SMART ENGINEERING SYSTEM DESIGN:

NEURAL NETWORKS, FUZZY LOGIC,

EVOLUTIONARY PROGRAMMING,

COMPLEX SYSTEMS AND ARTIFICIAL LIFE

VOLUME 13

ASME PRESS SERIES ON INTELLIGENT ENGINEERING SYSTEMS THROUGH ARTIFICIAL NEURAL NETWORKS

EDITOR

C.H. Dagli, University of Missouri-Rolla, Rolla, Missouri, USA

Intelligent Engineering Systems Through Artificial Neural Networks, Volume 1, edited by Cihan H. Dagli, Soundar R.T. Kumara, and Yung C. Shin, 1991

Intelligent Engineering Systems Through Artificial Neural Networks, Volume 2, edited by Cihan H. Dagli, Laura I. Burke, and Yung C. Shin, 1992

Intelligent Engineering Systems Through Artificial Neural Networks, Volume 3, edited by Cihan H. Dagli, Laura I. Burke, Benito Fernandez, and Joydeep Ghosh, 1993

Intelligent Engineering Systems Through Artificial Neural Networks, Volume 4, edited by Cihan H. Dagli, Benito Fernandez, Joydeep Ghosh, and R.T. Soundar Kumara, 1994

Intelligent Engineering Systems Through Artificial Neural Networks, Volume 5, edited by Cihan H. Dagli, Metin Akay, C.L. Phillip Chen, Benito Fernandez, and Joydeep Ghosh, 1995

Intelligent Engineering Systems Through Artificial Neural Networks, Volume 6, edited by Cihan H. Dagli, Metin Akay, C.L. Phillip Chen, Benito Fernandez, and Joydeep Ghosh, 1996

Intelligent Engineering Systems Through Artificial Neural Networks: Smart Engineering Systems: Neural Networks, Fuzzy Logic, Data Mining and Evolutionary Programming, Volume 7, edited by Cihan H. Dagli, Metin Akay, Okan Ersoy, Benito Fernandez, and Alice Smith, 1997

Intelligent Engineering Systems Through Artificial Neural Networks: Smart Engineering Systems: Neural Networks, Fuzzy Logic, Evolutionary Programming, Data Mining and Rough Sets, Volume 8, edited by Cihan H. Dagli, Metin Akay, Anna L. Buczak, Okan Ersoy, and Benito Fernandez, 1998

Intelligent Engineering Systems Through Artificial Neural Networks: Smart Engineering System Design: Neural Networks, Fuzzy Logic, Evolutionary Programming, Data Mining and Complex Systems, Volume 9, edited by Cihan H. Dagli, Anna L. Buczak, Joydeep Ghosh, Mark Embrechts and Okan Ersoy, 1999

Intelligent Engineering Systems Through Artificial Neural Networks: Smart Engineering System Design: Neural Networks, Fuzzy Logic, Evolutionary Programming, Data Mining and Complex Systems, Volume 10, edited by Cihan H. Dagli, Anna L. Buczak, Joydeep Ghosh, Mark Embrechts, Okan Ersoy and Stephen Kercel, 2000

Intelligent Engineering Systems Through Artificial Neural Networks: Smart Engineering System Design: Neural Networks, Fuzzy Logic, Evolutionary Programming, Data Mining and Complex Systems, Volume 11, edited by Cihan H. Dagli, Anna L. Buczak, Joydeep Ghosh, Mark Embrechts, Okan Ersoy and Stephen Kercel, 2001

Intelligent Engineering Systems Through Artificial Neural Networks: Smart Engineering System Design: Neural Networks, Fuzzy Logic, Evolutionary Programming, Complex Systems and Artificial Life, Volume 12, edited by Cihan H. Dagli, Anna L. Buczak, Joydeep Ghosh, Mark Embrechts, Okan Ersoy and Stephen Kercel, 2002

Intelligent Engineering Systems Through Artificial Neural Networks: Smart Engineering System Design: Neural Networks, Fuzzy Logic, Evolutionary Programming, Complex Systems and Artificial Life, Volume 13, edited by Cihan H. Dagli, Anna L. Buczak, Joydeep Ghosh, Mark Embrechts and Okan Ersoy, 2003

SMART ENGINEERING SYSTEM DESIGN:

Neural Networks, Fuzzy Logic, Evolutionary Programming, Complex Systems and Artificial Life

VOLUME 13

Proceedings of the Artificial Neural Networks in Engineering Conference (ANNIE 2003) held November 2-5, 2003, in St. Louis, Missouri, U.S.A.

EDITORS
Cihan H. Dagli
Department of Engineering Management
University of Missouri-Rolla

Anna L. Buczak
Lockheed Martin
Camden, New Jersey

Joydeep Ghosh
University of Texas at Austin
Austin, Texas

Mark Embrechts
Rensselaer Polytechnic Institute, RPI
Troy, New York

Okan Ersoy
Purdue University
West Lafayette, Indiana

ASME PRESS NEW YORK 2003

Library of Congress 92-30949
ISBN No. 0-7918-0204-3

CONTENTS

PREFACE

As a follow up to the previous twelfth volumes of Intelligent Engineering Systems Through Artificial Neural Networks, edited by Dagli, C.H. et. al., this volume contains the edited versions of the technical presentations of ANNIE 2003. The thirteenth international gathering of researchers interested in Smart Engineering System Design using neural networks, fuzzy logic, evolutionary programming, data mining and artificial life, was held on November 2-5, 2003 in St. Louis, Missouri, USA. The conference covered the theory of neural networks, fuzzy logic, evolutionary programming, data mining, and artificial life; and their applications in the engineering domain. The papers included in this volume provide a forum for researchers in the field to exchange ideas on smart engineering systems design.

An extended version of each paper selected for inclusion was reviewed by two referees, then revised, edited, and condensed to the format herein. The papers in this edited book are grouped into eight categories:

- Architectures
- Evolutionary Programming
- Fuzzy Systems
- Adaptive Control
- Pattern Recognition
- Prediction
- Data Mining and Decision Support Systems
- Smart Engineering Systems

There were six plenary sessions scheduled for ANNIE 2003. Dr. Paul Werbos presented "Unified Neural Network Designs: The Key to Large-Scale Applications and Understanding the Brain" for the Monday morning plenary session. At noon on Monday, Dr. Piero Bonissone offered "Soft Computing: a Common Framework to Represent Meta-Heuristics. In Tuesday's morning plenary session, "Dynamics of Respiratory Neural Networks During Maturation" was presented by Dr. Metin Akay. Tuesday's luncheon plenary "A View on the State of AI and New Computing Paradigms" was offered by Toshinori Munakata. Dr. Robert Kozma offered "Nonconvergent Neural Memories for Robust Encoding of Noisy Sensory Data" for the Wednesday morning plenary and Martin Pelikan presented "Hierarchical Bayesian Optimization Algorithm: Toward a New Generation of Evolutionary Algorithms" for the luncheon plenary.

Our Banquet Plenary Speaker for this year's ANNIE was Dr. James Gantt and his talk was entitled "Net Centric Systems: Lessions Learned and Directions" Dr. Gantt was born in Murray, Kentucky. After graduation from high school, he attended Murray State University and received a Bachelor of Science with honors (Cum Laude) in 1971. In 1972, he received a Master of

Science degree in Computer Science, majoring in Operations Research, from the University of Missouri at Rolla. After graduation, he entered the United States Army and served as an instructor at the Army Computer Science School at Fort Harrison, Indiana.

In 1978, he received an Education Specialist degree in Management from Ball State University. He moved to Atlanta, Georgia and was employed as an Operations Research Analyst with the Army Institute for Research in Management Information, Communications, and Computer Sciences (AIRMICS), which is now part of the U.S. Army Research Laboratory (ARL). He was selected to spend the 1981-1982 school year at the Massachusetts Institute of Technology as a Fellow in the Center for Advanced Engineering Studies.

From 1983 to 1992, he served as the Chief of the Management Information Systems Division at AIRMICS. He completed his Ph.D in Industrial and Systems Engineering at the Georgia Institute of Technology in 1989.

When ARL was formed in 1992, he became the Chief of the Software Technology Branch and then served as the Chief of the Software and Intelligent Systems Division.
In September 1996, he became the Acting Director of the Information Science and Technology (IS&T) Directorate. He was selected to become a member of the Senior Executive Service and appointed as Director of IS&T in June 1998. In May 2000,
Dr. Gantt became the Deputy Director of ARL's new Computational and Information Sciences Directorate (CISD), which was formed by merging the Computational Sciences with the Tactical Information Processing and Communications Research programs. In July 2003 he became the Acting Director of CISD and Acting Chief Information Officer (CIO) of ARL.

The CISD of ARL is responsible for conducting research for the Army in Computational Science, Computer Science, Communications, and Weather. The current budget for the Directorate is approximately $175M. About 350 government employees and another 350 on-site contractors execute this budget.

The editors would like to once again thank the plenary speakers, and the authors for their contributions. We would also like to recognize the organizing committee members of ANNIE 2003 for their excellent support in promoting the conference internationally. Further, we wish to express our gratitude to all the referees for their excellent and timely review efforts, which made this edited book possible within a short period of time.

Although space will not permit for the recognition of all contributors special thanks goes to the session chairs, whom contributed to the quality of the discussions in the presentations. We would also like to mention our appreciation to the staff of the Engineering Management Department for their timely and continuous support and particularly to Cindy Harper. Her ownership of ANNIE conferences and her excellent ability to communicate with authors all over the world and organizing abilities along with putting this volume together made her an integral part of the ANNIE conferences. One last appreciation goes out to Debbie Singer, although she came in late without her the final touches would never have come together.

Lastly, but most importantly, we would like to thank all of our families for their patience and sustenance during the many long hours that it took to create yet another ANNIE, ANNIE 2003, and subsequently, this book.

Cihan H. Dagli
University of Missouri-Rolla

Anna Buczak
Lockheed Martin

Joydeep Ghosh
University of Texas at Austin

Mark Embrechts
Rensselaer Polytechnic Institute, RPI

Okan Ersoy
Purdue University

PART I: ARCHITECTURES

Multiple-Labeled Instance Classifier in Functional Genomics

Jack Y. Yang, Ph.D.; Mary Qu Yang and Okan K. Ersoy, Ph.D.
School of Electrical and Computer Engineering
Purdue University West Lafayette, IN 47907-1285 USA
Contact: purduejk@purdue.edu, purduexy@purdue.edu, ersoy@purdue.edu

ABSTRACT

With the advent of high-throughput genome sequencing technology, more and more genomes are available; consequently our ability to sequence genomes has outstripped our ability to analyze the resulting data in order to determine the functions and structures of proteins encoded in the genomes. In reponse, Prof. Erosy introduced rare event rule extraction by Neural Networks and Decision Trees in the Human Genome Projects and developed Erosy's PHSNN. We noticed that disordered proteins carry functions but do not have 3-D structures and complex protein-protein and protein-DNA interactions enabled multiple functional natures of a proteins, consequently that functions of most proteins are either completely unknown or not completely known and no machine learning based method could well classify the complex functions of a protein. We introduced a new machine learning classifier called Multiple-Labeled Instance Classifier (MLIC) that handles instances belonging to many classes corresponding to the multiple functionalities of a protein, which has not been studied in previous computational intelligence approaches. We developed an automated machine learning based approach to provide information concerning multiple functional relations among a large group of proteins simultaneously. We used MLIC to build a most comprehensive protein phylogenetic profile library based on 61 genomes; it is an improvement from the results of other protein phylogenetic profiles based on 24 genomes. Experimental results show MLICs outperform other computational intelligence methods such as Support Vector Machines and Decision Trees, and provide a viable alternative to the supervised or unsupervised methods alone.

1. INTRODUCTION

The development of high-throughput genome sequencing technology has helped biology enter a new era in which one is seeking an understanding of biological processes from complete genomes. We have utilized a comprehensive data at the present time, with at least 61 completely sequenced genomes, including 12 archaebacterial, 45 eubacterial, and 4 eukaryotic species. Our data contains information of a total of more than 500,000 genes. We introduce a new classifier to handle the complex nature of protein functionalities and try to avoid issues such as inaccuracies in protein structures, missing information about disordered proteins in PDB (Protein Data Bank) and missing information in SVM or Decision Tree based classifiers. Recently, several authors have begun to apply supervised learning techniques to genomic data; Vert (2002) proposed a Support Vector Machine (SVM) tree kernel, and showed that it outperforms the "heterogeneous" SVM kernels that Pavlidis et al (2002) used. Ersoy et al. developed rare event rule extraction by Decision Trees and Neural Networks in the Human Genome Project (Ersoy 1999,2000). Yang and Ersoy (2002) developed gene finding and protein functional determination methods by using Bayesian trees and physical trees. So far, the Support Vector Machine is the only machine learning method that has been used in a protein phylogenetic profile data outside Prof. Ersoy's group. Experiments that Pavlidis et al. (2002) and Vert (2002) have performed were based on 24 species augmented with 79 dimensional DNA microarray vectors (heterogeneous data). Specifically, the dataset consists of protein phylogenetic profiles (computed using E-values) of their selected 2465 genes (they called "learnable" genes) from a yeast genome, along with the corresponding microarray data, consisting of 79 element gene expression vectors. For each selected protein functional class, a Support Vector Machine classifier is trained to distinguish proteins of that class from all other proteins. The report by Pavlidis et al., indicated, that 22 out of 27 classifiers have error rates greater than 72% (5 best learnable classes are reported with one having 72.1% error rate). Vert extended this work using the same dataset by considering a Support Vector Machine in which the traditional kernel is replaced by a tree kernel using a different set of 16 protein functional classes. He concluded that the tree kernel performed better than the traditional kernel on the 12 out of16 protein functional classes, and worse on the 4 out of 16 functional classes, which, he pointed out, tended to be larger and more general than the other classes. Proteins (genes) that belong to the rare event classes (less than 10 occurrences) are ignored in their work due to insufficient instances for training a Support Vector Machine. Our approach differs in several aspects from the earlier work in that it uses a multi-stage decomposition that makes use of both unsupervised and supervised machine learning techniques; referred this as Unsupervised-Supervised Tree (UST) algorithm (Yang and Ersoy 2002). The UST structure performs such a hierarchical decomposition of the feature space in a manner that groups proteins (genes) with similar profiles at each stage, thus transforming a complex problem into simpler subproblems that improve the generalization. In order to use UST as a supervised classifier to handle the instance belonging to many classes, we developed a new algorithm called MLIC (Multiple-Labeled Instance Classifier). The MLIC approach yields better results than SVMs and Decision Trees and provides improved performance on cases such as when instances belonging to even more than 10 classes actually occur. MLICs may provide a viable alternative to supervised or unsupervised methods.

2. MLIC

The core of UST is the MCT (Maximum Contrast Tree). MCTs are constructed so that at each stage the two training instances with maximum contrast (that is, the two training instances with maximum separation according to some distance measure) are used as seeds from which a partition of the feature space (into two sets) is grown (this is called splitting a node). The nearest neighbor instances should always go to the same node. The construction of the tree begins with a root node N that contains all the training data. The procedure split(N) splits the root node into two child nodes N_L and N_R, which are then recursively split using the same procedure.

Procedure split(N)

Let T be the set of training instances in node N. First we find the two instances $\overrightarrow{X_L}, \overrightarrow{X_R}$ in T that have maximum separation according to some distance measure $d(\cdot,\cdot)$, i.e. $d(\overrightarrow{X_L}, \overrightarrow{X_R}) = \max_{i,j} d(\overrightarrow{X_i}, \overrightarrow{X_j})$

Let A_N^L be a set of training instances that initially contains only the element $\overrightarrow{X_L}$, and A_N^R be a set of training instances that initially contains only the element $\overrightarrow{X_R}$. Let $d(\vec{X}, A_N^L)$ be the minimum distance of instance $\overrightarrow{X}$ to any instance in A_N^L, and similarly let $d(\vec{X}, A_N^R)$ be the minimum distance of instance $\overrightarrow{X}$ to any instance in A_N^R. We grow the sets A_N^L and A_N^R until they include all the training instances in T as follows:

Repeat until $A_N^L \cup A_N^R = T$:

- Let $\overrightarrow{X_1}$ be the instance in the set $T-(A_N^L \cup A_N^R)$ that minimizes the distance to any instance in A_N^L, i.e.

  $$\vec{X}_1 = \underset{\vec{X} \in T-(A_N^L \cup A_N^R)}{\arg\min} d(\vec{X}, A_N^L),$$

and let $\overrightarrow{X_2}$ be the instance in the set $T-(A_N^L \cup A_N^R)$ that minimizes the distance to any instance in A_N^R, i.e. $\vec{X}_2 = \underset{\vec{X} \in T-(A_N^L \cup A_N^R)}{\arg\min} d(\vec{X}, A_N^R)$

- If $d(\vec{X}_1, A_N^L) < d(\vec{X}_2, A_N^R)$ then add $\overrightarrow{X_1}$ to A_N^L

  else add $\overrightarrow{X_2}$ to A_N^R.

Create two new nodes N_L and N_R, which are, respectively, the left and right child of node N. Initialize node N_L so that it contains all the training instances in the set A_N^L, and similarly initialize node N_R so that it contains all the training instances in the set A_N^R.

While we have used a Euclidean distance measure in our experiments, other distance measures could be used as well; we list two distance measures:

Euclidean distance:

$$d(\overrightarrow{X}, \vec{Y}) = \sqrt{\sum_{i=1}^{m} (X_i - Y_i)^2}$$

Correlation coefficient:

$$d(\vec{X},\vec{Y}) = \frac{\sum_{i=1}^{m}(X_i - \overline{X}_i)(Y_i - \overline{Y}_i)}{\sqrt{\sum_{i=1}^{m}(X_i - \overline{X}_i)^2}\sqrt{\sum_{i=1}^{m}(Y_i - \overline{Y}_i)^2}}$$

Clustering algorithms can often be derived by first defining an appropriate energy function over the parameter space. Choosing the parameters that minimize the energy can serve as the basis for defining clusters. Therefore if the new coming instance approaches the nearest neighbor, the energy is minimized. The new instances should enter the sub-notes of the tree according to the nearest neighbor rules. This serves two purposes. First two child nodes will be formed as the maximum separation between each other. It maximizes boundaries that separate these two nodes. Secondly, instances inside a child note are maximally similar to each other. Therefore, MCT is constructed to minimize the energy function at each local step. The split at each node of the MCT can be viewed as minimizing an energy function. Let T be the set of training instances at a node. For a given partition of the set T into two sets A^L and A^R, we define the energy of this partition to be: $E(\{A^L, A^R\}) = -\min_{X \in A^L, Y \in A^R} d(X,Y)$ Then, the goal is to find the two sets A^L, A^R that minimize $E(\{A^L, A^R\})$ subject to the constraint that $A^L \cup A^R = T$. A plausible algorithm for constructing the maximal partition is as follows. Define a third set, A^U, the set of undecided instances, and initially let this set consist of all instances in T. Let A^L and A^R initially be empty. To start off the algorithm, we want to choose two instances from A^U; one will be added to A^L, and the other will be added to A^R (both instances will then be deleted from A^U). Which two instances should we choose to give the minimum value to $E(\{A^L, A^R\}) = -\min_{X \in A^L, Y \in A^R} d(X,Y)$? The answer is that we should choose the two most widely separated instances, just as we do in the maximum contrast algorithm. The algorithm is iterative: on each iteration, we move one instance from A^U into either A^L or A^R. We choose to move the instance that will result in the smallest value of $E(\{A^L, A^R\}) = -\min_{X \in A^L, Y \in A^R} d(X,Y)$. This instance turns out to be the instance in A^U that is closest to either A^L or A^R. If this instance is closest to A^L, then we move this instance from A^U to A^L, otherwise if it is closest to A^R then we move this instance from A^U to A^R. Since on each iteration, we remove one instance from A^U, the algorithm will eventually terminate when A^U is empty, at which point we will have $A^L \cup A^R = T$. Since at each step

we have chosen the action that minimize:
$E(\{A^L, A^R\}) = -\min_{X \in A^L, Y \in A^R} d(X,Y)$, it is plausible that when the algorithm terminates, the resulting partition will be optimal. Further note that the steps above correspond exactly to the steps involved in splitting a node in the natural growing MCT. The MCTs belong to the family of step-wise greedy algorithms. We also referred it as Sequential Bifurcation Tree (SBT). All the 6357 proteins (derived from open reading frames) inside the yeast genome were mapped onto a functionally related tree structure by the UST, so that protein functional relationships can be viewed at any layer and at any node (Figure 1). Such tree structure preserves all complex protein functional information, including proteins belonging to the rare event classes (classes of proteins with fewer occurrences inside a genome. The MCT allowed us to establish many links between proteins of related functions in the yeast *Saccharomyces cerevisiae* (See Figure 1.Overall process). In fact, many proteins are multi-functional, and this poses a challenge for machine learning based classifiers. We introduce a new machine learning classifier that can handle such situations called Multiple-Labeled Instance Classifier (MLIC). We found out that a good approach for classification is to use the MCT as a Tree-based bottom-up K Nearest Neighbor classifier. For example, we choose K equals to 3. We divide all the known proteins (genes) into training set and testing set. We tested MCTs by cross validation methods. The training set is used to generate the MCT. For each test instance, we determine the nearest protein (gene) on a leaf terminal node and move up one or two layers until the upper node contains at least 3 proteins (genes). In case there are more than three proteins (genes), three closest proteins (genes) are chosen based on closeness to the test protein (gene) on the node. Then the 3-Nearest Neighbor classifier with a majority voting scheme is performed. The output of such a classifier is the set of class labels shared by any two of the three nearest neighbors, where the 3 nearest neighbors are defined by the tree structure. This classifier is illustrated in Figure 2 and an example is given in Figure 3. MLICs are multi-functional themselves. Our experimental results showed the reject rate tended to be low, and even if we treat rejecting rate as error rate to give an upper bound on error rate, the results from UST were still better than the results of SVMs and Decision Trees. When we use the MLIC as a two-class classifier, there is definitely a result of either correct classification or misclassification from the majority voting at the node. Then, there is no reject rate. In this way, performance can be compared numerically in detail to the Support Vector Machines and to the Decision Trees using the same dataset. The advantage of MLICs is that it can be used as either a 2-class classifier or a complex classifier involving instances belongs to more than one labels or a regular multi-class classifier. In all cases, results are better than SVMs and Decision Trees. MCT can be used as a classifier in two different modes, which we refer to as one-level-up and two-level-up. In one-level-up mode, given a test instance, we first search the leaves of the tree to find the leaf node that contains the closest training instance. We then look at the set of training instances in the parent of that leaf node, find the K instances that are nearest to the test instance

from this set, and use the majority class label of these K training instances as the classification for the test instance. If the parent node does not contain K training instances, then we look to the parent of that parent node (grandparent node), and apply the voting strategy just described. Two-level-up mode is similar, but skips the step of looking at the parent of the leaf node, and looks directly at the parent of the parent (grandparent) of the leaf node to find the K nearest training instances. We also constructed a library of yeast protein phylogenetic profiles based on 60 genomes from complete 6357 genes (ORFs) and compared our results to the results by Pavlidis et al (2002) and Vert (2002), both based on 24 species from the selected "learnable" 2465 genes (ORFs) from the same yeast genome. Our library is an improvement over results from others. Our algorithms compare favorably against existing classification algorithms, and, more broadly, compare common classification algorithms over a more comprehensive protein phylogenetic profiles library than has been used previously.

3 DATA GENERATION

For each of the 6357 ORFs in the yeast *S cerevisiae* genome, we constructed a protein phylogenetic profile over 60 complete genomes. The ith element of the phylogenetic profile for a given ORF A is obtained by using the BLASTp program (Altschul 1997) to search the ith genome for the presence of ORF A. The score reported by BLASTp (means blast algorithm on proteins) is given by:

$$\max_{ORF\ \text{B in ith genome}} Evalue(A,B)$$

where Evalue(A,B) is a matching score that lies between 0 and infinity. The Evalue is obtained from another quantity called the p-value via the transformation: Evalue(A,B) = - ln (1 – Pvalue(A,B)) The Pvalue is matching score (probability) that lies between 0 and 1; Pvalue(A,B) equals one if internal amino-acid sequences of ORFs A and B are identical, and equals zero if they are completely different. The phylogenetic profiles were then normalized by subtracting off the expectation value of each attribute and dividing by the standard deviation of each attribute. Protein function descriptions and class labels were assigned according to the MIPS Comprehensive Yeast Genome Database (CYGD) using the Pedant (Protein Extraction, Description and Analysis Tools) program. Of the 6357 ORFs, 3887 are functionally known, and 2470 are functionally unknown or questionable; only 1084 carry a unique label. It is important to be aware of the hierarchical labeling scheme corresponding to the different levels of biological pathways that proteins participate. The overall process resulting in the dataset to be fed into the MLIC and subsequent processing is shown in Figure 1.

4 EXPERIMENTAL RESULTS

We compared the MLIC classifier to the Decision Tree algorithms C4.5 and C5, and to Support Vector Machines, both in the complex multi-class case and in the two class cases. The MLICs have the flexibility to be used in the complex multi-class case in which instances can carry multiple class labels. In the complex multi-class case, in which instances can carry multiple class labels, we used the classification strategy described in Figures 2-3. The Decision Tree algorithms C4.5 and C5 Decision Trees cannot handle instances that belong to more than one class, and therefore we had to select only proteins belonging to a single functional class, which reduced the number of available proteins (genes) from 6357 to 1804. Although the training errors of both C4.5 and C5 were always less than 10%, both C4.5 and C5 had difficulty learning this data set, and the testing errors were always greater than 70%. In the case of Support Vector Machines, for which detectors were constructed for each protein functional class, 22 out 27 proteins gave testing error rates greater than 72% using the data from Pavlidis et al (Pavlidis 2002). The results of the SVMs using our data are shown in Tables 1-2, in which 16/18 proteins had error rates ranging between 25 to 50%. This means our data is better than Pavlidis's data (in such usage as protein detectors). Because most proteins are multi-functional, we randomly draw multi-functional proteins (genes) for testing. The testing error rates for the UST-based MLICs ranged from 8% to 19% over 100 runs (on each experiment, a different set of training and testing samples were drawn). On the basis of these results (Tables 1-5 & Fig. 4) we conclude that the UST-based MLIC outperforms both Decision Trees and SVMs. For each protein functional class C, we defined a two class problem by defining all instances that include C as one of the labels to be of class 0, and all other instances to be of class 1. After training, the resulting two-class classifier functions as a detector for instances that belong to that class. This procedure is repeated for each protein functional class. We constructed more than 200 protein detectors, examples for the classes used to evaluate the performance of the tree when used as a classifier are given in Tables 3-5. Figures 4 plot the some examples in comparing to Support Vector Machines and Decision Trees. Those results are from the library of protein phylogenetic profiles for the yeast *S. cerevisiae* we built. MLICs not only handle complex multiple classes, but also can be used as a two-class classifier using the UST-based MLIC algorithm. There are two modes on a MLIC, we named them as one-level and two-level. For one-level method, we first examine the node just above the identified nearest terminal node (parent node) of the testing protein (gene). If such l-level up node already contain at least 3 proteins (genes), then perform MLIC on this node, otherwise, perform the MLIC on the 2-level up node (grandparent node). For the two-level method, we perform MLIC test directly on the node that is two-level above the terminal node (grandparent) of the nearest protein (gene) without exam the 1-level up node. On both one-level-up and two-level-up methods, we performed 3-fold cross validation and repeated the test many times with different randomly drawn data to obtain reliable average classification rate. " > " is used to denote the number of times UST is better than the compared classifier, and " <" is used to denote the number of times UST is worse than the compared classifier during the testing. From the tables of results for the various classes, it is evident that Decision Trees do not perform well on this dataset. A possible explanation is that Decision Trees only split on one feature at a time, that is, on one bit position in the phylogenetic profiles. However, the fact that two phlylogenetic profiles either agree or disagree in one bit position says almost nothing about the overall similarity of one phlylogenetic profile to another. By contrast, a MLIC takes into account the overall similarity of one whole phylogenetic profile to another. This may explain why MLIC methods appear to outperform Decision Trees on this dataset. The philosophy of a Support Vector Machine is to transform original (low dimensional) data into higher dimensional space, hoping that inseparable original data may be separable in the high dimensional space. A Support Vector Machine relies on a hyperplane in the transformed high dimensional space to separate the data and such hyperplane can not accommodate to the complexity of data if a real boundary to separate the data is not anything similar to a hyperplane in the transformed space as well. Therefore, Support Vector Machine may still yield high misclassification rates.

5 CONCLUSIONS

Our ultimate goal is to utilize computational intelligence to establish theory of evolution at the genomic level which is the key to the mystery of life on the earth. It is expected that computational intelligence techniques such as MLICs and Ersoy's rare event rule extraction by Neural Networks and Decision Trees in the Human Genome Project shall further develop to learn more valuable information from the human genome and other genomes efficiently. High dimensional feature space, in relation to the number of available training instances, as well as the fact that an instance can belong to more than one class make the classification a challenging problem. Based on our experiments, MLICs perform considerably better than Decision Tree algorithms C4.5 and C5, and Support Vector Machines, and can provide a viable alternative to supervised or unsupervised methods alone. In addition, MLIC classifiers are capable of handling protein functional classes with a small number of proteins (rare events), and also handle multifunctional proteins. The abilities of MLICs to handle such cases means that a larger dataset can be used, which may provide deeper insight into protein functional relationships and thus may lead to a better understanding of evolution at a higher molecular and genomic level.

Table 1 Performance comparisons of UST-based MLICs with Decision Trees and SVMs.

Algorithms	Error Rate on Testing	Multiple Hierarchical Protein Classes?	Dataset
USTs	8%-19%	Yes, Complex multi-class system	3887 Genes (use a number of different training / test instances on 60 genomes
Decision Tree C45	72-85%	No, must be single class for each gene	1500 training / 304 test instances on 60 genomes
Decision Tree C5	71-84%	No, must be single class for each gene	1500 training / 304 test instances on 60 genomes
SVMs (Pavlidis et al Data)	22/27 proteins have error rates greater than 72%	Yes, after converting into two classes yes or no SVM protein detectors	2465 genes mixed with protein phylogenetic profile data and DNA microarray data on 24 genomes
SVMs (Our data)	16/18 proteins have error rates 25% - 50% see Table.2	Yes, after converting into two classes yes or no SVM protein detectors	3887 genes (use hold a number of different training / test instances on 60 genomes (Table 2)

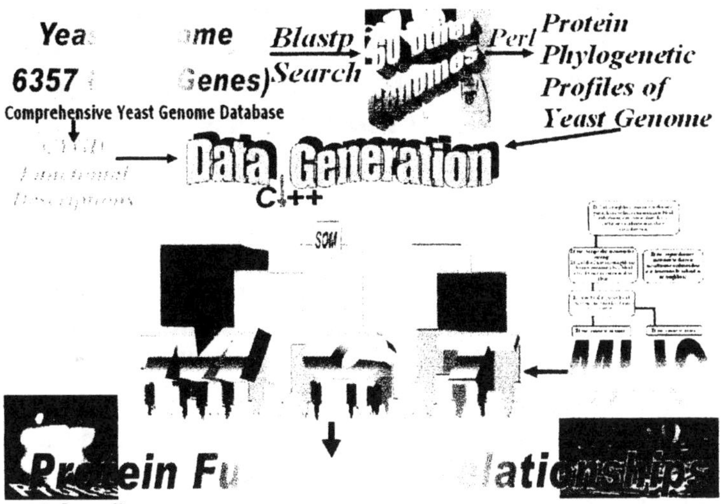

Figure 1.Overall process

Table 2. Results of Support Vector Machines using our data

Protein functional categories	Class	Correct	Error	Total	Error rate
METABOLISM	01	424	228	652	34.97 %
ENERGY	02	110	46	156	29.49 %
CELL CYCLE AND DNA PROCESSING	03	230	158	388	46.75%
TRANSCRIPTION	04	267	221	478	46.23%
PROTEIN SYNDISSERTATION	05	142	78	220	35.45%
PROTEIN FATE	06	224	142	366	42.26%
CELLULAR TRAN & TRANSPORTMECHANISM	08	186	132	318	41.51%
CELL COMMUN / SIGNAL TRANSDUCT MECH.	10	23	23	46	50.00%
CELL RESCUE, DEFENSE AND VIRULENCE	11	94	80	174	45.98%
REGULAT / INTERACTION CELL ENVIRONMT	13	64	56	120	46.67%
CELL FATE	14	157	103	260	39.62%
DEVELOPMENT (Systemic)	25	1	3	4	75%
TRANSPOS ELEM, VIRAL&PLASMIDPROTEINS	29	6	4	10	40%
CONTROL OF CELLULAR ORGANIZATION	30	72	60	132	45.5%
SUBCELLULAR LOCALISATION	40	698	470	1168	40.24%
PROTEIN ACTIVITY REGULATION	62	8	0	8	0%
POTEIN W/ BIND FUNCT OR COFACTOR REQUI	63	3	1	4	25%
TRANSPORT FACILITATION	67	117	79	196	44.89%

Table 3: Protein 02.13 Respiration with UST based MLIC versus others results.

Average classification accuracy				UST	
Label: 2.13	C45	C5	SVM	1 Level	2 Level
AVERAGE	0.556992	0.567444	0.580289	0.653173	0.652118
STD	0.067531	0.074686	0.068411	0.060416	0.059277
VAR	0.004560	0.005578	0.004680	0.003650	0.003514
Comparisons of methods					
1 Lev > C45	1 Level < C45	1 Lev > C5	1 Lev< C5	1 Level > SVM	1L<SVM
111	14	103	20	107	14
2 Lev> C45	2 Lev < C45	2 Lev > C5	2 Lev < C5	2 Level > SVM	2L <SVM
108	15	98	22	103	14

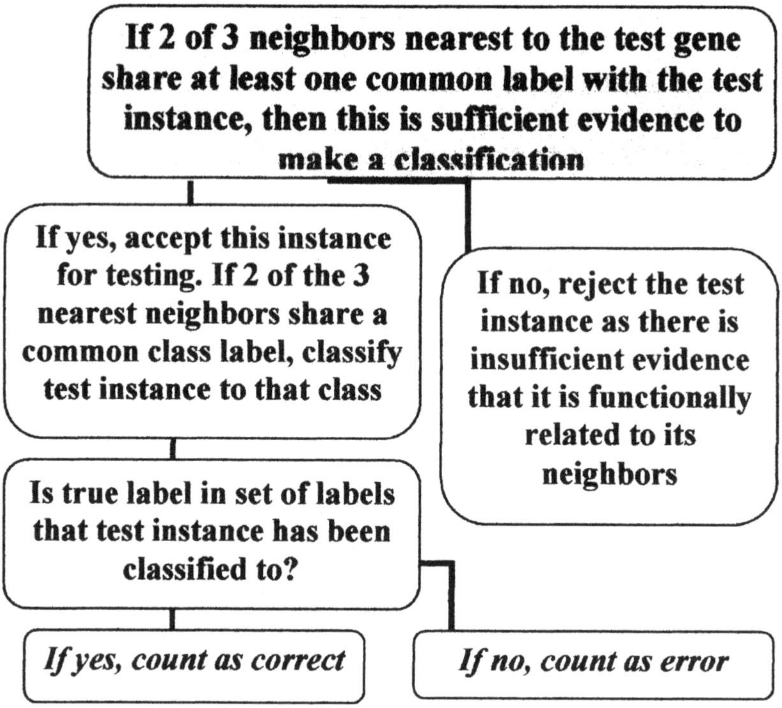

Figure 2.Block diagram of the Multiple-Labeled Instance Classifier.

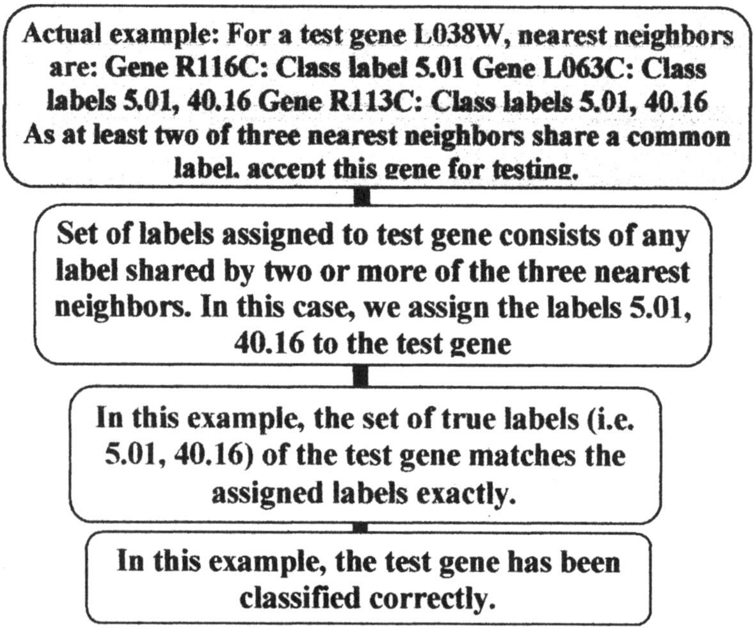

Figure 3: An example of using MLIC

Table4: Protein 02.13 Respiration with UST based MLIC versus others results.

Average classification accuracy				UST	
Label: 2.13	C45	C5	SVM	1 Level	2 Level
AVERAGE	0.556992	0.567444	0.580289	0.653173	0.652118
STD	0.067531	0.074686	0.068411	0.060416	0.059277
VAR	0.004560	0.005578	0.004680	0.003650	0.003514
Comparisons of methods					
1 Lev > C45	1 Level < C45	1 Lev > C5	1 Lev< C5	1 Level > SVM	1L<SVM
111	14	103	20	107	14
2 Lev> C45	2 Lev < C45	2 Lev > C5	2 Lev < C5	2 Level > SVM	2L <SVM
108	15	98	22	103	14

Table 5: Protein 01.20.01 Metabolism of primary metabolic sugars derivatives

Average classification accuracy				UST	
1.20.01	C45	C5	SVM	1 Level	2 Level
AVE	0.647228	0.651643	0.685763	0.711537	0.717163
STD	0.093500	0.092932	0.089360	0.100229	0.105281
VAR	0.008742	0.008636	0.007985	0.010046	0.011084
Comparisons of methods					
1L>C45	1 Level < C45	1 Level > C5	1Level < C5	1 Level > SVM	1 L<SVM
207	87	210	84	142	83
2L>C45	2 Level < C45	2 Level > C5	2 Lev < C5	2 Level > SVM	2L <SVM
212	81	210	83	158	77

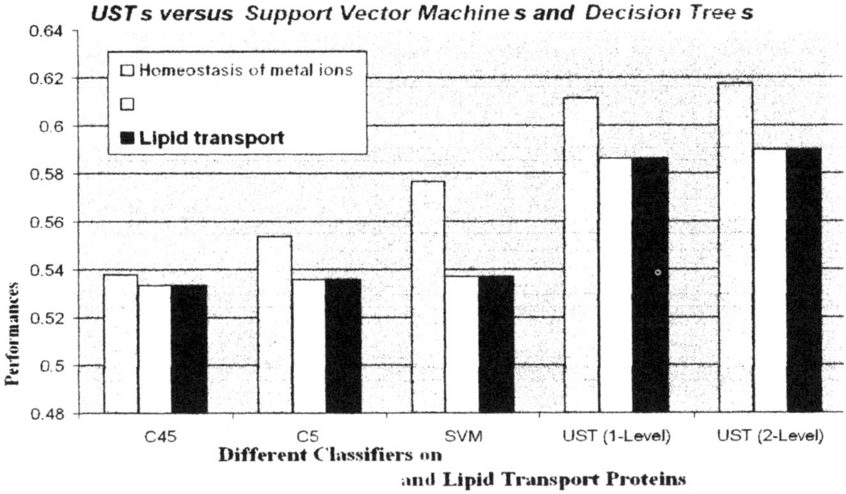

Figure 4.Results on proteins of homeostasis of ions, cell import & lipid transport

Reference:

Altschul, S.F., et al. (1997) "Gapped BLAST and PSI-BLAST: " Nucleic Acids Res. 25:3389-3402.

Codrington, Craig W. (2001) "Boosting with Confidence Information", ICML, June 29 2001

Ersoy,O.K, Choe W, Bina M (2000) "Neural network schemes for detecting rare events in human genomic DNA" Bioinformatics Vol. 16 no 12 Pages 1062-1072

Pavlidis et al (2002) "Learning Gene Functional Classifications..". J of Com Biology,Vol.9 401-411

Pellegrini et al (99) "Assigning protein functions by compar. genome analysis….", PNAS p.4285-88

Yang, Jack; Yang, Mary and Ersoy,O.K. (2002) "Gene finding and protein functional determination by protein phylogenetic profile and computational intelligence". Intelligent Engineering Systems Vol 12. Page 733-740 ANNIE 2002 ASME Press

Vert J.(2002) "A tree kernel to analyze phylogenetic profiles" Bioinformatics Vol 18 Sup.1. P. S276

AN ANALYSIS OF MCMC SAMPLING METHODS FOR ESTIMATING WEIGHTED SUMS IN WINNOW

QINGPING TAO AND STEPHEN D. SCOTT
Department of Computer Science & Engineering
University of Nebraska
Lincoln, Nebraska

ABSTRACT
Chawla et al. introduced a way to use the Markov chain Monte Carlo method to estimate weighted sums in multiplicative weight update algorithms when the number of inputs is exponential. But their algorithm still required extensive simulation of the Markov chain in order to get accurate estimates of the weighted sums. We propose an optimized version of Chawla et al.'s algorithm, which produces exactly the same classifications while often using fewer Markov chain simulations. We also apply two other sampling techniques and empirically compare them with Chawla et al.'s Metropolis sampler to determine how effective each is in drawing good samples in terms of accuracy of weighted sum estimates and total time. We perform our analyses in the context of learning DNF formulas using littlestone's Winnow algorithm.

INTRODUCTION

In this paper, we consider as our application area the problem of learning general DNF formulas, which has been heavily studied in a learning-theoretic framework (e.g. Bshouty et al. (1999); Khardon et al. (2001)). Let $f = P_1 \vee P_2 \vee \ldots \vee P_K$ be the target function, where $P_i = c_{i1} \wedge c_{i2} \wedge \ldots \wedge c_{in}$ is a term and c_{ij} is a constraint on the value of attribute j. If attribute j takes on values from $\{1, \ldots, k_j\}$, then $c_{ij} = \ell \in \{1, \ldots, k_j\}$ means that for an example $\mathbf{x}$ to satisfy constraint c_{ij}, $x_j = \ell$. If $c_{ij} = 0$, then x_j can be any value from $\{1, \ldots, k_j\}$ and still satisfy the constraint. If $x_j = 0$, then this attribute value is unspecified and only satisfies when $c_{ij} = 0$. In other words, $\mathbf{x}$ satisfies P_i iff for all j, either $x_j = c_{ij}$ or $c_{ij} = 0$. The set of terms available for f and the instance space are both $\Omega = \prod_{j=0}^{n-1} \{0, \ldots, k_j\}$. If example $\mathbf{x}$ has $n_{\mathbf{x}}$ specified values, then there are exactly $2^{n_{\mathbf{x}}}$ terms satisfied by it.

One useful method for solving this problem is using multiplicative weight update (MWU) algorithms such as Winnow (Littlestone, 1988). The learning of Winnow proceeds in *trials*. At trial t, Winnow receives an input vector $\mathbf{x}'_t$ and makes a prediction $\hat{y}_t = 1$ if $W_t = \mathbf{w}_t \cdot \mathbf{x}'_t \geq \theta$ and 0 otherwise, where $\mathbf{w}_t$ is its weight vector and θ is the threshold. Then Winnow updates its weight vector as follows: $w_{t+1,i} = w_{t,i}\alpha^{x'_{t,i}(y_t - \hat{y}_t)}$ for some learning rate $\alpha > 1$, where y_t is the true label. We call it a *promotion* if $w_{t+1,i} > w_{t,i}$, and *demotion* if $w_{t+1,i} < w_{t,i}$. Winnow can be used to learn DNF formulas by using all possible terms as its inputs: e.g. if $k_i = 2$ for all i, there will be $N = |\Omega| = 3^n$ possible terms. Input x'_i to Winnow is 1 if $\mathbf{x}$ satisfies term i and 0 otherwise.

The advantage of using Winnow is that the number of mistakes that Winnow will make on any sequence of examples is at most $O(K \log N) = O(Kn)$, where K is the number of terms in the target disjunction (Littlestone, 1988). But the disadvantage is that brute-force computation of $W_t = \mathbf{w}_t \cdot \mathbf{x}'_t$ is takes time $\Omega(N)$, which is exponential in n. Thus another approach is needed. One possibility is to use kernels, as illustrated by Khardon et al. (2001) for the Perceptron algorithm (i.e. using additive weight updates). However, while they showed that it is possible to efficiently compute the weighted sum for Perceptron when learning DNF, in the worst case, their kernel-based algorithm makes $2^{\Omega(n)}$ prediction mistakes. They also argued that unless P = #P, it is impossible to efficiently exactly simulate Winnow for learning DNF. Thus we look to Chawla et al. (2003), who use MCMC methods to *estimate* W_t for Winnow *with high probability*, as opposed to Khardon et al.'s hardness result that says no *deterministic* simulation of Winnow is possible for DNF.

Although Chawla et al.'s preliminary empirical results are much stronger than what their theoretical results imply, they still required extensive simulation of the Markov chain in order to get accurate estimates of the weighted sums. This significantly slowed their algorithm. Here we propose an optimized version of Chawla et al.'s algorithm, which often uses less computation time without any loss in classification accuracy. We also empirically compare two MCMC techniques (Gibbs and Metropolized Gibbs) to Chawla et al.'s Metropolis sampler in terms of accuracy of weighted sum estimates.

MARKOV CHAIN MONTE CARLO METHODS

Markov chain Monte Carlo methods have been widely used to solve problems in e.g. statistical physics and Bayesian statistical inference. They can provide approximations to quantities by performing statistical sampling experiments. Generally, to apply MCMC methods, a Markov chain M with state space Ω and stationary distribution π is designed to be *ergodic*, that is, the probability distribution over Ω converges asymptotically to π regardless of the initial state. Then M is repeatedly simulated for T steps and generates S samples almost according to π. These S samples are then used to estimate the quantity of interest. Usually we discard states met in the first T_0 steps and assume there would be a rapid convergence to π during these steps. This T_0-step procedure is called *burn-in*. After burn-in, M would be simulated for another S steps and each additional step would generate a new sample. The most commonly used MCMC method is the Metropolis sampler (Metropolis et al., 1953). In each step, the Metropolis sampler repeatedly considers randomly generated changes to one of the components of the current state P and accepts the new state Q with probability $\min\{1, \frac{\pi(Q)}{\pi(P)}\}$.

CHAWLA ET AL.'S MCMC SOLUTION FOR WINNOW

We now describe Chawla et al.'s solution for estimating $W_t(\alpha) = \sum_{P \in \Omega_t} w_{t,P}(\alpha)$, where $w_{t,P}(\alpha) = \alpha^{\varpi_t(P)}$ is term P's weight of training Winnow with learning rate α, and $\varpi_t(P) = u_t(P) - v_t(P)$, where $u_t(P)$ is the total number of promotions of term P at time t and $v_t(P)$ is the total number of demotions.

Let $\Omega_t \subset \Omega$ be the set of 2^{n_t} terms that are satisfied by example $\mathbf{x}_t$ ($n_t \leq n$ is the number of non-zero values in $\mathbf{x}_t$). For each term $P = (p_1, \ldots, p_{n_t}) \in \Omega_t$, $p_i = 1$ if constraint $c_{p_i} > 0$ and otherwise $p_i = 0$. Then a set of Markov chains $\mathcal{M}_t$ is built on the state space Ω_t. Each chain $M_t(\alpha') \in \mathcal{M}_t$ has a specific learning rate α' and a stationary distribution $\pi_{\alpha',t}(P) = \alpha'^{\varpi_t(P)} / W_t(\alpha')$.

Chawla et al. then define $f_{i,t}(P) = \frac{w_{t,P}(\alpha_{i-1,t})}{w_{t,P}(\alpha_{i,t})}$, where $\alpha_{i,t} = (1 + \frac{1}{m_t})^{i-1}$ for $1 \leq i < r_t$, $\alpha_{r_t,t} = \alpha$, r_t is the smallest integer such that $(1 + \frac{1}{m_t})^{r_t - 1} \geq \alpha$, and $m_t = u_t(P_e) + v_t(P_e)$ where $P_e = (0, 0, \ldots, 0)$. Because $\mathrm{E}[f_{i,t}(P)] = \frac{W_t(\alpha_{i-1,t})}{W_t(\alpha_{i,t})}$, $\frac{W_t(\alpha_{i-1,t})}{W_t(\alpha_{i,t})}$ can be estimated by computing the sample mean of $f_{i,t}(P)$, which allows $W_t(\alpha)$ to be computed since $W_t(\alpha) = \left(\frac{W_t(\alpha_{r,t})}{W_t(\alpha_{r-1,t})}\right) \cdots \left(\frac{W_t(\alpha_{2,t})}{W_t(\alpha_{1,t})}\right) W_t(\alpha_{1,t})$ and $W_t(\alpha_{1,t}) = W(1) = |\Omega_t| = 2^{n_t}$. Therefore, for each value $\alpha_{2,t}, \ldots, \alpha_{r,t}$, S_t samples are drawn through $M_t(\alpha_{i,t})$. Let $X_{i,t}$ be the sample mean of $f_{i,t}(P)$ and $|\mathcal{M}_t| = r_t - 1$, then the estimate of $W_t(\alpha)$ is $\hat{W}_t(\alpha) = 2^{n_t} \prod_{i=2}^{r_t} 1/X_{i,t}$.

OUR OPTIMIZED MCMC SOLUTION

In Chawla et al.'s MCMC solution, $r_t - 1$ Markov chains need to be simulated. Here we give an optimized solution that is based on the idea that to exactly simulate Winnow, we only need to know what Winnow's prediction is going to be (i.e. on what side of the threshold θ that W will fall on), not what the weighted sum exactly is. So it is possible that we could stop computing our estimate after a subset of the chains in $\mathcal{M}_t$ have been run.

Let $\wp_t^{max} = \max_{P\in\Omega_t}\{u_t(P) - v_t(P)\}$, $\wp_t^{min} = \min_{P\in\Omega_t}\{u_t(P) - v_t(P)\}$, i.e. the maximum and minimum number of net promotions, and $\Psi_t = \{2, \cdots, r_t\}$. We define the following two constraints:

$$\mathcal{C}\prod_{i\in\Psi'} \frac{W_t(\alpha_{i,t})}{W_t(\alpha_{i-1,t})}\left(\frac{\alpha_{i,t}}{\alpha_{i-1,t}}\right)^{-\wp_t^{min}} > \theta, \tag{1}$$

$$\mathcal{D}\prod_{i\in\Psi'} \frac{W_t(\alpha_{i,t})}{W_t(\alpha_{i-1,t})}\left(\frac{\alpha_{i,t}}{\alpha_{i-1,t}}\right)^{-\wp_t^{max}} < \theta, \tag{2}$$

where $\mathcal{C} = 2^{n_t}\prod_{i=2}^{r_t}\left(\frac{\alpha_{i,t}}{\alpha_{i-1,t}}\right)^{\wp_t^{min}}$ and $\mathcal{D} = 2^{n_t}\prod_{i=2}^{r_t}\left(\frac{\alpha_{i,t}}{\alpha_{i-1,t}}\right)^{\wp_t^{max}}$. Then we can prove the following theorem.

Theorem 1 *If $\exists\Psi' \subseteq \Psi_t$ that satisfies constraint (1), then $W_t(\alpha) > \theta$; If $\exists\Psi' \subseteq \Psi_t$ that satisfies constraint (2), then $W_t(\alpha) < \theta$.*

Proof Sketch: Because $\alpha_{i,t} > \alpha_{i-1,t} > 0$ and $\varpi_t(P) - \wp_t^{min} \geq 0$ for all $P \in \Omega_t$, $\sum_{P\in\Omega_t} \alpha_{i,t}^{\varpi_t(P)-\wp_t^{min}} \geq \sum_{P\in\Omega_t'} \alpha_{i-1,t}^{\varpi_t(P)-\wp_t^{min}}$. So $\frac{W_t(\alpha_{i,t})}{W_t(\alpha_{i-1,t})}\left(\frac{\alpha_{i,t}}{\alpha_{i-1,t}}\right)^{-\wp_t^{min}} \geq 1$. Then $W_t(\alpha) = \mathcal{C}\prod_{i=2}^{r_t}\frac{W_t(\alpha_{i,t})}{W_t(\alpha_{i-1,t})}\left(\frac{\alpha_{i,t}}{\alpha_{i-1,t}}\right)^{-\wp_t^{min}} \geq \mathcal{C}\prod_{i\in\Psi'}\frac{W_t(\alpha_{i,t})}{W_t(\alpha_{i-1,t})}\left(\frac{\alpha_{i,t}}{\alpha_{i-1,t}}\right)^{-\wp_t^{min}} > \theta$. Similarly we can prove the second statement. □

Theorem 1 tells us that it would not always be necessary to run all $r_t - 1$ Markov chains if we were only interested in Winnow's predictions. Instead, we can sometimes limit our simulations to a subset of Markov chains. So what we want is to find such a subset with the smallest size.

Let $\Gamma_1(\Psi_t)$ be the set of all Ψ' that satisfy constraint (1), and $\Gamma_0(\Psi_t)$ be the set of all Ψ' that satisfy constraint (2). We define $\Psi_1^{min} \in \Gamma_1(\Psi_t)$ as a *minimum 1-prediction set* if $|\Psi_1^{min}| \leq |\Psi'|$ for all $\Psi' \in \Gamma_1(\Psi_t)$, and $\Psi_0^{min} \in \Gamma_0(\Psi_t)$ as a *minimum 0-prediction set* if $|\Psi_0^{min}| \leq |\Psi'|$ for all $\Psi' \in \Gamma_0(\Psi_t)$. Then we give Theorem 2.

Theorem 2 *If Ψ_1^{min} exists, $\{r_t, r_t - 1, \cdots, r_t - |\Psi_1^{min}| + 1\}$ is a minimum 1-prediction set, and if Ψ_0^{min} exists, $\{2, 3, \cdots, |\Psi_0^{min}| + 1\}$ is a minimum 0-prediction set.*

Proof Sketch: Using Cauchy's inequality, we can prove that $\frac{W_t(\alpha_{i+1,t})}{W_t(\alpha_{i,t})} \geq \frac{W_t(\alpha_{i,t})}{W_t(\alpha_{i-1,t})}$ for $\forall i, 2 \leq i \leq r_t - 1$. Then let $\Psi' = \{r_t, r_t - 1, \cdots, r_t - |\Psi_1^{min}| + 1\}$,

$$\mathcal{C}\prod_{i\in\Psi'}\frac{W_t(\alpha_{i,t})}{W_t(\alpha_{i-1,t})}\left(\frac{\alpha_{i,t}}{\alpha_{i-1,t}}\right)^{-\wp_t^{min}} \geq \mathcal{C}\prod_{i\in\Psi_1^{min}}\frac{W_t(\alpha_{i,t})}{W_t(\alpha_{i-1,t})}\left(\frac{\alpha_{i,t}}{\alpha_{i-1,t}}\right)^{-\wp_t^{min}} > \theta.$$

So $\{r_t, r_t - 1, \cdots, r_t - |\Psi_1^{min}| + 1\}$ is a minimum 1-prediction set. Similarly we can prove $\{2, 3, \cdots, |\Psi_0^{min}| + 1\}$ is a minimum 0-prediction set. □

According to Theorem 2, if $W_t(\alpha) > \theta$ and we simulate Markov chains in the order of $r_t, r_t - 1, \cdots, 2$, and halt when constraint (1) is satisfied, we need no more computation than any other sequence of Markov chains in $\mathcal{M}_t$. Similarly we use no more chains than any other sequence if $W_t(\alpha) < \theta$ and we simulate them in the order of $2, 3, \cdots, r_t$ and check if constraint (2) is satisfied. Then we get an optimized MCMC solution for Winnow, which chooses one of the two orders ($\{r_t, \cdots, 2\}$ or $\{2, \cdots, r_t\}$) by guessing the most likely prediction y'_t. When we use Winnow to predict, we could just assume y'_t is 1. When we are training Winnow, we can set y'_t as the class label of training example $\mathbf{x}$. But a better way is that at the t-th training iteration let $y'_t = \hat{y}_{t-1}(\mathbf{x})$, where $\hat{y}_{t-1}$ is the prediction of $\mathbf{x}$ at the $(t-1)$-th iteration. The heuristic is that the weighted sum of $\mathbf{x}$ might not change too much after the last time Winnow met $\mathbf{x}$. At the beginning of training, all weights of Winnow are 1. So $W_1(\alpha) = 2^{n_t}$ for all examples. If $2^{n_t} \geq \theta$, $y'_1 = 1$, otherwise 0.

SAMPLING FROM $\pi_{\alpha,t}$

In Chawla et al.'s MCMC solution, the computation time of estimating $W_t(\alpha)$ depends on the number of chains $r_t - 1$, the number of burn-in steps T_0 and the sample size S. Besides our optimized solution, another way to reduce the computation time is choosing relatively small T_0 and S. This could be achieved by using a good sampler that can efficiently draw samples from the desired distribution. Chawla et al. (2003) applied the Metropolis sampler to their MCMC solution. Here we adapt two other MCMC sampling techniques, Gibbs sampler and Metropolized Gibbs sampler.

Gibbs Sampler for Winnow

The Gibbs sampler (Geman & Geman, 1984) has a number of distinct features. In a single step of the Gibbs sampler, each component is replaced with a value picked from its distribution conditional on the current values of all other components. The conditional distributions are constructed on prior knowledge of π. Furthermore, the Gibbs sampler is, by construction, multidimensional. It generates new values for all components and only after that it outputs a sample.

For any state $P \in \Omega_t$, each variable p_i only has two possible values, 0 and 1. If $P_0 = (p_1, \cdots, p_i = 0, \cdots, p_{n_t})$ and $P_1 = (p_1, \cdots, p_i = 1, \cdots, p_{n_t})$, the conditional distribution is $\pi_{\alpha,t}(p_i|P\backslash\{p_i\}) = \frac{\pi_{\alpha,t}(P)}{\pi_{\alpha,t}(P_0)+\pi_{\alpha,t}(P_1)}$. Then we get the conditional distribution of the Gibbs sampler for $M_{DNF,t}(\alpha)$, that is,

$$\pi_{\alpha,t}(0|P\backslash\{p_i\}) = 1/(1 + \alpha^{\varpi_t(P_1)-\varpi_t(P_0)}) \quad (3)$$

$$\pi_{\alpha,t}(1|P\backslash\{p_i\}) = 1/(1 + \alpha^{\varpi_t(P_0)-\varpi_t(P_1)}). \quad (4)$$

Metropolized Gibbs Sampler for Winnow

The Metropolized Gibbs sampler (Liu, 1996) is a modification of the Gibbs sampler, which has been proven to be statistically more efficient than the Gibbs sampler. The sampler draws a new value p'_i with probability $\frac{\pi_{\alpha,t}(p'_i|P\backslash\{p_i\})}{1-\pi_{\alpha,t}(p_i|P\backslash\{p_i\})}$, and accepted with the Metropolis-Hasting acceptance probability $\min\{1, \frac{1-\pi_{\alpha,t}(p_i|P\backslash\{p_i\})}{1-\pi_{\alpha,t}(p'_i|P\backslash\{p_i\})}\}$.

As mentioned before, each component in state P only has two possible values. So the Metropolized Gibbs sampler becomes a Metropolis sampler that repeatedly updates all components in a fixed order. We then build the Metropolized Gibbs sampler for $M_t(\alpha)$ that updates each component p_i in P one by one with a value p'_i randomly drawn from 0 and 1, and accept it with the acceptance probability

$$\min\{1, \alpha^{\varpi_t(P\backslash\{p_i\}\cup\{p'_i\})-\varpi_t(P)}\}. \quad (5)$$

EXPERIMENTAL RESULTS

In our experiments, we compared three MCMC sampling techniques (Metropolis, Gibbs and Metropolized Gibbs) on two types of data (simulated data used by Chawla et al. (2003) and UCI data sets). We then compared the computation costs of our optimized solution with Chawla et al.'s algorithm in term of the total number of Markov chain simulations. In all experiments, we set the burn-in time $T_0 = n^2$ because it takes about T^2 steps to move to a state T steps away due to the random walk nature of MCMC samplers (Neal, 1995). Our experiments showed that this burn-in time worked well. In each experiment, we also counted each update of a single component of current state as a single sampling step. We used the same number of sampling steps T_s for all three samplers.

Comparisons on Estimating Weighted Sums

To evaluate how well the weight estimation procedures with different samplers guessed the weighted sums, we compared them using the measure *Guess Error* given in Chawla et al. (2003), which is the average error of the estimates ($|\hat{W} - W|/W$). We trained Winnow for 20 rounds on 10 partitions of Voting data while varying the sampling time T_s. Figure 1(a) shows averages over more than 70000 estimates obtained from the experiment. For all T_s, Guess Errors of Gibbs and Metropolized Gibbs are always lower than Metropolis. Although Metropolized Gibbs has a lower Guess Error than Gibbs, the difference between these two samplers is very small, especially when $T_s = 9600, 12800, 16000$. To evaluate the effect of varying n, we measured Guess Error of all three samplers on the simulated data. In Figure 1(b), T_s is fixed at 10000. Gibbs and Metropolized Gibbs are still always better than Metropolis. The Guess Error of Metropolized Gibbs is lower than Gibbs when $n \leq 25$. But when $n > 25$, Gibbs becomes the best sampler. Considering results in Figures 1, we can say that Gibbs sampler is the best choice on average in terms of the accuracy of estimating weighted sums.

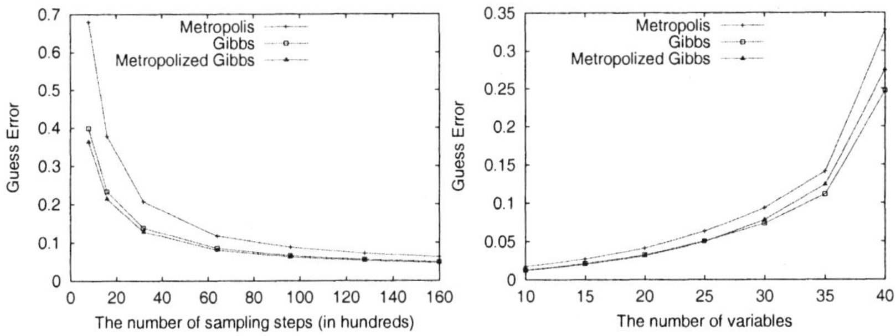

(a) Guess Error vs. T_s on Voting **(b) Guess Error vs. n on simulated data.**

Figure 1: Comparison on Guess Error.

Comparisons on Computation Cost

Here we report speedups of our optimized algorithm over Chawla et al.'s solution in terms of the total numbers of Markov chains that are used. Table 3 summarizes average results of 10-fold cross-validation on simulated data with 20 training rounds and five UCI data sets with 100 training rounds. In Table 3, "MCMC" is the number of chains used by Chawla et al.'s solution. "Opt MCMC" is the number of chains used by our algorithm. "Savings" is the percentage of chains our algorithm saved. Results in Table 3 confirms that we do not need to run all chains sometimes. However, for different data sets, our algorithm has different performance. A possible reason is that weighted sums of Winnow are much less or greater than threshold for some data sets. Thus, there are probably more chances for our algorithm to stop early.

Table 3: Comparisons on the total number of Markov chains (in thousands).

(a) Simulated data

n	MCMC	Opt MCMC	Savings(%)
10	4.6	4.1	11.3
15	11.9	11.2	6.3
20	20.8	18.7	10.0
25	23.5	21.2	9.7
30	40.8	37.5	8.1
35	58.7	53.0	9.7
40	60.9	53.2	12.7

(b) UCI data sets

Data Sets	MCMC	Opt MCMC	Savings(%)
iris	217.7	200.1	8.0
car	14272.7	14032.0	1.7
breast cancer	45176.2	45111.5	0.2
voting	1185.5	1132.0	4.5
auto	13822.8	13751.7	0.5

CONCLUSIONS

We proposed an optimized MCMC solution for estimating weighted sums in Winnow, which often uses less computation time than Chawla et al.'s solution without any loss of classification accuracy. Our experimental results confirmed that our algorithm only needs to use a subset of all Markov chains implied by the original solution. We showed how to get such a subset with the smallest size. We also empirically applied two MCMC sampling techniques: Gibbs and Metropolized Gibbs. They all showed better performance than Chawla et al.'s Metropolis sampler in terms of accuracy of weighted sum estimates.

ACKNOWLEDGMENTS

This work was funded in part by NSF grants CCR-0092761 and EPS-0091900 and a grant from the University of Nebraska Foundation. It was also supported in part by NIH Grant Number RR-P20 RR17675 from the IDeA program of the National Center for Research Resources. This work was completed in part utilizing the Research Computing Facility of the University of Nebraska.

REFERENCES

Bshouty, N. H., Jackson, J., & Tamon, C., 1999, "More efficient PAC-learning of DNF with membership queries under the uniform distribution," *Proc. of the 12th Annual Conf. on Computational Learning Theory*, pp. 286–295.

Chawla, D., Li, L., & Scott, S. D., 2003, *"On approximating weighted sums with exponentially many terms,"* (Technical Report TR03-02-02). Univ. of Nebraska-Lincoln. Early version in COLT'01.

Geman, S. & Geman, D., 1984, "Stochastic relaxation, Gibbs distributions and the Bayesian restoration of images," *IEEE Trans. on Pattern Analysis and Machine Intelligence*, vol. 6, pp. 721-741.

Khardon, R., Roth, D., & Servedio, R., 2001, "Efficiency versus convergence of boolean kernels for online learning algorithms," *Advances in Neural Information Processing Systems 14*.

Littlestone, N., 1988, "Learning quickly when irrelevant attributes abound: A new linear-threshold algorithm," *Machine Learning*, vol 2, pp. 285–318.

Liu, J.S., 1996, "Peskun's Theorem and A Modified Discrete-State Gibbs Sampler," *Biometrika*, vol. 83, pp. 681-682.

Metropolis N., Rosenbluth A.W., Rosenbluth M.N., Teller A.H. and Teller E., 1953, "Equations of state calculations by fast computing machines," *J. of Chemical Physics*, vol. 21, pp. 1087-1091.

Neal, R. M., 1995, *"Suppressing random walks in Markov chain Monte Carlo using ordered overrelaxation,"* (Technical Report No. 9508). Dept. of Statistics, University of Toronto.

THE ALGORITHM OF SYNTHESIS OF LOGICALLY TRANSPARENT MULTILAYER NEURAL NETWORKS

VALERIY I. DUBROVIN
Dep. of Software, Zaporozhye National Technical University, Zhukovskiy str., 64, Zaporozhye, 69063, Ukraine
e-mail: vdubrov@zntu.edu.ua

SERGEY A. SUBBOTIN
Dep. of Software, Zaporozhye National Technical University, Zhukovskiy str., 64, Zaporozhye, 69063, Ukraine
e-mail: subbotin@zntu.edu.ua

YURIY N. VNUKOV
Zaporozhye National Technical University, Zhukovskiy str., 64, Zaporozhye, 69063, Ukraine

ABSTRACT

The novel algorithm of a neural network synthesis is considered. It makes possible to get a logically transparent neural network models of different objects and processes. Such models are very useful for a knowledge extraction from the data and analysis of a data topology. The algorithm based on a cluster-regression approximation and allows to form a network topology and evaluate neural network weights in a non-iterative mode.

INTRODUCTION

The artificial neural networks are an effective mean for a construction of recognition and diagnostic systems based on black-box models [1,2]. However, such a models are not useful for a knowledge extraction because their structure is a difficult for a human analysis. Nowadays the development of algorithms allowing to built a logically transparent neural network models is a very actual.

This paper is devoted to the development of a new algorithm of dependence approximation on a dot data and it's interpretation for a synthesis and training of logically transparent neural networks.

Let we have a training set $x = \{x^s_j\}$, where x^s_j is a value of j-th feature of s-th sample, y^s is a value of target parameter of s-th sample, $s = 1, 2, ..., S$; S - number of samples in a training set. Then the dependence y from x we can present as: $y = f(w,x)+Err$, where f(w,x) is a function, which type is determined by the approximation algorithm or by user, w is a set of parameters (weights), which gives possible to adjust function to solve of a decided problem, Err is an error, which appears because the approximating function is not complectly corresponds to a real dependence and approximating function parameters evaluated with error.

For approximating function parameters evaluation it is possible to use regression analysis. When the one-dimensional linear regression models are used for approximation of many-dimensional nonlinear dependence, as a rule it fails to ensure a required exactitude of prediction. But the parameters of approximating function can be calculated in a uniterated condition. On the other hand in separate areas in a feature space the value of a target parameter can be determined by someone feature, which most correlated with it in a given area.

So, in general we can concludes that piece-linear regression for some features can supply the best approximation, than the linear regression model for a one feature. The construction of a piece-regression model requires a selection of a cluster - compactly grouped samples areas in a feature space with close values of a target parameter.

THE ALGORITHM OF A CLUSTER-REGRESSION APPROXIMATION

The algorithm of a cluster-regression approximation consists from the next steps.

Step 1. Initialization: form a set of samples x and a set of target parameter values y.

Step 2. On the basis of a training set form a clusters C^q, q = 1,2,...,Q, where Q is a number of formed clusters. For clusters forming we propose to use a below-described algorithm of a cluster selection.

Step 3. Separate samples of training set on groups for each cluster by the minimal distances from a sample to cluster center in a feature space.

Step 4. For samples of each group find values of coefficients of correlation of their features and values of target parameter:

$$r(x_j, y) = \frac{\sum_{s=1}^{S}\left(x_j^s - \frac{1}{S^q}\sum_{s=1} x_j^s\right)\left(y^s - \frac{1}{S^q}\sum_{s=1} y^s\right)}{\sqrt{\sum_{s=1}^{S}\left(x_j^s - \frac{1}{S^q}\sum_{s=1} x_j^s\right)^2 \sum_{s=1}^{S}\left(y^s - \frac{1}{S^q}\sum_{s=1} y^s\right)^2}}, x^s \in C^q,$$

j=1,2,...,N.

Step 5. For each group of samples find a number of a feature, which has a maximal absolute value of correlation coefficient with a target parameter:

$$g^q = \arg\min_j |r(x_j, y)|,$$

q =1,2,...,Q; j=1,2,...,N.

Step 6. For each group of samples evaluate values of parameters of linear regression equation for target parameter by the feature with number g^q.

If the linear regression equation of a target parameter on a feature g^q for samples of a q-th cluster have a view:

$$y_q = b_{q0} x_{g^q} + b_{q1},$$

than the values of coefficients b_{q0} and b_{q1} will be determined by the formulas [3]:

$$b_{q0} = (a^q_1 - a^q_2 a^q_3)/(a^q_4 - a^q_3 a^q_3),$$

$$b_{q1} = (a^q_4 a^q_2 - a^q_3 a^q_1)/(a^q_4 - a^q_3 a^q_3),$$

$$a^q_1 = \frac{1}{S^q}\sum_{s=1}^{S} y^s x^s_{g^q}, x^s \in C^q,$$

$$a_2^q = \frac{1}{S^q}\sum_{s=1}^{S} y^s, x^s \in C^q,$$

$$a_3^q = \frac{1}{S^q}\sum_{s=1}^{S} x_{g^q}^s, x^s \in C^q,$$

$$a_4^q = \frac{1}{S^q}\sum_{s=1}^{S} \left(x_{g^q}^s\right)^2, x^s \in C^q,$$

where S^q is a number of samples related to the q-th cluster.

The algorithm of output parameter evaluation on the basis of cluster-regression approximation can be presented as following steps.

Step 1. Initialization: give a set of samples $x=\{x^s_j\}$ for which the output parameter values must be evaluated and set a number of clusters Q and cluster centers coordinates $C=\{C^q_j\}$. For each cluster set a feature number g^q and values b_{q0} and b_{q1} of parameters of regression equation.

Step 2. For each sample find distances from sample to centers of all clusters $R(x^s,C^p)$, s=1,...,S; p=1,...,Q.

Step 3. For each sample find

$$q = \arg\min_p R(x^s, C^p),\ s=1,...,S;\ p=1,...,Q.$$

Step 4. The value of output parameter for each s-th sample can be defined by formula:

$$y^s = y_q^s = b_{q0} x_{g^q}^s + b_{q1},\ s=1,...,S.$$

THE ALGORITHM OF A CLUSTER SELECTION

For selection of clusters (compact regions of samples grouping in a feature space) we propose to use a following algorithm [4].

Step 1. Initialization of parameters of algorithm of cluster selection. Give an initial set of samples $x_{init.}$ and associated with them values of target parameter $y_{init.}$, and also L - a number of separations of initial set. Set variable N equal to the number of features of samples, set the variable S equal to the number of samples in initial set. Install the width of permissible interval of variation of target parameter: $dy = |\max(y_{исх.}) - \min(y_{исх.})| / L$, where min(g) and max(g) are the minimal and maximal values of a vector g, accordingly. Set newind=1.

Step 2. Evaluate distances between samples of initial set:

$$R(p,q) = \begin{cases} \sqrt{\sum_{j=1}^{N}\left(x_j^p - x_j^q\right)^2}, & p \neq q; \\ \text{RealMax}, & p = q, \end{cases}$$

p=1,...,S, q=1,..,S

where RealMax is maximal number represented in computer, x^p_j is a value of j-th feature of p-th sample.

Step 3. Find in the matrix of distances R the minimal element minx and it's indexes q and p, and also the maximal element maxx, in case, when the minimum and maximum are determined hear and below we ignore the elements equal to RealMax.

Step 4. Set: a=| maxx-minx | / (2L).

Step 5. If minx<Realmax, then go to the step 6, else go to the step 13.

Step 6. Set: $C^{(newind)} = x_{init.}(q)$, $y^*(newind) = y_{init.}(q)$, where C and y* are dimensions of standard samples of clusters and associated with them target values, accordingly. Set: newind=newind+1, the value of current minimal element in row: yteck=R(q,p). Find the minimal element mminx and it's indexes mqmin and mpmin from elements of q-th row of matrix R.

Step 7. If mminx<Realmax, then go to the step 8, else go to the step 11.

Step 8. Set the value of the pointer of deleted sample from the column deleted = 0 (in the matrix R the numbers of rows and columns must started from 1).

Step 9. If | yteck - R(q,mpmin) | $\leq$ a, then go to the step 10,

else set: deleted = mpmin, R(q,mpmin) = Realmax, R(mpmin,q) = Realmax and goto the step 11.

Step 10. If | $(y_{init.}(mpmin)-y_{init.}(q)| \leq dy$,

then set: R(v, mpmin) = realmax, R(mpmin, v) = realmax, v=1,..,M,

else set: $C^{(newind)}=x_{init.}(mpmin)$, $y^*(newind) = y_{init.}(mpmin)$, newind = newind+1, R(v, mpmin) = Realmax, R(mpmin,v)=Realmax, v=1,...,S.

Step 11. Find the minimal element mminx and it's indexes mqmin and mpmin from the elements of q-th row in matrix R.

Step 12. Go to the step 7.

Step 13. Set: R(v, q) = Realmax, R(q, v) = Realmax, v = 1,...,S, the pointer to deleted sample from the row: dstr = q. Find in matrix of distances R the minimal element minx and it's indexes q and p.

Step 14. Go to the step 5.

Step 15. If (deleted $\neq$ dstr) and (deleted > 0), then set:
$C^{(newind)} = x_{init.}(deleted)$, $y^*(deleted) = y_{init.}(deleted)$.

Step 16. End.

As a result of executing of this algorithm for initial set $x_{init.}$ and associated set of target parameter $y_{init.}$ we obtain a set of standard samples for each cluster C and associated with them set of values y*. After that it is needed to separate on clusters of all samples of initial set by the minimal distance to standard sample of clusters, and than for each cluster set coordinates of a cluster center:

$$C^q_j = \frac{1}{S^q}\sum_{s=1}^{S} x^s_j, x^s \in C^q,$$

$q = 1,..., Q; j = 1,...,N;$

where S^q is a number of initial set samples that lies in q-th cluster, Q is a number of selected clusters.

THE NEURAL NETWORK INTERPRETATION OF CLUSTER-REGRESSION APPROXIMATION ALGORITHM

After a clusters forming the algorithm of a cluster-regression approximation may be used for a synthesis and weights evaluation of multi-layer neural network.

If the Q clusters are formed than the neural network will be consist from 5 layers of neurons with following discriminant functions:

$$\varphi^{(1,i)}(x,w)=\sum_{j=1}^{N} w_j x_j + w_0 \text{ , i=1,2,...,Q;}$$

$$\varphi^{(1,i)}(x,w)=\sum_{j=1}^{N} (x_j - w_j)^2 \text{ , i = Q+1, Q+2,...,2Q;}$$

$$\varphi^{(2,i)}(x,w)=\sum_{j=1}^{2} w_j x_j + w_0 \text{ , i=1,2,...,} Q^2;$$

$$\varphi^{(3,i)}(x,w)=\sum_{j=1}^{Q} w_j x_j + w_0 \text{ , i=1,2,...,Q;}$$

$$\varphi^{(4,i)}(x,w)=\prod_{j=1}^{2} w_j x_j \text{ , i=1,2,...,Q;}$$

$$\varphi^{(5,i)}(x,w)=\sum_{j=1}^{Q} w_j x_j + w_0 \text{ , i=1;}$$

where $\varphi^{(\mu,i)}$ - is a discriminant function of i-th neuron of a μ-th layer.
The transfer functions of neurons will be determined by equations:

$$\psi^{(\mu)}(\alpha)=\begin{cases}0, \alpha \le 0, \\ 1, \alpha > 0;\end{cases} \quad \mu=2,3;$$

$$\psi^{(\mu)}(\alpha) = \alpha, \ \mu=1,4,5.$$

The weight coefficient of a j-th input of i-th neuron of μ-th layer of network will be evaluated by the formulas:

$$w_j^{(\mu,i)}=\begin{cases}0, \mu=5, i=1, j=0; \\ 1, \mu=5, i=1, j>0; \\ 1, \mu=4, \forall i, j>0; \\ 1-Q, \mu=3, i=1, j=0; \\ 1, \mu=3, i=1, j=0; \\ 0, \mu=2, \forall i, j=0;\end{cases} \qquad w_j^{(\mu,i)}=\begin{cases}1, \mu=2, \forall i, j=2; \\ -1, \mu=2, \forall i, j=1; \\ b_{q0}, \mu=1, i=q, j=g^q, q=1,...,Q; \\ b_{q1}, \mu=1, i=q, j=0, q=1,...,Q; \\ 0, \mu=1, i=q, 0<j\ne g^q, q=1,...,Q; \\ C_j^q, \mu=1, i=Q+1, Q+2,...,2Q, j=1,2,...,N.\end{cases}$$

The scheme of the neural network synthesed and trained by the proposed algorithm is shown on a figure.

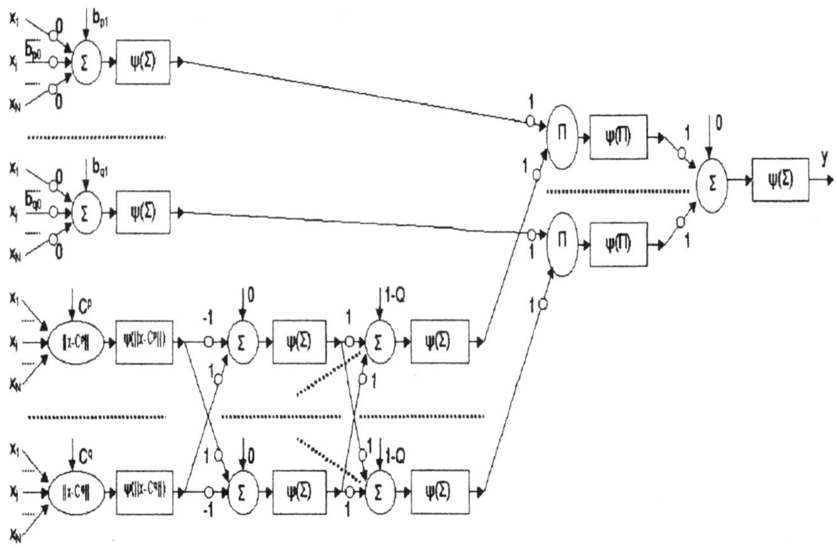

Fig. - Cluster-regression approximation neural network scheme.

CONCLUSIONS

The cluster-regression approximation algorithm discussed in this paper and it's neural network interpretation was realized in a software package Matlab 6.0. The developed program was used for a solution of several practical problems in technical and bio-medical diagnostics [2,5]. The results of conducted experiments shows that proposed algorithm gives possible to obtain better results than known linear regression methods, and it have a such simple and effective computing procedure.

The neural network syntesed by the proposed algorithm is a logically transparent and may be simple transformed to a decision rules tree, which gives possible to use it for knowledge extraction from the data. The correctness of a model and a training speed of neural network designed by the developed algorithm are sufficiently high for most applications in diagnostics.

REFERENCES

1. Beale M., Demuth H. Neural Network Toolbox User's Guide for Use with MATLAB 5.0.- Natick: Mathworks, 1997.- 700 p.
2. Dubrovin V., Subbotin S., Morshchavka S., Piza D. The plant recognition on remote sensing results by the feed-forward neural networks // International Journal of Smart Engineering System Design, 2001, № 3, P. 251-256.
3. Tennant-Smith J., Basic statistics, London, Butterworth and Co Publishers, 1985.
4. Dubrovin V.I., Subbotin S.A. Methods of increasing of effectiveness of neural network diagnostic procedures // Neurocomputers: design and application, 2002, vol.3.- Pp. 3-9. (in Russian)
5. Dubrovin V.I., Subbotin S.A., Yatzenko V.K. Neural network model of hardening coefficient of airengine details // Smart Engineering System Design: Neural Networks, Fuzzy Logic, Evolutionary Programming, Data Mining, and Complex Systems / ed.: C.H. Dagli, et al.- New York: ASME press, 2001, vol. 11, P. 939-944.

IMPROVING KERNEL-BASED CLASSIFIERS BY GUIDED DYNAMIC SAMPLE SELECTION

INMA MORA-JIMÉNEZ
Dept. of Signal Processing and Communications, Univ. Carlos III de Madrid, 28911 Madrid, SPAIN

REBECA GARCÍA-MARCIEL
Dept. of Signal Processing and Communications, Univ. Carlos III de Madrid, 28911 Madrid, SPAIN

ANIBAL R. FIGUEIRAS-VIDAL
Dept. of Signal Processing and Communications, Univ. Carlos III de Madrid, 28911 Madrid, SPAIN

ABSTRACT

There is a growing body of evidence which supports the view that learning processes improve when focused on the more difficult examples. This principle can be applied to pattern recognition by means of selecting those samples which are critical to define the decision boundary. In this paper, we present an efficient learning strategy for kernel-based classifiers, consisting on training with those samples which provide an output whose magnitude is under a given threshold plus those producing an output under the estimate provided by a voting *k*-Nearest Neighbour (*k*NN) rule. Simulation results for some synthetic and real problems show that this approach leads to both a reduced computational burden and/or better performance when compared with classical and decision-based procedures.

INTRODUCTION

In disciplines like artificial intelligence or psychology, a growing body of research supports the view that the learning process improves when the subject focuses in those issues more difficult to learn. Based on these works, some authors (Munro, 1992) have analyzed the effects that the harder to learn samples have on neural networks, and different pattern selection strategies have been proposed (Cachin, 1994). Within this context, the philosophy of Munro represents a valid framework to carry out a pattern selection strategy, and so it will remain in the background of our work.

When dealing with a pattern recognition problem, it should be remarked that we are mainly concerned about the allocation of a pattern to the proper class, while the certainty of the classification can be pushed into the background. However, in the classical training of neural classifiers, the network is forced to reach the desired values (or targets) for every sample, trying to achieve the maximum certainty on every decision it makes. As the purpose of the training should be to avoid missclassifications, it can be considered that classical training creates unnecessary compromises, what leads to a waste of computational resources and time, as well as to a possible loss in accuracy. It seems therefore convenient to "relax the targets" along the training process, not forcing the network to reach the ideal values when it is able to easily classify without mistake. This strategy will allow the training to focus on the "key" samples, essentially those patterns lying near the division boundary between classes, while forgetting those which have been already mastered by the network.

Following this line, we have developed a method to train kernel-based classifiers with a classical architecture. Our contribution is basically founded on the dynamic selection of the training samples used to update the network parameters: the most influential patterns will be automatically identified by the network, and its number will evolve according to the training process. This technique can be applied to any kind of neural network, as long as it is based on local approximators.

Our proposal can be seen as a mixture of two type of methods. On the one hand, we have the so-called Decision Based Neural Networks (DBNN) by Kung and Taur (1995), where a sample is only considered for learning if it is missclasified by the net. Its main disadvantage lies in its poor generalization capabilities and bad convergence properties. In the second family of methods, all patterns are considered along the training process. This classical approach implies high computational resources when large databases are used.

The rest of the paper is organized as follows. In next Section and to set the background of our approach, we revise classical kernel-based classifiers. Then, the learning strategy we propose is described and analysed, presenting experimental results and discussing the advantages of our scheme. Finally, some concluding remarks and suggestions for future lines of research are provided.

CLASSICAL LOCAL METHODS FOR CLASSIFICATION

Without loss of generality and for simplicity, let us assume a binary classification problem, where both classes are represented by symbols C_0 and C_1. In order to train the network we have a collection S of data (called training set), which is composed of N patterns $\mathbf{x}_i$ with their corresponding targets t_i: $S = \{\mathbf{x}_i, t_i\}$, i =1, ..., N with $t_i \in \{-1, +1\}$. From now on, we will consider that a datum belongs to class C_1 if its target is "t=1" and the opposite for C_0.

In this paper we are interested in classifiers with local properties, since the understanding and interpretability of the training process allows to monitor the influence of different parameters over certain regions (Bishop, 1995). Among all local classifiers, the neural network with Radial Basis Functions (RBF) has been chosen, due to its proved universal approximation capabilities (Park and Sandberg, 1991) and well-studied architecture.

A RBF network is a feedforward network with a single layer of centroids (or basis functions), linearly combined to produce the output $o(\mathbf{x})$. The most typical basis function is the gaussian one: for a given a sample $\mathbf{x}$, a RBF with N_C spherical gaussian centroids provides the following output:

$$o(\mathbf{x}) = \sum_{s=1}^{N_c} w_s \exp\left(\frac{-\|\mathbf{x} - \mathbf{c}_s\|^2}{2\sigma_s^2}\right) \tag{1}$$

where $\mathbf{c}_s$ corresponds to the s'th centroid, σ_s is its dispersion parameter and w_s is a weightening factor that connects $\mathbf{c}_s$ with the output layer.

The values for the parameters $\{\mathbf{c}_s, w_s\}$, s=1, ..., N_C will be set in a supervised manner (Bishop, 1995) by the minimization of a cost function C

through a gradient descent algorithm. We discard the training of the dispersion parameters to preserve the network local properties.

$$\min_{\mathbf{c},\mathbf{w}} C(\mathrm{t},\mathrm{o}) = \sum_{i=1}^{N} (\mathrm{t}_i - \mathrm{o}_i)^2 \tag{2}$$

In our approach, the standard *k*-means clustering algorithm (Bishop, 1995) has been used to find a set of initial *k* centroids. The dispersion parameters σ_s are determined from the *k* clusters previously obtained by finding the nearest samples to each centroid: σ_s is calculated through the sample variance of the patterns assigned to each centroid. A factor *F* is applied to σ_s to ensure that all regions are represented by any center. In this way, the network outputs can be interpreted as the difference between the *a posteriori* probabilities of class membership:

$$\mathrm{o}(\mathbf{x}) = \hat{\mathrm{P}}(\mathrm{C}_1 \mid \mathbf{x}) - \hat{\mathrm{P}}(\mathrm{C}_0 \mid \mathbf{x}) = \Delta\hat{\mathrm{P}}(\mathbf{x}) \tag{3}$$

However, even when the RBF is able to estimate Eq.(3), it is not always essential. In most practical cases, it suffices if the net provides a good class prediction: this idea is the basis of our approach.

STRATEGY OF TRAINING USING DYNAMIC PATTERN SELECTION

The philosophy of our proposal is implicitly based on a pattern selection strategy: a sample will be used in the training process unless the network classifies it correctly and with enough accurate margin. Although Eq.(2) is still the goal, we assume that a sample $\mathbf{x}_i$ has been learned even when its output o_i has not reached the target value t_i. Therefore our approach is based on a pattern selection strategy, where patterns already mastered by the network are discarded as the training proceeds. The effect of this learning strategy is double: saving computational resources and improving performance.

To carry out such a training strategy, a selection criterion must be specified first. As Eq.(3) is progressively achieved along training, we should use another complementary method which provides a reliable mesasure about the classification accuracy. In order to deal with this task, we have turned to the *k*-Nearest Neighbour (*k*NN) classifier (Ripley, 1996).

Classification with *k*NN is based on the search of the *k* most similar patterns to a certain sample **x**, using to this purpose all the training set *S*. Once the *k* nearest neighbours have been found, the sample is assigned to the class with the predominant target among its *k* nearest patterns. This memory-based method estimates the difference between the *a posteriori* probabilities according to:

$$\Delta\hat{\mathrm{P}}_{k\,\mathrm{NN}}(\mathbf{x}) \cong \frac{k_1 - k_0}{k_1 + k_0} = \frac{\Delta k}{k} \tag{4}$$

where k_1 and k_0 are the number of patterns for classes C_1 and C_0, respectively, among the *k* nearest neighbors ($k = k_1 + k_0$).

This way, our approach can be seen as a hybrid classifier: we use Eq.(2) to train the network, but only with those samples lying on areas where the class estimate is doubtful. This corresponds to regions of high overlap and therefore near the decision boundary. The training criterium we propose can be finally specified as follows: an input pattern ($\mathbf{x}_i$) will be used to train the network only if it fulfils any of these conditions:

Condition 1: It is missclassified by the network: sign (o_i) $\neq t_i$
Condition 2: Its output is not sufficiently close to its target value:
$|o_i| < T$ and $|\Delta \hat{P}_{kNN}(\mathbf{x}_i)| \neq 1$
Condition 3: The estimate provided by *k*NN is doubtful: $|\Delta \hat{P}_{kNN}(\mathbf{x}_i)| < T$

where *T*, specified as $T \cong u_k/k$, is a parameter named threshold value; *k* is the parameter used in Eq.(4) and u_k is an integer value in the range [0,*k*]. Let us briefly comment each requisite:

- Condition 1 coincides with the DBNN technique, where a correct classification implies the sample will not be used in training next epoch.
- Condition 2 allows the selection of those samples near the division boundary; i.e. patterns which the network does not classify with an accurate margin: condition $|o_i| < T$ avoids the selection of well-learned samples for training. As in our kernel-based classifier the area of influence for every kernel function is limited, regions far from any centroid can lead to output values close to zero. It is important to note that, among the samples with low output are those located far away from the decision border, which do not add any relevant information and so should be discarded: the constraint $|\Delta \hat{P}_{kNN}(\mathbf{x}_i)| \neq 1$ is then necessary to fulfil this requirement.

  As the output values tend to increase along the training and the threshold *T* is constant, the number of samples which satisfy $|o_i|<T$ usually decreases during training: at the beginning, patterns located in a wide region near the boundary are selected; as the training proceeds, this region is progressively reduced so that only the closest samples to the divisory line are selected. Condition 2 may be therefore seen as a "dynamic" criterion, which adapts the selection of patterns to the training process.
- Condition 3 adds reliability and robustness to the selection rule: samples selected by this condition are independent of the training stage, and so it guarantees that the most difficult to learn patterns are always included.

It is worthy to remark that, despite the net learns from a subset of the original dataset, all regions are considered thanks to the initial clustering with the whole training set.

EXPERIMENTAL WORK

In this section we present experimental work to show the effect of selecting training samples in a guided manner. We have considered a range of binary benchmark problems: *Ripley* is a synthetic bidimensional dataset (Ripley, 1996) with a Bayes accuracy of 92% over the test partition; the rest, named *Pima Indian*, *Ionosphere* and *Abalone*, are from the UCI Machine Learning Repository (Blake and Merz, 1998). As in *Pima Indian* there is no test partition, we have created it choosing 40% of the whole dataset for this purpose. Regarding *Abalone*, we use its binary version, provided by (Ruiz and López-de-Teruel, 2001), where the best reported rate is lower than 80%. All data have been first preprocessed to have zero mean and unit deviation and a validation set with 20% of the whole dataset has been randomly chosen.

We have experienced with three learning methods: *Classical* training, that uses all examples to update the network parameters; *Decision-based* training, where only the misclassified patterns are used for the parameters updating; and our approach, *k-NN guided* training, which can be considered as a hybrid between the previous ones. In order to make a fair comparison, we have established the same architecture and initial parameters for all of them. Specifically, we have considered a RBF network with spherical gaussian kernels and one non-linear output (hyperbolic tangent). We have considered 10 values for the numer N_C of centroids (N_C around 5 % of the training size). To have a different covering field for every kernel, several values F for the dispersion parameter σ have been tested: F={1, 1.5, 2, 2.5, 3}. A gradient descent algorithm has been applied to update weights and centroids using a constant learning rate (0.001) and applying cross-validation as stopping criterion.

For the k-NN guided scheme, it is necessary to set two additional parameters: k and u_k. Note that, even when T is the same for different k and u_k, parameter k affects to $\Delta\hat{P}_{kNN}$ and so it has influence on Conditions 2 and 3. Although the exact value for k is not critical in this technique, it should be as lower as possible to preserve the kNN local behaviour.

TABLE 1. Comparison of average test classification rates (in %) and standard deviation (in brackets) for different problems and training schemes, according to a cross-validation procedure. Results in boldface correspond to the best accuracy for each problem.

Database	*Classical* Test accuracy % (standard deviation) (% kernels, F)	*Decision-based* Test accuracy % (standard deviation) (% kernels, F)	*k-NN guided* Test accuracy % (standard deviation) (% kernels, F, k, u_k)
Ripley	**90.7** (0.4) (8, 1)	87.0 (2.3) (6, 1)	90.6 (0.5) (6, 1, 7, 3)
Ionosphere	83.1 (6.6) (8.7, 1)	94.1 (3.7) (11.2, 1.5)	**94.7** (4.2) (8.7, 2, 3, 0)
Pima Indian	76.7 (1.8) (7.6, 2)	74.5 (2.9) (5.4, 2.5)	**77.6** (2.3) (6.5, 3, 7, 3)
Abalone	79.8 (0.4) (1, 2.5)	77.8 (0.6) (1, 2)	**80.2** (0.3) (1,2.5, 15, 9)

Table 1 summarizes the results of our experiments in terms of test accuracy. It shows the average rate (in %) and standard deviation (in brackets) for the test set over 10 different runs. From these figures we conclude that the k-NN guided scheme outperforms classical and DBNN classifiers. This advantage is specially remarkable in *Pima Indian* and *Abalone* (real-world problems). Additionally, *Abalone* is the dataset with more training samples (2507), what also implies a significant reduction (around 30%) in the number of patterns processed.

Although for *Ionosphere* our scheme also presents lower error rate, the corresponding standard deviation increases in comparison with DBNN, which implies in this case both techniques perform in a similar way. We can conclude the *Ionosphere* dataset is too small (201 training samples and 34 features) to be effectively split for cross-validation purposes. Regarding *Ripley*, the advantage of our approach lies in the reduction of samples used to update parameters, since the accuracy is comparable with the one obtained with the classical method.

We finish the discussion by saying that the k-NN guided scheme usually needs less kernels than the classical one, what also contributes to reduce computational effort. Although not presented in this paper, we consider important to remark that k-NN guided also outperforms voting kNN classifier.

A final comment on the T value should be made: except for the *Ionosphere* problem, it can be checked it is close to 0.5. This characteristic could be used to avoid the search of a good relation between k and u_k, and it is also a consequence of using a hybrid learning scheme.

CONCLUSIONS

A new learning technique has been introduced in this paper. The proposed scheme is implicitly based on a dynamic pattern selection strategy, where the most influential patterns are automatically identified by the network and used to update its parameters. This pattern selection is guided by the information given by a kNN rule, which provides statistical consistency to the method. Techniques as the proposed in (Fukunaga and Narendra, 1975) can be used to reduce considerably the computational burden inherent to the kNN rule.

The proposed k-NN guided algorithm has been compared with other existing learning methods: classical training and DBNN technique. Experiments reveal that the k-NN guided algorithm shows both better performance and reduced computational burden, specially when applied to large databases.

Although at first sight our technique requires to establish some extra parameters (u_k, k), we have stated through experimentation that most problems show best performance when the relation $T = u_k/k$ is close to 0.5. This result can be used as a shortcut to reduce the parameters search. Connected with this idea, a variable T which adapts to the training stage is proposed as a future work.

REFERENCES

Blake, C. L., and Merz, C.J., 1998, UCI Repository of machine learning databases, Irvine, CA, University of California, Department of Information and Computer Science. [http://www.ics.uci.edu/~mlearn/MLRepository.html].

Bishop, C. M., 1995, *Neural Networks for Pattern Recognition*, Oxford University Press.

Cachin, C. , 1994, "Pedagogical pattern selection strategies", *Neural Networks*, vol. 7, pp. 171-181.

Fukunaga K., and Narendra P.M., 1975, "A Branch and Bound Algorithm for Computing the k-Nearest Neighbors", *IEEE Trans. on Computers*, pp.750-753.

Kung S. Y., and Taur J. S., 1995, "Decision-Based Neural Networks with Signal/Image Classification Applications", *IEEE Trans. on Neural Networks*, Vol. 6, no. 1, pp.170-181.

Munro, P. W., 1992, "Repeat until bored: a pattern selection strategy", Advances in Neural Information Proc. Sys. Vol.4, pp. 1001-1008, in J.E. Moody et al. (Eds.): Morgan Kaufmann.

Park, J., and Sandberg, I.W., 1991, "Universal Approximation using Radial Basis Function Networks", *Neural Computation*, vol. 3, no. 2, pp.246-257.

Ripley, B. D. , 1996, *Pattern Recognition and Neural Networks*, Cambridge University Press.

Ruiz, A., and López-de-Teruel, P. E., 2001, "Nonlinear Kernel-based Statistical Pattern Analysis", *IEEE Trans. on Neural Networks*, vol. 12, no. 1, pp.16-32.

PARAMETERIZATION OF SCATTERED SURFACE POINTS USING A SOFM

GEORGE K. KNOPF, ARCHANA SANGOLE AND PHILIP C. IGWE
Department of Mechanical & Materials Engineering,
Faculty of Engineering, The University of Western Ontario,
London, Ontario, Canada, N6A 5B9

ABSTRACT
Computer aided design (CAD) and computer graphic applications require smooth surfaces to be constructed from irregularly spaced and unorganized digitized data. Parametric surfaces are often the preferred mathematical representations in freeform modeling because they permit simple shape modification by changing only a small number of parameters such as control points, knot values, or weight values. In this paper, a 2D self-organizing feature map (SOFM) is used to assign parametric values to the measured data prior to Bézier, B-spline or NURBS surface fitting. The biparametric function defines a path where the parameter pair (u,v) represent the normalized distance of a data point along the reconstructed surface. Common parameterization algorithms such as the chord length and centripetal methods require additional information about the connectivity between measured points. Due to the predefined lattice of the SOFM it is possible to track neighbouring surface points along lines of linked nodes and assign parametric pairs (u, v) to the original scattered data. Illustrative examples are presented to demonstrate how the SOFM can be used to parameterize scattered, range data prior to Bézier surface fitting.

INTRODUCTION

The mathematical reconstruction of freeform surfaces from non-uniformly spaced, or scattered, data is a common problem found in a variety of scientific and engineering disciplines such as geology, cartography, virtual reality modeling, computer-aided design, and reverse engineering. Digitized data acquired by range sensing technology are an enormous volume of irregularly spaced coordinate points and therefore cannot be directly incorporated into commercially available computer graphics and CAD software. Consequently, a great deal of post-processing is required to reduce the number of points prior to fitting continuous surface patches to the data.

Although ordinary first-degree analytical expressions can be used to define planar facets in a polygonal surface mesh, continuous curved surfaces require at least a three-degree polynomial expression. Parametric curves and surfaces are used in most design software because the mathematical tool allows the user to specify various smooth freeform shapes based on data interpolation or approximations of the designer's intent. The parametric surfaces are constructed using tensor product methods (Hoschek and Lasser, 1993) that multiply the basis functions of two parametric curves in the u and v directions. The parametric function defines a path where the parameter pair (u, v) represents the normalized distance of a data point along the reconstructed surface in the two directions. The combined basis function is then applied to control points in the 3D data space in order to create a smooth continuous tensor product surface, Figure 1.

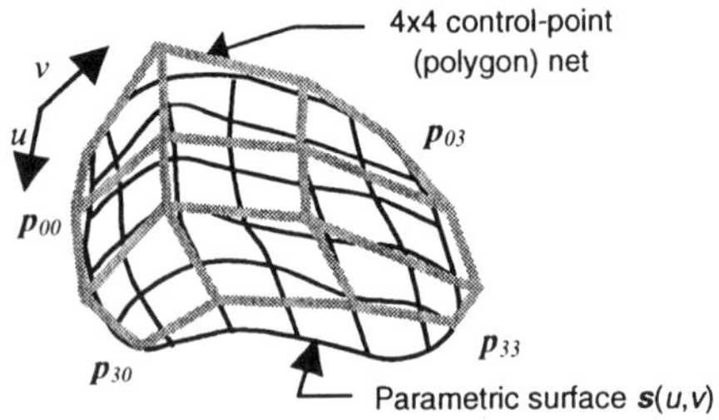

Figure 1. Synthetic surface patch s(*u*,*v*) created by Bernstein polynomials.

The control polygon for the tensor product surface is given by a mesh with vertices at p_{nm}. The combined basis function for the surface patch, $\Phi_{nm}(u,v)$, is the product of the curve basis functions given by

$$\Phi_{nm}(u,v) = \phi_{nN}(u)\ \phi_{mM}(v) \quad , \tag{1}$$

where $\phi_{nN}(u)$ and $\phi_{mM}(v)$ are basis polynomials and the point on the surface at (u,v) is

$$s(u,v) = \sum_{n=0}^{N} \sum_{m=0}^{M} \Phi_{nm}(u,v)\ p_{nm} \quad , \tag{2}$$

and $p_{nm} = [\,p_{x_{nm}}, p_{y_{nm}}, p_{z_{nm}}\,]^{T}$ is the spatial vector of the (n^{th}, m^{th}) control point for $0 \leq n \leq N$ and $0 \leq m \leq M$. A basic property of the combined basis function is that it must sum to one at any (u,v) pair.

Parametric forms such as Bézier, B-spline, and non-uniform rational B-splines (NURBS) permit simple shape modification by changing only a small number of parameters such as control points, knot values, or basis function weights. Creating continuous biparameteric surfaces from measured data involves two principle steps: parameterization and surface approximation. *Parameterization* is necessary to determine the topology, geometry and boundary of the final surface. Different parameterization strategies can be used to assign parametric values (u,v) to the individual data points: equally spaced, chordal, centripal, geometric, and affinity invariant parameters (Hoschek and Lasser, 1993). Gu and Yan (1995) describe a novel parameterization scheme that exploits the learning ability of a supervised neural network to map parameters (u, v) onto corresponding coordinate points (x, y, z). However, these techniques are based on the assumption that the data points are arranged on a grid and are therefore not suitable for applications where the data is unorganized or scattered.

Methods that deal with scattered points often project the original 3D coordinate data onto a 2D parametric plane and then calculate the corresponding (u,v) values with respect to the projected boundary. Unfortunately, the topology of the reconstructed surface is not always preserved because the parameterization scheme does not guarantee that the relative position between original 3D points is maintained. Ma and Kruth (1995) avoid this problem by utilizing a nonplanar surface, which is a rough approximation of the final fitted form, as the initial base surface that is iteratively adjusted until it converges to the final surface. Although, projecting the data points onto an adaptive 3D base surface produces a better assignment of (u,v) values to the original 3D data the technique may produce a base surface with concave boundaries that can lead to undesirable results such as a self-intersecting surface (Barhak and Fischer, 2001a).

Barhak and Fischer (2001) describe an alternative parameterization method based on the self-organizing feature map (SOFM) and PDE (Partial Differential Equation) techniques. They examined two cases; one that involves training the nodes along the boundary of the map prior to adjusting any interior nodes, and the second is the reverse. In both cases however, the SOFM is trained on a 2D projection of the coordinate points and the primary focus is to generate a parametric base surface that 'best' represents the object surface. Várady *et al.* (1998, 1999) apply the SOFM directly to the sample coordinate points instead of 2D projections of the points.

This paper discusses a coordinate parameterization technique for free-form surface fitting similar to the adaptive method described by Várady *et al.* (1998, 1999) and Barhak and Fischer (2001). A 2D SOFM is used to establish a topological order to scattered data and, thereby, provide connectivity information about the measured surface points. However, all the nodes are trained simultaneously and no constraints are placed on the boundary and internal nodes of the SOFM. Furthermore the size of the network is kept constant throughout the training process. The objective is to determine the connectivity between 3D scattered coordinate points using the self-organizing capability of the SOFM without making any prior assumptions about the data and utilize the resulting mesh representation, with as the facetted base to "parametrize" the scattered data prior to generating a smooth Bézier surface. Illustrative examples using range data are provided to demonstrate data parameterization and Bézier surface reconstruction.

DATA PARAMETERIZATION

The polygonal mesh with quadrilateral elements created by a SOFM is used to assign the parametric values (u, v) to each measured point prior to Bézier surface fitting. The SOFM is an unsupervised clustering algorithm to learn the topological order of scattered data by grouping similar inputs into common neighbourhoods (Kohonen, 1997). The algorithm arranges the weights of nodes, $\boldsymbol{w}_{ij}$, on a predefined lattice such that they reflect some physical characteristic of the external input $\boldsymbol{X}^d = [\, x_1^d \,,\; x_2^d \,,\; x_3^d \,]$, where $d = 1, 2, \ldots D$, and D is the total number of sample coordinate data points used for training. Each input data vector $\boldsymbol{X}^d$ corresponds to (x, y, z) coordinates in 3D measurement space.

The procedure for creating the base surface for parameterization begins by initially setting the node weights to small values distributed throughout the input space. For each randomly selected coordinate point an error or difference, ξ_{ij}^d, using,

$$\xi_{ij}^d = \sum_{k=1}^{3} \left(x_k^d - w_{nij} \right)^2 \tag{3}$$

where x_k^d is the kth input, w_{nij} is the weight from the kth input to the $(i, j)^{\text{th}}$ node, $i = 1, 2, \ldots$ I, and $j = 1, 2, \ldots$ J. The winning node $(i, j)^*$ is selected as the unit with the minimum distance ξ_{ij}^d. During the adaptation phase, weights of the node $(i, j)^*$ and all the nodes residing within the specified neighborhood, NE_{ij}, are updated using

$$\boldsymbol{w}_{ij*}\,(\text{new}) = \boldsymbol{w}_{ij*}\,(\text{old}) + \alpha\,[\,\boldsymbol{X}^d - \boldsymbol{w}_{ij*}\,(\text{old})\,] \tag{4a}$$

where,

$$\alpha = 0.001 + \mu\,(NE_{ij*} \,/\, NE_{\text{initial}}) \;, \tag{4b}$$

μ is a predefined learning rate, and NE_{initial} is the initial neighborhood size in terms of number of nodes. The size of the neighbourhood is decreased as training progresses until it consists of only the winning unit.

On completion of the adaptation process, the connected of the nodes along the grid lines of the lattice represents a non-planar base surface for determining the parametric values of each measured coordinate point X^d and provides the sequence of coordinates along lines of constant u and v in the parametric space, Figure 2. Each node is assigned a parametric pair (u_{ij}, v_{ij}) using centripetal approximation (Hoschek and Lasser, 1993). If the number of nodes in the lattice is approximately equal to the number of data points, then each coordinate point is given the parameter pair of the nearest $(i, j)^{th}$ node. Alternatively, greater accuracy can be achieved by interpolating the (u, v) value based on the relative distances of surrounding nodes.

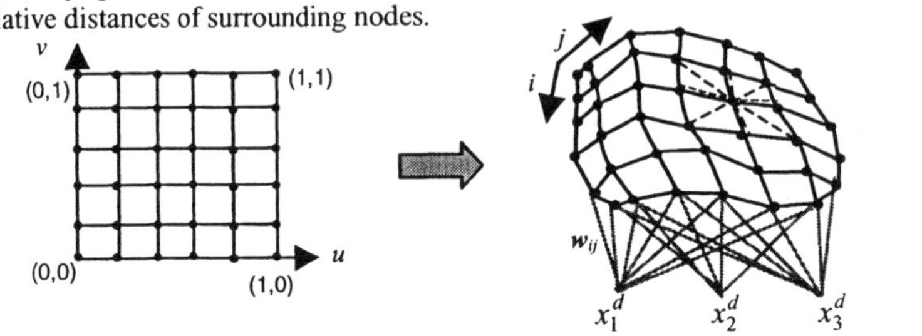

(a) Parametric grid. (b) SOFM lattice fitted to measured data.

Figure 2. Parameterization of coordinate points X^d using a 2D SOFM lattice which is topologically equivalent to the mesh in the parameter space.

The proposed scattered data parameterization method is essentially an iterative clustering technique and, therefore, will take as much computational time to reach an acceptable solution as any other unsupervised clustering program. The number of nodes (IxJ), number of data points (D), dimensionality of the input data vectors (e), and the number of training cycles (N_{cycle}) all influence the computation time. For every input vector presented to the network, (IxJ) sequential operations are performed to search for the winning node. Therefore, if $T = D \times N_{cycle} \times e \times (I \times J)$ then the computation time can be approximated as a linear operation O(T) (Standish, 1994) - with some constant allowance for addition and multiplication operations required to compute the error measure during learning. The size of the neighbourhood NE_{ij} during weight adaptation is another key parameter. Significant efficiencies in computing time can be achieved if an efficient neighbourhood reduction strategy is used during weight training.

SURFACE RECONSTRUCTION

The order of a bi-parametric surface patch is directly related to the number of control points. The greater the number of control points the higher the surface order and interpolation will require high-order polynomials that produce undesirable oscillatory behaviour around the points. For dense data, it is therefore necessary to reduce the volume of the measured points prior to fitting. The surface geometry is then approximated in a "best-fit" sense using a least-mean square approach. A variety of techniques can be used to fit smooth parametric functions to parameterized coordinate data. The Bernstein Basis Function (BBF) network (Knopf and Kofman, 2001) is one adaptive approach to determine the control points, $\mathbf{p}_{nm}$, necessary to construct a Bézier function that approximates the underlying surface. Individual units in the first layer of the BBF network transform the parametric inputs, (u_{ij}, v_{ij}), into nonlinear outputs, $\phi_{nN}(u_{ij})$ and $\phi_{mM}(v_{ij})$, using Bernstein polynomials. The second layer transforms the outputs of the first layer (ϕ_{nN} , ϕ_{mM}) into the surface coordinates $\mathbf{s}(u_{ij}, v_{ij})$ with weight vectors $\mathbf{p}_{nm}$, that represent Bézier control points. The position of a point, $\mathbf{s}(u_{ij}, v_{ij})$, on the Bézier surface is given by Eqn. (2). The shape of a Bézier surface patch is determined by the location of the control points that make up the characteristic polygon.

EXPERIMENTAL STUDY

Examples include coordinate data of a man's face and nose (Campbell and Flynn, 1998). Figure 3 shows the partitioned regions. The data set for the face contains 7257 coordinate points while the nose data set contains 6377 points. In both examples, a 2D SOFM with approximately the same number of nodes as data vectors was used to assign (u,v) parametric values to each measured coordinate point.

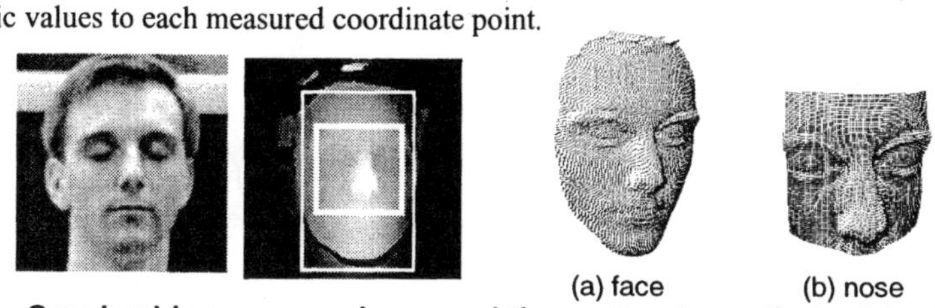

Figure 3. Grey level image, range image and the scattered coordinate data of the man's face (7257 points) (a), nose (6377 points) (b). Each point is given by vector X^d.

For the man's face, a 85x85 SOFM lattice with an initial neighbourhood of 9 units (19x19 units) was used to parameterize the face data set, Figure 4a. The coordinate points and their corresponding parametric pairs were then fitted by a 13x13 Bézier patch, Figures 4b, with a maximum number of 100 cycles through the data set.

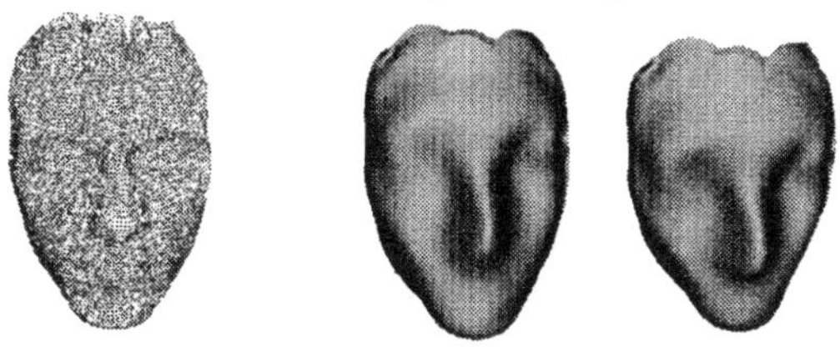

(a) SOFM grid of face. (b) Rendered Bézier surface patch.

Figure 4. The polygon mesh (85x85 SOFM) (a) and reconstructed 13x13 control net Bézier surface (b) of the man's face.

The resultant patch is a smooth function that approximates the underlying surface given by the cloud of points. The blurring effect is expected since the entire face is represented by a single Bézier function. In many practical applications, the dense data is first segmented into contiguous regions with common surface curvature properties and the points within these partitioned regions are then fitted with low-order surface patches. In a similar fashion, for the nose data, a 79x79 SOFM network with an initial neighbourhood of 9 units (19x19 units) was used and the parameterized data was then represented by a 13x13 Bézier patch, Figures 5. Small distortions exist in the final reconstructed surface because of the global modification characteristics of the Bernstein polynomials used for the approximation. In both examples, the SOFM algorithm enabled a faithful representation of the freeform surface for parameterizing the coordinate points.

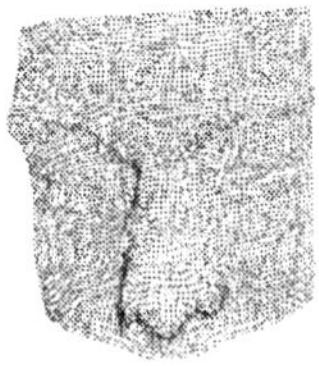

(a) SOFM grid of nose region.

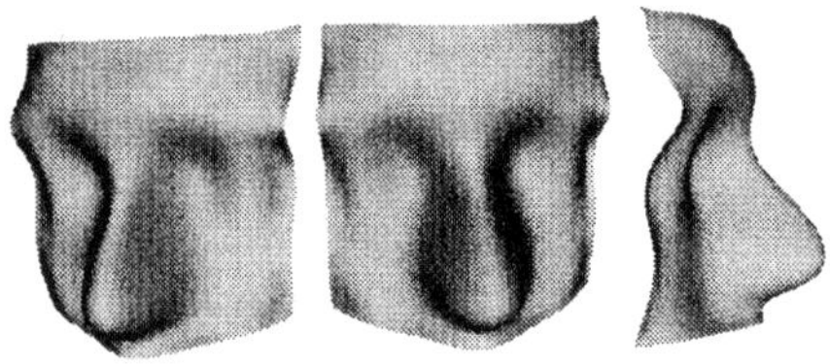

(b) Rendered Bézier surface patch.

Figure 5. The polygon mesh (79x79 SOFM) (a) and reconstructed 13x13 control net Bézier surface (b) of the man's nose.

CONCLUSIONS

A data parameterization method using a 2D SOFM was used introduced in this paper to determine the underlying surface topology of numerous scattered digitized data and help assign (u,v) parametric values to the measured 3D points. The technique is based on reconstructing a nonplanar parametric base surface with the SOFM, calculating the parameter pairs using centripetal approximation on the generated grid, and fitting the parameterized data values with a Bézier surface. The feasibility of the parameterization method is demonstrated on range images captured by a laser scanning system.

ACKNOWLEDGEMENTS

This work has been supported, in part, by the Natural Sciences and Engineering Research Council of Canada.

REFERENCES

Barhak J. and Fischer A., (2001), "Parameterization and reconstruction from 3D scattered points based on neural network and PDE techniques", IEEE Trans on Visualization and Computer Graphics, vol. 7, no. 1, pp. 1-16.

Campbell R. and Flynn P., (1998), "A WWW-accessible 3D image and model database for computer vision research", Empirical Evaluation Methods in Computer Vision, K.W. Bowyer and P.J. Phillips (Eds.), IEEE Computer Society Press, pp. 148-154.

Gu P. and Yan X., (1995),"Neural network approach to the reconstruction of freeform surfaces for reverse engineering", Computer-Aided Design, vol. 27, no. 1, pp. 59-64.

Hoschek J. and Lasser D., Fundamentals of Computer Aided Geometric Design. A.K. Peters Ltd.: Wellesley MA, 1993.

Knopf G.K. and Kofman J., (2001), "Adaptive reconstruction of free-form surfaces using Bernstein Basis function networks", Eng. Appl. of Artificial Intelligence, vol. 14, pp. 577 – 588.

Kohonen T., (1997), Self-Organizing Maps. New York: Springer.

Ma W. and Kruth J. P, (1995), "Parameterization of randomly measured points for least squares fitting of B-spline curves and surface", Journal of Computer-Aided Design, vol. 27, no. 9, pp. 663-675.

Standish T.A., (1994), Data Structures, Algorithms, and Software Principles, Addison-Wesley Longman.

Várady L. and Hoffmann M., (1998), "Free-form surface for scattered data by neural networks", Journal for Geometry and Graphics, vol. 2, no. 1, pp. 1-6.

Várady L., Hoffmann M., and Kovács E., (1999), "Improved free-form modelling of scattered data by dynamic neural networks", Journal for Geometry and Graphics, vol. 3, no. 2, pp. 177-181.

AN ADAPTIVE SAMPLING SCHEME FOR GENETIC ALGORITHMS ON THE SAMPLED ONEMAX PROBLEM

Tian-Li Yu[1], Ying-Ping Chen[2], David E. Goldberg[3], and Jian-Hung Chen[4]

[1,2,3] Illinois Genetic Algorithm Laboratory (IlliGAL)
University of Illinois at Urbana-Champaign
Urbana, IL 61801
[4] Department of Information Engineering and Computer Science
Feng Chia University, Taichung 407, Taiwan

Email: {tianliyu[1], ypchen[2], deg[3]}@illgal.uiuc.edu and jh.chen@ieee.org[4]

ABSTRACT
This paper proposes an adaptive sampling scheme for genetic algorithms. The adaptive sampling scheme is tested on the sampled OneMax problem. The results suggest that through this scheme, speed-up is obtained for problems with non-uniformly scaled building blocks (BBs). For problems with uniformly scaled BBs, the proposed adaptive sampling scheme does not give speed-up but still maintains the same performance with respect to the number of function evaluations when the adaptive sampling scheme is not adopted.

1 INTRODUCTION

Over the last few decades, significant progress has been made in the theory, design and application of genetic and evolutionary algorithms. A decomposition design theory has been proposed and several *competent* genetic algorithms (GAs) have been developed [1]. By competent GAs, we mean GAs that can solve hard problems quickly, reliably, and accurately. Competent GAs solve difficult problems within only a polynomial (usually subquadratic) number of function evaluations.

However, in real-world problems, even the subquadratic number of function evaluations can be intractable. Therefore, a number of efficiency enhancement techniques (EETs) have been proposed to alleviate the computational burden. One such technique is evaluation relaxation [2], in which the accurate, costly fitness evaluation is replaced by a cheap, less accurate evaluation. Partial evaluation through sampling is an example of evaluation relaxation scheme, which has been empirically shown to yield significant speed-up [3]. Evaluation relaxation through sampling has also been analyzed by developing facetwise and dimensional models [4, 5, 6]. Specifically, it has been shown that sampling does not yields speed-up for the sampled OneMax problems (SOM) with uniformly-scaled building blocks (BBs) [6]. This study extends the theoretical analyses and investigates the utility of sampling as an evaluation-relaxation technique for problems with non-uniformly scaled BBs.

The objective of this paper is to exploit sampling techniques on problems with exponentially scaled BBs to yield speed-up, which is an extend the work of [6]. Specifically, an adaptive sampling scheme is proposed and tested on the SOM. This paper is composed of four primary parts: (1) background knowledge including previous work and an introduction to the SOM, (2) descriptions of the adaptive sampling scheme, (3) empirical results, and (4) extensions and conclusions of this work.

2 Previous Work

Grefenstette and Fitzpatrick achieved a great success in applying sampling techniques to the image registration problem [3]. Their success motivated the work of [7], which gave theoretical analyses of a related problem. They derived and empirically verified the theoretical optimal sampling size when the fitness function is clouded by an additive Gaussian noise. Giguère and Goldberg investigated a uniformly-scaled building blocks (BBs) sampled OneMax (SOM) problem where the fitness value is calculated only based on sampled bits [5]. Their results empirically showed that sampling is not really useful when the gambler's ruin population-sizing model [8, 9] is adopted. More recently, Yu, Goldberg, and Sastry [6] gave analytical models that support the results in [5] based on the convergence time model proposed in [2].

The previous work has shown that sampling does not yield speed-up on the SOM with uniformly scaled BBs when the population is properly sized. This paper makes a further step to investigate the SOM with non-uniformly scaled BBs.

3 The Sampled OneMax Problem (SOM)

The sampled OneMax problem (SOM) is like an ordinary OneMax problem except the fitness value is calculated through sampling. For simplicity, call the SOM with uniformly-scaled building blocks (BBs) the uniform SOM, and call the SOM with exponentially-scaled BBs the exponential SOM.

The fitness of an individual $\vec{x} = x_1x_2x_3..x_l$ (l is the chromosome length) in the uniform OneMax problem and the exponential OneMax problem is $\Sigma_{i=1}^{l} x_i$ and $\Sigma_{i=1}^{l} x_i 2^{l-i}$ respectively. When sampling is taken into consideration, given a sampling set S, a subset of $\{1, 2, \cdots, l\}$, only x_i where $i \in S$ is known. Like the messy GA [10], we assume the partial string, partial evaluation (PSPE) is possible for now. In other words, the fitness of an individual $\vec{x}$ in the uniform SOM and the exponential SOM is $f(\vec{x}) = \Sigma_{i \in S} x_i$ and $f(\vec{x}) = \Sigma_{i \in S} x_i 2^{l-i}$ respectively. A template-filling method is proposed to solve the PSPE problem in the next section when the adaptive sampling scheme is described in detail. Note that in this paper, the number of function evaluations means the number of bits that GAs visited.

4 The Adaptive Sampling Scheme

It has been shown that sampling does not give speed-up for the uniform SOM [5, 6]. However, for the nonuniform SOM, obtaining speed-up through sampling is possible. Here the speed-up means the decrease of number of function evaluations under the same predefined desired solution quality (the average fitness of the population). For example, for a 10-bit exponential SOM, the fitness of the individual 1111100000 is 992 (about 97% of the highest fitness 1023), while only 5 bits (half of all 10 bits) are correctly converged. The key to obtain speed-up for the exponential SOM is to be able to identify significant bits. This section describes the adaptive sampling scheme which identifies significant bits in a GA manner and gives speed-up to nonuniform SOM problems.

4.1 Chromosomes in the Adaptive Sampling Scheme

Based on the above observation, an adaptive sampling scheme is proposed. Each individual is attached with a binary sampling string, where an one means the corresponding bit

solution string	x_1	x_2	x_3	x_4
sampling string	1	0	1	0

Figure 1. A chromosome is composed of a solution string and a sampling string. In this example, x_1 and x_3 are sampled.

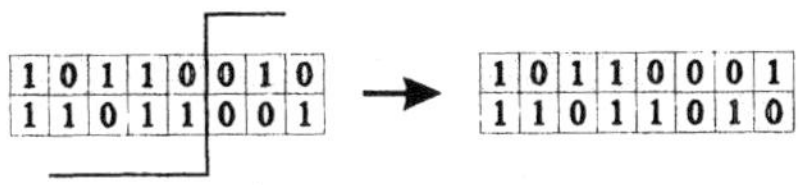

Figure 2. One-point crossover acting on a chromosome in adaptive sampling scheme. The crossing site is the same on both the solution string and the sampling string.

is sampled, and a zero means the corresponding bit is not sampled. Therefore, a chromosome is composed of two binary strings, named the solution string and the sampling string. In Fig. 1, only x_1 and x_3 are sampled, and hence the values of x_2 and x_4 are unknown. The crossover operator performs as usual but now it acts on both the solution string and the sampling string (Fig. 2). The mutation operator acts only on the solution string. Note that if all sampling strings in the population are all 1's, the GA with the adaptive sampling scheme reduces to an ordinary GA. We initialize sampling strings as follows: At the beginning, for each sampling string, one bit is randomly chosen to 1, and all the other bits are set to 0. The reason of such initialization will become clearer in the next subsection.

4.2 Fitness Function Modification

Now the task is to modify the fitness function so that the fitness value reflects both the solution quality and the sampling quality. A sampling string has a higher quality if it gives a considerable speed-up with little or no loss of solution quality. For example, in a 4-bit exponential SOM defined in section 3, a sampling string 1000 outperforms 0001. The sampled bits should be as few as possible so that maximal speed-up is obtained. However, at the end of the GA run, every bit should be sampled to ensure a high-quality convergence. Based on the argument, in the 4-bit exponential SOM, the ideal sampling string should be 1000 at the beginning of the GA run; with the GA progressing, the ideal sampling string becomes 1100, 1110, and finally 1111.

For the exponential SOM, define $f(\vec{x}) = \Sigma_{i \in S} x_i 2^{l-i}$. If the GA adopts $fitness = f$ directly, the selection operator would prefer the sampling string with all 1's since f is a positive summation. As a result, only little or no speed-up is obtained. The first modification of the fitness function is $fitness = \frac{f}{|S|}$ (the sampled fitness value divided by the number of sampled bits), which reflects the efficiency of sampling. With this modification, however, the selection always prefers the sampling string $1000\cdots$ because the first bit is most significant. Finally, the fitness function is modified as following:

$$fitness = \frac{f(\vec{x}) - \bar{f}}{|S|}, \tag{1}$$

where $\bar{f}$ is the average f-value of the population. At the beginning, for each sampling string, one bit is randomly chosen to 1, and all the other bits are set to 0. Therefore, initially $\bar{f}$ is small. Thinking $\bar{f}$ is small enough to be omitted, the fitness function acts like $\frac{f}{|S|}$, and hence the selection operator prefers the sampling string $1000\cdots$. Take a 4-bit exponential SOM as an example. When the most significant bit is almost converged to 1, $\bar{f}$ is roughly 8. Every chromosome with an f-value less than 8 results in a negative fitness value according to Eq. (1). Among those chromosomes with positive fitness values, the selection prefers the sampling string 1100 because it is most efficient now. Based on the argument, the modified fitness function in Eq. (1) acts similarly to the ideal case discussed above.

4.3 *check-and-set* and *lift-up*

Two operators, *check-and-set* and *lift-up* are introduced to make the adaptive sampling scheme more practical. Define $A_i \in [0,1]$ as the proportion of 1's of the i-th bits of all sampling strings in the current population. Initially, the expectation of A_i is $1/l$. The *check-and-set* operator does two tasks: (1) if a sampling string is all 0's, randomly set one bit to 1, and (2) if A_i is less than $1/2l$, randomly set the i-th bit to 1 among all sampling strings until A_i is greater than or equal to $1/2l$. The first task is to avoid a sampling string with all 0's, and the second task is to prevent that some bit is never sampled. The *lift-up* operator sets the i-th bit of every sampling string to 1, when A_i is greater than $2/l$. The purpose of the *lift-up* operator is to speed up the convergence of the sampling string. When A_i grows more than twice of its initial expectation $(1/l)$, it is of high confidence that the i-th bit should be always sampled. Note that both *check-and-set* and *lift-up* act only on the sampling strings and only change 0 to 1. As indicated before, a GA with the adaptive sampling scheme where all sampling strings are all 1's reduces to an ordinary GA. Therefore, these two operators do not bias the solution given by the GA.

4.4 Partial String Partial Evaluation Problem

Similar to [10], a fixed template is chosen to fill those un-sampled bits to solve the partial string partial evaluation (PSPE) problem. Unlike [10], a local minimal template is preferred in a maximization problem. The reason is that we want the fitness value to reflect the efficiency of sampling. Therefore, those un-sampled bits should contribute as little as possible to the fitness value. For the OneMax problem, the best template is a string with all 0's. In implementation, the template is chosen by a bit-wise local searcher, which costs $2l$ function evaluations. Note that we do not actually replace the un-sampled bits with the template. The template is only needed for the calculation of f.

4.5 Putting It All Together

To summarize, a GA with the adaptive sampling scheme acts as follows:

1. Initialize chromosomes.
   (a) Initialize solution strings.
   (b) Initialize sampling strings (section 4.1).
2. *check-and-set* (section 4.3).
3. Calculate fitness values (section 4.2).
4. Perform *selection*.
5. Perform *crossover* (section 4.1).
6. Perform *mutation*.
7. *lift-up* (section 4.3).
8. Go to step 2 until the terminate condition is satisfied.

Next section shows experiments on a uniform SOM and an exponential SOM.

5 EMPIRICAL RESULTS

Experiments are done on two test functions: a 100-bit uniform SOM and a 20-bit exponential SOM. In both cases, a simple GA (SGA) with binary tournament selection, uniform crossover, and mutation was used. Crossover probability is 1 and mutation probability is $1/l$, where l is the chromosome length. All experiments are averaged over 100 independent runs. For a better comparison, the solution quality showed in below figures is the average f-value, not *fitness* in Eq.(1), of the population without sampling.

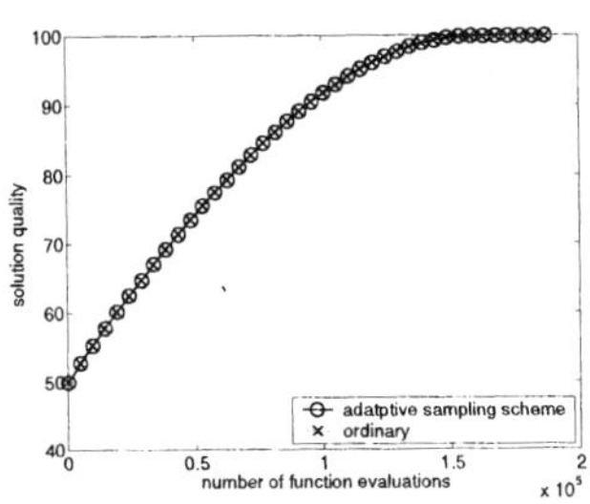

Figure 3. Solution quality versus the number of function evaluations of a SGA with and without the adaptive sample scheme on a uniform SOM. The adaptive sampling scheme does not yield any speed-up.

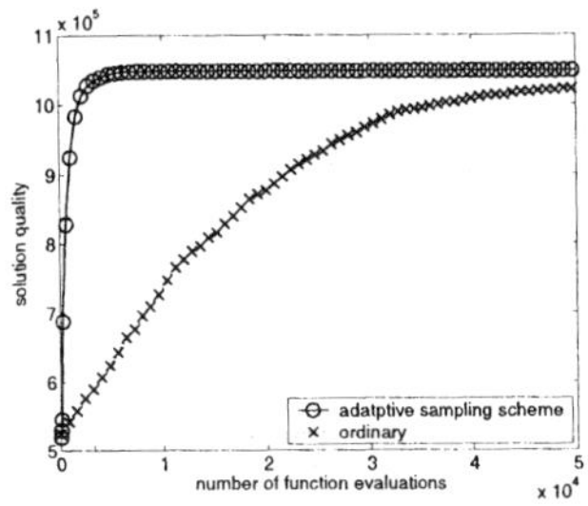

Figure 4. Solution quality versus the number of function evaluations of a SGA with and without the adaptive sample scheme on a exponential SOM. The adaptive sampling scheme gives a significant speed-up.

Given the failure rate 1/100, the population size of a 100-bit OneMax problem is about 48 according to the gambler's ruin model [8, 11]. When the adaptive sampling scheme is adopted, the population size should be larger because sampling introduces uncertainty when fitness values are calculated. However, when doing the experiments, we observed that the sampling strings becomes all 1's within the second or the third generation because of the *lift-up* operator. Therefore, the SGA with the adaptive sampling scheme acts the same as the ordinary SGA after two or three generations, and hence the population size of the SGA with the adaptive sampling scheme was set the same. Figure 3 indicates that the adaptive sampling scheme does not give speed-up for the uniform SOM. This is consistent with the results in [5, 6].

According to [12], the population size of an exponential OneMax is $8.89\sqrt{l(1+r)}$, where $r = \frac{\sigma_e^2}{\sigma_f^2}$. σ_f^2 is the fitness variance and σ_e^2 is the error variance. Based on the model, the population size of a 20-bit exponential OneMax is set to 40. Again, when the adaptive sampling scheme is adopted, the population size should be larger. However, according to our observations, significant bits were identified very quickly, which makes r small. Therefore, the population size was set the same when the adaptive sampling scheme is adopted. Figure 4 illustrates that a significant speed-up is obtained by the adaptive sampling scheme. It is because the adaptive sampling scheme enhances the domino convergence [13]. For example, when the sampling string is converged to $10000\cdots$, the GA only looks at the most significant bit, and the 20-bit SOM is now like a 1-bit OneMax problem. After the most significant bit converged, the most selection-favored sampling string is $11000\cdots$. Since the most significant bit is already converged, the problem now is still like a 1-bit OneMax problem. As a result, the 20-bit SOM is like 20 independent 1-bit OneMax problems. Moreover, solving the most significant seven 1-bit OneMax problems already yields a 99% solution quality.

6 Conclusions

This paper proposed an adaptive sampling scheme. Empirically, the adaptive sampling scheme is shown to yield a significant speed-up for the exponential SOM and maintains the same performance with respect to the number of function evaluations and solution quality for the uniform SOM. In general, we believe that the adaptive sampling scheme always gives speed-up for nonuniform SOM problems.

The OneMax problem provides a good starting point but somehow lacks of analogy to real-world problems, where deceptions and noises exist. With the help of many linkage learning algorithms [14, 15, 16], and analogizing each BB to a bit, the adaptive sampling scheme should be able to apply in a straight forward manner. However, further investigations and experiments are needed to check the argument.

ACKNOWLEDGEMENT

This work was sponsored by the Air Force Office of Scientific Research, Air Force Materiel Command, USAF, under grant F49620-03-1-0129, and the National Science Foundation under grant DMI-9908252. The U.S. Government is authorized to reproduce and distribute reprints for government purposes notwithstanding any copyright notation thereon.

REFERENCES

[1] Goldberg, D.E.: The design of innovation: Lessons from and for competent genetic algorithms. Kluwer Academic Publishers, Norwell, MA (2002)

[2] Sastry, K., Goldberg, D.E.: Genetic algorithms, efficiency enhancement, and deciding well with differing fitness variances. Proceedings of the Genetic and Evolutionary Computation Conference (2002) 528–535 (Also IlliGAL Report No. 2002002).

[3] Grefenstette, J.J., Fitzpatrick, J.M.: Genetic search with approximate function evaluations. (1985) 112–120

[4] Miller, B.L., Goldberg, D.E.: Genetic algorithms, selection schemes, and the varying effects of noise. Evolutionary Computation **4** (1996) 113–131

[5] Giguère, P., Goldberg, D.E.: Population sizing for optimum sampling with genetic algorithms: A case study of the Onemax problem. (1998) 496–503

[6] Yu, T.L., Goldberg, D.E., Sastry, K.: Opitmal sampling and speed-up for genetic algorithms on the sampled onemax problem. Proceedings of the Genetic and Evolutionary Computation Conference 2003 (to appear)

[7] Miller, B.L., Goldberg, D.E.: Optimal sampling for genetic algorithms. Proceedings of the Artificial Neural Networks in Engineering (ANNIE '96) conference **6** (1996) 291–297

[8] Harik, G., Cantú-Paz, E., Goldberg, D.E., Miller, B.L.: The gambler's ruin problem, genetic algorithms, and the sizing of populations. (1997) 7–12

[9] Miller, B.L.: Noise, sampling, and efficient genetic algorithms. doctoral dissertation, University of Illinois at Urbana-Champaign, Urbana (1997) Also IlliGAL Report No. 97001.

[10] Goldberg, D.E., Korb, B., Deb, K.: Messy genetic algorithms: Motivation, analysis, and first results. Complex Systems **3** (1989) 493–530 (Also TCGA Report 89003).

[11] Miller, B.L., Goldberg, D.E.: Genetic algorithms, selection schemes, and the varying effects of noise. Evolutionary Computation **4** (1996) 113–131

[12] Albert, L.A., Goldberg, D.E.: Efficient discretization scheduling in multiple dimensions. Proceedings of the Genetic and Evolutionary Computation Conference 2002 (2002) 271–278

[13] Thierens, D., Goldberg, D.E., Pereira, Â.G.: Domino convergence, drift, and the temporal-salience structure of problems. Proceedings of 1998 IEEE Iternational Conference on Evolutionary Computation (1998) 535–540

[14] Smith, J., Fogarty, T.C.: An adaptive poly-parental recombination strategy. In: Evolutionary Computing, AISB Workshop. (1995) 48–61

[15] Harik, G.: Linkage learning via probabilistic modeling in the ECGA. IlliGAL Report No. 99010, University of Illinois at Urbana-Champaign, Urbana, IL (1999)

[16] Yu, T.L., Yassine, A.A., Goldberg, D.E.: A genetic algorithm for developing modular product architectures. 15th International Conference on Design Theory and Methodology (DTM) (to appear)

ADAPTATION OF REPRESENTATION IN GP

CEZARY Z. JANIKOW
University of Missouri – St. Louis
Department of Mathematics and
Computer Science
St Louis, Missouri

RAHUL A DESHPANDE
University of Missouri – St. Louis
Department of Mathematics and
Computer Science
St Louis, Missouri

ABSTRACT
This paper discusses our initial work on automatically adapting Genetic Programming (GP) representation. We present here two independent techniques: AMS and ACE. Both techniques are based on Constrained GP (CGP), which uses mutation set methodology to prune the representation space according to some context-specific constraints. The ASM technique monitors the performance of local context heuristics when used in mutation/crossover, during GP evolution, and dynamically modifies the heuristics. The ACE technique iterates complete CGP runs and then uses the distribution information from the best solutions to adjust the heuristics for the next iteration. As the results indicate, GP is able to gain substantial performance improvements as well as learn qualitative heuristics.

INTRODUCTION

Genetic Programming (GP), proposed by Koza (1992), follows the evolutionary algorithms approach to problem solving by utilizing a population of solutions evolving under limited resources. The solutions, called chromosomes, are evaluated by a user-defined objective fitness evaluation. They compete for survival based on this fitness, and they undergo evolution by means of simulated crossover and mutation operators.

GP differs from other evolutionary computation methods by using trees as the solution representation. Trees provide a rich representation, one that is sufficient to represent computer programs, analytical functions, variable length structures, even computer hardware. The user defines the representation space by defining labels for the trees: functions of specific arities to label the internal nodes, and terminals (functions of no arguments and ephemeral constants) to label the leaves. The arities define the syntactic constraints of the trees. The fitness of a tree is evaluated by recursively applying function/terminal semantics, as defined by the user.

Two important principles need to be satisfied for GP to be successful (Koza, 1992). First, the principle of *sufficency* requires that the set of functions and terminals, along with their semantics, be rich enough to allow building solution trees of desired quality. Unfortunately, the user is often unaware of the minimal set and thus provides an enlarged set to be on the safe side. This enlarges the search space and thus often reduces search efficiency. Second, the principle of *closure* requires that no context constraints be imposed on labeling a tree except for labeling with proper arity. This was required in the absence of methodologies to enforce any such constraints.

Because of the requirements of both sufficiency and closure, the actual representation space that is explored during a GP simulated evolution is much larger than the desired or the actual solution space for the problem in hand. In the late nineties, two similar generic methods for automatic processing of additional constraints were proposed: Strongly Typed GP (STGP) and Constrained GP (CGP) (Janikow, 1996). Both follow the idea of enclosing the search in the feasible (or desired) space. That is, the initial population is generated so that each chromosome satisfies all constraints. Subsequently, every operator (mutation and crossover) is guaranteed to produce constraint-valid offspring if constraint-valid parents are used. Similar ideas have been later incorporated in Context Free Grammar-based GP (CF-GP), where the constraints are in the form of BNF production rules for program construction.

These methodologies allow explicitly expressing a set of local context rules to prune the actual trees that are evolved – only those satisfying the rules will be evolved. For example, an *if* function may now require that the test subtree produce a valid condition.

However, in either of the above methodologies, the user is still responsible for presenting the constraints to the system. Unfortunately, in many cases the user may not be aware of the best heuristics to solve a particular problem.

This paper presents two different techniques, based on CGP, which allow GP to monitor its performance and adjust the heuristics. The heuristics are in the form of *weak* constraints, that is to say they do not prohibit a given local structure to evolve in GP, but rather they allow different local structures to have different probabilities of labeling the tree nodes. However, they can be used as *hard* constraints as well, *e.g.*, when a given heuristic falls below certain probability threshold. The first technique, called AMS, is very simple and uses very limited information. It is based on comparing the fitness of parents to that of offpring in determing the utility of a given local heuristic when used in mutation/crossover. This is expected to work better for crossover, where one offspring is generated by one heuristic. The actual results validate this observation. The other technique, called automatic constraint extraction (ACE), is more complex in terms of run complexity. It iterates CGP runs, with fixed heuristics for each iteration. At the end of each complete CGP run (iteration), the best solutions are evaluated for distribution of local labeling contexts, and these contexts are then applied to modify the heuristics for the next iteration.

In the following, we describe both techniques and then present some experimental results using the 11-multiplexer problem.

CGP – CONSTRAINED GENETIC PROGRAMMING

CGP extends GP to allow mutation and crossover to generate only those offspring that are valid with respect to user provided constraints. The contsraints are in the form of local contexts (parent-child node relationship), and are precompiled into lookup tables - Mutations Sets (MS) (Janikow, 1996). CGP2.1 uses Typed MS structures to accommodate different data types and hence overloaded function instances (Janikow, 1999). For the experiments reported here, we used untyped MS. CGP also allows local heuristics, which are in the form of probabilities of labeling specific context structures. CGP allows the user to specify individual weights for different elements of every mutation set (Janikow, 1996). A weight twice as high as another weight indicates that the corresponding function or terminal should be twice as likely to be used in a given context. A weight set to zero in fact results in removing the corresponding function or terminal from consideration during evolution (only in the specific context, and not globally). For example, the user may specify that a given argument of a particular function be quite unlikely to use recursion on itself, without prohibiting recursion completely. It is this capability that is used in our techniques – both use some runtime information to adjust these probabilities.

ASM - ADAPTATION OF MUTATION SETS BY JUDGING OFFSPRING

The AMS algorithm, presented below, compares the fitness of a parent with its offspring. It then either increases or decreases the probability of the local heuristic that was used to generate the offpring.

1. Run CGP to solve a particular problem, while starting with the user provided constraints and heuristics (none if not given).
2. If crossover is selected to create an offspring (we consider both offspring individually), perform the following:
    a. Select a subtree to be inserted according to the current heuristics for the local context where the subtree is being inserted. This heuristic is denoted with h_i.
    b. Compute Δf_i, the difference in fitness values of the offspring with that of the receiving parent. Positive difference implies that a better offspring is generated.

c. Modify the specific heuristic h_i, expressed as weight (before normalized as probability), according to: w_i += Δw_c. We use a reward technique such that if Δf_i >0 then Δw_c is a positive constant, otherwise Δw_c is a negative constant (we have experimented with Δw_c proportional to Δf_i but the results were similar).

3. If mutation is selected to create an offspring, perform the following:
   a. Select a subtree to be removed.
   b. Grow the new subtree according to the current heuristics for the local contexts. Note each applied heuristic is denoted at h_i (multiple heuristics are required and a given heuristic can be used multiple times c_i).
   c. Compute Δf_i, the difference in fitness values of the offspring with that of the receiving parent. A positive difference implies that a better offspring is generated.
   d. Modify each used heuristic h_i, expressed as weight (before normalized as probability) according to: w_i += $c_i \Delta w_m$. Again, if Δf_i >0 then Δw_m is a positive constant, otherwise Δw_m is a negative constant (again we experimented with Δw_m proportional to Δf_i but the results were similar).
   e. Observe that we used Δw_m much smaller than Δw_c simply because single mutation involved using many heuristics at once and the reward is shared among all heuristics.

ACE – AUTOMATIC CONSTRAINT EXTRACTION

This technique runs on the top of CGP. In other words, it utilizes complete runs of CGP. To avoid statistical anomalies, ACE in fact can utilize a number of independent CGP runs simultanbeously . The algorithm has already been extended, but the results and descriptions will be presented separately.

ACE takes input parameters which determine the number of indepedent runs, the number of CGP generations per run, and the number of iterations that ACE is to run. The iterations are also stopped if no further significant distribution change is detected. Other parameters determine the number of best quality trees to be investigated from each CGP run, the total number of trees selected from all the independent runs, as according to user-specific objectives, and the objectives guiding the selection. At present, the selection can be based on quality or size in any combination. The selected trees are investigated for the distribution of functions and terminals, and the discovered distribution is fed back into CGP to adjust the constraint weights. Weights falling below specific threshold can also be dropped to 0 (thus becoming *hard* constraints).

For example, if ACE runs 10 independent runs, and it extracts the 15 best quality trees from each run, it ends up with 150 trees at the end of the iteration. If 20% of the trees are to be selected for investigation, and the objective is to minimize size, the 30 smallest trees from the 150 quality trees are selected. Distribution statistics from the 30 trees is computed, and it is fed back into the next ACE iteration in the form of CGP weights (or just adjustments of the weights).

ACE can start with no constraints or it can start with user-defined constraints and initial heuristics.

EXPERIMENTS AND RESULTS

To illustrate both techniques, we selected the well known and studied 11-multiplexer problem [1][3]. The 11-multiplexer problem is to discover the boolean function that passes the correct data bit (out of eight bits *d0…d7*) when fed three addresses (*a0… a2*). There are 2048 (2^{11}) possible combinations. Koza [3] and others have used a set of four atomic functions to solve the problem: *if/else*, *and*, *or*, and *not*, in addition to the data and address bits. This set is not only sufficient but is also redundant. In (Janikow, 1996), we have shown that operating with only *if* (sufficient by itself) dramatically improves the performance by simply reducing the search space. Moreover, we have shown that the performance is further enhanced when we restrict the *if*'s condition argument to test addresses only, while restricting the two action arguments to select only data bits or recursive *if*.

Name	Experiment	Crossover/Reproduction/Mutation
E1	No AMS w/all operators	85/10/5%
E2	AMS w/crossover+rep	90/10/0 %
E3	AMS w/mutation+rep	0/10/90%
E4	AMS all operators	85/10/5%

Table 1: AMS experiment setup.

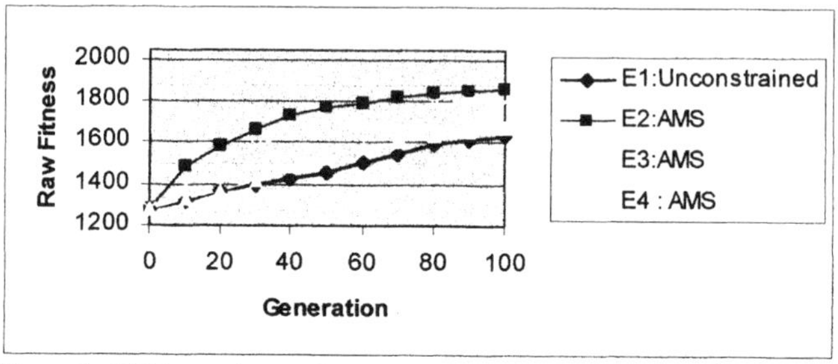

Figure 1: AMS experiments in practice.

For AMS, our objective was to determine if CGP can improve its run by adjusting the heuristic weights. In addition, another objective was to evaluate the utility of mutation and crossover for the technique. We used $\Delta w_c = 0.01$ and $\Delta w_m = 0.0005$ for $\Delta f_i > 0$, and $\Delta w_c = -0.01$ and $\Delta w_m = -0.0005$ otherwise. We performed two sets of different experiments. The first set, as defined in Table 1, started with no constraints or heuristics and different operators while running with and without AMS. The results are illustrated in Fig. 1. They indicate that AMS is able to adapt the heuristics while running CGP so that the overalll performance improves very significantly. The only case when AMS does not show significant improvement is when it is run with mutation only (reproduction accesses only global fitness as opposed to adapting local heuristics). This effect was expected – a single mutation requires many heuristics yet there is only one offspring. In crossover, on the other hand, one offspring is generated by one heuristic.

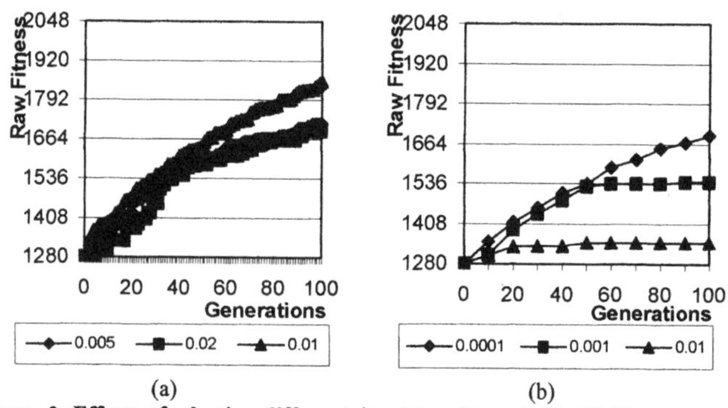

Figure 2: Effects of selecting different Δw_c (a) and Δw_m (b) in AMS.

The second set of experiments was designed to determine the influence of some of the AMS parameters, namely Δw_m and Δw_c. The results are presented in Fig. 2. As seen, the magnitude of crossover adjustment has little impact on the resulting improvement in evolution speed. However, the magnitude of mutation adjustment proves to be very important. For some values, the adjustments produce no improvements over multiple generations, while for others, the effect is always positive (0.0001 in this case). This verifies our previous intuition that in this case of very limited information, when only the parent-child relationship is observed, it is much more difficult to derive any heuristic information from mutation due to its necessity to apply multiple heuristics before any feedback is received. Some more sophisticated bucket brigade algorithms may prove more beneficial.

To test the ACE technique, we conducted three different experiments with three different extraction objectives: quality, size, and a mix of the two. All experiments were conducted with 10 independent CGP runs, 10 iterations, and all remaining parameters being the dafult CGP parameters (Janikow 1996).

Fig. 3 (left) illustrates the best quality tree from among the trees selected for computing the distribution statistics at the end of every ACE iteration. As can be seen, we gain perfect performance just after two iterations. When optimizing for size or mix (size+quality), we may see a dip after iteration 6 – size optimization may comprimize quality. Except for the dip, there seem to be no apparent improvement after the saturation. However, this is not so. As illustrated in Fig. 3 (right), the number of required generations for an iteration to be successful keeps decreasing after that saturation (ieration 2).

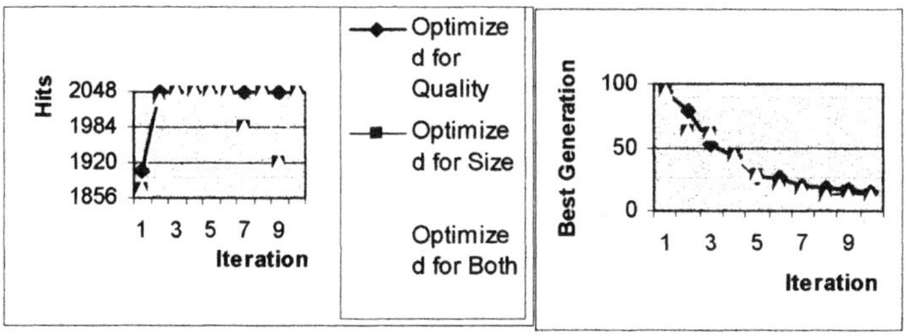

Figure 3: Improvements due to ACE: best of iteration (left) and the number of generations needed to solev the problem, per iteration (right).

The next question is that of the interpretation of the evolve heuristics. Fig. 4 illustrates the evolving distribution of the condition part of *if*. As stated before, the best prior results were obtained while testing only addresses. Fig. 4 left indicates that the evolved heuristics was to use address *a1* and the function *not* in the condition part. However, the same figure on the right indicates that the remaining two addresses, *a0* and *a2*, were discovered to be used indirectly through the argument of the *not* function. Thus, all three addresses are tested by *if* either directly or indirectly. The remaining functions are terminals have been automatically degraded to zero total weight and thus removed from consideration.

CONCLUSION AND FUTURE RESEARCH

We have presented a new methodology for adapting the representation of solutions in GP. The methodology is based on CGP, which allows heuristics to be used for biasing the representation space. The heuristics are local, that is they differentiate function/terminals to be used in specific parent-child contexts. In the AMS technique, these heuristics are adapted by observing the utility of specific heuristics when used to create offspring in crossover and mutation and when judged by the relative fitness of the offspring as related to that of the parent. This particular technique is very simple yet still able to generate significant improvements on the speed of evolution of solutions. In the ACE technique, the heuristics are adapted by evaluating the distribution of labels in the best trees selected from a pool of independent CGP runs. This technique uses more reliable information, and it seems to produce much better results yet at a higher computational cost. Further improvements and results will be presented separately.

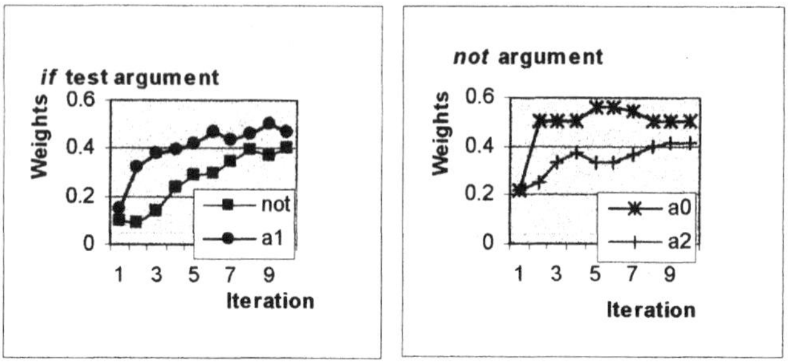

Figure 4: Direct (left) and indirect distribution on the condition part of *if*.

REFERENCES

Janikow, C.Z., 1996. "A Methodology for Processing Problem Constraints in Genetic Programming", Computers and Mathematics with Applications, Vol. 32, No. 8, pp. 97-113, 1996.

Janikow, C.Z., 1999. "CGP 2.1 User's Manual".

Koza, J.R., 1992. Genetic Programming: On the Programming of Computers by Natural Selection. The MIT Press, Cambridge, MA.

Bonzaf, et al. 1998. Genetic Programming: an Introduction. Morgan Kaufmann.

A HYBRID GA/PSO TO EVOLVE ARTIFICIAL RECURRENT NEURAL NETWORKS[1]

MATTHEW L. SETTLES
University of Idaho
Department of Computer Science
Moscow, Idaho 83844

TERENCE SOULE
University of Idaho
Department of Computer Science
Moscow, Idaho 83844

ABSTRACT
In this paper we propose a novel hybrid algorithm (GA/PSO) combining the strengths of particle swarm optimization with genetic algorithms to evolve the weights of recurrent neural networks. The hybrid algorithm combines the standard velocity and update rules of PSOs with the ideas of selection and crossover from GAs. We compare the hybrid algorithm to both the standard GA and PSO models, when evolving weights to a recurrent neural network problem. This paper presents results using the new hybrid algorithm and describes the hybrid's benefits[2].

INTRODUCTION

Genetic algorithms (GA) and particles swarm optimization (PSO) are both population based algorithms that are successful in solving very difficult problems, including recurrent artificial neural networks (RANN). However, both models have strengths and weaknesses. Comparisons between GA and PSOs have been performed by both Eberhart (1998) and Angeline (1998) and both conclude that a hybrid of the standard GA and PSO models could lead to further advances. In this paper we present a novel hybrid algorithm, combining the strengths of GAs with those of PSO. The hybrid algorithm is compared to the standard GA and PSO models in evolving weights to a RANN problem.

Traditional gradient search neural network (NN) techniques, such as back propagation, perform poorly on RANNs. Due to the complex search spaces of RANN theses techniques generally get stuck on local optima's and flat regions. Furthermore, gradient descent training requires 'unfolding' the network (Hammer, 2000). In contrast, population based searches are not restricted to a local search and can easily move across flat regions.

GAs and PSO are performance (fitness) based, no unfolding is required and are much less likely to become stuck at local optima, or in a flat region of the search space. There have been several successful examples of using a GA to evolving RANN architectures and weights (Angeline et al, 1994, Saunders et al, 1994, Mandischer, 1995, Settles and Soule, 2002). Salerno (1997), Settles and Soule (2002) used a PSO to evolve the weights of RANNs.

---

[1] This work is supported by NSF EPSCoR EPS-0132626. The experiments were performed on a Beowulf cluster built with funds from NSF grant EPS-80935 and a generous hardware donation form Micron Technologies.
[2] Thanks to Dr. Hiromoto of the University of Idaho for a conversation that sparked many of these ideas.

GENETIC ALGORITHM AND PARTICLE SWARM OPTIMIZER

Genetic Algorithms were first introduced by John Holland in the 1960s. In this problem, the GA uses a chromosome consisting of real values. Each real value corresponds to the weight between one pair of nodes (or between a node and itself).

The genetic algorithm is generational. The five best individuals are copied into the next generation (elitism). Tournament selection is used, with a tournament of size 3. The mutation rate is 1% and can change a weight by up to ±25% of the weight's original value.

When crossover is applied to two individuals the same random (non-input) node is chosen in both individuals. The weights of all connections leading *into* the chosen node are exchanged between the two individuals. The crossover rate used is 0.8.

The PSO algorithm was first proposed by Kennedy and Eberhart (2001). PSO consists of a population of particles, where each particle keeps track of its current position, velocity, best position thus far, best fitness thus far and current fitness. In this problem, a particles position and velocity are real valued vectors, with one value for each network weight. Thus the movement of each particle through the search space is governed by Equations (1) and (2).

$$\vec{v}_i(t) = I * \vec{v}_i(t-1) + \rho_1(\vec{p}_g - \vec{x}_i(t-1)) + \rho_2(\vec{p}_i - \vec{x}_i(t-1)) \quad (1)$$

$$\vec{x}_i(t) = \vec{x}_i(t-1) + \vec{v}_i(t) \quad (2)$$

Where $\vec{x}_i(t)$ is the position of the particle at time t, $\vec{v}_i(t)$ is the velocity of the particle at time t, I is the *inertia* weight, reduced linearly from 0.9 to 0.2 by epoch, $\vec{p}_g$ is the position of the global best particle, $\vec{p}_i$ is the particles previous best position and ρ_1 and ρ_2 are random values in the range [0.0,1.0].

To prevent the particle from venturing outside the search space a limit, called V_{max}, is applied to the velocity equation.

The PSO algorithm is conceptually simple and can be implemented in a few lines of code. PSOs also have memory, whereas in a GA if an individual is not selected, the information contained by that individual is lost. However, without a selection operator PSOs may waste resources on a poor individual stuck in a poor region of the search space.

A PSO's group interactions enhance the search for an optimal solution, whereas GAs have trouble finding an exact solution and are best at reaching a global region (Eberhart, 1998). Hybrid GA/Gradient Searches are often used to overcome this problem.

THE HYBRID (GA/PSO) MODEL

The hybrid algorithm combines the standard velocity and position update rules of PSOs with the ideas of selection and crossover from GAs. The algorithm is designed so that the GA performs a global search and the PSO performs a local search. Other hybrid GA/PSO algorithms have been proposed, and have been tested on function minimization problems, but not NN or RANN (Lvbjerg et al, 2001, Robinson et al., 2002).

In previous work, a GA and PSO were tested and compared when evolving the weights for a fixed topology RANN (Settles and Soule, 2003). Both the GA and PSO were successful in evolving RANNs. However, the algorithms produced different results for different network sizes, demonstrating that they implement different search strategies. We believe that the correct combination of both models has the potential to achieve better results faster.

Two parameters are added to the hybrid to allow balancing of the GA and PSO. The PSO's velocity vector is multiplied by an *influence* term, with a range of [0.0, 1.0], this term can reduce the size of the steps taken by the particles. The GAs selection operator has a *replacement* term, with a range of [0.0, 1.0], which determines the percentage of individuals in the population who get replaced and crossed over in the current generation. The GA mutation operator was not used in the hybrid algorithm.

First, the hybrid algorithm performs the standard PSO velocity and position update rules, with the *influence* term. GA Selection and Crossover then occur on the appropriate number of individuals determined by the *replacement* term, the top (*population size – replacement term*population size)* individuals, based on fitness, are copied into the new population. Tournament selection, with a tournament size of 3, is used to select parents to fill the remainder of the population.

Two crossover operators for the hybrid were used in these experiments, node crossover, as described above, and a new variation of crossover: node weighted average crossover. Node weighted average crossover is a hybrid crossover that incorporates the PSO velocity vector. The goal is to create two new child particles whose position is between the parents position, but accelerated away from the current direction to increase diversity. Node weighted average crossover is governed by equations (3) and (4).

$$child_1(x_i) = -parent_1(v_i) * (parent_1(x_i) + parent_2(x_i)) / 2.0 \tag{3}$$

$$child_2(x_i) = -parent_2(v_i) * (parent_1(x_i) + parent_2(x_i)) / 2.0 \tag{4}$$

The children's velocity vectors are exchanged at the same node and the previous best vector is set to the new position vector, effectively restarting the children's memory. Fitness is evaluated on the new population and the process is repeated.

Four tests of the hybrid algorithm, using different *influence* and *replacement* parameters and crossover operators, were tried (see Table 1).

All tests are performed with a population size of 500 and each test is run 50 times.

The rows in Table 1 are as follows:

- Hybrid-1 - This test was designed with the expectation that the best results would be with the GA performing an initial global search with the PSO finishing as a local search.
- Hybrid-2 - This test was designed with the expectation that the PSO performing an initial global search with the GA finishing as a local search would produce poor results.

Table 1: Test algorithm parameters and crossover type. $0.0 \rightarrow 1.0$ means that the parameter begins at 0.0 in the first generation and is linearly increased, by generation, to 1.0 in the last generation.

Model	Influence	Replacement	Crossover Type
GA	0.0 → 0.0	0.99 → 0.99	Node crossover
PSO	1.0 → 1.0	0.0 → 0.0	Node crossover
Hybrid-1	0.0 → 1.0	0.99 → 0.0	Node crossover
Hybrid-2	1.0 → 0.0	0.0 → 0.99	Node crossover
Hybrid-3	1.0 → 1.0	0.5 → 0.5	Node crossover
Hybrid-4	1.0 → 1.0	0.5 → 0.5	Node weighted average crossover

The results (described below) ran counter to our expectations, so two additional test are included.

- Hybrid-3 - The goal of this test was to allow the PSO to naturally converge (local search) on a solution, while allowing half the population to "spread out" through crossover.
- Hybrid-4 - Here the PSO is again allowed to naturally converge on a solution, while half the population, due to the averaged weights and negative velocity, actively "disperses", increasing diversity.

SIMULATING A BIOLOGICAL NEURON

The problem was to evolve a network that produces a simple pulsed output when an activation 'voltage' is applied to the network's input (Soule et al, 2002, Settles and Soule, 2003). The error is the sum of the absolute value of the difference between the desired output and the actual output at each time step plus a penalty of 0.5 if the slope of the desired output differs in the direction from the slope of the actual output.

The neural network used in this experiment is a 2 layer and 3 nodes per layer strongly connected recurrent neural network, consisting of 49 weights. It has a single input node and a single output node. All nodes in the network used the symmetric sigmoid activation function, with an output in the range [-1.0,1.0]. The initial weights were randomly chosen in the range [-2.0,2.0].

At each time step the input node's activation level (0 or 1) is set. Then for each hidden layer the activation level of the nodes in that layer are calculated. Effectively the activation levels of the neurons in the same hidden layer are updated simultaneously. The activation levels of the neurons in different layer are updated sequentially (Soule et al, 2002, Settles and Soule, 2003).

Training was allowed to continue for 250 generations. A successful run was defined as a periodic output of the desired frequency.

Table 2 summarizes the results for each experiment. Figures 3 and 4 show the best and average results for the GA, PSO, Hybrid-1 and Hybrid-4, respectively.

Hybrid-4 and the standard PSO algorithm outperform all other models on this problem, with Hybrid-4 marginally outperforming the PSO algorithm. Hybrid-4 is able to find a better solution faster than any other algorithm and Hybrid-4 has the best overall run with an output more closely matching that of the desired

Table 2: Experimental results from simulating a biological neuron.

Model	# of Successes	Best Fitness Average	Best Fitness Std-dev	Average Fitness Average	Average Fitness Std-dev
GA	21	22.7	9.2	27.3	8.3
PSO	28	22.0	9.6	30.1	10.4
Hybrid-1	8	31.1	4.7	31.1	4.7
Hybrid-2	14	27.7	8.6	27.7	8.7
Hybrid-3	24	23.7	8.9	24.2	9.8
Hybrid-4	30	21.5	9.1	34.5	7.3

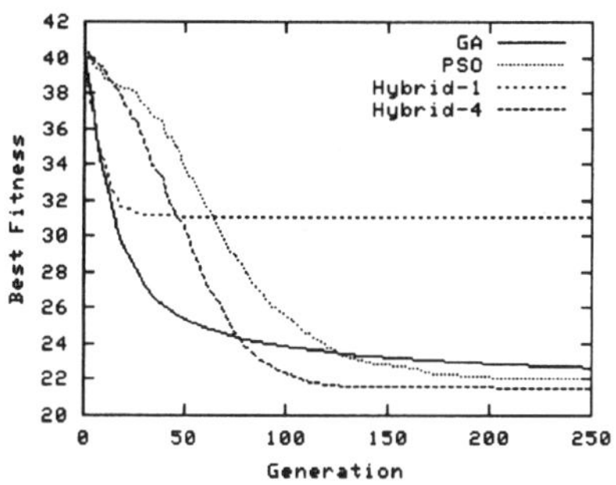

Figure 2: Best Fitness values of each algorithm over 50 runs

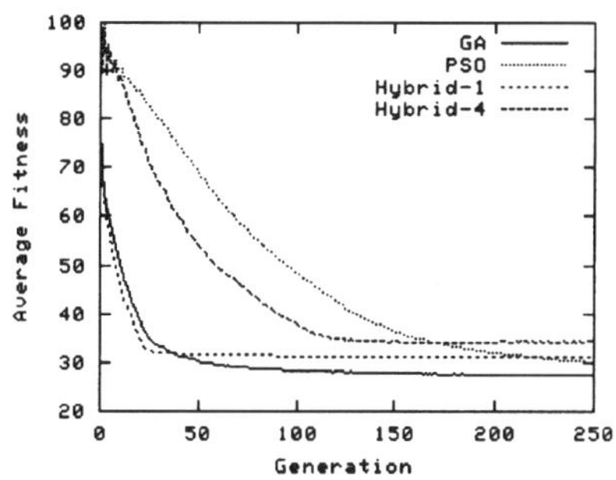

Figure 3: Average fitness values of each algorithm over 50 runs

output. Hybrid-4 has relatively poor average results due to the modified crossover operation. Hybrid-3 and the standard GA algorithm perform the next best with 24 and 21 successes, respectively. Hybrid-1 and Hybrid-2 performs poorly on this problem, Hybrid-1 performs extremely poorly, with the population converging on a suboptimal solution after 30 generations.

As a further test each of the algorithms were also run on a finite state machine problem. The problem was to learn a minimal six state finite state machine. Similar results were observed, with Hybrid-4 outperforming the other algorithms.

CONCLUSIONS

Our initial hypothesis was that the best hybrid would use the GA to perform an initial global search and the PSO would perform a local search at the end of the run (Hybrid-1). In fact, Hybrid-1 performed very poorly. We hypothesis that the combination of GA selection and PSO convergences produces too much pressure towards premature convergence, often in 30 generations or less.

Two additional hybrids (Hybrid-2 and Hybrid-3) that varied the relative influence of the GA and the PSO were then introduced. They perform marginally better than the initial hybrid, but it appeared that convergence was still a problem.

Hybrid-4 was an extension to Hybrid-3 using a new crossover operator. The choice of values for the *influence* and *replacement* parameters allow half the population, due to the PSO portion of the algorithm, to naturally converge on a solution. The other half the population is subject to selection and crossover. Our choice for the crossover operator, weighted averaged node crossover, actively "disperses" the population, resulting in increasing diversity throughout the run (see Figures 3). The increased diversity in the population did improve overall results.

It is important to note that these results suggest that any hybrid of PSO and GA will require some additional diversity maintenance. Future work will include testing this model in other problem domains.

REFERENCES

Angeline, P. J., Saunders, G. M., and Pollack, J. P., 1994. "An Evolutionary Algorithm that Constructs Recurrent Neural Networks," *IEEE Transactions on Neural Networks*, Vol.. 5, Num. 1, pp. 54-65.

Angeline, P.J., 1998. "Evolutionary Optimization Versus Particle Swarm Optimization: Philosophy and Performance Differences," *Evolutionary Programming VII, Lecture Notes in Computer Science* 1447, pp. 601-601.

Eberhart, R.C., and Shi, Y., 1998. "Comparison between Genetic Algorithms and Particle Swarm Optimization," *Evolutionary Programming VII, Lecture Notes in Computer Science* 1447, pp. 611-616.

Hammer, B., 2000. *Learning with Recurrent Neural Networks*. Vol. 254. Springer.

Horne, B. G., and Giles, C. L., 1995. "An Experimental Comparison of Recurrent Neural Networks," *In Advances in Neural Information Processing Systems*, G. Tesauro, D. Touretzky, and T. Leen, Eds., Vol. 7, The MIT Press, pp. 697-704.

Kennedy, J., and Eberhart, R., 2001. *Swarm Intelligence*. Morgan Kaufmann Publishers, Inc., San Francisco, CA.

Lvbjerg, M., Rasmussen, T., and Krink, T., 2001. "Hybrid Particle Swarm Optimizer with Breeding and Subpopulations," *In Proceedings of the Genetic and Evolutionary Computation Conference* (GECCO-2001).

Mandescher, M., 1995. "Evolving Recurrent Neural Networks with Non-binary Encoding," *In Proc. Second IEEE Int'l Conf. Evolutionary Computation*, Vol. 2, pp. 584-589, Perth, Australia. IEEE Press, Priscataway, NJ.

Robinson, J., Sinton, S., and Rahmat-Samii, Y.., 2002."Particle Swarm, Genetic Algorithm, and Their Hybrids: Optimization of a Profiled Corrugated Horn Antenna," 2002 IEEE Antennas and Propagation Society International Symposium and URSI National Radio Science Meeting, San Antonio, TX.

Salerno, J. 1997. "Using the Particle Swarm Optimization Technique to Train a Recurrent Neural Model," 9th *International Conference on Tools With Artificial Intelligence* (ICTAI '97), IEEE Press.

Saunders, G.M., Angeline, P.J., and Pollack, J.B., 1994. "Structural and Behavioral Evolution of Recurrent Networks," *In Advances in Neural Information Processing Systems*, J.D. Cowan, G. Tesauro, and J. Alspector, Eds., Vol. 6, Morgan Kaufmann Publishers, Inc., pp. 88-95.

Settles, M., Soule, T., 2003. "Comparison of Genetic Algorithm and Particle Swarm Optimizer When Evolving a Recurrent Neural Network," *In Proceeding of the Genetic and Evolutionary Computation Conference* (GECCO-2003), In Press.

Soule, T., Chen, Y., and Wells, R., 2002 "Evolving a Strongly Recurrent Neural Network to Simulate Biological Neurons," *In Proc. Of The 28th Annual Conference of the IEEE Industrial Electronics Society.*

ADAPTIVE LEARNING ALGORITHMS FOR NERNST POTENTIAL IN NERVE CELL MEMBRANE ION CHANNELS

Vikram Krishnamurthy
Department of Electrical Engineering,
University of British Columbia,
Vancouver, V6T 1Z4, Canada.
vikramk@ece.ubc.ca

ABSTRACT

We present discrete stochastic optimization algorithms that adaptively learn the Nernst potential in membrane ion channels. The proposed algorithms dynamically control both the ion channel experiment and the resulting Hidden Markov Model (HMM) signal processor and can adapt to time-varying behaviour of ion channels. One of the most important properties of the proposed algorithms are their its self-learning capability – they spends most of the computational effort at the global optimizer (Nernst potential). Numerical examples illustrate the performance of the algorithms on computer generated synthetic data.

I. INTRODUCTION

An ion channel is a hole or pore in a nerve cell membrane. In physical structure, an ion channel is a large molecule whose different configurations correspond to the ion channel being in a *closed* state or *open* state. The measurement of ionic currents flowing through single ion channels in cell membranes has been made possible by the giga-seal *patch-clamp* technique [7], [5]. This was a major breakthrough for which the authors of [7] won the 1991 Nobel prize in Medicine. Because all electrical activities in the nervous system, including communications between cells and the influence of hormones and drugs on cell function, are regulated by membrane ion channels, understanding their mechanisms at a molecular level is a fundamental problem in biology. Moreover, elucidation of how single ion channels work will ultimately help neuro-biologists find the causes of, and possibly cures for, a number of neurological and muscular disorders.

Ion channel currents are typically of the order of pico-amps (i.e. , 10^{-12} amps). In patch clamp experiments these minute ion channel currents are obfuscated by large amounts of thermal noise. Chung *et al.* [3] first introduced the powerful paradigm of Hidden Markov Models (HMMs) to characterize patch-clamp recordings of small ion channel currents contaminated by random and deterministic noise. By using sophisticated HMM signal processing methods, [3] demonstrated that the underlying parameters of the HMM could be obtained to a remarkable precision despite the extremely poor signal to noise ratio. These HMM parameter estimates yield important information into the dynamics of ion channels.

I-V Curve and Nernst Potential: A typical trace of the ion channel current measurement from a patch clamp experiment (after suitable anti-aliasing filtering and sampling) shows that the channel current is piecewise constant discrete time signal that randomly jumps between two values – zero amperes which denotes the *closed state* of the channel, and $I(v)$ amperes which denotes the *open* state. $I(v)$ is called the *open-state* current level. It depends on the voltage v that is applied by the experimenter to the ion channel. Let $\{i_n(v)\}$

denote the discrete-time ion channel current sequence with $n = 0, 1, \dots,$ denoting discrete time.

In characterizing different types of ion channels, neuro-biologists routinely construct current-voltage (I-V) curves. The curve represents the variation of the open state current level $I(v)$ of the ion channel as a function of the applied voltage value v. Such I-V curves yield a unique signature of a particular ion channel, revealing its operating characteristics. For example, the relationship may be linear or Ohmic in some ion channels, whereas it may be superlinear or sublinear for other ion channels. The magnitude of the current flowing in one direction at a given potential difference may be equal in some ion channels, whereas the two arms of the I-V curve are asymmetrical in other ion channels. The I-V curve is always monotonically increasing – i.e. , $I(v)$ is a monotonically increasing function of v.

The zero point of the I-V curve, i.e. , the voltage v^* at which the open state current level $I(v^*)$ is zero, is known as the *Nernst potential.* The Nernst potential gives information about the relative concentrations at the two faces of the ionic channel. The value of the open state current level $I(v)$ is described by the Nernst-Planck equation.

The aim of this paper is to propose algorithms that efficiently learns the I-V curve from the noisy observed channel current sequence $\{y_n(v)\}$ by dynamically controlling the applied voltage v. A schematic of our proposed methodology is summarized in Fig.1. The proposed algorithm dynamically controls (schedules) the choice of voltage v at which the ion channel operates in order to efficiently estimate the Nernst potential (zero current point) and deduce how the current increases or decreases as the applied voltage deviates from the Nernst potential. Thus at a given time instant, given the current estimate from a HMM estimator operating at a particular voltage, the aim is to devise a scheduling algorithm that dynamically decides which voltage value to pick at the next time instant to apply to the ion channel. The most important aspect of the resulting combined experiment scheduling/HMM estimation algorithm is its self learning capability – it is provably convergent to the Nernst potential estimate and is provably efficient – that is the algorithm spends more time running the ion channel at the Nernst potential than any other voltage.

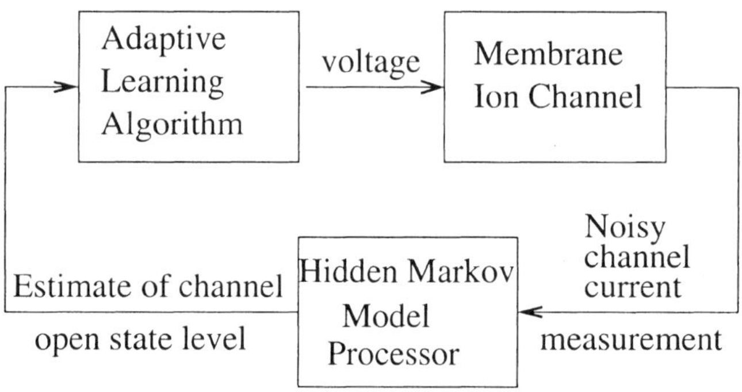

Fig. 1. Dynamic Control of Ion Channel for Learning Nernst Potential

II. Ion Channel Model and Current Voltage (I-V) Curve

A. *Measurements of Ionic Currents Flowing Across Single Ion Channels*

To record currents from single ion channels, the tip an electrode, with the diameter of about 1 μm, is pushed against the surface of a cell, and then a tight seal is formed between the rim of the electrode tip and the cell membrane. A patch of the membrane surrounded by the electrode tip contains one or more single ion channels. The current flowing from the inside of the cell to the tip of the electrode through a single ion channel is monitored. In a living cell, there is a potential difference between its interior and the outside environment, known as the membrane potential. Typically, the cell interior is about 60 mV more negative with respect to outside. Also, the ionic concentrations (mainly Na^+, Cl^- and K^+) inside of a cell is very different from outside of the cell. In the cell-attached configuration, the ionic strength in the electrode is usually made same as that in the outside of the cell. Let E_i and E_o, respectively, denote the resting membrane potential and the potential applied to the electrode. If E_o is identical to the membrane potential, there will be no potential gradient across the membrane patch confined by the tip of the electrode. Let c_i denote the intracellular ionic concentration and c_o the ionic concentration in the electrode. Here the intra-cellular concentration c_i inside the cell is unknown as is the resting membrane potential E_i. c_o and E_o are set by the experimenter and are known. Let $v = E_o - E_i$ denote the potential gradient. Both the potential gradient v and concentration gradient $c_o - c_i$ drive ions across an ion channel resulting in an ion channel current $\{i_n(v)\}$. This ion channel current is a piece-wise constant signal that jumps between the values of zero and $I(v)$, where $I(v)$ denotes the current when the ion channel is in the *open state*. The potential E_o (and hence potential difference v) is adjusted experimentally until the current $I(v)$ goes to zero. This voltage v^* at which the current $I(v^*)$ vanishes is called the *Nernst potential*

$$v^* = -\frac{kT}{e}\ln\frac{c_o}{c_i} = -59\log_{10}\frac{c_o}{c_i}(\text{mV}), \tag{1}$$

where $e = 1.6 \times 10^{-19}$ C denotes the charge of an electron, k denotes Boltzman's constant and T denotes the absolute temperature. The Nernst equation (1) gives the potential difference v required to maintain electrochemical equilibrium when the concentrations are different on the two faces of the membrane.

B. HMM for Ion Channel Current and Parameter Estimation of Burst Current State $I(v)$

We formally define the HMM for the ion channel current and briefly describe MLE algorithms for HMMs. Such probabilistic models for ion channels based on HMMs are now widely used [3], [8].

Markov Model for Ion Channel Current: Suppose a patch clamp experiment is conducted with a voltage v applied across the ion channel. Then, as described in [3], [8], the ion channel current $\{i_n(v)\}$, can be modelled as a three state homogeneous first order Markov chain. The state space of this Markov chain is $\{0_g, 0_b, I(v)\}$ corresponding to the physical states of *gap mode*, *burst-mode-closed* and *burst-mode-open*. For convenience, we will refer to the burst mode closed and burst-mode-open states as the *open* and *closed* states, respectively. In the gap mode and the closed state the ion channel current is zero. In the open state, the ion channel current has a value of $I(v)$. The (3×3) transition probability matrix $A(v)$ of the Markov chain $\{i_n(v)\}$, which governs the probabilistic behaviour of the channel current, is given by

$$A(v) = \begin{array}{c|c|c|c|} & 0_g & 0_b & I(v) \\ \hline 0_g & a_{11}(v) & a_{12}(v) & 0 \\ \hline 0_b & a_{21}(v) & a_{22}(v) & a_{23}(v) \\ \hline I(v) & 0 & a_{32}(v) & a_{33}(v) \\ \hline \end{array} \tag{2}$$

The elements of $A(v)$ are the transition probabilities $a_{ij}(v) = P(i_{n+1}(v) = j | i_n(v) = i)$ where $i, j \in \{0_g, 0_b, I(v)\}$.

HMM Observations: Let $\{y_n(v)\}$ denote the measured noisy ion channel current at the electrode when conducting a patch clamp experiment:

$$y_n(v) = i_n(v) + w_n(v), \quad n = 1, 2, \ldots \tag{3}$$

Here $\{w_n(v)\}$ is thermal noise and is modelled as zero mean white Gaussian noise with variance $\sigma^2(v)$. Thus the observation process $\{y_n(v)\}$ is a Hidden Markov model sequence parameterized by the model $\lambda(v) = \{A(v), I(v), \sigma^2(v)\}$ (see for example [3]), where v denotes the applied voltage.

HMM Parameter Estimation of Current Level $I(v)$: Given the HMM model, estimating $I(v)$ for a fixed voltage v, involves processing the noisy observation $\{y_n(v)\}$ through a Hidden Markov Model maximum likelihood parameter estimator. The most popular way of computing the maximum likelihood estimate (MLE) $I(v)$ is via the Expectation Maximization (EM) algorithm (Baum Welch equations) which is now fairly standard in the signal processing and neuro-biology literature – see [4], [3].

Let $\hat{I}_\Delta(v)$ denote MLE of $I(v)$ based on the Δ-point measured channel current sequence $(y_1(v), \ldots, y_\Delta(v))$. For sufficiently large batch size Δ of observations, due to the asymptotic normality of the MLE for a HMM [2],

$$\sqrt{\Delta}\left(\hat{I}_\Delta(v) - I(v)\right) \sim N(0, \Sigma(v)) \tag{4}$$

where $\Sigma^{-1}(v)$ is the Fisher information matrix. Thus asymptotically $\hat{I}_\Delta(v)$ is an unbiased estimator of $I(v)$, i.e. , $\mathbf{E}\{\hat{I}_\Delta(v)\} = I(v)$ where $\mathbf{E}\{\cdot\}$ denotes expectation.

III. Discrete Stochastic Optimization based HMM Algorithm

A. *Formulation as Discrete Stochastic Optimization Problem*

Determining the Nernst potential v^* requires conducting experiments at different values of voltage v. In patch clamp experiments, the applied voltage v is usually chosen from a finite set. Let

$$v \in V = \{\theta(1), \ldots, \theta(M)\}$$

denote the finite set of possible voltage values that the experimenter can pick. For example, in typical experiments, if one needs to determine the Nernst potential to a resolution of 4 mV, then $M = 80$ and $\theta(i)$ are uniformly spaced in 4 mV steps from $\theta(1) = -160$ mV and $\theta(M) = 160$ mV.

Note that the Nernst potential v^* (zero crossing point) does not necessarily belong to the discrete set V – instead we will find the point in V that is closest to v^* (with resolution $\theta(2) - \theta(1)$). For the rest of this paper with slight abuse of notation we will denote the element in V closest to the Nernst potential as v^*. Thus determining $v^* \in V$ can be formulated as a discrete optimization problem: $v^* = \arg\min_{v \in V} |I(v)|^2$. Due to the presence of large amounts of thermal noise, $I(v)$ cannot be exactly evaluated and only unbiased estimates $\hat{I}(v)$ are available. Thus computing the Nernst potential is equivalent to the following discrete stochastic optimization problem:

$$\text{Compute } v^* = \arg\min_{v \in V} \left[\mathbf{E}\{\hat{I}(v)\}\right]^2 \tag{5}$$

where $\hat{I}(v)$ is the MLE of the parameter $I(v)$ of the HMM. Since for a HMM, no closed form expression is available for $\Sigma^{-1}(v)$ in (4), the above expectation cannot be evaluated analytically. This motivates the need to develop a simulation based (stochastic approximation) algorithm.

HMM Brute Force Approach: The brute force approach for solving (5) involves an exhaustive enumeration as follows: For each $v \in V$, run an independent experiment to gather the sample path $\{y_1(v), y_2(v), \dots, y_\Delta(v)\}$ for a very large batch size Δ. Compute the MLE $\hat{I}(v)$ via a HMM parameter estimator. Finally pick $\hat{v}^* = \arg\min_{v\in V} |\hat{I}(v)|^2$. Since for any fixed $v \in V$, the MLE $\hat{I}(v)$ is strongly consistent [6], $\hat{I}(v) \to I(v)$ w.p.1, as the batch size $\Delta \to \infty$. The finiteness of V then implies that as $\Delta \to \infty$, $\arg\min_{v\in V} \left(\hat{I}(v)\right)^2 \to \arg\min_{v\in V} \left(I(v)\right)^2$ w.p.1.. Thus in principle, the above brute force simulation method can solve the discrete stochastic optimization problem (5) for large Δ and the estimate is *consistent*. However, the method is highly inefficient since $\hat{I}(v)$ needs to be evaluated for each $v \in V$. The evaluations of $\hat{I}(v)$ for $v \neq v^*$ are wasted because they contribute nothing to the estimation of the ion channel current $i(v^*)$ at the Nernst potential v^*. Also the brute force approach does not exploit the monotone nature of the I-V curve.

B. Discrete Stochastic Approximation Algorithm

The idea of discrete stochastic approximation [1] is to design a plan of experiments which provides more observations in areas where the Nernst potential is expected and less in other areas. More precisely what is needed is a dynamic resource allocation (control) algorithm that dynamically controls (schedules) the choice of voltage at which the HMM estimator operates in order to efficiently obtain the zero point and deduce how the current increases or decreases as the applied voltage deviates from the Nernst potential. We propose a discrete stochastic approximation algorithm that is both *consistent* and *attracted* to the Nernst potential. That is, the algorithm should spend more time gathering observations $\{y_n(v)\}$ at the Nernst potential $v = v^*$ and less time for other values of $v \in V$. There are several different classes of methods that can be used to solve the discrete stochastic optimization problem (5); see [1] for a recent survey.

Notation: Let $n = 1, 2, \dots$ denote discrete time. Since experiments will be conducted over batches of data, it is convenient to introduce the following notation. Group the discrete time into batches of length Δ – typically $\Delta = 10,000$ in experiments. We use the index $N = 1, 2, \dots$ to denote batch number. Thus batch N comprises of the Δ discrete time instants $n \in \{N\Delta, N\Delta + 1, \dots, (N + 1)\Delta - 1\}$.

Let $D_N = (D_N(1), \dots, D_N(M))'$ denote the vector of duration times the algorithm spends at the M possible potential values in V. For notational convenience define the M dimensional unit vectors, e_m, $m = 1, \dots, M$ as $e_m = \begin{bmatrix} 0 & \cdots & 0 & 1 & 0 & \cdots & 0 \end{bmatrix}'$ with 1 in the m-th position and zeros elsewhere.

The discrete stochastic approximation algorithm of [1] is not directly applicable to the cost function (5) – since it applies to optimization problems of the form $\min_{v\in V} \mathbf{E}\{C(v)\}$. However, (5) can easily be converted to this form as follows: Let $\hat{I}_1(v)$, $\hat{I}_2(v)$ be two statistically independent unbiased HMM estimates of $I(v)$. Then defining $\hat{C}(v) = \hat{I}_1(v)\hat{I}_2(v)$, it straightforwardly follows that $\mathbf{E}\{\hat{C}(v)\} = \left[\mathbf{E}\{\hat{I}(v)\}\right]^2 = |I(v)|^2$.

The discrete stochastic approximation algorithm we propose is as follows:

Algorithm 1: [Algorithm for Learning Nernst Potential]

- `Step 0`: (Initialization.) At batch $N = 0$, select starting point $X_0 \in \{1, \dots, M\}$ randomly. Set $D_0 = e_{X_0}$, Set initial solution estimate $\widehat{v}_0^* = \theta(X_0)$.
- `Step 1`: (Sampling.) At batch N, sample $\tilde{X}_N \in \{X_N - 1, X_N + 1\}$ with uniform distribution.
- `Step 2`: (Evaluation and Acceptance.) Conduct two independent experiments with voltage $\tilde{v} = \theta(\tilde{X}_N)$ to obtain two Δ length batches of HMM observations.

Let $\hat{I}_N^{(1)}(\tilde{v})$ and $\hat{I}_N^{(2)}(\tilde{v})$ denote the HMM-MLE estimates for these two batches obtained using the EM algorithm, see Appendix A. Set $\hat{C}_N(\tilde{v})) = \hat{I}_N^{(1)}(\tilde{v})\hat{I}_N^{(2)}(\tilde{v})$. Similarly conduct two independent experiments with voltage $v = \theta(X_N)$. Let $\hat{I}_N^{(1)}(v)$ and $\hat{I}_N^{(2)}(v)$ denote the HMM-MLE estimates for these two batches. Set $\hat{C}_N(v)) = \hat{I}_N^{(1)}(v)\hat{I}_N^{(2)}(v)$.
If $\hat{C}_N(\tilde{v}) < \hat{C}_N(v)$, set $X_{N+1} = \tilde{X}_N$, else, set $X_{N+1} = X_N$.

- `Step 3`: (Update occupation probabilities of X_N.) $D_{N+1} = D_N + e_{X_{N+1}}$.
- `Step 4`: (Update estimate of Nernst potential.) $\widehat{v}_N^* = \theta(m^*)$ where $m^* = \arg\max_{m\in\{1,\dots,M\}} D_{N+1}(m)$, set $N \to N+1$, go to Step 1.

The main idea behind the above algorithm is that the sequence $\{X_N\}$ (or equivalently $\{\theta(X_N)\}$) generated by Steps 1 and 2 is a homogeneous Markov chain with state space $\{1,\dots,M\}$ (respectively, V) that is designed to spend more time at the global maximizer v^* than any other state. In the above algorithm, $\hat{v}_N^*$ denotes the estimate of the Nernst potential at batch N. In Step 3, the vector D_{N+1} is a counter for the occupation times of X_N in the M possible states, i.e. , $D_N(m)$, $m \in \{1,\dots,M\}$ measures the number of times the Markov chain $\{X_N\}$ has visited the state m until batch N.

C. Convergence and Attraction of Algorithm 1

Throughout this section we assume:

(N) The batch size Δ is sufficiently large (e.g $\Delta = 10,000$) so that due to (4), the MLE estimates $\hat{I}_N^{(1)}(v)$, $\hat{I}_N^{(2)}(v)$ are $N(I(v),\Sigma(v))$ Gaussian random variables.

(O) For any $m \in \{1,\dots,M-1\}$,

$$I^2(\theta(m+1)) > I^2(\theta(m)) \implies P\Big(\hat{C}(\theta(m+1)) > \hat{C}(\theta(m))\Big) > 0.5 \qquad (6)$$

$$I^2(\theta(m+1)) < I^2(\theta(m)) \implies P\Big(\hat{C}(\theta(m+1)) > \hat{C}(\theta(m))\Big) < 0.5 \qquad (7)$$

Theorem 1: Under the condition (O) above, the sequence $\{\theta(X_N)\}$ generated by Algorithm 1 is a homogeneous, aperiodic, irreducible Markov chain with state space V. Furthermore, Algorithm 1 is attracted to the Nernst potential v^*, i.e. , for sufficiently large N, the sequence $\{\theta(X_N)\}$ spends more time at v^* than an other state. (Equivalently, if $\theta(m^*) = v^*$, then $D_N(m^*) > D_N(j)$ for $j \in \{1,\dots,M\} - \{m^*\}$.)

Remark: Complete proofs, numerical results and adaptive versions of Algorithm 1 will appear in a paper in IEEE transactions Nano-Bioscience.

REFERENCES

[1] S. Andradottir. Accelerating the convergence of random search methods for discrete stochastic optimization. *ACM Transactions on Modelling and Computer Simulation*, 9(4):349–380, Oct 1999.

[2] P.J. Bickel, Y. Ritov, and T. Rydén. Asymptotic normality of the maximum-likelihood estimator for general hidden Markov models. *Annals of Statistics*, 26:1614–1635, 1998.

[3] S.H. Chung, V. Krishnamurthy, and J.B. Moore. Adaptive processing techniques based on hidden Markov models for characterising very small channel currents buried in noise and deterministic interferences. *Proc. Phil. Trans. Roy. Soc. Lond. B*, 334:357–384, 1991.

[4] Y. Ephraim and N. Merhav. Hidden Markov processes. *IEEE Trans. Inform. Theory*, 48:1518–1569, June 2002.

[5] O.P. Hamill, A. Marty, E. Neher, B. Sakmann, and F.J. Sigworth. Improved patch-clamp techniques for high-resoluton current recodring from cells and cell-free membrane patches. *Pflüders Arch.–Eur. J. Physiol.*, 391:85–100, 1981.

[6] V. Krishnamurthy and T. Rydén. Consistent estimation of linear and non-linear autoregressive models with Markov regime. *Journal of Time Series Analysis*, 19:291–308, May 1998.

[7] E. Neher and B. Sakmann. Single-channel currents recorded from membrane of denervated frog muscle fibres. *Nature*, 260:799–802, 1976.

[8] L. Venkataramanan, R. Kuc, and F.J. Sigworth. Identification of hidden Markov models for ion channel currents – part ii: Bandlimited, sampled data. *IEEE Trans. Signal Proc.*, 48(2):376–385, Feb. 2000.

NON-LINEARLY SEPARABLE CLUSTER CLASSIFICATION: AN APPLICATION FOR A PULSE-CODED CMOS NEURON

SAI RAMESH J. RAO KONDURI
University of Idaho, MRC Institute,
Moscow, Idaho 83844-1024, USA

JAMES F. FRENZEL
University of Idaho, MRC Institute,
Moscow, Idaho 83844-1024, USA

ABSTRACT
Application of a biologically inspired neuron to cluster classification is presented. Pulse-coded neurons and the networks constructed from such have the ability to mimic behaviors found in biological neural systems. Pulse streams have the advantage that they are easy to generate, possess high noise immunity, and can be distributed over relatively large regions, the latter facilitating the construction of very large neural networks using CMOS VLSI circuit technology. Non-linear cluster classification involves differentiating between different clusters with a set of inputs and associated outputs, which are non-linearly separable. A 3-cluster classification problem is presented and solved using artificial neural network. The training mechanism implemented for the network proposes training the weights for the inputs of the neurons as well as changing the threshold of the neurons for sampling weak inputs. Simulation results in Microsim Pspice showed good agreement when compared to the theoretical results.

INTRODUCTION

This paper discusses an application of the pulsed CMOS neuron for VLSI circuits described in (Liu and Frenzel, 2002). Pulse-coded neurons and its associated networks function similar to a biological neuron with synaptic junctions (Ota and Wilamowski, 2000). The Neuro fuzzy techniques are attractive to approximate pattern recognition solutions when they are implemented as parallel systems with adaptive capability (Garcia-Lamont, et al., 2002). It is practical to build the pulse-coded neural networks as high speed electronic chips since the connectivity requirements for these networks are less (Johnson and Padgett, 1999).

Cluster analysis is one of the basic tools for exploring the underlying structure of a given data set and is being applied in a wide variety of engineering and scientific disciplines such as medicine, psychology, biology, society, pattern recognition and image processing (Su, et al., 1997). Cluster classification is to generate a nonlinear multidimensional decision boundary that partitions the pattern space into prescribed classes. Classification techniques are finding increasing application as tools for consistent and accurate interpretation of data (Polikar, et al., 1999).

Winner-take-all (WTA) network, also referred to as the Kohonen network is the best known neural network for data clustering in which samples are assigned to one of the K initially specified clusters detected in a competitive learning process (Eltoft and deFigueiredo, 1998). Competitive learning process modifies weights such that similar input feature vectors are clustered into distinct classes (Mann, et al., 1998). Learning non-linear multivariate mappings involves partitioning the input space into regions and fitting a different mapping within each region (Bishop, 2000). During this learning only the weight vector, which is most similar to the given input vector, is accepted for weight building. The training results in the weights converge toward the cluster centers.

The organization of this paper is as follows: Section II presents a brief review on the operation of the CMOS neuron. Section III describes the problem description in detail. Section IV describes the training mechanism implemented, scalability issues, simulation results and conclusions.

II. OPERATION OF THE NEURON

The basic structure of the neuron is shown in Fig. 1. The neuron has multiple synaptic inputs which are connected to a capacitive storage node. This storage node is connected to a threshold circuits block followed by a ring oscillator which provides the final output of the neuron.

Each synaptic input to the neuron, either from other neurons or fed back from the neuron output, can be set as excitatory, inhibitory, or both, based on the values of the synaptic weights, WE2-0 and WI2-0 as shown in Fig. 1. The input voltage to the synaptic input of the neuron is converted to current by using binary signals for controlling the synaptic weights.

The circuits composed by MOS transistors working as current-mode signals reduce active area and interconnection complexity (Garcia-Lamont, et al., 2002). These simple analog current-mode circuits can approximate complex arithmetic processes that appear in Neuro fuzzy systems. The current from a current-mode synaptic input is used to charge a capacitor internal to the neuron. Each neuron can accommodate multiple synaptic inputs. The voltage on this capacitive storage node represents the level of neuron activation. In order to ensure that the neuron returns to a relaxed state even in the absence of inhibitory inputs, a resistive load is attached to this capacitive storage node, forming a leaky integrator. The neuron output behavior is dependent on the level of activation, which is accomplished using multiple Schmitt triggers with different set points. A pulse generator circuit using a ring oscillator with multiple taps generates the neuron output.

III. PROBLEM DESCRIPTION

A 3-cluster classification problem is presented and is solved using an artificial neural network. The inputs and corresponding non-linearly separable output patterns of the 3-cluster classification problem are shown in Fig. 2. The inputs X and Y are 2-bit patterns and correspond to (X1 X0) and (Y1 Y0) respectively.

Straight lines, which represent neurons, can be used to separate the regions between the non-linearly separable output patterns. These regions or neurons can be combined to form different clusters. These separation neurons with their corresponding direction are shown in Fig. 2. Writing the equations for these separation lines will form the relation between the inputs and outputs for the neurons.

From Fig. 2, it can be seen that the region formed by taking N2 in upward direction and N1 in downward direction belongs to the first cluster. Using another neuron, which uses the outputs of N1 and N2 as inputs, the first cluster can be formed. In a similar manner, the regions formed by N3 and N2 with N2 in the opposite direction or N1 and N4 with N4 in the opposite direction form the second cluster. The regions formed by N1 and N4 or N3 with N3 in the opposite direction form the third cluster. These regions for the individual clusters are combined using other neurons. The final network structure for cluster classification is shown in Fig. 3. The output layer has the same dimensionality as the number of output bits. The input layer has the same dimensionality as the number of separation lines to be used.

The input patterns and the outputs for all the neurons in the network of Fig. 3 are shown in Table 1. The separation neurons of the first layer are formed with X and Y inputs. Taking the outputs of the first layer as inputs to the second layer form the regions required for the clusters. The second layer outputs are used as inputs to the third and final layer of the network for additional integration of regions. The final neural network structure is thus formed for this cluster classification problem.

The 10th neuron has an extra input (pulsed high) in order to ensure inversion in the case of a low input. This is applicable in logical operations when all the inputs are low and the expected output is high.

IV. TRAINING THE NETWORK

The initial weights are to be determined analytically from the striaght line equations. The input patterns resulting in true at the output of a neuron are identified. The excitatory or inhibitory or no effect of the input bits is decided by checking these patterns. The inputs with no effect need not be given as synaptic inputs to the neuron. For example, considering neuron N1, it has a true output when only Y0 and Y1 are true. The output does not depend on X0 or X1. Hence, Y0 and Y1 have excitatory effect on the output of N1 while X0 or X1 have no effect.

When the input has an excitatory effect or inhibitory effect, imposing the minimum value of these effects by taking the weights to be the least non-zero value performs the training of the neuron. The training of each neuron involves analysis of the output of the neuron for all combinations of inputs. The weights then changed by checking the charge on the storage node to meet the threshold and by checking the validity of the output for all the input combinations. This process is repeated until good convergence on the output is observed. Excessive training after reasonable convergence of output causes loosing some of the obtained clusters (Wilamowski and Vieira, 1995).

For training, changing the thresholds of neurons in the network by changing the Schmitt triggers inside the neurons can also be useful. It is beneficial in cases where the output of the neuron is used as an input to other neuron and vice-versa. The output of a neuron is not a good digital waveform and is of varying time period but requires to be sampled correctly even for a low pulse when used in an excitatory manner. In this case reducing the neuron threshold will be useful to make sure the sampling of the pulse is done.

SCALABILITY ISSUES

The number of neurons required for this cluster classification network depend on the clustering problem itself and the number of separation lines required for that problem. As the separation lines increases the number of neurons required in the first layer also increases. And more the regions required for a cluster, more neurons would be required for the other layers. Therefore, scalability is dependent on the problem complexity. Increase in size results in increase in power dissipation.

SIMULATION RESULTS AND CONCLUSIONS

Each input combination in Table 1 is applied for 250ns in the order mentioned. All inputs start with a low value and the duty cycle for each input is 0.5. The time periods for Y0, Y1, X0 and X1 are 500ns, 1000ns, 2000ns and 4000ns respectively. The simulation results for the third layer are shown in Fig. 4. Because of the propagation delay through the neuron, the output of any neuron exhibits some delay from its inputs. Therefore, some delay is obvious and is observed in the simulation results.

The proposed approach will work well for n-bit cluster classification. Simulation results in Microsim Pspice showed good agreement when compared to the theoretical results shown in Table 1. A test chip has been fabricated in CMOS process (AMI 1.2μ) and is undergoing tests. This test chip will enable verify the simulations results.

FIGURES AND TABLES

X1	X0	Y1	Y0	N1	N2	N3	N4	N5	N6	N7	N8	N9	N10
0	0	0	0	0	0	0	0	0	0	0	0	0	1
0	0	0	1	0	0	1	0	1	0	0	0	1	0
0	0	1	0	0	1	1	0	0	0	0	1	0	0
0	0	1	1	1	1	1	0	0	1	0	0	1	0
0	1	0	0	0	0	0	0	0	0	0	0	0	1
0	1	0	1	0	0	1	0	1	0	0	0	1	0
0	1	1	0	0	1	1	0	0	0	0	1	0	0
0	1	1	1	1	1	1	0	0	1	0	0	1	0
1	0	0	0	0	0	1	0	1	0	0	0	1	0
1	0	0	1	0	0	1	0	1	0	0	0	1	0
1	0	1	0	0	1	1	0	0	0	0	1	0	0
1	0	1	1	1	1	1	0	0	1	0	0	1	0
1	1	0	0	0	0	1	1	1	0	0	0	1	0
1	1	0	1	0	0	1	1	1	0	0	0	1	0
1	1	1	0	0	1	1	1	0	0	0	1	0	0
1	1	1	1	1	1	1	1	0	0	1	0	0	1

Table 1. Truth table showing desired Input / Output behavior.

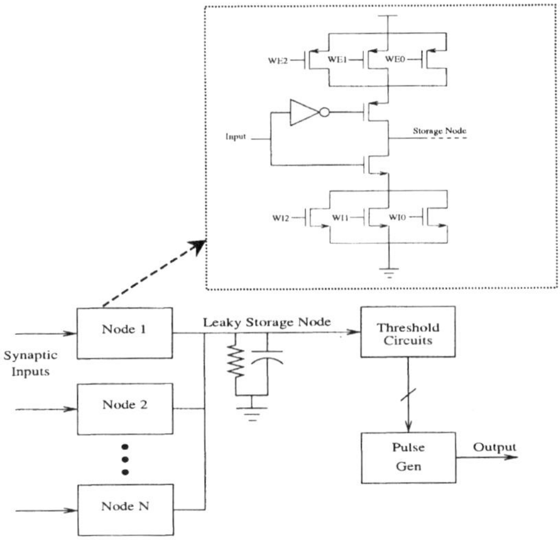

Fig. 1. Block diagram of a single neuron with multiple synaptic inputs.

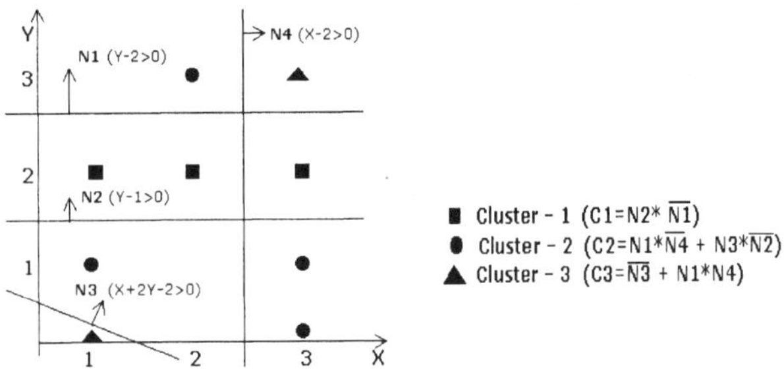

Fig. 2. Graph showing straight lines to separate regions between outputs.

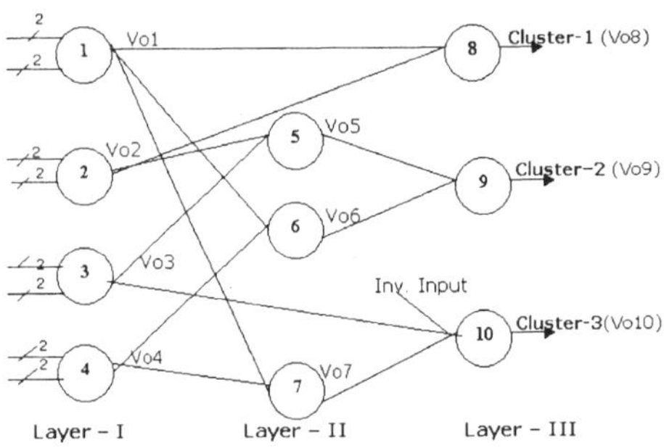

Figure 3. Neural network structure for 3-bit cluster classification problem.

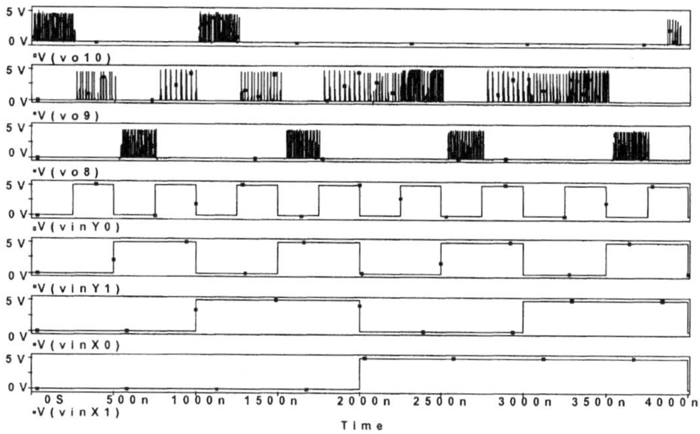

Figure 4. Simulation results showing inputs and final layer outputs.

ACKNOWLEDGMENTS

This work is supported by the NSF-Idaho EPSCoR Program and the National Science Foundation under award number EPS-0132626.

REFERENCES

Bishop, C. M., 2000, "Neural Networks for Pattern Recognition," Oxford University Press Inc., New York., pp. 136-137.

Eltoft, T., and deFigueiredo, R. J. P., 1998, "A New Neural Network for Cluster-Detection-and-Labeling," IEEE Transactions on Neural Networks, Vol. 9, no. 3, pp. 1021-1035.

Garcia-Lamont, J., Flores-Nava, L. M., Gomez-Castaneda, F., and Moreno-Cadenas, J. A., 2002, "CMOS Analog Integrated Circuit for Fuzzy C-means Clustering," Proceedings, 5th Biannual, World Automation Congress, Vol. 13, pp. 462-467.

Johnson, J. L., and Padgett, M. L., 1999, "PCNN Models and Applications," IEEE Transactions on Neural Networks, Vol. 10, no. 3, pp. 480-498.

Liu, B., and Frenzel, J. F., 2002, "A CMOS Neuron for VLSI Circuit Implementation of Pulsed Neural Networks," Proceedings, 28th Annual Conference of the IEEE Industrial Electronics Society, Seville, Spain, pp. 3182-3185.

Mann, J., Lippmann, R., Berger, B., and Raffel, J., 1998, "A Self-Organization Neural Net Chip," IEEE Custom Integrated Circuits Conference, pp. 10.3.1-10.3.5.

Ota, Y., and Wilamowski, B. M., 2000, "CMOS Architecture of Synchronous Pulse-Coupled Neural Network and its Application to Image Processing," Proceedings, 26th Annual Conference of the IEEE Industrial Electronics Society (IECON'00), Nagoya, Japan, pp. 1213-1218.

Polikar, R., Udpa, L., and Udpa, S. S., 1999, "Nonlinear Cluster Transformations for Increasing Pattern Separablity," Proceedings, IJCNN International Joint Conference on Neural Networks, Vol. 6, pp. 4006-4011.

Su, M.-C., DeClaris, N., and Liu, T.-K., 1997, "Application of Neural Networks in Cluster Analysis," IEEE international conference on Systems, Man and Cybernetics, Vol. 1, pp. 1-6.

Wilamowski, B. M., and Vieira, K., 1995, "Clustering of Patterns using Radial Base Function Networks," Proceedings, ANNIE'95 - Artificial Neural Networks in Engineering, St. Louis, Missouri, USA. pp. 109-115.

Line detection using a biologically-motivated, continuous-time, feedback system based on the visual cortex: Part I

Robert Woodley, and Levent Acar
Department of Electrical and Computer Engineering and
Intelligent Systems Center
University of Missouri-Rolla

ABSTRACT:
The visual cortex is a well studied structure in the brain. Numerous laboratory tests have produced a wealth of data concerning the input-output behavior of the cortex. One of the main functions of the cortex is to detect lines in the visual image. Many theories as to how the cortex performs this task have been proposed. What we have done is to take several of these theories and combine them with the concept of a continuous-time, feedback system. Our results show that the orientation and length of the lines in an image can be detected using our system. This article is part one of two and covers the main detection engine consisting of Gabor filters used as the kernel of an integral transform.

1 Introduction

In 1962, Hubel and Wiesel published their article about the functional architecture of the cat visual cortex (Hubel and Wiesel, 1962; Hubel, 1988). Their discoveries sparked a renewed interest in research of the cortex. Their most important discovery being the layers of orientation sensitive cells. For the first time a glimpse of how the brain actually processes an image was seen.

In 1993, Zeki published his book," A Vision of the Brain" (Zeki, 1993). In this book, Zeki describes various paths that the visual image takes while being processed by the cortex. In particular he describes the form, motion, color, and form from color paths. These four paths can be seen as independent, although they are all part of the cortex. Zeki shows through clinical examples how each path resides within different structures within the cortex. This concept was also published by Sereno (Sereno, 1993). Our cortex is modeled on this concept. This article focuses on the form path, but the other paths will soon be implemented. In addition to the separate path concept, Zeki also mentions that there is feedback and an integration process being performed by the cortex. Cortical feedback is also described in (Nielsen, Tanabe, Imaruoka, Sekiyama, Tashiro, and Miyauchi, 1999). We will be exploiting this in future work.

Several others have been used in development of our cortex system. Chang and Chatterjee (Chang and Chatterjee, 1993) describe that the orientation sensitivity can be realized using Gabor functions. Gabor and other similar type filters have also been used to describe the function of the cortex (Zakaria, Khondker, and Akers, 1997; Folsom and Pinter, 1998; Milanova, Elmaghraby, Wachowiak, and Campilho, 2000; Daugman, 1985). We use these filters embedded within a transform operation that has also been used to describe the cortex (Thomson, Cowan, Elder, Daviet, Soga, Zhang, Grant, and Zuckerman, 1995; Cavanagh, 1985; Amari, 1985). Finally, our cortex system, once it is complete, will be part of a larger hierarchical scheme to perform object recognition similar the concept found in (Hartmann, 1985). Additional resources include (Orban, 1985; Payne and Peters, 2002).

This article will proceed as follows. Section 2 describes the integral transform representation for finding the lines in an image using Gabor filters as the integral kernel. We then show our results in Section 3 and draw conclusions in Section 4.

2 An Integral Transform Using Gabor Filters

In this section we will show what is a Gabor filter, how it can be directionally tuned, then how it can be embedded in an integral transform to perform the operation of line detection.

2.1 The Gabor filter

The mathematical expression for the Gabor filter is shown in Equation 1 (Chang and Chatterjee, 1993)

$$g(x - x_0, y - y_0, u_0, v_0) = e^{-\left[\frac{(x-x_0)^2}{2\sigma_x^2} + \frac{(y-y_0)^2}{2\sigma_y^2}\right] + i2\pi[u_0(x-x_0)+v_0(y-y_0)]}, \tag{1}$$

where x and y are the spatial locations, x_0 and y_0 are the spatial centers of the filter, and u_0 and v_0 are the spatial frequencies in the x and y direction, respectively. From this formulation the size and directionality can be tuned.

Equation 1 is a two dimensional (2-D) case of the Gabor filter. It consists of a 2-D Gaussian modulated by sine waves in the x- and y- directions. The 2-D Gaussian marks the center of the filter in spatial location by means of the x_0 and y_0 terms. The spatial bandwidth of the filter is set by the σ_x and σ_y terms. The sine waves represented by the complex exponential term sets the directionality by means of the u_0 and v_0 terms.

2.2 Tuning the Gabor filter

Tuning the Gabor filters consists of three steps. Step one, the center location must be determined. Given the pixels from the image, the location where a line is to be detected are set by the terms x_0 and y_0 as stated above. Step two, the width of the filter must be determined. It was shown by Hubel and Wiesel (Hubel, 1988) that the surround field of the cortex cell is basically circular, however they prefer lines in particular directions. Therefore, we selected our Gaussian to have a circular shape since this represents the surround field of the filter. We selected σ_x and σ_y to both have a value of one. This made the spatial bandwidth circular, and also set the extent of which cells/pixels have any effect on the output. It basically causes a surround field of two pixels, with the most importance being given to the pixel directly next to the target pixel.

The most difficult part of the filter tuning was step three, determining the spatial frequency. The u_0 and v_0 terms while representing the spatial frequencies also perform the job of setting the directionality of the filter. The job of setting the direction is made easier by performing a variable transformation into polar coordinates, $(u_0, v_0) = (\omega_0\cos(\theta), \omega_0\sin(\theta))$, where ω_0 is the spatial frequency in the θ direction. Therefore, the directional sensitivity can be selected by simply choosing θ. This then leaves selecting ω_0. Selecting too high a frequency causes the image to blur and no lines are detected. Selecting too low a frequency leads to bias problem due to the center being at a value of one while neighboring pixels are set near zero. Thus the center is too strong and biases the output to the point where a threshold operation would be difficult. The frequency that we selected was $\omega_0 = 0.5$. This frequency causes the positive and negative parts of the filter to be about even.

2.3 Gabor filter as an integral transform kernel

The form of the Gabor filter is a reason we selected it for our work. In (Woodley and Acar, 2001; Woodley and Acar, 2002) we showed certain criterion that the kernel must have. The Gabor filter has these criterion and is already in the form needed to be used as an integral transform kernel. The transform expression is shown in Equation 2

$$y(X_0, \omega_0, \theta) = \int_{\partial X} g(X - X_0, \omega_0, \theta) u(X) dX, \tag{2}$$

where y is the output, X is the vector notation for (x, y), g is the Gabor filter, ∂X means the surface of X, and u is the input to be transformed.

The operation of the transform integral is then obvious. Given a location X_0 where line is to be detected, and the direction of interest, the integral transform performs a convolution operation that will produce a spike if a line exists and zero if no line exists, for the particular direction of interest, at that point. The intensity of the spike will be influenced by the contrast between the line and its surround as well as the correlation with the direction. A line that is a few degrees off the direction of the filter will have a smaller output than a line that is in the correct orientation. This type of operation is the same as that described by Hubel and Wiesel for the operation of the cortex. The cortex is composed of layers, where each layer is sensitive to a particular direction, while the position within the layer corresponds to a particular spatial location on the retina (Hubel, 1988).

3 Results

Two similar images were used to test the cortex. First, an image that was composed of three flat sections and two sections of large changes was transformed. Figure 1 shows the first image. The sections with the large changes are where the lines are located.

Three outputs were generated. The first output shown in Figure 2 is a filter set in the perpendicular direction to that of the lines. The transform detected lines in this perpendicular direction, even though it seems it should not have. However, upon further inspection, the lines are composed of small lines in this perpendicular direction. In fact the filter correctly found the lines, and we show later that it also detects the larger lines with the same ease, much the same way our own vision works. Figure 3 shows the output to the filter for a 45 degree filter direction.

The output shows activity all along the actual lines. This is exactly how it should react. Although the the sensitivity is in the 45 degree direction, given the pixel pattern it is possible to interpret the line as being at 45 degrees. The hierarchy that is to follow will determine the actual direction. The actual direction should be simply a matter of finding the maximum output. As shown in Figure 4 the output in the true direction is significantly larger than that of the 45 degree direction.

The second input was a slightly more noisy image. Instead of nice smooth sections and well defined edges, Figure 5 shows very jagged sections and narrow edges. Once again the image was transformed using three directional sensitivities.

Figure 6 shows the perpendicular direction output for the noisy image. The lines have been complete removed, and the peaks of the noise have been reduced. The noise haven't been completely eliminated since the noise may in fact represent surface texture. The transform preserves the texture such that higher levels of the hierarchy may be able to use the information to fill in the complete structure of the viewed object.

Figure 7 shows the 45 degree direction output for the noisy image. Figure 8 shows the 90 degree direction output for the noisy image. Similar to the smooth case, the strongest output is in the 90 degree direction. The noise has mostly been ignored, yet the texture still remains. The 45 degree case is also very near zero since the lines are less wide in this case compared to the smooth surface.

4 Conclusion

In this article we present the first part of two about the first stage of an image processing system based on the visual cortex. This first part covers the main processing element of our cortex system, the integral transform based on a Gabor filter kernel. Part two of the series covers the embedding of the transform into a continuous-time, feedback loop.

We give a brief description of the Gabor filter. The Gabor filter is a modulated Gaussian signal by a sine wave in both the real and imaginary domains. The spatial bandwidth and orientation sensitivity may be explicitly set with the Gabor filter which makes it a very useful tool in image processing. In addition, its particular mathematical form also makes it an excellent candidate as a transform kernel.

Our results show that the location and orientation of lines are easily determined using the transform. The transform accurately determined the location and orientation of the lines for both a smooth surface and a noisy surface. In addition, the texture of the noisy image was preserved for future processing, while not producing an erroneous output for the line detection.

References

Amari, S. (1985). Formation of retinopy and columnar microstructures by self-organization: A mathematical model. In D. Rose and V. G. Dobson (Eds.), *Models of the Visual Cortex.* John Wiley and Sons, Ltd.

Cavanagh, P. (1985). Local log polar frequency analysis in the striate cortex as a basis for size and orientation invariance. In D. Rose and V. G. Dobson (Eds.), *Models of the Visual Cortex.* John Wiley and Sons, Ltd.

Chang, C. and Chatterjee, S. (1993). Ranging through gabor logons-a consistent, hierarchical approach. *IEEE Transactions on Neural Networks 4*(5), 827–843.

Daugman, J. G. (1985). Representational issues and local filter models of 2-d spatial visual encoding. In D. Rose and V. G. Dobson (Eds.), *Models of the Visual Cortex.* John Wiley and Sons, Ltd.

Folsom, T. C. and Pinter, R. B. (1998). Primitve features by steering, quadrature, and scale. *IEEE Transactions on Pattern Analysis and machine Intelligence 20*(11),

1161–1173.

Hartmann, G. (1985). Hierarchical contour coding by the visual cortex. In D. Rose and V. G. Dobson (Eds.), *Models of the Visual Cortex.* John Wiley and Sons, Ltd.

Hubel, D. (1988). *Eye, brain, and vision.* Scientific American Library.

Hubel, D. and Wiesel, T. (1962). Receptive fields, binocular interaction, and functional architecture in the cat's visual cortex. *Journal of Physiology, London 160*, 106–154.

Milanova, M. G., Elmaghraby, A. S., Wachowiak, M. P., and Campilho, A. (2000). A computational model of visual cortex receptive fields using spatio-temporal filters. In *Proceeings of the 2000 IEEE/EMBS Conference on Information Technology Applications in Biomedicine.*

Nielsen, M. L., Tanabe, H., Imaruoka, T., Sekiyama, K., Tashiro, T., and Miyauchi, S. (1999). Reciprocal connectivity in visual cortex: Evidence from fmri. In *Proceeings of the 1999 IEEE/SMC Conference on Systems, Man, and Cybernetics.*

Orban, G. A. (1985). *Neuronal operations in the visual cortex.* Springer-Verlag Berlin Heidelberg NewYork Tokyo.

Payne, B. R. and Peters, A. (2002). *The cat primary visual cortex.* Academic Press.

Sereno, M. E. (1993). *Neural computation of pattern motion.* MIT Press.

Thomson, J. R., Cowan, W., Elder, K. R., Daviet, P., Soga, G., Zhang, Z., Grant, M., and Zuckerman, M. J. (1995). Modelling pattern formation on primate visual cortex. In P. D. E. Mosekilde and P. D. O. G. Mouritsen (Eds.), *Modelling the Dynamics of Biological Systems: Nonlinear Phenomena and Pattern Formation.* Springer-Verlag erlin Heidelberg.

Woodley, R. S. and Acar, L. (2001). A generalized unstructured artificial neural network architecture: A first study. In *Intelligent Engineering Systems Through Artificial Neural Networks.* New York, New York: ASME Press. Proceedings of the Artificial Neural Networks in Engineering (ANNIE 2001).

Woodley, R. S. and Acar, L. (2002). A generalized unstructured artificial neural network architecture for control: A first study. In *Proceedings of the 2002 American Control Conference.* IEEE Press. ACC 2002.

Zakaria, A., Khondker, T., and Akers, L. A. (1997). Analog implementation towards artificial post-retinal processing. In *Proceeings of the 1997 IEEE International Symposium on Circuits and Systems.* Hong Kong.

Zeki, S. (1993). *A vision of the brain.* Blackwell Scientific Publications.

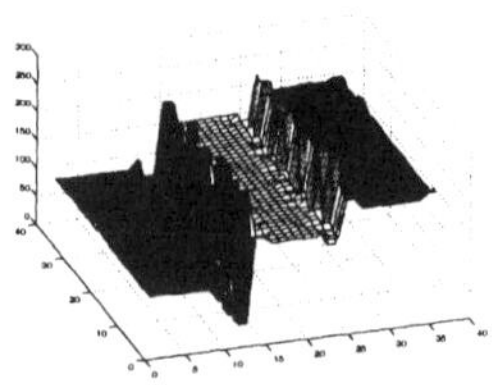

Figure 1: Smooth surface with two lines and three different intensities.

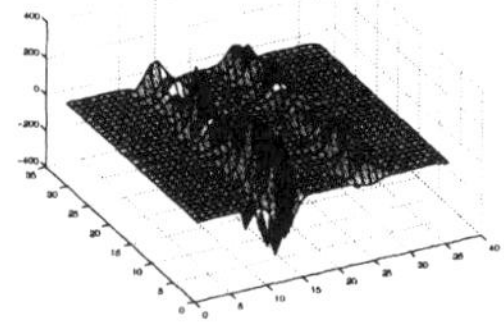

Figure 2: Filter output to the smooth surface set at 0 degrees, i.e. perpendicular to the lines.

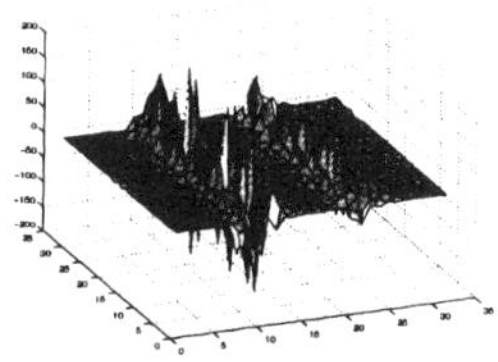

Figure 3: Filter output to the smooth surface set at 45 degrees.

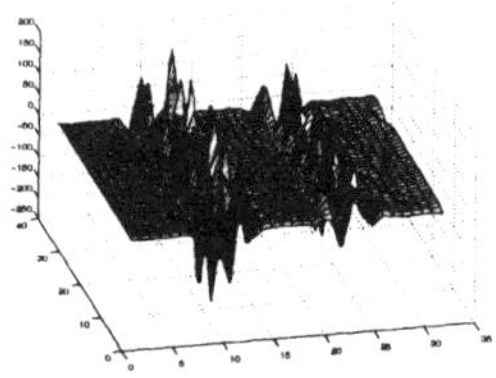

Figure 4: Filter output to the smooth surface set at 90 degrees.

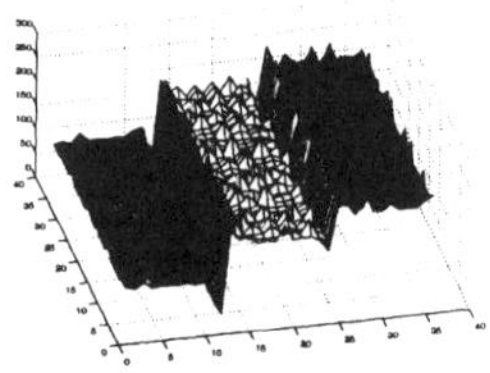

Figure 5: Noisy surface with two lines and three different intensities.

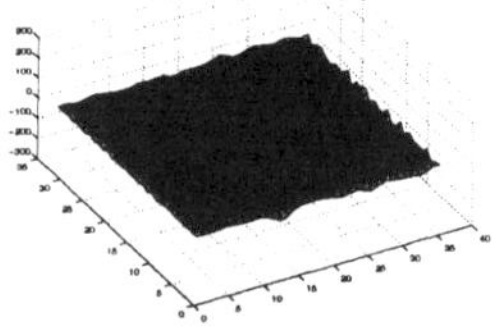

Figure 6: Filter output to the noisy surface set at 0 degrees, i.e. perpendicular to the lines.

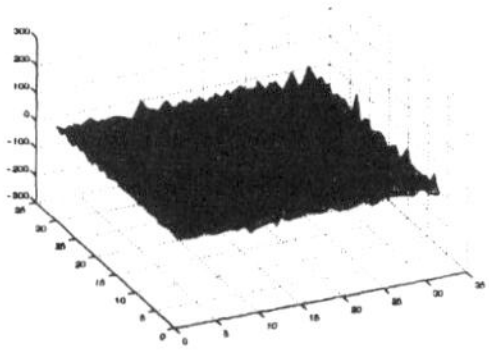

Figure 7: Filter output to the noisy surface set at 45 degrees.

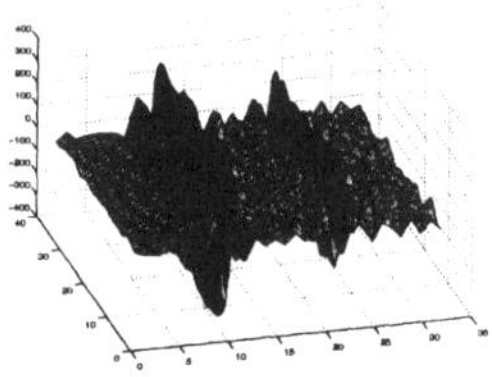

Figure 8: Filter output to the noisy surface set at 90 degrees.

Line detection using a biologically-motivated, continuous-time, feedback system based on the visual cortex: Part II

Robert Woodley, and Levent Acar
Department of Electrical and Computer Engineering and
Intelligent Systems Center
University of Missouri-Rolla

ABSTRACT:
The visual cortex is a well studied structure in the brain. Numerous laboratory tests have produced a wealth of data concerning the input-output behavior of the cortex. One of the main functions of the cortex is to detect lines in the visual image. Many theories as to how the cortex performs this task have been proposed. What we have done is to take several of these theories and combine them with the concept of a continuous-time, feedback system. Our results show that the orientation and length of the lines in an image can be detected using our system. This article is part two of two and covers the feedback system that is used to enhance the image.

1 Introduction

In 1962, Hubel and Wiesel published their article about the functional architecture of the cat visual cortex (Hubel and Wiesel, 1962; Hubel, 1988). Their most important discovery being the layers of orientation sensitive cells.

In 1993, Zeki published his book,"A Vision of the Brain" (Zeki, 1993). In this book, Zeki describes various paths that the visual image takes while being processed by the cortex. This concept was also published by Sereno (Sereno, 1993). In addition to the separate path concept, Zeki also mentions that there is feedback and an integration process being performed by the cortex. Cortical feedback is also described in (Nielsen, Tanabe, Imaruoka, Sekiyama, Tashiro, and Miyauchi, 1999).

We will be using a feedback mechanism similar to those found in (Thomson, Cowan, Elder, Daviet, Soga, Zhang, Grant, and Zuckerman, 1995; Cavanagh, 1985; Amari, 1985). The main difference in our system is the use of Gabor filters as the transformation kernel embedded in the feedback loop. Gabor and similar type filters have also been used to describe the function of the cortex (Zakaria, Khondker, and Akers, 1997; Folsom and Pinter, 1998; Milanova, Elmaghraby, Wachowiak, and Campilho, 2000; Daugman, 1985). Additional resources include (Hartmann, 1985; Orban, 1985; Payne and Peters, 2002).

2 An Integral Transform Using Gabor Filters

The mathematical expression for the Gabor filter is shown in Equation 1 (Chang and Chatterjee, 1993)

$$g(x-x_0, y-y_0, u_0, v_0) = e^{-\left[\frac{(x-x_0)^2}{2\sigma_x^2}+\frac{(y-y_0)^2}{2\sigma_y^2}\right]+i2\pi[u_0(x-x_0)+v_0(y-y_0)]}$$

The form of the Gabor filter is a reason we selected it for our work. In (Woodley and Acar, 2002) we showed certain criterion that the kernel must have. The Gabor filter has these criterion and is already in the form needed to be used as an integral transform kernel. The transform expression is shown in Equation 2

$$y(X_0, \omega_0, \theta) = \int_{\partial X} g(X - X_0, \omega_0, \theta) u(X) dX,$$

where y is the output, X is the vector notation for (x, y), g is the Gabor filter, ∂X means the surface of X, and u is the input to be transformed.

3 A Feedback Representation of the Cortex

Figure 1 shows the feedback system. This diagram is based on the mathematics found in (Thomson, Cowan, Elder, Daviet, Soga, Zhang, Grant, and Zuckerman, 1995). It consist of a nonlinear feedback in that the feedback term is not subtracted from the input to cause an error term, rather it multiplies the input. This type of feedback is based on Hebbian learning (Amari, 1985).

The signal that multiplies with the input, V, represents the excitation of the connection. The excitation signal, V, is a continuous time function. The time integral, $\frac{1}{s}$, is the memory unit that makes it possible to store information about the image from one moment to the next. The input to the integrator is also nonlinear. The input is composed of a normalized and scaled multiplication of the input signal and the output. The multiplication reinforces the sections where the lines occur, and inhibits the sections where there are no lines. The scale factor, A, sets the adaptation rate for the system. The normalization keeps the signal from becoming too excited and saturating.

Finally, the main processing element is the integral term as shown in Equation 2. As mentioned earlier, this part of the block diagram is performing the task of line detection. More importantly, it is performing a transformation from each pixel representing a point on the retina to representing lines.

4 Results

To test the system, a constant black surface was applied to the artificial cortex. Figure 2 shows the constant surface. At time 0.5s a square in the center of the image was raised to a gray color as shown in Figure 3.

The output figures shown are for a single orientation. Similar outputs are achieved for different orientations. Figure 4 shows the excitation just before the center square is applied. Figure 5 shows the output of the cortex.

Figure 6 shows the excitation immediately following the application of the center square. The excitation is very small at this point, therefore, sending the image directly to the filter. Figure 7 is the output. The filter detects the edges of the square, for the given orientation of the filter.

Figure 8 shows the excitation after 1.0s of time. The output, shown in Figure 9, is reduced by a large amount compared to the previous output. This is very similar to a phenomena with human vision. An image that is projected on the human retina will slowly fade unless the image can be made to shift to a new location on the retina.

To test how it performs with noisy images, a noisy black image is applied as shown in Figure 10. Figure 11 shows the output. The noise is greatly reduced, and a large

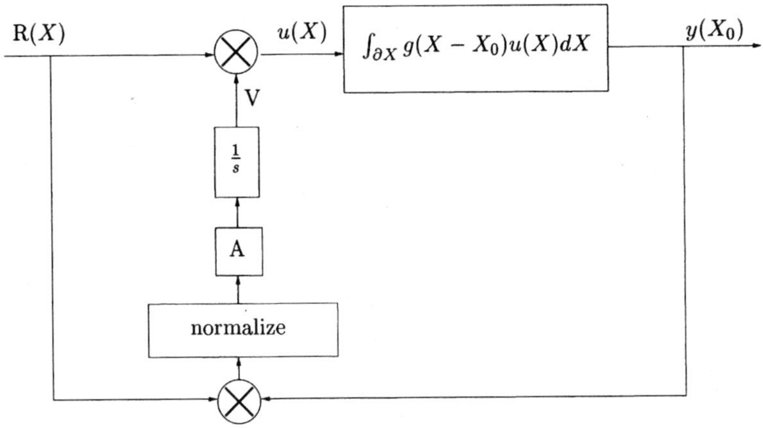

Figure 1: Block diagram of the feedback system.

threshold would interpret the surface as flat. However, the texture can be preserved if a lower threshold is used.

The input image is a real output from the previous stage of our vision system. The previous stage enhances edges via the Mach effect (Woodley and Acar, 2000). This causes the image to change slightly, hence, the edges become more pronounced as shown in Figure 12. The output, shown in Figure 13, shows a much more defined output of where the edges occur in the image.

5 Conclusion

In this article we present the second part of two about the first stage of an image processing system based on the visual cortex. This second part covers the feedback of our cortex system. Part one of the series covers the main precessing unit of the integral transform using a Gabor filter kernel.

We give a brief description of the Gabor filter. The spatial bandwidth and orientation sensitivity may be explicitly set with the Gabor filter which makes it a very useful tool in image processing. In addition, its particular mathematical form also makes it an excellent candidate as a transform kernel.

Our results show that the feedback system adapts to the input image. The ability to detect lines of particular orientation is preserved, it also has the ability to detect both a light edge and a dark edge just as a human. The feedback loop also reduces noise effects using the adaptation effect.

References

Amari, S. (1985). Formation of retinopy and columnar microstructures by self-organization: A mathematical model. In D. Rose and V. G. Dobson (Eds.), *Models of the Visual Cortex*. John Wiley and Sons, Ltd.

Cavanagh, P. (1985). Local log polar frequency analysis in the striate cortex as a basis for size and orientation invariance. In D. Rose and V. G. Dobson (Eds.), *Models of the Visual Cortex.* John Wiley and Sons, Ltd.

Chang, C. and Chatterjee, S. (1993). Ranging through gabor logons-a consistent, hierarchical approach. *IEEE Transactions on Neural Networks 4*(5), 827–843.

Daugman, J. G. (1985). Representational issues and local filter models of 2-d spatial visual encoding. In D. Rose and V. G. Dobson (Eds.), *Models of the Visual Cortex.* John Wiley and Sons, Ltd.

Folsom, T. C. and Pinter, R. B. (1998). Primitve features by steering, quadrature, and scale. *IEEE Transactions on Pattern Analysis and machine Intelligence 20*(11), 1161–1173.

Hartmann, G. (1985). Hierarchical contour coding by the visual cortex. In D. Rose and V. G. Dobson (Eds.), *Models of the Visual Cortex.* John Wiley and Sons, Ltd.

Hubel, D. (1988). *Eye, brain, and vision.* Scientific American Library.

Hubel, D. and Wiesel, T. (1962). Receptive fields, binocular interaction, and functional architecture in the cat's visual cortex. *Journal of Physiology, London 160*, 106–154.

Milanova, M. G., Elmaghraby, A. S., Wachowiak, M. P., and Campilho, A. (2000). A computational model of visual cortex receptive fields using spatio-temporal filters. In *Proceeings of the 2000 IEEE/EMBS Conference on Information Technology Applications in Biomedicine.*

Nielsen, M. L., Tanabe, H., Imaruoka, T., Sekiyama, K., Tashiro, T., and Miyauchi, S. (1999). Reciprocal connectivity in visual cortex: Evidence from fmri. In *Proceeings of the 1999 IEEE/SMC Conference on Systems, Man, and Cybernetics.*

Orban, G. A. (1985). *Neuronal operations in the visual cortex.* Springer-Verlag Berlin Heidelberg NewYork Tokyo.

Payne, B. R. and Peters, A. (2002). *The cat primary visual cortex.* Academic Press.

Sereno, M. E. (1993). *Neural computation of pattern motion.* MIT Press.

Thomson, J. R., Cowan, W., Elder, K. R., Daviet, P., Soga, G., Zhang, Z., Grant, M., and Zuckerman, M. J. (1995). Modelling pattern formation on primate visual cortex. In P. D. E. Mosekilde and P. D. O. G. Mouritsen (Eds.), *Modelling the Dynamics of Biological Systems: Nonlinear Phenomena and Pattern Formation.* Springer-Verlag erlin Heidelberg.

Woodley, R. S. and Acar, L. (2000). An analysis of the human vision process as a decentralized continuous-time dynamical system. In *Proceedings of the Sixth International Conference on Control, Automation, Robotics and Vision*, Singapore, December.

Woodley, R. S. and Acar, L. (2002). A generalized unstructured artificial neural network architecture for control: A first study. In *Proceedings of the 2002 American Control Conference.* IEEE Press. ACC 2002.

Zakaria, A., Khondker, T., and Akers, L. A. (1997). Analog implementation towards artificial post-retinal processing. In *Proceeings of the 1997 IEEE International Symposium on Circuits and Systems.* Hong Kong.

Zeki, S. (1993). *A vision of the brain.* Blackwell Scientific Publications.

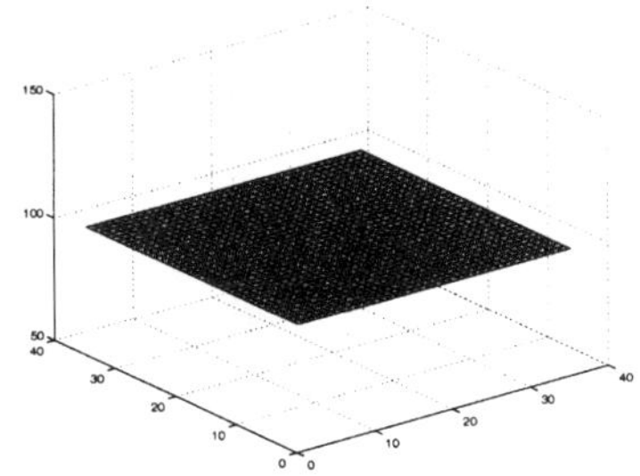

Figure 2: Test surface for time 0.0 to 0.5 second.

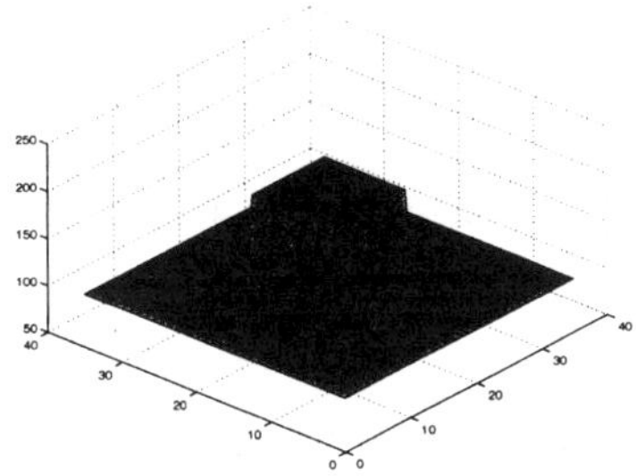

Figure 3: Test surface for time 0.5 to 1.0 second.

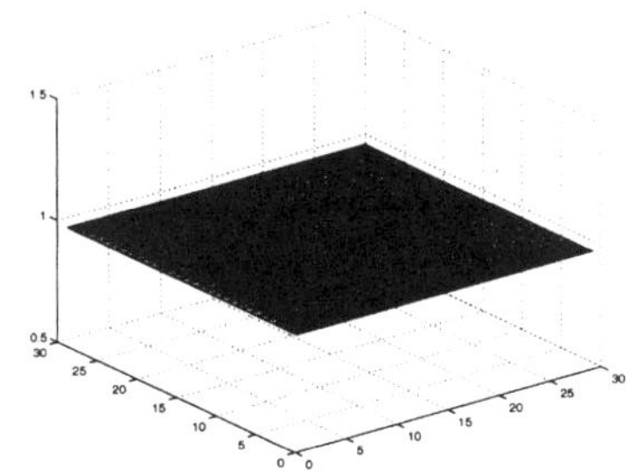

Figure 4: Excitation surface for time 0.4975 second.

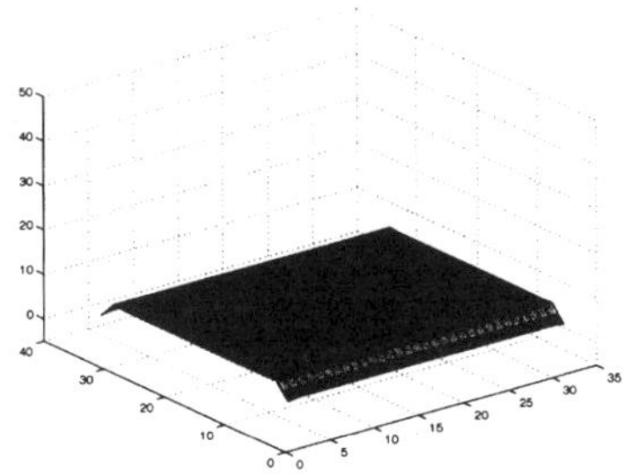

Figure 5: Output surface for time 0.4975 second.

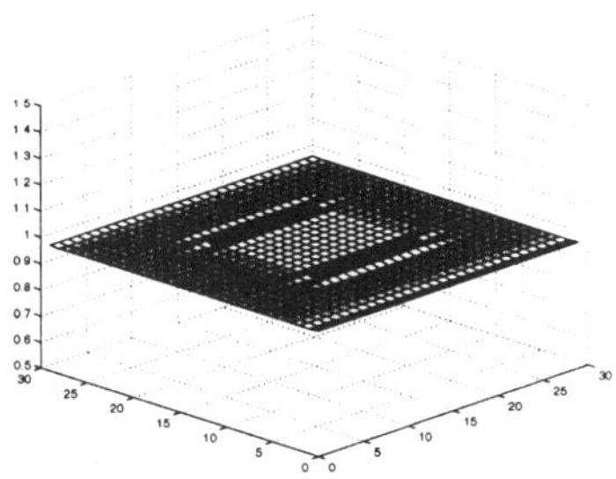

Figure 6: Excitation surface for time 0.5125 second.

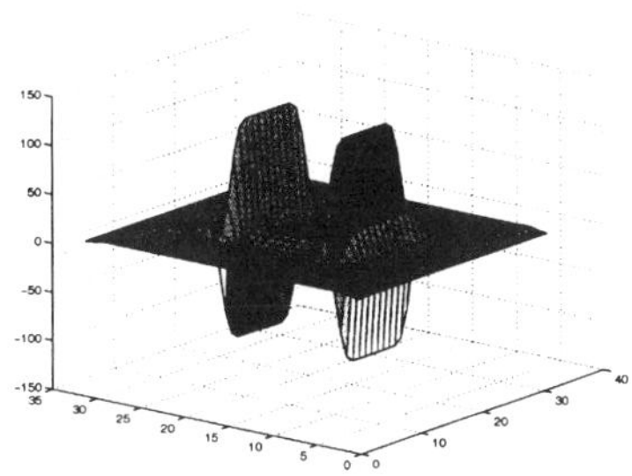

Figure 7: Output surface for time 0.5125 second.

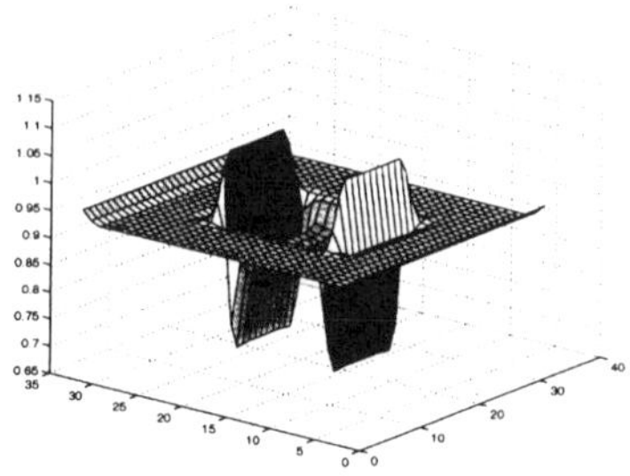

Figure 8: Excitation surface for time 1.0 second.

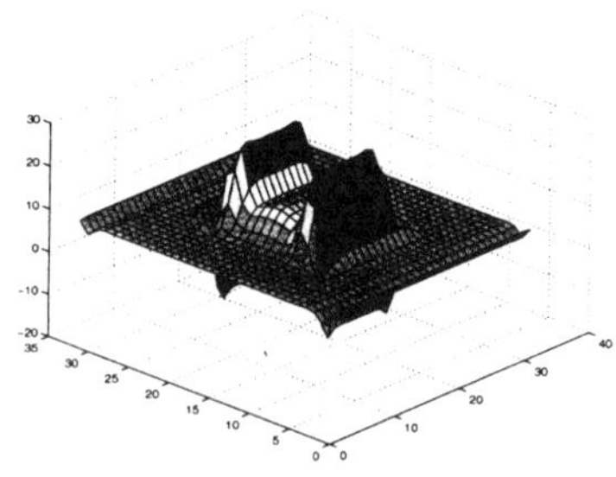

Figure 9: Output surface for time 1.0 second.

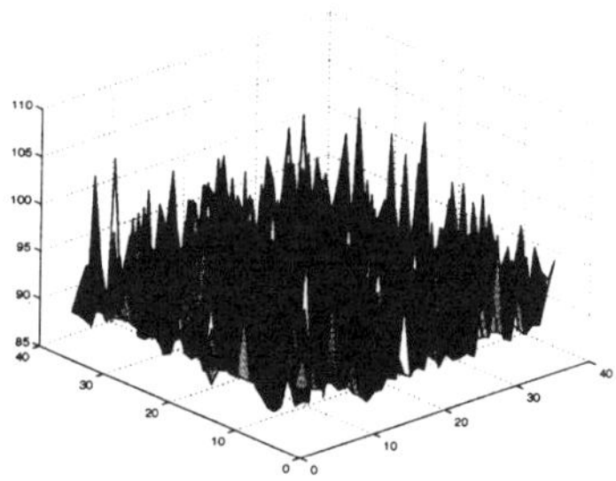

Figure 10: Noisy input surface for time 0.2625 second.

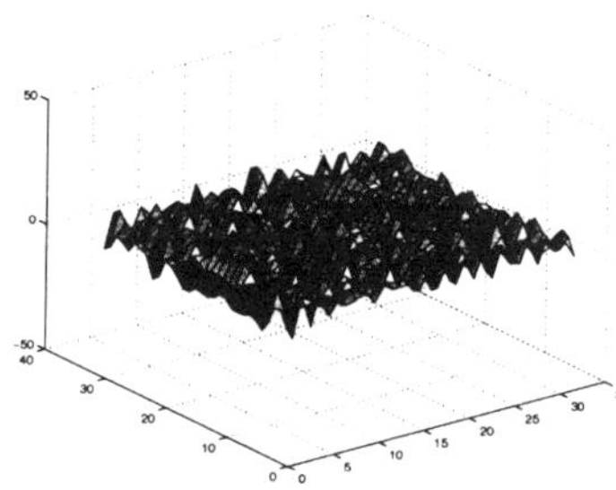

Figure 11: Output surface for time 0.2625 second.

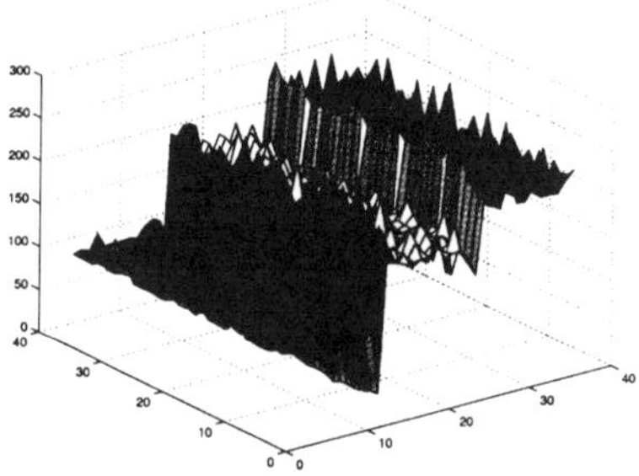

Figure 12: Noisy input surface for time 0.75 second.

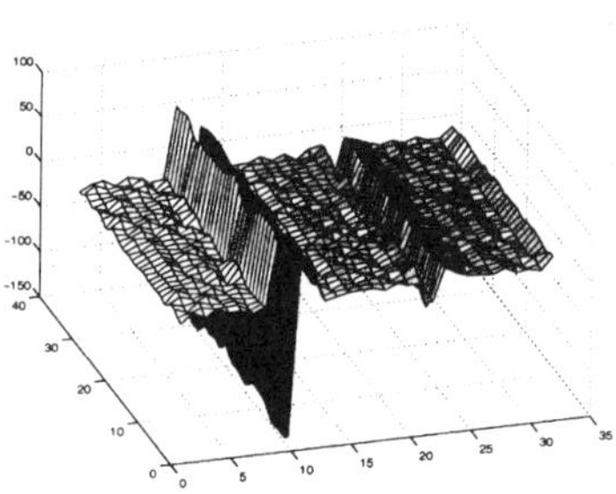

Figure 13: Output surface for time 0.75 second.

The Architecture of a Wisdom Machine (WM)

SYED V. AHAMED
City University of New York
365, Fifth Avenue, New York 10016

VICTOR B. LAWRENCE
Bell Labs Innovation
Holmdel, New Jersey 07733

ABSTRACT
In this paper, we propose that wisdom can be distilled from knowledge, knowledge can be derived from information, information can be gathered from data and finally data can be assembled, acquired, switched and channeled by the existing tools and techniques borrowed from computer science and network technology. A variety of data sensors activate the system and at the second level the traditional AI techniques will process the data through the information and the knowledge levels. The final pinnacle of wisdom is an all-human activity where the mind is the super-processor from information and knowledge derived from six computer systems (or subsystems). Each subsystem performs a step either in a sequential or a recursive mode of execution to either forward a new concept in the dictionary of Wisdom or it will link and classify the realm of current activity to an existing tree of knowledge or as minutia of known knowledge.

I. INTRODUCTION

The word *wisdom* dates back to Aristotle. The original use is to mean the capacity of judging rightly between the good and just life and, as we know, the word *philosophy* is rooted in the Greek word for wisdom. It has not been in philosophic discourse since the Middle Ages. The *Oxford English Dictionary* is not much help, either, the definition is basically circular. The early definitions of wisdom lean toward philosophy (Hughs, 2000). From modern perspective, wisdom points to knowledge, virtue and lasting truth. Here, we present an architectural arrangement of a group of computers that examine the flow of data and information through the networks and extract a first level of knowledge from data to evaluate it further to extract any wisdom that is embedded in the data or events that the machines are constantly sensing and processing. The circular definition of Wisdom in the Oxford English Dictionary is broken by tapping into the flow of information and distilling knowledge and then refining this knowledge to wisdom, if it is wisdom that meets the directional properties (based on the area of knowledge), its confidence level (of the verifiable truth) and virtue that is associated with social value.

This rational is derived from the classic works in science, arts and philanthropy. *First,* wisdom like truth is illusive; but that has not prevented scientists from writing equations and identities. *Second,* wisdom like beauty is also illusive; but that has not prevented architects, sculptors, and painters from creating masterpieces. *Third,* wisdom like virtue is illusive; but that has not prevented philanthropists from building churches, mosques and shrines for the peace of mind. The list continues. It is evident that even though we cannot define wisdom absolutely, but we can pursue it and we can instruct the machine(s) with every (AI) program to track data, information and knowledge upon which wisdom is based.

In the modern information society, knowledge drives the economy of minds. But the ground rules of the economics of knowledge and information do not completely follow the rules of microeconomics (Michael, 1999) or the rules of macroeconomics (Friedman, 2002). Emotion, passions and feelings get activated as sensitive information and knowledge is processed in the mind. Hence, the machines do the initial processing and reason and rationality prevail. In addition, the machine and human intelligence can tirelessly pursue these two attributes with utmost precision and accuracy. Wisdom has the highest quality of reason embedded in it with timelessness of the truth in the statement of the concept being derived from knowledge. The machines fail to pursue such lofty goals and the human being cannot sustain the stamina for the search covering the Internet space. Thus, a realm of cooperation and coexistence is sought between the machines that do not tire and the minds that do not give up. In this paper we explore the region of intersection and overlap between the capacity of the machines and the networks that work for us in an extremely disciplined way and the minds of humans that will not stop till a discovery is made or an invention has evolved or till a new frontier is crossed or a new barrier is broken. Further, the vigor or vitality of the human mind may last a lifetime, even if the goals are abstract. In the Wisdom Machine (WM), the human being commands a set of slave machines to permit a limited amount of programmed freedom for each machine to do its task but also to quickly abandon fruitless activity or alter the course of execution of programs.

II. ROLE OF DECISION SUPPORT SYSTEMS (DSS)

Decision Support Systems (Stuart, 1993a) are in use for the last fifty years. More recently these software systems have impacted all fields of endeavor ranging from business (The SAP Organization, 2002), medicine (Krol and Reich, 2000), education (Ramakumar, 2003), electronic governments (Ahamed and Lawrence, 2002), and even the judicial systems (Stuart, 1993b). Numerous applications of the knowledge engine are documented by the SAP Organization (2002), Krol and Reich (2000), Ramakumar (2003), and the Mathworks Corporation (2002). A typical framework of a modern DSS is shown in Figure 1. This framework provides an embryonic precursor to the WM. Two features derived from the DDS are *(a)* the man-machine interaction at interface A, and *(b)* the refinement of search for an optimal solution. Queries placed via the input bus (A1 in Figure 1), get recompiled at A, and the environmental variables are received from the Operational databases at B. The DSS also deploys the conventional models from the pure and applied sciences to find valid and acceptable solutions at C. These three segments of the DSS are presented as follows.

2.1 The I/O Subsystem (Interface A)

Problems launched at the input bus A1 get recompiled to a series of instructions that the machine is equipped to process. The usual steps in compilation, assembling, machine code generation and finally the execution follow. The results are reported at the output stations via the output bus A2. Any amount of processing is feasible in the I/O subsystem to suit the user needs. Both image and data processing are appropriate in this subsystem. Rudimentary form of pattern and object recognition, as they are used in context of Artificial Intelligence, are incorporated to further tune the overall DSS to suit the needs of any particular user such as a handicapped person or a genius interacting with the DSS. During this interaction, the machine is sometimes programmed to learn as much about the native intelligence of the user as much as the user is learning about the type of solutions that the machine will provide. Both *deductive* and *predictive* (rather than *inductive*) processing of information and knowledge takes place.

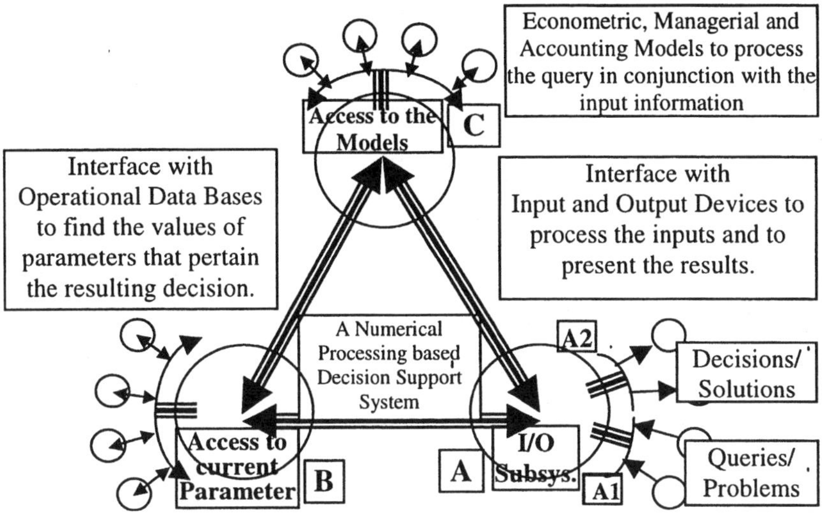

Figure 1. A typical decision support system and its implementation on a computer. Typical queries are received and processed by the (typical) mathematical models in light of the current set of external inputs from the operational databases. Results of possible decisions are presented in view of pros and cons of each decision. Medical, stock market, educational and business systems use the DSS approach effectively and extensively.

2.2 The Operational Parameters (Interface B)

Operational parameters influence the solution that the DSS provides. These parameters are stored in databases, retrieved from networks and/or the DSS may read them as contextual data. The database or network interface at B is shown in Figure 1. It provides the machine with the capacity to retrieve these parameters. This restriction simplifies the function and the decision is provided in the current time and social setting. The question remains unanswered if the question was asked in the earlier centuries or likewise if the question was asked in a different social/cultural environment. Such restrictions of a conventional DSS are removed from the WM by providing it total access to all databases. The machine thus processes information for knowledge or processes knowledge for wisdom from all sources.

2.3 The Scientific, Social and Mathematical Models (Interface C)

Once the parameters are retrieved at the interface B, the scientific, social, econometric/mathematical models stored in database C determine their influence on the solution to the problem. The models are invoked numerous times. At inception of the problem, they provide an initial solution. When the total solution is reviewed, the DSS also checks if the solution is consistent with the established scientific principles and concepts embedded in all the models. In the conventional DSS, these models are generally static and not dynamically updated by self-learning and self-healing algorithms.

III. ROLE OF SENSING NETWORKS

It is possible to make the DSS environment self-monitoring, self-sufficient and dynamically balanced. It can be made artificially intelligent by providing the AI software

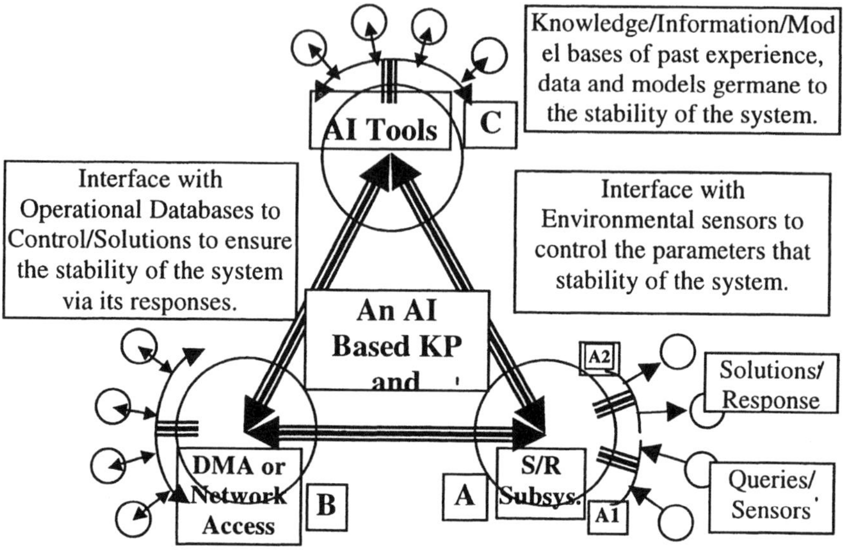

Figure 2. The modification of a typical decision support system and its implementation on a computer with three bus structures **A,** dedicated to I/O operations with sensing functions at the input side and control function on the output side, **B,** automated access to all the relevant information that would impact the action being taken to adjust the control parameters and **C,** the frame work for accessing the knowledge or ground rules that should govern the stabilizing response from the system. The knowledge processing (KP) associated with the AI tools permit the system to process the objects and their attributes.

system to analyze and recognize input parameters, run expert systems software and derive its own optimal solution to a dynamic environment. In a sensing network, shown in Figure 2, the three major building blocks and the interfaces are retained. These sensing networks offer stability of a given environment. If the sensors detect any threat to the overall stability or functionality, then elaborate feedback control systems make localized decisions. A typical example of such a system is found in the cruise control of a car or the autopilot of an aircraft.

3.1 The Sensor/Response System

If the I/O subsystem of the DSS is replaced by sensor/response system, then the machine is forced to function as a stable feed back processor that checks an array of inputs (via the Box A1, Figure 2) and adjusts an array of outputs (via the Response Box A2, Figure 2). The sensors are local or global, they provide current or derived information, or they preprocess the sensor data and act as filters. In the WM, the role of such a sensor system detects new and relevant knowledge from which a new statement of wisdom may be derived.

3.2 The Current Parameter Database

This is akin to the operational database of the DSS. In addition to accessing the parameters that have an impact on the course of action by the processor, the database predicts the changes of such parameters at the time when the corrective measure will be implemented. The predictive nature in the estimation of the parameters would impart a greater sense of stability to the system. It also eliminates possibly unexpected oscillations in the transitory behavior of the sensing systems/networks and of the entire system. Critically damped response or incrementally corrective measures are also invoked from these networks by additional processing at B and C interfaces (see Figure 2).

3.3 The Knowledge, Information and Models Databases

As the rules of stability of the system get complicated or it is delicately balanced, then the AI tools in handling objects (including personalities of people or the medical records of patients) embedded as the parameters become active. This approach that parallels the functions of the algebraic/numerical tools, which would are activated in handling symbols/numbers in the simpler DSS systems. The process of final implementation of the corrective strategy may be purely human activity. This approach is frequently seen in the managerial and political environments. If the machine is to accurately replace the human process (during the execution of medically tailored AI programs such as *Intern* or *Mycin*), then the confidence levels of the machine-based decision are computed along with the inductive reasoning to reach the decision. This approach is further accentuated in the statistically estimation in the WM.

IV. THE CONCEPTUAL FRAMEWORK FOR THE WISDOM MACHINE (WM)

If the execution of the routine function is the forte of the machine, then the creation of novelty is the challenge of the human being. Machines have already started to assist the human beings in almost all the activities ranging from medicine (Krol and Reich, 2000) to electronic governments (Ahamed and Lawrence, 2002). The limitations in the architecture of the hardware and the design of their software (at all the numerous levels) are documented for the computer and network scientists. For this reason, the division of tasks starts to become rational; the numerous approaches that the machine takes to execute its tasks are made available to the human who studies and interprets the validity of the steps the machine took to construct a new knowledge tree (as in *Mycin*) and then derive a new concept (such as an ailment). In the WM, this concept is further validated to check if the concept is an axiom or a universal truth. A framework for the new machine is proposed in Figure 3, with a group of machines for Level I functions, a group of machines interacting with humans for Level II, man-machine functions, and a human environment for Level III, all human functions. These functions are sequential, but they can be iterative and reentrant. Since some of these functions are human, human creativity is invoked. Both *inductive and deductive* thinking is implied at any iteration and at any point of reentrancy.

Parallel tasks can be programmed in the hardware of the WM since the operating system can have the parallel processing and multitasking features. But what is important to determine if the partial results of one task in one processor will not adversely affect the parallel process of the same task in next processor or the inference of the human being. This precaution is more stringent in the processing of knowledge for wisdom than in the processing of data. In addition, the machine environment can be as independent (group of machines) or networked together to perform as a system sharing the tasks. The human

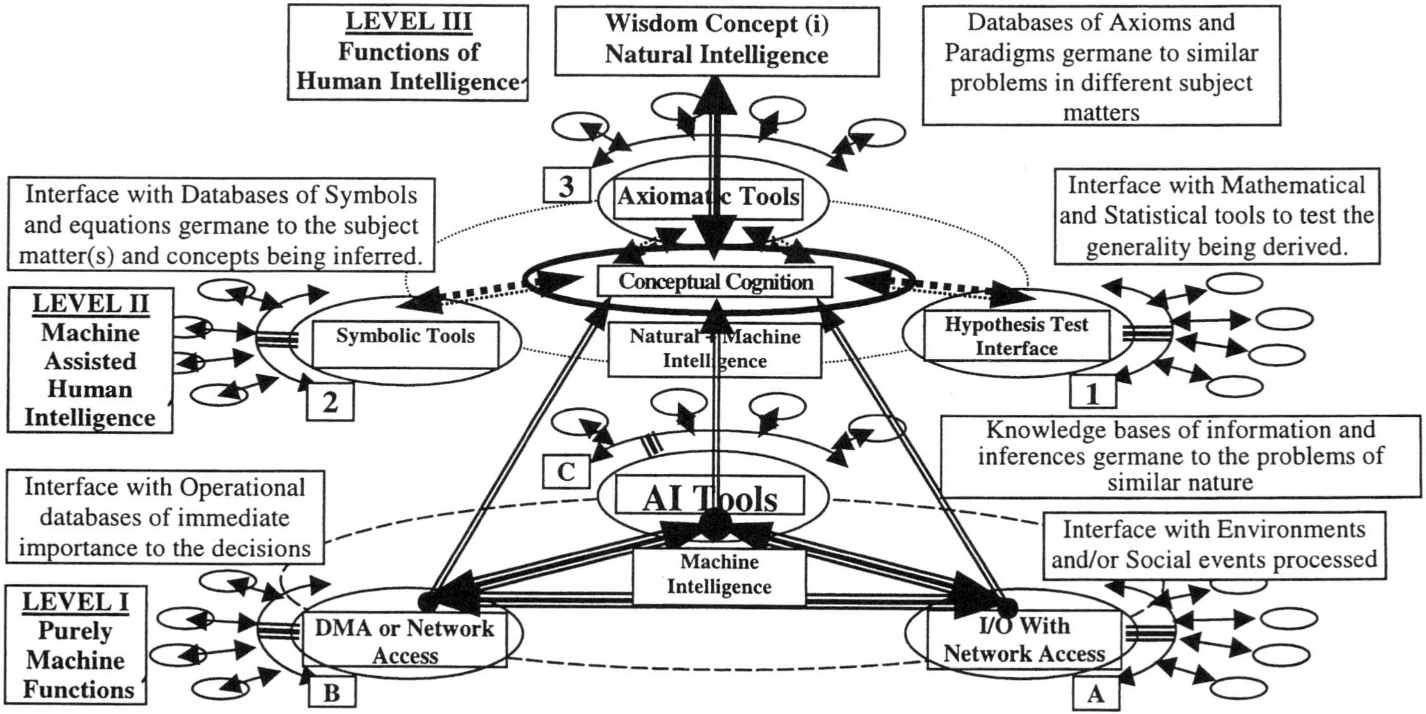

Figure 3. A Three Level Model for the extraction of wisdom from knowledge and decision-making activity. The proposed system needs at least six machines or interfaces to deal with the environmental and social setting (A), knowledge accumulated thus far (B), Cross referencing other disciplines (C) at Level I. Axiomatic tools (3), mathematical validity of the proposed concept (2) and hypothesis testing (1) at Level II permit generality of an axiom of wisdom. It may become logical and necessary to establish additional connectivity between these six machines depending on the generality of the concept being deduced by the Man-Machine system.

environment needs its own design considerations. However, we do not foresee a problem since human beings are more adept than machines.

4.1 Level I (All-Machine) Tasks

All machine tasks are essentially programmable. Even if the machine is capable of altering its own codes and sequence of codes, it is still instructed to do so. Total randomness is a close call to chaos. These routine functions that the machine can execute with great dependability and total accountability are made Level I tasks of the WM. The human mind with its almost non-programmable tasks has a short span of intense attention. All the information necessary for a conceptual breakthrough should be accurately and readily available during these moments of heightened activity (or excitement). Endurance to error and faults without sacrificing the persistence in reaching the goal is still a very human activity. Once documented enough it may even reach the methodology for becoming programmable. Technology now provides efficient computational, access, database, network and AI capabilities. Whereas, we may install independent machines to perform the core computational, access, database and AI services, the (inter and intra) networking services may be independent or shared. This aspect of the WM is not significant and we go on to classify all versions AI based function as Level I functions.

4.1.1 Blocking Trivia from Wisdom. Users can play unproductive games with the WM. The attempt may be voluntary or involuntary. For example if one start to search for established facts as wisdom, or forces the machine to evaluate the value of the Euler's Constant or of pi, beyond the limits of accuracy of the machine, the machine will go into an oscillatory mode hovering its best value that the machine can evaluate. Such a scenario can be handled by the interrupt generated by the Numeric Processor Unit (NPU) within the CPU in traditional computers. Such a trap also needs to be broken by the operating system of the WM. Further, the information or the knowledge equivalent of this scenario is not so easily identified by the AI process handler or the Knowledge Processing Unit (KPU) of the WM. For example, if the user instructed the machine to search for the cure for depression or leukemia, then the machine looks in its knowledge bases and offers a set of possible solutions with a confidence number and keep searching the current updates through out the globe. But at some stage the operating system assigns a lower priority to the open-ended problems. It keeps executing the tasks of routine pursuits that offer germane knowledge/wisdom. Strategies such as deadlock avoidance, Halting algorithms and virus removal are well known to the designers of software and operating systems. The nature of the operating system, the inner most layers of privilege instructions (PI), and the core layer of the software for the "bare WM," need sophistication far in excess of those of the normal computers to prevent silliness, abuse or any unethical pursuits of the users. Very powerful hardware, software and its architecture can indeed, be equally well used for counterproductive purposes. The routines of the WM to block and defend itself from an onslaught of socially destructive, morally corrupt and intellectually demeaning activity need a true interception of (entirely) human wisdom with an immortal sense of truth, beauty and virtue. Such pursuits are foreign to the hardware, software or the firmware of the WM, but the human counterpart is expected to intervene.

4.2 Level II (Man-Machine) Tasks

4.2.1 The Realm of Ingenuity. If ingenuity can be programmed, the machine can handle it and to the extent that it is non-programmable, the human being can handle it. For this reason, the greater adaptation to solving any particular problem is still that of a human. The highest level of programmed rationality of the machine may approach the lowest level of human creativity, since creativity can be spontaneous, it can beautifully true, truly beneficial and universally beautiful. These modes of functionality are still left in the human realm. The human being becomes as much a driver for the machine as the machine is a programmed slave to the human. In the synergy of tasks between the humans and the machines, a trained human may surpasses the purely predictive reasoning or deductive logic of the machine. For this reason, we propose an architecture of the WM that has human beings mentor and monitor a group of machines performing all machine (Level I) tasks. Such an approach is routine in space travel and orbit or landing control of space vehicles. We do believe it is possible to have an architecture where there is one centralized machine and a group of humanoids for Level II functions. The organization would be akin to that of an ant colony in which the queen would also be a robotic machine.

4.2.2 Synergy of Thought and Machine Intelligence. In the cooperative mode between the human and the machine, the burden is heavier on the human. The multiple processes of the machine need to be vectored (by defining the knowledge domain) and mentored (by specifying a confidence level) in the subconscious of the human. A harmony between the instantaneous inferencing based on the synergy between the machine process and the human thinking needs to be achieved with delicate (human) care. In other areas of human endeavor (such as music, painting, sculpting, etc.) this feat is mastered. The artist and the instrument get unified and lose their boundary. In the functionality of the WM, this later task is nontrivial but the human nature has a built in rescue mode and the machine can seize a core-dump. The human mind retains visual and derived information (of significant value) in its medium term memory and it (by itself) simmers over a period of time till it is recovered and reprocessed by the mind or rejected as irrelevant in the context of the solution. Frequently, the human realizes the complete picture/solution in a flash and thus an incremental step in the deriving the wisdom or its increment is achieved as an inspiration. Extending this process one layer higher (Level III), the many macro steps may yield the whole axiom of wisdom.

The delayed perception and sudden realization of the mind compensate the exact synergy between the human and the machine. For this reason the database capabilities of the WM should be able to freeze the process at any given instant of time. Such timings should be an option for the human being interacting with the machine. Sometimes, to restart the natural intelligence process from a new and fresh perspective offers a new and perhaps a more creative solution.

4.2.3 Post Processing of the Extracted Wisdom. When the man machine team infers any new concept or an axiom of wisdom, further processing of the axiom becomes essential to prove its validity. For the concept to have validity, it needs a certain confidence value. The machine discovers similar concepts and explores them from a database in light of the new axiom. The hypothesis embedded in the new concept is tested by hypothesis testing subroutines. These post processing functions (database searches, hypothesis testing and the evaluation of confidence interval) are feasible in the machine domain.

To accomplish an iterative refinement, the machine activity is invoked numerous times to sharpen the initial axiom of wisdom. This type of activity is pursued in most human environment when a resolution or a byelaw is reexamined many times before the final adaptation. The interaction between the members of the all-human team needs to be reincarnated in the man machine interaction within the WM. The Level II functions (see Figure 3) are thus iteratively and recursively invoked.

A range of all machine post processing functions start to emerge: *First,* consider the relaxation of the parameters within the range of their natural bounds to check if the axiom of wisdom proves to be false and thus generates a measure of confidence. Further constraints also need to be imposed to maintain a cooperative and controlled behavior of the machine. The search for absolute wisdom is futile for two reasons: *(a)* network access provides ever increasing information and knowledge bases and their contents around the world and the WM will lose and *(b)* total generality of wisdom is impossible because of the search becomes absolute over all areas of knowledge. This leads to the *second* of the post processing control by the human being on the machine; the search for wisdom is forced to be directional (a knowledge identifier, which may be relaxed later) and numeric (a confidence level, which may be tightened later).

Assigning numerical identifiers to knowledge domains were introduced by Dewey in 1873 as the Dewey Decimal Classification or DDC (OCLC, 2003). Later the Library of Congress classification (United States Government, 2003) further refined this classification and offered even more flexibility. The area of search for wisdom or its axioms is directional by defining a knowledge domain by its numerical identifier.

In studying human behavior in corporations, March and Simon (1993) have documented the directed and limited search strategy. Directed searches are deployed in finding specific solutions. The management team also looks for a feasible solution to the corporate problems rather than an absolute or a perfect solution. We import this idea to control the WM activity in two ways. Before the machine generates its own confidence level for its inference, the human being initially assigns an estimated number for the confidence and forces the machine to search for its validity from all the data and knowledge bases. Hence, two bounds are imposed on the all-machine activity in refining wisdom at Level II of WM shown in Figure 3. These two numbers also force the machine activity from becoming uncontrolled or chaotic.

The processing of the preliminary axioms of wisdom is forced to be vectorial: direction being defined by the knowledge domain identifier and numerical by the confidence level the axiom has to meet. During the normal operation of the WM, the vectored search is modified many times. It is either sharpened or relaxed to evaluate the validity of the axiom of wisdom in three dimensions (universality, truth and social benefit).

If the WM has access to the equations in Mathematics (DDS= 510) or to the equations of Physics (DDS of 530) or both, then it will offer mathematical and physical wisdom as equations. By broadening the search to look for the symbol ∇, the grad operator, the machine would indicate how these mathematical identities deploy ∇. If there were a false deployment of the operator in the current equation or axiom, the WM would provide an area of conflict or falsification. Searching for $\nabla X \nabla$ with a 100% confidence level in direction 510 and/or 530 will positively lead to Maxwell's Equations and so on. It will also validate the initial and the boundary conditions in which the symbol is used and relations pertaining to the use of such a symbol. The WM thus provides a framework of programmability in particular validity or falsification of any axiom with a given confidence level.

4.3 The All Human Functions

The human being keeps the overall solution and the contributions of several independent and interdependent machines in perspective. More than that the human being whose intuition is faster than the computation, forces each machine to function in the direction in which the wisdom is being perceived. False pursuits and trivial are soon eliminated. For example, the WM is perfectly in its bounds to pick up dictionary meaning of words and phases as classic wisdom. It can also rephrase and replace the statements from an encyclopedia and so on with any level of confidence. It can restate basic equations as words of wisdom such as the ratio of circumference to diameter, etc. or within the context of it knowledge bases, it can guide men to function better than machines etc.

The human interacting with this machine is much like a pilot flying a (wisdom) machine in the knowledge domain to explore new dimensions/manifestations of truth and social benefit. To this extent the human wisdom becomes superior to the machine inference, but the sum of the two can become significantly greater than either one. Over a period of time the performance of both is likely to better.

V. SPECIFIC EXAMPLES AND CONFIGURATIONS

The search for *total* wisdom becomes a quest for perfection in *every* direction. One of positive way to defeat the purpose of the WM is to force a DDS of 000.000 to 999.999 (i.e., the entire human knowledge) and a confidence level of 1.00 for its results. Sooner or later, it is bound for failure. On the one hand, if one starts to limit the direction to be highly specific (such as symbols, elements, weather, etc.), and starts to probe a certain context (such as use of symbols in mathematical equations, the atomic weight of elements, the pressure-velocity in weather patterns, etc.) then all the information is gathered by the Level I functions of the WM. Patterns of events and occurrences are documented in the knowledge bases using AI techniques, modern networks and information highways. Pattern recognition in computer graphics, medical diagnostic systems, etc. yield preliminary results. The distillation of the concept is attempted at the Level II functions and a few generalities (Level III functions) are derived and tested leading to a new discovery or an age-old wisdom.

If one starts to become quantitative and assigns high directionality (by defining the exact, precise and the entire Call Number of 621.3191'21-dc20 (including the DDS classification of 621.319) then the search becomes highly specific. Similarly if the user commands the WM with a call number TK5103.15.M38 1995 (including the LOC classification TK5103) and assign a confidence level of 100% then the WM can only produce a book ISBN 0-7803-1121-3). In search for the clues of wisdom, the human being should use the vector of wisdom (direction and confidence value) with due care. Significant results and comparison can be made to test the wisdom of any axiom that the man machine team may initially uncover. If one discovers with 98% of the time that the 523.5 DDS code (for Meteors, Solar winds and Zodiacal light) occurs with 523.75 DDS code (for Chromosphere and Corona) and then one may generalize "Meteors, solar winds and Zodiacal Light lead to Corona discharges." On the other hand in reality, a Library of Congress search under the most liberal definition of these words yields 0.01 confidence that corona and meteors are related. Only additional humanistic effort may shed more light. Such searches and generalizations can be non-trivial. For instance, it is known that diamonds are found where the conditions fitted specific sets of geological patterns (temperature, depth, pressure, soil chemistry, volcanic activity, etc.; X1, X2, X3, X4,

X5...Xn) that persisted for a given length of time (100 years, 1000 years, 10000 years, etc). Conversely, searching for similar geological patterns from the knowledge bases may lead to the diamond mines (e.g., recently discovered diamonds mines in Russia). Other examples of search for species, birds, plants, etc. also exist.

4.1 Other Possible Applications

Consider all the symbols that are used for the noble metals (Au, Ag, Pt, etc.) at Interface A, Level I in Figure 3. Level I functions of WM compiles all the equations of chemistry where they are deployed (Interface B, Level I, Figure 3), and the reason (Interface C) for the use. The dearth of chemical compounds derived from these metals (derived by machine intelligence Level - I) points to their higher atomic weight (derived from Interface B) by either Level I or Level II functions of the machine. This initial inference relating the high atomic weight to their chemical inactivity is submitted to Level II or Level III functions for human approval or denial. Initial inference may hardly be an axiom, but the WM forces the human to investigate further to explore the truth.

Other examples of early WM's also exist. When a group of IBM 7090 computers solved the Stone Henge mystery, or when a computer-imaging machine (purely a machine) was programmed to identify malignant tumors in the human body, then these machines in an artificially intelligent (wise) way have provided the answers that would are (almost) impossible for humans to pursue. The riddles of Stone Henge or the process for discovery of the tumors were resolved. In the present implementation, the WM would not only give these answers but also check of the intellectual heritage of the Early Bronze Age peoples to have had the access astronomical charts and their accuracy of the their sighting devices. All the supporting logic would be traced from the linkages of the builders of the Henge to their own sources of knowledge. To the extent that the WM is not able to document thoroughly, its confidence number would wane. At every stage, the machine offers a conclusion (or a derivation or an inference) and offers its confidence in the result of the micro-step. Such micro-steps are cascaded to draw a major conclusions based on scientific data and/or religious documents traced to the Stone Henge. Weather systems are also good examples of the cases of specifically tailored WM's. Numerical data from the pressure-velocity charts are tied in with the observations of the past and monitored on a continuous basis to affirm or negate any partial hypotheses derived from earlier observation to better estimate the current judgment. The results and their confidence are both specified in the inference of these systems.

When the Level I functions are confined to the search for words that rhyme, ideas that blend or concepts that flow by searching the current databases, the logical linkages are established between the poems of well known poets or the contributions of philosophers. Once a new database is derived from words, ideas and concepts, then the human being can so instruct the machine to represent the current events within the society in words that rhyme, ideas the blend or concepts that flow. The human freedom to play amusing and variation of cross-cultural cross-time games becomes enormous. One can force the ideas of Shakespeare to be recomposed as the Rubayiat of Omar Khayyam, just as well as Dover Beach (by Matthew Arnold) can be composed for the Middle East conflict, and so on. Thus, the WM faithfully generates poems and/or graphics like a computer that generates music. Whereas such music may be discernible by the masters of symphonies and operas, it may suffice for jazz bands of street corners.

WM can also solve a variety of problems. The machine can import art of Renascence in the paintings of Picasso at the Level I of Figure 3. Though unthinkable for human beings, the WM would struggle very hard to make sense of an impossible question or situation. To the extent that any possibility exists, it will be retrieved, derived, inferred or inducted. To the extent that the possibility does not exist, and falls out of the level of confidence imposed by the human being, it is offered to the human being (Level II of Figure 3) to think beyond what the machine was programmed to process (Level I of Figure 3).

A variety of roles for the WM are possible. The edge that WM has over the human being is that it is self-learning at a rate faster than ordinary humans. To this extent the machine can also assume the role of an instructor that learns from some humans and teaches others, much like an instructor would do so.

As another example of the role of the WM as a Labor-Management negotiator, consider the WM at Level I communicating with both Labor and Management to the fullest extent and evaluating the 'Cost of Agreement' and the 'Cost of Disagreement' for every 'Demand of the Labor', and for every 'Concession of the Management'. The first three or four sessions/meetings between the two parties are carried out diligently, thoroughly, and without any deception that the parties may be drawn into. The 'Hicks-Effect" can be reasonable accurately drawn out as the length of negotiations or strike/lockout are prolonged. The WM as a negotiator will make the 'Agreement Process' (if it is possible at all) a smooth and critically damped in moving the terms of contract from one contract period to another. Such a role of the WM is possible for other type of human and national negotiations.

The WM can be abused just as well as it can be used. If the user primes negative intelligence (with terror, anger and hate instead of truth, virtue and beauty), Level I and II functions will turn horror filled or destructive and the WM can be forced to serve as a war machine. In a positive sense, most of the major inventions are blended with the nature of the user to provide a social value or service. For any machine, abuse is also a possible use that it provides. To prevent such an abuse, the WM may be supported (with tamperproof intelligence or wisdom) that is not accessible to the user. Such judgment at the execution time is forced on the WM by forcing the WM to track the direction of the goal and to evaluate its social/human value. When such a mechanism (hw, sw, fw or chip) is inhibited by the abuser, the WM can work towards wicked goals with the same vehemence and provide negative wisdom/intelligence to its user.

VI. CONCLUSIONS

In this paper we present an architecture for building generic man-machine knowledge processing environment that leads to specified generality and to any given level of confidence. The pursuit of total generality like the pursuit of perfection is likely to fail. It is an indefinite run for the resources till the human beings (Levels 2 and 3) would collapse and/or machines would reach their limits of precision. The WM supplements the environment generated from its own knowledge banks of accumulated procedures and inferences about the topic under investigation. The human intelligence firmly drives the programmed AI of the machine in directions that validate or repudiate an axiom of wisdom.

Search and infer activity, at the first level, is purely machine based. With human interaction at the second level, the machine functions as a highly scientific tool with built in numeric validity for all its functions. Furthermore, the partitioning between the extents of the search for wisdom between the machine and human beings is adjusted and gradually tightened as the search proceeds by assigning increasing levels of confidence, as the direction of search gets narrower and narrower. The process is thus forced into a recursive mode yielding some results. In the traditionally man-machine systems, the search for wisdom or inference (as in research) is a human activity. Even though search engines exist, they do not compile the conditions when the results can occur and draw a conclusion based on the results from iterative and reentrant cycles where the machine also plays a significant role. In addition, the WM identifies the weakest link in reason and its validation. Its selectivity and confidence may be both reinforced by human beings. The WM is likely to alter the human perception of wisdom/values much like the information from space vehicles that has altered the perception of the universe.

REFERENCES

Ahamed, S.V. and Lawrence, V. L., 2002, "Evolving Network Architectures for Medicine, Education and Government", Pacific telecommunications Council 24th Annual Conference Paper W.2.2.1, Jan 13-17, , Honolulu, HI. , PTC Paper

Friedman, M., 1976, "Price theory", Aldin de Gruyter, see also Columbia University, "Columbia Encyclopedia," Columbia University Press, New York 2002

Hugh H. B., 2000, "Socratic Wisdom, The Model of Knowledge in Plato's Early Dialogues", Oxford University Press.

March J. G. and Simon H. A., 1993, "Organizations", Blackwell Publishers, 2nd Edition.

Mathworks and Simulink 2003, “New Real-time Embeded Coder 2”, Release 13. Refer to www.mathworks.com, for verification of concept, design and production code for devices and products, MathWorks Consulting Services, Novi, Michigan.

Michael M., 1999, "Dilemmas in Economic Theory: The Persisting Foundational Problems of Microeconomics" Oxford University Press.

OCLC- Dewey Decimal Classification, 2003, "Dewey Decimal Classification and Relative Index, Edition 22," ISBN 0-910608-70-9, OCLC, 6565 Frantz Road Dublin, OH 43017.

Krol M., and Reich D. L., 2000, "Development of a Decision Support System to assist anesthesiologists in operating room," Journal of Medical Systems, June, Vol. 24 No. 3, pp. 141-6.

Ramakumar, V., 2003, "eLearner, Sovler, Platform and Technology" Cloverworxs Inc., Suite 1806, 4411 Spicewood Springs Road, Austin Texas 78759 or visit www.cloverworxs.com. URL accessed July 2003.

SAP (UK) Limited, 2002, "SAP Solutions, and eLearning," Clockhouse Place, Bedfont Road, Feltham, Middlesex, TW148HD.

Stuart, N. S. 1993a, "Computer-Aided Decision Analysis," Quorum Books, Westport, Conn.

Stuart, N. S. 1993b, "Legal Scholarship, Microcomputers and Super-Optimizing Decision-Making," Quorum Books, Westport, Conn.

United States Government (2003), Library of Congress Classification, Refer to http://catalog.loc.gov for an explanation of the LOC classification, coding, and call numbers. URL accessed June 2003.

More Neuromorphic Artificial Neural Networks for Modeling General Neural Control

John R. Alexander
Computer and Information Sciences Department
Towson University
Towson, MD 21252
jalexander@towson.edu

Michael Williams
Northrup Grumond Corp.
Linthicum MD 21090
michael_m_williams@md.northgrum.com

Abstract Over the past nine years, the ability of small artificial neural networks, each consisting of two input neurons and one output neuron, to accomplish a variety of control tasks [1] has been studied. Such three-neuron nets are called Three-Neuron Controllers (TNCs). By 1999 the TNC had proved successful in solving toy control tasks, among them the truck backing problem, but no biological tasks had been attempted. To this end, the task chosen to begin testing the TNC's biological abilities was to determine this circuit's ability to act as a pacemaker [2], [3]. These studies too proved successful. Prompted by Daniel Gardner's [4] call for more neuromorphic ANNs, the truck backing task was re-visited. Using one of Ronald MacGregor's [5] biological neuron models, an attempt was made to use total spike count as a parameter to convey the necessary control information to back the truck. Again, this proved successful [6]. A yet more advanced model, which also generates spiking neurons, NEURON, [7] was tested in this same task. This paper discusses the biological inconsistency of the TNC which was noted in this study. We end our attempt to model toy control problems with TNCs, and begin our attempt to model the control abilities of very elementary biological creatures.

Background As mentioned in the **Abstract** the first author had been investigating the ability of a class of artificial neural networks to solve toy control problems. The origin of the artificial neural network in question was the first author's PhD dissertation [8]. However, the network used in the majority of the studies was a modification of the "IAC" equations of McClelland and Rummelhart [9] which these authors had written as difference equations. The first author cast these equations into differential equation form. In the two-input neuron, one-output neuron case, the resulting non-linear equations could be integrated in closed form [10].

The name of the original differential equations (from the dissertation) was the RX equations. Those equations inspired by McClellend and Rumelhart were called the ENIAC (<u>Ex</u>teNded <u>IAC</u>) equations. Both sets of equations describe a two layered net, i.e. only input and output layers. In the two-input, one-output case (Three-Neuron Controllers or TNCs), the two input nodes represent sensors and the output node represents a controller. Neither the RX or ENIAC version of the equations included a learning algorithm - in the RX equations, probability data served as weighted and parameters. In the ENIAC case, weights were simply assigned.

In the RX case, the investigations were general in nature. The main thrust of the effort was to compare the predictions of the RX network to the predictions of a backpropagation network. The performance of each of these network was similar [8]. In the early 90s control problems were attempted. These included controlling the height of water in a tank under conditions of random change in the inflow (and outflow), truck backing, truck and trailer backing, and the inverted pendulum problem. Wishing to introduce a biological perspective, the truck-backing algorithm was modified to allow control using MacGregor's [5] PTNRN10 model. All previous work is summarized in the journal article **Control: An Emergent Property of Biological Neurons** [1]. This article contains the definition and solutions of both RX and ENIAC equations. To make this article self contained we have included a brief discussion of the ENIAC equations and their solution in Appendix A.

Property of the ENIAC and RX Equations
Figure 1 displays the geometry of the "Truck Backing" problem. The task of the Truck Backing Problem is to back the truck to the center point of the loading dock (coordinate (50, 100)). Only backing is allowed. The ENIAC Equations are presented in Equations 1 and 2, Appendix A. The solutions to the ENIAC equations, Equations 7 and 8 in Appendix A, both contain the term $w_{11} * m_1 + w_{12} * m_2$. Also, given in Appendix A,

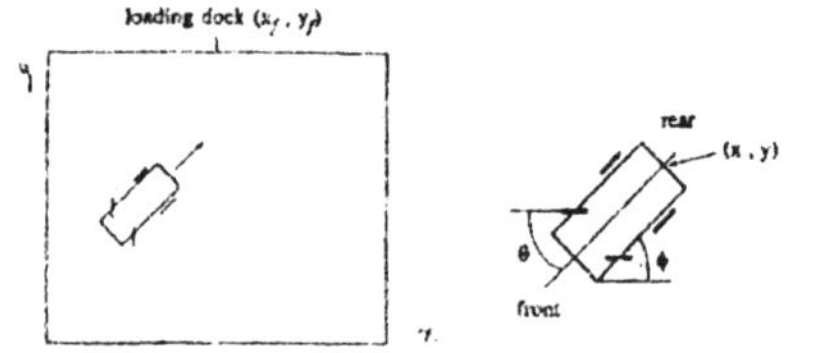

Figure 1 Geometry of the Truck Backing Problem
[Copied with permission.]

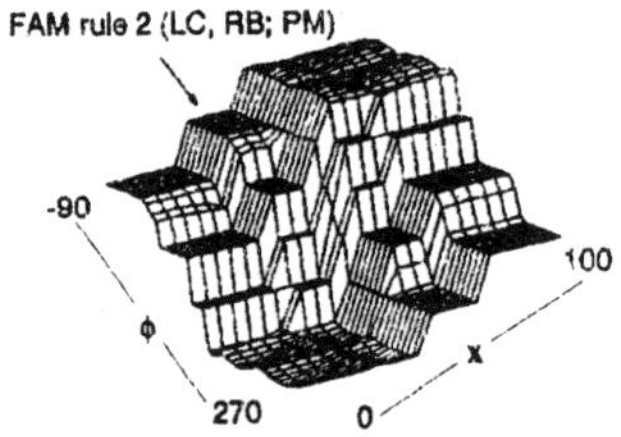

Figure 2 Control Surface for backing a truck
[Copied with permission.]

Equations 3 and 4, represent the mapping for calculating m_1 and m_2, from the inputs, neuron 1 and neuron 2. The angle ϕ which the centerline of the truck makes with the loading dock is the "observed" value for m_1 and the x coordinate of the center of the rear of the truck is the "observed" value for m_2. Figure 2 displays the (fuzzy logic) control surface for the backing of the truck. As can be seen, as ϕ increases the appropriate steering angle decreases, and as x increases the appropriate steering angle increases. For this reason the value of w_{11} was assigned a value of -1, and the value of w_{12} was assigned a value of +1. The rule which is used by our TNC to combine the net input impinging on the output neuron is to simply sum the magnitudes of the excitatory and inhibitory inputs, i.e. $w_{11} * m_1 + w_{12} * m_2$. A TNC assumes that it is possible to linearly add inhibition and excitation. As will be seen in the next section, this assumption appears to be a biological transgression.

NEURON NEURON is an extensive computation tool designed to assist in neuroscience research. It is a joint and continuing effort of Duke and Yale Universities [7]. This effort is funded by the NSF. The entire system, along with complete documentation, can be downloaded from the internet [7]. The individuals who manage this project at Yale and Duke offer seminars on how to use NEURON, normally twice a year, a one day seminar and a one week seminar. The first author was able to attend a one day seminar on the use of this tool during the Annual Neuroscience Conference in New Orleans in November of 2000 on his way to the ANNIE2000 Conference. For more information on NEURON go to http://www.neuron.yale.edu/neuron.

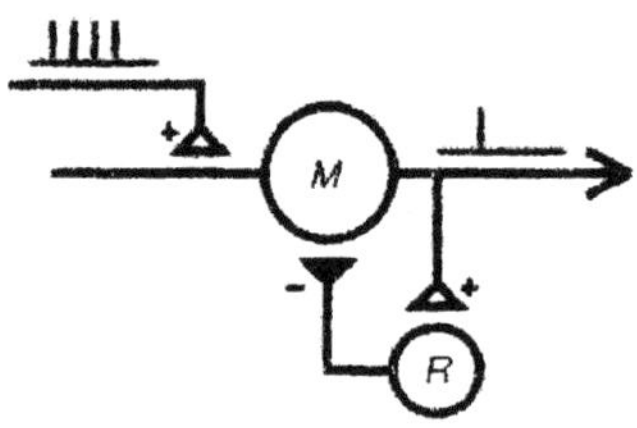

Figure 3 Simple neural circuit from NEURON Tutorial
[Copied with permission.]

Consider the neural circuit displayed in Figure 3. A solution to this circuit is presented in the Tutorial which accompanied the seminar. The neural net given in Figure 3 requires that we define the excitatory synapse (with the "+" sign in Figure 3) and the inhibitory synapse (with a "-" sign in Figure 3). The solution presented in the tutorial shows that the weight associated with the excitatory synapse is 30 times as great as the weight associated with the inhibitory synapse. What could cause this to be so? An explanation of synaptic inhibition is given in [11], on page 207. "The process of inhibition by ligand-activated [responding to extracellularly-applied chemicals of various kinds] channels involve the same principles as those underlying synaptic excitation; only the change in postsynaptic conductance is different. Excitation is achieved by driving the membrane towards threshold; whereas inhibition is achieved by holding the membrane below threshold." Thus, the excitatory and inhibitory excitations are achieved by the release of different neurotransmitter. Each neurotransmitter has its own distinct physical and chemical properties and the time courses of each will therefore be different. Thus, the assumption made by the TNC, that the effects of the excitatory and inhibitory are exactly equal, is a gross simplification.

Implications and Conclusions. It was stated in the abstract that we are proposing a major change to the direction of our research. The change from modeling toy control problems to attempting to model elementary biological control problems is such a change. The TNC had modeled toy control problems on a par with comparable fuzzy controllers. We had observed that simple biological problems appeared to be similar to the toy control problems we had solved, in that biological control problems also had monotone control surfaces. For example, the faster an animal runs, the more blood must flow to support the increased need for oxygen. So our TNC studies [1] suggested that neurons appeared first in the role of controllers, and (much) later as memories. In fact, the jellyfish of phylum Coelenterata, controls itself in swimming, mating, and feeding, but has little or no memory abilities. In aquaria they must be kept in circular tanks; otherwise, they will swim into any corners, and keep swimming into the corner until they starve to death. The question immediately arises - what was the evolutionary path which led from neural circuitry which supports the control function to neural circuitry which supports the memory function?

Rather than continuing experiments with TNCs, it is our intention to employ NEURON [or other more biologically oriented models] to attempt to simulate the neural circuitry of animals lsuch as leech, aplasia, or pehraps some avian species. These creatures have little or no memory capacity. Perhaps, by employing a more realistic modeling tool, we may begin to discover an answer to the question proposed at the end of the previous paragraph.

References

[1] Alexander, J. and Calif, S. (2000) **Control: An Emergent Property of Biological Neurons.** *International Journal of Systems Science.*

[2] Turk, R. P. and Alexander, J. (1999) **A Biologically Motivated Modification to Three-Neuron Controllers.** *Intelligent Engineering Systems Through Artificial Neural Networks.* **9**, November, 1999, ACME Press, New York.

[3] Alexander, J. and Lipinski, J. (2000) **On the Criticality of Timing Neuronal Firings.** *Intelligent Engineering Systems Through Artificial Neural Networks.* **10** November, 2000, ACME Press, New York.

[4] Gardner, D (1993) *The Neurobiology of Neural Networks* MIT Press. Cambridge, MA.

[5] MacGregor, R. (1987) *Neural and Brain Modeling* Academic Press. San Diego, CA.

[6] Sorkin, S.J. and Alexander, J.R. (1998) **A Three-Neuron Controllers: Shifting to the Biological Perspective.** *1998 IEEE Systems Man and Cybernetics Proceedings.*

[7] Carnevale, N. and Hines, M. (1998) *NEURON* http://www.neuron.yale.edu

[8] Alexander J.R. (1991). An Analysis of the General Properties and Convergence of a Connectionist Model Employing Probability Data as Weights and Parameters. PhD Dissertation. University of Maryland Baltimore County, Baltimore MD.

[9] McClelland J. and Rumelhart D. (1988) *Explorations in Parallel Distributed Processing.* MIT Press. Cambridge MA.

[10] Cox, J.T. and Alexander, J.R. (1996) **Backing Up a Truck and Trailer Using Sets of Three-Neuron Controllers.** *Intelligent Engineering Systems Through Artificial Neural Networks* **6**, ASME Press, New York, NY.

[11] Nichols, J.G., Martin, A.r., and Wallace, B.G. (1992) *From NEURON To Brain.* Sinaure. Sunderland, MA

[12] Carnevale, N.T. and Hines, M.I. (2001) *Tutorial 2: Making Hybrid Nets.* http://www.neuron.yale.edu/neuron/docs/netbuild

APPENDIX A. Brief Development of the ENIAC Equations Concerning the terminology, in Equation 1 and 2 below, w_{ij} is the weight representing the strength of the connection from the jth input to the ith output node, v_{ik}, is the weight representing the inhibition between $a_i(t)$ and $a_k(t)$, $\bar{a}_i$ is the average or resting value of $a_i(t)$, and d represents the decay factor or tendency of all $a_i(t)$ to settle back to their resting values. The equations are listed below.

$$\dot{a}_i(t) = \frac{[1-a_i(t)]}{(1-\bar{a}_i)} * \{\sum_{j=1}^{J} (a_i(t)-\bar{a}_i)w_{ij}m_j - \sum_{k\neq i}^{N} (a_i(t)-\bar{a}_i)v_{ij}[a_k(t)-\bar{a}_k] - d*(a_i(t) - \bar{a}_i)\} \quad \text{When } \sum_j w_{ij}*m_j > 0 \tag{1}$$

$$\dot{a}_i(t) = \frac{a_i(t)}{\bar{a}_i} * \{\sum_{j=1}^{J} (a_i(t)-\bar{a}_i)w_{ij}m_j - \sum_{k\neq i}^{N} (a_i(t)-\bar{a}_i)v_{ij}[a_k(t)-\bar{a}_k] - d*(a_i(t) - \bar{a}_i)\} \quad \text{When } \sum_j w_{ij} * m_j \leq 0 \tag{2}$$

We use a different equation when $w_{11} * m_1 + w_{21} * m_2$ (the weighted sum) is positive, and another when the weighted sum is negative. By doing this we limit $a_i(t)$ to be on (0,1). The reason for the factors $1/(1 - \bar{a}_i)$ in Equation 1, and $1/\bar{a}_i$ in Equation 2, is to keep the derivatives equal at the point $a_i(t) = \bar{a}_i$. The w_{ij} may be positive (excitatory) or negative (inhibitory).

The input nodes are represented by m_j, where, in the TNC case, j is 1 or 2. The input is mapped from its natural range to [-1,1]. Equations 3 and 4 give this mapping.

$$m_j = \frac{[Observ_j - Ave_j]}{[Max_j - Ave_j]} \quad \text{(3)}$$

$$If \;\; Observ_j > Ave_j$$

$$m_j = \frac{[Observ_j - Ave_j]}{[Ave_j - Min_j]} \quad Otherwise \quad \text{(4)}$$

Although the ENIAC equations are non-linear, when only one output neuron exists the equations may be integrated in closed form.

With only one output (or upper level) node and two input (or lower level) nodes, the ENIAC equations reduce to equations of the form, given as Equation 5 when the weighted sum is positive (note the i = 1 always, and we drop the subscript), and Equation 6 when it is negative.

$$\dot{a}(t) = C_1 a^2(t) + C_2 a(t) + C_3$$
$$when \sum_j w_{1j} * m_j > 0 \quad \text{(5)}$$

$$\dot{a}(t) = C_4 a^2(t) + C_5 a(t)$$
$$when \sum_j w_{1j} * m_j < 0 \quad \text{(6)}$$

The details of how the C_x terms contain the m_j, w_{ij}, d, and $\overline{a}$, and the actual integration of Equations 5 and 6 are omitted.

Equations 7 and 8 display the formula for a(t) which result from the integrations.

$$a(t) = \bar{a} + \frac{(1-\bar{a})}{1+(-1+\frac{(1-\bar{a})}{(a_0-\bar{a})})*\exp(-K_1*(1-\bar{a})*t)} \qquad (7)$$

$$\text{Where } K_1 = \frac{(w_{11}m_1 + w_{12}m_2 - d)}{(1-\bar{a})} > 0$$

$$a(t) = \frac{\bar{a}}{1 + (\frac{\bar{a}-a_0}{a_0})*\exp(-K_2 * \bar{a} * t)} \qquad (8)$$

$$\text{Where } K_2 = \frac{(w_{11}m_1 + w_{12}m_2 + d)}{\bar{a}} \leq 0$$

We have let d = zero in all our work thus far.

VISUALIZING SPECIATION IN EVOLVING NEURAL NETWORKS

RANDY L. RIBLER
CHRISTOPHER R. JORDAN
Department of Computer Science
Lynchburg College
Lynchburg, Virginia 24501
ribler@lynchburg.edu
jordan_c@students.lynchburg.edu

ABSTRACT

Speciation is generally considered a desirable phenomenon in evolving neural networks and genetic algorithms, both as a means of encouraging diversity and to aid in the development of multilayered network solutions. This paper describes a general tool for visualizing the similarities in both the genotypes and phenotypes present in a population. The visualizations employ a clustering algorithm that can provide important insights into the degree to which speciation is occurring, the number of evolving species, the growth of evolving species, and the degree to which the species are effecting a population

INTRODUCTION

Speciation is generally considered a desirable phenomenon in evolving neural networks and genetic algorithms, both as a means of encouraging diversity and to aid in the development of multilayered network solutions. This paper describes a general tool for visualizing the similarities in both the genotypes and phenotypes present in a population. The tool can be used to provide insights into how speciation occurs during the process of evolution, and how techniques such as fitness sharing effect the speciation. The tool also has the potential to be used to help steer the evolutionary process, adjusting parameters to affect speciation at runtime.

VISUALIZING GENOTYPE SIMILARITY

It would not be useful to simply display all the bits in all the genes and look for similarities. The number of bits in a gene may be large, and detecting matching patterns across genes would be very difficult. Instead, we use a visualization that shows which p-length schemata are common in the population.

Our *genotype view* compares a single member of the population, which we will call the *base*, with all other members in the population simultaneously. The visualization shows which p-length schemata contained in the base genotype also exist in other members of the population. The point (x, y) is plotted if the p-length schema that begins at x in the base, exists at location x in member y. Figure 1 illustrates how the display is constructed with a schema length $p = 4$. In the figure, the base gene sequence is 11010001. So the first p-length schema in the base is the first four bits, 1101. Of the four genes in the population, the only genes that start with 1101 are the base gene (Gene 1) and Gene 2. Therefore, points are plotted at $(1, 1)$ and $(1, 2)$. The schema used to determine the points plotted at $x = 2$ is the p-length sequence that begins at the second bit in the base gene. This sequence is 1010. Because genes 1, 3, and 4 all contain the schema 1010 starting at the second bit, the points $(2, 1)$, $(2, 3)$, and $(2, 4)$ are plotted.

The user of the visualization tool selects the schema length p, based on the desired granularity of similarity. Larger values of p will require larger sections of the genes to match to become visible. Smaller values of p will allow smaller matching sections to be visible. Generally, the user will select larger values for p when the genes in the population are more homogenous.

A user can use a mouse to designate any gene in the population as the base gene. The display quickly refreshes to show the new view. The user can alternatively use the return key to change the base gene to the next sequential member of the population. Thus the user can quickly browse the entire population, comparing each member with all the others in the population.

The genes that are present in the population may be the genes from one generation, from a subset of the population, or even from different generations. One effective approach is to compare a genetic population from an earlier generation to the dominant gene in the final population. This method can provide insight into the way in which the population converged.

VISUALIZING PHENOTYPE SIMILARITY

For visualization purposes, we treat the phenotype of a member of the population as a bit string where each bit corresponds to one case in the training set. A bit in the phenotype is set if the corresponding case yielded the desired result, and is cleared otherwise. Our *phenotype view* simply displays each member's performance in solving the training set. The point (x, y) is plotted if member y successfully solves problem x in the training set. We combine the genotype and phenotype views, by placing the phenotype view to the right of the genotype view. Thus each member's genotype is displayed immediately to the left of its phenotype.

CLUSTERING

Our goal is to identify species within the population. Unfortunately, the concept of species is ill-defined and neither biologists nor evolutionary programmers have been able to agree on a definition. Because we cannot identify any species with absolute certainty, we use a clustering algorithm to identify *candidate species.* It will be up to the user of the visualization to determine whether or not a cluster determines a unique species. This is why visualization is particularly well-suited to this problem. Whether or not a subgroup of the population constitutes a species requires a judgment. The visualization provides the researcher with accessible information upon which to base a judgment.

In order to better visualize candidate species, we use a k-means clustering algorithm[Fukunaga, 1990]. This algorithm partitions the population into k clusters that will constitute our candidate species. The algorithm needs a measure of similarity that it can apply to pairs of members in the population. The measure employed here is based on the Hamming distances between pairs of genotypes and pairs of phenotypes. In an attempt to give equal weight to similarities in the genotype and phenotype, we use the following formula

$$\text{distance}(m_0, m_1) = H(g_0, g_1)^2 + w^2 H(p_0, p_1)^2$$

where

$\text{distance}(m_0, m_1)$ is the distance between members m_0 and m_1,

$H(g_0, g_1)$ is the Hamming distance between the genotypes of m_0 and m_1,

$H(p_0, p_1)$ is the Hamming distance between the phenotypes of m_0 and m_1 and,

w is the number of bits in the genotype divided by the number of bits in the phenotype. This term compensates for differences in the lengths of the genotype and phenotype.

Depending on the objective of the visualization, alternative formulas might rely on the genotype only, the phenotype only, or on a different function of the two.

The k-means clustering algorithm that our visualization employs proceeds as follows.

- The population is partitioned into k equal-sized partitions. Membership in a particular partition is determined at random.
- While we have not encountered an iteration in which no member moves to a different partition
    - for each member in the population
        - Determine the member's average distance to the other members of the population. If a member's average distance to the members in another partition is smaller than the average distance to the members in its current partition, the member is moved to the partition with the smaller average distance.

This algorithm will yield k or fewer candidate species. To make the candidate species more easily identifiable in the visualization, the members are sorted by candidate species, and a unique color is used to display members of the same candidate species. Figure 2 shows the resulting visualization. The left half of the display shows the genotype view, and the right half shows the phenotype view. The candidate species are grouped into candidate species consisting of members 0-25, members 26-47, members 48-85, members 86-91, and members 92-99. The phenotype view shows clear distinctions between species. For example, it is easy to see that the phenotype of the candidate species at the bottom of the display is more or less complimentary to the phenotype of the candidate species that includes members 48-85. This indicates that one species of neural network generates correct outputs in cases where the other species generates incorrect outputs, and conversely. The genotype view shows that while there is clear speciation in the phenotype, the genotypes remain relatively diverse. In contrast, Figure 3 shows the same visualization without clustering. In this view the speciation that was obvious in Figure 2 is not at all apparent.

TRACKING SPECIES ACROSS GENERATIONS

In a genetic algorithm in which all or most of each generation is replaced by the next generation, it is not clear how to track species from one generation to the next. The approach we use attempts to match the clusters of the current generation with those of the previous generation. For each member in the current population, we find the best match from among the previous generation's clusters. If more than half of the members in a candidate species can be mapped back to the same cluster from the previous generation, we assume that this is the same species crossing generations. When we generate visualizations for sequential populations, we use the same colors to identify candidate species that crossed generations. Using this visualization tool, a researcher can view the current set of candidate species, advance the genetic algorithm by one or more generations, and see how the population of each candidate species has grown or diminished.

CONCLUSIONS

The process of evolving neural networks is generally considered to be opaque. However, the visualizations described here, when combined with a simple clustering algorithm, can provide detailed information about speciation in a population, and can show how that speciation changes across generations. This information can be used to document the effectiveness of mechanisms designed to effect speciation.

NOMENCLATURE

- *Schema:* A subsequence of bits in a gene specified using the characters "0", "1", and "*" for "don't care". The plural of schema is *schemata.*

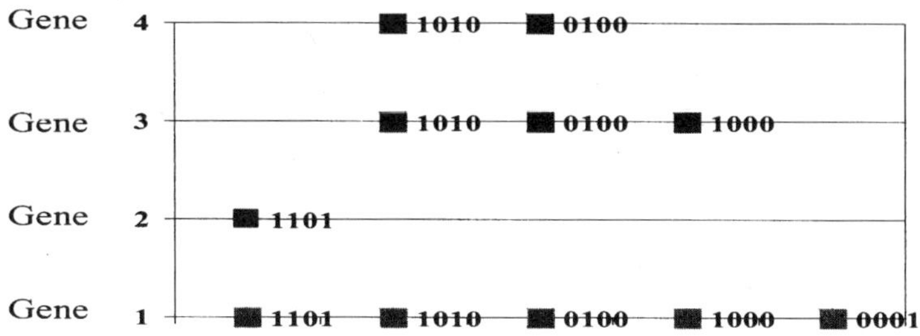

Gene 4: 0 1 0 1 0 0 1 1

Gene 3: 0 1 0 1 0 0 0 0

Gene 2: 1 1 0 1 1 0 1 1

Gene 1: 1 1 0 1 0 0 0 1 (base)

Figure 1: Comparing the Genotype of One Member to the Rest of the Population

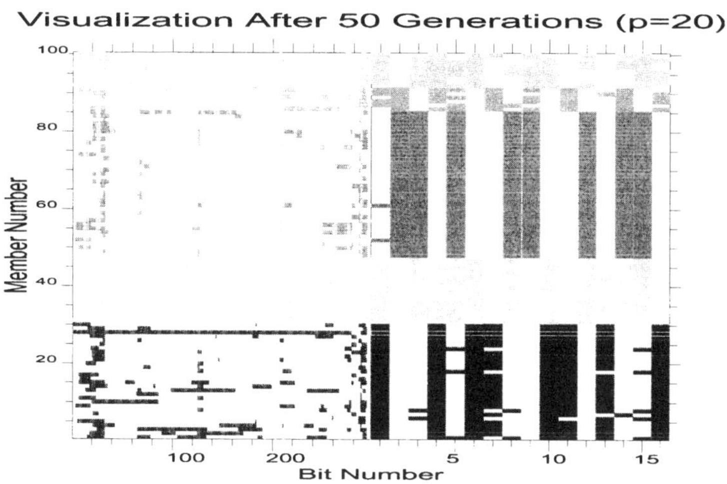

Figure 2: Genotype/Pheotype View

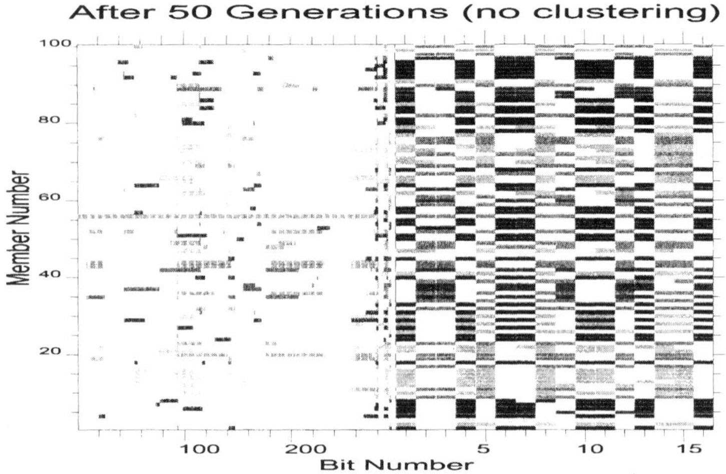

Figure 3: Genotype/Pheotype View Without Clustering

None of the schemata discussed in this paper contain any "*s."

- *Hamming Distance:* the number of bits in which two equal-length strings differ. For example, the Hamming distance between 1001 and 1100 is two, because the strings differ in the second and and fourth bits.

References

[Alsabti et al., 1998] Alsabti, Ranka, and Singh (1998). An efficient parallel algorithm for high dimensional similarity join. In *IPPS: 11th International Parallel Processing Symposium*. IEEE Computer Society Press.

[Fukunaga, 1990] Fukunaga, K. (1990). *Introduction to Statistical Pattern Matching*. Academic Press, New York.

[Ribler, 1997] Ribler, R. (1997). *Visualizing Categorical Time Series Data with Applications to Computer and Communications Network Traces*. PhD thesis, Virginia Polytechnic Institute and State University. <URL: http://scholar.lib.vt.edu/theses/public/etd-1711111139751001/etd-title.html.

[Ribler and Abrams, 2000] Ribler, R. and Abrams, M. (2000). Using visualization to detect plagiarism in computer science classes. In *IEEE Symposium on Information Visualization 2000*, Salt Lake City, UT.

[Smyth, 1996] Smyth, P. (1996). Clustering using monte carlo cross-validation. In *Proc. Second International Conference on Knowlege Discovery and Data Mining*, Portland, OR. <URL: http://www.hensa.ac.uk/www94>.

[Yao, 1999] Yao, X. (1999). Evolving artificial neural networks. *Proceedings of the IEEE*, 87(9):1423–1447.

CLUSTER VALIDITY AND THE SEMI-SUPERVISED CLASSIFICATION PROBLEM

NEIL HARVEY
Computing and Information Science
University of Guelph
Guelph, Canada

DEBORAH A. STACEY
Computing and Information Science
University of Guelph
Guelph, Canada

ABSTRACT
Training a classifier requires a supply of example problems and the correct classification (label) for each. In some practical situations examples are plentiful, but obtaining labels for them is costly. Several algorithms exist for learning classification when only a few examples are "labelled" at the outset and the rest are "unlabelled." This paper presents continued work on one such algorithm developed by Dara, Kremer, and Stacey. It investigates whether cluster validity indices can guide the choice of parameters for best performance on eight real-world data sets. An examination of a simple clustering problem points to explanations for the algorithm's behaviour, and tests of a variant algorithm that capitalizes on these observations are presented.

INTRODUCTION AND BACKGROUND

Semi-supervised learning methods are used when one has a small number of "labelled" example problems (for which the true answer is known) and a large number of "unlabelled" examples (for which the true answer is not known). These methods try to find groups or clusters of similar examples, so that labelled examples may be used to estimate an answer for unlabelled examples in the same cluster. The end goal is a larger set of labelled examples for supervised learning.

Shahshahani and Landgrebe (1992, 1994), Miller and Uyar (1996, 1997), and Nigam et al. (1998, 1999) developed successful semi-supervised learning methods that model data as a mixture of probability distributions and find a maximum-likelihood estimation of the model parameters.

In the neural network field, Huang and Lippmann (1988) used a Kohonen-like feature map algorithm to discover clusters. Miller and Uyar (1996) tried Radial Basis Function networks, and Dara, Kremer, and Stacey (2002) used Self-Organizing Maps. Other recent work has employed graph cuts (Blum and Chawla, 2001) and Support Vector Machines (Bennett and Demiriz, 1999; Fung and Mangasarian, 2001).

This paper uses Dara et al.'s SOM-based algorithm. They favoured the SOM because it groups similar clusters spatially. When *very* few labelled examples are available, the SOM makes it easy to examine labelling information in neighbourhoods of clusters, not just single clusters.

CHOOSING A NUMBER OF CLUSTERS

Whether you use a mixture model, an RBF network, a SOM, or even k-means to cluster data, you must choose the number of clusters.

Shahshahani and Landgrebe (1992) and Nigam et al. (1998) began by assigning one mixture component (cluster) to each class. In a follow-up paper, Shahshahani and Landgrebe (1994) allowed each class to be modelled by more than one component, although they had no systematic way to assign each example to a particular component. Nigam et al. hypothesized that representing each class by just one component in their research problem (text classification) was unrealistic, so they too split each class into multiple components, and suggested finding the ideal number of components by training several classifiers with different values and picking the best one by cross-validation.

Another possibility is using cluster validity to choose the best clustering solution to use for labelling examples.

There is no universally agreed-upon definition of what a cluster *is*, just an intuitive sense that two points from the same cluster should be somehow more similar to each other than to a point from a different cluster. Many researchers have translated this notion into numerical measures of the quality or "validity" of a clustering solution. Milligan and Cooper (1985) compared 30 measures from the literature and exposed many of them as very poor ones.

Weingessel et al. (1999) did a study of 14 measures with more difficult problems and a measure that wasn't available in 1985. However, they used only binary data. The data sets used in this paper contain a mix of real and binary attributes, so Milligan and Cooper's recommendations may be more relevant.

Of the ten top-ranked measures from Milligan and Cooper's study, we eliminated some immediately: $Je(2)/Je(1)$, Beale, and Mojena work only with merging or splitting clustering algorithms, and Gamma and $G(+)$ were eliminated based on running-time considerations.

That leaves (ordered with the highest-scoring measure first)

1. the Calinski and Harabasz measure, which uses between- and within-cluster sums of squares. It should be maximized.
2. the C-Index created by Hubert and Levin, which uses between-point distances. It should be minimized.
3. the cubic clustering criterion (CCC) created by Sarle, which compares the actual and expected proportion of variance accounted for by the clustering solution. It should be maximized.
4. a point-biserial correlation, which tests whether membership in the same cluster correlates with small between-point distances. It should be maximized.
5. the Davies and Bouldin index (DB-Index), which uses between- and within-cluster distances. It should be minimized.

EXPERIMENTS

The prescribed way to use cluster validity measures is to plot validity vs. number of clusters for a range of clustering solutions, then find a minimum or maximum in the plot. As a baseline, Fig. 1 shows how the measures rate clustering solutions found in *random* data. If they behave this way with real data, we should suspect that the data might not contain any meaningful groupings.

Figure 1: Quality of clusters found in random data. Validity measures that should be maximized are on the left, those that should be minimized are on the right. Each line's y-range has been scaled to [0,1] to highlight the shape.

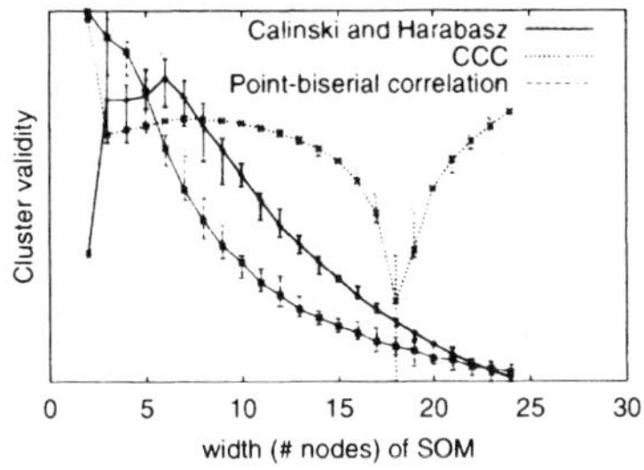

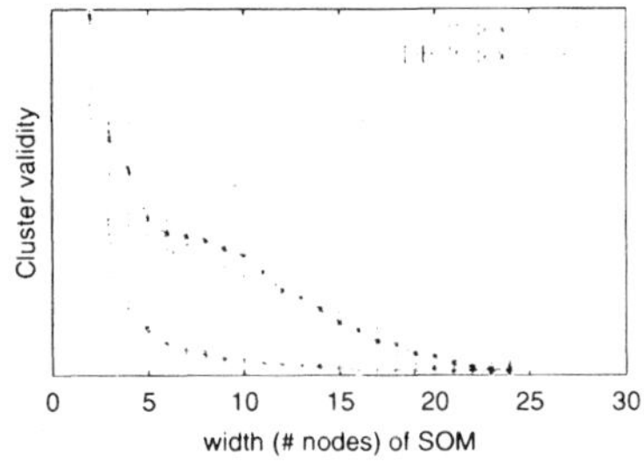

Table 1: Natural number of clusters.

Data set	Clusters	Based on
abalone	35-55	C&H, CCC, DB-Index
breast cancer	3	C&H, CCC, DB-Index
diabetes	3	C&H, DB-Index
heart disease	3-20	C&H, p-b
iris	3-8	C&H, *C*-Index, CCC, DB-Index
letter	3	C&H, DB-Index
mushroom	3-20	p-b, DB-Index
sick-euthyroid	3-12	C&H, p-b, DB-Index

("C&H" = Calinski and Harabasz measure; "p-b" = point-biserial correlation)

The data sets used in this study are from the UCI Machine Learning Repository: abalone, breast cancer, diabetes, heart disease, iris, letter recognition, mushroom, and sick-euthyroid. For the abalone and letter recognition data sets, the examples were grouped to create two-class problems.

The data sets were clustered by SOM, with the map size increasing from 2×1 to 24×18 (432 nodes) over 500 runs. For the larger maps, proportions of about 4×3 were maintained.

Table 1 shows the "natural" number of clusters that the measures appear to recommend, i.e., the number of SOM nodes in the clustering solution with the best score. Ranges are given because the different measures, even when they do clearly deviate from the behaviour shown in Fig. 1, do not always agree.

The clustering solutions for each data set were then used for semi-supervised learning. 7% of the training examples from each class were chosen to be labelled; labels were removed from the rest. The unlabelled examples were labelled using Dara, Kremer, and Stacey's rules, and the success of this "self-labelling" was measured by the labelling ability (the fraction of unlabelled examples that could be labelled) and labelling accuracy (the fraction of those that were labelled correctly). We assume that it is desirable to assign correct labels to the largest possible number of unlabelled examples, even at the cost of mis-labelling some. This was repeated with 15%, 30%, and 60% labelled examples.

Despite the varied origins of the data sets, the overall behaviour was con-

Figure 2: Ability to label heart disease data (starting with 7% labelled points).

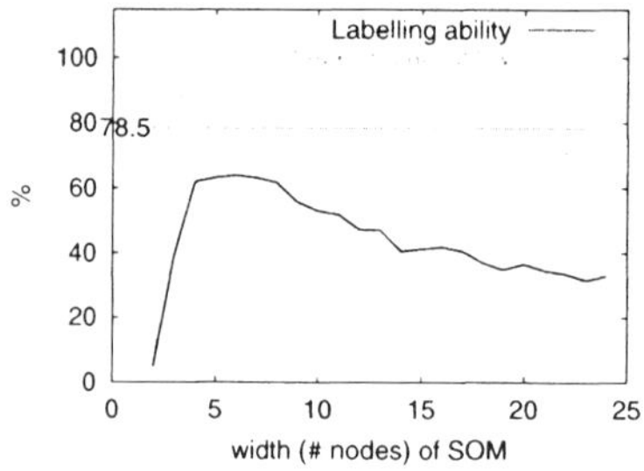

Figure 3: Effect of number of original labelled examples on best map size for breast cancer data.

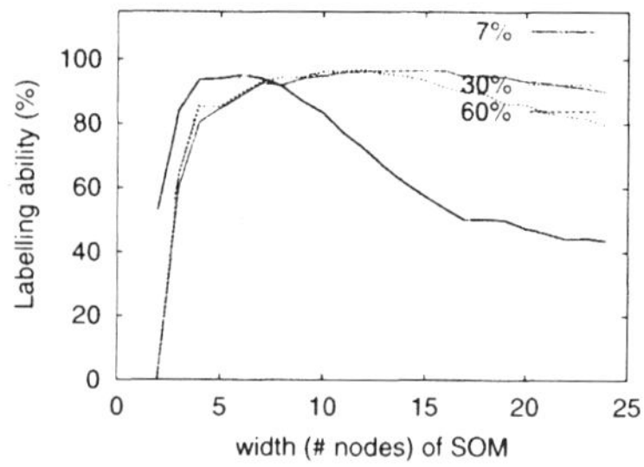

Figure 4: Two one-dimensional classes, arbitrarily divided into k clusters.

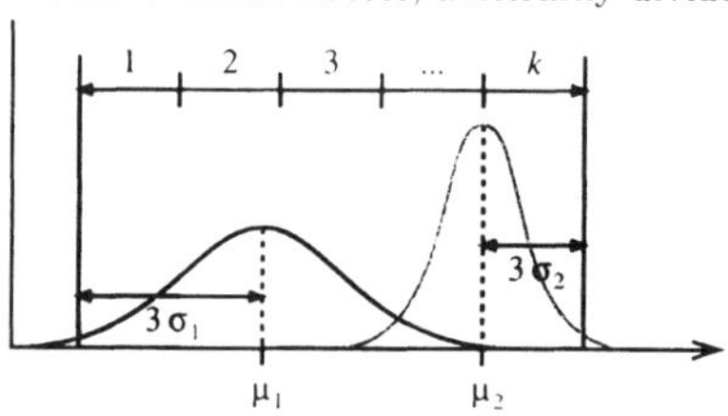

sistent: as the number of clusters increased, the labelling ability rose, peaked, then slowly fell off. Labelling accuracy peaked quickly and remained high. Fig. 2 shows an example.

The labelling ability peaked at smaller map sizes when fewer examples were initially labelled (Fig. 3 shows an example). In other words, the "best" map size for a data set is a moving target: it strongly depends on the number of original labelled examples.

The labelling rule used by Dara et al. states that unlabelled examples are labelled only if *all* labelled examples in their cluster have the same label. Some experimentation with a one-dimensional analogue of the problem (Fig. 4), which shows behaviour just like that in Figs. 2 and 3, suggests that this rule creates the sensitivity to the number of original labelled examples: it makes the probability that a cluster is unusable equal to the probability that *even one* example from the "wrong" class falls inside the cluster.

It might be worthwhile to relax the labelling rule, and insist just that a majority (or some other arbitrary proportion p_0) of the labelled points agree on a class label. A kind of "quality control" can be maintained even with this relaxation by insisting that the entire confidence interval of the calculated proportion lie above p_0.

With this new labelling rule asking for 50% agreement, labelling accuracy was roughly the same for the breast cancer, diabetes, heart disease, iris, and letter

Figure 5: Ability to label heart disease data with 50% cutoff (starting with 7% labelled points).

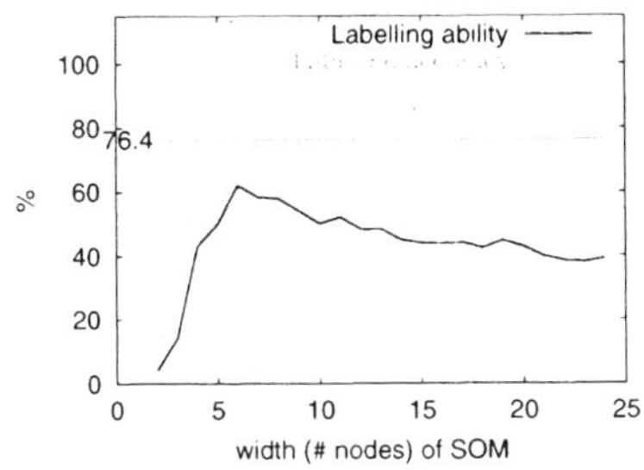

Figure 6: Effect of number of original labelled examples on best map size for heart disease data (modified labelling rule).

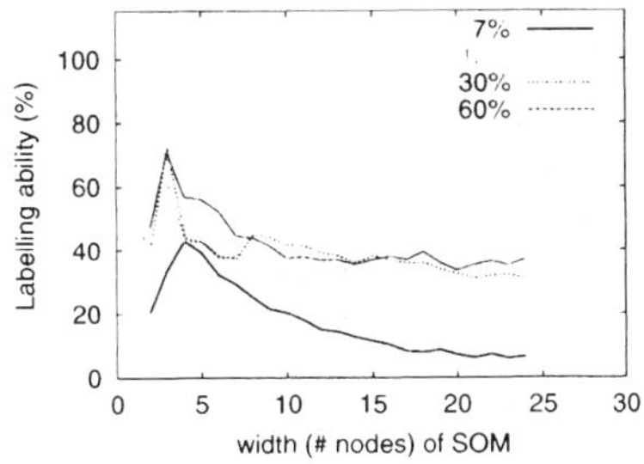

Table 2: Natural and optimal (w.r.t. labelling ability) number of clusters.

Data set	Natural # of clusters	Optimal # of clusters
abalone	35-55	2-3
breast cancer	3	2
diabetes	3	2-3, 24
heart disease	3-20	3-12
iris	3-8	6-8
letter	3	2-3
mushroom	3-20	20
sick-euthyroid	3-12	2-6

recognition data sets (Fig. 5), and slightly worse for abalone and sick-euthyroid. Requiring 66% agreement raised the accuracy for the sick-euthyroid data to the same level achieved originally, and boosted labelling accuracy for the diabetes, heart disease, and letter recognition data sets above the level achieved originally.

For the abalone, breast cancer, heart disease, letter recognition, and sick-euthyroid data, the optimal number of clusters was the same no matter whether 7, 15, 30, or 60% of the examples were labelled. Fig. 6 shows an example. For the diabetes data, the optimal number of clusters was lower when more examples were labelled. Only for the iris data did the labelling ability show behaviour like that encountered for the original labelling rule.

The optimal number of clusters was within the range recommended by the cluster validity measures (Table 2) for the breast cancer, heart disease, iris, letter-recognition, mushroom, and sick-euthyroid data. For the diabetes data, the labelling ability peaked at the "natural" number of clusters only when larger amounts of labelled examples (30%, 60%) were available.

CONCLUSIONS

Cluster validity has a long history in the literature, so using it as a way to choose a clustering solution in semi-supervised learning is an attractive notion.

In this study, we saw that the number of clusters chosen by validity measures coincided with the number of clusters that gave the best self-labelling performance for over half of the data sets. However, we also saw that different validity measures do not always agree, and that cluster validity may not be usable with a particular algorithm until dependencies that affect the relationship between number of clusters and self-labelling performance are identified and understood.

The last point may be a blessing: the fact that validity measures are independent of the clustering algorithm used (in this paper it was the SOM, but it did not have to be) could make them a useful tool for exposing such dependencies during algorithm design.

REFERENCES

Bennett, K. P. and Demiriz, A. 1999, Semi-Supervised Support Vector Machines, in *Advances in Neural Information Processing Systems 11*, ed. M. S. Kearns, S. A. Solla, and D. A. Cohn (MIT Press), pp. 368–374

Blum, A. and Chawla, S. 2001, Learning from Labeled and Unlabeled Data Using Graph Mincuts, in *Proc. 18th International Conference on Machine Learning* (San Francisco: Morgan Kaufmann), pp. 19–26

Dara, R., Kremer, S. C., and Stacey, D. A. 2002, Clustering Unlabeled Data with SOMs Improves Classification of Labeled Real-world Data, in *Proc. WCCI, Hawaii May 2002*

Fung, G. and Mangasarian, O. L. 2001, Semi-Supervised Support Vector Machines for Unlabeled Data Classification, *Optimization Methods and Software*, 15, pp. 29–44

Huang, W. Y. and Lippmann, R. P. 1988, Neural Net and Traditional Classifiers, in *Neural Information Processing Systems Denver, CO 1987*, ed. D. Z. Anderson (New York: American Institute of Physics), pp. 387–396

Miller, D. J. and Uyar, H. S. 1997, A Mixture of Experts Classifier with Learning Based on Both Labelled and Unlabelled Data, in *Advances in Neural Information Processing Systems 9*, ed. M. C. Mozer, M. I. Jordan, and T. Petsche (MIT Press), pp. 571–577

Miller, D. J. and Uyar, S. 1996, A Generalized, Gaussian Mixture Classifier with Learning Based on Both Labelled and Unlabelled Data, in *Proc. 30th Annual Conference on Information Science and Systems (CISS'96)*, ed. S. R. Kulkarni and M. T. Orchard, pp. 783–787

Milligan, G. W. and Cooper, M. C. 1985, An Examination of Procedures for Determining the Number of Clusters in a Data Set, *Psychometrika*, 50, pp. 159–179

Nigam, K., McCallum, A., Thrun, S., and Mitchell, T. 1998, Learning to Classify Text from Labeled and Unlabeled Documents, in *Proceedings of the Fifteenth National Conference on Artificial Intelligence* (AAAI Press)

Nigam, K., McCallum, A. K., Thrun, S., and Mitchell, T. 1999, Text Classification from Labeled and Unlabeled Documents using EM, *Machine Learning*

Shahshahani, B. M. and Landgrebe, D. A. 1992, On The Asymptotic Improvement Of Supervised Learning By Utilizing Additional Unlabeled Samples; Normal Mixture Density Case, in *Proc. SPIE Neural and Stochastic Methods in Image and Signal Processing*, ed. S.-S. Chen, Vol. 1766, pp. 143–155

Shahshahani, B. M. and Landgrebe, D. A. 1994, The Effect of Unlabeled Samples in Reducing the Small Sample Size Problem and Mitigating the Hughes Phenomenon, *IEEE Transactions on Geoscience and Remote Sensing*, 32, pp. 1087–1095

Weingessel, A., Dimitriadou, E., and Dolnicar, S. 1999, An Examination of Indexes For Determining The Number Of Clusters In Binary Data Sets

USING A CMAC NEURAL NETWORK IN NOISY ENVIRONMENTS

GARY DEMPSEY, MANFRED MEISSNER, CHRISTOPHER SPEVACEK
Department of Electrical and Computer Engineering
Bradley University, Peoria, Illinois 61625 U.S.A.

ABSTRACT
The cerebellar model articulation controller (CMAC) is commonly used as an adaptive feed-forward controller with a conventional PID-type controller. In this paper, results are presented where the CMAC is used in a noisy environment. The experimental setup for CMAC control and testing is a small robot arm system. The majority of noise is produced from the calculation of velocity from a position sensor. Degradation of the control system is discussed in terms of noise level and velocity calculation methods. Use of an adaptive filter (ADALINE) in conjunction with a bandwidth-limited differentiation proved to be the best method for velocity calculation.

INTRODUCTION

A block diagram of the experimental system is shown in Fig. 1. The plant consists of a power amp, DC motor, gear train, external load, and a position sensor. The load consists of an aluminum rod. The digital controllers are implemented in C-language with a sampling frequency of 50 Hertz. Because the plant exhibits a high degree of static and kinetic friction, steady-state accuracy is poor (Dempsey, 1999). Use of integral control produces a limit cycle. The CMAC neural network can be used to reduce steady-state error without integral control. The following sections will discuss the controller designs, velocity calculation methods, use of the CMAC in a noisy environment, and experimental results.

CONVENTIONAL SINGLE-LOOP AND TWO-LOOP CONTROL

Appropriate conventional controller design can significantly reduce the cost associated with the CMAC neural network and improve overall performance (Dempsey, 1999 and Miller et. al., 1990). The specifications for the position control system were a minimum phase margin (PM) specification of 90 degrees, maximum overshoot of 2 degrees, and steady-state error less than $\pm$ 2 degrees. Position commands were limited to a range of $\pm$90 degrees and rate-limited to assure the maximum overshoot did not exceed 2 degrees. Three loads were used: short arm (mass = 110 grams), medium arm (mass= 127 grams), and medium arm with a load of 27 grams attached at the end. These loads will be referred to as the light, medium, and heavy loads in the following discussion. A proportional controller cannot meet the required specifications. The second controller design consisted of a single-loop PD controller with a feed-forward compensator. Although improved, the steady-state error specification is not met with this design.

The third design consisted of a two-loop controller shown in Fig. 2. The transfer function A(s) is the model for the A/D converter's anti-alias filter and Hp(z) is a software first-order low-pass filter (pole at 150 rad/sec). Transfer function V(z) represents the velocity calculation method, Gcv(z) the forward velocity controller, Gcp(z) the forward position controller, and F(z) the feed-forward controller.

Position data is filtered by A(s) and Hp(z). A differentiation of the data to obtain velocity feedback reveals the noise in the system which consists of power supply noise and

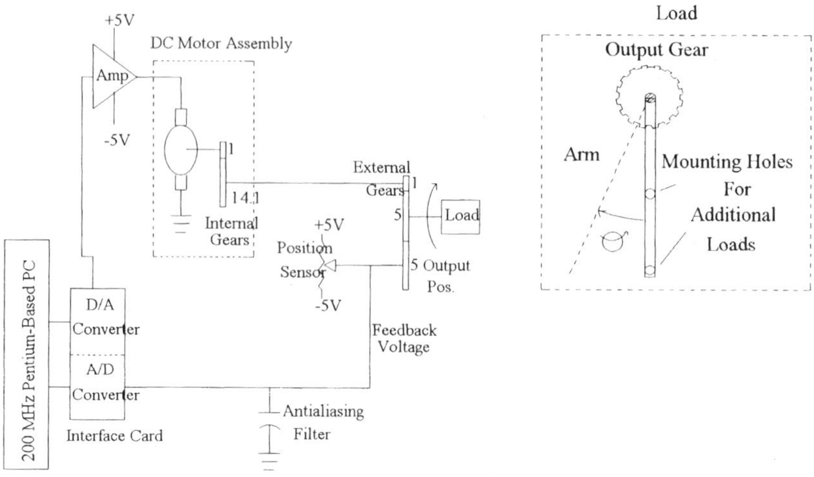

Figure 1: System Block Diagram of Small Robot Arm System.

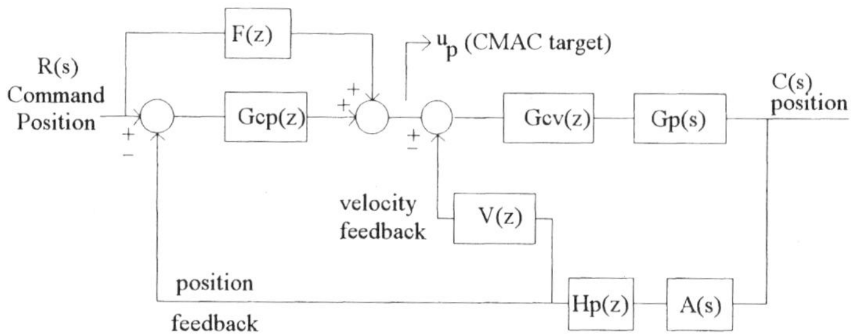

Figure 2. Two-loop Control System Block Diagram.

quantization error noise from the 12-bit A/D converter. This is illustrated in Fig. 3 where the command velocity (slope of position command) is 40 degrees/sec. All experimental data was uploaded to MATLAB for plotting. The first-order difference equation

$$y(n) = \left[x(n) - x(n-1)\right]\left(1/T\right) \qquad (1)$$

was used for the velocity calculation. Parameter T is the sampling period and x is the current and past samples from filter Hp(z). This is a common approach for systems with minimal noise. Note that the noise will depend significantly on the sample period T. A period of 0.02 seconds was the maximum for our system to meet the PM specification.

A bandwidth-limited differentiation calculation in the form of

$$V(s) = \frac{50s}{s+50} \qquad (2)$$

where 50 rad/sec is the low-pass filter pole is a better method for noisy systems. A digital

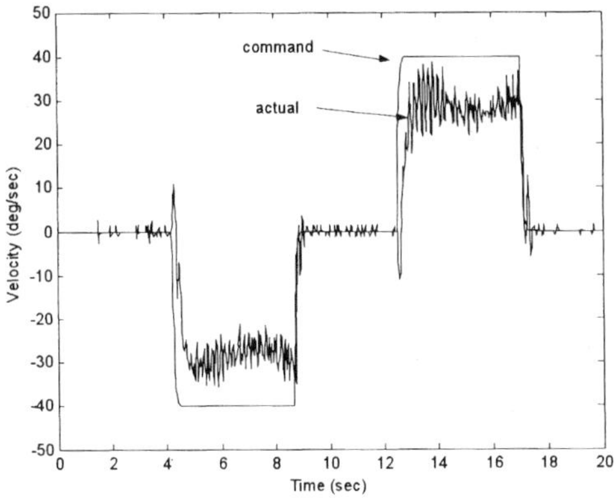

Figure 3. Command/Actual Velocity Versus Time (first-order difference).

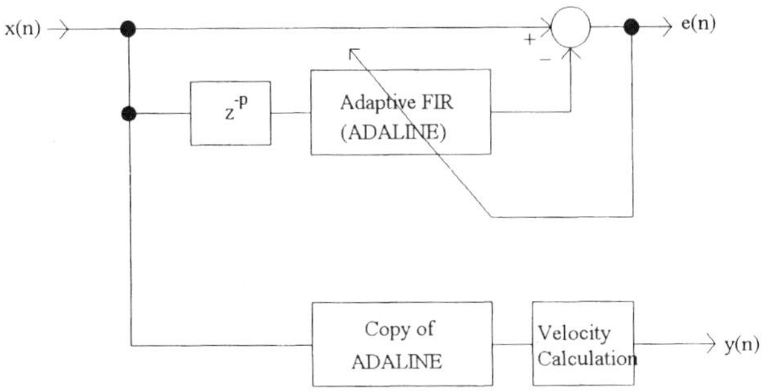

Figure 4. Predictor-based Differentiation.

filter equivalent can be obtained using the bilinear transformation. A lower value of the pole results in less noise but more distortion of the velocity signal from the ideal command velocity. The lowest pole for our application was determined to be 50 rad/sec. This resulted in a reduction of the peak to peak velocity noise from 17 to 16 degrees/sec.

The third method investigated for velocity calculation was an adaptive FIR filter approach suggested by Vainio (1999) which showed promising results for simulated noise. The adaptive filter method is shown in Fig. 4. Widrow-Hoff's (1960) least-mean-square algorithm was used for weight adaptation in our application. In our system, x(n) is the position data. The prediction step length is controlled by the delay parameter "p". A value of "p" equal to one was optimal for our application. Because of the closed-loop application, a copy of the converged ADALINE is used in our control scheme. Note that the copy of the ADALINE is presented with the input x(n) in Fig. 4 (no delay). Because it was trained with a one step delay, the copy will have a predictive characteristic.

Eight inputs for the ADALINE were determined to yield the best noise reduction and convergence characteristics. The synaptic weight modification rule for the ADALINE is

$$w(n+1) = w(n) + \eta \cdot e(n) \cdot x(i) \tag{3}$$

where η is the learning rate, e(n) is the error signal in Fig. 4, and x(i) is the synaptic weight input (x(n) or a delayed version). The output of the ADALINE copy is differentiated using a digital version of Eq. (2). The ADALINE converges in approximately 300 presentations (epochs) of the position data (6 seconds). After 300 epochs, the ADALINE weights remain relatively constant at different positions. The noise reduction is illustrated in Fig. 5. The maximum peak to peak noise is reduced from 16 to 3.3 degrees/sec.

The frequency response of the ADALINE must be considered in the velocity loop controller design because of the PM specification. The forward velocity controller Gcv(z) in Fig. 2 is a lag controller and is used to shape the frequency response of the velocity loop. The feed-forward controller F(z) is a first-order approximation of the inverse of the closed-loop velocity control. A proportional controller Gcp(z) is used for the outer position loop and is tuned to assure the overshoot is less than 2 degrees for command inputs for the three loads. The worse-case steady-state errors were 3, 4.5, and 5.2 degrees for the light, medium, and heavy loads.

CMAC NEURAL NETWORK

A block diagram of the CMAC is shown in Fig. 6. See Miller and Glanz (1996) for a discussion of configuring the CMAC for different applications and the initial selection of CMAC parameters. The amount of RAM memory for the Albus CMAC can be very large. One method for reduction is the use of hash memory as shown in Fig. 6 (Albus, 1975).

Referring to Fig. 6, the C hash memory cells activated by the input are summed to produce the CMAC output. This is referred to as the "controller phase". The inputs are the command position and the current output position. In the "learning phase", the CMAC inputs are the current output position, the past output position, and the target. The target is the signal u_p shown in Fig. 2. The CMAC is used as an adaptive feed-forward compensator and learns the closed-loop velocity inverse.

CMAC memory (weight) adaptation is performed using

$$\Delta w = \frac{\beta}{C} D[\text{target-output}], \qquad \begin{aligned} D[x] &= 0 && \text{if } |x| \leq d_0 \\ &= x - d_0 && \text{if } x \geq d_0 \\ &= x + d_0 && \text{if } x < -d_0 \end{aligned} \tag{4}$$

where d_0 is defined as the dead-zone size and parameter β is the learning rate. This modified least-mean squares algorithm prevents weight modification if the error is less than the dead-zone. The algorithm in Eq. (4) was presented by Chen and Chang (1996) and prevents a limit cycle due to memory adaptation. The parameter selection is critical in high noise environments to assure long term stability.

CMAC NEURAL NETWORK EXPERIMENTAL RESULTS

Figure 7 shows a comparison between the two-loop controller without feed-forward (No CMAC in the figure), a position CMAC, and a position/velocity CMAC. Both CMAC versions were used in conjunction with the conventional two-loop controller. The ADALINE was used for velocity noise reduction in all three controllers. The figure shows the result after 20 cycles of CMAC training. Each cycle requires 16.66 seconds. The case shown is for the medium load, a command velocity of 40 degrees/sec, and a 90-degree set-point. The CMAC parameters were optimized and kept constant for controller comparisons. A hash memory of 95,000 locations (32 bits per location) was used for the CMAC memory.

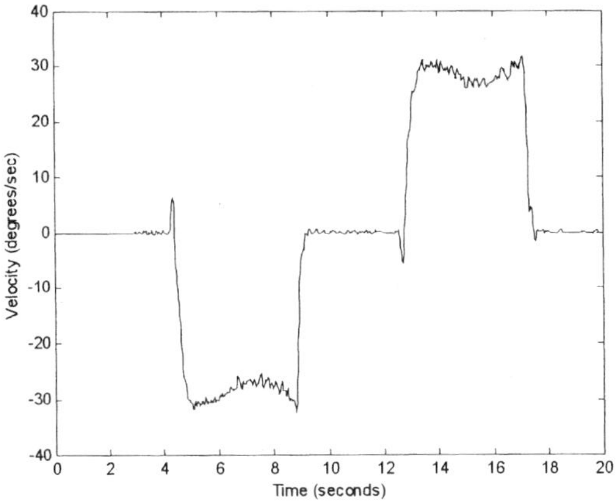

Figure 5. Differentiation of ADALINE Output.

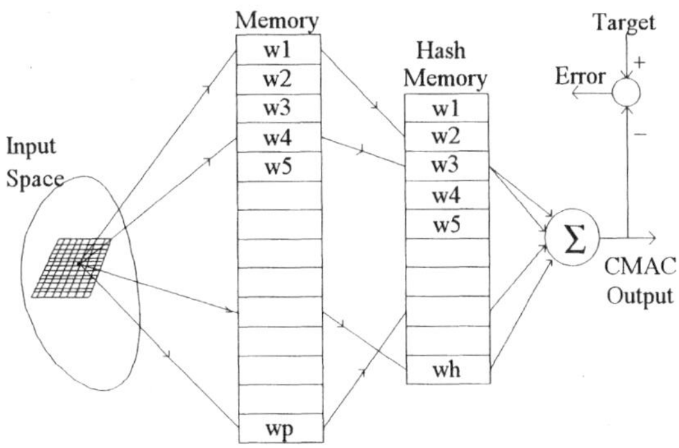

Figure 6. CMAC Block Diagram.

The position CMAC is a three-dimensional map. The learning inputs are the current and past position and target signal u_p in Fig. 2. The control inputs are the command position and current position. The hash memory accessed was 7,336 locations. An additional input is supplied for the position/velocity four-dimensional CMAC. The current velocity state along with the current and past position states and target signal are used for the learning phase. Command position, command velocity, and the current position are used for the control phase. Command velocity is calculated by taking the derivative of the position command. The hash memory accessed was 17,417 locations.

From Fig. 7, the steady-state errors in degrees are 19.4 (no CMAC), 7.9 (position CMAC) and 1.4 (position/velocity CMAC). The worse-case tracking errors in degrees are 41.2 (no CMAC), 39.0 (position CMAC) and 22.5 (position/velocity CMAC). The position/velocity CMAC with the ADALINE met the specifications with the three loads.

As discussed by Chen and Chang (1996), large training cycle experiments may be required to determine the existence of CMAC instability. In our application, limit cycles were observed when the CMAC was required to generate a steady-state output for a long

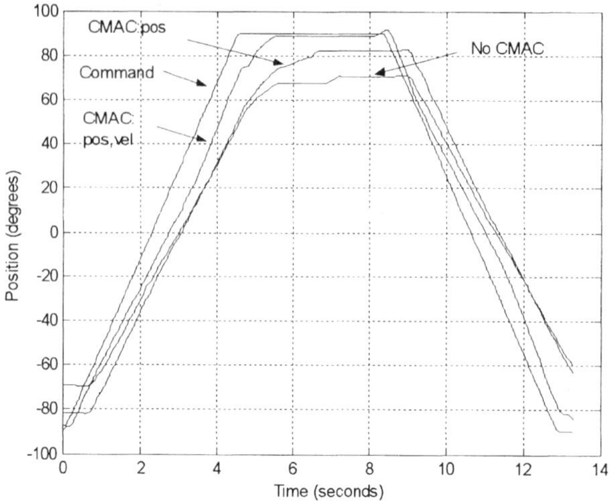

Figure 7. Controller Ramp/Set-Point Responses.

period of time. Only the position/velocity CMAC with and without the ADALINE were tested since they were demonstrated to be the best controllers. The limit cycle was first observed in the CMAC output signal and then eventually in the arm position. The CMAC without the ADALINE failed in 16 seconds. The CMAC with the ADALINE failed in 7.5 minutes. This illustrates the importance of reducing the noise in the CMAC's velocity input.

The solution for the limit cycle is adjustment of the dead-zone parameter d_0 in Eq. (4). A dead-zone of 0.6 was used for the results shown in Fig. 7. A dead-zone of 2.0 was required with the ADALINE and 3.5 without the ADALINE to prevent CMAC instability for long durations at the set-point. A larger noise level will require a larger dead-zone and in general degrade steady-state error and tracking performance. For example, with the dead-zone changes, after 20 cycles of the ramp/set-point training, the worse-case tracking error was 33.0 degrees with the ADALINE and 36.2 degrees without the ADALINE.

REFERENCES

Albus, J., 1975, "*A New Approach to Manipulator Control: The Cerebellar Model Articulation Controller*," ASME Journal of Dynamic Systems, Measurement, and Control, pp. 220-227.

Chen, F., and Chang, C., 1996, "*Practical Stability Issues in CMAC Neural Network Control Systems*," IEEE Transactions on Control Systems Technology, Vol.4, pp. 86-91.

Dempsey, G., 1999, "*Using Conventional Controllers with the CMAC Neural Network*," Proceedings of the Artificial Neural Networks in Engineering (ANNIE) Conference, St. Louis, Mo., pp. 125-130.

Miller, W., Hewes, R., Glanz, F., and Kraft, L., 1990, "*Real-Time Dynamic Control of an Industrial Manipulator Using a Neural-Network-Based Learning Controller*," IEEE Transactions on Robotics and Automation, Vol. 6, pp. 1-9.

Miller, W., and Glanz, F., 1996, "*Chapter 26: Cerebellar Model Arithmetic Computer*," Fuzzy Logic and Neural Network Handbook, editor C. Chen, McGraw-Hill, Inc.

Vainio, Olli, 1999, "*Adaptive Derivative Estimation for Delayed-Constrained Acceleration Measurement*," IEEE Transactions on Industrial Electronics, Vol. 46, No. 5, pp. 933-935.

Widrow, B. and Hoff, M., 1960, "*Adaptive switching circuits*", 1960 IRE WESCON Convention Record, New York: IRE, pp 96-104.

A BENCHMARK FOR THE RATE OF CONVERGENCE IN NEURAL NETWORK CLASSIFICATION ALGORITHMS

MASOUD GHAFFARI SOUMA M. ALHAJ ALI
XIAQUAN LIAO ERNEST HALL
Department of Mechanical, Industrial and Nuclear Eng.
University of Cincinnati
Cincinnati, Ohio

ABSTRACT
The purpose of this paper is to demonstrate a new benchmark for comparing the rate of convergence in neural network classification algorithms. The benchmark produces datasets with controllable complexity that can be used by an algorithm. The dataset generator uses the concept of random numbers and linear normalization to generate the data. In a case of a one-layer perceptron, the output datasets are sensitive to weight or bias of the perceptron. A Matlab implemented algorithm analyzed the sample datasets and the benchmark results. The results demonstrate that the convergence time varies based on some selected specifications of the generated dataset. This benchmark and the generated datasets can be used by researches that work on neural network algorithms and are looking for a straightforward and flexible dataset to examine and evaluate the efficiency of neural network classification algorithms.

INTRODUCTION

Artificial neural networks (ANN) have been an active area of research in the last decade and many achievements have been emerged. After any development in the techniques and algorithms, the question comes to the application. Any improvement in neural network (NN) algorithms needs a test bed and problem to show the results. Some authors have used the real data of their applied problems and others used the well-known problems of NN literature.

For example, Ng et al. (1999) developed a fast convergent generalized back-propagation algorithm. They have conducted a number of experiments on three different problems, including XOR, 3-bit parity and the 5-bit counting problems to illustrate the various aspects of the new algorithm. The network of XOR problem consists of two input nodes, two hidden nodes and one output node. In the 3-bit parity problem, the network consists of three input nodes, two hidden nodes and one output node.

Another popular benchmark problem is the two-spiral problem. The two-dimensional (2D) spiral data set was proposed by Alexis Wieland of MITRE Corporation and now forms one the important benchmarks at the Carnegie Mellon repository (Singh, 1998). The two-spiral problem is often used as a test for comparing the quality of different supervised learning algorithms and architectures (Alvarez, 1999). Many authors now include it in the benchmarks for speed and quality of learning for new algorithms and architecture types (Riedmiller and Braun, 1993; Treadgold and Gedeon, 1997).

There are many common datasets described in ANN literature that have been used as the applications of techniques. Some examples of those datasets include the IRIS, Fishers Iris Set (Fisher 1936), Cleveland Heart Diseases data (Murphy and Aha, 1994); IMOX and 80X, hand printed character sets (Jian and Ramaswami, 1988), Congressional voting dataset (UC-Irvin), Churn dataset (Feraud, 2002), BLOOD published by the American Statistical Association (Cox et al., 1982), Sonar (Gorman and Sejnowski, 1988), Glass, a collection of glass fragments (Ripley, 1994), Tremor, Parkinson's disease data (Spyer-Ashby, 1996), and Ionosphere, radar data (Sigillito et al., 1989).

Studies show that for almost any ANN a dataset can be constructed for which it solves it well (Duin, 1996). The performance of network is based on the class distribution and sample size and therefore of the application. Thus, an application domain has to be defined. The common way to do this is by selecting a collection of datasets.

This method of performance evaluation has some pitfalls. A collection of datasets may show the diversity but does not show the weight of a particular dataset in the overall performance. In addition, some classifiers have many user-adjustable parameters such as step sizes, momentum terms, weights and stopping procedures and therefore the results are user dependent, which makes the performance comparisons more difficult because different researchers may get different results for the same problem. The next problem of traditional datasets is the need for the training data and limitation of sample size. Generally, datasets are divided into three parts training, tuning and testing sections. The common pitfall of this procedure is that most researchers tend to adjust their algorithms after the testing, with the result that they may be using from the testing data, for the training and therefore the results are biased (Duin, 1996).

Since many researchers and students are developing new methods, a standard benchmark dataset is a necessity. The benchmark should be large enough and it should include a collection of diverse problems. In addition, it should be renewed from time to time. There have been some attempts to build such a standard benchmark (Duin, 1996; Michine et al., 1994; Prechelt, 1994). Also a workshop about NN benchmarking at NIPS*95 (Neural Information Processing Systems) addressed some issues regarding a standard benchmark.

The purpose of this study is to present a new benchmarking dataset that can be used for the test of algorithms associated with neural networks. The proposed data generator will provide a flexible dataset without the limitation of sample size. It also may provide a good tool for parameter adjustment of the network. By choosing different data each time, a specific behavior of a neural network can be studied. This benchmark is not a replacement for real application datasets, rather, it is a complementary benchmark that can be used in standard benchmark datasets.

THE BENCHMARK

The main idea of this benchmark based on generating random numbers from the normal distribution with different complexities for different purposes. What will be described in this paper is a special case of the benchmark data in two-dimensional space. The idea can be applied to higher dimensions.

The benchmark contains two types of datasets, Benchmark 1, shown in Fig. 1 shows the idea of a two-dimensional set of the benchmark data. The sample data has a random distribution in two sectors with the radius of one. Based on α, the angle between two lines of Fig. 1, two classes of data can be separated by a linear perceptron with different complexities. This benchmark is sensitive to weight parameter estimation of the perceptron.

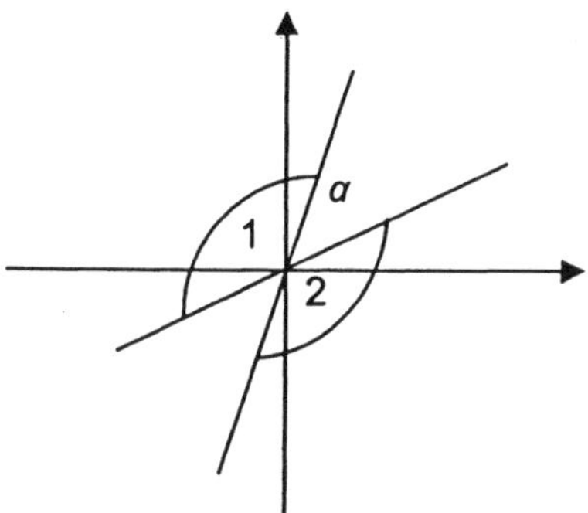

Figure 1. Random data in two sectors

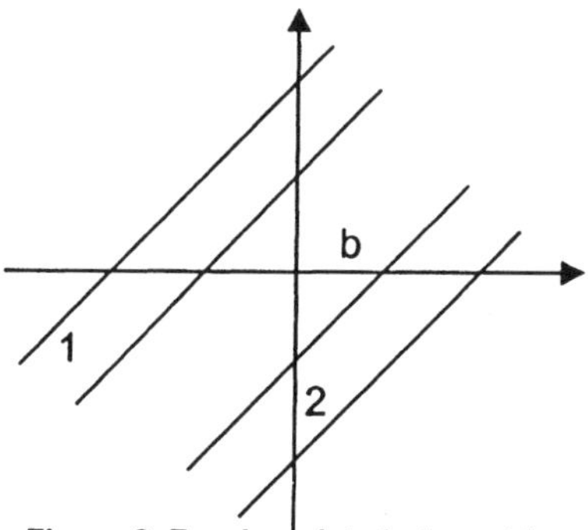

Figure 2. Random data in two strips

Benchmark 2, shown in Fig. 2, represents two classes of random data between two parallel strips. When these two classes are far from each other, separation is easier. By changing the b, different data can be obtained. This benchmark is sensitive to estimation of bias in the perceptron.

Research hypothesis states that 'when the angle of α and bias b decrease, the data set is more complex and execution time for classification algorithms increase'. If the research hypothesis is proved, it means the benchmark has the ability to generate different standardized datasets for benchmarking and comparing the rate of convergence, in classification algorithms.

The following is an example for benchmark 1:

In the polar coordinate system:

	Class 1	*Class2*
	$\rho = Rand(0,1)$	$\rho = Rand(0,1)$
(1)	$\theta = \pi/4 + \alpha/2 + (\pi - \alpha) * Rand(0,1)$	$\theta = \pi/4 - \alpha/2 + (-\pi + \alpha) * Rand(0,1)$

and in the cartesian coordinate system:

	Class 1	*Class2*
	$x = \rho \sin(\theta)$	$x = \rho \sin(\theta)$
(2)	$y = \rho \cos(\theta)$	$y = \rho \cos(\theta)$

Each time a new random number for ρ and θ should be used to generate independent (iid) x and y in any class. The above formulas generate m two-dimensional random sample datasets in two classes.
The following example is performed for benchmark 2:

Class1 | *Class 2*

$$x - y = b + \sqrt{2} * Rand(0,1) \qquad x - y = -b - \sqrt{2} * Rand(0,1)$$

$$(3) \quad x = -B + 2B * Rand(0,1) \qquad x = -B + 2B * Rand(0,1)$$

This formula generates different samples, by changing b for a specific B. These sets of data are sensitive to bias and can be used in comparing the learning speed in different algorithms.

RESULTS

A Microsoft Excel program produced several data sets for benchmark 1 and benchmark 2 based on α and b. Figure 3 shows a sample data set for α=30 and Fig. 4 represent the scatter plot of a sample data for $b = 8$.

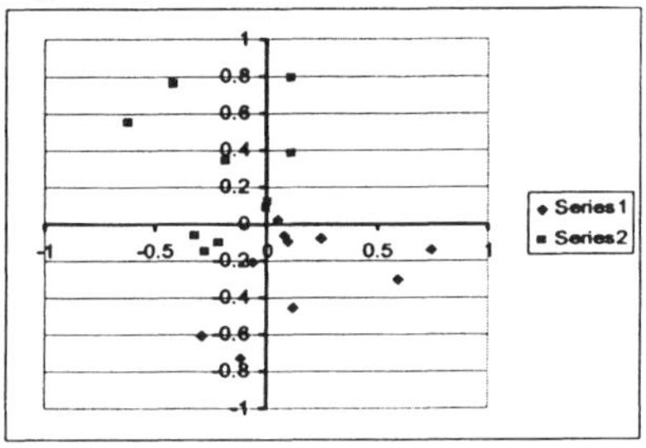

Figure 3. Data set scatter plot for α=30

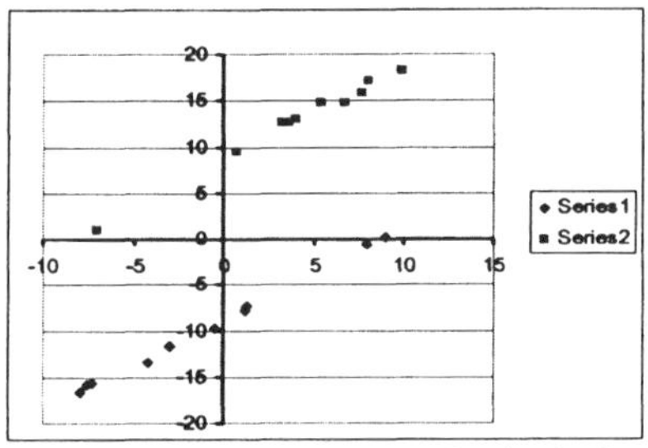

Figure 4. Data set scatter plot for b=8

These data were used in a MATLAB program and execution time for any dataset was measured. Figures 5 and 6 demonstrate the execution time verses angle and bias, which are two parameter to control the complexity of datasets. Some researches chose the number of iteration as an indicator for the learning speed of algorithm however; in this example, the execution time has been chosen that is better indicator for the computational work (Auda, 1998).

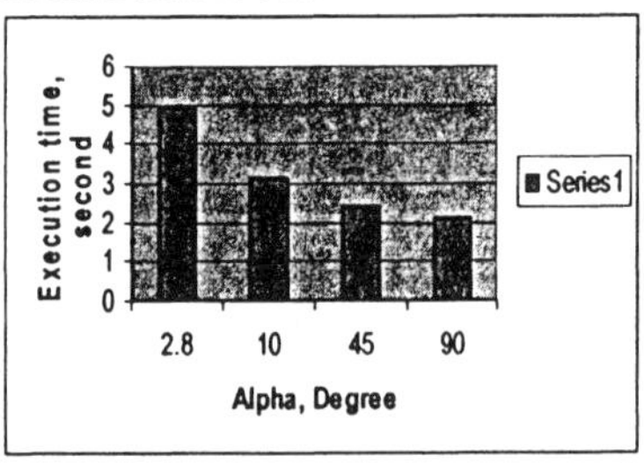

Figure 5. Execution time verses angle

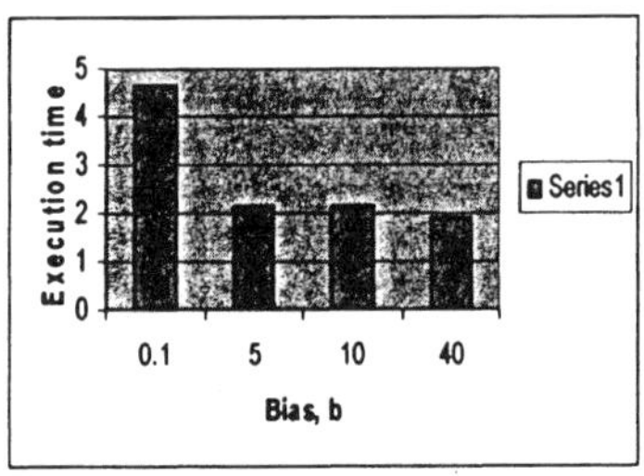

Figure 6. Execution time verses bias

The results confirm the hypothesis of the research. The benchmark is able to generate different data sets with various complexities. Figures 7 and 8 show two samples output of the MATLAB program.

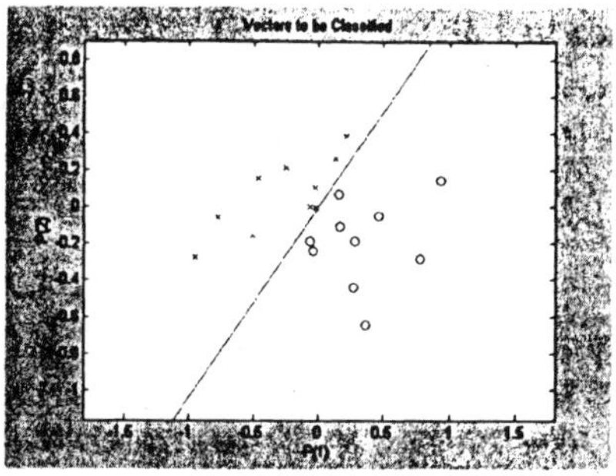

Figure 7. MATLAB output for a=2.8

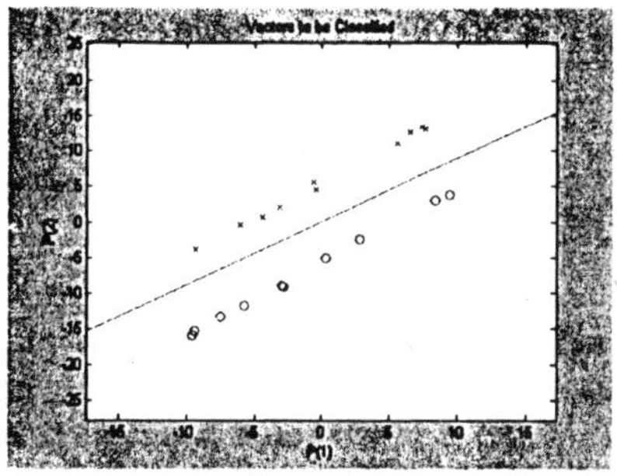

Figure 8. MATLAB output for $b=5$

GENERALIZED BENCHMARK

The demonstrated experiment used the following assumptions:

- Two-dimensional space
- A slope of 45° and $B=10$ (width of strips) for the benchmark 2
- Random numbers in Cartesian coordinates
- One perceptron classification
- Linear classification
- Ten samples for any class.

Relaxing the assumptions provide a more generalized benchmark. Generalizing from two-dimensional space to m dimension easily is possible. Also for slope and B different numbers can be chosen. The interesting generalization is from Cartesian coordinate to polar format. Figure 9 shows the 2D spiral problem in Cartesian coordinates and Fig. 10 represents the radius in function of angle for the 2D spiral problem. The points of a spiral obey the equation $r=\rho(\theta+2\pi n)+r_0$. Figure. 10 demonstrates that a 2D problem can be resulted from benchmark 2.

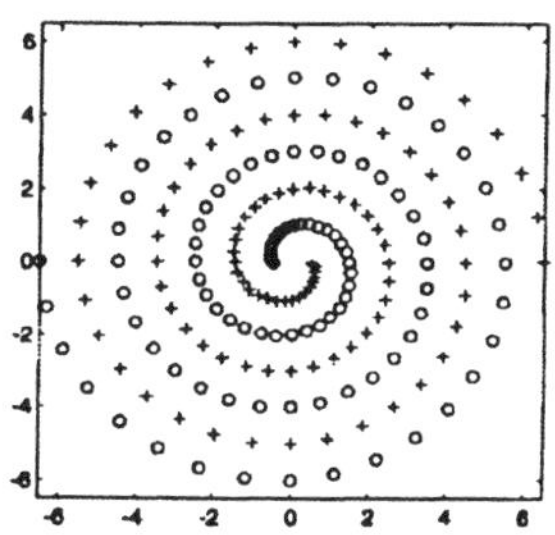

Figure 9. Cartesian plane representation of 2D

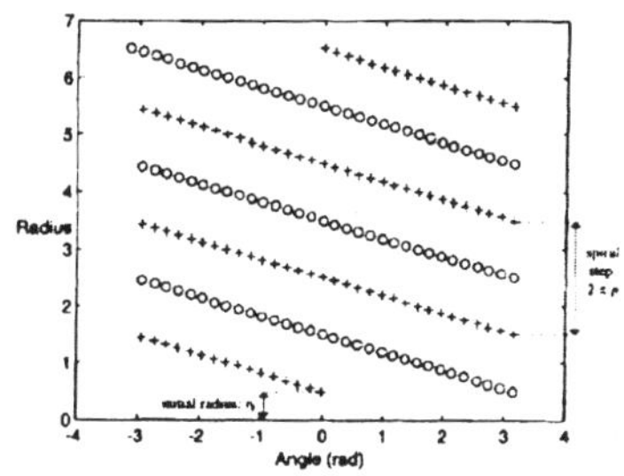

Figure 10. Representation of the radius in function of the angle for 2D (From Alvarez-Sanchez, 1999)

Relaxing the assumption of one perceptron is also possible. Figure 10 illustrates the idea. In the current format $G_1[\alpha,n,m]$ and $G_2[b,n,m]$ represent the benchmark when α=angle, b=bias, n= dimension and m= number of samples.

CONCLUSION

In this paper, a benchmark problem for the neural network classification algorithms was presented. The results demonstrate the flexibility and capability of the bearcat benchmark in generating the variety of data sets with different complexities for comparing the learning speed of algorithms. Using these two benchmarks can also be useful in estimating the initial weight factors in the algorithms. In addition, the generalized benchmark can stimulate ideas for further research in this field. The output of these researches can be installed in the University's website as a data repository.

REFERENCES

Alvarez-Sanchez J. R. , 1999, " Injecting Knowledge into the Solution of the Two-Spiral Problem," *Neural Comput & Applic*, Vol. 8, pp. 265-272.

Cox L.H., et al., 1982, "Exposition of statistical graphing technology," *ASA Proc. Statistical Computation Section*, pp. 55-56.

Duin R., 1996, "A note on comparing classifiers," *Pattern Recognition Letters*, Vol. 17, pp. 529-536.

Feraud R., Clerot R., 2002, "A methodology to explain neural network classification," *Neural Networks*, 15, pp. 237-246.

Fisher, R.A., 1936, "The use of multiple measurements in taxonomic problems," *Ann Eugenics*, Vol. 7, pp. 280-322.

Gorman, Sejnowski T.J., 1988, "Learned classification of sonar targets using massively parallel network," *IEEE Tran. Acous. Speech Signal Process.* , 36, 7, pp 1135-1140.

Jain, A.K., Ramaswami M.D., 1988, "Classifier design with Parzen windows in: E.S. Gelsema nad L.N. Kanal, Eds.," *Pattern Recognition and artificial intelligence*, Amsterdam, pp. 211-228.

Michie D., et al., 1994, 1994, *Machine Learning, Neural and Statistical Classification*, Ellis Horwood, New York.

Murphy P., Aha D., 1994, UCI repository of machine learning databases, Technical report, University of California, Irvin.

Ng, S.C., Leung, S.H., Luk, A., 1999, Fast Convergent Generalized Back-Propagation Algorithm with Constant Learning Rate, *Neural Processing Letters*, 9,1, pp. 13-23.

Prechelt L., 1994, "A study of experimental evaluations of neural network learning algorithms: current research practice," Technical report 19/94.

Riedmiller M, Braun H., 1993, "A direct adaptive method for faster backpropagation learning: the RPROP algorithm," Proceeding, *IEEE International Conference on Neural Networks*, San Francisco, 1993.

Riply B.D., 1994, "Neural networks and related methods for classification," *J. Roy. Statist. Soc. B*, 56, 3, pp. 409-456.

Sigillito V., et al., 1989, "Classification of radar returns from ionosphere using neural networks," *John Hopkins APL Technical Digest*, 10 pp. 262-266.

Singh S., 1998, " 2D spiral pattern recognition with possibilistic measures," *Pattern Recognition Letters*, Vol. 19, pp. 141-147.

Spyers-Ashby J., 1996, "The recording and analysis of tremor in neurological disorders," PhD dissertation, Imperial College, London.

Treadgold NK, Geden TD., 1997, "A cascade network algorithm employing progressive RPROP," *Biological and Artificial Computation*, Vol. 1240.

INTERACTIVE EVOLUTION FOR SYSTEMATIC EXPLORATION OF A PARAMETER SPACE

DARYL H. HEPTING
Department of Computer Science
University of Regina
Regina, SK, S4S 0A2, CANADA
dhh@cs.uregina.ca

ABSTRACT
Interactive evaluation of a fitness function for a genetic algorithm through direct manipulation is known as interactive evolution. Because it removes the need to specify a fitness function prior to exploration, the user can change evaluation criteria over time. This is especially important when the fitness function is unknown or not easily specified beforehand. This paper describes a system that allows the user to evaluate and evolve small collections of candidate solutions that represent the range of available solutions in order to explore parameter ranges of interest. Although users may be expert with respect to their particular tasks, some may be novices in relation to their software. For them, interactive evolution removes syntactic barriers to the specification of alternative candidate solutions and navigation amongst them. Expert users benefit from exposure to candidate solutions outside of their prior experience and so may find solutions that better meet their needs.

INTRODUCTION

Consider that every parameter under consideration has some permissible values, either discrete or continuous. The space defined by these parameters is the Cartesian product of all values for all parameters. For N parameters, each N-tuple of values is a distinct point in the parameter space, as follows:

$$\langle v_1, v_2, \ldots, v_N \rangle \in P_1 \times P_2 \times \ldots \times P_N$$

One might easily encounter what is called the "curse of dimensionality" or a "combinatorial explosion" when exploring this space. If each of 3 parameters have just 10 values, there are 1000 distinct tuples in this space, making an exhaustive search overwhelming. It is also fruitful to consider this parameter space as a solution space (Simon, 1977), since the expectation exists that a solution to the current problem can be found amongst the combinations of parameters values.

The computer has the promise to democratize exploration of these parameter spaces, as suggested by Jessup (1992), for example. The promise, however, will not be fully realized until software addresses the issue of access. In general, the relationship between humans and computers can be viewed as either automatic, manual, or augmented (Kochhar et al., 1991). Automatic systems, like Mackinlay's APT [A Presentation Tool] (1986), are generally prescriptive and work with very little input from the user to determine the visual representation for the user. Manual systems, like AVS [Advanced Visualization

System] described by Upson et al. (1989), provide the building blocks but require the user to take responsibility to assemble them. Augmented systems encompass a wide range of designs, including examples such as VISTA (Senay and Ignatius, 1994) and Design Galleries (Marks et al., 1997), and are intended to allow the computer to support the user's activities by working with the user.

Although many, including Foley and Ribarsky (1994) consider that automatic systems are the only feasible solution to the curse of dimensionality, these systems presuppose a complete, or at least adequate, articulation of the problem. Winograd and Flores (1985), however, contend that such a definitive articulation is not possible but it is rather an ongoing process. This paper describes an augmented system, called *cogito*, that employs interactive evolution to allow for an initially incomplete articulation of the problem to be refined as the interaction progresses. The rest of this paper is organized as follows. The next section details the need for a new approach to parameter space exploration. The following section details the design of the present software system, called *cogito*. The implementation section describes some details of the prototype version now in use and relates some of the experience gained thus far. Finally, the last section describes the conclusions drawn and areas for future work.

MOTIVATION

In all cases when a user of computer software wants to find a solution, either he or she knows what is needed or wants to explore what is available. If the user is familiar with computers in general and the software specifically, then it may be an easy matter specify a particular solution. However, this user may also be unaware of better solutions that exist outside of his or her experience, content with a local rather than global maximum. When exploration is indicated, it may involve considerable work even for an expert user. In the domain of data plotting, Bertin (1983) rightly indicated that it would take "more patience than imagination to generate 100 DIFFERENT FIGURES from the same data." That patience may give out before the best solution may be found is a serious issue.

If a complete articulation could be done for any one instance of the problem, that solution would not have to be located again. And, there would be an algorithmic mapping of the problem to the solution. However, there are many issues that work to create new instances of the problem and solution: task and individual experience being just two examples (Casner, 1991). If the solution must be adapted for the individual and the circumstances, there must be a means by which this adaptation can be achieved to support Hadamard's (1954) view that, "in the end, I know only one inventor, that is myself."

For users who do not have sufficient expertise to operate the particular software, the translation between the semantics of the problem and the syntax of the software can become more onerous. Norman (1988) uses the term "gulf of execution" to describe the distance betwen the user's concept of what is to be acheived and how to realize that goal using the software. A similar difficulty arises when evaluating the output of the software, in an effort to reconcile the computer's state with the original goal state. This is what Norman called the "gulf of evaluation." Most often, an iterative process is required to narrow these gulfs in order to reach a satisfactory outcome, the length of the process reflects the difficulty. From the perspective of solution spaces, Perkins (1995) classifies spaces depending on the amount of clues available: Klondike spaces are clue-poor and homing spaces are clue-rich. Ideally, tools would enable users to transform a Klondike space into a homing space for themselves.

Beyond the specifics of the commands, the translation of the user's internal representation to one that may be communicated externally may also involve translation (Paivio, 1986), specifically when the goal is visual in nature. A user's internal representation of his or her goal must be converted to the programmer's internal representation and then that converted to the software. Even if the user has a mental picture of what is to be acheived, it can still be difficult to communicate verbally. There is increasing evidence, in part reported by Schooler and Engstler-Schooler (1990), which indicates that attempts to verbalize descriptions of non-reportable phenomena may overshadow the original information. This is particularly important for people who do not have a strong graphical vocabulary and this extra difficulty can keep people from finding what they need. If a programmer is employed, translation errors between the user and programmer may also become a serious issue.

DESIGN

Although it is difficult for a user to generate hundreds of alternatives, it is not so for the computer. If it is possible to parameterize the form of the solution, all available solutions that fit that model can be computed automatically. Instead of quickly returning to the problem of the combinatorial explosion, a different approach is used. The parametric representation of the solution makes it well-suited to a genetic algorithm (Goldberg, 1989), but the standard implementations may be too automatic. Other automatic systems (APT), and some augmented ones (Design Galleries) also automate this initial review by having the computer first evaluate each solution. Such a specification can unwittingly eliminate desirable alteratives. Even though Design Galleries presents the user with a sampling of the full range of solutions, based on the results of the initial evaluation, specification of the metric *a priori* may represent a significant hurdle especially when the requirements are hard to specify, or even articulate.

The involvement of users is important from the perspective of personalization. Therefore, the design of *cogito* relies on the user beginning with the whole space and making decisions at the time of inspection instead of beforehand. Rather than attempt an exhaustive search for the optimal solution, the focus is to find satisficing solutions instead (Simon, 1977). Yet, if not all solutions are likely to be directly evaluated, the system must provide an effective means to choose the combinations that will be reviewed, since all will be available. This is done by allowing the sapce to be partitioned and having each partition sampled.

The paradigm of interactive evolution is well-suited to a system that is designed to encourage individual involvement. The user is be able to see alternatives without having the burden of directly specifying them. This sort of approach, which has the benefit of a low syntactic burden for the user by virtue of its direct manipulation style, has been used to great effect by Sims (1991), for example.

With respect to Perkins, the visual interface style of interactive evolution allows users to explore and locate landmarks within the solution space. In this regard, Boden (1996) was critical of Sims' system since its genetic programming component (that permitted the functionality to also evolve) made it insufficiently deterministic, compared to others like that of Todd and Latham (1992), which allowed more directed exploration. This determinism is also a goal of the *cogito* system.

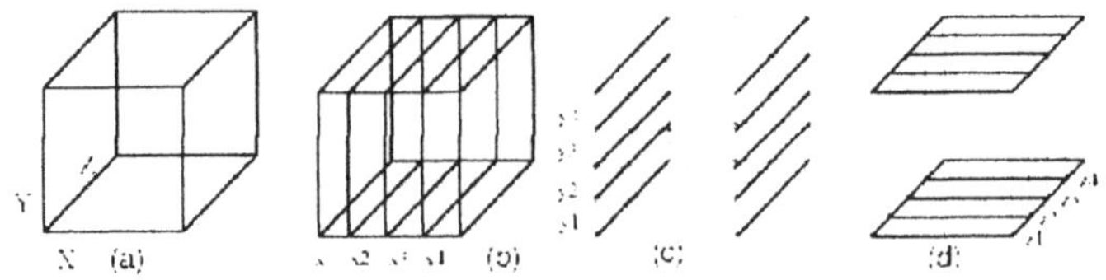

Figure 1: Consider a three-dimensional parameter space, depicted in (a), with *parameters* X, Y, *and* Z *(each with 4 values). Organizing the space in terms of any of* those 3 axes leads to other states shown (b-d). For example, as in (b), if space is partitioned according to values in X, those 4 can be shown sequentially while values from Y and Z can be chosen pseudo-randomly.

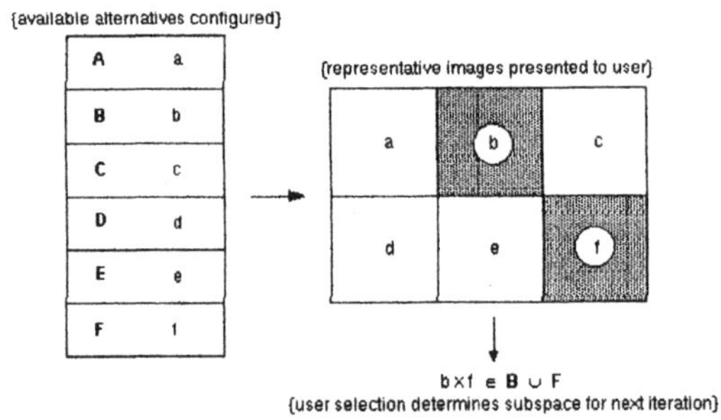

Figure 2: Schematic view of the interface: the space of available alternatives is grouped according to user-specified criteria (see Figure 1). Each group (A-F) has a representative element (a-f) that is displayed to the user. The subspace for the next iteration is based on the user selection (b and f).

As the solution space is explored and landmarks are found, the user is able to develop a mental model of the solution space just as one develops a model of an application's interface (Rosson and Carroll, 2002). This process helps the user to gradually articulate his or her needs in a particular situation. Despite the fact that a complete articulation, especially before any exploration, is not possible (Winograd and Flores, 1985), Wegner (1997) asserts that this interactive approach is also much more powerful.

Based on the preceding two figures, consider the following simple example. Imagine that the user has organized the space according to the X parameter, and subsequently chosen samples from the x1 and x3 partitions. If these selections were in fact <x1,y3,z2> and <x3,y4,z1>, the newly generated subspace consistent with those selections would also contain <x1,y3,z1>, <x1,y4,z2>, <x1,y4,z1>, <x3,y3,z2>, <x3,y3,z1>, and <x3,y4,z2>. The new subspace contains 8 of the original 64 available combinations.

The interaction is not unlike the notion of computational steering (Bullock et al., 2002), in the sense that the user can guide the computation and focus interest on areas of the solution space that he or she finds particularly important. Because users evaluate the computational results for a few points in the solution space without ever seeing the whole space, it is expected that users will not become overwhelmed by the total number of available alternatives.

The benefit of the genetic algorithm in searching the solution space is its global character that allows global extrema to be found over local ones. The common genetic operations of crossover (illustrated in the earlier example) and mutation provide very powerful means for exploring the space. Crossover, for example, allows similar solution-space points to be located. Mutation, which allows unexpected values to be added to combinations, could be used to balance the coverage of a space (for example, by periodically adding infrequently-visited values to the mix). This approach benefits the user by encouraging deeper understanding of the problem through involvement and exploration, and ultimately providing direct access to a variety of solutions. It provides better access to the solution space than the popular spreadsheet paradigm (Jankun-Kelly and Ma, 2001, and Chi, 1998) because there the user must still exercise great caution in choosing the parameters and values to explore, all done two at a time.

IMPLEMENTATION AND EXPERIENCE

This system was implemented in C++ under the IRIX operating system using X Windows for the interface and OpenInventor (Wernecke, 1994) for the graphics. It permitted a generic approach to interactive evolution. The first applications written for the system were visual, where each valid combination of values from the parameters was realizable as a graphic (see Figure 3 below). Compatibilities were defined to describe rules for combining values (code modules) from different parameters. The executable code to realize the graphics was primarily stored in dynamic shared objects that were loaded at run-time. A text file enabled a description of the application and the space in terms of parameters and values to the *cogito* system that managed the exploration of that solution space by the user.

Rather than complete particular solutions, this approach allowed the programmer to create specific modules to correspond with each value for particular parameters (but the number of modules may be less than the number of values, since the same code could be reused with different arguments.)

This interface allows the user to organize his or her view of the space according to different parameters, creating a hierarchical organization of the space best suited to the user. In addition to allowing the search space to be narrowed, it can also be expanded by re-introducing all values for a particular parameter, for example.

Figure 3 shows the first two screens of an interaction where the user moves towards a desired visual representation. From a very large initial space that could contain over 25 million points, a user can quickly select interesting visual representations and see the system display alternatives consistent with those selections, through application of a crossover operation.

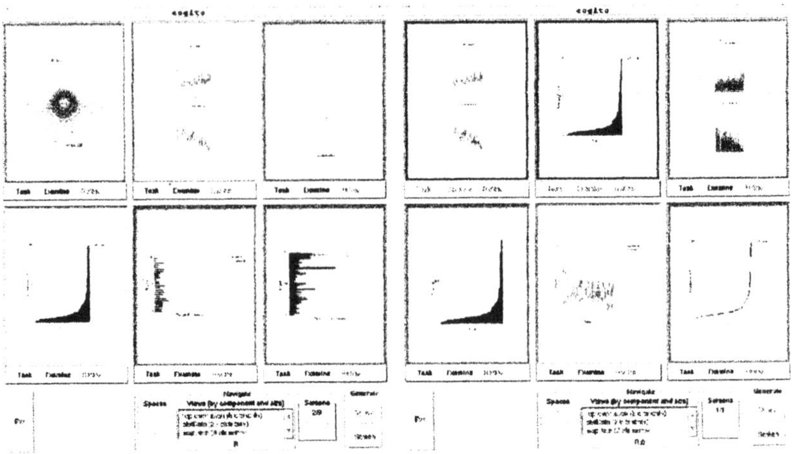

Figure 3: Two screens in an interaction with the *cogito* system. In the left screen, the user selects the top middle and bottom left cells. In the subsequent screen on the right, the system first displays the two "parent" selections then other alternative visual representations that are consistent with those selections, as determined by the application of a crossover operation. In this example, the user has chosen to proceed by selecting the bottom middle cell, although he or she may view more alternatives before making any selections. Through this system interface, the user is also able to direct the search by reintroducing values for specific parameters before continuing with selections.

A preliminary user study asked participants to choose plots to answer questions about a set of data. The *cogito* system (with the interactive evolution interface shown in Figure 3) was tested against the same system with an interface more typical of traditional visualization systems. The study, described elsewhere (Hepting 2002, Hepting 1999), found a preference for the interactive evolution interface, in particular when accessing alternatives.

CONCLUSIONS AND FUTURE WORK

This new paradigm for acess to a large, complex information space has proven effective for exploration. The user study indicated the need for both the global search provided by the interactive evolution interface and the local control available through the more typical interface. Therefore, the appropriate integration of those two access methods is now being examined. Although the user may have a conceptual map of the solution space, a graphical representation of this map is also being developed for the next version of the software. Such a map, and a more detailed history of explorations, will be valuable for individual users over time and also to facilitate collaboration between users. Overall, the system is being redesigned to accomodate a wider range of applications.

REFERENCES

Bertin, J. Semiology of graphics: diagrams, networks, maps. University of Wisconsin Press, 1983.

Boden, M. A., Creativity. In M. A. Boden, editor, *Artificial Intelligence: Handbook of Perception and Cognition*, Chapter 9. Academic Press, second edition, 1996.

Bullock, S. et al., Prospects for Computational Steering of Evolutionary Computation. E. Bilotta et al. (Eds.). *Workshop Proceedings of the Eighth International Conference on Artificial Life*, Sydney, Australia, December 8-13, pages 131-137, MIT Press, Cambridge, MA, 2002.

Casner, S. M. A task-analytic approach to the automated design of graphic presentations. ACM Transactions on Graphics, 10(2), April, 1991.

Chi, E. H., Principles for information visualization spreadsheets. IEEE Computer Graphics and Applications, July/August 1998.

Foley, J. & Ribarsky, W., Next-generation data visualization tools. In L. Rosenblum et al., editors, Scientific Visualization, Academic, 1994.

Goldberg, D. E., Genetic algorithms in search, optimization, and machine learning. Addison-Wesley, Reading, MA, 1989.

Hadamard, J., *The Psychology of Invention in the Mathematical Field.* Dover, 1954.

Hepting, D. H., A New Paradigm for Computer-Aided Exploration, Ph.D. Thesis, Simon Fraser University, Canada, 1999.

Hepting, D. H., Towards a visual interface for information visualization. Proceedings of the 6th International Conference on Information Visualization, IEEE Press, pp. 295–302, 2002.

Jankun-Kelly, T. J. & Ma, K.-L., Visualization Exploration and Encapsulation via a Spreadsheet-Like Interface, IEEE Transactions on Visualization and Computer Graphics, 7(3), pages 275-287, 2001.

Jessup, M.E., Scientific visualization: Viewpoint on collaborations of art, science, and engineering. SIGBIO Newsletter, February, pages 1-9, 1992.

Kochhar, S. et al., M., Interaction paradigms for human-computer cooperation in graphical-object modelling. In S. MacKay and E. M. Kidd, editors, Proceedings of Graphics Interface '91, 1991.

Mackinlay, J., Automating the design of graphical presentations of relational information. ACM Transactions on Graphics, 5(2), 1986.

Marks, J. et al. Design galleries: A general approach to setting parameters for computer graphics and animation. In *Computer Graphics: SIGGRAPH '97 Conference Proceedings*, 1997.

Norman, D.A. *The psychology of everyday things*. Basic Books, New York, 1988

Paivio, A. *Mental representations: A dual-coding approach*. Oxford University Press, 1986.

Perkins, D. N., Insight in minds and genes. In R. J. Sternberg and J. E. Davidson, editors, *The Nature of Insight*, pages 495-533. MIT Press, Cambridge, MA, 1995.

Rosson, M. B. & Carroll, J. M., Usability Engineering: Scenario-based development of human-computer interaction. Morgan-Kaufmann, 2002.

Schooler, J.W. & Engstler-Schooler, T.Y. Verbal overshadowing of visual memories: Some things are better left unsaid. *Cognitive Psychology*, volume 22, pages 36-71, 1990.

Senay, H., & Ignatius, E., A knowledge-based system for visualization design. *IEEE Computer Graphics and Applications*, 14(6), 1994.

Simon, H., *Models of Discovery*, Reidel, 1977.

Sims, K., Artificial evolution in computer graphics. In R. J. Beach, editor, Computer Graphics: SIGGRAPH '91 Conference Proceedings, pages 319-328. ACM Press, 1991.

Todd, S. and Latham, W., *Evolutionary art and computers*. Academic Press, London, 1992.

Upson, C. et al., The application visualization system: A computational environment for scientific visualization. *IEEE Computer Graphics and Applications*, 9(4), pages 30-42, 1989.

Wegner, P., Why Interaction Is More Powerful Than Algorithms, Communications of the ACM, 40(5), pages 80-91, 1997.

Wernecke, J., *The Inventor Mentor*. Addison-Wesley, 1994.

Winograd, T. and C. F. Flores. *Understanding Computers and Cognition*. Ablex, Norwood, New Jersey, USA, 1985.

A VISUALIZED INITIAL WEIGHTS SELECTING FOR MULTI-LAYER FEEDFORWARD NETWORKS

YEOCHANG YOON
Dept. of Computer Science and Statistics, Woosuk Univ., 565-701, Korea
yoonyc@woosuk.ac.kr

ABSTRACT
In this paper, we propose a new visualized initial weights selecting method for multi-layer feedforward networks. Most of the previous studies used the so called random number generation to deal with the initialization problems. Here we consider a visualized initialization. Our initial weights are simply selected through visualization box. The proposed method is based on the concept of Easton's controller box [1]. Although there are many difficulties with a multi-dimensional structure in the visual aspect and practical use, our methods provide the dynamic results of initial weights on the controller box to get the learning off to as fast a start as possible. Using this method, we demonstrate the visual effects of weight initialization on artificial series subject to $1 \times n \times 1$ network.

□. INTRODUCTION

Neural networks are usually trained by the backpropagation algorithm which is often too slow for most applications. One of the approaches to speed up the training rate is to estimate optimal initial weights [2]. In most cases, a brief summary on major methods determining initial weights is provided.

When the neural network is first set up, the weights must be given some initial values with random number generating from which the learning process may begin. Our goal in choosing these values is to get the learning off to as fast as start as possible. Generally speaking, we can get the quickest start by beginning with values that result in a weighted sum of inputs close to 0 for every node, regardless of the inputs. At this time, initial weights play an important role in the learning procedures.

The general method is based on the usual weight initialization. This is a uniform random initialization in the range [-0.05, 0.05].

Reference [3] simply uses the random number which is generated from $0.2 \times (X\text{-}0.5)$, $X \sim U(0,1)$. Reference [4] selects a initial weight on the interval ± 2.4 divided by number of hidden nodes. Reference [5] proposed a minimum bound for the weight initialization, which is improving speed in the backpropagation network. The initialization is still random, but satisfies the minimum. Reference [6] proposed another method quite different to the usual initialization. First, the multi-layer network is partitioned at the hidden layer into two simple perceptrons. Then the weights of the perceptrons are initialized with zero values and it is performed an independent pre-training of both perceptrons is performed by using the delta rule. Reference [7] proposed another method based on geometrical considerations. Reference [8] proposed to assign different initial connection strengths according to the importance of the inputs. A more important input should be initialized with larger weights. Reference [9] use the concept of paralyzed neuron percentage. This concept may be defined by testing how many times a neuron is in a saturated state and the magnitude of at least one output error is high. Reference [10] made another attempt to comparison the speed, generalization and the probability of convergence for several initialization methods. Reference [11] examined various weight initialization methods for multi-layer perceptrons. They applied various methods to eight real world benchmark problems. Their results show that a convenient weight initialization method for high order perceptrons with the logistic and the identity activation function is a random initialization.

A new approach for evaluating the optimal weights of neural network is presented in this paper. The methods based on visualized multi-dimensional geometrical approach. The concept is to prevent the network from getting saturation in the beginning training stage through proper weights initialization. In order to get the more proper weight initialization with a fast training network, we propose that a visualized initial weights selection using the controller box be included in this study.

The paper is organized as follows: Section 2 provides background on controller box based selection of weights and some examples. Section 3 includes a discussion of results.

II. VISUALIZED INITIAL WEIGHTS

In this work we consider a $n \times n \times 1$ neural network. Where n is the number of examples in the data set.

The equation for the outputs of the hidden nodes of a network with n hidden nodes is

$$y_j = g(a_{0j} + \sum_{i=1}^{n} a_{ij} z_i), \quad j = 1, \ldots, n,$$

and the output of a network is

$$z_{j+1} = g(b_0 + \sum_{k=1}^{n} b_k y_k), \quad j = 1, \ldots, n,$$

and activation function is the sigmoid,

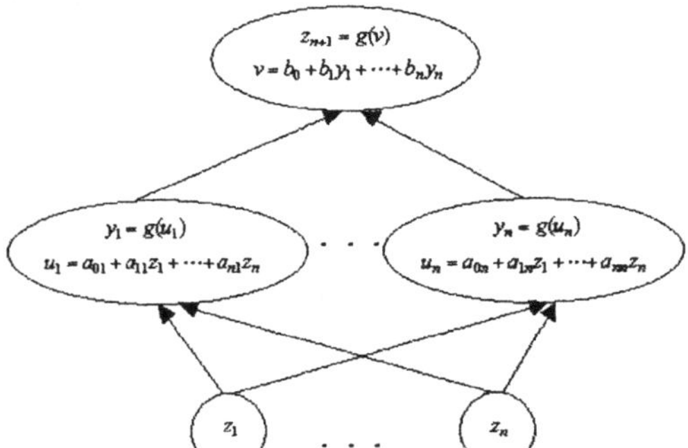

Fig. 1. $n \times n \times 1$ neural network.

$g(u) \equiv 1/\left(1+e^{-u}\right), \quad 0 < g(u) < 1$.

The learning process is generally designed to find the coefficients or weights that provided the best fit between the mapping function and data consisting of examples of the target function. Learning law is the adaptive learning rate method. This learning process using the criteria MSE is similar to [3].

2.1 Controller box

A controller box can mapping multi-dimensional points in cube to two-dimensional plane. Suppose the start vertex is A on Fig. 2. A set of 2^3 isosceles triangles is △OAB, △OBC, △OCD, △ODE, △OEF, △OFG, △OGH, △OHA that each edge of a cube, (A,B,C,D,E,F,G,H) is connected middle point O. If the position of each edge from the middle point O is positive sign, otherwise negative sign. So the $p = n$ dimensional figure is consist of 2^n isosceles triangles on the same two dimensional plane.

The point on the controller box is represented to the point on p-dimensional space is as follows. Suppose $p = 2$ dimensional, then the number of isosceles triangle are 4 and the middle point is relative coordinate (0,0). Let relative distance between center point and each vertex is k, a recurring path from starting vertex $(-k,-k)$ to $(-k,k)$, (k,k), $(k,-k)$, $(-k,-k)$ be the base of an isosceles triangle in regular sequence. Then we generate 2^2 isosceles triangles consist of the lengths of both side are k and the length of base is $2k$. The first isosceles triangle of controller box is generated as Fig. 2.

Each vertex (a_0, b_0), (a_i, b_i), (a_{i+1}, b_{i+1}) is given as an absolute coordinate on monitor and absolute location (a, b) of selected point is known, so we calculate the absolute coordinate (a^*, b^*) and the length of m, n, l', l as follows:

$$\|m\| = \sqrt{(a-a_0)^2 + (b-b_0)^2}\,, \quad \|n\| = \sqrt{(a-a^*)^2 + (b-b^*)^2}\,,$$

$$\|l'\| = \sqrt{(a_i-a^*)^2 + (b_i-b^*)^2}\,, \quad \|l\| = \sqrt{(a_{i+1}-a_i)^2 + (b_{i+1}-b_i)^2}\,.$$

Suppose relative distance between starting vertex (a_i, b_i) and absolute coordinate (a^*, b^*) is c, we can find the relative coordinate of absolute coordinate (a^*, b^*). So $l : l' = 2 : c$, $c = 2 \cdot l'/l$, $t = -k + c$.

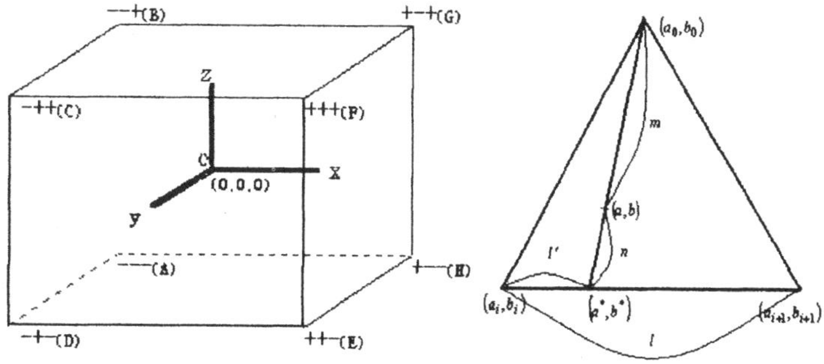

Fig. 2. 3-Dimensional figure and first isosceles triangle △OAB of controller box.

Therefore relative coordinate of absolute coordinate (a^*, b^*) is $(-k, t)$. So relative coordinate (x, y) of absolute coordinate (a, b) is a dividing point between a vertex and base of isosceles triangle.

$$(x, y) = \left(\frac{-m \cdot k}{m+n}, \frac{m \cdot k \cdot t}{m+n} \right).$$

As we expanding this procedure to $p = 3, 4, \ldots$, the location of absolute coordinate selected on the controller box can be recognized to weights that representing relative coordinate as follows:

$$\text{Case } p = 3: \quad (x, y, z) = \left(\frac{-m \cdot k}{m+n}, \frac{-m \cdot k}{m+n}, \frac{m \cdot k \cdot t}{m+n} \right).$$

$$\text{Case } p = 4: \quad (x, y, z, w) = \left(\frac{-m \cdot k}{m+n}, \frac{-m \cdot k}{m+n}, \frac{-m \cdot k}{m+n}, \frac{m \cdot k \cdot t}{m+n} \right).$$

Fig.3 is an visual examples of relative weights that represented by each dimensional relative coordinate on controller box at p=2,3,4. In this example for p=3, a relative coordinate of △ODE on controller box is (-3.98, 7.44, -7.44).

2.2 Initial value selection using controller box

In this section we consider $1 \times n \times 1$ network using controller box. First, for the simple case we consider $1 \times n \times 1$ network with one input and one hidden node. Then weight a is 2-dimentional of (a_0, a_1), weight $\underline{b}$ is 2-dimentional of (b_0, b_1). These weights are represented on the 2-dimentional controller box like Fig. 4(a) and 4(b).

Suppose the initial values of the learning are automatically given with 0.2× (X-0.5), X~U(0,1) at the beginning. And permitted limits of initial value on controller box setting up $|X| \leq 10$. And to investigate how each of the initial weights affects fitting, we intentionally select with small training sets. Specifically, we train the networks to fitting the following function:

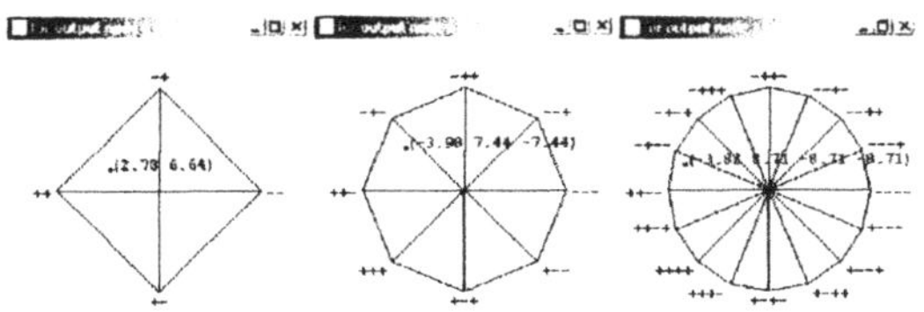

Fig.3. Weights on the controller box at p=2,3,4.

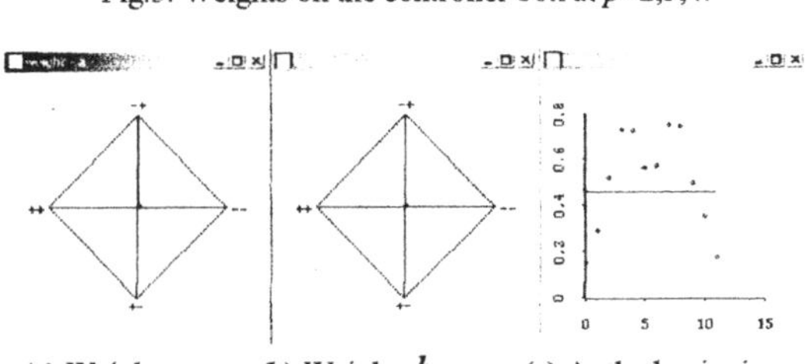

(a) Weight $\underline{a}$, (b) Weight $\underline{b}$, (c) At the beginning.

Fig. 4. Controller box at $1 \times 1 \times 1$ network and beginning of learning with generated random number.

$$f(x)=e^{-(x-1)^2}+e^{-(x+1)^2}\ ,\ x\in[-2.5,2.5].$$

For generating the training set, 12 training samples are uniformly spaced in x and target values of samples are corrupted with Gaussian noise with zero mean and variance 0.005. and for network topology, multi-layer perceptron with a single input(x) unit, single output(t) unit and one hidden unit. Hidden unit is logistic sigmoid activation.

Initial values that can be selected on controller box are represented at 2^p space of p degree initial value space. The solid lines are initial actual output with random number generation of [3] and disks are the target value at Fig. 4(c).

Fig. 5 indicates arbitrary initial value on the controller box and its results. If we start learning with that initial value, we can reduce the learning time and avoid the saturation. Therefore in this study, initial value selection at simple neural network does not use the $0.2\times(X\text{-}0.5), X\sim U(0,1)$ of [3] and $\pm(2.4/\text{number of hidden nodes})$ of [4] but real time selecting through controller box like Fig. 5.

Fig. 5 represents initial windows of initial values selected on the controller box using the generated data. It represents that visualized mapping effects for weights $\underline{b}$ subjected to select on the first, second, third and fourth quadrant given $\underline{a}$. If we move to other position on the fourth quadrant of controller box at Fig. 5, then we can meet an appropriate mapping status. Visually selected weights then become new initial values.

Fig. 6 is a final result after only 127 times learning and then MSE is 9.98E-4. The left window of Fig. 6 shows convergence processing and the right window shows the learning result. At this learning, tolerance is 10^{-3} and the learning time is only 127.

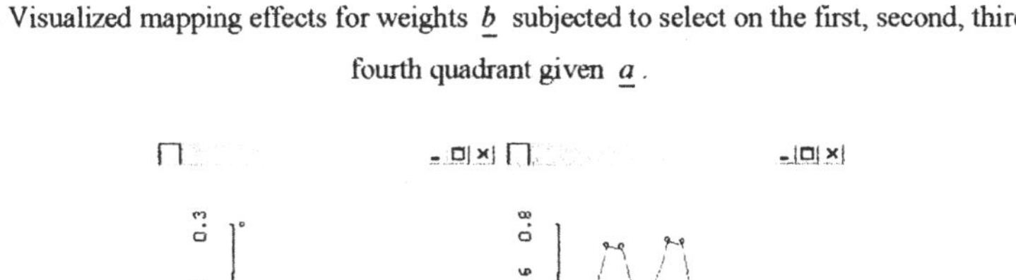

Fig. 5. Visualized mapping effects for weights $\underline{b}$ subjected to select on the first, second, third and fourth quadrant given $\underline{a}$.

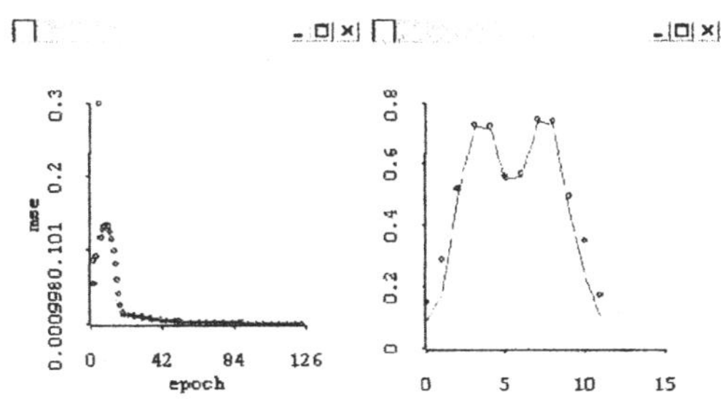

Fig.6. Learning result using the initial weights from right window of Fig.5.

Therefore, if the number of hidden nodes is enough for solving complex problem, a selection of initial weight at neural learning can be important problem. But if we can select on the visualized controller box under the suitable neural network, it can be one of the improved method.

III. DISCUSSION

We described a visually selection of initial value on the controller box for increasing efficiency of neural training at $1 \times n \times 1$ neural networks.

The initial values are visually selected appropriate weights within restricted range rather than simple random number generating. This is an alternative method that decreases training time and avoids a saturated point. We used small number of hidden node. The controller box represents multi-dimensional points at simple neural network and can be expended to other application fields.

There are under many restrictions about neural network modeling and selection of weights. Viewed in this visualization or application side, the controller box do not represent four or more multi-dimensional space. So this visualizes technique leaves much room for improvement. Therefore we need to improved selection methods on the multi-dimensional space.

REFERENCES

[1] Easton, G.S., 1994, "A simple dynamic graphical diagnostic method for almost any model," *Journal of the American Statistical Association*, 89, pp.201-207.

[2] Yam, J.Y.F. and Chow, T.W.S., *2001*, "Feedforward networks training speed enhancement by optimal initialization of the synaptic coefficients," *IEEE Transactions on Neural Networks*, Vol.12, No.2, pp. 430-434.

[3] Smith, M., 1993, *Neural Networks for Statistical Modeling*, New York, Van Nostrand Reinhold.

[4] Russo, A.P., 1991, "Neural networks for sonar signal processing," *IEEE Conference on Neural Networks for Ocean Engineering*,Tutorial Nn. 8.

[5] Kim, Y.K., and Ra, J.B., 1991, "Weight value initialization for improving training speed in the backpropagation network," *Proceedings of International Joint Conference On Neural Networks*, Vol.3, pp.2396-2401.

[6] Li, G., Alnuweiri, H., Wu, Y., 1993, "Acceleration of backpropagations through initial weight pre-training with delta rule," *Proceedings of the IEEE International Conference on Neural Networks*, ICNN'93, Vol.1, pp.580-585.

[7] Shimodaira, H., 1994, "A weight value initialization method for improved learning performance of the back propagation algorithm in neural networks," *Proceedings of the 6th International Conference On Tools with Artificial Intelligence*, pp.672-675.

[8] Yoon, Ho-Sub, Bae, Chang-Seok, Min, Byung-Woo, 1995, "Neural networks using modified initial connection strengths by the importance of feature elements," *International Joint Conference on System,Man and Cybernetics*, Vol.1, pp.458-461.

[9] Drago, G.P. and Ridella, S., 1992, "Statistically controlled activation weight initialization," *IEEE Transactions on Neural Networks*, Vol.3, No.4, pp.627-631.

[10] Espinosa, C.H. and Redondo, M.H., 2001, "Multilayer feedforward weight initialization," *Proceedings of International Joint Conference On Neural Networks*, pp.166-170.

[11] Thimm, G. and Fiesler, E., 1997, "High-order and multilayer perceptron initialization," *IEEE Transactions on Neural Networks*, Vol.8, pp.349-359.

GLIDE ALGORITHM WITH TUNNELING: A FAST, RELIABLY CONVERGENT ALGORITHM FOR NEURAL NETWORK TRAINING

VITIT KANTABUTRA and
BATSUKH TSENDJAV
Computer Science Program
College of Engineering
Idaho State University
Pocatello, Idaho 83209, U.S.A.

ELENA ZHELEVA
(Work done while at Idaho State University)
Department of CSEE
The University of Vermont
Burlington, Vermont 05405, U.S.A.

ABSTRACT
Lack of reliable convergence of training algorithms due to flat regions in the quadratic error curve is a most serious problem in neural networks. A new type of training algorithm presented here, collectively called the Glide Algorithm with Tunneling, deals successfully with this problem. This (type of) algorithm converges reliably in several hard test problems and appears to be applicable to general feed-forward, multilevel perceptrons. Tests show fast convergence 100% of the time for Parity-4, character recognition, and Two-Spiral problems. For comparison, Gradient Descent with or without momentum only converges at a 40%-80% rate on the same problems using comparable parameters. Additionally, the average CPU time used by the Glide Algorithm with Tunneling can be as little as 1/12 of that which is used by regular Gradient Descent. The new algorithm's fast, reliable convergence and robustness to parameter variations makes it attractive for general, practical applications.

INTRODUCTION

Multilevel artificial neural networks are capable of efficient universal computation, and are very useful for pattern recognition. Therefore they would see widespread use if only they could be trained reliably. Unfortunately the known algorithms, at least those that do not build the network itself, do not converge reliably for difficult problems. Some algorithms are also very sensitive to tuning.

Since the discovery of the gradient-descent based backpropagation training technique and its relatives starting in 1969 [BH69, P85, RHW86, We74], many improvements have been made. These improvements are often second-order methods, which work very well when the error curve is not too flat. The big problem is that there are vast areas in which the error curve is flat, sometimes with gradient magnitudes of a thousandth or less. In these areas, all derivatives are very small, so gradient descent moves very slowly since it moves in proportion to the gradient magnitude. Fancy, higher-order optimization algorithms are likewise ineffective, since most of them assume that the error curve is of a certain shape (not flat). For example, high-order algorithms work by assuming that the error is locally shaped like a paraboloid, and these algorithms work by jumping directly to the minimum of this paraboloid. Unfortunately a profoundly flat region is not approximated very well by a paraboloid (or by any non-flat shape that can easily be characterized mathematically). In fact, some of the most successful higher-order algorithms (like the Levenberg-Marquardt algorithm [HM94]) turn away from the shape-approximation scheme when far from the solution and resort back to the plain gradient descent idea when far away from the solution.

Flat region problem

Flat regions in the error curve have long been recognized as a problem by researchers [Kara96, Sl93, Y95]. However, solutions are few and far between. Perhaps the most prominent attempted solution that existed before the work presented in the current paper involves modifying backprop's formula for computing output-level error signal. There are at least two modified error signal formulae, one by Solla [So89], and the other by Fahlman [F88]. While both formulae help reduce the flatness problem, neither seems to really eliminate the problem according to the experiments performed by the authors of this paper. Fahlman's formula used with plain gradient descent (with momentum) gives very quick convergence where there is convergence, but sometimes there is no convergence using just gradient descent with momentum and Fahlman's formula. Fahlman's formula (with usual gradient descent backprop with momentum) is, however, still useful, and will be used here to help the two-spiral training move towards 0 error faster *initially*. However, if this scheme does not appear to converge, then a switch is made to the new algorithm to bring the training process to convergence reliably.

Previous algorithm by the authors of this paper

The first version of the Glide Algorithm was in [KZ02]. The idea in that paper was that flat regions appear to be 'safe' simply because derivatives are very close to 0. Thus, it appeared to be a good idea to move a constant distance per step (such as .1) instead of a constant proportion of the (norm of the) gradient. Moving a constant distance in itself is not a new idea, but here a very simple switching rule is used: move a constant distance if the gradient is below some cutoff point (such as .08), else perform regular gradient descent.

While it was possible to show in that paper that the simple Glide Algorithm converges significantly faster than regular gradient descent when it does converge, the fact is that the simple Glide Algorithm converges less frequently than regular gradient descent. For 'mild' parameter values, the simple Glide Algorithm is not much worse than gradient descent, so the former is certainly useful. However, since even plain gradient descent only converges 60% of the time or so (just citing a typical number; convergence rates vary widely in practice), one would hope to come up with a training algorithm that converges better, rather than worse, than gradient descent.

Algorithmic improvements presented here

More recently, the Glide Algorithm has been modified by a great degree into a second-generation version, which is called the Glide Algorithm with Tunneling. Tunneling refers to changing the weights less than the usual distance of gradient norm times the learning rate. Tunneling is very important where the error curve is rugged. In the two-spiral problem, tunneling can bring down an error of 3 untrained points to 0 in 1 step!

This new Glide Algorithm (with Tunneling) is the subject of this paper. The new algorithm converges very reliable, converging every time in several hundred. In fact for one of the test problems it converges even initialized with 'optimally bad' weights, that is, with weights that would compute the 'complement' of the correct function. 'Complement' here means that where the function value should be 0, those bad initial weights would give an output of 1, and vice versa.

The Glide Algorithm with Tunneling appears to be applicable wherever Gradient Descent is, without limitations. The new algorithm's fast, reliable convergence and robustness to parameter variations makes it attractive for general, practical applications.

The next section (Section II) explains the new type of Glide Algorithm. Section III explains the experiment setup and gives the experimental results on three hard problems.

NEW GLIDE ALGORITHM WITH TUNNELING

The work done so far only concentrates on the flat regions and the convergence problems they bring. This is not because the non-flat regions are not important. They are, but there already exist a multitude of excellent solutions for those regions. The flat regions, on the other hand, have not received nearly the same amount of effective attention.

New concepts

The new Glide Algorithm grew out of the Hairpin Observation, which is an observation that shows why this paper's authors' original Glide Algorithm did not converge as well as regular gradient descent. The Hairpin Observation says simply that in flat regions, regular gradient descent usually sends weights in a straight average trajectory (averaging allows zigzags to be smoothed out), but can take sharp, unexpected hairpin turns.

Several new concepts are introduced here. First of all, the new glide algorithm with tunneling does not move in the direction of the gradient when the gradient is small, but instead it moves in the average direction of two or more gradients to avoid severe slowdowns due to zigzagging.

Additionally, when the new algorithm gets stuck, it attempts a gentler type of randomization before the usual type of 'complete' randomization.

The new algorithm

Unlike most other algorithms in neural nets, this algorithm makes extensive use of the computer as a decision maker. Most algorithms in this field are simply iterations of mathematical formulae. (The formulae themselves may be complex, but the decision process regarding what to do in each iteration is usually simple.) Note also that many variants of our new algorithm are possible, so it is more important to convey the basic idea behind the algorithm than the precise details.

The algorithm first computes a few gradient descent moves and saves the resulting points without necessarily making the move permanently. One version of the algorithm that tested computes two gradient descent moves. More such moves could lead to fewer iterations overall, but slower ones as well. For this discussion, assume two gradient descent moves are computed. Call the original point $\mathbf{w}_0$ and the points after the two gradient descent moves $\mathbf{w}_1$ and $\mathbf{w}_2$, respectively. These 3 points are related by the equation $\mathbf{w}_{i+1} = \mathbf{w}_i - \eta \nabla E(\mathbf{w}_i)$, where E is the error function and η is the learning rate (less than 1) of each gradient descent move.[1]

- Based on the "tentative" gradient descent moves in part A., several possibilities exist, as will now be detailed. (Here is where the computer's power to make complex decisions is found useful.)
  - Most often, the error decreases or does not increase too much after the moves, for example, $E(\mathbf{w}_2) \le 1.001 * E(\mathbf{w}_0)$. (Note that even if the error decreases on average, it can not be the case that $E(\mathbf{w}_1) < E(\mathbf{w}_0)$ because

[1] Why compute two gradient descent moves instead of one? This is because gradient descent often takes a zigzag course as it descends through channel-like terrain on the error curve. As it takes the zigzag course, the error can alternate between rising and falling. By computing two gradient descent moves, it becomes more likely that if the error is going down on averages, then the error at the second weight point will be less than that at the zeroeth.

the weights can move in a zigzag fashion, where the error alternately increases and then decreases. Note also that the one per mill threshold is arbitrary. A slightly different threshold would probably also work.) In this case it now makes sense to figure out how much it is possible to go further in the same direction while keeping the error decreasing. In one version of the algorithm, a linear search in the direction $\mathbf{w}_2 - \mathbf{w}_0$ is performed. However, such action is time-consuming.

- Observe that when the gradient is not small, it is efficient enough to just take the two-step gradient descent and call it an iteration.
- However, when the gradient is small, for example, less than 0.08 (a rather arbitrary threshold that the authors actually used in the experiments), then just taking the two-step gradient descent does not lead the weight very far. Instead, the algorithm attempts to perform a *"glide,"* which means that the weight trajectory is force to go a *fixed, rather large (e.g., 0.1) distance* in the direction of $\mathbf{w}_2 - \mathbf{w}_0$. However, to be prudent the algorithm checks to see whether or not the weight resulting from the glide maps into an error that is smaller than or equal to the weight before the glide. If so, the algorithm performs the glide. Otherwise the algorithm tries a more sophisticated type of glide called a "two-step glide." This move is done by first making two glides, but only tentatively. Then, derive a direction from the two glides, and glide in that direction. This move may not be necessary, however, if not much slowdown happens in ordinary gliding. The 2-spiral program was written without this move and still works very efficiently.
- If a two-step glide increases error, then perform a *random glide* without minding whether or not such a move increases the error. This move involves going a rather large, random distance in the direction of $\mathbf{w}_2 - \mathbf{w}_0$. (Recall that $\mathbf{w}_2$ is the weight resulting from two tentative, or simulated, gradient descent moves.) The random gliding distance is randomly generated and should have the possibility of being significantly larger than the normal gliding distance, which was 0.1 or so. The random gliding distance norm was set to double itself to a certain value, then reset to a low value again.

o Sometimes the error increases more than a very small amount, i.e., $E(\mathbf{w}_2) > 1.001 * E(\mathbf{w}_0)$. This is when *tunneling* or *bracketing* is performed. Such error increase indicates ruggedness in the terrain, and warrants spending extra computational time moving very carefully. This kind of move involves first computing a few tentative or simulated gradient descent moves beyond $\mathbf{w}_2$. With two additional weight points, there would now be weight points $\mathbf{w}_0, \ldots, \mathbf{w}_4$ available, where each point is computed from the previous one by one gradient descent step. Next pick among these points the one that maps into the least error. Say $\mathbf{w}_i$ is that point. Perform a line search (by bracketing, for example) for the minimum weight point between $\mathbf{w}_i$ and $\mathbf{w}_{i+1}$. The actual destination weight taken is the weight found using this line search. Note that it is not a good idea to spend time finding the minimum to any great accuracy. Additionally, do not let the weight get too close to $\mathbf{w}_0$ since that would imply a very small weight change. Generally speaking it is better to have some increase in weight than a tiny step. There is also an empirical reason for performing a few gradient descent steps before the line search. Experience shows that if the algorithm only performs a line search from $\mathbf{w}_0$, then too often the answer would lie too close to $\mathbf{w}_0$ itself, hindering progress.

- The algorithm performs a *belt-tightening* step once in a while to reduce weight spread due to gliding. This step simply involves multiplying each weight by a constant, between .95 and .99. A check is made of the new set of weights to make sure the error doesn't increase by more than a very small amount, say 1 per mill.
- In one version of the algorithm, if there is no progress after 2000 iterations then the algorithm completely restarts. Note that some critics could say that the algorithm of course will converge surely since randomness has been introduced. However, in truth this algorithm converges not only surely but also quickly.

PROBLEMS THE ALGORITHM HAS SOLVED SO FAR

The algorithm was first tested on the parity-4 network with 4 hidden neurons, the Lang-Witbrock [LW88] network for the 2-spiral problem (2-5-5-5-1) with shortcuts, and a small character-recognition network.

To test the Glide-Tunnel algorithm on parity-4, a few hundred runs were performed with random starting weights, normally distributed with mean = 0 and S.D. = 1/3. All runs converged. Bogdan Wilamowski [Wi02] suggested a more difficult test. In the first (odd) run of each or 100 pair of runs, the starting weights would be generated randomly. But in the second (even) run of the pair, the algorithm used the *previous run's final weights* as the starting weights. The interpreted language Octave was used. The CPU times for odd and even runs stay within less than 1000s, *always* converging. The mean, SD, and median CPU times are 253s, 194s, and 211s, respectively. Runs with plain gradient descent, with or without momentum, converge at a rate no better than 60%, with CPU time allowed to reach 4000s before killing the run. The new algorithm also converged 100% of the time when we let some key parameters vary randomly.

The two-spiral problem is considered to be much harder problem than parity-4. Java was chosen as the programming language. Our algorithm converges reliably, with 100 runs "officially" tested and tabulated. Of those, the first 50 runs were done with just our algorithm (not shown here), and the other 50 runs, started with gradient descent with momentum and Fahlman's delta (output-layer error signals). The first 50 runs converged faster on average in terms of epochs, but the last 50 runs were probably faster in terms of CPU time because each epoch of regular gradient descent was approximately 4 times as fast as an epoch of our algorithm. Plain gradient descent, with or without momentum, converge at a rate of only about 60-70% on average.

The authors are not aware of any other reliably convergent training algorithm for this problem, except ones that construct their own networks such as Fahlman's (unpublished, but mentioned in the history of the two-spiral problem on CMU's archival site).

A relatively small character-recognition network was tried and the new algorithm (a relatively untuned version of it) was tested on this problem against gradient descent. A network for an 8-pixel x 6-pixel character recognition problem from an old classroom exercise was trained on a few hundred samples of the digits 0-9. Gradient descent often did not converge after 20,000 epochs whereas the new algorithm presented in this paper converged on all the 60 official tests in less than 5,000 epochs. The Glide Algorithm's average running time is 47s as opposed to regular gradient descent algorithm's average running time of 3.7 min.

CONCLUSIONS

This paper presented a new algorithm for training neural networks that converges reliably on 3 hard problems on which plain gradient descent does not converge. The results obtained are very good for parity-4 and character recognition problems. As for the 2-spiral problem, the average number of epochs needed is not so good, but at least the algorithm still converges reliably, unlike other non-network-building algorithms that the authors are aware of. The algorithm also seems quite robust, at least for parity-4, in the sense that it converges even when important parameters are randomized. The next major step in this line of research would be to improve this algorithm, if possible, to the point where it converges quickly on all problems up to a certain level of "difficulty." To be able to prove this sort of thing, it is probably necessary to characterize error curves in terms of their ruggedness. The authors are optimistic that this will eventually be done, but such work would probably entail the invention of a theory for describing error curves that goes far beyond quadratic curve fitting that is traditionally used. More specifically, this new theory would probably have to deal more effectively with flat regions.

REFERENCES

[BH69] Bryson, A. E., and Y-C. Ho (1969) *Applied Optimal Control.* New York: Blaisdell.
[F88] Fahlman, S. (1988), Fast learning variants on back-propagation, an empirical study. *Proceedings 1988 Connectionist Models Summer School,* Pittsburgh.
[HM94] Hagan, M. T. and Menhaj, M. (1994), Training feedforward networks with the Marquardt algorithm. *IEEE Transactions on Neural Networks,* vol. 5, no. 6,
[KZ02] Kantabutra, V., and E. Zheleva (2002), Gradient Descent with Fast Gliding over Flat Regions: A First Report, Proceedings of IECON, Sevilla, Spain.
[Kara96] Karayiannis, N. B. (1996), Accelerating the training of feedforward neural networks using generalized Hebbian rules for initializing the internal representations," *IEEE Trans. Neural Net.,* vol. 7, no. 2, 419-426.
[LW88] Lang and Witbrock, Learning to tell two spirals apart. Proceedings 1988 Connectionist Models Summer School, Pittsburgh.
[LC85] Le Cun, Y. (1985). Une procédure d'apprentissage pour réseau à seuil assymétrique. In *Cognitiva 85:* À la frontière de l'intelligence artificielle des sciences de la connaissance des neurosciences (Paris 1985), 599-604. Paris: CESTA.
[P85] Parker, D. B. (1985). Learning Logic. Tech. Rep. TR-47, Center for Computational Research in Economics and Management Sciences, M.I.T., Cambridge, MA.
[RHW86] Rumelhart, D. E., G. E. Hinton, and R. J. Williams (1986). Learning Representations by Back-Propagating Errors; *Nature* **323,** 533-536.
[Sl93] Slock, D. T. M. (1993), On the convergence behavior of the LMS and the normalized LMS algorithms," *IEEE Trans. Signal Processing,* vol. 41, no. 9, 2811-2825.
[So89] Solla, S. A. (1989), Learning and Generalization in Layered Neural Networks: the Contiguity Problem. In Neural Networks from Models to Applications (Paris), eds. L. Personnaz et G. Dreyfus, 168-177. Paris: I.D.S.E.T.
[We74] Werbos, P. (1974). Beyond Regression: New Tools for Prediction and Analysis in the Behavioral Sciences. Ph.D. Thesis, Harvard University.
[Wi02] B. M. Wilamowski, (2002). *Private Communication.*
[Y95] Yu, X.-H., et al. (1995). Dynamic Learning Rate Optimization of the Backpropagation Algorithm, *IEEE Trans. Neural Net.,* vol. 6, no. 3.

RECONFIGURABLE COMPUTING ARCHITECTURE BASED ON CELLULAR AUTOMATA WITH A FIXED RULE

IOS A CRUZ
Electrical Engineering Department
ITESM Campus Monterrey
Monterrey, Mexico

ALFONSO AVILA
Electrical Engineering Department
ITESM Campus Monterrey
Monterrey, Mexico

ABSTRACT
This paper introduces a general-purpose computer architecture based on cellular automata. The cellular automaton is a computing model based on physical constrains to exploit nature's parallelism. The use of a fixed rule is proposed to simplify the hardware and to facilitate the cellular automata implementation with current semiconductor technology. The paper describes the architecture specification, the methodology to select an efficient and universal cell's rule, and simulation results verifying the correct operation of the cellular automata after computing simple Boolean functions.

INTRODUCTION

Computer architectures based on reconfigurable hardware have become a promising choice to increase computing performance by exploiting hardware parallelism. Current computer architectures, capable of performing parallel computation, are based on both the fifty-year old Von Newman architecture and the sequential disposition of the common processors. Computer architectures, based on FPGAs [Ligon 1990] [Joao and Cardoso 1999,] perform data processing by embedding the algorithms in reconfigurable hardware instead of executing a sequential set of instructions.

Computer models based on nature physical constrains, as opposed to traditional constrains, usually increase the system's parallelism level and simplify the complexity of the system's elements. Table 1 lists the main characteristics of traditionally constrained models versus physically constrained models. A physically constrained computer model aims for exploiting the nature's parallelism efficiently. The cellular automaton is a computing model based on physical constrains that takes advantage of nature's concurrency, discreteness and space-time invariance. Cellular computing is the computing science that relies on the cellular automata [Sipper 1990.]

Table 1: Traditional vs Physical

Aspect	Traditional	Physical
Orientation	Human	Physical
Execution	Sequential	Parallel
Programming	Algorithmic	Behavioral
Processing	Centralized	Distributed
Memory	Centralized	Distributed
Hardware	Complex	Simple
Software	Complex	Highly complex
Scalability	Hard	Easy

This paper introduces a general purpose architecture based on cellular automata. The automata cellular uses a fixed rule to simplify the hardware and to ease the implementation by using available integrated circuit technology. Cellular automata based architecture is expected to offer a higher density compared to traditional computer architectures. It is also expected to transfer the hardware complexity to the software layer and to use genetic algorithms to manage the software complexity. This paper presents the completed work related to defining the architecture topology, specifying the cell behavior, choosing the fixed rule, and testing the architecture using simple Boolean functions.

RELATED WORK

Perrier and Sipper developed a cellular automaton capable of instantiating universal computers and performing self-reproduction; however this sequentially conceived system is unable to exploit massive parallelism [Snipper and Perrier 1996]. Other research efforts have focused on demonstrating the existence of cellular automata with universal computing capabilities [Wolfram, 2002], and on the study of cellular automata based on physical models [Margolus, 2002.] Studies have explored the use of genetic algorithms to find cellular automata rules to solve particular problems demonstrating the cellular automata capability to perform massively parallel computing [Mitchel and Crunchfield, 1996.] The most successful cellular automata hardware is found in the Cellular Automata Machines (CAM) [Toffoli, 1985] and the use of cellular automata in reconfigurable computing has been recently studied [Ogury, 1998.]

ARCHITECTURE

The proposed cellular automata architecture has a homogeneous neighborhood and uses a fixed rule. The cellular automaton has a finite number of cells arranged in two-dimension lattice. Each cell has communication with the neighbors surrounding the cell in the up, down, left, and right directions. The *communication among the cells is 1-bit wide to keep the cell simple and* increase the cell density. A cell is modeled by a 4-input and 4-output combinational function. Figure 1(a) shows the convention to represent a cell and Figure 1(b) shows the convention for a cellular automata. The I/O labels of the cellular automata are placed on the Cellular Automata boundary.

The cellular automata architecture can perform an execution cycle after the corresponding programming cycle. A set of instruction bits must be applied to all the inputs (i_0 to i_{11}) to program the cellular automata; a wait time is also required for the cellular automata to reach a target configuration state. The cellular automaton is able to process the data applied to inputs (i_0 to i_{11}) after reaching the target configuration. The cellular automata process the input data and place the results in the cellular automata outputs (o0 to o11.)

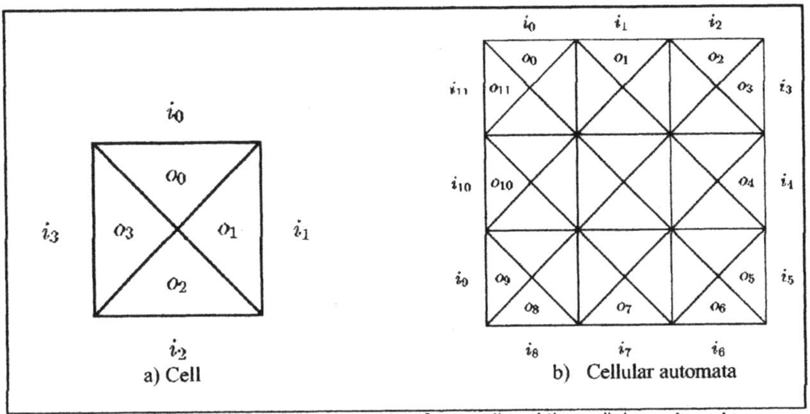

Figure 1: Representation convention for a cell and the cellular automata

SEARCHING FOR A FIXED RULE

The criteria presented in this section made possible to select one rule from the large number of rules available for a cell with n inputs and m outputs. The number k of possible rules can be obtained from the equation shown below. The total number of rules is 2^{64} for a cell with $m = 4$ outputs and $n = 4$ inputs.

$$k = 2^{m2^n} \tag{1}$$

The first criterion is to search for the rotation invariant rules. The function for O_0 also defines the rotation invariant function assignment for the remaining outputs as illustrated in equations 2, 3, 4 and 5. A 16-bit number, known as seed, is assigned to output to O_0 representing a rule in the rotational invariant rule space. The truth table shown in Table 2 illustrates the seed assignment for each cell output and the definition of the other outputs using rotation invariant mapping.

$$o_0 = f_r(i_0, i_1, i_2, i_3) \tag{2}$$
$$o_1 = f_r(i_1, i_2, i_3, i_0) \tag{3}$$
$$o_2 = f_r(i_2, i_3, i_0, i_1) \tag{4}$$
$$o_3 = f_r(i_3, i_0, i_1, i_2) \tag{5}$$

The total number of rules, calculated with the equation 6, shows a significant reduction after applying the first criterion. The number of rules is 2^{16} for n = 4 inputs and m = 4 outputs.

Table 2: Rotation invariant mapping

i_3	i_2	i_1	i_0	o_3	o_2	o_1	o_0
0	0	0	0	s_0	s_0	s_0	s_0
0	0	0	1	s_2	s_4	s_8	s_1
0	0	1	0	s_4	s_8	s_1	s_2
0	0	1	1	s_6	s_{12}	s_9	s_3
0	1	0	0	s_8	s_1	s_2	s_4
0	1	0	1	s_{10}	s_5	s_{10}	s_5
0	1	1	0	s_{12}	s_9	s_3	s_6
0	1	1	1	s_{14}	s_{13}	s_{11}	s_7
1	0	0	0	s_1	s_2	s_4	s_8
1	0	0	1	s_3	s_6	s_{12}	s_9
1	0	1	0	s_5	s_{10}	s_5	s_{10}
1	0	1	1	s_7	s_{14}	s_{13}	s_{11}
1	1	0	0	s_9	s_3	s_6	s_{12}
1	1	0	1	s_{11}	s_7	s_{14}	s_{13}
1	1	1	0	s_{13}	s_{11}	s_7	s_{14}
1	1	1	1	s_{15}	s_{15}	s_{15}	s_{15}

$$k = 2^{2^n} \quad (6)$$

Choosing the rules satisfying two or more cell's local behaviors is the second selection criterion. The local behaviors considered for this criterion are: reversible, value symmetric, value conservative and complement invariant. A reversible rule relates an input state to a unique output resulting in a cellular automata configuration with the ability to go backwards in time. A value conservative rule keeps the output values equal to the input values for all the possible input values. Equation 7 defines a value symmetric function with n inputs and m outputs. Equation 8 defines a complement invariant function with n inputs and m outputs.

$$f_j(i_{n-1},\ldots,i_1,i_0) = \overline{f_j(i_{n-1},\ldots,i_1,i_0)} \text{ for each } j = 0,1,\ldots,m\text{-}1 \quad (7)$$

$$f_j(i_{n-1},\ldots,i_1,i_0) = f_j(\overline{i_{n-1}},\ldots,\overline{i_1},\overline{i_0}) \text{ for each } j = 0,1,\ldots,m\text{-}1 \quad (8)$$

The analysis of the rotational invariant rule space and the identification of rules with two or more local behaviors reduced the number of rules to 244. A program written to calculate the rule's count in the rotational invariant space for each local behavior generated the results shown in Figure 2(b). Figure 2(a) shows the mix of local behaviors in the rotational invariant rule space. Only the 3.6% of the tested rule space had four of the local behaviors: reversible, value symmetry, value conservative and axial symmetry. Rules with complement invariant behavior (0.4%) were discarded; they had no match with the other behaviors included in the second criterion.

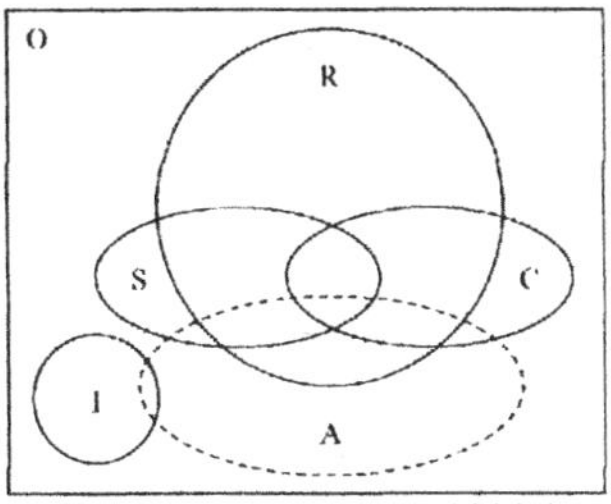

O : Rotation invariant space (100 %)
R : Reversible rules (2.3%)
C : Value conservative rules (0.3%)
S : Value symmetric rules (0.4%)
A : Axial symmetric rules (6.3%)
I : Complement invariant rules (0.4%)

a) Rotation invariant rule space

Local Behavior	Count	Tested
Reversible	1536	224
Value Conservative	192	128
Value Symmetric	256	128
Complement Invariant	256	0
Axial Symmetric	4096	40

b) Total rule count

Figure 2: Rotation invariant rule space

The last criterion uses a rule score to finally determine the fixed rule associated to each cell of the cellular automata. The rule with the highest score is selected. The rule score is the sum of the maximum fitness of the last generation of each test program. The rule score is obtained after running a set of test programs on the cellular automata for each of the 224 rules. The test programs validate the following global behaviors: program retention, data mobility and data independence. Program retention occurs when the programmed behavior has no change every time new data is entered. Data mobility is the ability to move within the cellular automata without change. Data independence indicates no data streams must interact unless interaction is required. A genetic algorithm is run twice for every rule. The genetic algorithm reads in the rule and test program and outputs a fitness result, using the parameters such as: population is 100 individuals, tournament is 5 individuals, finish criteria is 20 generations, crossover rate is 90%, resizing rate is 30% and mutation rate is 7%. The rules 62658 and 48336 had the highest scores. It can be proved that these two rules are equivalent; one is the complement of the other. so any one can be selected. The rule 48336 was selected after following the previously described criterion.

TESTING THE ARCHITECTURE WITH SIMPLE PROGRAMS

The selected rule was tested with programs focused on simple Boolean functions. These programs include the implementation of 2-input Boolean functions: AND, OR, XOR, NAND and XOR. The genetic algorithm had difficulties to find a solution with fixed input/output locations; input/output locations should no be restricted to find solutions. Table 3 shows the obtained results. The time delay of the solution is associated with the maximum Manhattan distances between the inputs cells and the output cell. The fitness values obtained are near the maximum possible value (100) indicating the cellular automata behavior is close to target behaviors. The NOR function was the only function with unexpected behavior. This was determined after applying random input data and tracing the cellular automata state manually. A 2-bit counter program was used to test memory related global behaviors in the cellular automata. The genetic algorithm required modifications to handle the memory behavior and the lack of inputs. The target behavior simulator was configured as a 2-bit counter with a two-step period.

Table 3: Boolean logic functions on rule 48336.

Function	Fitness	a	b	c	Machine Code
$c = ab$	92.4	i_3	i_5	o_1	0001101001000100001 0001000000111101110
$c = a + b$	91.8	i_0	i_1	o_{14}	0000110100001001111 1111000000000111111
$c = a \oplus b$	93	i_5	i_3	o_{11}	1101110000010100101 1111110000000000101
$c = \overline{ab}$	91.4	i_{18}	i_{14}	o_5	1000000000100111101
$c = \overline{a + b}$	84.8	i_3	i_{10}	o_{19}	0110111110010000011 0111011110110100000 0011011111110000000 0011011111110000001

CONCLUSIONS

Essential behaviors for the proposed programming and execution scheme were identified in some rules. These essential behaviors are: program retention, data mobility and data independence. The research results also proved that a compiler based on genetic algorithms can make a cellular automata behave as desired by applying stimulus on the cellular automata boundary. It was possible to select a unique rule that could evolve simple Boolean functions and simple memory behaviors. Cellular automata input/output data boundary locations should not be forced to specific locations. No solutions may be available to some I/O placement configurations. The genetic algorithm requires further work to obtain better solutions in less time and to maximize the reuse of previously related solutions.

REFERENCES

Ligon III, W.B., 1990, "Exploration of reconfigurable architectures: An empirical approach," Technical Report, Georgia Institute of Technology, Georgia

Joao, M. P. V. Joao and Cardoso, M.P., 1999," Architectures and compilers to support reconfigurable computing," *ACM Crossroads.*

Sipper, M., 1990, " The emergence of cellular computing," *IEEE Computer*, pp. 18-25.

Goldberg, D. E., 1989, "Genetic Algorithms in Search, Optimization and Machine Learning," Addison Wesley Longman, Inc.

Perrier, J. Z. J. Y. and Sipper, M., 1996, "Toward a viable, self-reproducing universal computer," Swiss Federal Institute of Technology.

Wolfram S., 2002, "A New Kind of Science," Wolfram Media, Inc.

Margolus, N., 2002, "Universal cellular automata based on the collision of soft spheres," New Constructions in Cellular Automata.

Mitchel, M. and Crunchfield, J. P., 1996, "Evolving cellular automata with genetic algorithms: A review of recent work," *Proceedings of the First International Conference on Evolutionary Computation and Its Applications.*

Mazumder P., 1999, "Genetic Algorithms for VLSI Design Layout and Test Automation," Prentice-Hall, Inc.

Toffoli, T., 1985, "Cellular Automata Machines, A new environment for modeling," MIT Press Series in Scientific Computation.

Oguri, K, 1998, "The plastic cell architecture for dynamic reconfigurable computing," NTT Optical Network Systems Laboratories.

CLASSIFICATION OF QRS COMPLEXES USING A MIXTURE OF EXPERTS

JUAN CARDELINO
Institute of Electrical Engineering, School of Engineering, Universidad de la República, Montevideo, Uruguay

NICOLÁS LOEFF
Institute of Mathematics and Statistics, School of Engineering, Universidad de la República, Montevideo, Uruguay

ABSTRACT
The QRS complex is one of the most important signals found in the ECG as far as the detection of cardiac anomalies is concerned. The presented algorithm decomposes each complex into its first principal components and adds temporal information. Two classifiers were developed using a supervised variant of the Competitive Selective Learning (CSL) algorithm, which is based on an optimal result of vector quantization, and a Gaussian mixture model. One of them was trained with a general ECG database and the other with patient specific data. Both classifiers were combined using a mixture of experts approach. This procedure presented a significantly higher classification rate than other results published when classifying beats in the MIT/BIH database.

INTRODUCTION

Several decades of investigation have turned electrocardiography into a basic discipline for the diagnosis of cardiac disorders. Due to its simplicity, low cost and being a non-invasive technique, it is still widely used in spite of the development of more modern techniques. Electrocardiographic signal (ECG) classification algorithms are usually developed and tested over large signal databases. These are extracted from different individuals and cover a wide range of possible heartbeats. ECG analyzers proposed in the literature present different performances (Lagerholm et al., 1998, Barro et al. 1998a, 1998b, Hu et al., 1997, Silipo and Marchesi, 1998, Watrous et al., 1994, 1995), but the main problem they face is the great variability among morphologies, mainly between patients but also due to movements of the patient, changes in the characteristics of electrical conduction in the body, etc. Therefore, a training set that takes all cases into account is impossible to build.

Algorithms trained in comprehensive sets may present significant degradation of performance when tested on other patients, impeding their application in clinical environments. Moreover, the complexity of the classifier increases with the training set, and also the number of exceptions that must be taken into account.

One possible solution to this problem is the development of techniques that can achieve certain adaptation to each patient. Some work has been carried out along this line. Watrous and Towell (1995) present a model based on feed-forward networks, where a number of neurons are left untrained when using the generic ECG database. These units are tuned with patient-specific data. One problem this approach faces is that the update of these weights (due to back-propagation) depends on the rest of the network, thus limiting the adaptability of the algorithm to each patient.

The algorithm presented in this article is based on (Hu et al., 1997) where a patient-adaptable scheme combing two different classifiers (or "experts") is proposed. One (*Global Expert* or GE) is trained as described before, using a broad, general ECG database. The other (*Local Expert* or LE) is trained with patient-specific records. Both experts are combined using a mixture-of-experts (MOE) algorithm, a decision function which weights each expert's classification based on the vector presented to the experts in order to reach a final decision.

This article will be focused on the study of the QRS complex[1], which is associated with electrical ventricular activity. The complex contains important clinical information, such as the morphological changes it undergoes in ventricular heartbeats. It is also found in most of the heartbeats and the signal-to-noise ratio is the highest of all the waves present in the signal.

SCSL ALGORITHM

Let $Q(x)$ be a vector quantifier, a function which maps a d-dimensional input space onto a finite number of representative vectors (code-vectors). A suitable measure of the quantization error is the distortion D,

$$D = \frac{1}{d} E\{d(x, Q(x))\} = \frac{1}{d}\int_{R^d} d(x, Q(x)) p(x) dx \tag{1}$$

Gerscho (1979) suggested the equidistortion principle, which unified and generalized the existing results of the asymptotic optimality of vectorial quantifiers. His work shows that for a smooth probability density $p(x)$, a large number N of code-vectors and an Euclidean measure of distortion, the minimum of the distortion is reached only when the probability density of the code-vectors is proportional to $p(x)^{d/(d+2)}$. Gerscho shows in this case all the sub-distortions, D_i, $i=1,2,\ldots,N$, are equal in an optimal partition. The sub-distortion is defined as:

$$D_i = \frac{1}{d} P(x \in S_i) E\{d(x, Q(x)) | x \in S_i\} = \frac{1}{d}\int_{S_i} d(x, Q(x)) p(x) dx \tag{2}$$

where S_i is the region associated to each code-vector by the nearest neighborhood rule.
It was assumed above that the underlying density of probability is continuous. Ueda and Nakano (1994) extended Gerscho's result to the multiple disjoint clusters case. According to the equidistortion principle, clusters with greater sub-distortion should be divided into more regions, and there should be fewer regions in clusters with lower sub-distortion. Therefore, Ueda and Nakano added a selection mechanism to the traditional competitive learning algorithm to achieve this (hence the name, Competitive Selective Learning, CSL).
This mechanism is based on the removal of code-vectors from zones with lower distortion, moving them to zones of higher distortion. The number of code-vectors to add (eliminate) is computed with a fitness function computed from the sub-distortion. Thus, the number of code-vectors in the regions can be adjusted by their sub-distortions without changing the total number of code-vectors. As learning progresses, the sub-distortions tend to equalize until the optimum is reached.
In addition to the usual competitive learning adjustment, a selection step in each epoch is performed. In order to assure the convergence of the selection process, the number of code-vectors to select must decrease monotonically after each step of the selection. Therefore, this process is always finished in a finite number of steps.

---

[1] The cardiac heartbeat has a bioelectric expression, constituted by several waves, named P, Q, R, S and T respectively. In this work, the expression QRS complex is employed when referring to the signal composed of the three central waves of each heartbeat

Traditional CSL asymptotically minimizes distortion as an unsupervised problem. In order to deal with several classes ($c=1,2,...,K$), a suitable choice of error function is the global distortion per class. That is, the distortion of all classes is minimized subject to a constant total number of neurons (allowing variations in the number of neurons per class). This is equivalent to designing a vector-quantizer per class $Q_c(x)$. The region S_i associated to each code-vector is defined by nearest neighborhood among the code-vectors of its class. Therefore Eq. (1) should be changed to reflect this,

$$D_c = \frac{1}{d} P(x \in \text{class } c) E\{d(x, Q_c(x)) | x \in \text{class } c\} = \frac{1}{d} \int_{S_i} d(x, Q(x)) p(x) dx \tag{3}$$

where D_c is the distortion per class. Sub-distortion (per neuron) is defined analogously, but contrary to other competitive learning algorithms (such as learning vector quantization, LVQ), a neuron in the proposed Supervised Competitive Selective Learning (SCSL) algorithm ignores input vectors not belonging to its class during training. Thus, global distortion is $D^{global}=\Sigma_c D_c=\Sigma_i D_i$. The overall selection mechanism is the same as in CSL, but the class of a neuron is preserved in reproduction. Hence, the SCSL algorithm asymptotically solves the problem of optimal labeling of neurons.

GAUSSIAN MIXTURE MODEL

Acording to (Gersho, 1979), as d is high, the code-vectors of class c probability density roughly approximates $p(x|c)$. Thus, we consider a model $\phi(x|c)$ for the density $\phi(x|c)$ as a linear combination of component densities $\phi((x|c)|j)$ in the form

$$\phi(x|c) = \sum_{i \in I_c} \phi((x|i)|c) P(i|c), \qquad \phi((x|i)|c) = \frac{1}{(2\pi\sigma^2)^{\frac{d}{2}}} \exp\left\{-\frac{\|x_n - \mu_i\|^2}{2\sigma^2}\right\} \tag{4}$$

where $P(i|c)$ are the mixing parameters and $I_c=\{i$: the i-th gaussian component belongs to class $c\}$. The GMM was initialized using the code-vectors produced by the SCSL algorithm as the centers of the gaussians. Likewise, the i-th gaussian component is assigned to the same class as the i-th code-vector. The GMM was trained using a simple case of the EM algorithm following (Bishop, 1996)

In order to train the MOE algorithm the output of each expert must be the posterior probability densities $\phi(c|x)$. Therefore, this density is computed using Bayes rule from the previously computed $\phi(x|c)$.

MIXTURE OF EXPERTS

The probabilistic output of a mixture of M experts is the weighted mean of the outputs of each expert,

$$p(c|x) = \sum_{j=1}^{M} g_j(x) \phi_j(c|x) \tag{5}$$

where $\phi_j(c|x)$ are the probabilistic expert outputs explained in the previous section and $g_j(x)$ are the weights (or gating network outputs). These are chosen to be soft-max functions. A standard error function used to train a mixture is based on the maximum likelihood principle (see (Bishop, 1996)). Thus, following (Moerland, 1997), a gradient descent update of the gating network parameters can be devised.

IMPLEMENTATION

The signal was prefiltered to attenuate noise. In order to generate feature vectors, the first nine coefficients of the projection onto the space of the Karhunen-Loève transform were used, and temporal information of the signal was added. The following features were considered: the instantaneous *RR* interval (time between *R* peaks, denotes the cardiac rhythm), the ratio of the instantaneous *RR* to the average *RR* (the average is of the last 8 *RR*s); this measures how premature or delayed the heartbeat is), and the width of the QRS complex. The vector thus generated was normalized over the MIT-BIH database.

The MIT-BIH ECG database consists of 48 records, each approximately 30 minutes long. The records are numbered as from 100 to 124 (1XX) and 200 to 234 (2XX) with some missing; it is generally considered the 2XX records present more complex and interesting cases. In order to compare this algorithm to (Hu et al., 1997), the 1XX records where used to train the global expert, and the 2XX records to test the algorithms.

AAMI guidelines (Mark and Wallen, 1987) allow the use of the first 5 minutes of each record for fine-tuning. Following Hu et al., the first 2.5 minutes where used to train the local expert, the other 2.5 to train the mixture and the rest (25 minutes) for testing purposes.

RESULTS

According to the AAMI recommended practice, the 16 available heartbeat types should be divided into 4 clases. Following (Moody, 1989) a finer partition was used, consisting of 6 classes, as shown in table I. Even though paced beats (/) are exempted from evaluation requirements in the AAMI guidelines, they were included in order to devise a more general classifier. As described above, the algorithm was tested over last 25 minutes of the 2XX records.

Two magnitudes were employed to measure the performance: sensitivity (*Se*) and positive predictivity (*+P*). Per class sensitivity (Se_i) and positive predictivity ($+P_i$) are defined and as

- Se_i : Ratio of heartbeats of class *i* correctly classified to the total of heartbeats assigned to class *i* by the algorithm. It measures the degree of certainty the heartbeat belongs to the group assigned by the algorithm and not to any other.
- $+P_i$: Ratio of heartbeats of class *i* properly classified to the total of heartbeats of class *i*. It measures the degree of certainty that the heartbeats belonging to class *i* are properly assigned.

Global performances are defined in the same way, averaging over all clases.

Table II shows the performances (per class) achieved by the algorithm over the MIT-BIH database. Maximum values are marked in boldface.

Local Expert: Table II shows global sensitivity was 96.15% and global positive predictivity was 96.20%. Normal beats are the most common class (81%) and present a higher sensitivity compared to the other classes. The exception are paced beats which are present only in record 217 and therefore exhibit similar morphologies.

Global Expert: The performance achieved by this classifier was significantly lower than that of the LE (*Se*: 87.06%, *+P*: 88.52%), as shown in Table II. The main reason is the inadecuate composition (in number and diversity of beats) of the training set for high variability data sets such as those in clinical environments. In addition the choice of 1XX

Table I: Classes for Evaluation

Class	Beat type (MIT-BIH Annotation codes)
N	Normal QRS beats (N, L, R, B, e, j, n)
S	Premature beats (S, A, a, J)
V	Ventricular beats (V, r, E, !)
P	Paced beats (/)
F	Fusion beats (F, f)
Q	Unclassifiable beats (Q, ?)

Table II: Global Performance (Per Class)

		S (%)			+P (%)		
Class	Beats	LE	GE	MOE	LE	GE	MOE
N	29356	99,10	91,84	**99,45**	96,77	95,02	**97,38**
S	1035	**64,23**	44,80	62,99	**98,30**	45,98	97,29
V	3872	88,49	88,99	**92,58**	94,08	70,14	**94,83**
F	830	**80,74**	0,36	76,81	71,03	0,28	**73,33**
/	1240	99,84	55,65	**99,92**	**99,84**	87,12	98,88
Q	8	0	**12,50**	0	X	**0,35**	0
Total	36341	96,15	87,06	**96,73**	96,20	88,52	**96,74**

records for training hampers the performance when evaluating the morphologically more complex 2XX records. The poor performance achieved by the GE in class F (lower than 1%) is caused by the presence of different types of F beats in the training and testing period.

Mixture of Experts: Global *Se* was 96.73% and +*P* was 96.74%, 0.58% and 0.54% higher than both experts. Overall, the results of the mixture show an improvement; this can be seen in table II, where the performance of the mixture is higher in most classes than that of LE and GE. Behavior per record is similar, the performance of the MOE is similar or slightly higher than the best expert in most records. Even when one expert classifies poorly, the mixture does not. It acts as a regulator; its performance is never lower than both experts, excepting record *209*. This results demonstrate the importance of this approach. If one expert is employed the performance is not only lower but more variable.

Comparison to other algorithms

(Hu et al., 1997) make use of a manually edited combination of SOM and LVQ algorithms to train 150 neurons for the LE and GE. The MOE algorithm was trained using a sum of squares error function and gradient descent. During training phase all 4 AAMI recommended classes where used but in the evaluation where divided in ventricular and non-ventricular beats.

In order to compare results with Hu et al.'s the following magnitudes should be defined: Sensitivity (*SE*), is the ratio of correctly classified V beats to the total number of V beats; Specificity (*SP*), is the ratio of non-V beats assigned to a non-V class to the total number of non-V beats; Positive Predictivity (*PP*), is the ratio of correctly classified V beats to the number of beats classified as V and Classification Rate (*CR*), is the ratio of beats correctly discriminated between V and non-V to the total number of beats.

The performance obtained is *SE*: 92.5%, *SP*: 99.3%, *PP*: 95.1% and *CR*: 98.5%, in this work and *SE*: 82.6%, *SP*: 97.1%, *PP*: 77.7% and *CR*: 94.0% in (Hu et al., 1997). This significantly superior performance can also be observed in most records, with the notable exception of record 201 (4.4% drop in classification rate).

Also, the performance of Hu et al.'s LE is superior to that of its mixture of experts, diminishing the reasons behind this approach. This behavior is not present in this work, as the mixture has superior performance compared to the best expert.

CONCLUSIONS

The present article studies the problem of classifying QRS complexes, and the use of a patient adaptable algorithm following (Hu et al., 1997). Two experts were trained using the SCSL algorithm. This supervised variant of the CSL algorithm was developed and tested in this problem. It proved to be adequate for this task and presented superior performances in both experts compared to the consulted bibliography. The conditional probabilities of the experts were estimated using a gaussian mixture model. Combining the output of the two experts, a mixture algorithm was developed and evaluated over the database obtaining an improvement in performance. Thus, the main contributions of this work are the introduction of the SCSL algorithm and the revision and enhancement of Hu et al.'s work. The developed algorithm presents a sensitivity of 96.73% and a positive predictive value of 96.74% discriminating the signals in the MIT-BIH database into six classes, significantly higher than the reported by Hu et al. The MOE algorithm acted as a regulator of both experts, and improved their performance. The performance of the experts is promising but the algorithm still has several drawbacks. A proper, representative training period must be used to train the local expert and the mixture algorithm. Also, the global expert showed the necessity of a large and diversified database.

ACKNOWLEDGEMENTS

The authors would like to thank Prof. Franco Simini and the Biomedical Engineering group at Universidad de la República for their help and financial support.

REFERENCES

Barro, S., Fernandez-Delgado, M., Vila-Sobrino, J., Regueiro, C. V., , and Sanchez, E., 1998a. "Classifying multichannel ECG patterns with an adaptive neural network". *IEEE Eng. in Med. and Bio.* [Jan./Feb.], pp. 45–55.

Barro, S., and Fernandez-Delgado, M., 1998b. "MART: A multichannel ART-based neural network". *IEEE Trans. on Neural Networks,* **9** [Jan.], pp. 139–150.

Bishop, C. M., 1996. *Neural Networks for Pattern Recognition*, Second ed. Clarendon Press, Oxford, ch. 2, pp. 59–73.

Gerscho, A., 1979. "Asymptotically optimal block quantization". IEEE Trans. on Information Theory, IT-25 (4) [July], pp. 1211–1227.

Hu, Y. H., Palreddy, S., , and Tompkins, W. J., 1997. "A patient-adaptable ECG beat classifier using a mixture of experts approach". *IEEE Trans. on Biomed. Eng.*, Vol. 44, No. 9, pp. 891–900.

Lagerholm, M., Peterson, C., Braccini, G., Edenbrandt, L., and Sornmo, L., 2000. "Clustering ECG complexes using hermite functions and self-organizing maps". *IEEE Trans. on Biomedical Engineering*, vol. 47, pp. 838–848.

Mark, R., and Wallen, R., 1987. AAMI-recommended practice: Testing and reporting performance results of ventricular arrhythmia detection algorithms. Association for the Advancement of Medical Instrumentation, Arrhytmia Monitoring Subcomitee, Tech. Rep. AAMI ECAR-1987.

Moerland, P., 1997. "Some methods for training mixtures of experts". IDIAP-COM (97-05).

Moody, G. B., 1989. ECG database programmer's guide. Harvard-Massachusetts Institute of Technology (MIT) Division of Health Science and Technlogy.

Silipo, R., , and Marchesi, C., 1998. "Artificial neural networks for automatic ECG analysis". *IEEE Trans. Signal Processing,* Vol. 46, no. 5, pp. 1417–1425.

Ueda, N., and Nakano, R., 1994. "A new competitive learning approach based on an equidistortion principle for designing optimal vector quantizers". *Neural Networks*, 7 (8) [Jan.], pp. 1211–1227.

Watrous, R. L., Towell, G., Glassman, M. S., Shahraray, M. S., and Theivanayagam, D., 1994. "Synthesize, optimize, analyze, repeat (SOAR): Application of neural network tools to ECG patient monitoring". *Proceedings of the Third International Congress on Air- and Structure-Borne Sound and Vibration, Montreal*, pp. 997–1004.

Watrous, R. L., and Towell, G., September 10-13, 1995. "A patient-adaptive neural network ECG patient monitoring algorithm". *Computers in Cardiology, Vienna, Austria*, pp. 229–232.

SYNTHESIS OF A COMPLEX CONTROL SYSTEM ON THE BASIS OF NEURAL NETWORKS

ALEXANDER KALININ
Voronezh State Technical
University, Russia

EUGENE SOLDATOV
Voronezh State Technical
University, Russia

SYMON PODVALNY
Voronezh State Technical
University, Russia

ABSTRACT
Traditional synthesis of a control system is carried out in the following different directions – creation of mathematical models, synthesis of algorithms by managements by dynamic modes and the expert analysis, adjustment of model by the skilled expert. The first aspect is solved within the traditional theory of control systems, but the second aspect concerns to intellectual tasks *and it is hardly gives into algorithmization. In the given paper the integrated* approach of synthesis of a complex adaptive control system on a basis of neural net technologies is offered.

INTRODUCTION

Up-to-date approach to the elaboration of complex control systems suggests the following tendencies:

- creation and perfection of mathematical models for production processes being the essential element required to solve management problems by nonlinear dynamic systems;
- construction of the optimum system using object simulator for control purposes, adapting to its behavior;
- control algorithm synthesis by dynamic process modes.

At the same time, the analysis of the latest surveys in the field of artificial neural network theory indicates that this mathematical tool can be effectively implemented related to management problems connected with complicated chemical-engineering processes, combining in itself the elements of the 3 above outlined approaches.

Currently, a full range of architectures of neural network control systems has been worked out, which has displayed wide practical application. Thus, the most widely spread networks are: direct control network – where neural network serves as a controller (fig. 1) and indirect control network – where neural network provides characteristic model identification (fig. 2)

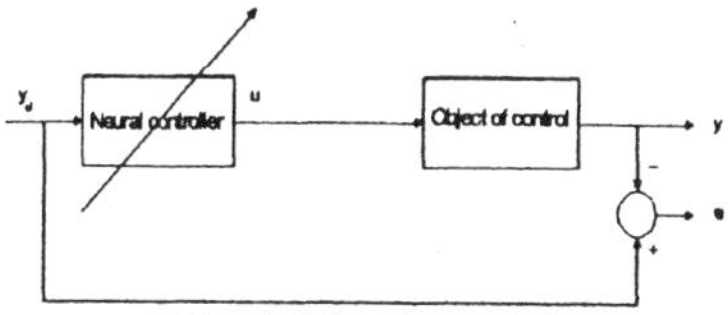

Fig. 1. Direct control

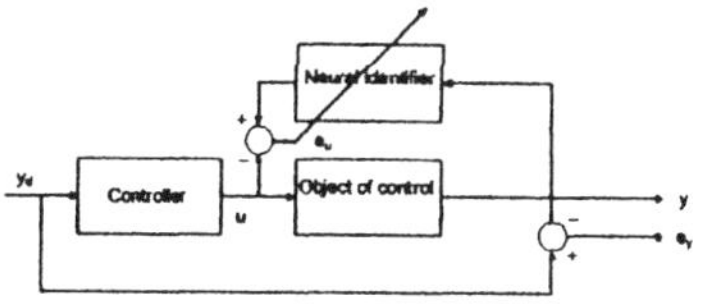

Fig. 2. Indirect control

In this connection, we cannot avoid mentioning some obvious flaws concerning controller construction in neural network technology. Firstly, it is a fairly slow convergence and, secondly, in comparison to classic methods, a lower feedback steadiness. However, on the other hand, neural networks possess a higher flexibility: they can automatically adjust to real conditions, and, potentially, due to constitutively appropriate concurrency can reach a much higher productivity level (up to in-line processing scale performance).

Basically, our investigation is dedicated to viewing the integrated construction opportunity of neural network controller for management problem of nonlinear process of divinyl-styrene rubber dissolving copolymerization.

MATHEMATICAL MODEL OF THE PROCESS

The most significant stage in synthetic rubber production can be regarded as monomers polymerization generation, within which physical-mechanical transformation of raw materials and formation of qualitative rubber activities occur. Divinyl-styrene synthetic rubber is a result of styrene and dissolving divinyl copolymerization in the reactors cascade.

Structural-level principle is assumed as a basis for mathematical model construction of the under review process. This principle comprises sequential modules accretion scrutinizing into kinetics, hydrodynamics and process heat transfer. Taking into account the community of kinetic regularity, the mathematical formulation is generally viewed in the light of butadiene and styrene copolymerization example proceeding in the grout of catalytic complex on basis of lithiumbytyl.

We are gravely interested in such a model construction which would ensure that we determine technological parameters of the process, numerical characteristics of the molecular weight rate distribution and major quality indices of the terminal product.

Mathematical model of the nonlinear process of the dissolving polymerization (for k-cascade reactor there has been acquired in the form of differential equation system):

$$\begin{cases}
\Theta^{(k)}\dfrac{dP^{(k)}}{dt} = P^{(k-1)} - P^{(k)} - \Theta^{(k)}\left(k_b^j P^{(k)}B^{(k)} + k_c^j P^{(k)}C^{(k)}\right) \\
\Theta^{(k)}\dfrac{dB^{(k)}}{dt} = B^{(k-1)} - B^{(k)} - \Theta^{(k)}\left(k_h^j P^{(k)}B^{(k)} + k_h^p B^{(k)}\mu_{0h}^{(k)} + k_{bc}^p B^{(k)}\mu_0^{(k)}\right) \\
\Theta^{(k)}\dfrac{d\mu_h^j}{dt} = \mu_{jb}^{(k-1)} - \mu_{jh}^k + \Theta^{(k)}k_j^h P^{(k)}B^{(k)} + \Theta^{(k)}k_h^p B^{(k)}\left(\sum_{n=0}^{j}\binom{n}{j}\mu_{nh}^{(k)} - \mu_{jh}^{(k)}\right) + \Theta^{(k)}k_{bc}^p B^{(k)}\left(\sum_{n=0}^{j}\binom{n}{j}\mu_{nc}^{(k)} - \mu_{jc}^{(k)}\right) \\
\Theta^{(k)}\dfrac{dC^{(k)}}{dt} = C^{(k-1)} - C^{(k)} - \Theta^{(k)}\left(k_c^j P^{(k)}C^{(k)} + k_c^p C^{(k)}\mu_o^k + k_{bc}^p C^{(k)}\mu_{0h}^{(k)}\right) \\
\Theta^{(k)}\dfrac{d\mu_c^j}{dt} = \mu_{jc}^{(k-1)} - \mu_{jc}^{(k)} + \Theta^{(k)}k_c^j P^{(k)}C^{(k)} + \Theta^{(k)}k_c^p C^{(k)}\left(\sum_{n=0}^{j}\binom{n}{j}\mu_{nc}^k - \mu_{jc}^k\right) + \Theta^{(k)}k_{bc}^p C^{(k)}\left(\sum_{n=0}^{j}\binom{n}{j}\mu_{nh}^{(k)} - \mu_{jh}^k\right) \\
\Theta\dfrac{dT^{(k)}}{dt} = T^{(k-1)} - T^{(k)} + \Theta\phi\left[(B^{(k-1)} - B^{(k)})q_b + (C^{(k-1)} - C^{(k)})q_c\right] - k_{xa}^{(k)}(T^{(k)} - T^{(h)})
\end{cases} \quad (1)$$

$$k_{xa}^{(k)} = \frac{k_t G_x^{(k)} S C_x}{G_n C_n (k_t S + G_x^{(k)} C_x)},\ \varphi = \frac{G_m C_m}{G_n C_n},\ \phi = \frac{G_m}{G_n C_n \rho_m},\ \Theta = \frac{\rho_m V}{G_m},\ \binom{n}{j} = \frac{j!}{n!(j-n)!},$$

$$P^{(k)}(0) = P_0^{(k)},\ B^{(k)}(0) = B_0^{(k)},\ C^{(k)}(0) = C_0^{(k)},\ \mu_h^{(k)}(0) = \mu_{ho}^{(k)},\ \mu_c^{(k)}(0) = \mu_{co}^{(k)}$$

where $B^{(k)}$, $B^{(k-1)}$, $C^{(k)}$, $C^{(k-1)}$, $P^{(k)}$, $P^{(k-1)}$ – the concentration of the butadiene, styrene and the catalyst on an input and an output from a reactor accordingly; q_b, q_c – the thermal effect of the butadiene and styrene polymerization; $\mu_{jb}^{(k)}$, $\mu_{jc}^{(k)}$ – the *j-th* order moments of the chain-length distribution; G_m – the consumption of the charge; C_m – the thermal capacity of the charge; ρ_m – the density of the charge; G_n – the consumtion of the polymerizate on an output from a reactor; C_n – the thermal capacity of the polymerizate; G_x – the consumption of the refrigerant in the *k*-th reactor lagging; C_x – the thermal capacity of the refrigerant; $T^{(k)}$ – the temperature in a *k*-th reactor; $T^{(k-1)}$ – the temperature of the mass flow acting in *k*-th reactor, k_t – the constant of the heat transfer; k_b^i, k_c^i – the speeds of butadiene and styrene initiation; k_b^p, k_c^p – the speeds of butadiene and styrene growing; k_{bc}^p – the speed of copolymer joint growing; Θ – the time of stay of reagents in a reactor; S – the surface of the heat pick-up; V – the volume of a reactor.

In consequence of the identification, there have been gained dependences of constants initiation of the polymeric chains, emerging on active polymerization cores of diverse nature and constants of polymer speed growth arising from comonomers concentration, catalyst and reaction temperature:

$$\begin{aligned} k_b^i &= \alpha_1(1+\beta_1 P_{in}^2)\cdot\exp(-E_i/R\cdot T);\\ k_c^i &= \alpha(1+\beta P_{in}^2)\cdot\exp(-E_{ci}/R\cdot T);\\ k_c^p &= \gamma_c/(1+f_c C_{in}^2)\cdot\exp(-E_{cp}/R\cdot T);\\ k_{bc}^p &= d_1\cdot\exp(-E_r/R\cdot T). \end{aligned} \tag{2}$$

where $E_i,\ldots,E_r$ – the activation energy of the appropriate reactions of initiation and growth; R – absolute gas constant; T – the reaction temperature

It has been actually proved that the elaborated mathematical model can be adequate to the real object in accordance with the mathematical statistics approaches gained by building up links within residual dispersion model and experiment reproduction dispersion for each coordinate: monomers concentration and reaction temperature.

Nonetheless, the unequivocal specific feature of this problem holds that the dependence of quality indices on input parameters variability (non-measurable included) leads to the necessity of adaptation mechanism model application within in-line processing, permitting the system to get artificially adjusted to the altering conditions at the expense of maintenance, handling and analysis of deficient information about the process together with its management.

In general outline, the succession of operations can be conceived in the following way:

1. Structural synthesis of the mathematical model process.
2. The mathematical model parameter identification.
3. Differential equation system solving by numerical computations.
4. Adaptive control action making.

Hereby, the foremost problem lies low in the fact that the obtained differential equation system is relatively complicated in terms of its solution, which demands considerable time outlay, in some cases making on-line control impossible. Therefore, application of neural network machinery perfectly enables us to cope with the set problem.

NEURAL NETWORK CONTROL

To begin with, we will attempt to investigate into indirect control. In this case every single mathematical model parameter B_i, C_i, P_i, ..., G_x is quantized with the specified pitch and differential equation system is elected and solved in advance for each unit of the input parameter, adequate to the technological process. Neural network, within such approach, acts as the intellectual model classifier and allows us to opt for the most imminent in-line process solved equation system in accordance with the B_i, C_i, P_i, ..., G_x (fig. 3) value parameters. Thus, as the input sample we leave quantum value of non-identified parameters, as outputs sample the class code of the corresponding model:

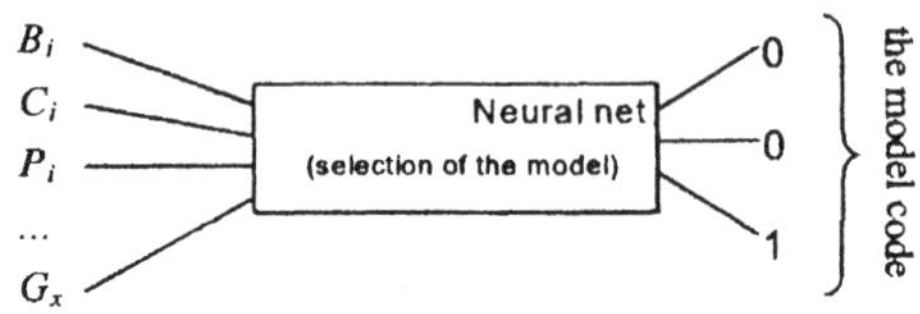

Fig. 3. The neural net selection of the model

On average, this approach is fairly basic in terms of its realization since it is theoretically grounded and vastly approved. Unfortunately, it is not universal. Limitations in memory volume make it unviable to establish a small quantization pitch. Moreover, solutions generated for intermediate parameter meanings can differ from those in quantization point disguising the possibility of management impact inadequacy. Thus, it is necessary to build up a non-linear control function in accordance with discrete parameter value, which gradually forms the problem of approximation in generalized sense.

It is theoretically well known that neural network of MLP class possesses universal approximation opportunities, affording ground to introduce a specifically accurate arbitrary multidimensional function. So, there is a theoretical possibility to create neural network controller of direct control. Unlike indirect control technology, neural network must independently realize both model option and control synthesis (fig. 4):

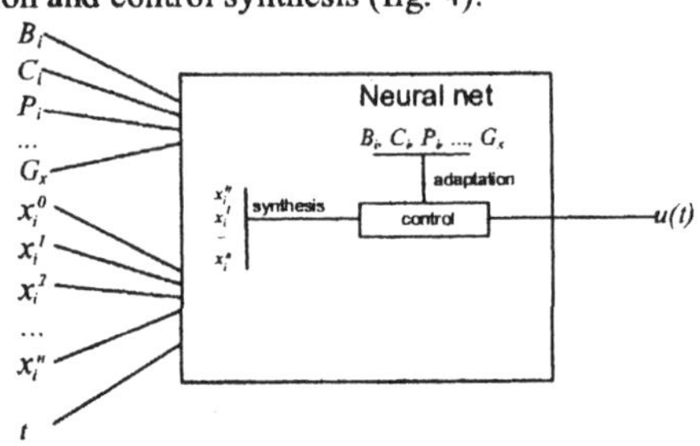

Fig. 4. The complex direct neural net control

Input parameters for such neural network are framed within the vector of identified computer model parameters and vector of non-identified parameters. To the group of identified parameters we can refer such indices as input butadiene, styrene and catalyst concentration; consumption, heat capacity and batch density; polymerization heat capacity; refrigerant consumption in reactors jacket; temperature of material flow getting into the reactor; cooler surface; reactors size and time. However, control function is the neural network output. The learning set makes the array of all dimensional vectors of the definition field of each input parameter.

Yet, within this approach appears a complex problem typical of neural networks. The fact is that neural network must synthesize the solution directed to unlimited time span $t \in [0; \infty]$. But in the classical interpretation such problem is insolvable for neural network since in this case it is impossible to specify the gradual learning set, which would seize the entire solution range. The suggested practical way out can be, for instance, to establish the solution in the form of finite series and to consider neural network output as coefficient within these series. In this context functions, subjected to approximation, belong to space C_m (*m* times differentiated functions zone), where to set its coefficient it is compulsory to know *m*-th derivatives of the sought function.

The introduction of differential equation solution in the form of Taylor's series does not require to fulfill the procedure of its numerical integration, since for *n*-th equation the first *(n-1)* coefficient is a primary condition, *n* coefficient is ingenuously obtained from differential equation: the rest coefficients are received by initial equation differentiation.

All in all, the algorithm of neural network controller (fig. 5) build up can be systematized as follows:

1. Input learning set creation as the sum total of the parameters unit, each of which is uniformly distributed within data domain.
2. Differential equation system solution for each set member.
3. Differential equation solution expansion into Taylor's series omitting the procedure of its numerical integration.
4. Solution comparison, obtained in 2-3 points where, in case of unsatisfactory result, it is urged to increase series remainders.
5. Output learning sample formation as a set of Taylor's series coefficients.
6. Neural network training on the basis of input and output learning sample.

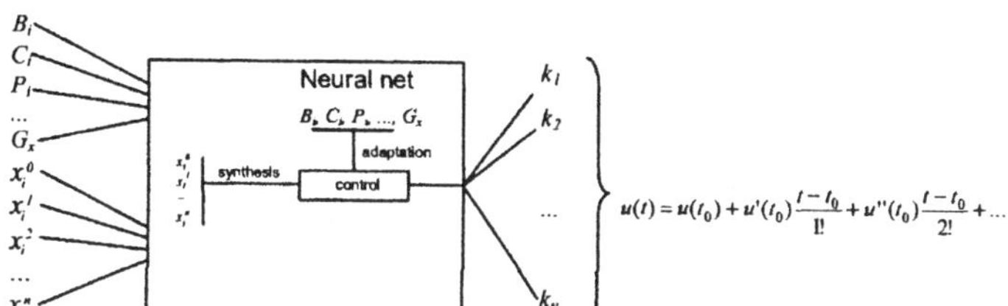

Fig. 5. The synthesis of the control by the neural net

To conclude, we have launched neural network controller in accordance with the above-established algorithm for rubber polymerization process. Presently, in the neural network theoretical interpretation it is hardly possible to expose analytical methods of exact net topology, whereas in practical realm we can trace some functional tips. So, to determine the quantity of neurons of the buried layer it is recommended to single out the next quality:

$$N_{hid} = \left[\frac{(N_p - 1)\cdot N_{out}}{N_{imp} + N_{out} + 1}\right], \tag{3}$$

where N_{imp}, N_{out} – input and output net dimension, N_p – training examples number.

Neural network, assigned to control neural network, comprises about 1000 links. Since the above-described controller construction technology requests a huge amount of calculating experiments, we have worked out algorithms of parallel neural network training.

As for the net on-line high speed, detecting direct signal makes a certain totality of fundamental operations of scalar product and non-linear function activation calculation which, despite net capacity, does not demand considerable calculation resources. Besides, neural networks embody structural concurrency; consequently even complex nets of direct distribution located in parallel processors are capable of functioning in in-line processing scale.

Another advantage of the elaborated neural network controller is the fact that it is open to be adjusted by the expert. While testing the controller, the expert is able to make correctional alterations, which can be recorded by neural network controller.

First experiments have revealed that equation synthesis by this controller is quite satisfactory through the entire parameters data domain, and the finite series of Taylor's series expansion is 2-3 remainders more than degree of the differential equation system, representing the computer process model.

CONCLUSIONS

From the perspective of complex adaptation controller construction, neural network apparatus application is a relatively effective mechanism, which basis is formed by structural computer model of the chemical-technological process. Neural network is trained to synthesis control impact during parametric model identification. Control can be performed as Taylor's series expansion, whereas neural network output is introduced by coefficients within these series remainders. Hereby, neural network synthesizes control of the entire parameter model data domain.

To sum up, one of the most significant advantages of neural network controller is that it provides the chance of maintaining extra flexible adjustment at the stage of practical testing – neural network output correction for specified parameters makes learning set precise and after some extra iteration neural network is able to record this modification.

ON VIRTUES AND VICES OF SECOND-ORDER MEASURES OF QUALITY FOR BINARY CLASSIFICATION

Fahed Abdallah, Cédric Richard, Régis Lengellé
Laboratoire de Modélisation et Sûreté des Systèmes (LM2S),
Université de Technologie de Troyes (UTT), France

ABSTRACT
When deriving a detector, we are often led to consider design criteria such as second-order measures of quality. The aim of this paper is to provide a critical overview of these criteria. We first consider the case of deriving unconstrained detectors. We show that second-order criteria must satisfy a non-trivial condition to yield Bayes-optimal receivers, and thus to be considered as relevant criteria for detector design. Next, we address the case where constraints are imposed on the detection structure, leading us to consider some set $\mathcal{C}$ of admissible detectors. In these conditions we prove that even if it exists a monotonic function of the likelihood ratio in $\mathcal{C}$, obtaining this statistic via the optimization of a second-order criterion, relevant or not, is not guaranteed. Finally, results are illustrated with simulation examples.

Introduction

Let (X, Y) be a pair of random variables taking their respective values from $\mathbb{R}^n$ and $\{0, 1\}$, where X is the observation and Y is the class. The purpose of detection is to determine to which of two classes $Y = 0$ or $Y = 1$ a given observation X belongs. According to classical statistical detection theories, comparing any strictly monotonic function of the likelihood ratio $L(X)$ with a threshold value is the optimum test [7]. In practical applications, implementing such a test may be impossible because of incomplete specification of the conditional probability densities $p(X|Y = 0)$ and $p(X|Y = 1)$, denoted by the standard notations $p_0(X)$ and $p_1(X)$, respectively. Therefore we are often led to consider alternative design criteria such as second-order measures of quality. These criteria are easy to use since they only depend on first and second-order moments of the statistics S to be sought [3, 4]. A wide variety of second-order measures of performance have been proposed and several contributions have been presented to prove their efficiency, e.g., [5] and references therein. In particular, some of these criteria guarantee the best solution in the Bayes sense since there optimization leads to a monotonic function of the likelihood ratio, as has been shown for well-known criteria such as Fisher crterion, mean-square error and signal-to-noise ratio.

The aim of this paper is to provide an overview of virtues and vices of second-order criteria. First, we consider the case of deriving unconstrained detectors. We show that second-order criteria must satisfy a non-trivial condition to provide Bayes-optimal receivers, and thus to be considered as relevant second-order criteria for detector design. Next, we address the case where constraints are imposed on the structure of the detector, leading us to restrict our attention to some set $\mathcal{C}$ of admissible detectors. In these conditions, we prove that even if it exists a monotonic function of the likelihood ratio in $\mathcal{C}$, obtaining this statistic via the optimization of a relevant criterion is not guaranteed. Finally, results are illustrated with simulations.

Characterization of relevant second-order criteria

Background and notations

Let $S(X) : \mathbb{R}^n \longrightarrow \mathbb{R}$ be an arbitrary measurable function and let $g : \mathbb{R}^n \longrightarrow \{0,1\}$ be the decision function based on $S(X)$:

$$g(X) = \begin{cases} 1 & \text{if } S(X) > \nu \\ 0 & \text{otherwise,} \end{cases} \tag{1}$$

which errs on X if $g(X) \neq Y$. Bayes decision theory leads to the result of major importance that the optimum detector g^* in the sense that it minimizes the risk consists in comparing the likelihood ratio $L(X) \triangleq p_1(X)/p_0(X)$ with a given threshold ν in order to make a decision [7]. Note that g is equivalent[1] to the Bayes-optimal detector g^* if $S(X) = \Phi\{L(X)\}$, where Φ is any monotonic function. Since the implementation Bayes test may be impossible in many practical applications, we are often led to consider simpler procedures for designing (1). In particular, one can use alternative design criteria such as second-order measures of performance. These criteria are defined in terms of first and second-order moments of the statistics $S(X)$, namely $m_i \triangleq \mathrm{E}\{S \mid Y = i\}$, $\sigma_i^2 \triangleq \mathrm{Var}\{S \mid Y = i\}$, with $i \in \{0,1\}$. There have been many contributions to explore, individually, the relevancy of well-known second-order criteria such as Fisher, mean-square error and signal-to-noise ratio (see, e.g., [5] and references therein). We show in the next subsection that any $\Psi(m_0, m_1, \sigma_0^2, \sigma_1^2)$ must satisfy a non-trivial condition to provide Bayes-optimal detectors, and thus to be considered as a *relevant second-order criterion* for detector design.

Relevant second-order criteria

Let Ψ be any function of $m_i \triangleq \int S(X)\, p_i(X)\, dX$ and $\sigma_i^2 \triangleq \int (S(X) - m_i)^2\, p_i(X)\, dX$, where $S(X)$ denotes any decision statistic. We first have to characterize statistics $S(X)$ which optimize Ψ. Operating on Ψ with a variational calculus, we obtain $\delta\Psi = \frac{\partial\Psi}{\partial m_0}\,\delta m_0 + \frac{\partial\Psi}{\partial m_1}\,\delta m_1 + \frac{\partial\Psi}{\partial\sigma_0^2}\,\delta\sigma_0^2 + \frac{\partial\Psi}{\partial\sigma_1^2}\,\delta\sigma_1^2$. Since $\delta m_i = \int \delta S(X)\, p_i(X)\, dX$ and $\delta\sigma_i^2 = 2\int (S(X) - m_i)\,\delta S(X)\, p_i(X)\, dX$ with $i \in \{0,1\}$, we obtain

$$\delta\Psi = \int \Bigg[\frac{\partial\Psi}{\partial m_0}\, p_0(X) + \frac{\partial\Psi}{\partial m_1}\, p_1(X) + 2\,(S(X) - m_0)\,\frac{\partial\Psi}{\partial\sigma_0^2}\, p_0(X) + 2\,(S(X) - m_1)\,\frac{\partial\Psi}{\partial\sigma_1^2}\, p_1(X) \Bigg]\, \delta S(X)\, dX. \tag{2}$$

To make $\delta\Psi = 0$ regardless of $\delta S(X)$, the $[\cdot]$ term given above must be equal to 0. We finally get the expression of the statistic $S(X)$ optimizing Ψ as a function of the likelihood ratio

$$S(X) = -\frac{1}{2}\,\frac{\frac{\partial\Psi}{\partial m_0} + \frac{\partial\Psi}{\partial m_1}\, L(X)}{\frac{\partial\Psi}{\partial\sigma_0^2} + \frac{\partial\Psi}{\partial\sigma_1^2}\, L(X)} + \frac{m_0\,\frac{\partial\Psi}{\partial\sigma_0^2} + m_1\,\frac{\partial\Psi}{\partial\sigma_1^2}\, L(X)}{\frac{\partial\Psi}{\partial\sigma_0^2} + \frac{\partial\Psi}{\partial\sigma_1^2}\, L(X)}. \tag{3}$$

The above statistic $S(X)$ leads to a Bayes-optimal detector if, and only if, it is a strictly monotonic function of $L(X)$. Therefore, the first order derivative of $S(X)$ with respect to $L(X)$ must be $\neq 0$. This result leads directly to the following proposition [8].

Proposition 1. *$\Psi(m_0, m_1, \sigma_0^2, \sigma_1^2)$ is a relevant second-order criterion Ψ_R, i.e., it guarantees the best solution in the Bayes sense, if and only if*

$$(m_1 - m_0)\,\frac{\partial\Psi}{\partial\sigma_0^2}\,\frac{\partial\Psi}{\partial\sigma_1^2} + \frac{1}{2}\left(\frac{\partial\Psi}{\partial\sigma_1^2}\,\frac{\partial\Psi}{\partial m_0} - \frac{\partial\Psi}{\partial\sigma_0^2}\,\frac{\partial\Psi}{\partial m_1}\right) \neq 0. \tag{4}$$

[1]Throughout this paper, two detectors are said to be equivalent if their receiver operating characteristic are the same.

Since it is very difficult, if not impossible, to find the solutions of (4), the above property can only be used to test the relevance of any criteria $\Psi(m_0, m_1, \sigma_0^2, \sigma_1^2)$. However, note that (4) is a non-restrictive condition, i.e., there exists a broad class of second-order criteria that lead to Bayes-optimal detectors [1]. This result is one of the most interesting virtues of second-order criteria.

Constrained detector design using second-order criteria

As shown in the previous section, there exists a broad class of second-order criteria that lead to a detection statistic $S(X)$ equivalent to the likelihood ratio $L(X)$. However, implementing $S(X)$ remains an unsolved problem since it depends on the probability densities $p_0(X)$ and $p_1(X)$ via $L(X)$, which are unknown. Therefore, we are often led to consider the following strategy for deriving receivers [2]:

1. selecting a class $\mathcal{C}$ of detection statistics;
2. picking the statistic of $\mathcal{C}$ that optimizes a given measure of performance, e.g., a relevant second-order criterion.

Unfortunately, this approach does not necessarily provide a Bayes-optimal detector since it generally requires the optimum statistic (3) to be a member of $\mathcal{C}$. We shall now discuss this drawback in the case where $\mathcal{C}$ denotes the class of linear statistics.

Linear detectors design

Linear classifiers are the simplest one as far as implementation is concerned, and are directly related to many known techniques such as correlations and Euclidean distances [4]. We shall now show how second-order criteria can be used for designing linear detectors, which are defined as follows:

$$g(X) = \begin{cases} 1 & \text{if } S(X) = K^T X - \nu > 0 \\ 0 & \text{otherwise.} \end{cases} \tag{5}$$

Here K denotes the direction onto which any n-dimensional observation X is projected, and ν is the detector threshold. The conditional expected values and variances of $S(X)$ are given by $m_i = \mathrm{E}\{S \mid Y = i\} = K^T M_i - \nu$ and $\sigma_i^2 = \mathrm{Var}\{S \mid Y = i\} = K^T \Sigma_i K$, where M_i and Σ_i are the conditional expected vectors and covariance matrices of X. Let Ψ be any second-order criterion. The optimal statistic $S(X)$ is given by equating to zero the partial derivatives of Ψ with respect to K and ν. As shown in [4, pp. 133-4] and [6], solving this linear system leads directly to the following proposition.

Proposition 2. *Let $S(X) \triangleq K^T X + \nu$ be any linear decision statistic. The optimum projection vector K under which the maximum value of any second-order criteria Ψ is reached satisfies*

$$K_\rho = [\rho\Sigma_0 + (1-\rho)\Sigma_1]^{-1}[M_1 - M_0], \tag{6}$$

where M_i and Σ_i are the conditional expected vectors and covariance matrices of X. The parameter ρ depends on the criterion Ψ as follows:

$$\rho = \frac{\frac{\partial \Psi}{\partial \sigma_0^2}}{\frac{\partial \Psi}{\partial \sigma_0^2} + \frac{\partial \Psi}{\partial \sigma_1^2}}. \tag{7}$$

The optimum projection direction K_ρ depends on Ψ through a single parameter $\rho \in]-\infty, +\infty[$. Then the latter can chosen so as to optimize the performance of the detector. Note that $\rho \in [0, 1]$ if, and only if, $\partial\Psi/\partial\sigma_0^2$ and $\partial\Psi/\partial\sigma_1^2$ are of the same sign (**Property 1**). This condition means that Ψ varies in the same way with σ_0^2 and σ_1^2, which is a desirable but non-mandatory requirement for design criteria. Let us now concentrate on $K_{-\infty}$ and $K_{+\infty}$. They are both proportional to $[\Sigma_0 - \Sigma_1]^{-1}[M_1 - M_0]$ since we have

$$K_{\pm\infty} \propto \lim_{\rho \to \pm\infty} \frac{1}{\rho}[\Sigma_0 - \Sigma_1]^{-1}[M_1 - M_0]. \tag{8}$$

up to a scaling factor such that $\|K_{\pm\infty}\| = 1$. The projection directions $K_{-\infty}$ and $K_{+\infty}$ then lead to equivalent detection structures (**Property 2**).

In the following, Proposition 2 and the above properties are illustrated through some classic detection problems. We also show that (relevant) second-order criteria does not guaranty an optimal detector in the Bayes sense if we restrict the solution space to a specific class $\mathcal{C}$ of detectors. Here this drawback is illustrated through the following situations:

Scenario 1: the optimization of any second-order criteria in $\mathcal{C}$ leads to a Bayes-optimal detector,

Scenario 2: there exist second-order criteria that provide Bayes-optimal receivers in $\mathcal{C}$,

Scenario 3: there exists a Bayes-optimal detector in $\mathcal{C}$ but it cannot be reached by optimizing any second-order criteria,

where $\mathcal{C}$ denotes the class of linear detectors (5).

Scenario 1: case of normal distributions with equal covariances

When $p_0(X)$ and $p_1(X)$ are normal with expected vectors M_0 and M_1 and covariance matrices Σ_0 and Σ_1, it is well-known that the Bayes-optimal statistic is given by:

$$S(X) = \frac{1}{2}(X - M_0)^T \Sigma_0^{-1}(X - M_0) - \frac{1}{2}(X - M_1)^T \Sigma_1^{-1}(X - M_1). \tag{9}$$

This equation shows that the decision boundary is a quadratic form in X. When $\Sigma_0 = \Sigma_1 = \Sigma$, the boundary becomes a linear function of X as

$$S(X) = (M_1 - M_0)^T \Sigma^{-1} X. \tag{10}$$

Eq. (10) indicates that the direction onto which any X is projected is given by $K^* = \Sigma^{-1}(M_1 - M_0)$. Let us now determine the projection direction under which the optimal value of any second-order criterion is reached. Eq. (6) gives:

$$K_\rho = \Sigma^{-1}(M_1 - M_0). \tag{11}$$

Comparing the projection directions K^* and K_ρ, we immediately conclude that they both correspond to equivalent detection structures. This very simple example shows that any second-order criterion, relevant or not, can sometimes lead to a Bayes-optimal receiver. However, such a success is not always guaranteed as illustrated in the next subsection, even if it exists a Bayes-equivalent receiver in $\mathcal{C}$.

Scenarios 2 and 3: case of exponential distributions

Let us consider that the components X_j of X are exponentially distributed and mutually independent. Then we have:

$$p_i(X) = \prod_{j=1}^{n} \frac{1}{\lambda_{ij}} \exp\left(-\frac{1}{\lambda_{ij}} X_j\right) u(X_j), \qquad i \in \{0, 1\}, \tag{12}$$

with λ_{ij} the parameter of the exponential distribution of the random variable X_j, and $u(.)$ the step function. It can easily be shown that the linear function given below is the Bayes-optimal detection statistic:

$$S(X) = \sum_{j=1}^{n} \left(\frac{1}{\lambda_{0j}} - \frac{1}{\lambda_{1j}}\right) X_j. \tag{13}$$

Then it is associated with the following Bayes-optimal projection direction:

$$K^* = \left(\frac{\lambda_{11} - \lambda_{01}}{\lambda_{11}\lambda_{01}}, \ldots, \frac{\lambda_{1j} - \lambda_{0j}}{\lambda_{1j}\lambda_{0j}}, \ldots, \frac{\lambda_{1n} - \lambda_{0n}}{\lambda_{1n}\lambda_{0n}} \right). \tag{14}$$

The expected vector M_i and the covariance matrix Σ_i of X, which is exponentially distributed according to (12), are given by $M_i = (\lambda_{i1}, \ldots, \lambda_{ij}, \ldots, \lambda_{in})^T$ and $\Sigma_i = \text{diag}(\lambda_{i1}^2, \ldots, \lambda_{ij}^2, \ldots, \lambda_{in}^2)$. Applying (6) to determine the projection direction under which the optimal value of any second-order criterion is reached, we obtain:

$$K_\rho = \left(\frac{\lambda_{11} - \lambda_{01}}{\lambda_{11}^2 - \rho(\lambda_{11}^2 - \lambda_{01}^2)}, \ldots, \frac{\lambda_{1j} - \lambda_{0j}}{\lambda_{1j}^2 - \rho(\lambda_{1j}^2 - \lambda_{0j}^2)}, \ldots, \frac{\lambda_{1n} - \lambda_{0n}}{\lambda_{1n}^2 - \rho(\lambda_{1n}^2 - \lambda_{0n}^2)} \right). \tag{15}$$

Comparing (14) and (15) shows that the collinearity of K^* and K_ρ depends on ρ. We shall now illustrate Scenarios 2 and 3.

Consider the case of two-dimensional observations X with $\lambda_{11} \neq \lambda_{01}$ and $\lambda_{12} \neq \lambda_{02}$. Vectors K^* and K_ρ are collinear if, and only if,

$$\rho = \frac{\lambda_{02}\lambda_{12}\lambda_{11}^2 - \lambda_{01}\lambda_{11}\lambda_{11}^2}{\lambda_{02}\lambda_{12}(\lambda_{11}^2 - \lambda_{01}^2) - \lambda_{01}\lambda_{11}(\lambda_{12}^2 - \lambda_{02}^2)}. \tag{16}$$

This means that any second-order criterion Ψ guarantees the best solution in the Bayes sense if, and only if, it satisfies:

$$\frac{\frac{\partial \Psi}{\partial \sigma_0^2}}{\frac{\partial \Psi}{\partial \sigma_0^2} + \frac{\partial \Psi}{\partial \sigma_1^2}} = \frac{\lambda_{02}\lambda_{12}\lambda_{11}^2 - \lambda_{01}\lambda_{11}\lambda_{11}^2}{\lambda_{02}\lambda_{12}(\lambda_{11}^2 - \lambda_{01}^2) - \lambda_{01}\lambda_{11}(\lambda_{12}^2 - \lambda_{02}^2)}, \tag{17}$$

an example of which is the generalized signal-to-noise ratio Ψ_α with α given by (16):

$$\Psi_\alpha(S) = \frac{(m_1 - m_0)^2}{(1-\alpha)\sigma_1^2 + \alpha\sigma_0^2}. \tag{18}$$

Here m_i and σ_i^2 denote the conditional expected values and variances of S. This illustrates Scenario 2.

Consider now that X is a n-dimensional observation, with $(n > 2)$. Except for very particular cases, we can notice that it does not exist any ρ ensuring collinearity of K^* and K_ρ. This means that optimization of second-order criteria, relevant or not, does not necessarily lead to a Bayes-equivalent detector, even if there is one that is a member of $\mathcal{C}$. Figures 1-(a) and 1-(b) illustrate this situation, called Scenario 3, for three-dimensional observations X. The projection direction K_ρ is represented on the unit sphere as a function of $\rho \in]-\infty, +\infty[$. One can observe the collinearity of projection directions $K_{-\infty}$ and $K_{+\infty}$ (see Property 2). Finally, one can notice that there is no value ρ_0 such that K_{ρ_0} and K^* are collinear, which is one of the weaknesses of second-order criteria.

Conclusion

The theoretical results reported in this paper are concerned with the virtues and vices of second-order criteria used for detector design. First, we have given a necessary and sufficient condition for these measures of performance to guaranty the best solution in the Bayes sense when deriving unconstrained detectors. Next we have considered the case where constrained are imposed on the structure of detectors, leading us to restrict our attention to a class $\mathcal{C}$ of admissible detectors. We have shown that any second-order criterion, relevant or not, can sometimes lead to a Bayes-optimal receiver. However, such a success is far from assured in the general case, even if exists a Bayes-equivalent detector in $\mathcal{C}$.

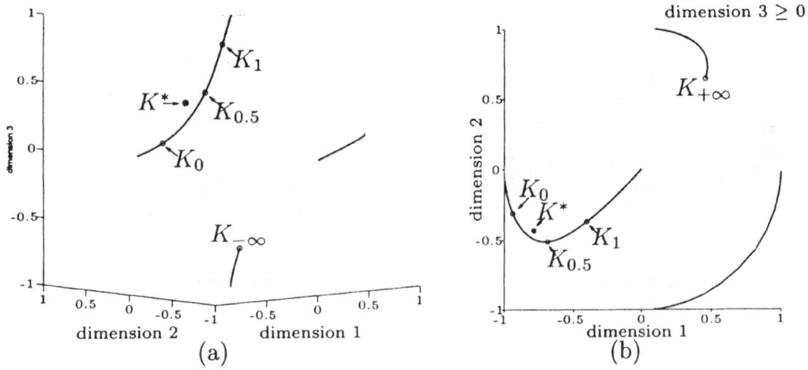

Figure 1: (a) Evolution of the projection direction K_ρ on the unit sphere as a function of $\rho \in]-\infty, +\infty[$, in the case of three-dimensional exponential distributions ($\lambda_{01} = 5$, $\lambda_{11} = 2$, $\lambda_{02} = 3$, $\lambda_{12} = 2$, $\lambda_{03} = 2$, $\lambda_{13} = 3$). The projection direction associated with the Bayes-optimal detector is referred to as K^* in this figure. (b) Overhead view of the unit sphere, i.e., dimension $3 \geq 0$.

References

[1] F. Abdallah, C. Richard and R. Lengell. On equivalence between detectors obtained from second-order measures of performance. *XI European Signal Processing Conference*, September 3-6, 2002, Toulouse, France.

[2] L. Devroye, L. Gyorfi and G. Lugosi. *A probabilistic theory of pattern recognition*, Springer-Verlag, 1996.

[3] R. O. Duda and P. E. Hart. *Pattern Classification and Scene Analysis*, Wiley and Sons, 1973.

[4] K. Fukunaga. *Statistical Pattern Recognition.* Academic Press, 1990.

[5] W. A. Gardner. A unifying view of second-order measures of quality for signal classification. *IEEE Transactions on Communications*, vol. 28, no. 6, pp. 807–816, 1980.

[6] D. W. Patterson and R. L. Mattson. A method of finding linear discriminant functions for a class of performance criteria. *IEEE Transactions on Information Theory*, vol. 12, no. 3, pp. 380-387, 1966.

[7] H. V. Poor. *An introduction to signal detection and estimation.* Springer-Verlag, 1994.

[8] C. Richard, R. Lengell and F. Abdallah. Bayes-optimal detectors design using relevant second-order criteria. *IEEE Signal Processing Letters*, vol. 9, no. 1, 2002.

GRANULAR SETS AND GRANULAR RELATIONS FOR ALGEBRAIC KNOWLEDGE MANAGEMENT

ANTONI LIGĘZA
AGH – University of Science and Technology
Kraków, Poland

ABSTRACT
This paper presents a generic concept of granularity for efficient and flexible knowledge representation and manipulation in Knowledge Management Systems. The basic idea is that the notion of set may consist of basic elements grouped into bigger granules. In this way the set is structured into a semi-partition. A semi-partition of a set is any collection of (disjoint) subsets of it. Then, concept of granular set is defined. A granular set is formed from a universe set and a semi-partition defining granules of its elements. Further, the concept of granular relation is also defined and its use for knowledge representation is proposed. A granular relation can be used to implement granular relational database systems and tabular knowledge based systems. The proposed formalism can be used for knowledge representation in the form of Extended Attributive Decision Tables.

INTRODUCTION

Knowledge management becomes an active research area and domain of industrial applications. However, current approaches are based mostly on extensive use of Relational Database Systems (RDBS) (Conolly, 1999) while knowledge engineering techniques (Torsun, 1995) are still not in common use. This is because neither really efficient tools for knowledge management, i.e. comparable to RDBS, have been developed, nor standards for knowledge representation and manipulation language have been widely accepted.

Efficient knowledge representation for dealing with complex, realistic tasks, where symbolic reasoning can be mixed with use of numerical values seems to be still a hard task for knowledge engineers. Although a number of knowledge representation formalisms were developed (Torsun, 1995) it seems that no single formalism is capable of adequate knowledge representation in more sophisticated systems. One of the inherent problems is the lack of ability of dealing with different levels of details and allowing for abstraction of knowledge within the same formalism. Moreover, many knowledge representation formalisms offer poor and incomplete knowledge manipulation formalisms, at least in comparison to the relational algebra and SQL (Conolly, 1999), (Torsun, 1995).

In order to try to overcome this issue a wider study of hierarchical and granular knowledge representation formalisms has recently been undertaken. Some current research include object oriented and relational-object databases, various clustering algorithms, approaches based on rough sets (Pawlak, 1991) and indiscernibility relation (Peters, 2002a, 2002b), and various ideas of the so-called *qualitative approaches*.

The main aim of the paper is to put forward yet another concept of granularity of data and knowledge no matter of their particular form. A general definition of *granular sets* is proposed; it allows to consider arbitrary *granules* of the elements of such sets instead of too numerous and too detailed atomic elements. Using such sets the notion of *granular relation* can be defined. Further, an extended version of well-known and widely accepted *Relational Algebra Systems*, applied in *Relational Database Systems* can be developed.

Relational Algebra (Conolly, 1999) is a simple yet powerful knowledge representation and manipulation tool; however, in order to fit practical applications it needs some extensions. The proposed extensions are reduced to some minimal set of innovations, so the formalism is kept as close as possible to the original relational algebra. The results are called an *Extended Attributive Decision Table* (tabular systems) and *Granular Relational Algebra*. The paper is based on some former presentations (Ligęza, 2002a, 2002b, 2003); details on tabular systems can be found in (Ligęza, 2001).

MATHEMATICAL FOUNDATIONS

Consider some universe U of atomic objects, symbolic or numerical values of interest or even more abstract objects such as logical formulae. For example, this can be both a set of numbers or vectors, such as $\mathbf{R}$ or $\mathbf{R}^n$ or a finite or infinite set of symbolic names, such as list of all students of some university. The idea of granular knowledge representation put forward in this paper consists in considering some distinguished *subsets* of the universe instead of atomic elements.

Consider an example concerned with time-table development for a university or a school. First, there is certainly a finite set of students, say S. Instead of specifying for each student his personal schedule, the university authorities consider 'granules' of them, i.e. years, groups, etc. If S_1, S_2, and S_3 are the groups of the first year, then a granular structure $G(S) = (S, \{S_1, S_2, S_3\})$ can be considered useful when assigning classes to the students of the first year. Further, time is also considered granular – instead of precise exact time one would rather consider traditional intervals, such as lessons (e.g. 45 or 55 minutes each) or periods of the length 1h30min which form a frame for constructing the schedule. Let T be the discrete set of time values from 7:00 to 21:00, and let T_1=[8:00,9:30], T_2=[9:30,11:00], T_3=[11:00,12:30], T_4=[12:30,14:00], T_5=[14:00,15:30], T_6=[15:30,17:00], T_7=[17:00,18:30] and T_8=[18:30,20:00]. Some other sets, such as the set of professors P, the set B of rooms or the set of classes (subjects) C are considered here at the level of single items. To conclude, the basic idea is that certain elements of a set can be 'glued' to form groups, clusters or granules. This allows to refer to such more abstract objects at a higher level.

Now, consider a set V and several subsets of it, say $V_1, V_2, \ldots, V_k$.

Definition 1 *A* semi-partition *of V is any collection of its subsets $V_1, V_2, \ldots, V_k$. A semi-partition is normalized (in normal form) iff $V_i \cap V_j = \emptyset$ for all $i \neq j$.*

A semi-partition will be also called an *s-partition* or a *sigma-partition*, for short. An s-partition of V (or maybe in V) will be denoted as $\sigma(V)$. If not stated explicitly, all the considerations will concern normalized s-partitions.

In case of an ordered set R any s-partition of it is a set of disjoint intervals covered by R (the intervals are ordered), and in case of a nominal set N any s-partition of it is just a collection of disjoint subsets of N. The case of a lattice is specific; an s-partition of the partially ordered set L is $\sigma(L) = \{L_1, L_2, \ldots, L_k\}$, such that $L_1, L_2, \ldots, L_k \in L$ and for any $L_i \neq L_j$, $L_i \cap L_j = \bot$. For intuition, in case of lattice a semi-partition is formed by incomparable, separate elements of the set L.

For example, if K is any finite set, a lattice can be formed as $L = 2^K$ with the partial order relation being the set inclusion. In such a case an s-partition of L can be defined as any collection of mutually disjoint subsets of K.

If $\sigma(V)$ is an s-partition, $\sigma(V) = \{V_1, V_2, \ldots, V_k\}$, then the set of all the elements of V occurring in the s-partition $\sigma(V)$ will be denoted as $[\![\sigma(V)]\!]$ and called the *support* of the s-partition.

Definition 2 *The support of an s-partition $\sigma(V)$ is defined as $[\![\sigma(V)]\!] = V_1 \cup V_2 \cup \ldots \cup V_k$.*

A nice thing about s-partitions is that any family of subsets (i.e. any unnormalized s-partition) of some set V can be transformed into an equivalent (covering the same elements) s-partition of V (a normalized one). Let us consider an arbitrary collection of subsets of a finite set V, say $V' = \{V'_1, V'_2, \ldots, V'_m\}$ (not necessarily disjoint ones). By subsequent replacing any two sets V'_i and V'_j of V' (ones having nonempty intersection) with three sets: $V'_i \setminus V'_j$, $V'_j \setminus V_i$ and $V'_i \cap V'_j$ one can generate an s-partition $\sigma(V) = \{V_1, V_2, \ldots, V_k\}$ of V as the fixed point of this replacement operation.

More formally, for any two sets $V'_i, V'_j \subseteq V'$ one may define the following partition generation operation as $V'_i \sqcap V'_j = \{V'_i \setminus V'_j, V'_j \setminus V_i$ and $V'_i \cap V'_j\}$. Further, $\Pi(V')$ can be defined as $\Pi(V') = \{V_1, V_2, \ldots, V_k\}$, where $V_i \cap V_j = \emptyset$ for $i \neq j$, $V_1 \cup V_2 \cup \ldots \cup V_k = V'_1 \cup V_2 \cup \ldots \cup V'_m$, and $\Pi(V')$ is the fixed point of the $V'_i \sqcap V'_j$ operation over the subsets of V' and derived from them sets.

GRANULAR SETS AND GRANULAR RELATIONS

For a given set S a granular set over S can be defined as follows.

Definition 3 *A granular set $G(S)$ is a pair $G(S) = (S, \sigma(S))$, where $\sigma(S)$ is any s-partition defined on S. If the s-partition $\sigma(V)$ is unnormalized, then the granular set is also determined as unnormalized one.*

The set S is called the domain *of the granular set, while the s-partition $\sigma(S)$ defines the so-called* signature of granularity.

Using the idea of granular set a granular relation can be defined in a straightforward way. Consider some collection of sets $D_1, D_2, \ldots, D_n$. Let there be defined some granular sets on them, i.e. $G_1 = (D_1, \sigma_1(D_1)), G_2 = (D_2, \sigma_2(D_2)), \ldots, G_n = (D_n, \sigma_n(D_n))$.

Definition 4 *A* granular relation *$R(G_1, G_2, \ldots, G_n)$ is any set R_G such that $R_G \subseteq U_G$ where*

$$U_G = \sigma_1(D_1) \times \sigma_2(D_2) \times \ldots \times \sigma_n(D_n). \tag{1}$$

The set U_G will be referred to as granular universe *or* granular space. *If at least one of the granular subsets was unnormalized, the relation is also said to be unnormalized one.*

The elements (rows) of a granular relation will be called *boxes*. Note that in fact a granular relation defines a kind of meta-relation, i.e. one based on sets instead of single elements. In fact, if R is a relation defined as $R \subseteq D_1 \times D_2 \times \ldots \times D_n$, then any tuple of R is like a thread in comparison to elements of R_G which are like a cord or a pipe.

Return for a while to the example with designing a time-table for a university or school. For simplicity, we focus on the schedule for some specific day, so the problem is to assign each group a professor, room and subject for any legal time interval. If $P = \{p_1, p_2, p_3\}$ is the set of professors, $B = \{b_1, b_2\}$ is the set of rooms and $C = \{c_1, c_2, c_3\}$ is the set of subjects, the relation representing the schedule can be as shown in Fig. 1.

For example, in Fig. 1 one can see a granular relation defined by the following tuples: $\{(S_1, T_2, p_1, b_1, c_1), (S_2, T_6, p_2, b_2, c_2), (S_3, T_8, p_3, b_2, c_3)\}$. In the tuples of this relation only the first two elements of each tuple are granules; the other three are basic items.

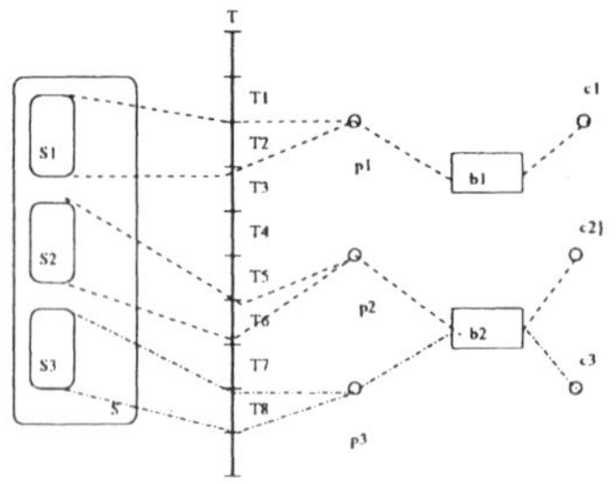

Figure 1: Two examples of granular sets.

EXTENDED ATTRIBUTIVE DECISION TABLES

Granular sets and granular relations can be applied to develop simple but powerful knowledge specification tool, i.e. the so-called *Extended Tabular Systems* or *Extended Attributive Decision Tables* (Ligęza, 1991). The scheme of a table consists of a sequence of (distinct) attributes, say $A_1, A_2, \ldots, A_n$, selected as common characteristics of data items. For any attribute A_i there is defined its domain D_j, for $j = 1, 2, \ldots, n$. The basic data and knowledge representation structure is just a table with the columns labeled with the attributes and the rows being records, as below:

$$\mathbf{T} = \begin{array}{|c|c|c|c|c|c|} \hline A_1 & A_2 & \ldots & A_j & \ldots & A_n \\ \hline t_{11} & t_{12} & \ldots & t_{1j} & \ldots & t_{1n} \\ \hline t_{21} & t_{22} & \ldots & t_{2j} & \ldots & t_{2n} \\ \hline \vdots & \vdots & \vdots & \vdots & \vdots & \vdots \\ \hline t_{i1} & t_{i2} & \ldots & t_{ij} & \ldots & t_{in} \\ \hline \vdots & \vdots & \vdots & \vdots & \vdots & \vdots \\ \hline t_{m1} & t_{m2} & \ldots & t_{mj} & \ldots & t_{mn} \\ \hline \end{array} \quad (2)$$

The above table **T** represents m uniformly described objects (entities) through their characteristics expressed with m *records*, one for one object. From mathematical point of view any such table represents a *granular relation*, as defined above. A constructed *ad hoc* table will probably be equivalent to some unnormalized granular relation. Such a relation can be transformed into a normalized, equivalent form with appropriate use of the Pi operation (Ligęza, 2002a,2002b).

In case of classical RDBs each $t_{ij} \in D_j$ is an *atomic value* of the appropriate attribute. This constraint defines the so-called *First Normal Form* (1NF) (Conolly, 1999). In classical RDBs the 1NF is required to establish the possibility to apply operations of relational algebra in a direct way; further, high computational efficiency can be assured. In the example below it is shown how an Extended Tabular Systems can be used to reduce knowledge specification.

AN EXAMPLE

After Pawlak (1991) let us consider the following example system tabular system. The attributes and their domains are as follows:

- A_1 := age; $D_1 = \{y, p, q\}$, where: y – young, p – pre-presbyotic, q – presbyotic,
- A_2 := spectacle; $D_2 = \{m, h\}$, where: m – myope, h – hypermyope,
- A_3 := astigmatic; $D_3 = \{n, y\}$, where: n – no, y – yes,

- A_4 := tear production rate; $D_4 = \{r, n\}$, where: r – reduced, n – normal,
- D := type of contact lenses (decision attribute); $D_D = \{H, S, N\}$, where: H – Hard contact lenses, S – Soft contact lenses, N – No contact lenses.

Now, consider the following Optician's Decision Table, being a perfect example of a atomic tabular system.

Number	Age	Spectacle	Astigmatic	Tear p.r.	Decision
1	y	m	y	n	H
2	y	n	y	n	H
3	p	m	y	n	H
4	q	m	y	n	H
5	y	m	n	n	S
6	y	n	n	n	S
7	p	m	n	n	S
8	p	n	n	n	S
9	q	n	n	n	S
10	y	m	n	r	N
11	y	m	y	r	N
12	y	n	n	r	N
13	y	n	y	r	N
14	p	m	n	r	N
15	p	m	y	r	N
16	p	n	n	r	N
17	p	n	y	r	N
18	p	n	y	n	N
19	q	m	n	r	N
20	q	m	n	n	N
21	q	m	y	r	N
22	q	n	n	r	N
23	q	n	y	r	N
24	q	n	y	n	N

This table can be regarded as a complete decision table (ready for use) or as a specification of some tabular rule based system. A brief analysis assures us that there are no redundant or subsumed rules. The system is deterministic and complete.

The table can be reduced to a form using granular relations – instead of atomic values of attributes subsets of domains are used. In order to perform reduction one has to search for certain rows identical apart values of a single attribute. For example, rows no. 1 and 2 of the initial table differs only on the second position, and the given values are the only ones, i.e. the completeness condition is satisfied. The two rows can be reduced to a single row of the table below (row no. 1). Reduction to complete domains (total reduction in which no subsets of the domains are admissible) leads to:

Number	Age	Spectacle	Astigmatic	Tear p.r.	Decision
1	y	_	y	n	H
2	_	m	y	n	H
3	y	_	n	n	S
4	p	_	n	n	S
5	_	n	n	n	S
6	_	_	_	r	N
7	p	n	y	_	N
8	q	m	n	_	N
9	q	n	y	_	N

If more specific granular tables are admitted, a further total reduction is possible; as the result the following table is obtained:

Number	Age	Spectacle	Astigmatic	Tear p.r.	Decision
1	y	_	y	n	H
2	_	m	y	n	H
3-4	{ y, p }	_	n	n	S
5	_	n	n	n	S
6	_	_	_	r	N
7-9	{ p, q }	n	y	_	N
8	q	m	n	_	N

The output table is in its minimal form (maximally reduced one). It is still complete (logically equivalent to its input form) and deterministic.

CONCLUSIONS

Extended Tabular Systems based on the idea of granular relations can constitute a tool for efficient knowledge representation and management. In order to complete this task algebraic operations, analogous to the ones of RDB algebra should be defined; some first proposals are given in (Ligęza, 2002a, 2002b). Further, analysis, verification, synthesis and inference issues are discussed in (Ligęza, 2001).

ACKNOWLEDGMENTS

Research carried out within AGH-UST internal grant No.: 10.10.120.105.

REFERENCES

Conolly, T., Begg, C., and Strachan, A., 1999, "Database Systems. A Practical Approach to Design, Implementation and Management", Addison-Wesley, Harlow, England.

Ligęza, A., 2001, "Toward logical analysis of tabular rule-based systems", *Int. J. of Intelligent Systems*, Vol. 16, No. 3, pp. 333-360.

Ligęza, A., 2002a, "Granular sets and granular relations: Towards a higher abstraction level in knowledge representation", *Intelligent Information Systems 2002*, M.A. Kłopotek, S.T. Wierzchoń and M. Michalewicz (Eds.), Physica-Verlag, A Springer-Verlag Company, Heidelberg, New York, pp. 331-340.

Ligęza, A., 2002b, "Granular sets and granular relations: An algebraic approach to knowledge representation and reasoning", *AI-Meth 2002 – Methods of Artificial Intelligence*, T.Burczyński, W. Cholewa and W. Moczulski (Eds.), Silesian University of Technology, Gliwice, Poland, pp. 47-54. Available at:
`http://www.ai-meth.polsl.gliwice.pl/AI2002.pliki/AIMETH_2002.htm`

Ligęza, A., 2003, "Dual resolution for logical reduction of granulartables", *Intelligent Information Processing and Web Mining 2003*, M.A. Kłopotek, S.T. Wierzchoń and K. Trojanowski (Eds.), Physica-Verlag, A Springer-Verlag Company, Heidelberg, New York, pp. 363-372.

Pawlak, Z., 1991, "Rough Sets. Theoretical Aspects of Reasoning about Data", *Kluwer Academic Publishers*, London.

Peters, J.F., Pawlak, Z., and Skowron, A., 2002a, "Clustering: A Rough Set Approach to Measuring Information Granules" *IEEE COMPSAC'02 Workshop on Data Mining and Rough Set Computing*, Oxford.

Peters, J.F., A. Skowron, Z. Suraj, W. Rząsa and M. Borkowski: Clustering: A Rough Set Approach to Constructing Information Granules. *SCDP'02*, Rzeszów, Poland, 2002b.

Torsun, I.S., 1995, "Foundations of Knowledge-Based Systems", *Academic Press*, London.

DISTRIBUTED EVOLUTIONARY PROGRAMMING FORMULATION OF SUPPORT VECTOR MACHINES (SVMS) FOR DETECTING AND CLASSIFYING ORGANOPHOSPHATE NERVE AGENTS

WALKER H. LAND, JR.,
Research Professor,
Dept. of Computer Science
Binghamton University
Binghamton, New York

OMOWUNMI SADIK
Associate Professor
Dept of Chemistry
Binghamton University
Binghamton, New York

MARK J. EMBRECHTS
Associate Professor
Dept. of Decision Sciences and Engineering Systems
Rensselaer Polytechnic Institute
Troy, New York

LUT WONG
Graduate Student
Dept. of Computer Science
Binghamton University
Binghamton, New York

ADAM WANEKAYA
Graduate Student
Dept. of Chemistry
Binghamton University
Binghamton, New York

MICHIKO UEMATSU
Graduate Student
Dept. of Chemistry
Binghamton University
Binghamton, New York

ABSTRACT

This paper extends the research described in Land et al. (2002) to include the following: (1) establishing SVM classification performance in the detection of organophosphates nerve agents using four separate model stimulant compounds, and (2) upgrading the architecture of EP SVMs to run on multiple machines simultaneously by using a client-server architecture (Distributed EP). Due to the increased threats of chemical and biological weapons of mass destruction (WMD) by international terrorist organizations, a significant effort is underway to develop tools that can be used to detect and effectively combat biochemical warfare. Furthermore, recent events have highlighted awareness that chemical and biological agents (CBAs) may become the preferred, cheap alternative WMD, because these agents can effectively attack large populations while leaving infrastructures intact. Despite the availability of numerous sensing devices, intelligent hybrid sensors that can detect and degrade CBAs are virtually nonexistent.

INTRODUCTION

The US is experiencing increasing threats of chemical and biological WMDs from international terrorist organizations. The new Office of Homeland Security is seeking new technologies to combat these threats. Consequently, there is great interest in developing tools that can be used not only to detect but also to effectively combat biochemical warfare. Recent events in the country have increased the awareness that CBA agents may become the preferred, cheap alternatives to weapons of mass destruction as these agents can effectively attack large populations while leaving infrastructures intact.

This paper examines the use of multiarray sensors coupled with support vector machines for the identification, detection, and classification of chemical warfare agents.

We demonstrate the effective use of the electronic nose coupled with SVM for the detection and classification of support organophosphate nerve agent stimulants, parathion and dichlorvos, dichlorvos and trichlorfon, and parathion and paraoxon, all of which are organophosphate (OP) nerve agents. These chemicals were paired based on their structural similarity. For example, the difference between paraoxon and parathion compounds is only the presence of the sulfur atom attached to the phosphorous for parathion, and the oxygen atom attached to the phosphorous for the paraoxon. Consequently, these to chemicals should be very difficult to correctly classify and will provide a good test for the SVM classification process.

AROMOSCAN ELECTRONIC NOSE

In recent years, a family of chemical sensors commonly referred to as electronic nose (EN) has been widely reported. A typical EN system consists of an array of chemical sensing elements coupled to headspace sampling, pattern recognition modules and appropriate transducer Pearce et al. (1993), Rose-Pehrsson (1988), Zellers (1995) and Grate (1993). The system employs metal oxides, quartz crystal arrays, surface acoustic wave devices, electrochemical cells and conducting polymers, or a combination of these sensors to mimic the human sense of smell.

The AromaScan EN consists of an array of 32 semi conductible polypyrrole sensors. Electricity is passed over this sensor array to provide a local environmental baseline reading. An analyte is exposed to the sensor array using a vapor generator, which controls concentration, exposure time as well as flow rate. The resultant DC resistance across each of the films constituting the array is sampled every 0.01 seconds. The values are then passed through an A to D converter and the result is then processed by Artificial Neural Network (ANN) classification software. The polypyrrole used by the AromaScan is capable of rapidly absorbing substances in addition to being electrically conductive. Consequently, absorption of any gas molecule will cause a physical "swelling", which leads to a linear increase in capacitance, thereby increasing the intensity of the signal when a current is "passed" over it.

The AromaScan software uses a conventional Artificial 3-layer Neural Network to perform classification. This is trained by backpropagation using the stochastic gradient ascent method. Elements of the validation set are then classified by this trained ANN. Such an ANN-based trained classifier has serious drawbacks, such as: (i) without a infinitely large training data set, the ANN is almost universally trained to a local minimum thereby providing substandard classification accuracy, and (ii) no known theory exists which will guarantee that the ANN is trained to a global minimum. This ANN will be used to establish the baseline classification accuracy of the existing AromaScan system in the experiments to be subsequently described. This ANN software is replaced by software implementing the distributed EP SVM architecture.

SUPPORT VECTOR MACHINES

Several references are available that provide extensive information about development of the mathematical foundation of SVMs, for example Vapnik (1998) and Schölkopf and Smola (2002). The key concepts of SVMs are generally covered in these references and will not be summarized here, except to explain the theoretical concept of why SVMs will provide a global minimum, whereas neural networks cannot.

Assume there exist N observations from the data set. Each observation (or training example) consists of a vector $\mathbf{x}_i$ containing the input pattern and a corresponding known classification y_i. The objective of the learning machine would be to formulate a mapping $\mathbf{x}_i \rightarrow y_i$. Now consider a set of functions $f(x, \alpha)$ with adjustable parameters α, that define a set of possible mappings $\mathbf{x} \rightarrow f(\mathbf{x}, \alpha)$. Here, $\mathbf{x}$ is given and α is chosen. In the case of a traditional neural network of fixed architecture, the α values would correspond to the weights and biases. The quantity R(α), known as the expected (or true) risk, associated with learning machines is defined as:

$$R(\alpha) = \int \frac{1}{2} \left| y - f(\mathbf{x},\alpha) \right| p(\mathbf{x},y)\, dx\, dy \tag{1}$$

where $p(\mathbf{x}, y)$ is an unknown probability density function from which the examples were drawn. This risk function is the expected (or true) value of the test (or validation) error for a trained learning machine. It may be shown that the best possible generalization ability of a learning machine is achieved by minimizing $R(\alpha)$, the expected (or true) risk. This generalization bound, for binary classification, holds with the probability of at least $1 - \eta$ ($0 \leq \eta \leq 1$) for all approximating functions that minimize the expected (or true) risk.

$$R(\alpha) \leq R_{\text{emp}}(\alpha) + \sqrt{\frac{h\left(\log(2N/h)+1\right) - \log(\eta/4)}{N}} \tag{2}$$

The first term on the right hand side of Eq. 2 is known as the "empirical risk". The empirical risk $R_{\text{emp}}(\alpha)$ is defined as

$$R_{\text{emp}}(\alpha) = \frac{1}{2N} \sum_{i=1}^{N} \left| y_i - f(\mathbf{x}_i, \alpha) \right| \tag{3}$$

This function is a measure of the error rate for the training set for a fixed, finite number of observations. This value is fixed for a particular choice of α and a given training set ($\mathbf{x}_i$, y_i). The second term in Eq. 2 is the *Vapnik-Chervonenkis (VC) confidence interval*, and is a function of the number of training samples N, the probability value η and the VC dimension h. The VC dimension is the *maximum* number of training samples that can be learned by a learning machine without error for *all* possible labelings of the classification functions $f(\mathbf{x}, \alpha)$, and hence is a measure of the capacity of the learning machine. In traditional neural-network implementations, the confidence interval is fixed by choosing a network architecture *a priori*. The function is minimized by generally obtaining a local minimum from minimizing the empirical risk through adjustment of weights and biases. Consequently, neural networks are trained based on the Empirical Risk Minimization (ERM) principle. In a SVM design and implementation, not only is the empirical risk minimized, the VC confidence interval also is minimized by using the principles of Structural Risk Minimization (SRM). Therefore, SVM implementations simultaneously minimize the empirical risk as well as the risk associated with the VC confidence interval, as defined in Eq. 2. Eq. 2 also shows that as $N \rightarrow \infty$, the empirical risk approaches the true risk because the VC confidence interval risk approaches zero. The reader may recall that obtaining larger and larger sets of valid training data would sometimes produce (with a great deal of training experience) a better performing neural network (NN), which resulted from classical training methods. This restriction is not incumbent on the SRM principle and is the fundamental difference between training NNs and training SVMs. Finally, because SVMs minimize the true risk, they provide a global minimum.

DISTRIBUTED EP SVM ARCHITECTURE

The architecture of EP SVM has been modified to run on multiple machines simultaneously by using a client-server architecture (distributed EP), because of the excessive time generally required to train the families of SVMs for these types of problems. The EP process used is described in Land et al. (2002). In this architecture, tasks are assigned by a central computer, called a server, to a group of computers connected to it, called clients. In the case of the EP configured SVM, each job refers to an incomplete population element. Therefore, when a client requests a job from the server, it will receive a population element from the current generation that it will work on and then will relay the results of its execution back to the server.

The execution of the system is as follows: First, a client connects to the server and requests a job. The server finds a population element that has not already been completed and sends the set of parameters that make up that specific population element to the client, who then disconnects. The client program then runs a copy of the SVM with the parameters obtained from the server. When the SVM outputs its results, the client program reconnects to the server and sends these results, along with what population elements the results belong to. The server then gives the client another job to process. When all population elements for a given generation are complete, the server "tells" any requesting clients to wait. While the clients wait, the server performs tournament selection and establishes the working population for the next generation. When this is complete, the next generation begins, and clients receive new jobs.

It is important to note that during the execution of this program care must be taken to ensure that each client possesses the exact same SVM. Training and validation data as are found in the server because: (1) the use of different SVMs could cause discrepancies in the results obtained, and (2) if a client is running its SVM on incorrect data, performance of the EP process will be degraded, as the process will attempt to find a single globally optimum solution for two different data sets.

EXPERIMENTAL RESULTS

Organophosphates (OPs) are among the most toxic substances and are thus commonly used as pesticides and nerve gases. Because of the inherent toxicity of these compounds, there is a considerable interest in the development of highly sensitive, selective, rapid and reliable devices for their detection. As stated, this research uses four OP pesticides as nerve agent stimulants. These were parathion, paraoxon, dichlorvos and trichlorfon (Fig. 1), which were evaluated by pair wise testing. The last two OPs to undergo pair wise comparison testing, and most difficult to classify, were paraoxon and parathion. These OPs were paired because of their structural similarity. As previously stated, the only difference in the structures of the 2 compounds is that the P=O bond in paraoxon is replaced by P=S bond in parathion. Trichlorfon and dichlorvos were also paired because of their structural similarities.

Parathion vs. Dichlorvos

Table 1 depicts the classification accuracy of these three kernels when compared to the AromaScan system, which was considered as the baseline performance system.

Table 1 shows that, when compared to the AromaScan baseline system: (1) a 23% improvement in the overall ROC Az index using the Schölkopf (S2000) kernel, with similar improvements with the Gaussian and polynomial (of degree 2) kernels, (2) a significant 173 % improvement in specificity with the S2000 kernel. This means that the

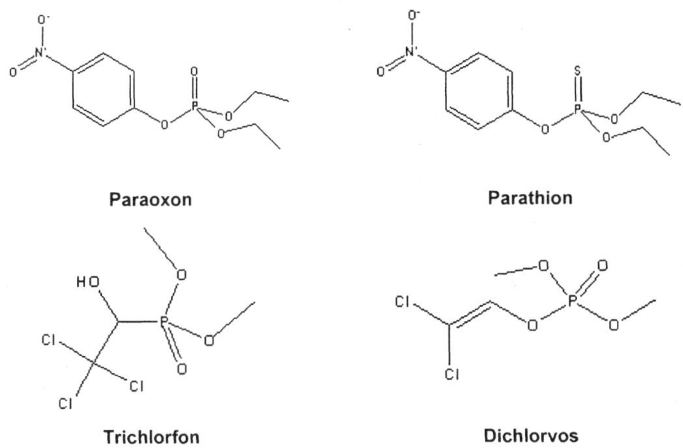

Figure 1. Organophosphate molecules

number of false negative errors were reduced by 173%, while making no false positive errors, when compared to the AromaScan base line performance. (3) The Gaussian and polynomial kernels demonstrated similar specificity at 100% sensitivity. (4) One would expect slightly better performance at 98% sensitivity, and this generally happens. However, because of the small validation sample size, the results were the same. These results will improve as larger sample sizes are processed.

Table 1. Classification Accuracy for the Parathion vs. Dichlorvos Pair

	AromaScan	s2000	%improv	rbf	%improv	Poly	%improv
Az	0.8021	0.9853	22.84	0.9444	17.75	0.9759	21.68
spec 100%	0.3280	0.8963	173.27	0.6834	108.36	0.7977	143.20
Ppv 100%	0.6460	0.9027	39.73	0.8685	34.44	0.8834	36.75
spec 98%	0.3280	0.8963	173.27	0.6834	108.36	0.7977	143.20
Ppv 98%	0.6460	0.9027	39.73	0.8685	34.44	0.8834	36.75

Dichlorvos vs. Trichlorfon

Table 2 depicts the SVM classification Measures of Performance (MOPs) for the dichlorvos vs trichlorfon pair. The results in this table show: (1) almost perfect (> 99%) ROC AZ values form all three kernels, (2) almost perfect partial Az values (Az values over the top 90% of the ROC curve), which are also > 99%, and (3) comparable (almost perfect) values for the specificity and PPV from 95-100% sensitivity. In summary, all three SVMs are essentially perfect classifiers for this chemical pair.

Table 2. Classification Accuracy for Dichlorvos and Trichlorfon

kernel	Az	Az90	100% spec	100% ppv	98% spec	98% ppv
rbf	0.9982	0.9943	0.9934	0.9934	0.9934	0.9933
S2000	0.9988	0.9959	0.9922	0.9942	0.9965	0.9968
poly	0.9999	0.9989	0.9965	0.9969	0.9965	0.9968

Parathion vs. Paraoxon

The reader will recall that the difference between paraoxon and parathion compounds is only the presence of the sulfur atom attached to the phosphorous for parathion, and the oxygen atom attached to the phosphorous for the paraoxon. Consequently, these to chemicals should be very difficult to correctly classify and will provide a good test for the SVM classification process. Table 3 depicts the classification results for this organophosphate pair. Given that only one atom differentiates these nerve agents, the results (using the three SVM kernels) depicted in Table 3 are very encouraging and show: (1) ROC Az indices from approximately 93% to greater than 99%, (2) partial Az values from ≈ 79% to 93%, (3) specificities from 76% to ≈ 84% at 100 and 98% sensitivity, and (4) PPVs from 73% to ≈ 84% at 100% and 98% sensitivities.

Table 3. SVM Classification Performance for the Parathion vs. Paraoxon Pair

kernel	Az	Az90	100% spec	100% ppv	98% spec	98% ppv
rbf	0.9275	0.7881	0.7633	0.7304	0.7633	0.7304
S2000	0.9844	0.9002	0.8701	0.8359	0.8701	0.8359
poly	0.9916	0.9344	0.8739	0.8366	0.8739	0.8366

CONCLUSIONS

This research demonstrates that for the *Parathion vs. Dichlorvos* pair, in comparison to the AromaScan baseline system: (1) a 23% improvement in the overall ROC Az index using the Schölkopf (S2000) kernel, with similar improvements with the Gaussian and polynomial (of degree 2) kernels, (2) a significant 173% improvement in specificity with the S2000 kernel. This means that the number of false negative errors were reduced by 173%, while making **no** false positive errors, when compared to the AromaScan base line performance. (3) The Gaussian and polynomial kernels demonstrated similar specificity at 100% sensitivity. All SVM classifiers provided essentially perfect classification performance for the *Dichlorvos vs. Trichlorfon* pair. For the most difficult classification task, the *Parathion vs. Paraoxon* pair, the following results were achieved (using the three SVM kernels): (1) ROC Az indices from approximately 93% to greater than 99%, (2) partial Az values from ≈ 79% to 93%, (3) specificities from 76% to ≈84% at 100 and 98% sensitivity, and (4) PPVs from 73% to ≈84% at 100% and 98% sensitivities. These are excellent results, considering only one atom differentiates these nerve agents.

Finally, the distributed architecture training times are a maximum of 60 times faster, when compared to standard iterative training time methods used in Land et al. 2002.

ACKNOWLEDGEMENTS

This work was performed under NSF grant no. CHE-0210968. The contributions of M. J. Embrechts were supported under NSF grant IIS-9979860. The authors also acknowledge the distributed architecture work of R. Velasquez, R. Gonzalez and R. Lewin.

REFERENCES

Grate J. W., Rose-Pehrsson S. L., Venezky D. L., Klusty M., Wohltjen H., 1993, *Anal. Chem.*, 65, p.1868.

Land, W.H. Jr., Sadik, O.A., Leibensperger, D. and Breimer, M., 2002, "Using Support Vector machines for Development and Testing Intelligent Sensors to Combat Terrorism," *Intelligent Engineering Systems through Artificial Neural Networks*, Eds, C. H. Dagli, et.al., pp. 811-816, ANNIE 2002

Pearce T. C., Gardner J.W., Friel S., Bartlett P. N., Blair N., 1993, *Analyst*, 118, p.371.

Rose-Pehrsson S. L., Grate J. W., Ballantine D. S. Jr., Jurs P. C., 1988, *Anal. Chem.*, 60, p.2801.

Schölkopf, B. and Smola, A.J., 2002, *Learning with Kernels*, MIT Press.

Zellers E. T., Batterman S. A., Han M., Patrash S. J., 1995, *Anal. Chem.*, 67, p.1092.

Vapnik, V.N., 1998, *Statistical Learning Theory*, Wiley, New York.

ROBUST SUPPORT VECTOR REGRESSION AND APPLICATIONS

THEODORE B.TRAFALIS
Laboratory of Optimization and Intelligent Systems
School of Industrial Engineering
University of Oklahoma
202 West Boyd, Room 124,
Norman, Oklahoma 73019
ttrafalis@ou.edu

SAMIR A. ALWAZZI
Laboratory of Optimization and Intelligent Systems
School of Industrial Engineering
University of Oklahoma
202 West Boyd, Room 124,
Norman, Oklahoma 73019
al7941@hotmail.com

Abstract
A robust convex optimization approach is proposed for Support Vector Regression with noisy input data. This approach deals with a "decision environment" which is characterized by crude knowledge of the uncertainty: all that is known about the uncertain data vector v, referring to a perturbation of an input data point, is that it belongs to a given uncertainty set U. We show that in the case that the set U is a hyper sphere; the Support Vector Regression is equivalent to a convex quadratic optimization problem with conic quadratic constraints. Preliminary computational results are also presented.

INTRODUCTION

Estimation of parameters in the presence of data uncertainties is a problem of significant practical importance and many estimators have been proposed in the optimization and statistics literature with the objective of handling modeling errors and measurement noise. Uncertainty is due to data collection, soft constraints (partial knowledge), model reduction (approximate modeling), and uncertainty about future (e.g in finance). Several optimization criteria have been also used like linear least–squares, regularized least-squares, ridge regression, total least squares and robust estimation (Lawson and Hanson, 1995; Golub and Van Loan, 1980, 1997; Ghaoui and Lebret, 1997; Huffel and Vandewalle, 1991). More recently, robust optimization techniques have been investigated by several authors (Ben-Tal and Nemirovski, 1996, 1998; Boyd, et al., 1998). These techniques are more meaningful in formulations with prior bounds on the size of the uncertainties on the data. The solutions coming from robust optimization models are more stable. Robust estimation is a problem that has been given a lot of attention in the statistics literature (Smola and Schhölkopf, 1998) with recent developments in robust regression models. These models usually assume noise in the functional values and not in the input data.

In this paper we consider the Support Vector Regression (SVR) problem with noise or uncertainty in the input data. Support Vector Machine (SVM) is a new type of learning that keeps the training error fixed (i.e. within given limits) while minimizing the confidence interval (it matches the machine capacity to data complexity). The resulting learning problem is a quadratic programming problem. Recently, a lot of work has also been done on formulating the SVM learning as a linear programming problem (Vapnik, 1998). Specifically, we consider the regression problem in which a response y is modeled using functions of input vectors x in R^d in the form $f(x)=\sum_j b_j \phi_j(x)+b$, where $\phi_1(x)$, $\phi_2(x)$ is a collection of possible nonlinear features. The traditional SVM solution takes b_j of the form $b_j=\sum_{i=1}^{l}\alpha_i \phi_j(x_i)$, where $x_1,\ldots, x_l$ are the observed input vectors in the data set of size l. Then the regression estimator takes the form:

$f(x) = \sum_{i=1}^{l} \alpha_i k(x, x_i) + b$, where the kernel $k(x, x_i) = \sum_j \phi_j(x)\phi_j(x_i)$, and the $\alpha_1, \ldots \alpha_l$ solve a quadratic optimization problem. Seeking sparsity, the l_1 norm may be used in place of the l_2 norm. Assuming the "noise" of each input data point to be constrained to a given sphere S, a robust optimization approach is proposed with the objective of developing a robust regressor with good generalization properties.

This paper is organized as follows. In section 2, the basic idea of support vector regression (SVR) is developed. Section 3 focuses on SVR with bounded error in the input data. Section 4 develops the robust of SVR for the nonlinear case. In section 5 preliminary computational results are provided and section 6 concludes the paper.

2 FUNDAMENTALS OF SUPPORT VECTOR REGRESSION

Consider a set of given training data $\{(x_1,y_1)\ldots(x_l,y_l)\}$, where l represents the number of samples or the training set size, where $x_i \in \Re^d$ and $y_i \in \Re$. Support Vector Regression uses the ε-insensitive loss function, (Vapnik, 1995, 1998), as a tool to find a function f (x) that has at most ε deviation from the original or actual obtained targets y_i for all the training data, and at the same time, is as flat as possible. It is assumed that the function f is linear in the feature space (ϕ space). Specifically $f(x) = \langle w, x \rangle + b$, where $\langle w, x \rangle$ denotes the dot product of the vectors *w* and *x*. A kernel function, $k(x, x')$, can be used to substitute $\langle x, x' \rangle$ in the nonlinear case. Then each data point x in the input space is mapped into a higher dimensional feature space using a feature map φ. In the new space, $\langle x, x' \rangle$ becomes $\langle \varphi(x), \varphi(x)' \rangle$. A nonlinear kernel function, $k(x,x')$, can be used to substitute $\langle \varphi(x), \varphi(x)' \rangle$. In the higher dimensional feature space, we can construct a linear regression function that represents a nonlinear function in the input space. In order to avoid any unnecessarily complicated notation, we will abbreviate φ(x) by **x** for the rest of the paper. Given a kernel K: $\mathbb{R}^d \times \mathbb{R}^d \rightarrow \mathbb{R}$ and a set of l points $x_1, \ldots, x_l$ in $\mathbb{R}^d$, we call the $l \times l$ matrix with, $K_{ij} = \langle \varphi(x_i), \varphi(x_j) \rangle = \langle \mathbf{x}_i, \mathbf{x}_j \rangle$, the kernel matrix of k at $x_1, \ldots, x_l$. There are many types of kernels to be used. The best one to substitute the dot product depends on the data. One has to do some experiments to find the best kernel function that fits the data.

Our objective is to find a w, small in terms of length. This can be accomplished by minimizing the norm of w (Smola and Schhölkopf, 1998). Thus, if the deviation between the actual and predicted value is smaller than ε, then the regression function is acceptable. Mathematically, we need to find a w in $\mathbb{R}^d$ such that

$$-\varepsilon \le \langle w, x_i \rangle - b - y_i \le \varepsilon,\ \forall i = 1, \ldots, l \text{ or } |y_i - \langle w, x_i \rangle - b| \le \varepsilon,\ \forall i = 1, \ldots, l.$$

Geometrically, this is a band or a tube of 2ε around the regression function points inside the tube are without error. Points outside the band are infeasible.

In Vapnik (1995), it can be seen that the SVR formulation, is described by the following primal optimization problem: $Min\ \frac{1}{2}\|w\|^2$ $\quad Min\ \frac{1}{2}\|w\|^2 + C\sum_{i=1}^{l}(z_i + \hat{z}_i)$

$$\text{st} -\langle w, x_i \rangle - b + y_i \le \varepsilon \quad (1) \quad \Rightarrow \quad \text{st } y_i - \langle w, x_i \rangle - b - z_i \le \varepsilon \quad (2)$$

$$\langle w, x_i \rangle + b - y_i \le \varepsilon \quad \forall i = 1, \ldots, l \qquad -y_i + \langle w, x_i \rangle + b - \hat{z}_i \le \varepsilon, \ , \ z_i, \hat{z}_i \ge 0.$$

(1) has been assumed to be feasible, but sometimes it is not, thus slack variables z and $\hat{z}$ are introduced for two kinds of training errors. They are different from zero for points outside the tube, and progressively decrease, until they reach the zero value for points inside the tube according to the loss function used (Smola and Schhölkopf, 1998). The above description is known as the ε-SV regression. (1) can be modified to cope with the infeasible constraints and it becomes as in (2). The above formulation can be used for the linear ε-insensitive loss function (Smola and Schhölkopf, 1998). The regression function f (x) corresponding to problem (2) is as follows:

$$f(x) = \langle w, x \rangle + b, \tag{3}$$

where x is in the φ space. One can compute the Lagrangian dual and by using kernel functions (Smola and Schhölkopf, 1998), a nonlinear regressor can then be constructed. The corresponding regression function to the nonlinear case is as follows:

$$f(x) = \sum \alpha_i k(x_i, x) + b \tag{4}$$

3. SUPPORT VECTOR MACHINE REGRESSION WITH BOUNDED ERRORS IN INPUT DATA

Using the above problem (2), we will follow the same logic as in the case of classification (Trafalis and Alwazzi, 2003), where the data points $\{x_i\}_{i=1,\dots,l}$ have some noise (uncertainty). Specifically, let $x_i = \tilde{x}_i + u_i$, *such that* $\|u_i\| \leq \sqrt{\eta}$, where u_i represents a bounded uncertainty and $\eta > 0$. By substituting this equality in the constraints of (2), then:

$$\begin{aligned} &y_i - \langle w, \tilde{x}_i + u_i \rangle - b - z_i \leq \varepsilon \\ &-y_i + \langle w, \tilde{x}_i + u_i \rangle + b - \hat{z}_i \leq \varepsilon \\ &z_i, \hat{z}_i \geq 0, \quad i = 1,\dots,l \end{aligned} \tag{5}$$

$$\Rightarrow \begin{aligned} &y_i - \langle w, \tilde{x}_i \rangle - \langle w, u_i \rangle - b - z_i \leq \varepsilon \\ &\langle w, \tilde{x}_i \rangle + \langle w, u_i \rangle + b - y_i - \hat{z}_i \leq \varepsilon \\ &z_i, \hat{z}_i \geq 0, \quad i = 1,\dots,l, \end{aligned} \tag{6}$$

$\forall\, u_i \in \Re^d$, such that $\|u_i\| \leq \sqrt{\eta}$. $\forall\, u_i \in \Re^d$, such that $\|u_i\| \leq \sqrt{\eta}$.

In order to satisfy inequalities (6) in a "robust way", we need to minimize the dot product of w and u_i subject to $\|u_i\| \leq \sqrt{\eta}$. Thus, we need to solve the following problem: $Min\langle w, u_i \rangle$ st $\|u_i\| \leq \sqrt{\eta}$, which provides the worst possible scenario for satisfaction of the inequalities in (6). Replacing the minimum of $\langle w, u_i \rangle$ by its value $-\sqrt{\eta}\|w\|$ in (6) then the robust problem of (2) is:

$$Min \frac{1}{2}\|w\|^2 + C\sum_{i=1}^{l}(z_i + \hat{z}_i)$$

$$\text{st} \quad \begin{aligned} &\sqrt{\eta}\|w\| - \langle w, \tilde{x}_i \rangle - b + y_i - z_i \leq \varepsilon \\ &-\sqrt{\eta}\|w\| + \langle w, \tilde{x}_i \rangle + b - y_i - \hat{z}_i \leq \varepsilon \\ &z_i, \hat{z}_i \geq 0, \quad i = 1,\dots,l,\ \forall u_i \in \Re^d, with \|u_i\| \leq \sqrt{\eta}. \end{aligned} \tag{7}$$

4. KERNELIZATION

In the previous section, the linear SVR problem was provided. In this section, the above problem can be reformulated to deal with non-linear regression problems. A general formulation will be given. As in the classification case (Trafalis and Alwazzi, 2003), the weight vector $\mathbf{w}=\sum_{i=1}^{l}\alpha_i \mathrm{x}_i$, and $\|\mathbf{w}\| = \sqrt{\alpha^t k \alpha}$. Now since then:

$\langle \mathrm{w}, \tilde{\mathrm{x}}_i \rangle = \sum_{j=1}^{l} \alpha_j k\langle \mathrm{x_j}, \tilde{\mathrm{x}}_i\rangle$. By setting $\mathrm{x_j} = \tilde{x}_j + u_j$, so $\langle \mathrm{x_j}, \tilde{\mathrm{x}}_i\rangle = \langle \tilde{\mathrm{x}}_j, \tilde{\mathrm{x}}_i\rangle + \langle \mathrm{u_j}, \tilde{\mathrm{x}}_i\rangle$.

Now $\langle \tilde{\mathrm{x}}_j, \tilde{\mathrm{x}}_i\rangle = \mathrm{k}(\tilde{x}_j, \tilde{x}_i)$, $\langle \mathrm{u_j}, \tilde{\mathrm{x}}_i\rangle = -\sqrt{\eta}\,\|\tilde{\mathrm{x}}_i\|$ and $\|\mathrm{u_j}\| \le \sqrt{\eta}$. Given all these, (7) can then be stated as:

$$
\begin{aligned}
& Min \frac{1}{2}\alpha^t k\alpha + C\sum_{i=1}^{l}(z_i + \hat{z}_i) \\
& st \;\; \sqrt{\eta}\sqrt{\alpha^t k\alpha} - \sum_{j=1}^{l}\alpha_j k(\tilde{x}_j, \tilde{x}_i) + (\sum_{j=1}^{l}\alpha_j\sqrt{\eta})\|\tilde{\mathrm{x}}_i\| - b + y_i - z_i \le \varepsilon \\
& -\sqrt{\eta}\sqrt{\alpha^t k\alpha} + \sum_{j=1}^{l}\alpha_j k(\tilde{x}_j, \tilde{x}_i) - (\sum_{j=1}^{l}\alpha_j\sqrt{\eta})\|\tilde{\mathrm{x}}_i\| + b - y_i - \hat{z}_i \le \varepsilon \\
& z_i, \hat{z}_i \ge 0,\; i = 1,\dots,l,\; j = 1,\dots,l,\; \alpha \text{ is unrestricted in sign.}
\end{aligned}
\tag{8}
$$

This model is the robust SVR with a non-linear kernel and its regression function is: $f(x,\alpha,b,u) = \sum_{i=1}^{l}\alpha_i\langle \tilde{\mathrm{x}}_i, \mathrm{x}\rangle + \sum_{i=1}^{l}\alpha_i\langle \mathrm{u}_i, \mathrm{x}\rangle + b.$ Note that $\langle \tilde{\mathrm{x}}_i, \mathrm{x}\rangle = k(\tilde{x}_i, x)$, and by Cauchy-Schwarz inequality we have: $|\langle \mathrm{u}_i, \mathrm{x}\rangle| \le \|\mathrm{u}_i\|.\|\mathrm{x}\|$. Replace $\langle \mathbf{u}_i, \mathrm{x}\rangle$ by its minimum value $-\sqrt{\eta}\,\|\mathrm{x}\|$ with $\|\mathrm{x}\| = \sqrt{k(x,x)}$ its robust estimator is

$$
\hat{f}(x,\alpha,b) = \sum_{i=1}^{l}\alpha_i k(\tilde{x}_i, x) - \sqrt{\eta}\,\|\mathrm{x}\|(\sum_{i=1}^{l}\alpha_i) + b, \tag{9}
$$

If $\eta=0$ is substituted in (8) and (9), then a precise non-linear SVR optimization problem with its regressor are obtained. If L_1 norm is used, then (8) will be transformed into a robust linear (precise If $\eta=0$), optimization problem but with the following different objective function: $Min\sum_{i=1}^{l}\alpha_i + C\sum_{i=1}^{l}(z_i + \hat{z}_i)$ and its regressor is as (9).

5. COMPUTATIONAL RESULTS

In our experimentations a Matlab code developed by the authors is used. The kernel is a polynomial of the form of $k(x,y) = (x^T y)^2$.

Synthetic Toy Problem, Description and Analysis

The data of this problem was generated according to the following equation: $f(x) = x^2_1 + x^2_2$. We randomly select 11 data sets each consists of 100 points. The uncertainty η describes perturbations of the input data by vectors of length of at most $\sqrt{\eta}$ .Table 1 shows the results.

Table 1. The Average Mean Square Error with different values for the Synthetic Problem

The Uncertainty Eita (η)	Average of Testing Mean square Error (mse) 70&30 pointsfor training &testing
0	1.69E-18
0.000001	2.00E-6
0.00001	2.00E-5
0.0001	2.00E-4
0.1	1.67E-01
0.5	5.00E-01
1	6.667E-01

We run 77 experiments (7 experiments for each data set). When the uncertainty value increases, the value of the MSE also increases in a continuous way as shown in Fig 1.

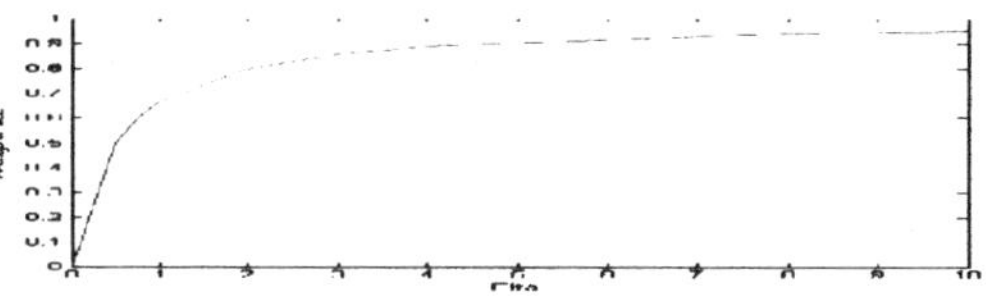

Figure 1. The Average testing mean square error versus the uncertainty eita.

Lynx Real World Problem - Description and Analysis

In this application, the total number of data points used is 85 (UCI Machine Learning Database).We performed five experiments with different values of η and C as shown in Table 2. Figure 2 provides the graphs of the predicted values, when η = 0 and C = 10. The testing MSE increases proportionally to changes of η .

Table 2. - Tests on the Lynx Regression Problem.

C	Epsilon	Eita	Training Error (mse) 51points	Testing Error (mse) 34points
10	0.5	0	0.6029	0.7604
100	0.5	0.0001	2.6153	5.1858
10	0.5	0.001	34.4571	36.3809
10	0.5	0.1	53.6306	128.5312

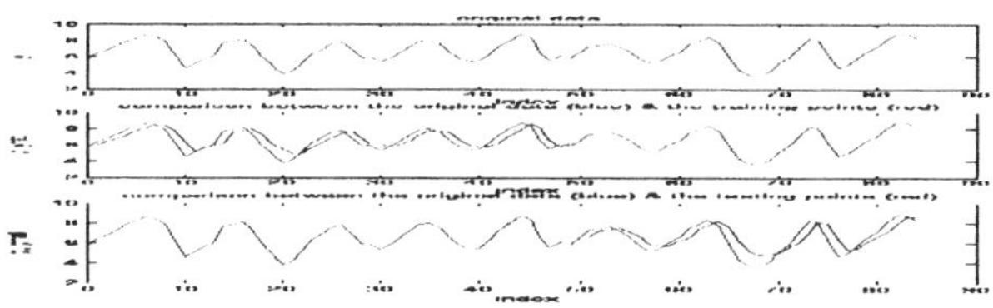

Figure 2. Comparison of Predicted ($\hat{y}$) and Target (y) values for the Lynx regression

6. CONCLUSION

In this work, we have developed a new approach for solving problems in regression analysis with spherical uncertainties in the input data. The behavior of the solutions was examined under uncertainties η . The contribution of this research was to investigate the robustness and stability of the SVR model under bounded perturbations of the input data. Preliminary experimentations have shown that robust techniques possess a good stability behavior and good potential generalization for predicting solutions under bounded perturbed data.

Acknowledgments: The present work has been partially supported by the NSF grants ECS-9978813, ECS-0099378.

Nomenclature

b: threshold
C: regularization/trade off
f(x) : regression estimator,
$\hat{f}$ (x) : robust regression estimator
k: kernel
l: number of samples or training set size
w: weight vector
$\tilde{x}$: center of the uncertainty sphere
α: auxiliary variable
ε :error/deviation from the actual target
η: uncertainty (eita)
$z_i, \hat{z}_i$: error variables/nonnegative slack
$\phi_j(x_i)$: mapping
u_i : radius of the uncertainty sphere
x_l: input parameters
y_l: output parameters
$\langle .,. \rangle$: dot product

REFERENCES

Ben-Tal, A. and A. Nemirovski, "Robust solutions to uncertain linear programs via convex programming," *Operations Research Letters*, vol.25:1, pp.1-17, 1996.

Ben-Tal, A. and A. Nemirovski, "Robust convex optimization," *Mathematics of Operations Research*, vol.23:4, pp. 769-805, 1998.

Boyd, S., M.S.Lobo, and L. Vandenberghe, "Applications of second-order cone programming," *Linear Algebra and its Applications*, vol.284, pp. 193-226, 1998.

Ghaoui, L.E. and H. Lebret, "Robust solutions to least- square problems with uncertain data," *SIAM J.Matrix Anal. Appl.*, 18, pp. 1035-1064, 1997.

Golub, G.H., "Some modified matrix eigenvalue problems," *SIAM Rev.*, 15, pp.318-344, 1973.

Golub, G.H. and C.F. Van Loan, "An analysis of the total least squares problem," *SIAM J. Numer. Anal.*, 17, pp.883-893, 1980.

Golub, G.H. and C.F. Van Loan, *Matrix Computations*, 3rd ed., Baltimore, MD, The Johns Hopkins University Press, 1997.

Huffel, S.V. and J. Vandewalle, *The Total Least Squares Problem: Computational Aspects and Analysis*, Philadelphia, PA, SIAM, 1991.

Lawson, C.L. and R.J. Hanson, *Solving Least-Squares Problems*, Philadelphia, PA, SIAM, 1995.

Smola, A. and B. Schhölkopf, A Tutorial on Support Vector Regression. *Statistics and Computing*. Also: Neuro COLT Technical Report NC- TR - 98-030, 1998.

Trafalis, T.B. and S. A. Alwazzi, "Robust Optimization in Support Vector Machine Training with Bounded Errors", Accepted Paper (in print), Proceedings IJCNN2003, Portland, Oregon, USA, July 2003.

UCI Machine Learning Data base ftp. ics. uci. edu /pub/machine-learning databases.

Vapnik, V., *The Nature of Statistical Learning Theory*, Springer Verlag, 1995.

Vapnik, V., *Statistical Learning Theory*, Wiley, 1998.

EXPLOITING MOBILE MULTI-AGENT SYSTEMS FOR MULTICLASS PROXIMAL SUPPORT VECTOR CLASSIFICATION

VIVEK MENON
Research Associate, Systems
Amrita Institute of Management
Amrita Vishwa Vidyapeetham
Coimbatore 641105
India
vivek_menon@amrita.edu

SHYAM DIWAKAR M.
Research Assistant, Centre for
Excellence in Computational
Engineering and Networking
Amrita Vishwa Vidyapeetham
Coimbatore 641105
India
m_shyam@amrita.edu

ABSTRACT
The concept of Proximal Support Vector Machines (PSVM) has redefined the functioning of support vector machine classifiers primarily with respect to the speed of classification of data. In one such technique, a data set of size n with r categories of a response variable is divided into r dyadic sets, with each dyadic set G comprising of x elements of a particular category and (n-x) elements from the remaining categories. Complexity and time required for data classification is directly proportional to the number of categories of response variable and size of data sets. We propose a Mobile Multi-Agent System (MMAS) approach to reduce the turnaround time by distributing the classification process to r agents spread across a network and working in parallel.

INTRODUCTION

A. Proximal Support Vector Machines

Support Vector Machines (SVM) are powerful tools for data classification (Mangasarian, 2001), which classify 2-category points by assigning them to one of the two disjoint half-spaces in either the original input space of the problem for linear classifiers or in a higher dimensional feature space for nonlinear classifiers. SVMs (Burbidge and Buxton, 2001) lie at the intersection of learning theory and practice. In other words, they are learning systems that use a hypothesis space of linear functions in a high dimensional feature space. Trained with a learning algorithm from optimization theory, SVMs implement a learning bias derived from statistical learning theory. They construct complex models that are easily amenable to mathematical analysis. Recently, much simpler classifiers such as the Proximal Support Vector Machines (Fung and Mangasarian , 2001) were implemented wherein each class of points is assigned to the closest of two parallel planes (in input or feature space) and are pushed apart as far as possible. This formulation, which can also be interpreted as regularized least squares or in the more general context of regularized networks, leads to an extremely fast and simple algorithm for generating a linear or nonlinear classifier. The classifier is obtained by solving a single system of linear equations.

Consider now the optimization function,

$$\min_{(\omega,\gamma,y)\in R^{n+1+m}} \gamma e'y + \frac{1}{2}\omega'\omega \tag{1}$$

Here (1/2) ω' ω is the reciprocal of the distance between the two hyperplanes

$$\begin{aligned} x'\omega &= \gamma + 1 \\ x'\omega &= \gamma - 1 \end{aligned} \tag{2}$$

And γ **e'y** is γ times the total error i.e. γ ($y_1 + y_2 + y_3 \ldots + y_m$). **ω** is the normal to the bounding planes:

The Eq. (1) bounds most of the sets as A+ and A- respectively. The constant determines their location relative to the origin. To generate a non-linear classifier, replace data A by K, where K is the non-linear kernel K (A, A').

The arena of multiclass PSVM algorithms (Shigeo, 2003) has seen various styles, the most popular being the splitting of a n-class problem into n (n-1)/2 two-class problems (Kreßel, 1999). A one-against-all method has been adopted to distribute the classification problem as r two-class cases G $\{g_1, g_2, g_3 \ldots g_r\}$ as needed by the PSVM classifier.

B. Software Agents

The concept of a software agent is straightforward and natural. An agent is a computer system that is capable of independent action on behalf of its user or owner. An agent can solve for itself what it needs to do in order to satisfy its design objectives, rather than having to be told explicitly what to do at any given moment. A multi-agent system as the name suggests is a collection of agents that work together for accomplishing the goals or objectives set by the owner of the agent (Wooldridge, 2002). These multi-agents may be located on the same network or spread across different geographically separated networks. They collaborate with each other by exchanging messages amongst themselves across the network. In the world of software agents, although mobility of code and state is not necessarily a required property, it is more of an advantage than an overhead in majority of the systems. These mobile agents are a class of software agents that have the distinctive capability to transport themselves from one network to another and thus move to a node that contains other agents or services like a database with which they can interact locally. They operate asynchronously, autonomously, reduce network traffic and provide an effective means of overcoming network latency (Lange and Oshima, 1998).

SCOPE AND MOTIVATION

Traditional distributed applications are designed as a set of processes assigned to a given execution environment (acting as managers of local resources) that cooperate in a (mostly) network unaware fashion (Adler, 1995). Instead, the mobile agent paradigm defines applications composed by network-aware entities (agents) capable of actively changing their execution environment by transferring themselves while executing (mobility) (Fuggetta et.al, 1998). Of the multiple approaches to multi-class prediction using PSVM, one technique stands out in terms of efficiency and ingenuity of operation. In this technique, a data set of size n with r categories of a response variable is divided into r dyadic sets, with each dyadic set G comprising of x elements of a particular category and (n-x) elements from the remaining categories. In this setting, complexity and hence turnaround time increases as the number of response variable categories and size of data sets increase. Most real world classification problems fall in the above category. An efficient solution to such problems can focus on either optimizing the speed of execution or the classification algorithm. It is in this context thtat we propose a Mobile Multi-Agent System (MMAS) approach to boost the speed of execution. The MMAS approach distributes the classification process to r agents working in parallel and spread across a network, thus effectively reducing the turnaround time. The case of multi-class PSVM is a data-intensive application where the data is remotely located, is owned by the remote service provider, and the user has specialized needs (Milojicic, 1999). Here, the user sends r agents to r nodes hosting redundant copies of the same data file. The price paid for redundancy of data files (memory) is offset by a reduced turnaround time for classification (speed).

APPROACH

The normal method of implementation for a multi-class PSVM classification problem works sequentially on a data set, implying that each of the r dyadic sets of G $\{g_1, g_2, g_3, \ldots g_r\}$ is probed one after the other to yield r different classifications. Assuming that each classification run on the data set lasts $t_1, t_2, t_3 \ldots t_r$ seconds respectively, the total time (T) for classification would be

$$T = \sum_{i=1}^{r} t_i \tag{3}$$

Each of the above classification runs over the data set is in effect the application of the same algorithm with generic modifications for each run, which are independent of the previous runs. Thus, we can construe that an r response variable classification problem is an inherently parallel task that can distribute across r nodes in parallel. Theoretically, the best possible overall turnaround time T_p should be

$$T_p = max\,(t_1, t_2, t_3 \ldots t_r) + \chi \tag{4}$$

where χ is the network turnaround time associated with the agent that takes the maximum time to communicate a classification result. Although in practice this may not be realized due to overheads involved in distributing the task and network delays in communication of results, the savings in turnaround time are immense and of the order r. This is a typical problem that fits into the domain of Cooperative Distributed Problem Solving (CDPS) (Durfee, 1999), as the agents in the system implicitly share a common goal. This generous assumption implies that the primary concern is the overall system objectives whilst the individual objectives of agents are secondary. Since the agents in the approach would be invariably 'owned' by a single entity, the overheads in design and implementation are cut down. If we can assume that all the agents need to be concerned about is the overall utility of the system, then we can design the overall system so as to optimize this (Wooldridge, 2002). Although it is also important to distinguish CDPS from parallel problems solving (Bond and

Gasser, 1988) the distinction blurs in certain problem domains, especially when agents are involved. With respect to the three-tier architecture of CDPS viz., Problem Decomposition, Sub-problem solution and Answer synthesis (Smith and Davis, 1980), this problem can be viewed wholly as a task sharing activity. Hence the intermediate sub-problem solution stage does not suffer from any time delays on account of any agent communications as results sharing amongst the remote agents is absent. Task sharing in this case is unambiguous, and the problem of multi class PSVM can be decomposed into sub-tasks that can be assigned to homogenous individual agents (slaves), which follow a centralized plan developed by the 'Master' agent. Under the Master-Slave pattern, the 'Master' agent distributes the plan (classification task) to the 'Slaves' who execute their part of the plan (Lange and Oshima, 1998) i.e., the corresponding classification run over the data set and get back with the classification results to their 'Master' agent. Thus the 'Master' agent plays an important role in initial problem decomposition and also in the final answer synthesis by collating the classification results from the 'Slave' agents executing on the remote nodes.

IMPLEMENTATION

From a range of possible options (Dantas et al, 2002), we chose to implement our agent model using the Java based IBM Aglets framework (ASDK) initially developed by the IBM Tokyo Research Laboratory and currently listed in Sourceforge.net. These Java Aglets are lightweight, easily deployable mobile agents that enable the autonomous execution of program on remote heterogeneous hosts. The Aglets version 2.0.2, listed in Sourceforge.net was chosen as opposed to the earlier versions owing to its compatibility with Java versions 1.2 and beyond. The 'Master' agent identifies the r categories of data based on the copy of a data set available on the Master node and serially spawns r 'Slave' agents. Each of the agents contains similar code to classify the data set based on one of the r response variables assigned to it. Each of the 'Slave' agents is dispatched to one of the r remote nodes in the network as soon as they are created. Once loaded in the memory of the remote nodes (also running the Aglets framework), these 'Slave' agents utilize the node's processor cycles and memory to complete the classification process in parallel. They are programmed such that they communicate their results to the 'Master' agent and dispose of themselves at the remote nodes. The order of arrival of the results from the 'Slaves' need not be in the order of dispatch due of the nature of the classification problem, possible network delays or other overheads at the remote site that can interfere with the execution of the 'Slave' agent.

A Sample Pseudo-code of the implementation

On Main Node

a) Create and Initialize the 'Master' Agent

b) Read the Data file and identify the categories of response variable

c) While more response variables

  i. Create a 'Slave' agent

  ii. Link a 'Slave' agent with a category of response variable

  iii. Dispatch 'Slave' agent to a remote node

On Remote Nodes after arrival of slave agents

d) Slave' agent executes in parallel and classifies local copy of data

e) Communicates execution results back to 'Master'

On Main Node

f) While more results awaited

i. Receive the results from the 'Slaves' and collate them.

CONCLUSIONS

In this paper, we have presented our research on how to enhance the turnaround time of a multi-class PSVM problem using a mobile multi-agent system. Our approach uses mobile agents that work in parallel, distributing the sub-tasks across the nodes in the network. In using Aglets the overheads of start-up time, code and state mobility increased the turnaround time for small data sets with few response variables when compared to a normal standalone serial Java application. The real benefits of a MMAS can be observed on large data sets with substantial categories of response variables.

We have made a static binding with respect to the number of response variables by chosing a pre-determined data set. Thus we know the number of agents deployed upfront. However, we can add dynamism to the system by detecting the number of response variables for any data set and thereby spawning the required number of agents. Another approach to bring parallelism to this classification problem is by storing the data set in a single node with multiple processors thus avoiding redundant copies of the data set. But this approach may open up intricate issues in parallel programming.

ACKNOWLEDGEMENTS

We would like to thank Dr. K.P. Soman, Head, Centre for Excellence in Compuatational Engineering and Networking (CEN), Amrita Vishwa Vidyapeetham and Dr. Srinivasan Rangan, Associate Professor, Amrita Institute of Management, Amrita Vishwa Vidyapeetham for their valuable comments and suggestions.

NOMENCLATURE

G : set of all dyadic sets g_i

g_i : a dyadic set

n : size of data set

r : number of dyadic sets

T : total time for classification

t_i : time to classify a data set based on a response variable

T_p : overall turnaround time

x : number of elements in a particular category of response variables

χ : Network turnaround time

REFERENCES

Adler, R. M., 1995, "Distributed Coordination Models for Client-Server Computing," *IEEE Computer*, Vol. 29, No. 4, pp. 14-22.

Bond, A.H., and Gasser, L. (eds), 1988, *Readings in Distributed Artificial Intelligence*, Morgan Kaufmann, San Mateo, CA., pp. 3-35.

Burbidge, R., and Buxton, B., 2001, "An Introduction to Support Vector Machines for Data Mining," *Keynote Papers, Young OR 12*, 3-15, M. Sheppee, ed., University of Nottingham, Operational Research Society.

Dantas, M. A. R., et al, 2002, "An Enhanced Scheduling Approach in a Distributed Parallel Environment Using Mobile Agents", *16th Annual International Symposium on High Performance Computing Systems and Applications, (HPCS 2002)*, Moncton, NB, Canada.

Durfee, E.H., 1999, "Distributed Problem Solving and Planning," *Multiagent Systems: A Modern Approach to Distributed Artificial Intelligence*, G. Weiss, ed., MIT Press, Cambridge, MA, pp. 121-164.

Fuggetta, A. et al, 1998, "Understanding Code Mobility", *IEEE TSE*, Vol 24, No. 5, pp. 343-346.

Fung, G., and Mangasarian O., 2001, "Proximal Support Vector Machine Classifiers," *Proceedings of Knowledge Discovery and Data mining*, F. Provost and R. Srikant, eds, ACM Press, NY, USA, pp.77-86.

Gray R., et al, 1996, "Mobile Agent: The Next Generation in Distributed Computing", Dept. of Computer. Science. - Dartmouth College – Hanover, *IEEE - Proceedings of the 2nd AIZU International Symposium on Parallel Algorithms / Architecture Synthesis (pAs'97)*, NH, USA. pp. 8-24.

Kreßel, U. H. G., 1999, "Pairwise classification and support vector machines," *Advances in Kernel Methods: Support Vector Learning*, B. Scholkopf et.al, ed., MIT Press, Cambridge, MA, pp. 255–268.

Lange, D. B. and Oshima, M, 1998, *Programming and Deploying Java Mobile Agents with Aglets.* Addison-Wesley, Reading, MA, pp. 2-5.

Ibid, pp.138-140.

Mangasarian, O., 2001, "Data Mining with Support Vector Machines," *Proceedings of IFIP Conference on System Modelling and Optimization*, Data mining Institute Report 01-05, University of Wisconsin, Madison.

Milojicic, D., 1999, "Mobile Agent applications", *IEEE Concurrency*, Vol. 7, No. 3, pp. 80-90

Shigeo, A., 2003, "Analysis of Multiclass Support Vector Machines," *Proceedings of International Conference on Computational Intelligence for Modelling Control and Automation*, Vienna, Austria.

Smith, R.G. and Davis, R., 1980 "Frameworks for Cooperation in Distributed Problem Solving", *IEEE Transactions on Systems, Man and Cybernetics*, Vol. 11, No.1, pp. 61-70

Wooldridge, M., 2002, *An Introduction to Multiagent Systems*, John Wiley, West Sussex, England, p. 3.

Ibid, pp.189-220.

SELF ORGANIZING MAP AND LINEAR SUPPORT VECTOR MACHINE DECISION TREE WITH OPTIMIZED CLASS SEPARABILITY

MEHMET IZZET SAGLAM
Istanbul Technical University, Informatics Institute, Advanced Technologies in Engineering, Satellite Communications and Remote Sensing Program, Maslak, Istanbul, Turkey

BINGUL YAZGAN
Istanbul Technical University, Faculty of Electrical and Electronic Engineering, Electronics and Communication Dept., Maslak, Istanbul, Turkey

OKAN ERSOY
Purdue University, School of Electrical and Computer Engineering, West Lafayette, Indiana,USA

ABSTRACT
Multispectral remote sensing data are obtained by using high resolution imaging sensors. Such data involves some problems. For example, the large number of classes, complex statistical distribution of these classes, and the smallness of the training data are among these. A new learning algorithm called self-organizing map, linear support vector machine decision tree (SOM-LSVMDT) with optimized class separability was developed to help solve complex remote sensing classifications problems. The method consists of a clustering part and a binary tree with linear support vector machine at all tree nodes. The proposed algorithm also generates a given number of clusters such that the classes chosen in each cluster are optimized in terms of a separability measure. The SOM -LSVMDT can also minimize problems associated with rare events. The performance of the SOM-LSVMDT algorithm was evaluated on a comparative basis with linear support vector machine decision tree (LSVMDT) algorithm, using three remote sensing data sets from the Colorado region. The SOM-LSVMDT was observed to yield better performance than the LSVMDT.

INTRODUCTION

High resolution imaging sensors are very important in modern remote sensing technology. These sensors produce multispectral data, and thereby result in one image per wavelength. However, they also involve problems of a large number of classes, complex statistical distribution of classes, and smallness of the size of the training data. In order to achieve better performance than what is possible with existing algorithms such as support vector machines and decision trees, the SOM -LSVMDT with optimized class separability has been designed.

The rare event problem corresponds to having a class or classes with small probability of occurrence. This problem occurs both naturally and during the design of the branches of a classification tree. The proposed algorithm is also more robust in handling the rare event problem.

The SOM-LSVMDT includes two parts. The first part involves clustering input data in to several subsets, and the second part is a decision tree with a linear support vector machine at each node (LSVMDT). In the clustering part, the data is first partitioned in to many clusters with the SOM in the current algorithm. Then, these clusters are grouped together in to a given number of clusters. In each cluster, the present classes have maximum separability according to a class separability measure. The data in each cluster is further classified by the LSVMDT. The experiments show that the proposed algorithm obtains better performance than the SOM, LSVMDT, and SOM-LSVMDT in which optimized class separability is not included [Saglam, Yazgan, and Ersoy, 2003].

The paper is organized in to six sections. Section 1 is introduction. Section 2 is background research, describing a complex remote sensing classification problem. In Section 3, the architecture and the operation of the linear support vector machine decision tree are discussed. Section 4 discusses the proposed algorithm. In Section 5, the experimental results are described. Section 6 covers conclusions.

2. BACKGROUND RESEARCH

In remote sensing applications, the classification problem is usually characterized as a highly complex problem with a multidimensional and highly nonlinear problem space. The Colorado data sets used in the experiments are representative of such a space.

The ten-class Colorado data set consists of 7 data channels obtained from four data sources [Valafar, 1994]. The area, which has 10 ground cover classes, is a mountainous area in Colorado. Each channel comprises an image of 135 rows and 131 columns, all of which are co-registered. Ground reference data were compiled for the area by comparing a cartographic map to a color composite of the Landsat data and also to a line printer output of each Landsat channel. By this method, 2019 ground reference points (11.4% of the area) were selected. Ground reference consists of two or more homogeneous fields in the imagery for each class.

3. LINEAR SUPPORT VECTOR MACHINE DECISION TREE (LSVMDT) ALGORITHM

First, the LSVMDT algorithm is discussed with the two-class problem for the sake of ease of understanding. The multiclass problem is described at the end of the section. The LSVMDT algorithm involves a binary classification tree at each node of which there is a linear SVM [Ersoy, Chi, 2002]. These linear SVMs separate their input space into two regions during the training process. In this process, they generate linear decision hyperplanes with their input data. The data, which is in the negative decision region, is further fed into linear SVM in the left child of the current node for training. The data, which is in the positive decision region, is further fed into linear SVM in the right child of the current node for training. This process stops, when all data vectors are classified correctly and a leaf node is reached. The label in the leaf node is matched with the class label.

When the input data vectors are linearly separable, the LSVMDT needs only one node to classify them correctly. But, when the data vectors are linearly nonseparable, the

LSVMDT classifies them as best it can by dividing into subsets with linear decision hyperplanes. It is very normal that there are misclassification errors in these subsets. Therefore, the classifying process continues by the other SVMs in later nodes. Finally, they are classified correctly at the leaves.

If there are unproportionally represented classes, such as rare classes, most data vectors lies on only one side of the hyperplane after training at the current node. To solve this problem, we use a heuristic algorithm. That is, we add randomly rare class vectors to the data set for making the class ratio equal to 1 [Ersoy, Chi, 2002]. In neural networks, this method is also used successfully. However, there is a limit on the random addition especially in the deeper nodes of the tree. Therefore, we add fewer data vectors in this part of the tree. In testing, we traverse down the tree until a leaf node is reached to classify a test vector.

Up to now, we assumed that there are two classes, and a binary tree is used. If we have m classes, each node in the tree can have a maximum of m children and $\frac{m(m-1)}{2}$ linear SVMs [Friedman, 1996]. Then, each of these SVMs is trained with two classes. Thus, each SVM produces a linear decision hyperplane. This hyperplane divides two classes from each other, and gives a vote in the testing procedure. The class label which has the maximum number of votes is selected as the output of the current node. This process is repeated until a leaf node is reached in testing [Ersoy, and Chi, 2002].

4. THE SOM-LSVMDT ALGORITHM WITH OPTIMIZED CLASS SEPARABILITY

The SOM-LSVMDT algorithm is has two-parts. The first part utilizes self organizing map and a given number of clusters with optimized class separability, and the second part is LSVMDT.

In the first part, there are four design issues. These are the number of neurons, the dimension of the map grid, map lattice and shape. The neighborhood size controls the smoothness and generalization ability of the mapping. It is desirable to make the number of neurons as big as possible. The shape of the map grid can be chosen as rectangles, because the map can orientate itself properly. We prefer hexagonal lattice, because all 6 neighbors of a neuron are at the same distance (as opposed to the 8 neighbors in a rectangular lattice). The procedure for learning a SOM is as follows:

- **Initialization:** Initialization is used to obtain a faster algorithm. There are three kinds of initialization:
  - Random initialization
  - Sample initialization
  - Linear initialization

  Before training with SOM, the weight vectors are initialized and the weight vectors are filled with the initialized values.

- **Training:** There are two stages in the training:
  - First, a sample vector is chosen from the input data vectors randomly. Then, the similarity between it and all the weight vectors of the map is calculated and a winner is chosen according to

$$\|x - m_c\| = \min_i \{\|x - m_i\|\} \tag{0.1}$$

- In the adaptation stage, the winning weight vector is updated as follows:

$$m_i(t+1) = \begin{cases} m_i(t) + \alpha(t)[x(t) - m_i(t)] & \in N_c \\ m_i(t) & \text{otherwise} \end{cases} \tag{0.2}$$

This adaptation procedure moves the prototypes of the best matching unit (BMU) and its topological neighbors towards the sample vector. This process is repeated many times until convergence. Then, the clusters that correspond to characteristic features are formed into the map automatically.

The number of output nodes in the SOM is chosen quite large, such as 150. The resulting clusters are further grouped in to a smaller number of clusters, each of which contains classes with optimized class separability in the proposed algorithm. For this purpose, two matrix measures of separability are used to create clusters. After that, a separability metric in terms of the ratio of the norms of those two matrices is chosen. These matrices are *within-class scatter matrix* and *between-class scatter matrix* [Balakrishnama, Ganapathiraju, and Picone, 98]. The within-class scatter matrix S_w is given by

$$S_w = \sum_{i=1}^{M} p(w_i) C_i \tag{0.3}$$

where $C_i = E\left((x - m_i)(x - m_i)^T\right)$ is the covariance matrix, and $p(w_i)$ is the a priori probability of class w_i, given by

$$p(w_i) = \frac{N_i}{N}$$

where N_i is the number of samples from class w_i, and N is the total number of samples. $trace(S_w)$, the sum of the diagonal elements of S_w, is a measure of the average variance of the features over all classes. To optimize the separability of classes, the norm of within-scatter matrix should be as small as possible.

The between-class scatter matrix S_b is given by

$$S_b = \sum_{i=1}^{M} p(w_i)(m_i - m_0)(m_i - m_0)^T \tag{1.4}$$

where m_0 is the global mean vector:

$$m_0 = \sum_{i=1}^{M} p(w_i) m_i \tag{0.4}$$

$trace(S_b)$ is a measure of the average distance of the mean of each class from the global mean value over all classes. Intuitively, for good class separability, the size of the between scatter matrix should be large. The ratio

$$J = \frac{trace\{S_b\}}{trace\{S_w\}} \tag{0.5}$$

is used to choose which classes should be grouped in a cluster This value is large if S_w is small and S_b is large. That is,

- Samples in the n dimensional space are well clustered around their mean within each class (small $trace(S_w)$).
- The clusters of the different classes are well separated (large $trace(S_b)$).

J is a good indicator of more separable classes. When clusters are chosen with a random combination of classes instead, the performance of the algorithm decreases.

In the second part of the SOM-LSVMDT, the classes in each cluster are classified with the LSVMDT algorithm. By using a small decision tree as a result of fewer classes in each cluster, the rare class problem is also reduced to a large degree.

For testing, we use the SOM algorithm together with the optimized combination of classes to divide the testing data into clusters. This is followed by classification in each cluster by the LSVMDT. The total test error is calculated by recording the number of vectors which are misclassified.

EXPERIMENTAL RESULTS

The classification of eight, ten, and thirteen class Colorado data sets was investigated. These data sets are known to represent very complex remote sensing classification problems. The SOM-LSVMDT and LSVMDT algorithms were performed on a comparative basis.

Each linear SVM in the LSVMDT and SOM -LSVMDT was implemented using OSU SVM Classifier Matlab Toolbox available at http://eewww.eng.ohio-state.edu/~maj/osu_svm. In addition, the SOM Toolbox by Esa Alhoniemi, Johan Himberg, Juha Parhankangas and Juha Vesanto available at http://www.cis.hut.fi/projects/somtoolbox/download/ was used.

Tables 1 and 2 show the results of LSVMDT and SOM-LSVMDT, respectively. For example, for 8 class Colorado data set, the minimum classification error of LSVMDT is equal to % 7.17. In SOM-LSVMDT, two clusters were generated. The first combination of classes in the first cluster, namely, classes 1, 6, 7 and 8 achieves % 2.67 error for the same data. Similar results are obtained with the other data sets, as observed in Table 2.

Table 1. Performance of the LSVMDT with the Colorado data sets.

LSVMDT	% ERROR
8 Class Colorado	7,17
10 Class Colorado	48,62
13 Class Colorado	29,48

Table 2. Performance of the SOM-LSVMDT using the Colorado data sets with clusters and optimized class separability.

CSS-LSVMDT	Class Labels	J	% ERROR
8 Class Colorado	1,6,7,8	6,8977	2,97
10 Class Colorado	1,2,3,4,5	0.3616	35.02
13 Class Colorado	7,8,9,10,11,12,13	0.6999	22.64

CONCLUSIONS

In this paper, a new classifier called the Self Organizing Map Linear Support Vector Machine Decision Tree with Optimized Class Separability was introduced. The SOM-LSVMDT is a combination of a clustering method and a linear support vector machine decision tree. This combination is shaped into a new architecture, which is designed to divide the complex remote sensing data into smaller parts and thereby make classification easier.

Computer experiments shows that COM-LSVMDT can achieve a better performance than LSVMDT. The relationship between the class separability and the performance of the proposed algorithm is also shown in this paper. The error of the SOM-LSVMDT decreases when the class separability increases. If clusters with a random choices of classes are used instead, the performance of the proposed algorithm decreases

REFERENCES

Balakrishnama, S., Ganapathiraju, A., Picone, J., Linear Discriminat Analysis for Signal Processing Problems, Institute for Signal and Information Processing

Chi, H., Ersoy, O.K., H-C. Chi, O. K. Ersoy, "Support Vector Machine Decision Trees with Rare Event Detection," *Int. Journal for Smart Engineering System Design*, Vol. 4, pp 225-242, 2002.

Friedman, J., Another Approach to Polychotomous Classification, Technical Report, Stanford University, 1996.

Saglam, M.I., Yazgan, B., Ersoy, O.K., 2003. Classification of Satellite Images by using Self Organizing Map and Linear Support Vector Machine Decision Tree, MS Thesis, Istanbul Technical University, May 2003.

Valafar, F., 1994. Parallel Probabilistic Self Organizing Hierarchical Neural Network, PhD thesis, Purdue University, and ECE Technical Report No. 1025-5818/02.

PART II: EVOLUTIONARY PROGRAMMING

EVOLUTIONARY PLATFORM FOR AGENT LEARNING

ANNA L. BUCZAK, DAVID G. COOPER, AND MARTIN O. HOFMANN
Lockheed Martin Advanced Technology Laboratories
Cherry Hill, New Jersey

ABSTRACT
This paper describes a novel methodology for agent learning. This method allows agents to learn how to better operate given the external conditions, in order not to repeat their previous mistakes. Evolutionary Platform for Agent Learning (EPAL) allows agents to change their behavior by changing the parameters of the tasks they are executing and by adding new programming constructs and tasks to their workflow. The method was especially constructed for EMAA agents developed in Lockheed Martin Advanced Technology Laboratories and proven in many military and DARPA programs. EPAL presents a general approach to agent learning, based on an extension to Genetic Programming that we developed. Our approach is suitable to learning by any agent that can be represented as a workflow that can be further decomposed into building blocks, such as operators, tasks, and parameters.

1. INTRODUCTION

Learning plays a fundamental role in most human activities. Humans learn from their experiences, both successes and mistakes. Babies learn to walk and talk, adults learn foreign languages, new skills and concepts. Learning seems to be the fundamental property that allows humans to adapt to changes in the environment and to be successful. The skill of learning is of dynamic nature in humans who continuously acquire and modify representations of the world. On the other hand, artificial systems such as software agents are usually static. For software agents to work properly, humans need to envision all contingencies and code them at the agent development time. Most agents react the same to the similar external stimuli, even when the previous reaction was a failure. We want the agent to learn different behavior in response to those stimuli.

If software agents could "just" learn from their successes and mistakes, they would be more rapidly deployable. However developing a general learning method for software agents is a difficult proposition. Several attempts have been made in the past. Most approaches use reinforcement learning. Stone [1998] proposes a layered learning approach where hierarchical task decomposition is performed that allows the learning to proceed at each level of the hierarchy. Learning performed at a lower level directly affects the learning at the higher level. A new reinforcement learning based method is introduced and applied to robotic-soccer applications.

ABLE agents from IBM [Bigus, 2002] have a library of components such as learning beans that can be used to build an agent. The learning beans implement different learning algorithms (e.g. temporal difference learning, neural networks) and a human can compose an agent that will use any of the learning beans. However this form of agent learning is limited to what a given learning algorithm does and how the human used it when configuring the agent. Conceptually in the ABLE framework there is a place for agent's learned behavior, however the only method that exists for continuous learning is the reinforcement-learning algorithm.

A so-called "conscious learning" mechanism for software agents was proposed in [Ramamurthy, 1998]. Their agent CMattie takes a role of departmental seminar organizer. She understands past events using her built-in domain knowledge such as a description of the features of a seminar. Should a new situation occur (e.g. dissertation defense) she first treats it as a seminar concept but through e-mail interaction with the organizer learns that a dissertation is a new concept. The "conversation" between the organizer and CMattie is stored in case-based memory. CMattie performs analogy matching and learns through case-based reasoning. The new concepts learned are added to her domain knowledge and the next time she encounters them she recognizes the past experience. CMattie must also learn new behaviors when faced with a dissertation defense and unfortunately this difficult subject was not addressed in the paper.

2. AGENTS AND EXTENDABLE MOBILE AGENT ARCHITECTURE

Lockheed Martin Advanced Technology Laboratories (ATL) developed the Extendable Mobile Agent Architecture (EMAA) over the last seven years. EMAA is used in about two dozen projects covering a full range of intelligent systems, including information management for time-sensitive strike [Hofmann, 2001], situation awareness for small military units, and executing user requests entered via spoken language dialogue. Starting with the Domain Adaptive Information System [Hofmann, 1998], experimentation with prototypes in military exercises has guided our research and development towards the adaptable EMAA architecture. EMAA was used in the US Navy Fleet Battle Experiment (FBE) series as a human aiding tool. From these experiments, an Interoperable Intelligent Agent Toolkit (I2AT) was created under DARPA Control of Agent Based Systems (CoABS) funding to allow people to construct agent systems without coding the agents from scratch, but by generating a workflow of tasks and parameters in a graphical fashion. The toolkit was used to support the Expeditionary Sensor Grid Enabling Experiments where it was used for the first time by non-Java developers to create agents.

EMAA agents are designed with a workflow model for agent construction. An EMAA agent's workflow (Fig. 1) is a list of tasks, linked by execution paths that can be conditional or unconditional. Each task can take a certain number of task dependant parameters. The work-flow architecture enables a new approach to agent development called agent composition, characterized by choosing, configuring, and assembling elementary agent tasks into workflows. Typical tasks include queries to relational databases, retrieval of content from Web pages, reading and sending e-mail, etc. When a user composes an agent workflow in I2AT, the tool produces an XML mark-up of the workflow that specifies agent tasks, their interconnections, and their parameters. Another component of I2AT creates on-the-fly agent instances ready to be executed. EMAA is an environment conducive to learning applications. In the past

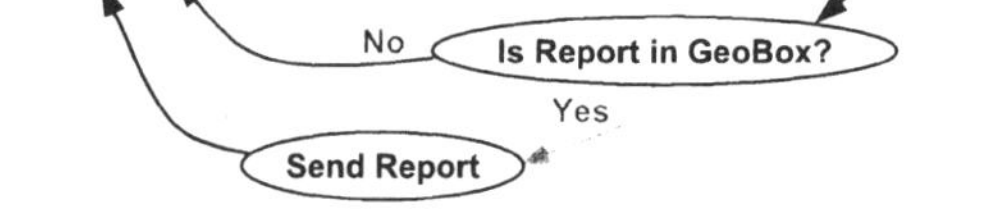

Figure 1. Example of an EMAA Agent Workflow.

years key features have been added to EMAA that constitute the support on which we are basing agent learning. The goal of EMAA's new learning capability is allowing agents to adapt to the environment in which they operate.

3. AGENT LEARNING APPROACH BASED ON GENETIC PROGRAMMING

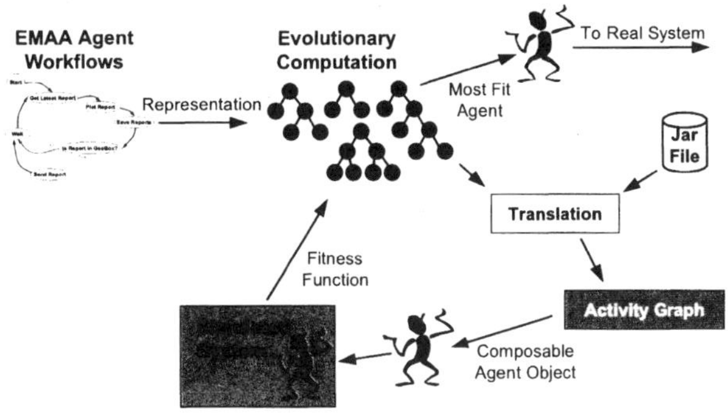

Figure 2. Evolutionary Platform for Agent Learning.

Our novel approach to agent learning is graphically described in Fig. 2. The first step is to represent agent workflows in a Genetic Programming (GP) tree form. This new representation is described in detail in Section 4. GP individuals are translated from a tree format into an Activity Graph, using their jar file descriptions from I2AT toolkit. The Activity Graph is then used to instantiate a composable agent, which is injected into the simulated system. Fitness is collected while the agent is run in the simulated environment.

GP operators of reproduction, crossover, and mutation work on agents' genetic material to generate new agents that have learned to overcome certain problems. Once the agents have learned, the most fit are injected into the real system and take the place of their parents who did not know how to overcome the problems. The method proposed is a general method that can generate completely new agent workflows, as well as related workflows but with new parameters. As long as the tasks to add to an agent's workflow are on a task list and related jar files exist, these tasks can be added to an agent, augmenting its workflow and changing how the agent behaves.

4. GENETIC PROGRAMMING EXTENSIONS FOR AGENT LEARNING

Koza [1993] invented GP as a means of program synthesis by genetically breeding a population of computer programs using the principles of Darwinian evolution. In GP each individual is represented as a tree constructed from functions and terminal symbols and often the trees represent programs in LISP. Our EMAA software agents are much more complicated than the small software programs that GP usually evolves. In order to evolve meaningful agents in a realistic time frame our representation of GP agents needs to be at a higher level than simple Java instructions and their parameters. We have chosen as our main GP building blocks the tasks that an agent can do. The tasks in an agent's workflow can be combined by several programming constructs such as *execute-sequentially*, *loop*, *if-then-else*. These constructs define the set of our operators (functions). Each of the tasks mentioned previously has a name and an associated list of parameters that represent our terminal symbols. Therefore our GP tree has three types of nodes, not two as usual GP trees have. The highest level is the operator level, the next level is the task level and the lowest level is the terminal level. Each GP tree can be of any depth allowing both simple and sophisticated agents (see an example on Fig. 4).

Each of the operators takes always a certain number of parameters and those parameters need to be of a predefined type. For example *execute-sequentially* takes always two parameters, each being an operator or a task. The set of tasks contains all the tasks that an agent can perform. In general this set is quite large but at this point we are allowing only a small subset of tasks (about twenty tasks dealing with acquiring, filtering and reporting information, computing network statistics). The set of terminals contains all the possible parameters of the tasks in the task set. Since our tasks are diverse, their parameters vary widely in number and type. We are not in a position to obey the GP closure property that requires each function in the function set to be able to accept as its arguments, any value that may possibly be returned by any function in the function set and any value that may be assumed by any terminal in the terminal set. Since the closure property does not prevail, special steps are needed to generate feasible offspring.

Our GP agents have several syntactic restrictions similarly to syntactic restrictions existing in programming languages. For example in the GP individual tree the root can be an operator or a task but it cannot be a terminal. The largest number of restrictions concerns the task parameters. Tasks have different numbers of parameters and those parameters are of different types (e.g. latitude, proximity, e-mail address, file name). The parameter restrictions heavily affect the design of our GP tree structure and operators.

Our GP agents explore the search space by performing crossover and mutation. We use a one–point crossover that we restricted in comparison to a regular one-point crossover. The restricted crossover does not allow the formation of infeasible individuals. In regular crossover entire subtrees from the parents are swapped and because of the closure property, offsprings created are always feasible. The restricted crossover allows only swapping portions of the trees that will lead to feasible individuals. It will allow swapping two subtrees that have a task at their head but it will not allow swapping two parameters unless they are the same type.

In GP there are usually two types of mutation, each starting by randomly choosing a mutation point in the tree. The first type chooses randomly a mutation point, the subtree rooted at that point is deleted, and a new subtree is grown there using the random growth process. In the second type of mutation, a single terminal is exchanged into another terminal or a single function is exchanged into another function (with its subtrees staying the same as previously). For agent learning we use the first type of mutation described above but since it is very disruptive we use it with a low probability of occurrence. We are using mainly three more restrictive types of mutation that we defied:

a. Parameter mutation. It deals exclusively with parameters and it performs a mutation of one parameter value into another of the same type (e.g. "3" into a "7" or emaa@lmco.com into agent@lmco.com). This is a restrictive mutation operator since it cannot simply mutate one terminal symbol into another one but it needs to keep track of the range of values that a given parameter in a given task can take.
b. Task mutation. It performs the mutation of tasks and is capable of substituting at the mutation point one task for another. It differs from the usual GP function mutation because it does not keep subtrees (parameters in the case of agent learning) intact. It assigns standard or random parameters to each new task generated.
c. Operator mutation. It mutates one operator into another and it keeps most of the subtrees intact. However there are situations when the subtrees need to be changed (e.g. when mutating *if* into *execute-sequentially*, the conditional part is discarded).

An important part of GP is the fitness of an individual. Our ultimate goal is to make agents learn everything, similarly to what humans do. Consequently we do not want to give too much guidance to the GP individuals and our fitness function should be of a general nature. The fitness should be some type of self-awareness function, describing how well an individual is performing. At this stage we want the agents to learn what to do when the network slows down/speeds up considerably. Our fitness function describes how slow/fast the network is in comparison to how it was previously.

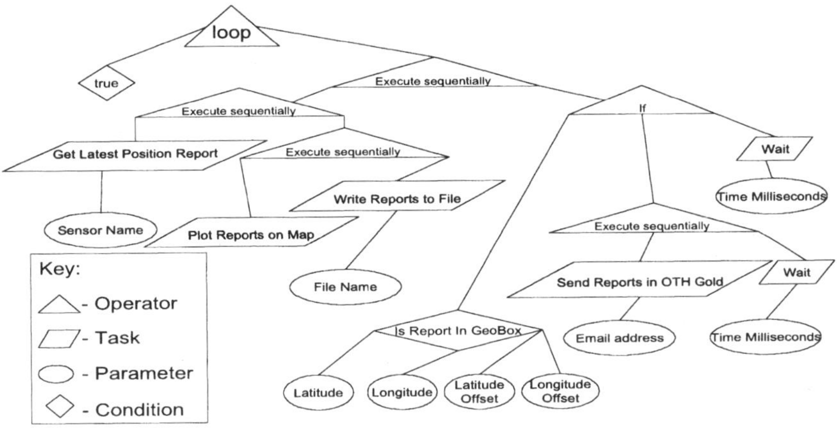

Figure 4. GP tree representation of an example agent from FBE experiments.

Our GP implementation is based on Evolutionary Computation in Java (ECJ) from University of Maryland. ECJ [ECJ] is a very flexible genetic algorithm and genetic programming freeware, written by Sean Luke et al. Its source code is readily available which allowed us to make all the modifications necessary for agent learning.

5. AGENT LEARNING EXAMPLE

One example of where learning is desirable is in the FBEs that used EMAA agents on the CoABS grid. One of the biggest challenges of this system is making sure that the network is not overloaded with data. One way to do this is to filter data, but the current filtering methods are static. In FBEs agents were used to pass relevant data to the correct participants; they stored data and kept track of who was online so that data could be refreshed when needed. While performing real-life FBE experiments, EMAA agents successfully performed all of their tasks that they were supposed to do. One type of agents was tasked to report all the tracks in a Geo Box with specific latitude-longitude coordinates. When the network was working at usual speed, the agents performed this task well but when the network became congested, the latency of reported tracks was too long. Humans observing the experiment wished that the agents decrease the size of Geo Box in order to provide tracks faster. It never happened since the agents could not learn. This is one of the simplest forms of learning that we want our agents to perform: adjust the task parameters. A more sophisticated form of learning is the one that requires agents to add/delete tasks and whole branches from their workflow. Again, while performing FBE experiments, a need for this type of learning became evident. Some of our agents were querying a database for relevant information and sending this information to the human. When unexpectedly a new database came on-line that had similar data, that database was not queried. Humans wished again that agents could figure on their own that the new database needs to be queried as well. This type of learning would translate into adding a new branch to the workflow with a few tasks in it. The agent learning methodology that we devised will be able to address both types of learning.

Our system starts by generating an initial population of agents. Certain individuals in this population are the agents that humans generated while other members are generated randomly. The fitness function is a measure of the latency of the network in comparison to previous latency. The premise is that if the agent is getting/sending more information it gets rewarded, and if it is getting less information it gets punished. We are not telling in the fitness function what the agent should learn or how it should learn.

6. CONCLUSIONS

EPAL is a framework for software agents to automatically learn how to better perform. It is especially well suited for ATL EMAA agents but the methodology that we developed can be applied to any software agents, as long as they are described by a workflow that can be further decomposed into simple building blocks. Our method allows agents to perform two forms of learning: simple learning where only the task parameters are changed and advanced learning, where the agent's workflow can be completely changed, allowing agents to find and query new databases, send reports to new locations, or really to do anything for which a task exists in our task database. As our task database grows, agents should be able to "learn anything". We developed an extension of GP representation that allows representing agents as GP trees. This representation has three types of nodes (operators, tasks and parameters) rather than two types present in standard GP trees. For this representation, specific mutation and crossover operators have been developed in order to efficiently find agent workflows with increased performance.

The learning method developed is unique because it generates real software agents that can be executed in real military scenarios. We are not talking about toy problems and toy applications. To our knowledge the solution we formulated is the first platform in which real software agents can learn from their experience to perform standard tasks better. In the future we need to augment the set of tasks that agents can build their workflows from and to design a general self-awareness module. This self-awareness module will replace the fitness function that we are presently using.

7. REFERENCES

Bigus, J.P., Schlosnagle, D.A., Pilgrim, J.R., Mills III, W.N., Diao, Y., 2002, "ABLE: A Toolkit for Building Multiagent Autonomic Systems," *IBM Systems Journal,* Vol. 41, No. 3.

ECJ, www.cs.umd.edu/projects/plus/ec/ecj/.

Hofmann, M.O., Chacón, D., Mayer, G., Whitebread, K.R., Hendler, J., 2001, "CAST Agents: Network-Centric Fires Unleashed," *Proceedings of the 2001 National Fire Control Symposium*, Lihue, HI, August 12-30.

Hofmann, M.O., McGovern, A., and Whitebread, K.R., 1998, "Mobile Agents on the Digital Battlefield," *Proceedings of the Second International Conference on Autonomous Agents* Minneapolis/St. Paul, May 10-13, 1, pp. 219-225.

Koza, J., 1993, "Genetic Programming – On the Programming of Computers by Means of Natural Selection," Cambridge, Massachusetts, MIT Press.

Ramamurthy, U., Bogner, M., Franklin, S., 1998, "Conscious Learning in an Adaptive Software Agent," *Proceedings of Second Asia Pacific Conference on Simulated Evolution and Learning,* Canberra, Australia.

Stone, P., 1998, "Layered Learning in Multi-Agent Systems," PhD Thesis, CMU.

INCREMENTAL BROADCASTING AS A STRATEGY FOR MULTI AGENT COMMUNICATION

IREN VALOVA
Computer and Information Sciences
University of Massachusetts
Dartmouth
285 Old Westport Rd,
North Dartmouth, MA 02747

NATACHA GUEORGUIEVA
Computer Science
CSI/City University of New York
2800 Victory Blvd.
Staten Island, NY 10314

VALENTIN ZACHARIAS
Computer and Information Sciences
University of Massachusetts
Dartmouth
285 Old Westport Rd,
North Dartmouth, MA 02747

ABSTRACT

"Direct routing", "routing by signal propagation" and "public notice routing" [3] are seen as the possible natures of communication between distributed agents. This paper will introduce a forth one: incremental broadcasting. It could be used when a broadcast protocol is needed, but the communication medium only allows point-to-point connections. Given a number of agents and certain environmental restrictions, different algorithms can be used to build the network underlying broadcast routing, therefore quality measures for such a network are proposed and some algorithms are compared.

INTRODUCTION

Broadcasting can be realized with point-to-point connections either by relaying agents or by requiring each agent to send messages to all other agents. Then, why is there a need for "incremental broadcasting"? Further, isn't incremental broadcasting just another protocol that should be subsumed with the broadcast communication mode (as a kind of "simulated broadcasting")?

In both cases we need to have agents that know every agent and are able to send a point-to-point message to everyone. In the case of the relaying agent, we are imposing a centralization on the system that may otherwise not be necessary for the accomplishment of the task. In addition, we are introducing a "single point of failure" - the breakdown of the relaying agent effectively destroys the system.

In the case of agents sending messages to all other agents we are imposing the burden of keeping track of the whole system on each client, a burden which becomes especially large in the case of coming, going and disappearing agents [4], i.e. going without having a chance to notify anyone of this, ex. communication failure.

Secondly, if this is another kind of simulated broadcasting and should be in the same category as broadcasting, is much harder to answer. The following features make it worthwhile to treat this approach separately:

Natural approach to complete broadcasting: The broadcasting mechanisms proposed by Ferber [2] fall into two different categories: "incomplete broadcasting", realized through message propagation will usually not reach all other agents and "Complete broadcasting", realized by messages send out by one agent and reaching everyone. Note, that it's very hard to come up with an example of an natural multi agent system for the second mechanism, while it's easy to think of one for incremental broadcasting (think for example of rumors).

Technical feasibility: The incremental broadcasting model is very well suited for the environment of today's omnipresent internet, where broadcast mechanisms exist only for relatively small subnets. A multi agent system based on this communication mode could therefore easily take advantage of existing standardized, easy to use and cheap things like wireless modems, internet and tcp/ip libraries, hereby lowering both developing time and cost.

Specific problems: Multi agent systems based on this communication mode face problems and performance issues unlike systems based on any of the other communication modes. These are discussed later in this paper.

The algorithm works as follows: each agent knows few other agents. When an agent whishes to send a broadcast message, it will send messages to all agents he can. These agents receive the message and send it to their neighboring agents, and so on. In order to prevent infinite circulation of messages, all agents will propagate a particular message only once. This will allow to build the system in a way, that the agents gather information about other agents by observing addresses appended to the broadcast messages [1].

TERMINOLOGY AND COMMUNICATION STRUCTURE

The property that each client can only send messages to a small number of other agents leads to a representation of the communication structure as a graph:

$$G = (V,E)$$

where V is the set of agents and each element of E is of the form (x,y), $x \in V$, $y \in V$ and represents the possibility for agent x to send a message to agent y.

To simplify the analysis, to stick closer with the metaphor of rumors and to stay close to existing connection oriented protocols, we will assume an undirected graph,

$$(x,y) \in E \leftrightarrow (y,x) \in E.$$

And by saying that agents are able to communicate with only a small number of agents, we're saying that the degree of any $v \in V$ is much smaller than the number of nodes, the number of elements in V.

Quality Of The Structure

To asses what quality means in such a communication structure, we assume the connections depicted in Graphs 1-3 and that agent 1 sends a message and we are looking to see when agent 10 receives this message. Most of the example graphs discussed here have certain properties (especially being k-regular [5]) which is not necessary for “incremental broadcasting” but makes them easier to compare.

With the communication structure shown in Graph 1, the message will never reach agent 10. The trivial first conclusion concerning the quality of the structure is therefore: it has to be connected.

In graph two the message from agent 1 to agent 10 has to be received and retransmitted four times, while in Graph 3 this is only needed once. Note that in Graph 3 this holds for every combination of two agents, i.e. a message send by any agent reaches all agents after being retransmitted only once.

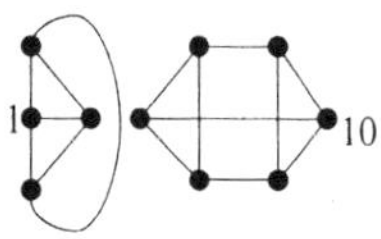

Graph 1

Graph 2

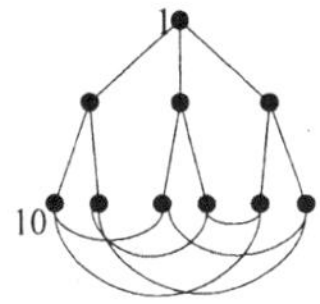

Graph 3

If we assume that each client can only have a maximum of 3 connections, then this graph could be called perfect in relation to this communication structure. Because assuming a maximum of k connections and with a maximum of r retransmissions not more than s(k,r) clients can be connected, where

$$s(k,0) = k + 1$$

$$s(k,r) = (k-1)^r * k + s(k,r-1)$$

and s(3,1) = 10. But although the construction of this graph looks simple, it does not work with larger graphs.

Not taking into account different communication speeds or costs, the third graph represents therefore a much better communication structure. To assess this kind of differences and their big impact on the efficiency of the communication structure, the analysis of such an structure should include the diameter, i.e. the longest path between two nodes in the communication graph. Because the communication graphs are not without cycles, a path could potentially by infinite, therefore we have to calculate the shortest path between all possible pairs of nodes, and take the longest found. For the last measurement of such a communication structure, we have to use the Graphs 4 and 5.

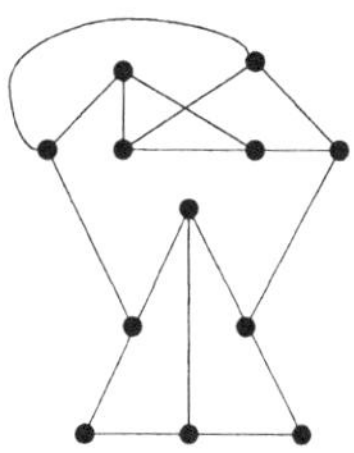

Graph 4

Both are connected and the diameter of Graph 4 is by one larger than the diameter of Graph 5 (5 vs 4). Graph 4 simply looks as if the nodes are better connected. If we calculate the average number of clients reached after any number of retransmissions (Fig. 1). in very few cases there are 4 retransmissions necessary and if we lock at how many agents are reached after 2 and 3 steps, Graph 4 is better, i.e. in most cases if we lock at the global situation after a given number of retransmissions, more agents in Graph 4 will know the message. This calculation is done via a depth first traversal of the tree. It is broken down into steps, i.e. all clients reached directly from the beginning node are added to the counter for step 1, afterwards the nodes reachable from these nodes and so on. After this calculation was

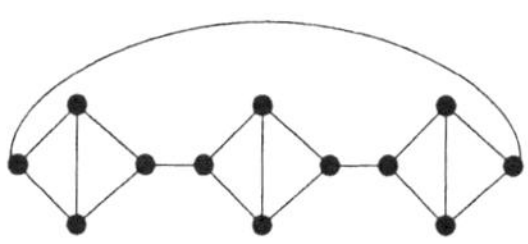

Graph 5

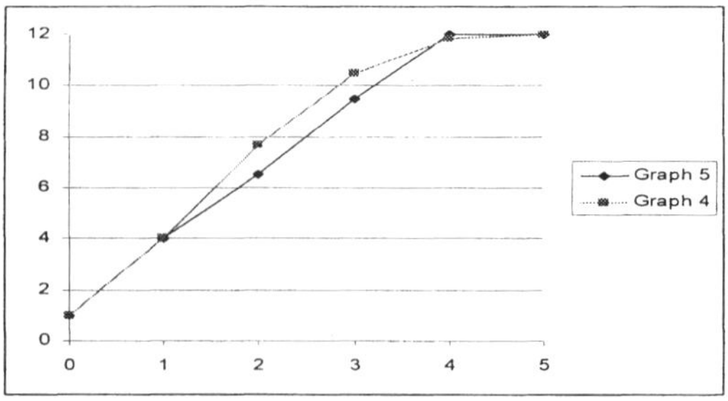

Figure 1 The number of agents reached after a given number of steps

performed for every possible start node, the values of the counter for each step are divided by the number of agents/nodes.

Robustness - one of the advantages of incremental broadcasting is its egalitarian structure that should still function after some agents dropped out. In order to assess the performance of a communication structure, tests after random agents dropped out are needed.

Costs and time - so far we have mostly implicitly assumed, that the cost for each connection and the time that a message needs to travel among agents that are directly connected, is constant. The last assumption is, of course, quite unrealistic. But given identical agents with identical communication devices, the time should not differ too much and the costs for communication are very small, so there is a value in studying incremental broadcasting under this simple conditions. However, if we have different communication devices, costs and communication speeds that depend for example on the geographical distance, these must be included in the assessment of such a communication structure.

The structure is determined by the following factors: How many other agents the agent should know as first connections, and is the agent able to acquire more knowledge about other agents by observing sender addresses appended to the messages? It should be noted that a connection change normally involves a termination of connection and choosing another agent to connect to.

SIMULATION

In all cases we simulated 485 agents, having totally random communications. Each agent was able to maintain 4 connections to other agents (was able to send messages to four other agents), and tried to get at least 3. We used this number of agents, because it's the largest possible 4-regular graph with diameter 5, but finding this graph is NP complete Weifa Lang et al. and was not done here. For some agents maintaining at least 3 connections was not possible, because all the agents they knew had already four connections. All connection dropping and establishing happened at the same time. For the strategies, where we assume that the agents obtained additional knowledge through looking at the sender addresses of the broadcast messages, we assumed that every agent knows half of the other agents. For the strategies where the agents in the beginning know only very few other agents, the agents where put into a grid and the agents tried to connect to randomly selected neighbors. When the agents had only limited knowledge for their initial connections, it sometimes happened, that the communication graph was not connected (1-4 agents where not able to connect to anyone), these cases where excluded.

Limited Knowledge

Four strategies where tested and compared using the "number of agents reached after x steps" graph introduced earlier. The strategies where:

Neighbors: The agents make connections to their neighbors in the grid, nothing more was done.

Dublicates: Starts with the same connections like the strategy above, but whenever a message arrived at an agent that it already saw, it increments a counter for the agent that send this message. After some time it is checked if the number of dublicates divided by the total number of messages coming from agent is higher than twice the average for the other agents. We did not want to use a threshold, because the numbers encountered here depend largely on the total number of agents. So using a threshold would require additional knowledge.

If this is true, this connection is dropped and a new connection is build to the agent known to be the furthest away. Note that this strategy requires that a message has an appended sender message as well as an retransmission count.

Dublicates with distance: Like the dublicates strategy, just when a message arrived twice the counter is incremented not by one, but by 1 divided by the transmission count. Like the pure dublicates strategy these strategies converge very quick and won't drop any connections after a short time.

Random drops: With a probability of 5% a agent drops a random connection and connects to the agent known to be furthest away.

Results. Fig.2 shows the results for these strategies. It can be seen that the neighboring strategy (as expected) performs badly and that any of the three proposed strategies improves the connections by a large margin.

However this better performance comes at a price, not only are the messages larger (because of the appended sender address and the retransmission count) and more bookkeeping is needed, but connections are dropped and build up again. In the case of the random drop agent, this occured on average 24 times, with the dublicates strategy 41 times and with the "dublicates with distance strategy" 55 times.

The performance of the other three strategies is surprisingly similar, and the two more difficult strategies are most probably not worth the effort. The "dublicates with distance" strategy reaches a little bit more agents after the first steps, the dublicates strategy is afterwards the best.

Complete Knowledge

For the complete knowledge condition we compared two strategies:

Random: The agents connect randomly to three other agents. Agents that already have four connections will refuse any attempts to connect to them.

Systematic: A strategy designed to come close to the smallest possible diameter, in this case 5. It starts out building a tree, beginning with a random node. After all agents are included in this tree, all nodes but the leafs already have four connections. Now the algorithm takes the first leaf, computes the distance to all other nodes and connects this node with the node furthest away with less than four connections. This is continued until all agents have four connections.

As can be easily seen, this algorithm runs in $O(n^2k^2)$ and is therefore feasible even for large number of agents. Here n is the number of agents and k the maximum number of connections per agent.

Results: Fig.3 shows the result for these two strategies. It can be seen that the systematic strategy outperforms the random strategy, although the diameter is in both cases almost identical. If we compare these strategies to those working with limited knowledge we also see a very large difference in favor of the latter two.

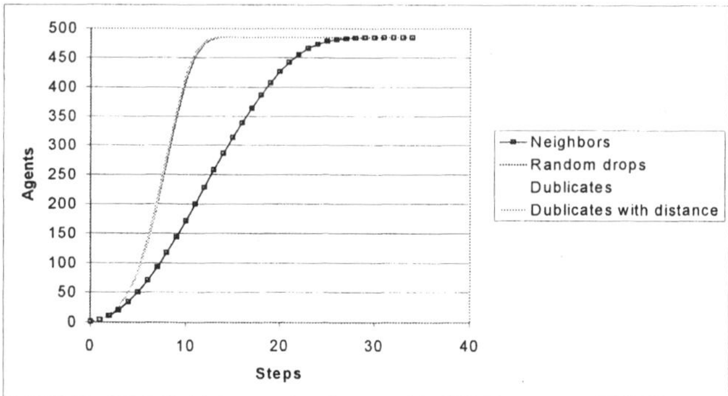

Figure 2 The number of agents reached after a given number of steps

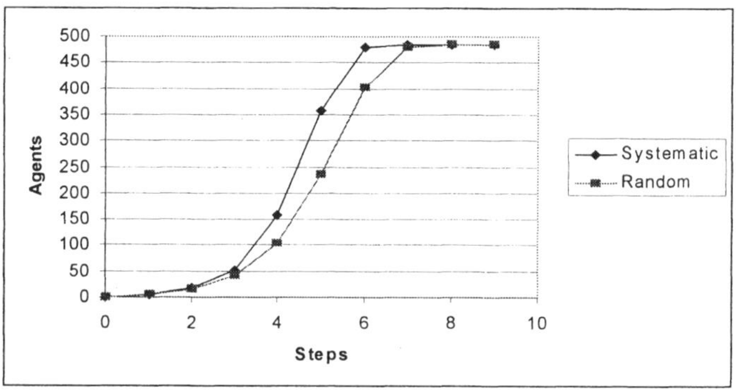

Figure 3 The number of agents reached after a given number of steps

REFERENCES

[1] Gnutella Specification: can be found for example at: www.clip2.com/GnutellaProtocol04.pdf

[2] Ferber, J., "Multi-Agent Systems - an Introduction to Distributed Artificial Intelligence", Addison-Wesley, 1999

[3] Kornfeld, W., "ETHER: A Parallel Problem Solving System". In Proc. of 6th IJCAL, 1979

[4] Sperber, D. and Wilson D., "Relevance: Communication and Cognition" Blackwell, 1986

[5] Liang, W., Havas, G. and Street A., " Finding a Low-Diameter and Low-Weight k-Connected Subgraph" Congressus Numerantium 136, pp.161-175, 1999

GROUP FORMATION OF EMOTION AGENTS

KEITARO NARUSE
HOKKAIDO UNIVERSITY

MASAHIRO KINOSHITA
HOKKAIDO UNIVERSITY

HIROSHI YOKOI
HOKKAIDO UNIVERSITY

YUKINORI KAKAZU
HOKKAIDO UNIVERSITY

ABSTRACT
In the field of multiple robot control and artificial life, group formation of agents, such as aggregation, distribution, lining and so on, is one of the important problems. Particularly, we focus on a group such that the shape of the group is almost same but its characteristic is different, for example, a group that is easy to distribute and a group which is robust to a noise. The objective of this paper is to develop a mechanism of forming such an agent group. For making such the group, in this paper, we set two-dimensional inner state in the agent, as well as subjective one-to-one evaluation. Then we investigate the characteristic of the proposed agent system.

INTRODUCTION

In most of multiple agent systems, how to achieve a team play or group behavior among agents is one of the central topics. It is achieved, for example, by learning, competition and cooperation among them. However, if we will control a great number of agents, we will need a new scheme for the control, because the analysis of each agent becomes almost impossible. We cannot cope with the agents as an individual, but as a group, mob, or society.

Many features of creatures and human being have been already introduced into the agent system. However, emotion has been employed in the agent system a little. The emotion can not be only an inner state of the agent, but it can be a non-verbal communication method. The creatures can survive themselves by watching the emotion of the others and by showing the emotion to the others.

The objective of this paper is to introduce a function of the emotion to the agent system. In this paper, we look at two dimensional emotion of pleasure-sadness and anger-fear, and affection to the other agents as one-to-one subjective relation. It is known in the psychology that a group of persons who share a same experience at the same time tend to have higher affection each other, because it is reported that the shared experience impose them shared emotion, which increases their affection. In other words, the emotion and affection influence each other. Changing the viewpoint to the agent system, the emotion and affection corresponds to an inner state and one-to-one subjective evaluation of the other agent, respectively.

In this paper, we propose an agent system including emotion-like two-dimensional inner states and affection-like subjective evaluation of the others. The system is expected to form more variety of groups or societies than the conventional one. We investigate the characteristics of the agent system.

This paper is organized as follows: In the next section, multiple agent system and artificial life systems are summarized from the viewpoint of the number of agents and an inner state. Then, the proposed agent system is defined in the following section. For verifying the applicability of the proposed system to the engineering problem, in the following section, we apply the system to a block removing problem, and we discuss the characteristics of the system. The last section gives conclusion of the paper.

RELATED WORK

Boid was proposed by Reynolds (1987) for simulating animal motion such as bird flocks and fish schools. A basic flocking model consists of three simple steering behaviors which describe how an individual boid maneuvers based on the positions and velocities its nearby flockmates, and it has no inner state. Since the interest of the system was placed on simulated motion of the bird flocks and fish schools, the system showed us only aggregation and obstacle avoidance.

Colorni, Dorigo and Maniezzo proposed the ant algorithm (1992), which is inspired by collective behavior of ants. A simulated ant is supposed to put a chemical material of pheromone in its trail. The ant moves towards the direction of the maximum pheromone in each step. If no pheromone is around, the ant moves randomly. Putting a large number of the ants to the field, it can perform massive parallel search. The ant agent has no inner state, as same as the boid agent does not.

Tierra was developed by Ray (1992), which demonstrates a society of computer programs. Each program reproduces itself as a creature does. While the reproduction, miss copy of the program can happen sometimes, like DNA copy error. In a limited resource area of the programs, it emerges the interaction among the programs such as parasitism and hyper-parasitism. It can be viewed as the programs develop a social hierarchy. In the process of the reproduction, the agent requires some byte of memories, but it is not the inner state that we have been discussing.

Lindgren (1992) simulated evolution of agents in the iterated prisoner's dilemma (IPD) game. Each agent has a history of the last several games, and it is used for deciding a move in the next game. The agent learns a strategy through the evolution. Simulating the evolution for thousands of games, several groups of the agents with a same strategy are formed. The agent has the history memories for the strategy evolution, but it is not the inner state considered here.

From the viewpoint of the inner state and the subjective evaluation, the above agent systems are summarized as Table 1.

Table 1: Summary of the artificial life systems.

	State	Evaluation	Agent	Density
Boid	none	local	many	middle
Ant algorithm	none	global	many	middle
Tierra	none	global	middle	high
IPD	none	local	middle	high
Proposed	two	subjective	middle	middle

PROPOSED AGENT SYSTEM

Emotion

In the psychology, several emotion models have been already proposed, although the unified model has not been established yet. In this paper, we employ two sorts of the emotion: pleasure-sadness, and anger-fear, because these two are commonly introduced in most of the models. In addition to the emotion, we introduce affection, which represents a value between like and hate, as one-to-one subjective evaluation.

In the field of the psychology, the following three tendencies are well known: 1) If a group of persons shares emotion by experiencing a same thing, their affection increases each other. 2) If a group of persons have same affection each other, the magnitude of their emotion increases. 3) If a person is affected by the other person, his affection to the other person increases, which is the retribution of the affection. We introduce the above nature and mobility to a proposed agent.

Agent definition

In the following definition, the emotion is represented as two-dimensional inner states and the affection is represented as a scalar value (Naruse, 2003).

The agent i is defined as the following 3- tuple.

$$Agent_i = (A_{ij}, E_i^k, \mathbf{P}_i), \qquad (1)$$

where

A_{ij}: an affection value of the agent i to the agent j. A positive and negative value mean like and dislike, respectively. The magnitude of the value represents the one of the affection. It takes the range of (-1,1).

E_i^k : an emotion value of the agent i to the agent j. The emotion value is two-dimensional (k=1,2), and we call the first dimension (k=1) as the pleasure-sadness axis and the second one (k=2) as the anger-fear axis. The values take the range of (-1,1).

$\mathbf{P}_i$: a position of the agent i.

Agent dynamics

The agent is supposed to interact with all of the other agents in a virtual world, it changes the affection and emotion values, and it determines the direction and magnitude of motion according to the values. The magnitude of the interaction is large if the distance of the agents is small. The following equations are given in the discrete time steps, t and t+1.

The emotion value of the agent i in the k axis follows the next equations.

$$E_i^k(t+1) = E_i^k(t) + \Delta E_i^k(t), \qquad (2)$$

$$\Delta E_i^k(t) = a_1(1+E_i^k(t))(1-E_i^k(t))\{A_{ij}(t)A_{ji}(t)\exp(-lr_{ij})\} - a_2\tau E_i^k(t), \qquad (3)$$

where a_1 is a coefficient of an emotion changing ratio, r_{ij} is a distance from the agent i to j, l is a decay factor with respect to the distance, a_2 is a coefficient of a natural decay, and τ is a time constant of the natural.

The emotion value change of each axis adopts the same equation, which is k=1,2 in Eq. (2) and (3). The change follows a logistic curve with the range of (-1,1), which is represented in the first two terms in Eq. (3). The third and fourth terms in Eq. (3) represents that the emotion increases if the two agent have same affection (the sign of the affection value of the two agents are same). The last term represents the natural decay, so that the emotion goes to neutral (zero) if there is no interaction.

On the other hand, the affection value follows the next equations.

$$A_{ij}(t+1) = A_{ij}(t) + \Delta A_{ij}(t) \quad (4),$$

$$\Delta A_{ij}(t) = b(1+A_{ij}(t))(1-A_{ij}(t)) \\ (c_1 E_i^{\,1}(t) E_j^{\,1}(t) + c_2 E_i^{\,2}(t) E_j^{\,2}(t) + c_3 A_{ji}(t)), \quad (5)$$

where b, c_1, c_2, c_3 are coefficients of each term.

The affection value changes the logistic curve with the range of (-1,1). Also it changes by emotion sharing and the retribution of the affection.

The agent i moves toward the agent j if the affection value is positive. We adopt a repulsive force for preventing collision of the agents. Therefore, the equation becomes

$$\mathbf{P}_i(t+1) = \mathbf{P}_i(t) + \sum_j (d_1 A_{ij} - d_2 \frac{1}{r_{ij}^{\,2}})\mathbf{e}_{ij}, \quad (6)$$

where $\mathbf{e}_{ij}$ is a unit vector from the agent i to j, d_1 and d_2 are coefficients of the terms, respectively.

Figure 1 illustrates the schematic diagram of interaction between two agents.

APPLICATION TO THE BOX REMOVING PROBLEM

Box removing problem

For verifying if the proposed agent system is useful to the engineering problem, we apply it to a box removing problems.

The problem is as follows: We have a workspace filled with a large number of boxes, and we have less free space to maneuver the boxes. An exit where the boxes should go out through is supposed to be narrow, so the boxes are often jammed around the exit. In this situation, we must get all the boxes of the workspace as fast as possible.

Implementation

We need to implement the problem to the affection and emotion of the agent. First, we assign the agent to each of the boxes. The motion of the agent is directly translated into the one of the box. Then, we put task-solving performance into the affection value of the agent as follows: If a box collides with another box and it cannot move toward the exit, the corresponding affection value is decreased. On the other hand, if the box can move smoothly towards the exit, the emotion of the pleasure-sadness axis of the corresponding agent increases.

This implementation is expected to work as follows: If the box is stacked, the affection values of a corresponding agent decreases and it turns to a negative one, which keeps the box away from the stacked position. Therefore, we put the following two equations when the agent collides with the other agent.

$$E_i^{\,2}(t+1) = 1.0, \quad (7)$$

$$A_{ij}(t+1) = -1.0. \quad (8)$$

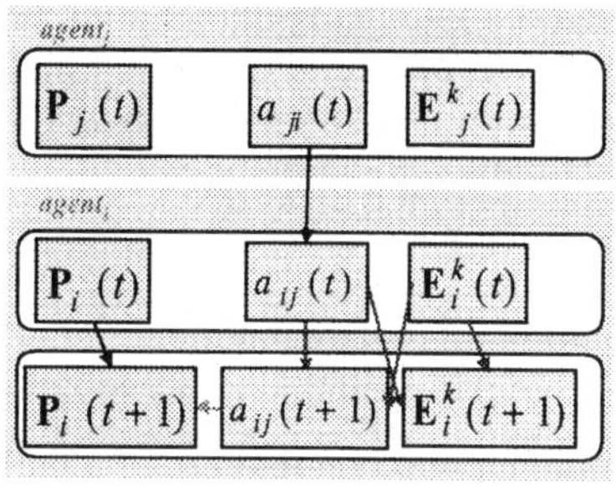

Figure 1: Interaction between agent i and j.

The equations mean the agent gets angry and dislikes an opponent agent. Furthermore, for approaching to the goal, we put the goal potential for the agent motion as follows:

$$\mathbf{P}_i(t+1)' = \mathbf{P}_i(t+1) + \frac{1}{r_i}\mathbf{g}_{ij}, \tag{9}$$

where $1 / r_i$ is a distance from the agent i to the goal, and $\mathbf{g}_{ij}$ is the unit vector from the agent i to the goal.

Experiment

Figure 2 illustrates an inner state of the agents represented by a color and the initial box configuration and the goal location. The boxes can move any directions. When the box is placed on the goal, it is removed from the workspace.

For investigating the effect of the two dimensional innter state, we conduct experiments with an agent with one and two dimensional inner state. Figure 3 shows the change of the number of goal arrival boxes. In the two dimensional one, which includes Eq. (7) and (8), the agents go through the goal more quickly than the one dimensional one.

Figure 4 shows the snapshot of the agents. In the one dimensional inner state, the agents are stacked around the goal and they are falled into a deadlock. An attractive force to the goal and a repulsive force by the other agents getare almost equal, and the agents cannot move anoymore. On the other hand, in the two dimensional case, the anger keeps the agents away from each other when they get close too much. From these results, it seems that the two dimensioanl emotion can mitigate a deadlock around the goal.

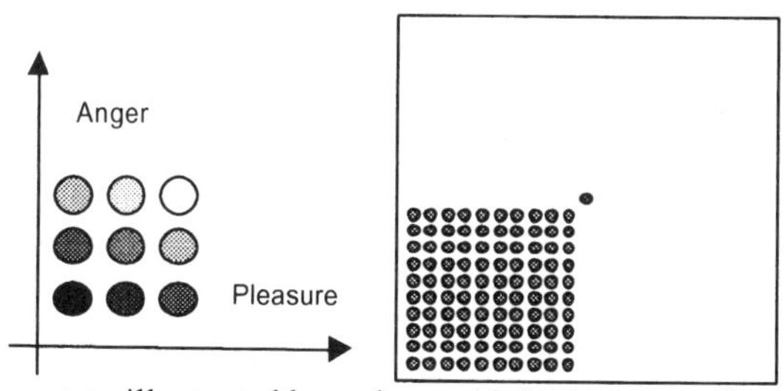

Figure 2: Inner state illustrated by color and initial box configuration.

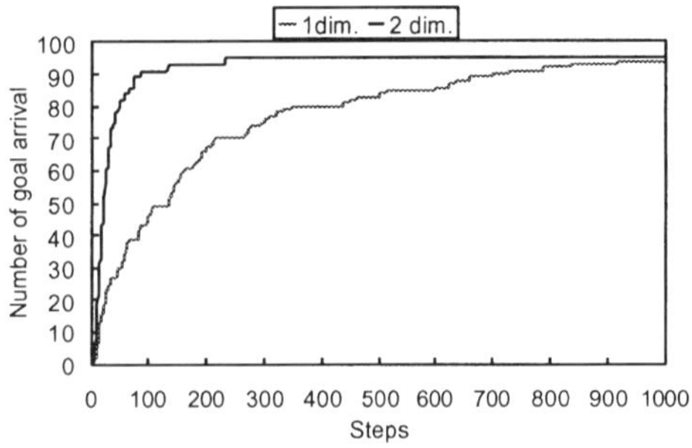

Figure 3: The number of goal arrival boxes.

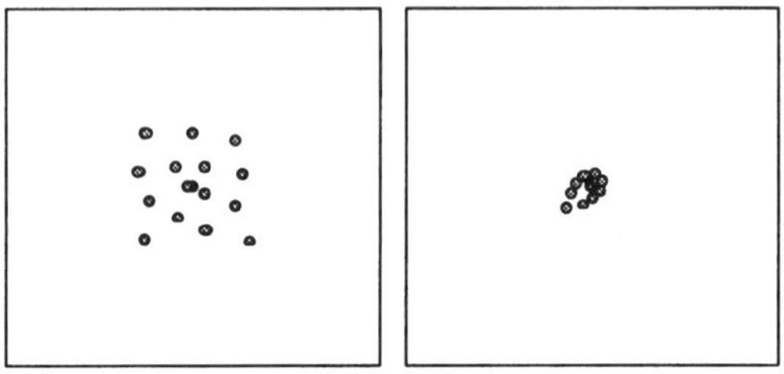

Figure 4: Snapshot of the agents. left: 1-dimensional inner state at the 500th step, right: 2-dimensional one at the 100th step.

CONCLUSION

In this paper, we have proposed the agent system including the emotion-like two-dimensional inner states and affection-like subjective evaluation of the others. For demonstrating the applicability of the proposed system to the engineering problems, we have applied the system to the box-removing problem. We have obtained good performance for the problem.

REFERENCES

Colorni, A., Dorigo, M., and Maniezzo, V., 1992, "An Investigation of Some Properties of an Ant Algorithm," Proceedings of the Parallel Problem Solving from Nature Conference (PPSN 92) , R. Mnner and B. Manderick (Eds.), Elsevier Publishing, pp. 509-520.

Lindgren, K., 1992, "Evolutionary Phenomena in Simple Dynamics," Langton, C., C. Taylor, J. D. Farmer, & S. Rasmussen (Eds.), Artificial Life II, Santa Fe Institute Studies in the Sciences of Complexity, vol. XI, pp. 295-312, Addison-Wesley.

Naruse, K., Kinoshita, M., Yokoi, H. and Kakazu, Y., 2003, "Group Formation of Agents with Two-dimensional Inner State and One-to-one Subjective Evaluation", Proc. of the 5th IEEE International Symposium on Computational Intelligence in Robotics and Automation (CIRA2003), CD-ROM.

Ray, T. S., 1992, "An Approach to the Synthesis of Life," Langton, C., C. Taylor, J. D. Farmer, & S. Rasmussen (Eds.), Artificial Life II, Santa Fe Institute Studies in the Sciences of Complexity, vol. XI, pp. 371-408, Addison-Wesley.

Reynolds, C. W., 1987, "Flocks, Herds, and Schools: A Distributed Behavioral Model," in Computer Graphics, 21(4), pp. 25-34.

THE EFFECT OF IRRELEVANT FEATURES ON AIRS, AN ARTIFICIAL IMMUNE-BASED CLASSIFIER

LOIS BOGGESS
Department of Computer Science and Engineering
Mississippi State University

JANNA SHAFFER HAMAKER
Department of Computer Science and Engineering
Mississippi State University

ABSTRACT
The AIRS classifier, based on principles derived from resource limited artificial immune systems, performs consistently well over a broad range of classification problems. AIRS is a learning system which responds to a training instance by evolving "cells" which are similar to the training instance, as the body's immune system evolves lymphocytes which "match" invading pathogens. Just as in nature some of the lymphocytes are retained indefinitely, in AIRS evolved "memory cells" are retained to be the basis of a k-Nearest-Neighbor classification system. One advantage that AIRS has is that it performs well without extensive experimentation to find an appropriate architecture. Some classifiers are able to ignore or at least not be adversely affected by irrelevant features. This paper explores the effects of irrelevant features on AIRS' performance and compares the performance of AIRS to a similar classifier, LVQ, on classification problems to which irrelevant features have been introduced.

INTRODUCTION

A new classifier, AIRS (Artificial Immune Recognition System), has been shown to perform well over a broad range of classification problems. These include the Iris, Sonar, and Ionosphere classification problems from the UCI machine learning repository (Blake and Merz 1988), which have real-valued features. See (Watkins 2001) for the original AIRS algorithm and its performance on these data sets and (Watkins and Timmis 2002) for the revised AIRS algorithm and its performance on these classification problems. AIRS has also performed competitively with the best classifiers listed in (Duch 2000) on data with nominal features: AIRS's performance on the Cleveland Heart Disease dataset using tenfold cross validation over three runs would place it in eighth place on the Duch list for that data set (Goodman et al. 2003). Two other classification problems with many nominal features are the credit-crx problem, for which AIRS has achieved the highest accuracy of any single classifier of which we are aware (Goodman et al. 2002), and the Wisconsin Breast Cancer data set (Goodman et al. 2003). Marwah and Boggess (2002) report results from a variation of the original AIRS algorithm which achieves an accuracy 5 percent higher on the e. coli data set than any recorded at the UCI repository. AIRS has been tested with as many as hundreds of features, and with few to many classes.

Based on principles derived from resource limited artificial immune systems, AIRS responds to a training instance by evolving "cells" which are similar to the training instances, just as a body's immune system evolves cells which have increasingly higher affinity to an invading pathogen, where affinity is a biological form of matching. In nature a few lymphocytes are retained indefinitely, quickly matching later invasions by the same pathogen. AIRS models this behavior by retaining evolved "memory cells" that become the basis of a k-Nearest Neighbor (kNN) classification system. AIRS typically has higher accuracy than kNN for most of the public classification problems on which it has been tested showing that it is capable of generalizing from the training data.

AIRS has a number of parameters which can be set by the user to fine tune performance for a given classification problem. The most important element of the classifier, however, is self-derived: During training, AIRS determines how many memory cells it should generate to characterize the problem space, within constraints placed upon it by the user's specification of its operating parameters. When a user defaults to the parameter values supplied with the sample problem included with the algorithm, the classifier normally performs within a few percentage points of its optimal performance. Contrast this to many general-purpose classifiers which require much experimentation before an appropriate architecture is determined. On the other hand, some classifiers are able to ignore redundant or irrelevant features. Given that AIRS currently uses Euclidean distance and kNN to classify from the memory cells that it generates during training in reaction to the training data, we surmised that irrelevant features would negatively affect the AIRS classifier. We set out to test this conjecture, and to compare AIRS with Kohonen's Learning Vector Quantization (LVQ) on the same data, since LVQ is widely used and develops output classification vectors which are somewhat analogous to AIRS' memory cells.

EXPERIMENTAL DESIGN

We chose three data sets for the experiment. The Iris data set is well known, uses real-valued features, has a small number of features and is a set for which AIRS has a high accuracy rate (96.0% for the current version of AIRS, as reported in Watkins and Timmis (2002)). Since there are only four features for the Iris data, adding a fifth, irrelevant feature is expected to have a serious negative impact, especially since AIRS uses Euclidean distance and normalizes the data so that all features occupy the same range in feature space. We created the fifth feature by randomly reordering the values which occur in the fourth feature. The values did apply to the Iris classification problem, in proportions which occur in nature, but they were applied randomly without regard to class.

The second data set chosen was the Wisconsin Breast Cancer data set. The original data set has nine nominal features, represented as integer values. We introduced a tenth feature which was randomly generated from the integers from 0 to 9. We chose the Wisconsin data set because it is a nominal data set with a larger number of features, in contrast to the Iris set, but it is also a classification problem for which the AIRS classifier performs well (an average of 96.2% over three runs of 10-fold cross validation (Goodman et al. 2003). In the latter respect it is closely comparable to the Iris data.

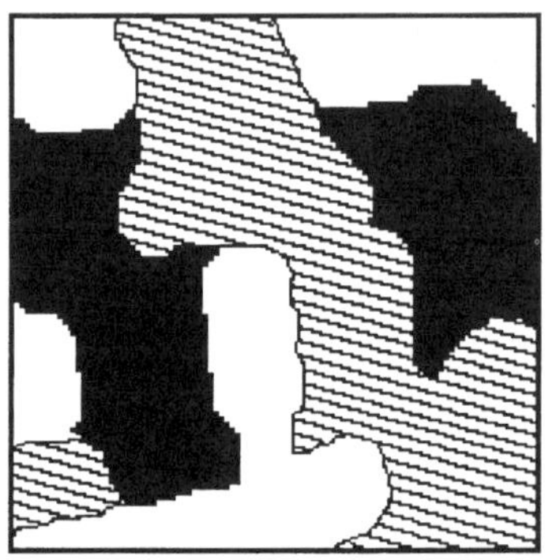

Figure 1. Artificial three-class problem. All white areas belong to one class, both lined areas to another, and both black areas to a third class.

Our third data set was chosen because in some respects it is similar to the Iris problem, yet it is a somewhat more difficult classification task. The data are 1,000 points chosen randomly from the real plane R^2, between (0.0, 0.0) and (1.0, 1,0), where each point belongs to one of three classes, as indicated by the white, gray, and black areas of Figure 1. All points within the white areas belong to the first class, all points within the lined areas belong to the second class, and all points within the black areas belong to the third class. The problem is unlike many naturally occurring problems in that there is no clustering of the most prototypical members of a class - the members of the class are smoothly distributed. The Iris data are limited to one digit before and one digit after the decimal place. In this artificial 3-class problem, we similarly discretized the x and y features to two decimal places, so that in effect a vector in this problem space can have one of at most 10,000 possible values. Choosing vectors for training and testing means that more than 90% of the problem space for this problem is unknown to the classifier. For this experiment a third feature was added - a random floating point number in the range of [0.0, 1.0). Hence one-third of the available data for each vector will be irrelevant.

Given that the accuracy of a classifier can be affected by the random assignment of the data set members to the training and test sets during cross validation, care was taken to assure that LVQ and AIRS received identical training and test subsets during the 10-fold cross validation runs, with the exception of required removal of data labels for LVQ.

OPTIMIZATION

Several kinds of "optimization" were performed to assure that performances for both AIRS and LVQ are not substantially below what they can achieve.. There are several user-set parameters for AIRS. We are highly interested in the behavior of AIRS when it determines the number of memory cells on its own. Therefore we always set the number of seed cells to zero. We ran 10 runs of 5-fold cross-validation for all variations while optimizing AIRS' parameters. First, ATS is varied, leaving other parameters at default values. After a best value for ATS is determined, it is used for the rest of the experiment with that data set. StimT is then varied. After adopting the best StimT found, we determine the optimum number of resources. Finally, the best k to be used for kNN is found.

Once the optimal parameters were determined, we then executed three runs of 10-way cross validation for the Iris and artificial 3-class problem. We report the average of these three runs. For the Wisconsin Breast Cancer data, we noted that variation among the runs while we were modifying parameters was greater than for the other data sets even in the later stages of optimizing. We therefore chose to be more conservative and use five runs of 10-fold cross validation for the final test runs for that data set, and we report the average of those five runs.

LVQ requires the user to specify how many output vectors must be created for the final classifier. To optimize LVQ we started with a small number of vectors (5, 10, 15, and so on), progressing until the average accuracy over the set of three (or five) runs of 10-fold cross validation leveled off or decreased.

EFFECT OF IRRELEVANT FEATURES ON AIRS

Figure 2 shows the effect of adding an irrelevant feature on the accuracy of AIRS. Adding an irrelevant feature to the Iris data caused a drop in accuracy from 96.0% to 95.5%. Adding an irrele-vant feature to the Wisconsin data caused a drop from 97.2% to 96.43%. Adding an irrelevant feature to the artificial 3-class problem caused a drop from 93.8% to 88.0%. In each case the impact of the irrelevant feature was less than expected. In the case of the artificial classification problem, the irrelevant data introduced constituted one-third of all available information to the classifier.

Data reduction is an important aspect of AIRS. Typically, it gives high accuracy with a memory pool size that is 20-50% of the size of the training set. In early stages of optimization for Iris, we noted that increasing the StimT parameter is accompanied by both an increase in memory cells and an increase in accuracy, after the StimT parameter's value rises above 0.5. (See Figure 3, showing averages over 10 runs of 5-way cross validation for the Iris set.) The other classification problems show similar behavior. The default value for StimT as shipped is 0.9. We advise against using low StimT values.

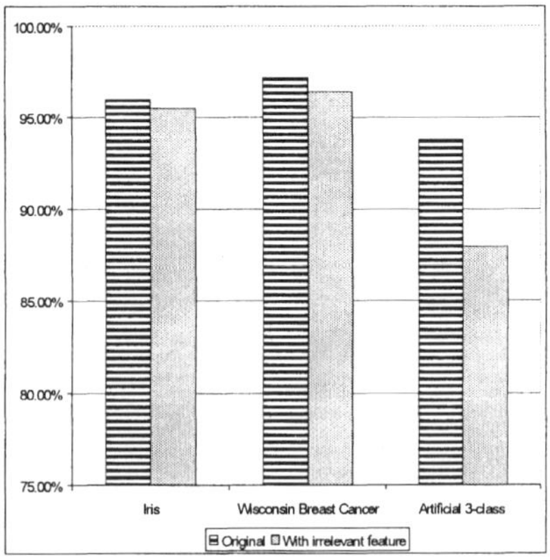

Figure 2. Comparison of accuracy of AIRS on original data and on data with an added irrelevant feature. Values are averages of three runs of 10-fold cross validation except for second column of Wisconsin data, which came from five runs of 10-fold cross validation.

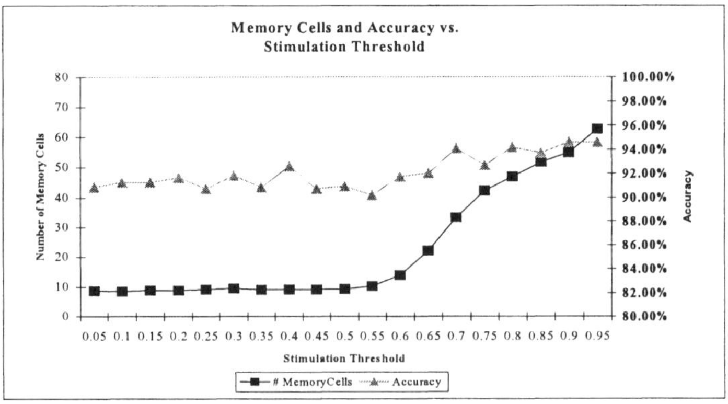

Figure 3. As Stimulation Threshold increases, number of memory cells holds steady until about 0.6. Then number of memory cells climbs quickly, and accuracy slowly increases to ~95%.

LVQ COMPARED TO AIRS ON DATA WITH IRRELEVANT FEATURES

For comparison, we tested Kohonen's LVQ classifier on the data with irrelevant features. LVQ requires that the user stipulate the number of output vectors to be used for the final classifier. To determine an "optimal" size for the LVQ classifier for each data set, we began by stipulating a small classifier, with five output vectors, and increased the number of output vectors until the performance of the classifier was not improved by adding output vectors. Performance was measured as the average of three runs of 10-fold cross validation. In order for the comparison to be fair, the subsets for the 10-fold cross validation were identical to those for the AIRS runs which were reported above except for stripping vector labels, as required by LVQ. Figures 4, 5, and 6 show the performance of LVQ on these data as the number of specified output vectors is increased until performance levels out or decreases. Like AIRS, LVQ performs well on the Iris and Wisconsin Breast Cancer data with added irrelevant features. It does not perform quite so well on the artificial 3-class data set, for which the proportion of irrelevant information is higher (see figure 7).

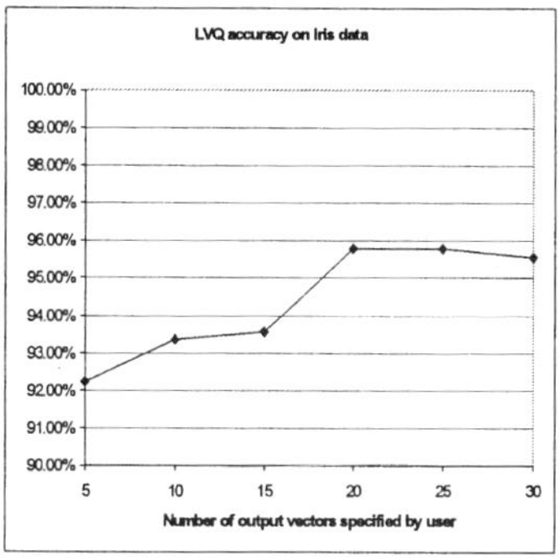

Figure 4. LVQ accuracy on Iris data, as number of output vectors specified by user increases from 5 to 30.

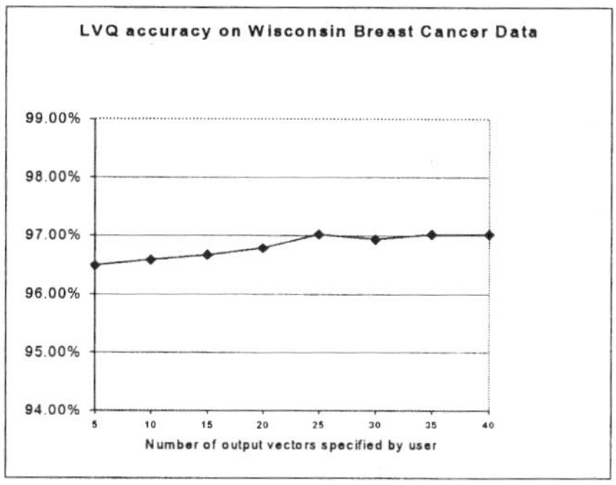

Figure 5. LVQ accuracy on Wisconsin Breast Cancer data, as number of output vectors specified by user increases from 5 to 40

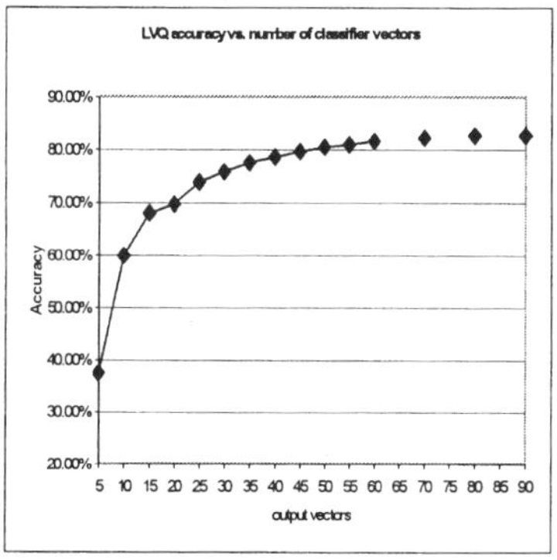

Figure 6. LVQ accuracy on artificial 3-class data, as number of output vectors specified by user increases from 5 to 90.

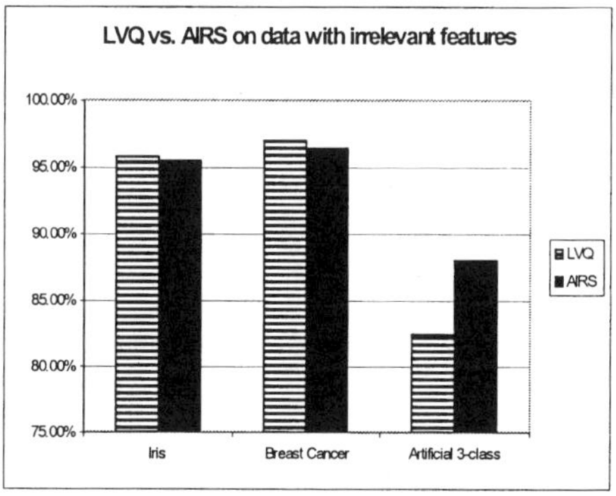

Figure 7. Comparison of AIRS-recommended LVQ with AIRS on data sets with added irrelevant features.

Table 1 shows the results when LVQ was required to use approximately the same number of output vectors as AIRS derived on the data with irrelevant features. For the Iris and Wisconsin data, the resulting performance would be remarkable even on the original data *without* irrelevant features. For example, AIRS achieves a 97.2% on the unmodified Wisconsin data. That accuracy would place in second position of the list of classifiers for that data published by Duch (2000). LVQ, using the number of output vectors that AIRS finds appropriate for its own memory cell pool has a slightly higher accuracy than AIRS on the original data, even though LVQ is dealing with an irrelevant feature.

Table 1: Improved performance of LVQ using AIRS "recommended" number of output vectors.

	LVQ	"Big" LVQ	AIRS recommendation
Iris	95.78%	97.11%	60 vectors
Wisconsin Breast Cancer	97.01%	97.25%	180 vectors
Artificial 3-class	82.50%	86.27%	800 vectors

As previously observed (Goodman et al. 2002), AIRS is a good predictor of optimal size for LVQ classifiers. Conversely, simply increasing LVQ size until performance levels off does not provide LVQ's best performance.

CONCLUSIONS

The experiment results indicate that the presence of irrelevant features has much less impact on AIRS than might be expected. The worst case was a drop of 5% in the case of a complex classification task with only two relevant features, after adding a third irrelevant feature. Similarly, LVQ performed well on data to which irrelevant features were added. The same data set that occasioned the sharpest decrease in performance for AIRS presented an even greater obstacle for LVQ. On the other hand, using the number of memory cells generated by AIRS as an estimate for an appropriate number of output vectors to be specified to LVQ led to superlative results for that classifier on the two real-world problems of the experiment. Finally, our results combined with others strongly suggest that classifiers using LVQ and basing their choice of size on increasing the number of output vectors until accuracy tapers off are probably under-sizing their classifier and unnecessarily hurting accuracy.

REFERENCES

Blake, C.L. and Merz, C. J., 1998, UCI Repository of Machine Learning Databases. http://www.ics.uci.edu/~mlearn/MLRepository.html. University of California at Irvine, Department of Computer Science.

de Castro, L. N. and Von Zuben, F. J., 2002, "Learning and Optimization Using the Clonal Selection Principle," *IEEE Transactions on Evolutionary Computation, Special issue on Artificial Immune Systems*. ftp://ftp.dca.fee.unicamp.br/pub/docs/vonzuben/lnunes/ieee_tec01.pdf

Duch, W., 2000, "Datasets Used for Classification: Comparison of Results," http://www.phys.uni.torun.pl/kmk/projects/datasets.html.

Goodman, D. E., Boggess, L. and Watkins, A., 2002, "Artificial Immune System Classification of Multiple-class Problems," *Artificial Neural Networks in Engineering (ANNIE 2002)*.

Goodman, D. E., Boggess, L. and Watkins, A., 2003, "An Investigation into the Source of Power for AIRS, an Artificial Immune Classification System," *Proceedings of the 2003 International Joint Conference on Neural Networks*, July 2003 (accepted).

Marwah, G. and Boggess, L., 2002, "Artificial Immune Systems for Classification: Some Issues," *1st International Conference on Artificial Immune Systems*, J. Timmis and P.J. Bentley, eds., pp. 149-153.

Timmis, J. and Neal, M. 2001. "A Resource Limited Artificial Immune System for Data Analysis," *Knowledge Based Systems*, vol. 14(3-4), pp. 121-130.

Watkins, A. 2001. "AIRS: A Resource Limited Artificial Immune Classifier," M.S. thesis, Department of Computer Science. Mississippi State University. http://nt.library.msstate.edu/etd/show.as;?etd=etd-11052001-102048

Watkins, A. and Boggess, L., 2002a, "A New Classifier Based on Resource Limited Artificial Immune Systems," *Proceedings of the 2002 Congress on Evolutionary Computation (CEC2002)*, IEEE Press, pp. 1546-1551.

Watkins, A. and Boggess, L., 2002b, "A Resource Limited Artificial Immune Classifier," *Proceedings of the 2002 Congress on Evolutionary Computation (CEC2002)*. Special Session on Artificial Immune Systems. IEEE Press, pp. 926-931.

Watkins, A. and Timmis, J., 2002, "Artificial Immune Recognition System (AIRS): Revisions and Refinements," *1st International Conference on Artificial Immune Systems*, J. Timmis and P. J. Bentley, eds., pp. 173-181

PARALLELIZING AN IMMUNE-INSPIRED ALGORITHM FOR EFFICIENT PATTERN RECOGNITION

ANDREW WATKINS[§], XINTONG BI, AND AMIT PHADKE
{andrew,xb7, aap50}@cse.msstate.edu
Department of Computer Science and Engineering,
Mississippi State University, USA
[§]Also: Computing Laboratory,
University of Kent, Canterbury, UK

ABSTRACT
In recent years, there has been a growth of investigation into the use of mammalian immune systems as a source of inspiration and metaphor for computational tasks. One avenue of this investigation has been the exploration of the learning capabilities demonstrated by these biological systems. A second appealing aspect of biological immune systems is their inherent distributedness. This paper examines these two strains by proposing modifications to an existing immune-inspired algorithm to provide a more efficient, parallel version. This investigation is offered by way of a case-study in the gains paralleization can provide to current Artificial Immune Systems and as a proof-of-concept in the field of parallel immune learning.

INTRODUCTION

In recent years, there has been a growth of investigation into the use of mammalian immune systems as a source of inspiration and metaphor for computational tasks (deCastro and Timmis 2002). One avenue of this investgation has been the exploration of the learning capabilities demonstrated by these biological systems (Watkins, Timmis, and Boggess 2003; Timmis and Neal 2001; deCastro and Von Zuben 2002). A second appealing aspect of biological immune systems is their inherent distributedness. This paper examines these two strains by proposing modifications to an existing immune-inspired algorithm to provide a more efficient, parallel version. Since very little work has been undertaken, to date, in the parallelization of immune-inspired **learning** algorithms, this investigation is offered by way of a case-study in the gains paralleization can provide to current Artificial Immune Systems (AIS) and as a proof-of-concept in the field of parallel immune learning.

The algorithm under investigation, CLONALG, utilizes the immunological concepts of memory cells, free antibodies, clonal selection, and affinity maturation to provide solutions for pattern recognition and function optimization tasks. This algorithm was developed by Leandro de Castro and offers us an ideal test bed for our initial explorations of immune learing in parallel. In (deCastro and Von Zuben 2002), the authors introduce an artificial immune system algorithm inspired primarily by the clonal selection theory first popularized by Burnet (1959). For a complete specification of this algorithm, please see (deCastro and Von Zuben 2002).

CLONALG's primary goal is in the development of memory cells which provide some representation of the data set. Similar to many evolutionary algorithms, CLONALG proceeds through a number of generations in the development of this representation. What really intrigues us about CLOANALG is the almost embarrassingly parallel nature of the algorithm. Since the algorithm is one of the more simplistic immune-inspired models, this seemed to be an excellent starting place for our exploration of parallel computing techniques and Artificial Immune Systems.

DESCRIPTION OF CLONALG AND ITS PARALLELIZATION

CLONALG, as presented in (deCastro and Von Zuben 2002), has two incarnations: one for pattern recognition tasks and one for multi-modal function optimization. We have decided to explore the pattern recognition version more thoroughly. This version of the algorithm proceeds in a similar fashion to many evolutionary algorithms; however, the inspiration for the algorithm is clearly from the realm of immune systems. The input (or antigens) to the algorithm consists of a collection of ***S*** patterns to be recognized. The algorithm begins by generating a random population (***P***) of *N* cells (or antibodies), where $|P| \geq |S|$. For the current application, these cells simply consist of a binary feature string that represents a given binary pattern. A pattern from ***S*** is presented to each cell in ***P***, and the affinity of each cell to the pattern is calculated. Affinity, for the current application, is based on hamming distance and is measured simply as the number of incorrect bits in the antibody when compared to the input pattern. Therefore, a lower affinity number implies a closer match to the input pattern. Once the affinity is calcluated for each antibody in ***P*** as compared to a given input pattern, the best *n1* antibodies are selected to undergo somatic hypermutation as a mechanism for providing affinity maturation. The authors of (deCastro and Von Zuben 2002) specify that the mutation should be dependent on the affinity value of the antibody in two ways: 1) the number of mutated offspring a given cell is allowed to produce increases as the hamming distance decreases and 2) the lower the hamming distance of an antibody the less mutation of the features should occur. CLONALG encourages the exploitation of the good solutions it has found by allowing high affinity (low hamming distance) cells to produce more offspring. It encourages exploration by increasing the mutation rate for low-affinity cells. The input pattern is then presented to each of the newly created offspring, and their affinity levels are calculated. The single best antibody is then compared to the memory cell for the given input pattern. If this antibody's affinity is better than that of the memory cell, the newly generated antibody replaces the established memory cell. Finally, *n2* random antibodies are created and replace existing cells in the population ***P*** of antibodies. This process continues until all the input patterns have been presented to the population. One cycle through the input patterns is considered one generation. The algorithm can continue for a fixed number of generations or until a certain convergence criterion is reached. For the current experiments, the algorithm is only halted when absolute convegence occurs. By absolute convergence, we mean that the memory cell set which has evolved exactly matches the input pattern set. While in real-world learning situations this convergence criterion would not be as useful as some fuzzier threshold which would allow for greater generalization by the system, for our current experiments that examine the behavior of the algorithm as various parameters are adjusted we felt this was a sufficient stopping criterion. The algorithm is presented below, in terms of the immune processes employed.

1) *Initialization*: Create a random population of *N* antibodies ***P*** and memory cells ***M***.
2) *Antigenic Presentation*: for each antigenic pattern in ***S*** do:
    a) *Clonal Expansion*: Determine the affinity of each antiboody in ***P*** to the given antigenic pattern. Select the best *n1* antibodies to produce clones.
    b) *Affinity Maturation*: Mutate clones based on affinities.
    c) *Clonal Selection*: Choose the best clone and replace the memory cell in ***M*** if the clone has a better affinity.
    d) *Metadynamics*: Replace *n2* antibodies in ***P*** with randomly generated cells.
3) *Cycle*: Repeat step 2 for a fixed number of generations or until the memory cells in ***M*** have converged to within some threshold.

Since, essentially, there is no connection among the cells in the CLONALG algorithm (as there would be in a network model of the immune system (Timmis and Neal 2001)), there is no real need for all of this process to occur on a single processor. The goal of the algorithm is simply to discover a set of memory cells and the development of a single memory cell can occur independently of the development of the other cells. For the parallelization of CLONALG, the input set is simply divided by the number of processes involved in the parallel job. Each process then evolves that number of memory cells. Once this finishes, a root process gathers all of these memory cells together in order to present a final result to the user of the evolved pattern recognizing cells. There are three parameters that we examined in our experiments: *N*, the size of the population ***P***; *n1*, the number of antibodies selected for mutation and cloning during each step of the antigenic presentation; and *n2*, the number of random cells generated to replace cells in ***P*** after each antigenic presentation. With the parallelization we need to decide how to handle these user-defined parameters. Currently, we have investigated two options: we can divide the parameter by the number of processes and let that be the setting for each process, or we can let each parameter be the full amount on each process. That is, for example, if the user specified 80 as the parameter for *N* and there are 4 processes in the parallel job, we can let the value of *N* be 20 at each of the processes or 80. Either way should have an impact on the performance of the algorithm.

EXPERIMENTAL DESIGN, RESULTS AND DISCUSSION

For the experiments described below, the binary character recognition data set discussed in (deCastro and Von Zuben 2002) were used. This data set consisted of 8 binary patterns representing a 10x12 graphical representation of the arabic digits 0-4, 6, 7, and 9. The length of each pattern to be recognized, therefore, was 120. We investigated the time to convrgence as well as the number of generations the algorithm took to converge (where convergence is as described above). In order to parallelize the algorithm, we simply divided the input data set accross the processes which each then proceeded to use this divided data set as its input. Once all of the processes completed, the resulting memory cells were gathered back to a root process, and the time and number of generations to convergence (max across the processes) were recorded. With most of the results presented below, we divided these parameter values across the number of processes as well. All results presented represent an average of three runs.

Table 1 shows the effects of varying the parameter *N* on the behavior of the algorithm both in terms of time to convergence and number of generations to convergence. This data is shown for runs across 1, 2, 4, and 8 processors and employs the division of parameters technique. One thing this table shows is that the *N* parameter tends to have a fairly significant effect on the time and number of generations to convergence. What is interesting is that there seems to be a definite trade-off between time and number of generations. That is, the amount of time to convergence increased as the value of *N* increases; however, the number of generations it took to reach this convergence seems to decrease as *N* increases. Also of note is that while increasing the number of processes definitely decreases the overall running time of the algorithm, this increase in processes increases the number of generations needed to converge to a solution. Practically speaking, the decrease in run-time is probably more important to us than the actual number of generations the algorithm took to reach this convergence. Additionally, this increase in number of generations to convergence should not be entirely surprising. With our current scheme of dividing the resources across the various processors, we decrease the number of items the algorithm has to cycle through during a given generation. Naturally, this decrease means that we can then process more cycles in the a given amount of time. The rate of convergence for larger values of our parameters seems to indicate that convergence is encouraged through the diversity of cells being processed which is increased as the various parameters (well *N*, at least) are increased.

Table 2 shows the effects of varying the parameter *n1* on the behavior of the algorithm both in terms of time to convergence and number of generations to convergence. This data is shown for runs across 1, 2, 4, and 8 processors and employs the division of parameter technique. Again we see that an increase in the parameter's value leads to an increase in the running time of the algorithm, and again, we see that the number of processors employed decreases the running time. However, the *n1* parameter has little distinguishable affect on the number of generations needed for convergence. Table 3 shows the effects of varying the parameter *n2* on the behavior of the algorithm both in terms of time to convergence and number of generations to convergence. This data is shown for runs across 1, 2, 4, and 8 processors and employs the division of parameter technique. The *n2* parameter seems to have little overall affect on the behavior of the algorithm, whether in run time or number of generations to convergence. It is possible that larger values for this parameter might produce an effect or that in more complex data sets this parameter might have more significance.

One metric for determining what is gained through the parallelization of an algorithm is speed-up. Speed-up is simply a ratio of the serial run time over the parallel execution time. Ideally, speed-up exhibits linear behavior such that with an increase in the number of processors one sees a proportional decrease in the run time. Figure 1 shows the speed-up obtained from a given run (*N*=40,*n1*=8, *n2*=2) for both styles of parameter division in the parallelization. What is striking about this curve is that we see superlinear speed-up when we employ our parameter divsion method of parallelizing CLONALG. We also see the almost ideal curve when we do not employ this parameter division across the processes. This implies that the parameter division method employed for the parallelization of CLONALG has a significant impact on its overall performance. From Table 1 and Table 2, we suspect this has to do with the reduction of the *N* and *n1* parameters more than anything else. As evidenced from the descriptions of the algorithm in the previous section, these two parameters have the greatest effect on the size of the antibody populations being generated and the interaction of this popualation with the input patterns. A reduction in one or both of these would reduce the number of cells that need to be manipulated as the input patterns are learned.

CONCLUSIONS AND SOME REMARKS CONCERNING PARALLEL AIS

This paper has shown that even simple parallelization techniques can have a role in the field of Artficial Immune Systems. Since often the goal of machine learning systems is to efficiently assist humans in the finding of interesting patterns in large amounts of data, any techniques that can speed up this process should have value. From our experiments with CLONALG, we were able to achieve a great deal of reduction in processing time through some basic parallelization. This seems encouraging.

Beyond just increased computational efficiency, parallelizing immune-inspired algorithms could also enable us to explore the immune metaphors employed more fully. Biological immune systems are inherently decentralized and distributed. Yet, there has been a lack of exploration of the decentralized nature of the immune system particularly in immune-inspired *learning* algorithms. One question that needs to be addressed is what, if anything, will be gained through the use of parallel/distributed computing techniques when applied to these algorithms. Perhaps it will be something as simple as the time-savings as seen with our experiments with CLONALG. However, perhaps it will allow us to build more complex models of the immune system that can solve more complex problems. There is still much left to be explored in this direction including some of the obvious quality of population discussions presented in (Cantú-Paz 1998).

REFERENCES

Burnet, F. 1959. *The Clonal Selection Theory of Acquired Immunity*. Cambridge University Press.

Cantú-Paz, E. 1998. A Survey of Parallel Genetic Algorithms. *Calculators Parallels*, 10 (2): 141-71.

deCastro, L.N. and J. Timmis. 2002. *Artificial Immune Systems: A New Computational Intelligence Approach*. Springer-Verlag.

de Castro, L. N. and Von Zuben, F. J. 2002. Learning and Optimization Using the Clonal Selection Principle. *IEEE Transactions on Evolutionary Computation*, 6(3): 239-51.

Timmis, J. and Neal, M. 2001. A Resource Limited Artificial Immune System for Data Analysis. *Knowledge Based Systems*, 14(3-4):121-30.

Watkins, A., J. Timmis, and L. Boggess, 2003. Artificial Immune Recognition System (AIRS): An Immune-Inspired Supervised Learning Algorithm. *Genetic Programming and Evolvable Machines*, In press.

Table 1: Convergence Times and Number of Generations when Varying the *N* Parameter

	1 proc		2 procs		4 procs		8 procs	
N	Conv. T(s)	Conv. Gen.	Conv. T(s)	Conv. Gen.	Conv. T(s)	Conv. Gen.	Conv. T(s)	Conv. Gen.
10	11.2	72	3.7	149	0.8	293	0.2	587
20	18.4	44	3.3	63	1.2	162	0.3	341
40	31.6	32	6.4	49	1.1	73	0.3	134
80	44.7	22	9.8	33	1.5	39	0.4	79
160	78.1	18	13.7	20	3.0	31	0.5	42

Table 2: Convergence Times and Number of Generations when Varying the *n1* Parameter

	1 proc		2 procs		4 procs		8 procs	
n1	Conv. T(s)	Conv. Gen.	Conv. T(s)	Conv. Gen.	Conv. T(s)	Conv. Gen.	Conv. T(s)	Conv. Gen.
2	11.1	31	1.8	44	0.7	75	0.3	134
4	19.3	33	3.4	48	0.7	75	0.3	134
8	31.6	32	6.4	49	1.1	73	0.3	134
16	48.3	31	9.2	42	2.4	91	0.5	133
32	89.3	33	16.7	46	2.9	69	0.8	134

Table 3: Convergence Times and Number of Generations when Varying the *n2* Parameter

	1 proc		2 procs		4 procs		8 procs	
n2	Conv. T(s)	Conv. Gen.	Conv. T(s)	Conv. Gen.	Conv. T(s)	Conv. Gen.	Conv. T(s)	Conv. Gen.
0	38.3	38	6.2	50	1.3	74	0.3	135
1	31.5	32	6.4	49	1.1	73	0.3	134
2	31.6	32	6.4	49	1.1	73	0.3	134
4	30.6	31	5.6	43	1.1	73	0.3	134
8	33.5	34	5.6	43	1.2	73	0.3	134

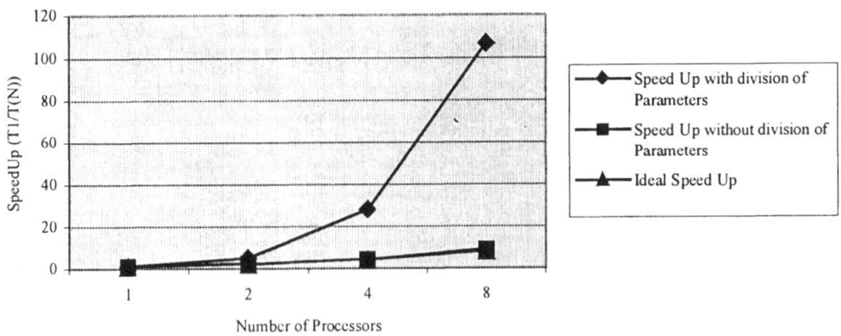

Figure 1: SpeedUp of CLONALG provided through Parallelization

EMERGENT SHORTEST PATHS

CHRIS COULSTON
Computer Engineering
Penn State Erie
Erie, PA 16563

LEE A. MITCHELL
Computer Engineering
Penn State Erie
Erie, PA 16563

ABSTRACT
Ant Colony Optimization (ACO) is an optimization paradigm based on the emergent computational properties of simple ant agents. In order to enhance their ability to explore many solutions simultaneously, discrete ants are replaced with an ant density function. The dynamics of the ant density function are governed by two interacting behaviors utilizing only local information. The resulting algorithm is applied to the problem of finding a shortest path between pairs of points. When the heights of the cells are all equal the algorithm quickly converges to an optimal shortest path.

INTRODUCTION

Ants have the ability to find the shortest path from their colony to sources of food (Dorigo 99). As they travel, ants deposit pheromone trials that are detected and followed by other ants. A computational paradigm built on this theme is embodied in ant colony optimization (ACO) algorithms. ACO algorithms have been applied to a variety NP-hard problems such as the traveling salesman problem and bin-packing (Dorigo 96).

The recent interest in ACOs has motivated examining their application to a variety of geometric problems. Finding a shortest path between two ant colonies in a hexagonally partitioned plane is a simple task which provides a good testbench for a new algorithm. The solutions to shortest path problems generally have a huge number of feasible solutions only a few of which are optimal. In addition, an optimal algorithm already exists for this problem (Coulston 02). Hence, the quality of the derived solutions can be readily computed. It must be stressed that this work is not motivated by the inherent difficulty of finding a shortest path in a hexagonal lattice, rather it is motivated by the desire to explore how two simple local behaviors interact to form global emergent structure, a shortest path.

PRELIMINARIES

Problem instances are created in a lattice of packed hexagons called *cells*. A hexagonal lattice was chosen because it tiles the plane and the distance between the centers of adjacent cells are equal. A cell A is adjacent to cell B if they share a common edge. P is a connected path between cells A and B if starting at A, one can reach B by moving between adjacent cells in P. Each cell is assigned a cost which can be interpreted as the cells height. The goal in this paper is to find a minimal cost path connecting a pair of points.

The thesis of this work states that true collective behavior can only be actualized when the identity of the individual is no longer discernible. This thesis is realized by replacing the representation of an individual ant on a cell with the density of ants in a cell. Mathematically, density is defined as a mass per unit volume. In this application the volume is the constant sized hexagon. Thus, the term mass will be used interchangeably with density. At the beginning of each time step the mass of ants in each cell is completely distributed among the 6 neighboring hexagons. This dividing up of the mass is accomplished using a *distribution function* F(x), defining the proportion of mass to be distributed to neighboring cell x. In order to be a distribution function $\mathcal{F}$ must have the following properties:

- $x \in \{1,2,3,4,5,6\}$
- $0 \leq F(x) \leq 1$
- $\Sigma F(x) = 1$

If the mass of ants in a cell is m then mF(x) mass of ants is moved into neighboring cell x at the beginning of the next time step. From an individual ants perspective the distribution function is a properly defined probability function (Bain 89). The distribution function represents the aggregate behavior of an uncountable number of individual ants via the law of large numbers.

Ants wander around looking for other colonies. While doing this ants will tend towards cells containing the other colonies scent and to low terrain. In order to model an ants propensity to focus on things ahead of it, a weight vector is introduced. The weight vector assigns a scalar to each neighboring hexagon representing the percentage of the ants attention paid to that cell. Cells in front of the ant have a higher weight than those to the rear.

ALGORITHM

The algorithm used to build a path between a pair of colonies in the hexagonal lattice consists of two phases; path formation and path extraction. During path formation the distribution function is invoked to distribute the mass of ants in each cell to their neighbors. During this movement of ant mass scent will be deposited leaving trails for subsequent ants to follow. Later, the path extraction process is invoked, teasing out a shortest path from the scent information.

PATH FORMATION

Since ants are drawn to areas where there is a lot of scent each hexagonal cell must have an associated scent. However, a single scent value for each cell does not afford an ant the opportunity to differentiate its own colonies scent from that of other colonies. Without the ability to differentiate between scents an ant would wander forever around its home colony because this area is generally saturated with its own colonies scent. Hence, each cell stores a scent vector, the ith element denoting the amount of scent from the ith colony. The probability that an ant from colony **c** moves to neighboring cell x is equal to the weighted proportion of scent coming from cell x over the entire weighted neighborhood scent. Scents from an ants home colony are ignored. Thus we have equation 1 describing the distribution function for scent.

$$F_s(x) = \frac{w^x s_c^x}{S} \qquad (1)$$

s^x_c is the sum of the scent in cell x, ignoring any contribution from colony c, w^x is the xth element of the weight vector, and $S = \Sigma\, w^x\, s^x_c$ is the weighted sum of the neighboring scents. Since equation 1 defines F in terms of a part to the whole, F is a distribution function. In addition to scent, ants use height to guide their movement. Since ants favor lower terrain then there must be an inverse relationship between the height of a cell and an ants propensity to move to it. Hence, the probability that an ant moves to neighboring cell x is equal to the weighed inverse of its height over the weighted total of the neighbors. Thus, we have equation 2.

$$F_h(x) = \frac{w^x}{Hh^x} \qquad (2)$$

Where h^x is the height of cell x and $H=\Sigma w^x/h^x$ is the weighted sum of the neighboring heights. As in equation 1, equation 2 defines a proper distribution function. A linear combination of equations 1 and 2 is used to insure that both scent and height guide ant movement. The resulting equation has two non-negative coefficients α and β, determining the relative emphasis placed on scent and height. The sum of these two coefficients is constrained to add to one, hence the linear combination is also a proper distribution function. Thus, we have Equation 3, the distribution function for the mass of ants that move to cell x.

$$F(x) = \alpha F_s(x) + \beta F_h(x) = \alpha \frac{w^x s_c^x}{S} + \beta \frac{w^x}{Hh^x} \qquad (3)$$

At the beginning of each time step all the mass of ants in all the cells is moved according to the distribution function given in equation 3. This movement of mass can result in 1 of 3 outcomes. Either the ants reach a colony or they wander off the lattice or they go to a normal hexagonal cell. In the first two cases the ants are reincarnated back at their original home colony. In the later case the ants deposit scent and then are moved in the next time step.

In order to make the path formation process more efficient, each colony is given a *scoreboard* on which the colony records which directions have lead to which colonies. This scoreboard is used to give ants reincarnated back at their home colony a productive initial direction. If a mass m of ants from a colony *src* arrives at colony *dest* heading in direction d then the scoreboard associated with colony *src* is increased by m k_1 in entry *dest*,d. If this mass of ants instead wandered off the world's edge then the scoreboard entry associated with colony *src* is decreased by m k_2 in entry *des*,d. In order to determine the initial direction of ants at the home colony the scoreboard is summed over all the colony entries for each of the 6 directions. Thus, 6 values are obtained each specifying the utility of a specific direction. A distribution function is then created determining the proportion of ants sent to which neighboring cells.

The scent evolution equations for a cell is based on the premise that the amount of scent deposited in a cell is proportional to the mass of ants in the cell. In addition, the amount of scent in a cell will decrease, via evaporation, by a constant proportion every time unit. These two observations lead to Equation 4 describing the change in a cells scent associated with colony c per unit time.

$$\Delta s^c = E(s^c + D^c) \qquad (3)$$

Where E is the evaporation rate, s^c is the amount of scent of colony c, and D^c is the density of colony c ants in the cell.

PATH EXTRACTION

Path extraction is subdivided into two subphases; extracting a path based on scent and refining the scent path according to height. Path extraction based on scent starts at the source colony. This cell is marked as SCENT_PATH and is assigned the *current cell.* All the other cells are marked as UNVISITED. At each time step the extraction algorithm ranks all UNVISITED neighbors of the current cell according to their *total scent*, the sum of the scent vector elements. The cell with the largest total scent becomes the new current cell and the previous cell is marked as SCENT_PATH. The procedure is halted when the destination colony is reached. The first (and second) phase of the extraction process is shown in Figure 1.

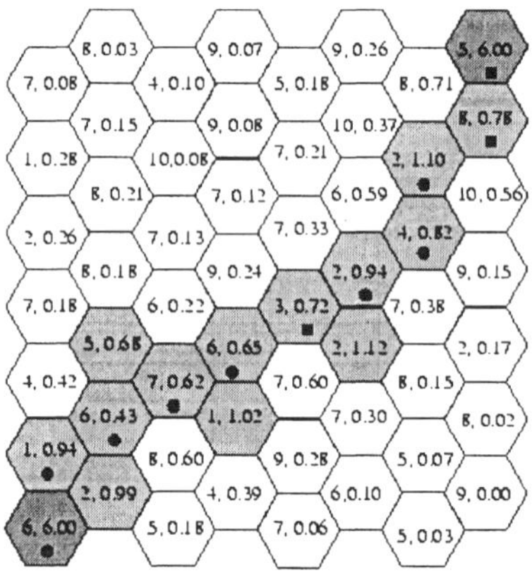

Figure 1: The extraction of a path from scent and height information. The entries in the cells are (height,scent). The source colony is lower left and the destination colony is in the upper right. Cells choosen as part of the SCENT_PATH are shaded and cells choosen as part of the HEIGHT_PATH have black dots in them.

Unfortunately, the first extraction phase generates sub-optimal paths, characterized by self intersecting twists and turns. In order to remove cells not making a necessary contribution to the shortest path a second path extraction pass is performed on the SCENT_PATH cells. Initially, the cell containing the source colony is marked as HEIGHT_PATH and is assigned the current cell. At each time step all SCENT_PATH neighbors of the current cell are ranked according to their height. The cell with the lowest height is picked as the new current cell and the previous cell is marked as HEIGHT_PATH. While ranking the SCENT_PATH neighbors of the current cell if the number of HEIGHT_PATH cells is greater than 1 then the current cell is on a path which intersects itself. Since loops are never present on a minimal path, the search backtracks to the former current cell and chooses the next lowest SCENT_PATH cell. The procedure is halted when the destination colony is reached.

EXPERIMENTS

When all the cells were assigned the same height the algorithm performed optimally. In addition, the system exhibited a strong stable attraction to the global optimum, and was fairly insensitive to changes in the systems parameters. To illustrate this phenomena, the evolution of the scent trail between two colonies is shown in Figure 2. In this figure there are two colonies positioned in opposite corners. Most of the initial salvo of ants miss the opposite colony from time 0 to 80. However, after this first salvo, enough scent was deposited by the few successful ants that a strong path forms from time 100 to 160. Note, the scoreboard is very effective in situations like the one pictured. For example, sending ants to the right from the upper right colony is useless. The scoreboard for the upper right colony would quickly develop a negative value for "right", hence ants would be forbidden to leave the colony in that direction.

CONCLUSIONS

In this paper we show how an ACO can be adapted to solve a geometric problem. The algorithm is based on the premise that collective behaviors are only realized when the identity of the individual is no longer discernible.

While the problem solved, shortest path, is not computationally difficult the methodology employed is novel in that it is a completely local interaction of two behaviors which generates a globally sound structure.

Future work needs to focus on two areas; improving the results presented and extending the results. The simulation parameters are currently set in an ad-hoc manner. A systematic analysis should be undertaken to establish relationships between parameters and their effect on the solution. Second, the work presented should be extended to solve other interesting geometric problems which are intractable.

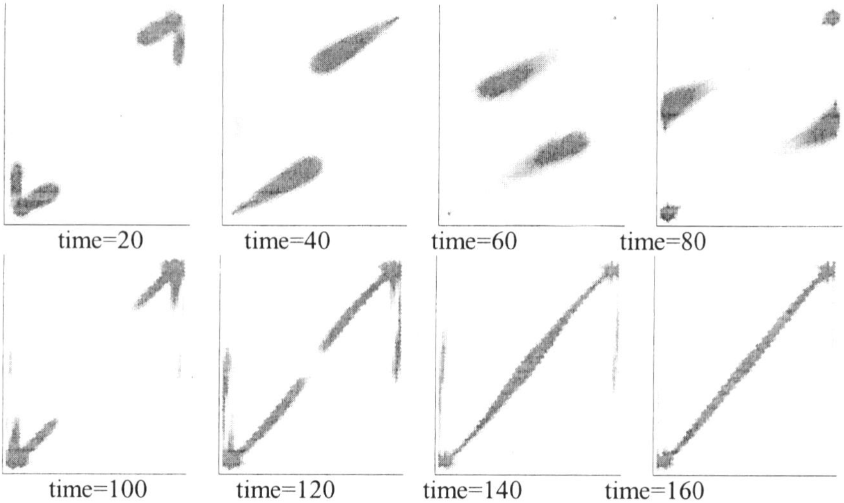

Figure 2: A sequence of images showing the total scent of each cell as a path evolves between two colonies in opposite corners. Darker cells have higher total scent.

REFERENCES

L.J. Bain, M. Engelhardt, *Introduction to Probability and Mathematical Statistics*. PWS-Kent, Boston, 1989.

C. Coulston, M. Smith and K. Werger, "Ants and Evolution: the non-uniform Steiner minimal tree problem", *ANNIE '02*, 7pp. 384-390, 2002.7

M. Dorgio and L.M. Gambardella, "Ant Colony systems: A Cooperative Learning Approach to the Traveling Salesman Problem", *IEEE Trans. on Evolutionary Computing*, Vol. 1, No. 1, pp. 73-81, 1996.

M. Dorigo, G. DiCaro and L.M. Gambardella. "Ant Algorithms for Discrete Optimization", *Artificial Life*, Vol. 5, No. 2, pp. 137-172, 1999.

FOLLOW THE LEADER MOTION STRATEGY IN DOUBLE-OCTAHEDRAL PARALLEL TRUSSES

IRA DUNN
Engineering Management Department
University of Missouri – Rolla
Rolla, Missouri

CIHAN DAGLI
Engineering Management Department
University of Missouri – Rolla
Rolla, Missouri

ABSTRACT

Variable Geometry Truss robot manipulators (VGTs) using parallel structure show many characteristics superior to serially-connected manipulators, including low weight, high structural strength, and a high number of degrees of freedom. The characteristics of one particular VGT, the Double-Octahedral Parallel Module (DOPM), allow the truss structure to move and compress with three degrees of freedom. This paper describes a new motion method for the DOPM-VGT, called Follow the Leader, in which serially-connected VGT modules move single file in a path to reach a goal trajectory. This strategy relieves the system from the necessity of determining complex obstacle avoidance strategies. Moreover, Follow the Leader motion allows the VGT manipulator to work in a decentralized manner, minimizing interconnectivity and central processing requirements. This paper improves upon existing control techniques for DOPM chains by implementing an on-board neural network.

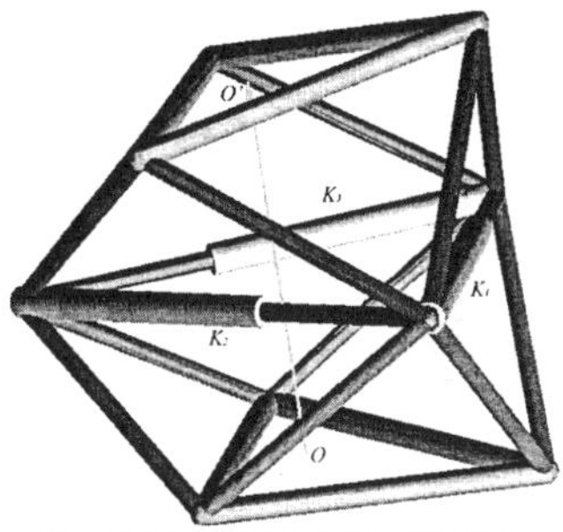

Figure 1. Double Octahedral Parallel Module

INTRODUCTION

A Variable Geometry Truss manipulator (VGT) is any of a number of configurations of structures with variable-length members, typically arranged in a parallel, truss-like structure as seen in Fig (1). Forms of this type of robotic manipulator have been in use for almost 50 years (Stewart, 1965), but the modern concept of the VGT was originally conceived in 1985 by Crawford and Mikulas (1985) and Miura et al (1985). In this paper, a new methodology of motion and path planning for serially-connected VGT chains, called the Follow the Leader Strategy is demonstrated. This strategy exploits the VGT's ability to flex in two axes as well as to shorten/extend in length.

The VGT that this paper is concerned with is the Double-octahedral Parallel Module (DOPM), as developed by NASA in the 80's (Rhodes and Mikulas, 1985). Williams (1994) and others have shown that this module can be modeled as the serial mechanism of a spherical joint connecting both faces by extensible translational joints of equal length, as seen in Fig (2). The length of the actuators corresponds in a highly non-linear fashion to the coordinates of the serial model parameters, θ, ϕ, and L.

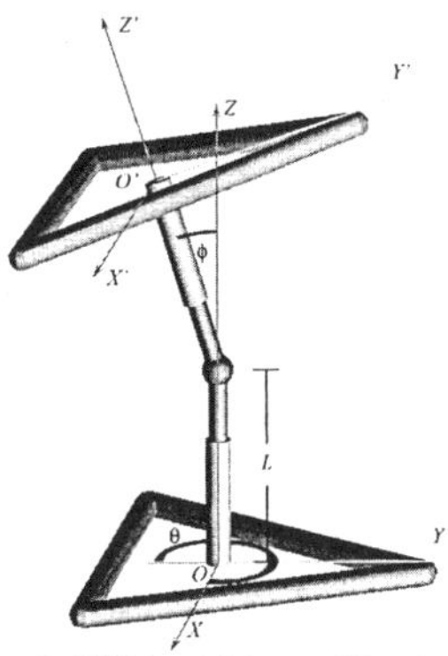

Figure 2. DOPM modeled as a serial mechanism

FOLLOW THE LEADER STRATEGY SUMMARY

The focus of this paper is a strategy in which the individual VGT modules in a chain "follow" each other during motion, giving the chain the appearance of remaining on the same path in space. This is dubbed "Follow the Leader," or FtL in this research. Follow the Leader avoids complex obstacle avoidance strategies, because only two space-paths must be generated that are obstacle-free. The chain moves from its initial position to its terminal position by "recessing" and "retracting" backwards, then "precessing" and "extending" forwards. The motions of recession and precession are such that the modules attempt to cover a constant length of the space curve, whereas the motions of retraction and extension are such that the modules cover less or more curve length, respectively.

There are many benefits to FtL motion, including simple obstacle avoidance, simple distributed computation, and less interconnectivity. The last benefit comes from the fact that a module only communicates with neighboring modules during motion. It can be inferred, also, that if optimization has been performed for the path, such as minimized path length, then the same optimization applies to the individual modules.

The precepts of FtL are implemented here by dividing the curve into discrete vectors $\boldsymbol{S}$ of constant length ρ. The vector shown in Figure 1 from O to O' is the characteristic vector describing the VGT's orientation, and is the sum of N discrete vectors, as seen in Eq (1).

$$\mathbf{V}_{VGT} = \sum_{j=1}^{N} \mathbf{S}_j \tag{1}$$

When the module moves forward or reverse the $\boldsymbol{S}$ information is passed forward or reverse to neighboring modules and the vector is redefined. The order of operations in FtL motion are listed here: 1) Communication of new $\boldsymbol{S}$-vector to end module 2) Redefinition of V_{VGT} in all modules 3) Recalculation of $\boldsymbol{Q}$ 4) Application of $\boldsymbol{Q}$ to on-board neural network to find K-vector of actuator lengths 5) Change in actuator lengths.

Calculation of New Q-vector.

The most algebraically complex calculation performed during motion is the determination of the new $\boldsymbol{Q}$-vector. From the geometry of the serial model, we have the relationships in the following equations.

$$\tan\left(\cos^{-1}\left(\frac{V_z}{\|\mathbf{V}\|}\right)\right) = \frac{\sin\phi_G}{1+\cos\phi_G} \tag{2}$$

$$L = \frac{\|V\|}{2\cos\frac{\phi_G}{2}} \tag{3}$$

$$\theta_G = \theta_V = \arccos\left(\frac{V_x}{\sqrt{V_x^2 + V_y^2}}\right) \tag{4}$$

Forward Kinematics of the DOPM.

The fundamental control problem concerning parallel mechanisms is the non-linear transformation from the [x, y, z] space to the [x', y', z'] space seen in Fig (2). In the forward kinematic problem, the actuator lengths are known and the transformation must be gained from the DOPM parameters given. For the VGT in Fig (1), the kinematic parameters used in this research are given in Table 1. (C.f. Williams, 1998)

Table 1. List of VGT parameters for the DOPM

Symbol	Parameter Name	Description	Range
L_0	Lower face batten length	The length of each of the lower face members	114mm
L	Lower longeron length	The length of each of the 6 longeron connecting the lower face with the midplane face	100mm
K_1, K_2, K_3	Actuator length (variable)	The length of each of the 3 linear actuators in the midplane face	105<K_i< 175mm
S	Offset length	An offset length used to compensate for actual VGT components at the actuators	S=0 (S<<L typically)

The forward kinematic problem is NP, and we apply an Evolutionary Strategy (ES) to first solve for a set of intermediate variables, Θ_1, Θ_2, and Θ_3 seen in Fig (3), given a single sample of K-values. The Euclidean distances between the tips of the lower three faces represent the actuator lengths $K_1 \ldots K_3$. The ES uses the maximum error in distances as the fitness for each trial solution and evolves accordingly. Subsequently, the spherical variables θ_G, ϕ_G, and L_G for each solution are found deterministically from the $\Theta_1 \ldots \Theta_3$ provided by the ES. (Sofge and Chiang)

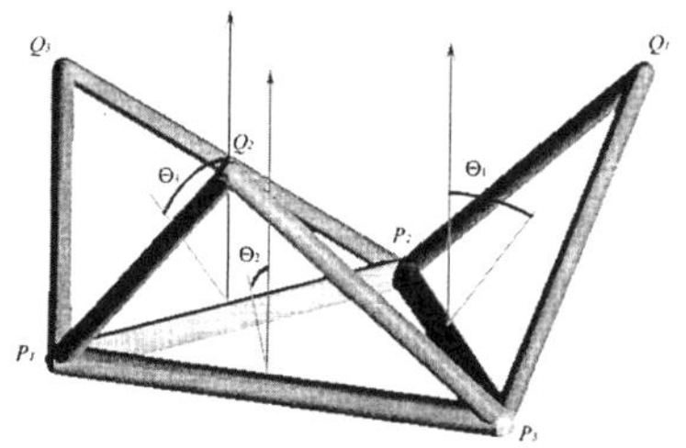

Figure 3. Lower half of DOPM-VGT, showing intermediate Θ-variables

We are interested in the distances between the Q-points, so we define the distance set in Eq (5)

$$\mathbf{K}' = [\ \|\mathbf{Q}_2 - \mathbf{Q}_3\|, \|\mathbf{Q}_3 - \mathbf{Q}_1\|, \|\mathbf{Q}_1 - \mathbf{Q}_2\|\ ] \tag{5}$$

Basic Evolutionary Strategy Architecture.

Since the equations defining $\boldsymbol{Q}$ are non-linear and non-separable for $\boldsymbol{\Theta}$, a numerical method employing an evolutionary strategy is used to solve the transformation $\boldsymbol{K} \rightarrow \boldsymbol{\Theta}$. The ES architecture is similar to a steady-state evolutionary computation algorithm, commonly known as a $(\mu + \lambda)$ algorithm. The population members in this work consist of 3 values: The 3-dimensional Θ-vector sample, a 3-dimensional vector σ specifying a variance for each dimension of Θ, and the fitness evaluation for the Θ-vector. When being mutated, the members use a Gaussian normal perturbation with zero-mean and sigma equal to the associated dimension's variance. The ES uses the delta-sigma variance update routine to evolve the standard deviations along the way. Some ES parameters are listed in Table 2.

Table 2. Evolutionary strategy parameters.

Parameter	Description	Value
Generations	Strategy iterations	4000
Population size P	Sample solutions in population	100
Mutation probability	Probability of being mutated in reproduction	10%
Crossover probability	Probability of two parents contributing to offspring	5%
$\Delta\sigma$	Stochastic variance of member variance values	0.1
Number of parents	Members to be chosen each generation for reproduction	8

The strategy begins with a single sample vector $\boldsymbol{K}$ and a set of random vectors $\Theta_1 \ldots \Theta_P$, where P is the population size. It then proceeds to solve for the distance set $\boldsymbol{K}'$ in Eq (5) for each of the Θ-vectors population members. Each member is assigned a fitness term equal to the maximum absolute distance error between the elements of $\boldsymbol{K}'$ and its respective goal vector $\boldsymbol{K}$.

$$Fitness = \max(abs(\mathbf{K}'(i) - \mathbf{K}(i))),\ i = 1 \ldots 3 \tag{6}$$

Each population member is ranked by a summed percentage method. That is, every member's fitness contributes some percentage to the total fitness of the population. In this case, fitness is a function to be minimized, and the individual fitness contribution is described in Eq (7).

$$Contribution = \frac{1 - Fitness / \sum_{i=1}^{P} Fitness(i)}{P - 1} \tag{7}$$

The difference between two consecutive members' ranks is the percentage contribution of the second. In this way, a roulette-wheel selection is used without a sorting routine by using a normally distributed random variable. Two parents at a time are chosen and stochastic crossover and mutation operators are applied to generate new offspring. The offspring are injected into the population via a highly deterministic algorithm. That is, the offspring are compared to the least-fit existing population member. If they are more fit, they replace the existing population member. If an offspring fares worse than the worst population member, that offspring is discarded and the next offspring is compared until the offspring list is exhausted.

We sampled the K-space bounded by the limits in Table 1 (K_{min} = 105mm, K_{max} = 175mm) with a sample spacing for each dimension of 5mm. This yielded a total of $15^3 = 3375$ data points to solve for. The ES program was run on each sample point until a solution within a specified tolerance was found. An error tolerance of 0.7mm was chosen since the actuator swing is 70mm, making the error be <1% of the relative change in actuator length. Because some populations did not converge on the correct solution in the first execution, the algorithm was run again on the failed solutions only until all sample solutions were collated. The overall program was executed 30 times, with an average in-tolerance count of 2239 (σ = 47.83) and mean population fitness of 1.409 (σ = 0.054).

Inverse Kinematics of the DOPM.

The final step in offline programming is to generate the K-values given the coordinates of the upper face, or "end-effector," in terms of the gimbal parameters. Again, this is an NP-problem, and proves to be even more complex than the forward kinematics. However, since we know a sampled relationship between $\boldsymbol{K} = [K_1, K_2, K_3]$ and $\boldsymbol{Q} = [\theta_G, \phi_G, L_G]$ from the ES, we train a neural network to learn the inverse kinematics for us, allowing us to interpolate for any $\boldsymbol{Q}$. We implement a feed-forward neural network to map $\boldsymbol{Q}$ back onto $\boldsymbol{K}$. This is similar to that developed by Zanganeh and Williams (1998), in that it is a well-studied, simple architecture trained using back-propagation of error.

The particular architecture to use is a four-layer feed-forward network with an input layer, two non-linear hidden layers, and a linear output layer. The input is the three dimension $\boldsymbol{Q}$-vector and the output is the three dimension $\boldsymbol{K}$-vector. The first hidden layer is composed of 12 neurons with tangent-sigmoid transfer functions. The second hidden layer contains 10 neurons, also with tangent-sigmoid transfer functions. The output layer contains three saturating linear neurons, chosen to prevent the actuators from going beyond their physical limits (K_{min}=105, K_{max}=175).

Neural Network Training.

The network is weighted and biased randomly upon initial startup. The network training is performed over the 1687 data points spaced 10mm apart per dimension using the Levenberg-Marquardt modification of Newton's Method. The training was performed 14 times on a 2.4GHz PC. The Mean Squared Error goal was 1.0E-05 and the learning rate was .01. The average MSE after 500 epochs was 8.279E-6 with standard deviation of .776e-6. Statistical analysis showed that there was a 95% confidence that the network would train to within [7.9, 8.6]*e-6 MSE after 500 epochs.

SIMULATION AND RESULTS

This software simulation was performed as a composite of programs using C++, Microsoft Excel, Matlab v6.5, and the kinematic modeling package ADAMS v12.

The offline applications are used solely for training purposes. The forward kinematic problem $\boldsymbol{K} \rightarrow \Theta$ is solved for a sample population in advance of simulation execution. This is done with the author's C++ evolutionary algorithm executable. The conversion of $\Theta \rightarrow \boldsymbol{Q}$ data was performed offline in a spreadsheet. Matlab performed supervised neural network training with the $\boldsymbol{K}$ and $\boldsymbol{Q}$ data and saved the weights for online implementation.

Matlab was then used to determine the parameterization of the VGT modules online, calculating all tabular data and handling module memory. This information was handed to ADAMS via a file of commands that represent sample data derived in Matlab. ADAMS then simulated the VGT in a modeling environment with primitive objects.

Simulation Examples

A series of simulations were executed to test for error, range of motion, and, most importantly, total path maintenance. The VGT chain in test was made up of 10 modules, which begin in a semi-circular spooled position. The vectors describing the loci of the end-effectors are listed in Table 3.

(1) The VGT modules are positioned in a point in space by fitting themselves to a straight line. The modules then apply FtL techniques to reach a different point, also by a straight line. This motion is a good exercise in aligning all modules into the same path, since deviation from this path will be clear.

(2) The VGT move along a sinusoidal curve in space with fixed magnitude and frequency. The chain then moves to a new position in space by a different sinusoid with a second frequency and magnitude. The first curve has a period of 1000 steps and the second has a period of 1400 steps.

(3) The VGT chain is extended to a point in space by fitting the modules to an arc. The chain then fits the modules to an arc in an orthogonal direction, reaching a new goal point. The first arc is in the –z direction, making the VGT chain lie in the y-z plane, while the second arc is in the +x direction, making the chain lie in the x-z plane. Both arcs have radii equal to the spool radius. This is a good test of the VGT's ability to synthesize symmetry, important because the VGT structure is made of triangular faces and non-orthogonal members and, thus, is non symmetric about the Cartesian axes

(4) This complex shape is formed by adding a third-dimension component to an arc. The first helix is a right-handed screw along the +x-axis, whereas the second helix is a right-handed screw along the –x-axis. This simulation will examine the overall motion capabilities of the chain. It will be the most complex motion described thus far.

Table 3. End-Effector locations for simulation

Simulation	EEx	EEy	EEz
L1	0.0	0.0	-1232.3
L2	-271.1	-400.8	-1135.9
S1	0.0	760.6	-871.3
S2	0.0	852.1	-792.9
A1	0.0	-784.5	0.0
A2	784.5	0.0	0.0
H1	200.0	912.2	-223.4
H2	-200.0	-912.2	-223.

Experimental Results

The end effector was moved to the first point, brought back onto the spool, then moved to the second goal point. The error points are described as the simulated position relative to the theoretical position of the tenth module's distal platforms' center of mass. Table 4 charts the errors of the terminal end-effector locations.

(1) The straight path errors are seen in Table 4 entries L1 and L2. Both goals were reached with reasonable accuracy (1.5% and 0.51%). Several "humps" of increasing magnitude could be seen along the path trace.

(2) The sinusoids behaved better than the linear paths from the above section. No "humps" were witnessed, but the simulations had the largest errors (1.9% and 2.2%) of any of the simulations.

(3) The arcs showed the least error (0.17% and 0.5%), despite being characteristically unique from the spool arc. The paths also did not show any "jerking" motion.

(4) The helix showed some error, but as a complex motion, it proved effective in reaching the goal point using FtL strategies within acceptable tolerance (1.6% and 1.3%).

Table 4. Component error for simulation

Simulation	Δx	Δy	Δz	Magnitude	% error
L1	12.9	-12.8	2.50	18.3	1.5
L2	1.7	-5.9	1.2	6.3	.51
S1	1.94	-18.4	-14.6	23.6	1.9
S2	4.22	-15.0	-23.4	28.1	2.2
A1	-0.91	0.0	-1.9	2.1	.17
A2	1.1	-5.5	-2.5	6.1	.5
H1	3.8	0.2	-19.5	19.9	1.6
H2	1.7	0.7	16.7	16.8	1.3

Discussion of Error.

The simulations of the VGT in ADAMS were run several (3 or 4) times apiece, all yielding results similar to the respective results noted above. The maximum error seen in Table 4.1 was 28.1mm in the secondary goal of the sinusoidal simulation. This corresponds to 2.28% error in overall length traversed by the VGT chain. In angular terms, if the error vector was orthogonal to the end effector, this would mean an angular difference of 1.3 degrees at the VGT base since the truss structure is approximately 1.2m in length.

During the linear simulations, the effect of small errors near the spool on the end effector could be seen in a trace line of the center of mass of the end effector. These errors increase in magnitude as the VGT extends. The immediate cause of these humps is not known. However, analysis of these actuators reveals that one or more actuators neared a limit. Successive simulations showed that when the limit is reached, larger, more pronounced jerking of the truss could be seen.

CONCLUSIONS

This paper outlines the theoretical motion methodology of Follow the Leader. This strategy was implemented on a 30 degree of freedom variable geometry truss structure using double octahedral parallel modules. The conclusions of the simulation are two-fold. First, the VGT has shown considerable versatility in its movements when maintaining a single path of motion according to the precepts of Follow the Leader. Second, the VGT has attained successive goal points with reasonable error.

The first conclusion is drawn from the qualitative nature of motion of the VGT chain. The simulations show that the chain moves off of the spool in an orderly and pursuant path, with each module "following" the module ahead of it. Little translational motion can be observed throughout the simulation, as is dictated by FtL. The second conclusion can be gained by the errors gathered from ADAMS at simulated goal points. Some of this error can be accounted for in small errors introduced by the neural network and the simulator. Further investigations will be needed to indicate how pronounced the network or simulator errors are, or how profound their effects are.

Of special note are the results of the two tools implemented in this research—the evolutionary strategy and the artificial neural network. The ES proved effective in generating test and training points for later use within the specified tolerances. It was able to overcome the problem of solving the three non-linear, non-separable equations defining the VGT's shape in a reasonable and acceptable amount of time. The neural network proved itself an indispensable tool in its ability to map the serial parameters into the parallel actuator lengths quickly and easily. It was trained and tested in an acceptable amount of time, and performed within desired error limits.

ACKNOWLEDGMENTS

I would like to acknowledge the guidance and inspiration of Dr. Cihan Dagli, as well as the help and suggestions received from Drs. David Enke and Daniel Tauritz.

REFERENCES

Stewart, D., "A Platform with Six Degrees of Freedom." Proc. of Institute of Mechanical Engineers, Vol. 180, Part 1, No.15, 1965, pp 371-38

Crawford, R. F., and Mikulas, M. M., "Sequentially Deployable Maneuverable Tetrahedral Beam." United States Patent, Patent NR. 4,557,097. 1985

Miura, K., Furuya, H., and Suzuki, K., "Variable Geometry Truss and Its Application to Deployable Truss and Space Crane Arm." Acta Astronautica Vol. 12. No.7/8. 1985. pp 599-607

Rhodes, M.D., and Mikulas, M.M., Jr, "Deployable Controllable Geometry Truss Beam." NASA Technical Memorandum 86366, NASA Langley Research Center, Hampton, VA 1985

Williams, R.L., II, "Kinematic Modeling of a Double Octahedral Variable Geometry Truss (VGT) as an Extensible Gimbal." NASA Technical Memorandum 109127, NASA Langley Research Center, Hampton, VA. 1994

Zanganeh, K. E., and Hughes, P. C. "Fast Estimation of the Kinematics of the Parallel Modules of the Variable Geometry Truss manipulator using Neural Networks." IEEE International Conference on Robotics and Automation. 1998

Williams, R.L., II, 1998, "Maximizing Kinematic Motion for a 3-DOF VGT Module." Journal of Mechanical Design, Vol. 120, No. 2, 1998, pp 333-336

THE ROLE OF HYPOTHESIS FILTERING IN AIRS, AN ARTIFICIAL IMMUNE CLASSIFIER

DONALD E. GOODMAN, JR.
Mississippi State University
Department of Psychology
Mississippi State, Mississippi

LOIS C. BOGGESS
Mississippi State University
Department of Computer Science
Mississippi State, Mississippi

ABSTRACT
The recently introduced AIRS classifier, based on principles of Artificial Immune Systems, has been shown to be a highly effective general-purpose classifier when tested on a broad range of publicly available classification problems with both real-valued and discrete-valued features, large numbers of features and/or large numbers of classes. Central to the classification process of this new classifier is the pool of "memory cells" analogous to lymphocytes in a natural immune system which have been successful in reacting and adapting to an invader and now are permanently part of the immune system, ready to recognize these invaders if they are ever encountered again. This paper empirically examines the hypothesis that the power of the AIRS classifier is not so much in the evolution of the memory cells themselves but in the decision process by which a potential candidate for the memory cell pool is accepted into that pool as well as the determination of whether a new memory cell replaces an existing member of the pool.

INTRODUCTION

In this paper we explore the characteristics of the Artificial Immune Recognition System (AIRS), a recently introduced classifier based on the principles of resource-limited artificial immune systems. AIRS is a resource-limited, supervised artificial immune machine learning system primarily inspired by work with resource limited unsupervised systems (AINE) by (Timmis and Neal, 2001).When compared against a listing of the 10-30 best classifiers on publicly available classification problem sets (Duchs, 2000a,b), AIRS's performance was found to be competitive with the top five to eight classifiers for every task but one (Watkins, 2001; Watkins and Boggess, 2002a; Duch, 2000a). The details of the AIRS algorithm are given in (Watkins, 2001) and (Watkins and Boggess, 2002b). Since its introduction AIRS has been tested on additional well-known classification problems and has been found to have the best accuracy of any single classifier (i.e., exclusive of committee machines, bagging and boosting) of which we are aware, on the E. coli data set (Marwah and Boggess, 2002) and on the credit-crx problem (Goodman, Boggess, and Watkins, 2002). The latter result was obtained with the revised AIRS algorithm which began distribution in 2002 (Watkins and Timmis, 2002).

Artificial immune networks have been shown to be robust platforms for machine learning. In the case of AIRS, the system is self-tuning, to a degree: For a given set of user-specified parameters, it determines an appropriate number of memory cells for the resulting classifier. Furthermore, the accuracy of AIRS when run with the default parameters tends to be within a few percentage points of its accuracy with optimal parameters. Hence it requires no *a priori* knowledge of the mathematical properties of the dataset under scrutiny, unlike some classifiers. Moreover, the search for an appropriate architecture, which is so important for neural networks such as backpropagation and LVQ, is of much less significance for AIRS. AIRS uses Euclidean distance in feature space to perform classification. Initially it was demonstrated to perform well on classification tasks using real-valued features (the Iris data, Ionosphere data, Pima Indian Diabetes data, and Sonar data from the UCI repository (Blake and Merz, 1998)) (Watkins and Boggess, 2002a). Since then it has been shown to compete with the top classifiers of data sets with nominal data, such as the Cleveland Heart Disease data (Watkins, Timmis, and Boggess, 2003) and the Wisconsin Breast Cancer data (Goodman, Boggess, and Watkins, 2003).

The remainder of this paper will be divided into two parts. The first part will briefly review the AIRS algorithm. The second part will then detail the experimental procedure and results for this investigation.

THE AIRS LEARNING ALGORITHM

The AIRS training algorithm is aimed primarily at developing a pool of memory cells. These memory cells are used during testing to classify novel vectors using k-nearest neighbor, an idea first developed by (de Castro and Von Zuben, 2002),. Unlike ordinary k-nearest neighbor, the vectors used during testing are not identical to the training set, but instead are refined from it through the training algorithm. This training algorithm proceeds as follows.

The memory cell pool is first seeded with the first training instance from each class; if the problem is a two-class problem, the memory cell pool will be seeded with one training vector from class 1 and one from class 2. The remaining training vectors are then presented to AIRS one at a time. For each training vector, the nearest memory cell in feature space with the same class as the training vector is chosen. This memory cell undergoes hypermutation, producing many offspring, who in turn compete for access to resources. Resources are allocated based on similarity to the presented training vector. Those most similar receive a larger proportion of the resources. Those offspring without enough resources die off, and the remaining offspring are allowed to mutate. This process repeats until the average distance of all surviving offspring is within a pre-specified threshold of the training vector. At this point, the closest offspring is chosen as a memory cell candidate.

The memory cell candidate is then submitted to the memory cell selection process. This process is the focus of our investigation, as it filters the memory cell candidates, keeping the size of the memory cell pool small. The candidate memory cell is admitted to the pool if it responds to the training vector more strongly than any other existing memory cell. The candidate memory cell will replace the previous best responder (its ancestor) if the distance between the candidate and its ancestor is within a specified threshold: it is this function specifically which prevents the memory cell pool from growing without bound. Usually, the memory cell pool will grow quickly with the first few training vectors, and then shift around once enough memory cells have been found to more or less cover the feature space.

Typically, the number of cells that AIRS creates in the memory cell pool is about half the number of training cells presented to it. Once training is complete, the memory cells are used as a k-nearest neighbor classification system for test data.

MEMORY CELL SELECTION

In this investigation, we are interested in understanding the role that the memory cell selection routine plays in AIRS's overall performance. We believe that one of the strengths of AIRS is its ability to maintain classification accuracy while maintaining a small pool of memory cells to use for classification.

We constructed a number of alternatives to AIRS's original memory cell selection routine. Each of these replacement routines was designed to eliminate one or more features of the original that we believed were crucial to proper functioning, high accuracy, and high generalizing power (i.e. data compression). The first revision (mod1) is like the regular AIRS selection routine in that it requires a candidate memory cell to be closer to the training cell than any existing memory cell in order to be added to the memory cell pool. However, it never replaces an existing memory cell; it is simply added to the pool. The second routine (mod2) always replaces: that is, a candidate cell is always admitted, and always replaces its ancestor. Finally, the third replacement routine (mod3) simply admits the candidate without test, but never replaces an existing candidate. In the figures, we refer to the original, unmodified AIRS algorithm as mod0; in this algorithm, a candidate must pass an admitance test (it must be closer to the training vector than any memory cell) and may replace its ancestor (if it is close enough to its ancestor to be within a specified threshold).

To recap: in regular AIRS, each training cell is the basis for the development of a potential new memory cell. The candidate memory cell is likely to be very similar to, but not identical with the training cell. If the candidate is closer to the training cell than the previously closest memory cell, then it is admitted to the memory cell pool, and if it is sufficiently close to that previously closest memory cell, it will replace it. Mod1 never replaces existing memory cells, so it ends with many more memory cells. Mod3 is like mod1 except that it makes no admittance tests, either, so that it ends with even more memory cells. Mod2, on the other hand, because it always replaces the previous best memory cell for a given class, ends with the same number of memory cells that it started with. AIRS allows a classifier to be seeded with an initial number of memory cells greater than one per class. However, most of our recent explorations of AIRS have been associated with its properties when it is allowed to determine an appropriate sized classifier without explicit seeding by the user. Hence, for the experiments reported here, mod2 ends with one representative per class. Please note that the memory cells which are used for classification at the end of the training process are *not* elements of the training set, in the typical case. While it is possible for a user to set certain parameters so stringently that successful candidates are forced to be virtually identical to the training instances which gave rise to them, this is not the general case.

There is one parameter, the stimulation threshold, which would affect the performance of all of the revised routines. The stimulation threshold determines how close a candidate memory cell must be to the original training vector in order for the candidate-producing process to stop. During training, after the closest memory cell is located and it undergoes hypermutation, the average distances to the original training vector of all the mutated offspring is measured. If this average distance is less than the stimulation threshold, that phase of training stops, and the closest offspring is nominated as the memory cell candidate. If the average distance is not within the stimulation threshold, the process continues, with under-resourced offsrping dying off, and offspring with access to many resources undergoing hypermutation. We varied the stimulation threshold over four values for this experiment, from 0.3 to 0.9 (the legal range is [0.0, 1.0]).

We report three data sets in our evaluation. Two of these were taken from the UCI Machine Learning Repository (Blake & Merz, 1998): the credit-crx dataset and the iris dataset. Additionally, we used the 3-class artificial dataset from (Goodman, Boggess and Watkins, 2002). Credit-crx was chosen because to date, AIRS has been shown to be the best performing single classifier on this dataset; it is therefore a dataset that AIRS should be able to perform relatively well on even when the algorithm has been deliberately degraded. The iris dataset was chosen because it is one of the best-known benchmarks of classification. Lastly, we had previously found (Goodman, Boggess and Watkins, submitted) that the 3-class artificial dataset shows some fundamental differences from datasets taken from real-world data. It is somewhat more challenging, however, so might show problems from the modifications that would not be evident in "easier" classification problems.

EFFECT OF DIFFERENT MEMORY CELL SELECTION POLICIES

Figures 1 through 3 show the results of running AIRS with each of the four memory cell selection routines, at four levels of stimulation threshold, on three different datasets. Each data point presented represents 20 runs of 5-fold cross-validation, to ensure maximal significance; the error bars represent the 95% confidence interval about each data point.

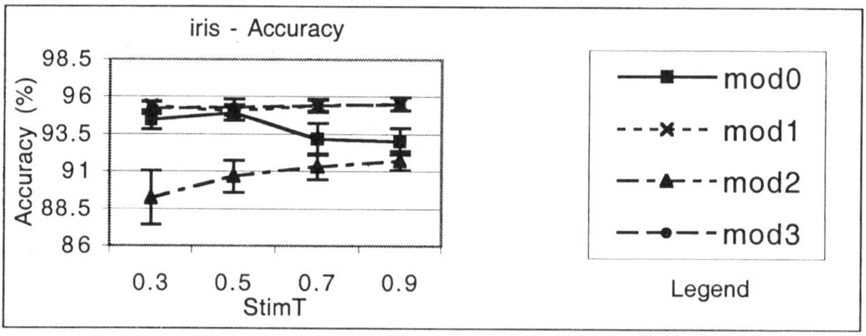

Figure 1.

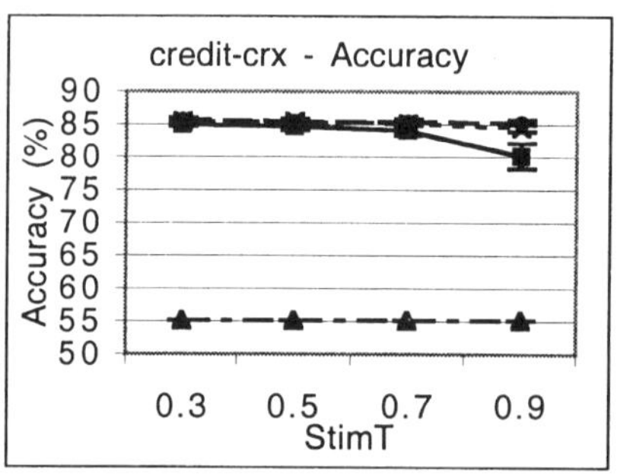

Figure 2.

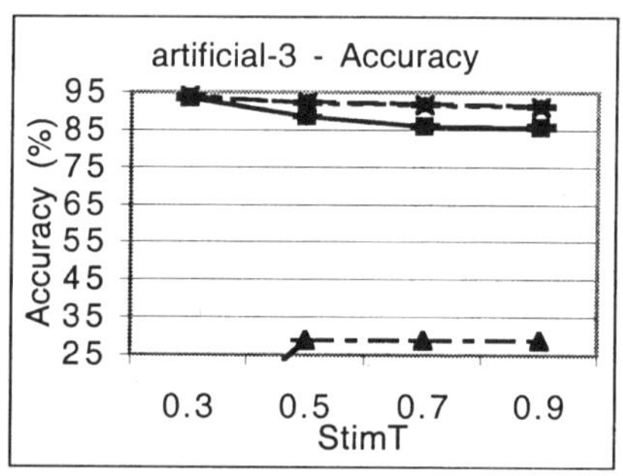

Figure 3.

The first and most obvious point to make is that mod2 (always admit, always replace) ends with one exemplar from each class, which is highly similar to the last training instance of that class presented during training. To a certain extent it is surprising that mod2 is as successful as it is. Secondly, it is clear that mod1 and mod3 (never replace, admit if pass test and never replace, always admit, respectively) stay on par with the original, unmodified selection routine (labeled "mod0") with respect to accuracy. To understand the full implication of this observation, we must examine the size of the final memory cell pool - that is, the amount that the training data has been reduced.

An examination of the number of memory cells retained at the end of the training period reveals that mod1 and mod3 in essence have one memory cell for every training instance. For mod3 this was expected - there should be the initial seeds (one per class) plus one new memory cell for each training cell. (Recall that these cells do not match the training data exactly, because the training data are pre-processed: the candidate memory cells will probabilistically fall within a certain, non-zero, distance of the original training vector. There is still a one-to-one correspondence, however.) We had expected mod1 to have fewer memory cells than mod3 but more than regular AIRS, since it never replaces a memory cell but does require that a new memory cell be closer to the training instance than the progenitor which gave rise to it. However, modifications to the original AIRS algorithm which were intended to improve run-time efficiency and model somatic hypermutation (Watkins and Timmis, 2002) have had the effect that in the version of AIRS currently being distributed, in practice the candidate memory cell is virtually always closer than its progenitor. Hence it appears that the admission test of regular AIRS and mod1 has extremely little effect. As the table shows, the noticeable difference between mod1 and mod3 is that mod3 has the same number of memory cells as mod1, plus the initial seed cells (one per class) that began the memory cell generation process. Mod2, starting out with one seed memory cell per class, and consistently replacing them with new candidate cells, accumulates no additional cells. As shown in Table1, however, Mod0, the regular AIRS selection scheme, substantially reduces the data while retaining essentially the same high accuracy rates as the modifications which retain one cell for every training instance.

Table 1.
Average Memory Cell Pool Size

dataset	Mod0	Mod1	Mod2	Mod3
Iris	17.74	119.84	3.00	123.00
Credit-crx	100.75	532.13	2.00	534.80
Artificial-3	244.32	800.00	3.00	803.00

CONCLUSIONS

From the data presented in this investigation, and the data from previous investigations (Goodman, Boggess and Watkins, 2003; Watkins and Boggess, 2002b), we may draw some conclusions about the source of the computational power of AIRS. AIRS has two features that make it an attractive classifier. The first is high accuracy on a broad range of datasets. The second is strong data reduction. This paper investigated the effect of modifying several parts of the memory cell selection algorithm. In addition to discovering that one aspect of that algorithm no longer serves a practical purpose, we have shown that the parts of AIRS which perform data reduction allow for it to perform at essentially the same accuracy as it would with no data reduction.

We note in passing that the accuracy attainable with a single representative of each class in the well-known Iris data set is high - a fact which might serve as a baseline for all classifiers tested on this data set.

We have begun a comparison of AIRS with the performance of regular k-nearest neighbor applied to the actual training vectors themselves, to determine the degree to which the genearalization incorporated into memory candidate cell production affects accuracy. We expect to report on the comparison in the very near future.

REFERENCES

Blake, C.L. and Merz, C. J., 1998, UCI Repository of Machine Learning Databases. http://www.ics.uci.edu/~mlearn/MLRepository.html. University of California at Irvine, Department of Computer Science.

de Castro, L. N. and Von Zuben, F. J., 2002, "Learning and Optimization Using the Clonal Selection Principle," *IEEE Transactions on Evolutionary Computation, Special issue on Artificial Immune Systems.* ftp://ftp.dca.fee.unicamp.br/pub/docs/vonzuben/lnunes/ieee_tec01.pdf

Duch, W., 2000a, "Datasets Used for Classification: Comparison of Results," http://www.phys.uni.torun.pl/kmk/projects/datasets.html.

Duch, W., 2000b, "Logical Rules Extracted from Data," http://www.phys.uni.torun.pl/kmk/projects/rules.html.

Goodman, D. E., Boggess, L. and Watkins, A., 2002, "Artificial Immune System Classification of Multiple-class Problems," *Artificial Neural Networks in Engineering (ANNIE 2002).*

Goodman, D. E., Boggess, L. and Watkins, A., 2003, "An Investigation into the Source of Power for AIRS, an Artificial Immune Classification System," *Proceedings of the 2003 International Joint Conference on Neural Networks*, July 2003.

Marwah, G. and Boggess, L., 2002, "Artificial Immune Systems for Classification: Some Issues," *1st International Conference on Artificial Immune Systems*, J. Timmis and P.J. Bentley, eds., pp. 149-153.

Timmis, J. and Neal, M. 2001. "A Resource Limited Artificial Immune System for Data Analysis," *Knowledge Based Systems*, vol. 14(3-4), pp. 121-130.

Watkins, A. 2001. "AIRS: A Resource Limited Artificial Immune Classifier," M.S. thesis, Department of Computer Science. Mississippi State University.

Watkins, A. and Boggess, L., 2002a, "A New Classifier Based on Resource Limited Artificial Immune Systems," *Proceedings of the 2002 Congress on Evolutionary Computation (CEC2002)*, IEEE Press, pp. 1546-1551.

Watkins, A. and Boggess, L., 2002b, "A Resource Limited Artificial Immune Classifier," *Proceedings of the 2002 Congress on Evolutionary Computation(CEC2002).* Special Session on Artificial Immune Systems. IEEE Press, pp. 926-931.

Watkins, A. and Timmis, J., 2002, "Artificial Immune Recognition System (AIRS): Revisions and Refinements," *1st International Conference on Artificial Immune Systems*, J. Timmis and P. J. Bentley, eds., pp. 173-181

Watkins, A. , Timmis, J. and Boggess, L., 2003, "Artificial Immune Recognition System (AIRS) Revisited," *Genetic Programming and Evolvable Machines,* Klewer, (accepted for publication in November, 2003)

A NOVEL GENETIC ALGORITHM WITH KING STRATEGY

HUI WANG[a], ANNA L. BUCZAK[b*], HONGAN WANG[a]
[a]Institute of Software, Chinese Academy of Sciences, Beijing, China, 100080
[b]Advanced Technology Laboratories, Lockheed Martin, Cherry Hill, NJ 08002

ABSTRACT
In this paper, a new reproduction strategy for Genetic Algorithms (GA) is proposed, and a novel GA with King reproduction strategy (King GA) is established. A new mutation operator is also proposed that helps to deal with the multi-modality of the fitness functions and performs search more efficiently than mutation operators from the literature. Experimental results show that King GA with the new mutation yields much faster convergence to the optimum than the traditional GA on a set of multi-modal and deceptive test functions.

1. INTRODUCTION
During the last four decades there has been a growing interest in algorithms that rely on analogies to natural phenomena. One type of such algorithms is the Genetic Algorithm (GA), which imitates the principles of natural evolution. In this paper we are approaching the problem of GA convergence from a practical standpoint. In the majority of practical problems, the most important goal is to obtain a quasi-optimal solution in the shortest possible time. In all but the simplest GA applications, fitness function evaluation takes the highest percentage of GA run time. This is why CPU time spent by GA on finding the optimum is proportional to the number of fitness function evaluations performed. As such the smaller search space that can be covered and still a quasi-optimal solution found, the less time consuming the GA search will be.

Our objective is to develop a GA that ensures faster convergence to the optimum than canonical GAs. Wolpert et al. (1996) presented a number of "no free lunch" theorems and established that for any algorithm, any high performance over one class of problems means underperformance over another class. The purpose of our study is not to propose a new GA that works well on all problems, but to develop a GA that gives superior performance for massively multi-modal and deceptive functions, that are difficult and time consuming to optimize due to large search spaces and multiple optima that the functions exhibit. Our interest in optimization of such functions is due to the fact that they describe well real-world phenomena such as choice of routers in computer networks, organization of transistors on chips, RNA folding, evolvable hardware, tracking of targets by sensor networks, etc.

2. GENETIC ALGORITHM WITH KING STRATEGY
The GA that we designed imitates somewhat the reproduction mechanism of bees. Bees are social insects, living together in highly organized colonies. There are three kinds of bees [www-1]: workers, drones and the queen. Worker bees are the most numerous members of the colony; they do not mate and therefore

---

[*] This work was performed when Anna L. Buczak was with Honeywell Laboratories.

cannot produce fertile eggs. Drones are male bees, and their only function is to mate with new queens. The queen is the only sexually productive female in the colony, only she can lay eggs and as such she is the mother of all drones, workers, and future queens.

Inspired by the bees' reproduction process, a novel GA, called King GA, is proposed. In King GA, a special individual, the fittest individual in the population, is always selected in the reproduction process to be one of the parents, the second parent is chosen by roulette-wheel selection or by tournament selection. This is quite different from canonical GAs that generally utilize the same type of method to select both parents. King GA reproduction process is shown in Fig. 1 a.

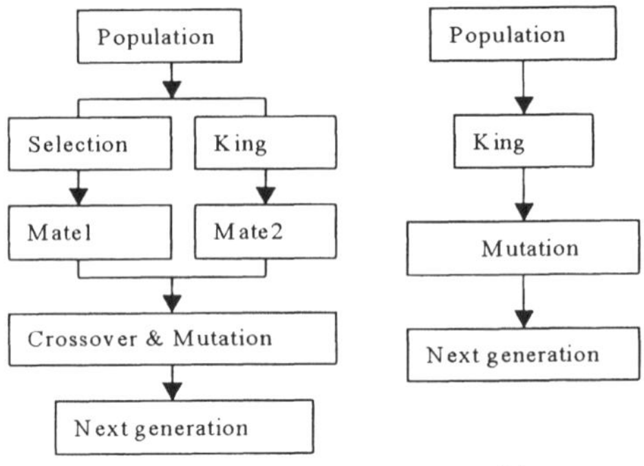

Fig. 1: The reproduction in a) King GA; b) King Mutation.

3. SPECIAL MUTATION OPERATOR

In GAs, mutation is generally thought of as disruptive and important only for ensuring that useful genes are not lost. Some researchers (Aguirre, 2002; Srinivas, 1995) however pointed out the importance of mutation for performing the actual search. The mutation methods they proposed were very similar, the difference merely being the value of the mutation rate, or whether the rate was constant or adaptive. Our previous experiments with GAs (Buczak, 2001) showed that when mutation performs a strong enough search, crossover is not necessary for finding the optimum of multi-modal functions with non-ordered parameters. Based on those results, we are proposing King Mutation, a version of King GA in which only mutation takes place. The reproduction process of King Mutation is shown in Fig. 1b.

A GA individual is composed of several chromosomes that have many genes (bits) each. The individual on Fig. 2 has 6 chromosomes, and 8 genes in each chromosome. During regular mutation each gene is chosen for mutation with a probability P_m. Therefore multiple genes in multiple chromosomes may be mutated in one event; the number of chromosomes or genes to be mutated is not pre-specified but each gene in each chromosome undergoes mutation with probability P_m. Consequently the number of chromosomes that can undergo mutation ranges from 1 to n, where n is the number of chromosomes in an individual.

In the new mutation operator that we are proposing, called MutationC_2, exactly two chromosomes of an individual are randomly selected to be mutated. Any number of genes in chosen chromosomes may undergo mutation, each gene with probability P_m. Fig. 2 shows the process of MutationC_2. Exactly two chromosomes are randomly selected to undergo mutation: the first and third chromosome. On Fig. 2 there is a mutation of one bit in the first chromosome, and a mutation of three bits in the third chromosome. King Mutation algorithm with only the mutation of type Mutation C_2 is called King MutationC_2.

Individual before mutationC_2: ***10010110*** 11001111 ***00001010*** 10001011

Individual after mutationC_2: ***10110110*** 11001111 ***11001000*** 10001011

Fig. 2. MutationC_2 occurs on exactly two chromosomes (in this case 1 and 3).

4. EXPERIMENT DESCRIPTION

In an attempt to examine the quality of the GAs proposed, we performed a set of experiments that compared the performance of KingGA, KingMutationC_2, and canonical GA on a set of test functions. We use the Effectiveness to compare the performance of different GAs. To decrease the effect of randomness, several runs of each experiment were performed; the runs that reached the optimum are called "optimal runs". We define Effectiveness as the ratio of the number of optimal runs to the number of total runs:

$$\text{Effectiveness} = \text{Number of Optimal Runs} / \text{Total Runs} \qquad (4\text{-}1)$$

Another important issue in GAs is how fast the algorithm reaches the optimal solution. In our experiments, all runs were done on a fixed Search Space (SS), which is the total number of fitness evaluations performed. The size of SS is important because of the nominal amount of time that each fitness calculation takes. An algorithm is desirable if high Effectiveness can be obtained when covering a small SS.

Deb et al. (1999) investigated the performance of GAs on a set of test functions ranging from unimodal to massively multimodal and deceptive. The functions described below were used for checking canonical GA performance by Deb and were tested by us with King Mutation algorithm and its variations.

Test 1: Rastringin's function

$$F1 = 100 + \sum_{i=1}^{10} (x_i^2 - 10\cos(2\pi x_i)) \quad x_i \in [-6,6] \qquad (4\text{-}2)$$

Rastringin's function was defined with 10 independent variables, and in this form it is considered massively-multimodal. To find its optimum using a GA each independent variable is coded as a separate chromosome and each individual is made up of ten chromosomes.

Test 2: Deceptive function

The deceptive function optimized (see Eq. 4-3) was the sum of ten unitation functions. The unitation function *u* computes the number of ones in a string. F2 is a difficult function to find the optimum of, since the low-order building blocks corresponding to the deceptive attractor (string of all zeros) are better than those of the global attractor (string of all ones).

$$F2 = \sum_{i=1}^{10} g(u_i) \qquad (4\text{-}3)$$

u, number of 1s in the 4-bits string	0	1	2	3	4
g(u)	3	2	1	0	4

5. RESULTS AND DISCUSSION

<u>5.1 Test1: Results for Rastringin's function</u>

The experimental results for different GA algorithms for F1 are shown in Table 1. The population size, and the number of generations were constant across the algorithms (100 and 450 respectively). We used 16 bits to code each variable, the length of a chromosome is 16, and the length of an individual is 10*16. The mutation rate is set to 1/IndividualLength, which is 0.00625. The mutation rate for King MutationC_2 is 1/ChromosomeLength, which is 0.0625.

Table 1. Performance of Different GAs in Optimizing Rastringin's Function. Where Opl#: Number of runs that found the optimum; Runs: Number of experiments. Ef: Effectiveness

GA method	Pc	Pm	Op#	Runs	Ef
GA	0.9	0.00625	6	30	0.20
King GA	0.9	0.00625	12	30	0.40
King Mutation	0.0	0.00625	1	30	0.03
King Mutation	0.0	0.0625	0	30	0.00
KingMutationC_2	0.0	0.0625	30	30	1.00

King GA gave a better result than the canonical GA, increasing the effectiveness from 0.2 to 0.4. King MutationC_2 gave the best results of all the GAs studied and when compared to a canonical GA, the Effectiveness was increased fivefold.

In order to evaluate how mutation rate, population size and number of generations affect the results of King MutationC_2, we performed experiments described in Table 2. The results show that King MutationC_2 is a robust algorithm effective across a wide spectrum of population sizes and mutation rates.

Table 2. Further experiments with King MutationC_2. Where Pop: Population size; Gen#: Number of generations.

Pc	Pm	Pop	Gen#	Op #	Runs	Ef
0	0.3	100	450	30	30	1.00
0	0.0625	10	4500	30	30	1.00
0	0.0625	1000	45	28	30	0.93
0	0.1	10	2000	30	30	1.00

The best results obtained by King MutationC_2 and the best results obtained by Deb for F1 are given in Table 3. The important factor when comparing the results is the number of fitness function evaluations performed by each algorithm i.e. the size of SS.

Table 3. Performance of King MutationC_2 and Best of Deb GAs.

Method	Pc	Pm	Pop	Gen#	Op#	Runs	Ef	SS	Saving
KingMutationC_2	0	0.1	10	1000	24	30	0.80	10000	64.2%
KingMutationC_2	0	0.1	10	2000	30	30	1.00	20000	28.3%
Deb's best GA	0.9	0	1000	45	45	50	0.90	27900	0.00%

In his thorough experiments, Deb (1999) tested numerous GAs with population size ranging from 2 to 2000, probability of crossover ranging from 0.0 to 0.9, and probability of mutation ranging from 0.0 to 1/L (L is the length of an individual). GAs tested performed well only with large populations - the best performance was obtained for a population size of 1000, a high crossover rate, and no mutation. His conclusion was that adding mutation even with low probability drastically reduces Effectiveness. In contrast, the population size for the King MutationC_2 was only 10 for both experiments and in second experiment Effectiveness of 1 was obtained (as compared to Deb's 0.9) while reducing the SS. This result shows how much more effective King MutationC_2 is than traditional GAs for optimization of a massively multi-modal function.

5.2 Test2: Results for Deceptive Function

The results for deceptive function F2 are shown in Table 4. The population size, and the number of generations are constant across the methods (100 and 450 respectively). The length of a chromosome is 4, and the length of an individual is 10*4. The mutation rate is set to 1/IndividualLength, which is 0.025. The mutation rate for King MutationC_2 is 1/ChromosomeLength, which is 0.25.

Table 4. Performance of Proposed GAs in Optimization of F2.

GA method	Pc	Pm	Op #	Runs	Ef
GA	0.9	0.025	0	30	0
King GA	0.9	0.025	0	30	0
King Mutation	0	0.025	0	30	0
King Mutation	0	0.25	0	30	0
KingMutationC_2	0	0.25	29	30	0.97

King MutationC_2 achieves a very high Effectiveness of 0.97 compared with the canonical GA result of 0.0; King GA doesn't perform well on this deceptive function. King Mutation with traditional mutation mechanism doesn't perform well either, no matter how large or small the mutation rate is.

Table 5 shows again that King MutationC_2 is robust and gives good results even with small populations. Larger mutation rate gave better results than smaller rate for F2.

Table 5. Further experiments with King MutationC_2.

Pc	Pm	Pop	Gen#	Op #	Runs	Ef
0	0.25	10	800	20	30	0.67
0	0.25	10	1000	22	30	0.73
0	0.5	10	400	30	30	1
0	0.5	10	800	30	30	1

The GAs tested by Deb(1999) performed well only with large populations, with the best results obtained for population sizes ranging from 500 to 750 and a SS of 15,000. In contrast, King MutationC_2 gave very good results even with small (10) population size while covering a very small SS of 8,000. Again King MutationC_2 was more effective in optimizing the deceptive function than the best of traditional GAs.

Considering the performance of King Mutation and King MutationC_2 on both test functions we think that both the concept and the mechanism of mutation are important. The concept of mutation defines the purpose of mutation: Is mutation just used to introduce new genes to serve crossover, or to generate better offspring as a power search operator ? This question is critically important. The mechanism of mutation deals with the implementation of mutation: it describes how the mutation is implemented. Both King Mutation and KingMutationC_2 have the same mutation purpose that consists of using mutation to generate better offspring; but they have quite different mutation mechanisms: KingMutationC_2 has a much more powerful mechanism.

When researchers test the role of mutation in GAs, they usually set the crossover rate constant, vary the mutation rate, and then compare the results; if no good result is available, they come to the conclusion that mutation is incapable of performing search effectively and thus is a useless operator. This is not the right conclusion, because the concept of such kind of mutation is only to introduce new genes for crossover, not to generate better offspring. Under such conditions, it is only safe to conclude that the mutation mechanism used doesn't work. If the concept of mutation used is to generate better offspring rather than introduce new genes for crossover, new mutation mechanisms should be used. For mutation to become a power search operator, without crossover being used, mutation needs to change genes with higher probability than that when both crossover and mutation are used. Otherwise the search by mutation is too slow and thus ineffective. MutationC_2 proposed in this paper performed well because its mutation mechanism correctly reflects its mutation purpose.

6. CONCLUSIONS

We proposed a novel GA with King selection strategy which somewhat imitates the reproduction process of bees. The new selection strategy, especially when coupled with a new mutation operator (MutationC_2) significantly improves the performance of GA for massively multi-modal and deceptive functions. The new algorithm is very robust, giving good results for a wide range of population sizes and mutation rates. This is in contrast with traditional GAs where it is very difficult to set the value of population size, crossover and mutation rates. King MutationC_2 reliably finds the optimum of massively multi-modal and deceptive functions tested while covering a small search space. Compared to traditional GAs tested by Deb, it needs considerably smaller search space to achieve the same Effectiveness.

In the future we plan to assess the performance of our new GA on a full set of test functions ranging from unimodal to deceptive, to describe classes of functions for which it works exceptionally well, and describe theoretic reasons for its performance.

7. REFERENCES

H. Aguirre, K. Tanaka, 2002, "Parallel Varying Mutation Genetic Algorithms", Proceedings of the Congress on Evolutionary Computation, pp. 795-800, Hawaii.

A. L. Buczak, H. Wang, 2001b, "Optimization of Fitness Functions with Non-Ordered Parameters by Genetic Algorithms", Congress on Evolutionary Computation '2001, Korea.

K. Deb, S. Agrawal, 1999, "Understanding Interactions Among Genetic Algorithm Parameters", Foundations of Genetic Algorithms5, Morgan Kaufmann Publishers, San Francisco, CA.

M. Srinivas, L.M. Patnaik, 1994, "Adaptive Probabilities of Crossover and Mutation in Genetic Algorithms", IEEE Transactions on Systems, Man and Cybernetics. Vol. 24, No.4

D. H. Wolpert and W. G. MacReady. 1996, "No free lunch theorems for optimization". IEEE Transactions on Evolutionary Computation.

[WWW-1] http://www.satweb.co.za/bees/bees.htm

MULTIPROCESSOR SCHEDULING ALGORITHMS BASED ON CELLULAR AUTOMATA TRAINING

FRANCISZEK SEREDYNSKI
Wyższa Szkoła Umiejętności,
Wesola 52, 25-353 Kielce, Poland

PASCAL BOUVRY
University of Luxembourg,
Luxembourg

ABSTRACT
This article introduces multiprocessor scheduling algorithms based upon cellular automata. To design cellular automata corresponding to a given program graph a generic definition of program graph neighborhood is used, transparent to the various kinds, sizes and shapes of program graphs. The cellular automata-based scheduler works in two modes. In learning mode a genetic algorithm (GA) is applied to discover rules of cellular automata (CAs) suitable for solving instances of a scheduling problem. In operation mode discovered rules of cellular automata are able to find automatically an optimal or suboptimal solution of the scheduling problem for any initial allocation of a program graph in two-processor system graph. Discovered rules are typically suitable for sequential cellular Automata working as a scheduler. Experimental results concerning scheduling algorithms discovered in the context of cellular automata - based scheduling system are presented.

INTRODUCTION

Multiprocessor scheduling belongs to a special category of computational problems. On one side it is closely related to the issue of practical performance of current and future computers. On the other side, the problem even limited to the simplest case considered in the paper when we have to do with the two processor system but any parallel program is an example of computationally difficult unsolved research problem, known to be as an NP-complete problem [2]. Therefore heuristics based in particular on genetic algorithms (GA) , neural networks and simulated annealing are effectively used today (see, e.g. [7]) to solve scheduling problems.

One of the problems which still remains is a high computational cost of a scheduler. Among sources of scheduling overhead is neglecting the potential knowledge about the scheduling problem which could be gained during solving instances of the scheduling problem. The motivation of our work is to develop a framework for designing scheduling algorithms where knowledge about scheduling process can be extracted and potentially used for solving new instances of scheduling problem.

For this purpose we propose to use a recently emerged and very promising hybrid technique combining evolutionary computation and computation with cellular automata (CA). The remainder of the paper is organized as follows. The next section presents the scheduling problem. Section 3 gives an overview of CA. Section 4 presents the concept of multiprocessor scheduling with use of CA. Section 5 contains experimental results concerning sequential CA applied to scheduling and introduces a coevolutionary GA - based engine for discovering parallel CA scheduler. Last section contains conclusions and discusses future works.

MULTIPROCESSOR SCHEDULING

Both a multiprocessor system and a parallel program are represented by corresponding graphs. A multiprocessor system is represented by an undirected unweighted graph $G_s=(V_s,E_s)$ called a *system graph*. V_s is the set of N_s nodes representing processors and E_s is the set of edges representing bidirectional channels between processors. A parallel program is represented by a weighted directed acyclic graph $G_p=<V_p,E_p>$, with a set V_p of N_p nodes and a set E_p of edges, called a *precedence task graph* or a *program graph*. V_p is the set of N_p nodes of the graph representing elementary tasks. The weight b_k of the node k describes the processing time needed to execute a task k on any processor of the system. E_p is the set of edges of the precedence task graph describing the communication patterns between the tasks. The weight a_{kl} of the edge (k,l) describes a communication time between the pair of tasks k and l when they are located in neighboring processors.

Figure 1 (upper) shows examples of the program graph and the system graph. The program represented by the graph consists of 4 tasks with $b_0=1$, $b_1=b_3=2$, $b_2=4$ (numbers on the left side of nodes), and $a_{01}=a_{02}=a_{13}=1$ (numbers on the left side of edges). The system graph represents a multiprocessor system consisting of two processors *P0* and *P1*.

The purpose of *scheduling* is to distribute the tasks among the processors in such a way that the precedence constraints are preserved, and the response time T (the total execution time) is minimized. The response time T depends on the *allocation* of tasks in the multiprocessor topology and on *scheduling policy* applied in individual processors: T=f(allocation, scheduling_policy).

We assume that the scheduling policy is a user-defined parameter, the same for all processors, but the allocation is a subject to change by a scheduling algorithm.

CELLULAR AUTOMATA

One dimensional CA [6] is a collection of two-state cells arranged in a lattice of the length *N*, and interacted locally in a discrete time *t*, usually in a parallel and synchronous way. For each cell *i* called the central cell, a neighborhood of a radius *r* is defined. It is assumed that a state q_i^{t+1} of the cell *i* at the time *t+1* depends only on states of its neighborhood at the time t, i.e. $q_i^{t+1} = f(q_i^t, q_{i1}^t, q_{i2}^t, \ldots, q_{in}^t)$. A transition function f defines a rule of updating the cell i. A length L_g of a rule and a number of neighborhood states for a binary uniform CA is $L_g=2^n$, where $n=n_i$ is a number of cells of a given neighborhood, and a number of such rules can be expressed as 2^{Lg}. For a CA with e.g. r=2 the length of the rule is equal to $L_g=32$, and a number of such rules is 2^{32} and grows very fast with L_g.

A CONCEPT OF CELLULAR AUTOMATA-BASED SCHEDULER

An idea of CA-based scheduler is presented in Figure 1. With each task of a program graph from Figure 1 (upper) an elementary cell of CAs is associated. An initial state of the CAs correspond to an initial allocation of tasks in the two-processor system (Figure 1 (lower-left)). Next, the CAs start to evolve in time according to some rule. Changing states of the CA cells corresponds to changing the allocation of tasks in the system graph, what results in changing the response time *T*. A final state of the CAs correspond to a final allocation of tasks in the system (Figure 1 (lower-right)).

To construct the CA-based scheduler one must solve several problems:(a) what is the topological structure of proposed CAs: linear as shown in Figure 1 (lower-left) or nonlinear, related in some way to the topological structure of a program graph, (b) what kind of a local neighborhood of a program graph is the most appropriate to design corresponding CAs, and (c) how to find in a huge space of CA rules, the rule capable to solve the scheduling problem.

In the approach we adopt, (see [5]) the structure of the CA is nonlinear and corresponds to the topology of the program graph. Elementary cells are associated with tasks of a program graph and a some neighborhood of a central task is created. The central cell takes only values 0 or 1, what results in considering the scheduling problem only for the 2-processor topology: the state 0 or 1 of a cell means that a corresponding task is allocated either in the processor *P0* or *P1*, respectively. The scheduler operates in two modes: a mode of learning CA rules and a mode of normal operating. The purpose of the learning mode is to discover effective rules for scheduling. Searching rules is conducted with use of GA in the following way:

GA to discover CA rules:
t=0
create an initial population P() of size n_{pop} of rules
WHILE *termination_condition* **NOT TRUE**
BEGIN
  create a set of a size *n_test* of test problems
  IF *hillclimbing_condition* **TRUE**
    THEN hillclimbing
  FOR i=1 **TO** n_{pop}
    BEGIN
      $T^*_{ij} = 0$
      FOR j=1 **TO** n_{test}
        $T^*_{ij} = T^*_{ij} + CA(rule_i, test_j, seq/par, CA_{steps})$
    END
  sort P() according to T^*_i
  move E of the best individuals from P(t) to P(t+1)
  FOR k=1 **TO** n_{pop} - *E*
    REPEAT
      rule1^{parent}=select()
      rule2^{parent}=select() $\neq$ rule1^{parent}
      (rule1^{child}, rule2^{child})=
        crossover(rule1^{parent}, rule2^{parent})
      mutation(rule1$_{child}$, rule2$_{child}$)

UNTIL Hamming($rule1_{child}$, rules)>=H
AND
Hamming ($rule2_{child}$, rules)>=H

t=t+1

END

Problem_solution = the best rules from P().

For this purpose an initial random population of rules is created. Also a set of random initial allocations of a program graph into the system graph is created. States of the CAs are initiated according to a given allocation of a program graph. The CAs equipped with a rule from the population of rules start to run during predefined number of time steps. Changing states of the CAs corresponds to changing an allocation of task of a program graph. The response time T for a final allocation is evaluated. This procedure of evaluation of a rule is repeated for each allocation from the set of initial allocations. Eventually, a fitness value T* for the rule is evaluated as the sum of final values of T corresponding to each initial allocation of tasks. After evaluation in a similar way of all rules from the population, genetic operators are applied. *Selection* with *elitism* transfers some percent of the best rules, called *elite* to the new population which will processed in the next generation. The remaining part of the population is created by *crossover* between members of the elite, and next *mutation* is applied to new members (children) of the population. Also a new set of initial allocations is randomly created at the beginning of each generations. Evolutionary process is continued a predefined number of generations. In the mode of normal operating, when a program graph is randomly allocated, the CA is initiated and equipped with a rule taken from the set of discovered rules. We expect in this mode, that for any initial allocation of tasks of a given program graph, the CAs will be able to find in a finite number of time steps, an allocation of tasks, providing an optimal or suboptimal value of T.

EXPERIMENTS

In the experiments reported in this section it is assumed that the CAs work asynchronously, i.e. at a given moment of time only one cell updates its state.

A number of experiments with program graphs available in the literature has been conducted (for details, see [5]). Figures 2a,b and 3 show results of experiment with a program graph called g40 consisting of 40 tasks.

A population of rules of a size 200 was used in the learning mode in this experiment. Figure 2a shows that rules of CAs providing an optimal scheduling with a response time T=80 was found after 160 generations. Figure 2b shows the performance of found rules evaluated in the operation mode. One can see that the best rules found in the learning mode provide in the average near optimal solutions in the operation mode.

Figure 3 shows a run of CA-based scheduler with the best found rule. One can see that the CA finds a steady-state corresponding to an allocation providing an optimal response time T=80 in the step 14.

A parallel version has been proposed in [5] based on using *coevolutionary GA* [3] based on a *predator-prey* paradigm for discovering rules. Results of the experimental study show that the coevolutionary GA is able to find rules for parallel CAs, which provide significantly faster scheduling in comparison with a sequential CA-based scheduler.

CONCLUSIONS

We have presented in the paper a novel approach to design CA-based scheduling algorithms. We used GA in learning mode of work of the scheduler to discover scheduling rules of CA. In this mode a knowledge about solving a given instance of the scheduling problem is extracted and coded into CA rules.

A number of questions in this area are still open. The most important of them is how to use the knowledge extracted during learning process and coded into CA rules. Some results of study show that rules discovered for different instances of the scheduling problem may be ranked according to their possibility to solve automatically, in the operation mode, other instances of the scheduling problem. It leads to the concept of artificial immune system [1] for scheduling problem [4] in which discovered rules are reused to solve new instances of the scheduling problem.

FIGURES

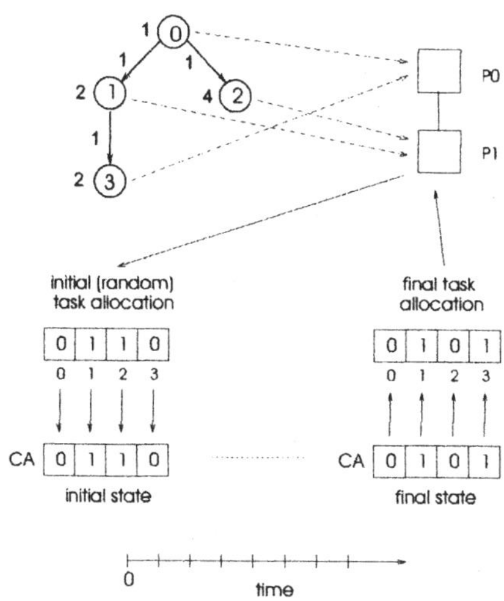

Figure 1. *An idea of CA-based scheduler: an example of a program graph and a* system graph (upper), corresponding CA performing scheduling (lower).

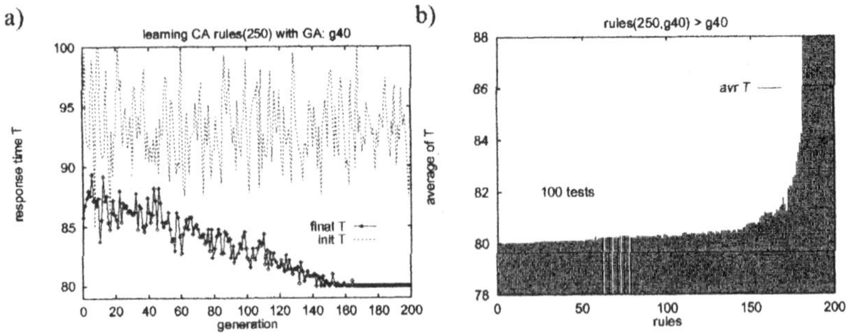

Figure 2. Sequential CA-based scheduler for g40: learning mode (a) and operation mode (b)

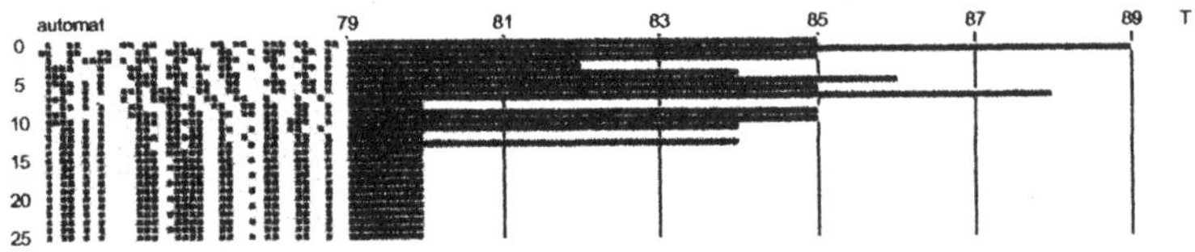

Figure 3. Space-time diagram of CA-based scheduler for g40

REFERENCES

[1] D. Dasgupta (ed.), *Artificial Immune Systems and Their Applications*, Springer, 1999.

[2] H. El-Rewini and T. G. Lewis, Scheduling Parallel Program Tasks ontoArbitrary Target Machines, *Journal of Parallel and Distributed Computing* 9, 1990, pp. 138-153.

[3] J. Paredis, Coevolutionary Life-Time Learning, in H. -M. Voigt *et al.* (eds.), *Parallel Problem Solving from Nature -- PPSN IV*}, LNCS 1141, Springer, 1996, pp. 72-80.

[4] F. Seredynski, A. Swiecicka, Immune-like System Approach to Cellular Automata - based Scheduling, in R. Wyrzykowski *et al.* (eds.), *Parallel Processing and Applied Mathematics*, Springer, LNCS 2328, pp. 626-633.

[5] F. Seredynski and A. Zomaya, Sequential and Parallel Cellular Automata-based Scheduling Algorithms, *IEEE Transactions on Parallel and Distributed Systems*, Volume: 13 Issue: 10 , Oct. 2002.

[6] S. Wolfram, Universality and Complexity in Cellular Automata, *Physica D* 10, 1984, pp. 1-35.

[7] A. Y. Zomaya, C. Ward, and B. Macey, Genetic Scheduling for Parallel Processor Systems: Comparative Studies and Performance Issues, IEEE *Trans. on Parallel and Distributed Systems*, Vol.10, No. 8, 1999, pp. 795-812.

RATE OF INFORMATION TRANSFER IN GRAPH BASED EVOLUTIONARY ALGORITHMS

STEVEN M. CORNS
Department of Mechanical Engineering
Iowa State University
Ames, Iowa

KENNETH M. BRYDEN
Department of Mechanical Engineering
Iowa State University
Ames, Iowa

DANIEL A. ASHLOCK
Department of Mathematics
Iowa State University
Ames, Iowa

ABSTRACT
Combinatorial graphs have recently been used to control the rate of spread of information in evolutionary algorithms. This allows for the preservation of diversity found necessary as the fitness landscape grows in complexity. Another factor that impacts the rate of spread of information and diversity is the population size of the graph. This paper examines the combined impact of graph type and population size on the spread of a superior solution using graph based evolutionary algorithms. This study identifies a correlation between population size, graph, and time for a good solution to spread. While no numerical relationships are introduced here, it is readily apparent that the required number of mating events for a solution to spread across an entire graph is proportional to the graph diameter, population size, and the fitness difference of the individuals.

INTRODUCTION

Evolutionary Algorithms (EAs) (Banzhaf et al., 1998) have been used to emulate the ideas of biological evolution in computer code. This has been used as an approach to solve many different math, physics and engineering problems. Populations of designs are evolved to search a solutions space for an acceptable answer. These algorithms can avoid early convergence to sub-optimal solutions, a common problem with gradient search methods (Bryden et al., 2003). Recently, combinatorial graphs have been used to limit the spread of information in EAs, helping preserve diversity. One theory as to why diversity has not disappeared in nature is Sewall Wright's theory of isolation by distance (Wright, 1986), where mating restrictions are imposed in nature by geographical obstacles. Graph Based Evolutionary Algorithms (GBEAs) emulate these obstacles, preserving diversity and slowing time to convergence. This helps keep the algorithm from getting stuck in a local optima for deceptive problems (Ackley et al., 1992; Mühlenbein, 1991). Initial research with has indicated that similar types of graphs (families) have roughly the same number of mating events solution for the tested problems (Bryden et al., 2003). It also showed that for problems with more difficult landscapes there was a significant reduction in time to convergence when a diversity-preserving graph was used. This study examines the rate of information spread within a GBEA.

GRAPH BASED EVOLUTIONARY ALGORITHMS

A *combinatorial graph* or *graph* (G), is a collection of vertices (V(G)) and edges (E(G)) where E(G) is a set of unordered pairs from V(G). Two vertices of the graph are *neighbors* if they are members of the same edge. GBEAs are evolutionary algorithms that impose a geography to restrict the number of possible mates a population member can interact with. Each vertex V(G) of the graph holds a population member, which can only interact with an individual whose vertex shares an edge E(G) with it. Evolution occurs one mating event at a time (steady state [Reynolds, 1992; Syswerda, 1991; Whitley, 1989]). The rate of information spread can be controlled by changing the graph type, thereby controlling diversity loss. For this study, the general strategy is to use the graph (population structure) to constrain mating by applying the graph as a geography on which the population lives, permitting mating only between neighbors and preserving diversity without sacrificing progress due to heterogeneous crossover.

LIST OF GRAPHS

In this section, the random graphs used in this study are defined. All non-random graphs can be found in Graph Based Genetic Algorithms (Ashlock et al., 1999). The randomized graphs were generated using edge moves (Bryden et al., 2003) in a randomized algorithm applied to the toroidal, Petersen and hypercube graphs. A fourth type of random graph was also generated by randomly placing points on a unit torus and creating an edge between two points if they were within a certain distance of each other. Three different instances of each random graph class were produced for this experiment. Complete information on how the random regular graphs were generated can be found in Bryden et al. (2003). Table 1 lists the graphs used in the study

Graph	**Index**	**Regularity**	**Random**	**32**	**64**	**128**	**256**	**512**	**1024**	**2048**	**4096**
C_n	0	2	No	16	32	64	128	256	512	1024	2048
H_n	1	$Log_2(n)$	No	5	6	7	8	9	10	11	12
K_n	2	n-1	No	1	1	1	1	1	1	1	1
$P_{n,1}$	3	3	No	9	17	33	65	129	257	513	1025
$P_{n,17}$	4	3	No	n/a	9	10	10	18	25	34	67
$P_{n,3}$	5	3	No	6	8	14	24	46	88	174	344
$P_{n,7}$	6	3	No	5	6	10	14	22	42	78	150
R(n,3,1)	7	3	Yes	6	7	8	9	11	12	12	15
R(n,3,2)	8	3	Yes	6	7	8	9	10	12	13	14
R(n,3,3)	9	3	Yes	6	7	8	10	10	11	13	15
R(n,4,1)	10	4	Yes	4	5	6	6	8	8	9	10
R(n,4,2)	11	4	Yes	4	5	6	7	7	8	9	9
R(n,4,3)	12	4	Yes	4	5	6	6	7	8	9	10
R(n,9,1)	13	$Log_2(n)$	Yes	3	4	4	4	4	5	5	5
R(n,9,2)	14	$Log_2(n)$	Yes	3	4	4	4	4	4	5	5
R(n,9,3)	15	$Log_2(n)$	Yes	3	4	4	4	4	4	5	5
R(r,1)	16	No	Yes	4	8	10	13	19	23	38	30
R(r,2)	17	No	Yes	4	6	9	26	20	29	47	50
R(r,3)	18	No	Yes	6	5	8	17	16	25	29	40
$T_{16,m}$	19	4	No	n/a	n/a	n/a	16	24	40	72	136
$T_{4,m}$	20	4	No	6	8	18	34	66	130	258	514
$T_{8,m}$	21	4	No	n/a	10	12	20	36	68	132	260

Table 1. Graphs used, their index numbers, and diameters by population size.

EXPERIMENTAL DESIGN

In this study, a series of numerical experiments were performed to examine the information mixing rates within various combinatorial graphs used in GBEAs. Each of the mixing experiments involved 1000 runs for each population size and graph type used. For each run, the graph is populated with a creature at each vertex, assigned a fitness of 1. The run is started by injecting an individual with a higher fitness, and performing a mating event. A mating event consists of selecting of a population member (randomly after the introduction of the high fitness individual), and then selecting a mate using roulette selection of the available neighbors. If one individual has a lower fitness, it is then replaced by a copy of the higher fitness individual. After initialization, mating continues until the entire population has this higher fitness, recording the number of mating events after initialization required to reach whole number percentages. Letting α be the fitness ratio between the superior individual and the initial population, runs were performed for α values of 1.5, 1.75, 2, and 3. In this way, the rate at which the superiorsolution spread across the graph can be tracked. This is similar to work by Syswerda (1991) and Goldberg (1993), but is purely a measure of information spread by selection. The data is reported as the number of mating events required as a function of percentage spread, and then as the number of mating events required divided by the population size as a function of percentage spread, in an attempt to normalize the results.

RESULTS

As described above, the experiment was designed to investigate the spread of information in a GBEA, as well as the effect the difference in fitness has on this rate of spread. Figure 1 shows the information spread (% of vertices with higher fitness value)

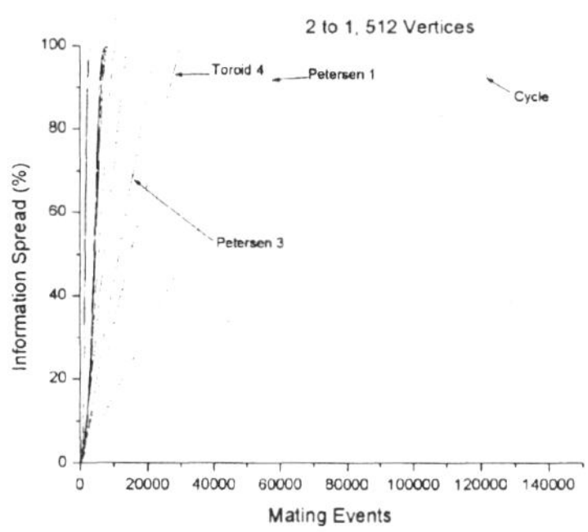

Fig. 1. Information spread as a function of mating events and graph, 512 vertices, α= 2.

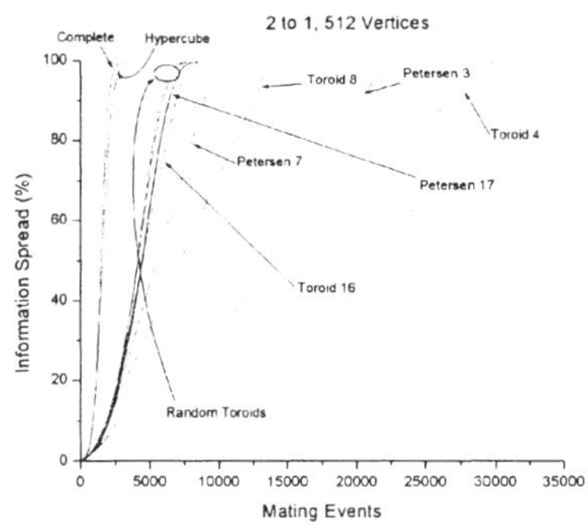

Fig. 2. Information spread as a function of mating events and graph, 512 vertices, α= 2.

as a function of mating event for the cycle; Petersen n,1; Petersen n,3; and toroid graphs with a population size of 512. Because of the scaling of Fig. 1, the performance of the additional graphs is shown, but not identified. These additional graphs are shown in Fig. 2. As shown in Figs. 1 and 2, the general relationship between the number of mating events and the extent of the information spread is initially slow. This occurs because the number of high fitness members in the population is small and the randomly selected individual and its neighbors will not include a high fitness member each time. As the number of high fitness members grows, the extent of the information spread grows nearly linearly, until the population is composed of primarily high fitness members. At this point, the rate of information spread slows, because of the difficulty of finding and selecting low fitness members to replace. In the case of the cycle graph, the rate of information spread is constant for the entire experiment until only one low fitness individual is left, as there are always two edges where change can occur. There is a curve at the end of the run, when the final low fitness member must be found. For all other graphs, the rate of information spread slow to begin with, increases to a nearly linear rate, and then slows again as the last low fitness members are found. As the rate of information spread within the graph increases, the S-shape of the mating event curves becomes more pronounced. As expected, the ranking of the graphs by

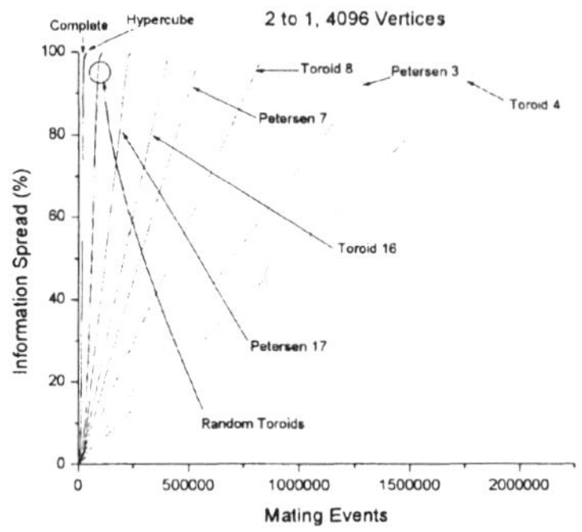

Fig. 3. Information spread as a function of mating events and graph, 4096 vertices, α= 2.

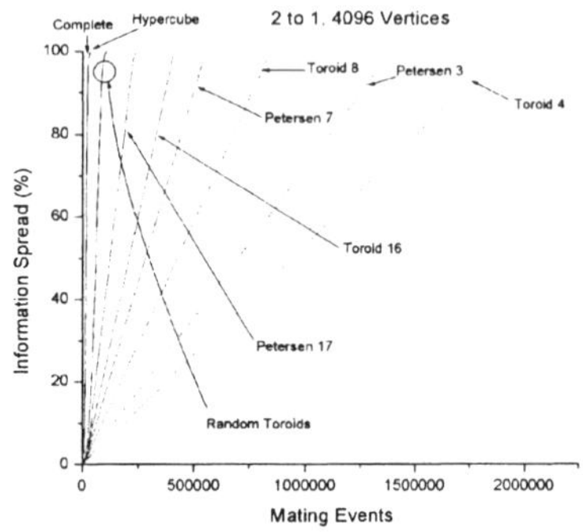

Fig. 4. Information spread as a function of mating events and graph, 4096 vertices, α=2.

slowest to fastest rate of information spread is cycle, Petersen and toroid, random toroid, hypercube, and complete (Fig. 2.) Figures 3 and 4 shows the rate of information spread for a population of 4096. As shown, the general shape of the curves and ranking of the graphs remains the same for all population sizes. While for smaller populations there are step increases in the percentage spread, this is due to the change in fitness of a single individual causing a more than one percent increase in the percentage spread; the underlying relationship is still taken to be linear. The toroid and Petersen graphs were interspersed between the other graph families, with higher diameter graphs (Table 1) having a slower spread rate, and lower diameter graphs having faster spread

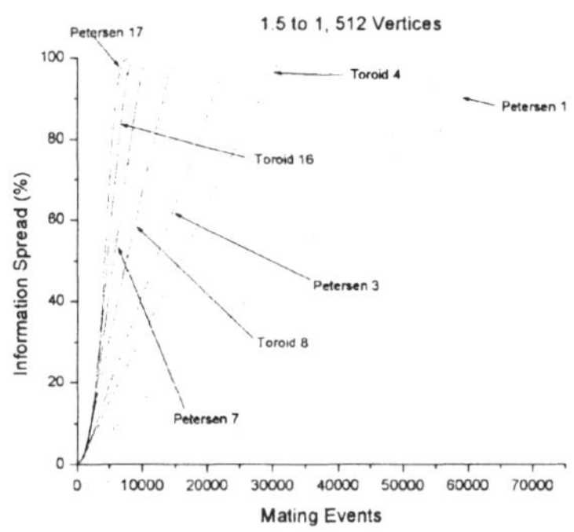

Fig. 5. Information spread as a function of mating events and graph, 512 vertices, α= 1.5.

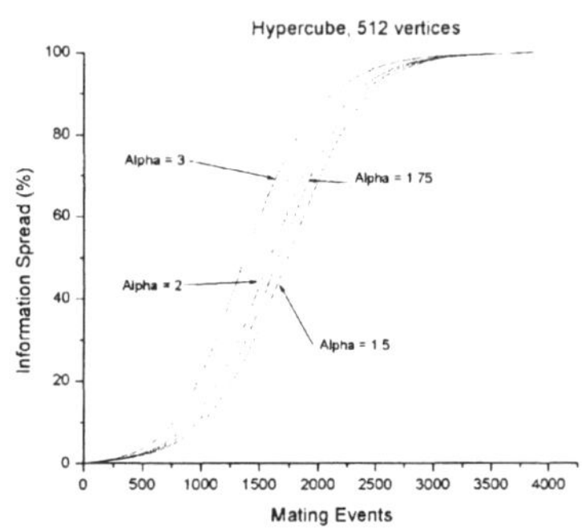

Fig. 6. Information spread as a function of mating events and α for the hypercube graph, 512 vertices.

rates (Fig. 5.) Another expected result was that for all population sizes and graph types, the rate of spread was proportional to α, with the effect appearing to diminish as α increased (Figs. 6 and 7.) When the number of mating events is divided by the

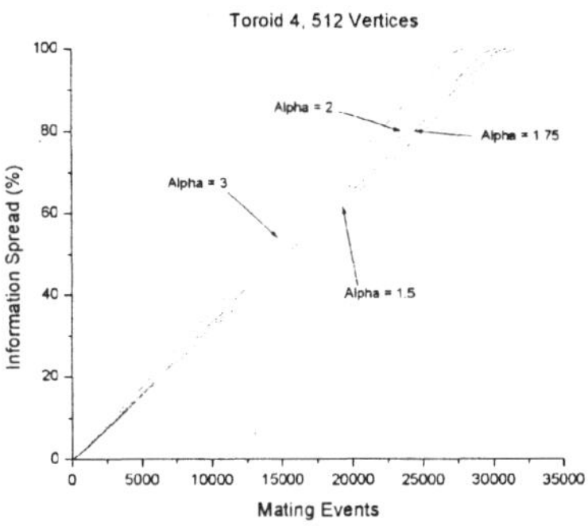

Fig. 7. Information spread as a function of mating events and α for the toroid 4 graph, 512 vertices.

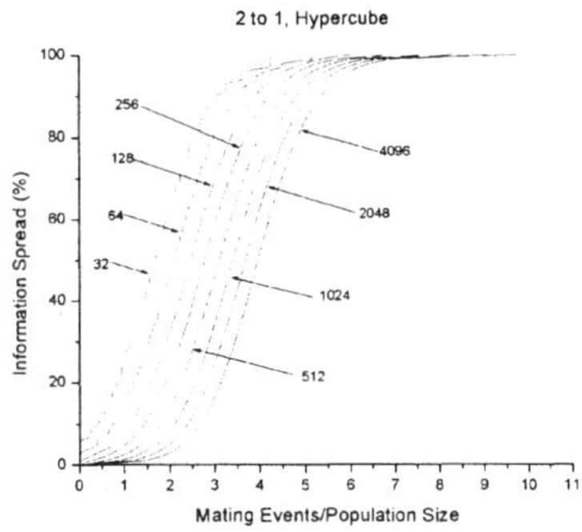

Fig. 8. Information spread as a function of mating events/population size for the hypercube, α=2.

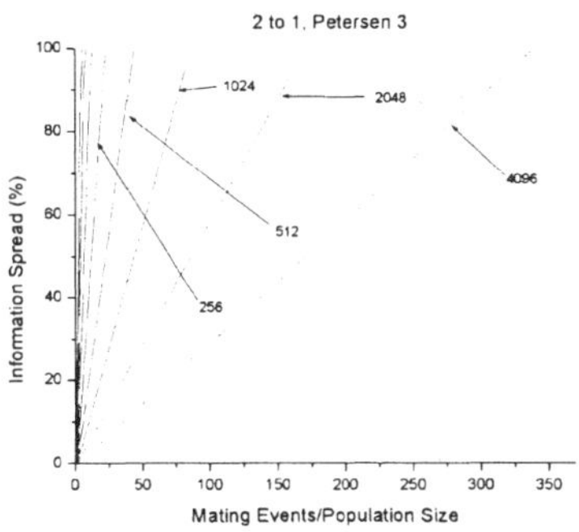

Fig. 9. Information spread as a function of mating events/population size for the Petersen 3 graph, α=2.

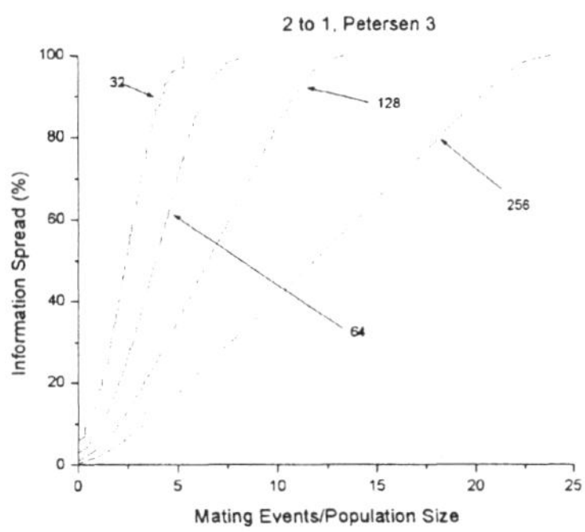

Fig. 10. Information spread as a function of mating events/population size for the Petersen 3 graph, α=2.

population size, it can be seen that, for the hypercube and complete graphs, the difference increases in discrete amounts as the population size is doubled (Fig. 8.) For the other types of graphs, this appears to be more of a doubling of the difference per increment (Figs. 9 and 10.) When the number of mating events divided by the population size at 100% information spread is compared to the graph diameters found in Table 1, they are found to be roughly equal for all graphs but the complete.

CONCLUSIONS AND FUTURE WORK

As shown, when the graph diameter and population size increase, the rate of information spread decreases, while increasing α has the opposite effect. The number of mating events appears to be directly proportional to the population size and graph diameter, and roughly inversely proportional to alpha. This study provides additional evidence that graphs can be used to tune the rate of information spread and population to maintain and control diversity. Earlier studies have indicated that diversity is a key parameter in determining graph performance. In addition, this type of study provides a relatively quick and easy methodology for comparing graph performance. Based on this, a test for non-standard graphs, such as connected cycles, could be developed. This would enable a much broader number of graphs to be investigated for their potential in use in a GBEA. In previous studies (Bryden et al.,2003), results have indicated that graph families perform so similarly that it is difficult to obtain statistically significant performance differences between family members. This study shows that, while the rate of information spread is similar, they are different between members. This would indicate that there is some other dynamic at work, possibly more than one superior solution contesting the spread of the others. Further analysis is required to obtain a numerical understanding of the relationship. Further work could involve using different selection methods, which may change the dynamics of the information spread. Also of interest would be to introduce multiple numbers of superior individuals to determine how they would interact. This could also be applied to genetic programming methods to test different methods of representation in an effort to determine robustness.

REFERENCES

Ackley, D.L. and Littman M.L., 1992, "A Case for Distributed Lamarckian Evolution," Working Paper, Cognitive Science Research Group, Bellcore, New Jersey.

Ashlock, D.A., Walker, J., and Smucker, M., 1999, "Graph Based Genetic Algorithms," *Proceedings of the 1999 Congress on Evolutionary Computation*, pp. 1362-1368.

Banzhaf, W., Nordin, P., Keller, R.E., and Francone, F.D., 1998, *Genetic Programming: An Introduction*. Morgan Kaufmann, San Francisco, CA.

Bryden, K.M., Ashlock, D.A., and Corns, S.M.,2003, "Graph Based Evolutionary Algorithms," submitted to *IEEE Transactions on Evolutionary Computations*.

Bryden, K.M., Ashlock, D.A, McCorkle, D.S., and Urban, G.L., 2003, "Optimization of Heat Transfer Utilizing Graph Based Evolutionary Algorithms," *International Journal of Heat and Fluid Flow* Vol. 24, pp. 267-277.

Goldberg, D.E. 1989, *Genetic Algorithms in Search, Optimization, and Machine Learning*. Addison-Wesley Publishing Company, Inc., Reading, MA.

Goldberg, D.E., Deb, K, and Clark, J.H., 1992, "Genetic Algorithms, Noise and the Sizing of Populations," *Complex Systems*, Vol 6, Number 4. 333-365)

Goldberg, D.E., Deb, K, and Thierens, D., 1993, "Towards a Better Understanding of Mixing in Genetic Algorithms," *Journal of the Society of Instruments and Control Engineers* 32(1), pp. 10-16.

Grefenstette, J.J., 1986, "Optimization for Control Parameters for Genetic Algorithms," *IEEE Transactions on Systems, Man, and Cybernetics*, Vol. SMC-16, No. 1.

Kimura, M. and Crow, J., 1963, "On the Maximum Avoidance of Inbreeding," *Genetic Research*, Vol. 4, pp. 399-415.

Mühlenbein, H., 1991, "Darwin's Continent Cycle Theory and Its Simulation by the Prisoner's dilemma," *Complex Systems*, Vol. 5, pp. 459-478.

Reynolds, C., 1992, "An Evolved, Vision-based Behavioral Model of Coordinated Group Motion," *From Animals to Animats 2*, pp. 384-392.

Syswerda, G., 1991, "A Study of Reproduction in Generational and Steady State Genetic Algorithms," *Foundations of Genetic Algorithms*, pp. 94-101.

Urban, G.L., K.M. Bryden, and Ashlock, D.A., 2002, "Engineering Optimization of an Improved Plancha Stove," *Energy for Sustainable Development*, 6(2), pp. 5-15.

Whitley, D., 1989, "The Genitor Algorithm and Selection Pressure: Why Rank Based Allocation of Reproductive Trials is Best," *Proceedings of the 3rd ICGA*, pp. 116-121.

Wright, S., 1986, *Evolution*. University of Chicago Press, Chicago, IL.

COMBINATORIAL OPTIMIZATION OF MUTUAL INFORMATION ESTIMATES USING TIME-DELAYED INPUT COORDINATES

P.B. DEIGNAN JR.
G.B. KING
P.H. MECKL
School of Mechanical Engineering
Purdue University
West Lafayette, Indiana, USA 47907-1288

ABSTRACT
A method of selecting time-delayed input coordinates based on an information-theoretic estimate of dependency to a single output is introduced. The problem of dimensionality is addressed through upper bounding the search space by the embedding dimension of the individual inputs, the set of candidate delays of which are identified by the first stage of a two-stage combinatorial optimization process utilizing a branch-and-bound search of joint mutual information estimates. Additionally, a method of selecting a useful binning interval for nonparametric probability density estimation in reference to mutual information estimates is presented for finite data.

INTRODUCTION

A fundamental problem in the reconstruction of dynamical manifolds from data is the selection of the best space in which to attempt the reconstruction. Ideally, this space is chosen from a set of measurable, related variables so that the trajectory of the variable or variables of interest lies in a coherent subspace of small enough dimensionality as to provide good estimates with limited data. Locally, the dynamical manifold is a coordinate map in Euclidean space that defines a functional relationship between inputs and an output as in the prototypical system identification problem. Globally, however, the manifold defines a multi-value relation between variables so a measure of central tendency is inappropriate as a criterion of merit for the global input selection problem; a more suitable measure of dependence is average mutual information (MI).

Another problem is concerned with the time-evolution of the system. The simplest case is to reconstruct a manifold which is statistically static over any significant time period. These manifolds are practical as models of fading memory processes that are deterministic and which may vary in memory time constant by input or locale in the dynamic subspace. The most interesting systems interact with their environment so that the trajectories along the manifold do not form closed orbits. It is rare outside of simulations to have enough control of the sampling interval and access to repeatable measurements to estimate the instantaneous, local derivatives of the variable of interest in a possibly high dimensional space to any degree of confidence. A utilitarian approach to introducing system dynamics into an otherwise static model is to expand the input space into time-delayed coordinates. While straightforward, this approach does increase the dimensionality of the candidate set vastly increasing the computational challenge. We will address this challenge by a multi-stage branch and bounding strategy.

If the output is dependent on the candidate input, the output trajectory can be embedded in the time-delayed input space with relatively few time delays. The embedding dimension is the effective upper bound of the number of relevant time delays of a particular input to the output. Unfortunately, the embedding dimension does not specify which time delays are most relevant nor is the estimated embedding dimension invariant in respect to the operating regime or sampling time of a particular set of measurements – the data set must be preconditioned to incorporate these factors [1]. The false nearest neighbors (FNN) algorithm is one method of estimating the useful embedding dimension that is both computationally efficient and compatible with the nonparametric estimation of mutual information between inputs and output.

After a review that presents details pertinent to the methodology, we introduce a binning algorithm for MI estimates that directly addresses the limitations of finite data sets. Next, a two-stage process of combinatorial optimization of inputs for modeling the dynamics of one particular output by time-delayed coordinates is described. Both results are demonstrated on a data set derived from measurements of diesel engine operation with net engine torque as the object output signal.

MUTUAL INFORMATION

When restricted to finite probability spaces, normalized mutual information (MI) satisfies all of the Rényi postulates for measures of dependence [2]. Due to this and the property that mutual information may be estimated uniformly to an arbitrary degree of resolution in a continuum of measured data, an optimal input set may be selected with this measure as the criterion of merit

Mutual information quantifies the capacity to map the system output, $y \in Y$, given the system input, $\mathbf{x} \in \mathbf{X}$. Specifically, average mutual information is the expected value of the Kullback-Leibler distance between the conditional probability distribution, $p(y \mid \mathbf{x})$, and the output marginal probability distribution, $p(y)$. For discrete data, it is given by the formula,

$$I(\mathbf{X};Y) = \sum_{\mathbf{x}\in\mathbf{X}} \sum_{y\in Y} p(\mathbf{x}, y) \log\left(\frac{p(y \mid \mathbf{x})}{p(y)}\right). \tag{1}$$

With $\log_e$ as the logarithm, the average mutual information is reported in natural units (nats). Note that there is no implicit or explicit presumption of a functional relationship between random variables in the interpretation of the measure. Thus, the degree of MI between input sets and an output effectively quantifies the ability to support a one-to-one input-output correspondence.

MI is the entropy difference between the marginal output and conditional distributions,

$$I(\mathbf{X};Y) = H(Y) - H(Y \mid \mathbf{X}) \tag{2}$$

where the entropy measure is defined as

$$H(Z) = -\sum_{\zeta\in Z} p(\zeta) \log\left(p(\zeta)\right) \tag{3}$$

Alternatively,

$$I(\mathbf{X};Y) = H(Y) + H(\mathbf{X}) - H(\mathbf{X},Y) \tag{4}$$

or in terms of a measure on sets,

$$\begin{aligned} I(\mathbf{X};Y) &= H(Y) + H(\mathbf{X}) - H(\mathbf{X} \cup Y) \\ &= H(Y \cap \mathbf{X}). \end{aligned} \tag{5}$$

It should be clear from the last equation that input subset selection is a set-covering optimization problem on $H(Y)$. The entropy of the output signal is the maximum information content to be covered and provides an upper limit on the useful descriptive entropy of any input signal set for the particular method of estimating the joint and marginal probabilities.

BRANCH AND BOUND ALGORITHM

Optimal input set selection with Shannon's joint mutual information as the criterion of merit evaluated over the candidate input-output data set is a combinatorial optimization problem in integer programming. Specifically, it is a set-covering problem of the output entropy in which the difference in conditional entropies of various input combinations are not necessarily disjoint (5). This caution holds as well for estimates of mutual information calculated here by uniformly binning the data over its domain.

Monotonicity of the estimated mutual information follows from the convexity of the natural logarithm over $[a,b] \in \mathbb{R}$. By monotonic, we mean that

$$\hat{I}(X_1;Y) \le \hat{I}(X_1,X_2;Y) \le \hat{I}(X_1,X_2,X_3;Y)\ldots \tag{6}$$

for any $\{X_1, X_2, X_3, \ldots, X_n\} \in \mathbf{X}$. This is easily seen from Jensen's discrete inequality,

$$\log\left(\sum_{i=1}^{m} p_i \mathbf{x}_i\right) \le \sum_{i=1}^{m} p_i \log(\mathbf{x}_i) \tag{7}$$

where $p_i \ge 0$ and $\sum_{i=1}^{m} p_i = 1$. The estimate of mutual information is nondecreasing with bin division so that on average with infinite data, the MI estimates increase by a log-linear trend per total number of uniform bins [3].

The algorithm enumerates integer combinations via a spanning tree, the branches of which are evaluated in top to bottom and right to left order. If the current optimal MI value lies within a bound set by an inequality based on the monotonicity of (6), subordinate branches are evaluated and ordered by decreasing MI (Figure 1). In case the inequality is not satisfied for the current optimal MI value, the algorithm backtracks and selects the next unexplored branch under the current node until the set is exhausted. The computational cost of evaluating MI increases with dimensionality at a rate that justifies the calculation of intermediate joint MI values in order to possibly avoid searching through all branches of a node.

The simplest bound is derived from the entropy inequality,

$$H(\mathbf{X},Y) \le H(Y) + H(\mathbf{X}). \tag{8}$$

Therefore,

$$I(X_1,X_2;Y) \le I(X_1;Y) + I(X_2;Y) \tag{9}$$

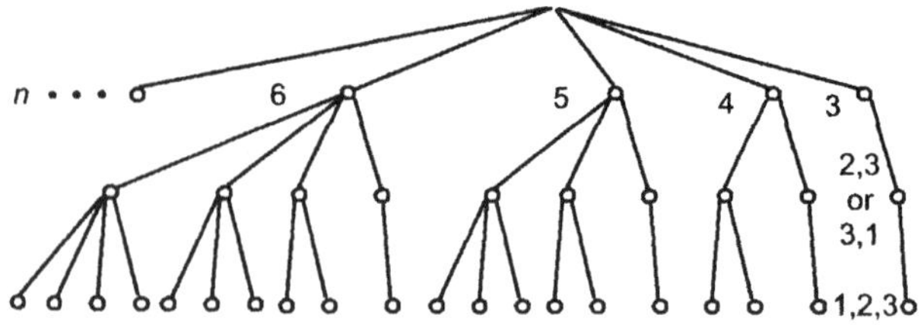

Figure 1: Branch and bound enumeration tree for three inputs out of a pool of *n* candidate inputs ordered by decreasing values of joint MI.

and in general ,

$$I\left(X_1,X_2,\ldots,X_n;Y\right) \le I\left(X_1,\ldots,X_{n-1};Y\right) + I\left(X_n;Y\right) \tag{10}$$

To the first degree of interaction, the bound on a branch of dimensionality k in a problem of dimensionality n is

$$I\left(\mathbf{X}^{(n)};Y\right) \le I\left(\mathbf{X}^{(n-k)};Y\right) + \max_{\Omega} \sum_{i=1}^{k} \left\{ I\left(\mathbf{X}^{(n-k)};Y\right) - I\left(\mathbf{X}^{(n-k)} \setminus X_i;Y\right) \right\}, \quad \left\{X_i \in \Omega \mid i = 1,2,\ldots,M\right\} \tag{11}$$

Since the average mutual information of a null input set is zero, this inequality also imposes an implicit bound to reject irrelevant inputs in this ordered search.

MI BINNING RULES FOR FINITE DATA SETS

While experience has shown that the relative ranking of inputs by joint average mutual information is not particularly sensitive to binning interval, this choice is nonetheless fundamental to the accuracy of any mapping and computational feasibility of the analysis [3]. Once the candidate data set is conditioned for outliers and noise, the binning interval is first determined for the output and then for the inputs. The detrended entropy reveals two distinct partitioning regions: one dominated by statistical structure and the other by quantization effects (Figure 2.). The range over which the trend is calculated is a function of data sufficiency.

There are two distinguishing features characterizing the structure and quantization noise regions: regularity of variability and skewness. Variability of detrended entropy estimates in the structure region is related to data sufficiency and smoothness of the population distribution. It is difficult to formulate a general rule applicable to arbitrarily irregular populations. While it is likely that the regions may be discriminated by the variation of a local measure of smoothness, this characteristic is readily perceived by direct examination.

The effect of finite data is to reduce the rate of increase of estimated MI with binning interval in the limit so that a detrended graph which extends into the structure region exhibits a negative skewness. When this feature coincides with the quantization region, we call this the depletion region. A limit of the admissible skewness may be used to define the searchable binning range. The minimum binning interval can based on the record length provided that the interval is significantly within the structure region.

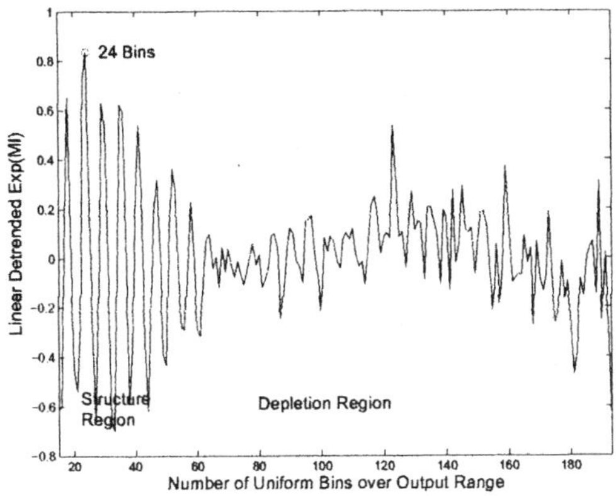

Figure 2: Analysis of output entropy: *I* (*Torque*; *Torque*) v. number of uniform bins over output range. The minimum skewness is −0.10.

The binning interval for MI estimation is a compromise between resolution of structure and computational practicality. In this illustration, we choose 24 bins for the output; however, any binning interval yielding a positive peak within the structure region is a feasible choice. The binning intervals for the inputs are chosen likewise, but based on MI relative to the predetermined output binning.

SEARCHING FOR MI OPTIMAL INPUT TIME DELAYS

Time-delayed coordinates are a ready means of introducing dynamics into an otherwise static analysis — especially for sampled data systems. For example, an MI search through time-delayed coordinates of an input/output may identify transport lags through a salient peak in MI over number of delays. Alternatively, the first minima in MI may be inferred to mark the delay after which the data streams are nonlinearly independent [4], [5]. An obvious drawback of time-delayed coordinates is the concurrent increase in dimensionality of the candidate input set.

If the inputs are not mutually dependent functions of time, the time delays of individual inputs may be searched separately by the aforementioned technique for optimal MI combinations up to the lesser of the embedding dimension or dimensionality of the reconstruction space. While an upper bound of usable joint MI may be found directly, the embedding dimension analysis by FNN maintains the temporal ordering of the data streams, does not impose quantization limits as in binning by dimension value, and is computationally much faster. Thus, the embedding analysis inherently introduces a temporal regularity constraint in the process. The domain of the search space can be defined *a priori* by considerations of causality or by a one dimensional MI search through the admissible range of time delays of that input. The resulting set of time-delayed inputs is the candidate set for the global branch and bounding search among all remaining combinations of time-delayed inputs.

Table I: Pairwise MI Optimal Estimates

Output: Torque (t)	*Embed. Dim.*	*Optim Bins/k*	*MI(t−.2k)max (nats)*
1: Speed	5	71/7	0.8073
2: Throttle	4	48/0	1.0412
3: Air Flow	4	57/2	0.9989
4: IMP	5	27/4	1.0262
5: Rail	5	23/10	0.6002

Table II: Joint MI by Input Dimensions

	2-Stage Estimation		*Full Combinatorial*	
	MI (nats) / min.	Input Indices : (Delays)	MI (nats) / min.	Input Indices : (Delays)
2	1.895 5.452	2: (2) 3: (8)	2.055 6.153	1: (0) 2: (2)
3	2.270 5.859	1: (0,8) 2: (4)	2.316 18.66	1: (0,7) 2: (2)
4	2.388 53.99	1: (0,2,8) 2: (2)	2.396 1161	1: (0,3,7) 2: (3)

The two-stage process was tested on a data set from a diesel engine undergoing a test cycle of 20 minutes. Data was sampled synchronously every 0.2 seconds from crank speed, throttle position, intake air flow, intake manifold pressure, fuel rail pressure, and brake torque. Table I gives an overview of MI estimates and parameters for individual inputs to the output, torque, using the method for bin selection as described previously.

The reduction in computation of the two stage process over a global search is directly related to the reduction in number of time-delayed inputs to be calculated. For comparison, the results of a full MI evaluation of up to eight consecutive time-delayed inputs in reference to the output from synchronicity are compared to the two-stage results in terms of inputs chosen, optimal MI reached, and minutes of computation. Note that the computational improvements of the two stage approach will increase exponentially as dimensionality is increased beyond the embedding dimension. Accuracy depends directly upon the separability of the contributions of the various inputs in time.

CONCLUSIONS

The two-stage branch and bound strategy performs surprisingly well in finding near optimal joint MI combinations while avoiding the combinatorial explosion of a full search for our arbitrarily chosen data set of a real physical process. This methodology shows promise as an input set selection preprocessor for dynamic neural network models.

REFERENCES

[1] Deignan, P.B. Jr., 1999, "Virtual Sensing: The Development of a Methodology for Internal Combustion Engine Torque Estimation," Masters Thesis, Purdue University, West Lafayette, IN.

[2] Bell, C.B., 1962, "Mutual Information and Maximal Correlation as Measures of Dependence," *Annals of Mathematical Statistics*, vol. 33, pp. 587-595.

[3] Deignan Jr., P.B., Meckl, P.H., and Franchek, M.A., 2002, "The MI-RBFN: Mapping for Generalization," *Proceedings of the American Control Conference*, pp. 3840-3845.

[4] Fraser, A.M. and Swinney, H.L., 1986, "Independent coordinates for strange attractors from mutual information," *Physics Review A*, Vol. 33, No. 2, pp. 1134-1140.

[5] Arbarbanel, H.D.I., Masuda, N., Rabinovich, M.I., and Evren, T., 2001, "Distribution of mutual information," *Physics Letters A*, Vol. 281, pp. 368-373.

L-GRAMMAR INDUCTION USING GENETIC ALGORITHMS

DAVID CALVERT
Computing and Information Science
University of Guelph
Guelph, ON
CANADA

NARENDRA PERSHAD
Computing and Information Science
University of Guelph
Guelph, ON
CANADA

ABSTRACT
In this study two technologies, Genetic Algorithms (GA) and Lidenmayer-Systems (L-System), are combined to produce a system which evolves a grammar that can recreate a target image. A hierarchical encoding is used to represent the grammar. The encoding allows the GA operations to manipulate the grammatical structure such that a correct grammar is always produced. A target image of a plant is used to calculate the fitness for the GA.

INTRODUCTION

The objective of this work is to demonstrate that genetic algorithms can be used to induce an L-Grammar which closely models a target image. The grammars associated with L-Systems are very terse. Their encoding makes it difficult to predict the appearance of the image without interpreting the grammar and rendering the image. In this work, an L-System is used to generate a given target image of vegetation previously generated by an L-System. A GA is used to evolve a grammar which can recreate the target image. The GA is therefore searching the set of possible L-Grammars for one which models the target image. The effects of mutation and crossover rates, and population size on the GA's performance are evaluated for target images of different complexity.

IMPLEMENTATION DETAILS

L-Systems are described using five parameters. These are the Length in pixels of the shortest branch in the image, the Iteration which represents the number of times to apply the rewriting rules, Angle whish represents the number of degrees between two adjacent branches in the image, an Axiom which is string representing the seed of the L-System, and a Rule set contains all the production and rewriting rules. Only deterministic L-Systems were used in this work.

The L-Grammar is represented as a rooted hierarchical collection of nodes, an example of which is seen in Figure 1. Each node has a single parent except the root node. All nodes have one of seven coded values F, f, G, O, 1, 2 or 3 which translate into instructions for drawing the grammar. The first three symbols are traditional symbols where f represents a forward step, F is a forward step while drawing a line, and G represents the seed grammer. The other symbols are decoded as follows: O is replaced by a null operation, 1 is replaced with +F, 2 is replaced with -F, 3 is replaced with FF. The + and – indicate an

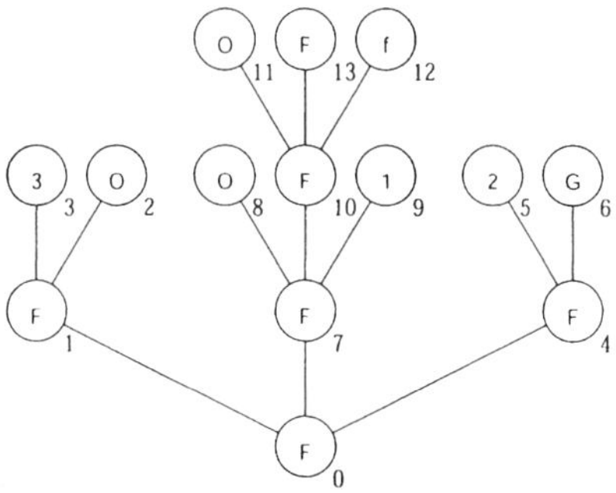

Figure 1: The Structure of a Chromosome

angle between two adjacent lines. These contractions were used because they are commonly occuring structures found in L-Grammars. A node may be directly linked with up to three child nodes. The sequence of these links will determine if the child node will be a left branch, a right branch or an extension of the current branch.

The algorithm for traversing the tree visits the nodes in the order of parent node, left child, right child, centre child. Each of the child branches are evaluated until leaf nodes are reached and the decoded results of the branches are returned to the parent nodes. The structure in Figure 1 is numbered with the sequence in which it would be traversed.

The grammars generated by this system contains symbols that are not present in a regular L-Grammar, so they must be decoded after the grammar has been constructed from the tree. The structure in Figure 1 would be translated to an L-Grammar expression via the following stages:

Node0 ⇒ F[-Node1][+Node4]Node7
Node1 ⇒ F[+O]3
Node4 ⇒ F[-2]G
Node7 ⇒ F[-O][+1]F[-O][+f]F
Node0 ⇒ F[-F[+O]3][+F[-2]G]F[-O][+1]F[-O][+f]F
Node0 ⇒ F[-FFF][+F[--F]G]F[++F]F[+f]F

This hierarchical representation was chosen over a fixed-length binary string representation which is commonly used in GAs because a balanced bracketed expression will always be produced and it may be extended to represent an arbitrary number of branches. This also allows there to be no restriction on the size of shape of the structure and to facilitate crossover, mutation, and conversion to a bracketed L-Grammar.

The initial value of a node are is assigned randomly. Links to the node are also determined randomly. Fitness is calculated using a bitmap measure where a target image and the images created from the L-Grammars generated by the GA are compared. Fitness proportionate selection and single point crossover are used. Crossover points are found by randomly choosing a node in each tree. Crossover is performed by swapping the entire subtree attached to these nodes. This guarantees that a correct structure is maintained and a correct bracketed L-Grammar will be produced.

Figure 2 shows two grammars before crossover, the X's indicate the crossover points. Figure 3 shows the same two chromosomes after crossover.

Three types of mutation are used in this work. These include changing the value of a selected node, creating a new node, and removing a node. If the selected node is null then a new node is created. If a node already exist then it is either removed or the value changed with equal chance.

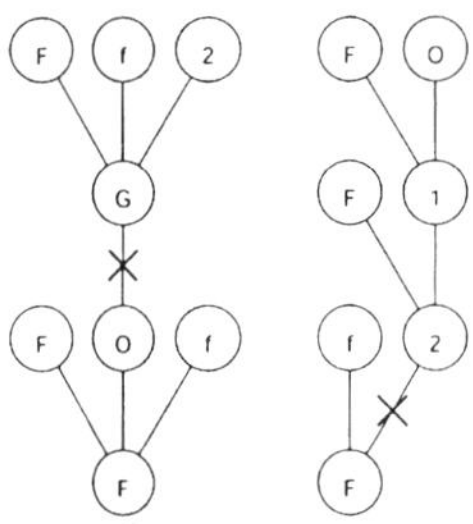

Figure 2: Before Crossover

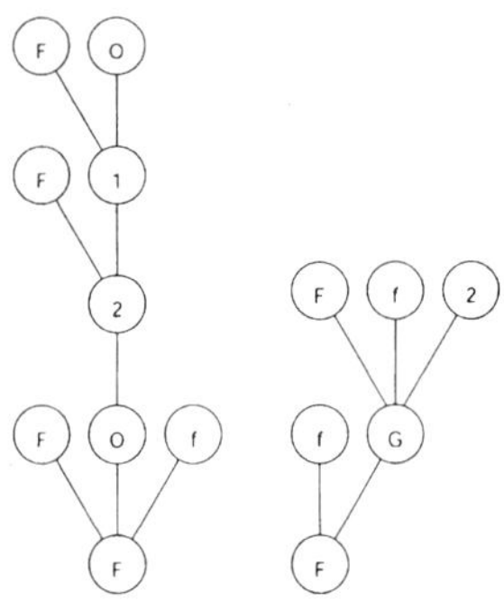

Figure 3: After Crossover

RESULTS

A simple grammar was chosen to investigate the effects of crossover and mutation on the ability of the GA to generate a solution within a given number of iterations. In many cases, the solution generated by this system was identical to the target solution. For each epoch the average fitness, the variance, and the best individual fitness were recorded.

The target grammar chosen was F[-F]F[+F]F. This is a very simple grammar and it will be referred to as Grammar A. The population size was set to 100. Crossover rate varied from 0.45 to 0.85 in 0.05 increments while the mutation rate varied from 0.05 to 0.85 also in 0.05 increments. Using these limitations all possible combinations of crossover and mutation rates were used. Each combination was run ten times until the correct solution was found or the number of iterations exceeded 500.

The average fitness of the best individual with the same crossover rate is shown in Figure 4. Each point in this figure is the average of all tests using the same crossover rate and differing mutation rates. This appears to indicate that the GA is relatively unaffected by changes in the crossover rate.

The variation of the average fitness of the best individual with the same mutation rate regardless of crossover rate is shown in Figure 5. This appears to indicate that the GA performs better at higher mutation rates.

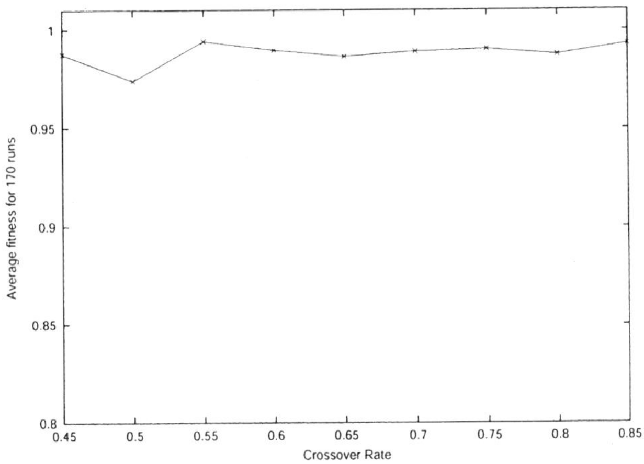

Figure 4: Crossover Rate and its Effect on Average Fitness for Grammar A

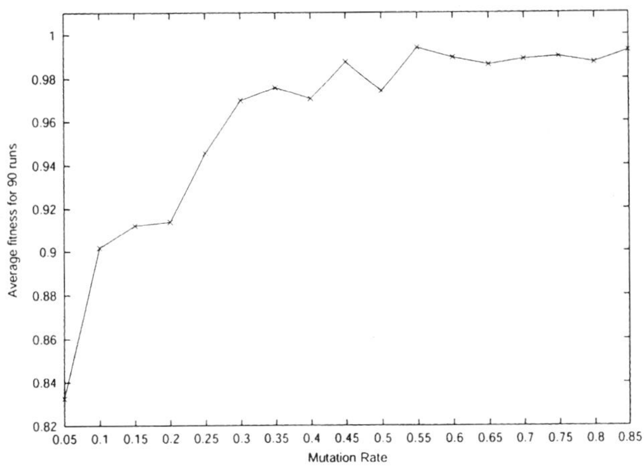

Figure 5: Mutation Rate and its Effect on Average Fitness for Grammar A

GENETIC ALGORITHM FITNESS CHARACTERISTICS

In 90% of the tests the best individual found had a fitness value of more than 0.95 out of 1.0.The changes in average fitness of the population for each cycle for all the runs is shown in Figure 6.

Out of the 1530 tests, the GA found a solution which exactly recreated the target image 391 times within the 500 cycle tests. It found the exact solution 33 times but it also found 274 different variations of the solution that will correctly reproduce the target image.

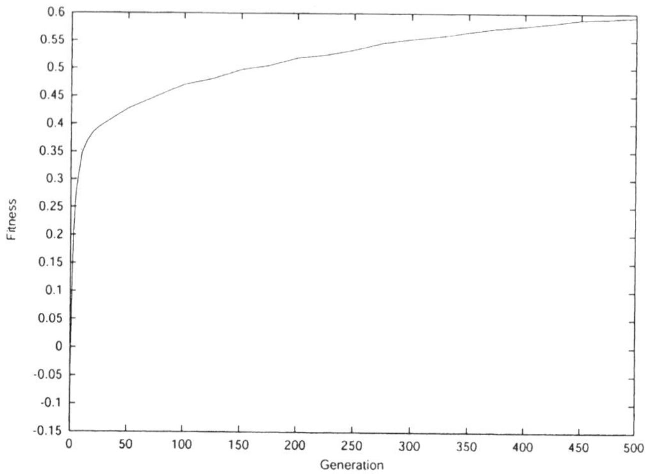

Figure 6: Variation in the Average Fitness and Generation for Grammar A

EFFECT OF COMPLEXITY ON PERFORMANCE

An attempt was made to measure the performance of the system by comparing the fitness of generated solutions to various complexity measures for the grammars. Using the same values of mutation and crossover, tests were performed with 18 grammars of varying complexities. Because these grammars constitute harder problems, the GA was allowed to run for a maximum of 5000 cycles for 5 sets of tests. Complexity was computed based on several common metrics. These included height, width, out degree, number of nodes in the tree. Additional measures of complexity which were tested include the number of pixels required to draw the image, and a weighted node count. The weighted node count is calculated by assigning a weight based on the height of the node in the tree. The further away it is from the root the more heavily it weighted. The weight of a node is calculated as $3^{level-1}$ so nodes on the 1st, 2nd, 3rd level on a tree are 1, 3 and 9 respectively. When this fitness was compared to the complexity it was hoped that a correlation between the GA's ability to find a solution and the complexity would be evident. There does not appear to be a strong correlation between the fitness and any of the complexity measures.

There does appear to be a correlation between the complexity measures themselves. The similar results from the different measures suggests that they are likely measuring the same characteristics of the tree.

RESULTS FOR MORE COMPLEX L-SYSTEMS

Two other grammars were tested with the same combinations of mutation and crossover values that were used to test Grammar A. The moderately complex grammar F[-F+F][++F-F]FF was used and will be referred to as Grammar B. The very complex grammar F[- -[F]][+F[+F]]F[-F][++F]F was also used, and this will be referred to as Grammar C for the rest of this study. Compared to tests with Grammar A both of the more complex grammars show a significant increase in range of fitness and required more time to find the solution. This illustrates that the system has more difficulty in finding the target image when the grammar is more complex.

CONCLUSIONS

A system was developed which will induce an L-Grammar that can recreate a target image. The system starts with a randomize population of L-Grammars and uses a genetic algorithm to evolve a grammar which can recreate the target image. Fitness for the GA is calculated by comparing the target image to the images create by the grammars in the GA population.

Various crossover and mutation rates where tested while evolving L-Grammars. The crossover parameter had little effect on the systems ability to find the desired solution. Within the range of 45% to 85% there was little change in the GA performance. The GA performs better at higher mutation rates. Within the range of 5% to 85% the performance increased steadily and then plateaus at around 55%.

Several complexity measures where tested with the system in an attempt to predict the computational cost of inducing a grammar for the target image. No apparent correlation between the complexity measures and the system's performance was found.

The results of testing more complex grammars indicate that the GA converged more slowly than it did with the simple grammar. The average fitness rates were lower and there was a much higher variance in fitness values.

REFERENCES

Goldberg, David. Genetic Algorithms in Search, Optimization, and Machine Learning, Addison-Wesley Publishing Company, 1989.

Herman, Gabor T., Grzegorz Rosenberg, and Aristid Lindenmayer. Developmental Systems and Languages, North-Holland Publishing Company, 1975.

Koza, J.R.. Genetic Programming: A Paradigm for Genetically Breeding Populations of Computer Progams to Solve Problems, MIT Press, 1990.

Przemuslaw Prusinkiewicz and Aristid Lindenmayer. The Algorithmic Beauty of Plants, Springer-Verlag, 1996.

Smith, A.R. Plants, Fractals, and Formal Languages, Computer Graphics, SIGGRAPH 84 Conference Proceedings, Association of Computer Machinery, 1984.

DATABASE-DRIVEN NEURAL NETWORKS AND EVOLUTIONARY COMPUTATION

BRIAN BLAHA
Department of Computer Engineering
University of Missouri-Rolla
Email: bblaha@umr.edu

DON WUNSCH
Department of Computer Engineering
University of Missouri-Rolla
dwunsch@ece.umr.edu

ABSTRACT
More complex problems call for more complex solutions, but these solutions require either more programmers to encode, or more time. More programmers have more difficulty coordinating and end up wasting effort. Spending more time means waiting longer before reaping any benefits, and, if taken to the extreme, can completely prevent a project from being finished, as regular turnover replaces those working on it as fast as new programmers can be brought up to speed. When development time is valuable enough to sacrifice speed in the finished application, an entirely new set of programming languages becomes viable. In particular, a neural network can be cast as a pattern of database queries to a simple set of tables. This approach yields safer programs sooner, but the programs run rather slowly.

EXPLANATION OF THE PROBLEM

As computers have increased in power, more complicated problems have become tractable, and more complicated programs have been written to solve them. However, more complicated programs take longer to write and debug. Therefore, the problems must be split into simpler, more easily implemented problems. Historically, this has been evident in the development of programming languages, from machine to assembly to compiled languages to object-oriented languages. Each successive language further facilitates solving difficult problems (Prechelt, 1997 and Lindvall, 1998).

On expensive, slow computers, each program needed to be optimized for its exact task. With faster, cheaper computers, programs can be more flexible, at the cost of some speed. Evolutionary computation is a very flexible, but time-consuming solution to a wide variety of problems. Neural networks are similarly flexible, though less time-consuming, and a great many varieties have been implemented for various purposes.

However, some of the problems that these techniques could be applied to require time-consuming programming to get the raw data to the algorithm, especially in the case of dynamic problems, where data is arriving in real time and must be incorporated into the solution. Chaotic system prediction, including weather forecasting and stock market prediction, is the most dependent on

The authors gratefully acknowledge the support of the National Science Foundation and the Mary K. Finley endowment

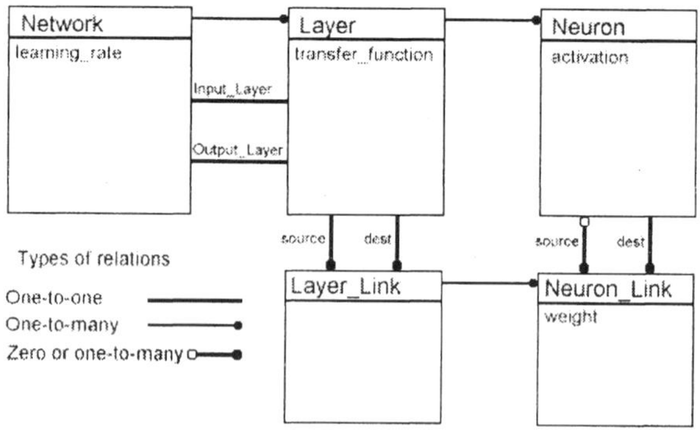

Figure 1: **Neural Network Organization**

current data, given chaotic systems' short windows of predictability. In other cases, such as (Putten, 1999), the data required to evaluate potential solutions is already in a database.

For this, SQL, an ANSI-standardized relational database query language, offers rapid programming and debugging. SQL easily supports altering the data used by a currently executing program, and it provides functions to naturally handle sets, such as a population of potential solutions. Recent C++ programming environments have just begun to offer some of the debugging assistance SQL provides, such as writing, testing, correcting, and recording code sequentially, so that errors immediately become evident (Gat, 2000 and Agrawal et al, 1993).

EXPLANATION OF SQL

A relational database is organized as a set of tables, each with columns, each potentially with relations to other columns in other tables. Each row in a table contains information which corresponds to the columns of the table. A database can be described concisely by a UML diagram.

Fig. 1 represents a generic neural network, in UML. Each network can contain any number of layers, and each layer contains any number of neurons. Each layer can be directionally connected to any number of other layers, although some additional mechanism would need to decide which connection to use and when. Each layer has its own transfer function associated with it, which can be defined much like in MATLAB. Each neuron can be connected to any number of other neurons, although such connections would be constrained to be between connected layers. Furthermore, as shown, a neuron can be the destination of a link with no source. This represents a bias.

Standard SQL does not support defining functions as sets of SQL statements. For this, PHP can be used. PHP is a language designed both for use on hypertext pages and for communicating with SQL databases. Large programs can be

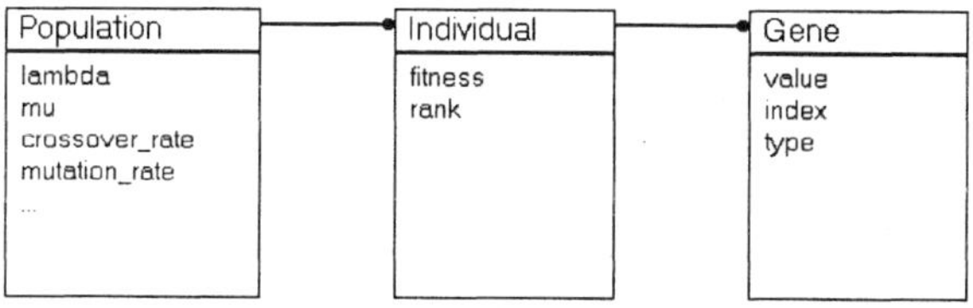

Figure 2: **Genetic Algorithm Organization**

written as a mixture of PHP and SQL statements. Learning techniques, such as backpropagation, can be implemented by adding some SQL tables and PHP functions.

Translating the UML in Fig. 1 into pseudocode SQL yields a set of table definitions and constraints:

create table network (ID integer primary key, learning_rate double not null);

create table layer (ID integer primary key, function varchar(20) not null, network integer not null);

alter table layer add constraint layer_network_fk foreign key (network) references network on delete cascade;

create table neuron (ID integer primary key, activation double, layer integer not null);

alter table neuron add constraint neuron_layer_fk foreign key (layer) references layer on delete cascade;

create table layer_link (ID integer primary key, source integer not null, dest integer not null);

alter table layer_link add constraint layer_link_source_fk foreign key (source) references layer on delete cascade;

alter table layer_link add constraint layer_link_dest_fk foreign key (dest) references layer on delete cascade;

create table neuron_link (ID integer primary key, weight double, source integer, dest integer not null, layer_link integer not null);

alter table neuron_link add constraint neuron_link_source_fk foreign key (source) references neuron on delete cascade;

alter table neuron_link add constraint neuron_link_dest_fk foreign key (dest) references neuron on delete cascade;

alter table neuron_link add constraint neuron_link_layer_link_fk foreign key (layer_link) references layer_link on delete cascade;

Because SQL inherently supports multiple clients on a single server, altering a network on the fly is straightforward, and the state of a network can be easily displayed on a web page, using a database-related language such as PHP.

Just as neural networks can be elegantly implemented in SQL, so can be other group-oriented algorithms, such as genetic algorithms, used to train neural networks in (Fogel, 2002). The UML in Fig. 2 describes both an SQL implementation and an object-oriented C++ implementation of one genetic algorithm.

The UML diagram in Fig. 2 has meaning as a set of C++ classes:

```
struct Gene{
    double value, stddev;
};
class Individual{
public:
    double fitness;
private:
    Gene* genes;
    size_t genes_count;
};
class Population{
public:
    size_t lambda, mu, individual_size;
    double crossover_rate, mutation_rate, gene_mutation_rate, stddev_mutation;
private:
    Individual* individuals;
    size_t individuals_count;
};
```

Of course, it also describes a database schema:

```
create table population ( ID integer primary key, crossover_rate double, mutation_rate double, mu integer, lambda integer, genotype_length integer, stddev_mutation double );

create table individual ( ID integer primary key, fitness double, rank integer, population integer not null );

alter table individual add constraint individual_population_fk foreign key (population) references population on delete cascade;

create table gene ( ID integer primary key, value double, type text not null, index integer not null, individual integer not null );

alter table gene add constraint gene_individual_fk foreign key (individual) references individual on delete cascade;
```

C++ classes are equivalent to SQL tables, and C++ objects are equivalent to SQL table rows. C++ methods are implemented as either sets of SQL statements, or functions in PHP. SQL constraints concisely describe and enforce relationships among data, whereas C++ requires explicit function calls to check relationships.

SQL statements handle columns and entire tables as primitives, much as MATLAB handles vectors and matrices. Just as multiplying two matrices is a brief statement in MATLAB, extracting useful information from a database can often be done with a short, legible statement, whereas the same operation in C++ would be a complicated function, prone to subtle bugs.

COMPARISON TO C++ ETC

A PHP function to perform mutation on one individual is longer than equivalent C++ code, but the C++ code uses symbols in place of words, and lends itself to inflexibility. If, for example, a programmer wants to add another set of genes corresponding to the standard deviation of the standard deviation, neither mutation function would need extensive modifications, but either the C++ class definition for the individual would need to be changed or most of the member functions which access the genes array would need to be changed. The SQL table definition wouldn't need a single change, and very few other functions would need alterations.

SQL and PHP code work with tables, columns, and relations, whereas C++ works with functions, integers, and pointers. What would require a complicated class structure in C++ is already provided with SQL. SQL works at a more abstract level than C++, and is as a result easier to use, write, and maintain. In C++, objects have member data and member functions, which can be called to modify the object. SQL revolves around relational database structure: no table row owns another; rather, certain rows share "keys," or columns used specifically for identification. in which certain columns' elements are intentional copies of elements in other tables. This defines the relationships between tables.

In C++ functions, most pieces of information must be explicitly passed, with the exception of class members used in member functions. A PHP function would perform a search for the population row with the same identifier that is stored in the population column of the current individual, because the relationship between individual and population is inherently adirectional, instead of the unidirectional relationship in C++. Although SQL/PHP suffers a loss of speed for searching, searches run in logarithmic if not constant time, limiting or preventing an increase in run-time complexity.

DISADVANTAGES OF SQL

In order to provide all its functionality, search-based relationships, security, and data integrity, SQL sacrifices some speed. Depending on just how the data is managed, an SQL/PHP implementation of a program could be anywhere from slightly slower to several orders of magnitude slower than the equivalent C++ program. However, if data required to solve the problem starts in a database, then any algorithm must lose speed in retrieving it, reducing the speed penalty for using SQL/PHP for the rest of the algorithm.

For some applications, the speed of the resulting program is much less important than the time it takes to develop the program. SQL/PHP programs contain relatively few bugs because the structure imposed by the relational database paradigm limits the potential for obscure errors. Those bugs they do contain can be removed swiftly because being able to program interactively allows spurious statements to be carefully tested with ease (Gat, 2000 and Prechelt, 1997). Since MATLAB, an inherently slow language, is commonly used to run neural networks, even less speed would be lost in converting to SQL/PHP.

Whereas, in C++, most code can be compiled in any compiler, SQL statements frequently differ between implementations. Most of these differences occur in proprietary extensions, but these extensions are unfortunately often necessary. Furthermore, the syntax required for PHP to communicate with the SQL server differs significantly, to the point that automatically converting from one format to another would require quite a complicated program. Manual conversions are required, and unfortunately become a barrier to switching SQL server programs.

SQL/PHP running time (seconds)						
	Hidden layer size					
Epochs	0	10	30	100	300	1000
10	9	10	10	13	16	35
30	26	28	28	32	37	67
100	84	90	91	97	104	153
300	254	268	273	285	302	391
1000	840	889	905	946	1011	1235

MATLAB running time (seconds)					
	Hidden layer size				
Epochs	10	1000	3000	10000	30000
10	1.1	1.2	1.3	1.5	2.0
30	3.0	3.0	3.5	3.8	5.6
100	9.6	9.6	11	12	24
300	28	28	32	36	110

C++ running time (seconds)					
	Hidden layer size				
Epochs	30	300	1000	3000	10000
10	2.6e-4	2.9e-3	3.3e-2	.11	.37
30	7.1e-4	9.0e-3	.10	.33	1.1
100	2.4e-3	3.2e-2	.35	1.2	3.7
300	7.1e-3	8.5e-2	.97	3.2	9.8
1000	2.4e-2	.23	3.1	10	34
3000	7.1e-2	.62	9.3	30	100
10000	.23	2.0	31	99	330

TEST RESULTS

Each implementation of a neural network was designed to have one input neuron, a hidden layer of arbitrary size, and one output neuron in a simple feed-forward structure. Each network was trained on a simple sinusoid function for a variable number of epochs. First, this network was implemented in a combination of SQL and PHP, and accessed via a local HTTP server and a loopback connection.

As demonstrated by the running times for a hidden layer size of zero, much of the algorithm's time was spent in overhead, including communication among the SQL engine, PHP, and the HTTP server. Batch training could substantially reduce this overhead, but since this paper focuses more on the potential of SQL, its performance was compared to MATLAB's performance, when also using incremental training (through the adapt function). It should be noted that MATLAB was run on a 2.2 GHz computer, whereas the C++ and SQL/PHP programs were run on a .95 GHz computer.

MATLAB's neural network toolkit spent about a second per ten training epochs in function call overhead, which is only somewhat better than SQL's performance. However, MATLAB spent much less time per neuron apart from overhead. This is largely due to the lower-level implementation of core neural network training routines. SQL routines could likewise be replaced by C++ implementations if speed is crucial. For reference, both neural network implementations were compared against a C++ implementation designed to be as similar in structure to the SQL implementation as possible.

Under these conditions, the C++ code of course outperforms MATLAB, but when non-scaling overhead is neglected, the speed difference is fairly small. Of course, this is no surprise when most of MATLAB's time is spent in non-MATLAB code. Therefore, the two elements holding back the speed of the SQL implementation are function call overhead and slow data processing. The first can be overcome with an implementation of batch training, while the second can be lessened by implementing C++ helper functions to manipulate large amounts of data (such as layers with many neurons) without invoking the SQL engine frequently.

Furthermore, despite the slowness of the SQL implementation, it served an important debugging role. The C++ implementation was virtually bug-free because all of the algorithm's logical bugs had been worked out interactively, with the verbose errors SQL and PHP provide. In particular, PHP is grammatically extremely similar to C++, and thus very simple to translate. Similarly, the C++ helper functions would be much easier to debug, since they would be based on extant SQL functions.

CONCLUSION

For problems in which new data is arriving relatively slowly, SQL can still keep up, even unoptimized. When this is the case, the benefits of SQL can be exploited. The resulting code is abstracted from machine code, allowing high-level operations to be written easily. Similarly, high-level changes in the problem, such as a change in the training method, can be easily integrated into existing code. Even if the speed penalty of SQL becomes intolerable, the time invested in developing code for it pays off when reimplementing the algorithms in other languages

MATLAB has the advantage of being already written for reasonable speed, but has limited ability to interface with a network, be run remotely, or allow data to be accessed by multiple users. Furthermore, MATLAB is not a viable option if the desired program cannot take advantage of a toolkit.

REFERENCES

Agrawal, H., DeMillo, R. A., and Spafford, E. H., 1993, "Debugging with Dynamic Slicing and Backtracking," Software - Practice and Experience. Volume 23, number 6, p. 589-616.

Fogel, D., 2002, *Blondie24: Playing at the Edge of AI,* Morgan Kaufmann Publishers.

Gat, E., 2000, "Lisp as an alternative to Java," Intelligence. p. 21-24.

Lindvall, M., 1998, "Monitoring and measuring the change-prediction process at different granularity levels: an empirical study," Software Process: Improvement and Practice. Volume 4, number 1, p. 3-10.

Prechelt, L., 1997, "An Empirical Comparison of C, C++, Java, Perl, Pythod, Rexx, and Tcl for a Search/String-Processing Program," http://citeseer.nj.nec.com/547856.html

Putten, P., 1999, "Data Mining in Direct Marketing Databases," World Scientific.

A HIERARCHICAL GENETIC ALGORITHM APPROACH FOR MODULAR NEURAL NETWORKS EVOLUTION

PATRICIA MELIN AND OSCAR CASTILLO
Department of Computer Science
Tijuana Institute of Technology
P.O. Box 4207, Chula Vista CA 91909, USA
pmelin@tectijuana.mx

ABSTRACT

We describe in this paper the evolution of modular neural networks using hierarchical genetic algorithms. Modular Neural Networks (MNN) have shown significant learning improvement over single Neural Networks (NN). For this reason, the use of MNN for complex system identification is well justified. However, network topology design of MNN is at least an order of magnitude more difficult than for classical NNs. We describe in this paper the use of a Hierarchical Genetic Algorithm (HGA) for optimizing the topology of each of the neural network modules of the MNN. The HGA is clearly needed due to the fact that topology optimization requires that we are able to manage both the layer and node information for each of the MNN modules. Simulation results shown in this paper prove the feasibility and advantages of the proposed approach.

INTRODUCTION

The results of the different applications involving MNN lead to the general evidence that the use of modular neural networks implies a significant learning improvement comparatively to a single NN and especially to the backpropagation NN [2]. Indeed, to constrain the network topology on connectivity, increases the learning capacity of NN and allow us to apply them to large-scale problems [6]. This is highly confirmed by the experience carried out by [1], which shows that a random pruning of connections before any learning improves significantly the network's performance. Also, we can note that [4] and [14] argue that a complex behavior requires bringing together several different kinds of knowledge and processing, which is, of course, not possible without structure (modularity).

Usually the MNN implementations are based on the "divide and conquer" principle, which is well known in computer science. This principle consists first in breaking down a task into smaller and less complex subtasks, to make learn each task by different experts (i.e. NN modules) and then, to reuse the learning of each subtask to solve the whole problem. For example, [8] has shown that NN training can be greatly simplified by identifying subtasks in the problem and embedding them into the network structure.

This method can only be applied if the a-priori knowledge concerning the task is sufficiently precise to enable a split up of the task into subtasks. If one can separate a task in distinct subtasks, each task can be trained off-line and later integrate in the global architecture [12]. This enables an acceleration of the learning [5].

To make an efficient use of modular neural networks, however, we need to optimize their topology for a specific problem. Here is where the evolutionary approach to modular neural network design comes into place. We propose the use of a hierarchical genetic algorithm for evolving the topology design of the complete modular network, which means optimizing each and all of the modules of the MNN. An HGA is needed because the MNN topology information is hierarchical, i.e. we need to manage information about the number of modules, number of layers and nodes of the network. As a consequence the chromosome of an HGA contains genetic information about the different levels of the MNN topology.

The paper is structured as follows. First, we describe some basic concepts about MNNs and evolution. Second, we describe the basic theory of HGAs for neural network design. Then, we describe the proposed approach for evolving an MNN using HGAs. We also show simulation results for the case of complex function identification. Finally, we conclude the paper with discussion and future work.

EVOLVING NEURAL NETWORKS

Neural Networks are biologically motivated approaches to machine learning, inspired by ideas from neuroscience. Recently, some efforts have been made to use genetic algorithms to evolve aspects of neural networks [11].

In its simplest feedforward form, a neural network is a collection of connected neurons in which the connections are weighted, usually with real-valued weights. The network is presented with an activation pattern on its input units, such as a set of numbers representing features of an image to be classified. Activation spreads in a forward direction from the input units through one or more layers of middle units to the output units over the weighted connections. This process is meant to roughly mimic the way activation spreads through networks of neurons in the brain. In a feedforward network, activation spreads only in a forward direction, from the input layer through the hidden layers to the output layer. Many people have also experimented with "recurrent" networks, in which there are feedback connections between layers.

In most applications, the neural network learns a correct mapping between input and output patterns via a learning algorithm. Typically the weights are initially set to small random values. Then a set of training inputs is presented sequentially to the network. In the backpropagation learning procedure, after each input has propagated through the network and an output has been produced, a "teacher" compares the activation value at each output unit with the correct values, and the weights in the network are adjusted in order to reduce the difference between the network's output and the correct output. This type of procedure is known as "supervised learning", since a teacher supervises the learning by providing correct output values to guide the learning process.

There are many ways to apply genetic algorithms to neural networks. Some aspects that can be evolved are the weights in a fixed network, the network architecture (i.e., the number of neurons and their interconnections can change), and the learning rule used by the network. David Montana and Lawrence Davis (1989) took the first approach of evolving the weights in a fixed network [13]. That is, Montana and Davis were using the genetic algorithm instead of backpropagation as a way of finding a good set of weights for a fixed set of connections. Several problems associated with the backpropagation algorithm (e.g., the tendency to get stuck at local minima, or the unavailability of a "teacher" to supervise learning in some tasks) often make it desirable to find alternative weight training schemes.

Neural network researchers know all too well that the particular architecture chosen can determine the success or failure of the application, so they would like very much to be able to automatically optimize the procedure of designing an architecture for a particular application. Many believe that genetic algorithms are well suited for this task [11].

MODULAR NEURAL NETWORKS

This section describes a particular class of "modular neural networks", which have a hierarchical organization comprising multiple neural networks; the architecture basically consists of two principal components: local experts and an integration unit, as illustrated in Figure 1. A variety of modular connectionist architectures have been discussed, and thus such diverse names as "committees of networks", "adaptive mixtures", and "hierarchical mixtures of experts" have all been mentioned.

In general, the basic concept resides in the idea that combined (or averaged) estimators may be able to exceed the limitation of a single estimator. The idea also shares conceptual links with the "divide and conquer" methodology. Divide-and-conquer algorithms attack a complex problem by dividing it into simpler problems whose solutions can be combined to yield a solution to the complex problem. In other words, the central idea is task decomposition. When using a modular network, a given task is split up among several local experts NNs. The average load on each NN is reduced in comparison with a single NN that must learn the entire original task, and thus the combined model may be able to surpass the limitation of a single NN. The outputs of a certain number of local experts (O_i) are mediated by an integration unit. The integrating unit puts those outputs together using estimated combination weights (g_i). The overall output Y of the modular network is given by

$$Y_i = \Sigma g_i O_I \qquad (1)$$

Nowlan, Jacobs, Hinton, and Jordan [7] described modular networks from a competitive mixture perspective. That is, in the gating network, they used the "softmax" activation function, which was introduced by McCullagh and Nelder [9]. More precisely, the gating network uses a softmax activation g_i of the ith output unit given by

$$G_i = \exp(ku_i) / \Sigma_j \exp(ku_j) \qquad (2)$$

Where u_i is the weighted sum of the inputs flowing to the ith output neuron of the gating network. Use of the softmax activation function in modular networks provides a sort of "competitive" mixing perspective because the ith local expert's output O_i with a minor activation u_i does not have a great impact on the overall output Y_i.

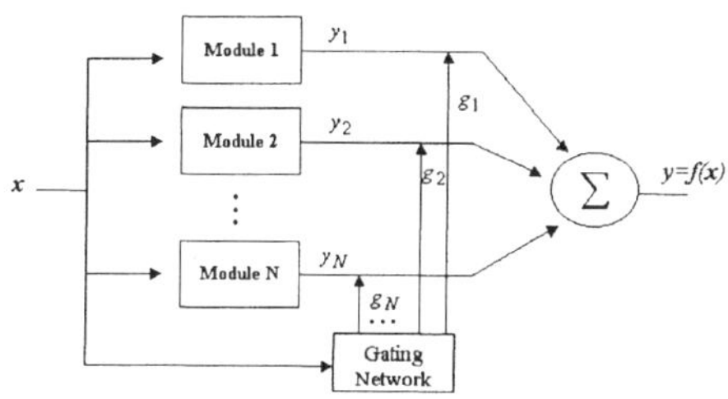

Figure 1 Architecture of a modular neural network.

HIERARCHICAL GENETIC ALGORITHMS FOR NEURAL NETWORKS

The bottleneck problem for NN application lies within the optimization procedures that are used to obtain an optimal NN topology. Hence, the formulation of the Hierarchical Genetic Algorithm (HGA) is applied for this purpose [10]. The HGA differs from the standard GA with a hierarchy structure in that each chromosome consists of multilevel genes. Each chromosome consists of two types of genes, i.e. control genes and connection genes. The control genes in the form of bits, are the genes for layers and neurons for activation. The connection genes, a real value representation, are the genes for connection weightings and neuron bias.

With such a specific treatment, a structural chromosome incorporates both active and inactive genes. It should be noted that the inactive genes remain in the chromosome structure and can be carried forward for further generations. Such an inherent genetic variation in the chromosome avoids any trapping at local optima, which has the potential to cause premature convergence. Thus it maintains a balance between exploiting its accumulated knowledge and exploring the new areas of the search space. This structure also allows larger genetic variations in chromosome while maintaining high viability by permitting multiple simultaneous genetic changes. As a result, a single change in high level genes will cause multiple changes (activation or deactivation in the whole level) in lower level genes. In the case of the traditional GA, this is only possible when a sequence of many random changes takes place. Hence the computational power is greatly improved.

EXPERIMENTAL RESULTS

We show the feasibility of the proposed approach with a problem of system identification. We consider the non-linear function given by the following equation:

$$y = 2e^{(-0.2X)}\left|\sin(x)\right| \tag{3}$$

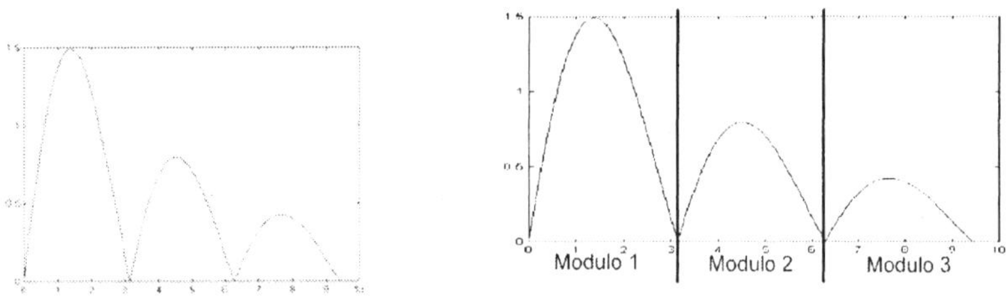

Figure 2 (a) Graph of the function eq. (3) (b) Modular approach to function identification

The graph of this function is shown in Figure 2 (a). From this figure, we can appreciate that this function has three different regions. For this reason, we can use this function to test the modular approach, i.e. one neural network module for each region can be used. This idea is illustrated in Figure 2 (b).

We used as range for the function the [0, 9.45] interval with a 0.01 step size, which gives us 945 points. The data is divided in the three modules as follows:

Module 1: from 0 to 3.15, Module 2: from 3.15 to 6.30, Module 3: from 6.30 to 9.45

The idea behind this data partition is that learning will be easier in each of the modules, i.e. a simple NN can learn more easily the behavior of the function in one of the regions. We used three-layer feed-forward NNs for each of the modules with the Levenberg-Marquardt training algorithm. We show in Figure 3 the general architecture of the MNN used in this paper.

Regarding the genetic algorithm for MNN evolution, we used a hierarchical chromosome for representing the relevant information of the network. First, we have the bits for representing the number of layers of each of the three modules and then we have the bits for representing the number of nodes of each layer. This scheme basically gives us the chromosome shown in Figure 4.

The fitness function used in this work combines the information of the error objective and also the information about the number of nodes as a second objective. This is shown in the following equation.

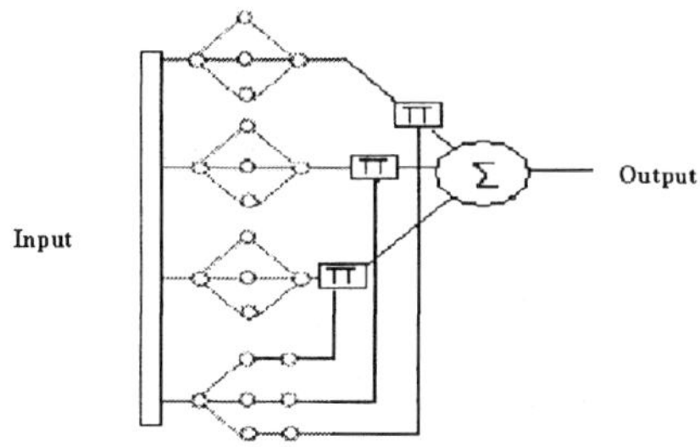

Figure 3 General architecture of the modular neural network.

CM1	CM2	CM3	NC1M1	NC2M1	NC3M1	NC1M2	NC2M2	NC3M2	NC1M3	NC1M3	NC1M3
3 BITS	3 BITS	3 BITS	8 BITS	8 BITS	8 BITS	8 BITS	8 BITS	8 BITS	8 BITS	8 BITS	8 BITS

Figure 4 Basic structure of the chromosome containing the information of the MNN.

$$f(z) = \left(\frac{1}{\alpha * Ranking \quad (ObjV \;\; 1) + \beta * ObjV \;\; 2} \right) * 10 \tag{4}$$

The first objective is basically the average sum of squared of errors as calculated by the predicted outputs of the MNN compared with real values of the function. This is given by the following equation.

$$f_1 = \frac{1}{N} \sum_{i=1}^{N} (Y_i - y_i) \tag{5}$$

The parameters of the genetic algorithm are as follows:

Type of crossover operator: Two-point crossover, Crossover rate: 0.8
Type of mutation operator: Binary mutation, Mutation rate: 0.05
Population size per generation: 10, Total number of generations: 100

We show in Figure 5 the topology of the final evolved modules of the neural network for the problem of function identification. As we can appreciate, from this Figure, module 2 is the smallest one and module 3 is the largest one. The result of MNN evolution is a particular architecture with different size of the modules (neural networks).

The MNN architecture shown in Figure 5 is the best one for the specific problem of system identification. It is worthwhile noting that this network topology is difficult to design manually, for this reason the HGA approach is a good choice for neural network design and optimization.

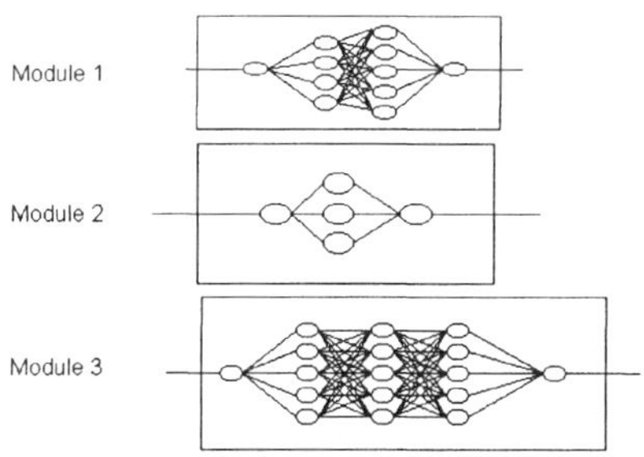

Figure 5 Topology of the final evolved MNN for function identification.

CONCLUSIONS

We described in this paper our hierarchical genetic algorithm approach for modular neural network topology design and optimization. The proposed approach was illustrated with a specific problem of function identification. The best MNN is obtained by evolving the modules (single NNs) according to the error of identification and also the complexity of the modules. The results for the problem of function identification are very good and show the feasibility of the HGA approach for MNN topology optimization. The future work will consist in the application of HGA to MNN in more complex problems, like in control and time series forecasting. Also, parallel processing of the HGA is being considered for speeding up the calculations.

ACKNOWLEGMENTS

We would like to thank the Research Grant Committee of COSNET for their financial support (under Grant 487.02-P). We would also like to thank the Department of Computer Science of Tijuana Institute of Technology for the time and resources given to this research project.

REFERENCES

[1] G. Barna and K. Kaski, "Choosing optimal network structure", Proceedings of the International Neural Network Conference (INNC90), pp. 890-893, 1990.

[2] O. Castillo and P. Melin,"Hybrid Intelligent Systems for Time Series Prediction using Neural Networks, Fuzzy Logic and Fractal Theory", IEEE Transactions on Neural Networks, Vol. 13, no. 6, pp. 1395-1408, 2002.

[3] D. J. Chalmers, "The Evolution of Learning: An Experiment in Genetic Connectionism", Proceedings of the 1990 Connectionist Models Summer School, Morgan Kauffman, 1990.

[4] J. Feldman, "Neural representation of conceptual knowledge", in Neural connections, mental computation (Nadel and et al., eds.), MIT Press, 1989.

[5] F. Fogelman-Soulie, "Multi-modular neural network-hybrid architectures: a review", Proceedings of 1993 International Joint Conference on Neural Networks, 1993.

[6] B. Happel and J. Murre, "Design and evolution of modular neural network architectures", Neural Networks, vol. 7, pp. 985-1004, 1994.

[7] R. A. Jacobs, M. I. Jordan, S. J. Nowlan and G. E. Hinton, "Adaptive Mixtures of Local Experts", Neural Computation, vol. 3, pp. 79-87, 1991.

[8] R. Jenkins and B. Yuhas, "A simplified neural network solution through problem decomposition: The case of the truck backer-upper", IEEE Transactions on Neural Networks, vol. 4, no. 4, pp. 718-722, 1993.

[9] M. I. Jordan and R. A. Jacobs, "Hierarchical Mixtures of Experts and the EM Algorithm", Neural Computation, vol. 6, pp. 181-214, 1994.

[10] K. F. Man, K. S. Tang and S. Kwong, "Genetic Algorithms: Concepts and Design", Springer-Verlag, 1999.

[11] M. Mitchell, "An Introduction to Genetic Algorithms", MIT Press, 1996.

[12] C. Monrocq, "A probabilistic approach which provides and adaptive neural network architecture for discrimination", Proceedings of the International Conference on Artificial Neural Networks, vol. 372, pp. 252-256, 1993.

[13] D. J. Montana and L. D. Davis, "Training Feedforward Networks using Genetic Algorithms", Proceedings of the International Joint Conference on Artificial Intelligence, Morgan Kauffman, 1989.

[14] H. Simon, The sciences of the artificial, MIT Press, 1981.

SEARCH OF GLOBAL OPTIMA WITH EVOLUTION STRATEGIES

FERNANDO M. ORSATTI
Escola Politécnica da USP – Departamento de Engenharia de Telecomunicações e Controle – São Paulo, Brazil.

RICARDO C. SOUZA
Escola Politécnica da USP – Departamento de Engenharia Mecatrônica – São Paulo, Brazil.

JOSÉ R. C. PIQUEIRA
Escola Politécnica da USP – Departamento de Engenharia de Telecomunicações e Controle – São Paulo, Brazil.

AGENOR T. FLEURY
Instituto de Pesquisas Tecnológicas – Mechanical and Electrical Engineering Division. Control System Group – São Paulo, Brazil.

ABSTRACT
Use of evolutionary algorithms to solve optimization problems faces the same difficulty found in the majority of the other search techniques: convergence to local optima instead of global ones. Lack of convergence is increased when the quality function of the optimization problem is too "thin" in the region of the global optimum. Having this situation in mind we propose a new strategy able to find other quality functions, called search functions that are "thicker" in the region of the global optimum even if their optima are not exactly the same of the original quality function. The problem of designing a controller to minimize vibrations of a very light flexible robot arm performing a rotational maneuver was addressed to test the proposed strategy.

INTRODUCTION

Evolution Strategies (ES) together with Genetic Algorithms (GA) and Evolutionary Programming (EP) are the most important paradigms of the so called field Evolutionary Computation.

The Evolution Strategies are the result of the german development of the idea of using the evolution model for solving engineering problems. The greatest differences from the Genetic Algorithms are the coding of individuals with real numbers, and the self adaptation of the strategic parameters during the evolution. For descriptions of ES see Rechenberg (1994) and Bäck et al (1991) and for comparisons with GA, Bäck (1992).

The Evolution Strategies faces the same problem of other optimization technics, the difficulty of finding global maxima instead of local ones, and for this reason we propose an improvement on the method in order to try to surpass this difficulty.

The proposed procedure has been tested in the problem of designing a controller for a flexible robot arm during a rotational manuever with the objective of minimizing arm vibrations induced by the movement.

Use of ES to design controllers seems to be very reasonable, since the coding of individuals is made with real numbers and so are the parameters of controllers. An example of this type of application can be found in Richter and Hofmann (2000).

THE SEARCH METHOD

The first step of any implementation of Evolution Strategies is to generate an inicial population of feasible solutions for the optimization problem. After this inicial population is settled, the evolution operators start to act: mutation leads to small changes in the genotip of the individuals and so small changes in their phenotyp are expected in a sense that they can "move" in the space of solutions, searching for places where the fitness function has higher values, or, in other words, searching for a maximum (Holland, 1975).

Thinking just about this small step evolution, it is clear that if only mutation is employed – and that means that no information is exchanged between individuals – all individuals will end the evolution process in a maximum of the fitness function that is closer to them when they were generated.

The implementation of recombination, as a way of imitating the sexual reproduction, is of great importance to increase variability in the population. The exchange of information between individuals make possible for new individuals to overcome which are seemly distant from the parents in the space of possible solutions of the problem.

The recombination operator is a way of speeding up the evolution process, as it allows feasible bigger steps to occur, and then, the fine tunning towards the maximum should be obtained as a consequence of mutations. For details on the implementation of recombination in evolution strategies see Bäck (1996).

DIFFICULTIES ON FINDING GLOBAL OPTIMA

As a result of the search strategy of an usual implementation of the Evolution Strategies one can see that a maximum for the optimization problem may always be achivied when there is an individual that is in a neighbourhood of this maximum. The existence of an individual near a maximum should be a consequence of the generation of the initial population or a result of recombination, and convergence to the maximum is a result of the mutation process, when small steps are made.

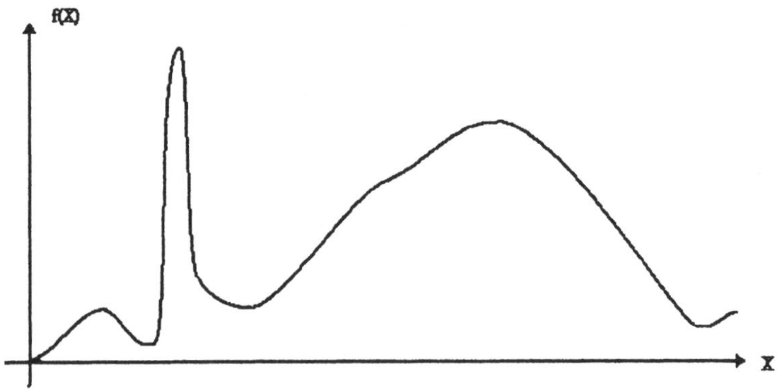

Figure 1: Example of fitness function with "thin" global maximum

A very difficult situation for an Evolution Strategy used for optimization occurs when the global maximum of the fitness function is very "thin" when compared to other local maximums. Existence of individuals in the neighbourhood of this "thin" maximum may demand an implementation with very large number of individuals, and finding the global maximum starts to be a matter of chance, rather than the result of a search strategy. This situation is represented in Fig. 1 for a simple one dimensional problem.

A NEW SEARCH STRATEGY

We now propose a new strategy for dealing with problems like the one resumed in Fig. 1. The main idea is to find another function, that will be called search function, so that this function has a global maximum near the global maximum of the fitness function and that this maximum is "thicker" than the maximum of the fitness function. A possible search function for the example of Fig. 1 is shown in Fig. 2.

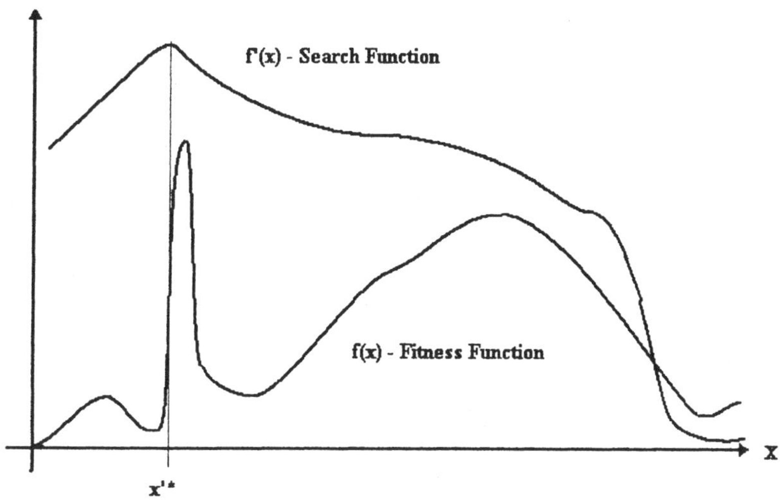

Figure 2: Possible search function for the given fitness function

The main purpose of finding this search function is that it could be used as the fitness function for an auxiliary implementation of an Evolution Strategy. With this implementation the value **x'*** should be determined and then introduced in the initial population for the evolution process with the original fitness function.

The ideia is not to use a fitness function that varies according to the search evolution as in Petridis and Kazarlis (1994), but to try other fitness function with maxima near the maximum of the orginal function, to verify if there is no "thin" hidden global maximum in the problem.

DETERMINING SEARCH FUNCTION

The complicated problem for the proposed implementation is to find a suitable search function. For this issue we suggest to use again Evolution Strategies. In this case, the objective of the evolution process is to find coefficients and parameters that compose a search function. To evaluate the quality of a search function, we use it as fitness function for an evolution strategy with very few iteractions and the result of this short evolution process is evaluated with the original fitness function.

In other words, we try to use a search function as the fitness function for an evolution strategy. If, in few iteractions, a good result, evaluated in the original fitness function, is obtained, one can say that this search function has a maximum near a maximum of the original fitness function, and also that this maximum is easily achievable.

In this way, a good search function is the one with which we can obtain a solution in the neighbourhood of the global maximum of the original fitness function and this solution can be obtained with very few interactions, meaning that the search function is "thick" in the region of the maximum, satisfying the two conditions expected for a good search function. Doing so, we have a way to evaluate search functions, and so we implement an evolution strategy that search for search functions.

CONTROLLER DESIGN FOR A FLEXIBLE ROBOT ARM PERFORMING A ROTATIONAL MANUEVER

The problem used to test the method was the problem of controlling a thin robot arm during a rotational maneuver. Due to its thin profile, the robot arm is very flexible, what leads to great vibrations on its tip. So the focus was the designing of a controller for the maneuver in order to minimize the vibration. Beyond the vibration we also had fixed maneuver time of 2 seconds, range of 15° and overshoot of 10%.

The fitness function for the problem was determined as follows:

$$f = 2 \cdot |M_p - 0.1| + |t_s - 2| + 10 \cdot v_p + 100 \cdot |\theta_f - 15°| + 100 \cdot v_m \quad (1)$$

The first attempt to design a suitable controller has been made through classical Pole Assignment techniques. Results obtained for this controller are shown in Fig. 3.

The first graph in Fig. 3 shows the angular position of the arm and the second shows the arm tip displacements resulting from the movement. This result was evaluated with the fitness function, Eq. (1), and the achieved value is 0.75.

After that a normal implementation of an Evolution Strategy was used to design the controller and the results are shown in Fig. 4. The resulting value for the fitness function in this case is 0.33, showing an improvement when compared to the results of classical Pole Assignment techniques.

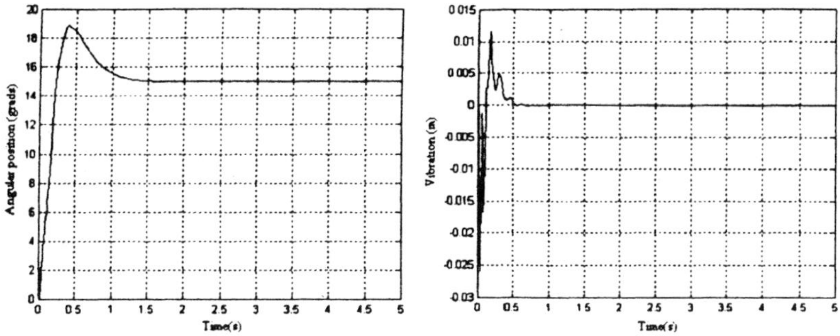

Figure 3: System with controller designed using evolutionary strategies

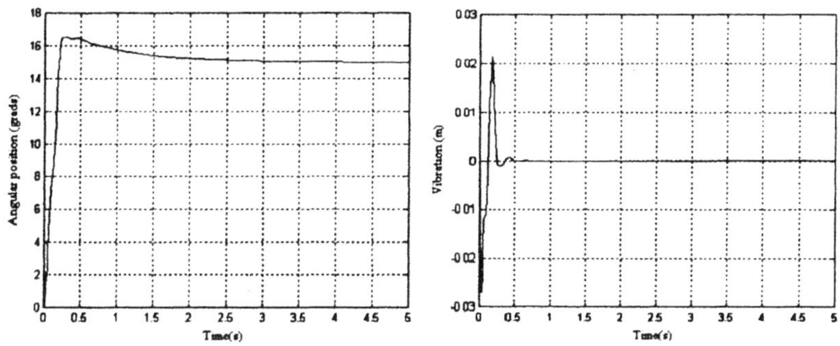

Figure 4: Pole Assignment controller

The last controller has been designed according to the method proposed in this article, using the ideas of search functions to help the search for a better solution for the problem. The results obtained for this implementation are shown in Fig. 5.

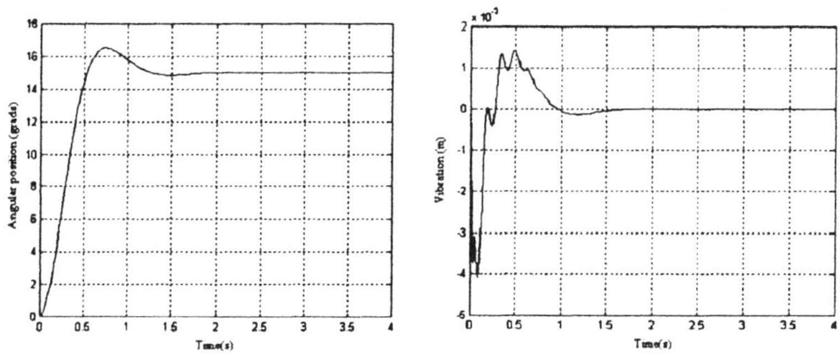

Figure 5: System with controller designed using ES with search functions

From Fig. 5 it is possible to verify that the results for the last implementation are much better than the results obtained so far. The value for the fitness function in this case is 0.064.

CONCLUSIONS

The results obtained with the proposed implementation of evolution strategies have been very satisfactory for the problem tested. Further works must follow to assure the robustness of the technique.

A problem to consider on this implementation is the large computation effort taken for finding the search function. A good strategy is to try search functions that are similar to the original fitness function. Doing so, the probability of finding search functions with maxima near the global maximum of the fitness function is higher. If other approachs are taken, finding the search functions would face the same problem of obtaining individuals for the initial population in the region of the global maximum, that is, it would become just a matter of chance.

It seems also reasonable that other techniques may be proposed for the determination of the search function, or even a sequence of search functions depending on the stage of evolution.

NOMENCLATURE

M_p – Overshoot.
t_s – Settling time.
v_m – Displacement mean value of the arm tip.
v_p – Peak value of vibration of the arm tip.
θ_f - Final angular position of the arm.

REFERENCES

Bäck, T., Hoffmeister, F. and Schwefel, H. P., 1991, "A survey of evolution strategies", in Belew and Booker [5], pp 2-9.

Bäck, T., 1992, "Evolutionary Algorithms", *ACM SIGBIO Newsletter*, pp. 26-31.

Bäck, T., 1993, "Optimal mutation rates in genetic search," in *Proc. 5th Int. Conf. on Genetic Algorithms,* S. Forrest, Ed. San Mateo, CA: Morgan Kaufmann, pp. 2–8.

Bäck, T, 1996, *Evolutionary Algorithms in Theory and Practice.* New York: Oxford Univ. Press, pp 82-83.

Holland , J. H, 1975, *Adaptation in Natural and Artificial Systems.* Ann Arbor, MI: Univ. of Michigan Press, pp 109-111.

Petridis, V. and Kazarlis, S., 1994, "Varying Quality Function in Genetic Algorithms and the Cutting Problem", *Proceedings of the First IEEE Conference on Evolutionary Computation,* IEEE Service Center, 1994, Vol. 1, pp. 166-169.

Rechenberg, I.,1994, *Evolutionsstrategie '94.* Fromman-Holzboog Verlag, Stuttgart.

Richter, R., Hofmann, W, 2000, "Optimized control on a two axis-robot by means of Evolution Strategies", *AMC 2000 Nagoya. Proceedings,* pp. 276-280.

VISUALIZING INFORMATION FLOW IN EVOLVING GRAPH-BASED POPULATION

BALU KARTHIKEYAN
Dept. of Mechanical Engineering
Iowa State University
Ames, Iowa, 50011

KENNETH M. BRYDEN
Dept. of Mechanical Engineering
Iowa State University
Ames, Iowa, 50011

DANIEL A. ASHLOCK
Dept. of Mathematics
Iowa State University
Ames, Iowa, 50011

ABSTRACT
Evolutionary algorithms mimic sexual reproduction in nature by blending pairs of solutions to a problem in hopes of creating novel or better solutions. As the evolutionary process proceeds this blending of characteristics tends to lead to a loss of diversity. This loss of diversity can narrow the search space for the solution, resulting in premature convergence to the sub-optimal solution and trapping between solutions. There are several approaches to maintaining diversity. These include careful choice of mutation rate, niche specialization, the use of island evolutionary algorithms, and graph based evolutionary algorithms. Graph based evolutionary algorithms manage diversity loss by imposing geography upon the population in the form of a combinatorial graph. Mating is only permitted between adjacent neighbors. Although graph based evolutionary algorithms have been found to increase diversity and speed of convergence in some cases, the choice of a graph is primarily a trial and error process. This paper presents a new methodology for visualizing the flow of information in a graph-based population within a virtual environment. Visualizing the flow of information across a graph can provide new insights to the evolutionary algorithm designer. This can reduce the time needed to choose a good graph for a particular problem.

INTRODUCTION

Evolutionary algorithms (EAs) mimic the nature's sexual reproduction by blending pairs of solutions in hopes of creating new or better solutions. EAs represent a robust method for optimization in high dimensional spaces (Holland, 1975; Forrest, 1993). EAs have been successful when applied to difficult optimization problems: for example, in job shop scheduling, high performance network design, DNA fragment assembly and models of the earth's mantle viscosity. In a simple genetic algorithm, the only constraint on reproduction is that more fit individuals have a higher probability of being selected for reproduction (Goldberg, 1989). When reproduction happens, individuals in the population are replaced with individuals composed of parts copied from a stochastically restricted subset of the population (Bryden et al., submitted). This results in a population predominantly with the gene of the fittest individual. Hence, the population diversity is lost as the algorithm proceeds.

Various approaches to manage diversity loss include using a high mutation rate, reducing the fitness of organisms in proportion to the number of organisms representing similar solutions (niche specialization), directly rejecting duplicate solutions (e.g. taboo search), and imposing a geography upon the population (Ackley and Littman, 1992; Mühlenbein, 1991). Imposing a combinatorial graph on an evolving population, mimics the behavior of many constraints on mating observed in biology (Ashlock et al., 1999). The biological mating constraints include separation by distance, cultural constraints, unique mating rituals, pheromones, and many other species dependent characteristics. This work focuses on visualizing the information flow in a graph based population. By visualizing the information flow within a GBEA in real time the EA designer can more quickly identify the optimum graph, population size, and issues such as trapping.

BACKGROUND

EAs work in an algorithmically simple manner but produce a vast amount of data. The evolved outcome of the GBEA depends on various factors such as initial fitness function of the mating creatures, type of crossover, mutation and replacement methods. Each of these factors influences the evolution process depending on the nature of the problem in-hand. Because EAs are non-linear solvers in which many complex interactions work together to create the final solution locating useful information and using this to understand the progress of the EA is challenging. Hence, a tool is needed for the EA designer to ensure the evolution always proceeds in the way that is intended to. Visualization techniques have been used to study EAs using both online and off-line systems. Several off-line visualization systems have been developed. Generally these provide a variety of information at varying levels of resolution. These include the individual creatures, family histories, and can examine the population and the runs.

Online systems allow users to closely follow and evaluate the progress of an EA and, in some cases may allow users to interactively influence or guide the directions of evolution of the system (Collins, 1998; Jones, 1993). One method of online visualization is to present the variables of the best creature and all other creatures in every generation utilizing representative diagrams for the progress of an optimization run (Wu et al., 1999). In this system the user can actually see how the variables change and which combinations of variables are successful. These types of visualization tool can help EA designer do "on-the-fly" modifications to the evolving creatures and observe the occurrence of interesting phenomenon. However depending on the population size information overload can be a problem. Off-line systems use visualization techniques to display information about an EA run after it is complete and may allow the display of data from multiples times of a run and movement both forwards and backwards in the evolutionary process (Shine and Eick, 1994).

One area where visualization system may provide essential information is in the design of GBEAs. By tuning the information flow within an evolving population GBEAs can provide substantial benefit. However the choice of the optimum graph is problem specific and is not known a priori. Because of this, insight into the flow of information within a GBEA can help the EA designer choose the optimum graph.

GRAPH BASED EVOLUTIONARY ALGORITHMS

A combinatorial graph or graph, G, is a set of V(G) of vertices and E(G) of edges where E(G) is a subset of the unordered pairs that can be drawn from V(G). Two vertices of the graph are neighbors if they are members of the same edge (West, 1996). Some examples of graphs are shown in Fig. 1. A GBEA imposes geography in the form of combinatorial graphs that restricts the rate of information spread within the evolving population. Within this population structure the creatures in the population are held on the vertices and reproduction is only permitted between the creatures that share a common edge. The restrictions imposed by GBEAs increase diversity and prevent premature convergence to a sub-optimal solution space. Converting traditional EAs to a GBEA has a low additional run-time cost but requires preliminary studies to determine which graphs will perform well for a specific problem being optimized. On highly disconnected graphs little or no loss of diversity was observed. In contrast there was significant loss of diversity in highly connected graphs (Ashlock et al., submitted).

GBEA becomes equivalent to standard EA when the complete graph is used. The choice of graph depends on the fitness landscape of the problem studied. The fitness landscape of the problem considered by the EA designer is not known a priori. Rather trial runs need to be conducted to determine the optimum graph for the problem being studied. Bryden et al. (submitted) have presented results that suggest that simple, unimodal problems are solved more quickly (up to 10 X faster) using graphs with a high degree of connectivity that enables information to be exchanged quickly. The extensive search on the unimodal fitness landscape provides no advantage to the solution process. Problems with complex polymodal fitness landscapes perform better (up to 12 X faster) with less connected graphs because a slower rate of information transmittal allows for a more thorough exploration of the fitness landscape.

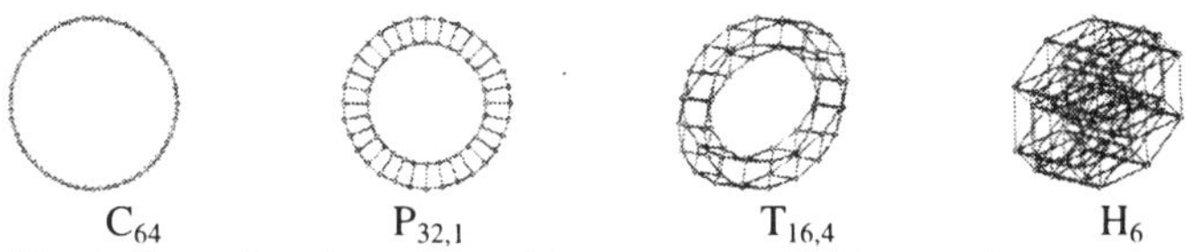

Fig. 1. Examples of complete, Petersen, torus and hypercube graphs

Additionally, a third group of problems was identified in which the global optimum was composed of pieces of local optima. These problems involve a two-step process to find the global optima-identify the building blocks and then assemble the building blocks correctly. In these problems the impact of graph selection was small.

VISUALIZATION TOOLS DEVELOPED

For this work an interactive online visualization system was developed to study the information flow in an evolving graph-based population. This tool couples the GBEA with interactive visual environment. This code utilizes VR Juggler as the virtual environment manager and hence can be used in a variety of computational/visualization environments. Including non-stereo and stereo based desktop environments, and multi-wall virtual reality environments. Because of the multidimensional complexity of the graphs used in GBEAs and the importance of understanding information spread within the graph, fully immersive virtual reality provides an ideal visualization platform.

The primary goal in the initial implementation at this tool was to combine a three-dimensional representaton of the graph with a visual measure of fitness in a real time (online) viewing system. The fitness functions of the individual in the population are visually distinguished by colors placed on the vertices of the graph. Black represents the lowest fitness and the white represents the highest fitness. In addition to the online visualization of the fitness function of the creatures in the population, an output file is generated by each problem case that lists the fitness function of all the creatures in the population after a specified number of mating events. The output consists of two geometrically identical graphs. The left graph shows the initial or any previously computed population, the right graph shows how the fitness is updated during the evolutionary process. Because this is a virtual environment the user can interact with evolving solution by scaling or rotation as well. Figs. 2-3 show the visual output for the torus and Petersen graphs respectively, in black and white.

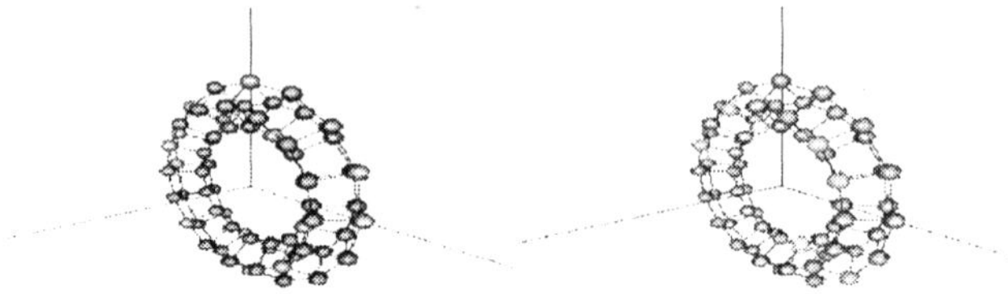

Fig 2. PORS-16 problem using torus graph

Fig. 3. PORS-15 problem using Petersen graph

TESTING AND RESULTS

To evaluate the initial implementation of this visualization tool, one simulation run was performed on three different test problems. These where the one-max problem, plus-one-recall-store (PORS) with n = 15 and n = 16. Four basic graphs that were used in each of the test problems include C_{64} cycle graph, $P_{32,1}$ Petersen graph, $T_{16,4}$ toroidal graph and H_6 hypercube.

The one-max problem uses a string of bits for a chromosome. In this study a 20-bit string was used. The fitness of a string is the number of 1's in it. Local roulette mating, two-point crossover, and absolute replacement were used. The mutation operator flipped one bit selected uniformly at random. The choice not to use elite replacement reflects the essentially trivial nature of the one-max problem. The search problem is harder without elite replacement and so more likely to yield information about the relative merits of graphs. The PORS efficient node use problem is described in detail in (Ashlock and Lathrop, 1998). PORS is a type of maximum problem within the domain of genetic programming (Koza, 1992; Kinnear, 1994; Banzhaf et al., 1998) with a small operation set and a calculator-style memory. The goal of the test problem is to find parse trees that, when executed, generate the largest integer result possible given a fixed maximum number of parse tree nodes. The language has two operations, integer addition and a store operation that places its argument in an external memory location. The language also has two terminals, the integer 1 and recall from an external memory. Visualization was done on two different PORS problems n = 15 and n = 16, the one of the hardest and the one of the easiest of the PORS problems respectively. The fitness for a given parse tree was the size of the number it produced when evaluated. In PORS problems the initial population composed of randomly generated trees with exactly n nodes. A successful individual was defined to a tree that produces the largest possible number. Crossover was performed by the usual sub-tree exchange (Koza, 1992). If this produced a tree with more than n nodes, then a sub-tree of the root node iteratively replaced the node until the tree had less than n nodes. This operation is called chopping. Mutation was performed by replacing a sub-tree picked uniformly at random with a new random sub-tree of the same size for each new tree produced. For the PORS problems local elite roulette mating was used (Bryden et al., submitted).

For the initial testing of the visualization tools the three test problems each with four basic graphs was run to visually study the evolution process as it occurs. Visualizing the EA process helps the designer to readily check the influence of various parameters such as parent selection, mutation, crossover and replacement. Due to non-linear influence of these parameters the analytical prediction of EA results can become tedious and time-consuming. Using this tool it was found that the following phenomenon could be readily identified:

1. *Information trapping* was visualized in the one-max problem using cycle graph. Due to the absolute replacement, occasionally the creatures with higher fitness got replaced with the less fit creatures. This process deprived the population from having creatures that could have converged to a solution in fewer mating events. The occurrence of this phenomenon was attributed to the replacement method. Another form of trapping that prevented rapid convergence was observed, in this, the fitness function oscillated back and forth. This was visualized in some cases of one-max problem with Petersen and torus graphs. For an unknown problem, before finalizing the EA design, a trial run like this would help the EA designer to identify the pitfalls and change the replacement operator "on-the-fly".
2. PORS problems were observed to have low initial fitness values, 0 to 6 in a scale of 36. But, once the evolution started, the rate of increase of fitness was higher than that in the one-max problems. The quick increase in the fitness values was attributed to the non-aligned crossover operator used in the parse trees. Additionally in highly connected graphs the benefits of the visual interface was the ability to see that information flow was much faster than the loosely connected ones. But for complex problems like PORS 15, loosely connected cycle graphs worked well. This is due to the gradual and thorough progress in the search that eliminated in landing drastically from a good solution space to a faulty space. On the other hand, for trivial problems no intensive search was required and the spread of information in the neighborhood was faster with higher degree graphs.
3. In PORS 16 problem, the global optima are composed of a group of smaller local optima. The local optima served as building block for the global optima. This phenomenon was observed with one creature in the population getting higher fitness followed by its neighbors gradually. The characteristics of the good gene extended to the nearby creatures in a population.

Based on these types of observations it is apparent that this type of visualization could be used by the designer to gain incite in the EA process. Visualization tools such as this could also help the to test the EA performance and to design suitable experiments for non-familiar or unknown problems.

CONCLUSION AND FUTURE WORK

The visualization showed things that would have been hard to see and enable the user to understand information flow in a GBEA. Trial experimentation needs to be done to get the approximate fitness landscape with basic graphs. Visualizing the evolution of problems with unknown fitness landscape may help reduce the initial run-time. Visualizing can also reduce the user's time to analysis the EA process and will reduce the lead-time to select the suitable graph for the problem under consideration. EA designer can also ensure that no "information trap" occurs. The information trap is said to occur in two ways - if the current best creature is not getting picked for future mating events till the end of the run, in this case, the good "traits" present in the best creature that could have potentially reduced the time to solution will go un-utilized. On the other hand, if a replacement operator is selected such that the less fit new creature replaces fitter creature in the current population. Visualizing the information trap and rapid change of the same EA remains a challenge for the designer due to the absence of an effective and interactive online visualization tool.

The current work on online visualization can be extended by more/better testing process and by enhancing the offline capabilities such as "History Files" to keep track of information such as parent, co-parent, type of crossover etc, for each mating event. The problems with known fitness landscapes will have a set of stored musical octave notes, the problems with unknown fitness landscapes can be made to match with anyone of the known stored octave to approximately identify the fitness landscape.

REFERENCES

Ackley, D.L. and Littman, M.L., 1992, "A Case for Distributed Lamarckian Evolution," Working Paper, Cognitive Science Research Group, Bellcore, NJ.

Ashlock, D.A., Bryden, K.M., Johnson, P.E., and McCorkle, D.S., "A Data Segregation Strategy Using Graph Based Evolutionary Algorithms," submitted to the *International Journal of Smart Engineering System Design.*

Ashlock, D.A. and Lathrop, J.I., 1998, "A Fully Characterized Test Suite for Genetic Programming," *Evolutionary Programming VII,* Springer-Verlag, New York.

Ashlock, D., Walker, J., and Smucker, M., 1999, "Graph Based Genetic Algorithms," *Proceedings of the Congress on Evolutionary Computation,* IEEE Press, Vol. 2, pp. 1362-1368.

Banzhaf, W., Nordin, P., Keller, R.E., and Francone, F.D., 1998, *Genetic Programming: An Introduction,* Morgan Kaufmann, San Francisco, CA.

Bryden, K.M., Ashlock, D.A., and Corns, S.M., "Graph Based Evolutionary Algorithms," submitted to the IEEE Transactions on Evolutionary Computation.

Collins, T.D., 1998, "Understanding Evolutionary Computer: A Hands-on Approach," in WCCI-98.

Forrest, S., 1993, "Genetic Algorithms: Principles of Natural Selection Applied to Computation," *Science,* Vol. 261, pp. 872-878.

Goldberg, D.E., 1989, *Genetic Algorithms in Search, Optimization, and Machine Learning,* Addison-Wesley Publishing Company, Inc., Reading, MA.

Holland, J.H., 1975, *Adaptation in Natural and Artificial Systems,* University of Michigan Press, Ann Arbor, MI.

Jones, T., 1993, "An Introduction to SFI Echo," Santa Fe Institute Working Paper #93-12-074.

Kinnear, K., 1994, *Advances in Genetic Programming,* MIT Press, Cambridge, MA.

Koza, J.R., 1992, *Genetic Programming,* MIT Press, Cambridge, MA.

Muhlenbein, H., 1991, "Darwin's Continent Cycle Theory and Its Simulation by the Prisoner's Dilemma," *Complex Systems,* Vol. 5, pp. 459-478.

West, D.B., 1996, *Introduction to Graph Theory,* Prentice Hall, Upper Saddle River, NJ.

Wu, A.S., De Jong, A., Burke, D.S., Grefenstette, J.J., and Ramsey, C.L., 1999, "A Visual Analysis of Evolutionary Algorithms," *Proceedings of the Congress on Evolutionary Computation,* IEEE Press, Vol. 2, pp. 1419-1425.

OPTIMIZING TRAFFIC CONTROL PROBLEMS WITH MICRO-GENETIC ALGORITHMS

GHASSAN ABU-LEBDEH
Department of Civil &
Environmental Engineering
Michigan State University
E. Lansing, MI 48824-1226

YACOUB NAJJAR
Department of Civil Engineering
Kansas State University
Manhattan, KS 66506-5000

ABSTRACT
This paper describes an experimental approach to test the suitability of micro-Genetic Algorithms (m-GAs) to solve very large combinatorial traffic control problems. It establishes relationships between time to convergence and problem size. A discrete time dynamical traffic control problem with different sizes and levels of complexity was used as a test bed. The results showed that m-GAs can solve computationally demanding traffic control problems and produce control with desirable system aftereffects. Upon appropriately sizing the m-GA population, it converged to a near-optimal solution in a number of generations approximately equal to the string length. The results also demonstrated that with the selection of appropriate number of generations, it is possible to get most of the worth of the theoretically optimal solution but with only a fraction of the computation cost.

INTRODUCTION

The goal of intelligent transportation systems (ITS) is to optimize use of various system components. One critical component is traffic signals at facilities like isolated intersections, corridors and networks. Traffic signals control traffic on these facilities to ensure optimal overall system operations. Because traffic conditions change with time, and due to the large number of control parameters and time limitation, optimizing control in real time is a computationally demanding problem specially since near-optimal solutions have to be generated rapidly for on-line implementation. Genetic Algorithms (GAs) are particularly suited for such optimization problems.

This paper describes an experimental approach to evaluate the feasibility of using m-GAs to optimize large scale traffic control problems for real-time applications. Because of the time limitations within which a solution must be found, the paper also addresses the related issue of tradeoffs between solution quality and computation time. As a point of clarification, GAs and m-GAs are fundamentally similar but differ tactically. More on this later. The terms GA and m-GA are used interchangeably in this paper, the paper pertains exclusively to m-GAs.

BACKGROUND

Successful applications of GAs in traffic are numerous (Sadek et al., 1997; 2000; Duerr, 2000; Abu-Lebdeh & Benekohal, 1997a). Most of those applications involved optimizing combinatorial complex problems, but time limitation was not an issue. GAs, however, fell out of favor soon after users in the traffic engineering field realized that much work is needed to configure a GA before it performs optimally. In many cases the good side of GAs was overlooked and they were only noted for their slow convergence and difficulty to configure.

GAs are still a powerful tool if properly configured. Real-time traffic control involves the optimization of difficult and large scale problems, and GAs possess the right properties to solve such problems. This paper shows how m-GAs were able to solve large control problems in relatively short time. The resulting solutions have most of the desirable attributes. m-GAs were used because, due to their structure and functionality, they can produce better near-optimal solutions when computation time is limited.

PROBLEM STATEMENT

Optimizing traffic control in real time in ITS is a necessity. In most realistic conditions, such optimization problem is both complex and combinatorial. While GAs have the inherent capability of solving such problems successfully, several issues are still unclear, at least to practitioners in the transportation field: 1) is it feasible to use GAs to optimize real-time control, 2) how to best configure the GA; particularly, how long should the GA run, and 3) what are the tradeoffs between computation time and quality of solutions.

GENETIC ALGORITHMS FOR OPTIMIZATION

Figure 1 *without* the black blocks shows the functional structure of a GA. GAs differ from conventional iterative improvement algorithms in at least two respects: 1) GAs use a population of points instead of a single design point to start the search. And since several points are used as candidate solutions, GAs are less likely to get trapped at a local optimum, and 2) GAs do their search and optimization in parallel. The power of a GA derives from its ability to exploit, in a near-optimal fashion, information about the utility of a large number of structural configurations (i.e., parts of the solution) without the computational burden of explicit calculation and storage. This makes GAs particularly suited for real-time traffic control optimization. More on GAs is available elsewhere (Goldberg, 1989; Michalewcz, 1994.)

GAs time-to-convergence requirements is another attractive feature. Based on application of GAs to "laboratory" problems of known complexity and optimal solution value, it was shown that depending on scaling and selection type, GAs-time-to-convergence grows only as O(L), $(O\sqrt{L})$ or O(L log L), where O(•) is an expression of time complexity of the algorithm (Thierens et al., 1999; Muhlenbein, 1992). This property of GAs makes them particularly suited to solve optimization of traffic control problems since the search space of those problems was found to grow exponentially as the size of the transportation system grows linearly (Abu-Lebdeh & Benekohal, 2000). This property is even more valuable since many of the traffic control problems will have to be optimized on line for real-time implementation.

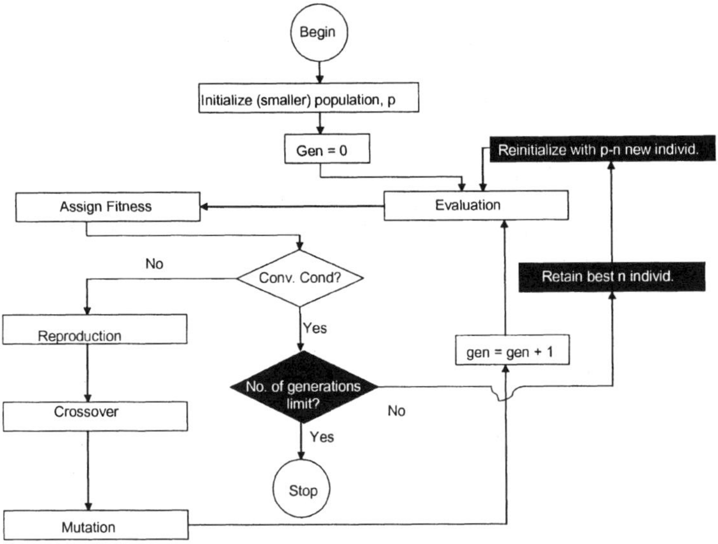

Figure 1: A flow chart of the functional components of a GA (Black boxes are components of m-GA only)

M-GAs differ from regular simple GAs in the population size choice and the way genetic diversity is introduced. A m-GA takes a small population n, converges it, keeps the best individual and randomly generates n-1 individuals, converges the new population and so on. In that regard, the difference between regular GAs and m-GAs is tactical, not fundamental. M-GAs have been successfully used in several applications (Abu-Lebdeh & Benekohal, 1997; Krishnakumar, 1989; Roy et al., 1996). Figure 1 *with* the black blocks shows the basic functional components of a m-GA.

EXPERIMENTAL DESIGN

The following experimental design was used to achieve the objectives of the paper. A complex large scale traffic control problem is selected and then a m-GA was used to optimize it. The complexity of the problem was then changed by varying the size of its search space and number of constraints. Three different system sizes were tested. For each case: 1) the traffic quality of the solution, and 2) the time needed to convergence was examined and then related to the size of the problem. One of the objectives of the paper is to determine if and what type of relation exists between problem size and time to convergence. First, we describe the traffic control problem and after that describe the specifics of the three systems that were tested.

Optimization Problem

The problem used is a dynamical traffic control problem of a signalized arterial system. The control maximizes system output while ensuring efficient use of green time, preventing blockage, and accounting for the non-steady state of traffic flow in the system. The problem is structured as a discrete-time dynamical system optimization problem (see Abu-Lebdeh & Benekohal, 1997b for detailed description). The number of intersections and cycles over which output is to be maximized were varied to obtain different system sizes. The optimization was designed such that green times are the only decision variables. The objective function of the optimization problem maximizes system output for a given system over space and time (i.e., for all intersections and for all cycles simultaneously). A linear system (single arterial) was used. This problem is highly constrained. The definitions of the symbols used in the formulation are as follows:

si adjusted saturation flow rate per lane of approach i, veh/sec
LV average effective vehicle length when stopped, ft.
Li length of link i, ft.
t(k)i+1 time for acceleration wave to reach tail of queue when signal turns green for approach i+1 at the start of cycle k, sec.
ts i+1 time for stopping wave to reach tail of queue when signal turns red for approach i+1when link i+1 is full, sec.
tq(k)i,i+1 time for front of queue i to join rear of queue i+1, sec.
υi speed of an acceleration wave (a shock wave) on link i, ft/sec
λi speed of a stopping wave (a shock wave) on link i, ft/sec
Cyc(k)i length of cycle k at intersection i, sec.
g(k)i effective green time for arterial approach i during cycle k, sec.
gc(k)i effective green time for cross approach i during cycle k, sec.
DV(k)i discharged vehicles from approach i during cycle k, veh.
AV(k)i arriving vehicles into approach i during cycle k, veh.
q(k)i length of queue on approach i at the beginning of cycle k, veh.
off(k)i,i+1 relative offset between approaches i and i+1, for cycle k, sec.
exoff(k)i,i+1 extended offset between intersections i and i+1 for cycle k, sec.
d(k)i mean departure rate from approach i during cycle k, veh/sec.

The complete control algorithm formulation is to

$$\text{Maximize}\Big(\sum_{k=1}^{n}\sum_{i=1}^{m} gc_{(k)i}d_{(k)i} + \{j \times \sum_{k=1}^{n} \min$$

$$[(q_{(k)j} + \frac{AV_{(k)j}}{g_{(k)j-1}} \times (g_{(k)j} - \frac{q_{(k)j}}{s_j}), AV_{(k)j} + q_{(k)j})]\} + \varpi + \omega\Big) \quad (1)$$

where n is the number of cycles, m is the number of cross street approaches, and j is the number of arterial links. All other terms are as defined above.

The first term of the objective function measures the output of the crossing streets. The second term is the output of the arterial street as measured by the number of link-vehicles. ϖ and ω are to adjust the output for vehicles that did not travel all system links. A vehicle is considered to have traveled a link only after it has exited that link. The above function was maximized subject to the following constraints on control and state variables:

1. $DV_{(k)i} \leq q_{(k)i} + DV_{(k)i-1}$ i=1,...j; k=1,..n
2. $off_{(k)i,i+1} = tq_{(k)i,i+1} - t_{(k)i+1}$ i=1,...j-1; k=1,.. n

$$t_{(k)i+1} = \frac{LV \times q_{(k)i+1}}{\upsilon_{i+1}}$$

3. $exoff_{(k)i,\, i+1} = Cyc_{(k)i+1} - off_{(k+1)i,\, i+1}$ i=1,...j-1; k=1,..n
4. $(g_{(k)i+1} + off_{(k)i,i+1} + t_{s\ i+1}) \geq g_{(k)i}$ i=1,...j-1; k=1,..n

$$t_{s\ i+1} = \frac{L_{i+1}}{\lambda_{i+1}}$$

5. $q_{(k+1)i} = q_{(k)i} - DV_{(k)i} + DV_{(k)i-1}$ i=1,...j; k=1,..n

$$DV_{(k)i} = \min[(q_{(k)i} + \frac{AV_{(k)i}}{g_{(k)i-1}} \times (g_{(k)i} - \frac{q_{(k)i}}{s_i}), AV_{(k)i} + q_{(k)i})]$$ i=1,...j; k=1,..n

6. $\varpi = -\sum_{i=2}^{j} q_{(1)i} \times (i-1)$ i=1,...j; k=1,..n
7. $\omega = \sum_{i=2}^{j} q_{(n+1)i} \times (i-1)$ i=1,...j; k=1,..n
8. $g_{(k)i} \leq$ maximum value i=1,...j; k=1,..n
9. $g_{(k)i} \geq$ minimum value i=1,...j; k=1,..n
10. $gc_{(k)i} \leq$ maximum value (key cross approach) i= m-1, m; k=1,..n
11. $gc_{(k)i} \geq$ minimum value (key cross approach) i= m-1, m; k=1,..n
12. $q_{(1)i}$ are given i=1,..j

The control algorithm was applied to congested systems as described in the following section.

Systems with Different Sizes

The m-GA was tested on three system sizes. System size here refers to the number of decision variables that are to be optimized. For the traffic control problem, these are the green times for the arterial system and the critical cross-street approach over the control period. That is, the variables are optimized over time and space simultaneously. Traffic flow was assumed one-way and turns were ignored. Simple signal phasing was used. Traffic input to each system was assumed known at the boundaries of the system but was variable as it enters and travels through the system. Intersection spacing was set at 320 meters. Typical values of vehicle acceleration and shock wave speeds were used. A more realistic, two-way system with complex phasing differs from the simple one only in the size of the solution set and number of constraints.

The large system is composed of 10 intersections and 34 cycles. Based on the selected values and ranges of green times the number of decision variables for this system is 408. This is determined as follows:

[1 (dummy intersection) + 10 (regular intersection)] × 34 (cycles) + 1 (cross-approach) × 34 (cycles) = 408 variables. The med-size system has 10 intersections and 17 cycles (204 variables) and the small system has 5 intersections and 8 cycles (56 variables.)

For all system sizes, the green time search was done in 5-second steps. The range of the green time for the arterial greens was 30 to 80 seconds, and for the critical cross street was 20 to 40 seconds. This means there were 11 possible green values to choose from for the arterial approaches and 5 possible values for the critical cross street approach. Given the above values, the search space sizes for the small system is:

$[(11 \text{ values for arterial})^{6 \text{ intersections}}]^{8 \text{ cycles}} \times [(5 \text{ values for cross-street})^{1 \text{ intersection}}]^{8 \text{ cycles}} = 3.8 \text{x} 10^{55}$. Similarly, the sizes for the med-size and large systems are $4.2 \text{ x} 10^{206}$, and 3.8×10^{413}, respectively.

Control for each of the 3 systems was optimized and convergence observed. Population was sized around $\sqrt{L}$ based on an earlier study (Abu-Lebdeh & Benekohal, 1999). L is the string length. Single point cross-over, no mutation, elitism and tournament selection were used. No attempt was made to optimize other parameter values or operator types.

RESULTS AND DISCUSSION

Figure 2 demonstrates one aspect of the appropriateness of the optimized solution in managing traffic within the systems. The Figure shows time plot of traffic queues on selected links within the small system where each link can physically hold up to 40 vehicles at one time. The control plan reduced queues over time, although there were occasional increases in queue length. However, in no case the number of vehicles exceeded the physical capacity of the link. This is critical, for if spillback occurs then gridlock would develop thus impacting more intersection over space and time. What is significant but not quite evident, is that these results are for systems that are congested. It is remarkable for example that not a single gridlock incident occurred in any of the three systems.

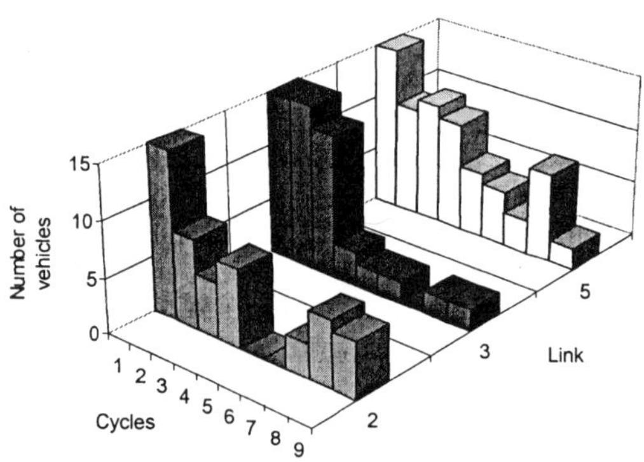

Figure 2: Change over time in traffic content on selected links for small system. Similar trends were observed for other system sizes.

The following subsections focus on the computational side of the results.

Establishing a Bench Mark Near-Optimal Solution

Since the global optimum is not known and the search space is enormous, an exhaustive search is impossible. It is necessary to establish a benchmark that marks a "practical" point of convergence. This is a point where, for practical reasons, one may consider the algorithm to have converged close enough to the optimal solution and thus terminate the run.

The benchmark was established based on inspection of the convergence behavior of the algorithm. A number of generations around 4L was found to be sufficient to ensure that the m-GA is close enough to the optimal value, that the run can be terminated without much loss in fitness value. The fitness value associated with 4L was then benchmarked as a "practical" near-optimal-value (NOV). It must be emphasized that this value is a relative one whicht may or may not be, time-wise, close to the true optimal value. From the pattern of convergence however, the NOV would not be statistically different from later generations' near-optimal values. Following results are based on multiple independent runs.

Convergence

For practical applications, the m-GA can be considered to have converged once the gain in fitness per computation step (i.e., generation) becomes marginal. This is the point of maximum curvature or bend on the fitness-time plot (see Figure 3). The Figure shows the progress of the m-GA towards the NOV for the smal system. It is seen that the most gain in the fitness value occurs within generations that are O(L), or around L. Similar trends were observed with the other systems. Beyond that point, the gain in the fitness value per generation is marginal. The point remains that running the m-GA for a number of generations that is at least O(L) is sufficient to ensure that the convergence has begun and that most of the worth of the optimal value has been attained.

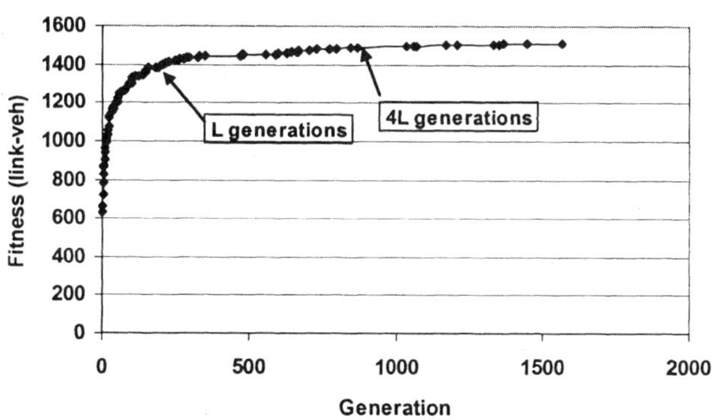

Figure 3: Progress of the m-GA towards the optimal solution – small system. Similar trends observed for other systmes

Computation Cost

Figure 4 compares the relationship between gain in fitness value and computation cost at generation O(L) to those at the NOV. For all three system sizes, the Figure shows clearly that a number of generations of O(L) is sufficient to attain most of the worth of the fitness at NOV but at only a fraction of the computation cost. According to Figure 4, ninety percent (90%) of the fitness at NOV can be attained for only 10% of the computation cost. Depending on the exact application at hand, this is an excellent tradeoff. This information has valuable implications to traffic application such as on-line real time control where, because of time constraints, "best-solutions-so far" are used.

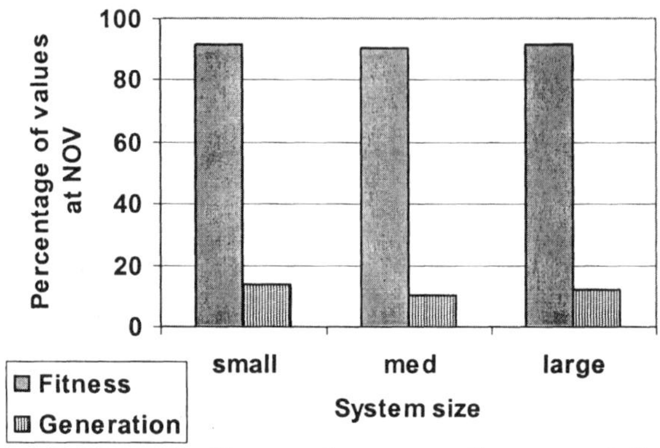

Figure 4: Change in fitness and corresponding computation time for the three system sizes

CONCLUSIONS

This paper described an experimental approach to examine the suitability of m-GAs for optimizing large scale traffic control problems. The paper also examined the convergence behavior of m-GAs and addressed the issue of computation cost in relation to gain in fitness. Different system sizes were optimized. Their convergence was observed over multiple independent runs.

The results showed that the micro-GA was able to generate traffic control solutions with the desired system aftereffects. For all system sizes, a number of generations equal to the length of string seems to be a sufficient lower-bound to ensure that the micro-GA has converged to a near-optimal solution. A tradeoff between computation cost and fitness improvement exists and can be used as a practical guideline to determine a cut-off point for the number of generations needed. Because there are numerous combinations of operator types and parameter values, it is recommended that the findings be tested with a variety of problems with other operators and parameter values.

REFERENCES

Abu-Lebdeh, G. & Benekohal, R.F. (1997a): "Dynamic Signal Coordination along Oversaturated Arterials". Proceedings of the ASCE Conference: Traffic Congestion and Traffic Safety in the 21st Century, Chicago, June, pp. 521-527

Abu-Lebdeh, G. & Benekohal, R.F.(1997b): "Development of a Traffic Control and Queue Management Procedure for Oversaturated Arterials". *Transportation Research Record 1603,* TRB, National Research Council, Washington, DC, 119-127

Abu-Lebdeh, G. & Benekohal, R.F. (1999) : "Convergence Variability and Population Sizing in Micro-Genetic Algorithms". *Computer-Aided Civil and Infrastructure Engineering (14)*, 321-334

Duerr, P.A. (2000): "Dynamic Right-of-Way for Transit Vehicles: Integrated Modeling Approach for Optimizing Signal Control on Mixed Traffic Arterials." *Transportation Research Record 1731*, TRB, National Research Council, Washington, DC, 31-37

Golderg, D.E.(1989b): "Genetic Algorithms in Search, Optimization, and Machine Learning." Addison Wesley

Krishnakumar, K.(1989): "Micro-Genetic algorithms for stationary and non-stationary function optimization." Proceedings of SPIE, vol. 1196, Intelligent Control and Adaptive Systems, 289-296

Michalewicz, M.(1994): "Genetic Algorithms + Data Structure=Evolution Programs." Springer-Verlag

Muhlenbein, H (1992): "How genetic algorithms really work I: Mutation and Hill-climbing." *Parallel Problem Solving from Nature, 2*, 15-26

Roy, S., Ghosh, S., & Shivipur, R.(1996): "Optimal Design of Process Variables in Multi-Pass Wire Drawing by Genetic Algorithms." Transactions of the ASME, 118, 244-251

Sadek, A.W.; Smith, B.L.; and Demetsky, M.J. (1997): "Dynamic Traffic Assignment: A Genetic Algorithm Approach." *Transportation Research Record 1558,* Transportation Research Board, Washington, D.C., 95-103.

Thierens, D., Goldberg, D.E., & Pereira, A. (1999): "Domino Convergence, Drift, and the temporal Salience Structure of Problems." 1998 IEEE International Conference on Evolutionary Computation Proceedings, 535-540.

SOLUTION OF A 2-D INVERSE HEAT CONDUCTION PROBLEM USING EVOLUTIONARY DATA SEGREGATION TECHNIQUES

PETER E. JOHNSON
Department of Mechanical Engineering
Iowa State University
Ames, Iowa

KENNETH M. BRYDEN
Department of Mechanical Engineering
Iowa State University
Ames, Iowa

DANIEL A. ASHLOCK
Department of Mathematics
Iowa State University
Ames, Iowa

ABSTRACT

Numerical solutions of differential equations are typically performed directly. That is, the boundary conditions and physical properties of the domain are given and the dependent variable is numerically computed throughout the domain. In contrast, in inverse solutions of differential equations the dependent variable is known at select locations throughout the domain. However, the material properties and/or the boundary conditions are unknown. This paper presents a novel technique for solving an inverse heat conduction problem. In the problem examined, the temperature profile within a two-dimensional material and the available materials are known. However, the placement of these materials and the heat flux at the boundaries are unknown. The proposed technique uses the Adaptive Modeling by Evolving Blocks Algorithm (AMoEBA) to optimize material configurations via an evolutionary algorithm. The binary trees of AMoEBA describe segregation schemes for the different materials available. The domains described by each tree are solved directly. Using the least squares fit between these candidate solutions and the known profile, a population evolves until the fitness criterion is met. At this point, the material placement is found, and the boundary heat fluxes are calculated.

INTRODUCTION

Inverse solution of engineering problems involves the calculation of the properties and/or the boundary conditions based on knowledge of the dependent variable throughout the domain. In contrast, in direct solutions the dependent variable is calculated throughout the domain based on known physical properties and boundary conditions. Although direct solutions are most common in engineering problem solving, situations in which severe conditions make it impossible to use the proper measurement tools or when making these measurements affects the properties being measured require inverse techniques to be used. Inverse problems can be more difficult to solve than direct problems because often, they are ill-posed and may have multiple solutions.

Tervola (1989) and Huang and Chin (2000) have each solved two-dimensional (2-D) inverse heat conduction problems (IHCPs) based on known temperatures with the goal of finding the thermal conductivity of the system. Tervola (1989) used a piecewise polynomial to model the thermal conductivity and then solved the resulting optimization problem using the Davidon-Fletcher-Powell method. Huang and Chin (2000) used the conjugate gradient method in their optimization process and were able to successfully estimate functions for predicting non-homogeneous thermal conductivities with a very fast rate of convergence. Others have also used inverse techniques in a similar manner to solve problems in heat transfer (Ponzini et al., 1989; Sridhar and Narh, 1998; Shenefelt et al., 2002; Drezet et al., 2000) These range from the identification of rock thermal conductivities through different layers (Ponzini et al., 1989) to the determination of physical properties and boundary conditions in casting situations (Drezet et al., 2000). Fan (1998) has used a combination of genetic algorithms and neural networks to solve an inverse engineering problem. Direct methods for designing diffuser blades use a time-consuming, iterative procedure of creating a temporary blade, computing and analyzing the resulting flowfield, and making changes to improve the blade's performance. An inverse technique can be used when expected conditions (surface velocity or pressure distribution) exist for the ideal blade. Fan's solution uses a genetic algorithm to create candidate blade profiles that are analyzed by a back-propagation neural net. By comparing the resulting flowfields to the target solution, an ideal, evolved blade is found.

This paper presents the Adaptive Modeling by Evolving Blocks Algorithm (AMoEBA) (Ashlock et al., submitted; Ashlock et al., 2002; Johnson et al., 2002) for the inverse solution of a heat transfer engineering design problem. In this problem, the temperature profile and materials involved in a 2-D heat conduction problem are known, but the placement of these materials and the boundary heat fluxes are unknown. AMoEBA uses evolving binary trees to define the geometry and properties of candidate material configurations. The candidate solutions are solved directly and compared with the known temperature profile using the least squared error. When this fitness reaches a defined level, the material placement scheme of the real system is found and boundary conditions matching the problem definition are identified. This process of adaptive blocks modeling applies to many inverse types of engineering problems because it can simultaneously optimize the domain decomposition and the equations describing the physics within the decomposed blocks.

PROBLEM DESCRIPTION

The application of AMoEBA to an IHCP is demonstrated on 2-D, steady state, heat conduction through three different materials. The heat flux, sizes, and placements of the materials are unknown. The exact temperature distribution and placement are shown in Figure 1. A square domain, normalized to the range [0,1] in both directions, is used with constant temperatures of 650 K and 400 K at the left (x = 0 m) and right faces (x = 1.0 m), respectively, and adiabatic boundary conditions on the top and bottom faces. The material placement is divided in two along the y = 0.3 m line. On top, the domain consists of Material 1 ($k = 80\ W\ m^{-1}\ K^{-1}$) from x = 0 m to x = 0.25 m and from x = 0.75 m to x = 1.0 m with Material 2 ($k = 5\ W\ m^{-1}\ K^{-1}$) placed in between from x = 0.25 m to x = 0.75 m. Below the dividing line of y = 0.3 m is Material 3 ($k = 30\ W\ m^{-1}\ K^{-1}$). The exact solution was used to generate a 200 X 200

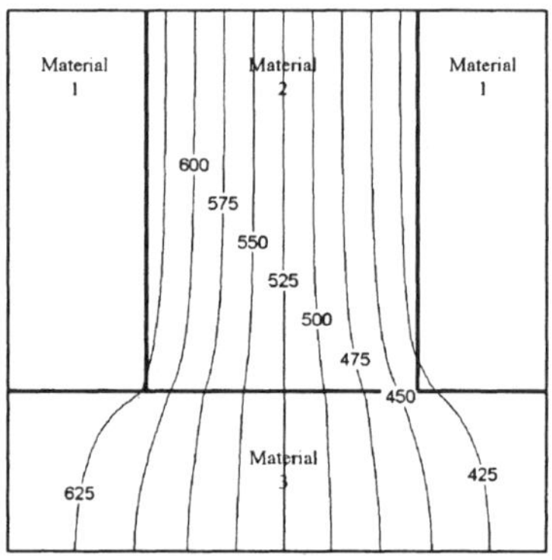

Fig. 1. The exact solution's temperature profile and material placement.

grid of 40,000 temperatures as input data for the IHCP.

At the boundaries between the materials a discontinuity occurs due to the property change. This can present problems depending on the solution method used. To determine the proper placement of the materials an inverse engineering problem must be solved. As the solution to this inverse problem is found, the following residual goes toward zero.

$$\text{Residual} = \sum_{j=0}^{M}\sum_{i=0}^{N}\left(T_{i,j} - \theta_{i,j}\right)^2 \tag{1}$$

where $T_{i,j}$ is the exact temperature profile, $\theta_{i,j}$ is the candidate profile, and M and N are the number of points in the x and y directions. The approach taken here is to minimize this residual by evolving a population of candidate material configurations.

AMOEBA DESCRIPTION

The technique described in this paper is termed the Adaptive Modeling by Evolving Blocks Algorithm (AMoEBA). Other applications of AMoEBA have included geometric decomposition of a diffusion data set [8, 9] and segregation and solver assignment for a multi-solver solution of a laminar, steady-state, fully-developed pipe flow problem.

AMoEBA partitions a data set into different regions and assigns solvers to each region through the use of a generational evolutionary algorithm. The data segregation and solver assignments for each creature in the evolutionary algorithm are encoded into a binary tree. Nodes are defined by both a floating-point value describing the location of a partition and an integer value that determines the direction of the segregation, either horizontal or vertical. The leaves are integers that assign materials to the region defined by traversing the tree to that leaf. Figure 2 shows a sample of the tree encoding structure.

Initially, a population consists of 104 randomly generated trees that define candidate solutions to the IHCP. Reproduction occurs by randomly dividing the population into tournament groups, each consisting of eight trees. In each tournament, the four trees that are more fit become the parents and duplicates (children) of these parents replace the least fit. Each pair of children have a 50% probability of being selected for crossover. During crossover, a node in each of the two duplicated trees is randomly selected. These nodes are then exchanged, transferring information between the trees (Fig. 3). The *chopping* process is then applied. Crossover between trees can result in geometric growth of the trees, creating partitions that contain no points. The chopping operation is used when a new tree contains more than eight total internal nodes and leaves. A subtree of the root node of the offending tree is randomly selected and promoted to be the new root, i.e.

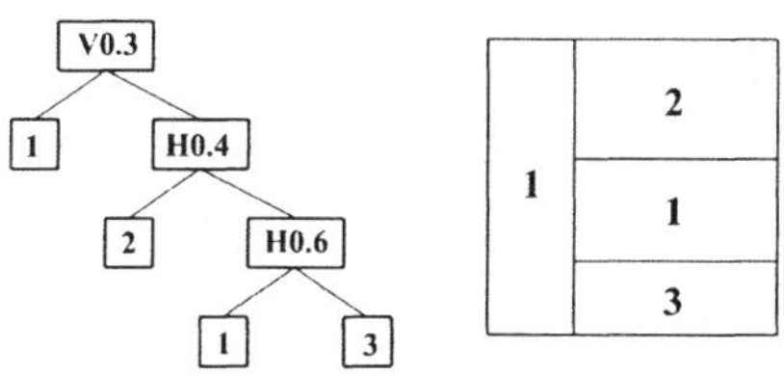

Fig. 2. A material segregation tree and the resulting material configuration. This example shows a two-dimensional tree making vertical (V) and horizontal (H) slices in the domain and placing Materials 1, 2, and 3 in the resulting partitions.

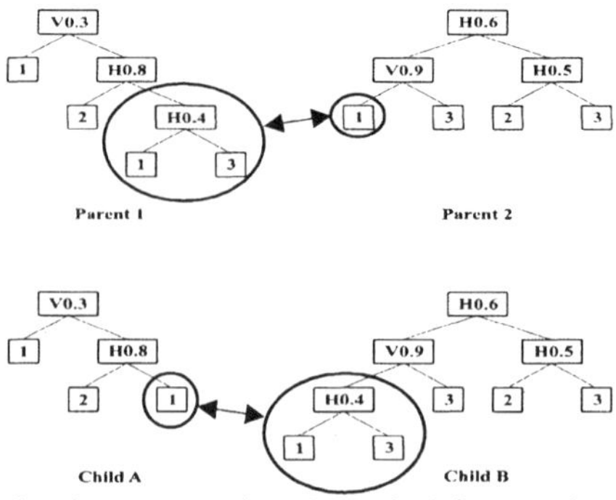

Fig. 3. An example of crossover. A node or leaf from each parent is selected and swapped resulting in two new children.

it becomes the new tree. This operation continues iteratively until the tree contains less than the eight node limit, a value chosen during preliminary studies.

Before reproduction is complete, the children are mutated. Each child has a 6/7 probability of being mutated by one of four different operators. In *random subtree mutation*, a randomly selected node in the tree is removed and replaced by a randomly generated subtree. This is similar to crossover in that it can be very disruptive and there fore has a 1/7 chance of occurrence. In *flip mutation*, the direction of a nodal partition is flipped, either from vertical to horizontal or vice versa. Flip mutation is also disruptive and occurs 1/7 of the time. *Boundary value mutation* adds a value selected uniformly in the range [-0.05, 0.05] to the node value. This operation moves the segregation line a random distance with reflection at the boundaries. This type of mutation operation is useful for "dialing in" the proper locations of the partitionsand therefore has a 3/7 chance of occurring. In *leaf mutation,* the material assigned to a randomly selected region is changed. This operation has a 1/7 chance of occurrence.

Forty-five separate trials of 104 randomly generated creatures (tree/solver combinations) were made. A fitness criteria of 0.5 was used to determine when an exact answer to the 2-D IHCP had been found. Finding an exact answer depends both on obtaining the correct temperature profile and on the conservation of energy for the candidate solution to ensure physically realistic results. Based on this, the fitness of a creature is the product of an energy balance penalty and the sum of the squared error between the creature's data set and the actual temperature profile. This is:

$$Fitness = (1+P_e)*\left[\sum_{i=0}^{M}\sum_{j=0}^{N}\left(T_{i,j}-\theta_{i,j}\right)^2\right] \tag{2}$$

where P_e is the *energy balance penalty.* The expression in the brackets is the residual from Eq. 1. For the energy to be balanced, the heat flux entering the domain from the x = 0 m end (q_0") and the heat flux leaving the domain from the x = 1 m end (q_1") must be equal. For the penalty function, these values are calculated for the candidate solution and their difference is normalized, first by the q_0" value and then by the q_1" value. The absolute values of these normalizations are then compared, and the maximum of these two normalizations is the penalty function:

$$P_e = \max\left(abs\left(\frac{q_1'' - q_0''}{q_1''}\right), abs\left(\frac{q_1'' - q_0''}{q_0''}\right)\right) \tag{3}$$

RESULTS

Of the 45 trials of AMoEBA on the IHCP, 30 were evolved to solutions that met the fitness criteria of 0.5 and 15 did not meet the criteria with the 2500-generation run limit that was enforced. Of the 30 that found a solution, the fastest (Trials 19 and 25) completed the search in 75 generations and the slowest (Trial 20) took 1905 generations. The average time to solution of the cases that converged was 549 generations with a standard deviation of 513. Looking at the 20 cases that converged in less than 500 generations, the average is 237 generations. The 10 cases that took more than 500 generations had an average of 1173 generations.

The most fit of the final results (Trial 34) obtained a fitness of 0.369 and the least fit of the final solutions (Trial 41) was 0.499. The average fitness of the final solutions was 0.453 with a standard deviation of 0.037. Of the 30 cases that found a solution, 24 found solutions that almost exactly matched the proper material configuration. However, both the most fit (Trial 34) and the least fit (Trial 41) of the final solutions obtained results that did not match the proper material configuration. Of the 24 trials that did find the proper material placement, the most fit was Trial 30 and the least fit was Trial 35. None of the cases that did not converge were able to find the proper material configuration. In other words, trials that did not reach the fitness criteria did not find the proper material configuration. The temperature profiles of the four trials (Trials 30, 34, 35, and 41 – shown in Figure 4) show the accuracy of the fitness criteria. There is very little difference between the temperature profiles of the improper material placements and the proper placements. The improper placements show that for this problem there are multiple answers based on the way the problem is solved. In this case the solution of this problem depends greatly on the accuracy of the direct solutions of the candidate placements.

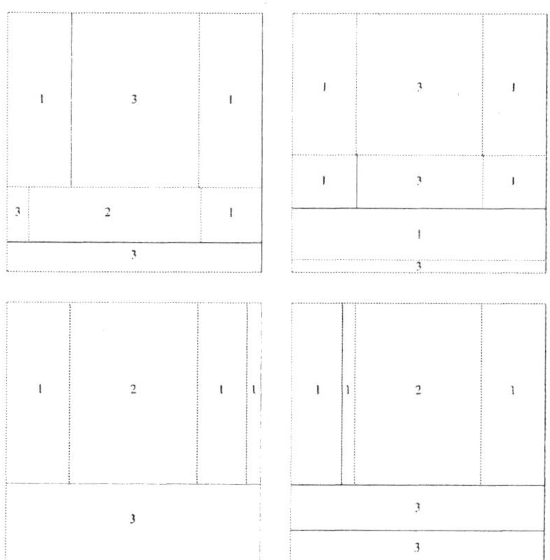

Fig. 4. Material placements of the most fit solution (Trial 34 – upper left) and the least fit solution (Trial 41 – upper right). Note that these are improper placements. Also shown are the most fit of the proper placements (Trial 30 – lower left) and the least fit of the proper placements (Trial 35 – lower right).

CONCLUSIONS/FUTURE WORK

Solution of a 2-D IHCP using evolutionary data segregation techniques has been demonstrated. The algorithm evolves until a suitable partitioning scheme is found based on the known thermal profile and the known materials involved. The algorithm needs to be extended to more complicated and/or three-dimensional geometries. In addition, more investigation is needed on the impact of the fitness criterion on the selection of the correct solution. The primary advantages of this algorithm are its ability to handle discrete rather than continuous data and geometries and its potential ability to handle complex geometries in the same manner as simple geometries.

NOMENCLATURE

i	index for the x location in the grid
j	index for the y location in the grid
k	thermal conductivity (W/m-K)
P_0	energy balance penalty function
q"	heat flux (W/m^2)
T	temperature of exact solution (K)
θ	temperature of candidate solution

REFERENCES

Ashlock, D.A., Bryden, K.M., Johnson, P.E., and McCorkle, D.S., 2002, "Improving Data Segregation with a Graph Based Evolutionary Algorithm," *Intelligent Engineering Systems Through Artificial Neural Networks,* Proceedings of the Artificial Neural Networks in Engineering Conference (ANNIE 2002), ASME Press, pp. 417-422.

Ashlock, D.A., Bryden, K.M., Johnson, P.E., and McCorkle, D.S., "Diversity Control of a Data Segregation Routine Using Graph Based Evolutionary Algorithms," submitted to the *International Journal of Smart Engineering System Design.*

Drezet, J.-M., Rappaz, M., Grun, G.-U., and Gremaud, M., 2000, "Determination of Thermophysical Properties and Boundary Conditions of Direct Chill-Cast Aluminum Alloys Using Inverse Methods," *Metallurgical and Materials Transactions A: Physical Metallurgy and Materials Science,* Vol. 31, pp. 1627-1634.

Fan, H.-Y., 1998, "An Inverse Design Method of Diffuser Blades by Genetic Algorithms," *Proceedings of the Institute of Mechanical Engineers,* Vol. 212, pp. 261-268.

Huang, C.-H. and Chin, S.-C., 2000, "A Two-Dimensional Inverse Problem in Imagining the Thermal Conductivity of a Non-Homogeneous Medium," *International Journal of Heat and Mass Transfer,* Vol. 43, pp. 4061-4071.

Johnson, P.E., Bryden, K.M., Ashlock, D.A., and Vasquez, E., 2001, "Evolving Cooperative Partial Functions for Data Summary," *Intelligent Engineering Systems through Artificial Neural Networks,* Proceedings of the Artificial Neural Networks in Engineering Conference (ANNIE 2001), ASME Press, pp. 405-410.

Ponzini, G., Crosta, G., and Giudici, M., 1989, "Identification of Thermal Conductivities by Temperature Gradient Profiles: One-Dimensional Steady Flow," *Geophysics,* Vol. 54, pp. 643-653.

Shenefelt, J.R., Luck, R., Taylor, R.P., and Berry, J.T., 2002, "Solution to Inverse Heat Conduction Problems Employing Singular Value Decomposition and Model-Reduction," *International Journal of Heat and Mass Transfer,* Vol. 45, pp. 67-74.

Sridhar, L., and Narh, K.A., 1998, "An Inverse Method for the Determination of Thermal Conductivity and Thermal Contact Resistance: Simulated Results," *Proceedings of the 7th AIAA/ASME Joint Thermophysics and Heat Transfer Conference—Part 3 (of 4),* pp. 267-273.

Tervola, P., 1989, "A Method to Determine the Thermal Conductivity from Measured Temperature Profiles," *International Journal of Heat and Mass Transfer,* Vol. 32, pp. 1425-1430.

EVOLVABLE HARDWARE APPLIED TO VOICE RECOGNITION

SUELY CUNHA AMARO MANTOVANI
Departamento de Engenharia Elétrica-
Universidade Estadual Paulista –
UNESP - Campus de Ilha Solteira
Fax:+551837431150
CP - 31 CEP - 15385-000 Ilha Solteira – SP, Brasil
suely@dee.feis.unesp.br

JOSÉ RAIMUNDO DE OLIVEIRA
Departamento de Computação e Automacão – DCA - FEEC
Universidade Estadual de Campinas - UNICAMP
Fax:+551937883845
CP -6101 CEP - 13081-970 - Campinas – SP, Brasil
jro@unicamp.br

ABSTRACT
This paper presents some results of the application on Evolvable Hardware (EHW) in the area of voice recognition. Evolvable Hardware is able to change inner connections, using genetic learning techniques, adapting its own functionality to external condition changing. This technique became feasible by the improvement of the Programmable Logic Devices. Nowadays, it is possible to have, in a single device, the ability to change, on-line and in real-time, part of its own circuit. This work proposes a reconfigurable architecture of a system that is able to receive voice commands to execute special tasks as, to help handicapped persons in their daily home routines. The idea is to collect several voice samples, process them through algorithms based on Mel - Ceptrais theory to obtain their numerical coefficients for each sample, which. compose the universe of search used by genetic algorithm. The voice patterns considered, are limited to seven sustained Portuguese vowel phonemes (a, eh, e, i, oh, o, u).
Keywords: Evolvable Hardware, Reconfigurable Hardware, Genetic Algorithms, Voice Digital Processing.

1 – INTRODUCTION

The Evolutionary Algorithms, widely used in system optimization, among the reconfigurable devices designed in the eighties, allowed the development of a new technology of circuits design – Evolvable Hardware (EHW). EHW is referred to a particular type of hardware, which architecture, structure and function change in a dynamical and autonomous way for adaptation to changes on the environment, while the traditional hardware has the structure and functions fixed by the manufacturer. Nevertheless most of the problems of the real world, are dynamical, they change with the time, therefore, different hardware structure are necessary. EHW provides the idea of adapting hardware for systems with dynamical behavior. EHW become feasible with the development of large scale reconfigurable devices, Programmable Logic Devices(PLDs). In the design and reconfiguration of the EHW are used the reconfigurable devices, that architecture and function are determined by evolutionary algorithms. This new technology is approached in several papers and conferences around the world, (Kitano, 1996),(Iwata, 1996). As showed by Zebulum et al (2002) EHW is a powerful tool of electronic and digital circuits synthesis .

In this paper, it is presented a proposal of an application of EHW in voice recognition, i.e., a system for recognizing sustained vowels for different persons. It is searched a device that allows recognize voice commands with the best possible rate of certainly and size(portable).

The automatic voice recognition is a promising research area, very developed in the last years, (Martins, 1997). The researches in this area aim to create systems able to process a voice signal and automatically determine the content, i.e., the information contained in the signal. As the voice is an individual characteristic of each person, it allows the safe access to determined information, places or services, system control, only with the voice command. The main interest in developing this work was the base of oral commands emitted by disabled persons, that become able to execute certain diary tasks(open and close doors, to set in motion devices, etc) without locomotion, as an example those with cerebral paralysis that have mobility and/or communication problems. Considering this user target and in addition the constraints of the physical dimension of the device to be designed, this work deals only with the voice command of sustained vowels.

The text is divided in five sections as follows. Section 2 presents an overview of EHW subject, the principal characteristics. Section 3 describes the application including the voice recognition and the technique adopted for the solution of the problem. Section 4 presents the preliminary results of the simulations, and the final considerations are in the section 5.

2 – STATE OF THE ART

2.1 - EHW Characterization

EHW is based on matching a reconfigurable device and evolutionary algorithms. However, it is noted by different proposals of works in the literature that the EHW circuits presents some divergences in its denomination according to the rule used for classification. Considering the several works and the proposal presented by Zebulum(1997) and other researchers, the EHW are characterized by the following properties: (1)Types of Evolutionary Algorithms; (2)Evolving Plataform (Hardware); (3)Hardware Evaluation Process; (4)Conection Basic Units; (5)Target Application Areas.

Types of Evolutionary Algorithms.

The EHW can be evolved by four different algorithms that are: Evolutionary Strategic(ES), Genetic Algorithms(GA), Genetic Programming(GP), Evolutionary Programming(EPs), (Michalewics, 1996). All of them are used mainly in system optimization. The difference are in the way representation and on used operators. The

GAs and GPs have similar characteristics to the evolutionary methodology used in this work, but they differ on the chromosome representation. The GA organizes the genes in a binary string, while the GP uses a gene tree and both the algorithms use crossover and mutation. EP and ES have no difference on representation, but use the mutation as the main operator.

Evolving.Plataform (Hardware)

The EHW needs a circuit for evaluate (genetically). Nowadays, the two great alternatives suitable for evaluating the EWH circuits are the programmable devices, even the logic and analogic ones and dedicated hardware. Among the programmable logic devices there are the PLA(Programmable Logic Arrays) and the more used FPGA(Field Programmable Gate Arrays), that present internally a quantity of digital gates that are connected by a binary string, or the FPAA(Field Programmable Analog Arrays) that reconfigure analogic components.

Custom Hardware or Dedicated Hardware, that can be obtained with the evolutionary process. For instance, a neural network dedicated architecture can be evolved using artificial neurons as a basic element of the evolutionary process (Zebulum, 1997).

Hardware Evalution Process

Considering the development of the different published works, there are two pattern classification adopted by the most of authors about the evaluation process or the environmental where the chromosomes are evaluated during the genetic learning: Intrinsic and Extrinsic EHW.

In the Extrinsic EHW approach or off-line, evolution is simulated by software. At the end of evolutionary process, the final solution (best chromosome) is downloaded into the reconfigurable device. This approach helps to draw theoretical results, used mainly for evaluating the fitness of a number of individuals.

In the intrinsic EHW or on line, evolution takes place in the reconfigurable device. The chromosome (individuals) are downloaded into the reconfigurable device and evaluated by the hardware. There are two advantages when is used this approach, the speed of the evolution process and the fact that no simulation model to evaluate the circuit. The evolutionary algorithm can be implemented in a special hardware - complete evolution or performed on a separate processor and incorporated into the chip containing the target EHW - evolution on - chip.

Conection Basic Units

The types of platform used on the EHW evolution differ on the basic units inner connections that form the circuits, i.e., the types of building blocks in circuits. Simple gates AND and OR available in PLA devices are not enough. The applications are needing hardware inner connections more complex for evolution. Three levels of complexity are found on the EHW works: analogic components (transistors, resistors, inductors and capacitors) (Zebulum, 2002); logic gates (OR and AND) (Kajitani, 1999),(Keymeulen, 1997); function (sine, cosine, addition, multiplication) (Murakawa, 1996).

Application Area

The application of EHW are mainly on the Recognition or Pattern Classification (Iwata, 1996), (Mantovani,2000); Control and Robotic (Thompson, 1995); Synthesis of Digital and Analogic Circuit design (VLSI Circuits) (Zebulum, 2002),(DeGaris, 1993); Tolerance to faults (Higuchi,1996). The major of these applications use logic gates and GA as technology. Promising results in analogic circuit design are found, where are searched using the evolution , optimized parameters for the components.

2.2 Genetic Algorithm

GA is the base of development and conception of evolutionary hardware. It is an iterative proceeding based on the natural selection used in optimization problems, presenting a good computational performance (Tomassini, 1997). By iterations the GA can eventually find acceptable solutions, i.e., a solution with a higher fitness. The best configuration found by the GA, is the necessary architecture of bits for programming the reconfigurable device.

3 DESCRIPTION OF PATTERN RECOGNITION SYSTEM

In a pattern recognition system there are two distinct phases: training and recognition. These phases present the same sequence of tasks, which consists of extracting the voice parameters and storage of the reference patterns (or test patterns) for each speaker. Based on a comparison with the reference and the test voice patterns, the GA provides the decision rule and recognition.

Following it is described the necessary steps for implementing the pattern voice recognition system using EHW.

3.1.Speech Processing

Generically, a speech recognition process consists of the steps illustrated in Fig.1.

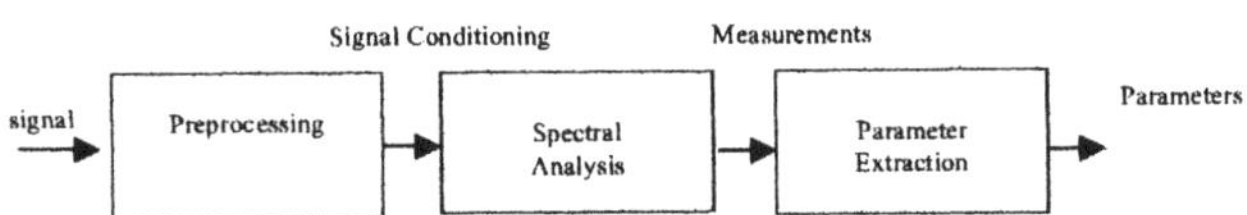

Fig .1 Block Diagram of a Speech Recognition Process

The signal conditioning and the measurements steps were implemented using: Sound Blaster board; Microphone; Wave format (without heading); Sample rate-22050 Hz and resolution of 16 bits; Preemphasis factor of the signal is 0.95; Wide Speech segment (NA), 512 samples.

3.2 Features Extraction

The parameters extracting is used to obtain the informations concerned to the differentiation of several speech events. For each vowel it is obtained a vector of characteristics containing every relevant information to the pattern classification, which should be small for minimizing the computational time. This recognition system uses the Mel - Cepstrais coefficients which provides a reliable speech signal, necessary for each vowel and the respective speaker. The usual speech recognition methods are available for stationary signals in an interval of milliseconds. This way, the input signal is divided in 40 frames (23 ms), without superposition due to the sustained vowels. For each speech segment, 12 coefficients mel – Cepstrais are extracted.

The Mel-Cepstrais coefficients are the most used ones, for describing the speech signal spectral shape. They are obtained, firstly, calculating the smooth spectrum of the digitized speech signal by FFT, then is filtered by a triangular filters, equally spaced according to the mel scale that simulates the frequency answer of the human audition (Picone, 1993).

3.3 GA Implementation

Considering the numeric nature of the problem, the simulations for the GA are executed for numeric digits. The chromosome wide of the bit architecture used for programming the reconfigurable circuit, is defined by the characteristic vector for each speaker, composed of 40 frames and 12 coefficients/frames, forming the string of 480 numbers (40x12=480).

The implemented GA is a simple genetic algorithm using selection by tournament, and the genetic operators, crossover in a single point and inductive mutation (position change). For adaptation function or evaluation criterion, it is used the Euclidian Distance technique, that identifies the minimum error with reference pattern and a test input signal for the system.

4. ANALYSIS OF THE RESULTS

The data base was formed by the samples of sustained vowel phonemes a, eh, e ,i, oh, o ,u, in Portuguese, considered as stationary signals able to be emitted by speakers that have or no , communication problems. It consists of 22 individuals (male and female) for training and 10 for test (the system is evaluated in the independent way of the speaker) . Each individuals recorded the seven samples vowels , add up 154 locution for the training and 70 for the test. The acquisition was effectuated in a natural environment (an office) without treating noises.

The recognition percentage is given by ratio by locution number correctly recognized and total number of test locution. Based on simulations performed on Mathlab environment are showed in the graphics some preliminaries results (Fig.2) that denotes reasonable behavior of propose recognition system.

5.FINAL CONSIDERATIONS

In this paper was presented one new methodology and preliminary results of EHW application in the Voice Recognition Pattern. This application has as main advantage the hardware flexibility, provide small size and speed process available the integrated circuits. The architecture propose in Fig 3 shows the basic units of the configuration required for performance the EHW for the Voice Recognition Pattern. There are two memories for storage the reference pattern and test pattern. The fitness evaluation is stored in separated memories. The CPU unit is used for interface communications with the external world. Process EHW is a promising idea that need more search to be developed.

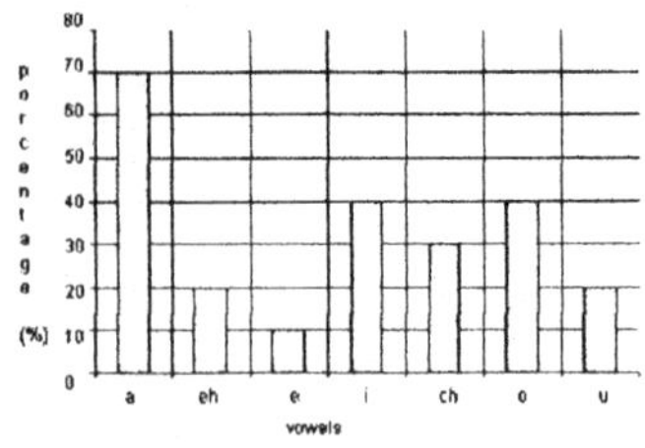

Fig .2 Preliminaries results

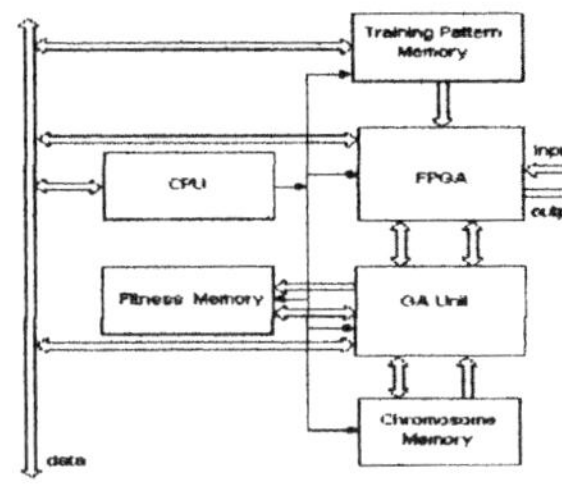

Fig .3 EHW Architecture

REFERENCES

DeGaris, H., 1993, "Evolvable Hardware Genetic Programming of a Darviw Machine", *International Conference On Artificial Neural Networks and Genetic Algorithms-ICANNGA93*, Innsbruck,Austria,1993.

Higuchi, T. et al,1996, "Evolvable Hardware and its Application to Pattern Recognition and Fault - Tolerant Systems", *Lecture Notes in Computer Science*, Vol. 1062, pp. 118-135.

Iwata, M. et al, 1996, "A Pattern Recognition System Using Evolvable Hardware", *Lecture Notes in Computer. Science*, Vol. 1141, pp. 761-770.

Kajitani, I., et al, 1999, "An Evolvable Hardware Chip for Prosthetic Hand Controller", *Proceedings, Seventh International Conference on Microelectronics for Neural, Fuzzy and Bio-inspired Syst.(MicroNeuro99)*, pp. 179-186.

Keymeulen, D., et al, 1997, "An Evolutionary Robot Navigation System Using a Gate-Level Evolvable Hardware", *Lecture Notes in Computer. Science*, Vol. 1259, pp. 195-209.

Kitano, H., 1996, "Morphogenesis for Evolvable Systems", *Lecture Notes in Computer Science*, Vol. 1062, pp. 19-47.

Mantovani,S.C.A., et al,2000, "Evolvable Hardware Applied to Voice Recognition for Carried Cerebral Palsy, Anais do Congreso Iberoamericano - 3º de Comunicación Alternativa y Aumentativa, y 1º de Tecnologías de Apoyo para la Discapacidad , Madrid - Espanha, 18,19 , 20, Outubro, 2000.

Martins, J. A ., 1997," Different Techniques Avaluation in Speech Recognition ", Doctor Thesis , DECOM/FEEC/UNICAMP, pp.82.(in Portuguese).

Michalewicz, Z., 1996 *Genetic Algorithms +Data Strutures = Evolution Programs*, 3rd ed. Heidelberg, Germany: Springer - Verlag, 1996.

Murakawa, M., et al, 1996, "Hardware Evolution at Function Level", *Lecture Notes in Computer Science*, Vol. 1062, pp. 62-71.

Picone, J.W., 1993, "Signal Modeling Techniques in Speech Recognition", *Proceedings of the IEEE*, vol.81, nº9, september 1993.

Tomassini, M., 1997, "Evolutionary Algorithms", *Lecture Notes in Computer Science*, vol. 1259, pp.195-209.

Thompson, A.,1995, "Evolving Electronic Robot Controllers that Exploit Hardware Resources", *Lecture Notes in Artificial Intelligence*, vol. 929, pp. 640-656.

Zebulum, R.S., et al,1997, "Evolvable Systems in Hardware Design: Taxonomy, Survey and Applications", *Lecture Notes in Computer Science*, vol. 1259, pp 344-358.

Zebulum, R. S., et al, 2002, *Evolutionary eElectronics, Automatic Design of Electronic Circuits and Systems by Genetic Algoritms*, The CRC Press International Series on Computational Intelligence., Boca Raton, Florida333431,pp.299.

GENETIC ALGORITHM DESIGN INSPIRED BY ORGANIZATIONAL THEORY: PILOT STUDY OF DEPENDENCY STRUCTURE MATRIX DRIVEN GENETIC ALGORITHM

Tian-Li Yu, David E. Goldberg, Ali A. Yassine, and Ying-Ping Chen

Department of General Engineering
University of Illinois at Urbana-Champaign
Urbana, IL 61801

Email: {tianliyu, deg, yassine, ypchen21}@uiuc.edu

ABSTRACT
This paper proposes a dependency structure matrix driven genetic algorithm (DSMDGA) which utilizes the dependency structure matrix clustering to extract building block (BB) information and use the information to accomplish BB-wise crossover. Three cases: tight, loose, and random linkage, are tested on both a DSMDGA and a simple genetic algorithm (SGA). Experiments showed that the DSMDGA is able to correctly identify BBs and outperforms a SGA by using the extracted BB information.

1 INTRODUCTION

Genetic algorithms (GAs) [1, 2] are stochastic black-box optimization methods. Holland [1] suggested that operators learning linkage information to recombine alleles might be necessary for GA success. Many such methods have now been developed to solve this including perturbation [3, 4, 5, 6], linkage learning schemes [7, 8], and model building schemes [9, 10, 11]. This paper adds to the literature of competent GAs with a method inspired by organizational theory.

In particular, this paper develops a technique for extracting building block (BB) information by using a dependency structure matrix (DSM) borrowed from the literature of project management and corporate organization, called a *dependency structure matrix driven genetic algorithm* (DSMDGA). The DSMDGA is able to identify BBs with the help of a DSM and accomplish BB-wise crossover. As the experimental results suggest, compared to a simple genetic algorithm, the DSMDGA is able to maintain a reliable solution quality under tight, loose, and random linkage with the same amount of function evaluations.

This paper consists four main parts: the DSM clustering, the description of the DSMDGA, the experimental results, and conclusions. In the DSM clustering part, an introduction to DSM is given and a minimum description length (MDL) clustering metric is described. Then the details of the design and implementation of the DSMDGA are revealed. The experiments part shows parameter settings, test functions, and results. Finally, conclusions and future work end this paper.

Figure 1. A DSM where clustering is not so obviously at the first glance. "x" means that dependency exists; the blank means no dependency.

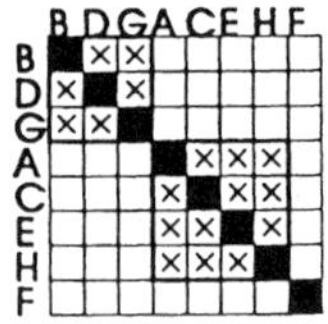

Figure 2. The same DSM but after re-ordered. The DSM can be cleanly clustered as ((B,D,G)(A,C,E,H),(F)).

2 THE DEPENDENCY STRUCTURE MATRIX CLUSTERING PROBLEM

This section describes the DSM decomposition problem which arises when the dependency structure matrix driven genetic algorithm (DSMDGA) is constructed in the next section. This section starts with an introduction to the DSM followed by the DSM clustering problem. At the end of this section, a minimal description length (MDL) clustering metric is described.

A DSM is a matrix where each entry d_{ij} represents the dependency between node i and node j [12, 13]. Entries d_{ij} can be real numbers or integers. The larger the d_{ij} is, the higher the dependency is between node i and node j. This paper concentrates on the 0-1 domain, where $d_{ij} = 0$ means that there is no dependency between node i and node j, and 1 means that node i and node j are dependent. The diagonal entries (d_{ii}) have no significance and are usually set to zero or blacked-out. Figures 1, 2 show examples of DSMs where an "x" symbol stands for 1, and a blank square stands for 0. From Fig. 1, for example, we know that A and C are dependent; A and F are independent. For elaborate exposition of DSM, please see MIT DSM web site: `http://web.mit.edu/dsm/`.

The goal of DSM clustering is to find subsets of DSM elements (i.e., clusters) so that nodes within a cluster are maximally dependent and clusters are minimally interacting. In Fig. 2, the DSM is clustered as ((B,D,G),(A,C,E,H),(F)). In a typical DSM clustering problem, overlapping clusters (clusters that share same nodes) are permissable. However, this paper focus on non-overlapping clusters.

DSM clustering is a sophisticated task which requires expertise. For example, not many people know how to cluster the DSM in Fig. 1. However, after nodes are reordered, it is easily seen that the DSM can be cleanly clustered into three parts (Fig. 2). In addition, A criterion that evaluates clustering arrangement is essential for automated clustering algorithms.

Previous DSM clustering algorithms can be found in [14, 15]. Their results showed that the objective function used was short of accurately predicting "good" clustering because the oversimplified objective function.

This paper adopts the minimum description length (MDL) clustering metric proposed in [16]. The only difference is that we use an ordinary summation instead of the weighted summation (α=β=1/3 in [16]). The metric can be written as following:

$$f_{DSM}(M) = n_c \log n_c + n_c \log n_n + \log n_n \Sigma_{i=1}^{n_c} cl_i + |S|(2\log n_n + 1), \quad (1)$$

where n_c is the number of clusters, n_n is the number of nodes in the DSM, cl_i is the size of the i-th cluster, and S is a mismatch set described in [16]. The objective is to find a model M that minimizes f_{DSM}.

3 THE DEPENDENCY STRUCTURE MATRIX DRIVEN GENETIC ALGORITHM

This section constructs the dependency structure matrix driven genetic algorithm (DSMDGA). The DSMDGA consists of three main tasks: constructing a DSM representing interactions between pairs of genes, clustering the DSM to obtain BB information, and perform BB-wise crossover by exploiting the BB information. This section describes the framework of the DSMDGA, and how to accomplish the three tasks mentioned above.

3.1 The Framework of the DSMDGA

There are two levels of evolutionary algorithms in the DSMDGA: one is to solve the given problem, called the *primary GA*, and other one is to solve the building block (BB) identification problem, called the *auxiliary evolutionary strategy* (ES) [17]. The basic idea of the DSMDGA is to use the auxiliary ES to identify BBs, then the primary GA solves the problem by utilizing the BB information.

Basically, the auxiliary ES receive a DSM which represents the dependencies between genes from the primary GA. After solving the DSM cluster problem, the auxiliary ES returns the BB information back to the primary GA, and the primary GA utilizes the BB information to accomplish BB-wise crossover. More details about the DSM construction and the auxiliary ES can be found in following subsections. Additionally, unlike a meta-GA [18], the auxiliary ES in the DSMDGA does not consume function evaluations of the given problem. Instead, the auxiliary ES uses only the MDL clustering metric, f_{DSM}, as described in section 2, which is much less computationally expensive than many real-world objective functions.

3.2 DSM Construction

The way that the DSMDGA detects the dependency of gene i and gene j is similar to LINC and LIMD [6]. Define $f_{a_i=x,a_j=y}$ as the fitness value of the schema where the i-th gene is x, the j-th gene is y, and the rest are * (wild card). For example, for $i=2$ and $j=5$ in a 5-bit problem, $f_{a_i=0,a_j=1} = f(*0**1)$. If the i-th gene and the the j-th gene are independent (linear), $f_{a_i=0,a_j=1} - f_{a_i=0,a_j=0}$ and $f_{a_i=1,a_j=1} - f_{a_i=1,a_j=0}$ should be the same. Therefore, the interaction (nonlinearity) between the i-th gene and the the j-th gene is defined as following:

$$\Delta_{ij} = |f_{a_i=0,a_j=1} - f_{a_i=0,a_j=0} - f_{a_i=1,a_j=1} + f_{a_i=1,a_j=0}|. \tag{2}$$

However, the fitness value of schemata cannot be computed unless every possible combinations are visited. Now the task is to approximate Δ_{ij} with s_{ij} which is computed based on the individuals seen so far. First, define the sampled fitness of a schema in the population of t-th generation as $f^t_{a_i=x,a_j=y} = \frac{1}{n_a}\Sigma_{a\in P_t,a_i=x,a_j=y} f(a)$, where t is the generation, P_t is the the population of the t-th generation, f is the fitness function, a is an individual where its i-th gene is x and j-th gene is y, and n_a is the number of such a in the population. We call that $f^t_{a_i=x,a_j=y}$ is *undefined* if n_a is zero. The information of interactions gathered from the population is $s^t_{ij} = |f^t_{a_i=0,a_j=1} - f^t_{a_i=0,a_j=0} - f^t_{a_i=1,a_j=1} + f^t_{a_i=1,a_j=0}|$. s^t_{ij} is *undefined* if *any* of the $f^t_{a_i,a_j}$ is *undefined*.

To utilize all individuals of all populations seen so far, s^t_{ij} is then averaged over generations. Defining a set $D = \{s^t_{ij} | s^t_{ij} \text{ is defined.}\}$, the average of s^t_{ij} is expressed as $s_{ij} = \frac{1}{|D|}\Sigma^T_{t=1,s^t_{ij}\in D} s^t_{ij}$, where T is the current generation. With a threshold θ, s_{ij} is then transferred into 0-1 domain, and a DSM is constructed:

$$d_{ij} = \begin{cases} 0 & ,\text{if } s_{ij} \le \theta. \\ 1 & ,\text{if } s_{ij} > \theta. \end{cases} \tag{3}$$

In this paper, the threshold is calculated by a two-mean algorithm (a special case of the k-mean algorithm [19], where $k = 2$). The threshold can also be set according to some prior knowledge if available.

3.3 DSM Clustering

An ES is used to solve the DSM clustering problem. Given a predefined n_{cmax}, the maximal number of clusters, and the number of nodes n_n, the chromosome is a $n_{cmax}+1$-ary string of n_n genes. If the i-th genes has a value $j \in [1, n_{cmax}]$, that means the i-th nodes belongs to the j-th cluster. A zero means that the node does not belongs to any cluster.

A simple mutation and $(\lambda + \mu)$ selection are performed. With a small probability p_m, the mutation randomly assigns a gene with a value between 0 and n_{cmax}. The population is initialized to all zeros, which is a known local minimum. A chromosome with all zeros means that all nodes are independent and no cluster exists at all. This is a local minimum because no clustering is often better a wrong clustering.

3.4 BB-wise Crossover

After the BB information is obtained from the auxiliary ES, BB-wise crossover can be achieved in the primary GA level. The BB-wise crossover is like an ordinary allele-wise crossover, but instead of mixing genes, the BB-wise crossover mixes BBs and will not disrupt BBs. The order of BBs is not significant because BBs have no (or little) interaction with each other by definition. For more details about BB-wise crossover, refer to [10].

4 EMPIRICAL RESULTS

The experiments were done by using a simple GA (SGA) as a test platform. The DSMDGA tested here is a simple GA with the the DSM construction and an auxiliary ES with the MDL clustering metric.

The test function is a 30-bit MaxTrap problem composed of 10, 3-bit trap functions. The 3-bit trap is given by $f^3_{trap}(u=0) = 0.9$, $f^3_{trap}(u=1) = 0.45$, $f^3_{trap}(u=2) = 0.0$, and $f^3_{trap}(u=3) = 1.0$, where u is the number of 1's.

Three linkage cases were tested: tight linkage, loose linkage, and random linkage. Define $U(x)$ as a counting function which counts the number of 1's in x. In the tight linkage test, genes are arranged as $fitness = f^3_{trap}(U(x_1, x_2, x_3)) + f^3_{trap}(U(x_4, x_5, x_6)) + f^3_{trap}(U(x_7, x_8, x_9)) + \cdots$. In the loose linkage test case, alleles are arranged as $fitness = f^3_{trap}(U(x_1, x_{11}, x_{21})) + f^3_{trap}(U(x_2, x_{12}, x_{22})) + f^3_{trap}(U(x_3, x_{13}, x_{23})) + \cdots$. For more details about this test scheme, see [3].

Given the failure rate to be 1/10, the population size is set as 182 by the gambler's ruin model [20, 21]. In the primary GA, binary tournament selection was adopted, and no mutation was used. In the auxiliary ES, the maximal number of cluster n_{cmax} is 10 (equal to m), and the mutation probability p_m was set to be $1/30$. A $(\lambda + \mu)$ selection was adopted, where $\lambda = 5$ and $\mu = 500$.

Figure 3 shows the performance of the SGA using two-point crossover. The SGA worked only for the tight linkage case. For loose and random linkage cases, SGA did not work because of BB disruption. Correspondingly, Figure 4 illustrates the performance of the DSMDGA using BB-wise two-point crossover. The DSMDGA converged for all three tests. Even in the tight linkage test, the DSMDGA (converged at the 22th generation) outperformed the SGA (converged at the 40th generation) because the DSMDGA disrupted fewer BBs. This argument can be verified by Fig. 5, which shows the DSM created by the DSMDGA for the tight linkage case. The prefect result is 10, 3-bit clusters located on the diagonal. As the figure indicates, at the fifth generation, the DSMDGA were able to identify eight BBs; the DSMDGA has successfully identified all BBs at the tenth generation.

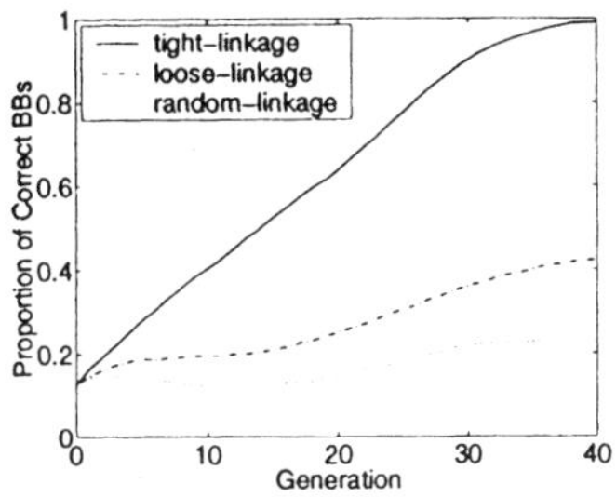

Figure 3. The performance of the SGA with two-point crossover.

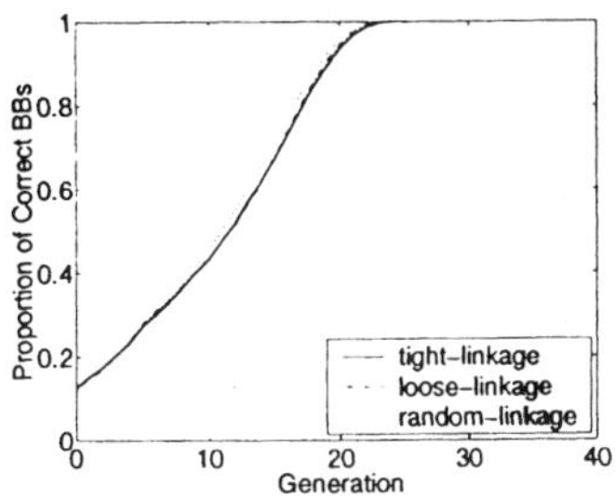

Figure 4. The performance of the DSMDGA with BB-wise two-point crossover.

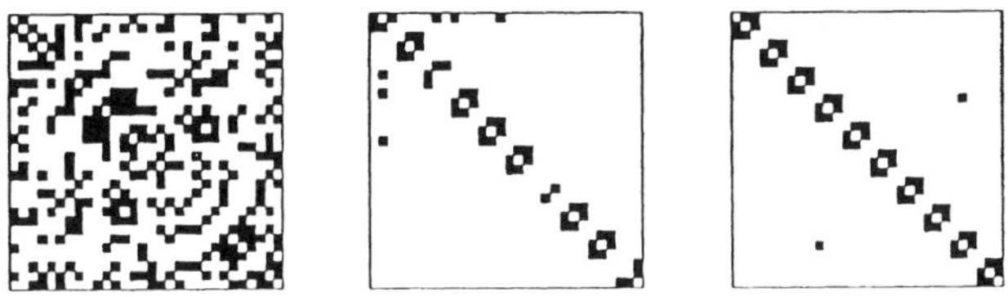

Figure 5. The DSMs created by the DSMDGA in the tight linkage test. From the left to the right, the DSMs are created at generation 0, 5, and 10, respectively. The prefect result should be 10, 3-bit clusters on the diagonal.

5 Conclusions

This paper proposed the dependency structure matrix driven genetic algorithm (DSMDGA), which utilizes the DSM clustering technique to identify BBs. A MDL clustering metric used by the auxiliary ES was constructed for the DSM clustering problem. Pilot empirical results have showed that using the BB information obtained by the DSM clustering technique can help the convergence of GAs on tight, loose, and random linkage problems.

This paper shows successful pilot results on the DSMDGA and there is no reason to believe that they will not scale. However, that needs to be checked. A short-term future work is to test scalability of the method by applying problems of varying size, complexity, noise. and scaling. There is an interesting circularity while we are doing this research. Originally started work is to cluster DSMs using GAs, but this has led to using DSMs to drive design of a competent GA. This circularity is somewhat puzzling and in longer term goal of this work is to understand why and when simple methods will work (or fail) in the search for key dependencies.

ACKNOWLEDGEMENT

This work was sponsored by the Air Force Office of Scientific Research, Air Force Materiel Command, USAF, under grant F49620-03-1-0129, and the National Science Foundation under grant DMI-9908252. The U.S. Government is authorized to reproduce and distribute reprints for government purposes notwithstanding any copyright notation thereon.

REFERENCES

[1] Holland, J.H.: *Adaptation in natural and artificial systems*. University of Michigan Press, Ann Arbor, MI (1975)

[2] Goldberg, D.E.: *Genetic algorithms in search, optimization, and machine learning*. Addison-Wesley, Reading, MA (1989)

[3] Goldberg, D.E., Korb, B., Deb, K.: Messy genetic algorithms: Motivation, analysis, and first results. *Complex Systems* 3 (1989) 493–530

[4] Goldberg, D.E., Deb, K., Kargupta, H., Harik, G.: Rapid, accurate optimization of difficult problems using fast messy genetic algorithms. *Proceedings of the Fifth International Conference on Genetic Algorithms* (1993) 56–64

[5] Kargupta, H.: The performance of the gene expression messy genetic algorithm on real test functions. *Proceedings of 1996 IEEE International Conference on Evolutionary Computation* (1996) 631–636

[6] Munetomo, M., Goldberg, D.E.: Identifying linkage groups by nonlinearity/non-monotonicity detection. *Proceedings of the Genetic and Evolutionary Computation Conference 1999: Volume 1* (1999) 433–440

[7] Harik, G.R., Goldberg, D.E.: Learning linkage. *Foundations of Genetic Algorithms 4* (1996) 247–262

[8] Smith, J.: On appropriate adaptation levels for the learning of gene linkage. *Journal of Genetic Programming and Evolvable Machines* 3 (2002) 129–155

[9] Bosman, P.A.N., Thierens, D.: Linkage information processing in distribution estimation algorithms. *Proceedings of the Genetic and Evolutionary Computation Conference 1999: Volume 1* (1999) 60–67

[10] Harik, G.: *Linkage Learning via Probabilistic Modeling in the ECGA*. IlliGAL Report No. 99010, University of Illinois at Urbana-Champaign, Urbana, IL (1999)

[11] Pelikan, M., Goldberg, D.E., Cantú-Paz, E.: BOA: The Bayesian optimization algorithm. *Proceedings of the Genetic and Evolutionary Computation Conference GECCO-99* I (1999) 525–532

[12] Steward, D.V.: The design structure system: A method for managing the design of complex systems. *IEEE Transactions on Engineering Managment* 28 (1981) 77–74

[13] Yassine, A., Falkenburg, D.R., Chelst, K.: Engineering design management: An informatoin structure approach. *International Journal of production research* 37 (1999) 2957–2975

[14] Fernandez, C.: *Integration Analysis of Product Architecture to Support Effective Team Co-location*. Master thesis, Massachusetts Institute of Technology (1998)

[15] Sharman, D., Yassine, A., Carlile, P.: Characterizing modular architectures. *ASME 14th International Conference, DTM-34024* (2002)

[16] Yu, T.L., Yassine, A., Goldberg, D.E.: A genetic algorithm for developing modular product architectures. *Proceedings of the 15th international Conference on Design Theory and Methodology (DTM)* (to appear)

[17] Rechenberg, I.: *Evolutions strategie optimierung technischer systeme nach prinzipien der biologischen evolution*. Friedrich Frommann Verlag, Stuttgart-Bad Cannstatt (1973)

[18] Mercer, R.E., Sampson, J.R.: Adaptive search using a reproductive meta-plan. *Kybernetes* 7 (1978) 215–228

[19] Hartigen, J.A., Wong, M.A.: A *k*-means clustering algorithm. *Applied Statistics* 28 (1979) 10–108

[20] Harik, G., Cantú-Paz, E., Goldberg, D.E., Miller, B.L.: The gambler's ruin problem, genetic algorithms, and the sizing of populations. *Proceedings of 1997 IEEE International Conference on Evolutionary Computation* (1997) 7–12

[21] Miller, B.L.: *Noise, sampling, and efficient genetic algorithms*. Doctoral dissertation, University of Illinois at Urbana-Champaign, Urbana (1997)

GENETIC ALGORITHM BASED DESIGN AND IMPLEMENTATION OF MULTIPLIERLESS TWO-DIMENSIONAL IMAGE FILTERS

DOUGLAS J. LOCKETT CHRISTOPHER D. ROBLEE MICHAEL RUDKO

Department of Electrical and Computer Engineering
Union College
Schenectady, New York 12308, U.S.A.

ABSTRACT

This paper outlines the use of a genetic algorithm to design multiplierless IIR filters for applications in hardware-based image processing. A unique genetic algorithm is developed to optimize filter coefficients such that the corresponding filter's frequency response matches that of an ideal system with the constraint that all coefficients are powers-of-two and the resulting filter is stable. The motivation for using power-of-two filter coefficients is to reduce the overall arithmetic complexity in any hardware based implementation by replacing digital multipliers with simpler shift operators. This approach is highly beneficial for computationally intensive image filtering applications.

The cases considered comprise Canny's edge detection filter as well as an image blur operator. It is shown that in many cases the multiplierless systems have a definite advantage in terms of their efficiency while maintaining a desired response, making them a viable alternative as image filters. It is demonstrated that custom genetic optimization is a reliable, and efficient means for designing such filters.

1. INTRODUCTION

The need for faster, embedded imaging systems has motivated research into various hardware-based imaging solutions. Currently, there exist numerous approaches for the performance improvement of digital image processors. Our objective is to explore a new and effective means to accomplish the optimization of hardware-based, IIR image filters.

We propose the use of multiplierless arithmetic reduction to replace conventional digital multiplication operations with less complex shift operations within the filter difference equation. This approach greatly improves the speed of existing real-time and non real-time image filters. We develop a custom genetic algorithm in software capable of approximating various IIR filter responses by generating multiplierless coefficients. We select this course based upon the proven ability of genetic algorithms to effectively and rapidly isolate a sufficient solution for a given optimization problem. Within, we assess a genetic algorithm's ability to realize multiplierless filters for certain image processing functions. We furthermore examine qualitatively the application of these multiplierless filters to a set of test images.

2. THEORY OF MULTIPLIERLESS ARITHMETIC

Binary multiplications consist of a series of shifts and additions, which in turn require additional carry and overflow-handling logic. This additional complexity and delay can greatly hamper the performance of processors that are required to perform arithmetically intensive computations. Multiplying, instead, by a power-of-two number saves execution clock cycles and entirely eliminates the need for adder logic.

A recursive difference equation is used as an implementation of our IIR filtering operation. When applied to a full 2D image, this operation requires a substantial number of multiplications and additions. This is the motivation behind multiplierless implementation enhancement for our application.

3. APPLICATION OF GENETIC ALGORITHMS

In general, genetic algorithms (GAs) rely upon the law of fittest member survival to optimize a set of possible outcomes or results. Loosely based on the laws of Darwinian genetics, GAs have many unique characteristics that make them ideal for many optimization problems. As Goldberg states (1989), "They [GAs] combine survival of the fittest among string structures with a structured, yet randomized information exchange to form a search algorithm with some of the innovative flair of human search."

Since our goal is to optimize IIR filters to contain only power-of-two coefficients, traditional design techniques are unsuitable. Restricting coefficients to power-of-two numbers forces us to explore other methods. The proposed technique employs a genetic algorithm to search for power-of-two filter coefficients.

3.1 Overview of the Designed Genetic Algorithm

The algorithm designed is based upon conventional GA principles and framework: it starts with a random population, evaluates member fitnesses, breeds randomly selected fittest members, occasionally mutates, and then repeats. The GA is designed in a modular fashion through the use of variable inputs in order to make important changes regarding the operation of the algorithm in a relatively straightforward manner.

The devised algorithm also makes certain major departures from conventional GA design. These different approaches in design are due to the precise nature of multiplierless optimization and the notion of representing IIR filters in terms of coefficient sets. The flowchart detailing the designed GA is shown in Figure 3.1 (page 4).

The most significant difference in the designed GA is that each member of the population is comprised of power-of-two, decimal values. These values represent the coefficients of a particular multiplierless IIR filter. Furthermore, each member coefficient set is divided into two separate numerator and denominator coefficient sets, which crossover separately. Fitness evaluation is based on the set of coefficients as a whole. This is a departure from conventional GA design in that each member of the population is traditionally represented as a single binary string. Coefficients remain intact as they are exchanged in their entirety throughout crossover. The implemented crossover scheme is a two-point crossover, as two distinct ranges of crossover are selected at random for each member of the population (Mitchell, 1996). This form of crossover is utilized in order to genetically enhance the numerator and denominator coefficient sets separately.

An additional deviation from general GA design is the designed algorithm's preemptive removal of unstable members (Figure 3.1). Since filter stability is the initial criterion for filter design, the GA is configured to remove unstable members before fitness evaluation and after mutation. The population is subsequently replenished with new random coefficient sets to fill vacancies left by unstable members that were removed during fitness evaluation.

It is noted that due to the restrictive nature of power-of-two coefficients and the devised crossover scheme, it is extremely difficult for our population to approach a point of convergence in which all members represent a uniform fittest coefficient set. Instead, the fittest member in the population is selected after the exhaustion of a specified number of iterations. This member represents the closest power-of-two filter approximation.

The designed GA also incorporates a form of elitism, a technique implemented in order to prevent fitness regression of the population. It prevents the loss of the fittest members due to crossover and mutation. Elitism ensures that the fittest members are retained by creating a copy of them before crossover and using them to replace the least fit members in the subsequent generation. The number of members held over is an input parameter of the GA.

3.2 Characteristics of GA Implementation

The fitness function in the designed GA compares the candidate multiplierless coefficient system responses from the population to a specified ideal filter's response. In order to do this, the absolute difference is taken between the magnitude frequency responses of the ideal system and the candidate multiplierless systems. This difference is then summed over all samples in the frequency response representation. Fitness is defined as being inversely proportional to this squared sum of differences between the ideal and candidate systems. This inverse relationship allows us to establish a basis of comparison between the members in the population. Members with lower squares of summed difference values are assigned higher probabilities of crossover since they represent the fitter members.

Each member is assigned a probability of crossover based on its relative fitness. A randomly generated number determines which element will be selected for breeding. This number falls within a particular range of crossover probability that corresponds to an individual coefficient set subsequently chosen as a breeding member. Consequently, members with higher fitnesses are more likely to be selected.

The magnitude frequency response is the basis for comparison between the candidate multiplierless coefficient sets and the specified ideal filter. A system of relative error tracks the progression of the GA population. Relative error is defined as the difference between magnitude frequency responses of the ideal and fittest multiplierless systems summed over all samples. This value allows us to track how the fittest member converges to a fittest solution over the total number of GA iterations.

The GA is designed to be able to optimize several types of filters as well as to adapt and modify its population in different ways. To do this, the GA incorporates several parameters that can be altered depending on the application needs. The implemented parameters comprise: number of genetic iterations, filter order, member mutation probability (rate), word length (bits) and sampling frequency.

4. RESULTS

Throughout the development phase of the GA, we utilize frequency analysis in order to compare the approximated and ideal magnitude responses of three distinct filters. The first is an ideal high-pass filter (HPF), displayed in Figure 4.1.

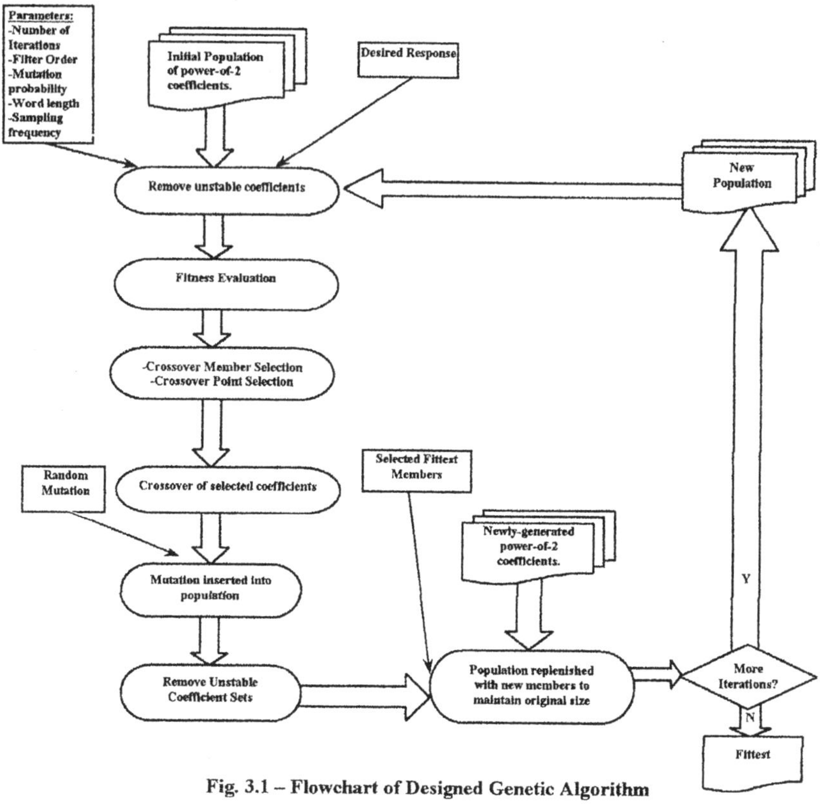

Fig. 3.1 – Flowchart of Designed Genetic Algorithm

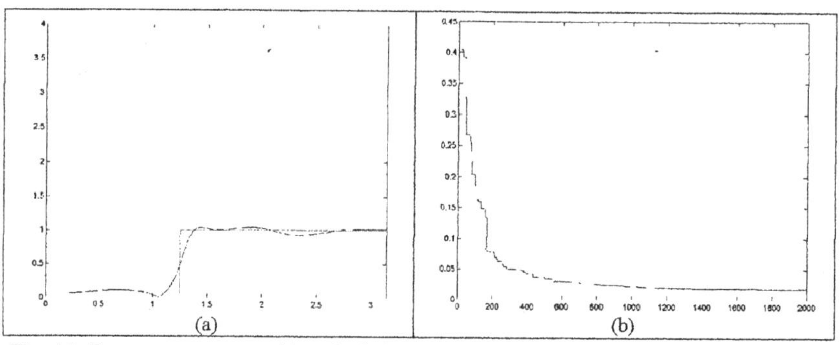

Fig. 4.1: Frequency responses (a) and relative error plot (b) of ideal and approximated filters *after 2000 GA iterations with mutation probability = 0.001.*

Given the specified parameters, the GA produces an accurate and sharp approximation of the ideal HPF with minimal relative error after 2000 iterations.

The second application attempted is Canny derivative edge detection (Parker, 1997). Canny's edge detector was designed as a convolution filter in to smooth the image noise (represented as a Gaussian), and to locate the edge. The optimal edge location filter is approximated as the first derivative of a Gaussian function in two directions, *G'(x,y)*. The edge detector convolves the input image with *G'* in both the *x* and *y* directions and then computes the magnitude between the two resultant vectors at each pixel. Smoothing (noise reduction) is performed in a similar manner using the original Gaussian function, *G(x,y)*, (Sarkar and Boyer, *1989*).

To test the multiplierless approximation of the Canny operator, the GA is executed with the Gaussian and derivative-Gaussian masks as the desired (ideal) responses. Results from the derivative Gaussian case are shown in Figure 4.2. In both trials, the relative error is minimized after the 200th GA iteration.

The third case considered is that of an image-blurring system, represented as an ideal LPF, with degree of blurring inversely proportional to cutoff frequency. The plots shown in Figure 4.4 are characteristic of those achieved throughout experimentation.

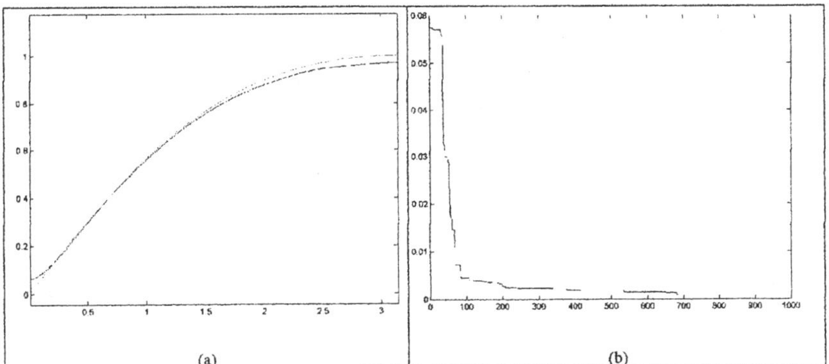

(a) (b)

Fig. 4.2: (a) Magnitude frequency responses of ideal and approximated derivative-Gaussian filters, after 1000 GA iterations, $\sigma = 1$ (b) Relative error plot.

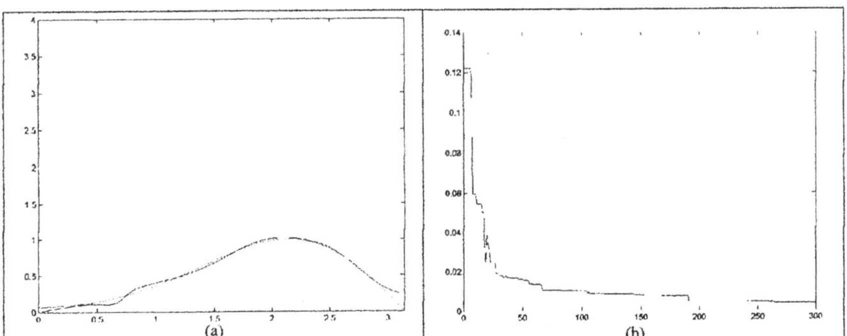

(a) (b)

Fig. 4.3: Magnitude frequency response (a) and relative error plot (b) of ideal and approximated Gaussian filter, after 300 iterations, $\sigma = 2$.

As a representative example of filter performance, the results from certain trials of the multiplierless Canny operator are displayed in Figure 4.5 and Figure 4.6.

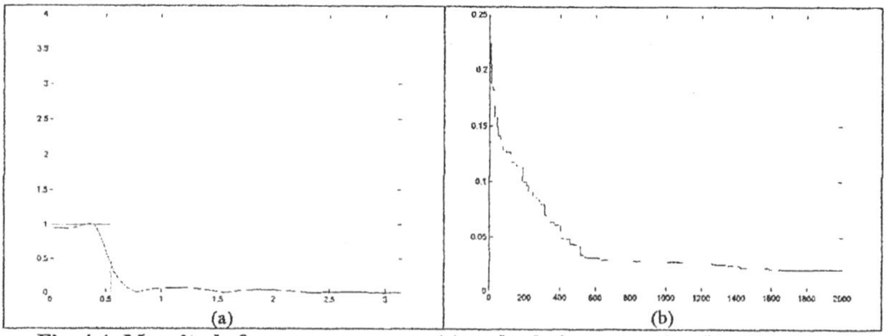

Fig. 4.4: Magnitude frequency response (a) and relative error plot (b) of ideal and approximated blurring filter, after 2000 iterations.

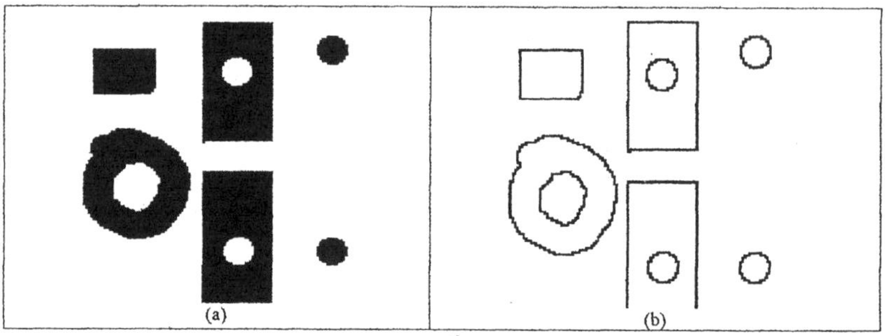

Fig. 4.5: Sample input (a) and output (b) of 10th order, multiplierless Canny edge detector.

We see how the Canny filter accurately detects all of the edges within the black and white bitmap image above. The edges are precise and very narrow, accentuated as single-pixel contours. The next trial involves a grayscale photograph (Figure 4.6).

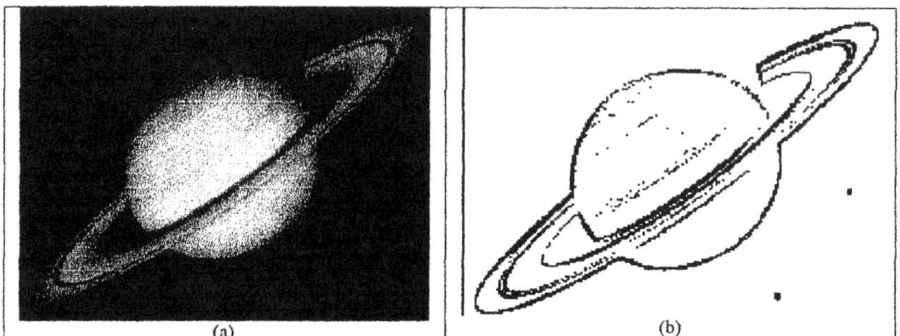

Fig. 4.6: Sample input (a) and output (b) of 10th order, multiplierless Canny edge detector.

5. COMPARATIVE ANALYSIS OF COMPUTATIONAL COMPLEXITIES

To quantify differences in complexity between the multiplier and multiplierless-based filtering algorithms discussed in section 2, two systems are compared through hypothetical implementations on identical target devices. We base all calculations upon the requirements of a 10th-order IIR filter throughout this procedure (Figure 5.1). We consider a generic field programmable gate array device (FPGA) implementation of the filter.

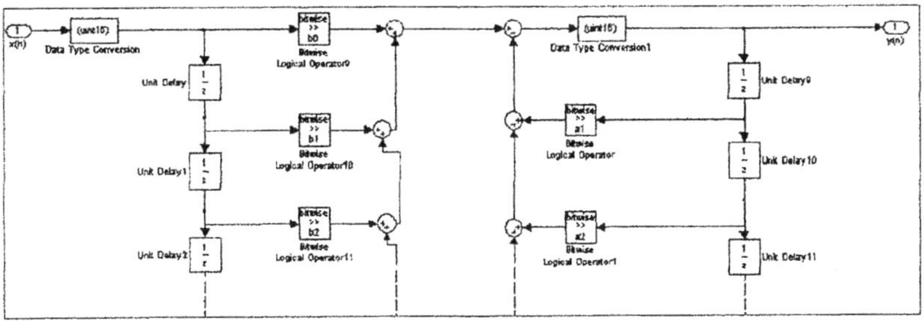

Fig. 5.1: Block diagram of multiplierless IIR system (cascade model).

Since they represent granular logic units within an FPGA, our primary analysis is conducted at the macro cell level. Hardware costs in number of macro cells for different variants of this IIR system are recorded in table 5.1. Here we see that a 10th order multiplierless system still outperforms a lower order (3) multiplier-based system.

Table 5.1: Hardware costs (in macro cell counts) for different implementations of IIR system.

Word Length	Multiplier-based				Multiplierless	
	Total Multiplier Cost		Total Cost		Total Shifter Cost	Total Cost
	3rd Order	10th Order	3rd Order	10th Order	10th Order	
n = 8	640	2,432	708	2,738	152	458
n = 16	2,560	9,728	2,689	10,322	304	898
n = 24	5,760	21,888	**5,892**	**22,770**	456	**1,338**

The multiplierless system cost is a linear function (order of n) of word length, whereas that of the multiplier-based approach is geometric (order of n^2). Hence, a multiplierless implementation represents an even better alternative for applications that require larger operands or higher levels of precision.

6. CONCLUSIONS

Our work saw the successful development of a genetic algorithm with several unique, application-oriented attributes, which is capable of optimizing filter coefficients such that the corresponding filter frequency response matches that of a desired (ideal) system with the constraint that all coefficients are powers-of-two and the resulting filter is stable. It has been shown how the genetically optimized multiplierless filters consistently yield image results comparable to those of their ideal counterparts. We may conclude that multiplierless-based image filtering is a viable design alternative, which is reliably implemented through the application of genetic algorithms.

REFERENCES

Goldberg, D. E., 1989, "Genetic Algorithms in Search Optimization and Machine Learning," Addison-Wesley, Boston.

Mitchell, M., 1996, "An Introduction to Genetic Algorithms," M.I.T. Press, Cambridge, MA.

Parker, J. R., 1997, "Algorithms For Image Processing And Computer Vision," John Wiley and Sons Inc., New York.

Proakis, J. G. and Manolakis, D. G., 1996, "Digital Signal Processing: Principles, Algorithms, and Applications", Third Edition, Prentice Hall Inc., Upper Saddle River, New Jersey.

Sarkar, S. and Boyer, K. L. 1989, "On Optimal Infinite Impulse Response Edge Detection Filters," *IEEE Transactions on Pattern Analysis and Machine Intelligence*, Vol. 13, No. 11, pp. 1154–1164.

(Full Paper Available at *www.vu.union.edu/~robleec/capstone*)

A Genetic Algorithm Search for Improved Halftone Masks

Peter G. Anderson[1], Jonathan S. Arney, Samuel A. Inverso, Daniel R. Kunkle, Timothy M. Lebo, and Chadd Merrigan
Rochester Institute of Technology
Rochester, New York, USA

Abstract

We present a genetic algorithm that automatically generates halftone masks optimized for use in specific printing systems. The search is guided by a single figure of merit based on a detailed model of the printing process and the human visual system. Our experiments show that genetic algorithms are effective in finding improved halftone masks and that two methods of reducing the search space to particular subsets of possible halftone masks greatly enhance the performance of the GA.

1. Introduction

Techniques for rendering excellent black and white approximations of grayscale images, called digital halftoning, are already abundant [8]. However, the reproduction of these halftone images by electrophotographic printers is not absolutely faithful, and the quality of each halftoned image is affected nonlinearly and in a manner specific to the particular printer used. The development of halftoning algorithms and masks has traditionally been a manual process, however, and thus it is too expensive to afford the creation of a specific mask for every printing system.

A number of efforts have already made use of genetic algorithms in designing halftone masks, including [1, 2, 7]. These evolve better masks by the reproduction, recombination, and modification over time of the best masks, as determined by a measure of their fitnesses. We use a fitness measure that can determine the quality of a given halftone mask relative to a specific printing system, given a few specific parameters obtained through a print sample analysis. Our present work focuses on the behavior of electrophotographic (laser) printers. The other significant improvements presented here are two methods of reducing the search space to a smaller subset of masks, all with above average fitness.

Digital halftone masks are represented as $H \times W$ grids containing each of the numbers $1 \ldots H \times W$ exactly once, though in any order. For example, the following spiral-dot mask:

$$\begin{bmatrix} 13 & 12 & 11 & 10 \\ 14 & 3 & 2 & 9 \\ 15 & 4 & 1 & 8 \\ 16 & 5 & 6 & 7 \end{bmatrix}$$

To generate a halftone image using a mask, the following process is executed, at least conceptually.

1. The mask grid is tiled to completely cover the image.
2. Pixels in the continuous image with gray levels less than or equal to the corresponding number in the mask which covers it become 0 in the halftone image; those greater become 1.

In practice, the mask or the image must first be scaled to the range of the other (usually $[0, 1]$ or $[0, 255]$), and step one can be simulated by a modulo operation in step two.

The best halftone images present the least noticeable disturbance in the continuity of a flat gray image to the eye, visible as such artifacts as worms, tears, and checkerboards. Due to the combined effects of neighboring pixels, random toner scattering, transmittance of light through ink, and other nonlinear effects, many halftone masks which appear to deliver very smooth images in their ideal form deliver very poor output in reality. Our fitness model as described in section 3 accounts for these factors to ensure that the masks selected for reproduction will provide the highest quality end result.

Section 4 presents two methods of creating halftone masks from a genotype, the *gear wheel* method and the *hybrid* method. Using either of these methods limits the search space to a subset containing only individuals with relatively high fitnesses and significantly improves the performance of the GA, though the hybrid method clearly provides the greatest enhancement.

2. The Genetic Algorithm

The GA we use is generational, based in part on Goldberg [5].

[1] Send correspondence to pga@cs.rit.edu. Internet: www.cs.rit.edu/~pga

1. A population of n random possible solutions is created.
2. The fitness of each individual (mask) is evaluated.
3. Repeat the following steps $n/2$ times to create a new generation.
   (a) Choose two parents using tournament selection.
   (b) With probability p_c, crossover the parents to create two children; otherwise clone parents to create children.
   (c) With probability p_m for each child, mutate that child.
   (d) Place the two new children into the next generation.
4. Repeat new generation creation until a solution is found or the process times out.

Some typical parameters for this genetic algorithm are $n \in \{20, 30, ..., 160\}$, $p_c = 0.8$, and $p_m = 0.3$. In this case, a mutation refers to the randomization of one location of the individual's genome.

The individuals in this case are permutations which map to halftone masks by any of our three methods detailed in section 4. A permutation must contain each member of a set in all cases, however, and a simple mutation by randomizing one element in a permutation is guaranteed to affect this complete inclusion. Thus they are represented in the genome by a signature which describes a sequence of reorderings of a simpler sequence of ordered integers. The signature can thus be mutated without affecting the inclusion of each member of the set in the final permutation.

These masks are evaluated according to the fitness function described in section 3. By continually selecting masks with higher fitness and creating new masks through the genetic operators of crossover and mutation the population of masks increases in quality. This procedure is usually given a specific amount of time and is then halted, at which time the best mask found is reported.

3. A Unified Figure of Merit

An earlier iteration of this project employed a multi-objective fitness function to optimize the printer noise level and the visual response separately. As with most multi-objective problems, it is difficult to find an ideal balance between the importance of each objective. Furthermore, variations which will increase the fitness on one objective will often decrease that on the other, as was the case with our previous work.

The former two-part function has been replaced by a single figure of merit which accounts for the significant effects of the printing process on the image. As before, the fitness is based on the noise in a uniform gray image after the appropriate transformations of the printing system and the human visual system are performed.

To calculate the noise level perceived by the eye after printing using an electrophotographic printer, we begin with a continuous grayscale image of some value in the range $[0, 1]$. This image is then halftoned using the mask to be evaluated to produce our basic image M.

The image M is super-sampled by a factor of 5 to produce a new image C with pixels small enough to represent the effects within one addressable printer pixel:

$$C_{i,j} = M_{\lfloor i/5 \rfloor, \lfloor j/5 \rfloor} \tag{1}$$

for $0 \leq i < 5H$ and $0 \leq j < 5W$, where H and W are the height and width of the original image M.

3.1. Dot Gain

Dot gain is a phenomenon by which an idealized image composed of black and white squares is altered by the facts that (1) printers produce round dots, not square, which must be larger in area than the ideal square in order to ensure full coverage of printer toner when the image is completely black, and (2) electric field interactions modify the charge delivered to the printer drum, especially toward the edges of each dot. Our method models this phenomenon by convolving the super-sampled image C with an experimentally derived point spread function PSF_m

$$C = M * PSF_m \tag{2}$$

where

$$\sigma = .034 \times (printer\ dots\ per\ millimeter) \times 5$$

and

$$PSF_m \propto \frac{1}{1 + (\frac{r}{\sigma})^4}$$

where r represents the distance from the center point of the PSF. The actual PSF_m is normalized to sum to 1, as are all of the point spread functions described here.

3.2. Toner Transfer

The transfer of toner to the paper is also a nonlinear phenomenon. Generally, there is a minimum charge on the photoconductive drum which is required to pick up and deposit any toner at all. Our model approximates this relationship by the equations

$$C_f = C^2 \tag{3}$$

and

$$C_n = C^2 + 2Ca \quad (4)$$

where a is a Gaussian random number near 0 with standard deviation 1.

Equation 3 calculates a simple relationship between charge level and actual toner transferred at any one point (i, j), recalling that the gray levels may range from 0 to 1. Equation 4, for C_n, represents the same relationship but adding the noise due to the intrinsic random toner particle scattering that is ubiquitous in the printing process. The two equations are necessary to be able to cache the random noise which results from this printer phenomenon and to add it to our final noise figure.

3.3. Paper and Ink Effects

With two separate images, C_n and C_f, we then calculate the actual reflectance from the printed image. Both calculations are performed as follows, here represented by only the noisy image C_n. First the transmittance of light through the ink is determined:

$$T_k = e^{-\epsilon \cdot n} \quad (5)$$

where $R_k = 0.05$ (reflectance of the ink), $R_g = 0.8$ (reflectance of the paper), and $\epsilon = -log(\sqrt{\frac{R_k}{R_g}})$. The result is convolved with the paper point spread function, PSF_p, and the overall reflectance R is calculated by the standard equation.

$$T_p = T_k * PSF_p \quad (6)$$

$$R = R_g \cdot T \cdot T_p \quad (7)$$

Two standard deviations are taken as a measure of the noise of each image, σ_n and σ_f. The difference of these two represents the noise attributable to the printer's random scattering effects.

$$\sigma_{printer} = max(\sigma_n - \sigma_f, 0) \quad (8)$$

3.4. Visual Effects

The human visual system can only perceive frequencies between alternations of light and dark with a minimum wavelength of $\frac{1}{150}$ of an inch [4]. Therefore, wavelengths of lesser lengths must be blurred out of the image if a true prediction of the eye's response is to be made. In our model, we use a simple Gaussian point spread function PSF_v (VTF in its familiar frequency-space form) to perform this blurring on the image without printer noise.

$$R_v = R_0 * PSF_v \quad (9)$$

Another standard deviation σ_v is extracted from this final image R_v.

3.5. Merit

The printer noise $\sigma_{printer}$ which was calculated in equation 8 is added to σ_v to provide a final noise estimation. Recalling that σ is a square root of the variance, the noises are combined:

$$\sigma = \sqrt{\sigma_v^2 + \sigma_{printer}^2} \quad (10)$$

The final figure of merit produces a number conveniently between 0 and 1, with 1 representing the highest fitness.

$$fig = \frac{\sigma_{max} - \sigma}{\sigma_{max}} \quad (11)$$

where σ_{max} represents the noise level resulting from an unblurred image, $(R_g - R_k) \cdot \sqrt{g(1-g)}$ where g is the testing gray level.

Because this maximum value varies with the gray level of the testing image, the fitness value is partially dependent upon the testing gray level. Early experiments using a different gray level for each generation used the iterative value $g_{next} = (\Phi g)\ mod\ 1$, where Φ is the golden ratio $\frac{\sqrt{5}-1}{2}$. Though this guaranteed an even coverage of all gray values over time, the fitness values varied wildly between generations.

Instead, the algorithm averages the fitnesses calculated from five test images of representative gray values chosen from the harmonic series $\frac{1}{2}, \frac{1}{3}, \frac{1}{4}, ...$ but alternating between those fractions and one minus those fractions, that is $\{.5, .33, .75, .2, .833\}$.

For further details see [3].

4. Methods of Reducing the Search Space

A number of methods for representing and creating masks were developed. Three methods, full permutation, gear wheel, and hybrid, are presented here. The gear wheel and hybrid methods reduce the search space by limiting the genetic algorithm to masks that are of better than average fitness; the full permutation method is provided as an unoptimized baseline.

4.1. Full Permutation

The full permutation method simply evolves masks over the full search space of permutations. For masks of size $H \times W$ the search space contains $(HW)!$ possible halftone masks. Normal mask sizes for the presented results are 64×64, resulting in an enormous search space of size $4096! \approx 1.2 \times 10^{13019}$.

This naive method is not efficient; it is merely described here as a contrast to the more intelligent methods described below. Figure 1 shows the evolution of a GA population over 12 hours of run time on a 440MHz Sun Ultra-10 workstation, or about 13,000 fitness evaluations.

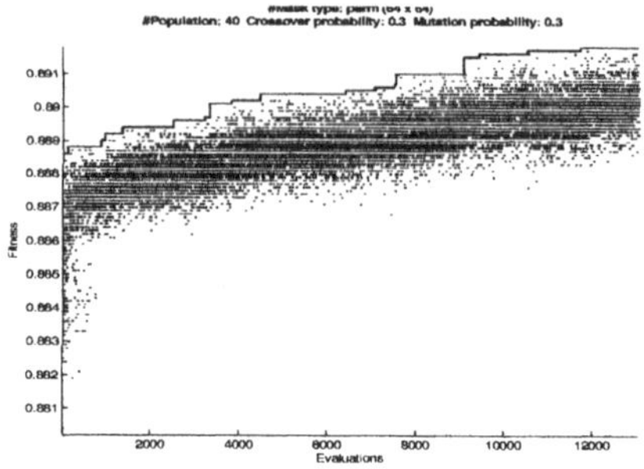

Figure 1: Full Permutation Method results - Every fitness value plotted against the point at which it occurred during GA evolution. The solid line represents best fitness found as of a given evaluation.

4.2. Gear Wheel

The gear wheel method evolves orderings of the two sets $\{1, 2, 3, ..., H\}$ and $\{1, 2, 3, ..., W\}$. This reduces the search space to only $H!W!$ possibilities, approximately 2.5×10^{176} times smaller than the space searched by the full permutation method for a similarly sized mask.

The two permutations are used to create one permutation of size $H \times W$ that represents a halftone mask. Conceptually, the two permutations are placed on separate gear wheels with tooth-counts H and W, and the wheels are turned one tooth at a time. The two numbers that meet where the two wheels touch describe the row and column in the mask where the next number in the permutation is placed. For this method to cover every coordinate pair of rows and columns, the height and width of the mask must be relatively prime (i.e., there is no number, other than 1, that is a factor of both H and W). By turning the H gear wheel W full rotations, the whole mask is filled. The table below shows the process of creating a 3×4 mask. The teeth on the hypothetical H and W gears are numbered in this instance by the trivial permutations $\langle 1, 2, 3\rangle$ and $\langle 1, 2, 3, 4\rangle$. Each numbered tooth in the H gear represents a row in the mask, and each in the W gear represents a column. Note that by the end of this table, each possible combination of H and W has appeared exactly once.

H Gear (row)	W Gear (column)	Mask value
1	1	1
2	2	2
3	3	3
1	4	4
2	1	5
3	2	6
1	3	7
2	4	8
3	1	9
1	2	10
2	3	11
3	4	12

The resulting mask has the following structure:

$$\begin{bmatrix} 1 & 10 & 7 & 4 \\ 5 & 2 & 11 & 8 \\ 9 & 6 & 3 & 12 \end{bmatrix}$$

Although gear wheels give a good conceptual visualization of this method, it is computationally impractical in our implementation language, MATLAB. An equivalent but more efficient algorithm is to initialize an $H \times W$ matrix using the two trivial permutations, as shown above, then permute the rows and columns of this matrix using the two chosen permutations. This allows us to create a trivial permutation mask like the one in the table above at the beginning of a GA run and use the row and column permutations when calculating the fitness of each possible mask.

Figure 2 shows the evolution of the GA population over 12 hours and approximately 8000 fitness evaluations. The best fitness values found using the gear wheel permutation method were 0.9428.

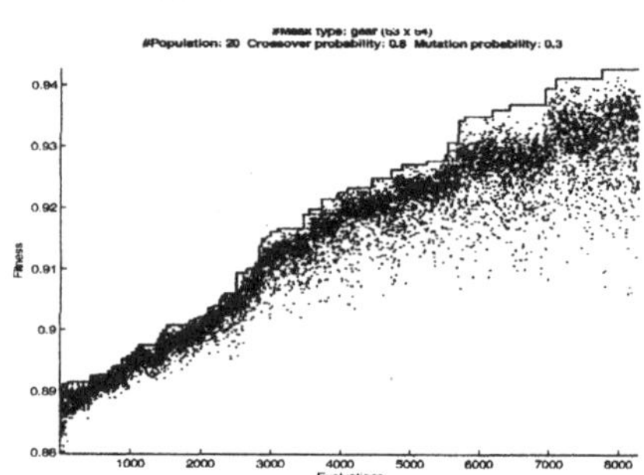

Figure 2: Gear Wheel Method results - Every fitness value plotted against the point at which it occurred during GA evolution.

4.3. Hybrid

The hybrid method simulates the construction of hybrid masks [6], which tend to be successful, using recursive tesellations. In this construction, two or more masks are combined recursively to produce the final mask which has dimensions equal to the product of the component masks' dimensions.

The combination of component masks M and N of size $h_1 \times w_1$ and $h_2 \times w_2$ respectively will result in a mask of size $h_1 h_2 \times w_1 w_2$ as shown below.

$$R = M \star N \tag{12}$$

$$R = \begin{bmatrix} (hw)M + N_{11} & \cdots & (hw)M + N_{1w} \\ \vdots & \ddots & \vdots \\ (hw)M + N_{h1} & \cdots & (hw)M + N_{hw} \end{bmatrix} \tag{13}$$

where h and w are the height and width of N, and N_{pq} is the value of N at row p and column q.

Visually, the matrix combination can be seen as tiling $h_2 \times w_2$ copies of the multiple $h_2 w_2 M$, where the (p, q) copy is offset by N_{pq}.

This process can be repeated with any number of matrices to produce the final mask whose fitness is evaluated as described earlier, using the visual and printer fitness measures.

For the results listed here, two component matrices, both of size 8×8 were used, producing a mask of size 64×64. The method ran for 12 hours computing approximately 12,000 fitness evaluations. Figure 4 shows the results of a GA using this method. GAs using this method consistently provide better results than either the full permutation or the gear wheel methods. (This search space has approximately 1.6×10^{178} elements.)

Using the hybrid tiling method described above, any number of component matrices of any size can be used to create a mask. Tests were conducted with all possible combinations of powers of two whose product is 64. There are 31 such combinations, arising from the 10 sets of dimensions and the permutations of those sets. The best performing combination has been demonstrated to be $(8 \times 8) \star (8 \times 8)$.

Figure 5 compares the results of the full permutation, gear wheel, and hybrid methods. The hybrid method produces masks with significantly higher fitness than the other methods. Specifically, hybrid methods whose first component matrix is 8×8 perform the best.

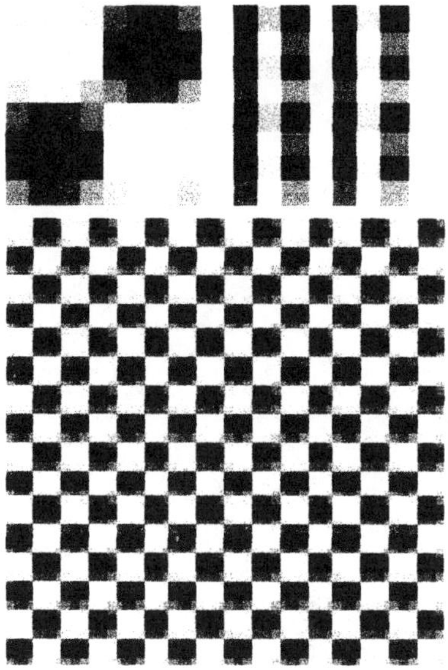

Figure 3: A hand designed micro-mask (top-left), an evolved macro-mask (top-right), and the resulting hybrid mask (bottom).

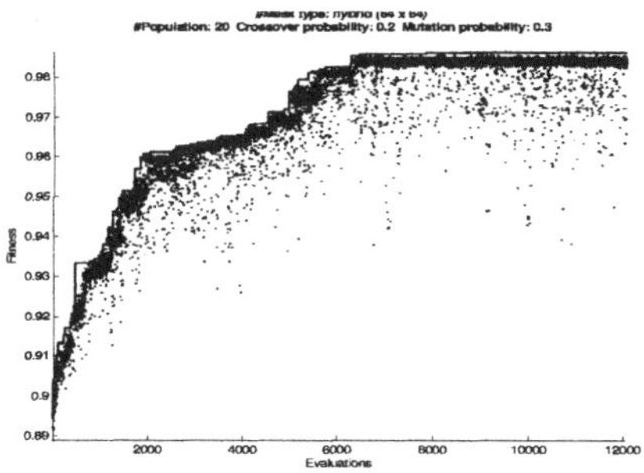

Figure 4: Hybrid Method results - Every fitness value plotted against the point at which it occurred during GA evolution.

5. Results

The images below show selected halftone ramp images for some of the highest fitness masks created by the three methods presented.

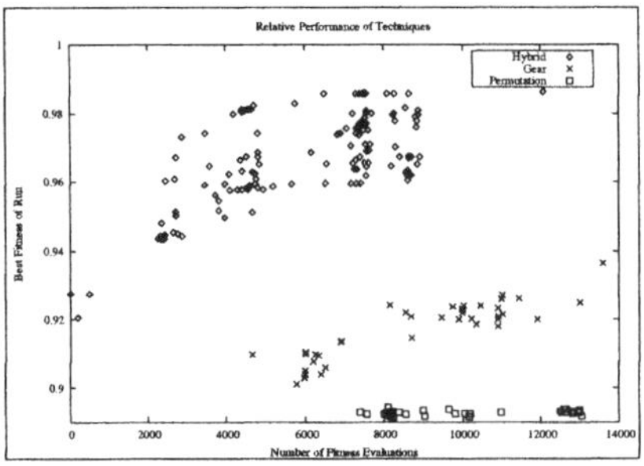

Figure 5: Comparison of best fitnesses using full permutation, gear wheel, and hybrid methods including power of 2 products. Masks with greater fitness appear in upper right of graph. The hybrid method consistently outperforms the other methods.

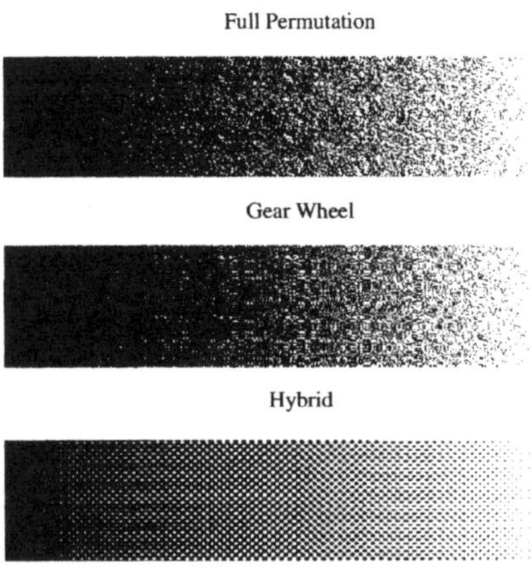

Figure 6: Horizontal gray ramps halftoned with masks found using the genetic algorithm.

It should be remembered that at the printed level of magnification, and without the effects of the printing process shown, it is difficult to distinguish a successful halftone mask from a casual glance at the images in figure 6.

6. Conclusion

Genetic algorithms are a valuable tool in designing halftone masks for specific printing systems. Their effectiveness is significantly increased by reducing the search space to masks with above average fitness. Ongoing work focuses on the development of even more accurate printer models.

7. Acknowledgments

We gratefully acknowledge the financial support of Hewlett Packard for this research.

References

[1] Hernán E. Aguirre, Kiyoshi Tanaka, Tatsuo Sugimura, and Shinjiro Oshita. Halftone image generation with improved multiobjective genetic algorithm. In Eckart Zitzler, Kalyanmoy Deb, Lothar Thiele, Carlos A. Coello Coello, and David Corne, editors, *First International Conference on Evolutionary Multi-Criterion Optimization*, pages 501–515. Springer-Verlag. Lecture Notes in Computer Science No. 1993, 2001.

[2] Jarmo T. Alander, Timo Mantere, and Tero Pyylampi. Threshold matrix generation for digital halftoning by genetic algorithm optimization. In David P. Casasent, editor, *Intelligent Systems and Advanced Manufacturing: Intelligent Robots and Computer Vision XVII: Algorithms, Techniques, and Active Vision*, pages 204–212, 1998.

[3] Peter G. Anderson. Optimizing halftone masks with genetic algorithms and a printer models. In *Proceedings of the IS&T Digital Printing Conference (NIP-19)*, September 2003.

[4] Peter Engeldrum. *Introduction to Image Microstructure*. Imcotek, Inc., Walworth, NY, 1986.

[5] D. E. Goldberg. *Genetic Algorithms in Search, Optimization & Machine Learning*. Addison-Wesley, Reading, MA, 1989.

[6] H. R. Kang. Frequency analysis of microcluster halftoning. In R. Eschbach, editor, *Recent progress in digital halftoning, II*. IS&T, 1999.

[7] Chih-Ching Lai and Din-Chang Tseng. Printer model and least-square halftoning using genetic algorithms. *Journal of Imaging Science and Technology*, 42(3):241–249, 1998.

[8] Robert Ulichney. *Digital Halftoning*. MIT Press, Cambridge, MA, USA, 1987.

A TABU SEARCH BASED APPROACH TO ROBUST STATE ESTIMATION IN RADIAL POWER DISTRIBUTION SYSTEMS

NORIYUKI KOSEMURA
Tokyo Electric Power Co., Inc
Chiyoda-ku, Tokyo 100-8560
Japan

HIROYUKI MORI
Dept. of Electrical and Electronics Eng.
Meiji University
Tama-ku, Kawasaki 214-8571
Japan

ABSTRACT
This paper proposes a robust static state estimation method for radial electric power distribution systems. In recent years distribution system operators are interested in improving the security function in distribution systems. In this paper the nested structure of the state variable is used to estimate the state variable of the distribution system power flow calculation. The problem of state estimation may be expressed by the minimization of noise of state variables at the substation node. That allows us to carry out static state estimation with a limited set of measurements. This paper presents a new robust algorithm for reducing noise and suppressing bad data. The conventional optimization is not applicable due to the complicated formulation. Meta-heuristics are very efficient for solving a combinatorial optimization problem. As one of meta-heuristics, this paper makes use of tabu search to solve the formulation. The proposed method is successfully applied to the 69-node distribution system. A comparison between the proposed and conventional methods is made to demonstrate the effectiveness of the proposed method.

INTRODUCTION

This paper proposes a robust static state estimation algorithm in radial distribution systems. Recently it is important to improve the security function in power distribution systems with the progess of distribution automation. In particular state estimation plays a key role to obtain reliable state variables. Static state estimation in distribution systems is quite different from that in transmission systems in a way that available measurements are extremely limited in distribution systerms. That is because the power flow calculation of distribution system called DistFlow has a feature that it calculates the state variables from the top node to terminal nodes successively (Baran and Wu, 1989). So far, many methods on state estimation have been deveropped in disribution systems. A model with the Kirchhoff's law was proposed to reduce nosise (Roytelman and Shahidehpour, 1993). Afterwards a method was developed to improve the quality of load data with state estimation and measured values (Baran and Kelly, 1994). Also the algorithm with the zero-injection constraints was presented for three-phase distribution systems (Lin and Teng, 1995). The methods above are based on the least squared method corresponding to the the ℓ_2 -norm based state estimation. However, the ℓ_2 -norm does not work so well in case where the measurements include bad data that implies measurements with gross errors. As a result, static state estimation with the ℓ_2-norm needs additional tasks to detect and reject bad data. On the other hand, it is known that the ℓ_1-norm is not significantly affected by bad data (Barrodale and Roberts, 1973; Kotiuga and Vidyasagar, 1982). Thus the ℓ_1-norm based static state estimation results in a robust state estimator.

In this paper static state estimation with the ℓ_1-norm is introduced to DistFlow based state estimation. The problem to be solved is to determine only the state variables at the substation node through DistFlow based state estimation(Mori and Ogita, 1999). It is necessary to evaluate the global optimal solution in minimizing the noise at the sustation node. The porposed method uses tabu search (TS) (Glover,1989) of meta-heuristics to estimate the state variables at the subsustation node. Meta-heuristics is useful for evaluating a globally optimum solution or its high approximate solutions(Osman and Kelly, 1996). TS is compared with other meta-heuristics methods such as simulated annealing (SA)(Kirkpatrick, et al., 1983) and genetic algorithm (GA) (Goldberg,1989) in terms of solution quality and computational effort. The proposed method is successfully to the 69-node sample system.

STATE ESTIMATION WITH ℓ_1-NORM

This section describes the ℓ_1-norm based static state estimation. Mathematically, static state estimation may be defined as a problem that determines the state variables from a given measurements. The objective is to minimize the errors between mesurements and estimates. Traditionally, most of the conventional methods make use of the least squared method (ℓ_2-norm). The cost function may be expressed as:

$$J_2 = \sum_{i=1}^{m} (z_i - h_i(x))^2 \qquad (1)$$

where,

J_2: cost function of ℓ_2-norm

z_i: *i-th* measurement

$h_i(\cdot)$: *i-th* measurement equation

x: state variable

m: number of measurements

zi: *i-th* measurement

Recently, distribution automation becomes more popular to carry out secure distribution system operation and planning. Thus, it is neccessary to improve the quality of state estimation and evaluate reliable estimate. To make state estimation more robust, the ℓ_1-norm is useful to obtain more reliable estimates. The cost function may be written as

$$J_1 = \sum_{i=1}^{m} |z_i - h_i(x)| \qquad (2)$$

where,

J_1: cost function of ℓ_1-norm

$|\cdot|$: absolute value of $\cdot$

It should be noted that the ℓ_1-norm and ℓ_2-norm correspond to the absolute error and squared error, respectively. The advantage of the ℓ_1-norm over the ℓ_2-

norm is to detect and reject bad data. Therefore, the ℓ_1-norm allow to realize a robust static state estimation including bad data. On the other hand, the ℓ_2-norm often deteriorates estimates for such a case.

DISTFLOW

This section outlines DistFlow that means power flow calculation tailored for radial distribution systems. Nonlinear equation $h_i(x)$ in Eqs.(1) and (2) correspond to the power flow equation of DistFlow. It determines only the state vectors with the top node of main feeder and laterals. Fig. 1 shows an example of a radial distribution system. Now, consider the main feeder in the figure. The relationship between nodes i and i+1 may be expressed by the following equation:

$$P_{0(i+1)} = P_{0i} - r_{0(i+1)} \frac{P_{0i}^2 + Q_{0i}^2}{|V_{0i}|^2} - PL_{0(i+1)} \tag{3}$$

$$Q_{0(i+1)} = Q_{0i} - x_{0(i+1)} \frac{P_{0i}^2 + Q_{0i}^2}{|V_{0i}|^2} - QL_{0(i+1)} \tag{4}$$

$$|V_{0(i+1)}|^2 = |V_{0i}|^2 - 2\left(r_{0(i+1)}P_{0i} + x_{0(i+1)}Q_{0i}\right) + \left(r_{0(i+1)}^2 + x_{0(i+1)}^2\right)\frac{P_{0i}^2 + Q_{0i}^2}{|V_{0i}|^2} \tag{5}$$

where,
$P_{0(i+1)}$ $(Q_{0(i+1)})$: active (reactive) power flow at node i+1
P_{0i} (Q_{0i}): active (reactive) power flow at node i
$r_{0(i+1)}$ $(x_{0(i+1)})$: resistance(reactance) between nodes i and i+1
$|V_{0(i+1)}|$: nodal voltage magnitude at node i+1
$|V_{0i}|$: nodal voltage magnitude at node I
$PL_{0(i+1)}$ $(QL_{0(i+1)})$: active (reactive) power load at node i+1
$_0$: denoting main feeder

The state variables are successively calculated from the top node of main feeder and the terminal nodes by Eqs. (3), (4) and (5). The power flow node becomes zero at the terminal. This constraint is called termnimal conditions. Once DistFlow determines the state variables at the top node, the state variables at the neighboring node are successivelly determined.

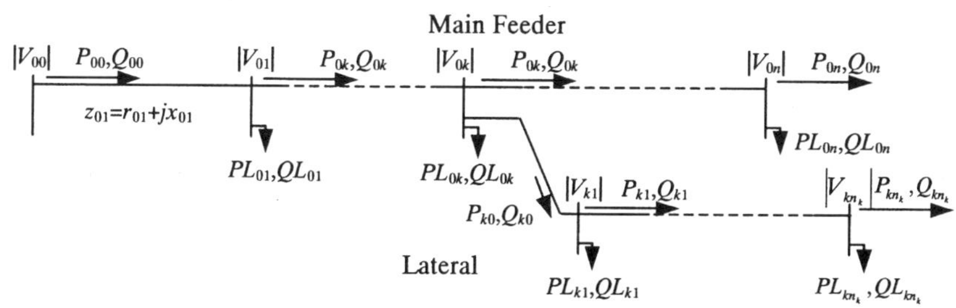

Fig. 1 Radial Distribution System

TABU SEARCH

This section describes the ouline of tabu search (TS). It is one of meta-heuristic methods for solving a combinatorial optimization problem. The basic mechanism of TS introduces an adaptive memory called tabu list into the neighborhood search on the hill-climbing method (HCM). HCM iteratively carries out local search in a neighborhood around the solution. However, HCM easily gets stuck in a local minimum. On the other hand, TS can continue the search process to escape from a local minimum with the function of the adaptive memory. The tabu list plays an important role to prevent the search process from the short-term circulation of solutions through storing some attributes and making them unchangeable for a while. The attributes are updated in the tabu list at each iteration. Once a new attribute enters into the tabu list and becomes fixed, the oldest one is released from it and becomes free. The period of keeping an attribute in the tabu list is refered to as the tabu length. It is necessary to tune up the parameter so that better results are obtained.

PROPOSED METHOD

This paper proposes an intergation method of the ℓ_1-norm and DistFlow for robust static state estimation in distribution systems. The proposed method does not need the additional tasks on the bad data detection and suppression. The ℓ_1-norm is useful for handling the bad data detection and suppression. DistFlow determines state variables with only state variables at the substation through the nested structure. The relationship between the substation node and other nodes may be written as follows:

$$\boldsymbol{x}_{i1} = \boldsymbol{h}_{i1}(\boldsymbol{x}_{i0}+\boldsymbol{\varepsilon}_i)=\boldsymbol{g}_{i1}(\boldsymbol{x}_{i0}+\boldsymbol{\varepsilon}_i) \tag{7}$$

$$\begin{aligned}\boldsymbol{x}_{i2} &=\boldsymbol{h}_{i2}(\boldsymbol{x}_{i1}) \\ &=\boldsymbol{h}_{i2}(\boldsymbol{h}_{i1}(\boldsymbol{x}_{i0}+\boldsymbol{\varepsilon}_i))=\boldsymbol{g}_{i2}(\boldsymbol{x}_{i0}+\boldsymbol{\varepsilon}_i)\end{aligned} \tag{8}$$

$$\vdots$$

$$\boldsymbol{x}_{in} = \boldsymbol{h}_{in}(\boldsymbol{x}_{i(n-1)})=\boldsymbol{g}_{in}(\boldsymbol{x}_{i0}+\boldsymbol{\varepsilon}_i) \tag{9}$$

where,

$\boldsymbol{x}_{ik}$: state vector at node k

$\boldsymbol{g}_{ij}(\cdot)$: nonlinear function vector mapping x_{i0} to x_{ij}

$h(\cdot)$: nonlinear function corresponding to DistFlow

The observation equation at node i may be written as

$$\boldsymbol{z}_{ij}=\boldsymbol{g}_{ij}(\boldsymbol{x}_{i0}+\boldsymbol{\varepsilon}_i)+\boldsymbol{r}_{ij} \tag{10}$$

where,

z_{ij} : measurement vector at node j

$\boldsymbol{r}_{ij}$: noise vector of z_{ij}

The state veariables at the sustation node and the source node of laterals are determined to minimize noise vector $\boldsymbol{r}_{ij}$ in Eq. (10). Thus, the cost function may be wriiten as

$$f_1 = \sum_{i=1}^{m}|r_i| + \sum_{i=1}^{k}\left(|P_{in_i}| + |Q_{in_i}|\right) \rightarrow \min \tag{11}$$

The first term means the minimization of the noise vector. Also, the second term implies the boundary conditions of the terminal conditions. The proposed method makes use of the ℓ_1-norm to remove effects of bad data. Therefore, it gives accurate estimates even if the measurement vector has bad data. To improve the solution quality, this paper minmizes Eq. (11) with TS so that more reliable state variables are obtained.

SIMULATION

Simulation Conditions

(i) Fig. 2 shows the 69-node sample system, where the black and gray squares imply meters without and with bad data.

(ii) Thirty sets of measurements noises following the normal distribution with the mean of zero and the standard deviation of 0.1 are prepared to add to the true value obtained by the power flow calculation, where bad data occurs at node 3 in Fig. 2. For convenience, the following cases are defined:

Case 1: measurements without bad data

Case 2: measurements with bad data

(iii) This paper makes a comparison between the ℓ_1-norm and ℓ_2-norm. In contrast with Eq. (11), the cost function with ℓ_2-norm may be given as follows:

$$f_2 = \sum_{i=1}^{m} r_i^T r_i + \sum_{j=1}^{k}\left(|P_{jn}| + |Q_{jn}|\right) \tag{12}$$

(iv) Eqs. (11) and (12) are minimized with TS in comparison with simulated annealing(SA) and genetic algorithm(GA).

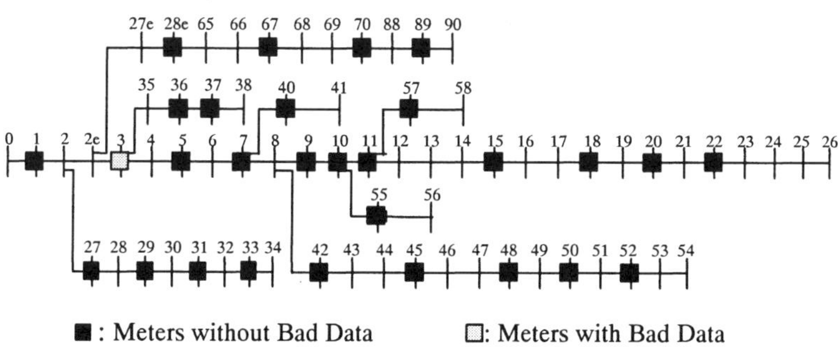

Fig. 2 69-node Distribution Systems

Table 1 Average Errors of Each Method

Methods		Case 1			Case 2		
		P	Q	|V|	P	Q	|V|
TS	ℓ_1-norm	0.64	0.94	1.02	0.56	1.06	1.31
	ℓ_2-norm	1.92	1.90	0.62	9.77	6.98	4.25
GA	ℓ_1-norm	1.07	1.04	0.96	1.11	1.18	1.27
	ℓ_2-norm	1.99	2.04	0.64	9.37	6.20	4.21
SA	ℓ_1-norm	2.82	2.66	1.57	3.54	3.01	1.92
	ℓ_2-norm	5.37	5.78	1.21	12.6	18.0	4.35

Table 2 Maximum Errors of Each Method

Methods		Case 1			Case 2		
		P	Q	|V|	P	Q	|V|
TS	ℓ_1-norm	1.64	2.88	3.13	2.14	2.88	3.52
	ℓ_2-norm	9.27	7.07	2.34	11.6	7.07	7.03
GA	ℓ_1-norm	2.28	2.99	3.16	3.29	3.69	3.94
	ℓ_2-norm	7.52	7.30	2.72	23.0	11.2	7.64
SA	ℓ_1-norm	8.61	8.97	4.33	15.1	10.7	4.14
	ℓ_2-norm	13.9	14.9	3.54	25.5	29.4	7.35

Table 3 Standard Deviations of Each Method

Methods		Case 1			Case 2		
		P	Q	|V|	P	Q	|V|
TS	ℓ_1-norm	0.63	0.57	1.24	0.66	0.69	1.36
	ℓ_2-norm	1.62	1.89	0.80	3.12	0.33	1.71
GA	ℓ_1-norm	2.09	0.68	1.13	2.13	0.83	1.34
	ℓ_2-norm	3.26	2.47	0.86	5.17	2.82	1.87
SA	ℓ_1-norm	4.80	4.44	2.24	4.72	3.92	2.42
	ℓ_2-norm	8.33	8.72	1.97	6.24	6.54	1.65

Table 4 Computational Time of Each Method

Methods	Computational Time (ms)
TS	1513
GA	4558
SA	5143

(v) All the calculations were performed on the Fujitsu Workstation S-7/400Uim270D(Ultra-5, SPECint 95: 9.1, SPECfp 95: 10.1).

Simulation Results

Table 1 shows the average and maximum errors of state variables at the substation node of main feeder. Looking over Case 1, it can be seen that the ℓ_1-norm has much smaller average and maximum errors than the ℓ_2-norm. Regarding Case2, the ℓ_2-norm gives much worse errors because of the existence of bad data. For instance, the average errors of real power (P) for the ℓ_1-norm and ℓ_2-norm are 0.64 and 1.92 in Case 1, respectively. In Case 2, the average erros of ℓ_1-norm and ℓ_2-norm are 0.56 and 9.77, respectively. Table 2 gives the maximum errors of each method. The difference between the ℓ_1-norm and ℓ_2-norm is that the ℓ_1-norm is more robust than the ℓ_2-norm in Case 2 with bad data. Table 3 shows the standard deviation for each method. It can be observed that the standard deviation of the ℓ_1-norm are smaller than that of ℓ_2-norm. In particular, the large difference is recognized in Case 2. Thus, It has been shown that the ℓ_1-norm state estimator is more robust although the ℓ_2-norm is affected by the bad data easily.

Next, TS is compared with SA and GA. Looking over the errors of the average and maximum errors for the ℓ_1-norm, TS is bettern than GA and SA. In particular, SA is much worse than others. The results of Table 3 show that TS gives more reliable state etimates since it has the smallest standard deviation than others. Namely, TS is more robust for the measurement noise than GA and SA. Table 4 gives the computational time of TS, GA and SA in case of the ℓ_1-norm. It can be seen that TS is 3.01-times and 3.40-times faster than GA and SA, respectively. Hence, it can be seen that TS is not affected by the initial conditions and more efficient than SA and GA. Therefore, the simulation results have shown that the proposed method with the ℓ_1-norm and TS is superior to other methods in terms of the solution quality and computational time.

CONCLUSIONS

This paper has proposed the ℓ_1-norm based state estimator with DistFlow in radial electric power distribution systems. The proposed method has indicated the following:

(i) In this paper, the ℓ_1-norm based estimator with DistFlow has been developed to minimize the measurement noise with TS. The poroposed method has a feature to determine the state variables at the substation node of main feeder and laterals through the nested structure of DistFlow. It has been confirmed that the poroposed method was more effective for rejecting bad data. Thus, the proposed allows to realize a robust state estimator in radial distribution systems.

(ii) The proposed method was applied to the 69-node distribution system. The simulation results have shown that the ℓ_1-norm was better than the ℓ_2-norm in terms of the average and maximum errors as well as the standard deviation for thirty initial conditions. For example, the average errors of the ℓ_1-norm and ℓ_2-norm for real power are 0.64 and 1.92 in case without bad data while the average erros of the ℓ_1-norm and ℓ_2-norm are 0.56 and 9.77 in case with bad data. Also, the simulation results have shown that TS was better than GA and SA in terms of robustness and computational time. TS is 3.01-times and 3.40-times faster than GA and SA, respectively.

REFERENCES

Roytelman, I., and Shahidehpour, S. M., 1993, "State Estimation for Electric Power Distribution Systems in Quasi Real-time Conditions," *IEEE Trans. on Power Delivery*, Vol. 8, No. 4, pp. 2009-2015.

Baran, M. E., and Kelly, A. W., 1994, "State Estimation for Real-time Monitoring of Distribution Systems," *IEEE Trans. on Power Systems*, Vol. 9, No. 3, pp. 1601-1609.

Lin, W. M., and Teng, J. H., 1995, "State Estimation for Distribution Systems with Zero-Injection Constraints," *Proc. of PICA'95*, pp. 523-529.

Mori, H., and Ogita, Y., 1999, "Tabu Search Based Static State Estimation for DistFlow in Radial Distribution Systems," in Dagli, C.H., *et al.* (Eds.), 1999, "Intelligent Engineering Systems through Artificial Neural Networks," Vol. 9, pp.1033-1039, *ASME Press*, New York, U.S.A..

Barrodale, I., and Roberts, F. D. K., 1973, "An Improved Algorithm for Discrete ℓ_1 Linear Approximation," *SIAM J. Numer. Anal.*, Vol. 10, No. 5, pp. 839-848.

Kotiuga, W. W., and Vidyasagar, M., 1982, "Bad Data Rejection Properties of Weighted Least Absolute Value Techniques Applied to Static State Estimation," *IEEE Trans. on Power Apparatus and Systems*, Vol. PAS-101, No. 4, pp. 844-853.

Baran, M. E. and Wu, F. F., 1989, "Optimal Sizing of Capacitors Placed on a Radial Distribution System," *IEEE Trans. on Power Delivery*, Vol. 4, No. 1, pp. 735-743.

Glover, F., 1989, "Tabu Search, Part I," *ORSA Journal on Computing*, Vol. 1, No. 3, pp. 190-206.

Osman, I.H. and Kelly, J.P. (Eds.), 1996, "Meta-Heuristics: Theory & Applications," *Kluwer Academic Publishers*, Netherlands

Goldberg, D.E., 1989, "Genetic Algorithm in Search, Optimization and Machine Learning," *Addison Wesley Publishing Company.*

Kirkpatrick, S., Gelatto, C. D., and Vecchi, M. P., 1983, "Optimization by Simulated Annealing," *Science*, Vol. 220, No. 4598, pp. 671-680.

A PROPOSAL AND EVALUATION OF A PARALLELIZED IMMUNE CO-EVOLUTIONARY ALGORITHM FOR DIVISION-OF-LABOR PROBLEMS

NARUAKI TOMA
Okinawa Industry Promotion Public Corporation
Okinawa, Japan.

SATOSHI ENDO, KOJI YAMADA
University of the Ryukyus Department of Information Engineering, Faculty of Engineering
Okinawa 903-0213, Japan

ABSTRACT
Recently, we can get ready for large-scale computing environments such as PC-cluster as a low price. Hereby, some parallelized optimization algorithms are proposed. By now, we have proposed a co-evolutionary immune algorithm (IA) to solve the division-of-labor problems, in particular n-th agent's Travelling Salesman Problem (*n*-TSP). In this paper, we extend the co-evolutionary IA for large-scale *n*-TSP with (1) the improvement for the search speed through parallelized search on the PC-cluster, and (2) the introduction of a new division-processing ``pre-estimated division processing" to improve the search ability.

1 INTRODUCTION

Recently, we can get ready for large-scale computing environments such as PC-cluster as a low price. Hereby, some parallelized optimization algorithms [Theodore 1995, Starkweather, 1991, et al] are proposed: for example, parallel-distributed genetic algorithm (PDGA) and PDGA with distributed environment scheme (PDGA/DE). In PDGA, population was parallelized to improve not only the search cost but also the search ability. And PDGA/DE could the cost-saving for parameter setting.

By now, we have proposed a co-evolutionary immune algorithm (IA) to solve the division-of-labor problems, in particular n-th agent's Travelling Salesman Problem (*n*-TSP) [Toma2000]. The purposes of this paper are to construct a parallelized optimization system for large-scale combinatorial problems, which the system is based on an immune co-evolutionary algorithm, and to introduce a new search operator ``pre-estimated division processing". The co-evolutionary IA searches a best solution through the interactions between two kinds of agents, one of the agent is called immune agents, which optimize the cost of its own work. And our target problem is an *n*-TSP.

We extend the algorithm so as to solve large-scale problem by using parallel processing on PC cluster. Our parallelized system solves problems through the following three steps. Firstly, the target problem is divided to the sub problems depending on the area, which is a problem domain in the *n*-TSP. Secondary, each sub problem area is processed by the different PCs with the immune co-evolutionary algorithm. Thirdly, a main PC gathers the sub results from the other PCs, and then it adjust the optimality of the sub results as a best solution by using escape processing. To investigate the validity, we verify through some computer simulations.

2 ISSUES OF DIVISION-OF-LABOR PROBLEMS

The key subject of division-of-labor problems is to focus on the issues of distribution of works for agents. The solvers of the problems are defined as follows: the domain that each agent covers is defined as work domains (WD), and the aggregate total of the work domains is defined as problem domain (PD). Thère are two aims in the division-of-labor problems. They are: (a) each WDi that is assigned by agenti must be divided evenly and, (b) the system performs effective division-of-labor by optimizing each WDi. As a typical case problem of the division-of-labor, we deal with n-TSP to investigate an adaptation ability of our model. The objective is finding the minimal tour by division among salesman. Note that one condition that all salesmen use the same city as the starting point was set in our experiments.

3 PROPOSED METHOD

The proposed algorithm consists of the co-evolutionary IA [Toma2000] and the following extensions:

- improvement of the search speed through a parallelization of problem area
- improvement of the search ability through a introduction of a pre-estimated division-processing.

3.1 Overall procedure

In order to perform the improvement of the search speed, our algorithm solves the problem through the following procedure:

(1) separate the problem into some sub-problems based on the problem space,
(2) solve each and all sub-problem on the each node (PC) with the co-evolutionary IA, and
(3) adjust the balance among sub-solutions through the escape-processing on the main node.

Of course, to search any tours striding plural sub-problems will be prevented under the parallelization. We revel this aspect by (3), which gathers the sub-solutions from all nodes and adjusts the cost balance among them through the above-mentioned escape-processing.

3.2 Parallelization for problem area

Our algorithm separates the given problem (problem domain, which consists of allocations for all cities in n-TSP) and processes the separated sub-problems are on each node. The problem domain separates to 2x2 matrix whose each lattice consists of same size, so there are four uniform lattices at the first stage. When any lattice has cities more than MaxCity, which is a reserved word as a limitation number for cities on node, the lattice is separated to 2x2 matrix again until the number of cities is lower then MaxCity. There are many separation methods: clustering for cities, distance based separation, and so on. However, we used a simple separation way (above-mentioned) for problem domain in n-TSP because we want to make an problem independent algorithm.

3.3 The co-evolutionary IA [Toma 2000]

The co-evolutionary IA solves the problems through two searching ways, (1) division-and-integration processing by salesman agents and (2) escape processing by city agents in the environment. The procedures of the proposed algorithm against an n-TSP are described as below.

[Step1. Definition of problems and immune functions.] Cities and Salesmen as the problem must be defined. Salesman's unique ID and division-and-integration processing for tours of each salesman must be defined. At this point, each city has the ID for expression of salesman visiting the city.
[Step2. Calculation of objective function.] The cost of salesman is calculated by following function.

$$Cost(S_i) = \sum distance(tour_{S_i}) \qquad (1)$$

- S_i: Salesman i.
- $tour_{Si}$: the tour of Salesman i.

[Step3. pre-estimated division processing.] Each salesman divides own tour into two subtours to optimize (decrease) the tour cost. Figure 1-(A) shows an just image, the detail procedure is described on subsection '3.4 Pre-estimated division-processing'.
[Step4. integration processing.] Two salesmen try to integrate the tours for searching lower costs by following steps (see also figure 1-(B)): decide of any two tours, make new tour by binding, calculate cost of the new tour and then, if the cost is improved, the integration performed (as a consequence, a original salesman is deleted).
[Step5. Mutation.] Swap any two cities. If the cost, after swap, is improved, the swap will be performed.
[Step6. Calculation of objective function.] The city agent's cost (same as a cost of visited salesman) is calculated by the function (1).
[Step7. Escape processing.] This approach changes ID of a city depending on neighbors cost to process following steps. First, check costs of salesmen that visited neighbor cities of target city. Second, if the other salesman's cost lower than its cost, the city changes its ID into the other. As a consequence, in this example figure 2, the ID of target city is changed to $S2$, and then, the target city is visited by $S2$ like the right side of figure 2.

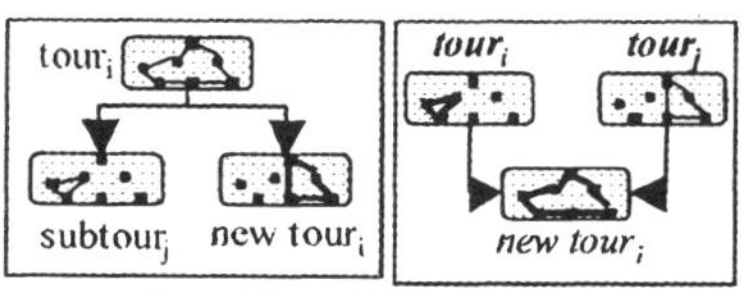

(A) Division (B) Integration

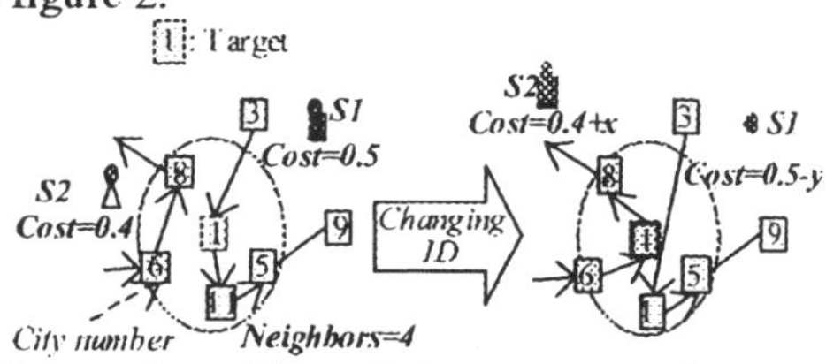

Figure 1: Examples of Division and Integration. Figure 2: Escape processing

3.4 Pre-estimated division-processing

The objective of division-processing, which is on step 3 in our algorithm, optimizes (decreases or minimizes) the cost through division of own tour into two subtours. For the purpose, we employed a kind of random way that decides a divided point at random and the judgment whether the division is achieved or not depends on a simulated annealing (a kind of climbing method with temperature, which the temperature denotes a possibility to go down; and the possibility is decreased step by step to converge the search). This old fashioned division-processing found better solutions, but it was a lack of a search ability for cost optimization, especially the number of cities on one salesman is over 100.

In this paper, we improve this point by introduce a new way ``pre-estimated division-processing" (an example shows on figure 3).

1. decides a part of own tour sub based on any two points.

At this time, the tour consists of *sub* and the back-and-force two cities call *Before1*, and other tour consists of a remained tour except for sub from *Before1* call *After1*.

2. finds the nearest city nearest from the head of sub.

3. calculates two candidate-tours' costs and finds the lower cost tour.

One of the tour consists of the back of *nearest* + *sub* + *nearest* itself, and the other tour consists of *nearest* itself + *sub* + the front of *nearest*. From the candidates, we choose a lower cost tour as *After2*, and the tour consists of a remained tour expect for *sub* from *After2* as *Before2*.

4. judges whether do division-processing or not depends on the estimated cost.

We estimated that *After1* + *After2* as one of the optimized possibilities for better solutions, and *Before1* + *Before2* as a current cost. To decide the judgment to do division-processing, we compare the estimated costs as followings:

If { Cost(A) > Cost(B) } then,
True → division
False → no division,

where *A* = *Before1+Before2*, *B* = *After1+After2*. ``No division" means the current tour doesn't divide into two subtours on this estimated step.

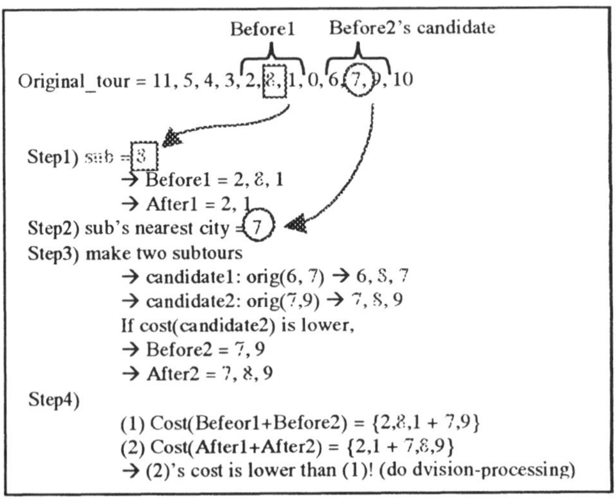

Figure 3: An example of pre-estimated division-processing.

3.5 Gather and adjust the cost balance among sub-solutions

After the terminal steps for the co-evolutionary IA, the sub-solutions on each node are gathered into main node to adjust the cost balance. The adjustment is achieved by escape-processing on the main node, which has all cities at this time.

4 EXPERIMENT

4.1 Simulation design

In order to evaluate the performance of our method, we compare the results between the co-evolutionary IA with one salesman on each node and our method. Both algorithms are applied to two following n-TSPs: (1) dual circle 49-city n-TSP with two peaks deceptive problem and (2) 1000-city n-TSP made at random (Table 1).

Table 1: Design for the first and the second n-TSP.

	First problem	Second problem
# of cities	49	1000
# of salesmen	4 (1 salesman on each sub-problem)	12 (3 salesmen on each sub-problem)
Arrangement	Dual circle	Random
Start city	Center of circle	Random
Radius (out, in)	Out=0.5, In=0.384	-
# of neighbors for escape-processing	5	5
Terminated time step	200	10000

4.2 Results and discussions:

Figure 4 shows a separation example on the first n-TSP, and each separated area is processed on each node as one independent problem.

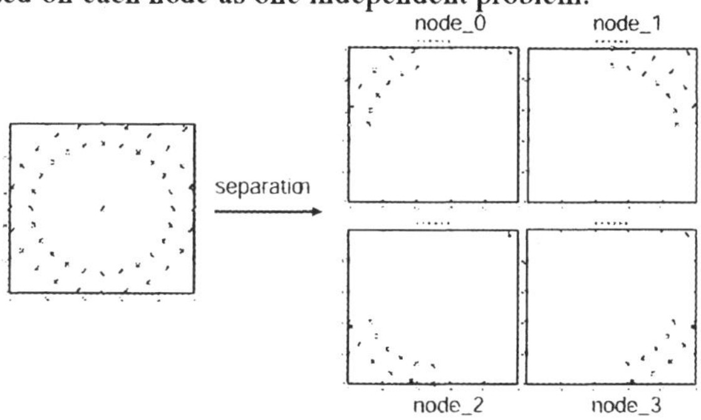

Figure 4: Separation example.

Table 2 for the first problem and Table 3 for the second problem show the obtained costs on each node. A "sum_error" on Table 3 denotes an error degree for equal division, is formulated by equation 2. Both of the current (proposed method) costs on two problems are smaller than or equal to the co-evolutionary IA, even though the sum_errors are higher sometimes. Especially, the decreace (improvement) degree on the costs is so huge: from 10% to 20% in average.

Table 2: The costs with each node on the first problem.

Node ID	the co-evolutionary IA	proposed method
Node_0	2.200276	2.053606
Node_1	2.569807	2.379430
Node_2	2.200277	2.148661
Node_3	1.822837	1.822837

Table 3: The costs with each node on the second problem. A parenthetic value shows "sum_error".

Node ID	the co-evolutionary IA	proposed method
Node_0	22.804396 (0.250477)	17.391262 (1.870297)
Node_1	16.375703 (0.080690)	12.025132 (0.014744)
Node_2	14.097906 (0.076036)	12.331801 (3.573724)
Node_3	14.758087 (0.041723)	14.672518 (0.043995)

$$sum\_error = \sum_i^N \left| \overline{\cos t} - x_i \right| \quad (2)$$

$$\overline{\cos t} = (\sum_i^N x_i)/N \quad (3)$$

5 CONCLUSIONS

In this paper, we accelerated the search speed through a parallelization of problem area, and improved the search ability through a introduction of a pre-estimated division-processing into the co-evolutionary IA. However the current simulator was implemented up to step 2 on overall procedure, the simulation showed pretty good results except for sum_error. We will improve the height of sum_error through an implementation of step 3. In addition, we have an interest in the search ability of the pre-estimated division-processing as genetic operator on GA, the results will be shown later.

REFERENCES

S. Forrest, A.A. Perelson (1990) Genetic algorithm and the Immune system. Proc. of 1st Workshop on Parallel Problem Solving from Nature, Dortmund, Federal Republic of Germany, 1-3, October, 1990.

Theodore C.Belding (1995) The Distributed Genetic Algorithm Revised, Proc. of the Sixth International Conference on Genetic Algorithms, pp.114-121.

T.Starkweather,D.Whitley and K.Mathimas (1991). Optimization Using Distributed Genetic Algorithms, Parallel Problem Solving from Nature, Springer Verlag, pp.176-183.

N. Toma, S. Endo, K. Yamada (2000). The Proposal and Evaluation of an Adaptive Memorizing Immune Algorithm with Two Memory Mechanisms. Journal of JSAI, Vol.15, No.6, pp.1097-1106.

N. Toma, S. Endo, K. Yamada, H. Miyagi (2000). The Immune Distributed Competitive Problem Solver Using Major Histocompatibility Complex and Immune Network. OPERATIONS RESEARCH / MANAGEMENT SCIENCE AT WORK, Kluwer's INTERNATIONAL SERIES, pp.129-147. (ISBN 0-7923-7588-2)

TRIANGULAR PERMUTATION SCAN ALGORITHM FOR ROBOT TOUR PROBLEM

JAYA SANKAR, K.,
Assistant Professor, Department of ECE, Vasavi College of Engineering, Ibrahimbagh, Hyderabad- 500 031, INDIA.

PANDHARIPANDE, V.M.,
Director,
Visvesvaraya National Institute of Technology,
Nagpur – 440 011, INDIA.

Moharir, P.S., Scientist 'F',
National Geophysical Research Institute,
Uppal Road,
Hyderabad – 500 007, INDIA.

ABSTRACT
This paper considers an optimisation problem arising in the automated manufacture of printed circuit boards (PCBs). An essential part of this problem is to determine robot tour sequence of placing the components on a PCB. This helps in reducing the overall assembly time significantly. This can be modelled as travelling salesperson problem (TSPP). Two heuristic algorithms are proposed in this paper to find optimal robot tour sequence. To test the efficacy of each of the algorithms, test problem with known optimal solution has been considered.

INTRODUCTION

Printed Circuit Boards (PCBs) are considered to be an integral part of three major industrial sectors namely Consumer Electronics, Computers and Telecommunications. A PCB consists of one or more layers of conductors and insulating material that allow for electronic components to be electronically interconnected and mechanically supported. During the last two decades the assembly of PCBs has generated a huge industrial activity [Golding, 1995]. From a labour intensive activity, PCB industry has evolved to a highly automated one. Use of robotics will help the agility and reliability required in PCB assembly lines [Holcomb, 1995]. Robots are capable of producing jobs of consistent quality, at a steady rate, with least error probability. Robotic assembly also allows quick changes from one product to another [Mair, 1988].

In PCB manufacturing, either 'plated-through-hole' technology or 'surface mount technology' is commonly employed. The conventional method is plated-through-hole technology, which is still popular in consumer electronic industry, since competitive product price is a key factor. But, a small portion of consumer electronic industry (for example cellular phone industry) uses surface mount technology, which is well suited to automated assembly process [Wang, 1998]. The assembly of PCBs on a production line can be divided into three process planning problems. These are: (i) The allocation of component feeders to machines, (ii) For each machine group, a sequence of the PCB types, indicating in which order the board types will be produced on these machines and (iii) For each pair consisting of a machine and a PCB type, a component placement sequence. Number of researchers proposed various models and techniques while addressing this complex optimization problem. Foulds and Hamacher [1993], Ball and Magazine [1988], Brad et al. [1994], Moyer and Gupta [1997], Leon and Peters [1996] are a few to name.

In this paper, a single component placement machine problem is considered, assuming that the first and second problems have already been solved. A robot tour to place (or solder) the components at specific locations on a PCB can be modelled as a Travelling Salesperson Problem [Burkard, 1998], in which robot is considered to be a travelling salesperson. Finding the minimal distance robot tour is complicated and time consuming. This problem is NP-hard and hence solving to optimality in reasonable computational time is difficult. It is therefore of interest to design heuristic algorithms which are more effective and can generate good solutions at a reasonable computational cost. A set of randomly selected feasible solutions can be iteratively improved by employing heuristic procedures. A stopping rule is required to indicate that there can be no further improvement possible and the solution on hand is optimal. Therefore, it would be good to have test problems with increasing complexity having known optimal solutions.

In this paper, ternary Gray code sequence generation, with the distance measure as the Hamming distance between natural ternary representations of the numbers, has been considered as a test problem. In this test problem, the optimum total distance of the tour is $N=3^n$, where 'n' is the number of gits (generalized bits) [Jaya Sankar, 2003] in the ternary Gray code. For a 4- git ternary Gray code, using the heuristic algorithms reported in this paper, optimum distance of N=81 (i.e.,$N= 3^4$) is obtained, which is the minimum distance between the code words with Hamming distance measure.

The paper is organized as follows: brief introduction about Hamming Scan and Backtracking algorithms is included in the following section. In the next section, the proposed heuristic algorithms are presented. In optimization problems, local minima are quite common, and therefore, techniques for avoiding or escaping local minima are needed. One such technique that uses perturbation method is also presented in this section. This is followed by a section that addresses test problem generation. Results and conclusions are included in the last section.

HAMMING SCAN AND BACKTRACKING

When an objective function defined on binary or ternary sequences is to be optimized, Hamming Scan Algorithm (HSA) is useful [Moharir, 1996]. During search, HSA changes sequentially every element in the binary sequence and compares its effect on the objective function. After performing a scan over all the first order Hamming neighbours of the binary sequence, the best Hamming neighbour is chosen and is considered to be the new starting point for another Hamming scan, which proceeds recursively till no further improvement is noticed in the objective function. For further improvement in the objective function, a second order algorithm, referred to as 'backtracking', which works even after the recursive HSA fails, is to be used. With the help of sub-optimal closest stored solutions, backtracking algorithm initiates HSA from each one of them. When improved solution is obtained from the chosen sub-optimal, the former replaces the latter and HSA is initiated at this stage till it fails. Backtracking is stopped when a combination of HSA and BT fails in that sequence.

TRIANGULAR PERMUTATION SCAN

For the TSPP, a triangular permutation scan (TPS) algorithm works on a square matrix in which the supra-diagonal and sub-diagonal elements are considered to be ordered pairs of cities. A binary option of no interchange or interchange replaces respectively a binary option of no mutation or mutation in HSA. After examining the effect of all binary interchanges on the objective function of TSPP, the best interchange is performed. The improvement obtained due to this on the objective function is accepted and no further interchanges are performed. This process of accepting the improvement due to the best interchange is referred to as First Grab (FG). The second version of TPS creates a pool of solutions without accepting the improved solution as soon as it is spotted. In this version, a maximum pool size is specified and iterations are continued either till the pool is completely filled up, or till the iterations stop without the pool getting filled up. In either case, the best solution in the pool is the one that is accepted. In either case, when TPS does not yield an improvement a larger maximum pool size can be tried. Thus, this algorithm offers the advantage of permitting variation of maximum pool size in an interactive manner.

BACKTRACKING AND PERTURBATION SEARCH

When all the interchanges in the ordered pairs yield a tour inferior to the staring tour, a triangular permutation backtracking (TPBT) algorithm helps in improving recursively the chosen sub-optimal solution. An option of increasing the backtrack pool size, or what is referred to as the span of TPBT, can be considered when a combination of TPS and TPBT fails.

Even with a larger span, when TPS and TPBT fail, perturbation search method can be employed. In this method by adding random noise, and initiating the search methods described above, the best solution obtained earlier can be further improved. This demonstrates that stochastic systematic search techniques (randomized constructive search + backtracking) combined with perturbation search method achieve competitive performance on hard combinational problem. By the addition of random noise, the available best sub-optimal solution can be further improved.

TERNARY GRAY SEQUENCE GENERATION AS TEST PROBLEM

With the help of a family of problems of increasing size with known optimal solutions, algorithms discussed in the previous sections are tested to assess their performance. Gray code sequence generation provides such an opportunity. Ternary Gray code sequence is an arrangement of the numbers from 0 to $(3^n - 1)$, where, 'n' denotes the number of gits in a codeword, such that when we move from one number to the next in their natural ternary representation (NTR), only one git changes [Jaya Sankar, 2003]. Though there are many Gray code sequences for non-trivial values of 'n', one of them, called the reflected Gray code sequence, is the most popular. In [Savage, 1997] it is reported that the Gray code is cyclic if the first and last elements on the list differ in the same way pre-specified for successive elements of the list by the adjacency criterion. For radices other than *even*, if reflection method is used to generate the Gray code words, cyclic property is not observed. Generalised (r, n) Gray code (where 'r' represents the radix) [Jaya Sankar, 2003], has the cyclic property, irrespective of whether 'r' is *even* or *odd*. In Table-1, a two git generalized Ternary Gray code is shown. Cyclic property can be noticed for the ternary Gray code shown in Table – 1. If the distance measure is the Hamming distance between the NTRs of the numbers, the problem of ternary Gray code sequence generation is a TSPP. The optimum total distance of the tour is $N = 3^n$, and one of the optimum tours is the reflected Gray code sequence.

RESULTS AND CONCLUSIONS

Most of the heuristic algorithms reported in the literature are for Euclidean TSPP [Durbin, 1987], [Burkard, 1998]. They work with distances that are derivable from the coordinates of the points. This is unrealistic (in practical instances of TSPP) because the actual distance is different from the one derived using the co-ordinates of the starting point and the destination. Any proposed algorithm should work with an arbitrary matrix of distances. TPS is one such algorithm. A four-git ternary Gray code sequence generation is considered as a test problem. The optimum total distance of the tour is N = 3^4 (i.e., 81) with the distance measure as the Hamming distance between Natural Ternary representations of the numbers. By getting the Gray code sequence for four-git ternary representation, the efficacy of each of the algorithms reported (i.e., TPS and TPBT) has been investigated and optimal solution has been obtained in a reasonable time.

TABLE-1: Generalised Gray Code for Radix – 3 (i.e., Ternary)

Gray code for single git	Gray code for Two gits
0	0 0
1	2 0
2	1 0
	1 2
	0 2
	2 2
	2 1
	1 1
	0 1

The programs developed are in MATLAB and the computational times reported are in seconds. Results in Table-2 are for TPS with pool size = 20 and for TPBT with a pool size of 50.

TABLE – 2 : Results of TPS & TPBT algorithms taking the distance measure as the Hamming distance between the NTRs with length n=4; poolsize in TPS = 20 and maximum backtrack poolsize=50 in TPBT

Sl. No.	Pool size		Cost	Time in Sec.
	TPS	TPBT		
1	20	---	107	106.18
2	---	50	98	399.36
			92	317.47
			90	317.42
			88	321.43
			87	328.01
			86	335.16
			85	349.10
				Total: 2471.13

When the combination of TPS and TPBT failed to improve the objective function, further improvement in the objective function was achieved by adding random noise to the available best sub-optimal solution and TPBT was performed again to obtain the optimal solution. Results are reported in Table – 3(A) to 3(D). This paper has considered an optimization problem encountered in PCB assembly process that can be modelled as TSPP. To solve this problem two algorithms are proposed. The efficacy of each of the proposed heuristic algorithms has been proved, by considering a test problem with a known optimal solution.

TABLE – 3(A) : Results of TPBT algorithm after Perturbation by addition of random noise and taking the distance measure as the Hamming distance between the NTRs with length n=4 ; Initial cost = 85; maximum backtrack poolsize = 50 in TPBT

Sl. No.	Pool size TPBT	Cost + Noise	Cost - Noise	Time in Sec.
1	50	122.1839		191.25
		119.9678		208.66
		118.8203		204.98
		118.3114		190.82
		116.9492		190.82
		116.8875		190.82
		115.8461		187.08
			83	174.94
			82	181.91
				Total: 1721.28

TABLE – 3(B) : Results of TPBT algorithm after Perturbation by addition of random noise and taking the distance measure as the Hamming distance between the NTRs with length n=4 ; Initial cost = 82; maximum backtrack poolsize = 50 in TPBT

Sl. No.	Pool size TPBT	Cost + Noise2	Cost – Noise2	Time in Sec.
1	50	117.1827		216.46
		116.7526		190.54
			83	181.85
				Total: 588.85

TABLE – 3(c) : Results of TPBT algorithm after Perturbation by addition of random noise and taking the distance measure as the Hamming distance between the NTRs with length n=4 ; Initial cost = 83; maximum backtrack poolsize = 50 in TPBT

Sl. No.	Poolsize TPBT	Cost + Noise	Cost - Noise	Time in Sec.
1	50	114.4020		187.08
		114.3637		186.80
			83	171.15
			82	174.67
				Total: 719.70

TABLE – 3(D) : Results of TPBT algorithm after Perturbation by addition of random noise and taking the distance measure as the Hamming distance between the NTRs with length n=4 ; Initial cost = 82; maximum backtrack poolsize = 50 in TPBT

Sl. No.	Poolsize TPBT	Cost + Noise3	Cost – Noise3	Time in Sec.
1	50	117.4510		197.68
		116.5214		236.29
		115.7815		186.58
		115.0984		193.88
		115.0069		193.72
		113.1067		204.59
			82	192.90
			81	185.42
				Total: 1591.06

ACKNOWLEDGEMENTS

The first author would like to thank Prof. P. Ramakrishna Rao, Principal, Vasavi College of Engineering for his support and Dr. V. P. Kodali for his encouragement. He also wishes to thank Prof. M. Satyam, Professor (Retd.), IISc., Bangalore and Prof. D. S. Venkateswarlu, Professor(Retd.), IT, BHU, Varanasi for helpful discussions.

REFERENCES

Ball, M. and Magazine, M., 1988, " Sequencing of Insertion in Printed Circuit Board Assembly", Operations Research, 36(2), pp 192 – 210.

Brad, J., Clayton, R. and Feo, T., 1994, " Machine setup and Component placement in Printed Circuit Board Assembly", International Journal of Flexible Manufacturing Systems, 6(1), pp 5 – 31.

Burkard, R.E., Vladimir G. Dieneko, Rene Vandal, Jack A.A. Vander Veen, and Gerhard J. Woeginger, 1998, "Well Solvable special cases of the Travelling Salesman Problem: A survey", SIAM Rev., Vol. 40, No.3, pp. 496-546.

Durbin, R., and Willshaw, D., 1987, "An analogue approach to the travelling salesman problem using an elastic net method," Nature, Vol. 326 (16), pp 689-691.

Foulds, L. and Hamacher, H., 1993, " Optimal bin location and sequencing in Printed Circuit Board Assembly", European Journal of Operational Research, 66(3), pp 279 – 290.

Foulds, L.R., 1981, "Optimization Techniques: An Introduction," Springer – Verlag, NY.

Golding, D., 1995, " PCB Assembly," Assembly Automation, Vol. 15(2), pp 10-13.

Holcomb, G.W., 1995, " Justifying Flexible Automation for PCB Assembly, " Assembly Automation, Vol. 15(2), pp 14-16.

Jaya Sankar, K., Moharir, P.S. and Pandharipande, V.M., 2003, " Generalised Combinatorial Cyclic Gray Codes", under review.

Leon, V. and Peters, B., 1996, "Replanning and Analysis of Partial setup strategies in Printed Circuit Board Assembly systems", International Journal of Flexible Manufacturing Systems, 8(4), pp 289 – 311.

Mair, G.M., 1988, " Industrial Robotics," PHI Ltd., U.K., pp 250-253.

Moharir, P. S., Singh R. and Maru, V.M., 1996, "S-K-H algorithm for Signal Design," Electronics Letters, 32, 18, pp. 1648-1649.

Moyer, L. and Gupta, S., 1997, " An efficient assembly sequencing heuristic for printed circuit board configurations", Journal of Electronic Manufacturing, 7(2), pp 143 – 160.

Savage, C., 1997, "A Survey of Combinatorial Gray Codes", SIAM Rev. Vol. 39, No. 4, pp 605 – 629.

Wang, C., 1998, " Layout Designs for Robotic PCB Assembly," Soldering and Surface Mount Technology, Vol. 10(2), pp 36-48.

CHEMICAL REACTION MODEL FOR PROBABILISTIC AND EVOLUTIONARY COMPUTATION

VIKRAM KRISHNAMURTHY
Department of Electrical and Computer Engineering,
University of British Columbia, V6T 1Z4,Vancouver, Canada
vikramk@ece.ubc.ca

V.K.MURTHY
School of IT and EE,
UNSW@ADFA,
University of New South Wales,
Canberra, ACT 2600, AUSTRALIA
murthy@cs.adfa.edu.au

ABSTRACT:
The GAMMA parallel programming model is based on the multiset datastructure . Here, a succession of chemical reactions consume the elements of the multiset and produce new elements according to specific rules, This paper extends GAMMA model to its probabilistic version - called P-GAMMA model to realise evolutionary computations -namely probabilistic, classifier, bucket - brigade learning , genetic algorithms, Evolutionary Optimization, Active -Walker model for nonequilibrium sytem and Swarm paradigms. We also explain how to support such computations through randomized choices and concurrent transformations on persistent globally accessible tuplespaces..

INTRODUCTION

Banatre et al [1] proposed a new paradigm for parallel and concurrent programming. This model, called GAMMA has the following features:

(i) Based on the chemical reaction model

It is based on the chemical reaction model, in which the datastructure is a multiset. The computations are interpreted as a succession of chemical reactions consuming the elements of the multiset and producing new elements according to specific rules. Since the transformation process is based on chemical reaction rules that are inherently parallel, any number of reactions can be performed in parallel provided they involve disjoint set of elements. If the attributes of its elements are known in advance, we may carry out intelligent search for those elements with the required attributes; if the attributes of its elements are not known in advance, these attributes may be permitted to evolve recursively, competitively or cooperatively toward an equilibrium state; hence, we can model a deterministic or nondeterministic or probabilistic evolutionary system that cover a wide range of paradigms - including getic algorithms ,swarm paradigm and ant colony optimization.

(ii) Sound method of program derivation

We derive the reaction condition as a negation of the termination condition so that it can be checked locally. The reaction conditions are then formulated as rule schemata.

(iii) Simplicity in implementation

The implementation of Gamma program is simple involving two basic tasks.

a.Searching for elements of the multiset satisfying the reaction condition

b.The application of action to these elements.

The above features make the Gamma model an important programming paradigm.

THE PROBABILISTIC GAMMA MODEL

The main feature of the GAMMA model [1] is the function:

$G(R,A)(M) =$ **if** $\exists\, x_1,x_2,...x_n \in M$ **such that** $R(x_1,x_2,...x_n)$ **then** $G(R,A)((M-\{x_1,x_2,...x_n\}) + A(x_1,x_2,...x_n))$ **else** M; here M denotes the initial multiset, the - sign denotes the multiset difference and the sign + denotes the multiset union.

The function R is the reaction condition (or condition text); it is a boolean function (or a predicate) indicating when some of the elements of the multiset M can react. The function A (action text) describes the result of this reaction. (Note that the function R corresponds to the

query evaluation function in a database, while the function A corresponds to the updating function for a set of database instances). Hence, if one or several reaction conditions hold for several non-disjoint subsets at the same time, the choice made among them can be nondeterministic. This aspect provides for competitive parallelism.
However, if the reaction condition holds for several disjoint subsets at the same time, the reactions can take place independently and simultaneously; this aspect provides for cooperative parallelism.
A useful variant of the GAMMA model is obtained by introducing probabilities for selection when one or more reaction conditions hold for several non-disjoint subsets at the same time. In this case the choice made among these subsets is determined by a random number generator that selects the ith possible subset with a probability p_i to perform the required actions - thus providing for probabilistic competition among the different choices. This results in the probabilistic version of the GAMMA model denoted by P-GAMMA model and is defined by the function:

PG (R /p_i, A) (M) = **if** $\exists\, x_1,x_2,\ldots x_n \; \varepsilon$ M **such that** R($x_1,x_2,\ldots x_n$) **then**
PG(R/p_i ,A)((M-{$x_1,x_2,\ldots x_n$ }) + A($x_1,x_2,\ldots x_n$)) **else** M.

where each of the possible number of subsets i that satisfy the conditions R is chosen with a probability pi and the corresponding text of action A is implemented. Note that the sum of p_i equals unity. Also when p_i is not specified the choice can be deterministic or nondeterministic. Thus P-GAMMA model can contain within itself the deterministic, nondeterministic and probabilistic components.
The P- GAMMA model can be useful in many ways:
1. It can be used to realize many different paradigms, e.g. genetic algorithms, swarm paradigm - that are based on nondeterministic or probabilistic selection and interaction.
2. It can support the design of a wide variety of programs to seek answers to questions such as :
a. Which is the most likely state a system will reach if supplied with a given input sequence ?
b. What is the average survival time of a population that is subject to reproduction and death of its members?

2.1.Properties of P-GAMMA programs

The nondeterministic and probabilistic computations are organized in two phases:
Phase 1: Guessing/tossing (random number)
Phase 2: Verifying validity of the result.
These two phases work interactively; thus nondeterministic and probabilistic computations are no more than guess-check and toss- check actions. Thus in both these computations two fundamental questions arise:
a. How much computational resource ***we must use*** to eliminate nondeterminism or probabilistic chance?
b. How much computational resource ***we can save*** if we are granted nondeterminism or probabilistic chance?
To prove the speed of convergence and termination of the P-GAMMA model, we need to use probabilistic arguments . In practice, to detect a fixed point (or equilibrium), we need to use some acceptance criteria and at the end of each trial evaluate an individual element or a selected subset or the whole multiset to decide whether to repeat the trial or to halt. That is, the evaluation of the multiset can take place at different levels of granularity depending upon the problem domain. Also, the acceptance criteria may be chosen dependent or independent of the number of previous trials and the choice of probabilities can remain static or can vary dynamically with each trial. Hence, depending upon the evaluation granularity, the acceptance criteria and the manner of assigning probabilities, we can devise various strategies. However, the termination of the resulting P-GAMMA program is not guaranteed and requires convergence analysis.
For example, the Monte Carlo method is realised in P-GAMMA MODEL thus:

the condition text (say R $/p_i$) prescribes the nature of the toss to be used for the selection of the elements of the multiset and also a set of **deterministic** criteria for the acceptance or rejection of the elements; the action text (A) implements these conditions and accepts or rejects the elements, as and when they are randomly generated (on- the- fly) [6-9].
The simulated annealing method is realized in P-GAMMA model thus:
the first condition text (say R /pi) prescribes the nature of the toss to be used for the selection of the elements of the multiset and also a set of **deterministic criteria** for the acceptance or rejection of the elements. The corresponding action text (A) implements these conditions and accepts or rejects the elements as and when they are randomly generated (on the fly). Following the rejection of elements, the second condition text (say R* /p*i) prescribes a **probabilistic criterion** to accept (or reject) some of the earlier rejected elements; the corresponding action text A* implements these conditions. The condition text R* /p*i is then varied as a function of the current number of trials and a parameter called "temperature". The effect of this variation is such that the probability of accepting a new solution that is worse than the current solution decreases with the degree of the deterioration of the solution, and more significantly with the run time of the method.
In both the Monte-carlo and simulated annealing methods the granularity of the evaluation takes place at the elemental level and hence they permit on - the- fly acceptance of elements thereby providing high concurrency in the implementation. Both the Monte-Carlo and simulated annealing methods permit on -the- fly acceptance of elements of the multiset involving only purely local actions and hence can have high concurrency in implementation.

SUPPORTING GAMMA PARADIGM

The GAMMA paradigm can be supported by a database transaction processing system if we identify the multiset M with a database, the condition text with a database query evaluation function (to find those elements or subsets of elements of the multiset satisfying particular conditions), and the action text with the updating operation in the database. Such an identification relates the Gamma programming style and the deductive database programming style .
When one or more reaction conditions hold for several disjoint subsets at the same time, the query or Read (R) operation and the update (Write or W) operation can take place concurrently. This parallelism corresponds to 'cooperative parallelism'.
If, however, one or more conditions hold for non-disjoint subsets of the multiset, then a transaction is chosen among the alternatives either nondeterministically, or probabilistically as dictated by a random number generator. The actions on the chosen subset are executed atomically and committed. In other words, the chosen subset undergoes an 'asynchronous atomic update' which ensures that the process of matching and the follow-up actions satisfy the atomicity, consistency, isolation and durability conditions . Once all the actions are carried out and committed the next set of conditions are considered. Concurrency can also be introduced among the actions on the nondeterministically or probabilistically chosen subset using the well- known concurrency control techniques . Thus, the transaction notion is useful for supporting the GAMMA programming paradigm.
As a result of the actions followed by commitment, we derive a new multiset; this may satisfy new conditions of the text and the actions are repeated by initiating a new set of transactions. These set of transformations halt when there are no more transactions executable or the multiset does not undergo a change for two consecutive steps indicating a new consistent state of the database. Such a scheme would correspond to the 'competitive parallelism' . From the above discussion it is clear that to support the GAMMA paradigm , we require the following facilities :
(i) breaking the multiset into chunks ,
(ii) handing over each chunk to an appropriate local processor, and
(iii) techniques to handle query and transaction processing .
Thus this paradigm will be suitable for cluster and grid computing.

EXAMPLES

1. Classifier / bucket brigade Systems

The classifier system [3],[4],[7],[9],[11],[14],[16],[17] is a parallel rule-based production system based on two-valued logic. Each classifier can be regarded as a separate instruction that takes messages as input and produces messages as output. Thus, like a production system there is a match cycle in which each rule is matched against the state of the short term memory containing a message list M. If its preconditions are satisfied by at least one message then the classifier is activated and an execute cycle carries out the required actions and posts an external message. All external communications are via the message list. Thus all internal control and external communication resides in the same datastructure. The classifier system [6] can therefore be realised by the GAMMA model provided the messages are represented by a multiset of appropriate type. When the classifier system is used deterministically, it iterates for a fixed number of times or until the message list does not change for two successive atomic steps. It has been shown in [6] that :

(i) Any finite function can be computed by some classifier system in a single match and execute step with an arbitrary amount of parallelism ,by distributing the representation of the function over enough number of processors and

(ii) A classifier system can be made to behave completely sequentially.

These two properties permit the classifier system to behave with maximal parallelism or completely sequentiality. Thus a classifier system offers a flexible formalism that permits optimization of the three parameters, namely, the number of processors (classifiers), the length of computation, and the amount of inter-processor communication.

A classifier system can be generalized by introducing two factors, called the strength and the support for the activation of each rule :

(1)the strength of each rule or reaction is a measure of its past success and

(2) the support for each rule is a measure of its likely relevance to the current situation.

These two factors, namely, the strength and the support provide for competition and cooperation among the different rules. When a message posted matches against a rule each classifier makes a bid proportional to its strength; hence highly fit rules are given preference. The bid made by an activated classifier is then proportionately divided and sent as a reward to other classifiers earlier responsible for activating it. Thus if a rule is instrumental to permit other rules to fire **favourably**, it receives payoff and hence its strength will be increased ; otherwise, its strength is decreased due to bidding. In the steady state the strength remains invariant. This classifier system is called a "Bucket brigade system" (BBS) .Thus BBS is useful to construct a multiset whose attributes are not known in advance, but are adaptive or evolving probabilistically .

The P-GAMMA model can realise the BBS, if the probabilities p_i are replaced by strengths and at the end of each atomic update, the actions are evaluated and the strengths are reassigned depending upon downstream and upstream pay off , [3,4,6,7,8,9,14,16].

2. Genetic algorithm

The genetic algorithm, goes a step further than the BBS ; after selecting the most successful rules, these rules are combined by selecting pairs and performing cross over and mutation. The cross over may be thought of as a combination of independent rules, while the mutation can be thought of as an error in input or an approximation.

The genetic algorithm [8,14] chooses an initial population of objects called chromosomes, represented by a multiset M of binary strings of length of m bits. Then it performs on M three different probabilistic operations that mimic the operations found in nature, namely, selective reproduction, crossover and mutation.These operations can be represented in the P-GAMMA model using four transactions:

Transaction 1 denoted by PG(R/q,A) evaluates the given generation for assigning the selection probabilities q to initiate the actions for the creation of a new generation.

Transaction 2 denoted by $PG(R1/ p_1, A1)$ performs reproduction; the condition text $R1/p_1$ enables us to select probabilistically, those strings which are fit and the action text(A1) replaces the original multiset with the fittest strings.
Transaction 3 denoted by $PG(R2/p_2,A2)$ performs crossover; $R2/p_2$ gives the probabilistic conditions for picking up any two elements and the crossover site; the corresponding action text A2 performs crossover and returns the two offspring back to the multiset.
Transaction 4 denoted by $PG(R3/p_3,A3)$ performs mutation; $R3/p_3$ gives the probabilistic conditions to select a string as well as the site for mutation; action text A3 performs mutation and returns the string back to the multiset. These sequence of transactions are then repeated until the multiset is stable or does not undergo a change.

If the operations of reproduction, crossover and mutation are performed sequentially, one after another, then the P-GAMMA program is given by:
GENE M = $PG(R3/p_3,A3)$ $PG(R2/p_2,A2)$ $PG(R1/p_1,A1)$ PG(R/q, A) M.
If we want to introduce concurrency among these three transactions 2,3 and 4 they must be interleaved arbitarily, subject to the constraint they act on different strings concurrently. Such a genetic algorithm called an "elitist model" [8,14,16] is denoted by the P-GAMMA program thus:
GENE M= ($PG(R3/p_3,A3)$,$PG(R2/p_2,A2)$, $PG(R1/p_1,A1)$) PG(R/q, A) M
In the elitist model the parents and offspring together can undergo selective reproduction, crossover and mutation [16].
Note that the evaluation transaction cannot be interleaved with others. In implementing GAMMA programs, therefore, we must decide how to interleave tasks of different types and size (granularity); this problem is called 'multilevel atomicity'.
It is possible to interleave the simulated annealing algorithm with the genetic algorithm; the accepted elements in the annealing algorithm can undergo selection, mutation, and cross-over or other new operations. Such a combination provides for a greater subspace search.

3. Swarm and Ant-colony Paradigms[2,5,12,13,17]
A swarm is a population of interacting elements that is able to optimize some global objective through collaborative search of a space. Interactions that are relatively local (topologically) are often emphasized. There is a general stochastic tendency in a swarm for individuals to move toward a center of mass in the population on critical dimensions , resulting in convergence on an optimum. Using the P-gamma paradigm, the particle swarm paradigm is easy to implement and can be made computationally efficient. The particle swarms are also related to cellular automata (CA). These have the following three properties:
1. Individual cell updates are done in parallel
2.Each new cell value depends only on the old values of the cell and its neighbours.
3. All cells are updated using the same rules.
Individuals in swarm can be conceptualized as cells in CA whose states change in many dimensions simultaneously. Thus the Gamma paradigm is very useful.

CONCLUDING REMARKS
The P-GAMMA paradigm can provide a new programming environment for probabilistic, classifier , bucket - brigade, genetic and learning algorithms and Swarm paradigm in AI.The programming paradigm outlined here can be implemented using MPI Gropp et al[10].
A complex system [17] consists of a large number of simple elements or intelligent agents interacting with each other and the environment. Two important concepts used

for modelling such complex systems are: self organized criticality [15] and the principle of active walks.[13] The former asserts that a large dynamical system tends to drive itself to a critical state . The latter principle describes how elements in a complex system communicate with the environment and with each other through the interaction with the landscape they share. The Gamma paradigm will be useful to simulate such systems and the rules of interaction can be revised or the elements can be updated .

REFERENCES

[1] J.-P.Banatre and D.L.Me'tayer, Programming by Multiset transformation, Comm.ACM **36** (1993) 98 -111.

[2]E. Bonabeau., M.Dorigo and G. Theraulaz,G: Swarm Itelligence: From natural to artificial systems, Oxford University Press, 1999.

[3] R.K.Belew and S. Forrest, Learning and Programming in classifier Systems, Machine Learning **3** (1988) 193-223.

[4] L.B.Booker, D.E.Goldberg, and J.H.Holland, Classifier systems and Genetic algorithms, Artificial Intelligence **40** (1989) 235-282.

[5]M.Dorigo,G.D. Caro and M.Sampels, Ant Algorithms, Lecture Notes in Computer Science, Vol.2463, Springer Verlag, New York, 2002.

[6]S.Forrest, Parallelism and Programming in classifier systems .Morgan Kauffman,San Mateo, California. 1991.

[7] S.Forrest, Emergent Computation (M.I.T Press, Cambridge, Ma. 1991).

[8] D.E.Goldberg, Genetic algorithms in search, optimization and machine learning,Addison Wesley, Reading, Ma. 1989 .

[9]J.J.Grefenstett, Credit Assignment in rule discovery systems based on genetic algorithms, Machine Learning **3** (1988) 225-245.

[10] W. Gropp,et al., Using MPI, M.I.T Press, Cambridge, Mass,1995 .

[11] J.H.Holland et al., Induction ,M.I.T.Press,Cambridge,Ma.1987 .

[12] J.Kennedy and R.C.Eberhart, Swarm Intelligence, Morgan Kauffman, London, 2001.

[13] L.Lam, Nonlinear Physics for beginners, World Scientific, singapore,1998.

[14] Z.Michalewicz, Genetic Algorithms + data structures = Evolution Programs, Springer Verlag, New York, 2000 .

[15] V.K.Murthy, V.Krishnamurthy, and E.V.Krishnamurthy, Can Interacting agents in WWW reach Self-organized criticality, Vol 12, Smart enginneringSytem Design, Pp.165-170, ASME Press, New york,2002.

[16] W.M.Spears and K.A.De Jong, An overview of evolutionary computation, Machine Learning Lecture Notes in Computer Science,Vol.667, pp. 442-459, (Springer Verlag, New York, 1993.

[17] S.Wolfram, A New kind of Science, Wolfram Media Inc.,Champaign, Ill.2002.

USING THE EVOLUTION STRATEGIES' SELF-ADAPTATION MECHANISM AND TOURNAMENT SELECTION FOR GLOBAL OPTIMIZATION

EFRÉN MEZURA-MONTES AND CARLOS A. COELLO COELLO
Evolutionary Computation Group at CINVESTAV-IPN (EVOCINV)
Depto. de Ingeniería Eléctrica
Sección de Computación
Av. IPN No. 2508,
Col. San Pedro Zacatenco,
México D.F. 07300, MEXICO,
emezura@computacion.cs.cinvestav.mx
ccoello@cs.cinvestav.mx

ABSTRACT
An approach based on a (μ+1)-ES and three simple tournament rules is proposed to solve global optimization problems. The proposed approach does not use a penalty function and does not require any extra parameters other than the original parameters of an evolution strategy. This approach is validated with respect to the state-of-the-art techniques in evolutionary constrained optimization using a well-known benchmark. The results obtained are very competitive with respect to the approaches against which our approach was compared.

INTRODUCTION

Having a strong theoretical background (Schwefel, 1995; Bäck, 1996; Beyer, 2000), Evolution Strategies (ES) have been found efficient in solving a wide variety of optimization problems (Asselmeyer, 1997). However, as other evolutionary Algorithms (EAs), ES lack an explicit mechanism to deal with constrained search spaces.

Several approaches to incorporate constraints into the fitness function of an EA (Michalewicz, 1996; Coello, 2002) have been proposed. The most common approach used to incorporate the constraints of the problem to the fitness function of an EA is the use of penalty functions, where the amount of constraint violation is used to punish or "penalize" an infeasible solution so that feasible solutions are favored by the selection process. Without concerning their simplicity, they have many drawbacks from which the main one is that they require a careful fine tuning of the penalty factors that accurately estimates the degree of penalization to be applied so that we can approach efficiently the feasible region (Smith, 1997; Coello 2002).

In this paper, we argue that the self-adaptation mechanism of a conventional evolution strategy combined with some (very simple) tournament rules based on feasibility can provide us with a highly competitive evolutionary algorithm for constrained optimization.

COMPARING DIFFERENT TYPES OF ES

The motivation of this work is based in the fact that some of the most recent and competitive approaches to incorporate constraints into an evolutionary algorithm use an ES (see for example (Runarsson, 2000; Hamida, 2002)). We then hypothesized that the self-adaptation mechanis of an ES plays an important role when solving numerical optimization problems. In order to verify our hypothesis, we performed a small comparative study among different types of ES. In order to determine the most useful type of ES to deal with constrained spaces, we performed a numerical comparison using five different types of ES. Both a $(\mu+\lambda)$-ES and a (μ,λ)-ES were implemented with and without correlated mutation. Additionally, we also implemented a $(\mu+1)$-ES using the "1/5-success rule". We decided to adopt a very simple constraint-handling approach which shares similarities with some previous proposals (e.g., (Jimenez, 1999; Deb, 2000)). Note however, that all of these previous proposals require an additional mechanism to maintain diversity in the population. In our proposal, however, no extra mechanisms are provided to maintain diversity. The tournament rules adopted in the five types of ES implemented are the following:

- Between 2 feasible solutions, the one with the highest fitness value wins.
- If one solution is feasible and the other one is infeasible, the feasible solution wins.
- If both solutions are infeasible, the one with the lowest sum of constraint violation is preferred.

We decided to use ten (out of 13) of the test functions described in (Runarsson, 2000) to evaluate our five types of ES. The test functions chosen contain characteristics that are representative of what can be considered "difficult" global optimization problems for an evolutionary algorithm. The expressions of the test functions can be seen in (Mezura-Montes, 2003).

TF	optimal	Best	Mean	Median	Worst	St. Dev.
g01	-15	-15	-14.8486	-14.998	-12.9999	0.410082
g02	0.803619	0.793083	0.698932	0.708804	0.576079	0.062927
g03	1	1	1	1	1	0.000014
g04	-30665.539	-30665.539	-30665.442	-30665.539	-30663.496	0.393918
g06	-6961.814	-6961.814	-6961.814	-6961.814	-6961.814	0.00
g07	24.306	24.3681	24.7025	24.7307	25.5167	0.242956
g08	0.095825	0.095825	0.095825	0.095825	0.095825	0.00
g09	680.63	680.6317	680.6736	680.6593	680.9151	0.052483
g11	0.75	0.75	0.7844	0.776296	0.8795	0.037345
g12	1	1	1	1	1	0.00

Table 1. Statistical Results of the $(\mu+1)$-ES after the first set of experiments.

We set a total of 350000 fitness function evaluations and performed 30 runs for each problem and for each type of ES. Equality constraints were transformed into inequalities using a tolerance value of 0.0001. For the $(\mu+1)$-ES the initial values are: $\sigma=4.0$, $C=0.99$, $\mu=5$, and maximum number of generations = 350000. For the $(\mu+\lambda)$-ES and (μ,λ)-ES, we adopted panmictic discrete recombination both for the strategy parameters and for the decision variables. The learning rates values were calculated as indicated in (Schwefel, 1995). The initial values for the standard deviations were 3.0 for all the decision variables. The initial values for the remaining ES are: $\mu=100$, $\lambda=300$, and maximum number of generations = 1166.

Due to space limitations we only show the statistical results of the (μ+1)-ES in Table 1 (the complete statistical results can be found in (Mezura-Montes, 2003)), which had the best overall performance, both in terms of the best solution found and in terms of its statistical measures. We found this a little bit surprising, because this is the most simple type of ES implemented in our comparative study. Based on the encouraging results that we obtained from this study, we decided to compare our (μ+1)-ES with respect to other approaches.

The details of the selected approach (we decided to call it Simple Evolution Strategy, or SES) are in the pseudo-code shown in Figure 1. We adopted the "1/5-success rule" to self-adapt the only sigma value used in the algorithm. At each generation μ individuals are created but none of them is evaluated. In fact, we only evaluate the offspring created by the union of the mutated individuals produced. The low computational cost of the approach is owing to the fact that it requires only one sigma value to adapt and one fitness function evaluation per generation.

```
Begin
  t=0
  Create a random initial solution x^0
  Evaluate f(x^0)
  For t=1 to MAX_GENERATIONS Do
    Produce μ mutations of x^(t-1) using:
          x_i^j=x_i^(t-1)+σ [t]•N_i(0,1)
    forall i ∈ n, j=1,2,…, μ
          Generate one child x^c by the combination of the μ mutations using
          m=randint(1, μ)
          x_i^c= x_i^m, forall i ∈ n
    Evaluate f(x^c)
    Apply comparison criteria to select the best individual x^t between x^(t-1) and x^c
    t=t+1
    If (t mod n = 0)  Then
          If (p_s>1/5) Then
                    σ [t]= σ[t-n]/c
          Else
                    If  (p_s<1/5) Then
                              σ [t]= σ[t-n]•c
                    Else
                              If  (p_s=1/5) Then
                                        σ [t]= σ[t-n]
                              End If
                    End If
          End If
    End If
  End For
End
```

Figure 1. SES algorithm (**n** is the number of decision variables of the problem, **ps** is the success rate of mutations in order to apply the "1/5 success rule")

It is worth emphasizing that our algorithm is based on two mechanisms: (1) the natural self-adaptative mechanism of the ES, which helps the approach to sample the search space well enough as to reach the feasible region reasonably fast and (2) the comparison criteria of our tournaments, which select the most promising solution in order to be used as a new starting point for the search.

COMPARING SES AGAINST OTHER APPROACHES

In the second part of our experiment, we compared the performance of our approach with respect to three techniques that are representative of the state-of-the-art in the area: the homomorphous maps "HM"(Koziel, 1999), the stochastic ranking "SR" (Runarsson, 2000) and the adaptive segregational constraint handling evolutionary algorithm "ASCHEA" (Hamida, 2002). In Tables 2, 3 and 4, we show the results of the comparison.

		Best		Mean		Worst	
TF	**optimal**	**SES**	**HM**	**SES**	**HM**	**SES**	**HM**
g01	-15	-15	-14.7886	-14.8486	-14.7082	-12.9999	-14.6154
g02	0.803619	0.793083	0.79953	0.698932	0.79671	0.576079	0.79119
g03	1	1	0.9997	1	0.9989	1	0.9978
g04	-30665.539	-30665.539	-30664.5	-30665.442	-30655.3	-30663.496	-30645.9
g06	-6961.814	-6961.814	-6952.1	-6961.814	-6342.6	-6961.814	-5473.9
g07	24.306	24.3681	24.62	24.7025	24.826	25.5167	25.069
g08	0.095825	0.095825	0.095825	0.095825	0.08916	0.095825	0.0291438
g09	680.63	680.6317	680.91	680.6736	681.16	680.9151	683.18
g11	0.75	0.75	0.75	0.7844	0.75	0.8795	0.75
g12	1	1	0.99999	1	0.99913	1	0.99195

Table 2. Comparison of results between our approach (SES) and the Homomorphous Maps (HM)

		Best		Mean		Worst	
TF	**optimal**	**SES**	**SR**	**SES**	**SR**	**SES**	**SR**
g01	-15	-15	-15	-14.8486	-15	-12.9999	-15
g02	0.803619	0.793083	0.803515	0.698932	0.781975	0.576079	0.726288
g03	1	1	1	1	1	1	1
g04	-30665.539	-30665.539	-30665.539	-30665.442	-30665.539	-30663.496	-30665.539
g06	-6961.814	-6961.814	-6961.814	-6961.814	-6875.940	-6961.814	-6350.262
g07	24.306	24.3681	24.307	24.7025	24.374	25.5167	24.642
g08	0.095825	0.095825	0.095825	0.095825	0.095825	0.095825	0.095825
g09	680.63	680.6317	680.63	680.6736	680.656	680.9151	680.763
g11	0.75	0.75	0.75	0.7844	0.75	0.8795	0.75
g12	1	1	1	1	1	1	1

Table 3. Comparison of results between our approach (SES) and the Stochastic Ranking (SR)

		Best		Mean		Worst	
TF	optimal	SES	ASCHEA	SES	ASCHEA	SES	ASCHEA
g01	-15	-15	-15	-14.8486	-14.84	-12.9999	NA
g02	0.803619	0.793083	0.785	0.698932	0.59	0.576079	NA
g03	1	1	1	1	0.9999	1	NA
g04	-30665.539	-30665.539	-30665.5	-30665.442	-30665.5	-30663.496	NA
g06	-6961.814	-6961.814	-6961.81	-6961.814	-6961.81	-6961.814	NA
g07	24.306	24.3681	24.3323	24.7025	24.66	25.5167	NA
g08	0.095825	0.095825	0.095825	0.095825	0.095825	0.095825	NA
g09	680.63	680.6317	680.63	680.6736	680.641	680.9151	NA
g11	0.75	0.75	0.75	0.7844	0.75	0.8795	NA
g12	1	1	NA	1	NA	1	NA

Table 4. Comparison of results between our approach (SES) and the Adaptive Segregational Constraint Handling Evolutionary Algorithm (ASCHEA) (NA means not available).

There are several issues derived from this comparison that deserve some discussion:

Our SES was able to converge to the global optimum in 8 of the test 10 functions used (g01, g03, g04, g06, g08, g09, g11 and g12), and it was able to converge very closely to the optimum in the other two.Also, with respect to the homomorphous maps (see Table 2), SES converged to a better "best" solution in 7 problems, and it obtained better average solutions in 8 problems. It also found a better "worst" solution in 6 problems. Thus, it should be clear that SES had a highly competitive performance, even improving the results of the homomorphous maps in some test functions. With respect to stochastic ranking (see Table 3), SES was able to converge to similar "best" solutions in 8 problems and it obtained similar average and worst results in 2 problems. SES found a better average and worst result only in one problem (g06). Finally, with respect to ASCHEA (see Table 4), the SES converged to similar "best" results in 7 problems and found better "best" results in problem g02. SES also obtained better average results in 3 problems and matched ASCHEA's average results in three more.

DISCUSSION OF RESULTS

The results described before indicate a competitive performance of our approach with respect to three techniques representative of the state-of-the-art in constrained evolutionary optimization. Besides being a very simple approach, it is worth reminding that SES does not require any extra parameters (besides those used with an evolution strategy). In contrast, the homomorphous maps require an additional parameter (called v) which has to be found empirically (Koziel, 1999). Stochastic ranking requires the definition of a parameter called P_f, whose value has an important impact on the performance of the approach (Runarsson, 2000). ASCHEA also requires the definition of several extra parameters and in its latest version, it uses niching, which is a process that also has at least one additional parameter (Hamida, 2002). Finally, the number of fitness funtion evaluations of our approach is less or equal to that requiered by the compared approaches: Stochastic ranking performed the same number of evaluations as SES (350000), but the homomorphous maps performed 1,400,000 fitness function evaluations, and ASCHEA performed 1,500,000 fitness function evaluations.

CONCLUSIONS AND FUTURE WORK

The self-adaptation mechanism of an evolution strategy combined with a set of simple tournament rules was used to deal with constrained search spaces. Based on the results obtained, we claim that the proposed approach is a viable alternative for constrained optimization. Its main advantage is that it does not require a penalty function or any extra parameters (other than the original parameters of an evolution strategy) to bias the search towards the feasible region of a problem. Additionally, our approach has a low computational cost and it is easy to implement. Our future work consists on experimenting with a different random numbers distribution (a Gaussian distribution was used in this paper) to avoid convergence to local optima, since this happens in some cases.

ACKNOWLEDGMENTS

The first author acknowledges support from the mexican Consejo Nacional de Ciencia y Tecnología (CONACyT) through a scholarship to pursue graduate studies at CINVESTAV-IPN's Electrical Engineering Department. The second author acknowledges support from CONACyT through project No. 32999-A.

REFERENCES

Asselmayer, T., Ebeling, W., and Rosé, H., 1997, Evolutionary strategies of optimization., *Phys. Rev. E*, 56(1):1171--1180, July.

Bäck, T., 1996, Evolutionary Algorithms in Theory and Practice., Oxford University Press, New York.

Beyer, H.G., 2001, The theory of Evolution Strategies., Springer, Berlin.

Coello Coello, C.A., 2002, Theoretical and Numerical Constraint Handling Techniques used with Evolutionary Algorithms: A Survey of the State of the Art., *Computer Methods in Applied Mechanics and Engineering*, 191(11-12):1245--1287, January.

Deb, K., 2000, An Efficient Constraint Handling Method for Genetic Algorithms., *Computer Methods in Applied Mechanics and Engineering*, 186(2/4):311—338.

Hamida, S.B. and Schoenauer, M., 2002, (ASCHEA): New Results Using Adaptive Segregational Constraint Handling., In *Proceedings of the Congress on Evolutionary Computation 2002 (CEC'2002)*, volume 1, pages 884--889, Piscataway, New Jersey, May, IEEE Service Center.

Jiménez, F. and Verdegay, J.L., 1999, Evolutionary techniques for constrained optimization problems., In Hans-Jurgen Zimmermann, editor, *7th European Congress on Intelligent Techniques and Soft Computing (EUFIT'99)*, Aachen, Germany, Verlag Mainz. ISBN 3-89653-808-X.

Koziel, S. and Michalewicz, Z., 1999, Evolutionary Algorithms, Homomorphous Mappings, and Constrained Parameter Optimization., *Evolutionary Computation*, 7(1):19--44.

Mezura-Montes, E. and Coello Coello, C.A., 2003, An Empirical Study about the Usefulness of the Evolution Strategies Natural Self-Adaptive Mechanism to Handle Constraints in Global Optimization., Technical Report EVOCINV-01-2003, Evolutionary Computation Group at CINVESTAV, Sección de Computación, Departamento de Ingeniería Eléctrica, CINVESTAV-IPN, México D.F., México. Available at http://www.cs.cinvestav.mx/~constraint

Michalewicz, Z. and Schoenauer, M., 1996, Evolutionary Algorithms for Constrained Parameter Optimization Problems. *Evolutionary Computation*, 4(1):1--32.

Runarsson, T.P. and Yao, X., 2000, Stochastic Ranking for Constrained Evolutionary Optimization. *IEEE Transactions on Evolutionary Computation*, 4(3):284--294, September.

Schwefel, H.P., 1995, Evolution and Optimal Seeking, John Wiley & Sons Inc., New York.

Smith, A.E. and Coit, D.W., 1997, Constraint Handling Techniques---Penalty Functions. In Thomas Back, David B. Fogel, and Zbigniew Michalewicz, editors, *Handbook of Evolutionary Computation*, chapter C 5.2. Oxford University Press and Institute of Physics Publishing.

ON THE EXPECTED CONVERGENCE TIME OF A GENETIC ALGORITHM WITH MINIMUM PARAMETERS

MARGARITA REYES-SIERRA AND CARLOS A. COELLO COELLO
Evolutionary Computation Group at CINVESTAV-IPN (EVOCINV)
Departamento de Ing. Eléctrica Sección Computación
Av. IPN No. 2508
Col. San Pedro Zacatenco
México, D.F. 07300, MÉXICO
mreyes@computacion.cs.cinvestav.mx
ccoello@cs.cinvestav.mx

ABSTRACT
In general, the problem of characterizing the behavior of a genetic algorithm is rather complex because it varies depending on the application domain and on the parameters adopted for the genetic algorithm. In this paper, we briefly discuss some of the most representative approaches that have been previously proposed in the literature to estimate convergence times of a genetic algorithm. Then, we introduce our own approach whose analysis is based on Markov Chains. Our model is empirically validated using a genetic algorithm with minimum parameters. Our results indicate the correctness of the proposed model.

INTRODUCTION

Genetic algorithms (GAs) are a heuristic search technique inspired on natural evolution (i.e., the survival of the fittest). In its origins, the genetic algorithm (now called "classical") was applied to single-objective optimization problems (Goldberg,1989). Rudolph proved in the mid-1990s convergence of this simple GA to the global optimum of a given function, under certain conditions (Rudolph, 1994). This work gave some of the desired theoretical foundations to the behavior of GAs. This paper extends Rudolph's model so that we can estimate the expected convergence time of a genetic algorithm with minimum parameters. We also detail the general features of the transition matrix of the elitist genetic algorithm discussed in (Rudolph, 1994). The results of our theoretical model are validated with a simple example. Our empirical results indicate the validity of our theoretical model.

PREVIOUS WORK

Ankenbrandt (1991) obtained a bound for the execution time to achive convergence of a genetic algorithm using a model based on allele convergence and a relatively simple proof by induction. On the other hand, taking advantage of the fact that natural selection uses the diversity of a population to give rise to adaptation, Louis & Rawlins (Louis, 1993) developed a model based on Hamming distances between individuals in a population. They argued that, if we ignore the effects of mutation, the lack of diversity makes useless the role of natural selection. Thus, they used a measure of diversity to estimate an upper bound on the time beyond which progress is unlikely (because of the lack of diversity). (Goldberg, 1987) presented a study similar to the work of (Louis, 1993). In this case, they estimated the convergence time of a genetic algorithm whose population consists of individuals of length 1 using a Markov chain model. Note that in the current paper we use a population consisting of individuals of length *n*.

MARKOV CHAINS

The basic concepts regarding Markov chains can be consulted in (Reyes, 2003). Next, we will discuss a special type of Markov chain on which the current work is based. The material of this section was obtained from (Rudolph 2000a, Iosifescu 1980).

Absorbing Chains

Consider a Markov chain with finite state space $S \neq \emptyset$. In this section the states of a Markov chain are classified based on whether it is possible to go from a given state to another state.

Definition 1. We say that the state *i* **leads to** state *j* and write $i \rightarrow j$ if and only if $p_{ij}^{k} > 0$ for an $k > 1$. If $i \rightarrow j$ and $j \rightarrow i$ we say that state *i* **communicates with** state *j* and write $i \leftrightarrow j$.

Beginning with the previous definition, the states are classified into "equivalence classes". Two states are in the same class if they are "communicated", i.e. if the process can go from one state to the another and viceversa. The equivalence classes are classified as **ergodic** sets (so called **recurrent**) or **transient** sets.This way, the corresponding states in those classes are called ergodic states and transient states, respectively. For each finite Markov chain there must be always at least one ergodic set; however, it is unnecessary to have transient sets. Once a chain leaves a transient set, it can never come back to it; conversely, once a chain enters into an ergodic set, it can never leave it. In particular, if an ergodic set contains just one state, this state is called an **absorbing** state, because once into it, the Markov chain will stay there forever. From the previous, we have that a state *i* is absorbing if and only if $p_{ii}=1$ (Kemeny, 1960). Now, a chain with transient states has a behavior such that it moves towards ergodic sets. The probability that the process lies within an ergodic set tends to 1. When all the ergodic sets are unitary, these chains are called **absorbing chains**, because they will eventually get trapped in an absorbing state. It is important to consider the canonical form of the transition matrix of a Markov chain. Let's assume that we have *s* transient states and *r-s* ergodic states, and that we cluster all the transient sets and all the ergodic sets together, the resulting form is:

$$P = \begin{pmatrix} S & O \\ R & Q \end{pmatrix}$$

The region O consists completely of zeros. The matrix $Q_{s\times s}$ represents the chain while it is in transient states, the matrix $R_{s\times r-s}$ represents the transition from the transient states to ergodic states and the matrix $S_{r-s\times r-s}$ represents the chain once that it is into an ergodic state.If we consider an absorbing chain, we have that by definition $S=I_{r-s\times r-s}$, so its canonical form is:

$$P = \begin{pmatrix} I & O \\ R & Q \end{pmatrix}$$

Definition 2.For an absorbing Markov chain we define the **fundamental matrix** to be $N = (I - Q)^{-1}$.

Definition 3. We define $\mathbf{n}_j$ to be a function whose value is the total number of times that the process is in a transient state s_j

Let **T** the set of transient states of the Markov chain. If we denote with $\mathbf{E}_i[\mathbf{n}_j]$ the expected value of $\mathbf{n}_j$ assuming that the process starts in the state s_i, we have the next result:

Theorem 1.(Kemeny, 1960) $\{\mathbf{E}_i[\mathbf{n}_j]\}=N$

Definition 4. Let **t** be a function whose value is given by the number of steps (including the initial state) in wich the process is in a transient state.

If the process starts in an ergodic state then **t**=0. If the process starts in a transient state, then **t** gives us the total number of necessary steps for reaching an ergodic state. In an absorbing chain, this is the *time to absorption*.

Let ξ a column vector with all entries equal to 1.

Theorem 2.(Kemeny, 1960)$\{\mathbf{E}_i[\mathbf{t}]\}=N\xi$.

EXPECTED CONVERGENCE TIME

In (Rudolph, 1994), we can study the mathematical models of a Simple Genetic Algorithm (SGA) and an Elitist Genetic Algorithm (EGA). It is required to consult this reference to get the preliminaries required to understand the work presented here. In the previous section, we showed some results on the fundamental matrix of a Markov chain. As we saw, such a matrix can be used to compute the expected convergence time of the chain. In this section, we will apply such results to the corresponding transition matrix of the EGA. Since in Rudolph's model the matrix **P** corresponds with the populations whose super individual is the global optimum, we can consider that when the chain is in one of those states, the search process has finished. Any further change in the population can be ignored because the super individual will be no longer modified. Therefore, we can rewrite the matrix like:

$$P^{+} = \begin{pmatrix} I & O \\ R & T \end{pmatrix}$$

We can see clearly now that the Markov chain corresponding to the EGA is absorbent. According to Definition 2, the fundamental matrix that interests us in this case is: $N=(\mathbf{I}-\mathbf{T})^{-1}$. Since our objective is to know the fundamental matrix N, we will start by studying the structure of the block **T**.

Matrix P

In this section, we will show the elements of the **P** matrix. As we know, this matrix is the result of the product: **P**=**CMS**. This is the reason why the elements of each of the corresponding matrices will be specified next. The details of the previous expressions may be found in (Reyes, 2003).

Elements of the matrix of crossover. The elements of uniform crossover were modeled. For that sake, the $\oplus$ operator was defined as the negation of an *or-exclusive*. Using this fact the following formula was developed (assuming a crossover percentage of 0.5):

$$c_{ij} = \prod_{q=1}^{n}(0.5)^{l}\prod_{r=1}^{l}\sum_{s=\phi(q)-1}^{\phi(q)}\pi_{s}^{r}(i)\oplus\pi_{q}^{r}(j)$$

where $\phi(q) = 2\left\lfloor \frac{q+1}{2} \right\rfloor$

Elements of the Mutation Matrix. The corresponding elements are: $m_{ij}=p_m^{H_{ij}}(1-p_m^{N-H_{ij}})$ where p_m is the mutation probability and H_{ij} is the Hamming distance between populations *i* and *j*.

Elements of the Selection Matrix.

The selection operator adopted is the well-known proportional selection:

$$s_{ij} = \begin{cases} \dfrac{\prod_{k=1}^{n} f(\pi_k(j))}{(\sum_{k=1}^{n} f(\pi_k(i)))^{n}} & \pi_k(j)\in\{\pi_k(i)\mid r=1,\ldots,n\}\forall k=1,\ldots,n. \\ 0 & otherwise. \end{cases}$$

where *f* is the objective function (fitness).

Elements of the **P** *Matrix.* Since the matrix **P** is the product of the crossover, mutation and selection matrices, we have that:

$$p_{ij} = \sum_{p=1}^{n}(\sum_{q=1}^{n} c_{ij} m_{qp}) s_{pj}$$

such that, given the previous expressions for each of the operators, it is possible to compute all the elements p_{ij}.

Matrix P

From (Rudolph, 1994) we conclude that this matrix has exactly one 1 per row. Given a fixed population is clear that its super individual can only improve. Then, $\mathbf{E}_{11}$ is an identity matrix and the matrices $\mathbf{E}_{aa}$ (a>2) are identity matrices with some zeros in the diagonal. It should also be clear that the **E** matrix has columns of zeros: those corresponding to populations whose super individual is not the best individual contained within. Conversely, in those columns corresponding to populations whose super individual is the best in the population, it is possible to find more than one 1. Then, as we descend throughout the rows of blocks, the populations get distributed along the row, depending on the quality of their maximum individual. We will assume that within each block the populations are sorted (by sets within which order is irrelevant) based on the quality of their maximum individual.

individual	fitness	rank
00	0.5	4
01	1.0	3
10	1.5	2
11	2.0	1

Table 1: The four possible individuals.

Experiments

Further details can be consulted in (Reyes, 2003). Let us consider the simplest case that has become commonly associated with the basic (minimum) conditions of a GA. Let l=2 and n=2. In this case, each population consists of 2 individuals and each individual is of length 2. This gives us a total of $2^{2\times2}=2^4=16$ populations to consider by the GA. With the aim of being able to rank the four individuals, we defined the following function: $f(x_1x_2)= x_1+0.5x_2+0.5$. As the function f is strictly positive for any individual, this same function was used as fitness function. In Table 1, we show the four possible individuals, their corresponding fitnesses and their rank. According to the model described in a previous section, the fundamental matrix N was obtained. The package MATHEMATICA 4.0 was used to carry out the computations corresponding to the fundamental matrix N, where $N = (I - \mathbf{T})^{-1}$. On the other hand, a GA with the conditions imposed by the model was run. The fixed parameters were: population size = 2, chromosome length = 2 and crossover probability = 1.0. Thus, the results only depend on the mutation probability (p_m). Note that the super individual (elitist individual) does not have to take part in the evolutionary process. Let g be the random variable whose value is the number of necessary iterations for the convergence of the GA. Next, we show the expected value of the variable g (E[g]) and the corresponding standard deviation (D[g]) obtained (using MATHEMATICA 4.0) by the theoretical model (TM) developed and the results obtained by the GA. In our experiments, we performed 100 runs with different random seeds. The results are shown in Table 2. In Figure 1, we show the graph of the values obtained by both methods.

p_m	GA		TM	
	E[g]	D[g]	E[g]	D[g]
0.001	345.05	540.001	514.137	660.781
0.005	90.98	131.461	103.782	132.47
0.01	40.89	54.019	52.491	66.433
0.03	13.0	16.836	18.307	22.415
0.07	5.08	7.036	8.561	9.852
0.1	4.69	6.5	6.380	7.03
0.2	2.69	3.47	3.872	3.780
0.5	1.32	1.847	2.527	1.964

Table 2: Results.

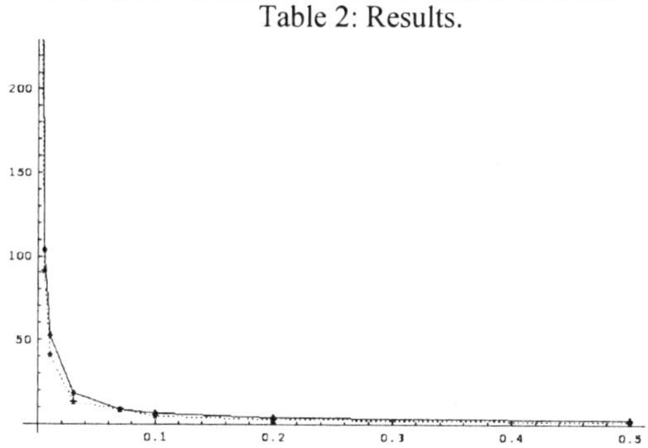

Figure 1:Graph of the values obtained by the GA and the theoretical model developed for the variable g, in terms of the probability of mutation. The continuous line corresponds with the values generated by the theoretical model and the dotted line corresponds to the values generated by the GA.

CONCLUSIONS

The current models to estimate the convergence time of a GA are quite simple and, therefore, very distant from the observed behavior of such type of algorithm. We developed a mechanism to estimate the convergence time of a GA by means of a model based on Markov chains. Our results for the most simple case that can be considered (perhaps argueably) as realistic, led us to conclude that the proposed model is correct. Nevertheless, it is clear that in such case it was relatively easy to obtain the corresponding matrix **P**, but this is a process that will generally get more complicated as we increase the size of the population and the chromosomic length. In general, the size of the matrix has an exponential growth. Therefore, we can conclude that, from a practical point of view, Markov chains are not a recommended theoretical tool for this sort of analysis (i.e., estimation of expected convergence time). Nevertheless, other alternatives exist to which it would be possible to resort (e.g., statistical mechanics (Prugel, 1994) and geometric interpretation approaches (Vose, 1991)). Another possible alternative is to focus the analysis in a different way. For example, in (Coffey, 1999), the author presents a derivation of an upper bound on the convergence of an elitist genetic algorithm which is based on the largest eigenvalue of the corresponding transition matrix. This author also presents an analysis in which he proves that GA convergence can be accelerated when its corresponding Markov chain is of the type called "rapidly mixing chains".

ACKNOWLEDGEMENTS

The first author acknowledges support from CONACyT through a scholarship to pursue graduate studies at the Computer Science Section of the Electrical Engineering Department at CINVESTAV-IPN. The second author gratefully acknowledges support from CONACyT through project 34201-A.

REFERENCES

Ankenbrandt, C. 1991. An extension to the theory of convergence and a proof of the time complexity of genetic algorithms. In *FOGAs*, pages 53–68. Morgan Kaufmann Publishers.

Coffey, S. 1999. An Applied Probabilist's Guide to Genetic Algorithms. Master's thesis, Department of Mathematics, University of Dublin, Trinity College, Dublin, Ireland, September.

Goldberg D. E. and Segrest, P. 1987. Finite Markov chains analysis of genetic algorithms. In *Proceedings of the Second ICGA*, pages 1—8. Lawrence Erlbaum Associates.

Goldberg D. E. 1989. *Genetic Algorithms in Search, Optimization and Machine Learning*. Addison-Wesley Publishing Co., Reading, Massachusetts.

Iosifescu M. 1980. *Finite Markov Processes and Their Applications*. Wiley, Chichester.

Kemeny J. G. and Snell J. L. 1960. *Finite Markov Chains*. D. Van Nostrand Company, Inc., Princeton, New Jersey.

Louis S. J.and Rawlins G. 1993. Syntactic analysis of convergence in genetic algorithms. In L. Darrell Whitley, editor, *FOGA*, pages 141– 151, Morgan Kaufmann Publishers.

Prügel-Bennett A. and Shapiro J. L. 1994. An analysis of genetic algorithms using statistical mechanics. *Physical Review Letters*, 72(9):1305–1309.

Rudolph G. 1994. Convergence properties of canonical genetic algorithms. *IEEE Transactions on Neural Networks*, 1(5):96–101.

Rudolph G. and Agapie A. 2000. Convergence properties of some multi-objective evolutionary algorithms. In *Proceedings of the CEC'2000*, pp. 1010–1016, Piscataway, New Jersey, IEEE.

Reyes M. and Coello C. 2003. On the Study of some theoretical aspects of genetic algorithms. Technical Report 02-2003 (available at http://delta.cs.cinvestav.mx/~ccoello/2003.html).

Vose M. D. 1991. Modeling simple genetic algorithms. In L. D. Whitley, editor, *FOGA 2*, pages 63–73, San Mateo California. Morgan Kaufmann Publishers

PLANNED TOURNAMENT SELECTION

DOUGLAS S. MCCORKLE
Dept. of Mechanical Engineering
Iowa State University
Ames, Iowa 50011

KENNETH M. BRYDEN
Dept. of Mechanical Engineering
Iowa State University
Ames, Iowa 50011

DANIEL A. ASHLOCK
Dept. of Mathematics
Iowa State University
Ames, Iowa 50011

ABSTRACT
Tournament selection is a model of evolution that permits an adjustable fitness gradient via choice of tournament size. Typically tournaments are chosen at random. This study explores the advantage of planning the selection of tournaments. Planned tournament selection can be used to restrict or enhance the rate of spread of information in a manner similar to graph based evolutionary algorithms, and can be used to create island-like models of evolution. Planned tournament selection also enables an efficient parallel schedule for fitness evaluation. (More complex versions of this technique permit multi-deme algorithms with multiple fitness functions.) In this study experiments elucidate the characteristics of planned tournament selection with the following problems: plus-one-recall-store (PORS) 15, PORS 16, location of self-avoiding walks, and the one-max problem for strings. Each problem represents a different class of search space and has different "building blocks" that are necessary to generate a solution. The results from this study are compared to results obtained with graph based evolutionary algorithms on the same problems. Initial results on approximate speed-up that can be gained by implementing planned tournament selection are given. For problems that require expensive fitness evaluations planned tournament selection can significantly reduce the wall clock time required to obtain a solution.

INTRODUCTION

An evolutionary computation system requires a method of selecting parents and placing children back into the population. Tournament selection performs both of these tasks in a simple manner. A tournament is a small group of population members from which good structures are selected for reproduction and bad structures are selected for replacement. A plethora of related techniques, differing in details of tournament size and method of selection, are all called tournament selection. Planned tournament selection (PTS) is any tournament selection scheme in which the individual tournaments are planned rather than chosen at random. There are a number of reasons to use PTS. Island genetic algorithms (Whitley et al., 1997) and graph based evolutionary algorithms (GBEAs) (Bryden et al., 2003a) are both examples of schemes where performance can be enhanced by restricting the flow of information by limiting the potential mate choice with a form of geography. PTS can create these and other restrictions to mating. In addition, if the tournaments are known in advance then the work of performing fitness evaluations can be sent to available processors in an efficient fashion.

BACKGROUND

There are two broad categories of evolutionary algorithms—steady-state and generational. A steady-state evolutionary algorithm replaces a few creatures with an operation called a mating event while a generational evolutionary algorithm replaces the whole population in one larger operation called a generation. These two techniques have been shown to control the information flow in a population differently (Syswerda, 1991). A benefit of steady-state evolutionary algorithms is that selection has much more control over the rate of information flow and how information flows between creatures in the population. A benefit of generational evolutionary algorithms is their naturally parallel nature. Steady-state evolutionary algorithms are not natuarlly parallelized because of the random selection processes that are traditionally executed in a serial fashion.

Graph based evolutionary algorithms (GBEAs) (Ashlock et al., 1999; Bryden et al., 2003a, Bryden et al., 2003b) provide a means to directly control the information flow between creatures. GBEAs utilize a combinatorial graph to impose a topology or geography on the evolving population. GBEAs are similar to island genetic algorithms (Whitley et al., 1997) in that they provide potentially good solutions time to mature before facing competing good solutions. Previous studies (Ashlock et al., 1999; Bryden et al., 2003b) have found that in addition to preserving population diversity the choice of graph in a GBEA can affect the number of mating events required to solve a problem. Problems with simpler fitness landscapes were solved up to 10 times faster and problems with difficult fitness landscapes up to 12 times faster by varying the choice of graph. This paper will present techniques that enable parallel evolutionary algorithms (PEAs) in both steady-state and generational situations while retaining the ability to control information flow as in a GBEA. The current work of parallelization techniques has been focused on generational evolutionary algorithms.

PLANNED TOURNAMENT SELECTION

In contrast to GBEAs, PTS is a naturally parallelizable algorithm and provides greater control over the rate and direction of information transfer. The primary difference between GBEAs and PTS is that in GBEAs order of choosing the tournament groups is random and the tournament groups are fixed by the imposed combinatorial graph. In PTS the order of selection is preplanned and the tournament groups may or may not be static.

In this study we use a type of tournament selection called *(a,b)-planned tournament selection*. If a is relatively prime to the population size n then the function $f(x)=(ax+b)(\mathrm{mod}\ n)$ is a permutation of the numbers $0,1,\ldots,n-1$. For planned tournament selection we must assign each member of the population to a tournament in each generation of the evolutionary algorithm. If the tournaments are of size k then we take the population in index order $(0,1,\ldots,n-1)$ in groups of size k to create the tournaments for the first generation. The permutation given by $f(x)=ax+b$ is used to re-order the population indices in each subsequent generation and the tournaments are still taken in groups of size k. The choice of a and b permit fine control of the mixing rate of the population and even permit the user to select a single mixing population or multiple independent islands. Several values of a and b are used for each test problem. Example:

if $a = 1$ and $b = 1$ and the tournament size is 4 and the population is 16 then the planned tournaments that follow can be seen in Table 1.

Mating Event	Tournament Group Members			
1	0	1	2	3
2	4	5	6	7
3	8	9	10	11
4	12	13	14	15
5	1	2	3	4
6	5	6	7	8
7	9	10	11	12
8	13	14	15	0

Table 1, Sample PTS mating event schedule

CURRENT PARALLEL EVOLUTIONARY ALGORITHMS

Classical parallel evolutionary algorithms have primarily consisted of generational algorithms not steady-state. These generational algorithms have been parallelized in the following manners: master-slave (Cantu-Paz and Goldberg, 2000), multi-deme (Whitley et al., 1997), and hierarchical (Goldberg, 1989).

Parallelization Issues

The steady-state nature of a GBEA does not support an easy parallel implementation. Aside from parallelizing a multi-population steady-state evolutionary algorithm there is limited research in this area. A potential parallel implementation of a GBEA could execute multiple tournament groups across multiple processors. To do this all the creatures in the selected tournament group must be available. That is no one tournament group can have a creature that is dependent on the outcome of a tournament group being simultaneously computed. This excludes simultaneous computation of tournament groups in which the same creature is present and simultaneous computations of tournament groups in which one provides a result to an "upstream" creature in the other tournament group. A parallel GBEA can be constructed as follows:

1. The tournament groups would have to be known beforehand. This requires that a limit be set on the number of mating events that can take place in any run.
2. Random numbers must be generated that represent the parents of all the mating events that will take place in a given GBEA run.
3. Given that the parents were randomly generated, their respective neighbors would be found via the graph that is imposed on the population.
4. Once all of the tournament groups have been defined a schedule can be generated that defines the order in which mating events can take place.

Once the schedule is defined a schedule optimization routine can be used to define when mating events will take place. This is one way to implement a parallel GBEA. From this it can be seen that there is an optimum path through the GBEA when parallelization is available. However this path will be challenging to identify and will have substantial load balancing problems. One way to overcome these challenges is to directly define the tournament sequence. This provides a means to control diversity and the rate of information spread similar to a GBEA while enabling an efficient parallel algorithm. This process is PTS.

To implement PTS in a parallel format on a distributed memory computer the following methods were used. The master-slave model is the starting place for the PTS parallelization model. One feature that is different from the master-slave is that the compute processors will be responsible for not only the fitness evaluation but also the crossover and mutation operations. The master node is responsible for its portion of the population and the assembling of the master population. These two changes allow more work to be done on the compute nodes and reduce the amount of compute expense on the master, providing better load balancing.

IMPLEMENTATION

For the purposes of this study the one-max, PORS 15, PORS 16, and self-avoiding walks were explored to see what values of *a* and *b* solved problems in the most efficient manner. The one-max problem was explored using a parallel implementation of PTS. The number of islands that were generated with each was also explored. For the purposes of this study values used for *a* were (1, 3, 7, 15) and values used for *b* were (1, 2, 4, 8, 16). From these selections of *a* and *b* several families of planned tournaments can be derived. The first family consists of (1,1) and (1,2). These pairings contain one island and have a low rate of information flow. The second family consists of (1,4), (1,8), and (1,16). These pairings contain 128 islands with a high rate of information flow on each island. The third family consists of (3,1), (7,1), and (15,1) which contain four, two, and one island(s) respectively with a moderate rate of information flow. The fourth family consists of the remaining pairings and contains one island with a high rate of information flow. These values allowed for a broad initial investigation of how diversity would be maintained within the population. This study consisted of 10,000 simulations for each of the four different test problems for each of the *a* and *b* values. The number of mating events required to find a correct solution to the problem were saved for each simulation. For each planned tournament and problem the mean and standard deviation of the number of mating events to solution was computed. The means are displayed in Figures 1a though 1d with 95% confidence intervals. The test problems are:

- The one-max problem evolves a population of 20 bit strings to become an all ones string. Fitness is the sum of the bits in a string. The simulation used two-point crossover and single bit-flip mutation.

- The PORS problem is described in detail in (Ashlock, 1998a). It is a type of genetic programming maximum problem (Banzhaf et al., 1998). The goal of the PORS problem is to compute the largest possible number using a fixed number of nodes that can be integer addition, the constant 1, or store and recall operations like those of a pocket calculator. An example of an optimal solution for n = 6 nodes is: (+ (Sto (+ 1 1)) Rcl). The difficulty of PORS varies strongly according to the congruence class (mod 3) of the number of nodes. This study implemented experiments on n = 15 and n = 16 nodes, respectively the hardest and easiest of the three classes. Standard variation (crossover and mutation) operators from genetic programming were used.

- The self-avoiding walk test problem uses a string that specifies a sequence of moves (up, down, left, right) on a grid. Moves off the grid are ignored and a string is scored on the number of squares it visits at least once. A 5x5 grid which admits 824 optimal walks was used in this study. The walk starts in position (1,1) and is permitted 24 moves so optimal walks must be self avoiding. The evolutionary algorithm uses two point crossover and single character replacement mutation when evolving self avoiding walks.

RESULTS

In the PORS 15 (Fig. 1a), PORS 16 (Fig. 1b), one-max (Fig. 1c), and self-avoiding walks (Fig. 1d) the four families of planned tournaments, as discussed in the Implementation Section, can be seen. Note the substantial variation in time to solution depending on the choice of PTS. Each of these problems prefers a different family for optimal performance. This occurs because each problem's character requires a different rate of information flow and a different number of islands to acquire the necessary building blocks to find the optimal solution.

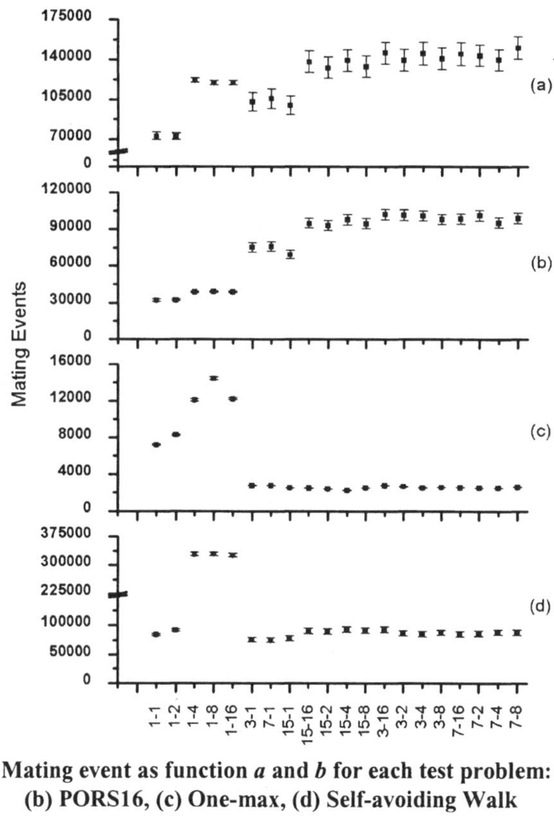

Figure 1, Mating event as function *a* and *b* for each test problem: (a) PORS15, (b) PORS16, (c) One-max, (d) Self-avoiding Walk

In the PORS 16 (Fig. 1b) the cyclic planned tournaments performed the best. The data is in agreement with previous work in GBEAs (Bryden et al., 2003a). The second family of planned tournaments performed better than the third family of planned tournaments. This is probably due to the PORS 16 problem favoring populations where the information flows at a lower rate. For the PORS 15 problem (Fig. 1a) the cyclic type planned tournament also performed the best. In the PORS 15 problem the inverse can be seen of that of the PORS 16 problem in that the third family performed better than the second family of planned tournaments. We conjecture this is due to PORS 15 not being able to locate the necessary building blocks in a planned tournament with 128 isolated islands.

The data indicate that the one-max (Fig. 1c) and self-avoiding-walks (Fig. 1d) prefer the planned tournaments in which there is a moderate rate of information flow, probably due to their simple fitness landscapes and a lack of dependence upon finding building blocks to find the optimal solutions to the problem respectively. In the case of the self-avoiding walks the problem does not exhibit behaviors that are conducive to multiple isolated islands. The problem does favor the third family of planned tournaments. This indicates that the self-avoiding walks benefit from a moderate rate of information flow and from multiple evolving populations. The one-max problem does not seem to favor any specific planned tournament characteristics other than that the problem benefits from moderate or better rates of information flow.

From the data discussed above it can be concluded that PTS does prevent problems associated with premature convergence to be maintained in the population and also employs a predefined schedule that allows the parallelization of a steady-state evolutionary algorithm. These results show that PTS is a viable selection method to incorporate in evolutionary computations. By choosing the appropriate P_a and P_b, PTS can create islands within the population. It is also demonstrated that different families of planned tournaments solve problems at different rates depending on the number of isolated islands created and the rate of information flow.

REFERENCES

Ashlock, D.A., Schmucker, M., Walker, J., 1999, "Graph Based Genetic Algorithms," *Proceedings of the 1999 Congress on Evolutionary Computation,* Washington, D.C., pp. 1362-1368.

Ashlock, D., Lathrop, J., 1998a, "A Fully Characterized Test Suite for Genetic Programming," *Evolutionary Programming VII,* Springer, NY, pp. 537-546.

Ashlock, D., 1998b, "ISAc Lists: A Different Program Induction Method," (with Mark Joenks), *Genetic Programming 1998, Proceedings of the Second Annual Conference on Genetic Programming,* pp. 18-26.

Banzhaf, W., Nordin, P., Keller, R., Francone, F., 1998, *Genetic Programming: An Introduction,* Morgan Kaufmann, San Francisco, CA.

Bryden, K.M., Ashlock, D.A., Corns, S.M., 2003a, "Graph Based Evolutionary Algorithms," submitted to the *IEEE Transactions on Evolutionary Computing.*

Bryden, K., Ashlock, D., McCorkle, D., Urban, G., 2003b, "Optimization of Heat Transfer Utilizing Graph Based Evolutionary Algorithms," *International Journal of Heat and Fluid Flow,* Vol. 24, pp. 267-277.

Cantu-Paz, E., Goldberg, D., 2000, "Efficient Parallel Genetic Algorithms: Theory and Practice," *Computer Methods in Applied Mechanics and Engineering,* Vol. 186, pp. 221-238.

Goldberg, D., 1989, *Genetic Algorithms in Search, Optimization, and Machine Learning,* Addison-Wesley, Reading, MA.

Syswerda, G., 1991, "A Study of Reproduction in Generational and Steady State Generic Algorithms," *Foundations of Genetic Algorithms,* pp. 94-101.

Whitley, D., Rana, S., Heckendorn, R., 1997, "Island Model Genetic Algorithms and Linearly Separable Problems," *Lecture Notes in Computer Science,* Vol. 1305, p. 109.

OPTIMIZING GENETIC ALGORITHMS TO TRAIN FULLY CONNECTED, FIXED FEEDFORWARD NEURAL NETWORKS

IREN VALOVA
University of Massachusetts
Dartmouth, MA
Computer Science Department

GARY MILANO
UMass Dartmouth, MA,
Raytheon Integrated Information
Systems, Falls Church, VA

KEVIN BOWEN
UMass Dartmouth, MA,
Raytheon Integrated Defense
Systems, Portsmouth, RI

ABSTRACT
Fully connected fixed feedforward neural networks possess pattern recognition properties that make them well suited for use in automatic target recognition systems. Their application is hindered, however, by the lack of a training algorithm, which reliably finds a nearly global optimal set of weights in a relatively short time. This paper examines the use of genetic algorithms as an efficient way to train a fully connected fixed feed-forward neural net. A set of experiments, using actual Laser Detection and Range data as training data, is described and the results presented. The effects of changing probabilities of mutation and crossover on neural net training, in the genetic algorithm are explored. The genetic algorithm is adjusted and the effects on training are analyzed from the standpoint of the number of generations required and the resulting correct and incorrect detections and classifications.

INTRODUCTION

Automatic Target Recognition (ATR) systems that utilize sophisticated sensors and advanced forms of machine intelligence to detect and classify objects are a major development area in the intelligence community [1]. The use of neural network algorithms as a way to recognize patterns continues to be a fertile area of development. Applications of this technology are wide ranging and include areas such as cancer tumor identification, Alzheimer pattern detection in brain scans, part identification in production pick and place robotic systems, etc. Essentially, any application that would benefit from expert pattern recognition where the orientation of the object being identified is unknown would be a candidate for this technology.

ATR SIGNAL PROCESSING

The block diagram depicted in Figure 1 provides an overview of the signal processing flow for the ATR system.

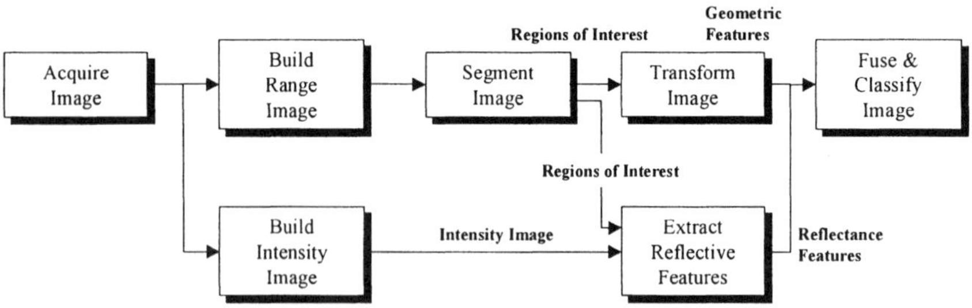

Figure 1. LADAR ATR Data Processing Flow

The LADAR sensor returns two channels of data; range and intensity. A LADAR sensor may be thought of as an array of very narrow radar beams, each beam creating a pixel of data which captures range (determined from round trip signal return time) and intensity (determined from peak signal strength). Therefore, for any given array scan, each pixel will have range and intensity data associated with it, for processing. The range data consists of a pixels image where each pixel has a scaled value representative of the range between the detector and the scene. The intensity image is the strength of the return signal. Both of these images are generated at the same resolution and at the same time, therefore they are registered, coherent images. During segmentation, range information is grouped into regions of interest (ROI) where pixels of the same or nearly the same distance are grouped and the groups identified as one of a tagged list of such ROIs. The tagged image is used to mask the intensity image providing a simple way to determine the foreground and background pixels in the intensity image.

Geometric features are used to classify ROI against tabulated geometries of known objects. Since target aspect angle is indeterminate from the sensor data, a Hough transform is used to generate geometric features at various poses. The transform returns both the dimensional and positional information. Due to the possibility of occlusion it is not possible to simply label the longer of the two horizontal length measurements the long axis of the object. Therefore two different aspects approximately 90 degrees apart are extracted from the histogram. Likewise, two more measurements are made approximately 180 degrees separated from the first two. All four of these measurements are presented to the classifier. The exact features are as follows:

Range: average distance between the detector and the object in a given ROI

Elevation: look-down angle of the ROI

Aspect angle: two long and two short axis measurements. Zero is looking down the estimated long axis. Each of the four aspects has a measurement for:

Length – long axis length

Height – length perpendicular to the long axis

Width – height above the estimated ground plane

Intensity data was used to extract two principal reflectance features: mean reflectance and contrast. Mean reflectance (of an ROI) is the average intensity of the pixels associated with an individual ROI and contrast is the relative reflectance between the ROI and that of background. Both of these parameters are used for classification.

APPLICATION OF A HYBRID FUZZY-NEURAL NETWORK CLASSIFIER

The object classifier consists of a fuzzy logic front end and a neural network as depicted in Figure 2. The neural network replaces a fuzzy-rules-table with a learning network.

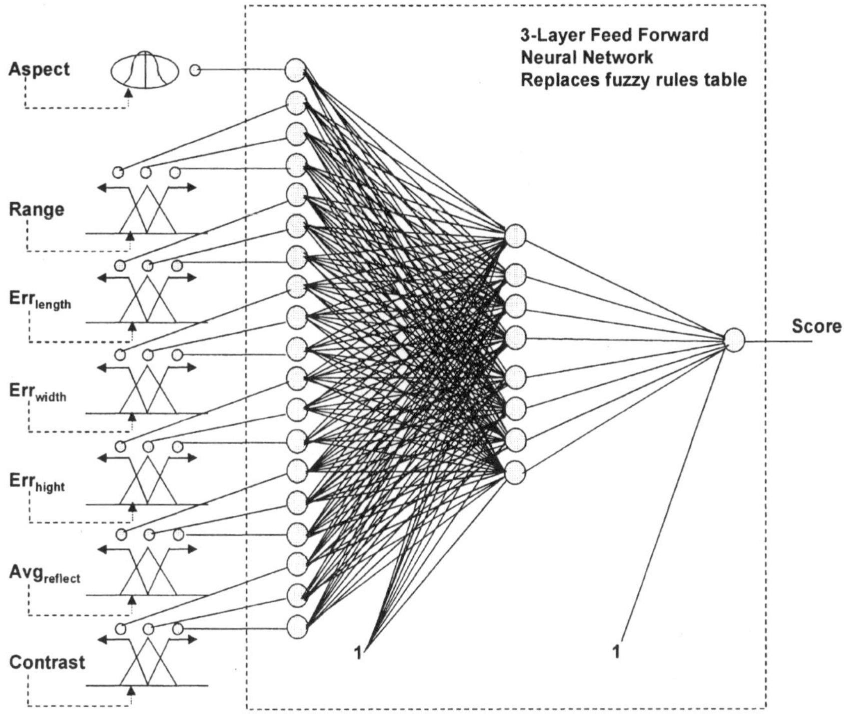

Figure 2. Hybrid Fuzzy-Neural Classifier

NEURAL NETWORK

The neural network consists of one input layer, one hidden layer and one output layer. The input layer accepts inputs from the fuzzy sets, 19 in all. The fuzzy logic front end transforms seven values (aspect, range, three geometric errors, reflectance and contrast) into 19 fuzzy memberships. These 19 inputs are processed by a three layer fully interconnected feed forward neural network. The number of input nodes is governed by the number of features being classified. The output is a single score. The number of hidden nodes was selected by using a rule of thumb that states the number of hidden neurons should be $\sqrt{n*m}$ where (n) is the number of inputs and (m) is the number of outputs [2]. The number was manually adjusted up and down to produce the desired classification. Eight was determined to be the optimal number while seven showed significant performance decrease and nine showed little improvement.

The authors in [3] showed that a genetic training algorithm works and outperforms backpropagation, the standard training algorithm, on complex examples. This work accepts Montana/Davis [3] result relative to the improvements gained by using genetic algorithms rather than backpropagation for network training, and leverages their work to evaluate aspects of optimizing the genetic algorithm to improve its overall performance.

The classifier with its topology defined, is fully defined by determining the adjustable parameters. Adjustable parameters include the thresholds for length, width, and height measurement errors. If the difference between an object table entry dimension and the measured dimension exceeds any one of the thresholds, the ROI is declared clutter and no further processing is made. There also is a score threshold for classifying an ROI as an object. Any classifier scores below this threshold are also identified as clutter. Fuzzy set parameters are also part of the adjustable parameters defining the classifier. The low and high fuzzy sets for each geometric and reflectance feature are defined statistically from the test data. The middle fuzzy set for each feature is also an adjustable parameter. Finally, the weights for the neural network are all adjustable parameters that need to be defined to fully complete the definition of the classifier. Using the genetic algorithm to determine the values for these adjustable parameters is the subject of this paper. In all, there are 179 adjustable parameters that fully define the behavior of the fuzzy-neural classifier.

PROCESSING SUMMARY

To summarize, the system captures images using a LADAR sensor. These images are processed as individual pixels of data. Pixels are combined to make Regions Of Interest (ROI) based on proximity to each other and separately based on range, and intensity (return signal strength). So then the image follows two separate process paths: one for regions of interest with common ranges and one for regions of interest with common intensities. Range ROIs are transformed into different aspect angles and evaluated against geometric thresholds (high and low) to determine if the ROI should be processed further. The same ROI is fused with reflectance data and evaluated against a reflectance threshold to determine if the signal strength (reflectance) is indicative of manmade objects. If it is, then the ROI is processed further. If the ROI has passed both the reflectance and geometric threshold tests it is processed through the classifier. The classifier will then evaluate the ROI against each entry in the target table. The first evaluation against a table entry returning a +1 score at the output of the classifier is called the classification. Failure of any evaluation against a table entry to return +1 results in the ROI being classified as clutter.

ATR TRAINING AND GENETIC ALGORITHMS (GA)

The ATR system obtains its object recognition memory using a combination of programmed-in specific object features and taught image recognition using sample data. The system assigns a classification to an object using the length, width, and height compensated for image range and resolved three geometric properties of about an aspect angle and, the two reflectance properties of intensity and contrast. Geometric properties, range, aspect angle and reflectance features are input to the fuzzy neural classifier which fuses the data and classifies the object. The neural network part of the classifier is trained in advance using actual data representative of what the system will see in operation to "teach" the network to recognize patterns in the fused data. Thus the classifier uses the features extracted from the LADAR image, pre-programmed geometric data, and taught information to assign the object one of the m designated classifications C1, C2, ..., Cm-1, Cm , or Cr where Cr denotes the reject class.

As explained above, the ATR design is incomplete (system does not contain all the necessary memory) until the neural network has been trained. Only after the network is trained and the weights established is the system fully configured and ready for use. Training of the network is done as a separate step in the design process and requires only the fuzzy-neural classifier portion of the ATR system. Training data is presented to the front end of the classifier and the GA training process is applied to determine the weights. Once the weights are selected and set in the system, the system can be put into operation.

APPLYING GENETIC ALGORITHMS TO NEURAL NETWORK TRAINING

Genetic algorithms apply population evolution using mutation and crossover with "survival of the fittest" to solve problems [4]. In the context of the neural network, population members are complete sets of weights (weight sets) for a defined neural network. One set of weights is considered one member of a population where the entire population may consist of many weight sets, 100, 200, 1,000, etc., whatever size population is selected as an initial population. GA uses the population to search the solution space using each member weight set for the neural network and running the training data through the network scoring the results [5]. Results are scored identifying the better performing sets in the population. From these results Tournament Selection is used to identify parents for next generation. Crossover and mutation are applied to the parents to form the next population. The cycle is repeated until scoring indicates that a weight set has achieved the desired performance threshold.

CONCLUSIONS

The objective of this study was to tune the system to optimize the clutter rejection and identification performance. This was done by:

- Modifying the probability of mutation
- Modifying the probability of crossover
- Use of weight Mutation with a Gaussian distribution

The paragraphs below summarize a set of experiments that were conduced using actual Laser Detection and Ranging data as training data, and present the results and conclusions:

All runs resulted in a Correct Identification for 53 of the 68 objects (Targets) in the data set.

The probability of mutation was varied from 10% to 50% and the probability of crossover was varied from 50% to 70%. Using the probability of mutation of 40% and the probability of crossover of 60% results in a much faster convergence on an optimum weight set which demonstrates superior pattern recognition and minimum mean square error. Clutter Called as Targets improved from 149 to 22. This seven-fold decrease in the number of false-positives greatly reduces unnecessary processing cost and allows for more sophisticated object recognition on the remaining 22 objects. Clutter Called as Clutter improved from 45 to 173, nearly eliminating the number of items of which we were uncertain. This improves the confidence in the overall classification scheme. The mean square error decreased from 0.4435 to 0.0027. Future studies could utilize this single parameter to dynamically assess performance and perhaps improve the convergence rate. The results are summarized in Table 1.

Table 1.Varying Probability of Mutation and Crossover Dramatically Improves Clutter Rejection, Clutter Recognition and Mean Square Error

Trial #	Probability of		Clutter classified as		
	Mutation	Crossover	Target	Clutter	MSE
21	0.1	0.8	149	46	0.4435
22	0.2	0.7	92	103	0.1407
23	0.3	0.7	72	123	0.0750
24	0.4	0.7	42	153	0.0158
27	0.4	0.6	22	173	0.0027
28	0.4	0.5	69	126	0.0669

These experiments proved the value of using genetic algorithms to train fully connected, fixed feedforward neural networks for object recognition. We also discovered that using biased weight mutation and using the probability of mutation set to 0.4 and probability of crossover set to 0.6 resulted in faster convergence with reduced error during training. The improved operational performance could be seen in the reduction of false positives and determination of Clutter-as-Clutter.

REFERENCES

[1] "Hybrid Fuzzy-Neural Classifier for Feature Level Data Fusion in LADAR Autonomous Target Recognition", Raytheon Company, 1999; S.W. Soliday, M.T. Perona, and D.G. McCauley

[2] "Neurocomputing", Addison Wesley, Reading, MA, 1990; R. Hecht-Nielsen

[3] "Training feedforward networks using genetic algorithms", In *Proceedings of the International Joint Conference on Artificial Intelligence*, 1989; D.J. Montana and L.D. Davis. Morgan Kaufmann

[4] "Mapping Classifier Systems into Neural Networks", *In Proceedings of the 1988 Conference on Neural Information Processing Systems*, 1988; L. D. Davis. Morgan Kaufmann

[5] "An Introduction to Genetic Algorithms", A Bradford book, Seventh printing 2001; Melanie Mitchell

A STUDY ABOUT AUTOMATIC DESIGN OF INTERSECTION MODEL USING CELLULAR AUTOMATA

1ST NOBUTO IJI
Master Course in Information Engineering, Graduate School of Science and Engineering

2ND SATOSHI ENDO
Department of Information Engineering, Faculty of Engineering

3RD KOJI YAMADA
Department of Information Engineering, Faculty of Engineering

4TH YUHEI AKAMINE
Doctoral Course of Complex Intelligent Systems Engineering, Graduate School of Science and Engineering

ABSTRACT
This paper describes about an automatic design method for an intersection that often causes a traffic congestion. We use *cellular automata* (CAs) for a traffic simulation, because CAs are able to reproduce a complicated phenomenon like traffic flow to decide local cell interaction. In recent research, acquisition methods of CA rule have been studied, but our point of view is to generate optimal intersection patterns by evolving cellular automata (EvCA). EvCA consists of two important elements. One is CA for reproducing of complex phenomenon, the other is genetic algorithm (GA) for acquiring the CA rules. However, in order to reproduce the traffic flow by CA, it is also important to design a suitable *initial configuration* that represent the intersection pattern. Then, we apply EvCA as the method of acquiring the optimal intersection pattern. The experiments show that some optimal intersection patterns are acquired by our method.

1. INTRODUCTION

In today, our living urban area is called the car-oriented society. Traffic congestion is one of the serious problems in city life. The first approach to solve this problem is to simulate traffic flow. By the simulation of the traffic network, we can find the point that we should improve and we can save a cost and time for this improvement[1]. *Cellular automata* (CAs) are able to reproduce a complicated phenomenon like traffic flow to decide local cell interaction. To simulate traffic flow by CA, we have to design *transition rules* and an *initial configuration*. The *transition rules* are the rule set for update from present *state* of each *cell* to next *state* synchronously. The *initial configuration* is the set of cellular states at initial time step $t = 0$. Since CA has typical characteristics homogeneity, locality and parallelism, it is applied to various fields, such as economic system, forest fire and traffic flow. Especially, development of traffic simulators are progressing. *MicroSim*[2] and *TRANSIMS*[3] are the representative examples of this simulator.

The intersection is mentioned as cause of traffic congestion, so we try to optimize intersection pattern. The method using *cellular automata* in combination with genetic algorithm (GA) is known as *Evolving Cellular Automata* (EvCA). EvCA is the framwork to generate the transition rule using evolutionary compuitng. The previous EvCA automatically designs the optimal transition rules. By contrust, we apply EvCA as the method of acquiring the optimal intersection pattern.

2. CELLULAR AUTOMATA

A *cellular automaton* (CA) consists of a regular lattice space of *cells*, each of which has a finite number of possible *states*. The states of all cells change synchronously in discrete time steps through *transition rules*. The transition rules depend only on states of cells in a certain local *neighborhood*, including the current cell. A CA has a *n-dimensional* (n-D) cellular space. Figure 1 shows an example of a 1-D CA, which has the neighborhood r = 3 and state s = 2. According to the transition rules of (a), the states of all cells in the next time step changes (b).

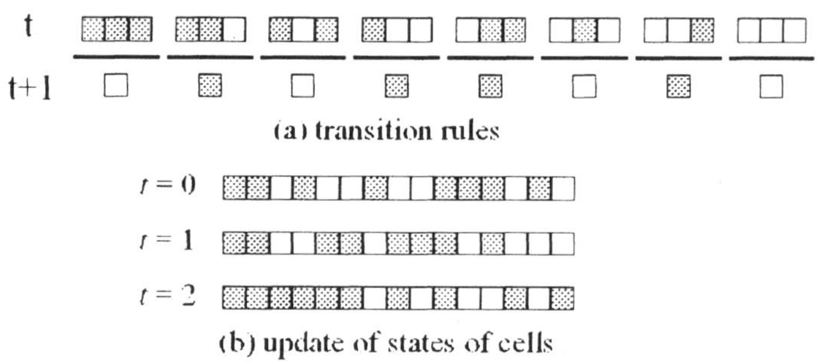

Figure 1: An example of a 1-D CA

3. DESIGN OF TRAFFIC MODEL BY EVCA

In this section, we determine the following two items to generate the intersection pattern by EvCA:

1. A traffic model using CA.
2. Design of GA for acquirement the intersection pattern.

3.1 THE TRAFFIC MODEL USING CA

The *traffic model* is exprssed as 2-D CA, which has four kinds of *cells*[4]:

- *Road cell* is the cell that vahicles can move on.
- *Generating cell* generates a vehicle. This cell tries to generate one vehicle for each time step.
- *Elimination cell* eliminates a vehicle on this cell.

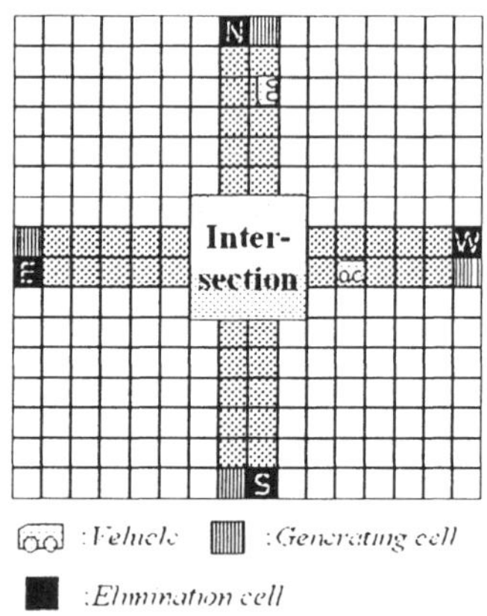

Figure 2: A traffic model

- *Intersection cell* indicates the direction where a vehicle can move. A intersection area is consists of 4x4 intersection cells.

Each cell has three *attributes*:

- *Attribute 1* expresses the existence of a vehicle and its current direction.
- *Attribute 2* expresses the speed of a vehicle. *Attribute 2* has two speeds: stop or move.
- *Attribute 3* expresses the target place of a vehicle. *Attribute 3* has four directions: north, south, west and east.

Only *intersection cell* has *attribute 4*:

- *Attribute 4* expresses the directions that a vehicle on the cell can move to.

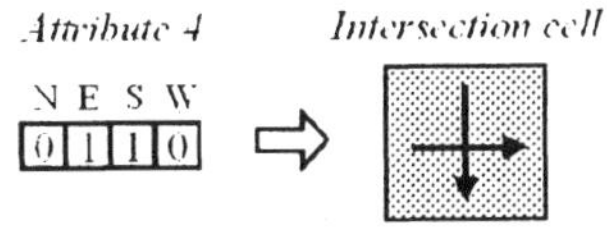

Figure 3: An example of intersection cell's pattern

Attribute 1, 2 and *3* are changed based on the following transition rules.

If (the front cell is free) *then* (the vehicle moves to the cell),

else (the vehicle stops).

Additionally, a vehicle on a intersection cell moves by *driver rules*. *Driver rules* is determine the moving direction of the vehicle. A vehicle tries to turn the direction basis of following priority.

1st. Direction for the target place of the vehicle (target direction).

2nd. The cuurent direction (*attribute 1*) of a vehicle.

3rd. Other direction.

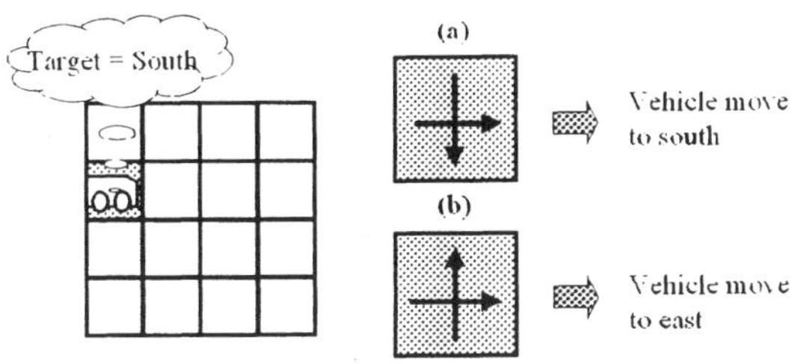

Figure 4: Two examples of determination of the moving direction

Two examples of determination of the moving direction are shown in Figure 4. If intersection cell's pattern is (a), vehicle move to right side of the current direction. And, if this pattern is (b), vehicle move to current direction.

However, if the target direction is same as current direction and a vehicle can move two other directions, the vehicle chooses the left side of the current direction. Here, if a vehicle can not move any directions, the vehicle stops.

In addition, the vehicle behaves following rules for collision avoidance:

- If the front left cell of the vehicle has other vehicle, the vehicle stops.
- If the front right cell of the vehicle has other vehicle, the vehicle moves.
- If the front cell of the vehicle has other vehicle, the vehicles stop.

3-2. AUTOMATIC DESIGN OF INTERSECTION PATTERN

(1) Coding

An *individual* is the array that the *attribute 4* of all *intersection cells* were connected. The *attribute 4* is coded by 4-bit string as shown in Figure.5 When a intersection model is condider a 4x4 lattice space, the bit length of a individual is set to 4x4x4 = 64 and search space is 2^{64}.

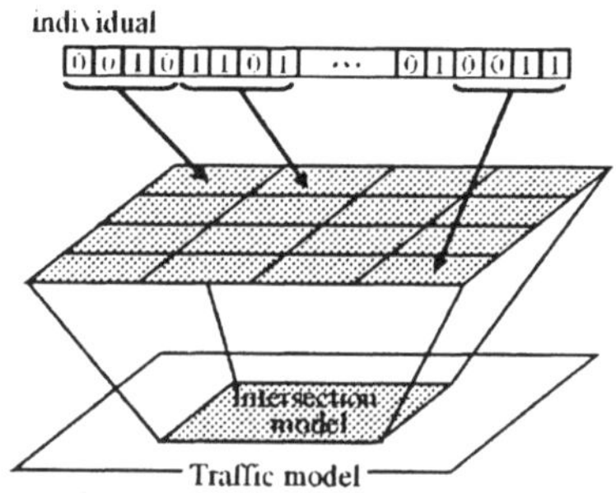

Figure 5: Relation between the intersection model and individual

(2) Fitness

The fitness function formulates as follows:

$$fitness = Weight \times f\_1 + (1 - Weight) \times f\_2 \qquad (1)$$

Where first term denotes the arrival rate for the vehicle, second term denotes the rate of *traffic congestion* on the traffic model. Each term can write the following formulas:

$$f\_1 = \frac{\text{The number of vehicles which arrive at the target place}}{\text{The number of vehicles which enter the intersection}} \qquad (2)$$

$$f\_2 = 1 - \frac{\text{The number of vehicles out of a traffic model}}{\text{The number of vehicles which are generated}} \qquad (3)$$

A new vehicle is not able to generate while the vehicle exists on generated cell. In this research, we defined this situation as traffic congestion, and its rate is computed by the number of times which has not generated a vehicle.

(3) Genetic operators

We use the operator in simple GA. Here, *one-point crossover* and *mutation* are applied.

4. A EXPERIMENTAL RESULT AND DISCUSSION

The parameters used in this experiment are shown in Table 1. Since the arrival rate is more important than the rate of traffic congestion, the *weight* is set up highly. The results are shown in Figure 6(a). Figure 6 on the top shows the intersection patterns, the actual routes on the bottom. As shown in Figure 6(a), We acquired the *crossroads* in this experiment.

Table 1: the parameters

population size	20
crossover rate	0. 6
mutation rate	0. 1
terminal generation	10000
weight	0. 7

However, in this experiment, the *rotary* that is one of the intersection pattern as a typical pattern was not generated. The cause of this result originate in the setup of *driver rules*. In the previous experiment, the vehicles tend to take a route to escape from the inersection in a short time. Under this tendency, the vehicles can not take a route like *rotary*. Then, we change the *driver rules* into what is adopted at the *rotary* intersection.

The result after change of *driver rules* is shown in Figure 6(b). As the result of this simulation, we got a intersection pattern like a *rotary*. From two results, we consider that optimal intersection pattern strongly depends on the determination of *driver rules*.

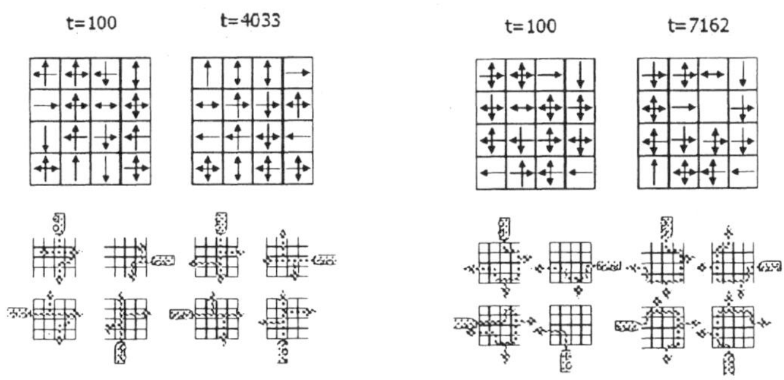

(a) (b)

Figure 6: (a) is the result of the experiment. (b) is the result of the experiment after change of *driver rules*.

5. INSTALLATION OF SIGNAL

We consider that installation of a signal is required for designing more actual intersection. In this research, we use the switching method of switch the intersection pattern. The set of patterns should be switched each intersection pattern is shown in Figure 7. As the parameters that control the timing of switch, we adopt the *cycle length* and *split*. A cycle length is the whole time one intersection pattern sequence take a round. The split is the interval for each intersection pattern.

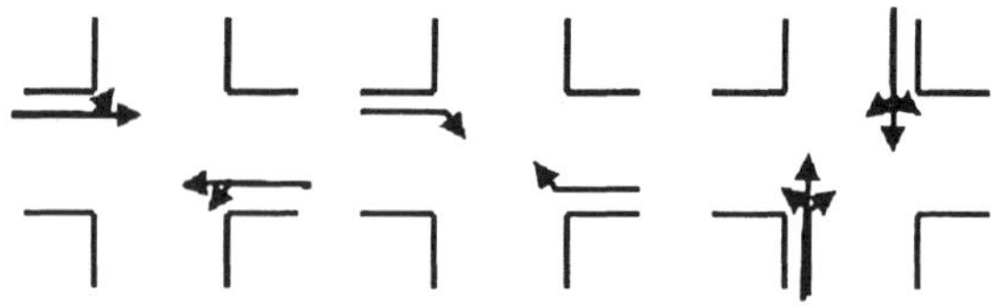

Figure 7: The set of patterns should be switched each intersection pattern

If a intersection pattern is switched roughly, the vehicles may stop in a intersection area. In order to avoid this situation, the time that the vehicles exit from a intersection area is required. In this experiment, we introduce this time as *yellow signal*.

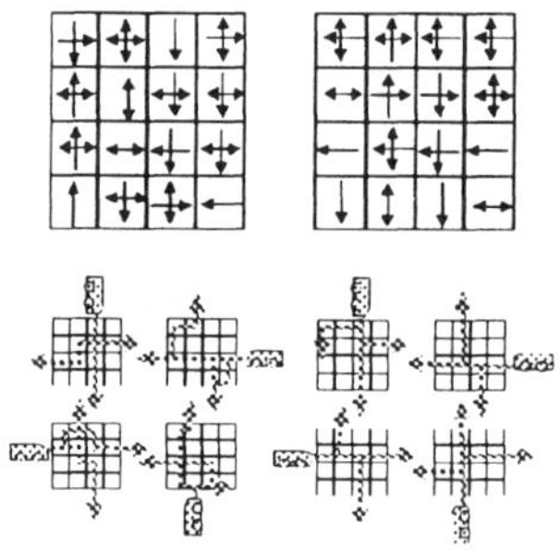

Figure 8: The experimental result after addition to signal control.

When this model applies th GA, each individual is coded as a string expression of two intersection patterns. Since a intersection area is the same as previous experiment, the bit length of an individual is 4x4x4x2=128, and the search space is 2^{128}

Figure 8 shows one of the results of the signal model. In this simulation, we can not acquire the intesection patterns as the signal intersection. We guess that the main cause of this result is setting low vehicle generating rate. In this case, no signal intersection has better traffic flow than signal intersection. We should investigate the cause and effect relation of the vehicle generating rate and the intersection pattern.

6. CONCLUSIONS

We described an automatic method of the intersection pattern using *EvCA*. Our EvCA, it generates not the *transition rules* but an *initial configuration*. From these results, we verified the validity in extension of *EvCA*. Furthermore, we found that optimal intersection pattern strongly depends on the determination of *driver rule*. As a future subject, we will research the coevolutionary learning method between the *initial configuration* and *driver rules*.

REFERENCES

[1] Y.Kato, "Traffic Flow Simulation by Cellular Automata Method", Journal of Japanese Society for Artifical Intelligence, Vol.15, No.2, pp.242-250, 2000.

[2] Rickert,M.and Wagner,P. Parallel Real-time Implementation of Large-scale, Route-plan-driven Traffic Simulation. Int.J.Mod.Phys.C, Vol.7, p.133-153, 1996.

[3] Nagel,K.and Barrett,C.L. using Microsimulation Feedback for Trip Adaptation for Realistic Traffic in Dallas. Int.J.Mod.Phys.C, Vol.8, p.505-525 1997.

[4] N.Iji, S.Endo, K.Yamada, Y.Akamine, "Discussion about Automated Design of Cross Junction Model using Cellular Automata", Fuzzy, Artificial Intelligence, Neural Networks and Computational Intelligence (FAN Symposium '02 in Saga), No.02-10, pp.335-338, 2002.

PART III: FUZZY SYSTEMS

POSSIBILITY REASONING WITH UNCERTAINTY

HENRY L. PFISTER, PHD.
Air Force Research Laboratory
Munitions Directorate
Eglin Air Force Base
email: pfister@eglin.af.mil

ABSTRACT
The problems of reasoning and uncertainty have been studied since the invention of probability three hundred and fifty years ago. This research presents a method for possibility reasoning with uncertainty developed for logical proposition inferences. The approach explicitly includes "uncertain" as a logic state along with "true" and "false". This leads to a model for possibility variables that can be solved as a linear program. The logical inferences from the proposition states can be computed in terms of the possibility variables using methods from fuzzy sets and logic. The 2000-year-old Epiminides paradox is solved as an example to illustrate the unique features of possibility reasoning with uncertainty. A data analysis application is developed for the well-studied Fisher Iris data classification problem. The result was a successful identification rate of 97.3% with low uncertainty and an identification of high uncertainty in the misclassified data samples.

INTRODUCTION

The task of reasoning with uncertainty has been studied throughout history with solutions ranging from astrology to artificial intelligence. Aristotle gave us a formal definition for reasoning when the propositions are certain to be either true or false. The oldest formalism for reasoning under uncertainty is probability theory developed by Pascal and Fermat in an exchange of letters in 1654 about a game of chance. George Boole introduced a formal language for making logical inference about propositions in 1847 with three laws of thought:

- Non-contradiction: No proposition is both true and false
- Excluded Middle: Every proposition is either true or false
- Identity: No proposition ever changes its truth value

They have been widely accepted in Boolean algebra but failed to address many classic reasoning paradoxes. The first recorded reasoning paradox is that of Epiminides the Cretan, who proclaimed that "All Cretans are liars." If we believe this proposition, then we should disbelieve it. But if we disbelieve it, then that is evidence that we should believe it. For centuries such conundrums were considered childish amusements. But around the turn of the twentieth century, they presented a formidable problem for philosophers and mathematicians trying to develop formal systems of reasoning. This paradox is analyzed as an illustration of the unique features of possibility reasoning with uncertainty. Zadeh [8] in 1965 relaxed the requirement for crisp logic required of probability and formal reasoning by introducing fuzzy sets with membership measures to formulate possibility theory. In essence he stated:

- Every fuzzy proposition is both true and false with a membership value
- A fuzzy proposition can change its truth membership value

From this he developed possibility theory and fuzzy logic [7]. Baldwin [9] formulated the use of possibility theory and fuzzy logic with logic programming in 1986 to evaluate uncertain propositions.

This research explores the application of possibility theory and fuzzy logic to reasoning with uncertainty along with a computational method to evaluate inferences. The premise is to explicitly include "uncertain" as a logic state along with "true" and "false." This leads to a formulation of possibility variables that can be solved as a linear optimization problem. Inferences on the proposition state can be then be computed in terms of the possibility values. Several example cases are presented for different classes of reasoning situations. This possibility reasoning with uncertainty process is then applied to the classic Fisher Iris data classification problem. This is one of the most cited statistics data sets with examples appearing in hundreds of papers over sixty years.

REASONING WITH UNCERTAINTY

An important point about the use of probability theory for reasoning is that it is not possible to precisely establish the probability of a combination of two or more logical propositions from the probabilities of the propositions alone. As a result it is only possible to establish the bounds on the probability of the combination. There are two approaches to this problem, either to change the representation from logic to something that is more natural from the point of view of probability theory or to use a numerical measure of uncertainty that is truth functional. The first approach leads to Bayes probability networks and decision trees. The second approach is to use a truth functional concept such as possibility as a measure of uncertainty. In this approach, possibilities associated with propositions are combined truth functionally, and may be used to infer the truth of the combined propositions within tighter bounds than with probability. Possibility theory is based around a measure, which quantifies the degree to which a proposition might have a particular property. For instance, the possibility that a person is tall is a measure of the degree to which they may be tall or are member in the set of tall things. One of the strengths of possibility theory is that it has a large number of different logical proposition combination operations, each of which assumes slightly different nuances of meaning for the reasoning situation. The possibility of a member of a fuzzy set X in a universe U is defined with membership functions m(x) as:

Poss(x in X) = sup min(m(x in U), m(x in X))

This defines limiting cases for the possibility of set intersections and unions by:

Poss(X union Y) = max(Poss(X), Poss(Y))
Poss(X intersect Y) = min(Poss(X), Poss(Y))

Another measure is a dual of the possibility called necessity N defined by:

N(X intersect Y) = min(N(X), N(Y))

Where N(X) = 1 indicates X is true and the dual relationship is

Poss(X) = 1 – N(NOT X) where min(N(X), N(NOT X)) = 0

That implies the following relationships:

Poss(X) => N(X) for all X in U
N(X) > 0 implies Poss(X) = 1
Poss(X) < 1 implies N(X) = 0

The support for a logical proposition A is given as a value pair [Necessity = Sn, Possibility = Sp] so that Sn <= Sp or there exists a necessity support measure less than or equal to a possible support measure and both have negations of 1 – Sn and 1 – Sp. When

Sn + (1 – Sp) <= 1

is satisfied and both Sn and Sp are constrained to the numerical range [0.0, 1.0]. Then the uncertainty of the support for the proposition can be defined as

(1 - Sn - Sp)

Given the support pairs for two propositions A and B as A:[SnA, SpA] and B:[SnB, SpB] and considering all logic combinations of the propositions in the following table defines the set of variables for the AND conjunction operation of the two propositions necessary support as Nij.

AND Support Variables - Nij	B True is SnB	NOT B is 1 - SpB	B Uncertain is SpB - SnB
A True is SnA	A AND B N11	A AND NOT B N12	A N13
NOT A is 1 – SpA	NOT A AND B N21	NOT A AND NOT B N22	NOT A N23
A Uncertain is SpA - SnA	B N31	NOT B N32	UNCERTAIN N33

The necessary support variable table entries Nij must satisfy the following seven equations obtained by summing rows, columns, and elements of the table.

A true implies:	**SnA = N11 + N12 + N13**	**(1)**
NOT A implies:	**1 – SpA = N21 + N22 + N23**	**(2)**
Uncertain A implies:	**SpA – SnA = N31 + N32 + N33**	**(3)**
B true implies:	**SnB = N11 + N21 + N31**	**(4)**
NOT B implies:	**1 – SpB = N12 + N22 + N32**	**(5)**
Uncertain B implies:	**SpB – SnB = N13 + N23 + N33**	**(6)**
Normalization:	**1.0 = SUM(All i and All j) of Nij**	**(7)**

These seven equations do not define nine unique values for the Nij variables but they do specify bounds that can be used to derive the following additional constraints. The argument to form the lower bound is that the minimum support that can be allocated to any variable is the maximum positive sum of the row and column support less 1.0 or zero. The argument to form the upper bound is that the maximum support that can be allocated to any variable is the minimum of the row and column supports leading to sixteen inequality constraints:

(8) Max(SnA + SnB – 1, 0.0) <= N11 <= Min(SnA, SnB)
(9) Max(1 - SpA - SpB, 0.0) <= N22 <= Min(1 - SpA, 1 - SpB)
(10) Max(SpA-SnA+SpB-SnB – 1, 0.0) <= N33 <= Min(SpA-SnA, SpB-SnB)
(11) Max(1 - SpA + SnB - 1, 0.0) <= N21 <= Min(1 - SnA, SnB)
(12) Max(SpA - SnA + SnB – 1, 0.0) <= N31 <= Min(SpA - SnA, SnB)
(13) Max(SnA + 1 - SpB - 1 , 0.0) <= N12 <= Min(SnA, 1 - SnB)
(14) Max(SnA + SpB - SnB – 1, 0.0) <= N13 <= Min(SnA, SpB - SnB)

(15) Max(SpA – SnA + 1 – SpB - 1, 0.0) <= N32 <= Min(SpA - SnA, 1 - SpB)
(16) Max(1 - SpA + SpB – SnB - 1, 0.0) <= N23 <= Min(1 - SpA, SpB - SnB)

A fuzzy logic based approach to solving this set of equations and constraints is to use the operations for the union of fuzzy sets (max) and the intersection of fuzzy sets (min) as operations on support functions as follows:

Sn(A AND B) = min(Sn(A), Sn(B))
Sn(NOT A AND NOT B) = min(Sn(NOT A), Sn(NOT B))
Sn(UNCERTAIN A AND UNCERTAIN B) =
min(Sn(UNCERTAIN A), Sn(UNCERTAIN B))
Sn(A OR B) = max(Sn(A), Sn(B))
Sn(NOT A) = 1 – SpA

Other entries in the table are computed from the row and column constraints.

Using equations and constraints (1) to (16), a linear programming algorithm can solve for the maximum values of the Nij variables as defined by the objective of the logical propositions. The following numerical example defines two first order propositions A and B for a target identification example as:

A: target(military): [Sn=.2, Sp=.5] AND B: target(tank): [Sn = .3, Sp = .67]

The following data is from a custom computer program implementing these computations where C(...) is the objective variable weight and X(...) is the solution value of the linear optimization program.

Input Data
Proposition A: Target is military with Possibility 0.50 and Necessity 0.20
Proposition B: Target is tank with Possibility 0.67 and Necessity 0.30
Solution for 25 equations 43 variables and iterations 21
Maximize -> Target is military AND Target is tank
C(1)=1.00 - X(1)=0.20 - Target is military AND Target is tank
C(2)=0.00 - X(2)=0.33 - NOT(Target is military) AND NOT(Target is tank)
C(3)=0.00 - X(3)=0.30 - UNCERTAIN(Target is military AND Target is tank)
C(4)=0.00 - X(4)=0.10 - NOT(Target is military) AND Target is tank
C(5)=0.00 - X(5)=0.00 - UNCERTAIN(Target is military) AND Target is tank
C(6)=0.00 - X(6)=0.00 - Target is military AND NOT(Target is tank)
C(7)=0.00 - X(7)=0.00 - Target is military AND UNCERTAIN(Target is tank)
C(8)=0.00 - X(8)=0.00 - UNCERTAIN(Target is military) AND NOT(Target is tank)
C(9)=0.00 - X(9)=0.07 - NOT(Target is military) AND UNCERTAIN(Target is tank)
The highest possibility is NOT(Target is military) AND NOT(Target is tank) with value 0.33
Inferences from the possibilities
Possibility that Target is military is true has support of 0.30
Possibility that Target is tank is true has support of 0.20
Possibility that Target is military is false has support of 0.50
Possibility that Target is tank is false has support of 0.33
Possibility that Target is military is uncertain has support of 0.07
Possibility that Target is tank is uncertain has support of 0.37
Max inference that Target is military is false has support of 0.50

The highest inference of 0.5 is that the target is not military with the possibility support of 0.33 that the target is both not military and not a tank. This illustrates explicit reasoning with uncertainty using fuzzy logic conditions and optimization. It is clear in this example that a number of other objective weights C(...) could be maximized to yield different reasoning possibility results.

Three major classes of reasoning situations can be stated for the two-proposition model and implemented with objective weights C(...). The first class is for conditional propositions so that if proposition B is true then proposition A is always true. The second class is for exclusive propositions where either A is true or B is true but not both and at least one is true. The third class is for independent propositions where A or B can be true or false in any combination.

Epiminides Paradox

The paradox of Epiminides the Cretan, who said, "All Cretans are liars" can be modeled by possibility reasoning with uncertainty. If we assert that both propositions are independent that A: Epiminides is a Cretan and B: "All Cretans are liars" with varying possibility and necessity values compute by the program:

A Sp	A Sn	B Sp	B Sn	Maximum Inference Conclusion
1.0	0.0	1.0	0.0	Epiminides is Cretan uncertain support = 1.00
0.5	0.2	0.5	0.2	Epiminides is Cretan false support = 0.50
0.6	0.1	0.5	0.2	All Cretan are Liars false support = 0.50
0.6	0.3	0.4	0.2	All Cretan are Liars false support = 0.60

Many more solutions can be constructed to investigate how this model solves simultaneous conflicting logical inferences. This is a feature not supported by classical reasoning using Boolean algebra with or without the use of probability. Of course there is always a danger of making serious reasoning errors by constructing illogical propositions, assigning unsupported possibilities, and drawing suspect inferences. However, the providing of insight to a 2000-year-old logic paradox is a unique feature of possibility reasoning with uncertainty.

THE FISHER IRIS DATA

The Fisher Iris data is a common benchmark used in object classification and pattern recognition studies and has been referenced by hundreds of papers over the last 67 years. It contains fifty measurements of four features from each of the three Iris species. The features are sepal width, sepal length, petal width, and petal length in centimeters. The species are *Iris setosa, Iris versicolor*, and *Iris virginica.* It was collected by Anderson [1] in 1935 and published by Fisher [2] in 1936 in one of the first statistical classification analysis studies for the genetic characterization of species. The *setosa* species is easily classified by simple methods and can be removed leaving only the problem of the confused species A: *versicolor* and B: *virginica.* The uncertain reasoning strategy is to create independent fuzzy set membership functions for each of the features from the sample data. These functions are used to estimate the support possibility Sp and support necessity Sn for each species using the sample data. The linear program is solved and the resulting supports and inferences are used to classify the sample as *versicolor* or as *virginica*. The results of the classification using various logical objective function weights are as follows:

Inference Objective Function	% Correct
[A true and B false] or [B true and A false]	95
[A true + B false, A false + B true] Triangular Membership	96
[A true + B false – A uncertain, B true + A false – B uncertain]	89
[A true + B false, A false + B true] Trapezoidal Membership	97.3

The 97.3% is a very good result for a general classification process that does not overly fit the training data or require any unusual logic inferences. At this point, it is prudent to ask if the algorithm is a better fit to the data or is the data suspect? An examination of the misclassified Iris sample cases reveals a high uncertainty in the classification. The samples have an inference possibility of being uncertain of 0.75 as compared to a truth possibility of 0.6 or about 30% higher possibility of an uncertain conclusion. The interesting question is what is the underlying source of this uncertainty? There has been conjecture that it is due to errors in the sample data measurement or transcription.

CONCLUSIONS

A possibility reasoning with uncertainty method was developed for evaluating logical proposition inferences. This lead to a model for a maximum possibility computation solved as a linear programming problem. The logical inferences of the proposition states are then computed in terms of the possibility variables. This model was used to analyze the classic Epiminides paradox and illustrates the unique capability of evaluating simultaneous conflicting logical statements. The model was then applied to the Fisher Iris data classification problem to achieve a successful performance of 97.3% and high uncertainty in the erroneous cases. The flexibility and complex inference reasoning ability of this model indicates a wide range of future application possibilities.

REFERENCES

[1] Anderson E., "The Irises of the Gaspe peninsula," *Bulletin of the Ameican. Iris Society*, vol. 59, pp. 2–5, 1935.

[2] Fisher R. A., "The use of multiple measurements in taxonomic problems," *Annals of Eugenics.*, vol. 7, no. 2, pp. 179–188, 1936.

[3] Chien B., Lin J., and Hong T., "Learning discriminant functions with fuzzy attributes for classification using genetic programming," *Expert Systems and Applications*, 2002.

[4] Wang C., Liu J., Hong T., and Tseng S., "A fuzzy inductive learning strategy for modular rules," *Fuzzy Sets and Systems 103*, 1999.

[5] Genther H., and Glesner M., "Advanced data processing using fuzzy clustering techniques," *Fuzzy Sets and Systems 85*, 1997.

[6] Bezdek J., Keller J., Krishnapuram R., Kuncheva L., and Pal N., "Will the Real Iris Data Please Stand Up?", *IEEE Transactions on Fuzzy Systems 7*, 1999.

[7] Zimmerman H., Fuzzy Set Theory and It's Applications - 2nd Edition, Kluwer, 1991.

[8] Zadeh L., "Fuzzy Sets," *Information and Control 8*, 1965.

[9] Baldwin J., "Support Logic Programming", *International Journal of Intelligent Systems 1*, 1986.

ASYMMETRIC KERNEL FUZZY CLUSTER LOADING

MIKA SATO-ILIC
Institute of Policy and Planning Sciences
University of Tsukuba
Tsukuba, Japan

ABSTRACT
We have proposed a model for obtaining the dynamic interpretation of the asymmetric property for 3-way asymmetric similarity data using the idea of fuzzy cluster loading (Sato-Ilic, 2002). This model can capture the quantities of relation between variables of the given observations and the obtained clusters. Since we can get the fuzzy cluster loading under the invariant clusters, we can compare the results of fuzzy cluster loading for the upper part or the lower part of the similarity data over the times and the result of dynamic fuzzy clustering. In this paper, we extend the fuzzy cluster loading for a higher dimension space considering asymmetric directions and times. We use the kernel method to extend the data space to a higher dimension space in order to extract the data structure more efficiently.

INTRODUCTION

In a model for capturing the dynamic change of asymmetry for 3-way asymmetric similarity data and the dynamic change for the interpretation of the obtained clusters over the asymmetric direction and the times (or situations) (Sato-Ilic, 2002), we use the technique of the dynamic clustering model for 3-way data (Sato-Ilic and Sato, 1998) and fuzzy cluster loading (Sato-Ilic, 2001a, 2001b). Due to a property of the dynamic fuzzy clustering model and fuzzy cluster loading model, we can get several results of fuzzy cluster loading under the same clusters and can compare them with respect to the times and directions of asymmetry of the asymmetric similarity data.

The essence of fuzzy clustering is to consider not only the belonging status to the assumed clusters, but also to consider how much the objects belong to the clusters. So, there is a merit to representing the complex data situations which real data almost always have.

However, the interpretation of such a fuzzy clustering causes us some confusion, because we sometimes think that objects which have a similar degree of belongingness can together form one more cluster. In order to obtain an interpretation of the fuzzy clustering result, we have proposed a model of fuzzy cluster loading which can show the degree of relationship between the obtained clusters and variables of a given data.

In this paper, we present a method to obtain the fuzzy cluster loading considering asymmetric direction and times in a higher dimension space than the data space. This model is for the nonlinear case of the fuzzy cluster loading model. We show how we can extend the model for the nonlinear case and obtain the estimate of the fuzzy cluster loading. Using the estimate and the kernel method, we can obtain the kernel based asymmetric fuzzy cluster loading for 3-way asymmetric similarity data. We demonstrate the numerical results of the proposed method.

DYNAMIC ADDITIVE FUZZY CLUSTERING MODEL FOR ASYMMETRIC SIMILARITY DATA

The 3-way asymmetric similarity data is observed by the values of similarity with respect to n objects for T times, and the similarity matrix of t-th time is shown by $S^{(t)} = (s_{ij}^{(t)})$, $s_{ij}^{(t)} \neq s_{ji}^{(t)}$, $i,j = 1,\cdots,n$, $t = 1,\cdots,T$. Suppose the upper triangular matrix of $S^{(t)}$ is denoted as $A^{(tU)}$ and the lower triangular matrix is $A^{(tL)}$. Then we create the following two matrixes:

$$S^{(tU)} = A^{(tU)} + (A^{(tU)})', \quad S^{(tL)} = A^{(tL)} + (A^{(tL)})',$$

where $(\cdot)'$ shows the transformed matrix of $\cdot$. That is,

$$S^{(tU)} = (s_{ij}^{(tU)}),\ s_{ij}^{(tU)} = \begin{cases} s_{ij}^{(t)}, & i \leq j \\ s_{ji}^{(t)}, & i > j \end{cases}, \quad S^{(tL)} = (s_{ij}^{(tL)}),\ s_{ij}^{(tL)} = \begin{cases} s_{ij}^{(t)}, & i \geq j \\ s_{ji}^{(t)}, & i < j \end{cases} \tag{1}$$

Then we assume the super-matrix as:

$$\tilde{S} = \begin{bmatrix} S^{(1U)} & S^{(1U)(1L)} & \cdots & S^{(1U)(TL)} \\ S^{(1L)(1U)} & S^{(1L)} & \cdots & S^{(1L)(TL)} \\ S^{(2U)(1U)} & S^{(2U)(1L)} & \cdots & S^{(2U)(TL)} \\ \vdots & \vdots & \vdots & \vdots \\ S^{(TL)(1U)} & S^{(TL)(1L)} & \cdots & S^{(TL)} \end{bmatrix}, \tag{2}$$

where

$$S^{(rU)(tL)} = (s_{ij}^{(rU)(tL)}),\ s_{ij}^{(rU)(tL)} \equiv m(s_{ij}^{(rU)}, s_{ij}^{(tL)}),\ r,t = 1,\cdots,T,\ i,j = 1,\cdots,n. \tag{3}$$

m is an average function from the product space $[0,1] \times [0,1]$ to $[0,1]$ and satisfies the following conditions:

(1) $\min\{x,y\} \leq m(x,y) \leq \max\{x,y\}$
(2) $m(x,y) = m(y,x)$:(symmetry)
(3) $m(x,y)$ is increasing and continuous

Moreover, from the condition (1), $m(x,y)$ satisfies the following condition:

(4) $m(x,x) = x$:(idempotency)

where $x,y \in [0,1]$.

Then a dynamic additive fuzzy clustering model is shown as

$$\tilde{s}_{ij} = \sum_{k=1}^{K} \rho(u_{i^{(c)}dk}, u_{j^{(c)}dk}) + \tilde{\varepsilon}_{ij}, \tag{4}$$

$$i,j = 1,\cdots,2Tn,\ i^{(c)} = 1,\cdots,n,\ c = 1,\cdots,2T,\ i = n(c-1) + i^{(c)}.$$

$\tilde{s}_{ij}$ is (i,j)-th element of $\tilde{S}$ and $u_{i^{(c)}dk}$ is a degree of belongingness of an object i to a cluster k at d-th direction (upper or lower parts) and satisfies the following condition:

$$u_{i^{(c)}dk} \geq 0, \quad \sum_{k=1}^{K} u_{i^{(c)}dk} = 1,\ \forall i^{(c)}, d. \tag{5}$$

c and d are suffixes for representing time and the direction of the asymmetric proximity data. The relationship is shown as follows:

$$d = \begin{cases} c \equiv 0 \pmod 2 &, \quad d = L \text{ and } t = \frac{c}{2} \\ c \equiv 1 \pmod 2 &, \quad d = U \text{ and } t = \frac{c+1}{2} \end{cases},$$

where $t = 1, \cdots, T$. L and U show the lower and upper triangular matrixes, respectively, that is, an asymmetric relationship in the same time. t shows t-th time, and $c \equiv 0 \pmod 2$ means that c is congruent to 0 modulo 2. $\tilde{\varepsilon}_{ij}$ is an error. $\rho(x, y)$ is assumed to be sharing a common property of x and y and ρ is a symmetric aggregation function which satisfies the following conditions for $x, y, z, w \in [0, 1]$:

$$0 \le \rho(x, y) \le 1, \ \ \rho(x, 0) = 0, \ \ \rho(x, 1) = x. \ \ \text{(Boundary Condition)}$$

$$\rho(x, y) \le \rho(z, w) \ \ \text{whenever} \ \ x \le z, \ \ y \le w. \ \text{(Monotonicity)}$$

$$\rho(x, y) = \rho(y, x). \ \text{(Symmetry)}$$

The merit of the model (4) is to obtain clustering results over asymmetric directions (i.e. upper or lower parts) and times under the same clusters. So, we can compare the results through the asymmetric directions and times. Then we can see dynamical changing situations over the asymmetric directions and times.

EVALUATION OF FUZZY CLUSTERING FOR 3-WAY ASYMMETRIC SIMILARITY

If we obtain the result of fuzzy clustering for 3-way asymmetric similarity data, $u_{i^{(c)}d_k}$, then using the model (4) and 3-way data for the same objects and times, we propose the following model:

$$u_{i^{(c)}d_k} = \sum_{a=1}^{p} x_{i^{(c)}a} z_{a^d k} + \varepsilon_{i^{(c)}d_k}, \ \ \forall i^{(c)}, k, \tag{6}$$

$$i, j = 1, \cdots, 2Tn, \ i^{(c)} = 1, \cdots, n, \ c = 1, \cdots, 2T, \ i = n(c-1) + i^{(c)},$$

$$d = \begin{cases} c \equiv 0 \pmod 2 & , \ \ d = L \ \text{and} \ t = \frac{c}{2} \\ c \equiv 1 \pmod 2 & , \ \ d = U \ \text{and} \ t = \frac{c+1}{2} \end{cases},$$

where $z_{a^d k}$ is fuzzy cluster loading which shows the fuzzy degree representing the amount of loading of cluster k to variable a at d direction (upper part or lower parts) and we call this fuzzy cluster loading. This parameter will show how each cluster can be explained by each variable. $\varepsilon_{i^{(c)}d_k}$ is an error.

The model (6) is rewritten as

$$\mathbf{1} = U_{dk}^{(c)} X^{(c)} \mathbf{z}_{dk} + \boldsymbol{\varepsilon}_{dk}^{(c)}, \tag{7}$$

using

$$U_{dk}^{(c)} = \begin{pmatrix} u_{1^{(c)}d_k}^{-1} & 0 & \cdots & \cdots & 0 \\ 0 & u_{2^{(c)}d_k}^{-1} & 0 & \cdots & \vdots \\ \vdots & \vdots & \vdots & \vdots & \vdots \\ 0 & \cdots & \cdots & \cdots & u_{n^{(c)}d_k}^{-1} \end{pmatrix},$$

$$X^{(c)} = \begin{pmatrix} x_{1^{(c)}1} & x_{1^{(c)}2} & \cdots & x_{1^{(c)}p} \\ x_{2^{(c)}1} & x_{2^{(c)}2} & \cdots & x_{2^{(c)}p} \\ \vdots & \vdots & \vdots & \vdots \\ x_{n^{(c)}1} & x_{n^{(c)}2} & \cdots & x_{n^{(c)}p} \end{pmatrix}, \ \ \mathbf{1} = \begin{pmatrix} 1 \\ 1 \\ \vdots \\ 1 \end{pmatrix}, \ \ \mathbf{z}_{dk} = \begin{pmatrix} z_{1^d k} \\ z_{2^d k} \\ \vdots \\ z_{p^d k} \end{pmatrix}.$$

Where $X^{(c)}(c = 1, \cdots, 2T)$ is a data matrix of n objects with respect to p variables. $u_{i^{(c)}d_k}$ satisfies the following condition:

$$u_{i^{(c)}d_k} \in (0.1), \quad \sum_{k=1}^{K} u_{i^{(c)}d_k} = 1, \quad \forall i^{(c)}, d,$$

which is a stricter condition than (5) in order to avoid $u_{i^{(c)}d_k} = 0$. From (7), we obtain the estimate of z_{dk} as

$$z_{dk} = ((X^{(c)})'(U_{dk}^{(c)})^2 X^{(c)})^{-1}(X^{(c)})' U_{dk}^{(c)}\mathbf{1}. \tag{8}$$

KERNEL METHOD

The kernel method was originally developed in the context of support vector machines (Cristianini and Shawe-Taylor, 2000), the efficient advantage of which is widely recognized in many areas. The essence of the kernel method is arbitrary mapping from a lower dimension space to a higher dimension space. Note that the mapping is an arbitrary mapping, so we do not need to find the mapping, this is referred to as the kernel trick. Suppose an arbitrary mapping Φ:

$$\Phi: R^n \to F, \tag{9}$$

where F is a higher dimension space than R^n. We assume

$$k(\boldsymbol{x}, \boldsymbol{y}) = \Phi(\boldsymbol{x})^t \Phi(\boldsymbol{y}),$$

where k is the kernel function which is defined in R^n and $\boldsymbol{x}$, $\boldsymbol{y} \in R^n$. The typical examples of the kernel function are as follows:

$$k(\boldsymbol{x}, \boldsymbol{y}) = \exp(-\frac{\| \boldsymbol{x} - \boldsymbol{y} \|}{2\sigma^2}). \tag{10}$$

$$k(\boldsymbol{x}, \boldsymbol{y}) = (\boldsymbol{x} \cdot \boldsymbol{y})^d. \tag{11}$$

$$k(\boldsymbol{x}, \boldsymbol{y}) = \tanh(\alpha(\boldsymbol{x} \cdot \boldsymbol{y}) + \beta). \tag{12}$$

(10) is a gaussian kernel, (11) shows polynomial kernel of degree d, and (13) is a sigmoid kernel. By the introduction of this kernel function, we can analyze the data in F without finding the mapping Φ explicitly.

KERNEL METHOD FOR FUZZY CLUSTER LOADING

From (8), z_{dk} can be shown as follows:

$$z_{dk} = ((G_{dk}^{(c)})^t G_{dk}^{(c)})^{-1}(G_{dk}^{(c)})^t \mathbf{1}, \tag{13}$$

where $G_{dk}^{(c)} = (g_{i^{(c)}d_{a(k)}})$, $g_{i^{(c)}d_{a(k)}} \equiv u_{i^{(c)}d_k}^{-1} x_{i^{(c)}a}$, $i = 1, \cdots, n$, $a = 1, \cdots, p$. We rewrite (13) as

$$z_{dk} = ((\boldsymbol{g}_{a(k)}^{(c)d})^t \boldsymbol{g}_{b(k)}^{(c)d})^{-1}((\boldsymbol{g}_{a(k)}^{(c)d})^t \mathbf{1}), \ a, b = 1, \cdots, p, \tag{14}$$

where

$$(\boldsymbol{g}_{a(k)}^{(c)d})^t = (g_{1^{(c)}d_{a(k)}}, \cdots, g_{n^{(c)}d_{a(k)}}),$$

$$(G_{dk}^{(c)})^t G_{dk}^{(c)} = ((\boldsymbol{g}_{a(k)}^{(c)d})^t \boldsymbol{g}_{b(k)}^{(c)d}), \ (G_{dk}^{(c)})^t \mathbf{1} = ((\boldsymbol{g}_{a(k)}^{(c)d})^t \mathbf{1}), \ a, b = 1, \cdots, p.$$

Then we introduce the mapping Φ in (9) for $\boldsymbol{g}_{a(k)}^{(c)d} \in R^n$. The kernel based fuzzy cluster loading in F for 3-way asymmetric data is defined as

$$\tilde{\boldsymbol{z}}_{dk} = (\Phi(\boldsymbol{g}_{a(k)}^{(c)d})^t \Phi(\boldsymbol{g}_{b(k)}^{(c)d}))^{-1} (\Phi(\boldsymbol{g}_{a(k)}^{(c)d})^t \Phi(\mathbf{1})), \ a, b = 1, \cdots, p.$$

Using the kernel representation, $\tilde{\boldsymbol{z}}_{dk}$ is obtained as

$$\tilde{\boldsymbol{z}}_{dk} = (k(\boldsymbol{g}_{a(k)}^{(c)d}, \boldsymbol{g}_{b(k)}^{(c)d}))^{-1} (k(\boldsymbol{g}_{a(k)}^{(c)d}, \mathbf{1})), \ a, b = 1, \cdots, p. \tag{15}$$

NUMERICAL EXAMPLE

We have two 3-way data, $X^{(t)} = (x_{ia}^{(t)})$, $i = 1, \cdots, n$, $a = 1, \cdots, p$, $t = 1, \cdots, T$ and $S^{(t)} = (s_{ij}^{(t)})$, $s_{ij}^{(t)} \neq s_{ji}^{(t)}$, $i, j = 1, \cdots, n$ for the same n objects and the same T times, we use telephone communication data. We show an example when $t = 1$. The data is observed as the number of telephone calls from one prefecture to another for 47 prefectures in Japan (NTT-east, 1998). For example, the number of telephone calls from A prefecture to B prefecture is different from the number of telephone calls from B prefecture to A prefecture, so we can treat this data as an asymmetric similarity data. For the same prefectures, we obtain the data of the sold number of telephone services which sends from each prefecture with respect to four leased circuit services (variables), 0.3kHz - 3.4kHz frequencies band service, voice transmission band service, code service (50bit/s) which is data transmission only, and code service (9600bit/s).

Using the asymmetric similarity data, we create the 94 $\times$ 94 super-matrix in (2). We use the arithmetic mean for the average function. In order to consider the difference of the population scale among the prefectures, we normalize the super-matrix data with respect to each prefecture. Using the model (4), we obtain the result of fuzzy clustering, $u_{i^{(c)d}k}$, when the number of clusters is assumed to be 2. In order to obtain the interpretations for the two obtained clusters, we use (8) and (15) and the data, $x_{i^{(c)}a}$, which is the sold number of telephone services with respect to the four variables. Fig. 1 shows the result of fuzzy cluster loadings in (8) for the upper part and the lower part, and Fig. 2 shows the result of kernel fuzzy cluster loadings in (15). We use polynomial kernel in (11) when $d = 2$.

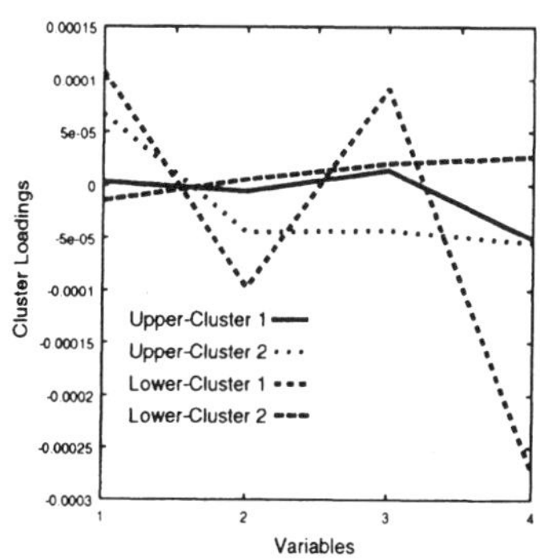

Figure 1 Result of Fuzzy Cluster Loading in Telephone Data

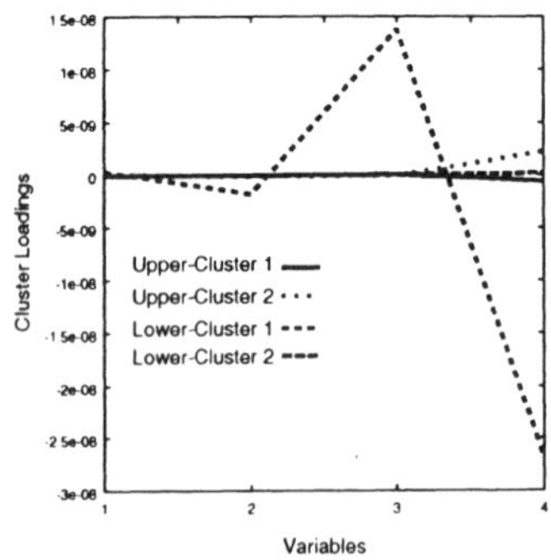

Figure 2 Result of Kernel Fuzzy Cluster Loading in Telephone Data
Variable 1: 0.3kHz-3.4kHz frequencies band service
Variable 2: Voice transmission band service
Variable 3: Code service (50bit/s)
Variable 4: Code service (9600bit/s)

From the results shown in Figs. 1 and 2, we can see a significant property of cluster 1 for receiving part with respect to variables 3 and 4, that is code service (50bit/s) and code service (9600bit/s). The reason seems to be that the prefectures which are in cluster 1 for lower part received a large number of telephone calls from the central area of Japan.

As the comparison between the two results of Figs. 1 and 2, we can see clearer result in Fig. 2, that is, we have large difference of cluster 1 for upper and lower parts, otherwise we cannot see any significant difference of cluster 2 for both parts, upper and lower parts. So, the result used the kernel fuzzy cluster loading may be able to avoid the noise compared with the result used the fuzzy cluster loading.

CONCLUSION

This paper presents extended fuzzy cluster loadings using the kernel method in a higher dimension space and the capability of the kernel fuzzy cluster loading for capturing the dynamic change of an asymmetric data structure over several times. The higher dimension space is related with the data space nonlinearly, so it seems to be adaptable for the dynamic spatial data in which the data performs irregularly over the variables and times.

REFERENCES

Cristianini, N., and Shawe-Taylor J., 2000, "*An Introduction to Support Vector Machines and Other Kernel-Based Learning Methods*", Cambridge University Press.

Sato-Ilic, M., and Sato, Y., 1998, "A Dynamic Additive Fuzzy Clustering Model", *Advances in Data Science and Classification*, A. Rizzi et al., ed., Springer-Verlag, pp. 117-124.

Sato-Ilic, M., 2001a, "Fuzzy Cluster Loadings for Weighted Regression Analysis", *Japanese Classification Society*, pp. 71-78.

Sato-Ilic, M., 2001b, "On Clustering based on Homogeneity", *Joint 9th IFSA World Congress and 20th NAFIPS International Conference*, pp. 2505-2510.

Sato-Ilic, M., 2002, "Dynamic Fuzzy Clustering using Fuzzy Cluster Loading", *Intelligent Engineering Systems through Artificial Neural Networks*, Vol. 12, pp. 309-314.

NTT-east, 1998, http://www.ntt-east.co.jp/info-st/network/traffic/

TUNING OF FUZZY INFERENCE SYSTEMS THROUGH UNCONSTRAINED OPTIMIZATION TECHNIQUES

ROGERIO A. FLAUZINO
UNESP/FE/DEE, CP 473
CEP 17033-360, Bauru-SP, Brazil

IVAN NUNES DA SILVA
UNESP/FE/DEE, CP 473
CEP 17033-360, Bauru-SP, Brazil

JOSE ALFREDO C. ULSON
UNESP/FE/DEE, CP 473
CEP 17033-360, Bauru-SP, Brazil

ABSTRACT

This paper presents a new methodology for the adjustment of fuzzy inference systems. A novel approach, which uses unconstrained optimization techniques, is developed in order to adjust the free parameters of the fuzzy inference system, such as its intrinsic parameters of the membership function and the weights of the inference rules. This methodology is interesting, not only for the results presented and obtained through computer simulations, but also for its generality concerning to the kind of fuzzy inference system used. Therefore, this methodology is expandable either to the Mandani architecture or also to that suggested by Takagi-Sugeno. The validation of the presented methodology is accomplished through an estimation of time series. More specifically, the Mackey-Glass chaotic time series estimation is used for the validation of the proposed methodology.

INTRODUCTION

The fuzzy inference systems design comes along with several decisions taken by the designers since it is necessary to determine, in a coherent way, the number of membership functions for the outputs and inputs, and also the specification of the fuzzy rules sets of the system, besides defining the strategies of rules aggregation and defuzzification of output sets. The need to develop systematic procedures to help the designers has been wide, since very often the trial and error technique is the only one available (Figueiredo and Gomide, 1997).

At present time, there are several researchers engaged in studies related to the design techniques involving fuzzy inference systems. A brief resume about the different approaches for tuning of fuzzy inference systems may be found in Guillaume (2001).

This paper presents a methodology of tuning fuzzy inference systems based on unconstrained optimization techniques. This methodology has the objective of minimizing an energy function associated to the fuzzy inference system. The definition of the energy function must obey the performance requirements of the fuzzy system. It may be thus observed that the correct definition of energy function is fundamental to the success of the desired adjustment (Kamimura et al., 1994).

Using this new approach from the definition point of view, the fuzzy system becomes defined as a three layers model. Each one of these layers represents the tasks performed by the fuzzy inference system, such as fuzzification tasks, fuzzy rules inference and defuzzification. The fuzzy inference system adjustment proposed in this paper is performed through the adaptation of the free parameters from each one of these layers, with the objective of minimizing the energy function previously defined.

To validate the proposed methodology, it is developed a fuzzy inference system to predict the Mackey-Glass time series. Due to its chaotic nature, this estimation problem offers an adequate application to validation of the hereby-underlined approach (Wan et al., 2001).

FUZZY INFERENCE SYSTEMS

The fuzzy inference systems may be treated as systems that use the concepts and operations defined by the fuzzy set theory and the by fuzzy reasoning methods, since they use the fuzzy inference process to perform their operational functions. Basically, these operational functions include the inputs fuzzification of the system, the inference rules associated to it, the aggregation of rules and the later defuzzification of the aggregation results, which represent the outputs of the fuzzy system.

Considering the operational functions performed by the fuzzy inference systems, it is convenient to represent them by a three layers model. Thus, a fuzzy inference system may be given by the sequential composition of an input layer, by an inference layer and by an output layer.

As illustration, Fig. 1 shows a multilayer model, with is constituted by two inputs and one output, having three fuzzy rules in its inference layer.

In the following subsections further details will be presented about how fuzzy inference systems can be represented through a three layers model.

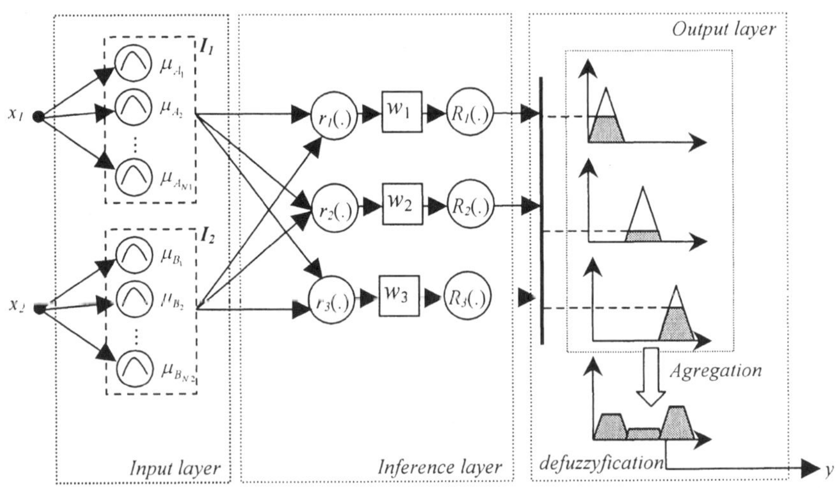

FIGURE 1. Fuzzy inference system composed by two inputs and one output.

Input Layer

The system inputs fuzzification has the purpose of determining the membership degree of each input related to the fuzzy sets associated to each input variable. To each input variable of the fuzzy system can be associated as many fuzzy sets as necessary. This way, given a fuzzy system with only one input and, to this input, associated with N functions, that is, N fuzzy sets which define that input, then the output of the input layer will be a column vector with N elements, which are representing the membership degrees of this input in relation to those fuzzy sets.

If we define the input of this fuzzy system with one only input x, then the input layer output of the fuzzy system will be the vector $\boldsymbol{I1}$, that is:

$$\boldsymbol{I}_1(x) = [\mu_{A_1}(x) \quad \mu_{A_2}(x) \quad \cdots \quad \mu_{A_N}(x)]^T \tag{1}$$

where $\mu_{Ak}(.)$ is the membership function defined to x input, which is referring to the k-th fuzzy set associated to this input.

The generalization of the input layer concept for a fuzzy system having p input variables can be achieved if we consider each input of this fuzzy system being modeled as a sub-layer of the input layer. Taking into account this consideration, the output vector of the input layer I(x) is then defined by:

$$\boldsymbol{I}(x) = [\boldsymbol{I}_1(x_1)^T \quad \boldsymbol{I}_2(x_2)^T \quad \cdots \quad \boldsymbol{I}_p(x_p)^T]^T \tag{2}$$

where x_i is the i-th input of the fuzzy system and $\boldsymbol{I}_k(.)$ is the k-th vector of membership functions associated to the x_k input. In Figure 1, it is illustrated the input layer for a fuzzy system composed by two inputs, which are mapped by the vector $\boldsymbol{I}_1$ and $\boldsymbol{I}_2$.

Inference Layer

The set of rules has fundamental importance to the correct functioning of the fuzzy inference system. There are several methods for the extraction of fuzzy rules from the tuning set.

In this paper, the fuzzy inference system has initially all the possible inferred rules. Therefore, the tuning algorithm has the task of weighting the inference rules. The weighting of the inference rules is an adequate way to represent the most important rules in the fuzzy system, or even to allow that conflicting rules are related to each other without any verbal completeness loss. Thus, it is possible to express the i-th fuzzy rule as in (3), that is:

$$R_i(\boldsymbol{I}(\boldsymbol{x})) = w_i \cdot r_i(\boldsymbol{I}(\boldsymbol{x})) \tag{3}$$

where $R_i(.)$ is the function representing the fuzzy weight value of the i-th fuzzy rule, w_i is the weight factor of the i-th fuzzy rule and $r_i(.)$ represents the fuzzy value of the i-th fuzzy rule. In Figure 1, it is shown the composition involving $r_i(.)$ and $R_i(.)$ for the three fuzzy rules belonging to inference layer.

Output Layer

The output layer of the fuzzy inference system aims to aggregate the inference rules, as well as the defuzzification of the fuzzy set generated by the aggregation of inference rules.

In the fuzzy inference systems design, the choice of not only the aggregation method but also the defuzzification method constitutes a very important decision. The aggregation method of the fuzzy inference rules must be in such a way that the fuzzy set resulting from aggregation is capable of adequately representing the knowledge explicited by this set of fuzzy rules. By analogy, the method chosen for the defuzzification must be capable of expressing, in a crisp value, the fuzzy set resulting from the fuzzy aggregation.

ADJUSTMENT OF THE FUZZY INFERENCE SYSTEM

The formalization of a fuzzy inference system in the form of a multilayer system, as presented in Section 2, can be justified not only by the different operational division of each one of these layers, but also by the presence in each of them of different free parameters.

This way, the mapping f between the input space $\boldsymbol{x}$ and the output space y may be defined by:

$$y = f\left(\boldsymbol{x}, \boldsymbol{mf}_{In}, \boldsymbol{w}, \boldsymbol{mf}_{Out}\right) \tag{4}$$

where $\boldsymbol{mf}_{In}$, $\boldsymbol{w}$ and $\boldsymbol{mf}_{Out}$ respectively represent the vectors of the input membership functions parameters, the weight of the inference rules and the output membership functions parameters. Therefore, $\boldsymbol{mf}_{In}$, $\boldsymbol{w}$ and $\boldsymbol{mf}_{Out}$ represents the free parameters of the fuzzy system and for this reason is better suitable rewrite the expression (4) as presented in (5).

$$y = f\left(\boldsymbol{x}, \boldsymbol{\Theta}\right) \tag{5}$$

where $\boldsymbol{\Theta}$ is the vector resulting from concatenation of the free parameters involved to system, that is,

$$\boldsymbol{\Theta} = \left[\boldsymbol{mf}_{In}{}^{T}, \boldsymbol{w}^{T}, \boldsymbol{mf}_{Out}{}^{T}\right] \tag{6}$$

The definition of the energy function to be minimized remains in function of the fuzzy mapping. Considering that the tuning set $\{\boldsymbol{x}, \boldsymbol{d}\}$ is fixed during the whole adjustment process, it may be written:

$$\xi_{(x,y)} = \xi_{(x,y)}(\boldsymbol{\Theta}) \tag{7}$$

where ξ represents the energy function associated to the fuzzy inference system f.

In problems like this, involving the minimization of energy functions, it is desired that to each iteration, the energy function value is lesser than that value obtained in the previous iteration. There are several techniques used to solve unconstrained optimization problems. A detailed description of the unconstrained optimization techniques may be found in (Bertsekas , 1995). The choice of the most adequate technique to be used is conditioned to the form by which the energy function is defined. For example, the Gauss-Newton method for the unconstrained optimization may be more applicable in problems where the energy function is defined as:

$$\xi(\boldsymbol{\Theta}) = \frac{1}{2}\sum_{i=1}^{m} e^{2}(i) \tag{8}$$

where $e(i)$ is the absolute error in relation to the i-th tuning pattern.

In this paper, a derivation of the Gauss-Newton method is used for the fuzzy inference system tuning. The Gauss-Newton expression to update the vector $\boldsymbol{\Theta}$ is defined by:

$$\boldsymbol{\Theta}_{next} = \boldsymbol{\Theta}_{now} - \frac{1}{2}\left(\boldsymbol{J}^T\boldsymbol{J}\right)^{-1}\boldsymbol{g} \tag{9}$$

where $\boldsymbol{g}$ is the gradient of ξ expressed in (9) and $\boldsymbol{J}$ is the Jacobian matrix of e presented in (9). The optimization algorithm used was the Levenberg-Marquardt method (Marquardt , 1963). The Levenberg-Marquardt method can handle well ill-conditioned matrices $\boldsymbol{J}^T\boldsymbol{J}$ by altering equation (9) to:

$$\boldsymbol{\Theta}_{next} = \boldsymbol{\Theta}_{now} - \frac{1}{2}\left(\boldsymbol{J}^T\boldsymbol{J} + \lambda\boldsymbol{I}\right)^{-1}\boldsymbol{g} \tag{10}$$

The calculation of the matrices $\boldsymbol{J}$ and the vectors $\boldsymbol{g}$ were performed through the finite differences method.

METHODOLOGY AND RESULTS

Using the methodology presented in this paper for fuzzy inference systems tuning, a fuzzy inference system of Mandani type was developed with objective to predict the Mackey-Glass time series (Mackey and Glass,1997), which is defined by:

$$\frac{dx(t)}{dt} = -b \cdot x(t) + \frac{a \cdot x(t-\tau)}{1 + x(t-\tau)^c} \tag{11}$$

where the values of the constants are often adopted as being $a = 0.2$, $b = 0.1$ and $c = 10$. The value for the delay constant τ is chosen as being 17. The tuning set was constituted by 500 patterns. The input variables of the fuzzy inference system were four, which correspond to values $x(t-18)$, $x(t-12)$, $x(t-6)$ and $x(t)$. As an output variable was adopted $x(t+6)$.

The fuzzy inference system was defined having four fuzzy sets attributed to each input variable and also to the output variable. A total of 64 inference rules have been used in the inference process.

The energy function of the system was defined as being the mean squared error between the desired values $x(t+6)$ and the values $\overline{x}(t+6)$, that is:

$$\xi(\boldsymbol{\Theta}) = \frac{1}{2}\sum_{i=1}^{L}\left[x_i(t+6) - \overline{x}_i(t+6)\right]^2 \tag{12}$$

where L is the number of data used in the tuning process (L=500).

The mean squared error of estimation for the proposed problem was 0.000598 with standard deviation of 0.02448. The prediction error for the 1000 points is shown in Fig. 2.

For comparison, it was developed a fuzzy inference system adjusted by the ANFIS (Adaptive Neural-Fuzzy Inference System) method. This fuzzy inference system was composed by 10 membership functions for each input, being the knowledge base constituted by 10 rules. The mean squared error of estimation for the proposed problem was 0.000165 with standard deviation of 0.0041.

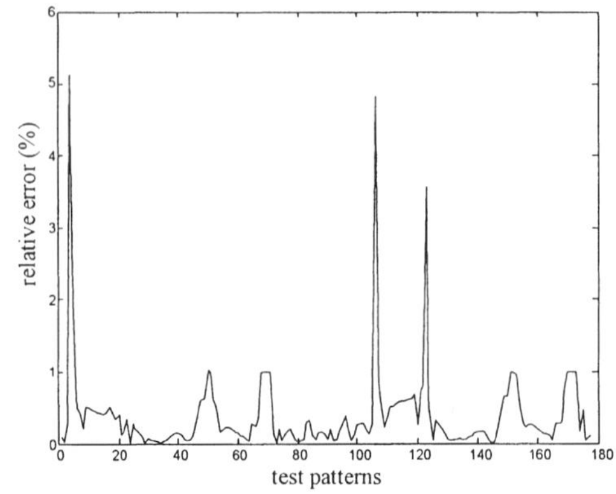

FIGURE 2. Prediction error for the Mackey-Glass series.

CONCLUSIONS

In this paper was underlined the basic foundations involved with the fuzzy inference systems tuning process from unconstrained optimization techniques. In order to become the more efficient tuning, it is necessary that the energy function is perfectly specified for the adjustment process.

To validation of the proposed methodology, the results obtained by the proposed approach were compared to those provided from the ANFIS methodology, and also through of a mathematical modeling problem. The results obtained from this methodology offers new perspectives of researches related to the fuzzy inference systems, allowing thus that problems previously treated only by artificial neural networks may also be treated through fuzzy systems.

REFERENCES

Bertsekas, D. P., 1995, "Nonlinear Programming", Athens Cientific, 1995.

Figueiredo, M., and Gomide F.,1997, "Adaptive Neuro Fuzzy Modeling", *Proceedings of the Sixth IEEE International Conference on Fuzzy Systems*, Vol. 3, pp. 1567-1572.

Guillaume, S., 2001, "Designing Fuzzy Inference Systems from Data: An Interpretability-Oriented Review", *IEEE Trans. Fuzzy Systems*, Vol. 9, No. 3, pp. 426-443.

Kamimura, R., Takagi, T., and Nakanishi, S., 1994, "Improving Generalization Performance by Information Minimization", *IEEE World Congress on Computational Intelligence*, Vol. 1, pp. 143-148.

Mackey, M. C., and Glass L., 1997, "Oscillations and Chaos in Physiological Control Sciences", *Science*, pp. 287-289.

Marquardt, D., 1963, "An Algorithm for Least Squares Estimamation of Nonlinear Parameters", *J. Soc. Ind. Appl. Math*, pp. 431-441.

Wan, W., Hirasawa, K., Hu J., and Murata, J., 2001, "Relation Between Weight Initialization of Neural Networks and Pruning Algorithms: Case Study on Mackey-Glass Time Series", *International Joint Conference on Neural Networks*, Vol. 3, pp. 1750-1755.

DEVELOPMENT OF NEUROFUZZY ARCHITECTURE FOR SOLVING THE N-QUEENS PROBLEM

IVAN NUNES DA SILVA
UNESP/FE/DEE, CP 473
CEP 17033-360, Bauru-SP, Brazil

ANDRE NUNES DE SOUZA
UNESP/FE/DEE, CP 473
CEP 17033-360, Bauru-SP, Brazil

JOSE ALFREDO C. ULSON
UNESP/FE/DEE, CP 473
CEP 17033-360, Bauru-SP, Brazil

ABSTRACT
A neural network model for solving the *N*-Queens problem is presented in this paper. More specifically, a modified Hopfield network is developed and its internal parameters are computed using the valid-subspace technique. These parameters guarantee the convergence of the network to the equilibrium points. The network is shown to be completely stable and globally convergent to the solutions of the *N*-Queens problem. A fuzzy logic controller is incorporated in the network to minimize convergence time. Simulation results are presented to validate the proposed approach.

INTRODUCTION

Artificial neural networks are nonlinear mapping systems whose structure is loosely based on principles observed in the nervous systems of humans and animals. These networks have been applied to several classes of optimization problems and have shown promise for solving such problems efficiently. In this paper, we present a modified Hopfield network designed for solving the *N*-Queens problem. As shown in the next sections, this problem can be treated as a typical problem of combinatorial optimization.

Many problems of both practical and theoretical importance concern themselves with the choice of a best configuration or set of parameters to achieve some goal. Combinatorial optimization problems have a fundamental role in many areas of sciences and engineering, where the objective is to optimize a cost function of a finite discrete system characterized by a large number of possible solutions (Papadimitriou and Steiglitz, 1982).

Basically, all of the neural networks used in combinatorial optimization problems contain some penalty parameters. The stable equilibrium points of these networks, corresponding to solutions of the optimization problem, are obtained only when the penalty parameters are sufficiently large. Moreover, the convergence process of the network depends on the correct adjustment of these parameters. Hence, it is proposed in this paper a modified Hopfield network not depending on the penalty parameters. Simulation results have shown that the network always converges to an equilibrium point representing a solution of the *N*-Queens problem.

THE MODIFIED HOPFIELD NETWORK

The mapping of combinatorial optimization problems by a Hopfield network (Hopfield, 1984), including the N-Queens problem, consists of determining the weight matrix $\boldsymbol{T}$ and the bias vector $\boldsymbol{i}^b$ to compute equilibrium points. A modified energy function $E^m(t)$ is used here, which is defined as follows:

$$E^m(t) = E^{conf}(t) + E^{op}(t) \tag{1}$$

where $E^{conf}(t)$ is a confinement term that groups all the constraints imposed by the problem, and $E^{op}(t)$ is an optimization term that conducts the network output to the equilibrium points. Thus, the minimization of $E^m(t)$ of the modified Hopfield network is conducted in two stages:
i) minimization of the term $E^{conf}(t)$:

$$E^{conf}(t) = -\frac{1}{2}\boldsymbol{v}(t)^T.\boldsymbol{T}^{conf}.\boldsymbol{v}(t) - \boldsymbol{v}(t)^T.\boldsymbol{i}^{conf} \tag{2}$$

where: $\boldsymbol{v}(t)$ is the network output, $\boldsymbol{T}^{conf}$ is weight matrix and $\boldsymbol{i}^{conf}$ is bias vector belonging to E^{conf}. This corresponds to confinement of $\boldsymbol{v}(t)$ into a valid subspace that confines the structural constraints imposed by the N-Queens problem.
ii) minimization of the term $E^{op}(t)$:

$$E^{op}(t) = -\frac{1}{2}\boldsymbol{v}(t)^T.\boldsymbol{T}^{op}.\boldsymbol{v}(t) - \boldsymbol{v}(t)^T.\boldsymbol{i}^{op} \tag{3}$$

where: T^{op} is weight matrix and i^{op} is bias vector belonging to E^{op}. This moves $v(t)$ towards an optimal solution (the equilibrium points). The operation of the modified Hopfield network consists of three main steps:

Step (I): Minimization of E^{conf}, corresponding to the projection of $\boldsymbol{v}(t)$ in the valid subspace (Aiyer et al., 1990; Silva et al, 1997) defined by:

$$\boldsymbol{v} = \boldsymbol{T}^{conf}.\boldsymbol{v} + \boldsymbol{s} \tag{4}$$

where: $\boldsymbol{T}^{conf}$ is a projection matrix ($\boldsymbol{T}^{conf}.\boldsymbol{T}^{conf}=\boldsymbol{T}^{conf}$) and $\boldsymbol{T}^{conf}.\boldsymbol{i}^{conf}=\boldsymbol{0}$. This operation corresponds to an indirect minimization of $E^{conf}(t)$.

Step (II): Application of a nonlinear 'symmetric ramp' activation function constraining $\boldsymbol{v}(t)$ in a hypercube:

$$g^r(v_i) = \begin{cases} 1 & \text{if} \quad v_i > 1 \\ v_i & \text{if} \quad 0 \le v_i \le 1 \\ 0 & \text{if} \quad v_i < 0 \end{cases} \quad ; \text{ where } v_i \in [0,1]. \tag{5}$$

Step (III): Determination of the size of the step Δt, related to network convergence time, by means of the Fuzzy Logic Controller (FLC). This computation is described in Section 4.

Step (IV): Minimization of E^{op}, which involves updating of $\boldsymbol{v}(t)$ in direction to an optimal solution (defined by $\boldsymbol{T}^{op}$ and $\boldsymbol{i}^{op}$) corresponding to network equilibrium points, which are the solutions for the combinatorial optimization problems, by applying the gradient in relation to the energy term E^{op}:

$$\frac{d\boldsymbol{v}(t)}{dt} = \dot{\boldsymbol{v}} = -\frac{\partial E^{op}(t)}{\partial \boldsymbol{v}}$$

$$\Delta\boldsymbol{v} = -\Delta t.\nabla E^{op}(\boldsymbol{v}) = \Delta t.(\boldsymbol{T}^{op}.\boldsymbol{v} + \boldsymbol{i}^{op}) \tag{6}$$

Therefore, minimization of E^{op} consists of updating $v(t)$ in the opposite direction to the gradient of E^{op}. These results are also valid when a 'hyperbolic tangent' activation function is used.

Each iteration represented by the above steps has two distinct stages. First, as described in Step (IV), $\boldsymbol{v}$ is updated using the gradient of the term E^{op} alone. Second, after each updating, $\boldsymbol{v}$ is projected directly in the valid subspace. This is an iterative process, in which $\boldsymbol{v}$ is first orthogonally projected in the valid subspace (Eq. 7) and then thresholded so that its elements lie in the range $[lim^{inf}, lim^{sup}]$. Thus, the mapping of constrained nonlinear optimization problems using a modified Hopfield network consists of determining the weight matrices $\boldsymbol{T}^{conf}$ and $\boldsymbol{T}^{op}$, and the vectors $\boldsymbol{i}^{conf}$ and $\boldsymbol{i}^{op}$.

MAPPING OF THE N-QUEENS PROBLEM

The N-Queens problem was studied by C. F. Gauss in 1850 and its goal is the following: N queens must be placed on a square n x n chessboard in such a way that they do not attack one another. Two chess queens attack each other if they are placed in the same row or in the same column or on the same diagonal of the chessboard. It is obvious that in a solution of the N-queens problem there must be exactly one queen in each row and exactly one in each column (structural constraints).

Formulation of the N-Queens Problem

Let vector $\boldsymbol{p} \in \Re^N$ represents the solution set of an N-Queens problem. The value p_i provides the position of a queen on the i-th row of the chessboard. Thus, the elements belonging to $\boldsymbol{p}$ have integer elements defined by:

$$p_i \in \{1,...,N\} \text{ where } i \in \{1..N\}$$

As a specific example, consider a N-Queens problem for $N = 4$, whose a final solution is given by one queen placed in the position 2 (row 1), one queen placed in the position 4 (row 2), one queen put on the position 1 (row 3), and one queen placed in the position 3 (row 4). The vector $\boldsymbol{p}$ representing this solution would be given by $\boldsymbol{p} = [2, 4, 1, 3]^T$.

The equations of $\boldsymbol{T}^{conf}$ and $\boldsymbol{i}^{conf}$ are designed to guarantee the validity of the structural constraints (there must be at most one queen placed in each row and at most one in each column) for the N-Queens problem. Such constraints limit neural activation to one and only one active state in each row and in each column. Thus, the parameters $\boldsymbol{T}^{conf}$ and $\boldsymbol{i}^{conf}$ can be expressed as follows:

$$\boldsymbol{T}^{conf} = (\boldsymbol{R}^N \otimes \boldsymbol{R}^N) \tag{7}$$

$$\boldsymbol{i}^{conf} = \frac{1}{N}(\boldsymbol{o}^N \otimes \boldsymbol{o}^N) \tag{8}$$

where $\boldsymbol{o}^N$ is an n-element vector of ones; the symbol "$\otimes$" denotes Kronecher product (Graham, 1981); and $\boldsymbol{R}^N$ is an NxN projection matrix defined by:

$$\boldsymbol{R}^N = \boldsymbol{I}^N - \frac{1}{N}\boldsymbol{O}^N$$

where $\boldsymbol{O}^N$ is an NxN matrix of ones. The sum of the elements of each row of a matrix is transformed to zero by post-multiplication with $\boldsymbol{R}^N$, while pre-multiplication by $\boldsymbol{R}^N$ has the effect of setting the sum of the elements of each column to zero.

The energy function E^{op} of the modified Hopfield network for the N-Queens problem is projected to satisfy the diagonal constraints (there must be at most one queen placed on each diagonal). When E^{op} is minimized, the optimal solution corresponds to the minimum energy state of the network. Therefore, the matrix $\boldsymbol{T}^{op}$ and vector $\boldsymbol{i}^{op}$ for the N-Queens problem can be expressed by:

$$\boldsymbol{T}^{op} = -\left[(\boldsymbol{Q}_1^{op} \otimes \boldsymbol{Q}_1^{op}) + (\boldsymbol{Q}_2^{op} \otimes \boldsymbol{Q}_2^{op}) + \ldots + (\boldsymbol{Q}_{(N-1)}^{op} \otimes \boldsymbol{Q}_{(N-1)}^{op})\right] \tag{9}$$

$$\boldsymbol{i}^{ot} = \boldsymbol{0} \tag{10}$$

where the matrix $\boldsymbol{Q}_k^{op} \in \Re^{N\text{x}N}$ is provided by:

$$[\boldsymbol{Q}_k^{op}]_{ij} = \delta_{j-k,i} + \delta_{j+k,i} \tag{11}$$

The term $\delta_{x,y}$ assumes value 1 if x is egual to y; otherwise, $\delta_{x,y}$ assumes value 0.

THE FUZZY LOGIC CONTROLLER

A fuzzy logic controller is associated with the modified Hopfield network and manages the trade-off between the network convergence speed and the accuracy of network equilibrium points. The controller input variables are the sum of squared errors (SSE) between two successive instants and the network energy variation (ΔE^m) defined by:

$$SSE(t) = \frac{1}{2}\sum_{i=1}^{n}[v_i(t) - v_i(t-1)]^2 \tag{12}$$

$$\Delta E^m(t) = |\, E^m(t) - E^m(t-1)\,| \tag{13}$$

The fuzzy logic controller output is given by the linguistic variable Δt, which represents the network convergence step. The universe of discourse associated with each input variable is given by {Very Small (*VS*), Small (*SM*), Medium (*ME*), Big (*BI*), Very Big (*VB*)}, and for the output variable is defined by {Small (*SM*), Medium (*ME*), Big (*BI*)}. The values of Δt are obtained by applying Zadeh's compositional inference to the decision-making rules "IF (SSE is Big) and (ΔE^m is Small) then (Δt is Medium)", as seen in Table 1.

TABLE 1. Set of fuzzy decision-making rules.

		$\Delta E^m(t)$				
	Δt	*VS*	*SM*	*ME*	*BI*	*VB*
	VS	*SM*	*SM*	*ME*	*ME*	*ME*
	SM	*SM*	*SM*	*ME*	*ME*	*BI*
$SSE(t)$	*ME*	*ME*	*ME*	*ME*	*BI*	*BI*
	BI	*ME*	*ME*	*BI*	*BI*	*BI*
	VB	*ME*	*BI*	*BI*	*BI*	*BI*

By using a methodology similar to that used in (Silva et al., 1997), one obtains the membership functions for the variables *SSE*, ΔE^m and Δt, as shown in Fig. 1. These functions are generalized after various simulation examples, and the parameters SSE_0 and ΔE_0 are obtained after network initialization.

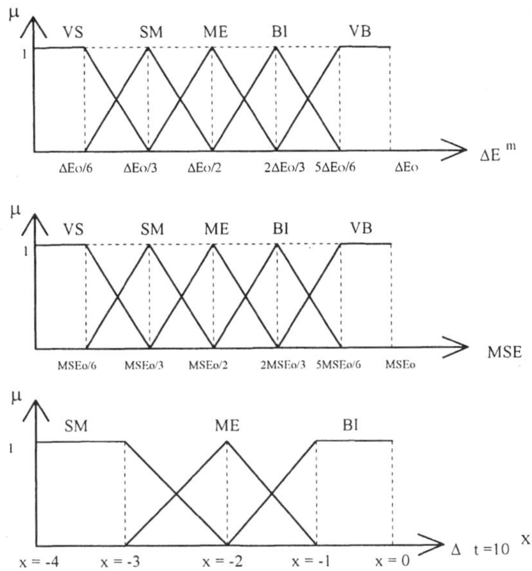

FIGURE 1. Membership functions.

For the rules in Table 1, all elements with non-zero truth values are activated, and the max-min inference procedure is applied to consider the consequents of each. The control action consists of the union of all of the calculated fuzzy functions. The final crisp value of Δt is determined using the method of the center of gravity (Lee, 1990).

SIMULATION RESULTS

To illustrate that the proposed network can be used efficiently, various *N*-queens problems for several values of *N* were simulated. As example, for *N*=5 the modified Hopfield network always converges after fifteen iterations. One of these solutions is given by: one queen placed on the field 1 in row 1, one queen placed on the field 4 in row 2, one queen placed on the field 2 in row 3, one queen placed on the field 5 in row 4, and one queen placed on the field 3 in row 5. In this case, the vectors $\boldsymbol{p}$ and $\boldsymbol{v}(\boldsymbol{p})$, and the matrix $\boldsymbol{V}(\boldsymbol{p})$ computed after convergence of the network are as follows:

$$\boldsymbol{p}^T = [1 \quad 4 \quad 2 \quad 5 \quad 3]$$

$$\boldsymbol{v}(\boldsymbol{p})^T = [\underbrace{1\,0\,0\,0\,0}_{\delta(p_1)^T} \quad \underbrace{0\,0\,0\,1\,0}_{\delta(p_2)^T} \quad \underbrace{0\,1\,0\,0\,0}_{\delta(p_3)^T} \quad \underbrace{0\,0\,0\,0\,1}_{\delta(p_4)^T} \quad \underbrace{0\,0\,1\,0\,0}_{\delta(p_5)^T}]$$

For N=8, the network reaches an equilibrium point representing a solution around forty iterations. An illustration representing two of theses solutions is shown in Fig. 2.

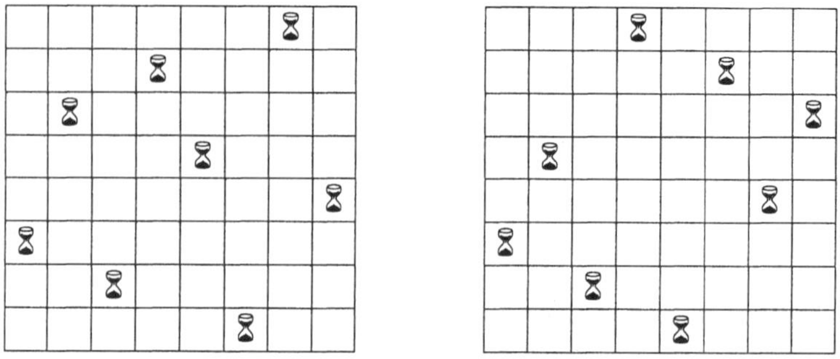

FIGURE 2. Illustration of solutions obtained by the proposed approach.

The main advantages of using a modified Hopfield network for solving the N-Queens problem are: i) consideration of optimization and constraint terms in distinct stages with no interference with each other, ii) use of the unique energy term (E^{conf}) to group all constraints imposed on the problem, and iii) lack of need for adjustment of weighting constants for initialization.

CONCLUSIONS

The development of a neurofuzzy architecture for solving the N-Queens problem is presented in this paper. A modified Hopfield network composes the neurofuzzy architecture and its internal parameters are computed using the valid-subspace technique. These parameters guarantee the convergence of the network to the equilibrium points. A fuzzy logic controller is incorporated in the network to minimize convergence time.

REFERENCES

Aiyer, S. V. B., Niranjan M., and Fallside F., 1990, "A Theoretical Investigation Into the Performance of the Hopfield Network", *IEEE Trans. on Neural Networks*, Vol. 1, pp. 53-60.

Graham, A., 1981, "Kronecher Products and Matrix Calculus," Ellis Horwood Ltd., Chichester, UK.

Hopfield, J. J., 1984, "Neurons With a Graded Response Have Collective Computational Properties Like Those of Two-State Neurons," *Proc. of the National Academy of Science*, Vol. 81, pp. 3088-3092.

Lee, C. C., 1990, "Fuzzy logic in control systems: fuzzy logic controller - Part I and II." *IEEE Trans. on Systems, Man and Cybernetics*, Vol. 20, pp. 404-435.

Papadimitriou, C. H., and Steiglitz, K., 1982, "Combinatorial Optimization – Algorithms and Complexity." Prentice Hall, Englewood Cliffs, New Jersey.

Silva, I. N., Arruda, L. V. R., and Amaral, W. C., 1997, "Robust Estimation of Parametric Membership Regions Using Artificial Neural Networks", *International Journal of Systems Science*, Vol. 28, No. 5, pp. 447-455.

A NEW APPROACH FOR FUZZY INFERENCE IN INTUITIONISTIC FUZZY SYSTEMS

OSCAR CASTILLO AND PATRICIA MELIN
Dept. of Computer Science, Tijuana Institute of Technology
P.O. Box 4207, Chula Vista CA 91909, USA
ocastillo@tectijuana.mx

ABSTRACT

We describe in this paper a proposed new approach for fuzzy inference in intuitionistic fuzzy systems. The new approach combines the outputs of two traditional fuzzy systems to obtain the final conclusion of the intuitionistic fuzzy system. The new method provides an efficient way of calculating the output of an intuitionistic fuzzy system, and as consequence can be applied to real-world problems in many areas of application. We illustrate the new approach with a simple example to motivate the ideas behind this work. We also illustrate the new approach for fuzzy inference with a more complicated example of monitoring a non-linear dynamic plant.

INTRODUCTION TO INTUITIONISTIC FUZZY LOGIC

The intuitionistic fuzzy sets where defined as an extension of the ordinary fuzzy sets [1]. As opposed to a fuzzy set in X [17], given by

$$B = \{(x, \mu_B(x))| \; x \in X\} \qquad (1)$$

where $\mu_B : X \to [0, 1]$ is the membership function of the fuzzy set B, an intuitionistic fuzzy set A is given by

$$A = \{(x, \mu_A(x), \nu_A(x))| \; x \in X\} \qquad (2)$$

where $\mu_A : X \to [0, 1]$ and $\nu_A : X \to [0, 1]$ are such that

$$0 \leq \mu_A + \nu_A \leq 1 \qquad (3)$$

and $\mu_A(x)$; $\nu_A(x) \in [0, 1]$ denote a degree of membership and a degree of non-membership of $x \in A$, respectively.

For each intuitionistic fuzzy set in X, we will call

$$\pi_A(x) = 1 - \mu_A(x) - \nu_A(x) \qquad (4)$$

a "hesitation margin" (or an "intuitionistic fuzzy index") of $x \in A$ and, it expresses a hesitation degree of whether x belongs to A or not. It is obvious that $0 \leq \pi_A(x) \leq 1$, for each $x \in X$.

On the other hand, for each fuzzy set B in X, we evidently have that

$$\pi_B(x) = 1 - \mu_B(x) - [1 - \mu_B(x)] = 0 \text{ for each } x \in X. \qquad (5)$$

Therefore, if we want to fully describe an intuitionistic fuzzy set, we must use any two functions from the triplet [16]:

- Membership function,
- Non-membership function,
- Hesitation margin.

In other words, the application of intuitionistic fuzzy sets instead of fuzzy sets means the introduction of another degree of freedom into a set description (i.e. in addition to μ_A we also have ν_A or π_A).

Since the intuitionistic fuzzy sets being a generalization of fuzzy sets give us an additional possibility to represent imperfect knowledge, they can make it possible to describe many real problems in a more adequate way.

INTUITIONISTIC FUZZY INFERENCE SYSTEMS

Fuzzy inference in intuitionistic systems has to consider the fact that we have the membership μ functions as well as the non-membership ν functions. In this case, we propose that the conclusion of an intuitionistic fuzzy system can be a linear combination of the results of two classical fuzzy systems, one for the membership functions and the other for the non-membership functions.

Assume that IFS is the output of an intuitionistic fuzzy system, then with the following equation we can calculate the total output as a linear combination:

$$IFS = (1 - \pi)\, FS_{\mu} + \pi\, FS_{\nu} \qquad (6)$$

where FS_{μ} is the traditional output of a fuzzy system using the membership function μ, and FS_{ν} is the output of a fuzzy system using the non-membership function ν. Of course, equation (6) for $\pi = 0$ will reduce to the output of a traditional fuzzy system, but for other values of π the result of IFS will be different as we are now taking into account the hesitation margin π.

The advantage of this method for computing the output IFS of an intuitionistic fuzzy system is that we can use our previous machinery of traditional fuzzy systems for computing FS_{μ} and FS_{ν}. Then, we only perform a weighted average of both results to obtain the final output IFS of the intuitionistic fuzzy inference system. We consider below a simple example to illustrate these ideas.

Example: Let us assume that we have a simple intuitionistic system of only two rules:

IF x is small THEN y is big
IF x is big THEN y is small

We will consider for simplicity uniform rectangular membership functions for both linguistic variables. We show in Figure 1 the membership functions for the linguistic values "small" and "big" of the input linguistic variable. We also show in Figure 2 the non-membership functions for the linguistic values of the output variable.

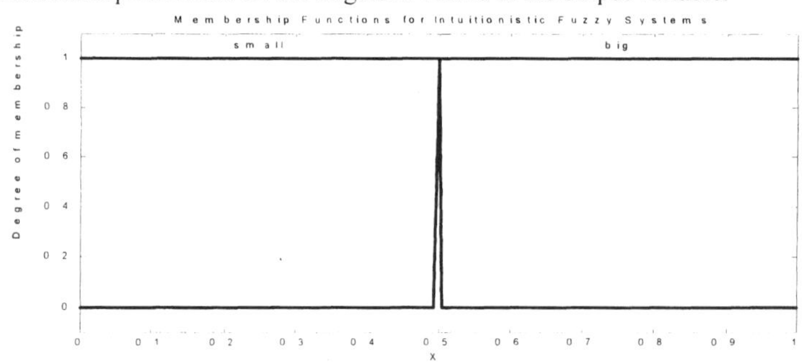

Figure 1 Membership functions for the "small" and "big" values of the input variable.

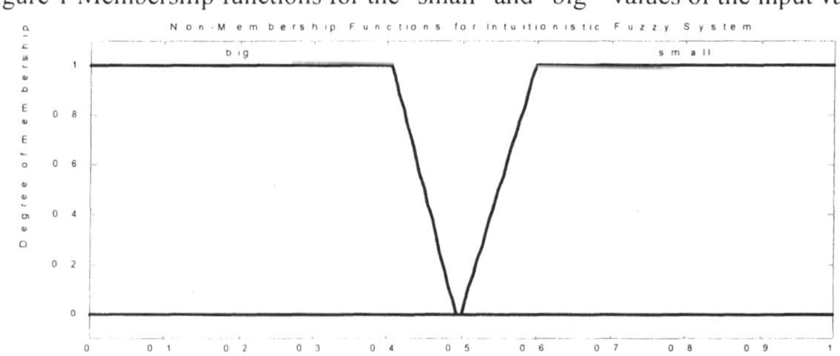

Figure 2 Non-membership for the "small" and "big" values of the output variable.

From Figure 2 we can clearly see that the hesitation margin π is 0.05 for both cases. As a consequence Equation (6) can be written for our example as follows:

$$IFS = 0.95FS_{\mu} + 0.05FS_{\nu} \qquad (7)$$

Now, let us assume that we want to calculate the output of the fuzzy system for a given input value of $x = 0.45$. In this case, we have that x is small with $\mu = 1$ and x is not small with $\nu = 0$, and x is big with $\mu = 0$ and x is not big with $\nu = 0.5$. As a consequence,

$$IFS = IFS_{small} + IFS_{big}$$
$$IFS = 0.95FS_{\mu small} + 0.05FS_{\nu small} + 0.95FS_{\mu big} + 0.05FS_{\nu big}$$
$$IFS = 0.95FS_{\mu small} + 0.05FS_{\nu big}$$
$$IFS = 0.95(0.75) + 0.05(0.765) = 0.74075$$

Of course, we can compare this intuitionistic output with the traditional one (of 0.75), the difference between these two output values is due to the hesitation margin. We have to mention that in this example the difference is small because the hesitation margin is small. We show in Table 1 the results of the intuitionistic system for several input values.

Table 1 Sample results of the intuitionistic fuzzy system for several input values.

Input values, x	Membership result	Non-membership result	Intuitionistic result
0.2500	0.7500	0.7766	0.741330
0.3500	0.7500	0.7766	0.741330
0.4500	0.7500	0.7650	0.740750
0.5500	0.2500	0.2359	0.249295
0.6500	0.2500	0.2250	0.248750
0.7500	0.2500	0.2250	0.248750

MONITORING AND DIAGNOSIS USING FUZZY LOGIC

Monitoring means checking or regulating the performance of a machine, a process, or a system [6]. Diagnosis, on the other hand, means deciding the nature and the cause of a diseased condition of a machine, a process, or a system by examining the symptoms. In other words, monitoring is detecting suspect symptoms, whereas diagnosis is determining the cause of the symptoms [4].

The importance of monitoring and diagnosis of plant processes now is widely recognized because it results in increased productivity, improved product quality and decreased production cost [5]. As a result, in the past decade, a large number of research and development projects have been carried and many monitoring and diagnosis methods have been developed [12, 13]. The commonly used monitoring and diagnosis methods include modeling-based methods [15], pattern recognition methods [14], fuzzy systems methods [2, 3], knowledge-based systems methods [13], and artificial neural networks [9]. It is interesting to note that even though these methods are rather different, they share a very similar structure as shown in Figure 3.

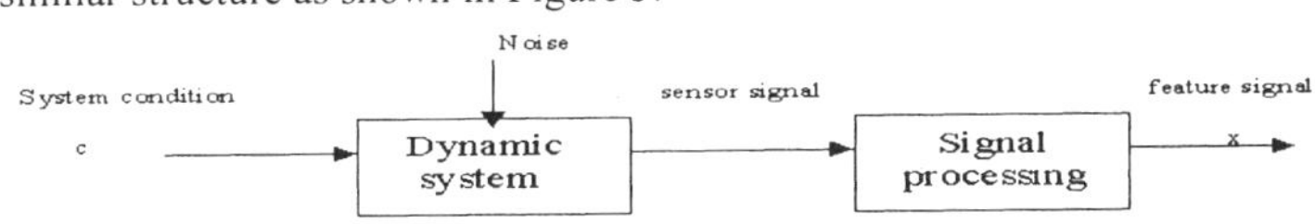

Figure 3 Plant Monitoring and Diagnosis.

In Figure 3 the engineering system is considered to be dynamic, i.e. changing its condition with time. For this reason, there is a block in Figure 5 called "dynamic system". Also, in Figure 3 there is block indicating that "signal processing" has to be performed on the sensor signals to extract the feature signals, which in turn are used to be able to monitor the dynamic process.

The "condition" of a machine, a process, or an engineering system (which will be referred to as system condition and denoted by $c \in \{c_1, c_2,..., c_m\}$) can be considered as the "input", the system working conditions and noises (including system noise and sampling noise) can be considered as the "noise", and the sensor signals are the "outputs" from the system. Typically, the sensor signals are processed by a computer, after which the signals are transformed into a set of features called feature signals, denoted as $\mathbf{x} = \{x_1, x_2,..., x_n\}$. In general, the systems conditions are predefined, such as normal, critical, etc. On the other hand, the features may be the mean of a temperature signal, the variance of a displacement signal, etc. [5, 6].

In particular, when a fuzzy system is used, the relationship is given by a set of rules as shown in Figure 4. The input to the fuzzy system is the feature signal and the output of the fuzzy system is the estimated plant condition(s) [i.e., $\mathbf{z} = (z_1, z_2,..., z_m)$ is an estimate of $\mathbf{c} = (c_1, c_2,..., c_m)$]. In other words, the fuzzy system models the inverse relationship between the system conditions and the feature signals.

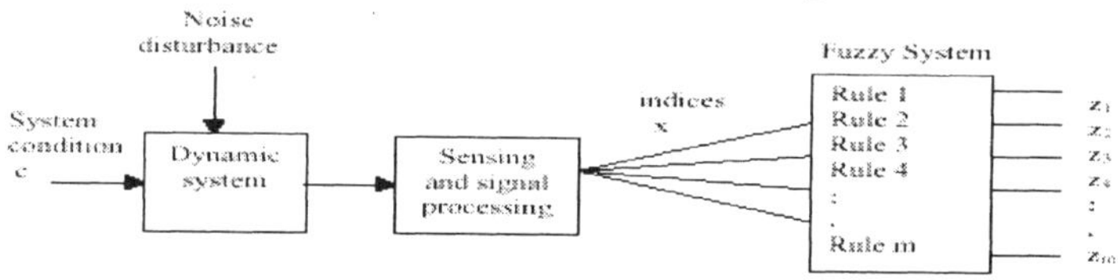

Figure 4 Fuzzy System for Plant Monitoring and Diagnosis.

MONITORING USING THE INTUITIONISTIC FUZZY LOGIC APPROACH

In this section, we show how to implement a fuzzy rule-based expert monitoring system with two basic sensors: temperature, and pressure. We also use as input the fuzzy fractal dimension of the time series of the measured variables [10, 11]. Individual sensors can identify two linguistic values (high and low). The three inputs can be combined to give 8 different scenarios. Let x_1 be the temperature, x_2 the pressure, x_3 the fuzzy fractal dimension, and y the diagnostic statement. Let L_i and H_i, represent the two sets of low range and high range for input data x_i, where i = 1, 2, or 3. Furthermore, let $C_1, C_2,..., C_8$ be the individual scenarios that could happen for each combination of the different data sets. The fuzzy rules have the general form:

$R^{(0)}$: IF x_1 is L_1 AND x_2 is L_2 AND x_3 is L_3 THEN y is C_1

... ... (8)

$R^{(i)}$: IF x_1 is V_1 AND x_2 is V_2 AND x_3 is V_3 THEN y is C_i

... ...

$R^{(26)}$: IF x_1 is H_1 AND x_2 is H_2 AND x_3 is H_3 THEN y is C_8

In this case, V_i represents L_i or H_i, depending on the condition for the plant. Experts have to provide their knowledge in plant monitoring to label the individual cases Ci for i = 1, 2,..., 8. Also, the membership functions for the linguistic values of variables have to be defined according to historical data of the problem and expert knowledge.

We can use the Fuzzy Logic Toolbox of MATLAB to implement the intuitionistic fuzzy monitoring system described above. In this case, we need to specify the particular fuzzy rules, and the corresponding membership functions and non-membership functions for the problem. Of course, in this case the fractal dimension is estimated using the box counting algorithm, which was implemented in MATLAB The actual 8 rules were defined according to expert knowledge on the process.

1. *If (temp is low) and (pressure is low) and (fracdim is low) then (condition is C1)*
2. *If (temp is low) and (pressure is high) and (fracdim is low) then (condition is C2)*
3. *If (temp is low) and (pressure is low) and (fracdim is high) then (condition is C3)*

4. If (temp is high) and (pressure is low) and (fracdim is low) then (condition is C4)
5. If (temp is low) and (pressure is high) and (fracdim is high) then (condition is C5)
6. If (temp is high) and (pressure is low) and (fracdim is high) then (condition is C6)
7. If (temp is high) and (pressure is high) and (fracdim is low) then (condition is C7)
8. If (temp is high) and (pressure is high) and (fracdim is high) then (condition is C8)

In Figure 5 the trapezoidal non-membership functions for the temperature variable are shown. On the other hand, Figure 6 shows the membership functions of the temperature variable. We can appreciate from these figures that the membership and non-membership are not complementary, which is expected in an intuitionistic fuzzy system.

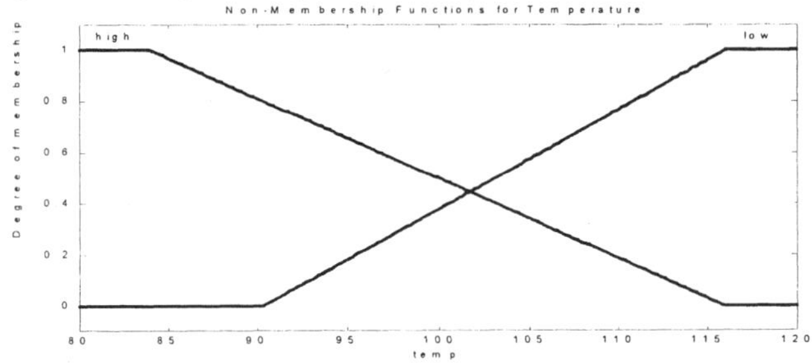

Figure 5 Trapezoidal non-membership functions of the temperature variable.

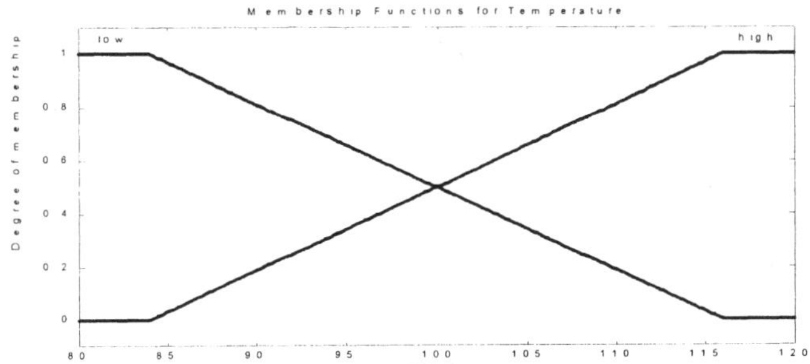

Figure 6 Trapezoidal membership functions of the temperature variable.

We also show in Figure 7 a plot of the outputs of the intuitionistic fuzzy system, and the membership and non-membership fuzzy systems. We can appreciate from this figure the difference between the outputs of the traditional fuzzy system and the intuitionistic fuzzy system.

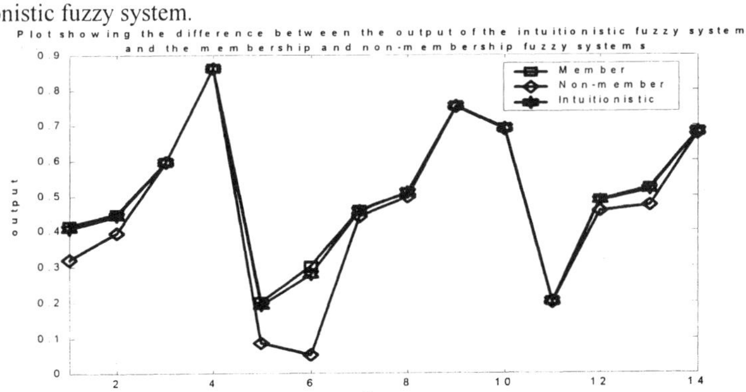

Figure 7 Comparison between the intuitionistic fuzzy system and the traditional one.

CONCLUSIONS

We described in this paper a new approach for fuzzy inference in intuitionistic fuzzy systems. The new approach is relatively easy to use and as consequence is reasonable to apply it in real-world problems. We have described the procedure and have shown a simple example to explain the ideas behind the new approach. We have also shown a more complete example of plant monitoring in which intuitionistic fuzzy logic has the potential of modeling the uncertainty in the identification of problems in a dynamic process. We think that this new approach for fuzzy inference can also be applied to other problems like in control and time series prediction.

ACKNOWLEDGMENTS

We would like to thank the Research Grant Committee of COSNET for the financial support given to this research project (under grant 488.02-P). We would also like to thank the Department of Computer Science of Tijuana Institute of Technology for the time and resources given to this project.

REFERENCES

1. Atanassov, K. (1999). "Intuitionistic Fuzzy Sets: Theory and Applications", Springer-Verlag, Heidelberg, Germany.
2. Castillo, O. and Melin, P. (1999). "Automated Mathematical Modelling for Financial Time Series Prediction Combining Fuzzy Logic and Fractal Theory", Edited Book " Soft Computing for Financial Engineering", Springer-Verlag, Heidelberg, Germany, pp. 93-106.
3. Castillo, O. and Melin, P. (2001). "Soft Computing for Control of Non-Linear Dynamical Systems", Springer-Verlag, Heidelberg, Germany.
4. Castillo, O. and Melin, P. (2003). "Soft Computing and Fractal Theory for Intelligent Manufacturing", Springer-Verlag, Heidelberg, Germany.
5. Chen, G. And Pham, T. T. (2001). "Introduction to Fuzzy Sets, Fuzzy Logic, and Fuzzy Control Systems", CRC Press, Boca Raton, Florida, USA.
6. Chiang, L.H., Russell, E.L., and Braatz, R.D. (2000). "Fault Detection and Diagnosis in Industrial Systems", Springer-Verlag, Heidelberg, Germany.
7. Du, R. (1998). "Engineering Monitoring and Diagnosis", Edited Book " Industrial and Manufacturing Systems", Academic Press, San Diego, California, USA, pp. 119-144.
8. Du, R., Elbestawi, M. A., and Wu, S.M. (1993). "Automated Monitoring of Manufacturing Processes", Transactions ASME J. Eng. Ind., pp. 245-252.
9. Jang, J.R., Sun, C.T. and Mizutani E. (1997). Neuro-Fuzzy and Soft Computing, Prentice Hall.
10. Mandelbrot, B. (1987). "The Fractal Geometry of Nature", W.H. Freeman and Company, New York, NY, USA.
11. Mandelbrot, B. (1997). "Fractals and Scaling in Finance", Springer-Verlag, Heidelberg, Germany.
12. Melin, P. and Castillo, O. (2001). "A New Method for Adaptive Model-Based Control of Non-Linear Dynamic Plants using a Neuro-Fuzzy-Fractal Approach", *Journal of Soft Computing*, Springer-Verlag, 5(2), pp. 171-177.
13. Melin, P. and Castillo, O. (2002). "Modelling, Simulation and Control of Non-Linear Dynamical Systems", Taylor and Francis, London, Great Britain.
14. Patton, R.J., Frank, P.M., and Clark, R.N. (2000). "Issues of Fault Diagnosis for Dynamic Systems", Springer-Verlag, Heidelberg, Germany.
15. Russell, E.L., Chiang, L.H., and Braatz, R.D. (2000). "Data-driven Methods for Fault Detection and Diagnosis in Chemical Processes", Springer-Verlag, Heidelberg, Germany.
16. Szmidt, E. and Kacprzyk, J. (2002). "Analysis of Agreement in a Group of experts via Distances between Intuitionistic Preferences", Proceedings of IPMU'2002, Universite Savoie, France, pp. 1859-1865.
17. Zadeh, L. A. (1965). "Fuzzy Sets", *Information and Control*, 8, pp. 338-353.

SPECIFYING THE SUBJECTIVE: FUZZY EVALUATION OF SYSTEM FEATURES

GORAN TRAJKOVSKI AND DARUSH DAVANI

Computer and Information Sciences Department
Towson University, 8000 York Road
Towson, MD 21252, USA
E-mail: {gtrajkovski, ddavani}@towson.edu

ABSTRACT
Often times we face problems of evaluation of system's features that are not easily expressible in numerical terms. Some of them are extremely dependent on the evaluator's subjectivity. In this paper we propose our original fuzzy evaluation environment to be used in systems evaluation that will ease the evaluation procedure. The procedure is more objective and founded upon a consistent criterion. The consistency of the criterion is congruent with the traditional evaluation procedure, since the fuzzy procedure is, eventually, expected to provide with an evaluation expressed in traditional terms. Prior to the evaluation process, the evaluation criteria are set in order to meet the personal expectations of the evaluator. We propose a neural network based algorithm towards establishing a standard evaluation criterion of an evaluator, and discuss the challenges within the multi-evaluator environments.

INTRODUCTION

Many systems deal with problems, whose specifications are complex, so that non-human methods can often be used to advantage, (Addis, 1986; Avritzer et al., 1996; Bastani et al., 1993). Complex problems usually lead to a waste number of possible solutions, each of which must be examined with care – an assessment task that is often well beyond normal human capability. Systems are constantly being developed commercially to solve nontraditional problems, (Perry, 1995). Taking into account the non-formal nature of many problems that involve human arbitrage, (Trajkovski et al., 1999), the exploration of possibilities within the fuzzy set theory can help, (Zadeh, 1996). The fuzzy logic is suitable for modeling human-like reasoning by the computing-with-words paradigm (Klir and Yuan, 1995; Trajkovski and Janeva, 1997).

Instead of assessing the attributes with a crisp number, an expert evaluator can choose to assess a system using fuzzy numbers, (Biswass, 1995; Trajkovski et al., 1997). A fuzzy number is a discrete or continuous function that has at most one local maximum or a maximal plateau. In other words, it is a possibility distribution, (Klir and Yuan, 1995). Therefore, the evaluating expert, imperfect as he/she is, distributes his belief over a given domain.

This paper proposes a fuzzy evaluating environment that will ease the evaluation procedure, will make it more objective, and is founded upon a consistent criterion. The consistency of the criterion is congruent with the traditional evaluation, since the fuzzy evaluation system is eventually expected to provide with an evaluation in traditional terms.

Prior to the evaluation process, the evaluation criteria is set in order to meet the personal expectation of the evaluator, as well as the possible expectations of an institution.

This paper is organized as follows. The next section formalizes the Traditional Evaluation Environment. The following section gives the frameworks of the proposed Fuzzy Evaluation Environment. Further, then, we discuss the issues that arise in a multi-evaluator environment. The section before the conclusion illustrate the process of deveoping a consistent fuzzy evaluation criterion of an evaluator, with the help of artificial neural networks. The last section concludes the work presented in the paper.

TRADITIONAL EVALUATION ENVIRONMENT

The traditional evaluation systems depend on the institution that performs the evaluation, and on the historically established practices in evaluation. The usual approach is assigning grades or numerical values according to some criteria learned with practice. In this section we formalize the traditional evaluationg environmnet and give a common example, form a perspective that would be congruent with the definition of the fuzzy counterpart evaluating environmnet.

Let us denote by G the set of all n grades that can be given by an evaluator $G=\{G_1>G_2>\ldots>G_n\}$ with the corresponding grade points (weight function w, defined on G): $w_i=w(G_i)$, $i=1,2,\ldots,n$. The pair (G,w) is called Traditional Evaluation Environment. The standard evaluation procedure within this environmnet consists generally of the following steps:

(i) Thorough examination of the product;
(ii) Marking the strengths and weaknesses to facilitate the overview, and
(iii) Deciding which grade to awarded to the product at hand. The level of performance is decided in qualitative terms, which are fuzzy in nature.

FUZZY EVALUATION ENVIRONMENT

The evaluation of an expert system, as stated above, is a problem that involves the concepts of uncertainty and subjectivity. The knowledge-based decision and evaluation performed during the process are complex and hard to deal with, unless with fuzzy tools. The fuzzy paradigm is suitable for dealing with tasks where the observed parameters and phenomena are not easily expressed in terms of exact numerical values. In the sequel of this section we present the theoretical framework of the fuzzy evaluation environmnet that we are proposing (Trajkovski, 1997; Trajkovski et al., 1997; Turski, 1998).

The non-empty set $U=\{u_1<u_2<\ldots<u_m\} \subseteq N_{100}=\{1,2,\ldots, 100\}$, is called universal set for the FEE. The elements are chosen in order to define the levels of evaluation on which the evaluator is to express his/her degrees of satisfaction by the aspect of the software product that is the subject of the evaluation. A fuzzy set on U is a mapping $M: U\rightarrow[0,1]$. For a given element $u\in U$, the value M(u) is referred as the membership degree or level of membership of u to the set M. The non-empty fuzzy set M defined on U is called fuzzy

evaluation grade (FEG) if its graph is of one of the following types: decreasing, increasing, or at first increasing, then decreasing.

The FEG is supposed to show the weight of belief $M_i = M(u_i)$ of the evaluator that the software product covers ui % of the evaluator's expectations.

Let M be a FEG. Then the vector (m-tuple) $M=(M_1,M_2,\ldots,M_m)$ is called FEG-vector of M. The set F of all FEGs that can be awarded is called the FEG-set of the FEE. The binary relation SS defined on F with Eq. (1),

$$SS(\mathbf{A},\mathbf{B}) = \frac{A \cdot B}{\max\{A \cdot A, B \cdot B\}} = \frac{\sum_{i=1}^{m} a_i b_i}{\max\left\{\sum_{i=1}^{m} a_i^2, \sum_{i=1}^{m} b_i^2\right\}}, \quad (1)$$

where A, B∈F ($A=(a_1,a_2,\ldots,a_m)$ and $B=(b_1,b_2,\ldots,b_m)$), and "." denotes the dot product of vectors, is called the measure of similarity. The value SS(A,B) from Eq. (1) is called the degree/coefficient of similarity between A and B. The fuzzy relation SS is a similarity (fuzzy equivalence) relation and is said to be a good choice as far as dilatation/concentration is concerned , (Biswass, 1995).

Let Si, i=1,…,n be FEGs. The matrix S, defined as follows: $S=[SS(S_i,S_j)]_{nxn}$ is called the similarity matrix of the sets S_i. The n-tuple $C=(S_1,S_2,\ldots, S_n)$ is called standard Fuzzy Criterion (SFC), if the values in the columns and the rows of the similarity matrix of the sets S_i, i=1,2,…,n, increase; decrease or at first increase then decrease as the row/column index increases/decreases. The sets Si are then called Standard Fuzzy Sets (SFSs) or components of C. This definition ensures that for any FEG M, one of the following inequalities will hold:

$$SS(M,S_1) \; O \; SS(M,S_2) \; O \ldots O \; SS(M,S_n) \quad (2)$$

where the n-1 circles (O) in Eq. (2) are substituted with either ≤ or ≥ signs, in that order that all the ≤ signs always precede the ≥ signs in the inequality. Thus, it is ensured that the similarity between M and one FEG is greatest, and the similarity grades decrease the further we go from the most similar Si to M.

Let M be a FEG and C a given SFC, with components S_i, and $s_i=SS(M,S_i)$, i=1,2,…,n. G_j is said to be the correspondent traditional grade (CTG) to the FEG M with respect to C if $s_j=\max\{s_i, i=1,2,\ldots n\}$. In case of having more than one maximal value between the coefficient s_i, it is left to the discretion of the evaluator to choose which grade to be awarded as the CTG to M wrt. C. Let G be the Grade set of TEE, C a SFC wrt. SS similarity measure, whose FEGs are defined on the universe U. Then, the quatriple (G,U,C,SS) is called Fuzzy Evaluation Environment (FEE).

MULTI-EVALUATOR ENVIRONMENTS

Evaluators typically have different degree of appropriateness for a given task. The appropriateness of a particular evaluator is a function of how well the agent's skill matches the expertise to do the task, the extent to which its limited knowledge is adequate for the task, its current processing resources and the quality of its communication with the other agents. In this section we discuss the challenges of a multi-evaluator fuzzy evaluation environment.

In real life, like in situations when we ask for a second opinion, it is not unlikely that two different physicians are two different answers. The same kind of mismatch may happen when we ask two (or more) experts to evaluate a tool/system. The reason why that happens is NOT the incompetence of the expert. The reason lies in casually–defined terms and notions in the domain knowledge.

Fuzzy Evaluation Grade (FEM)			Traditional Grade
M1	M2	M3	
0.0	0.8	1.0	A
0.2	0.7	1.0	A
0.1	0.3	1.0	A
0.0	0.3	0.9	A
0.0	0.4	0.8	A
0.1	0.4	0.9	A
0.1	0.5	0.4	B
0.3	0.5	0.7	B
0.1	0.4	0.1	C
0.3	0.9	0.3	C
0.2	0.9	0.1	C
0.2	0.7	0.2	C
0.5	0.5	0.1	D
0.9	0.7	0.0	D
0.6	1.0	0.2	D
0.7	1.0	0.0	D
1.0	0.3	0.0	D
0.1	0.3	0.0	D
1.0	0.5	0.0	F
1.0	0.0	0.0	F

Table 1. The training set of the NN

How to draw conclusions on the expert system features, when the traditional grades of the evaluators are different? The usual approach would be the use of voters. They can be majority, median, plurality and weighted voters, Lorczak et al., 1989), and we choose them so that they reflect the current situation with all the evaluating agent at hand with seemingly inevitable consequences on the reliability of the whole process, (Barbara and Garcia-Molina, 1987).

SOLVING THE PROBLEM OF STANDARDIZED CRITERION

In this section we give results of a practical approach towards a personalized SFC of an evaluator. It is a SFC obtained with the help of a backward propagation neural network (NN). Within this procedure we will try to reveal one of the four components of FEE, namely SFC, given the other three beforehand.

In order to standardize the criterion of an evaluator, we have chosen 20 expert systems written by different students. The systems that were already graded in a traditional manner. We asked the same evaluator to assign FEGs to each of them, using the grade set from Table 1 and U={0,50,100}, as a basis for the evaluation procedure.

We chose triplets because in our opinion it is not easy to evaluate in fuzzy terms if the evaluator is used to evaluating traditionally. On the other hand, we had left no possibility of subversion the evaluation process by just giving all-but-one 0's as membership degrees of the FEMs. We thought that the triplets, for an inexperienced fuzzy evaluators would be just the right transitional resolution in evaluation.

Traditional	FEG		
	M1	M2	M3
A	0.1	0.3	1.0
B	0.1	0.5	0.4
C	0.1	0.4	0.1
D	0.1	0.3	0.0
F	1.0	0.0	0.0

Table 2. Components of a SFC of the evaluator.

The evaluator was advised to use multiples of 0.1 as FEGs' membership degrees, for we think that beliefs cannot be more effectively expressed using the whole set of real numbers, due to the presence of the human factor, and its inability to evaluate with an ideal precision.

We trained a back propagation neural network by the means of the training set given in Table 1.

After the NN was trained and tested, it was given the task to evaluate all the possible FEMs in the given environment. We selected those that were evaluated as 4.00, 3.00, 2.00, 1.00 and 0.00, as "perfect" candidates for the SFC that we were to establish. They are given in Table 8.

In the next stage, of the standardization procedure, we combined the x.00 FEGs to see which ones will give a SFC. This process resulted with the six SFC, one of which is given in Table 2. Any one of these SFCs can be used in the fuzzy evaluation procedure.

CONCLUSIONS

In this paper we presented a fuzzy framework for evaluation of different aspects of expert systems. The approach is actually very general and can be applied in any evaluation-type of activities, when there is any expert/human-like knowledge involved in the assessment. Moreover, we illustrated an neural network approach towards standardizi, pp.ng the evaluator's criteria, in order to be able to evaluate in a consistent manner, even in a multi-evaluator environment.

REFERENCES

Addis, T.R., 1986, Designing Knowledge-Based Systems, Prentice Hall.

Avritzer, J.P. Ros, E.J. Weyuker, B. , 1996, "Reliability Testing of Rule-Based Systems," IEEE Software, September, pp. 76-82.

Barbara, D., Garcia-Molina, H., 1987, "The Reliability of Voting Mechanisms," IEEE trans. on Computers, C-36, pp. 1197-1208.

Bastani, F., Chen, I., Tsao, T., 1993 "Assessment of the Reliability of AI Programs," Int. J. Software Engineering and Knowledge Engineering, 3, pp. 99-114.

Biswass, R., 1995, "An Application of Fuzzy Sets in Students' Evaluation," Fuzzy Sets and Systems, North-Holland, Amsterdam, 72, pp. 187-194.

Klir, G., Juan, B., 1995, Fuzzy Sets and Fuzzy Logic, Prentice Hall, Upper Saddle River, NJ.

Lorczak, P., Caglayan, A.K., Eckhardt, D.E., 1989, "A Theoretical Investigation of Generalized Voters for Redundant Systems," 19th IEEE Int. Symposium on Fault-Tolerant Computing Digest of Papers, IEEE Computer Society Press, pp., 444-451.

Perry, W., 1995, Effective Methods for Software Testing, John Willey and Sons.

Trajkovski, G., Cukic, B., Stojanov, G., 1999, "Fuzzy Logic in Neurophysiology: A Case Study," Proc. Computational Intelligence: Methods and Applications (CIMA'99), Rochester, NY.

Trajkovski, G., 1997, Fuzzy Relations and Fuzzy Lattices, M.Sc. Thesis, University "St. Cyril and Methodius, Skopje, Macedonia, 1997 (in Macedonian).

Trajkovski, G., Stojanov, G., Bozinovski, S. Bozinovska, L., Janeva, B., 1997, "Fuzzy sets and Neural Networks in CNV detection," Proc. Interaction Technology Interfaces (ITI'97), Pula, Croatia, pp. 153-158.

Trajkovski, G., Janeva, B.. "Towards a Standardized Personal Fuzzy Criterion for Student Evaluation", 7th Int'l Fuzzy Systems Assoc. Congress, Prague, Czech Republic, pp. 62-67.

Turski, W.M., 1998, "Should/Could Software Be More Reliable than the 'World' in Which It Is Used? " Proc. ISSRE '98, Padeborn, Germany pp. 3-9.

Zadeh, L.A., 1996, "Fuzzy Logic = Computing with Words," IEEE Trans. Fuzzy Systems, 4, pp. 103-111.

A FUZZY FINITE-STATE AUTOMATON DEFINITION THAT UNIFIES A NUMBER OF OTHER POPULAR COMPUTATIONAL PARADIGMS

MANSOOR DOOSTFATEMEH and STEFAN C. KREMER
Dept. of Computing and Information Science, University of Guelph
Guelph, ON N1G 2W1, CANADA
{skremer|mdoostfa}@uoguelph.ca

ABSTRACT
This paper provides a unification of 4 important computational paradigms: deterministic finite-state automata (DFA), non-deterministic finite-state automata (NFA), second-order recurrent neural networks (SRNNs) and fuzzy finite-state automata (FFA). We present a definition of a *general fuzzy automaton* which, under various simple restrictions, degenerates to the paradigms mentioned above. This new definition provides a fuzzy finite state automaton that can be more readily applied to real-world problems than previous definitions. It also allows for comparisons between the automata of the differing paradigms and for the application of theoretical discoveries and practical algorithms across the computational models which previously had no operational relationships.

1 INTRODUCTION

We begin with some mathematical definitions to form the basis of our definition. First, we adopt the standard definitions of a fuzzy set and a fuzzy power set (as presented in, e.g. (Klir and Yuan, 1995)).

A *fuzzy set*, μ_A, is a function mapping each element of the set A to a unique element of the set $[0, 1]$:

$$\mu_A : A \rightarrow [0, 1] \tag{1}$$

$\widetilde{P}(A)$ represents the fuzzy power set of A. Specifically, $\widetilde{P}(A)$ is the set of all fuzzy subsets defined on the set A:

$$\widetilde{P}(A) = \{\mu_A \mid \mu_A : A \rightarrow [0, 1]\} \tag{2}$$

We can now define a general fuzzy automaton as follows:

Definition 1 *A general Fuzzy Finite-state Automaton (FFA),* $\widetilde{F}$ *is an 8-tuple denoted as* $\widetilde{F} = (Q, \Sigma, \delta, R, Z, \gamma, F_1, F_2)$,

where:

- Q is a finite (non-fuzzy) set of states, $Q = \{q_1, q_2, ..., q_n\}$;
- Σ is a finite (non-fuzzy) set of input symbols, $\Sigma = \{a_1, a_2, ..., a_m\}$;
- $\delta \in \widetilde{P}(Q \times \Sigma \times Q)$ is a fuzzy set of triples mapping all elements of $Q \times \Sigma \times Q$ to a membership value;
- $R \subseteq \mu_Q$ is a fuzzy set of start states with memberships (note that $R \in \widetilde{P}(Q)$);

- Z is a finite (non-fuzzy) set of output symbols, $Z = \{z_1, z_2, ..., z_p\}$;
- $\gamma : Q \rightarrow Z$ is an output function mapping each state in the active state set of an input string to a unique output symbol in Z;
- $F_1 : [0,1] \times [0,1] \rightarrow [0,1]$ is the membership assignment function; and,
- $F_2 : [0,1]^* \rightarrow [0,1]$ is the multi-membership resolution function.

2 OPERATION

The FFA operates by making transitions from one fuzzy state-set to another, starting with the fuzzy start state-set. It is convenient to specify the configuration[1] of an automaton at a specific point in time, t. To do this we use the fuzzy set notation introduced in Eq. (1) augmented with a superscript t as μ_Q^t. Thus, the μ_Q^t specifies the set of fuzzy active states at time t. Where the time in question is unambiguous or not applicable we will omit the superscript.

There are two important considerations that must be examined regarding the operation of a FFA. **(a)** For every transition, the relation between the source-state membership value, transition weight and the destination-state membership value must be defined; this will be done by F_1. **(b)** When two or more active states all have transitions to the same destination state under the same input symbol but with different weights, the destination state's activation value must be resolved from the different values obtained in step (a) for each of the transitions; this will be done by F_2.

We first consider the calculation of state membership in the simplest case. Suppose that there exists a transition from state q_i to state q_k under symbol a_j upon time t. Now suppose that the membership value of state q_i at time t, is $\mu_Q^t(q_i)$, and the transition weight is $\delta(q_i, a_j, q_k)$. Without loss of generality, we shall define the membership value in the state q_k at time $t+1$ in response to a_j as

$$\mu_Q^{t+1}(q_k) = F_1(\mu_Q^t(q_i), \delta(q_i, a_j, q_k)) \tag{3}$$

We will consider specific implementations of the function F_1 which serves to combine the state membership with the transition weight in Section 4.

The second issue that we must consider is the situation where more than one transition can lead to the same state under the same input symbol which we call *multi-membership*. Specifically, let us consider the multi-membership where the memberships in the states of the automaton are given by the fuzzy set μ_Q^t and the transition weights are given by the fuzzy function δ. While our previous definition tells us that for any state q_i, the membership in state q_k is given by Eq. (3), we must now consider the destination-state as a consequence of *all* states with activation values in μ_Q^t. Once again, we provide a completely general definition using an arbitrary function, in this case, F_2:

$$\mu_Q^{t+1}(q_k) = \underset{q_i}{F_2}[F_1(\mu_Q^t(q_i), \delta(q_i, a_j, q_k))] \ \forall q_i \in \mu_Q^t | (q_i, a_j, q_k) \in \delta \tag{4}$$

Now, we can ask the question "what will be the automaton's state memberships after the presentation of the string x?". We provide a recursive definition of the transitive closure of δ to answer this question:

$$\tilde{\delta}(\epsilon) \equiv R \tag{5}$$

[1]The term "configuration" is used in the same sense as in Hopcroft and Ullman's (1979) description of pushdown automata and Turing machines.

and,

$$\tilde{\delta}(xa_j) \equiv \underset{q_i}{F_2}[F_1(\tilde{\delta}(x), \delta(q_i, a_j, q_k))] \ \forall q_i \in \tilde{\delta}(x) \tag{6}$$

3 OUTPUT

Now that we have identified the state of the automaton in response to an arbitrary input string, we can consider the matter of output. Thus far, in our definition of the automaton, we have defined Z to be the set of output symbols and $\gamma : Q \rightarrow Z$ as the output function. It should be noted that we have associated output with states in our definition in a manner that is consistent with deterministic finite state automata and Moore machines (Moore, 1956). (A reformulation of output associated with transitions, in the manner of Mealy machines (Mealy, 1955), is also possible and the computational equivalence of the two paradigms persists in the fuzzy domain.) The definition of an output alphabet Z imparts a rich expressiveness to our FFA.

Now, we can describe the output of the FFA for a given input string. We make the assumption that output symbols are generated in direct correspondence with the fuzzy state-memberships[2]. That is:

$$\mu_Z^t(z_l) = \underset{q \in \gamma^{-1}(z_l)}{F_2} (\mu_Q^t(q)) \tag{7}$$

where γ^{-1} represents the relation that is the inverse of the γ function. That is, a relation that generates the list of all states which are associated with a given output (in this case, z_l), at time t. Of course, we are usually more interested in the output of an automaton in response to a given input string (rather than an arbitrary internal state), so we define the output of the automaton in response to string x as:

$$\tilde{\mu}_Z^t(x) = \underset{z_l \in \gamma(\tilde{\delta}(x))}{F_2} \left[\underset{q \in \gamma^{-1}(z_l)}{F_2} (\mu_Q^t(q)) \right] \tag{8}$$

In practice, we may be interested in the automaton's output *symbol* only after the presentation of an entire string (final output) like in a DFA or its output *string* computed over time as the inputs are successively presented like in a Moore machine.

4 EQUIVALENCES

We now show that our automaton is really a generalization of a number of other popular, formal automata.

4.1 FUZZY FINITE STATE AUTOMATA

Practical definitions of F_1 and F_2 are often missing from theoretical treatments of the subject at hand (see for example some parts of (Mordeson and Malik, 2002; Omlin et al., 1998)). One exception which does give a more practically realizable definition is that of (Omlin et al., 1999) which specifies the use of min and max, but does not formalize the operation of automata or their output. We define F_1 = max and F_2 = min. Together with our discussions on operation and output in Sections 2 and 3, these definitions provide practical specifications for the operation of automata.

---

[2]We have chosen to implement the output multi-membership resolution function in the same fashion as multi-membership in states, i.e. using F_2. In general, this need not be the case, and we can use a different function F_3.

4.2 DETERMINISTIC FINITE STATE AUTOMATA

A Deterministic Finite State Automaton (DFA) is defined by a 5-tuple, $M = (Q, \Sigma, \delta, q_0, F)$ where Q is the set of states, Σ the input alphabet, δ a transition function of the form $\delta : Q \times \Sigma \rightarrow Q$, $q_0 \in Q$ is the start state, and F is a set of final or accepting states (Hopcrotf and Ullman, 1979). Clearly the first two elements of the tuple already match the parameters of FFA in Definition 1.

The transition function in the DFA is both deterministic and absolute. That is for every current state and input symbol pair there is exactly one next state that becomes fully activated (i.e. with a membership value of 1). This form of operation can be easily imposed on a FFA by the following restrictions:

- **R 1** *All weight transitions have values of 0 or 1:*

$$\forall i, j, k \quad \delta(q_i, a_j, q_k) \in \{0, 1\} \tag{9}$$

- **R 2** *For any given source state q_i and input symbol a_j there is exactly one destination state, q_k:*

$$\forall q_i, a_j \ \exists\, q_k \ s.t.\ \delta(q_i, a_j, q_k) = 1 \wedge \forall\, q_l \in Q \mid q_l \neq q_k \Rightarrow \delta(q_i, a_j, q_l) = 0 \tag{10}$$

- **R 3** *The membership values of the start state, q_0, is 1 and all others are 0:*

$$R(q_0) = 1; \ \forall q_i \in Q \mid i \neq 0 \Rightarrow R(q_i) = 0 \tag{11}$$

- **R 4** *The output (decision) of the DFA is limited to accept and reject:*

$$Z = \{accept, reject\} \tag{12}$$

- **R 5** *All states in F are accepted, and all others rejected:*

$$\gamma(q_i) = \begin{cases} accept & if\, q_i \in F \\ reject & if\, q_i \notin F \end{cases} \tag{13}$$

- **R 6** *The degree of state membership in the destination state of a transition is the boolean AND of the originating state's membership and the transition weight:*

$$F_1(\mu_Q(q_i), \delta(q_i, a_j, q_k)) = \mu_Q(q_i) \wedge \delta(q_i, a_j, q_k) \tag{14}$$

We note that under these restrictions, the function $\widetilde{\mu}$ in Eq. (8) generates exactly the same classification decisions as the original DFA. Naturally, by defining the output over a more expressive set Z, the automaton becomes a Moore machine.

4.3 NONDETERMINISTIC FINITE STATE AUTOMATA

A Nondeterministic Finite State Automaton (NFA) is defined by the same 5-tuple, as a DFA with one exception: δ is a relation, rather than a function (Hopcrotf and Ullman, 1979). This implies two things: (1) not every state-input combination must have a δ value (and hence it need not have a defined transition), and (2) there may be more than one destination state for a given state-input combination. Fortunately, our definition of the FFA allows us to contend with both of these issues in a straightforward manner. For undefined transitions in the NFA, we assign a weight of 0 to the corresponding transition in the FFA. And, for multiple outcomes, we simply allow multiple transitions that differ only in their destination states (but not their origins and input symbols) to have a weight of 1. That is if we apply all restrictions defined for DFA in Section 4.2 except R 2, then the FFA becomes equivalent to a NFA; and, every NFA can be represented by a FFA with these restrictions. If NFA with a set of non-deterministic start states are required then restriction R 3 can be replaced by the more general:

- **R 7** *The membership values of the start state are restricted to 0 and 1:*

$$\forall q_i \;\; R(q_i) \in \{0, 1\} \tag{15}$$

If we consider ϵ as a member of the input alphabet, then we can represent NFA with empty transitions as well.

Before proceeding, however, we should be more careful to define in exactly what sense we mean "equivalent". Specifically, a NFA is considered non-deterministic in the sense that the resolution of the next-state ambiguity is not defined. In other words, the NFA is unrealizable because the next state a NFA will select for a given state-input combination is not uniquely defined. By contrast, the FFA we have defined here are deterministic but may not have unique state activations. In other words, the automaton's definition unambiguously specifies exactly which states will be selected and to what degree.

Instead of an operational definition of equivalence, we use a functional definition of equivalence. That is a NFA and a FFA are equivalent if they accept the same set of strings. This is analogous to the well established equivalence between NFA and DFA. However, while a DFA for a given NFA may have an exponentially increased number of states, the number of states in a corresponding FFA remains exactly the same. Using the restrictions above, it is clear that any NFA can be represented by an equivalent FFA.

4.4 SECOND-ORDER RECURRENT NEURAL NETWORKS

A Second-order Recurrent Neural Network (SRNN) (Giles et al., 1990) is a computational device consisting of a number of processing units whose activations, a_i fall in the interval of [0,1] or [-1,+1]. Their units are grouped into input and hidden units (a subset of which are also designated as outputs). Each hidden unit k receives a signal from each pair of one hidden unit i and one input unit j. When applied to symbolic problems their input units are restricted to the extreme values of the representational range and thus act as units that serve to gate the signal from hidden unit to hidden unit. The activation of any unit k, at time $t+1$, when the single input[3] associated with unit j is presented, is given by

$$a_k(t+1) = f(\sum_i w_{kij} \cdot a_i(t)) \tag{16}$$

If we equate internal unit activations ($a_i(t)$) with state membership values ($\mu_Q^t(q_i)$), second order weights (w_{kij}) with transitions ($\delta(q_i, a_j, q_k)$), and external unit activations with input presentations, then this equation reduces to

$$\mu_Q^{t+1}(q_k) = f(\sum_i \delta(q_i, a_j, q_k) \cdot \mu_Q^t(q_i)) \tag{17}$$

[3] As is obvious, in automata at any time, only one input is presented.

which is analogous in form to Eq. (4) where

$$\mathcal{F}_1(\mu_Q^t(q_i), \delta(q_i, a_j, q_k)) = \delta(q_i, a_j, q_k) \cdot \mu_Q^t(q_i) \quad (18)$$

and

$$\mathcal{F}_{k}{}_2(net_k) = f(net_k), \quad net_k = \sum_i w_{kij} \cdot a_i(t) \quad (19)$$

So, once again, with a specialized assignment of $\mathcal{F}_1$ and $\mathcal{F}_2$ this computational paradigm reduces to a special case of our FFA.

5 CONCLUSION

We have provided a general, yet practically applicable generalization of finite state automata using fuzzy logic. Our definition includes a specific *modus operandi* specifying both internal state changes and outputs which other treatises on the subject do not. Next, we have shown that many popular models of computation can be reduced to special cases of this new model. The results presented here set the stage for the cross-pollination of disciplines. In particular, we are interested in looking at applying recurrent neural network algorithms to general FFA models, applying results from automata theory to FFA and then to recurrent neural networks, and further extending this general framework to other models of computation.

6 ACKNOWLEDGEMENTS

Prof. Kremer is funded by grants from the NSERC, CFI, ORDCF and OIT.

REFERENCES

Giles, C., Sun, G., Chen, H., Lee, Y., and Chen, D. (1990). Higher order recurrent networks & grammatical inference. In Touretzky, D. S., editor, *Advances in Neural Information Processing Systems 2*, pages 380–387, San Mateo, CA. Morgan Kaufmann Publishers.

Hopcrotf, J. E. and Ullman, J. D. (1979). *Introduction to Automata Theory, Languages, and Computation*. Addison-Wesley, Reading, MA.

Klir, G. J. and Yuan, B. (1995). *Fuzzy Sets and Fuzzy Logic: Theory and Applications*. Prentice Hall, Upper Saddle River, NJ.

Mealy, G. (1955). A method for synthesizing sequential circuits. *Bell System Technical Journal*, 34(5):1045–1079.

Moore, E. (1956). Gedanken experiments on sequential machines. In *Automata Studies*, pages 129–153. Princeton University Press, Princeton NJ.

Mordeson, J. N. and Malik, D. S. (2002). *Fuzzy Automata and Languages*. Chapman & Hall/CRC, New York, NY.

Omlin, C. W., Giles, C. L., and Thornber, K. K. (1999). Equivalence in knowledge representation: automata, recurrent neural networks, and dynamical fuzzy systems. *Proc. of IEEE*, 87(9):1623–1640.

Omlin, C. W., Thornber, K. K., and Giles, C. L. (1998). Fuzzy finite-state automata can be deterministically encoded into recurrent neural networks. *IEEE Trans. on Fuzzy Systems*, 5(1):76–89.

A HARDWARE MODEL FOR FUZZY TEMPORAL KNOWLEDGE REPRESENTATION

KRESIMIR MIHIC
Cypress Semiconductor
Lexington, Kentucky

ABSTRACT
Fuzzy systems implemented in hardware operate with much higher performance than software implementations, and are more suitable for implementation in automated supervision systems. For systems with very strict temporal restrictions, those systems should be able to handle imprecise and uncertain information and should be capable of describing the evolution of the system in terms of a state change. This paper presents a hardware model capable of fuzzy temporal knowledge representation and reasoning. The model is based on the Fuzzy Temporal Knowledge Representation Scheme that uses the modified Petri Net theory, the possibility theory and a fuzzified version of Allen's temporal relations, and has been implemented in VHDL following the Register-Transfer-Level model approach.

INTRODUCTION

Hardware systems, from large industrial systems to small, distributed robots, require a form of supervision to handle unanticipated situations that do not match the expected behavior. For systems with very rigid temporal restrictions, supervision systems should be able to handle imprecise and uncertain information and should be capable of describing the evolution of the system in terms of a state change. This paper describes an attempt towards the development of a fully automated and independent supervising hardware system capable of fuzzy temporal knowledge representation and reasoning.

In many application domains of intelligent systems, human knowledge about time, which has to be implemented, is usually pervaded with uncertainty and vagueness (Ribaric, 2000). The first attempt (Dubois and Prade, 1989) to represent temporal data was done by Kahn in 1977, who proposed an ad hoc symbolic approach when dealing with linguistic time specifications. . The turning point, however, was the work on logic of temporal interval done by Allan (1983), followed by a fuzzy set based model of time intervals (Dutta 1988). Later, different models for processing fuzzy temporal knowledge were developed (Dubois and Prade 1989, Dalbelo Basic 1997, Kunzle et al. 1999).

The concept of fuzzy Time Petri Nets permits the modeling of dynamic behavior of discrete event systems, interleaving, concurrence, and conflict constraints between both states and events. It also allows handling explicit timing considerations by means of the fuzzy time intervals attached to and processed by transitions, as well as time intervals during which a place holds tokens. A number of Fuzzy Time Petri Nets models have been developed so far (Cardoso et. al. 1989, Valette et. al. 1989, Dalbelo Basic 1997).

The Hardware Model for Fuzzy Temporal Knowledge Representation gives a framework for building systems capable of monitoring and controlling situations in complex fuzzy temporal systems. The model is based on the Fuzzy Temporal Knowledge Representation Scheme that uses the modified Petri Net theory, the possibility theory, and a fuzzified version of Allen's temporal relations. The organization of this paper is as follows. First, the general information regarding fuzzy temporal reasoning is given. Second, the hardware model for temporal knowledge reasoning is proposed. Third, an example of the use of the proposed model is given, and finally, a summary of the future work concludes this paper.

BASIC NOTIONS OF FUZZY TEMPORAL KNOWLEDGE REASONING

The Fuzzy Time Petri Net (FTPN) (Dalabelo Basic 1997) is an 8-tuple $(P,T,I,O,m,\tau,\lambda,\Theta)$ where P, T, I and O are components of crisp Petri Nets; m is a marking function, $m: P \to [\Im(T) \times (\Im(T) \cup \varnothing)] \cup \varnothing$, where T denotes a time scale and $\Im$ is a set of all normalized and convex possibility distributions defined on T. A fuzzy time token, $m(p_i)$ is a pair $m(p_i) = m_i = ((\pi_{b(i)}, \varnothing), (\varnothing, \pi_{e(i)})) = (\pi_{b(i)}, \pi_{e(i)})$, i=1,2,…,n and π_b and π_e are the possibility distributions of beginning and end, respectively, of some action, state or event. The Function $\tau: P \to \Im(T) \cup \varnothing$ defines the time of fuzzy time token detainment in a place and the function $\lambda: T \to \Phi$ defines mapping from a set of transitions T to a set of fuzzy operations Φ defined on the fuzzy time token, $\lambda(t_j)=\lambda_j$. The function $\Theta: [\Im(T) \times \varnothing] \times \Im(T) \to \Im(T) \times \Im(T)$ defines the effects of fuzzy time token detainment in a place; it transforms a token $m_i = (\pi_{b(i)}, \varnothing)$ into $m_i = (\pi_{b(i)}, \pi_{e(i)})$ where $\pi_{e(i)} = \pi_{b(i)} \oplus \tau_i$. Firing of a transition in the FTPN is performed automatically and immediately after the transition has been enabled. A transition t of a FTPN $C=(P,T,I,O,m,\tau,\lambda,\Theta)$ is enabled for marking m if and only if $\forall p_i \in I(t_j), m(p_i) \neq \varnothing \wedge \forall p_i \in I(t_j), \Theta(m_i, \tau_i) = m_i$ and the fuzzy operation λ_j results in a convex and a normalized possibility distribution. Firing changes the initial state $m^{(k)}$ of an enabled transition t_j into $m^{(k+1)}$ by executing the *next-state* function $\delta(m^{(k)}, t_j) = m^{(k+1)}$. If $\forall p_i \in P, m^{(k+1)}(p_i) = m_i^{(k+1)}$ then $\forall p_i \in O(t_j)$, $m_i^{(k+1)} = \pi_{b(i)}^{(k+1)} =$ $= \lambda_j[(\pi_{b(1)}^{(k)}, \pi_{e(1)}^{(k)}), \ldots, (\pi_{b(1+s)}^{(k)}, \pi_{e(1+s)}^{(k)})], p_1, \ldots, p_{1+s} \in I(t_j)$. In the FTPN model, conflict states as described by Peterson (Peterson 1981) and potential conflict states are eliminated by multiplication of the fuzzy time token.

The Fuzzy Temporal Knowledge Representation Scheme (FTKRS) (Dalabelo Basic, 1997) is a 5-tuple $(FTPN, FTLM, \alpha, \beta, F)$ where *FTPN* is a Fuzzy Time Petri Net, *FTLM* is a Fuzzy Temporal Logic Module based on fuzzy intervals, α and β are elements of the Knowledge Representation Sheme, and F is a set of flags. The set of flags F, $f_i \in F$ represents a query posed on the knowledge base designed by FTKRS, where f_i is a 3-tuple: (p_i, p_j, ftr) for queries about a relation between temporal concepts assigned to places p_i and p_j; (p_i, a, ftr), $\mathbf{a} \in \Im(T) \times \Im(T)$ for queries about a relation between an interval a and a temporal concept assigned to a place p_i; (p_i, t, ftr), $t \in \Im(T)$ for queries that involve a relation between a time moment t and a temporal concept assigned to a place p_i. The query *ftr* is based on Allen's temporal relations. Table 1 lists basic concept relationships and associated fuzzy operations.

A HARDWARE MODEL FOR FUZZY TEMPORAL KNOWLEDGE REPRESENTATION (HMFTKR)

HMFTKR, based on the FTKRS theory described in the previous section, provides the means for describing situations in temporally rich domains. The model gives a designer an opportunity to construct and design a hardware based autonomous system either directly from a knowledge base based FTPN model or from a list of concepts, temporal data associated with the concepts, and relationships between these concepts.

The trapezoidal function $A(x;a,m,n,b)=max\{min[(x-a)/(m-a),\ 1,(b-x)/(b-m)],0\}$ is the function of choice used in this work for temporal datum representation.

The HMFTKR working principle relies on properties of a fuzzy temporal datum representation carried by a FTKM token. FTPN defines a token as a pair $m=(\pi_b,\ \pi_e)$, but in a hardware model, the token carries a single information, either about the beginning or the end of an action. A single token is a vector storing inputs and results of λ functions: $m_{HW}=(F1,I1,MF1,MI1,F2,I2,MF2,MI2)$ where $F1$ is a prime fuzzy number, and $I1$-$MI2$ are fuzzy numbers used at the output stage of a modeled system to reconstruct the accurate output of $\lambda \in \{\underline{max},\ \underline{min},\ \cap I,\ I\}$. This approach was used to keep the hardware model as simple as possible, thus saving resources and power consumption.

Hw_place and hw_transition, HMFTKR basic modules, perform FTPN place and transition functionality respectively. HMFTKR place consists of an addition module that evaluates Θ function and a pair of registers that store the beginning (π_b) and the end (π_e) of an action or event. HMFTKR transition is the 5-input arithmetical module performing λ functions. HMFTKR model supports:

(1) Addition, $\lambda \in \{\oplus\},\Theta$. Addition of a time period to a token is preformed both by places and transition instances as defined by Θ and λ respectively. A time period τ (or κ) is added to a token variable $F1$ (or I, $MF1$…) as follows: $F1\oplus\tau=(a_{F1},m_{F1},n_{F1},b_{F1}) \oplus \oplus (a_\tau,m_\tau,n_\tau,b_\tau)=(a_{F1+}\ a_{\tau},,m_{F1+}\ m_\tau,\ n_{F1+}\ n_\tau,\ b_{F1+}\ b_\tau)$.

(2) Minimum & Maximum, $\lambda \in \{\min,\max\}$. Both functions follow the same principle: If rising (falling) edges of compared time periods overlap, then M* variable elements store the result of the opposite function (for <u>min</u> it is maximum, and for <u>max</u> it is minimum), otherwise M* variables carry the same value(s) as prime numbers. The last enables the identification of intersected edges and thus the calculation of z_0 and $C(z_0)$ at the output. It should be noted that M* variables should be observed as rising/falling edge pairs *(a,m)* and *(n,b)* rather than trapezoidal numbers. Let A and B be two fuzzy time intervals as shown in Figure 1. Let <u>min</u> be a function of choice. The resultant token variables $F1$ and $MF1$ are then $F1=(a_A,m_B,n_B,b_B)$ and $MF1=(a_B,m_a,n_B,b_B)$. Min/max of other token variables (I1,F2,I2) follows the same principle.

(3) Interval, $\lambda \in \{\mathbf{I}\}$. By definition, **I** is the interval between the beginning and the end of an event A. Its membership function (Basic 1997) is an integral of input intervals, and input intervals can be reconstructed back from the function. This property enables an event to be further processed by other functions by modifying prime intervals rather than the membership function itself. It also simplifies the design, as all mathematical calculations are done at the output stage. If the duration of an event is bounded by two fuzzy numbers (Fig. 2), F_b and F_e, then token values assigned by **I** are $F1=AF1_b$ and $I1=AF1_e$. If the duration of an event is bounded by two **I** intervals, $(AF_b,\ AI_b)$ and $(AF_e,\ AI_e)$, then processing by another **I** function leads to $F1=AF1_b$ and $I1=AI1_e$ a. The process is repeated for M^* variables when defined.

(4) Interval intersection, $\lambda \in \{\cap\mathbf{I}\}$. The interval intersection is a complex function performing **I** and $\cap$ functionality. The **I** function constructs interval pairs that are then assigned to the token variables $F1$, $I1$, $F2$ and $I2$. The first interval pair is assigned to $F1$and $I1$ and the second to $F2$ and $I2$. The same principle is followed for M^* variables when defined. The exact membership function that involves intersection is reconstructed at the output stage. If input tokens A and B carry data in the form of trapezoidal fuzzy numbers then $F1=AF1_b$, $I1=AF1_e$, $F2=BF1_b$, and $I2=BF1_e$, but if input tokens carry **I** intervals then $F1=AF1_b$, $I1=AI1_e$, $F2=BF1_b$, and $I2=BI1_e$.

(5) Interval, $\lambda \in \{[]\}$. Interval [] is defined as a period between possible beginnings and possible ends of an action or event A (Dubois and Prade 1989). Evaluation of the

function is preformed by assigning beginning and end points of an interval to m_{HW} variables as follows: when an event is described by fuzzy numbers then $F1(a) = AF1_b(a)$, $F1(m) = AF1_b(m)$, $F1(n) = AF1_e(n)$, $F1(b) = AF1_e(b)$; when an event is described by an interval **I** then $F1 = AF1_b$, $I1 = AI1_e$, and when the event is described by an interval ∩**I** then $F1 = AF1_b$, $I1 = AI1_e$, $F2 = BF1_b$, $I2 = BI1_e$. The process is repeted for M^* variables.

In building a hardware model special attention was paid to the simplicity of design. All but the simplest addition is performed at the output level by either the output interface or user. Other functions are modeled at the transition level by fuzzy number points manipulation, which resulted in some limitations of the current model:

(a) min/max function limitations. In the current model, only one intersection can be stored, which may introduce an error by adding additional fuzziness to the system when the function is repeated. The function supports only the same variable input, i.e. only three input combinations are possible: min/max(A,B), min/max(**I**(A),**I**(B)) and min/max (∩**I**(A),∩**I**(B)). The function is limited to two inputs.

(b) **I** and ∩**I** function limitations. An event must be described by a pair of non-overlapping fuzzy numbers or intervals because the model supports only normalized fuzzy sets. The function ∩**I** cannot be repeated, i.e. ∩**I**(∩**I**(A, B),C) is not allowed.

EXAMPLE

This section gives an example of how HMFTKR can be used to process temporal knowledge. The knowledge base is built on a scenario that uses an every day event, but the very same approach can be followed for any system, including autonomous monitoring and control systems that HMFTKR targets:

Fred, John and Mark have a meeting as soon as they all arrive to work. Fred leaves home at about 7 o'clock., and gets to work about 20 minutes later. John comes to work a few minutes earlier than Fred. Mark leaves the house approximately at the same time as Fred. It takes him about 20 minutes to get to the bus station and few more minutes to get to work. As soon as Mark comes to work, he calls his wife Jane and goes to the meeting. Jane needs about 50 minutes to prepare and to come to the mall where she waits for Mark for at most 20 minutes. Mark's meeting lasts exactly 90 minutes. As soon as it ends, Mark is driving to meet Jane. It takes him near 10 minutes to get to the mall. What is the possible start time of the meeting? Will Mark and Jane meet? (Ribaric 2000).

Based on the data given, concepts and relationships between concepts (Table 2, 3) and an FTPN model (Fig. 3) were derived. Dummy place, p_7, is a work around to the limitation of the current model to two input transitions. The implementation was done in VHDL language following the Register-Transfer-Level approach.

The current model's universe of disclosure is given in minutes, so the initial marking (around 7 o'clock) is given as: $F1_{b(1)}$=(415, 420, 420, 425). The answer to the first question, what is the possible start of the meeting, is given by the input value of p_{10} token (Fig. 4). The output $\pi_{b(10)}$ (Fig. 5) is reconstructed to be: $F1_{b(10)}$ = (pobf1a, pobf1n, pobf1m, pobf1b) = (7:10, 7:30, 7:30, 7:55}, $MF1_{(10)}$ = (pobf1a, pobf1n, pobf1m, pobf1b) = (7:05, 7:20, 7:30, 7:50). The result can be interpreted as follows: The meeting starts anytime between 7:10 and 7:55 with the highest possibility for the start at 7:30. The height of the intersection of the output values of t_{o9} and t_{o10} gives the answer to the second question. The output π_{to9} is found to be $F1_{to9}$ = (7:45, 8:20, 8:40, 9:20) and π_{to10} to be $F1_{to10}$ = (8:44, 9:10, 9:10, 9:42) and $MF1_{to10}$ = (8:39, 9:00, 9:10, 9:42). The result (Fig. 5) can be interpreted as follows: A meeting between Mark and Jane may occur between 8:44 and 9:20 with the highest possibility for the start at 8:56. The meeting is not likely to take place as the height of the intersection is $hgt(\pi_{to9} \cap \pi_{to10}) = 0.54$.

CONCLUSIONS

The HWFTKR hardware implementation of the FTKR scheme developed in this work, successfully models any system that supports reasoning and planning in temporally rich domains. The model is based on two programmable blocks, hw_place and hw_transition, that model FTPN place and transition respectively. Each block is optimized to guarantee the fastest execution time with minimum hardware, which may make this model interesting for embedded real time systems. Although shown as useful and practical, the current model needs a number of improvements related to its limitation to two input functions, its support of only normal fuzzy sets and its time consuming placing scheme. The future work will include the removal of those limitations and a buildup of a programmable array of HWFTKR primitives able to adapt different FTKR models. Such a system should also incorporate better i/o interface.

REFERENCES

Allen, J.F. 1983. Maintaining Knowledge about Temporal Intervals. *Communications of ACM*, vol. 26, no. 11, pp. 832-43.

Cardoso, J. 1999. Time Fuzzy Petri Nets. In *Fuzzines in Petri Nets*. eds.: J. Cardoso and H. Camargo, Heidelberg, New York: Springer Verlag.

Cardoso, J., R. Valette, and D. Dubois 1989. Petri Nets with Uncertain Markings. *10th Int. Conf. Applications of Petri Nets*.

Cardoso, J., R. Valette, and D. Dubois 1996. Fuzzy Petri Net: An Overview. *Proc 13th IFAC World Congr*. San Francisco, CA, pp. 119-28.

Cardoso, J., R. Valette, and D. Dubois 1999. Possibilistic Petri nets. *IEEE Trans. Sys., Man.,and Cyber*, vol. 29, no.5, pp. 573-82.

Dalbelo Basic, B. 1997, Knowledge Representation Using Fuzzy and Fuzzy Time Petri Nets, Ph. D. dissertation. University of Zagreb.

Dubois, D., and H. Prade 1980. *Fuzzy Sets and Systems: Theory and Application*. NY: A. Press.

Dubois, D., and H. Prade 1989. Processing Fuzzy Temporal Knowledge. *IEEE Trans. Sys., Man.,and Cyber*, vol. 19, no.4, pp. 729-44.

Dutta, S. 1988. An event-based temporal fuzzy logic. *Proc. 18th IEEE Intl.Sym. Multiple-Valued Logic*. Spain

Hung, D.L. 1995. Dedicated digital fuzzy hardware. *IEEE Micro*, vol. 15 no. 4, pp. 31-39.

Jong, WT., YS. Shiau, YJ. Horng, HH. Chen, and SM. Chen 1999. Temporal knowledge representation and reasoning techniques using time Petri nets. *IEEE Trans. Sys., Man.,and Cyber*, vol. 29, no.4, pp. 541-45.

Pedrycz, W., and W. Gomide 1998. *An Introduction to Fuzzy Sets: Analysis and Design*. A Bradford Book, MA: MIT press.

Peters, L, and Sh. Guo 1998. Three generation of Fuzzy Hardware. *In Fuzzy Hardware: Architectures and Applications*, eds.: A. Kandel, G. Langholz, Kluwer A. Pub., pp.27-43.

Ribarić, S., B. Dalbelo Bašić, and D. Tomac 2000. Object-Oriented Implementation of a Model for Fuzzy Temporal Reasoning. *Proc. 8th Intl. Conf. on Inf. Proc. and Mng. of Uncertainty in Knowledge-based Systems*. IPMU 2000 Madrid, Spain, Vol. I, pp. 1247-54.

Valette, R, J. Cardoso and D. Dubois 1989. Monitoring manufacturing systems by means of Petri nets with imprecise markings. *Proc. IEEE Intl. Sym. on Int. Ctrl.*, 25-26 Sep 1989, pp. 233 –38.

Yakovlev, A.M L. Gomes and L. Lavango, eds. 2000. Hardware Design and Petri Nets. Kluwer Academic Publishers, MA.

Yao, J. 1994. A petri Net Model for Temporal Knowledge Representation and Reasoning. . *IEEE Trans. Sys., Man.,and Cyber*, vol. 24, no.9, pp. 1374-83.

APENDIX

Table 1. Concept relationships and associated fuzzy operations

λ	Operation	Concept
≡	Identity	As soon as
⊕	Addition	K units before/after
I	Fuzzy Time Interval	During
[ ]	Interval	Between possible beginnings and ends
I∩	I Intersection	At the same time
min_b	Minimum of beginnings	As soon as the last begins
min_e	Minimum of ends	As soon as the first finishes
max_b	Maximum of beginnings	As soon as the last begins
max_e	Maximum of ends	As soon as the last finishes

Table 2. Concepts derived from the example

p_i	Concept	τ
p_1	Fred is leaving home	(0,0,0,0)
p_2	Fred is coming to work	(0,0,0,0)
p_3	Mark is in the bus	~20 min (15,20,20,25)
p_4	Mark is arriving to work	(0,0,0,0)
p_5	Mark calls his wife Jane	few min (0,5,5,10)
p_6	John is arriving to work	(0,0,0,0)
p_7	Dummy place	(0,0,0,0)
p_8	Jane is preparing	~50 min (40,50,50,60)
p_9	Jane is waiting	at most 20 min (0,0,20,25)
p_{10}	Meeting	almost 90 min (87,90,90,93)
p_{11}	Mark is gets to the mall	near 10 min (7,10,10,13)

Table 3. Relationships between concepts

t_i	$I(t_i)$	$O(t_i)$	λ
t_1	p_1	p_2	⊕ (15,20,20,25) - later
t_2	p_1	p_3	⊕ (-5,0,0,5) - app. at the same time
t_3	p_3	p_4	⊕ (0,5,5,10) - later
t_4	p_2	p_6	⊕ (-10,-5,-5,0) - earlier
t_5	p_4	p_5	≡
t_6	p_2, p_5	p_7	max_e
t_7	p_7, p_6	p_{10}	max_e
t_8	p_5	p_8	≡
t_9	p_8	p_9	≡
t_{o9}	p_9	--	≡
t_{10}	p_{10}	p_{11}	⊕ (7,10,10,13) - near
t_{o10}	p_{11}	--	≡

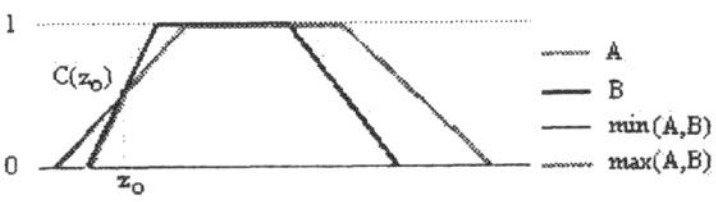

Figure 1. Min/max of two functions

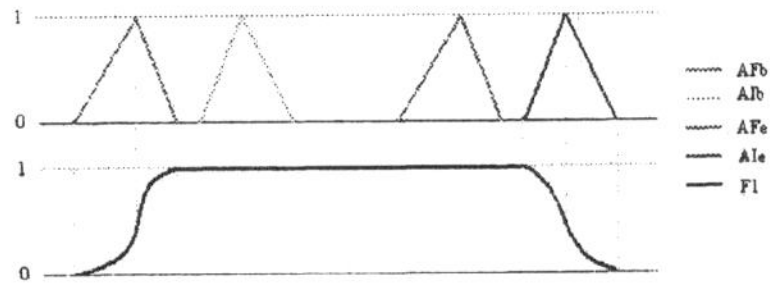

Figure 2. Interval I between two fuzzy numbers

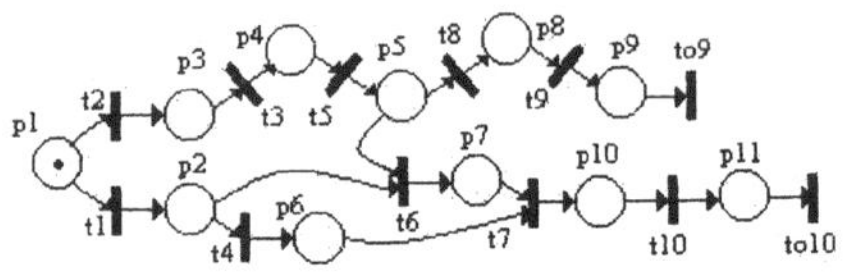

Figure 3. FTKR graph of the knowledge base for the example

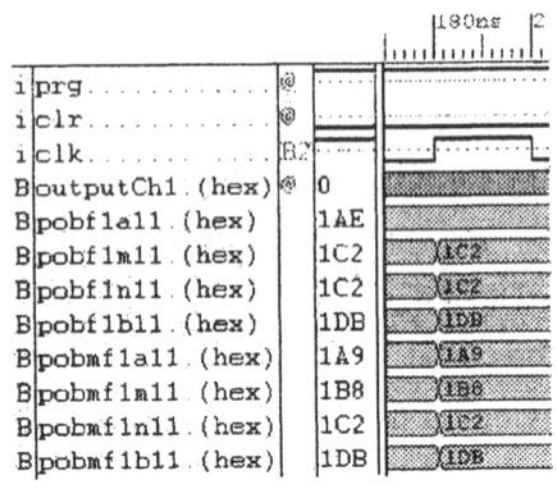

Figure 4. The input of p_{10}

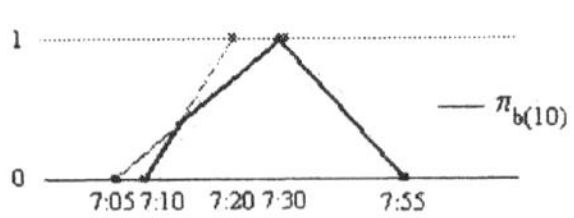

Figure 5. The possibility distribution $\pi_{b(10)}$

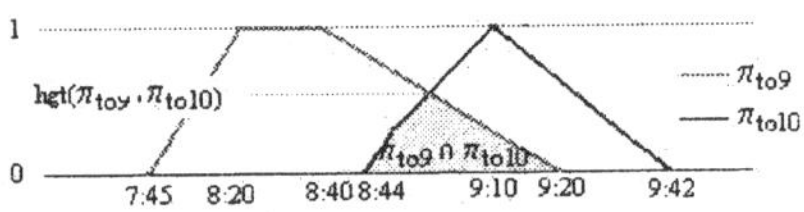

Figure 6. The intersection of possibility distributions π_{to9} and π_{to11}

FUZZY CORRELATION COEFFICIENT APPROXIMATE ESTIMATION

YONGSHEN NI
School of Electrical and Computer Engineering, University of Oklahoma , Norman, USA

JOHN Y. CHEUNG
School of Electrical and Computer Engineering, University of Oklahoma , Norman, USA

ABSTRACT
Correlation coefficient is an important measurement in data mining application. It reflects co-varying relationship between two random variables. Conventional methods assume the data set as crisp value, however many data in real world are ambiguous so it's not proper to define them as crisp value. In this paper, fuzzy set theory is introduced to deal with this problem. Since it's burdensome to compute fuzzy correlation coefficient precisely, we proposed an approximate method to estimate correlation coefficient for observations with fuzzy characteristics. Finally we did some simulation studies to verify this algorithm and compare the results with stochastic simulation method. Approximate estimation is very helpful for the purpose to have a preliminary knowledge of structure of data set in fuzzy environment.

INTRODUCTION

In the data research and analysis field, it's very useful to know whether one variable co-varies with another variable. Correlation coefficient plays one of the most important roles in gaining this end;

Pearson's correlation coefficient is the most widely used tool for assessing correlations. It measures the degree of linear relationship between two variables, one is called independent variable and the other called dependent variable,

First of all, let's take a short review about Pearson's correlation coefficient.

The below formula is his definition:

$$r_{xy} = \frac{\sum_{i=1}^{n}(x_i - \bar{x})(y_i - \bar{y})}{\sqrt{\sum_{i=1}^{n}(x - \bar{x})^2 \sum_{i=1}^{n}(y - \bar{y})^2}} \qquad (1)$$

In Eq.(1) x is independent variable, y is dependent variable and n is the size of dataset.

By definition, the correlation coefficient varies from -1 to +1. Zero means random relationship, while +1(or -1) shows perfect positive (or negative) linear relationship exists between variables.

Pearson's formula works well with the crisp data; however, in real world not every data has crisp characteristic. For example there are some subjective data which can't be represented by crisp value. So we have to develop more complicated way to better fit such cases.

RELATED WORK

Fuzzy correlation estimate has recently become a hot spot in data analysis area. So far, many valuable conceptions and algorithms have emerged.

Pedrycz(1999) proposed a method to analyze information granule in which case exploring local relationship in the dataset is more important than viewing the data globally. In his method data points have to be objective measurement, but people have to interpret those data subjectively. Pedrycz used fuzzy set theory to analyze data relationship at each α level and finally got a curve which is fuzzified correlation coefficient. Watanabe(1999) put the concept "fuzzy random variable (FRV)" in his method and computed the correlation coefficient at each α level. Both methods worked with crisp data and derived a membership function of correlation which is a fuzzy number ranging from -1 to 1.

Ding-An Chiang and Lin(1999) substituted data value in Pearson's formula with their membership function value. Chaudhuri and Bhattacharya (2001) also proposed a formula to measure subset hood between two intuitively fuzzy sets; their methods are used in pattern matching in the presence of ambiguity.

Liu(2002)'s paper provides a fuzzy number correlation coefficient ranging from -1 to 1. The model can be summarized into a nonlinear programming problem with bounded constraints; We know nonlinear optimization is an untraceable problem mathematically; to get the exact solution is computationally expensive sometimes impossible, especially when the size of dataset is big.

In next section, we suggest an approximate method to estimate correlation coefficient by means of fuzzy set theory. This approach has the virtue of fast computation and thus can be applied in applications which need to take a quick look at the relationship among fuzzy dataset.

ALGORITHMS DESCRIPTION

Many measurements in real physical environment are inaccurate in nature. In this section, we set up a model which can be applied to the uncertain environment.

A fuzzy measurement is defined as follows

$$Fx=\{ \mu M (x) \mid x\in R, \mu M : X\text{->}[0,1]\} \qquad (2)$$

We can get fuzzy measurements of a data set by fuzzifying (X, Y) to produce (Fx, Fy), μM is membership function which is designed according to the specific fuzzy environment.

Fuzzy number captures interval conceptions and plays an important role in many applications and research area. In this paper, we restrict the membership function to be a fuzzy number.

In an uncertain environment since the dataset is fuzzy, according to Extension Principle, the correlation coefficient which is computed with input data is also fuzzy. So we need to compute the membership function of this fuzzy measurement.

Suppose we define $r_{xy}(\alpha)$ as correlation coefficient at each α level .Because every fuzzy data has a pair of membership functions in both x and y directions, therefore two intervals are generated in both directions for each observation.

We will use these two intervals to compute correlation coefficient $r_{xy}(\alpha)$ with Pearson's formula. Therefore $r_{xy}(\alpha)$ is no longer a single value, but an interval. It's in fact a nonlinear optimization problem to find the data points which provide the minimum and maximum value of $r_{xy}(\alpha)$. There are lots of algorithms which can be applied to solve the problem, such as steepest gradient descent method, reduced gradient method etc. If we assume each fuzzy observation follows some kind of distribution in those two intervals, then we can pick up data point (X_i, Y_i) randomly to represent that observation, finally we should get one sample space to compute $r_{xy}(\alpha)$.

Suppose the number of fuzzy data of our dataset is N, and we have M random sampling for each α -cut level, then computational complexity of the above stochastic simulation algorithm is O(M*N). In general M can't be a small number if we want to get precise results. So this method (we call it stochastic simulation method) doesn't work very well with big size dataset.

Now let's consider an approximate method.

Firstly we set α value to 1 and pick those corresponding data points, then apply linear regression model to those data so we will obtain a straight line which represents linear relationship of the data set (α =1) in minimum mean square error sense. The idea of this method is to explore the impact of data fuzziness on the regression model. Here we have several cases:

i) Regression straight line goes through data area composed of $[X\alpha L, X\alpha U]_j$ $[Y\alpha L, Y\alpha U]_j$ (we will call it region in the later text). Depending on sign of the regression line slope, the point contributing to maximum (minimum)value of rxy(α) i.e. 1(-1) is any point on the straight line and in the region. As to the point contributing to minimum (maximum) value, we will use the corner point of the region which has the largest deviation from the regression line in y coordinate.

ii) Regression straight line doesn't pass through the region. As in case one we will use corner points of the region. We assume the point which has the smallest (largest) deviation from the line contributes to maximum (minimum) value of rxy(α), and the corner point with the largest(smallest) deviation from the line contributes to minimum(maximum) value.

In theory, with Pearson's formula correlation coefficient ranges from -1 to 1. +1(or -1) indicates perfect linear relationship exists between two random variables while zero means no relationship at all. So when searching the possible solution for this problem, approximately we can assume those points with the biggest deviation from the regression line will contribute to the minimum absolute value and those points with the smallest deviation will contribute to the maximum absolute value.

The algorithm is described as follows.

Initialize α=0.0, carrying out the following steps:

1. Fuzzifying each data with proper membership function.
2. Take α cut on membership functions (in both directions)
3. Apply approximate method on the data to obtain $r_{xy}(\alpha)$
4. $\alpha = \alpha + \Delta\alpha$, continue this process till α=1.0, go to (2)
5. Defuzzifying the correlation coefficient membership function.

Based on the above algorithm, $r_{xy}(\alpha)$ is computed with Pearson's formula, it varies in a closed interval $\in$ [-1, +1]. Considering we have restricted in this paper to discuss fuzzy number dataset, it's obvious the correlation coefficient which is computed with those fuzzy data is also a fuzzy number.

It's apparent that computational complexity of this method is only O (n) which is much faster than stochastic simulation method; however the solution is not local or global optimum.

After finishing the calculation at each α level, we will get the membership function of correlation coefficient. Then we can defuzzify this membership function to derive a crisp value which summarized the co-varying relationship of the fuzzy data set.

$$d_p(r) = \frac{\sum_{k=1}^{n} r(\alpha)\Delta\alpha}{\sum_{K=1}^{n} \Delta\alpha} \quad (3)$$

SIMULATION TESTS

In this section we will apply this approximate method to some fuzzy datasets, and see if it works by comparing with crisp case result. Suppose we have a data set with N observations, and all those observations are fuzzy data with triangle membership function

By implementing this algorithm, we get the following plots.

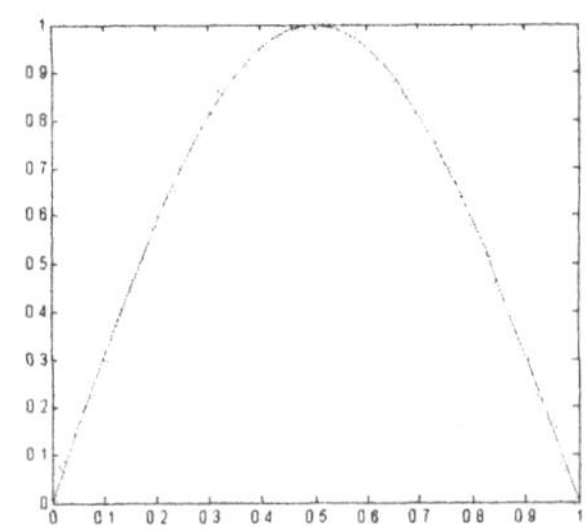

Fig. 1. Dataset 1

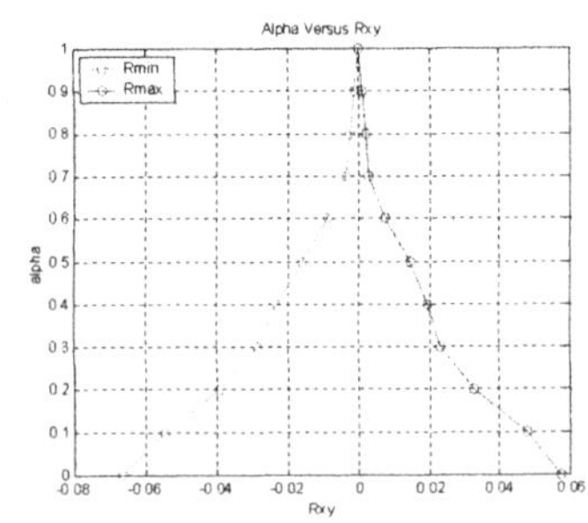

Fig. 2. Membership function of r_{xy}

Figure 1 is the first dataset we will test with the proposed algorithm. We fuzzify these data and apply the proposed algorithm. Figure 2 shows the simulation result, which is the membership function of correlation coefficient for dataset 1. From this figure we find correlation coefficient is equal to 0 when α=1 which means there is no uncertainty in data set. In that case fuzzy data is boiled down to crisp data, therefore fuzzy estimation is boiled down to Pearson's correlation coefficient.

Uncertainty increases with the decreasing of α value and when α equals to 0, we get the fuzziest case in which point correlation coefficient has biggest interval.

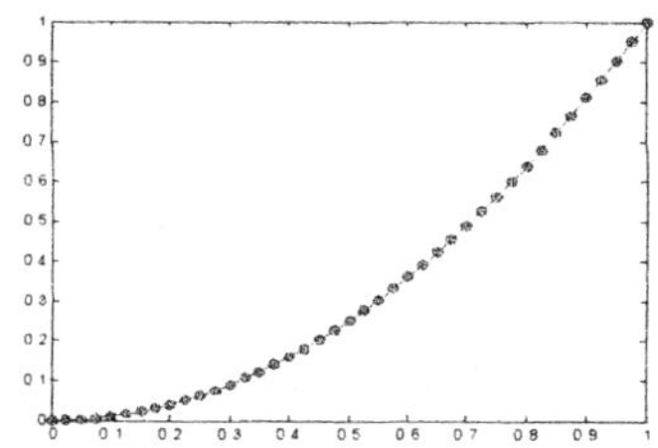

Fig. 3. **Dataset 2**

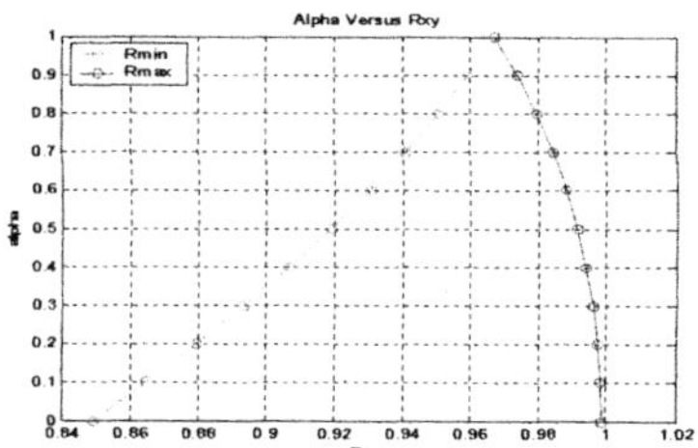

Fig.4. Membership function of r_{xy}

Figure 3 is the dataset we will test next, it's a squared curve. Just follow the same steps we used in the first simulation. Figure 4 shows the membership function of correlation coefficient for this data set.

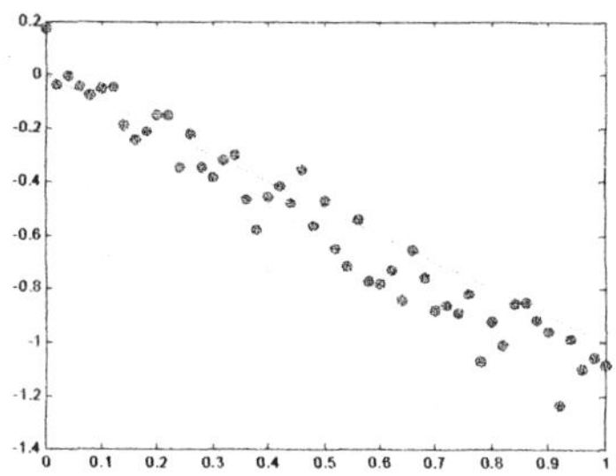

Fig. 5. Dataset 3

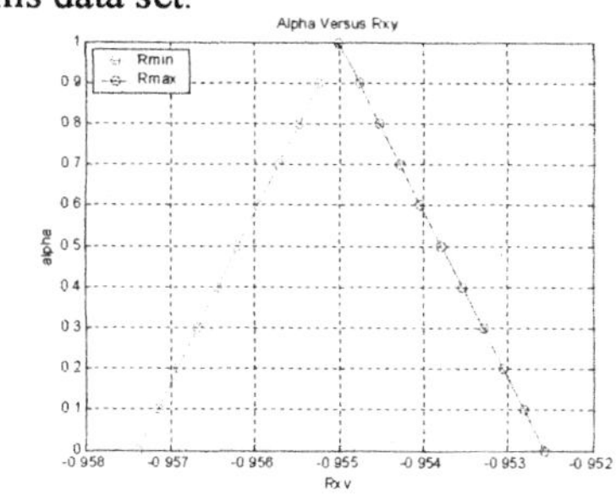

Fig. 6. Membership function of rxy

Figure 5 is a straight line dataset with some noise. Figure 6 shows the membership function of correlation coefficient for this data set. We find the resulting value is close to -1 which is not surprising.

Figure 7 describes a perfect straight line data set. Figure 8 displays the corresponding correlation coefficient.

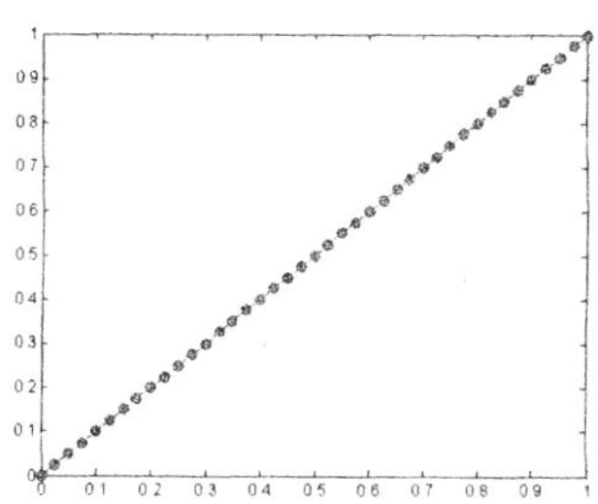

Fig. 7. Dataset 4

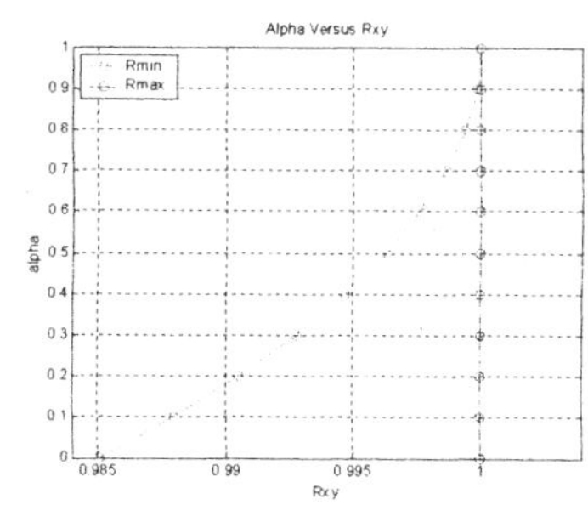

Fig. 8. Membership function of r_{xy}

Finally we summarize the simulation result as the following table.

Table 1 Simulation results of four datasets

Data set	Crisp case value	Defuzzified value	Defuzzified value
1	0	0.0120	-0.0213
2	0.9668	0.9580	0.9300
3	-0.961	-0.9544	-0.9501
4	1	0.9891	0.9908

The second column of table 1 is the result using Pearson's formula. The third column lists their defuzzified value according to stochastic simulation algorithm; the last column is the results computed with approximate method.

From the above table we find defuzzified correlation coefficient value of fuzzy data set computed with approximate method is close to their crisp observations counterparts and also close to the results got from stochastic simulation method. We can conclude this algorithm is intuitively reasonable; Observing fig. 2,4,6,8, Fuzzy correlation coefficient is in essence a nonlinear mapping from membership function of fuzzy data based on extension principle.

CONCLUSIONS

The method presented in this paper aims at data analysis in an uncertain environment. First we discussed Pearson's correlation solution and the present related works. Then we set up a new model to compute correlation coefficient of dataset within fuzzy context. Finally we test this model along with Pearson's method and stochastic simulation method on four data sets. From the simulation results, we can say that this model is efficient and reasonable.

REFERENCES

Daniel Ramot , Ron Milo, Menahem Friedman, and Abraham Kandel ,On Fuzzy Correlations, IEEE Trans. On Systems, Man, and Cybernetics VOL.31, NO.3,pp.381-390 (2001)

George J. Klir/Bo Yuan, Fuzzy Sets and Fuzzy Logic. Englewood Cliffs, NJ:Prentice-Hall, 1995

Witold Pedrycz, Michael H. Smith, Granular Correlation Analysis in Data Mining, 1999 IEEE

Norio Watanabe and Tadashi Imaizumi, A Fuzzy Correlation Coefficient for Fuzzy Random Variables, IEEE International Fuzzy Systems Conference Proceedings, pp.Π1035-Π1038(1999.8)

Ding-An Chiang, Nancy P. Lin, Correlation of fuzzy sets, Fuzzy Sets and Systems, 102(1999) 221-226

Ding-An Chiang, Nancy P. Lin, Partial correlation of fuzzy sets, Fuzzy Sets and Systems 110(2000) 209-215

B.B. Chaudhuri, A. Bhattacharya, On correlation between two fuzzy sets, Fuzzy Sets and Systems 118(2001) 447-456

Shiang-Tai Liu, Chiang Kao, Fuzzy measures for correlation coefficient of fuzzy numbers, Fuzzy Sets and Systems 128(2002) 267-275

Yongshen Ni, John Cheung, Correlation coefficient estimate for fuzzy data, in publishing.

MINING THE WORLD BANK DATA: THE FUZZY c-MEANS CLUSTERING APPROACH

IVETA MRÁZOVÁ
Dept. of Software Engineering
Charles University
Prague, Czech Republic

CIHAN H. DAGLI
Engineering Management Dept.
University of Missouri-Rolla
Rolla, MO, USA

ABSTRACT
The World Development Indicators (WDI) provide comparative socioeconomic data for state economies. With regard to imprecisions and incompleteness in the collected data, we use fuzzy c-means clustering algorithm (FCM) to group the economies according to their mutual similarities. To determine the appropriate number of clusters, various criteria are examined.

INTRODUCTION

World Development Indicators (WDI) reflect a comprehensive view of the development process in the countries. WDI are published annually by the World Bank (2001) and contain the data for the most recent year for which it is available. According to Deboeck (2000), most analyses of WDI performed so far were based on regression analysis. However, these models are incapable of discovering more complex non-linear relationships. From this point of view, other AI-based learning techniques like self-organizing maps (SOMs) seem to be more convenient for exploratory data analysis and data mining.

One of the first papers on detecting patterns of poverty in the World Bank data was published by Kohonen and Kaski (1996). Also G. Deboeck (2000) used SOMs to examine non-linear relationships between various WDI. But due to a relatively large amount of imprecisions in the data, it seems to be a quite natural choice to employ rather fuzzy-logic based methods for analyzing the World Bank data. Here, we applied fuzzy c-means clustering (FCM) algorithm proposed by Bezdek (1981) to group the country economies. In this context, we will examine various criteria for determining an appropriate number of clusters, too.

FUZZY c-MEANS CLUSTERING (FCM)

A fuzzy set A in the (feature) space X can be defined as a set of ordered pairs $A = \{(x, \mu_A(x)) | x \in X; \mu_A(x) \in [0,1]\}$ where $\mu_A(.)$ is called the membership function of A. A vector $\vec{x} \in R^n$ will be denoted as an input pattern. A finite set $X = \{ \vec{x}_1, \ldots, \vec{x}_P \}$ of P ($P > 1$) input patterns $\vec{x}_p \in R^n$ ($1 \leq p \leq P$) will be called the data set. In clustering, the main task consists in partitioning a data set into a number of more-or-less homogeneous subsets called clusters according to a chosen similarity measure. Patterns belonging to the same clusters should be mutually similar and patterns from different clusters should be as dissimilar as possible. A closely related problem called cluster validity refers to the identification of the most appropriate number, c; ($2 \leq c < P$), of data groups in X. In non-fuzzy "crisp" clustering, each input pattern is assigned to exactly one of c clusters. The end result of fuzzy clustering can be expressed in the form of the so-called fuzzy c-partition matrix U:

$$U = [u_{ip}] \; ; \quad u_{ip} \in [0,1] \quad \forall i,p, \quad 1 \le i \le c, \quad 1 \le p \le P \tag{1}$$

Each element u_{ip} of the matrix U expresses the membership degree of $\vec{x}_p$ to the i-th cluster. However, there are two additional constraints on U:

- total membership of $\vec{x}_p \in X$ to all the clusters is equal to 1 and
- every constructed cluster is non-empty and different from the entire set X.

The FCM-algorithm iteratively updates the cluster centers towards their "right" location in the feature space. Accordingly, the partition matrix is adjusted, too. The iteration steps are designed to minimize an objective function $J_s(U,\vec{v}) = \sum_{p=1}^{P} \sum_{i=1}^{c} (u_{ip})^s \, [\sum_{j=1}^{n} (x_{pj} - v_{ij})^2 \,]$ with $\vec{v}_i \in R^n$ being the i-th cluster center, $1 \le i \le c$. $s \in [1,\infty)$ stands for the fuzziness parameter. For larger values of s, the clustering will be "fuzzier". The FCM-algorithm can be summarized as it follows:

The FCM-Algorithm:

- **Step 1:** Specify the number of clusters c $(2 \le c \le P)$ and the fuzziness parameter s $(1 < s < \infty)$. Choose randomly an initial partition matrix $U^{(0)}$. Choose the termination criterion ε and set the iteration index t to 0: $t = 0$.
- **Step 2:** Calculate the fuzzy cluster centers $\{ \vec{v}_i^{(t)} \mid 1 \le i \le c\}$ by using $U^{(t)}$ according to:

$$\vec{v}_i^{(t)} = \frac{1}{\sum_{p=1}^{P} ({u_{ip}}^{(t)})^s} \sum_{p=1}^{P} ({u_{ip}}^{(t)})^s \, \vec{x}_p, \qquad 1 \le i \le c, \tag{2}$$

- **Step 3:** Calculate the new partition matrix $U^{(t+1)}$ by using $\{ \vec{v}_i^{(t)} \mid 1 \le i \le c \}$ according to:

$$ {u_{ip}}^{(t+1)} = \frac{(1 / \| \vec{x}_p - \vec{v}_i^{(t)} \|^2)^{1/s-1}}{\sum_{k=1}^{c} (1 / \| \vec{x}_p - \vec{v}_k^{(t)} \|^2)^{1/s-1}} , \qquad 1 \le i \le c; \; 1 \le p \le P. \tag{3}$$

- **Step 4:** Calculate $\Delta = \| U^{(t+1)} - U^{(t)} \| = \max_{i,p} | u_{ip}^{(t+1)} - u_{ip}^{(t)} |$. If $\Delta > \varepsilon$, then set $t = t+1$ and go to **Step 2**. If $\Delta \le \varepsilon$, then **Stop**.
- **END of FCM**

WHAT IS THE APPROPRIATE NUMBER OF CLUSTERS?

In this study, we used the following three cluster validity indicators to estimate an appropriate number of clusters present in the data set. For a (strictly) fuzzy c-partition U of P input patterns, these indicators are defined as it follows:

- partition coefficient $F(U;c)$:

$$F(U;c) = \frac{1}{P} \sum_{p=1}^{P} \sum_{i=1}^{c} (\, u_{ip} \,)^2 \tag{4}$$

- partition entropy $H(U;c)$:

$$H(U;c) = -\frac{1}{P} \sum_{p=1}^{P} \sum_{i=1}^{c} u_{ip} \ln(u_{ip}) \; ; \quad u_{ip} \ln(u_{ip}) = 0 \text{ for } u_{ip} = 0. \tag{5}$$

- proportion exponent $W(U;c)$ (Windham (1980)):

$$W(U;c) = -\sum_{p=1}^{P} \ln \left[\sum_{j=1}^{\lfloor \mu_p^{-1} \rfloor} (-1)^{j+1} \binom{c}{j} \left(1 - j \bigvee_{i=1}^{c} u_{ip} \right)^{c-1} \right] \tag{6}$$

$$\mu_p = \max_{1 \leq i \leq c} \{u_{ip}\} = \bigvee_{i=1}^{c} u_{ip} \text{ and } \lfloor \mu_p^{-1} \rfloor = \mathit{greatest\ integer} \leq (1/\mu_p)$$

The partition coefficient and the partition entropy seem to indicate reliably the right number of well-separated ("crisper") clusters. Unfortunately, both of these two (mutually complementary – Bezdek (1981)) criteria suffer from appearing to be monotonous for lower numbers of clusters. The proportion exponent $W(U;c)$ might be less reliable for values of s close to 1 but it is more capable of detecting the right number of mutually overlapping clusters for higher values of s. Unfortunately, then the following FCM-clustering often leads to almost identical cluster centers with membership values close to $1/c$.

When estimating an adequate number of clusters c according to the above indicators, we assume intuitively that the clustering structure is matched better when more points concentrate around the cluster centers. The found cluster centers should be located relatively far one from each other. This implies a crisper partitioning matrix (while considering a fixed fuzziness parameter s). According to this assumption, the heuristic rules for the best number of clusters c are:

- for the partitioning entropy: $\min_{c=2,\dots,P-1} \{ \min_{U \in \Omega_c} [H(U,c)] \}$, (7)
- for the partitioning coefficient: $\max_{c=2,\dots,P-1} \{ \max_{U \in \Omega_c} [F(U,c)] \}$, (8)
- and for the proportion exponent: $\max_{c=2,\dots,P-1} \{ \max_{U \in \Omega_c} [W(U,c)] \}$, (9)

where Ω_c denotes the set of all optimal solutions for a given number of clusters c.

SUPPORTING EXPERIMENTS

The tested data consist of 99 16-dimensional input patterns, each one corresponding to a state economy. These comprised the low- and middle-income countries with complete data records. For high-income countries, the missing data (on external debt only) was set to 0 because in comparison with low- and middle-income countries, their external debt is usually very low. 16 considered WDI-indicators listed below describe the economical and social potential of the countries, or their citizens. All indicators are relative to population.

1. Gross national product (GNP) per capita (in 1999 in USD)
2. Purchasing power parity (PPP) (in 1999, converted to USD)
3. Average annual growth rate of the gross domestic product (GDP) per capita (in 1998 – 1999)
4. GDP implicit deflator (average annual percentage growth during 1990 – 1999
5. External debt (present value % of GNP in 1998)
6. Total debt service (% of export of goods and services in 1999)
7. High-technology exports (% of manufactured exports in 1999)
8. Military expenditures (% of GNP in 1997)
9. Expenditures for research and development (% of GNP in 1987 – 1997)
10. Total expenditures on health (% of GDP in 1990 – 1998)
11. Public expenditure on education (% of GNP in 1997)
12. Male life expectancy at birth (in 1998)
13. Fertility rates (in 1998)

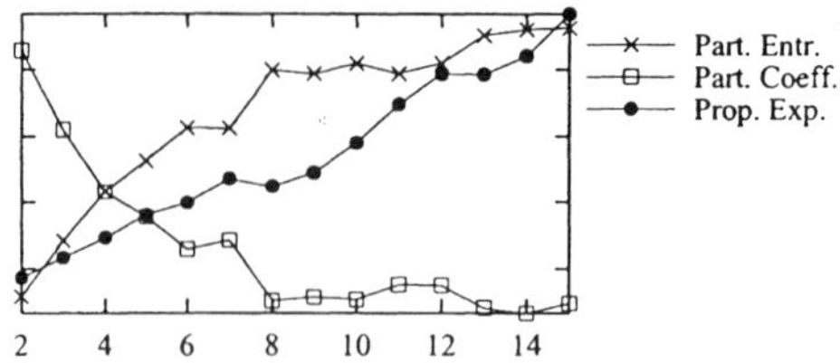

Figure 1: A graph of the trends for the partition entropy, partition coefficient and proportion exponent (for $s = 1.4$ and $\varepsilon = 0.05$). The x-axis indicates the number of clusters. Due to their different values, all three criteria were transformed (each one using a different but linear transformation) such that their trends would become more apparent. For 7 clusters, the graphs show a local minimum of the partition entropy and local maxima for the partition coefficient and the proportion exponent (cf. (7) – (9)).

14. GINI-index – the distribution of income / consumption (mostly for 1995)
15. Internet hosts per 10000 people (in January 2000)
16. Mobile phones per 1000 people (in 1998)

This data set was transformed element-wise to the interval $[0, 1]$ by means of:

$$x' = \frac{x - x_{min}}{x_{max} - x_{min}} \quad \text{and} \quad x'' = \frac{1}{1 + e^{-k(x' - 1/2)}} \qquad (10)$$

with $k = 4$. x_{max} and x_{min} denote the maximum and minimum value over all 99 input patterns in the considered dimension. To estimate the relevant number of clusters, we tested the above-mentioned cluster validity indicators. The trends in their evolution are depicted in Figure 1 for $s = 1.4$. We chose 7 to be an appropriate number of clusters. For this number of clusters, we can see a local minimum of the partition entropy and local maxima for both the partition coefficient and the proportion exponent in Figure 1.

For the World Bank data, the FCM-algorithm converged very quickly, in most cases within 15 iterations. The results of FCM-clustering the World Bank data (with $s = 1.4$ and $\varepsilon = 0.05$) are presented in the form of a "spread-sheet-like" table below. For some countries, almost a "crisp" clustering was achieved – see e.g. Guinea, Uzbekistan, or even Netherlands – while for others a fuzzier cluster belongingness occurs, see e.g. Azerbaijan, Korea, New Zealand, or Poland. There are even countries with much "fuzzier" partitions – see e.g. Brazil, Philippines, or Russia. This indicates rather a long distance from any of the formed clusters.

CONCLUSIONS

Experiments performed so far confirm that the FCM-algorithm is an efficient tool for clustering imprecise data. Together with reliable cluster validity indicators, FCM-clustering allows to estimate the appropriate number (7 in our case) of groups present in the data. To visualize the clustering results in a clear and convenient manner, we used the form of a "spread-sheet-like" table with higher cluster membership values indicated by darker background of the cells. In our opinion, this is a simpler and more illustrative approach than the complex one adopted by Kaski and Kohonen (1996) or Deboeck (2000)in SOM-based analyses of WDI.

The achieved results represent an important basis for solving other data mining tasks – like providing a set of simple rules explaining mutual relationships among the data, or extracting a set of descriptive landmarks characterizing best the found cluster regions.

ACKNOWLEDGEMENTS

This research was partly supported by the Fulbright commission and by the grants No. 300/2002/A INF/MFF (GA UK) and No. 201/02/1456 (GA ČR).

REFERENCES

Bezdek, J. C., 1981, *Pattern Recognition with Fuzzy Objective Function Algorithms*, Plenum Press, New York.

Deboeck, G., 2000, "Self-Organizing Patterns in World Poverty Using Multiple Indicators of Poverty, Repression and Corruption," *Neural Networks World*, 2000/1-2, pp. 239-254.

Kaski, S. and Kohonen, T., 1996, "Exploratory data analysis by the self-organizing map: structures of welfare and poverty in the world," *Refenes, A.P. et al. (eds): "Neural Networks in Financial Engineering"*, World Scientific, Singapore, pp. 498-507.

World Bank, 2001, *World Development Report 2000/2001: Attacking Poverty*, The World Bank Group, Washington, D.C.

Windham, M. P., 1980, "Cluster Validity for Fuzzy Clustering Algorithms," *J. Fuzzy Sets Syst.*, 3, pp. 1-9.

Partition table for the World Bank data:

Below are presented the results of FCM-clustering the World Bank data (with $s = 1.4$ and $\varepsilon = 0.05$). Each line of the table corresponds to one particular state economy. For each country, the 7 columns indicate the degree of its membership to that respective cluster. Higher membership values are indicated by darker background of the cells. Due to minor rounding errors, their sum might slightly differ from 1.

	Country	1	2	3	4	5	6	7
1	Algeria	0.01	0.82	0.01	0.10	0.01	0.00	0.04
2	Armenia	0.34	0.27	0.02	0.10	0.06	0.00	0.22
3	Australia	0.00	0.00	0.00	0.00	0.04	0.95	0.00
4	Austria	0.00	0.00	0.00	0.00	0.01	0.98	0.00
5	Azerbaijan	0.56	0.33	0.01	0.04	0.03	0.00	0.03
6	Bangladesh	0.02	0.72	0.04	0.17	0.02	0.00	0.04
7	Belarus	0.95	0.02	0.00	0.01	0.01	0.00	0.01
8	Belgium	0.01	0.01	0.00	0.01	0.11	0.86	0.01
9	Bolivia	0.00	0.03	0.03	0.27	0.00	0.00	0.67
10	Brazil	0.17	0.15	0.06	0.18	0.08	0.02	0.34
11	Bulgaria	0.06	0.83	0.00	0.03	0.05	0.00	0.02
12	Burkina Faso	0.00	0.00	0.97	0.02	0.00	0.00	0.00
13	Burundi	0.03	0.06	0.64	0.21	0.02	0.00	0.05
14	Cambodia	0.01	0.05	0.27	0.63	0.01	0.00	0.03
15	Canada	0.01	0.01	0.00	0.01	0.17	0.79	0.01
16	Chile	0.01	0.03	0.00	0.03	0.01	0.00	0.92
17	China	0.02	0.82	0.01	0.06	0.04	0.00	0.06
18	Colombia	0.01	0.04	0.01	0.07	0.02	0.00	0.85
19	Costa Rica	0.04	0.48	0.01	0.12	0.12	0.00	0.22
20	Cote d'Iv.	0.01	0.02	0.83	0.13	0.00	0.00	0.01
21	Croatia	0.14	0.26	0.02	0.11	0.35	0.02	0.12
22	Czech Rep.	0.01	0.03	0.00	0.01	0.95	0.00	0.01
23	Denmark	0.00	0.00	0.00	0.00	0.01	0.99	0.00
24	Dominican R.	0.03	0.55	0.03	0.15	0.04	0.00	0.20
25	Ecuador	0.02	0.13	0.03	0.15	0.02	0.00	0.66

	Country	1	2	3	4	5	6	7
26	Egypt A.R.	0.02	0.90	0.01	0.05	0.01	0.00	0.01
27	El Salvador	0.01	0.08	0.01	0.11	0.01	0.00	0.78
28	Estonia	0.04	0.13	0.01	0.06	0.67	0.01	0.09
29	Ethiopia	0.00	0.00	0.97	0.02	0.00	0.00	0.00
30	Finland	0.00	0.00	0.00	0.00	0.02	0.96	0.00
31	France	0.00	0.00	0.00	0.00	0.01	0.99	0.00
32	Georgia	0.96	0.02	0.00	0.01	0.01	0.00	0.01
33	Germany	0.00	0.00	0.00	0.00	0.03	0.97	0.00
34	Ghana	0.00	0.07	0.08	0.82	0.00	0.00	0.02
35	Greece	0.01	0.05	0.00	0.02	0.85	0.04	0.03
36	Guatemala	0.01	0.09	0.18	0.37	0.01	0.00	0.34
37	Guinea	0.00	0.00	0.99	0.01	0.00	0.00	0.00
38	Honduras	0.01	0.03	0.02	0.09	0.01	0.00	0.86
39	Hungary	0.03	0.24	0.01	0.04	0.65	0.01	0.02
40	India	0.01	0.85	0.01	0.11	0.01	0.00	0.02
41	Indonesia	0.06	0.43	0.10	0.20	0.05	0.01	0.16
42	Ireland	0.01	0.02	0.01	0.01	0.13	0.79	0.02
43	Italy	0.00	0.00	0.00	0.00	0.01	0.99	0.00
44	Jamaica	0.07	0.47	0.01	0.14	0.10	0.00	0.22
45	Japan	0.00	0.00	0.00	0.00	0.01	0.98	0.00
46	Jordan	0.09	0.24	0.06	0.26	0.14	0.02	0.20
47	Kazakhstan	0.84	0.10	0.00	0.03	0.01	0.00	0.01
48	Kenya	0.01	0.04	0.19	0.67	0.01	0.00	0.07
49	Korea	0.04	0.09	0.02	0.05	0.38	0.38	0.05
50	Kyrgyz R.	0.06	0.75	0.01	0.13	0.01	0.00	0.05
51	Lao PDR	0.01	0.03	0.87	0.09	0.00	0.00	0.01
52	Latvia	0.07	0.47	0.01	0.07	0.30	0.00	0.07
53	Lesotho	0.03	0.06	0.28	0.49	0.02	0.00	0.11
54	Lithuania	0.12	0.35	0.01	0.09	0.28	0.01	0.13
55	Madagascar	0.00	0.00	0.97	0.02	0.00	0.00	0.00
56	Malaysia	0.08	0.26	0.07	0.21	0.13	0.03	0.22
57	Mali	0.00	0.00	0.98	0.01	0.00	0.00	0.00
58	Mauritania	0.01	0.03	0.71	0.22	0.01	0.00	0.02
59	Mexico	0.02	0.22	0.01	0.15	0.03	0.00	0.57
60	Moldova	0.18	0.17	0.03	0.20	0.14	0.01	0.27
61	Mongolia	0.01	0.93	0.00	0.04	0.01	0.00	0.01
62	Morocco	0.02	0.41	0.01	0.27	0.01	0.00	0.28
63	Nepal	0.01	0.17	0.12	0.65	0.01	0.00	0.05
64	Netherlands	0.00	0.00	0.00	0.00	0.00	1.00	0.00
65	New Zealand	0.02	0.04	0.01	0.03	0.66	0.20	0.04
66	Nicaragua	0.06	0.16	0.12	0.25	0.08	0.02	0.30
67	Niger	0.00	0.01	0.90	0.06	0.00	0.00	0.02
68	Nigeria	0.00	0.01	0.91	0.06	0.00	0.00	0.02
69	Norway	0.00	0.00	0.00	0.00	0.02	0.97	0.00
70	Pakistan	0.04	0.31	0.16	0.35	0.04	0.00	0.09
71	Panama	0.01	0.12	0.00	0.07	0.02	0.00	0.77
72	Paraguay	0.01	0.03	0.01	0.08	0.01	0.00	0.86
73	Peru	0.00	0.03	0.00	0.04	0.00	0.00	0.93
74	Philippines	0.07	0.25	0.09	0.25	0.09	0.02	0.25
75	Poland	0.06	0.32	0.00	0.04	0.53	0.00	0.03
76	Portugal	0.01	0.02	0.00	0.01	0.92	0.03	0.01
77	Romania	0.11	0.59	0.01	0.08	0.09	0.00	0.11
78	Russian F.	0.14	0.34	0.04	0.23	0.06	0.01	0.19
79	Senegal	0.00	0.00	0.95	0.04	0.00	0.00	0.00
80	Slovak Rep.	0.02	0.09	0.00	0.01	0.86	0.00	0.01
81	Slovenia	0.00	0.01	0.00	0.00	0.98	0.01	0.00
82	South Africa	0.04	0.10	0.10	0.44	0.06	0.01	0.26
83	Spain	0.00	0.01	0.00	0.00	0.97	0.01	0.00
84	Sri Lanka	0.03	0.82	0.01	0.06	0.03	0.00	0.04
85	Sweden	0.00	0.00	0.00	0.00	0.02	0.97	0.00
86	Switzerland	0.00	0.00	0.00	0.00	0.00	0.99	0.00
87	Thailand	0.04	0.53	0.01	0.16	0.09	0.00	0.17
88	Tunisia	0.06	0.60	0.01	0.12	0.10	0.00	0.10
89	Turkey	0.02	0.12	0.01	0.10	0.03	0.00	0.72
90	Uganda	0.01	0.02	0.86	0.09	0.01	0.00	0.01
91	Ukraine	0.97	0.01	0.00	0.01	0.01	0.00	0.01
92	Un. Kingd.	0.00	0.00	0.00	0.00	0.01	0.98	0.00
93	Un. States	0.02	0.03	0.01	0.02	0.08	0.80	0.03
94	Uruguay	0.01	0.08	0.01	0.06	0.05	0.00	[illegible]
95	Uzbekistan	0.98	0.01	0.00	0.00	0.00	0.00	0.00
96	Venezuela	0.01	0.06	0.01	0.07	0.02	0.00	[illegible]
97	Vietnam	0.00	0.97	0.00	0.02	0.00	0.00	0.01
98	Yemen	0.07	0.15	0.26	0.37	0.05	0.01	0.09
99	Zambia	0.01	0.03	0.74	0.14	0.01	0.00	0.06

DYNAMIC NEURO-FUZZY SYSTEM FOR PROGNOSIS IN REHABILITATION

YU WANG
Dept. of Biomedical Engineering
Marquette University, WI 53233

JACK M. WINTERS
Chair and Professor
Dept. of Biomedical Engineering
Marquette University, WI 53233

ABSTRACT

This paper describes an event-driven dynamic recurrent neuro-fuzzy system that is designed to perform the process of prognosis (specifically, dynamic recovery procedure prediction) in stroke and cardiac rehabilitation. Four layers are implemented in this system, each of which consists of a cluster of neurons: input, rule, output, and outcome layers. Detected events are mapped as fuzzy variables in the input layer by nonlinear "receptive field" Gaussian-shaped membership functions and "hardness-variable" boundary membership functions. The rule layer is composed of dynamic neurons, which associate with given rules. The states of a rule-neuron are not only a function of the fuzzy rule, but also on a temporal dynamic process that depends on the time constants and weakly on connections with other rule-neurons that are complementary (excitatory connections) or competitive (inhibitory connections). The relationships of performance measures or decision-support information to the states of rule-neurons and/or input events are represented in the output layer. Outcomes are similar to optimization cost functions, which can be functions of input events, fuzzy states, and outputs. Reinforcement learning will be used to tune the system parameters.

I. INTRODUCTION

Rehabilitation exists as a field because the human system can change as a function of natural healing mechanisms and rehabilitative interventions [1]. Capturing such dynamic change with mathematical models has proven difficult. This paper presents a novel approach that integrates two modeling tools for complex systems. First, trained clinical experts often have a reasonably good capacitiy to predict changes as a function of intervention, based on clinical experience and knowldege of scientific liturature. A fuzzy inference system can be a good choice for capturing such expert knowledge of diagnosis-prognosis causality and of healing processes in the human body. Second, the human body can be viewed as a complex dynamic system, and a dynamic neural network [1,2] is an obvious choice for capturing key dynamic features of interest. Taking advantage of strengths of both, this paper uses a dynamic neural network to represent a fuzzy inference system, and then through simulation, predicts changes in key variables of interest that are attributed to the real human system.

II. SYSTEM DESCIPTION

This system is derived from the classic optimum control system [1,3], which is typically described by the following differential equations, where U is the input vector, X is the state vector, Y is the output vector.

$$\begin{cases} \dfrac{dX}{dt} = f(X,U) \\ Y = g(X,U) \end{cases} \tag{1}$$

The outcome J (sometimes called cost or performance critieria) of the system is a function of the input, state, and/or output, as Eq. (2) shows. Normally f is a function of integration (for analog models) or summation (for digital models).

$$J = f(U, X, Y) \tag{2}$$

Eq. (1) is an effectve way to represents dynamic systems. On the other hand, fuzzy logic is very powerful to simulate human knowledge. Also it is reasonable to construct the whole system with neural network since the strength of neuro-fuzzy is well documented [4].

This system is composed of four layer clusters of neurons, representing five categories of objects: input, rules, state, output, and outcome. These are shown in Figure 1.

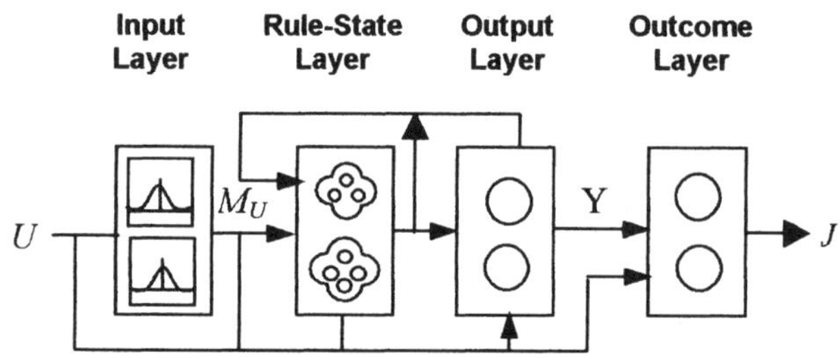

Figure 1. Layer structure of the system

A. Input Layer

Occurrence of specific events is detected by the input layer. Here an event is defined as a fuzzy event, which is a fuzzy variable with a degree of membership on the interval $<M_{thresh}, 1>$, where M_{thresh} is a specified constant on the interval $<0, 1>$. Within the input layer, neurons implement nonlinear mapping from input signal into fuzzy variables of a specific event. Two kinds of membership functions are available:

- "Receptive field," normally of Gaussian shape with mean and variance parameters:

$$M_u = \frac{1}{\sigma\sqrt{2\pi}} e^{-\frac{(u-m_0)^2}{2\sigma^2}} \tag{3}$$

  where m_0 is the mean and σ is the variance.

- "Soft boundary," defined by parameters of edge and hardness:

$$M_u = \begin{cases} 0 & u \le (b-l) \\ \dfrac{e^{\frac{k(u-b+l)}{l}}-1}{2(e^k-1)} & (b-l) < u \le b \\ 1-\dfrac{e^{\frac{k(-u+b+l)}{l}}-1}{2(e^k-1)} & b < u \le (b+l) \\ 1 & u > (b+l) \end{cases} \tag{4}$$

where $l = \dfrac{(1-h)(u_{\max} - u_{\min})}{2}$, k is the dimensionless shape factor and preset as 3, b is the edge parameter, h represents the extent of hardness (on interval <0, 1>), and $u_{\max}$ and $u_{\min}$ define the range of input signal.

An event is acknowledged as occurring if $M_u \ge M_{thresh}$.

For each of these two kinds of membership functions, crisp logic is a subset of fuzzy logic:

- If the variance of Gaussian membership function is zero, Eq. (3) becomes

$$M_u = \begin{cases} 0 & u \ne m_0 \\ 1 & u = m_0 \end{cases} \tag{5}$$

- If the hardness is one, the "soft boundary" membership function becomes a "hard boundary"

$$M_u = \begin{cases} 0 & u < b \\ 1 & u \ge b \end{cases} \tag{6}$$

Certain neurons within the nucleus may have second-generation effects such as "lateral inhibition" connections. Several fuzzy modifiers, such as the "Not" and "Very" operations, are also implemented.

B. Rule-State Layer

The expert implements fuzzy rules through an interactive GUI. The expert is required to add temporal dynamics in the implementation of the core logic of the rule. Experts have the option of directly entering time constant parameters, or providing temporal responses that are then fit. For example, in the case of a patient performing exercise of a certain intensity, if experts provide the strength level of the patient at several time points after the exercise intervention, the time constants will be estimated using a minimum square error (MSE) method. The resulting curve estimation will be shown through the GUI.

The states are included in this layer, which includes two kinds of neurons:

1. Rule-neurons that implement a rule from expert knowledge, using fuzzy mathematical operations,
2. State-neurons that represent a physical/physiological state of patient.

The rule-neurons can receive inputs from outputs of neurons in the input, rule-state and output layer. The output of a rule-neuron depends on both its input and on low-weight connections with other rule-neurons that are complementary (excitatory connections) or competitive (inhibitory connections). The logic for rules is normally associated with state, i.e., for each state, find all the factors that may have effect on this state, and then organize those factors into rules. The output of a rule is a function of the input signals, states, and/or outputs (see Figure 1), through a mapping process that implements fuzzy rule operations that include any combination of "and" and "or" operators or fuzzy variables:

$$R_i = f(M_U, M_X, M_Y) \tag{7}$$

R_i : Output of the ith rule-neuron (between <0, 1>).

M_U , M_X , M_Y : Mapping of inputs, states, and outputs that connect to this rule-neuron.

Each rule-neuron connects to a paired state-neuron (or to a small cluster of neurons providing sub-states). The state of a state-neuron is not only a function of the fuzzy rules, but also of a temporal dynamic process that depends on some time constants. An example of a "state" in rehabilitation is a physiological measure that can change over time, such as strength of a healing muscle. The general equation of the input to a state is:

$$f(x_k) = \sum_i (w_i R_i + F_i(\sum_j w_{ij} R_j)) \tag{8}$$

x_k : The kth state (between <0, 1>),

$f(x_k)$: Input to the kth state,

w_i : The weight of the ith rule-neuron (on <0, 1>),

w_{ij} : The weight of the connection between the ith rule-neuron and the jth rule-neuron (on <0, 1>),

F_i : A nonlinear squashing function.

In order to capture as many details of intrinsic dynamics of the states as possible, three types of possible dynamic functions are provided for $f(x_k)$.

1. A second-order dynamic function (or two first order dynamic function in series). There are two time constants that aim to capture both short-term and the long-term temporal dynamics.

$$\tau_1 \tau_2 \frac{d^2 x_k}{dt^2} + (\tau_1 + \tau_2)\frac{dx_k}{dt} + x_k = f(x_k) \tag{9}$$

where τ_1 and τ_2 have direct physical meaning that helps the experts associate meaning to the values. However, it can only describe the over-damped procedure, which is not the only case for dynamic healing processes in rehabilitation.

2. Two weighted first order dynamic function in parallel.

$$\begin{cases} \tau_1 \dfrac{dx_{k_1}}{dt} + x_{k_1} = f(x_k) \\ \tau_2 \dfrac{dx_{k_2}}{dt} + x_{k_2} = f(x_k) \end{cases} \qquad x_{k_1}, x_{k_2} : \text{Sub-states of } x_k$$

$$x_k = K_1 x_{k_1} + K_2 x_{k_2} \tag{10}$$

Obviously, this parallel representation can describe a greater variety of dynamic behavior than the serial one.

3. A state neuron "cluster" of sub-states that is a combination of 1 and 2. This is more capable of describing the dynamic procedures at the cost of the increased complexity, such as a sum of sub-states with dynamic behavior that can be similar to "phasic" and "tonic" interneurons.

C. Output Layer

Outputs typically perform an action (e.g., prompt user, initiate communication, store data in an electronic record) or provide decision-support information for the user. The output of a neuron in this layer is a function of the states of the rules and the input events (see figure 1).

$$y_i = f(U, X, M_U, M_X) \tag{11}$$

where y_i is the output of the ith output-neuron.

The function f can be a weighted sum (using standard or fuzzy mathematics) or a Sujeno fuzzy inference system [3].

D. Outcome Layer

Outcomes are defined as scalars or curves that relate to what in engineering optimization are called performance sub-criteria or cost functions, and can be a function of events, fuzzy states, and outputs. Examples of rehabilitative outcomes are numerical representations of terms such as impairment, disability, independence, quality of life, satisfaction, and cost. Normally, an outcome is a weighted sum, as well as their mapped results, with the "+" operation implemented using fuzzy mathematics:

$$\begin{aligned} J_i = & \sum w_j^U U + \sum w_j^X X + \sum w_j^Y Y \\ & + \sum w_j^{M_u} M_u + \sum w_j^{M_X} M_X + \sum w_j^{M_Y} M_Y \end{aligned} \tag{12}$$

where J_i is the value of the ith outcome.

Outcome metrics can be used to compare the effects of alternative intervention plans. Users may focus on different outcomes, and thus use results to select an intervention plan that best meets their preferred criteria.

E. Parameter Sensitivity Analysis and Optimization

Sensitivity analysis has been implemented for two reasons. First, it can be used by expert designers to better understand the sensitivity of global behaviors of interest to parameters associated with specific membership functions or rules.

Second, it can be used to help select candidate parameters for optimization, since the large number of parameters and non-linearity of the system makes it prohibitive to use gradient search algorithms with all parameters.

The above dynamic fuzzy inference system has been purposely implemented as a dynamic recurrent neural network so as to enable the system to improve its prognosis prediction capabilities for a given client through learning. There are two approaches to parameter optimization: supervised and reinforcement learning. When experts use the interactive GUI's to construct the rules and set the values of the parameters, they also may provide some reference (trial) data (e.g., states such as physiologic signals, outputs such as performance measures, outcomes such as actual measured outcomes). Supervised learning uses these data to adjust key parameters, selected by the expert user(s) based on sensitivity analysis or their degree of uncertainty in assigning parameters during the design phase. Because of the Health Insurance Portability and Accountability Act (HIPAA) [5], it is now more difficult to obtain large amounts of patient data, and in most cases supervised learning can be used only in the construction stage. The main purpose of reinforcement learning is to optimize system parameters during routine usage: a user (e.g., physical therapist) plays the role of an "adaptive critic," and feeds back a punishment/reward signal that is used to adjust parameters based on credit assignment.

CONCLUSIONS

This paper introduces a novel method to predict rehabilitative dynamics and evaluate the effect of different intervention plans. Its unique structure can take advantage of expert knowledge of system dynamics, and integrate with real data.

ACKNOWLEDGEMENT

The financial support from The Ralph and Marian Falk Medical Trust Foundation and the Rehabilitation Engineering Research Center on Telerehabilitation (U.S. Department of Education, NIDRR #H133A990008) is gratefully acknowledged. The opinions are those of the authors.

NOMENCLATURE

U: Vector of input signal
M_U: Vector of detected events
X: Vector of states
M_X: Vector of mapped states
Y: Vector of outputs
M_Y: Vector of mapped outputs
J: Vector of outcome
u: Input signal
M_u: Degree of membership of an event
x: State
M_x: Mapped state
y: Output
M_y: Mapped output

REFERENCES

[1] Winters, J.M., "Study posture/movement selection & synergies via a synthesized neuro-optimization framework," In: *Biomechanics and Neural Control of Movement*, Winters, J.M. and Crago, P.E., eds, Chapter 35, pp. 458-476, New York: Springer-Verlag, 2000.

[2] Draye, J.-P., Winters, J.M. and Cheron, G., *"Self-Selected Modular Recurrent Neural Netowrks with Postural and Inertial Subnetworks Applied to Complex Movements," Biol. Cybern.*, Vol. 82, pp. 82:27-39, 2002.

[3] Paraskevopoulos, P.N., *"Modern Control Engineering", Part II, New York: Marcel Dekker, 2002*

[4] Lin, C.T. and Lee, C.S., *"Neural Fuzzy Systems: A Neuro-Fuzzy Synergism to Intelligent Systems", New Jersey: Prentice Hall*, 1996.

[5] http://cms.hhs.gov/hipaa/

FRAMEWORK OF A FUZZY MULTIPLE CRITERIA MODEL FOR FACILITIES PLANNING

VIKAS KAPOOR
Dept. of Production & Industrial Engg.,
Jai Narain Vyas University Jodhpur,
Jodhpur, India.

SHYAM S. TAK
Member,
Raj. Public Service Commission,
Ajmer, India.

ABSTRACT

Facilities planning is a higher-level term, which encompasses both facilities location and facilities design at lower levels. Both these problems depend on quantitative as well as on qualitative factors, making it a time consuming & complex exercise. The paper investigates this long-standing problem and proposes a comprehensive framework for addressing it through a Fuzzy multiple criteria model. At the core of the proposed model is a Fuzzy variant of Saaty's (1980) Analytic Hierarchy Process (AHP), which employs 'linguistic variables' such as {very low, low, medium, high, very high} in place of numerical values 1-9. Thereafter, the paper demonstrates how this Fuzzy variant of AHP may be used either to replace the conventional AHP for solving these problems as has been suggested by Liang and Wang (1991), Dweiri and Meier (1996) respectively; or alternatively may be used in new algorithms for location and layout design.

INTRODUCTION

According to Tompkins and White (1984) 'facilities planning' is a higher-level term, which is a 'composite of facilities location and facilities design'. Facilities location would refer to the placement of a facility with respect to markets, resources, competitors, suppliers, labour etc. Facilities design, on the other hand, is concerned with how – by virtue of mutual correlation of positioning - the components of the facility can best support the objectives of the facility. Together, these two have engaged researchers for several decades and have emerged as complex issues requiring both quantitative as well as qualitative kinds of analysis. This paper addresses itself to both these elements of facilities planning and proposes a comprehensive mathematical model to deal with it.

The Background

There have been several efforts in the past few decades to deal with problems of facilities planning. However, no singular approach could promise a complete & error-free solution. The reason for this may be explained in the words of J.M. Apple (1974).

"...In carrying out the early research in modelling the layout process, the experimenters found the problem so overwhelming in its requirements that it was impossible to structure it in mathematical terminology. To facilitate their effort they made the necessary assumptions to cut it down to size. More often than not, such assumptions were false, unrealistic or deceptively simplified. Some misleading assumptions often made by early researchers are:

- *Material flows from the center of one department to the center of another*
- *Material handling cost is directly proportional to distance*
- *All data on material flow are known, and are deterministic in nature*

- *Waste or scrap material need not be considered.*
- *All travel is in two dimensions*

Obviously, all of these are likely to be false, and their use as bases for a model will result not only in a sub-optimum solution, but in a highly misleading representation of the actual situation...."

This pretty much sums the efforts of mathematicians and researchers engaged in this area for several decades. Given below is an overview of various attempts and techniques proposed by researchers in the field of Facilities Planning.

Facilities Location Problem. A majority of traditional techniques for location selection have been known to combine subjective considerations with objective requirements using a ranking and rating method. The evaluation of the critical factor measures (the ones for which absence or presence precludes the location of a plant at a site regardless of other conditions) is based upon the critical factor index (CFI) that may be assigned either a value of 0 or 1. In practice, however, the assignment of points and degrees to a prospective location for such an assessment are not sharply defined.

Facilities Design Problem. Traditional techniques for facilities design have ranged from the schematic approaches like operation process chart, flow diagrams, string diagrams etc., to the systematic approaches of travel charting, sequence analysis, systematic layout planning using Activity Relationship Chart and eventually computer assisted design algorithms. However, none of them could sustain for long, since they all relied too heavily on the subjective judgment of the evaluator and either became unstable when the problem involved a large number of facilities, or could generate a sub-optimal solution.

FUZZY APPLICATIONS IN FACILITIES PLANNING

In contrast to the conventional approach, fuzzy set theoretic approach was introduced by Zadeh (1965), (1975) in the mid 60s and developed through the 70s wherefrom it was immediately taken up for further research in a variety of Engineering Applications. In their excellent paper, Karwowski & Evans (1986) discuss the role of imprecision and subjective evaluation in conventional techniques of facilities planning. They were the first to suggest that distances between planned facilities may be expressed as being 'short', 'medium' or 'long' rather than by using 'crisp' values. This not only makes the approach more realistic but also reduces the complexity level immensely. The work of Karwowski et al. led to a spate of papers attempting to adopt this technique to the problems of facilities planning.

Given below is a brief description of two landmark papers dealing with the problems of facilities location and facilities design using fuzzy multi-criteria approach.

i) Liang and Wang (1991) proposed a fuzzy based multi-criteria decision-making method for facility site selection by developing an algorithm based on linguistic assessment of various attributes affecting the location. Their procedure operates at two distinct levels. At the first level the importance of various decision criteria being considered are evaluated using linguistic variables such as *very low, low, medium, high, very high*. And at the second level, the various sites are rated under each criterion using another set of linguistic variables viz., *very poor, poor, fair, good, very good*. The membership functions of the above terms assumed by the authors are of the form given in figure 1. Then using these values (and the concepts of operations on fuzzy numbers) the weightings of criteria and evaluation rating of alternatives are aggregated and 'fuzzy suitability indices' are obtained. The site that has the largest value of this index is the most suitable site.

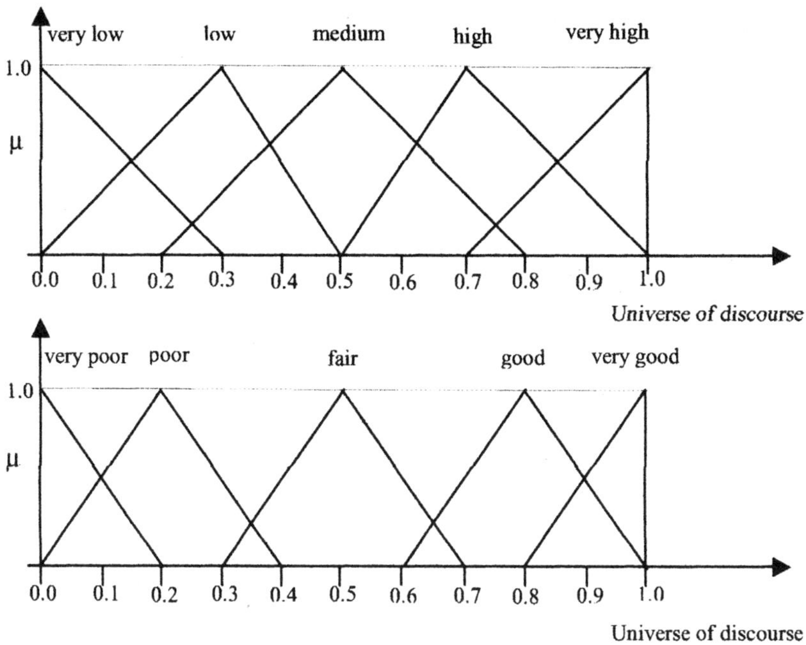

Figure 1 **Membership Functions used by Liang & Wang (1991)**

ii) Dweiri and Meier (1996) proposed a vigorous methodology for facilities design based on a popular multi-criteria decision making technique by Saaty (1980) – The Analytical Hierarchy Process (AHP). Dweiri and Meier argue that the closeness rating between departments in a plant depend upon both qualitative and quantitative factors. Their technique is based on the premise that some factor used in closeness rating may have greater effect on the designer's decisions than others. Thus, according to them the pairwise comparison of the analytical hierarchy process ensures the consistency of the designer's decisions when assigning the importance of one factor over another.

The procedure suggested by Dweiri and Meier consists of the following three distinct parts.

1. Generating the improved activity relationship chart using all possible factors that are believed to affect the layout using AHP and fuzzy set theory.
2. The generated activity relationship chart is then used to develop the layout on the pattern of CORELAP, using a computer program FZYCRLP (Fuzzy Computerized Relationship Layout Planning).
3. This layout is then evaluated using FELAP (Fuzzy Evaluation of the Layout Planning), a program in QBASIC.

Most papers using fuzzy approach, including the ones cited above, display a significant improvement over the traditional approaches in terms of time spent, level of complexity handled, reduction in arbitration of subjective judgment, adaptability to other applications etc.

AHP and its Fuzzy Variant

Listed below is stepwise description of the proposed methodology for generalized Fuzzy based Analytical Hierarchy Process developed by the authors:

1. Identify and select a problem that requires Multi Criteria Decision Making.
2. Choose the criteria on the basis of which evaluation/selection of factors/alternatives respectively need to be carried out. As far as possible those criteria must be chosen that can best be described fuzzily like 'educational facilities', 'climate', 'safety' etc.
3. To solve the selected problem, set aside the number of alternatives available (p) and identify the number of decision makers (q) for the evaluation.
4. Decide Term Sets of Fuzzy Linguistic Variables to be used like {Very Low, Low, Medium, High, Very High} for '*preference*' and {Very Poor, Poor, Fair, Good, Very Good} for evaluating the '*suitability*' of various alternatives against the identified criteria; and choose appropriate values for their Membership Functions.
5. Ask each of the 'q' decision makers to assign the relative weight values for each criterion using the linguistic variables from the chosen Term Set for '*preference*'.
6. From the assigned values, calculate aggregated *preference* weights using the Fuzzy Mean Operator as given in Equation (1). (Any other operator such as fuzzy median or fuzzy mode may also be used.)

   $$W_i = \frac{C_iD_1 + C_iD_2 + C_iD_3 \ldots + C_iD_q}{q} \quad (1)$$

   where C_i denotes the ith criterion and D_q the qth Decision Maker.
7. If any of the chosen criteria can be further sub-divided & evaluated using other sub-criteria then the overall influence of these sub criteria w.r.t. the criterion in question may also be calculated using the above aggregation formula. This stage is an optional one and may be skipped if such detailing is not desired.
8. Decide Fuzzy Linguistic Variables for these sub criteria also and get values for their Membership function like before. One may continue to employ a similar term set as used above for '*preference*' {Very Poor, Poor, Fair, Good, Very Good}.
9. Use aggregating formula $\alpha_{pq} = \Sigma\, [\, (A_pD_q)_m \times W_k\,]$ to get the final values for each sub criteria w.r.t. a criterion and assign linguistic variables accordingly from the term set decided in step 4.

   Where, p = number of alternatives, q = number of decision makers,
   A_p = p^{th} alternative, D_q = q^{th} decision maker,
   k = m = number of sub criteria, W_k=preference weight of kth sub criteria,
   C_i = i^{th} Criteria, Γ_p = fuzzy score for p^{th} alternative,
10. Calculate the suitability index to fit the numeric form (a,b,c,d) back to its relevant Linguistic Value using Equation (2). Suitability index for p^{th} alternative & i^{th} criteria is:

    $$S_{pi} = \frac{\sum_{\ell=1}^{\ell=q} (A_i D_\ell)_m}{q} \quad (2)$$

11. Fuzzy score for each alternative is calculated by using Equation (3).

    $$\Gamma_p = \frac{\sum_{\ell=1}^{\ell=q} (S_{p\ell} \times W_\ell)}{q} \quad (3)$$

    This is called the Process of Defuzzification by Centre of Area Method and is demonstrated graphically in Figure 2. For greater detail see Driankov et al. (1996).
12. The alternative giving highest value for Γ is taken to be the outcome of our methodology in the form of a chosen alternative or the selected department.

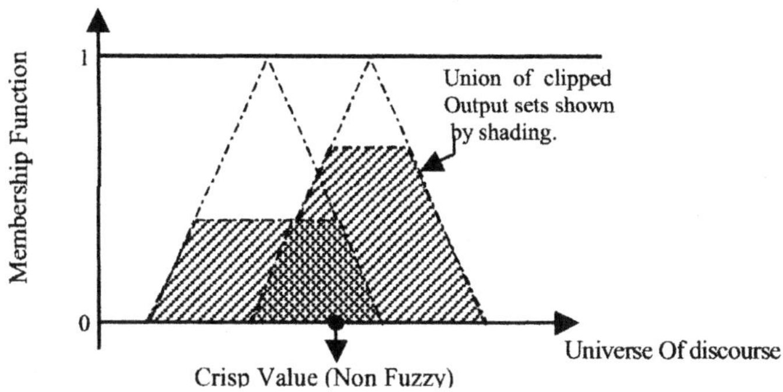

Figure 2 **Graphical Representation of Centre of Area Defuzzification**

Application of the Proposed Methodology

This variant of the AHP where the evaluator is expected to carry out a pair-wise comparison using words or phrases (instead of numbers) for first the myriad criteria of evaluation and then the alternatives is a more realistic and convenient manner of carrying out multi criteria assessment. It may be used to solve any Facilities Planning problem from the industry as follows:

Facilities Location Problem. It is suggested that the problem of Facilities Location be tackled through our proposed fuzzy multi criteria model. The most important aspect of the proposed methodology is the identification of selection criteria or factors affecting location selection. Though the choice of factors affecting selection of location is, in most cases, a context-dependent issue - Moore's (1962) enunciation of such factors still remains a fundamental basis for such an exercise. A table giving a brief listing of some important factors is being reproduced below.

Table 1 **General Factors for Location Selection**

Factors for choice of Location	Territory Selection	Selection of Site & Community
Nearness to Market	*	
Availability of Raw Material	*	
Transportation	*	*
Availability of Power	*	*
Climate and Fuel	*	
Labour and Wages	*	*
Laws and Taxation	*	*
Community Services & Attitude		*
Water and Waste		*

The decision regarding the final choice of the criteria of evaluation may be made after analyzing the problem at hand. If desired, choice of any or all factors from Table 1 may be made and their membership distributions may either be assumed (as most researchers prefer to do), or experimentally derived.

Facilities Design Problem. The proposed fuzzy variant of AHP may be used for solving the problem of layout development & design by either replacing the conventional AHP in each stage of the methodology suggested by Dweiri and Meier (1996); or alternatively by using them in new algorithms. Work is still on for developing an improved algorithm for allocating departments to various parts of the shop floor using the methodology in here.

CONCLUSION

The methodology developed and proposed here should be of great help in solving not only problems of Facilities Planning, but also of a variety of Multi Criteria Decision Making problems encountered in life. Merely by assigning linguistic values to the selection parameters, choosing the criteria and carrying out the evaluation one can obtain desired results using our work. Furthermore, the proposed technique has proved to be more powerful as compared to simple AHP and its variant: additive AHP as demonstrated in Kapoor & Banu 2002 (a). It has even proven itself to be amenable to computerization in Kapoor & Banu 2002 (b) where a software was developed in 'C' and problems such as car selection, robot selection were run on it using a desktop PC giving satisfactory results. A trade-off between the available memory space and execution time was maintained; though efforts are on for finding a still better method.

References

Apple, J.M., 1974, Plant Layout and Material Handling, *John Wiley & Sons*, U.S.A.

Driankov, D., Hellendoorn, H. and Reinfrank, M., 1996, An Introduction to Fuzzy Control, *Narosa Publishing House, India*. [Springer International Students Edition].

Dweiri, F. and Meier, F.A., 1996,"Application of Fuzzy Decision Making in Facilities Layout Planning," Int. J. of Prod. Res., Vol. 34(11), pp. 3207-3225.

Kapoor, V. and Banu, Z., 2002 (a) "Multi Criteria Decision Making and Fuzzy Analytic Hierarchy Process - A Comparative Study", Proceedings of National Seminar On "Soft Computing & Its Engineering Applications", February 15–17, 2002, Jodhpur.

Kapoor, V. and Banu, Z., 2002 (b), "Development Of A Software For 'Fuzzy' Analytic Hierarchy Process", Proceedings of International Conference on 'e-manufacturing: An Emerging Need for 21st Century World Class Enterprises', November 17-19, 2002, Maulana Azad College of Technology, The Institution of Engineers (India), Bhopal, India.

Karwowski, W. and Evans, G. W., 1986, "Fuzzy Concepts in Production Management Research: A Review," Int. J. of Prod. Res., Vol. 24(1), pp. 129 - 147.

Liang, G-S., and Wang, M-J., 1991, "A Fuzzy Multi-Criteria Decision-Making Method For Facility Site Selection", Int. J. Prod. Res., Vol. 29(11), pp. 2313 – 2330.

Moore, J. M., 1962, "Plant Layout and Design", *The Macmillan Com*pany, New York, U.S.A.

Saaty, T.L., 1980, The Analytic Hierarchy Process, *McGraw-Hill*, New York.

Suri, H. and Kapoor, V., 2001, "Application of Fuzzy Logic to the AHP for Robot Selection", Proceedings of XV National Convention of Prod. Engineers & National Seminar on "Emerging Convergence in Manufacturing Systems" The Institution of Engineers (India), March 3–4, 2001, Bhopal.

Tompkins, J. A., and White, J. A., 1984, Facilities Planning, *John Wiley & Sons*, New York.

Zadeh, L. A., 1965, "Fuzzy Sets," Information and Control, Vol. 8(3), pp. 338 – 353.

Zadeh, L. A., 1975, "The Concept of a Linguistic Variable and its Application to Approximate Reasoning", Parts 1 & 2, Information Sciences, Vol. 8, pp. 199 – 249, 301 – 357.

Intelligent Control of Nonlinear dynamic system Using Immune Fuzzy Fusion

DONG HWA KIM, WON PYO HONG

Dept. of Instrumentation and Control Eng.
Hanbat National University
16-1 San Duckmyong-Dong Yuseong-Gu, Daejon City, Korea, 305-719.
E-mail: kimdh@hanbat.ac.kr, Tel: +82-42-821-1170, Fax: +82-821-1164

ABSTRACT

This paper proposes non-linear control method using immune algorithm based fuzzy logic. Nonlinear dynamic system exist widely in many types of systems such as chemical processes, biomedical processes, and the inverted pendulum. Up to the present time, PID Controllers have been used to operate these systems. However, it is very difficult to achieve an optimal *PID* gain with no experience, because gain of the PID controller has to be manually tuned by trial and error. An inverted pendulum control problem is selected to illustrate the efficiency of the proposed method and defines relationship state variables $x, \dot{x}, \theta, \dot{\theta}$ using immune fuzzy.

INTRODUCTION

The Proportional-Integral-Derivative (PID) controller for non-linear system such as steam temperature control system of power plant, chemical process, an inverted pedulum has been widely used owing to its simplicity and robustness [1]. However, with only the P, I, D parameters, it can not effectively control a plant with complex dynamics, such as large dead time, inverse response, and highly nonlinear system [3].

It is a challenge in controller tuning technologies to explore novel control strategies and philosophies for complex industrial processes [2]. The application of intelligent system technologies in industrial control has been developing into an emerging intelligent control technology. Chronologically, fuzzy logic was the first technique of intelligent systems. Neural, neuro-fuzzy and evolutionary system and their derivatives followed later [4]. Each technique is offering new possibilities and making intelligent system even more versatile and applicable in an ever-increasing range of industrial applications [4]-[6].

On the other hand, biological information processing systems such as human beings have many interesting functions and are expected to provide various feasible ideas to engineering fields, especially intelligent control or robotics. Biological information in living organisms can be mainly classified into the following four systems: brain nervous, genetic system, endocrine system, and immune system [5].

This paper suggest control scheme to control effectively non-linear system using immune algorithm based fuzzy and simulate through computer.

II. DYNAMIC MODEL OF IMMUNE SYSTEM

A. Dynamics Of Immune System

In the immune system, the level to which a B cell is stimulated relates partly to how well its antibody binds the antigen. We take into account both the strength of the match between the antibody and the antigen and the B cell object's affinity to the other B cells as well as its enmity. Therefore, generally the concentration of i-*th* antibody, which is denoted by δ_i, is calculated as follows [7]:

$$\frac{dS_i(t)}{dt} = \left(\alpha \sum_{j=1}^{N} m_{ji} \delta_j(t) - \alpha \sum_{k=1}^{N} m_{ik} \delta_k(t) + \beta m_i - \gamma_i \right) \delta_i(t) \quad (1a)$$

$$\frac{d\delta_i(t)}{dt} = \frac{1}{1 + \exp\left(0.5 - \frac{dS_i(t)}{dt}\right)} \quad (1b)$$

where in Eq. (1), N is the number of antibodies, and α and β are positive constants. m_{ji} denotes affinities between antibody j and antibody i (i.e. the degree of interaction), m_i represents affinities between the detected antigens and antibody i, respectively.

$$m_{\alpha\beta} = \frac{1}{\{1 + \Omega(\alpha\beta)\}}, \quad \Omega(\alpha\beta) = H_s(x) = [f_1(x) + f_2(x) + f_3(x)] \quad (4)$$

where $\Omega(\alpha\beta)$ is an information which obtained by antibody α and antibody β. If $\Omega(\alpha\beta)$ =0, the antibody α and antibody β match completely. Generally $m_{\alpha\beta}$ is given by range of 0-1.

III. CONTROLLER DESIGN NON-LINEAR SYSTEM USING IMMUNE FUZZY FUSION

A. Sugeno Fuzzy Logic

Takagi, Sugeno, and Kang proposed the Sugeno fuzzy model [10] in an effort to develop a systematic approach to generating fuzzy rules from a given input-output data set. A typical fuzzy rule in that fuzzy model has the form

$$\text{If } x \text{ is } A \text{ and } y \text{ is } B \text{ then } z=f(x, y), \qquad (5)$$

where A and B are fuzzy sets in the antecedent, while $z=f(x, y)$ is a crisp function in the consequent.

$f(x, y)$ is a polynomial in the input variables x and y. When $f(x, y)$ is a first order polynomial, the resulting fuzzy inference system is called a first-order Sugeno fuzzy model.

B. Cart with Interted Pendulum

To demonstrate the availability, this paper takes an inverted pendulum system for simulation. The inverted pendulum system is composed of a rigid pole and a cart on which the pole is hinged as shown in Fig. 1. The cart can move on the rail tracks to its right or left, depending on the force exerted on the cart.

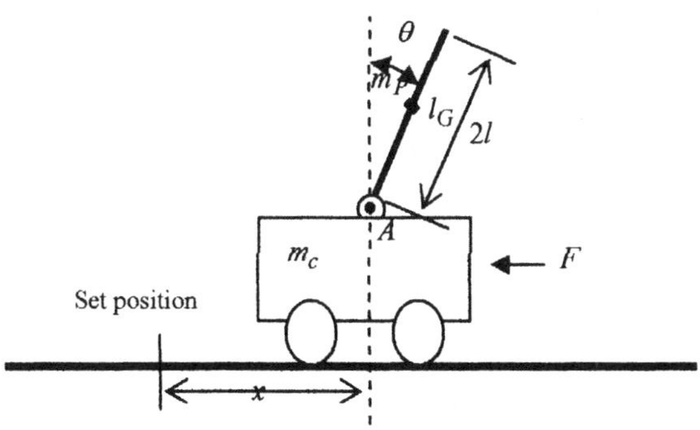

Fig. 1. The structure of cart with inverted pendulum.

The dynamics of the inverted pendulum system are characterized by four state variables: θ (angle of the pole with respect to the vertical axis), $\dot{\theta}$ (angular velocity of the pole), x (position of the cart on the track), $\dot{x}$ (velocity of the cart). The behavior of these four state variables can be expressed by the following two second-order differential equations [10].

C. Non-linear Controller Design Using Immune-Fuzzy Fusion

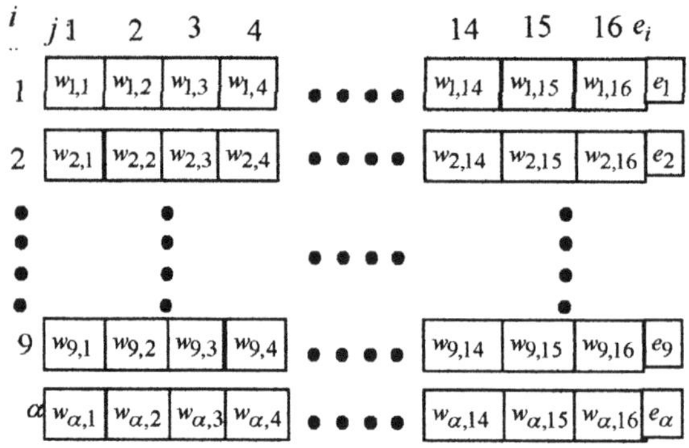

This paper expresses the distance between the target position (B) and the initial position (A) as α, and its resolution is regulated by:

Fig. 2. The tructure of antibody for memory cell.

$$\left|\min(A,B)+\left(\frac{\max(A,B)-\min(A,B)}{\alpha}\right)\times(i-1)\right|<|d_i|$$

$$|d_i|\le\left|\min(A,B)+\left(\frac{\max(A,B)-\min(A,B)}{\alpha}\right)\times i\right|,$$

$$i=\alpha,\alpha-1,\alpha-2,\cdots,1. \quad (8)$$

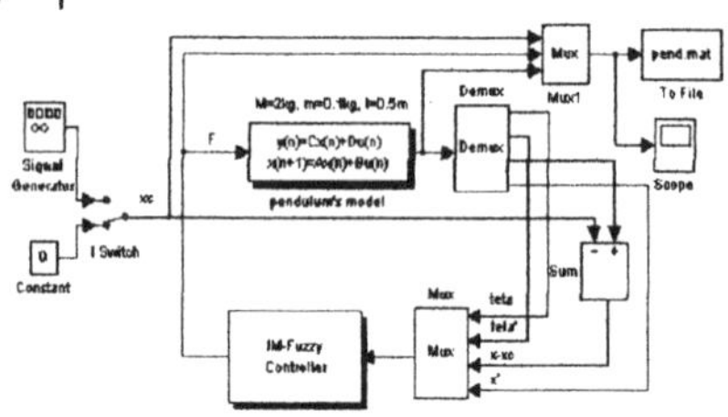

Where, d_i is the present position of the given cart. The antibody in immune network is produced as much as the number of the resolution α.

IV. SIMULATION AND

DISCUSSIONS

Fig. 3 Simulation structure for the immune algorithm-Fuzzy model.

Fig. 2 shows simulation structure by Simulink and Matlab S/W and the membership function for parameters of the pendulum is defined for Sugeno fuzzy model by 8-rules as the Table 1.[1]
Figs 4-5 illustrate the responses to position and angular when only fuzzy logic for the pendulum control is used. These responses are similar as response $\alpha = 1$.
Fig. 4 shows the graph on α =20 and Tables 1 are data for those.

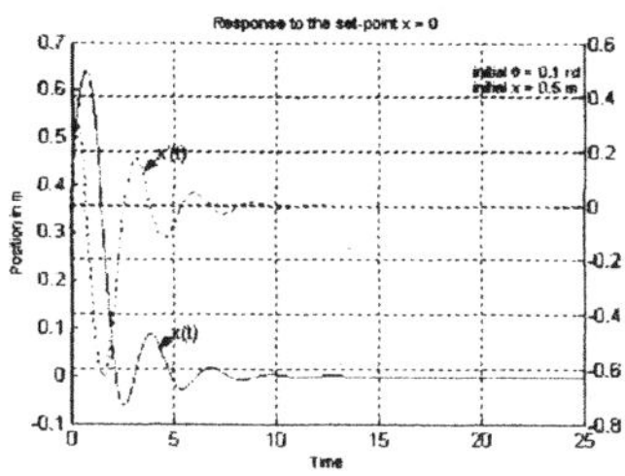

Fig. 4. The angular response on fuzzy controller Fig.

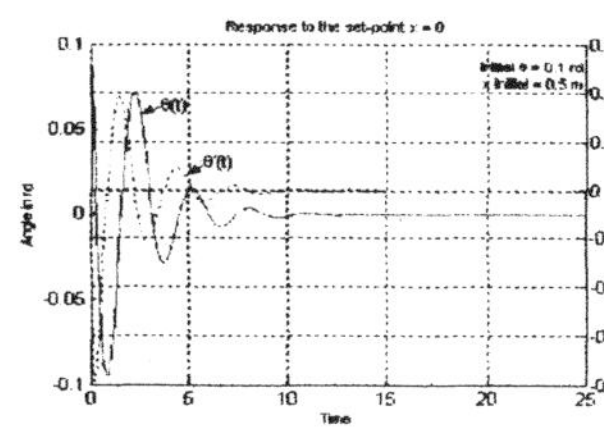

Fig. 5. The position response on fuzzy controller.

Table 2. Data on $\alpha = 5$.

	w_1	w_2	w_3	w_4	w_5	w_6	w_7	w_8	e_i
1	1.7092	1.1366	1.8172	1.8651	0.9971	0.3958	0.4252	0.1375	2747.3
2	1.6256	1.3709	1.864	1.8614	1.1668	1.67	1.7432	0.2961	22.206
3	0.2030	0.6786	0.0061	0.7707	0.0365	0.3096	1.9885	0.4221	4.2395
4	0.9361	0.4963	1.2069	0.4873	0.7342	1.9955	1.7916	0.7562	0.6114
5	1.2857	1.2261	1.9534	1.9603	0.9868	1.7432	1.8806	0.5040	78.815

Φ : 5.0299

V. CONCLUSIONS

This paper proposes control method for non-linear system such as power plant, chemical plant, inverted pendulum method using immune algorithm based fuzzy logic. This paper uses an inverted pendulum to illustrate the efficiency of the proposed method for non-linear system and defines relationship state variables $x, \dot{x}, \theta, \dot{\theta}$ using immune fuzzy, through simulation. The results represent satisfactory response.

REFERENCES

[1] D avid Lindsley, Boiler Control Systems, 1991, McGrawill.

[2] S. Matsummura, 1998, "Adaptive control for the steam temperature of thermal power plants," Proceedings the 1993 IEEE on Control applications, PP. 1105 - 1109.

[3] Teng Fong-Chwee, 1988, "Self-tuning PID controllers dor dead time process,"IEEE Trans., Vol. 35, No. 1, pp. 119-125.

[4] Yong Zai Lu ,1996, "Industrial intelligent control," John Wiley $Sons.

[5] J. D. Farmer, N. H. Packard and A. S. Perelson, 1986, "The immune system, adaptation, and machine learning," Vol. Physica. D, No. 22, pp. 187 – 204.

[6] C. V. Rao, An introduction to immunology, 2002, Alpha Science International Ltd.

[7] R. Brooks, 1986, "A robust layered control system for a mobile Robot," IEEE Jounal R&A, vol.2, no.& pp.14-23.

[8] Dong Hwa Kim, 2002, "Parameter tuning of fuzzy-neural networks by immune algorithm," 2002 IEEE international conference on fuzzy systems, Honolulu, Hawaii.

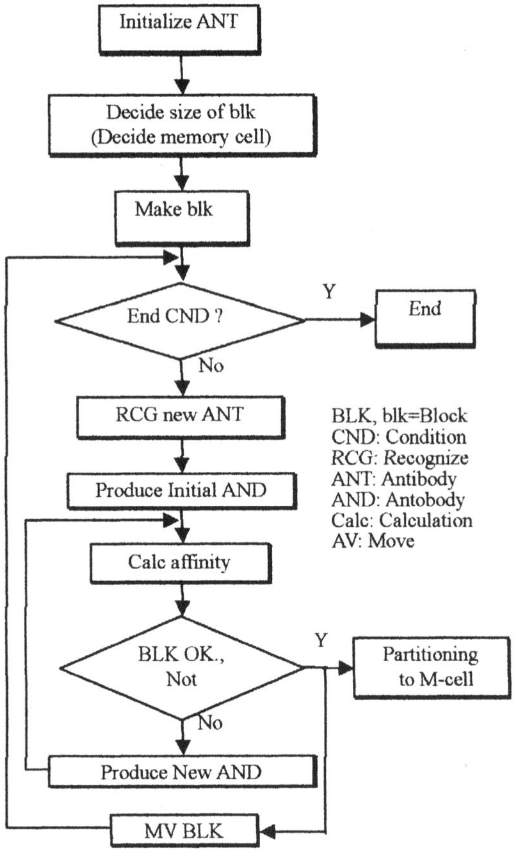

Fig. 6. Flowchart of immune algorithm.

AN ALGORITHM FOR FRACTAL IMAGE COMPRESSION USING FUZZY LOGIC

TANASANEE PHIENTHRAKUL
Technology of Information System Management, Faculty of Engineering, Mahidol University, Salaya, 73170
Thailand

RAWIN RAVIWONGSE, Ph.D.
Department of Industrial Engineering
Faculty of Engineering
Mahidol University, Salaya, 73170
Thailand

ABSTRACT

Graphic representations are widely used in various computer applications. Good quality graphics may require much storage space therefore several compression techniques have been proposed. Fractal image compression is a technique based on the concept of the self-similarity property of images, which offers a high compression ratio but consumes a lot of processing time. This research employs fuzzy logic with a traditional fractal image compression technique in order to reduce compression time. The proposed algorithm used fuzzy logic in two stages, (1) fuzzy classification homogeneity property of blocks, and (2) fuzzy identification similarity of domain and range blocks. Experimental results showed that the proposed algorithm yields a promising result. The efficiency of this algorithm was compared to Fisher's method. It proved to have a faster encoding time while the compression ratio and the quality of the reconstruction image did not differ from Fisher's algorithm.

INTRODUCTION

In recent times, graphic representations have been widely used in various applications. Though computer technologies have advanced, fast communication links and large memory storage have been developed. These are not able to handle efficiently the image now created. Image compression techniques are more important for reducing storage space and transmission times.

Fractal image compression is a new compression technique, based on the concept of the self-similarity property of an image. Instead of storing the whole image, it is enough to store the relevant parameters of the transformations [1]. This reduces the memory requirement. Although fractal image compression can produce a high compression ratio, it requires enormous computation and is very difficult to achieve in real time.

Fuzzy logic is a technology that has been successfully applied in many areas. It is a powerful problem-solving methodology, which provides a simple way to draw definite conclusions from vague, ambiguous or imprecise information [2, 3]. In this research, fuzzy logic is applied to fractal image compression in order to reduce compression time and give no loss in quality, which means the outcome requires less memory and has high resolution.

FRACTAL IMAGE COMPRESSION

Fractal image compression was first developed and implemented by Barnsley and Jacquin [4]; they had the idea of partitioning the image into non-overlapping ranges, and using Partitioned Iterated Function Systems (PIFS) for each range. The processing steps of PIFS can be described as follows:

1. Duplicate the image: one image is partitioned into non-overlapped blocks (called ranges), and another is partitioned into overlapped blocks (called domains) with double length of height and width to the corresponding range.

2. Compute the Euclidean distance between each domain and each range. Four pixels in each domain are computed with a pixel in the corresponding range. If the computed value of a domain is acceptable, then this corresponding domain index and the transformation are recorded. Otherwise, this domain and corresponding range are divided into ¼ (the height and width are divided into ½, respectively), until it reaches the smallest block. Then the domain and the corresponding transformation with the smallest computed value are recorded.

PROPOSED ALGORITHM

The proposed algorithm applies fuzzy logic to Fisher's algorithm [4] in order to classify domains and compare between domains and ranges. The main steps of the proposed method for fractal image compression can be summarized as follows:

Step 1: Allocate and initialize the image data and commutative image data.
Step 2: Partition the original image into a collection of non-overlapping range blocks.
Step 3: Partition the original image into a collection of overlapping of larger sub-images (domain blocks).
Step 4: The domain blocks are filtered and sub-sampled using pixel averaging so that it occupies the size of range blocks.
Step 5: Classify all domains and apply fuzzy logic to identify all the domains and ranges.
Step 6: Each range block identifies a class or partition to a smaller size. Then, domain pool is searched until a contracted domain block with the smallest mean square difference is found. This step applied fuzzy logic to find the appreciated domain.
Step 7: The suitable transformation parameters are determined.
Step 8: These mapped transformation coefficients are stored.
Step 9: Repeat step 6 for all other range blocks.
Step 10: Free all dynamically allocated memory.

Steps 5 and 6 of this algorithm apply fuzzy logic. Step 5 applies fuzzy logic to classify a domain and step 6 uses it to select a domain to compare with a range. Both steps decrease encoding time and their details are explained in next part.

Fuzzy Logic for Classification

The domain-range comparison step of the encoding is very computational intensive. A classification scheme is used in order to minimize the number of domains compared with a range. In the preliminary steps, all domains are classified to avoid reclassification of domains. During the encoding, a potential range is classified. Only those domains with the same (or near) classification are compared with the range. This significantly reduces the number of domain-range comparisons. Here this algorithm selects a schema presented in Fisher's book and uses fuzzy logic to classify the homogeneity properties of block. There are three fuzzy parameters.

1. *EdgeFreq* – Edge frequency of block. This value comes from counting pixels that are identified as edge in block. In this research, the Sobel edge detector is considered [5]. The ratio between edge pixels and all pixels in that block is computed as a percentage.

2. *DiffGrey* – Difference between maximum and minimum of grey levels in a block. This value reveals the range of grey levels in a block. If the result of the difference is high, there may be more than one object in this block.

3. *DiffVar* – Difference between the variance of block and average of variance in sub-blocks. This value is calculated by dividing a block into four sub-blocks. Then, the variance of a block and the variance of each sub-block are found. After that, the average value of these four variances is computed. Finally, the *DiffVar* is calculated by the difference between average variance and variance of a block.

For each of the above fuzzy parameters, there are three values: *'Low'*, *'Medium'*, and *'High'*. The output has five levels of *Homogeneity*: *'Very Low'*, *'Low*, *Medium'*, *'High'*, and *'Very High'*. The membership functions of these parameters and output are shown in Figure 1.

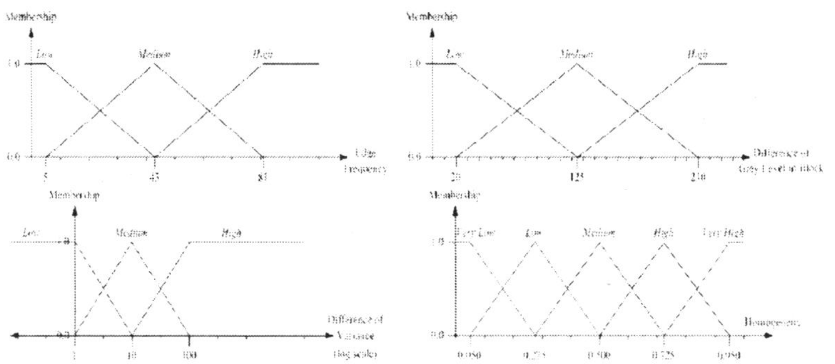

Figure 1. Membership Functions of *EdgeFreq*, *DiffGrey*, *DiffVar*, and Homogeneity

The examples of fuzzy rules are:

<u>Rule1</u>: IF EdgeFreq is *Low* AND DiffGrey is *Low* AND DiffVar is *Low* THEN Homogeneity is *Very High*.

<u>Rule2</u>: IF (EdgeFreq is *High* AND DiffGrey is *High*) OR (EdgeFreq is *High* AND DiffVar is *High*) OR (DiffGrey is *High* AND DiffVar is *High*) THEN Homogeneity is *Low*.

<u>Rule3</u>: IF (EdgeFreq is *Low* OR DiffGrey is *Low* OR DiffVar is *Low*) AND NOT (EdgeFreq is *High* OR DiffGrey is *High* OR DiffVar is *High*) THEN Homogeneity is *High*.

. . . etc. . . .

Fuzzy Logic for Domain Searching

The goal of searching is to get a domain block with a minimum of the root mean square difference (*rms*) between the transformation domain pixel values and the range pixel values. Generally, all domains are compared with a range. In this algorithm, fuzzy logic is applied to pre-compare between the range and domain. To reduce comparison between a range and domains, fuzzy logic is applied to find the *Similarity* between range and domain, which consists of three values: *'Differ'*, *'Similar'*, and *'Same'*. The best domain block for each range block is found by defining the fuzzy parameters as follows:

1. *delEdge* – The difference of edge frequency between range and domain.
2. *delMean* - The difference of average grey level between range and domain.
3. *delVar* - The difference of variance between range and domain.

For each of the above fuzzy parameters, there are three values: *'Small'*, *'Medium'*, and *'Large'*. The membership functions of these parameters and output are shown in Figure 2.

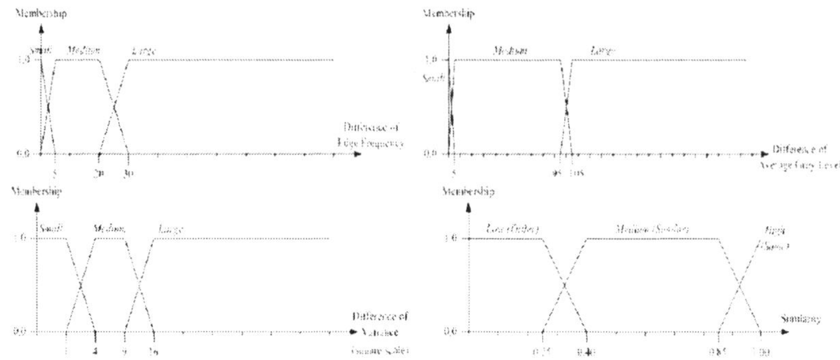

Figure 2. Membership Functions of *delEdge, delMean, delVar,* and Similarity

The fuzzy rules of similarity are:

Rule1: IF delEdge is *Small* AND delMean is *Small* AND delVar is *Small* THEN Similarity is *High*.

Rule2: IF delEdge is *Large* AND delMean is *Large* AND delVar is *Large* THEN Similarity is *Low*.

Otherwise: Similarity is *Medium*

EXPERIMENTAL RESULTS

The performances of the proposed algorithm are compared with Fisher's fractal encoding. Identical coding parameters are used in all experiments. Bitmap grayscale images of Lena, Baboon, and Balloon, shown in Figure 3, are tested. The image size is 512×512. The maximum range size is 32×32 (minimum quad tree depth 4), and the minimum is 8×8 (maximum quad tree depth 6). The performance comparisons of both algorithms are showed in Table 1.

Table 1. Performance Comparison of the Proposed Algorithm and Fisher's

Image	Fuzzy Fractal Encoding			Fractal Encoding		
	Encoding Time (Sec.)	Compression Ratio (%)	PSNR (dB)	Encoding Time (Sec.)	Compression Ratio (%)	PSNR (dB)
Lena	109.53	96.85	29.55	358.12	97.57	29.45
Baboon	111.61	94.57	21.47	466.92	94.58	21.69
Balloon	131.16	97.37	32.12	270.51	98.42	31.11

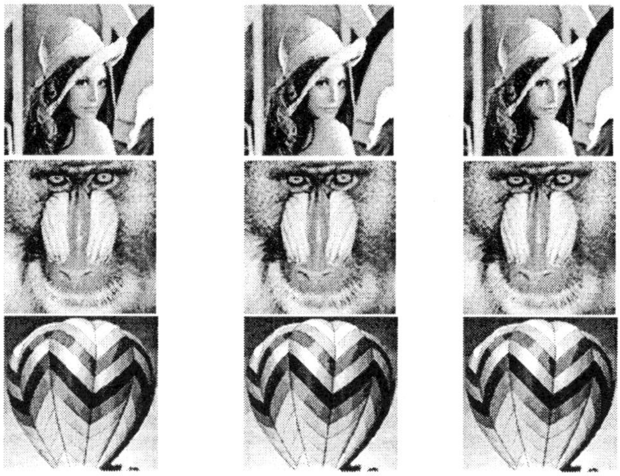

(a) (b) (c)

Figure 3. Test Images of Lena, Baboon, and Balloon with (a) Original Images, (b) Images Obtained from the Proposed Algorithm, and (c) Images Obtained from Fisher's

From Table 1, the encoding time of the proposed method is less than Fisher's method and the compression ratio is also slightly less than Fisher's. The quality of image reconstruction is greater than Fisher's method, except the Baboon, but the human eyes cannot notice the difference.

When we compare the results at the same compression ratio, the encoding time of the proposed algorithm is less than half of the encoding time of Fisher's algorithm, although the quality of a decoded image using the proposed algorithm is slightly less than Fisher's algorithm. The examples of these comparisons of Lena image are shown as graphs in Figure 4.

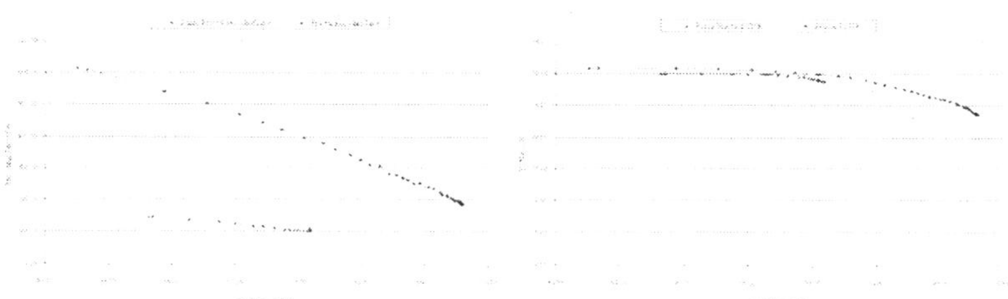

(a) (b)

Figure 4. Testing Results of Fuzzy Fractal and Fractal Algorithm for 512 × 512 Lena Image at the Same Compression Ratio (a) Encoding time (b) PSNR

These results indicate that the proposed algorithm has a better encoding time while maintaining the quality of the image. Figure 4(b) indicates that the graph of fuzzy fractal algorithm is shorter than the fractal algorithm. That is, the fuzzy fractal algorithm ensures the quality of the image. The quality of a reconstructed image using the fuzzy fractal algorithm will not be below than some values. Figure 5 compares the PSNR when the tolerance of encoding is changed. The PSNR of the proposed algorithm is decreased to a point and then the value of the PSNR is a constant, while the PSNR of Fisher's method still decreases.

These results clearly show that the proposed algorithm used significantly less encoding time than Fisher's algorithm, while the compression ratio of the proposed algorithm is decreased and the quality of a reconstructed image is more than Fisher's algorithm. These tests are compared at the same parameters. In practice, users can adjust parameters to get the expected compression ratio or PSNR. If users want a high compression ratio, the quality of a reconstructed image may decease. This is same in all methods including Fisher's algorithm and the fuzzy fractal algorithm.

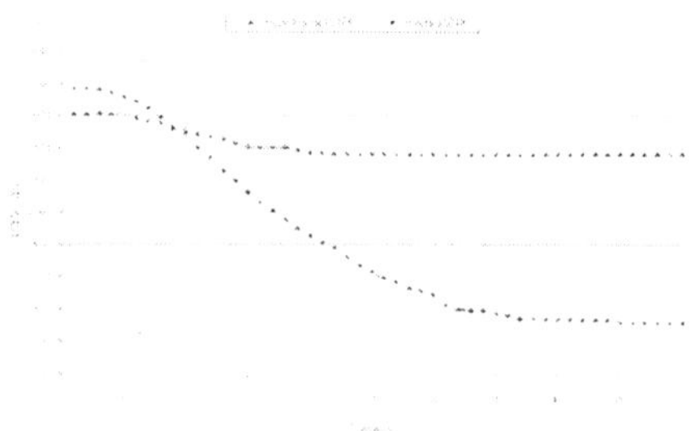

Figure 5. The PSNR of 512 x 512 Lena Image

CONCLUDING REMARKS

The proposed fuzzy fractal image compression algorithm uses fuzzy logic to reduce encoding time. It is applied at two positions, the first is to classify homogeneity of sub-images, and the other is to find a domain block that is similar to each range block.

The experimental results show that the encoding time is reduced by 40-75% in relation to Yuval Fisher's algorithm. The quality of the reconstructed image and the compression ratio of the proposed algorithm are close to Fisher's method.

The main benefit of the proposed algorithm is saving time, while the accuracy of the rebuilt image is comparable to Fisher's method. Moreover, the fuzzy logic in the comparison step ensures that the quality of the rebuilt image is not very low. However, the complexity of the proposed algorithm is greater than the conventional algorithm: the new algorithm must add fuzzification parts into processing steps. The future plans are to develop the fuzzy fractal image compression algorithm with other techniques, or to blend many techniques to improve time efficiency, quality of reconstructed image and compression ratio.

REFERENCES

[1] Nelson M., and Gailly, J. L., 1996, "The Data Compression Book," 2nd ed., M&T Books, NY.

[2] Ross T., 1995, "Fuzzy Logic with Engineering Applications," McGraw-Hill, NY.

[3] What is fuzzy logic? [Online] 15 April 1993. Available from: URL: http:// www.cs.cmu.edu/afs/cs.cmu.edu/project/ai-repository/ai/html/faqs/ai/fuzzy/part1/ faq-doc-2.html [Accessed 21 July 2001].

[4] Fisher Y., 1995, "Fractal Image Compression Theory and Application," Springer-Verlag, NY.

[5] Pitas I., 1993, "Digital Image Processing Algorithms," Prentice Hall, London.

APPLICATION OF FUZZY ADAPTIVE NETWORKS IN MANUFACTURING: PREDICTION OF MATERIAL REMOVAL RATE IN ROTARY ULTRASONIC MACHINING

YUE JIAO, PING HU, Z.J. PEI, SHUTING LEI, E. STANLEY LEE
Department of Industrial and Manufacturing Systems Engineering
Kansas State University, Manhattan, KS 66506

ABSTRACT
Rotary Ultrasonic Machining (RUM) is a hybrid machining process that combines the material removal mechanisms of diamond grinding and ultrasonic machining (USM), resulting in higher material removal rates (MRR) than those obtained by either diamond grinding or USM. There are various approaches proposed to model the machining processes. Fuzzy adaptive networks (FANs) provide alternative techniques to model the vague and not well-defined phenomena in machining processes. The knowledge and understanding thus obtained can be used to improve the efficiency of these processes. This paper presents the results of applying fuzzy adaptive networks to model material removal rates in rotary ultrasonic machining of ceramics. After a brief review on the background of FANs, the modeling methodology is outlined and the results of the pilot experiments are discussed to verify the inferred model.

INTRODUCTION

The modeling of machining processes plays an important role in modern manufacturing industry. Various modeling techniques have been proposed and comprehensive reviews have been given by Ehmann et al. [1997] and Merchant [1998]. Ehmann et al. classified the various approaches into three categories: analytical, experimental, and mechanistic and numerical. However, due to the vagueness and complexity, many manufacturing processes are too complex to be well understood or adequately modeled by the traditional approaches. Uncertainty and ambiguity due to complexity and vagueness are inherent characteristics in many manufacturing processes [Lee et al., 1999]. This is also true in the rotary ultrasonic machining process.

In order to handle this vague and complex phenomenon, the fuzzy adaptive networks (FANs) were developed [Cheng and Lee, 1999; Jiao, 2002]. The FANs networks are ideally suited for modeling the complex relationships between the various factors in manufacturing or machining processes. The resulting network representation can effectively predict process outputs when the process parameters are given. Due to the integration of fuzzy logic and neural network, fuzzy adaptive networks are capable of providing both learning ability and tolerance for imprecision, uncertainty and vagueness. This capability is desirable for the monitoring and control of complex processes. As an example, this paper presents the results of utilizing FANs to deal with the problem of modeling and prediction of MRR in RUM of ceramics.

The inception of RUM has been for more than several decades, and many investigators have reported their studies [Hards, 1966; Kubota et al., 1977; Markov and Ustinov, 1972; Prabhakar et al., 1992]. However, most of the studies are experimentally based, little attempt has been made to develop a model to describe the material removal mechnism [Pei et al., 1995]. Pei et al. [1995] studied the relations between the material removal rate and the controllable machining parameters. In that study, the main effects of process parameters were obtained . This study presents the results of applying neuro fuzzy approach to the modeling of MRR in RUM, which provides an alternative to identify the material removal mechanism in RUM. The comparison with theoretical results will also serve as further validation of the model.

The remainder of the paper is organized as follows. Section 2 introduces rotary ultrasonic machining process. In Section 3, a fuzzy adaptive network is formulated and utilized to model the relationships between material removal rate and the process parameters. Section 4 summarizes the results of pilot experiments. And, finally, some conclusions are given in Section 5.

ROTARY ULTRASONIC MACHINING

Fig. 1 shows the RUM process and Fig. 2 schematically illustrates the experimental setup. It consists of an ultrasonic spindle kit, a constant pressure feed system and a coolant system.

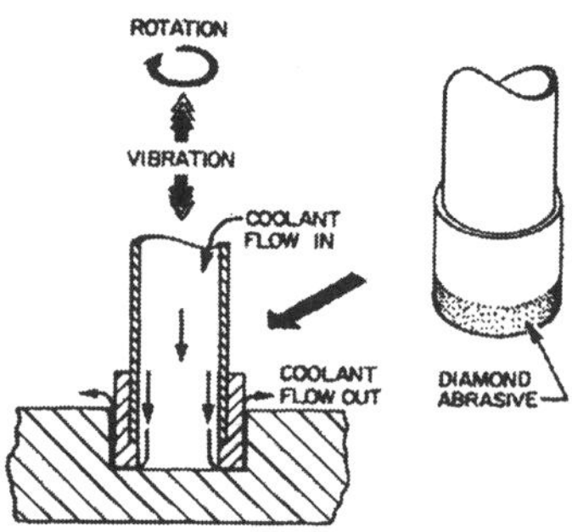

Fig. 1. Illustration of RUM (after [Stinton, 1988]).

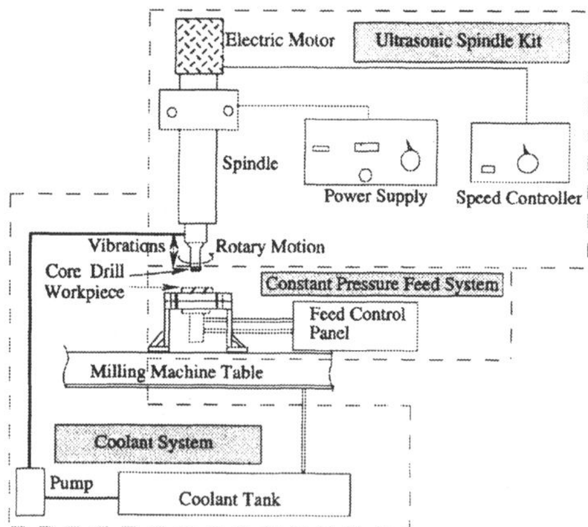

Fig. 2. Schematic illustration of the experimental setup for RUM.

The ultrasonic spindle kit comprises of an ultrasonic spindle, a power supply and a motor speed controller. The ultrasonic spindle is mounted on a milling machine (replacing the toolhead of the milling machine). The spindle contains an ultrasonic transducer. The power supply converts 50 Hz electrical supply to high frequency (20 kHz) AC output. This is fed to the piezoelectric transducer located in the spindle. The transducer converts electrical input into mechanical vibrations. By changing the setting of the output control of the power supply, the amplitude of the ultrasonic vibration can be adjusted. The rotational motion of the tool is supplied by the motor attached atop the spindle and different speeds can be obtained by adjusting the motor speed controller.

METHODOLOGY

Fuzzy adaptive networks (FANs) [Cheng and Lee, 1999; Jiao, 2002] essentially are a series of network-based fuzzy inference systems. To illustrate the approach, FAN III [Jiao, 2002], is used to model MRR of RUM for magnesia stabilized zirconia. The resulting model can be used to predict the influences of the five process parameters on MRR.

Background of FAN III Model

FAN III is based on the combined use of singleton fuzzifier, production-inference operation and Gaussian membership function [Wang, 1994; Jang et al., 1997]. Fig. 3 describes the general structure of FAN III. Each node of FAN III performs a particular node function on the incoming signals, which is characterized by a set of parameters. The adaptive nodes, whose parameters can be tuned by a learning procedure are represented by squares. And fixed nodes are represented by circles.

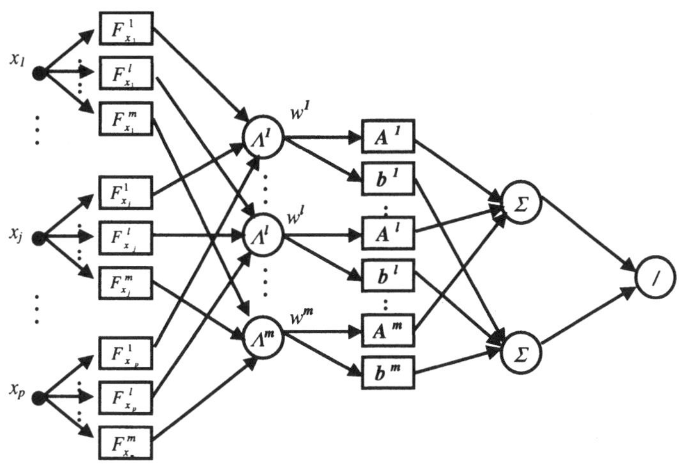

Fig. 3. General structure of FAN III.

Assume that a fuzzy inference system has p inputs, $X = (x_1, x_2, \cdots, x_p)^T$, using symmetric triangular fuzzy number for the output variable, we have:

$$\hat{Y} = (\hat{y}, \hat{e}) = \frac{\sum_{l=1}^{m} A^l \prod_{j=1}^{p} \exp[-\frac{(x_j - v_j^l)^2}{\sigma^2}]}{\sum_{l=1}^{m} b^l \prod_{j=1}^{p} \exp[-\frac{(x_j - v_j^l)^2}{\sigma^2}]} \quad (1)$$

where, A^l , b^l are symmetric triangular fuzzy numbers and numeric numbers, respectively.

The performance of FAN III is measured by the squared difference between the desired output Y_i and the network output $\hat{Y}_i$, called the error measure, E. In this investigation, both Y_i and $\hat{Y}_i$ are assumed as symmetric triangular fuzzy numbers, $Y_i = (y_i, e_i)$, $\hat{Y}_i = (\hat{y}_i, \hat{e}_i)$, where y_i and $\hat{y}_i$ are the modes of fuzzy numbers Y_i and $\hat{Y}_i$, respectively, and e_i and $\hat{e}_i$ are their spreads. Thus, the error measure, using the overall existence ranking index (OERI) concept (Chang and Lee, 1994), is defined as:

$$\mathrm{E} = \frac{1}{N}\sum_{i=1}^{N}\varepsilon_i = \frac{1}{N}\sum_{i=1}^{N}\frac{1}{2}[(\hat{y}_i - y_i)^2 + (\hat{e}_i - e_i)^2] \tag{2}$$

Generation of training dataset

The data used in this study is based on a previously published model [Pei et al., 1995]. The workpiece material investigated is magnesia stabilized zirconia with Young's modulus $E = 205{,}000$ MPa and Poisson's ratio $v = 0.31$.

The design of experiments involves five machining parameters. A 3^5 (five factors, three levels) full factorial design is used to generate the training dataset for fuzzy adaptive networks. Table 1 lists the three levels of these variables.

Table 1. Control variables and their levels

Control variable	Low level	Medium level	High level
Static Force (N)	100	225	350
Vibration Amplitude (mm)	0.02	0.03	0.04
Rotating Speed (rpm)	1000	2000	3000
Abrasive Grit Size (mm)	0.1	0.2	0.3
Abrasive Grit Number	300	600	900

Model Development

To determine the relationships between the dependent variable and the process parameters, a fuzzy inference system is constructed. This fuzzy inference system is then converted into FAN III. Thus, this five-input-one-output system include:

Output variable: material removal rate (MRR);

Input variables: static force (F_s),
ultrasonic vibration amplitude (A),
tool rotating speed (S),
abrasive grit size (d),
abrasive grit number (n).

The parameters F_s, A, S, d and n are treated as linguistic variables, whose membership functions are the following Gaussian membership functions: $\exp\left[-\left(\frac{F_s - v_{Fs}^l}{\sigma}\right)^2\right]$, $\exp\left[-\left(\frac{A - v_A^l}{\sigma}\right)^2\right]$, $\exp\left[-\left(\frac{S - v_S^l}{\sigma}\right)^2\right]$, $\exp\left[-\left(\frac{d - v_d^l}{\sigma}\right)^2\right]$ and $\exp\left[-\left(\frac{n - v_n^l}{\sigma}\right)^2\right]$.

Then the constructed FAN III is trained by the use of the training dataset.

EXPERIMENTAL VERIFICATION

In order to verify the inferred model, a pilot set of experiments was conducted, and the results are illustrated in Fig. 4 and Table 2. In Fig. 4, the circles represent the theoretical values vs. the predicted values of MMR under different machining conditions. For the convenience of comparison, a 45° line is placed on the scatter plots to display the relative agreement between the theoretical and the predicted values. It can be seen that the predicted values agree fairly well with the theoretical values for MMR.

Table 2. Results of pilot experiment verification

Test No.	Fs (N)	A (mm)	s (rpm)	d (mm)	n	MRR (mm^3/s) Theo.	MRR (mm^3/s) Pred.
1	110	0.022	1150	0.12	450	1.16	0.98
2	110	0.022	2850	0.28	850	1.78	1.61
3	190	0.038	2200	0.25	620	3.80	4.39
4	200	0.022	1850	0.12	400	2.17	2.24
5	200	0.03	2100	0.21	700	3.27	3.57
6	230	0.025	1500	0.17	500	3.09	3.08
7	230	0.028	1750	0.18	500	3.37	3.46
8	240	0.028	1750	0.19	350	3.61	3.53
9	290	0.033	3100	0.15	590	4.35	4.28
10	300	0.042	2900	0.12	300	4.59	4.60
11	320	0.035	2500	0.29	550	6.63	7.20
12	330	0.032	2700	0.31	600	6.76	6.89
13	340	0.041	3000	0.27	900	7.38	7.65
14	350	0.039	990	0.23	750	6.72	6.60
15	360	0.04	2400	0.27	650	7.69	7.60

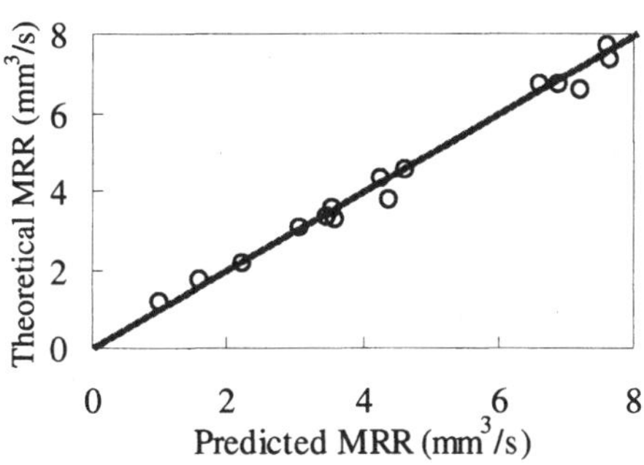

Fig. 4. Theoretical results vs. predicted results by FAN III model.

By using the trained FAN III model, the influences of the independent variables on the dependent variable, can be predicted as well. Due to the limited space, the prediction influences of process parameters (static force, ultrasonic vibration amplitude, tool rotating speed, abrasive grit size, abrasive grit number) on MRR in RUM will be presented in a follow-up paper.

CONCLUSIONS

FANs are network-based fuzzy inference systems, which have the capability of modeling complex and vague phenomena in real life systems.

Furthermore, the proposed approach can be used to continuously up-date and improve the initial established aproximate model by the use of regular operating data and the learning ability of the network. It should be emphasized that daily operating data are enough and not any special effort is need to obtain any special type of data such as the data obtained by statistical experimantal design.

As an example of application of this approach in manufacturing processes, this paper applies FANs based method to the modeling of the relationships between material removal rate (MRR) and the five controllable machining parameters in ultrasonic rotary machining process. Even from this limited results, we can conclude that the proposed FAN approach is a useful tool for modeling of the complex relationships of material removal rate in rotary ultrasonic machining process.

Although only RUM process is modeled, it is obvious that this powerful modeling tool can be applied for modeling and analysis of many other complex and difficult-to-define practical machining problems.

ACKNOWLEDGMENTS

Financial support for this work is provided by Society of Manufacturing Engineers through a Research Initiation grant and the Advanced Manufacturing Institute at Kansas State University.

REFERENCES

Cheng, C.-B., and Lee, E.S., 1999, "Applying fuzzy adaptive network to fuzzy regression analysis," *Computers and Mathematics with Applications*, 38, pp.123-140.

Chang, P.-T., and Lee, E.S., 1994, "Ranking of fuzzy sets based on the concept of existence," *Computers and Mathematics with Applications*, 27(9/10), pp.1-20.

Ehmann, K.F., Kapoor, S.G., DeVor, R.E., and Lazoglu, I., 1997, "Machining process modeling: a review," *Journal of Manufacturing Science and Engineering*, 119(4B), pp. 655-663.

Hards, K.W., 1966, "Ultrasonic speed diamond machining," *Ceramic Age*, 82(12), pp.34-36.

Jang, J.-S. R., Sun, C.-T., and Mizutani, E., 1997, *Neuro-fuzzy and Soft Computing: A Computational Approach to Learning and Machine Intelligence*, Prentice Hall, Upper Saddle River.

Jiao, Y., 2002, "Fuzzy adaptive networks and applications to humanistic systems," Ph.D. Thesis, Kansas State University, Manhattan, KS.

Kubota, M., Tamura, Y., and Shmanura, N., 1997, "Ultrasonic machining with a diamond impregnated tool," *Bulletin of the Japan Society of Precision Engineering*, 11(3), pp. 127-132.

Lee, Y.H., Yang, B.H., Moon, K.S., 1999, "An economic machining process model using fuzzy non-linear programming and neural network," *International Journal of Production Research*, 37(4), pp. 835-847.

Markov, A.I. , and Ustinov, I.D., 1872, "A study of the ultrasonic diamond drilling of non – metallic materials," *Industrial Diamond Review*, March, pp. 97-99.

Merchant, M.E., 1998, "Interpretive look at 20th century research on modeling of machining," *Machining Science and Technology*, 2(2), pp.157-163.

Pei, Z.J., Prabhakar, D., Ferreira, P.M. and Haselkorn, M., 1995, "A Mechanistic approach to the prediction of material removal rates in rotary ultrasonic machining," *Transaction of the ASME* 117, pp. 142-151.

Prabhakar, D., Ferreira, P.M. and Haselkorn, M., 1992, "An experimental investigation of material removal rates in rotary ultrasonic machining," *Transactions of the North American Manufacturing Research Institution of SME*, XX, pp. 211-218.

Stinton, D.P., 1988, Assessment of the state of the art in machining and surface preparation of ceramics, ORNL/TM – Rep. 10791, Oak Ridge National Laboratory, TN.

Wang, L.-X., 1994, *Adaptive Fuzzy Systems and Control: Design and Stability Analysis*, Prentice Hall, New Jersey.

ONE POSSIBLE MEMBERSHIP FUNCTION CONSTRUCTION MODEL

VIKAS KAPOOR
Dept. of Production & Industrial Engg.,
Jai Narain Vyas University Jodhpur,
Jodhpur, India.

SHYAM S. TAK
Member,
Raj. Public Service Commission,
Ajmer, India.

ABSTRACT

This paper reports a novel methodology for ascertaining one possible set of graphical representations of membership for common linguistic variables. A pilot survey was planned seeking to elicit responses from a large number of people, coming from different social & educational backgrounds, regarding their perception of the subjectivity attached with four linguistic values viz., DISTANCE, WEIGHTAGE, EVALUATION, & COMPATIBILITY. Based on the premise that the meaning carried by any linguistic term cannot accurately be reduced to precise mathematical equalities & relations, a questionnaire was designed to enable people to mark segments (and not specific points) within which an appropriate representation of the accompanying linguistic value could definitely be made. We report the design of our questionnaire, the assumptions that we made, the study that was conducted and the outcome of the study in this paper.

MEMBERSHIP FUNCTIONS

As is commonly understood, the most essential feature of any 'fuzzy' based algorithm is its membership function distribution. To paraphrase Timothy J. Ross (1995), "*Since the membership function essentially embodies all fuzziness for a particular fuzzy set, its description is the essence of a fuzzy property or operation.*" However, since there are no standard techniques or preferred shapes, it makes the task of a designer difficult.

There are countless ways available for generating suitable membership functions. Amongst them are concepts of Relative Preferences (Saaty, 1974), Deformable Prototypes (Bremermann, 1976), Intuition (Zadeh, 1965, 1975), Various Uses of Statistics (Dubois and Prade, 1980), Neural Networks (Hopfield, 1986), Genetic Algorithm (Forrest, 1993) etc., to name a few. However, the field is still open for newer methodologies that are less time consuming or less computation intensive.

The basic premise of Knowledge Engineering suggests that any form of membership distribution function may suffice as long as it follows the basic logic and gives valid results. Hence, we choose the basic trapezoidal shape, which is defined as follows. A fuzzy set representation A in R is characterized by a continuous mapping from R (set of real numbers) to a closed interval [0,1]; and is of a trapezoidal form (p,q,r,s) if its membership function $\mu_A : R \rightarrow [0,1]$ is given by (i) and illustrated in (ii) of figure 1.

(i)

$$\mu_A(x) = \begin{cases} (x - p) / (q - p) & \text{for } p \le x \le q \\ 1 & \text{for } q \le x \le r \\ (x - s) / (r - s) & \text{for } r \le x \le s \\ 0 & \text{otherwise} \end{cases}$$

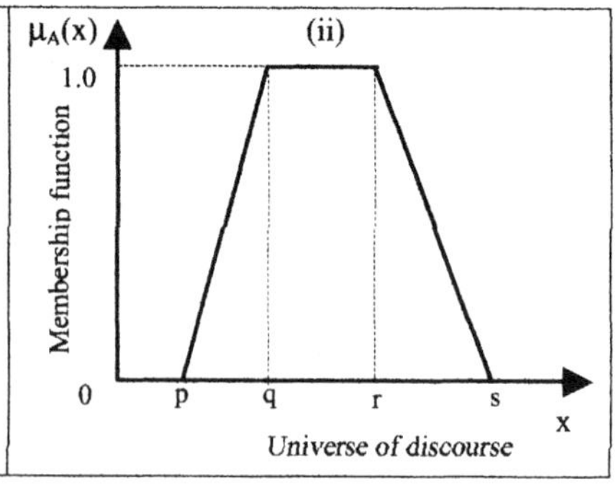

Figure 1 **Trapezoidal Fuzzy representation: Definition & Illustration**

The Methodology

The methodology reported in here outlines a convenient technique for eliciting responses from a large number of people, coming from different social & educational backgrounds, regarding their perception of the subjectivity attached with some common linguistic values. A questionnaire is proposed to enable people to capture some of this subjectivity and to portray it on a given length of line, with as little inaccuracy as they can manage. Since it is evident that the meaning carried by a linguistic term can not accurately be reduced to precise mathematical equalities & relations, the questionnaire has been designed to enable people to mark segments (and not specific points) within which an appropriate representation of the accompanying linguistic value could definitely be made. The only constraint, in the face of which respondents are required to express themselves is, that they must choose terms of description from a given set. For instance, they were asked to interpret the variable DISTANCE using only the terms from a given set of linguistic values viz., **Very Close, Close, Nearby, Far** during our study.

Since it is assumed, that regardless of the choice of context of reference and the scale of measurement, the segments marked by the evaluators would inadvertently, locate themselves in the same portion(s) of the line; the people interviewed may not be supplied with any specific reference or dimension to base their responses on. Further, to help them be free of any bias, (borne out of their 'crisp conditioning') they might be asked to mark each representation on a separate line. Of course, the fact must not be concealed from them that all terms pertaining to a specific variable would finally be superimposed and processed. Proceeding thus, we shall have, representation of DISTANCE by the term set **Very Close, Close, Nearby & Far** on separate line lengths and likewise for WEIGHTAGE, EVALUATION, & COMPATIBILITY. After obtaining the representation of at least 100 samples, their measurements may be integrated as follows:

1. Measure each marked segment and tabulate the 'limit inferior' and 'limit superior' of each representation. For convenience in further processing, make sure, for each term set that the first segment begins at the start and the last segment finishes at the end of the line being marked by the evaluators. If this is not so, use a suitable multiplying factor to neutralize the incongruity of proportion and bring them all to a uniform scale.
2. Scan the data obtained in this manner and for each set of segment lengths pertaining to a variable, identify that length, which is the **intersection of all segments** marked by the hundred odd respondents. Such a length would reflect every respondent's interpretation of the linguistic value in question.

3. Similarly, scan the data obtained for each set and identify that length pertaining to each variable, which is the **union of all segments**. Such a length would reflect the sum total of ambiguity attached to that particular linguistic value as expressed by all evaluators.
4. Carry out steps 2 & 3 for every set of measurements, (i.e., for each linguistic value) and obtain the intersection & union for each set for all the four linguistic values viz., DISTANCE, WEIGHTAGE, EVALUATION, & COMPATIBILITY.
5. Construct trapezoidal representations for each by taking the '**intersection segment**' to represent the '**core**' and the '**union segment**' to represent the '**support**' of the set. This is demonstrated in figure 2.

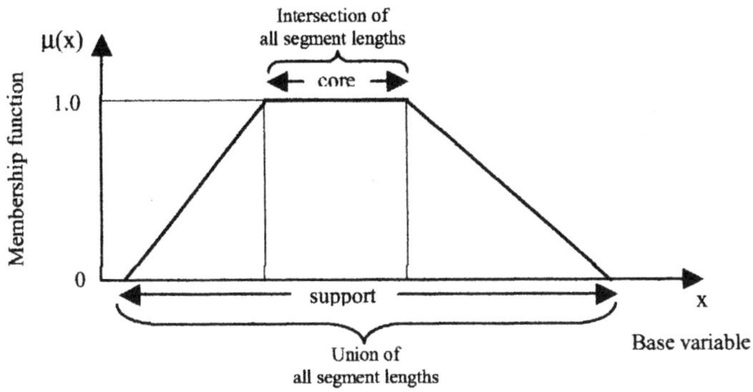

Figure 2 **Nomenclature of Membership Function Representation**

6. Allocate the base values of 'core' to a membership function of unity and those of limit inferior & limit superior of 'support' to a membership function of zero. Plot such representations for each set of readings and obtain values of membership functions for any number of intermittent points of the base variable.
7. Sketch or tabulate the final membership function distribution for the linguistic variable (such as DISTANCE) by superimposing such individual trapezoidal representations, (as shown above) for each value of its term set (such as **Very Close, Close, Nearby** etc.) on a common line length.

Many times it so happens that the 'support' of one representation overlaps that of another by a very great extent, as shown in the non-normalized representation of a fuzzy set (e,f,g,h) with reference to (a,b,c,d) in figure 3.

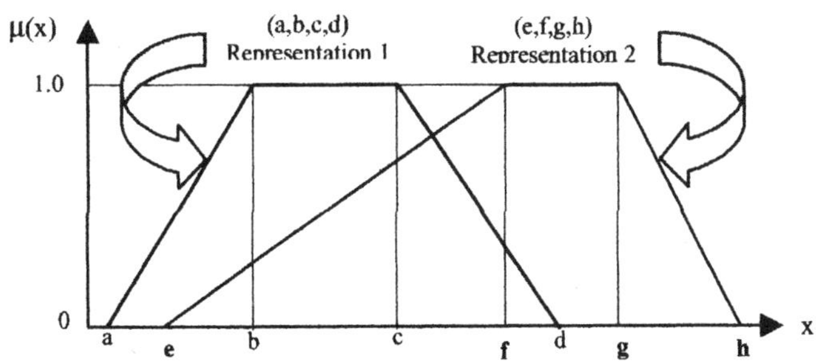

Figure 3 **A Non-normalized Representation of (e,f,g,h)**

The designer may proceed to normalize (e,f,g,h) by making it manifest only at (or after) the point where the 'core' of the previous representation ends, as shown in figure 4. This is sometimes useful for containing the spread and limiting the sensitivity of its terms.

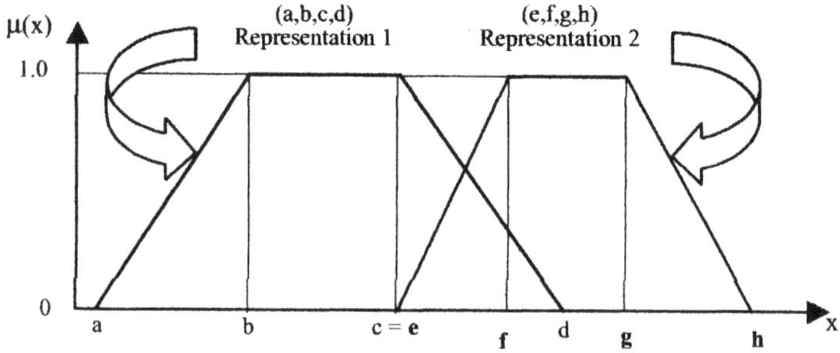

Figure 4 **A Normalized Representation of (e,f,g,h)**

Data Processing And Solutions

Given below is a stepwise description of the processing of our experimental data, authentication of its results and an account of their eventual use in our algorithms.

Membership Function Generation. The various questionnaires were obtained and the segments marked on them measured. The process of union – intersection of all the segments from the hundred odd questionnaires, as demonstrated in figure 2, generate the outcome after normalization, on an arbitrarily chosen scale of 10 as given in Table 1.

Table 1 **Measurements of 'Core' & 'Support' of the four Linguistic Variables**

Distance	Union of all	Intersection of all
Very close	0 - 3.5	0 - 0.4
Close	0.4 - 5.8	2.2 - 2.3
Nearby	2.3 - 8.3	4.2 - 4.4
Far	4.7 - 10.0	8.1 - 10.0
Weightage	Union of all	Intersection of all
Very low	0 - 3.1	0 - 0.5
Low	0.5 - 5.4	2.2 - 2.4
Medium	2.4 - 8.0	4.2 - 4.8
High	5.4 - 10.0	6.8 - 6.8
Very high	6.8 - 10.0	8.9 – 10.0
Evaluation	Union of all	Intersection of all
Very poor	0 - 3.4	0 - 0.5
Poor	0.5 - 4.6	1.8 - 1.9
Fair	1.9 - 7.2	4.1 - 4.3
Good	4.3 - 9.1	6.6 - 6.7
Very good	6.7 - 10.0	9.1 - 10.0
Compatibility	Union of all	Intersection of all
Negative	A continued segment 3.3 units long	
Very low	0 - 1.9	0 - 0.5
Low	0.5 - 3.2	1.5 - 1.5
Medium	1.5 - 5.0	3.0 - 3.2
High	3.2 - 6.0	4.6 - 4.6
Very high	4.6 - 6.7	6.0 - 6.7

Authentication of Derived Membership Functions. For the purpose of authentication of our experimentally derived membership functions, we deploy them on the problems solved by Liang & Wang (1991) and Raoot & Rakshit (1994) for site selection and layout planning respectively and report the results in the two parts of our paper, Kapoor & Tak (1999).

Using assumed membership spreads for WEIGHTAGE and EVALUATION, Liang & Wang (1991) have tackled a problem of selection of suitable site(s) from three alternatives, as evaluated by four decision-makers. Table 2 gives the final results of the problem of selection of suitable site(s) using both approaches - theirs and ours.

Table 2 **Comparison of Results for Site Selection Problem**

Alternative *to be* evaluated	**Solution by Liang & Wang (1991)**		**Solution by Kapoor & Tak (1999)**	
	Preference Rank	Suitability Index	Preference Rank	Suitability Index
A_1	Third	0.416	Third	0.324
A_2	**First**	**0.476**	**First**	**0.381**
A_3	Second	0.469	Second	0.364

Using assumed membership functions for all four variables viz., DISTANCE, WEIGHTAGE, EVALUATION, & COMPATIBILITY, Raoot and Rakshit (1994) have tackled a common pair of layout problems that had been solved by many researchers before them. One problem is of Single Objective Layout Development type while the other is of Multiple Objective Layout Development type. The summary of results obtained by us in Kapoor & Tak (1999) for both the problems is presented in Table 3.

Table 3 **Comparison of Results for Single & Multi Objective Facilities Layout Problems**

Single Objective Layout Development Problem									
Problem Size	**S. No.**	**Solution Given by**	**Location of Facilities**						**Cost**
			1	2	3	4	5	6	
6×6	1	Francis & White (1974)	5	1	4	6	3	2	98
	2	Golany & Rosenblatt (1989)	3	6	2	1	5	4	92
	3	Raoot & Rakshit (1994)	2	6	3	4	5	1	92
	4	**Kapoor & Tak (1999)**	**2**	**6**	**3**	**4**	**5**	**1**	**92**
Multiple Objective Layout Development Problem									
Problem Size	**S. No.**	**Solution Given by**	**Location of Facilities**						**Cost**
			1	2	3	4	5	6	
6×6	1	Rosenblatt (1983)	1	5	6	3	2	4	368
	2	Dutta & Sahu (1983)	1	5	6	3	2	4	368
	3	Fotenberry & Cox (1985)	3	1	4	2	5	6	368
	4	Raoot & Rakshit (1994)	1	5	6	3	2	4	368
	5	**Kapoor & Tak (1999)**	**2**	**5**	**6**	**3**	**1**	**4**	**368**

Application of Derived Membership Functions. As is clear from the above benchmark examples, our membership functions are capable of generating acceptable solutions even for problems solved by various researchers at different times over a period of nearly two decades. Subsequent to their authentication, our sets of membership functions have been used in a variety of fresh problems such as Robot Selection, Selection of an appropriate Materials handling Equipment, Ranking of alternatives, etc. with encouraging results.

A CRITIQUE OF RESULTS

The distribution of membership grades obtained through our experiment for various linguistic variables have been proven valid by the results. However, despite the results of both types of problems being comparable, there is a fair amount of difference between the experimentally derived and theoretically assumed membership functions. (For a detailed analysis, see Kapoor and Tak (1999).)

Our membership functions portray the actual representation of a large number of people and thus provide us with a much more realistic & widely acceptable basis for obtaining results. However, since our questionnaire leaves everything at the discretion of the respondent – right from the context of reference to the scale on which they are to mark the segments, the experimentally derived membership functions show fairly large segments of uncertainty. This tendency of the distribution to have an extended 'support' has partly been offset by the process of normalization. Any further attempt at capturing human subjectivity should take this into account.

Future Research

The proposed methodology needs to be tested further for optimality but it offers a fairly competent framework that may act as a launching pad for future investigations. Also, efforts need to be made to evaluate the results of any standard problem with regard to the sensitivity it exhibits to the changes in membership functions of its variables.

References

Bremermann, H., 1976, Pattern recognition, in H. Bossel, S. Klaszo, and N. Müller (eds.), *Systems theory in the social sciences*, Birkhaeuseur, Basel, pp. 116-159.

Dubois, D., and Prade, H., *1980*, Fuzzy sets and systems: Theory and applications, *Academic*, New York.

Dutta, K. N. and Sahu, S., 1983, Comments on A note on a multi-goal heuristic for facilities design problems: MUGHAL, Int. J. Prod. Res., Vol. 21, pp. 441 – 442.

Forrest, S., 1993, Genetic algorithms: principles of natural selection applied to computation, Science, vol. 261, pp. 872-878.

Fortenberry, J. C. and Cox, J. F., 1985, Multiple criteria approach to the facilities layout problem, Int. J. Prod. Res., Vol. 23, pp. 773 – 782.

Francis, R. L. and White, J. A., 1974, Facilities Layout and Location, *Prentice Hall*, Englewood Cliffs, New Jersey.

Golany, B. and Rosenblatt, M. J., 1989, A heuristic algorithm for the quadratic assignment formulation to the plant layout problem, Int. J. Prod. Res., Vol. 27, pp. 293 – 308.

Hopfield, J., and Tank, D., 1986, Computing with neural circuits: a model, Science, vol. 233, pp. 625-633.

Kapoor, Vikas and Tak, S.S., 1999, Application of Fuzzy Approach to Facilities Planning – A Case Study (Part 1 & 2) in International Conference On Advances in Materials and Processing Technology (AMPT-99) and 16th Annual Conference of 'The Irish Manufacturing Committee (IMC-16)', August 3 – 6, 1999, Dublin City University, Ireland.

Liang, G-S. and Wang, M-J., 1991, A Fuzzy Multi-Criteria Decision-Making Method For Facility Site Selection, Int. J. Prod. Res., Vol. 29(11), pp. 2313 – 2330.

Raoot, A. D. and Rakshit, A., 1994, A fuzzy heuristic for the quadratic assignment formulation to *the facility layout problem*, Int. J. Prod. Res., *Vol. 32, pp. 563 – 581.*

Rosenblatt, M. J., 1983, A note on 'A multi-goal heuristic for facilities design problems: MUGHAL', Int. J. Prod. Res., Vol. 21, pp. 437 – 439.

Ross, T. J., 1995, Fuzzy Logic with Engineering Applications, *McGraw Hills Inc.*, New York, pp. 474–475.

Saaty, T., 1974, Measuring the fuzziness of sets, J. Cybern., vol. 4. no. 4, pp. 53-61.

Zadeh, L. A., 1965, Fuzzy Sets, Information and Control, Vol. 8(3), pp. 338 – 353.

Zadeh, L. A., 1975, The Concept of a Linguistic Variable and its Application to Approximate Reasoning, Parts 1 & 2, Information Sciences, Vol. 8, pp. 199 – 249, 301 – 357.

PART IV: ADAPTIVE CONTROL

APPROXIMATE INPUT/STATE FEEDBACK LINEARIZATION FOR MULTIVARIABLE NONLINEAR SYSTEMS

KOU YAMADA
Department of Mechanical System Engineering, Gunma University
1-5-1 Tenjincho, Kiryu, Japan

KOUJI NITTA
Department of Mechanical System Engineering, Gunma University
1-5-1 Tenjincho, Kiryu, Japan

ATSUYUKI YUZAWA
Department of Mechanical System Engineering, Gunma University
1-5-1 Tenjincho, Kiryu, Japan

ABSTRACT

In this paper, we consider a design method for approximate input/state feedback linearization for multivariable nonlinear systems to which the exact linearization method is not applicable. We adopt a two-step procedure to solve the approximate input/output linearization for multivariable nonlinear systems.

Introduction

In this paper, we examine a design method for approximate input/state linearization for multivariable nonlinear systems. The linearization problem is one of the most important control problems for nonlinear systems. In general use, there is an approximate linearization method using a Taylor series approximation (Isidori, 1989, Vidyasagar, 1993, Slotine and Li, 1991). This method is effective for those nonlinear systems having a narrow range of action. However, this method is less effective for nonlinear systems with a wide range of action. Therefore, this method cannot be applied to robotic systems because, generally, robotic systems have a wide range of action. For robotic systems, exact feedback linearization using state feedback control is used (Isidori, 1989, Vidyasagar, 1993, Slotine and Li, 1991). Because this method does not use a Taylor approximation, it can be applied to nonlinear systems with a wide range of action. However, plants to which this method is applicable are restricted. For example, this method does not apply to a mobile robot. Isidori proposed exact linearization using nonlinear feedback control and coordinate transformation (Isidori, 1989, Vidyasagar, 1993, Slotine and Li, 1991). This method is a powerful tool for nonlinear systems to which the exact linearization method using nonlinear feedback control cannot be applied. It has been shown that this method can easily be applied to nonlinear systems of the first and second order (Isidori, 1989). However, it is difficult to apply to nonlinear systems of orders higher than the third. Yamada and Yuzawa (2002) proposed an approximate linearization method for single-input/multiple-output nonlinear systems to which the exact linearization method cannot be applied.

In the present paper, we expand the method by Yamada and Yuzawa (2002) and propose an approximate input/state linearization methodology for multivariable nonlinear systems to which the exact linearization method cannot be applied. We adopt a two-step procedure to solve the approximate linearization. First, a state transformation matrix is selected, so that the nonlinear system is approximately transformed into the controllable canonical form. Second, a standard nonlinear linearization method is used to transform the controllable canonical form into a stable linear system.

Problem formulation

Let us consider the nonlinear system written as

$$\left\{ \begin{array}{lcl} \frac{dx}{dt} & = & A(x)x + B(x)u \\ y & = & C(x)x \end{array} \right. , \tag{1}$$

where $x \in R^n$ is the state variable, $u \in R^m$ is the control input, $y \in R^p$, and $A(x)$ and $B(x)$ are analytic on x. Without loss of generality, the origin $x = 0$ is the equilibrium point. It is assumed that all the elements of $A(0)$ are finite and

$$\text{rank}\left[\begin{array}{cccc} B(x) & A(x)B(x) & \cdots & A^{n-1}(x)B(x) \end{array}\right] = n \tag{2}$$

holds in the neighborhood of $x = 0$.

The problem examined in the present paper is to propose a new approximate input/state linearization method for the nonlinear system in Eq.(1) using state feedback control and coordinate transformation.

Approximate input/state linearization

In this section, we propose an approximate input/state linearization for nonlinear systems in Eq.(1). We adopt a two-step procedure to solve the approximate linearization. First, a state transformation matrix is selected, so that the nonlinear system is approximately transformed into a controllable canonical form. Second, the standard nonlinear linearization method is used to transform the controllable canonical form into a linear system.

We adopt the coordinate transformation matrix written as

$$x = T(x)z. \tag{3}$$

In addition, we adopt the following approximation written as

$$\frac{dT(x)}{dt} \simeq 0 \ \ \forall t, x, \tag{4}$$

Under the assumption of $\det T(x) \neq 0$ and the assumption of Eq.(4), applying Eq.(3) to the system Eq.(1), the system Eq.(1) is rewritten as

$$\begin{aligned} \frac{dz}{dt} &= T^{-1}(x)\frac{dx}{dt} - T^{-1}(x)\frac{dT(x)}{dt}z \\ &= T^{-1}(x)A(x)T(x)z + T^{-1}(x)B(x)u \equiv A_z(x)z + B_z(x)u. \end{aligned} \tag{5}$$

A design method for the coordinate transformation matrix $T(x)$ is as follows. From the assumption of Eq.(2), there exists n of independent column vectors in $[\; B(x) \quad A(x)B(x) \quad \cdots \quad A^{n-1}(x)B(x) \;]$. Let the integer ν_i

$$\nu_i \equiv \min\left\{ j : A^j(x)b_i(x) \in \text{span}\left[B(x), A(x)B(x), \cdots, A^{j-1}(x)B(x), A^j(x)b_1(x), \cdots, A^j(x)b_{i-1}(x)\right]\right\}. \tag{6}$$

Then we have

$$\text{rank} \; [\; b_1(x) \quad A(x)b_1(x) \quad \cdots \quad A^{\nu_1-1}(x)b_1(x) \quad b_2(x) \quad \cdots \quad A^{\nu_2-1}(x)b_2(x) \quad \cdots \quad A^{\nu_m-1}(x)b_m(x) \;] = n, \tag{7}$$

since

$$\nu_1 + \nu_2 + \cdots + \nu_m = n \tag{8}$$

is satisfied. From the definition of ν_i, there exists $\alpha^i_{j,k}(x)$ satisfying

$$\begin{aligned} A^{\nu_i}(x)b_i(x) &= \alpha^i_{0,1}(x)b_1(x) + \cdots + \alpha^i_{0,m}(x)b_m(x) + \alpha^i_{1,1}(x)A(x)b_1(x) \\ &\quad + \cdots + \alpha^i_{1,m}(x)A(x)b_m(x) + \cdots + \alpha^i_{\nu_i-1,1}(x)A^{\nu_i-1}(x)b_1(x) \\ &\quad + \cdots + \alpha^i_{\nu_i-1,m}(x)A^{\nu_i-1}(x)b_m(x) + \alpha^i_{\nu_i,1}(x)A^{\nu_i}(x)b_1(x) \\ &\quad + \cdots + \alpha^i_{\nu_i,i-1}(x)A^{\nu_i}(x)b_{i-1}(x). \end{aligned} \tag{9}$$

We select

$$b_{i,j}(x) = \sum_{k=1}^{m} \alpha^i_{j,k}(x)b_k(x) \tag{10}$$

$$(i = 0, 1, \cdots, \nu_{i-1})$$

$$\tilde{b}_i(x) = \sum_{k=1}^{i-1} \alpha^i_{\nu_i,k}(x)b_k(x) \tag{11}$$

$$(j = 2, \cdots, m)$$

$$\tilde{b}_1(x) = 0.$$

Equation (9) is rewritten by

$$\begin{aligned} A^{\nu_i}(x)b_i(x) &= b_{i,0}(x) + A(x)b_{i,1}(x) + \cdots \\ &\quad + A^{\nu_i-1}(x)b_{i,\nu_i-1}(x) + A^{\nu_i}(x)\tilde{b}_i(x). \end{aligned} \tag{12}$$

$t_{i,j}(x)$ $(j = 1, \cdots, \nu_i)$ is defined by

$$\begin{aligned} t_{i,j}(x) &= A^{\nu_i-j}(x)(b_i(x) - \tilde{b}_i(x)) - b_{i,j}(x) \\ &\quad - A(x)b_{i,j+1}(x) - \cdots - A^{\nu_i-j-1}(x)b_{i,\nu_i-1}(x). \end{aligned} \tag{13}$$

From the definition of ν_i,

$$\text{rank} \; [\; t_{1,1}(x) \quad \cdots \quad t_{1,\nu_i}(x) \quad \cdots \quad t_{m,1}(x) \quad \cdots \quad t_{m,\nu_m}(x) \;] = n \tag{14}$$

is satisfied. Let $T(x)$

$$T(x) = [\; T_1(x) \quad T_2(x) \quad \cdots \quad T_m(x) \;], \tag{15}$$

where

$$T_i(x) = [\; t_{i,1}(x) \quad t_{i,2}(x) \quad \cdots \quad t_{i,\nu_i}(x) \;]. \tag{16}$$

Then

$$\begin{aligned} A_z(x) &= T^{-1}(x)A(x)T(x) \\ &= \begin{bmatrix} A_{1,1}(x) & A_{1,2}(x) & \cdots & A_{1,m}(x) \\ A_{2,1}(x) & A_{2,2}(x) & \cdots & A_{2,m}(x) \\ & \cdots & \cdots & \\ A_{m,1}(x) & A_{m,2}(x) & \cdots & A_{m,m}(x) \end{bmatrix} \end{aligned} \tag{17}$$

$$A_{i,i}(x) = \begin{bmatrix} 0 & 1 & 0 & & \mathbf{0} \\ 0 & 0 & 1 & & \\ \vdots & \vdots & \vdots & \ddots & \\ 0 & 0 & 0 & \cdots & 1 \\ \gamma_{i,i,1}(x) & \gamma_{i,i,2}(x) & \gamma_{i,i,3}(x) & \cdots & \gamma_{i,i,\nu_i}(x) \end{bmatrix} \in R^{\nu_i \times \nu_i} \tag{18}$$

$$A_{i,j}(x) = \begin{bmatrix} 0 & 0 & \cdots & 0 \\ & \cdots & \cdots & \\ 0 & 0 & \cdots & 0 \\ \gamma_{i,j,1}(x) & \gamma_{i,j,2}(x) & \cdots & \gamma_{i,j,\nu_j}(x) \end{bmatrix} \in R^{\nu_i \times \nu_j} \tag{19}$$

$$B_z(x) = T^{-1}(x)B(x) = \begin{bmatrix} B_{1,1}(x) & B_{1,2}(x) & \cdots & B_{1,m}(x) \\ B_{2,1}(x) & B_{2,2}(x) & \cdots & B_{2,m}(x) \\ & \cdots & \cdots & \\ B_{m,1}(x) & B_{m,2}(x) & \cdots & B_{m,m}(x) \end{bmatrix} \tag{20}$$

$$B_{i,i}(x) = \begin{bmatrix} 0 \\ 0 \\ \vdots \\ 0 \\ 1 \end{bmatrix} \in R^{\nu_i} \tag{21}$$

$$B_{i,j}(x) = \begin{bmatrix} 0 \\ 0 \\ \vdots \\ 0 \\ \beta_{i,j}(x) \end{bmatrix} \in R^{\nu_i} (i < j), B_{i,j}(x) = \begin{bmatrix} 0 \\ 0 \\ \vdots \\ 0 \\ 0 \end{bmatrix} \in R^{\nu_i} (i > j). \tag{22}$$

Next, we show that when the control input for the system in Eq.(5) is given by

$$u = -F_z(x)z + G(x)v, \tag{23}$$

then the system written as

$$\frac{dz}{dt} = \Big(A_z(x) - B_z(x)F_z(x)\Big)z + B_z(x)G(x)v \equiv A_f z + Hv, \tag{24}$$

can be transformed into stable linear state equation, where v denotes the new input.

We describe a design method for $F_z(x)$ such that $A_z(x) - B_z(x)F_z(x)$ is a linear stable matrix. Denote by $\tilde{a}_i$ the $(\nu_1 + \cdots + \nu_i)$-th column of $A_z(x)$. Vector $l_i(x) \in R^{1\times n}$ $(i = 1, 2, \cdots, m)$ and $q_1(x), q_2(x), \cdots, q_m(x)$ are selected sequentially as follows.

$$\left\{ \begin{array}{lcl} l_i(x) & = & \tilde{e}_i - \tilde{a}_i(x) - \sum_{j=i+1}^{m} \beta_{i,j}(x) l_j(x) \\ & & (i = 1, 2, \cdots, m-1) \\ l_m(x) & : & \text{arbitrary polynomial} \\ q_i(x) & = & -\sum_{j=i+1}^{m} \beta_{i,j}(x) q_j(x) \\ & & (i = 1, 2, \cdots, m-1) \\ q_m(x) & = & 1 \end{array} \right. , \tag{25}$$

where

$$\tilde{e}_i(x) = [\ 0 \ \cdots \ 0 \ \underset{\nu_1 + \cdots + \nu_i + 1}{\underset{\uparrow}{1}} \ 0 \ \cdots \ 0\ ]. \tag{26}$$

Using $l_i(x)$, when $L(x) \in R^{m \times n}$ and $q(x) \in R^m$ are selected by

$$L(x) = \begin{bmatrix} l_1(x) \\ l_2(x) \\ \vdots \\ l_m(x) \end{bmatrix}, q(x) = \begin{bmatrix} q_1(x) \\ q_2(x) \\ \vdots \\ q_m(x) \end{bmatrix}, \tag{27}$$

respectively, from Eq.(25), Eq.(17) and Eq.(20), we have

$$A_z(x) + B_z(x)L(x) = \begin{bmatrix} 0 & 1 & 0 & & \mathbf{0} \\ 0 & 0 & 1 & & \\ \vdots & \vdots & \vdots & \ddots & \\ 0 & 0 & 0 & \cdots & 1 \\ \xi_0(x) & \xi_1(x) & \xi_2(x) & \cdots & \xi_{n-1}(x) \end{bmatrix} \tag{28}$$

and

$$B_z(x)q(x) = \begin{bmatrix} 0 \\ 0 \\ \vdots \\ 0 \\ 1 \end{bmatrix}, \tag{29}$$

where $\xi_i(x) \quad (i = 0, 1, \cdots, n-1)$ are arbitrary functions of x.

Let $F_z(x)$ be

$$F_z(x) = -L(x) + q(x)k(x), \tag{30}$$

where $k(x)$ is written as

$$k(x) = [\ \xi_0(x) + \lambda_0 \quad \cdots \quad \xi_{n-1}(x) + \lambda_{n-1}\ ] \tag{31}$$

and $\lambda_i (i = 0, 1, \cdots, n-1)$ are arbitrary real numbers such that all roots of

$$s^n + \lambda_{n-1}s^{n-1} + \cdots + \lambda_1 s + \lambda_0 = 0 \tag{32}$$

are in the open left half plane.

From Eq.(30), since

$$\begin{aligned} A_z(x) - B_z(x)F_z(x) &= A_z(x) + B_z(x)L(x) - B_z(x)q(x)k(x) \\ &= \begin{bmatrix} 0 & 1 & 0 & & \mathbf{0} \\ 0 & 0 & 1 & & \\ \vdots & \vdots & \vdots & \ddots & \\ 0 & 0 & 0 & \cdots & 1 \\ -\lambda_0 & -\lambda_1 & -\lambda_2 & \cdots & -\lambda_{n-1} \end{bmatrix}, \end{aligned} \tag{33}$$

we can make $A_z(x) - B_z(x)F_z(x)$ be an linear stable matrix.

Next, we give a design method of $G(x)$ to make $B_z(x)G(x)$ be an linear matrix. $G(x)$ is rewritten in the block matrix form with the same form as $B_z(x)$ by

$$G(x) = \begin{bmatrix} G_{1,1}(x) & G_{1,2}(x) & \cdots & G_{1,m}(x) \\ G_{2,1}(x) & G_{2,2}(x) & \cdots & G_{2,m}(x) \\ & \cdots & \cdots & \\ G_{m,1}(x) & G_{m,2}(x) & \cdots & G_{m,m}(x) \end{bmatrix}. \tag{34}$$

This yields the i-th column and j-th row block matrix of $B_z(x)G(x)$, which is written as

$$\sum_{k=i}^{m} B_{i,k}(x)G_{k,j}(x).(i,j=1,2,\cdots,m) \tag{35}$$

$G_{i,j}(x)$ is selected sequentially by

$$\begin{aligned} G_{i,j}(x) &= -\sum_{k=i+1}^{m} \beta_{i,k}(x)G_{k,j}(x)+\epsilon_{i,j}(i=1,2,\cdots,m-1;j=1,2,\cdots,m) \\ & G_{m,j}(x)=\epsilon_{m,j}(j=1,2,\cdots,m) \end{aligned}$$

where $\epsilon_{i,j}$ are arbitrary real numbers. Then $B_z(x)G(x)$ is written as

$$B_z(x)G(x) = \begin{bmatrix} H_{1,1} & H_{1,2} & \cdots & H_{1,m} \\ H_{2,1} & H_{2,2} & \cdots & H_{2,m} \\ & \cdots & \cdots & \\ H_{m,1} & H_{m,2} & \cdots & H_{m,m} \end{bmatrix} \equiv H, \tag{36}$$

where

$$H_{i,j} = \begin{bmatrix} 0 \\ 0 \\ \vdots \\ 0 \\ \epsilon_{i,j} \end{bmatrix} \in R^{\nu_i}. \tag{37}$$

In this way, we can force $B_z(x)G(x)$ to be a linear matrix.

Conclusion

In this paper, we expand the method of Yamada and Yuzawa (2002) and propose an approximate input/state linearization methodology for multi-variable nonlinear systems to which the exact linearization method cannot be applied. If $\lambda_i(i=0,\cdots,n-1)$ in Eq.(31) is selected using Neural Networks, we can design the nonlinear control system to have Intelligences and linearity.

References

A. Isidori: Nonlinear Control Systems (Third Edition), Springer-Verlag, New York (1989)
M. Vidyasagar: Nonlinear Systems Analysis, Prentice Hall, New Jersey (1993)
J.J.E. Slotine and W. Li: Applied nonlinear control, Prentice Hall, New Jersey (1991)
K. Yamada and A. Yuzawa: Approximate feedback linearization for nonlinear systems and its application to the ACROBOT, American Control Conference 2002, 1672/1677(2002)

DYNAMIC SIMULATION OF COMPUTED-TORQUE CONTROLLERS FOR A WHEELED MOBILE ROBOT AUTONOMOUS NAVIGATION IN OUTDOOR ENVIRONMENTS

SOUMA M. ALHAJ ALI
University of Cincinnati
Department of Mechanical, Industrial and Nuclear Engineering
Cincinnati, Ohio

MASOUD GHAFFARI
University of Cincinnati
Department of Mechanical, Industrial and Nuclear Engineering
Cincinnati, Ohio

XIAQUAN LIAO
University of Cincinnati
Department of Mechanical, Industrial and Nuclear Engineering
Cincinnati, Ohio

ERNEST L. HALL
Director - Center for Robotics Research
Paul E. Geier Professor of Robotics
University of Cincinnati
Department of Mechanical, Industrial and Nuclear Engineering
Cincinnati, Ohio

ABSTRACT

This paper describes the development of PD and PID Computed-Torque (CT) controllers for autonomous navigation of a Wheeled Mobile Robot (WMR) in outdoor environments. WMRs have complex and nonlinear dynamics, which make their modeling and control difficult. The controllers select the suitable control torques, so that the WMR follows the desired path produced from a navigation algorithm described in a previous paper. PD CT and PID CT controllers were developed and tested under different motion trajectories. Software was developed to simulate the performance and efficiency of the controllers. Simulation results verified the effectiveness of the controllers. The significance of this work lies in the development of CT controllers for WMR navigation, instead of robot manipulators. These CT controllers will facilitate the use of WMRs in many applications.

INTRODUCTION

This paper presents the development of a proportional-plus-derivative (PD) CT and a proportional-plus-integral-plus-derivative (PID) CT controllers. A dynamic simulation, based on a framework developed by Lewis, et al. (1999), was conducted after modifying it to suit the navigation of a WMR. The simulation software takes, as input, the desired robot path from the navigation algorithm described in a previous paper by the authors (Alhaj Ali and Hall, 2002). The simulation software produced the suitable control torques. This simulation was developed by using matlab and C++.

Different controllers had been developed for the motion of robot manipulators. However, not until recently has there been an interest in moving the robot itself, not just its manipulators. Thus, WMR motion control is a new research area.

Based on empirical practice, Shim and Sung (2003) proposed a WMR asymptotic control with driftless constraints by using the WMR kinematic equations. The trajectory control of a wheeled inverse pendulum type robot had been discussed by Yun-Su and Yuta (1996).

Rajagopalan and Barakat (1997) developed a computed torque control scheme for Cartesian velocity control of WMRs. A discontinuous stabilizing controller for WMRs with nonholonomic constraints was presented by Zhang and Hirschorn (1997). A path tracking problem was formulated by Koh and Cho (1999) for a mobile robot, which follows a virtual target vehicle moving with a specified velocity exactly along the path. Zhang, et al. (2003) employed dynamic modeling to design a tracking controller for a differentially steered mobile robot subject to wheel slippage and external loads. A sliding mode control was used to develop a trajectory tracking control in the presence of bounded uncertainties (Corradini and Orlando, 2001). Jiang, et al. (2001) developed a model-based control design strategy dealing with global stabilization and global tracking control for the WMR kinematic model.

A fuzzy logic controller had also been tried for navigating WMRs. Montaner and Ramirez-Serrano (1998) developed a fuzzy logic controller, which can deal with the sensors inputs uncertainty and ambiguity for direction and velocity maneuvers. Toda, et al. (1999) employed a sonar-based mapping of crop rows and fuzzy logic control-based steering for navigating a WMR in an agricultural environment.

PD AND PID CT CONTROLLERS

This section describes the development of PD and PID CT controllers for WMR autonomous navigation in outdoor environments. To provide focus for the presented model, a particular configuration will be addressed: the Bearcat III robot, which has two driven fixed wheels and a caster. Bearcat III was developed at the Center for Robotics Research, at the University of Cincinnati.

A simplified kinematic and dynamic model is derived from the Newton-Euler method by considering the velocity along the x and y axes and the angular velocity, with the robot center of mass as a reference point.

$$M(q)\ddot{q} + J(q,\dot{q})\dot{q} + F = \tau$$

where:

$$q = \begin{bmatrix} x_c \\ y_e \\ \theta \end{bmatrix}, \quad F = \begin{bmatrix} \dfrac{-f_n er}{d} \\ \dfrac{-f_n er}{d} \end{bmatrix}, \quad \tau = \begin{bmatrix} \tau_r \\ \tau_l \end{bmatrix}$$

$$M(q) = \begin{bmatrix} \dfrac{(mr^2\cos\theta + 2J_0\cos\theta)}{2r} & \dfrac{(mr^2\sin\theta + 2J_0\sin\theta)}{2r} & \dfrac{(mr^2ed\sin^2\theta - mr^2ed\sin\theta\cos\theta + J_cr^2 + 2J_0d^2}{2rd} \\ \dfrac{(mr^2\cos\theta + 2J_0\cos\theta)}{2r} & \dfrac{(mr^2\sin\theta + 2J_0\sin\theta)}{2r} & \dfrac{(mr^2ed\sin^2\theta - mr^2ed\sin\theta\cos\theta - J_cr^2 - 2J_0d^2}{2rd} \end{bmatrix}$$

$$J(q,\dot{q}) = \begin{bmatrix} \dfrac{-J_0\dot{\theta}\sin\theta}{r} & \dfrac{J_0\dot{\theta}\cos\theta}{r} & \dfrac{-mre\dot{\theta}\cos\theta(\sin\theta + \cos\theta)}{2} \\ \dfrac{-J_0\dot{\theta}\sin\theta}{r} & \dfrac{J_0\dot{\theta}\cos\theta}{r} & \dfrac{-mre\dot{\theta}\cos\theta(\sin\theta + \cos\theta)}{2} \end{bmatrix}$$

x_c : is the x- coordinate of the robot's center of mass.

y_e : is the y- coordinate of the center of the wheel axle.

θ : is the orientation angle of the robot.

f_n : is the resultant normal force.

e : is the distance between the robot's center of mass and the center of the wheel axle.

r : is the radius of the fixed wheel.

d : is half the width of the robot.

τ_r : is the right motor torque, acting on the right wheel.

τ_l : is the left motor torque, acting on the left wheel.

m : is the robot's mass, excluding the wheels.

$J_0 : J_0 = J_w + m_w r^2$, where J_w is the inertia of the wheel, and m_w is the mass of the wheel.

J_c : is the robot's inertia, excluding the wheels.

(1)

For more details, please refer to Alhaj Ali (2003).

The robot dynamics can be written as follows (Lewis, et al., 1999):

$M(q)\ddot{q} + N(q,\dot{q}) = \tau$, where $N(q,\dot{q})$ represents the nonlinear terms, $N(q,\dot{q}) = J(q,\dot{q})\dot{q} + F$ (2)

The objective of a motion controller is to move the robot along a desired motion trajectory $q_d(t)$. The actual motion trajectory is defined as $q(t)$. The tracking error, in this case, is defined as (Lewis, et al., 2000):

$$e(t) = q_d(t) - q(t) \tag{3}$$

The Brunovsky canonical form is developed by differentiating $e(t)$ twice and writing it in terms of the state x (Lewis, et al., 2000):

$$\frac{d}{dt}\begin{bmatrix} e \\ \dot{e} \end{bmatrix} = \begin{bmatrix} 0 & I \\ 0 & 0 \end{bmatrix}\begin{bmatrix} e \\ \dot{e} \end{bmatrix} + \begin{bmatrix} 0 \\ I \end{bmatrix} u, \tag{4}$$

where $u \equiv \ddot{q}_d + M^{-1}(q)(N(q,\dot{q}) - \tau)$, $x = \begin{bmatrix} e^T & \dot{e}^T \end{bmatrix}^T$

To develop the CT controller, a linear system design procedure will be used to select a feedback control, $u(t)$, that stabilizes the tracking error equation. The torques needed for the motors will then be computed by using the inverse of the dynamic equation for the WMR:

$$\tau = M(q)(\ddot{q}_d - u) + N(q,\dot{q}) \tag{5}$$

A PD feedback for $u(t)$, with a derivative gain matrix, K_v, and a proportional gain matrix, K_p, produces the PD CT controller with the motor torques (Lewis, et al., 2000):

$$\tau = M(q)(\ddot{q}_d + K_v\dot{e} + K_p e) + N(q,\dot{q}), \tag{6}$$

which has the tracking error dynamics $\ddot{e} = -K_v\dot{e} - K_p e$

The PD CT controller described above can easily be adjusted to a PID CT controller by adding an integrator gain matrix, K_i, to $u(t)$ (Lewis et al., 2000):

$$\tau = M(q)(\ddot{q}_d + K_v\dot{e} + K_p e + K_i(\int e)) + N(q,\dot{q}), \tag{7}$$

which has the tracking error dynamics $\ddot{e} = -K_v\dot{e} - K_p e$

Simulation of the PD CT controller for WMR navigation

The simulation program has the following main components:

- The first component computes the desired WMR trajectory from the input of the navigation algorithm.
- The second component calculates the controller input by first calculating the tracking error between the desired trajectory and the actual trajectory and then computing the inertia term and the nonlinear term from the WMR dynamic model described in Eq. 1. Finally, the motor torques are calculated by using Eq. 6.
- The third component calculates the new position of the WMR, using the state-space equation, $\dot{x} = f(x,u)$, in which the state-space position/velocity form is used (Lewis, et al., 2000):

$$x \equiv \begin{bmatrix} q^T \\ \dot{q}^T \end{bmatrix}, \quad \dot{x} = \begin{bmatrix} \dot{q} \\ -M^{-1}(q)N(q,\dot{q}) \end{bmatrix} + \begin{bmatrix} 0 \\ M^{-1}(q) \end{bmatrix}\tau \tag{8}$$

This is used to update the WMR's actual position. Since $M(q)$ is not square, the Moore-Penrose generalized inverse is used.

Several experiments had been conducted on the PD CT simulation software; the robot parameters were from Bearcat III. Various controller parameters were tested by sinusoidal desired motion trajectories.

Experiments show that k_p and k_v need to be different for each component of the motion trajectory, q. Small positive values are required to obtain good results. The gain parameters, k_{p1}=2, k_{v1}=1, k_{p2}=0, k_{v2}=10, k_{p3}=2, and k_{v3}=1, give good results, as shown in the following figures. The tracking errors are in the range 0.00-0.04, as shown in Fig. 1. The desired versus the actual motion trajectory is shown in Fig. 2. As shown in the figures, the tracking errors are zero for both x and θ, with the actual motion trajectory following the desired motion trajectory for these two components. For y, the error oscillates around 0.02.

The selection process of proper parameters for the controller is a challenging task, since six parameters must be changed in each trial. Although x, y, and θ are independent of each other, they are, nevertheless, used to update the new values of the actual motion trajectories. Hence, they have an impact upon each other. Therefore, keeping the same controller parameters for a particular motion trajectory component does not imply that the actual motion trajectory will remain the same.

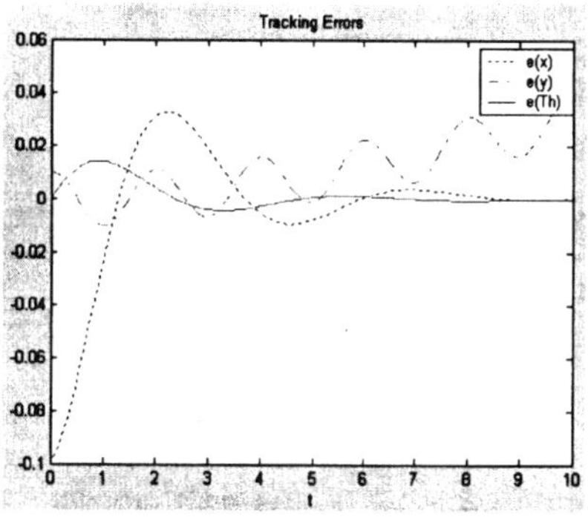

Figure 1: PD CT controller tracking errors versus time.

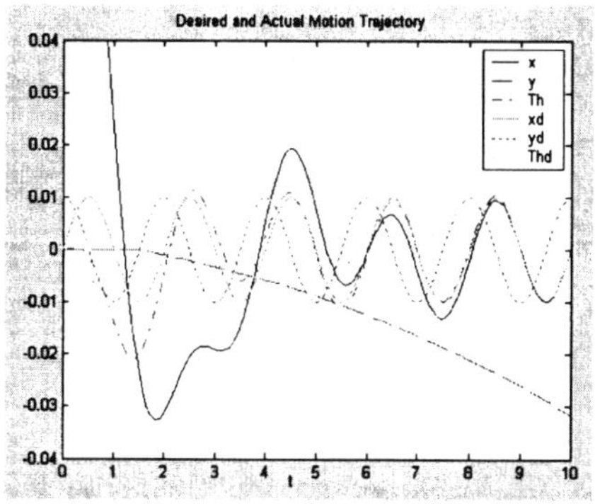

Figure 2: Desired versus actual motion trajectories.

Simulation of the PID CT controller for WMR navigation

The PID CT controller simulation program for WMR navigation was developed the same way as that of the PD CT controller, except for the equation for calculating the motor torques, where Eq. 7 was used.

Several experiments had been conducted on the PID CT simulation software, and different trajectories and controller parameters were tried.

The results of the experiments show that the controller parameters need to be small positive numbers to obtain good results. It is also noteworthy that increasing k_p and fixing k_v and k_i, or increasing k_v and fixing k_p and k_i, reduces the tracking errors of θ and x, while it increases the tracking error of y. However, there is a limit to this increase, which is about 10.

Using very large, or zero, values for k_p;k_v, or k_i is not recommended. Additionally, the value of k_i must not be too large, as a condition for having a stable tracking error.

k_p=2, k_v=1, and k_i=1 give very reasonable results, as shown in the following figures. The tracking error for θ is zero. For x, it oscillates around zero. For y, it starts at zero and increases to 0.35, as shown in Fig. 3. The desired versus the actual motion trajectories are shown in Fig. 4.

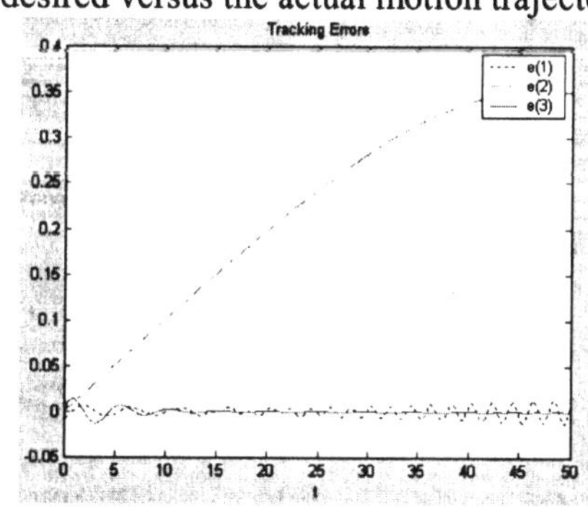

Figure 3: PID CT controller tracking errors versus time.

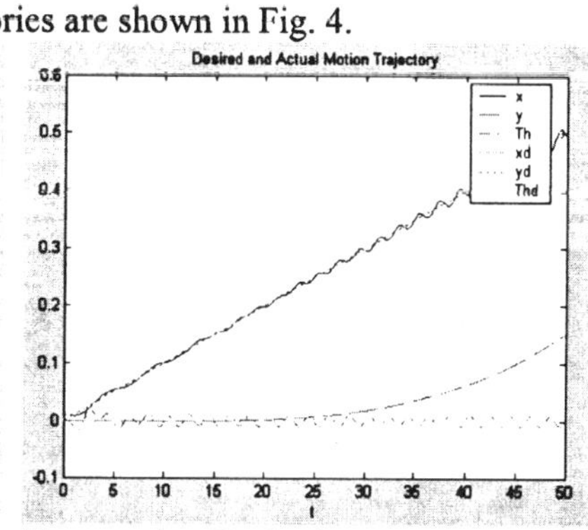

Figure 4: Desired versus actual motion trajectories.

CONCLUSIONS

In this paper, the development of a PD CT and PID CT controller for WMR autonomous navigation in an outdoor environment is presented. The controllers select suitable control torques, so that the WMR will follow the desired path from a navigation algorithm described in a previous paper. The controllers are tested under various control parameters and motion trajectories. Simulation results show that, for a PD CT controller, the gain matrices need to be different for each component of the motion trajectory; small positive values are needed to get good results. Simulation results for a PID CT controller show that the gain matrices also need to be small positive numbers. k_p=2, k_v=1, and k_i=1 produce reasonable results. Selecting the proper parameters for the controller is a challenging task, since there are too many parameters involved.

REFERENCES

Alhaj Ali, S. M., 2003, "Technologies for autonomous navigation in unstructured outdoor environments," Ph. D. dissertation, University of Cincinnati, Cincinnati, OH.

Alhaj Ali, S. M., and Hall, E. L., 2002, "Technologies for autonomous operation in unstructured outdoor environments, Part I: Navigation," Proceedings of the Artificial Neural Networks in Engineering Conference, Vol. 12, pp. 57-62.

Corradini, M. L., and Orlando, G., 2001, "Robust tracking control of mobile robots in the presence of uncertainties in the dynamical model," *Journal of Robotic Systems*, Vol. 18, Issue 6, pp. 317-323.

Jiang, Z.-P., Lefeber, E., and Nijmeijer, H., 2001, "Saturated stabilization and tracking of a nonholonomic mobile robot," *Systems and Control Letters*, Vol. 42, Issue 5, pp. 327-332.

Koh, K. C., and Cho, H. S., 1999, "A smooth path tracking algorithm for wheeled mobile robots with dynamic constraints," *Journal of Intelligent and Robotic Systems*, Vol. 24, Issue 4, pp. 367-385.

Lewis, F. L., Jagannathan, S., and Yesildirek, A., 1999, "Neural Network Control of Robot Manipulators and Nonlinear Systems," Taylor and Francis Ltd, T. J. International Ltd, Padstow, UK.

Montaner, M. B., and Ramirez-Serrano, A., 1998, "Fuzzy knowledge-based controller design for autonomous robot navigation," *Expert Systems with Applications*, Vol. 14, Issue 1-2, pp. 179-186.

Rajagopalan, R., and Barakat, N., 1997, "Velocity control of wheeled mobile robots using computed torque control and its performance for a differentially driven robot," *Journal of Robotic Systems*, Vol. 14, Issue 4, pp. 325 - 340.

Shim, H.-S. , and Sung, Y.-G., 2003, "Asymptotic control for wheeled mobile robots with driftless constraints," *Robotics and Autonomous Systems*, Vol. 43, Issue 1, pp. 29-37.

Toda, M., Kitani, O., Okamoto, T., and Torii, T., 1999, "Navigation method for a mobile robot via sonar-based crop row mapping and fuzzy logic control," *Journal of Agricultural Engineering Research*, Vol. 72, Issue 4, pp. 299 - 309.

Yun-Su, H., and Yuta, S., 1996, "Trajectory tracking control for navigation of the inverse pendulum type self-contained mobile robot," *Robotics and Autonomous Systems*, Vol. 17, Issue 1-2, pp. 65-80.

Zhang, M., and Hirschorn, R. M., 1997, "Discontinuous feedback stabilization of nonholonomic wheeled mobile robots," *Dynamics and Control*, Vol. 7, Issue 2, pp. 155-169.

Zhang, Y., Chung, J. H., and Velinsky, S. A., 2003, "Variable structure control of a differentially steered wheeled mobile robot," *Journal of Intelligent and Robotic Systems*, Vol. 36, Issue 3, pp. 301-314.

EVOLUTIONARY ACQUISITION OF BEHAVIOR USING ROBOT MODEL BASED ON SPRING-MASS SYSTEM

KAZUTOSHI UEHARA KEITARO NARUSE

HIROSHI YOKOI YUKINORI KAKAZU

Division of Systems and Information Engineering, Hokkaido University,
Kita-13, Nishi-8, Kita-ku, Sapporo, Japan 060-8628
{uehara | naruse | yokoi | kakazu} @complex.eng.hokudai.ac.jp
http://junji.complex.eng.hokudai.ac.jp/~uehara/

Abstract
In the field of evolutionary robotics, researches on the design of morphology and behavior have been advanced so far. However, the 'appropriate relation' between the morphology and behavior is not well understood. In this paper, as the basis of this research, a series of experiments that investigate a process of evolution in behavior are conducted. A robot employed here uses a spring-mass system as a fundamental model. This model has the feature that various behaviors and structure are possible, though its principle is simple. To evolve the behavior, genetic algorithms (GA) are applied with different designs of chromosomes and fitness functions. From the results, it is observed that individuals that move with various patterns of behavior (like walking, running, etc.) are emerged according to the settings of the fitness function.

INTRODUCTION

The virtual creatures created by Sims (1994a, 1994b) and Ventrella (1995) are one of the major achievement in the field of evolutionary robotics. Fascinating patterns of behavior are generated from interaction with environment, and these results have been attracting many researchers until now. It is a common feature in these researches that complex function and morphology are emerging, although the model of a creature is simple. This feature brings up some issue; What is the relation between the function and the morphology? Can we make it clear? Is it possible to control the emerging function and morphology through the use of the relation? Many works influenced by Sims and Ventrella have been carried out so far, analyses are not well advanced. This study aims at clarification of such relation.

In this paper, as the basis of this research, a series of experiments that investigate the process of evolution in behavior are conducted. A robot employed here uses a spring-mass system as a fundamental model. This model has a feature that various behaviors and structure are possible, though its principle is simple. Genetic algorithm (GA) is used for the computation of evolution. Several varieties of fitness functions are prepared from the combination of multiple performance functions. Under this condition, the evolution process of parameters relevant to behavior and morphology is observed. As a result, robots with different behavioral patterns are generated depending on the disign of fitness functions, and each parameter showed characteristic changes. Detailed analysis is shown to investigate the relationship between the behavior and the morphology.

THE MODEL

Overview

The model dealt with in this study is based on the model used by sodaplay (soda creature ltd., 2000), one of the works of Soda Creative Technologies. This model has a feature that various behaviors and structure are possible, though its principle is simple. In this paper, a robot based on the model acts in a virtual 2-dimensional environment with external force (gravity, viscosity, etc.). The robots shown in Fig.1 are examples of the model.

Basically, the model is a spring-mass system situated in a 2-dimensional space: a system made from a number of *mass points*, *springs* between them, and special springs called *muscle*. A muscle is a spring that oscillates its natural length with time. The model has a virtual *oscillator* that vibrates all muscles and makes a robot keep periodic motion. Each muscle has its own amplitude (called *relative amplitude*) and initial phase. A robot can behave in various way by tuning these two parameters. The spring constant is common in all springs and muscles.

Formulation

A *mass point* is an object that moves according to the equation of motion, $m\frac{d^2x}{dt^2} = \boldsymbol{F}$. $\boldsymbol{F}$ is the aggregation of the following forces: gravity, viscous resistance, frictional force, and repulsive force from spring and muscle.

A *spring* is used as the structural material of this model. The spring force $\boldsymbol{F}_{\text{Spring}}$ is applied to the mass points bonded to its both ends, and is calculated as follows (Hooke's law):

$$\boldsymbol{F}_{\text{Spring}} = -k(L_0 - l(t)), \tag{1}$$

where L_0 is its natural length; $l(t)$ is the length at the current time t; k is the spring constant.

A *muscle* is used as the actuator of this model. The muscle is a spring whose natural length isn't fixed but changes periodically; i.e. defined as a spring that has a periodic function $L(t)$ as a substitute for the L_0 in Eqn.(1). So, the force from a muscle $\boldsymbol{F}_{\text{Muscle}}$ is given by

$$\boldsymbol{F}_{\text{Muscle}} = -k(L(t) - l(t)), \tag{2}$$

where $L(t)$ is defined as follows:

$$L(t) \equiv L_0 + \Delta L_0, \ \ \Delta L_0 \equiv L_0 \cdot R \cdot A \sin\left(\frac{2\pi}{T}t - \phi\right). \tag{3}$$

The term $A \sin(\cdots)$ represents the effect of the virtual oscillator. L_0 in Eqn.(3) is a constant that each muscle has, and is used as its natural length in the case the oscillator is off. In Eqn.(3), there are two oscillator-specific parameters; $A \in [0, 1]$:Amplitude and $T \in [0, \infty)$:Period, and two muscle-specific parameters; $R \in [0, 1]$:Relative amplitude and $\phi \in [0, 2\pi)$:Phase.

The robot for experiments

The upper left of Fig.2 shows the robot for experiments of this paper. This is a robot assumed as a quadruped animal. There are eight muscles used as legs: four from mass point 1 to 6, 1 to 7, 2 to 6 and 2 to 7 (referred to as *rear leg* later); another four from 3 to 8, 3 to 9, 4 to 8 and 4 to 9. The body consists of six mass points (5, $\cdots$, 10) and these mass points are interconnected with springs except between 6 and 9, and between 7 and 8.

EXPERIMENT : ACQUISITION OF BEHAVIOR WITH GA

Overview

In this section, Genetic Algorithms (GAs) are investigated in four different ways of settings. These settings of GAs are concerning the fitness functions and the designs of chromosome. GAs are applied to the robot model shown in Fig.2, then the acquired behaviors are observed. Parameters for GAs are listed in Table 1. These values are common in all experiments. Detailed description of settings are given below.

The task and the environment

Fig.2 shows the task environment. The meshed part is called "goal area". The robot is located on the bottom center of the environment at start. The task for the robot is "to carry its center of mass toward the goal area".

Genetic Algorithm (GA)

■Encoding scheme In this paper, the real-coded genetic algorithm (RCGA) is employed. A variety of genetic operations (crossover, mutation) for RCGA have been proposed so far (Herrera, 1998). We choose BLX-α (BLX:Blend Crossover) (Eshelman, 1993) from among them for the reason that it is simple but provides better performance. BLX-α has a feature that it has possibility to produce offsprings into the whole of search space without using any kind of mutational operation.

With the use of RCGA, it is able to use a real vector as a chromosome, directly. In this experiment, a real vector is 19-dimensional one made from parameters listed in Table 2. Table 2 also shows the search width of each locus.

■Selection scheme As is the case with genetic operation, many kinds of selection methods have been proposed (Mitchell, 1996). In this paper, we use the tournament selection in combination with the elitist strategy.

■The fitnenss function and three evaluation functions Three types of evaluation function are provided for the computation of the fitness function.

Distance evaluation : f_{Distance} evaluates the moving distance.

$$f_{\text{Distance}} = \begin{cases} 1 & (x_m > x_{\text{Goal}}) \\ 1 - \dfrac{x_{\text{Goal}} - x_m}{x_{\text{Goal}} - x_{\text{Start}}} & (x_m \le x_{\text{Goal}}) \end{cases}, \tag{4}$$

where x_m is the x coordinate value of the center of mass; x_{Start} is the x coordinate value of the start line; x_{Goal} is the x coordinate value of the left edge of the goal area.

Altitude evaluation : f_{Altitude} evaluates the altitude that is recorded by the center of mass in one trial.

$$f_{\text{Altitude}} = \begin{cases} 1 & (y_{\text{Altitude}} < y_{\text{Start}}) \\ 1 - \dfrac{y_{\text{Altitude}} - y_{\text{Start}}}{y_{\text{fieldHeight}} - y_{\text{Start}}} & (y_{\text{Altitude}} \ge y_{\text{Start}}) \end{cases}, \tag{5}$$

where $y_{\text{fieldHeight}}$ is the height of the field; y_{Altitude} is the maximum center-of-mass altitude while in trial; y_{Start} is the y coordinate value of the center of mass at start.

Time evaluation : f_{Time} evaluates the elapsed time.

$$f_{\text{time}} = 1 - \frac{t_{\text{finish}}}{t_{\text{observe}}}, \tag{6}$$

where t_{observe} is the trial time for each individual; t_{finish} ($0 < t_{\text{finish}} \le t_{\text{observe}}$) is the time reached to the goal.

EXPERIMENTAL RESULTS

For each experiment, 10 cycles of GAs are applied with different random seeds. The following results are based on the 1,000 individuals (= 100 individuals from 1 GA × 10 times) of the final generation.

Experiment 1

Under the conditions described above, GAs are applied with the fitness function:

$$fitness \equiv f_{\text{Distance}}. \qquad (7)$$

Although the population is evolved up until the last generation, no tendency is found on any parameter (e.g. convergence on a certain value). Many individuals in the last generation jump up and down, randomly. There are few individuals which reach to the goal. The upper side of Fig.3 shows typical behavior of individuals in the last generation. The reason why some individuals can reach the goal is considered that those individuals happened to jump toward its goal direction. Such individuals tend to emerge even in the early stage of evolution, this is why the speed of evolution is slow.

Experiment 2 : Adding Altitude Evaluation

As a result of experiment 1, which generated random jumping, the altitude evaluation is added into the fitness function:

$$fitness \equiv f_{\text{Distance}} \times f_{\text{Altitude}}. \qquad (8)$$

Individuals in this experiment obtain high fitness by evolving toward the direction which makes the spring constant k low. These individuals tend to relax its posture just after the start, and crawl into the goal. Such behavior ensures individuals to obtain $f_{\text{Altitude}} = 1.0$ because the altitude of the center of mass never go beyond the altitude at start. The middle part of Fig.3 shows typical behavior of individuals in this experiment.

Experiment 3 : Reducing the Search Space

In experiment 1 and 2, the robot cannot acquire well-ordered behavior (like walking). So in this experiment, limits are set on oscillator amplitude A, oscillator period T and spring constant k. The values of A, T, k are fixed as; $A = 0.3$, $T = 6$, $k = 125$. The dimension of the chromosome vector is reduced from 19 to 16. The fitness function used here is the same as experiment 2.

Many individuals in this experiment tend to walk with short strides and keep its height of the center of mass as constant as possible. This behavior seems similar to common quadruped gaits of walk. The bottom of Fig.3 shows typical behavior of individuals in this experiment.

Experiment 4 : Adding Time Evaluation

In the previous experiment, the robot acquired behavior similar to quadruped gaits. Followed by the result, this experiment is conducted to investigate the effect on behavior when fitness function is changed. The time evaluation function is newly added to the fitness function as follows:

$$fitness \equiv f_{\text{Distance}} \times f_{\text{Altitude}} \times f_{\text{Time}}. \qquad (9)$$

Many individuals in this experiment tend to run, rather than walking. For example, some individual shows gaits of trot or pace, which are faster than that of walk. And many other individuals show gaits of gallop, which is the fastest gaits for quadruped animals. Fig.4 shows gallop behavior acquired by individuals in this experiment.

DISCUSSION

In experiment 1 and 2, parameters A, T and k are also contained in the search domain. We found that various random behavior (jumping up and down, etc.) were emerged in experiment 1. And in experiment 2, unexpected behavior has emerged due to adaptation of spring constant k, which is a parameter associated with morphology of the robot. From these results, we should say that it may be difficult to converge behavior patterns if a GA is approximately with large search domain including some morphology parameters. In experiment 3 and 4, we found that acquired behaviors were changed from walking to running, with the change of combination of evaluation functions. From above we should say that it is an effective method for evolutionary acquisition of behavior to evolve robots using the fitness function built with simple evaluation functions.

In this paper, we used "product" for the combination method of evaluation functions. It was intended to generate robots which achieve high value for all combined evaluation functions. However, we can think of other different method; "addition" is one of the methods which is considered reasonable and proper to apply. This method correspond to the weighted sum strategy ($fitness = w_1 f_1 + w_2 f_2 + \cdots$), a technique for multiobjective optimization. Although we could simultaneously achieve high value in f_{Distance}, f_{Altitude} and f_{Time}, there will be cases where simultaneous optimization is impossible if other kind of evaluation functions are added. By regarding this problem as that of multiobjective optimization, it is possible to get appropriate individuals as Pareto-optimum solutions.

CONCLUSION

A series of experiments that investigate the process of evolution in behavior are conducted. Through the results, we found that:

- it may be difficult to converge behavior patterns if a GA is applied with large search domain including some morphology parameters.
- to acquire well-ordered behavior by evolution, it is an effective method to evolve robots using the fitness function built with simple evaluation functions.

We need more detailed analysis and verification of availability in the case design of morphology is included in search domain.

References

Eshelman, L. and Schaffer, J., 1993, "Real-coded genetic algorithms and interval-schemata", *Foundations of Genetic Algorithms 2*, pp.187-202, Morgan Kaufman Publishers.

Herrera, M., Lozano, M., and Verdegay, J. L., 1998, "Tackling Real-Coded Genetic Algorithms: Operators and Tools for Behavioural Analysis" *Artificial Intelligence Review*, volume 12, number 4, pp.265-319.

Mitchell, M., 1996, "An Introduction to Genetic Algorithms" MIT Press.

Sims, K., 1994a, "Evolving Virtual Creatures" *Computer Graphics (Siggraph '94 Proceedings)*, pp.15-22.

Sims, K., 1994b, "Evolving 3D Morphology and Behavior by Competition" *Proceedings of the 4th International Workshop on the Synthesis and simulation of Living Systems (ALIFE IV)*, pp.28–39.

Soda Creative Technologies Limited, 2000, *sodaplay*. `http://sodaplay.com/`

Ventrella, J., 1995, "Disney Meets Darwin" *IEEE Computer Animation '95 Proceedings*.

Table 1: Setting of GA

maximum generation	100
population size	100
trial time for each individuals	100
crossover rate	0.5
winning percentage of the better individual	0.75
constant value for defining the search area of BLX-α	0.5

Table 2: Layout and description of the chromosome and search width of each locus

locus	parameter	search area
$1, \cdots, 8$	phase of each muscle $\phi_1, \cdots, \phi_8$	$[0, 2\pi)$
$9, \cdots, 16$	relative amplitude of each muscle $R_1, \cdots, R_8$	$[0, 1]$
17	spring constant k	$[0, 125]$
18	period of oscillator T	$[0, 60]$
19	amplitude of oscillator A	$[0, 1]$

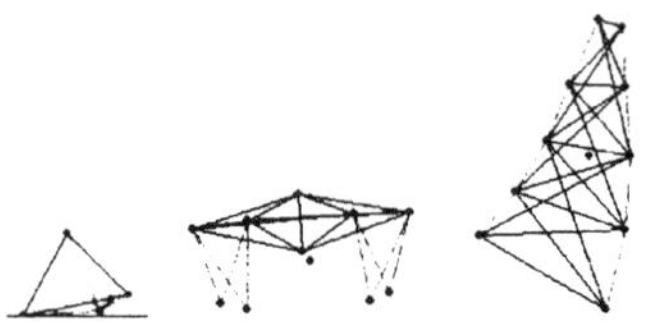

Figure 1: Robots based on spring-mass system

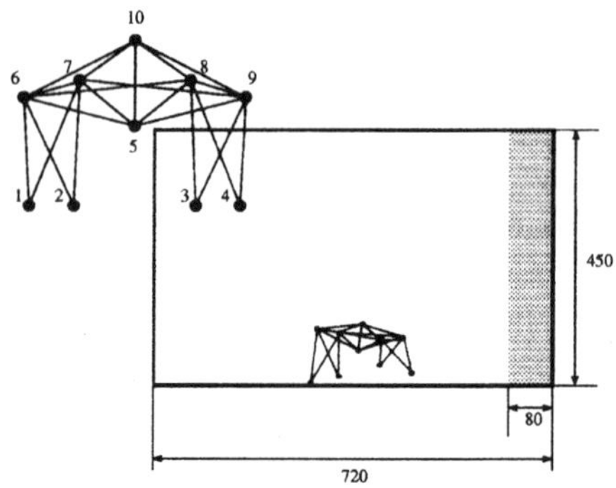

Figure 2: Robot, field and goal(shaded area)

Experiment 1 (*fitness* $\equiv f_{\text{Distance}}$) ; jump up and down, randomly.

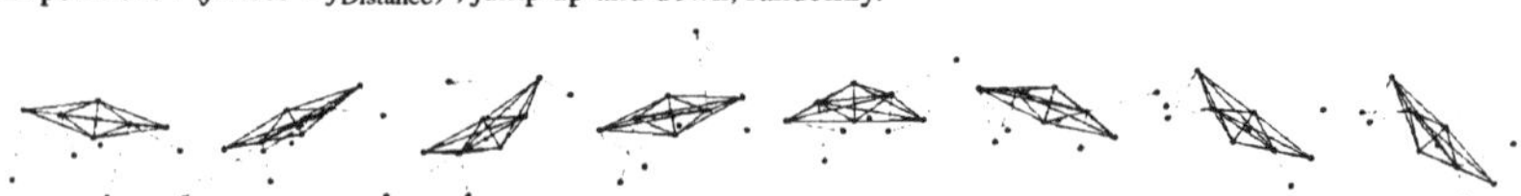

Experiment 2 (*fitness* $\equiv f_{\text{Distance}} \times f_{\text{Altitude}}$) ; relax its posture just after the start, and crawl into the goal.

Experiment 3 (*fitness* $\equiv f_{\text{Distance}} \times f_{\text{Altitude}}$, with reduced search area) ; walk with short strides.

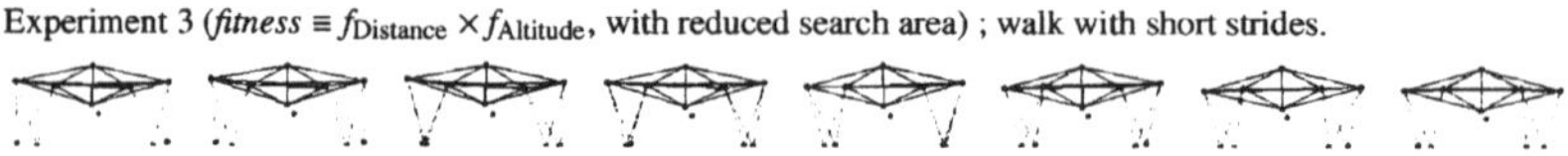

Figure 3: Typical behaviors in experiment 1-3

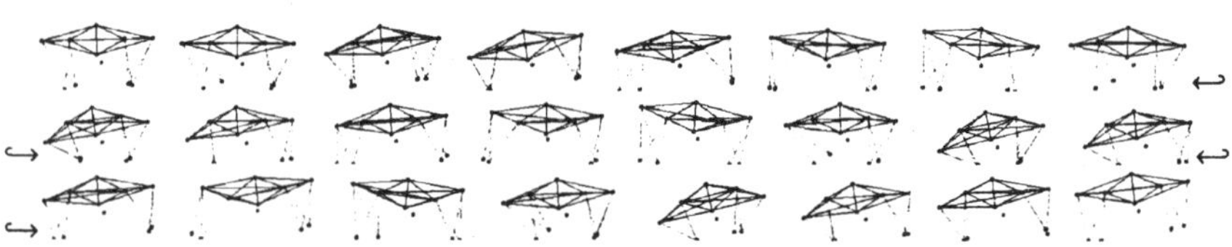

Figure 4: Typical behavior in experiment 4

CREATIVE CONTROL FOR INTELLIGENT AUTONOMOUS MOBILE ROBOTS

XIAOQUN LIAO MASOUD GHAFFARI
SOUMA M. ALHAJ ALI ERNEST L. HALL
Center for Robotics Research, University of Cincinnati, Cincinnati, Ohio

ABSTRACT
For intelligent robots to accomplish tasks in an unstructured environment, the adaptive critic algorithm has been shown to provide useful approximations or even optimal control policies to non-linear systems. The purpose of this paper is to explore the use of new learning control methods defined as Creative Learning or Creative Control that goes beyond the adaptive critic method for unstructured environments. The creative controller like the adaptive critic controller has information stored in a dynamic database (DB), plus a dynamic task control center (TCC) that functions as a command center to decompose tasks into sub-tasks with different dynamic models and multi-criteria functions. The simulation results based on adaptive critic learning are discussed in this paper. The significance of this paper is to better understand the adaptive critic learning theory and move forward to develop more human-intelligence-like components into the intelligent robot controller. Moreover, it should extend to other applications.

1. INTRODUCTION

Hall (1985) defines an intelligent robot as one that responds to changes to its environment through sensors connected to a controller. Dynamic Programming (DP) is perhaps the most general approach for solving optimal control problems. Adaptive Critics Design (ACD) offers a unified approach to deal with the controller's nonlinearity, robustness, and reconfiguration for a system whose dynamics can be modeled by a general ordinary differential equation. ANN and Backpropagation (BP) made it possible for ACD implementation (Werbos, 1994). However, in order to develop "brain-like intelligent control" (Werbos, 1999), it is not enough to just have the adaptive critic portion. Here we proposed a novel algorithm, called Creative Learning (CL) to fill this gap as discussed in Section 4.

2. NEUROCONTROLLER AND INTELLIGENT CONTROL THEORY

In order to design intelligent robot controllers, one must also provide the robot with a means of responding to problems in both temporal and spatial context. It is the goal of the robot researcher to design a neural learning controller to utilize the available data from the repetition in robot operations. The neural learning controller in Figure 1, based on the recurrent network architecture, has the time-variant feature that once a trajectory is learned, it should learn a second one in a shorter time.

ANN can be used to obtain the system model identification that can be used to design the appropriate intelligent robot controller. Once the real system model is available, they can also be used directly in design of the controller

(Narendra, etc., 1990). Neural network approaches to robot control can be classified as: supervised control, direct inverse control, neural adaptive control, backpropagation of utility involves information flowing backward through time, and adaptive critic method that uses a critic evaluating robot performance during training.

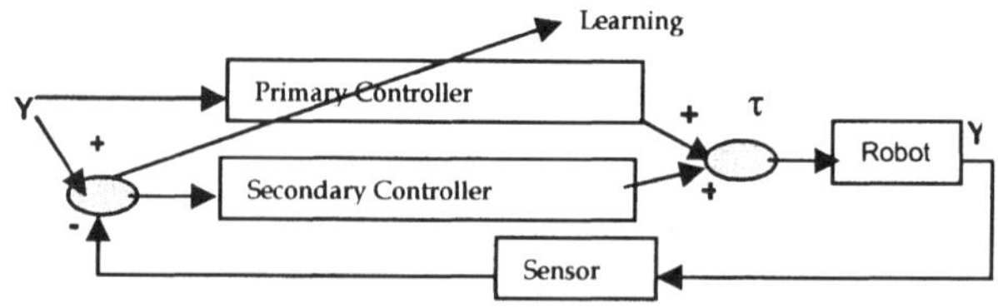

Figure 1 Recurrent neural learning controller

3. ADAPTIVE CRITIC LEARNING

Werbos (1995) summarized recent accomplishments in neurocontrol as a "brain-like" intelligent system. It should contain at least three major general-purpose adaptive components: (1) an Action or Motor system, (2) an "Emotional" or "Evaluation" system or "Critic" and (3) an "Expectations" or "System Identification" component.

"Critic" serves as a model of the external environment to be controlled, solving an optimal control problem over time may be classified as adaptive critic designs (ACD). ACD is a large family of designs which learn to perform utility maximization over time. In dynamic programming, normally the user provides the function $U(\underline{X}(t), \underline{u}(t))$, an interest rate r, and a stochastic model. Then the analyst tries to solve for another function $J(\underline{X}(t))$, so as to satisfy some form of the Bellman equation shown in Equation (1) that underlies dynamic programming (Werbos, 1999):

$$J(\underline{X}(t)) = \max_{\underline{u}(t)}(U(\underline{X}(t),\underline{u}(t)) + <J(\underline{X}(t+1))>/(1+r)) \qquad (1)$$

Where "<>" denotes expected value.

In principle, any problem in decision or control theory can be classified as an optimization problem. Many ACDs solve the problem by approximating the function J. The most popular methods to estimate J in ACDs are heuristic dynamic programming (HDP), Dual Heuristic Programming (DHP) and Globalized DHP (GDHP) (Werbos, 1995) (Prokhorov, 1997). HDP and its ACD form have a critic network that estimates the function J (cost-to-go or strategic utility function) in the Bellman equation of dynamic programming, presented as follows:

$$J(t) = \sum_{k=0}^{\infty} \gamma^k U(t+k) \qquad (2)$$

Where γ is a discount factor ($0<\gamma<1$), and U(.) is the utility function or local cost. An alternative approach referred to as Dual Heuristic Programming (DHP) has been proposed. Here, the critic network approximates the derivatives of the future cost with respect to the state variable. It has been proved that DHP is capable of generating smoother derivatives and has shown improved performance when compared to HDP (Lendaris, 1999) (Venayagamoorthy, 2002). Werbos (1992) first proposed the idea how to do GDHP. Training the critic network in GDHP utilizes an error measure which is a combination of the error measures of HDP and DHP.

4. Creative Learning and Control

"Creative Learning" provides a model, for understanding the kind of intelligence that exists in biological brains. A creative control architecture as shown in Figure 2 is proposed in this paper accoding to the creative learning theory (Liao, 2003). In this proposed diagram, there are three important components: task control center, criteria (critic) knowledge database, and learning system. Adaptive critic learning method is a part of the creative learning algorithm. However, creative learning with decision-making capabilities is beyond the adaptive critic learning. The most important characteristics of the creative learning structure are: (1) Brain-like decision-making task control center, entails the capability of human brain decision-making; (2) Dynamic criteria database integrated into the critic-action framework, makes the adaptive critic controller reconfigurable and enables the flexibility of the network framework; (3) Multiple criteria, multi-layered structure; (4) Modeled and forecasted critic modules result in faster training network. A creative controller is designed to integrate the domain knowledge and task control center into the adaptive critic controller. It needs to be a well-defined structure such as in the autonomous mobile robot application as the test-bed for the creative controller.

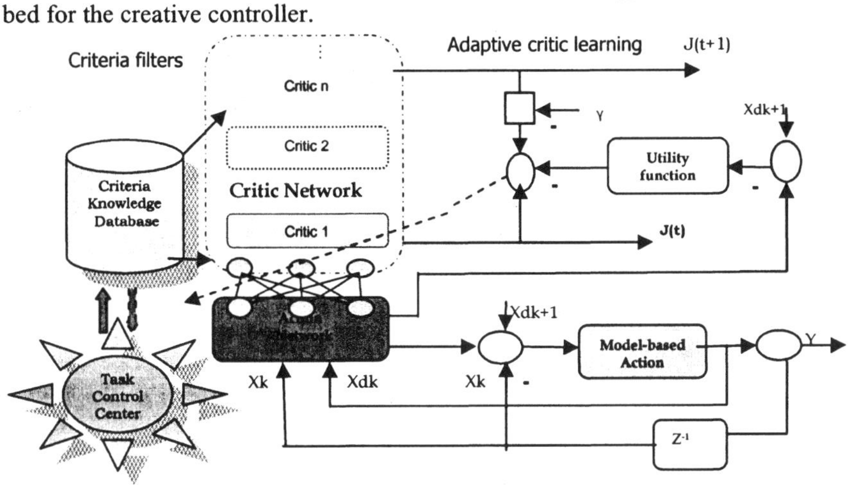

Figure 2 Proposed CL Algorithm Architecture

4.1 Task Control Center (TCC)

The task control center (TCC) can build task-level control systems for the creative learning system as shown in Figure 3. By "task-level", we mean the integration and coordination of perception, planning and real-time control to achieve a given set of goals (tasks). TCC provides a general task control framework, and it is intended to be used to control a wide variety of tasks and permits responsive actions based on mission commands, on interactions with other robots. Although TCC has no built-in control functions for particular tasks (such as robot path planning algorithms), it provides control functions, such as task decomposition, monitoring, and resource management, that are common to many applications. The particular task built-in rules in the dynamic database matches the constraints on particular control schemes or sub-tasks or environment allocated by TCC. The task control center acts as a decision-making system. It integrates domain knowledge or criteria into the database of the adaptive learning system. According to Simmons (1999), task control architecture for mobile robots provides a variety of control constructs that are commonly needed in mobile robot applications, and other autonomous mobile systems. Integrating TCC with adaptive critic learning system and interacting with the dynamic database, the creative learning system could provide both task-level and real-time control or learning within a single architectural framework.

4.2 Dynamic Knowledge Database (DKD)

Dynamic knowledge database serves as a "coupler" between a task control center and a nonlinear learning system. The purpose of the coupler is to provide the criteria functions for the adaptive critic learning system and filter the task strategies commanded by the task control center. The proposed dynamic database contains a copy of the model (or identification). Action and critic networks are utilized to control the plant under nominal operation, as well as make copies of a set of HDP or DHP parameters (or scenarios) previously adapted to deal with a plant in a known dynamic environment. It also stores copies of all the partial derivatives required when updating the neural networks using backpropagation through time (Yen, etc., 2002).

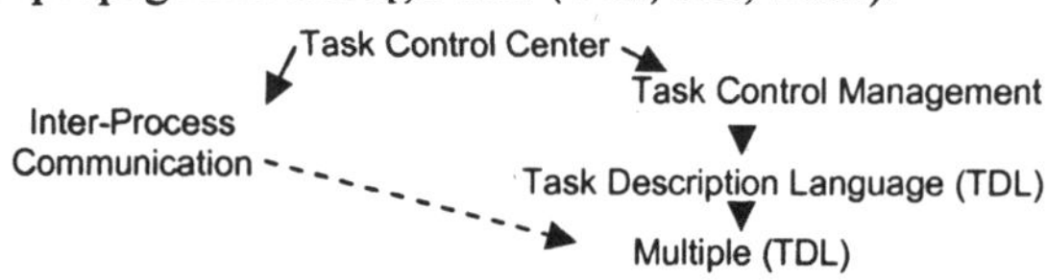

Figure 3 The structure of task control center

The data stored in the dynamic database can be uploaded to support offline or online training of the dynamic plant and provide a model for identification of nonlinear dynamic environment with its modeling function. Another function module of the database management is designed to analyze the data stored in the database including the sub-task optima, pre-existing models of the network and newly added models. The task program module is used to communicate with the task control center. The functional structure of the proposed dynamic database is shown in Figure 4 and 5.

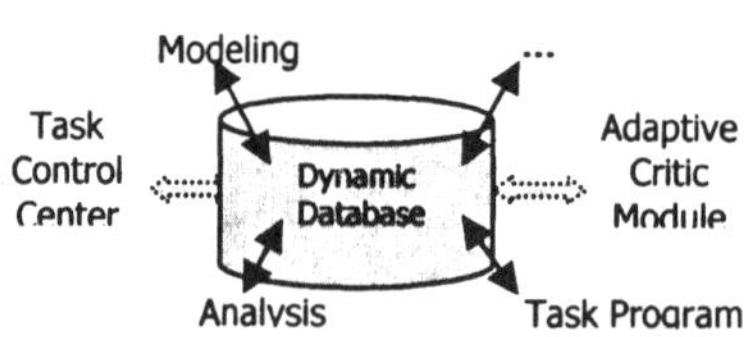

Database fields	
MODEL_ID	Action model ID
CRITERIA_FUN	Criteria function
Adaptive Critic Training Parameters	
INPUT_CRITIC	Input to critic network
DELT_J	J(t+1)-J(t)
...	...

Figure 4 Functional structure of dynamic database Figure 5 Semantic database structure

4.3 Creative Control Mobile Robot Example

Suppose a mobile robot is used for urban rescue as shown in Figure 6. Since it is in an urban environment, it must use the established roadways. Along the road ways it can follow pathways. However, at intersections, it can choose various paths to go to the next block. Therefore, it must use different criteria at the corners. The overall goal is to arrive at the rescue site with minimum distance or time in order to rescue people. This example requires the use of both continuous and discrete tracking, a database of known information and multiple criteria optimization. It is necessary to add a large number of real-world issues including position estimation, perception, obstacles avoidance, etc.

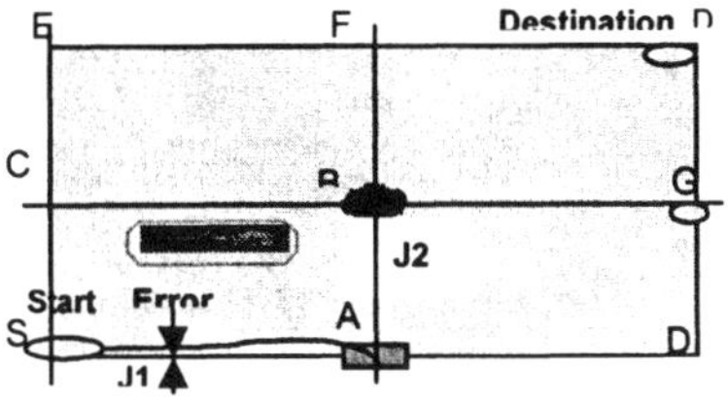

Figure 6 Simple urban rescue sites

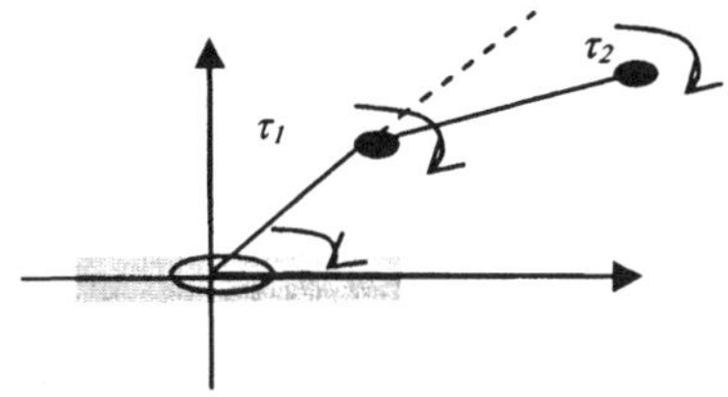

Figure 7 Two-link robot arm manipulator

As discussed in Section 4.1, the task control center (TCC) acts a decision-making command center. It takes perception information from sensors and other inputs to the creative controller and derives the criteria functions. It's appropriate to assume that J1 and J2 are the criteria functions that the task control center will transfer to the learning system at the beginning of the mission from the Start point (S) to Destination (D). The simplest style HDP in ACDs can be used. More discussion is presented in Liao's dissertation (Liao, 2003).

4.4 Creative Controller and Simulation Results

As an ANN robot controller, the block diagram of the creative controller can be presented as shown in Figure 8. A creative controller is integrated into the intelligent robot learning system. As an expermental study, we used a two-link robot arm maniputlator shown in Figure 7 as one simulation example for adaptive critic learning. The dynamics of the two-link robot arm manipulater is shown as Equation 3. The simulation results are shown in Figure 9. When training with the adaptive crtic system, it is more stable and the tracking errors are smaller. The furthur results present a comparison with the adaptive critic learning techiques and more case studies for the system shown in Eq. 3 (Liao, 2003).

$$M(q)\ddot{q} + V(q,\dot{q}) + F(q,\dot{q}) + G(q) + \tau_d = \tau \qquad (3)$$

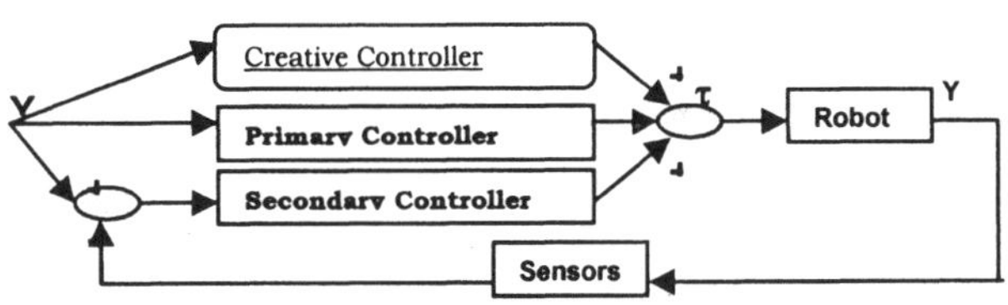

Figure 8 Creative controller structure

5. CONCLUSION

A creative learning theory proposed in this paper includes all the components in the adaptive critic family, which generalizes adaptive critic learning by modifying the learning rates and utilizing multiple critics and providing a model-based action database. A creative controller can be used to explore an unpredictable environment, and permit the discovery of unknown problems. By learning the domain knowledge, the system should be able to obtain the global optima and escape local optima. The creative controller for intelligent robots like the adaptive critic controller has information stored in a dynamic database, plus a dynamic task control center that functions as a command center to decompose tasks into sub-tasks with different dynamic models and criteria functions. Improvements such as those shown in Figure 9 have been achieved.

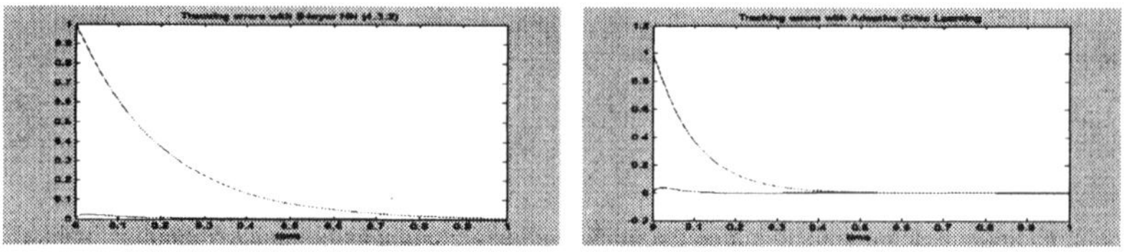

Figure 9 (a) Tracking error with NN (b)Tracking error with AC (t=1sec)

REFERENCES

Hall, E.L., and Hall, B. C. (1985), *Robotics: A User-Friendly Introduction*, pp. 1-8, Saunders College Publishing, Holt, Rienhart and Wilson, Orlando FL.

Liao, X., Creative Learning for Intelligent Autonomous Mobile Robots, PhD Dissertation, University of Cincinnati, 2003.

Lendaris, G.G., Shannon, T.T. and Rustan, A. (1999), A Comparison of Training Algorithms for DHP Adaptive Critic Neurocontrol, Neural Networks, 1999. IJCNN '99. International Joint Conference on, pp: 2265 -2270, v. 4.

Narendra, K.S., and Parthasarathy, K., (1990) Identification and control of dynamical systems using Neural networks, *IEEE Transactions on Neural Networks*, vol. **1(1)**, pp. 4-27.

Prokhorov, D., and Wunsch, D., (1997) Adaptive Critic Designs, *Neural Networks, IEEE Transactions on*, v. 8, n.5, pp.997-1007.

Simmons, R., Task Control Architecture, <http://www.cs.cmu.edu/afs/cs/project/TCA/www/TCA-history.html

Venayagamoorthy, G. K., Harley, R. G. and Wunsch, D. C., (2002) Comparison of Heuristic Dynamic Programming and Dual Heuristic Programming Adaptive Critics for Neurocontrol of a Turbogenerator, *IEEE Transactions on Neural Networks*, v. 13, no. 3, pp764-773.

Werbos, P. (1994), *The Roots of Backpropagation: From Ordered Derivatives to Neural Networks and Political Forecasting*, Wiley.

Werbos, P.J. (1999), Tutorial on Neurocontrol, Control Theory and Related Techniques: From Backpropagation to Brain-Like Intelligent Systems, the Twelth International Conference on Mathematical and Computer Modelling and Scientific Computing (12th ICMCM & SC).

Werbos, P., (1995) Optimal neurocontrol: practical benefits, new results and biological evidence, Wescon Conference Record, p.580-585.

Werbos, P.J. (1992), Approximate Dynamic Programming for Real-Time Control and Neural Modeling, In White D, A and D.A. Sofge, editors, *Handbook of Intelligent Control*, pages 493--525. Van Nostrand Reinhold.

Yen, G.G., and Lima, P. G., (2002) Dynamic Database Approach for Fault Tolerant Control Using Dual Heuristic Programming, Proceedings of the American Control Conference, Anchorage, pp.5080-5085.

MODAL ANALYSIS AND ACTIVE VIBRATION CONTROL OF A MULTI-DEGREE-OF -FREEDOM STRUCTURE

RANJEET AGARWALA
Texas A&M University-Kingsville
Mechanical Engineering
Kingsville, Texas.

ROBERT A. MCLAUCHLAN
Texas A&M University-Kingsville
Mechanical Engineering
Kingsville, Texas.

SELAHATTIN OZCELIK
Texas A&M University-Kingsville
Mechanical Engineering
Kingsville, Texas.

MOHAMMED FARUQI
Texas A&M University-Kingsville
Civil Engineering
Kingsville, Texas.

ABSTRACT

In this paper, the problem of active vibration control of multi-degree-of-freedom high-rise buildings is considered. The responses of each floor depend on the natural frequencies and the mode shapes of the structure. Each floor of the structure is modally analyzed and decoupled. The system is then subjected to earthquake vibrations and wind effects, which are treated as disturbances. Fuzzy gain scheduling of PID controller is implemented on the multi-degree-of-freedom structure, which is modeled as composing of spring-mass-damper subsystems to obtain satisfactory vibration suppression.

INTRODUCTION

Active vibration control has been closely studied and has been given a lot of importance since last several years. The theory of active vibration control has been extensively used for scientific research and commercial fields for the control of small as well as large structures (Wu et al., 1993). Large structures like high rise buildings, bridges, communication towers, space stations, etc are complex structures and hence require regulation and control of vibration due to disturbances, to ensure continued performance under adverse environmental uncertainties. Large structures are referred to as multi-degree-of-freedom structures due to their being composed of various subsystems and due to the complex nature of interconnectivity (Iwai et al., 1993) between the subsystems.

A high-rise building has been categorized as a multi-degree-of-freedom structure and many researchers have proposed techniques to minimize the vibrations, due to external disturbances that crop up during its performance. Designing a high-rise building for any type of usage requires careful consideration of all probable and possible environmental uncertainties and their intensity modes. Hence, the strategies used to actively control the vibration takes into account the above-mentioned design conditions to ensure safe and sound operation over a period of time. Active vibration control is one in which the controller is centralized and the energy supplied to it from an external source to suppress the vibrations due to external disturbances. LQ, H_∞ (Cui et al., 1992), etc. controls have been deployed as centralized controllers for multi-degree-of-freedom structures.

Hino, Iwai, Mizumoto and Kohzawa (Iwai et al., 1993) in their work have discussed the problem of active vibration control of a multi-degree-of-freedom structure, using robust decentralized simple adaptive control (RDSAC). The high-rise building was treated as a five-storied building and modeled as a spring- mass- damper system. The effect of the wind forces (Yoshida et al., 1988) treated as disturbances was considered on the second, third, fourth and the fifth floor. However, actual engineering standards require inclusion of the effect of the wind on the first floor, while designing any high rise building. In addition to that, they did not take into account shear force that the high rise building is subjected to, during an earthquake, distributed as lateral forces on every floor. Agarwala, Ozcelik and Faruqi (Agarwala et al., 2000) have further remedied the situation by considering the effect of earthquake shear force on each floor and the effect of the wind on the first floor. However in this paper a knowledge- based system is applied for vibration control.

The application of knowledge-based systems in process control is growing, especially in the fields of fuzzy control. In fuzzy control, linguistic description of human expertise in controlling a process are represented as fuzzy rules or relations. This knowledge base is used by an inference mechanism, in conjunction with some knowledge of the states of the process (say, of measured response variables) in order to determine control-actions. In this paper, fuzzy gain scheduling of PID (Zhao et al., 1993) controllers is utilized for vibration suppression. Although the controller is apparently not structured as PID controllers, fuzzy logic controllers is considered as a nonlinear controller whose parameters are determined on-line based on the error signal and their time derivative or difference. Modal Analysis is a computationally elegant technique for modeling structural dynamics. It is based on the use of eigenvalue and Eigenvector information of a system. The elegance and appeal of this technique is mainly due to its decoupling capability.

MODELING OF THE MULTI-DEGREE-OF FREEDOM-STRUCTURE (FIVE STORIED BUILDING)

The five-storied structure is shown in Fig 1. This structure consists of 5 floors (subsystems) and is subjected to wind disturbances on each floor, disturbances due to earthquake and shocks from the ground. An active control device (actuator), which supplies energy to suppress vibrations, is installed on first and the fifth floors respectively.

An important element of an active vibration control strategy is the actuator. As the name implies, these are active control devices (Iwai et al., 1993) that expend energy to attenuate disturbances at the corresponding subsystems or reduce the disturbances at the corresponding subsystems or reduce the vibration on floors at which they are installed.

Here we use two types of actuator as follows:

i) An active vibration isolator supported at a fixed point is used. It supplies control force directly to make the magnitude of undesirable vibration small. Since it requires support at a fixed point, it is installed on the first floor.

ii) Active dynamic vibration absorbers are used to supply inertia force to the floor. Here we realize such an effect by moving a mass on the floor.

The structural state-space model, based on (Iwai et al., 1993) of the building described in the Fig 1 will be considered in the following. The notations used here are as follows.

Subsystems that have actuators are denoted by the set 'N'.

Subsystems without actuators are denoted by the set 'M'.

Using the above notations, without loss of generality, the following set of equations can be obtained for each subsystem

Subsystem with actuator (S_i, $i \in N$)

$$\begin{cases} \dot{x}_i(t) = A_i x_i(t) + B_i u_i(t) + B_i L_i^T(t) x_{i-1}(t) + B_i L_{i+1}^T x_{i+1}(t) + T_i(t) \\ y_i(t) = C^T x_i(t) \end{cases} \tag{1}$$

Subsystem without actuator (S_i, $i \in M$)

$$\begin{cases} \dot{x}_i(t) = A_i x_i(t) + B_i L_i^T(t) x_{i-1}(t) + B_i L_{i+1}^T x_{i+1}(t) + T_i(t) \\ y_i(t) = C^T x_i(t) \end{cases} \tag{2}$$

A_i , B_i , C are the state space form of the subsystems of the plant. L_i is the interconnection factor affecting the adjacent subsystems. $T_i(t)$ is the disturbance that each subsystem is subjected to.

Here,

$$x_i(t) = \begin{bmatrix} y_i(t) \\ \dot{y}_i(t) \end{bmatrix}$$

$y_i(t)$ is the displacement of the i-th floor

The control force $F_i(t)$ generated by the i-th actuator is given by:

$$u_i(t) = \begin{cases} F_i(t), i = 1 \\ -F_i(t), i \neq 1 \end{cases} , \quad i \in N$$

A_i, B_i, and C matrices are given by:

$$A_i = \begin{bmatrix} 0 & 1 \\ -\dfrac{k_i + k_{i+1}}{m_i} & -\dfrac{c_i + c_{i+1}}{m_i} \end{bmatrix}$$

$$B_i = \begin{bmatrix} 0 \\ \dfrac{1}{m_i} \end{bmatrix}$$

$$C = \begin{bmatrix} 1 \\ 0 \end{bmatrix}$$

k_i : is the spring constant of the i-th floor

m_i : is the mass of the i-th floor

c_i : is the damping coefficient

$\hat{T}_i(t)$ represents the disturbances that the system is subjected to.
It represents earthquake vibrations, earthquake shear force and wind forces.
It is given by:

$$\hat{T}_i(t) = \begin{cases} L_1^T \left[e(t), \dot{e}(t) \right] , & i = 1 \\ H_i + S_i , & i = M, N \end{cases} \qquad L_i = \begin{bmatrix} k_i \\ c_i \end{bmatrix}$$

Here,

$T_i(t)$ is the disturbance factors entering the i-th floor.

L_i is the spring damper matrix augmenting the disturbance and interconnection of the i-th floor.

$e(t)$ is the earthquake disturbance on the first floor.

H_i is the wind force on the i-th floor

S_i is the earthquake shear force on the i-th floor

SIMULATION RESULTS

The system is simulated to demonstrate its ability to damp out vibrations, when subjected to earthquake shocks and wind load. The plant and actuator parameters are given in Tables 1 and 2. The system is subjected to earthquake vibration with an amplitude and frequency of 2 m and 1.6 Hz respectively.

Table 1: Plant parameters

Floor	1	2	3	4	5
Mass m_i (*kg*)	17500	9750	9750	9750	9750
Spring constant k_i (*kg*)	17500	9750	9750	9750	9750
Viscous damping coeff. C_i (*Ns/m*)	100	70	70	70	70

Table 2: Actuator parameters

Floor	Nominal Values
1st	k_a=12000(*N/m*), f_{a1}=0.567(1/s), f_{a2}=0.013(*m/s/v*)
5th	m_d=0.817(*Kg*), J_m=0.05(*kg.m²/s²*), K_m=10(*Nm/Amp*)

The initial states of the system are given by:

$$x_0(t),\ x_{n+1}(t) = 0\ k_{n+1} = 0,\ L_{n+1} = 0,\ c_{n+1} = 0 \tag{3}$$

The system is modally analyzed and simulated with the above given parameters. The natural frequency matrix of the system is given by:

$$\Omega = \begin{bmatrix} .0287 & 0 & 0 & 0 & 0 \\ 0 & 0.1264 & 0 & 0 & 0 \\ 0 & 0 & 0.1651 & 0 & 0 \\ 0 & 0 & 0 & 0.2659 & 0 \\ 0 & 0 & 0 & 0 & 0.4198 \end{bmatrix}$$

The eigenvector matrix, representative of mode shapes of the system is given by:

$$\Phi = \begin{bmatrix} 0.0021 & 0.5607 & 0.7602 & 0.4489 & 0.1056 \\ 0.0035 & 0.5803 & 0.4019 & -0.6376 & -0.6492 \\ 0.0045 & 0.4390 & -0.0192 & -0.6109 & 0.7515 \\ 0.0053 & 0.1810 & -0.4315 & 0.1348 & -0.0513 \\ 0.0058 & -0.3513 & 0.2720 & -0.0236 & 0.0033 \end{bmatrix}$$

Fig. 2 shows the mode shapes of the five-storied building. In Fig. 3, the inner curve indicates the controlled response of the system and the outer curve indicates uncontrolled response of the system. The fuzzy rules are applied taking into consideration the individual mode shape values of the system.

CONCLUSION

In this paper we implemented fuzzy gain scheduling of PID controllers to actively control the vibration of the multi-degree- of -freedom structure. The modeling of the system takes into account the wind effect on each floor and the earthquake shear force. The structure was modeled in terms of spring, mass, and damper subsystems. The simulation results exhibit that, the implementation of the control algorithm shows a good response in absorbing the vibration due to earthquake shear and wind force.

REFERENCES

Wu A.C., Fu L. C. and Hsu C. F., 1993, "A New Decentralized Model Reference Adaptive Control for a class of Interconnected Dynamic Systems Using Variable Structure Design, " *International Journal of Adaptive Control and Signal Processing*. Vol.7,pp. 33-43, 1993.

Agarwala R., Ozcelik S., Faruqi M., 2000, " Active vibration control of a multi-degree-of-freedom structure by the use of Direct Model Reference Adaptive Controller, " *American Control Conference*, Chicago, IL.

Yoshida K., Shimogo T., and Nishimura H., 1988, " Optimal control of Random Vibration by the use of an active Dynamic Vibration Absorber," *JSME International Journal*, Series III, Vol.31, No.2, pp.387-394.

Cui W., Nonami K., Nishimura H. and Shiba A., 1992, "Experimental study on H_∞ Control and H_2 Control for Vibration Systems with Multi-Degree-of-Freedom", *Proceedings of the 31st SICE Annual Conference*, DS55,pp-589-590.

Iwai Z., Hino M., Nakazano T. and Wakamiya R., 1993, "Active vibration control of a Multi-Degree-of-Freedom Structures Using Decentralized Simple Adaptive Control, "*Proceedings of the Asia-Pacific Vibration Conference*, Vol.3, pp.1078-1083.

Zhao, Tomizuka, Isaka, 1993," Fuzzy Gain Scheduling of PID Controllers",*IEEE Transactions on systems, Man, and Cybernetics*.

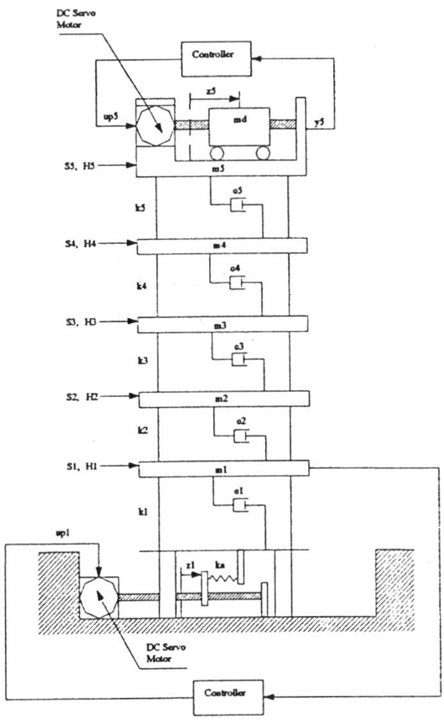

Figure. 1. Configuration of the System

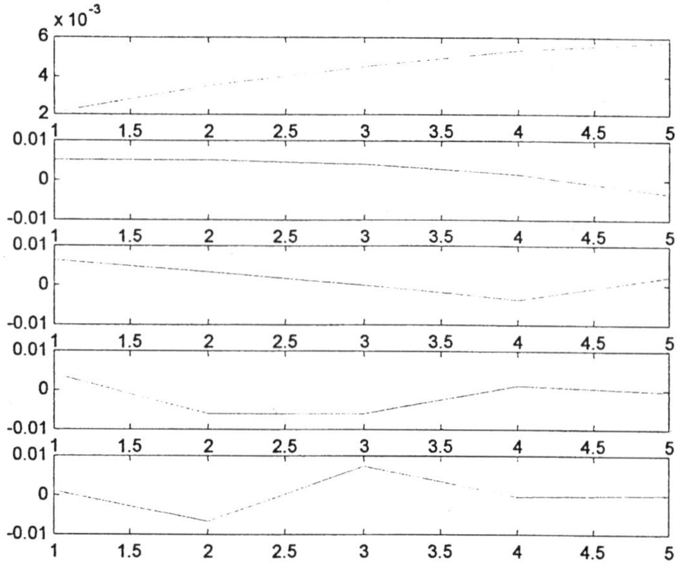

Figure. 2. Mode shapes of the System

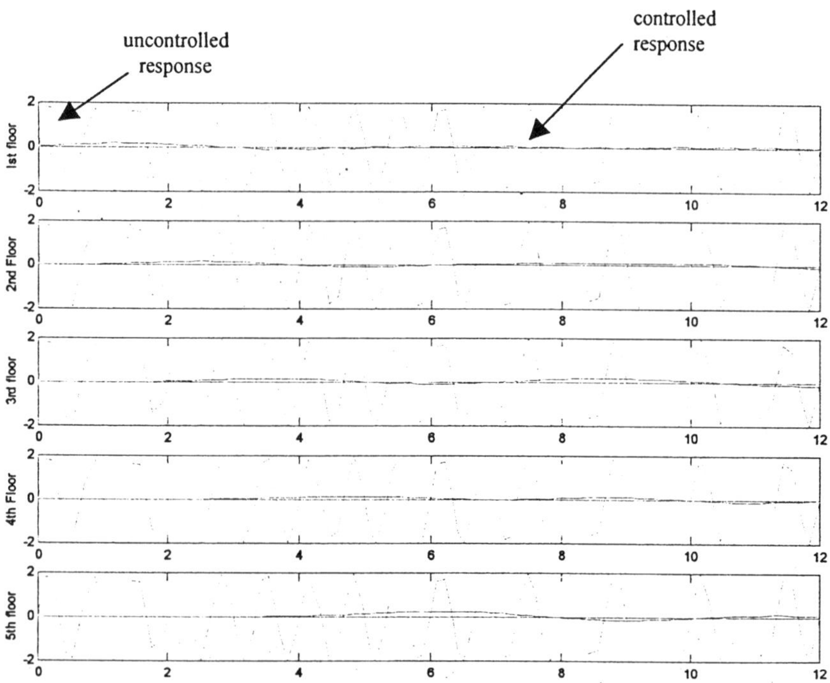

Figure. 3. Simulation results of the System

NONLINEAR FAULTS ISOLATION AND MULTIVARIATE PROCESS MONITORING – BASED ON NEURAL NETWORKS METHOD

FENG ZHANG
Department of Industrial Engineering
Texas A&M University
College Station, TX 77843-3131

ABSTRACT
As a linear method, Principal Component Analysis (PCA) is not accurate when nonlinear correlations are involved in the data structure. Principal curves method, which was proposed as the generalization of PCA, is shown to be an alternative to nonlinear feature extraction and dimension reduction. A nonlinear blind source separation method is proposed in this paper to separate multiple nonlinear spatial variation patterns in the manufacturing processes. Neural networks technique is incorporated into the nonlinear faults isolation and feature extraction for developing a multivariate process monitoring mechanism. A real example has demonstrated its utilization in complicated multivariate processes diagnosing and quality improvement.

INTRODUCTION

Though PCA has been extensively applied, its applicability is cumbered by its linearity (Hastie 1989; Chang 1998; Roweis 2000). In many applications, PCA is not accurate for nonlinear feature extraction and data description. For example, some minor components are not always unimportant or irrelevant, as the case in which they can be interpreted as the noise or lower variance. So discarding such minor components would cause the loss of some important information among the multivariate data. Consequently, nonlinear PCA methods have been shown to be alternatives to such situations (Dong 1995; Hsieh 2001).

Autobody assembly is one typical complex manufacturing system, where advanced in-process sensing stations are implemented among the assembly sages for measuring the deviations from nominal positions for each fixture location point. The measurements from car to car often exhibit some nonlinearities in the relationship between the coordinates. Therefore, given the measurement data for the assembled parts collected from the manufacturing processes, we would like to identify and investigate the underlying variation sources governing such variability among products. To achieve such tasks, it is natural to represent the nonliear variation pattern among the products first.

Principal Curves presented as the nonlinear generalization of principal component analysis, are smooth 1-D curves that pass through the middle of a *d*-D data set (Hastie 1989). Principal curves extend PCA by fitting a smooth curve f parameterized by arc length t_f, a projection index mapping from an observation to a point on curve f, i.e., $t_f(x) = \sup\{t : \|x - f(t)\| = \inf_s \|x - f(s)\|\}$. And a smooth curve is called a principal curve if $E[x \mid t_f(x) = t] = f(t)$. An iterative algorithm for finding such a principal curve alternates between estimating f with t fixed and estimating t with f fixed. When f is defined to be linear, the above principal curves algorithm is equivalent to extracting the first principal component line (Hastie 1989).

NONLINEAR VARIATION PATTERNS ISOLATION AND RECOGNITION

An appropriate model will be described to express the effects of the random sources on the measured product characteristics through some nonlinear variation patterns. This can be viewed as an extension of linear variation pattern analysis for manufacturing processes (Apley 2001).

Let $\boldsymbol{x} = [x_1, x_2, \ldots, x_d]^{\mathrm{T}}$ be a random vector that represents a set of d measured product characteristics. Model (1) below just presents such multiple nonlinear spatial variation patterns

$$\boldsymbol{x} = \boldsymbol{f}^1(v^1) + \boldsymbol{f}^2(v^2) + \ldots + \boldsymbol{f}^p(v^p) + \boldsymbol{w} \tag{1}$$

where $\boldsymbol{v} = [v^1, v^2, \ldots, v^p]^{\mathrm{T}}$ is a zero-mean random vector representing p distinct variation sources, and the variation pattern $\boldsymbol{f}^m(v^m)$ $(m = 1, 2, \ldots, p)$ are generated by scalar source v^m. Noise $\boldsymbol{w}$ is independent of $\boldsymbol{v}$ and represents the aggregated effects of measurement noise. For notation convenience, it will be assumed that the covariance matrix of $\boldsymbol{w}$ is $\boldsymbol{\Sigma}_w = \sigma^2\mathbf{I}$, where σ^2 is a scalar constant. The interpretation of model (1) is that the measurement vector $\boldsymbol{x}$ is affected by v_m, in which vector $\boldsymbol{f}_m$ indicate the nature of the spatial variation pattern

Among the collection of variation patterns, each $\boldsymbol{f}^m$ always exhibits a lower dimensional nonlinear feature embedded in the original data space. The implication that the signal components of $\boldsymbol{f}^m$ lie in a lower feature space allows us to represent $\boldsymbol{f}^m$ by principal curves method.

Blind source separation (BSS) is first proposed to solve a signal processing problem in which there is an array of spatially distributed sensors, each of which picks up signals from independent signal-emitting sources (Cardoso 1998). Considering the nonlinearity in data $\boldsymbol{x}$, the most challenging task is to separate the distinct variation patterns resulting from independent sources v_m.

If $\boldsymbol{f}^m$ lies in an r_m-dimensional ($r_m << d$ in manufacturing processes) linear variety for all v^m, and no other linear variety in which $\boldsymbol{f}_m$ lies has dimension smaller than r_m, then $\lambda^1 \geq \lambda^2 \geq \ldots \geq \lambda^{r_m} > 0 = \lambda^{r_m+1} = \lambda^d$ and each $\boldsymbol{f}^m$ can be represented by $\boldsymbol{f}^m = s^{m,1}\boldsymbol{a}^{m,1} + s^{m,2}\boldsymbol{a}^{m,2} + \ldots + s^{m,r_m}\boldsymbol{a}^{m,r_m}$. The dominating eigenvectors $\{\boldsymbol{a}^{m,r_k} : 1 \leq r_k \leq r_m\}$ are associated with those non-zero eigenvalues of covariance matrix for $\boldsymbol{f}^m$, and the PCA scores are given by $s^{m,r_k} = \boldsymbol{f}^{m,\mathrm{T}}\boldsymbol{a}^{m,r_k}$.

The PCA decompositions for $\boldsymbol{f}_i$ thus transform model (1) into a linear form

$$\begin{aligned}\boldsymbol{x} &= s^{1,1}\boldsymbol{a}^{1,1} + \ldots + s^{1,r_1}\boldsymbol{a}^{1,r_1} + \ldots + s^{p,1}\boldsymbol{a}^{p,1} + \ldots + s^{p,r_p}\boldsymbol{a}^{p,r_p} + \boldsymbol{w} \\ &\equiv \mathbf{A}\boldsymbol{s} + \boldsymbol{w}.\end{aligned} \tag{2}$$

It follows from the model assumption that random vectors $\boldsymbol{s}^m = [s^{m,1}, \ldots s^{m,r_m}]^{\mathrm{T}}$ are mutually independent. However, the elements within each vector $\boldsymbol{s}^m$ are uncorrelated PCA scores and not independent any more. Denote the length of random vector $\boldsymbol{s}$ as $r = \sum_{m=1}^{p} r_m$. Also, r is the number of dominating eigenvalues of covariance matrix Σ_x which are greater than σ^2.

Similar to linear BSS case, we first transform $\boldsymbol{x}$ into whitened data $\boldsymbol{y} = \mathbf{W}\boldsymbol{x} = \mathbf{W}(\mathbf{A}\boldsymbol{s} + \boldsymbol{w}) = \mathbf{H}\boldsymbol{s} + \mathbf{W}\boldsymbol{w}$, where $\mathbf{W}$ is the whiten matrix composed by the r dominating eigenvectors and associated eigenvalues of Σ_x.

As the linear source separation case, high order cumulants are shown to be a good alternative for nonlinear variation pattern separation.

For an arbitrary zero-mean r-dimensional random vector $\boldsymbol{s}$ with co-ordinate $s_1,\ldots, s_r$ and finite fourth-order moments, the second-order cumulant is $Cum^{i,j}(\boldsymbol{s}) = \mathrm{E}[s^i s^j]$, and the fourth-order cumulant of its ith, j^{th}, k^{th}, and l^{th} elements is

$$Cum^{i,j,k,l}(\boldsymbol{s}) = \mathrm{E}[s^i s^j s^k s^l] - \mathrm{E}[s^i s^j]\mathrm{E}[s^k s^l] - \mathrm{E}[s^j s^k]\mathrm{E}[s^i s^l] - \mathrm{E}[s^j s^l]\mathrm{E}[s^i s^k]. \quad (3)$$

A cumulant involving at least two different variables is called cross-cumulant. Note that the fourth-order cross-cumulants $Cum^{i,j,k,l}(\boldsymbol{s})$ $(i,j,k,l \neq i,i,i,i)$ are zero if the elements are mutually independent. Thus, the independence of elements of PCA score vector $\boldsymbol{y}$ can be examined by cross-cumulants. We introduced the cumulant matrices with simpler structures to offer nonlinear independent criterion by finding an optimal orthogonal rotation matrix $\mathbf{U}$ which minimizes the the off-diagonal elements of cumulant matrices (Cardoso 1998).

We decompose the r-dimensional vector $\boldsymbol{s}$ into a sum of p r-dimensional independent vectors $\{\tilde{\boldsymbol{s}}^1, \tilde{\boldsymbol{s}}^2, \ldots, \tilde{\boldsymbol{s}}^p\}$ such that different column vectors $\tilde{\boldsymbol{s}}^m$ are independent. However, the elements within each $\tilde{\boldsymbol{s}}_m$ are uncorrelated only. Using the linear formed model (2) and the cumulant properties (Gaussian rejection, additivity, multilinearity), it is straightforward to establish the structure of cumulant matrix $\mathbf{Q}^{\mathrm{Y}}(\mathbf{M})$ for vector $\boldsymbol{y}$ as

$$[\mathbf{Q}^{\mathrm{Y}}\mathbf{M})]^{ij} = \sum_m [\mathbf{Q}^{\mathrm{H}\tilde{s}^m}(\mathbf{M})]^{ij}\,.$$

And off-diagonal elements of $\mathbf{Q}^{\mathrm{H}\tilde{s}^m}(\mathbf{M})$ are zeros except for the diagonal block elements $[\mathbf{Q}^{\mathrm{H}\tilde{s}^m}(\mathbf{M})]^{ij}$ when $\sum_{n=1}^{m-1} r_n \leq i, j \leq \sum_{n=1}^{m} r_n$.

Since the elements in each vector $\tilde{\boldsymbol{s}}^m$ are not independent, the block structured cumulant matrix $\mathbf{Q}^{\mathrm{H}\tilde{s}^m}(\mathbf{M})$ will keep the same structure for any rotation. The idea of the proposed nonlinear independent criterion is to rotate the elements of $\boldsymbol{y}$ to approximate the independence. This diagonal block cumulant matrix thus allows us to isolate the nonlinear variation patterns using the diagonality independence criterion.

The nonlinear mappings $\boldsymbol{f}^m$ always lie in lower dimensional subspaces. Hence, it is reasonable to recover these nonlinear spatial features in the subspace spanned by the r_m dominating eigenvectors of covariance matrix.

For convenience, we denote the $r_m \times r_m$ square sub-matrix corresponding to nonlinear variation sources v^m as "nonlinear block". The rotation matrix $\mathbf{U}$ and "nonlinear blocks" provide the estimate of PCA scores. Note that the estimated scores will lie in the same feature space spanned by the r_m dominating eigenvectors $\{\boldsymbol{a}^{m,r_k}\}$. This allows us to extract the nonlinear variation feature $\boldsymbol{h}^m$ by principal curves method from the projection of PCA scores onto the identified feature space, and we can reveal the nonlinear variation pattern $\boldsymbol{f}^m$.

NONLINEAR PROCESS MONITORING USING NEURAL NETWORKS

The forms of principal curves change with respect to different kinds of processes. For such arbitrary nonlinear features, neural network is a good approach to model for its universal approximation property. The architecture of the multiple layer perceptron implementation is designed to be a 5-layer forward feedback neural network, as illustrated in Figure 1. The first 3-layer neural network sets the r_m-D nonlinear variation feature $\boldsymbol{h}_m$ as inputs, and the reduced outputs are the projection index. The nonlinearity is implied by the properly chosen activation function and trained weight parameters. For the second network, the inputs are the projection index, and the outputs are estimated r_m-dimensional principal curve. Actually the combination of these two networks is just a 5-layer neural network since the output layer of the first one is exactly the input layer of the second network. The advantage of such network architecture is that the two parts of the whole neural network could be trained separately.

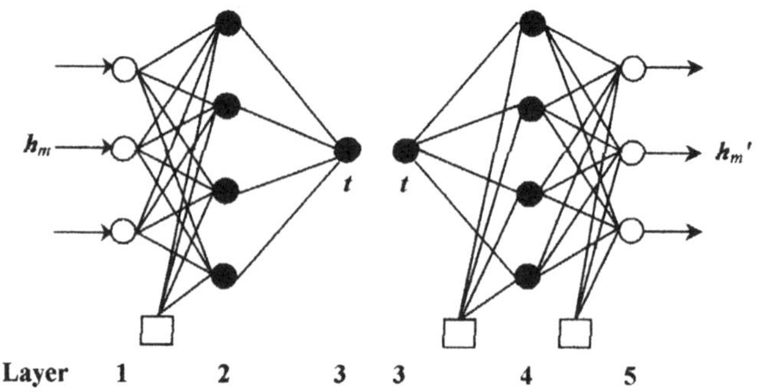

Figure 1. The architecture of MLP neural network for principal curves, where white circles denote the input and output layer and black circles for neurons in bottleneck layer and hidden layers.

For principal curves implementation with the above neural networks, conjugate gradient learning algorithm is a good and easy approach to train the weight parameters. The reasons for choosing conjugate gradient as a training algorithm lie on two folds. The first is that the learning speed of conjugate gradient is much faster than that of general back-propagation and other methods, and also the learning rate constants can be calculated automatically and adaptively rather than have to be specified before training.

The main feature distinguish the principal curves estimated by the neural network from those resulting from standard principal curve algorithms lie in the fact that the scores t_f are continuous, while the projection index given by the classical principal curves are discontinuous because of the structure of spatially distributed sample data. The neural network based principal curve estimation provides the data prediction for new identified nonlinear measurement data $\boldsymbol{x}$, which could be potentially used to process monitoring for charting new measurement data in some nonlinear multivariate processes.

As the linear case where PCA has been used in process monitoring, nonlinear PCA method can also capture the nonlinear relationship between the process variables. When this nonlinearity determined by the manufacturing process characteristics changes, the measurement data structures will follow this change. Hence, a nonlinear PCA model, e.g., principal curves, will act as a monitor mechanism for detect a fault process condition when it causes a deviation from the normal operation point.

Squared prediction error (SPE) plot is one appealing approach for monitoring such nonlinear multivariate process variables, defined as SPE = $\sum_{i=1}^{r_m}(h_{m,i} - h'_{m,i})^2$ where $\boldsymbol{h}_m$ is the mth nonlinear variation feature represented by a principal curve, and $\boldsymbol{h}'_m$ is the associated prediction by neural network. The magnitude of SPE is affected by the abnormalities in a process, where either some process variables deviates much while others remain the same, or the nonlinear relationship $\boldsymbol{h}_m$ changes. SPE chart thus is utilized to monitor the nonlinear process operations by reporting SPE for each measurement data, and is shown to be an effective multivariate process monitoring technique when linear PCA methods often fail to detect the faults (Dong 1995).

It is reasonable to assume that the prediction errors are normally distributed, so the SPE can be viewed as a $\tau \cdot \chi_{\upsilon}^{2}$ distribution where υ is the effective degrees of freedom and τ is a constant (Nomikos 1994). By matching the moments of $\tau \cdot \chi_{\upsilon}^{2}$ with the moments of the SPE for the normal operation data, υ and τ can then be determined.

In view of process monitoring, the system $\boldsymbol{x}$ is represented by a mixture of nonlinear variation patterns as model (1). For each of the nonlinear feature $\boldsymbol{h}_m$ embedded in $\boldsymbol{f}^m$, a particular decision function on incoming process measurement is performed. Therefore, the final decision function outputs $O^m(\boldsymbol{x})$ indicating the mth input process variable $\boldsymbol{x}$ falls into accepted region, can be expressed as $O^m(\boldsymbol{x})$ = accept if $SPE_m \leq \tau_m \cdot \chi^2_{m,\upsilon,\alpha}$; = reject if $SPE_m > \tau_m \cdot \chi^2_{m,\upsilon,\alpha}$, where SPE_m is the squared prediction error for the mth nonlinear feature $\boldsymbol{h}_m$ and $\tau_m \cdot \chi^2_{m,\upsilon,\alpha}$ is the control limit determined by sample data. To build a good monitoring system it is an important issue to establish a consistent and identify nonlinear BSS models with the historical data sets collected under normal working conditions. The monitoring decision function output based on the statistics is accepted when the corresponding system input patterns fall below the control limits. On the other hand, some specific input pattern that is *far away from all these ellipsoidal centers yields a rejection output signal.*

EXPERIMENTAL RESULTS

In the autobody assembly process, we collected a sample of data from d = 16 measurement locations in the liftgate opening, a subassembly part of sports utility vehicles fixed together with many tooling elements, such as clamps, and robotic weld guns, collectively referred to as process tooling. When a tooling element breaks, wears, malfunctions, or is not designed properly, it will generally result in a distinct variation pattern in the dimensional measurement data.

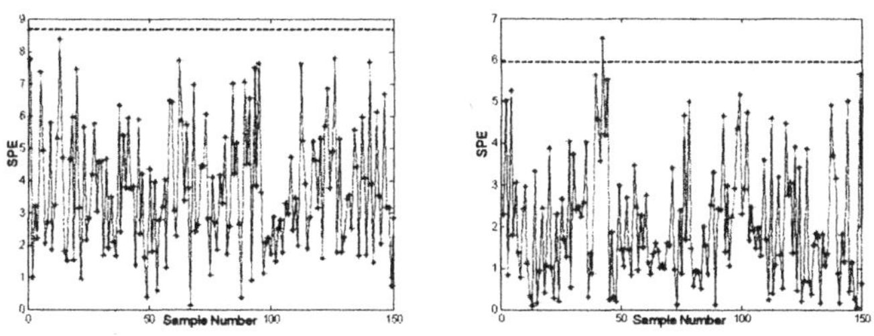

Figure 2. Multivariate SPE charts for monitoring deviation from nonlinear patterns $\boldsymbol{f}^m$ (m = 2, 3).

Based on N = 400 sample data, we recognized that there are p = 3 variation patterns existing in the multivariate data $\boldsymbol{x}$. In particular, the first variation pattern $\boldsymbol{f}_1$ is a linear function of variation source v_1, i.e., $\boldsymbol{f}^1 = \boldsymbol{c}^1 v^1$. The linear variation pattern vector $\boldsymbol{c}_1$ and source signals v_1 could be recovered from $\boldsymbol{f}_1$ using classical linear BSS algorithms. The BSS results are then used to isolate the remaining two nonlinear variation patterns which are estimated by nonlinear PCA method. To monitoring each single nonlinear variation fault caused by v^2 or v^3, we use neural networks to produce the prediction and squared prediction error for each new input measurement.

The 95% control limits are calculated based on sample data, as $SPE_{2,0.95}$ = 8.746 and $SPE_{3,0.95}$ =5.977. By moment matching with the sample mean and variance, we obtained that $\tau_2 = 0.79$, $\upsilon_2 = 5$ and $\tau_3 = 0.63$, $\upsilon_3 = 4$. Figure 2 shows the constructed SPC charts for 150 new measurement data. The monitoring decision is made by detecting any fault condition in chart from squared prediction errors. For example, around the 42th sample, the second chart detected a fault signal. Since the variation sources v^m are independent, the out-of-control signal in the separated faults detection charts can reflect the deviation from the specific underlying nonlinear variation feature.

This real application demonstrated that the multivariate process monitoring mechanism based on our nonlinear BSS algorithm and neural network is a potential systematic approach, which will be of importance for determining the characteristics of any variation patterns that are present, e.g., which can provide insight into the major root causes of product-to-product dimensional variability in manufacturing processes.

CONCLUSIONS

In this paper, we extended the linear BSS problem by considering the nonlinear variation patterns mapping from source to measurement data. Extra constrains are necessary for accurately achieving the solutions to source separation when using only the independence assumption. A special class of additive nonlinear mixture model was given special attention to propose such structure constraint.

A promising high order cumulants based separation approach was then derived to provide the source estimation and reveal the nonlinearity variation features through neural networks based principal curves method. The nonlinear variation faults could be isolated and monitored by a multivariate SPE charts for each separated variation pattern. The experimental result has illustrated the efficiency of nonlinear BSS algorithm combined with a multiple layer neural network implementation.

REFERENCES

Cardoso, J.F., 1998, "Blind Signal Separation: Statistical Principles," *Proceedings of the IEEE*, Vol. 86, pp. 2009-2025.

Hastie, T., Stuetzle, W., 1989, "Principal Curves," *Journal of the American Statistical Association*, Vol. 84, pp. 502-516.

Hsieh, W.W., 2001, "Nonlinear Principal Component Analysis by Neural Networks," *Tellus*, Vol. *53A*, pp. 599-615.

Dong, D., McAvoy, T.J., 1995, "Nonlinear Principal Component Analysis based on Principal Curves and Neural Networks," *Comput. Chem. Eng.*, Vol. 20, pp. 65-78.

Apley. D.W., Shi, J., 2001, "A Factor Analysis Method for Diagnosing Variability in Multivariate Manufacturing Processes," *Technometics*, Vol. 43, pp. 1-12.

Chang, K., Ghosh, J., 1998, "Principal Curves for Nonlinear Feature Extraction and Classification," *SPIE Applications of Artificial NN in Image Processing III*, Vol. 3307, pp. 120-129.

Roweis, S. T., Saul, L.K., 2000, "Nonlinear dimensionality reduction by locally linear embedding," *Science*, Vol. 290, pp. 2323-2326.

THE PARAMETRIZATION OF ALL STABILIZING CONTROLLERS FOR TIME-DELAY SYSTEMS

KOU YAMADA
Department of Mechanical System Engineering, Gunma University
1-5-1 Tenjincho, Kiryu, Japan

TOORU ARAKAWA
Department of Mechanical System Engineering, Gunma University
1-5-1 Tenjincho, Kiryu, Japan

N. MATSUSHIMA
Department of Mechanical System Engineering, Gunma University
1-5-1 Tenjincho, Kiryu, Japan

KEIJI SATOH
Department of Mechanical System Engineering, Gunma University
1-5-1 Tenjincho, Kiryu, Japan

ABSTRACT

In this paper, we give the parametrization of all stabilizing controllers, which are rational functions, for non-minimum phase time-delay systems. This basic idea is very simple. If the controller stabilizes the plant, then the controller must be included in the class of all stabilizing controllers for the plant. The parametrization of all stabilizing controllers for the plant were obtained by Youla et al. and Vidyasagar. The parametrization of all stabilizing controllers include a free parameter, which is stable. That is, stabilizing controllers for the non-minimum phase time-delay systems can be designed using the free parameter in the parametrization by Youla et al. and Vidyasagar. Using this idea, we obtain the parametrization of all stabilizing controllers, which are rational functions, for non-minimum phase time-delay systems.

Introduction

In this paper, we examine the parametrization of all stabilizing controllers, which are rational functions, for non-minimum phase time-delay systems. The parametrization problem is the problem finding all stabilizing controllers for the plant (Zames, 1981, Youla et al., 1976, Desoer, 1980, Zhou et al, 1996, Morari and Zafiriou, 1989, Glaria and Goodwin, 1994, Yamada, 2001a, Yamada, 2001b).

For an asymptotically stable plant, the parametrization of all stabilizing controllers has a structure identical to that of Internal Model Control (Morari and Zafiriou, 1989). For an unstable plant, the structure of the parametrization of all stabilizing controllers has full-order state feedback, including the full-order observer (Zhou et al., 1996). Glaria and Goodwin (1994) gave a simple parametrization for minimum phase plants and their control structure; however, one practical difficulty remains: for strictly proper systems, the parametrization of the stabilizing controllers given by Glaria and Goodwin generally includes improper controllers. In practice, the controller is required to be proper. If the parametrization of all proper stabilizing controllers for the minimum phase plant can be obtained, then the parametrization of all stabilizing controllers for minimum phase plants can be applied in several applications. Yamada (2001a) overcame the problem of the parametrization of Glaria and Goodwin and gave the parametrization for the class of all proper internally stabilizing controllers for minimum phase plants. On the other hand, the parametrization of all stabilizing controllers for time-delay systems has not been studied.

In this paper, we tackle the parametrization problem for time-delay systems and propose the parametrization of all stabilizing controllers, which are rational functions, for systems with input time-delay. The basic idea to obtain the parametrization of all stabilizing controllers for systems with input time-delay is very simple. If the controller stabilizes the plant, then the controller must be included in the class of all stabilizing controllers for the plant. The parametrization of all stabilizing controllers for the plant are obtained using the parametrization of all stabilizing controllers for the plant by Youla et al. (1976) and Vidyasagar (1985). The parametrization of all stabilizing controllers includes a free parameter, which is stable. That is, stabilizing controllers for the non-minimum phase time-delay systems can be designed using the free parameter in the parametrization by Youla et al. (1976) and Vidyasagar (1985). In practice, in many cases, we cannot obtain the exact time-delay, that is, the time-delay has uncertainty. If we can obtain the parametrization of all stabilizing controllers, which is independent of the time-delay, we can make the control system robustly stable, even with uncertainty in the time-delay. Next, using the parametrization of all stabilizing controllers, which are rational functions, for the time-delay systems, we can obtain the parametrization of all stabilizing controllers, which is independent of the time-delay, for time-delay systems.

Problem formulation

Let us consider the control system written as

$$\left\{ \begin{array}{lcl} y & = & G(s)e^{-sT}u \\ u & = & C(s)\left(r-y\right) \end{array} \right. , \tag{1}$$

where $G(s)e^{-sT}$ is the plant with input time-delay $T > 0$, $C(s) \in R(s)$ is the rational controller, $y \in R$ is the output, $u \in R$ is the input and r is the reference input.

The problem considered in this paper is to obtain the parametrization of all controllers $C(s)$, which are rational functions, to make the control system in Eq.(1) stable. The basic idea to obtain the parametrization of all stabilizing controllers $C(s)$ is very simple. From the assumption that the plant has input time-delay T, the plant $G(s)e^{-sT}$ can be factorized into the rational function part $G(s)$ and the time-delay part e^{-sT}. We will show later, if $C(s)$ is rational function, the parametrization of all stabilizing controllers $C(s)$ for $G(s)e^{-sT}$ is equivalent to that of all stabilizing controllers $C(s)e^{-sT}$ for $G(s)$. The parametrization of all stabilizing controllers for $G(s)$ was obtained by Youla et al. (1976) and Vidyasagar (1985). The parametrization of all stabilizing controllers by Youla et al. (1976) and Vidyasagar (1985) has a free parameter. If the free parameter is selected correctly and the controller can be written as $C(s)e^{-sT}$, then the controller $C(s)$ stabilizes $G(s)e^{-sT}$. Conversely, the controller $C(s)$ stabilizes the plant $G(s)e^{-sT}$, then the free parameter can be written in a certain form.

In the next section, using the above idea, we give the parametrization of all stabilizing controllers for the time-delay systems $G(s)e^{-sT}$.

The parametrization

In this section, using above idea, we give the parametrization of all controllers $C(s)$, which are rational functions, to make the control system in Eq.(1) stable.

The parametrization of all stabilizing controllers for the plant $G(s)e^{-sT}$ is summarized in the following Theorem.

Theorem 1 *Suppose that $C(s)$ is rational function. The control system in Eq.(1) is internally stable if and only if $C(s)$ can be written as*

$$C(s) = \frac{\tilde{X}(s) + D(s)Q(s)}{\tilde{Y}(s) - N(s)Q(s)} e^{sT}, \tag{2}$$

where $N(s) \in RH_\infty$ and $D(s) \in RH_\infty$ are coprime factors of $G(s)$ on RH_∞ satisfying

$$G(s) = N(s)D^{-1}(s) = \tilde{D}^{-1}(s)\tilde{N}(s) \tag{3}$$

$Q(s) \in H_\infty$ is written by

$$Q(s) = \frac{Q_{n1}(s) + Q_{n2}(s)e^{-sT}}{Q_{d1}(s) + Q_{d2}(s)e^{-sT}}, \tag{4}$$

where $Q_{n1}(s)$, $Q_{n2}(s)$, $Q_{d1}(s) \neq 0$, $Q_{d2}(s)$ are any RH_∞ functions satisfying

$$Q_{d2}(s)\tilde{Y}(s) - N(s)Q_{n2}(s) = 0, Q_{d1}(s)\tilde{X}(s) + D(s)Q_{n1}(s) = 0. \tag{5}$$

$\tilde{X}(s)$ and $\tilde{Y}(s)$ are RH_∞ functions satisfying

$$\begin{bmatrix} Y(s) & X(s) \\ -\tilde{N}(s) & \tilde{D}(s) \end{bmatrix} \begin{bmatrix} D(s) & -\tilde{X}(s) \\ N(s) & \tilde{Y}(s) \end{bmatrix} = I. \tag{6}$$

Proof of this theorem requires the following lemma.

Lemma 1 *Unity feedback control system in*

$$\begin{cases} y = G(s)u \\ u = -C(s)y \end{cases} \tag{7}$$

is internally stable if and only if $C(s)$ can be written as

$$C(s) = \frac{\tilde{X}(s) + D(s)Q(s)}{\tilde{Y}(s) - N(s)Q(s)}, \tag{8}$$

where $N(s) \in RH_\infty$ and $D(s) \in RH_\infty$ are coprime factors of $G(s)$ on RH_∞ satisfying

$$G(s) = N(s)D^{-1}(s) = \tilde{D}^{-1}(s)\tilde{N}(s) \tag{9}$$

$\tilde{X}(s)$ and $\tilde{Y}(s)$ are RH_∞ function satisfying

$$\begin{bmatrix} Y(s) & X(s) \\ -\tilde{N}(s) & \tilde{D}(s) \end{bmatrix} \begin{bmatrix} D(s) & -\tilde{X}(s) \\ N(s) & \tilde{Y}(s) \end{bmatrix} = I. \tag{10}$$

$Q(s)$ is arbitrary asymptotically stable function (Vidyasagar, 1985).

Theorem 2 *Suppose that $C(s)$ is rational function. $C(s)$ stabilizes $G(s)e^{-sT}$ if and only if $C(s)e^{-sT}$ stabilizes $G(s)$.*

(Proof) The proof is obvious because $C(s)$ is rational function. ∎

Using Lemma 1 and Theorem 2, we show the proof of Theorem 1.
(Proof) From Theorem 2, the parametrization of all stabilizing controllers $C(s)$ for $G(s)e^{-sT}$ is equivalent to that of all stabilizing controllers $C(s)e^{-sT}$ for $G(s)$.

First, necessity is shown. That is, we will show that if the controller $\hat{C}(s) = C(s)e^{-sT}$ stabilizes the plant $G(s)$ in Eq.(1), then $C(s) = \hat{C}(s)e^{sT}$ and $Q(s)$ are written as Eq.(2) and Eq.(4), respectively. In addition, $Q(s)$ is stable. From Lemma 1, the parametrization of all stabilizing controllers $\hat{C}(s) = C(s)e^{-sT}$ for $G(s)$ is written as

$$\hat{C}(s) = C(s)e^{-sT} = \frac{\tilde{X}(s) + D(s)Q(s)}{\tilde{Y}(s) - N(s)Q(s)}, \tag{11}$$

where this equation is equivalent to Eq.(2). Next, we show that if $C(s)e^{-sT}$ stabilizes the plant $G(s)$, then the free parameter $Q(s)$ in Eq.(2) is written as Eq.(4). From Eq.(11), we have

$$\begin{aligned} Q(s) &= \frac{-\tilde{X}(s)C_d(s) + \tilde{Y}(s)C_n(s)e^{-sT}}{C_d(s)D(s) + C_n(s)N(s)e^{-sT}} \\ &= \frac{Q_{n1}(s) + Q_{n2}(s)e^{-sT}}{Q_{d1}(s) + Q_{d2}(s)e^{-sT}}, \end{aligned} \tag{12}$$

where $C_n(s) \in RH_\infty$ and $C_d(s) \in RH_\infty$ are satisfying

$$C(s) = \frac{C_n(s)}{C_d(s)} \tag{13}$$

and $Q_{n1}(s)$, $Q_{n2}(s)$, $Q_{d1}(s)$ and $Q_{d2}(s)$ are

$$Q_{n1}(s) = -\tilde{X}(s)C_d(s), Q_{n2}(s) = C_n(s)\tilde{Y}(s),$$
$$Q_{d1}(s) = C_d(s)D(s), Q_{d2}(s) = C_n(s)N(s),$$

respectively. From Eq.(12), it has proved that if $C(s)$, written as Eq.(4), stabilizes $G(s)e^{-sT}$ then the free parameter $Q(s)$ in Eq.(2) is written as Eq.(4). Next it is shown that $Q(s)$ in Eq.(12) is stable. From the assumption that $C(s)$ written by Eq.(2) stabilizes $G(s)e^{-sT}$,

$$\begin{aligned} &1 + C(s)G(s)e^{-sT} \\ &= \frac{1}{C_d(s)D(s)}\left(C_d(s)D(s) + C_n(s)N(s)e^{-sT}\right) \end{aligned} \tag{14}$$

has no pole in the closed right half plane. This implies that $Q(s)$ in Eq.(12) is stable.

Next sufficiency is shown. That is, we will show that if $Q(s)$ can be written as Eq.(4), the controller $C(s)$ that is written as Eq.(2) stabilizes

$G(s)e^{-sT}$. Substituting Eq.(4) into Eq.(2), we have

$$C(s)e^{-sT} = \frac{\tilde{X}(s)+D(s)Q(s)}{\tilde{Y}(s)-N(s)Q(s)} = \frac{\left(Q_{d1}(s)\tilde{X}(s)+D(s)Q_{n1}(s)\right)+\left(Q_{d2}(s)\tilde{X}(s)+D(s)Q_{n2}(s)\right)e^{-sT}}{\left(Q_{d1}(s)\tilde{Y}(s)-N(s)Q_{n1}(s)\right)+\left(Q_{d2}(s)\tilde{Y}(s)-N(s)Q_{n2}(s)\right)e^{-sT}}. \tag{15}$$

From Eq.(15) and Eq.(5), $C(s)$ is written as

$$C(s) = \frac{Q_{d2}(s)\tilde{X}(s)+D(s)Q_{n2}(s)}{Q_{d1}(s)\tilde{Y}(s)-N(s)Q_{n1}(s)}. \tag{16}$$

$C(s)$ written as Eq.(16) is obviously a rational function.

We have proved Theorem 1. ∎

The parametrization of all stabilizing controllers which is independent of time-delay

In this section, using the result in the preceding section, we give the parametrization of all controllers that independently stabilize time-delay, for time-delay systems.

The parametrization of all controllers that independently stabilize time-delay, for time-delay systems is summarized by the following theorem.

Theorem 3 *Even if the time-delay T has uncertainty, the control system in Eq.(1) is internally stable if and only if $C(s)$ can be written as*

$$C(s) = \frac{\tilde{X}(s)+D(s)Q(s)}{\tilde{Y}(s)-N(s)Q(s)}e^{sT}, \tag{17}$$

where $N(s) \in RH_\infty$ and $D(s) \in RH_\infty$ are coprime factors of $G(s)$ on RH_∞ satisfying

$$G(s) = N(s)D^{-1}(s) = \tilde{D}^{-1}(s)\tilde{N}(s) \tag{18}$$

$Q(s) \in H_\infty$ is written as

$$Q(s) = \frac{Q_{n1}(s)+Q_{n2}(s)e^{-sT}}{Q_{d1}(s)+Q_{d2}(s)e^{-sT}}, \tag{19}$$

where $Q_{n1}(s)$, $Q_{n2}(s)$, $Q_{d1}(s) \neq 0$, $Q_{d2}(s)$ are any RH_∞ functions satisfying

$$Q_{d2}(s)\tilde{Y}(s) - N(s)Q_{n2}(s) = 0, Q_{d1}(s)\tilde{X}(s) + D(s)Q_{n1}(s) = 0. \tag{20}$$

$\tilde{X}(s)$ and $\tilde{Y}(s)$ are RH_∞ functions satisfying

$$\begin{bmatrix} Y(s) & X(s) \\ -\tilde{N}(s) & \tilde{D}(s) \end{bmatrix}\begin{bmatrix} D(s) & -\tilde{X}(s) \\ N(s) & \tilde{Y}(s) \end{bmatrix} = I. \tag{21}$$

(Proof) This theorem is proved using the same procedure as for Theorem 1. ∎

Even if $Q(s)$ is selected using Neural Networks, if $Q(s)$ satisfies Theorem 1, we can make the control system in Eq.(1) stable for time-delay systems, including uncertainty in the time-delay.

Conclusion

In this paper, we propose the parametrization of all stabilizing controllers, which are rational functions for systems with input time-delay. Finally, using the parametrization of all stabilizing controllers, which are rational functions for the time-delay systems, gave the parametrization of all stabilizing controllers, which is independent of time-delay, for time-delay systems. Using the result in this paper, even if $Q(s)$ is selected using Neural Networks, we can make the control system stable for the time-delay systems, including uncertainty in the time-delay.

References

C.A. Desoer, R.W. Liu, J. Murray and R. Saeks (1980). Feedback system design: The fractional representation approach to analysis and synthesis. *IEEE Trans. on AC* **25**, 399–412.

D.C. Youla, H. Jabr and J.J. Bongiorno (1976). Modern Wiener-Hopf design of optimal controllers. part i. *IEEE Trans. on AC* **21**, 3–13.

Glaria, J.J. and G.C. Goodwin (1994). A parametrization for the class of all stabilizing controllers for linear minimum phase systems. *IEEE Trans. on AC* **39**, 433–434.

K. Zhou, J.C. Doyle and K. Glover (1996). *Robust and optimal control.* Prentice Hall.

Morari, M. and E. Zafiriou (1989). *Robust Process Control.* Prentice Hall.

Vidyasagar, M. (1985). *Control System Synthesis–A factorization approach–.* MIT Press.

K. Yamada (2001). A parametrization for the class of all proper stabilizing controllers for linear minimum phase systems. *Preprints of the 9th IFAC/IFORS/IMACS/IFIP/ Symposium on Large Scale Systems: Theory and Applications*, pp.578-583. Zames, G. (1981). Feedback and optimal sensitivity: model reference transformations, multiplicative seminorms and approximate inverse. *IEEE Trans. on AC* **26**, 301–320.

A DESIGN METHOD FOR ROBUST REPETITIVE CONTROL SYSTEMS

KOU YAMADA
Department of Mechanical System Engineering, Gunma University
1-5-1 Tenjincho, Kiryu, Japan
KEIJI SATOH
Department of Mechanical System Engineering, Gunma University
1-5-1 Tenjincho, Kiryu, Japan

TOORU ARAKAWA
Department of Mechanical System Engineering, Gunma University
1-5-1 Tenjincho, Kiryu, Japan

ABSTRACT
We propose a design method for robust modified repetitive control systems. The basic idea of the design method is that if the modified repetitive control system is robustly stable, then the modified repetitive controller must satisfy the robust stability condition. The parameterization form of all robustly stabilizing controllers for the plant are obtained using an H_∞ problem based on either the Riccati equation or on a Linear Matrix Inequality. Using the free parameter of the parameterization, we can design robust modified repetitive control systems.

Introduction

In this paper, we investigate a design method for robust repetitive control systems. The repetitive control system is a type of servo mechanism designed for a periodic reference input (Inoue et al., 1980, Hara et al., 1988, Omata et al., 1987).

When modified repetitive control design methods are applied to real systems, the influence of uncertainties in the plant must be considered. In some cases, the uncertainties in the plant make the modified repetitive control system unstable, even though the controller was designed to stabilize the nominal plant. The stability problem with uncertainty is known as the robust stability problem (Doyle et al., 1989). The robust stability problem of modified repetitive control systems was considered by Hara et al. (1994). The robust stability condition for modified repetitive control systems was reduced to the μ synthesis problem (Hara et al., 1994), but the μ synthesis problem cannot be solved analytically. That is, in order to solve the μ synthesis problem, we must solve an H_∞ problem iteratively using the $D-K$ iteration method. Furthermore, the convergence of iterative methods to solve the μ synthesis problem is not guaranteed. Yamada et al. (2003) tackle this problem and propose a design method of robust modified repetitive control systems. The method by Yamada et al. (2003) is effective for the minimum phase systems, however the method by Yamada et al. is not so effective for the non-minimum phase systems, since the frequency range in which the output follows the periodic reference input is restricted.

In this paper, we expand the result by Yamada et al. (2003) and propose a two-step design method for robust modified repetitive control systems for non-minimum phase system such that the frequency range in which the output follows the periodic reference input is not restricted. The basic idea of the method is as follows. If the modified repetitive control system is robustly stable, then the modified repetitive controller must satisfy the robust stability condition. This implies that if the modified repetitive control system is robustly stable, then the modified repetitive controller is included in the parameterization of all robustly stabilizing controllers for the plant. The parameterization of all robustly stabilizing controllers for the plant is obtained using H_∞ control theory based on the Riccati equation (Doyle et al.,1989) and the Linear Matrix Inequality (LMI) (Iwasaki and Skelton, 1994, Gahinet and Apkarian, 1994). Robustly stabilizing controllers for the plant include a free parameter, which is designed to achieve desirable control characteristics. If the free parameter is chosen to give the control system robust servo characteristics for periodic reference input, then the controller operates as a robust modified repetitive controller. We present a design method for single-input/single-output systems so that the free parameter makes the control system have robust servo characteristics for periodic reference input.

Problem formulation

Let us consider the unity feedback control system shown in Figure 1 . Here, $G(s) \in R(s)$ is the single-input/single-output strictly proper plant,

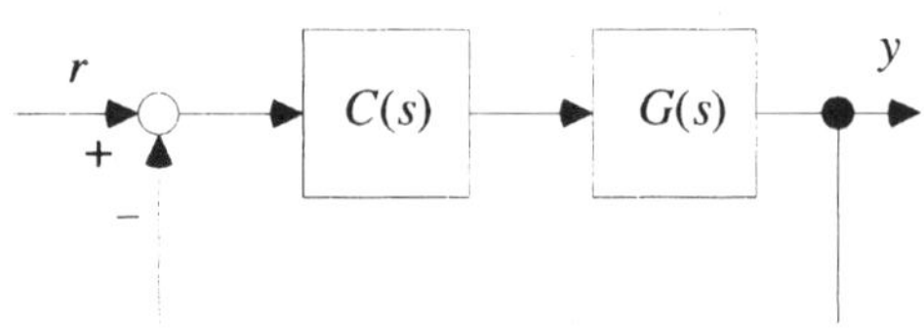

Figure 1: Unity feedback control system

$C(s) \in R(s)$ is the controller, $y \in R$ is the output, and $r \in R$ is the reference input with period T satisfying

$$r(t+T) = r(t) \quad \forall t \geq 0. \tag{1}$$

The nominal plant of $G(s)$ is denoted by $G_m(s) \in R(s)$. Both $G(s)$ and $G_m(s)$ are assumed to have no zero or pole on the imaginary axis. In addition, it is assumed that the number of poles of $G(s)$ in the closed right half plane is equal to the number of poles of $G_m(s)$. The relation between the plant $G(s)$ and the nominal plant $G_m(s)$ is written as

$$G(s) = G_m(s)(1 + \Delta(s)). \tag{2}$$

The set of $\Delta(s)$ is all rational functions satisfying

$$|\Delta(j\omega)| < |W_T(j\omega)| \quad \forall \omega \in R_+, \tag{3}$$

where $W_T(s)$ is an asymptotically stable rational function.

According to Inoue et al.(1980), Hara et al.(1988) and Omata et al.(1987), in order for the output y to follow the reference input r in Eq.(1) with small steady state error, the controller $C(s)$ must have the following structure

$$C(s) = \hat{C}(s) + \bar{C}(s)\frac{1}{1 - q(s)e^{-sT}}, \tag{4}$$

where $q(s)$ is an asymptotically stable strictly proper low pass filter satisfying $q(0) = 1$, $\hat{C}(s) \in R(s)$, and $\bar{C}(s) \in R(s)$. In the following, $1/\left(1 - q(s)e^{-sT}\right)$ defines the internal model for the periodic input with period T.

The problem considered in this paper is to design a modified repetitive control system such that the modified repetitive control system is robustly stable and the output y follows the periodic reference input r with small steady state error even in the presence of uncertainty $\Delta(s)$.

Robust modified repetitive control design

In order to design robust modified repetitive control systems such that the frequency range in which the output follows the periodic reference input is not restricted, a two-step design method is proposed.

The basic idea of a two-step design method is as follows. If the modified repetitive control system is robustly stable, then the modified repetitive controller must satisfy the robust stability condition. The parameterization of all robustly stabilizing controllers for the plant is obtained from H_∞ control theory based on the Riccati equation (Doyle et al.,1989) and on LMI (Iwasaki and Skelton, 1994, Gahinet and Apkarian, 1994). When the free parameter of the parameterization of all robust stabilizing controllers is correctly chosen, the structure of the controller is, equivalently, written as Eq.(4).

In the following, we describe in detail a two-step design method for robust modified repetitive control systems using the above idea. The robust stability condition for the plant $G(s)$ with uncertainty $\Delta(s)$ satisfying Eq.(3) is given by

$$\|T(s)W_T(s)\|_\infty < 1, \tag{5}$$

where $T(s)$ is the complementary sensitivity function given by

$$T(s) = \frac{G_m(s)C(s)}{1 + G_m(s)C(s)}. \tag{6}$$

The problem of obtaining the controller $C(s)$ satisfying Eq.(5) is equivalent to the following H_∞ problem. In order to obtain the controller $C(s)$ satisfying Eq.(5), consider the control system shown in Figure 2 . $P(s)$ is

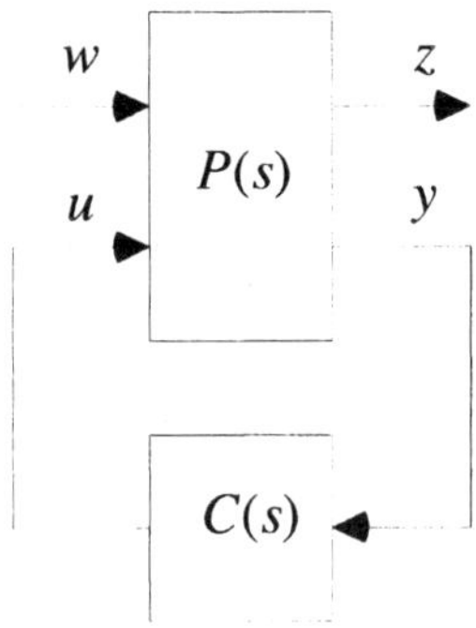

Figure 2: Block diagram of H_∞ control problem

selected such that the transfer function from w to z in Figure 2 is equal to $T(s)W_T(s)$. The state space description of $P(s)$ is, in general,

$$\left\{\begin{array}{lllll} \dot{x}(t) & = & Ax(t) & +B_1w(t) & +B_2u(t) \\ z(t) & = & C_1x(t) & & +D_{12}u(t) \\ y(t) & = & C_2x(t) & +D_{21}w(t) & \end{array}\right. , \tag{7}$$

where $A \in R^{n\times n}$, $B_1, B_2 \in R^n$, $C_1, C_2 \in R^{1\times n}$, $D_{12}, D_{21} \in R$. $P(s)$ is called the generalized plant. $P(s)$ is assumed to satisfy the standard assumption by Doyle et al.(1989).

With these assumptions, if controllers satisfying Eq.(5) exist, both

$$X\left(A - B_2D_{12}^{\dagger}C_1\right) + \left(A - B_2D_{12}^{\dagger}C_1\right)^T X + X\left(B_1B_1^T - B_2\left(D_{12}^TD_{12}\right)^{-1}B_2^T\right)X + \left(D_{12}^{\perp}C_1^T\right)^T D_{12}^{\perp}C_1^T = 0 \tag{8}$$

and

$$Y\left(A - B_1D_{21}^{\dagger}C_2\right)^T + \left(A - B_1D_{21}^{\dagger}C_2\right)Y + Y\left(C_1^TC_1 - C_2^T\left(D_{21}D_{21}^T\right)^{-1}C_2\right)Y + B_1D_{21}^{\perp}\left(B_1D_{21}^{\perp}\right)^T = 0 \tag{9}$$

have solutions $X \geq 0$ and $Y \geq 0$ such that

$$\rho(XY) < 1 \tag{10}$$

and both $A - B_2D_{12}^{\dagger}C_1 + \left(B_1B_1^T - B_2\left(D_{12}^TD_{12}\right)^{-1}B_2^T\right)X$ and $A - B_1D_{21}^{\dagger}C_2 + Y\left(C_1^TC_1 - C_2\left(D_{21}D_{21}^T\right)^{-1}C_2\right)$ have no eigenvalue in the closed right half plane. Using X and Y, the parameterization of all controllers satisfying Eq.(5) is given by

$$C(s) = C_{11}(s) + C_{12}(s)Q(s)(I - C_{22}(s)Q(s))^{-1}C_{21}(s), \tag{11}$$

where

$$\left[\begin{array}{cc} C_{11}(s) & C_{12}(s) \\ C_{21}(s) & C_{22}(s) \end{array}\right] = \left[\begin{array}{c|cc} A_c & B_{c1} & B_{c2} \\ \hline C_{c1} & D_{c11} & D_{c12} \\ C_{c2} & D_{c21} & D_{c22} \end{array}\right] \tag{12}$$

$$\begin{aligned}
A_c &= A + B_1B_1^TX - B_2\left(D_{12}^\dagger C_1 + E_{12}^{-1}B_2^TX\right) \\
&\quad - (I - XY)^{-1}\left(B_1D_{21}^\dagger + YC_2^TE_{21}^{-1}\right)(C_2 + D_{21}B_1^TX) \\
B_{c1} &= (I - XY)^{-1}\left(B_1D_{21}^\dagger + YC_2^TE_{21}^{-1}\right) \\
B_{c2} &= (I - XY)^{-1}(B_2 + YC_1^TD_{12})E_{12}^{-1/2} \\
C_{c1} &= -D_{12}^\dagger C_1 - E_{12}^{-1}B_2^TX \\
C_{c2} &= -E_{21}^{-1/2}(C_2 + D_{21}B_1^TX) \\
D_{c11} &= 0 \\
D_{c12} &= E_{12}^{-1/2} \\
D_{c21} &= E_{21}^{-1/2} \\
D_{c22} &= 0 \\
E_{12} &= D_{12}^TD_{12} \\
E_{21} &= D_{21}D_{21}^T
\end{aligned}$$

and the free parameter $Q(s)$ is any rational function satisfying $\|Q(s)\|_\infty < 1$.

Next, we present a design method for the controller $C(s)$ denoted by Eq.(11) that has the same structure as Eq.(4) and operates as a robust modified repetitive controller. So that $C(s)$, denoted by Eq.(11), has the same structure in Eq.(4), we use the free parameter $Q(s)$. When $Q(s)$ is selected as

$$Q(s) = \frac{1}{Q_1(s) + Q_2(s)e^{-sT}}, \tag{13}$$

then the controller $C(s)$ in Eq.(11) has the same structure as Eq.(4) and operates as a robust modified repetitive controller, where $Q_1(s) \in RH_\infty$ and $Q_2 \in RH_\infty$ are rational functions to hold $Q(s) \in H_\infty$. $Q_1(s)$ is settled to make $Q_1(s) - C_{22}(s)$ of minimum phase and $Q_2(s)$ is settled by

$$Q_2(s) = -\left(Q_1(s) - C_{22}(s)\right)\bar{q}(s). \tag{14}$$

Here, $\bar{q}(s) \in RH_\infty$ is a low pass filter satisfying $\bar{q}(0) = 1$, as

$$\bar{q}(s) = \frac{1}{(1 + s\tau)^\alpha} \tag{15}$$

is valid, where $\tau \in R$, and α is an arbitrary positive integer. If $\bar{q}(s)$ satisfies

$$\left\| \frac{1}{Q_1(s) + (Q_1(s) - C_{22}(s))\bar{q}(s)e^{-sT}} \right\|_\infty < 1, \tag{16}$$

then $Q(s)$ written as Eq.(13) satisfies $\|Q(s)\|_\infty < 1$, because

$$\begin{aligned}
\|Q(s)\|_\infty &= \left\| \frac{1}{Q_1(s) + Q_2(s)}e^{-sT} \right\|_\infty \\
&= \left\| \frac{1}{Q_1(s) + (Q_1(s) - C_{22}(s))\bar{q}(s)e^{-sT}} \right\|_\infty < 1. \qquad (17)
\end{aligned}$$

Substituting Eq.(13) into Eq.(11) yields

$$C(s) = C_{11}(s) + \frac{C_{12}(s)C_{21}(s)}{Q_1(s) - C_{22}(s)} \frac{1}{1 - \bar{q}(s)e^{-sT}}. \tag{18}$$

In this way, when $Q(s)$ is selected as Eq.(13), the controller Eq.(11) has the same structure as Eq.(4). From the above, when $C_{22}(s)$ is of non-minimum phase, if $Q(s)$ in Eq.(11) is chosen as Eq.(13), $Q_1(s)$ and $\bar{q}(s)$ holds Eq.(16), the controller $C(s)$ operates as a robust modified repetitive controller. In addition, since $\bar{q}(s)$ is settled by Eq.(15), it is obvious that the frequency range in which the output follows the periodic reference input is not restricted.

Conclusion

In this paper, we expand the result by Yamada et al. (2003) and proposed a two-step design method for robust modified repetitive control systems for non-minimum phase system such that the frequency range in which the output follows the periodic reference input is not restricted.

References

Inoue, T., et al., "High Accuracy Control Magnet Power Supply of Proton Synchrotron in Recurrent Operation," *The Trans. of The Institute of Electrical Engineers of Japan*, Vol.C100, No.7 (1980), pp.234-240.

Hara, S., Yamamoto, Y., Omata, T. and Nakano, M.,"Repetitive Control System: A New Type Servo System for Periodic Exogenous Signals," *IEEE Trans. on Automatic Control*, Vol.AC-33, No.7 (1988), pp.659-668.

Omata, T., Hara, S. and Nakano, M.,"Nonlinear Repetitive Control with Application to Trajectory Control of Manipulators," *J. of Robotic Systems*, Vol.4, No.5 (1987), pp.631-652.

Doyle, J.C., Glover, K., Khargonekar, P.P. and Francis, B.A., "State-space solution to standard H_2 and H_∞ control problems," *IEEE Trans. Automatic Control*, Vol.AC-34, No.8 (1989), pp.831-847.

Iwasaki, T. and Skelton, R.E., "All controllers for the general H_∞ control problem: LMI existence conditions and state space formulas," *Automatica*, Vol.30, No.8 (1994), pp.1307-1317.

Gahinet, P. and Apkarian, P., " A linear matrix inequality approach to H_∞ control," *International Journal of Robust and Nonlinear Control*, Vol.4 (1994), pp.421-448.

Hara, S., Trannitad P. and Chen, Y., "Robust stabilization for repetitive control systems," *Proceedings of the 1st Asian Control Conference*, pp.541-544, 1994.

Yamada, K., Arakawa, T, Hoshi, H, Okuyama, T,"Two-step design method of robust repetitive control systems," JSME International Journal Series C accepted for publication(2003)

A NEURAL CONTROL METHOD FOR IMPEDANCE MATCHING OF ANTENNA SYSTEMS

THOMAS L. HEMMINGER
School of Engineering and Engineering Technology
Penn State University, Behrend College, Erie, PA 16563 USA
e-mail: hemm@psu.edu

ABSTRACT
The primary intent of this paper is real-time compensation for changes in the driving point impedance of an antenna due to frequency deviations. In general, the driving point impedance of an antenna or antenna array is computed by numerical methods such as the method of moments or similar techniques. Some configurations do lend themselves to analytical solutions, which will be the primary focus of this work. This paper employs a neural control system to match antenna feed lines to two common antennas during frequency sweeps. In practice, impedance matching is performed off-line with Smith charts or relatively complex formulas but they rarely perform optimally over a large bandwidth. The approach proposed here employs three small neural networks to perform real-time impedance matching over a broad range of frequencies during transmitter operation. Double stub tuners are being explored in this paper but the approach can certainly be applied to other methodologies.

INTRODUCTION

Electrical engineers are usually introduced to the Smith chart in a first course in electromagnetic fields and waves (Kraus, and Fleisch, 1999), (Ulaby, 2001), (Cheng, 1993). A great deal of knowledge can be gleaned from a Smith chart, e.g., voltage standing wave ratio (VSWR), complex reflection coefficient, single and double stub tuner lengths, and much more. A Smith chart yields a relationship between input impedance along a transmission line and the corresponding reflection coefficient. This information is employed in the proposed controller. The work described here emphasizes transmission lines having zero resistance in the conductors and zero conductance in the insulators, e.g., lossless lines. Some researchers have experimented with Hopfield networks for impedance matching but these configurations can converge to local minima and are not always dependable (Prasad et al., 1992), (Prasad and Prasad, 1995). Others have applied neural networks to alternative electromagnetic problems, but to date these have been designated toward waveguides and microwave circuits, e.g., neural inverse space mapping (Bandler, et al. 2001), (Watson and Gupta, 1997). An important aspect of this control method is that the impedance trajectory in reflection coefficient space be a known function. For this reason, this work is primarily focused on the $\lambda/2$ dipole and the $\lambda/4$ vertical dipole over a ground plane. Controllers for Yagi-Uda antennas will be explored in the future.

The most commonly used Smith chart is a normalized rectangular plot that relates input impedance at a point, to reflection coefficient (Guru,and Hizirolu, 1998). Figure 1 illustrates a simple transmission line with a double stub tuner. The target VSWR is

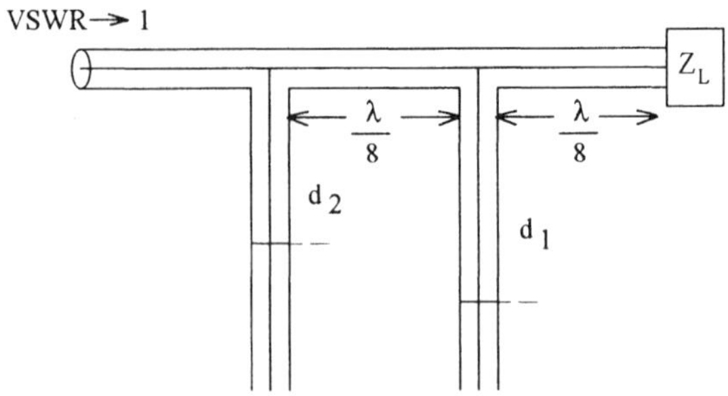

Figure 1. A double stub tuner with $\lambda/8$ stub spacing.

unity, as shown, although any value less than three is usually acceptable. In field theory, many transmission lines are considered ideal if field losses can be minimized, meaning that coaxial, parallel, and certain striplines tend to dominate. Generally, the objective of impedance matching is to connect loads (typically shorted stubs) across an antenna to modify its overall impedance. By way of careful adjustment, the line and antenna will appear to be matched resulting in minimal reflected waves. Through a somewhat lengthy but simple derivation from Maxwell's equations, the input impedance of a lossless transmission line can be expressed by the following equation:

$$Z_{in}(x) = Z_o \left[\frac{Z_L + j Z_o Tan(\beta x)}{Z_o + j Z_L Tan(\beta x)} \right] \tag{1}$$

where: $\beta = \frac{2\pi}{\lambda}$, Z_L is the load impedance, and Z_o is the line impedance. The objective is for $Z_{in}(x) = Z_o$. It can also be written in an alternative form relating input impedance to reflection coefficient [6].

$$Z_{in}(x) = Z_o \left[\frac{1 + \Gamma_v(x)}{1 - \Gamma_v(x)} \right] \tag{2}$$

where $\Gamma_v(x)$ is the voltage reflection coefficient. $\Gamma_v(x)$ is usually a complex number indicating $\Gamma_v(x) \rightarrow |\Gamma_v(x)| \angle\phi = \Gamma_r + j\Gamma_i$.

THE DIPOLE

Equations 3 and 4 can be employed to determine the driving point impedance of the common dipole antenna Z_{in}. The real part, R_r, is known as the radiation resistance and is directly related to the efficiency of the antenna.

$$R_r = \frac{\eta}{2\pi} \left\{ \begin{array}{l} C + \ln(k\,l) - C_i(k\,l) + \frac{1}{2} Sin(k\,l)[S_i(2k\,l) - 2S_i(k\,l)] + \\ \frac{1}{2} Cos(k\,l)[C + \ln(k\,l/2) + C_i(2k\,l) - 2C_i(k\,l)] \end{array} \right\} \tag{3}$$

Where $\eta \approx 120\Omega$ is the impedance of free space, $C = 0.5772$ (Euler's Constant), k is the wave number, and l is the length. S_i and C_i are the sine integral and cosine integral respectively. The imaginary component, X_m, can be computed with the following expression where a represents the wire radius:

$$X_m = \frac{\eta}{4\pi}\left\{\begin{array}{l} 2S_i(kl) + Cos(kl)[2S_i(kl) - S_i(2kl)] - \\ Sin(kl)\left[2C_i(kl) - C_i(2kl) - C_i\left(\frac{2ka^2}{l}\right)\right] \end{array}\right\} \quad (4)$$

These impedances can be mapped on the Smith chart yielding a very stable control design. For the half-wave dipole $Z_{in} \approx 73 + j42.5\Omega$. For proper matching, the antenna is trimmed to approximately 98% of $\lambda/2$, thus reducing the reactive component to zero without adversely affecting R_r. This is illustrated in Fig. 2. The VSWR can be computed easily from the values of R_r and X_m for a given transmission line impedance which in this case was chosen to be 52Ω as shown in Fig. 3.

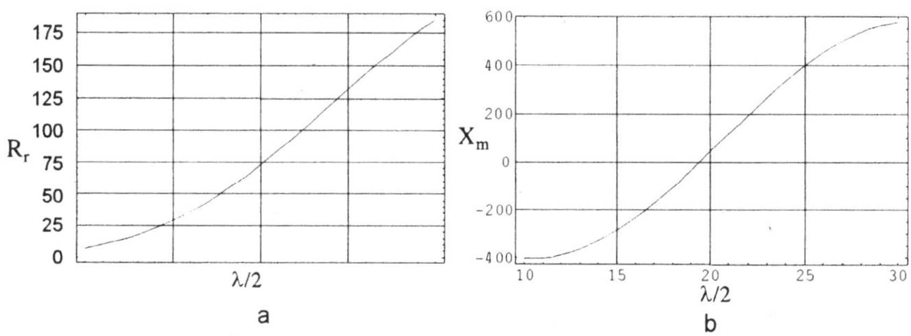

Figure 2. The resistive component of the antenna impedance (a). The reactive component (b) for the 40 meter band.

DEVELOPING THE SYSTEM

Through the use of equations 3 and 4 it is easy to find the VSWR, but the converse is not true. The value of l cannot be computed in closed form, and inductive or capacitive antennas will yield the same VSWR depending on length, i.e., the plot in Fig. 3 is not a one-to-one mapping. Higher frequencies yield VSWR values resulting from an inductive antenna while the antenna becomes capacitive at lower frequencies. To solve this problem two networks were trained, one for higher frequencies and one for lower values. The input was the VSWR and the targets were the appropriate antenna impedances. Fifty patterns were used on each network having six sigmoidal nodes in the hidden layer and two linear output nodes. Both networks trained within 200 epochs in MATLAB using the Levenburg-Marquardt algorithm. The outputs from these networks are input to another network called the tuner described below. Referring back to equation (1) it is easily distinguished that a shorted stub $(Z_L = 0)$ has the following impedance:

$$Z_{in}(x) = Z_o\left[\frac{0 + jZ_o Tan(\beta x)}{Z_o + j0Tan(\beta x)}\right] = jZ_o Tan(\beta x) \quad (5)$$

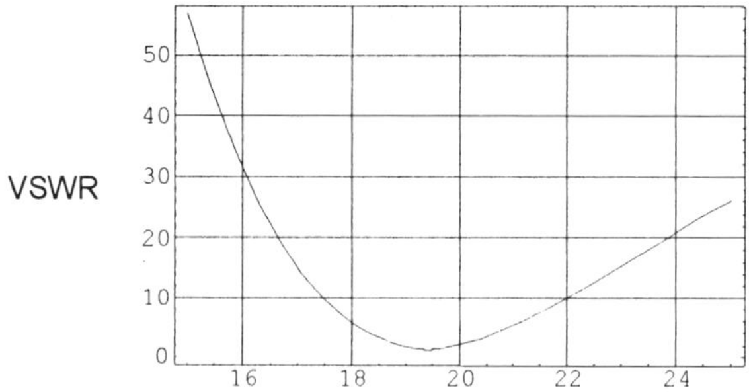

Figure 3. The VSWR of the antenna. The minimum is at point slightly less than one half wavelength.

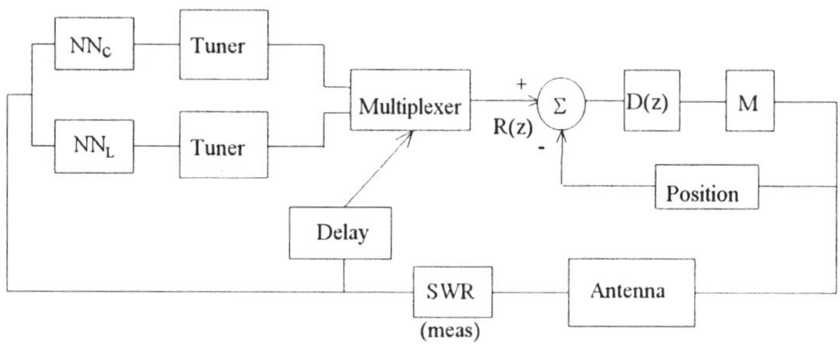

Figure 4. Overall control system block diagram. NN_C is a network that outputs the antenna characteristics of a capacitive system from impedance mismatch, and NN_L performs the same function for inductive systems. The tuner is a single network that adjusts the stub lengths.

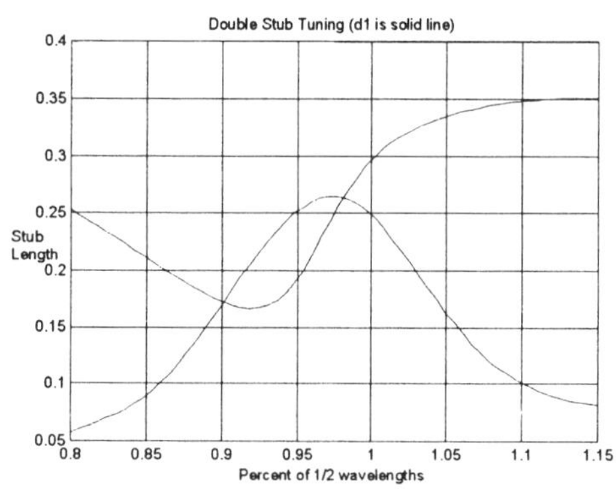

Figure 5. Stub length values shown against wavelength.

which is purely reactive for a lossless line. Furthermore, for the purposes of this paper Z_o is set to unity resulting in a stub reactance of $Z_{in}(x) = j\ Tan(\beta x)$ with $\beta = 2\pi$. This enables the data sets for the neural networks to be independent of the line impedance. A set of non-linear equations was solved numerically for the stub lengths d_1 and d_2 in terms of wavelength for antenna mismatches up to VSWRs of 20 due to capacitive and inductive reactance. This produced the data set for the tuner in Fig. 4.

It is a two input, two output network having 8 sigmoidal nodes in the hidden layer. Approximately 500 epochs were required to train it to an error of $10^{-3}\lambda$ for d_1 and d_2. Both tuners in Fig. 4 are identical but are shown separately for clarity. Figure 5 illustrates the values of d_1 and d_2 necessary to retain a low VSWR for the two types of antenna discussed in this paper. The main problem in realizing this system was the fact that the capacitive reactance and inductive reactance networks yield different results. One network will improve the match while the other worsens the effect. The tuner networks in turn produce differing values of stub length. This is the reason for the delay and the multiplexer. The multiplexer randomly chooses one of the tuner outputs and provides the set point $R(z)$ for a PI controller $D(z)$. M represents the set of motors for the stubs. The outer loop of Fig. 4 is the measured VSWR providing feedback to the system. If the match has degraded from the first correction the delay block switches the input of the multiplexer to the other tuner and the process is repeated. This allows for a very inexpensive VSWR bridge to be employed and the use of small networks that can be programmed onto a microcontroller.

SIMULATION RESULTS

At this point the system was simulated using MATLAB. The frequency was set at 50 MHz, the six-meter band, to enable it to be implemented with readily available components when time permits. The controller was able to compensate for frequencies of 40 MHz to 57 MHz while remaining stable as seen in Fig. 6.

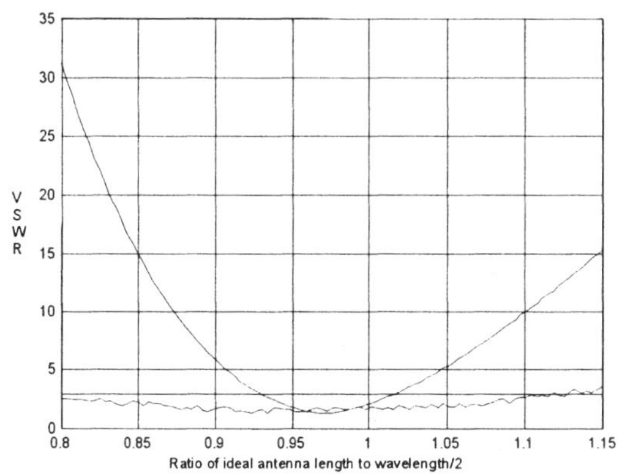

Figure 6. The system is able to keep the VSWR less than 3 (dotted line) for a 15% to 20 % change in frequency.

The minimum VSWR was 1.3 with an average of 2.1 across a 17 MHz sweep. The system could not stabilize outside those regions probably due to modeling discrepancies or controller settings. The values of the PI coefficients for the controller were set by trial and error to values of about 0.77 and 0.12 respectively. These values provided the fastest response with minimal settling time of about 1.5 seconds. The step response can be seen in Fig. 7 and it is worth noting that the system never completely settles due to the effects of the multiplexer and delay block (10 ms) in the controller. Once the system reaches the set point the multiplexer tends to oscillate between NN_C and NN_L indefinitely, but the overall error remains below 3%.

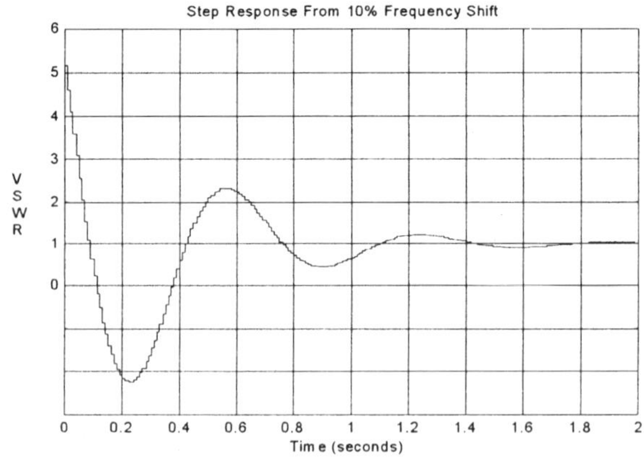

Figure 7. Step response for a 10% shift in frequency.

CONCLUSION

This controller is a hybrid between traditional (PI) control techniques and neural networks. In this case the networks were required to learn the impedance characteristics of a particular type of antenna through measurement of VSWR and determine the tuning parameters. The networks effectively provide a variable set point to a traditional controller. One of the main contributions of this work is that it should be very inexpensive to implement on a microcontroller due to the size of the networks and the relatively slow response speed yet the results indicate a very robust system. It is likely that other antenna models could be employed, e.g., Yagi-Uda, helical, or patch.

References

Kraus, J.D., and Fleisch, D.A., *"Electromagnetics with Applications,"* McGraw-Hill, Fifth Ed., 1999, pp. 137-155.

Ulaby, F.T., *"Fundamentals of Applied Electromagnetics,"* Prentice Hall, 2001, pp. 66-84.

Cheng, D.K., *"Fundamentals of Engineering Electromagnetics,"* Addison Wesley, 1993, pp. 366-380.

Prasad, M.V., Prasad, S, Wang, H., "A Smith Chart Represented by a Neural Network and its Applications," Microwave Symposium Digest, IEEE MTT-S International, 1992, pp. 1565-1568.

Prasad, M.V., Prasad, S, "Microwave Circuit Analysis and Design by a Massively Distributed Computing Network," IEEE Trans. On Microwave Theory and Techniques, Vol. 43, No. 5, May 1995, pp. 1087-1094.

Bandler, J.W., Ismail, M.A., Rayas-Sanchez, J.E., Zhang, Q.J., "Neural Inverse Space Mapping EM-Optimization," *2001 IEEE MTT-S Digest*, pp. 1007-1010.

Watson, P.M., Gupta, K.C., "Design and Optimization of CPW Circuits Using EM-ANN Models for CPW Components," *IEEE Trans. On Microwave Theory and Techniques*, Vol. 45, No. 12, Dec. 1997, pp. 2515-2523.

Guru, B.S., Hizirolu, H.R., *"Electromagnetic Field Theory,"* PWS Publishing, 1998, pp. 538-560.

ARTIFICIAL INTELLIGENCE FOR THE CONTROL OF UNDERWATER VEHICLES

PEPIJN W.J. VAN DE VEN, COLIN FLANAGAN and DANIEL TOAL
Control and Instrumentation Systems Group
Department of Electronic and Computer Engineering
University of Limerick, Limerick, Ireland

email: pepijn.vandeven@ul.ie, colin.flanagan@ul.ie, daniel.toal@ul.ie

ABSTRACT
In this paper recent research efforts in the field of the application of neural networks for the control of (semi-) autonomous underwater vehicles are reviewed. Based on a literature review a classification of neural network approaches to this control problem is proposed and the presented articles are categorized according to the identified categories. Based on practical results as described in the discussed literature a qualitative assessment regarding the performance of the control strategies is given. Per category, or control strategy, the major advantages and disadvantages are identified and discussed.

INTRODUCTION

In recent years, interest in underwater exploration has experienced a substantial increase. As this type of research is normally costly, labour-intensive and potentially hazardous, an increasing research effort in control of unmanned underwater craft can be noticed (Lorentz and Yuh 1996). However, due to the highly non-linear and dynamic characteristics of the underwater world, control of unmanned underwater vehicles (UUVs) is a far from trivial problem. This especially leads to big challenges for a special group of craft within the class of UUVs, the autonomous underwater vehicles (AUVs). These craft, which are deployed without any human supervision, rely on the availability of a highly robust control system. As neural networks are characterized by flexibility and an aptitude for dealing with non-linear problems, they are envisaged to be beneficial when used on AUVs. In this article, state of the art applications of neural networks for the control of AUVs are gathered and discussed. In order to create an overview, a framework of control strategies using neural networks is proposed and advantages and disadvantages for the several classes are identified.

CONTROL STRATEGIES

There are several strategies to apply neural networks (NNs) to the control of a system. The dynamics of the system can be controlled by training a NN to represent the inverse dynamics of the system. In another approach the forward dynamics are learned by the NN. Based on this model a controller is then constructed and consecutively used for control of the system. Or a NN can be used to replace the conventional system identifier in model based control schemes. Regardless of the chosen strategy, a NN will normally consist of at least one hidden layer with non-linear activation functions as the use of such layers enables the NN to solve non-linear classification problems. As underwater craft are highly non-linear systems (Akkizidis and Roberts 1998, Li *et al.* 2002), it is envisaged that neural networks will thus prove beneficial.

In this article, the techniques and strategies, as briefly discussed in the above, are analyzed and classified. The authors propose a classification into three major control strategies: (i) *combined control and learning, CCL* (closely related to direct control as found in literature), (ii) *separate control and learning, SCL* (closely related to indirect control as found in literature) and (iii) *augmented control, AC*.

Combined Control and Learning

In the field of (semi-)autonomous control of underwater vehicles numerous articles can be found presenting both theoretical and practical studies applying NNs that perform the tasks of control and online learning in the same loop. Learning and control are thus performed during consecutive moments in time. The popularity of this approach can be explained by the following. Often the conventional controller can be simply replaced by the NN controller, complexity of the NN controller is limited and the network can adapt to changing dynamics quickly.

An apparent drawback of this approach is that, due to the fact that learning is performed in between control actions, the time for learning is restricted by the desired update rate of the control action. As a result this strategy can be relatively demanding in terms of computational capacity.

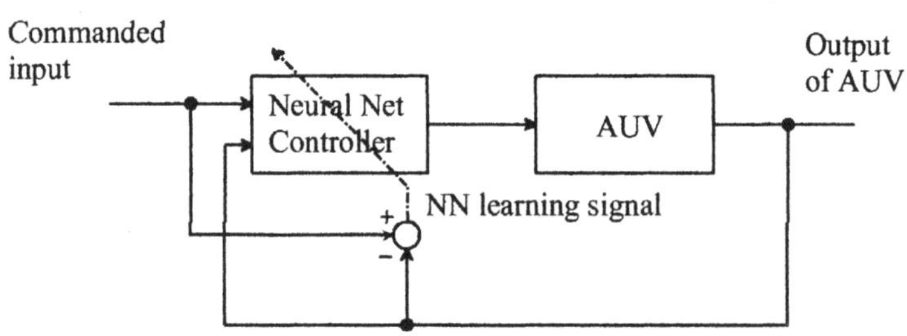

Figure 1: Block level representation of Combined Control and Learning

As previously mentioned, in combined control and learning (CCL) (see fig. 1), control of the craft and training of the NN are performed at consecutive moments in time (Akkizidis and Roberts 1998, Farrell *et al.* 1990, Guo *et al.* 1995, Kim and Yuh 2001, Labonte 2002, Seube 1991, Venugopal *et al.* 1992, Wang *et al.* 1999, Wang *et al.* 2000, Wang and Lee 2002, Yuh 1990, Yuh and Lakshmi 1993, Yuh 1994). Time sharing of the two actions is symbolized by the single feedback loop in fig. 1: the setpoint or desired system output is sent to the controller. Based on its representation of the craft, the controller determines the necessary control signal that is sent to the actuators. The output of the craft is then monitored and fed back to the NN controller. Based on the fed back output and the desired output, the NN is trained. As training is performed online and continuously, the NN adapts to changing dynamics.

Example of CCL. In (Guo *et al.* 1995) an architecture that can be classified as CCL is presented. A two layer NN (2 inputs, 5 hidden nodes, 1 output node) controller is used to control motion in the horizontal plain while feedback of the control parameters is obtained from a compass. Back propagation is used to update the NN in two phases. In the first phase, the initialization phase, the NN is trained off-line with a linear controller as its automated teacher. After off-line training the obtained weights are used as initial weights in the controller. During actual deployment of the craft the weights are further updated using an objective function incorporating tracking error and control rate requirement.

This controller is used to perform simulations and pool tests utilising a craft developed in the Department of Naval Architecture and Ocean Engineering of the National Taiwan University. The authors claim a good agreement between simulations and pool tests.

An interesting claim by the authors is that direct control (CCL) should be preferred to indirect control (SCL) as forward modelling is very problem specific. It usually requires a large network with recurrent connections.

Separate Control and Learning

In contrast to Combined Control and Learning, only a few examples of Separate Control and Learning (see fig. 2) can be found in literature. A reason might be that this strategy inherently needs a more extensive architecture, which is illustrated by (Ishii *et al.* 1994, Ishii *et al.* 1995, Ishii and Ura 2000). There are however several reasons to advocate this approach. Probably the most apparent difference between SCL and CCL is that in SCL learning and control are separate tasks that can be performed simultaneously. As a result, more time can be taken to train the NN and thus more extensive, and possibly more appropriate, network architectures can be applied. Recurrent networks for example, can be used to represent memory effects in the craft and are appropriate for filtering of noisy input data (Polycarpou and Ioannou 1993).

In SCL initial control is performed by a conventional controller or fuzzy network with *a priori* knowledge of the craft. During this period of initial control, which is sometimes referred to as the infancy of the craft, the NNs are trained. The first step in this training process is to train a NN to represent the forward dynamics of the craft. This is done by presenting the NN with control actions and the resulting craft's behaviour. Once the forward dynamics are identified, this NN is used to train a NN controller. After completion of this second training step (a copy of) the latter NN can be used for control of the real craft.

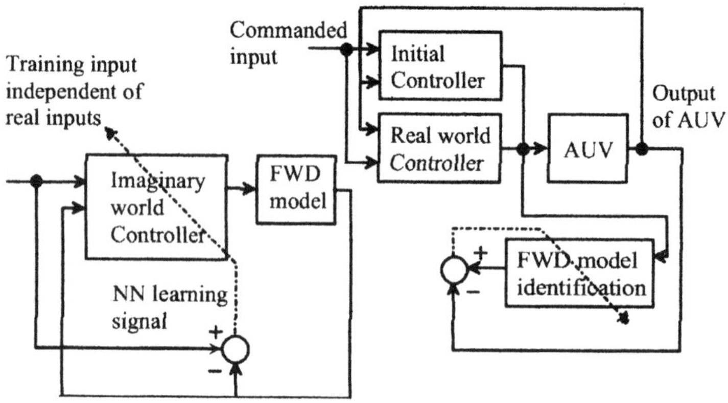

Figure 2: Block level representation of Separate Control and Learning

Example of SCL. A good example of this strategy is the controller called SONCS for Self-Organising Neural-net Control System (Fujii *et al.* 1993, Ishii *et al.* 1994, Ishii *et al.* 1995, Ishii and Ura 2000) that uses a so-called real-world and imaginary-world system. In the imaginary-world system, which is a copy of the real-world system (including the craft's dynamics), a controller is constantly trained using an online updated recurrent forward model of the craft dynamics. This model, in turn, is regularly updated using inputs and outputs from the actual craft. At specified intervals the real-world controller is updated with the learned imaginary-world controller weights. Learning is thus performed in a loop totally independent of the control loop. To control the craft during its infancy a fuzzy controller is used that implements some rudimentary behaviours and is overruled once the NN controller is trained. Figure 2 shows the architecture with the real-world and imaginary-world controllers. Back propagation is used to train the forward model and controller.

Tests are performed using the Twin-Burger test bed (Fujii *et al.* 1993). In these tests the Twin-Burger is required to follow a periodically changing reference signal (+0.5 rad to -0.5 rad and vice versa every 10s) while a constant disturbance of 2.0 Nm is applied to the craft. For this experiment the controller settles at the desired heading in about ten seconds with a small amount of overshoot.

Augmented Control

Augmented control relies on the assumption that a conventional controller can perform a significant portion of the control task under all circumstances. To aid the conventional controller, a NN is added, often in feed forward or feed backward configuration. It adds the necessary response to unmodelled or poorly modelled dynamics, disturbances and other uncertainties, to the total control action. The most apparent advantage of

Augmented Control is that it is normally possible to use an old architecture with only slight changes (Campa *et al.* 2000). As the NN can be trained outside the control loop (after all the conventional controller is assumed to yield some means of control) from a learning point of view there is no need for very fast processors. Most examples of augmented control use feed forward networks combined with back propagation algorithms (Li *et al.* 2002, Yamamoto 1995, Pollini *et al.* 1997).

CONCLUSIONS

In this article several examples of (semi-) autonomous control of underwater vehicles using neural networks were presented and categorized according to the proposed classification framework. As the chosen NN architecture, learning algorithms and test bed highly influence the presented practical results, it is far from trivial to draw conclusions as to which strategy, CCL, SCL or AC, should be regarded to be most favourable. However some general characteristics have been found for the defined classes. Future work will consist of simulations to verify the found results.

ACKNOWLEDGEMENT

This work was supported by the Irish Research Council for Science, Engineering and Technology: funded by the National Development Plan.

References

Akkizidis, I.S. and G.N. Roberts (1998). Fuzzy modelling and fuzzy-neuro motion control of an autonomous underwater robot. *5th International Workshop on Advanced Motion Control* pp. 641–46.

Campa, G., M. Sharma, A.J. Calise and M. Innocenti (2000). Neural network augmentation of linear controllers with application to underwater vehicles. *Proc. of the 2000 American Control Conference* **1**, pp. 75–79.

Farrell, J., B. Goldenthal and K. Govindarajan (1990). Connectionist learning control systems: submarine depth control. *Proc. of the 29th IEEE Conference on Decision and Control* **4**, pp. 2362–7.

Fujii, T. and T. Ura (1990). Development of motion control system for auv using neural nets. *Proceedings of the (1990) Symposium on Autonomous Underwater Vehicle Technology* pp. 81–6.

Fujii, T., T. Ura, Y. Kuroda, H. Chiba, Y. Nose and K. Aramaki (1993). Development of a versatile test-bed "twin-burger" toward realization of intelligent behaviors of autonomous underwater vehicles. *Proc. of IEEE OCEANS '93* pp. 186–91.

Guo, J., F.C. Chiu and Chieh-Chih Wang (1995). Adaptive control of an autonomous underwater vehicle testbed using neural networks. *Proc. OCEANS '95* **2**, pp. 1033–9.

Ishii, K. and T. Ura (2000). An adaptive neural-net controller system for an underwater vehicle. *Control Engineering Practice* **8**, pp. 177–84.

Ishii, K., T. Fujii and T. Ura (1995). An on-line adaptation method in a neural network based control system for auvs. *IEEE Journal of Oceanic Engineering* **2, issue 3**, pp. 221–8.

Ishii, K., T. Ura and T. Fujii (1994). A feedforward neural network for identification and adaptive control of autonomous underwater vehicles. *1994 IEEE International Conference on Neural Networks* **5**, pp. 3216–21.

Kim, T.W. and J. Yuh (2001). A novel neuro-fuzzy controller for autonomous underwater vehicles. *Proc. of the 2001 International conference on Robotics & Automation* pp. 2350–5.

Kodogiannis, P.J., G. Lisboa and J. Lucas (1996). Neural network modelling and control for underwater vehicles. *Artificial Intelligence in Engineering 1*, pp. 203–12.

Labonte, G. (2002). Fast adaptive control of a non-linear system by an adaline: motion in a fluid. *Proc. of the 2002 International Joint Conference on Neural Networks* **2**, pp. 1837–41.

Li, Ji-Hong, Pan-Mook Lee and Sang-Jeong Lee (2002). Neural net based nonlinear adaptive control for autonomous underwater vehicles. *Proc. of the 2002 IEEE International Conference on Robotics and Automation* **2**, pp. 1075–80.

Lorentz, J. and J. Yuh (1996). A survey and experimental study of neural network auv control. *Proc. of the 1996 Symposium on Autonomous Underwater Vehicle Technology* pp. 109–16.

Pollini, L., M. Innocenti and F. Nasuti (1997). Robust feedback linearization with neural network for underwater vehicle control. *Proc. of Oceans '97. MTS/IEEE* **11**, pp. 12–16.

Polycarpou, M. M. and P.A. Ioannou (1993). Stable nonlinear system identification using neural network models. *Neural Networks in Robotics* pp. 165–77.

Seube, N. (1991). Neural network learning rules for control: application to auv tracking control. *Proc. of the 1991 IEEE Conference on Neural Networks for Ocean Engineering* pp. 185–96.

Ura, T., T. Fujii, Y. Nose and Y. Kuroda (1990). Self-organizing control system for underwater vehicles. *Proc. of OCEANS '90* pp. 76–81.

Venugopal, K.P., R. Sudhakar and A.S. Pandya (1992). On-line learning control of autonomous underwater vehicles using feedforward neural networks. *IEEE Journal of Oceanic Engineering* **17**, pp. 308–319.

Wang, J.S. and C.S.G. Lee (2002). Self-adaptive recurrent neuro-fuzzy control for an autonomous underwater vehicles. *Proc. of the 2002 IEEE International conference on Robotics & Automation* pp. 1095–1100.

Wang, J.S., C.S.G. Lee and J. Yuh (1999). An on-line self-organizing neuro-fuzzy control for autonomous underwater vehicles. *Proc. of the 1999 International Conference on Robotics & Automation* pp. 2416–21.

Wang, J.S., C.S.G. Lee and J. Yuh (2000). Self-adaptive neuro-fuzzy systems with fast parameter learning for autonomous underwater vehicles. *Proc. of the 2000 IEEE International conference on Robotics & Automation* pp. 3861–6.

Yamamoto, I. (1995). Application of neural network to marine vehicle. *Proc. of the 1995 IEEE International Conference on Neural Networks* **1**, pp. 220–4.

Yuh, J. (1990). A neural net controller for underwater robotic vehicles. *IEEE Journal of Oceanic Engineering* **15, issue 3**, pp. 161–6.

Yuh, J. (1994). Learning control for underwater robotic vehicles. *IEEE Control Systems Magazine* **14, issue 2**, pp. 39–46.

Yuh, J. and R. Lakshmi (1993). An intelligent control system for remotely operated vehicles. *IEEE Journal of Oceanic Engineering* **18, issue 1**, pp. 55–62.

Fuzzy-Neuro System for Bridge Health Monitoring

Lakshmanan Meyyappan, Cihan Dagli
Smart Engineering Systems Lab, Department of Engineering Management
University of Missouri – Rolla, Rolla, Missouri 65409, USA

Abstract

Many civil and mechanical systems are in continuous use despite aging and associated risk for damage accumulation. Hence, the ability to monitor the structural health of these systems on a real-time basis is becoming very important. This paper describes a practical real-time structural health monitoring system using smart engineering tools and its application to the structural health monitoring of a steel bridge located in Missouri. Vibration data collected from this bridge was processed and fed to the fuzzy logic decision system. The fuzzy logic decision system makes use of fuzzy clustering to determine the possible presence of damage in the bridge. A neural network prediction system which makes use of backpropagation algorithm predicts the amount of actual damage in the members which were predicted damaged by the fuzzy logic.

INTRODUCTION

Over the past few decades structural health monitoring has received paramount importance. A need to develop reliable real-time health monitoring techniques is increasing. These types of health monitoring technique allow continuous monitoring of the systems, capable of detecting damage, locating the damage and estimating the severity of the damage. The real-time health monitoring system described in this paper meets the following requirements: it is a Non-Destructive Evaluation (NDE) technique, it is capable of monitoring the structure on a global basis, initial and operating costs are inexpensive, it is sensitive as to detect small damage, it does not cause any disruption to the normal operation of the structure, the equipment used for collecting data is small, it is easy to replace and is reliable. Real-time monitoring can be achieved through a smart health monitoring technique that uses fuzzy-neuro hybrid architecture. Damage is a state of change that affects the present or future performance of a system. Implicit in the above definition is the fact that damage detection involves comparison with some initial undamaged state. In this project the sensors were connected to the bridge, which was monitored. Wireless Bluetooth devices were used to collect vibration data and send it continuously to the laboratory where the data was analyzed. As a first step some pre-processing was done to feed the data to the fuzzy system. The fuzzy logic system was a decision-making system.

The fuzzy logic basically uses some relationships between the members of a bridge under normal undamaged operational condition to detect the possibility of damage in the bridge. If there is a possibility of damage, the fuzzy logic gives a list of members, which could possibly be damaged. This data is fed into the neural network system. The neural network takes speed as the input and tries to predict the vibration value that should occur if the bridge was undamaged. Based on this the neural networks gives out its result. Fuzzy logic and neural network were trained using the data collected from the Teardrop Bridge which is located at 20 miles from Rolla at Teardrop Road, Devils Elbow, MO 65457.

DATA ACQUISITION AND ANALYSIS

The fuzzy logic and neural network need to be trained using data collected from a system operating under normal undamaged condition. The sensors were mounted on various truss elements (both vertical and diagonal trusses) in one span of the bridge at a height of 5 feet above ground level, one sensor at a time. With the sensor placed in one of the members shown above, a test vehicle was driven from both directions at speeds of 10, 15 and 20 mph (16.1, 24.2 and 32.2 km/h) and the vibration data was collected.

The same procedure was followed with sensors in various members of the bridge. There are two sets of data acquisition system. The first data acquisition system is an in-field system. This consists of a number of sensors mounted on various members of the bridge. The sensors are wired to a micro-processor which is connected to a blue tooth device. The blue tooth device transmits the data continuously by wireless means to the second set of data acquisition system. The second system consists of a PC running on Windows and uses MATLAB for data analysis. The neural network and fuzzy logic is also run within the MATLAB. With the sensor mounted on one of the truss, data was collected continuously by driving the test vehicle (GMC pick-up) at the speeds previously described in both directions. Signal processing is performed on this data and the power spectrum values are deduced. The power spectrum values were unique for each of the different members at different speeds of the test vehicle.

DAMAGE DETECTION

This step involves detecting that damage has occurred, finding the location of the damage, and finally the severity of the damage. The data was analyzed to find a technique, which would uniquely identify the damage, its location and severity. From the analysis it was found that the power spectrum peak values are unique for every different location of the sensor and that it increases non-linearly with increase in speed. Even though it varies with speed, the pattern, that is, its relation with a different member remained the same. To put it more lucidly, if the power spectrum peak value at location 6 at a speed of 10mph was greater than the power spectrum peak value at location 5 at 10mph, the relation always holds at all different speed values. This fact was made use in detecting the presence of damage in any member. If there was some damage, then based on the various relationships the member that was actually faulty was determined and also the severity of the damage was determined.

This method is very good because it is independent of the speed and also the mass of the vehicle. At all speeds (10 mph, 15 mph and 20 mph) the power spectrum value at a particular location (say member 3) is smaller than at another location (say member 2) which will be smaller or greater than another member (say member 1). This relationship always holds. So if there is some discrepancy in this relationship, it can be concluded that there is some damage in the bridge and based on the various relationships the damaged member can be located. The first step of detecting a possible damage is done using fuzzy logic. If possible damage is predicted by fuzzy logic, then the members which may be possibly damaged are found out and the neural network is used to predict the amount of damage in each of those members. The neural network makes the prediction based on the speed of the vehicle. The speed has to be entered by the user. It can be collected using a speed gun in real time and fed to the neural network directly. The fuzzy logic system and the neural network architecture will be discussed in the following sections.

FUZZY LOGIC DECISION SYSTEM

A fuzzy logic program was written using MATLAB, which performs all the required operations. As described above the power spectrum peak values are given as input to the system. Initially fuzzy ranking was used to rank the members based on their vibration level. The system has been designed using 101 membership functions. The term set used is (0, 1, 2, 3 ... 99, 100). Numerical names are given to the term sets just for simplicity. In this system triangular fuzzy numbers are used as membership functions corresponding to the elements in term set. The reason of using triangular fuzzy number is that it is intuitively easy to be used by the user. The triangular fuzzy number is denoted as follows:

$$f_M(x) = \begin{cases} \dfrac{(x-a)}{(b-a)}, & a \le x \le b, \\[2ex] \dfrac{(x-c)}{(b-c)}, & b \le x \le c, \\[2ex] 0, & otherwise \end{cases}$$

Where a, b, and c are real numbers and ($a \le b \le c$)

The input (power spectrum value) was fuzzified and was converted to a triangular fuzzy number. An input value may belong to two membership functions with different degree of memberships. In such cases Zadeh operators listed below are commonly used:

- AND => use the minimum of the options.
- OR => use the maximum of the options
- NOT => use 1-option

According to Zadeh's interpretation, AND operator is used for combining two different input classes. OR operator is used when choosing the membership value that belongs to an output class. NOT operator is used while combining two incompatible input classes. AND operator was mainly used in this case. The fuzzy values are then ordered. The process of ordering fuzzy numbers is called ranking. There are several methods of ranking fuzzy numbers [8] [14] [16]. In this paper, the total integral value method [16] is used because it is very simple and also accurate. Let the total integral value for triangular fuzzy number $F = (a, b, c)$, be defined as

$$I_T^{\alpha}(F) = \left(\tfrac{1}{2}\right)\left[\alpha c + b + (1-\alpha)a\right]$$

Here, α is an index of optimism that represents the degree of optimism of the decision-maker it lays between 0 and 1. A larger value of α indicates a higher degree of optimism. For given fuzzy numbers F_i and F_j, if $I_T^{\alpha}(F_i) < I_T^{\alpha}(F_j)$, then $F_i < F_j$; if $I_T^{\alpha}(F_i) = I_T^{\alpha}(F_j)$, then $F_i = F_j$; and if $I_T^{\alpha}(F_i) > I_T^{\alpha}(F_j)$, then $F_i > F_j$. As described above the fuzzy system will give a ranked member relation output. This was compared with the general order member relation got as input from the user. If both these values matched perfectly then the bridge was proclaimed perfect. If there were some discrepancies then one or more member may be damaged. Based on the above calculations and comparisons some members were proclaimed as possibly damaged members. The fuzzy ranking model is a simple and straight forward tool and is easy to implement and use. But when the number of members under inspection is large, it is very difficult to use fuzzy ranking. So fuzzy clustering was used. Members with similar vibration levels were clustered together. If a member after clustering belonged to its original cluster, then it is perfect. If a member is grouped in a different clustering by the fuzzy clustering algorithm, then it may probably be damaged. The power spectrum value of the five members from which data was collected was fed to the fuzzy logic decision system.

The fuzzy logic decision system making use of FCM classifies the five members into three clusters. Members 1 and 5 have similar vibration levels; members 2 and 4 have similar vibration levels. Hence, if the bridge is perfect, cluster 1 consists of member 1 and 5, cluster 2 consists of member 2 and 4 and cluster 3 consists of member 3. FCM was performed on the 10mph N-S power spectrum values and the FCM graph is shown in figure 2. As can been seen in figure 2, the FCM clustering algorithm has grouped members 1 and 5 together, members 2 and 4 together and member 3 separately. If the entered data is classified as above, then the bridge is perfect. If there are some discrepancies in the relation, then the bridge may possibly be damaged. But this does not mean that the member is surely damaged. One set of data which was collected by driving the test vehicle at 10 mph from South to North yielded contradictory results. FCM analysis on this data is shown *below in figure* 3.

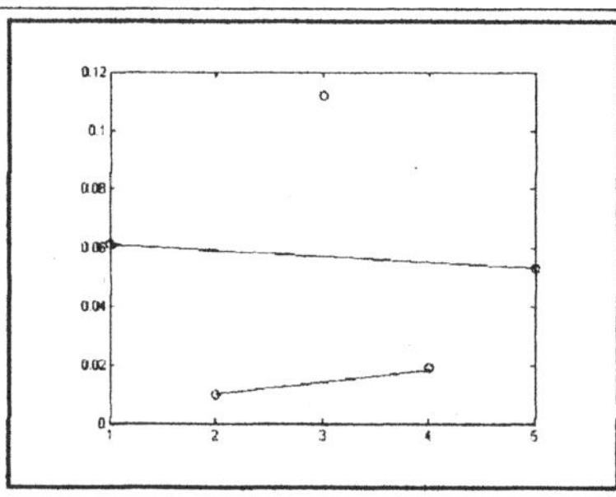

Figure 2. FCM on Data 1

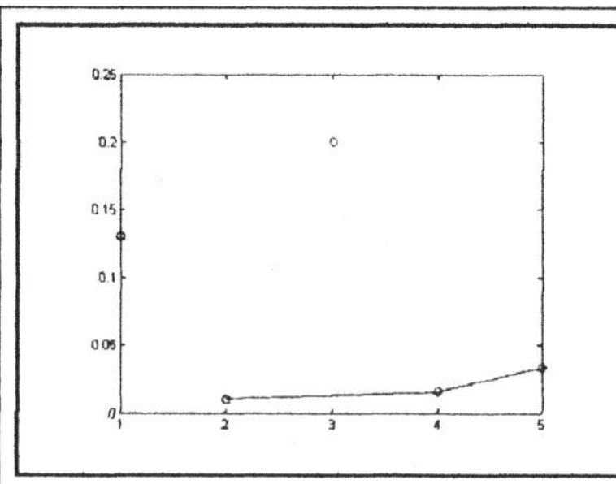

Figure 3. FCM on Data 2

As can be seen in figure 3, FCM classified members 2, 4 and 5 into one cluster and 1 and 3 separately. The fuzzy decision system based on the above system will declare that either member 1 or member 5 is possibly damaged even though both these members are perfect. This contradictory result may be due to a number of insignificant reasons such as the test vehicle hitting a pot hole during the test drive or may be because of some lose connection. So, further analysis is done using neural network.

NEURAL NETWORK PREDICTION SYSTEM

Neural network is a promising new generation information processing system that demonstrates the ability to learn, recall, and generalize from training patterns and data. They are deliberately constructed to make use of some organizational principle resembling human brain. When a car or any other vehicle moves on a bridge, the bridge vibration varies with respect to the speed of the vehicle. Therefore by training the neural network with different speeds and their corresponding vibration values (power spectrum values of vibration data was used) the neural network can be made to predict the power spectrum value of a particular vibration value at a particular speed for a particular member. A number of neural network architectures where tested for this application and it was found that back propagation network using Levenberg-Marquardt (LM) method yielded the best results. The next step involves selecting the number of neurons in the input, hidden and output layers. Different number of neurons in the hidden layer was used and tested, with the aim of having minimum possible number of neurons as this will not only ensure better performance and faster training but also will prevent any potential over fitting problems. So it finally boiled down to fifteen neurons. Different combinations of transfer functions in the hidden layer and the output layer were tested. Finally tansig in the hidden layer and purelin in the outer layer yielded the best results. The Backpropagation architecture is shown in Figure 4.

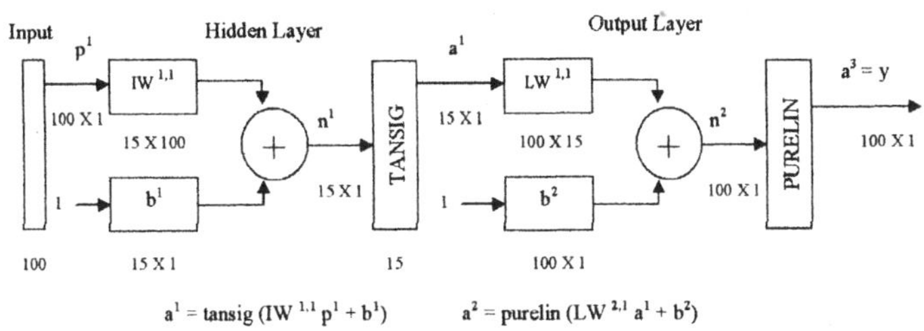

Figure 4. Neural Network Architecture

In order to overcome the over fitting problem of the BP network and improve its generalization, the early stopping technique was used. In this technique, the available data are divided into three subsets. The first subset was the training set, which was used for computing the gradient and updating the network weights and biases. The second subset was the validation set. The error on the validation set was monitored during the training process. The validation error will normally decrease during the initial phase of training, as also the training set error does. However, when the network begins to over fit the data, the error on the validation set will typically begin to rise. When the validation error increases for a specified number of iterations, the training was stopped, and the weights and biases at the minimum of the validation error were returned. The third set was the testing set. The user was asked to input the speed of the vehicle. The speed can be collected using a speed gun in real time. The neural network was run for different members (which were put in the possibly-damaged list by the fuzzy logic system) separately and it predicted a corresponding power spectrum peak value for each one of them. This predicted value was compared with the value entered by the user. Based on the difference between these two values the program outputs the result as: the member N is fine, the member N has small damage, the member N has medium damage, the member N has large damage. As can be seen above the output is a fuzzy linguistic value. The main reason for giving an output like this than giving values like 10 mm damage or so is because 10 mm damage may not give the user a good idea of the damage. 10 mm damage may be very small in the case of a very large bridge but may be significant in a small bridge. So, fuzzy linguistic variables are used which can be understood easily and corresponding actions taken.

CONCLUSIONS

An approach for monitoring the structural health of a bridge on a real-time basis based on vibration data using smart engineering systems and tools has been demonstrated. The demonstrated method has a number of advantages over the conventional methods. This system can monitor the bridge on a global and real-time basis. Data is collected and analyzed continuously. It is an NDE technique and does not cause disruption to the normal operation of the bridge. The sensitivity of the approach is high and it can detect small damages. The use of fuzzy logic and neural network makes it robust and helps in handling the non-linearities associated with the structural behavior. The initial and operating cost of this system is also less. It is a relatively simple and easy approach, yet it is accurate and reliable. Vibration based health monitoring has a number of advantages over the other conventional methods. But it does not come without a price. In order to detect damages efficiently and accurately data from damaged bridge is needed, which is generally not available. In such cases generally expert knowledge and experience is used in arriving at this data. The current system relies on this. A better way would be do simulation using finite element analysis to get the damaged data set. Currently FCM clustering is used to analyze the data. Sometimes FCM does not yield accurate results. Other fuzzy clustering techniques such as the subtractive clustering, adaptive fuzzy c-variety clustering or fuzzy c-rings algorithms may yield better results. SOM based neural network clustering may also prove efficient.

ACKNOWLEDGEMENTS

[1] Jose, Meta. Computer Engineering Department, UMR
[2] Silva, Pedro. Civil Engineering Department, UMR
[3] Pottinger, Hardy. Computer Engineering Department, UMR

REFERENCES

[1] Altrock, C. von, *Fuzzy Logic & Neurofuzzy Applications Explained*, Prentice-Hall Inc., Upper Saddle River, New Jersey Prentice-Hall, Inc. 1995

[2] Battani, M., Casciati, F., and Faravelli, L. "Fuzzy Control of Structural Vibration: An Active Mass System Driven by Fuzzy Controller", Earthquake Engr and Struct. Dyn., 27(11), 1998, pp. 1267-1276.

[3] Buckley, J. J. "Ranking Alternatives Using Fuzzy Numbers, Fuzzy Sets and Systems" 15, 1985, pp. 21.

[4] Chen, S. H. "Ranking Fuzzy Numbers With Maximizing Set And Minimizing Set, Fuzzy Sets And Systems", 1985, 17, pp. 113.

[5] Joghatatie, A., and Ghaboussi, J. "Neural Network and Fuzzy Logic in Structural Control." Proc. 1st World Conf. on struct. Control, Vol. 1, Pasadena, Calif., 1999, pp. 21-30.

[6] Liou, T. S. and Wang, M. J. J. "Ranking Fuzzy Numbers with Integral Value, Fuzzy Sets and Systems", 1992, 50, pp. 247.

[7] Rumelhart, D.E., Hinton, G.E., Williams, R.J. "Learning Representations by Backpropagation Errors," Nature, Vol. 323, 1986, pp. 533-6

A DESIGN METHOD FOR PI CONTROL FOR MINIMUM PHASE SYSTEMS

KOU YAMADA
Department of Mechanical System Engineering, Gunma University
1-5-1 Tenjincho, Kiryu, Japan

TAKAYUKI MOKI
Department of Mechanical System Engineering, Gunma University
1-5-1 Tenjincho, Kiryu, Japan

ABSTRACT
In the present paper, we examine a design method for PI (proportional-integral) control for minimum phase systems using the parametrization for the class of all proper stabilizing controllers for linear minimum phase systems. PI controller structure is the most widely used one in industrial applications. However no method has been published to guarantee the stability of PI control system. In this paper, we overcome this problem and propose a design method of PI control systems such that the stability of the PI control system is guaranteed.

Introduction

In this paper, we examine a design method for PI(proportional-integral) control for minimum phase systems. PI control is a control system of which the controller include the proportional and the integral (Minorsky, 1922, Callender et al., 1936, Astrom and Hagglund, 1955). PI controller structure is the most widely used one in industrial applications (Astrom and Hagglund, 1955). The controller has two tuning parameters which are often tuned by trial and error (Ziegler and Nichols, 1942, Astrom and Hagglund, 1955). Several papers on PI tuning methods have been considered for industrial use (Ziegler and Nichols, 1942, Astrom and Hagglund, 1955). However no method has been published to guarantee the stability of PI control system.

In this paper, we propose a design method of PI control systems for the minimum phase systems such that the PI control system is asymptotically stable. In order to design PI controller, we use the idea of fusion of the parametrization of all stabilizing controllers for the minimum phase systems (Yamada, 2001) and the model driven control (Kimura, 2000).

Problem formulation

Let us consider the control sytem written by

$$\left\{ \begin{array}{rcl} y & = & G(s)u \\ u & = & C(s)(r-y) \end{array} \right. , \tag{1}$$

where $G(s) \in R$ is the single-input/single-output minimum phase plant, that is, $G(s)$ has no zero in the closed right half plane. $C(s) \in R$ is the controller, $u \in R$ is the control input, $y \in R$ is the output and $r \in R$ is the reference input.

When the controller $C(s)$ is given by

$$C(s) = a_P + \frac{a_I}{s}, \tag{2}$$

then the controlle $C(s)$ is called PI controller, where $a_P \in R$ and $a_I \in R$. a_P and a_I are settled so that the control system in Eq.(1) has desirable control characteristics such as steady state characteristic and transient characteristic. The transfer function from r to y in Eq.(1) is written by

$$y = \frac{G(s)\left(a_P + \frac{a_I}{s}\right)}{1 + G(s)\left(a_P + \frac{a_I}{s}\right)} r. \tag{3}$$

It is obvious that when a_P and a_I are settled at random, the stability of the control system in Eq.(1) does not guaranteed. In addition, no design method of PI controller has been published such that the stability of control system in Eq.(1) is guaranteed.

The problem considered in this paper is to propose a design method of PI control system such that the stability of the control system in Eq.(1) is guaranteed. We propose a design method of PI control system using the idea of the fusion of the parametrization of all stabilizing controllers for the minimum phase systems (Yamada, 2001) and the model driven control systems (Kimura, 2000). In the next section, we describe the detail of a design method of PI control system using above-mentioned idea.

PI control

In this section, we propose a design method of PI control system such that the stability of the control system is guaranteed.

According to Yamada (2001), the parametrization for the class of all stabilizing controllers for $G(s)$ is given by

$$C(s) = \frac{C_f(s)}{1 + C_f(s)K(s)}. \tag{4}$$

$$\left(\lim_{\omega\to\infty} \left(1 + \bar{C}(j\omega)K(j\omega)\right) \neq 0\right)$$

Here $K(s)$ is any asymptotically stable biproper rational function that make $G(s) + K(s)$ of minimum phase. $C_f(s)$ is denoted by

$$C_f(s) = \frac{1}{Q(s)} - \frac{1}{G(s) + K(s)} \tag{5}$$

and $Q(s)$ is any non-zero asymptotically stable biproper rational function. $K(s)$ is settled satisfying

$$K(s)|_{s=0} = 0 \tag{6}$$

in the class of biproper rational function to make $G(s) + K(s)$ of minimum phase. The fact that $K(s)$ satisfying Eq.(6) exists is proved as following theorem.

Theorem 1 *If $G(s)$ is of minimum phase, there exists $K(s)$ that satisfies the following expressions:*

1. *$G(s) + K(s)$ is of minimum phase.*
2. *$K(s)$ is a biproper asymptotically stable rational transfer function.*
3. *$K(0) = 0$*

Proof of this theorem requires following lemma.
(Proof) Let $\bar{K}(s) \in RH_\infty$ and $A(s) = \frac{s}{s+\alpha}$, where $\alpha > 0$. $G(s)/A(s) + \bar{K}(s)$ is rewritten as

$$\frac{G(s)}{A(s)} + \bar{K}(s) = \frac{N(s) + \bar{K}(s)A(s)D(s)}{A(s)D(s)}, \tag{7}$$

where $N(s)$ and $D(s)$ are any right coprime factorization of $G(s)$ over RH_∞ given by

$$G(s) = D^{-1}(s)N(s). \tag{8}$$

The proof of Theorem 1 is obtained by showing that $\bar{K}(s) \in RH_\infty$ exists such that $N(s)+\bar{K}(s)A(s)D(s) \in \mathcal{U}$. This is equivalent to the interpolation problem satisfying

$$\frac{\mathrm{d}^j}{\mathrm{d}s^j}U(s_i) = \frac{\mathrm{d}^j}{\mathrm{d}s^j}N(s_i)(j = 0, \ldots, m_i - 1, i = 1, \ldots, l), \tag{9}$$

where $U(s)$ is defined by

$$U(s) = N(s) + \bar{K}(s)A(s)D(s), \tag{10}$$

$s_1, \cdots, s_l$ are distinct zeroes of $A(s)D(s)$ on the positive real axis and $m_1, \cdots, m_l$ are the corresponding multiplicities. Since $G(s)$ is of minimum phase, $N(s)$ is also of minimum phase. This implies that all of $N(s_i)$ are the same sign. From the result in the book (Vidyasagar, 1985), there exists $U(s) \in \mathcal{U}$ such that Eq.(9) holds.

Next, we will show that there exists $\bar{K}(s)$ which is biproper. That is, it will be shown that if $U(s) \in \mathcal{U}$ exist such that Eq.(9) holds, then $\bar{K}(s) = (U(s) - N(s))/(A(s)D(s))$ is biproper. $U(s)$ is written by

$$U(s) = N(s) + \bar{K}(s)A(s)D(s). \tag{11}$$

This equation is rewritten as

$$\bar{K}(s) = \frac{n(U(s))d(A(s))d(D(s))d(N(s)) - d(Q(s))d(A(s))d(D(s))n(N(s))}{d(U(s))n(A(s))n(D(s))d(N(s))}, \tag{12}$$

where $n(\cdot)$ denotes the numerator of $(\cdot)$ and $d(\cdot)$ denotes the denominator of $(\cdot)$. The assumption that $U(s)$ holds Eq.(9) implies that $K(s)$ written by

Eq.(12) is asymptotically stable. Since $U(s)$, $A(s)$ and $D(s)$ are biproper and $N(s)$ is strictly proper,

$$\lambda\{n(U(s))\} = \lambda\{d(U(s))\}, \quad \lambda\{n(N(s))\} < \lambda\{d(N(s))\}$$
$$\lambda\{n(D(s))\} = \lambda\{d(D(s))\}, \quad \lambda\{n(A(s))\} = \lambda\{d(A(s))\}$$

are satisfied, where $\lambda\{(\cdot)\}$ denotes the degree of $(\cdot)$. Therefore we have

$$\begin{aligned} & \lambda\{n(\bar{K}(s))\} \\ &= \lambda\left\{n\left(\frac{U(s)-N(s)}{A(s)D(s)}\right)\right\} \\ &= \lambda\{n(U(s))d(A(s))d(D(s))d(N(s)) - d(U(s))d(A(s))d(D(s))n(N(s))\} \\ &= \lambda\{n(U(s))d(A(s))d(D(s))d(N(s))\} \end{aligned} \tag{13}$$

and

$$\begin{aligned} \lambda\{d(\bar{K}(s))\} &= \lambda\left\{d\left(\frac{U(s)-N(s)}{A(s)D(s)}\right)\right\} \\ &= \lambda\{d(U(s))n(A(s))n(D(s))d(N(s))\} \\ &= \lambda\{n(U(s))d(A(s))d(D(s))d(N(s))\}. \end{aligned} \tag{14}$$

Thus we have

$$\lambda\left\{n\left(\frac{U(s)-N(s)}{A(s)D(s)}\right)\right\} = \lambda\left\{d\left(\frac{U(s)-N(s)}{A(s)D(s)}\right)\right\}. \tag{15}$$

Therefore, if there exists $U(s)$ satisfying Eq.(9), $\bar{K}(s)$ denoted by Eq.(12) is biproper.

The rest of proof of this theorem is to show that when $K(s)$ is settled by

$$K(s) = A(s)\bar{K}(s), \tag{16}$$

then $K(s)$ is biproper and satisfying $K(0) = 0$. The fact that $K(s)$ is biproper is obvious since both $A(s)$ and $\bar{K}(s)$ are biproper. Since $\bar{K}(s) \in RH_\infty$ and $A(0) = 0$, $K(s)$ written by Eq.(16) satisfies $K(0) = 0$. This completes the proof of this theorem. ∎

On the parametrization of all stabilizing controllers in Eq.(5) for $G(s)$, $G(s) + K(s)$ is considered as the biproper model of $G(s)$ since $G(s) + K(s)$ takes over the characteristics of $G(s)$ such as the poles of $G(s)$ and the minimum phase characteristics. There exists the idea of the model driven control (Kimura, 2000), which is to improve the control performance using the model of the plant. Using the idea of the model feedback control and the parametrization of all stabilizing controllers for $G(s)$, we propose a design method of PI control system such that the stability of the control system is guaranteed. In order to design the PI control system using the idea of model driven control, the free parameter $Q(s)$ in Eq.(5) of controller $C(s)$ in Eq.(4) is designed as

$$\frac{1}{Q(s)} = a_P + \frac{a_I}{s}. \tag{17}$$

According to Yamada(2001), the free parameter $Q(s)$ must be biproper asymptotically stable rational function. It is obvious that the necessary and sufficient condition that the $Q(s)$ in Eq.(17) is biproper is

$$a_P \neq 0. \tag{18}$$

The necessary and sufficient condition that the free parameter $Q(s)$ in Eq.(17) is asymptotically stable is summarized following theorem.

Theorem 2 *The necessary and sufficient condition that $Q(s)$ is asymptotically stable is that a_P holds*

$$a_P > 0, \qquad a_I > 0 \tag{19}$$

or

$$a_P < 0, \qquad a_I < 0. \tag{20}$$

(Proof) From Eq.(17), $Q(s)$ is rewritten by

$$Q(s) = \frac{s}{a_P s + a_I}. \tag{21}$$

From this equation, it is obvious that the necessary and sufficient condition that the free parameter $Q(s)$ is asymptotically stable is to hold Eq.(19) or Eq.(20). We have thus proved this theorem. ∎

Next, we show the role of a_P and a_I in Eq.(17) are equivalent to that of conventional PI parameters (a_P and a_I) in Eq.(2). First the role of a_P in Eq.(17) and that of a_P in Eq.(2) are considered. $C(s)$ in Eq.(2) satisfies

$$sC(s)|_{s=0} = a_I. \tag{22}$$

This implies that when a_I is made bigger, then the influence of the integrator is tendency bigger. Conversely when a_I is made smaller, then the influence of the integrator is tendency smaller. On the other hand, the propose PI controller in Eq.(4) satisfies

$$\begin{aligned} s\bar{C}(s)|_{s=0} &= s\left.\frac{G(s)+K(s)-Q(s)}{G(s)Q(s)+K(s)\left(G(s)+K(s)\right)}\right|_{s=0} \\ &= \frac{1}{\dfrac{1}{a_I}+k}, \end{aligned} \tag{23}$$

where $k = \left.\dfrac{K(s)}{s}\right|_{s=0} = \text{const}$. This implies that when a_I is made bigger, then the influence of the integrator is tendency bigger. Conversely when a_I is made smaller, then the influence of the integrator is tendency smaller. Therefore we can find that the role of a_I in the proposed PI controller in Eq.(4) is equivalent to that of a_I in the conventional PI controller in Eq.(2).

Next, we consider the role of a_P in Eq.(4) and that in Eq.(2). In order to make clear the role of a_P, let $a_I = 0$. The conventional PI controller $C(s)$ in Eq.(2) satisfies

$$C(s)|_{s=0} = a_P. \tag{24}$$

This implies that when a_P is made bigger, then the influence of the proportional is tendency bigger. Conversely when a_P is made smaller, then the influence of the proportional is tendency smaller. On the other hand, the propose PI controller in Eq.(4) satisfies

$$C(s)|_{s=0} = a_P - \frac{1}{G(0)}. \qquad (25)$$

This implies that when a_P is made bigger, then the influence of the proportional is tendency bigger. Conversely when a_P is made smaller, then the influence of the proportional is tendency smaller. Therefore we can find that the role of a_P in the proposed PI controller in Eq.(4) is equivalent to that of a_I in the conventional PI controller in Eq.(2).

In this way, we can design the PI control system such that the stability of the control system is guaranteed. Using the result in this paper, even if a_I, a_P and $K(s)$ are selected using Neural Networks, we can obtain high performance stable PI control systems.

Conclusion

In this paper, we proposed a design method of PI control systems for the minimum phase systems such that the PI control system is asymptotically stable. In order to design PI controller, we use the idea of fusion of the parametrization of all stabilizing controllers for the minimum phase systems (Yamada, 2001) and the model driven control (Kimura, 2000).

This research was partially supported by the Sasakawa Scientific Research Grant from The Japan Science Society.

References

N. Minorsky: Directional stability of automatically steered bodies, J. Amer. Coc., Naval Engineers, 42-2, pp.280-309(1922)

A. Callender, D.R. Hartree and A. Porter: Time-lag in a control system, Phil. Trans. Roy. Soc. London, A235-756(1932)

J.G.Zieglae and N.B. Nichols: Optimum settings for automatic controllers, Trans. ASME, 64, pp.759-768(1942)

K. Astrom and T. Hagglund: PID controllers: Theory esign, and tuning, Instrument Society of America, North Carolina(1995)

K. Yamada: A parametrization for the class of all proper stabilizing controllers for linear minimum phase systems, Preprints of the 9th IFAC/IFORS/IMACS/IFIP/ Symposium on Large Scale Systems: Theory and Applications, pp.578-583(2001)

H.Kimura: Is the Model a Good Controller? Perspectives on Brain Motor Control,CDC(2000)

M. Vidyasagar: Control System Synthesis–A factorization approach–, MIT Press, (1985)

PART V: PATTERN RECOGNITION

MODELING FOVEAL HUMAN IMAGE PROCESSING FOR ENHANCED CONTRAST AND EDGE DETECTION OF IMAGES USED IN ARTIFICIAL VISION SYSTEMS

PRAMOD K. VARMA [1]
Computer Science

DAVID ENKE [1]
Engineering Management

DANIEL ST. CLAIR [1]
Computer Science

ABSTRACT
The following paper describes an artificial neural network model that is inspired by human image processing. The proposed model attempts to integrate certain aspects of human image processing to incorporate the initial intelligent image processing necessary to allow artificial vision systems to function more efficiently in noisy and unknown environments. By incorporating the robust dynamic interactions between amacrine and bipolar cells, in addition to center-surround antagonism and photoreceptor blurring, the improved model demonstrates enhanced edge detection and contrast enhancement, in addition to providing gain control and noise reduction for the output response of the retina model. Real camera images were used to demonstrate the performance of the model. The model performance is compared with traditional image processing algorithms to illustrate the desired properties of the improved models.

INTRODUCTION

Modeling human vision has remained both a challenging and fascinating area of research. Modeling human visual processes for use in artificial vision systems may provide the necessary insights that will further advance the field of artificial vision. Using the knowledge gained from the primate vision system and earlier artificial vision systems, an improved retina model is developed and presented that can be used in artificial vision systems. The initial goal of the retina model is to provide gain controlled contrast enhanced edge detection. At the same time, this model will allow both high and low spatial frequency filtered signals to be sent to, and interact with, later vision systems.

To model retinal image processing, the proposed model is divided into two layers of processing, where visual information flows vertically through the layers and is reorganized laterally by interneuronal action. Each layer is modeled after cells and inter-cellular connections, with a special focus on the higher acuity fovea region. In addition, modeling of the antagonistic center surround receptive field structure is considered. The center surround antagonism provides the neural basis for the lateral inhibition and is the underlying mechanism for the enhancement of the image contrast, including reduction in the effects of intrinsic noise and spatial redundancy (Taylor, 1998; Dacey et. al., 2000). Similar to the human retina, the modeled ganglion and bipolar cells demonstrate an antagonistic center surround receptive field structure. Much of the modeling in Layer I is developed from the previous work of Enke and Dagli (1996). At Layer I, photoreceptors,

[1] University of Missouri-Rolla, Rolla (UMR), Missouri - 65409

bipolar cells, and horizontal cells are modeled, along with their interconnections. Layer I modeling can be found elsewhere (Varma, 2003). This paper will concentrate on the processing involved in Layer II. Most of the new modeling efforts involved interactions between the amacrine and ganglion cells in Layer II.

Layer II Processing

The resultant activation of bipolar cells generated at Layer I is given as input for the Layer II processing. Figure 1 illustrates the inter-neuron connections at Layer II. The difference from the interactions considered by Enke and Dagli (1996) is the inclusion of negative feedback from the amacrine cells to bipolar cells, which helps to remove uncorrelated noise. This feedback allows bipolar cells to give a spatially filtered response that enhances the resultant image response at the output of Layer II. The modified bipolar responses are again sent to amacrine cells to generate new amacrine activations. This feedback mechanism aids in the removal of uncorrelated noise generated at Layer I. This noise was neglected in the original model. In addition, only one type of amacrine cell is modeled in order to keep the retina model as simple as possible while maintaining the quality of resultant activations. Hence, the small field amacrine cells that were developed by Enke are not modeled here. The ganglion cells are modeled in a manner similar to bipolar cells, but with the addition of a push-pull effect that will be explained later.

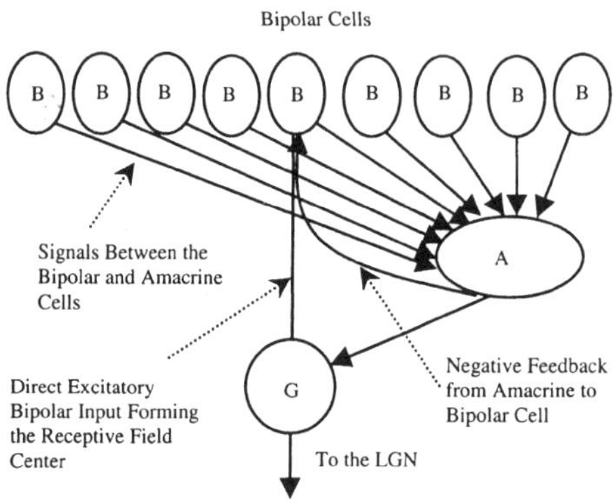

Figure 1: Connections Between Bipolar Cells, Amacrine Cells and Ganglion Cells

Modeling of Amacrine Cells

As suggested by Jacob and Werblin (1998), a robust dynamic interaction between bipolar and amacrine cells is responsible for sharp edge enhancement and noise filtering. The information coming from the bipolar cells will be summed over a neighborhood region that is defined by the span, n_a, of an amacrine cell. The activation of the amacrine cell, a(x,y), at the retina grid location *(x,y),* is represented as described by Equation 1. The term *b(i,j)* represents the bipolar cell activation at location *(i,j)* within the neighborhood grid, defined by span n_a. A new feedback mechanism is modeled from the amacrine cells to the bipolar cells. This feedback, net_feedback, is sent in the form of the difference between the amacrine cell activation and the bipolar cell activation, as given in Equation 2. Thus, the feedback consist of pure surrounding spatial information.

The bipolar cell will act upon the amacrine cell feedback in order to correct its signals. The new bipolar activation will be generated and is described in Equation 3, where K is a constant. Equation 4 gives the net amacrine cell activation, *neta(x,y)*, which is computed from the enhanced

$$a(x,y)= \sum_{i=x-n_a}^{x+n_a} \sum_{j=y-n_a}^{y+n_a} b(i,j) \tag{1}$$

$$net\_feedback = a(x,y) - b(x,y) \tag{2}$$

$$b_{t+1}(x,y) = \frac{b_t(x,y) * net\_feedback}{net\_feedback + K} \tag{3}$$

bipolar response by a spatial summation in order to perform a local enhancement.

$$neta(x,y) = R_a \sum_{i=x-n_a}^{x+n_a} \sum_{j=y-n_a}^{y+n_a} b_{t+1}(i,j) \quad (4)$$

R_a, as described by Equation 5, gives the number of neurons present within the neighborhood defined by the span n_a, while $b_{t+1}(i,j)$ represents the corrected bipolar response at location (i,j) within the neighborhood grid.

$$R_a = \frac{1}{(2n_a+1)^2} \quad (5)$$

$$a(x,y) = \frac{1}{1+e^{-S_a(neta(x,y)-T_a)}} \quad (6)$$

A sigmoid function will be used for the final amacrine cell response as explained by Enke (1997). Equation 6 gives the final activation of the amacrine cells, where S_a and T_a represents the slope and threshold of the amacrine cell, respectively, with a net activation, $neta(x,y)$, defined by Equation 4 at location (x,y).

By initially summing bipolar signals within its predefined dendritic field, the amacrine cells perform a local enhancement of the input signal. After each amacrine cell responds to this information, it will send its response to the ganglion cells. The resultant amacrine signals will enhance the regions of low contrast, while controlling the level of the much larger bipolar cell responses by firing action potentials. This process will perform a gain control for the retina, further limiting the range of bipolar cell response.

Modeling of Ganglion Cells

Ganglion cells are the final output neurons of the retina model. The initial modeling of ganglion cells is similar as proposed by Enke (1997). Both OFF-center/ON-surround and ON-center/OFF-surround ganglion cells are modeled. In addition to the model developed by Enke, the push-pull effect was modeled (Varma, 2003). This process is achieved by providing inhibitory input from the opposite pathway *via* amacrine cells. For example, the OFF-center ganglion cell's response to illumination is partly due to excitatory inputs from OFF-center bipolar cells and partly due to inhibitory input from amacrine cells, driven by the ON-center bipolar cells (Xing and Heeger, 2000). This will aid in improving the overall contrast gain at the ganglion cell level by incorporating visual information from both ON and OFF pathways. Since the ganglion cell fires an action potential, the final response g(x,y) of the cell can be modeled by placing the net activations for ON-center and OFF-center cells through a sigmoid function as given by Equation 7, where S_g represents the slope of sigmoid function, T_g represents the shift, and net_g is the net ganglion cell activation.

$$g(x,y) = \frac{1}{1+e^{-S_g(net_g(x,y)-T_g)}} \quad (7)$$

Response of the Improved Retina Model

The values of parameters for testing were set as follows: $n_a = 1.0$, $S_a = 7.0$, $T_a = 0.2$, $S_g = 50$, and $T_g = 0.1$, while n_g was initially set to 1 and varied later to different values in order to study its effect. The OFF-center responses of the ganglion cells were simulated as illustrated in Figure 2. It can be observed that ganglion cell activations are able to enhance contrast detecting sharp edges by retaining most of the high and low level features as shown in Figure 2b. Additional simulations were done by changing the span to $n_h = 2$ and $n_g = 2$. As observed in Figure 2c, the edges start becoming thicker, and at the same time the ganglion cell activations tend to refine the edges of the image, making the border appear thicker with contrast differences more pronounced. Of interest, it can also be observed that the edges are also filled to define a more connected image at the output response, but this is done at the expense of including low level noise. The ganglion threshold, slope of the sigmoidal function at the amacrine cell level, and threshold for the bipolar cell can be used to overcome this problem, but again, increasing the threshold for net activation is a tradeoff between the detection of edges and the filtering of noise.

a) Original Input Image b) $n_h = 1$ and $n_g = 1$ c) $n_h = 2$ and $n_g = 2$

Figure 2: OFF-center Responses of the Ganglion Cells

MODEL COMPARISONS

The development of a novel and simple retina model that demonstrates numerous image processing characteristics was described earlier. In this section, comparison of the proposed model with the Laplacian of Gaussian (LOG), Sobel Edge Detection, and the original connectionist model developed by Enke (1997) will be performed. Visual comparison, along with the three measures developed by Varma (2003), were used to apraise the performance of the improved model when compared to the above traditional algorithms. The three new measures were developed based on the most important characteristics often preferred to be present in the output response for high-level visual tasks involved in artificial vision systems. These important characteristics include: a measure of contrast, a measure of connectivity, and a measure of the uncorrelated noise present in the output.

Visual Comparison

The same image was provided to all four algorithms. Figure 3 shows the response of Sobel edge detection, LOG, Enke's model, and the new retina model when a human face is given as input. Clearly, one can note that Enke's connectionist model and the new retina model give responses on a wide range of contrast, ranging from low to high spatial frequency regions. LOG and Sobel edge detection do not perform well. Each is only successful in identifying low spatial frequencies, whereas regions of high spatial frequencies are to a greater extent lost. The new retina model consistently provides high contrast while simultaneously eliminating noise, which is still observed in Enke's model.

Sobel LOG Enke's Model New Retina Model

Figure 3: Response of the Four Algorithms for the Human Face Image

Measure of Contrast

A measure of contrast was developed in order to assess the quality of the image. An image with high contrast has higher mean luminance when compared with the background. Equation 8 describes the contrast computation when summed over the entire region of the image, where C_m is the measure for contrast when summed over the region of interest, l_o is the luminance of the pixel, and l_b is the luminance of the background.

$$C_m = \frac{\sum_{i=1}^{x}\sum_{j=1}^{y}[l_o(i,j) - l_b(i,j)]}{x * y} \tag{8}$$

Only the output of Enke's Connectionist model and the new retina model were evaluated on the basis of contrast, since both LOG and Sobel give binary outputs. When an average of the contrast was taken for both algorithms, there was found to be an improved efficiency of about 10% in overall contrast of the final output for the new model. The average intensity increased from $C_m = 210.65$ to $C_m = 231.98$.

Measure of Connectivity

The measure of connectivity was based on counting the number of correlated activations in the output response of the edge detector. From this measure, one can conclude that the image with the highest measure of connectivity has the higher number of correlated activations. In order to find the correlated activations, a neighborhood grid with different spans of 1 and 2 were used. All four algorithms were evaluated with the connectivity measure. The output response of Enke's connectionist model and the new retina model were converted to binary values by applying a threshold. Since object recognition was of interest, the image region that was extracted was the part of the face that had higher spatial frequencies. The region was cropped out of the output image and a ranking system was applied to compute the measure of connectivity.

Table 1 summarizes the results obtained after computation. It can be seen from the table that both biologically-inspired algorithms have higher measures of connectivity as compared to the traditional image processing algorithms. One difference between the two retina models was that the new retina model is successful in eliminating shot noise generated from the graded response, whereas Enke's connectionist model shows a very high level of shot noise. LOG and Sobel edge detector have very little noise, but are of less use due to their lower levels of connectivity. The slightly higher count of connectivity for Enke's model may be attributed to the presence of noise that could have been considered as correlated if the noise was close to other activated pixels.

Table 1. Measure of Connectivity and Noise

	Measure of Connectivity		**Measure of Noise**	
	$(n_g = 1)$	$(n_g = 2)$	$(n_g = 1)$	$(n_g = 2)$
New Retina Model	644	577	0	20
Enke's Connectionist Model	646	583	86	99
Laplacian Of Gaussian	329	308	5	13
Sobel	239	219	1	7

Measure of Uncorrelated Noise

All activation that is uncorrelated is classified as noise. Table 1 also gives a measure of uncorrelated noise for all four algorithms when a neighborhood span $n_g = 1$ and an uncorrelated noise of size 1 pixel is considered. It also gives a measure of uncorrelated noise when noise was classified as less than or equal to 2 uncorrelated activations. It can be observed that Enke's connectionist model has a high degree of uncorrelated noise while the LOG is a distant second when $n_g = 1$. The new retina model has the minimum count for noise but comes in third when the span is increased. This might be due to the pixels that were classified as connected as an edge of size 2 when $n_g = 1$ was considered, now get classified as noise since the edge size is less than 3 when $n_g = 2$ is considered. Sobel shows a minimum amount of uncorrelated noise, but the low level of connectivity comes as a disadvantage.

CONCLUSION

A retina model inspired by the human visual system has been developed for contrast enhancement and edge detection. This model is an extension in many aspects to models developed by Enke (1997). This model provides improved contrast enhancement, edge detection, gain control, and noise reduction as compared to traditional algorithms. It has been found to improve the image quality as well by reducing random noise. These characteristics are necessary for the efficient processing of later high-level visual tasks, such as boundary completion, visual attention, and object recognition.

This biologically-inspired retina model can act as a front end to higher intelligent artificial vision systems and can be applied in numerous industrial applications, such as the manufacturing industry, robotics, and diagnostics. The interactive nature of these systems will allow them to adapt in real-time to changing requirements and environmental conditions, even when specific learning algorithms are not incorporated. In addition, implementation of new intelligent visual processing, such as visual attention, can be naturally developed and utilized.

REFERENCES

Dacey, D., O.S. Packer, L. Diller, D. Brainard, B. Peterson, and B.B. Lee, "Center surround receptive field structure of cone bipolar cells in primate retina," *Vision Res.*, Vol. 40 (2000): 1801-1811.

Enke, D., "Biologically Inspired Neural Network Connectionist Models for use in Artificial Vision Systems," *Ph.D. Thesis*, UMR, 1997.

Enke, D.L., and C.H. Dagli, "Modeling Biological Visual Processes for Improved Contrast Enhancement and Edge Detection of Artificial Vision Systems," Applications and Science of Artificial Neural Networks II, SPIE Conf., Vol. 2760, (1996): 346-357.

Jacobs, A. and F. Werblin, "Spatiotemporal Patterns at the Retinal Output," *J. Neurophysiol.*,Vol. 80, No. 1 (1998): 447-451

Taylor, W.R., "TTX attenuates surround inhibition in rabbit retinal ganglion cells," *Visual Neuroscience*, Vol. 16 (1998): 285-290.

Varma, P., "Modeling Biological Visual Process in the Primate Retina for Enhanced Contrast and Edge Detection for use in Artificial Vision System," *Masters Thesis*, UMR, 2003.

Xing, J. and D.J., Heeger, "Center-Surround interactions in foveal and peripheral vision," *Vision Research*, Vol. 40 (2000): 3065-3072.

ROBUST FACIAL EXPRESSION RECOGNITION SYSTEM WITH FACE DETECTION USING CONVOLUTIONAL NEURAL NETWORKS

KATSUHIKO MORI
Canon Research Center
Atsugi, Japan
YUJI KANEDA
Canon Research Center
Atsugi, Japan
YUSUKE MITARAI
Canon Research Center
Atsugi, Japan
MASAKAZU MATSUGU
Canon Research Center
Atsugi, Japan

ABSTRACT
We developed a subject independent facial expression recognition system with translation, rotation, and scale invariance properties. This system recognized smiling faces for over 5600 still images of 13 persons of diverse ages, sexes and eye-shapes from images with varying facial expressions, including talking faces. The facial expression recognition algorithm used in that system is combined with robust face detection using a convolutional neural network. Features for facial expression recognition are obtained from the results of facial detection process. Using the normalized rate of changes in these features from normal face, the weighted score to recognize the facial expression is calculated. The weight values are determined based on the tendency of individualities of the features. Classification of facial expression is finally justified using the continuity of scores.

1. INTRODUCTION

Humans use non-verbal information as well as spoken language in smooth communication. Especially, facial expressions providing emotional states such as delight, anger, sorrow and pleasure are seemed to be one of the most important keys in communication. In order to be used for human-machine communication, facial expression recognition algorithm must have functional capabilities like (1)robustness to individuality, (2)robustness to head movement (pose, size, and translation invariance), (3)robustness to varying lighting conditions, (4)automatical processing, (5)real time processing.

There is a significant amount of research on facial expressions in computer vision. Some algorithms with practicability for human-machine-interaction are now being developed. As regards robustness to individuality in facial expression recognition, to the best of our knowledge, there are only a few algorithms (e.g., Ebine & Nakamura, 1999). Recognizing facial expression under rigid head movements was addressed by Black & Yacoob (1995). Most algorithms, so far, have addressed only a part of these functional capabilities (e.g., Földiák, 1991; Wallis & Rolls, 1997).

Convolutional neural network (CNN) models (LeCun & Bengio, 1995) as well as neocognitron (Fukushima, 1980) known as one of biologically inspired models, have been used for pattern recognition tasks such as face recognition and hand-written numeral recognition. The network includes feature detecting (FD) layers, each of which alternated with a sub-sampling layer (pooling layer) to obtain properties leading to translation and deformation invariance. Recently,

Fasel (2002) has proposed a model with two independent convolutional neural networks, one for facial expression the other for face identity recognition, which are combined by a MLP.

Proposed facial expression recognition system recognizes smiling faces as the facial expression to be recognized, based on difference in local features between a normal face and those not, from images with varying facial expressions, including talking faces, in almost real-time. Furthermore, the proposed model with the property of subject independence and that of pose, size, and translation invariance turns out to be much more efficient and compact than Fasel's model.

2. OVERVIEW OF FACIAL EXPRESSION RECOGNITION SYSTEM

This system consists of a video camera, a PC with a capture board and a monitor. Being input to the PC using the capture board, images from the video camera is displayed on the monitor and processed with our algorithm.

To recognize facial expressions independence of head motion, the proposed system at first detects a face in an image using a convolutional neural network with translation, rotation, and scale invariance properties. Secondly, by applying some detected features in facial detection process, it recognizes the facial expression in that face-region. If the result of facial expression recognition is the expected expression such as smiling face, the system saves that image, and if not, the system continuously captures image, detects face, recognizes facial expression.

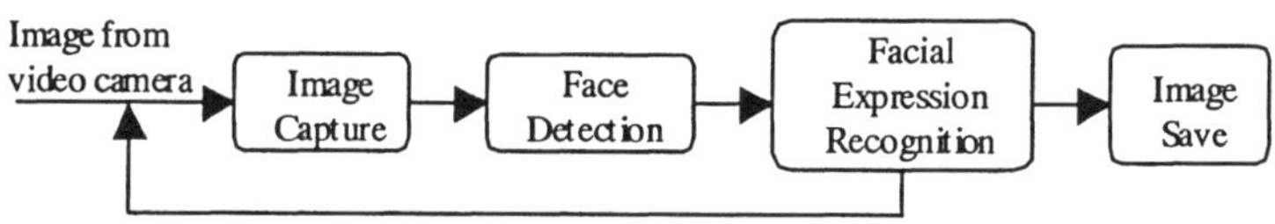

Figure 1: Overview of Facial Expression Recognition System

3. FACE DETECTION USING CONVOLUTIONAL NEURAL NETWORK

Figure 2 shows the architecture of our convolutional neural network for face

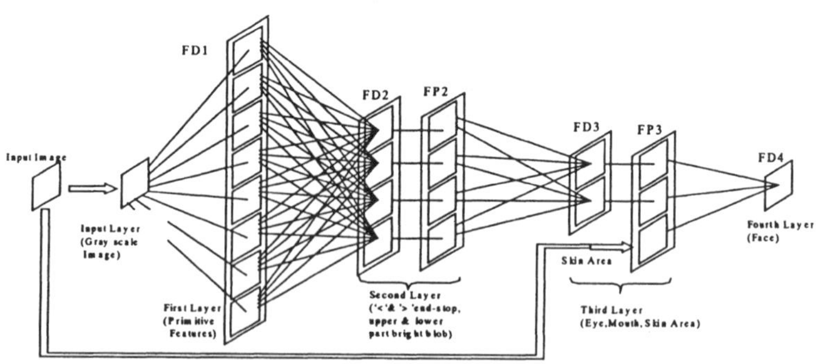

Figure 2 : Convolutional neural network architecture for face detection

detection. As in the previously proposed model (Matsugu et al., 2002), internal representation of face is provided by a hierarchically ordered set of convolutional kernels defined by the local receptive field of FD neurons. Face model is represented as a spatially ordered set of local features of intermediate complexity, such as eyes, mouth, nose, eyebrow, cheek, or else, and all of these features are represented in terms of a fixed set of lower and intermediate features.

In our network, local features to be detected in each layer are edge-like features in the first layer, ‘<’ and ‘>’ end-stop, upper part bright blob, and lower part bright blob in the second layer, eye and mouth in the third layer. Finally, using eye and mouth feature detected in the third layer and the skin area data defined by some restricted range of hue and saturation values, some face is detected in the forth layer.

In the training process, at first simple local feature detectors in FD1 and FD2 layer are trained using standard back-propagation with intermediate local features (e.g., ‘<’ end-stop) as positive trainig data sets and negative examples that do not constitute the corresponding feature category as false data. More complex local feature detectors in FD3 and FD4 layer are generated using the patterns extracted from face images as positive data sets and other features as negative data sets. We trained this network with varying size and rotation (up to 30 deg. for inplane axis as well as head axis), and lighting conditions.

Figure 3 shows the detected features in our face detection network. In this figure, the output value of neurons in the range of (-1.0 – 1.0) is converted to (0(black) – 255(white)).

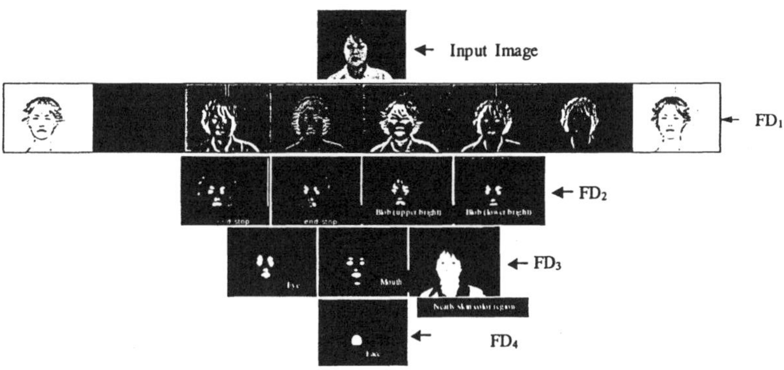

Figure 3 : Output of each module in each layer

When some faces are detected in the CNN, we also have some intermediate features including eye and mouth features available for facial expression recognition. So, we can use the result of facial detection process for facial expression recognition without detecting additional new features.

4. FACIAL EXPRESSION RECOGNITION

Our facial expression recognition algorithm has four distinct properties.

First, features used for facial expression recognition are obtained from the results of facial detection process in the CNN. Figure 4 shows processing stages to get these relevant local features. Based on the center of gravity in localized outputs of face detecting neurons with output values more than threshold, we decide the position of face. From that face location, the positions of eyes are pinned down using the output of the eye detecting neurons. Finally, we decide the position of the end-stops of both eyebrows and eyes using the output of primitive feature detecting neurons. All operations in the above are automatically processed.

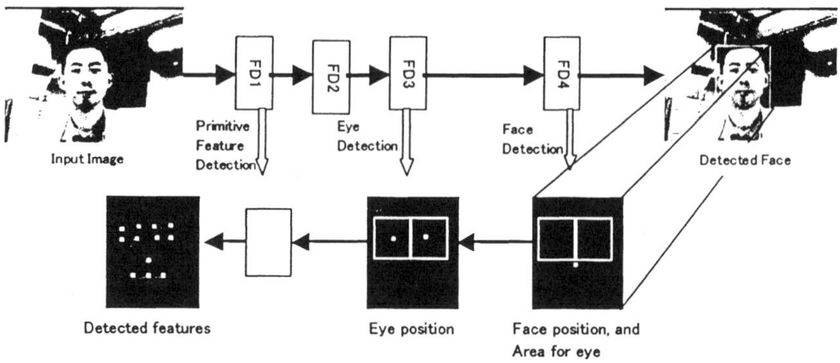

Figure 4 : Processes to obtain features for facial expression detection

Second, we use some normalized changes of these features between normal face and the input face. Figure 5 shows examples of visual features used to obtain the degrees of changes in the analysis of facial expressions.

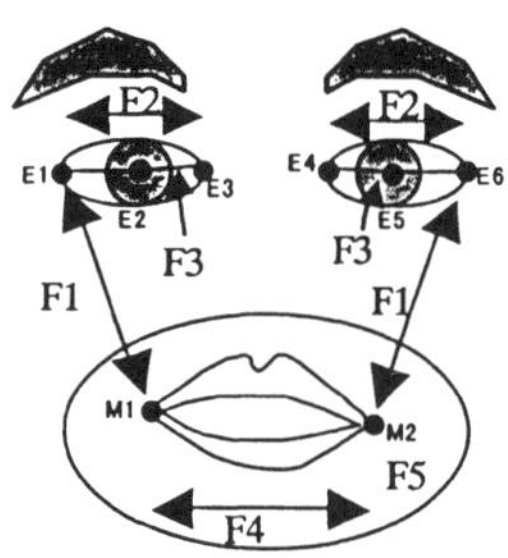

Figure 5 : Features for facial expression recognition

(1) F1 : distances of two end-stops in eye and mouth (E1-M1,E6-M2)
(2) F2 : widths of eye (E1-E3,E4-E6) in terms of distance between two end points of eye
(3) F3 : shapes of eyes (angle between E1-E2 and E2-E3, and between E4-E5 and E5-E6)
(4) F4 : the width of mouth
(5) F5 : the output value of the neurons around mouth in FD1 layer

To recognize facial expressions independent of differences among individuals, we use normalized changes of features in Figure 5 between normal face and processed face. Furthermore, to be robust against the size variability resulting from the head motion, those degrees of changes of these features are normalized with the distance between E3 and E4 feature.

Third, features to be used to recognize facial expression can be classified into two groups based on individuality concerning the degrees of changes. The degrees of changes in F1 and F4 feature are small among individuals. On the contrary, concerning the degrees of changes in F2 and F5 features, there are bigger differences among individuals. It shows that when smiling almost everyone opens their mouth broadwise and then the end-stops of mouth rise upward. And, for some people, some edges arise in cheek area in smiling and the length of eye varies. We calculate the saliency score, S, given by weighted sum of these degrees of change of features, s_k, using weight, p_k, depending on the individuality of respective features as follows.

$$S = \sum_k p_k s_k$$

Fourth, we use continuity measure to verify the classification of facial expression. It is well known that facial expressions as manifestations of emotinal states continue for a while, and on the contrary, facial expressions while talking varies quickly. So, we decide the input image as emotional faces if the score more than threshold conitnue over some consecutive image frames.

5. EXPERIMENTS AND RESULT

The proposed system has been evaluated with 13 subjects of different sex, ages and shape of eyes. The system is input images as a movie totally about 3 minutes (5600 still images) that includes normal faces, smiling faces to be recognized and other faces such as talking faces not to be recognized.

Figure 6 shows some recognized smiling faces among different people under varying image capturing conditions. The proposed system could recognize smiling faces with the recognition rate of 97.6% for 5600 still images.

Figure 6 : Detected smiling face

In this system, it takes 70msec for image caputure and initial setting, 49msec for face detection with CNN, 41msec for facial expression recognition per image with a 2.8GHz Xeon processor.

Figure 7 shows a sequence of normalized saliency scores in the scene smiling faces (A) and of talking faces (B). Figure 7(A) shows that in the scene of smiling faces the score grows up suddenly, stays with constant value, and

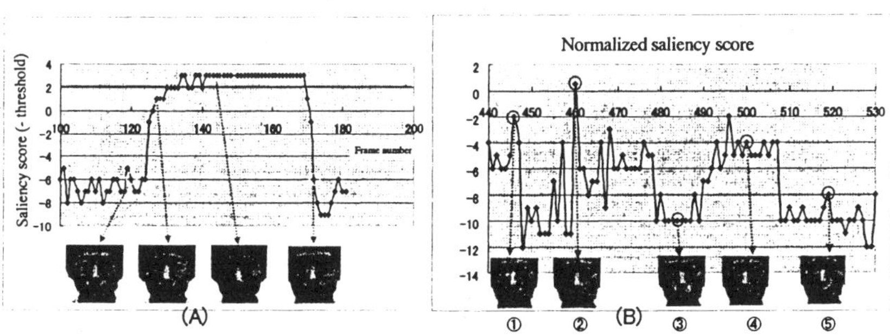

Figure 7 : Normalized saliency score of the scene

suddenly goes down. And Figure 7(B) shows that in the scene of talking faces the score changes frequently. In this figure, each image shows image of talking face for 'a' 'i' 'u' 'e' 'o' respectively. The system distinguished the scene of smiling faces from talking faces based on the condition that the score over threshold continues more than 3 frames or not.

6. CONCLUSIONS

We have shown that the proposed approach for recognizing facial expressions using face detecting convolutional neural network is very promising. Robust face detection with invariance properties in terms of translation, scale, and pose in our CNN brings some merits to robust facial expression recognition in terms of automated detection of useful features for recognizing facial expression with varying head motion, even under varying lighting conditions.

Facial expression recoginition algorihm in the proposed system has distinct properties, namely, using the degree of changes of features from normal face, using weighted score based on the tendency of individualities of the features, and using continuity of the score. These properties achive robustness to individuality and capability of making distinction between facial expression as manifestations of emotinal states and facial expression in talking.

As an extention of this work, we plan to recognize other facial expressions and automatical selection of normal face from images.

Acknowledgements : The facial data in this paper are used by permission of Softpia Japan, Reseach and Development Division, HOIP Laboratory. It is strictly prohibited to copy, use or distribute the facial data without permission.

REFERENCES

Black, M. & Yacoob, Y. (1995) Tracking and Recognizing Rigid and Non-rigid Facial Motions Using Local Parametric Models of Image Motion. Proceedings of IEEE Fifth International Conference on Computer Vision (pp. 374-381).

Ebine, H., & Nakamura, O. (1999) The recognition of facial expressions considering the difference between individuality. (in Japanese) Transaction on IEE of Japan, 119-C, 474-481.

Fasel, B. (2002) Robust face analysis using convolutional neural networks. Proceedings of International Conference on Pattern Recognition.

Földiák, P. (1991) Learning invariance from transformation sequences. Neural Computation, 3, 194-200.

Fukushima, K. (1980) Neocognitron: A self-organizing neural networks for a mechanism of pattern recognition unaffected by shift in position. Biological Cybernetics, 36, 193-202.

Le Cun, Y. & Bengio, T. (1995) Convolutional networks for images, speech, and time series. In: Arbib, M.A. (Ed.) The Handbook of Brain Theory and Neural Networks (pp. 255-258). Cambridge, MA: MIT Press.

Matsugu, M., Mori, K., Ishii, M., & Mitarai, Y. (2002) Convolutional spiking neural network model for robust face detection. Proceedings of The 9th International Conference on Neural Information Processing, (pp. 660-664) Singapore.

Wallis, G. & Rolls, E.T. (1997) Invariant face and object recognition in the visual system. Progress in Neurobiology, 51, 167-194.

HIERARCHICAL SEARCH AND BROWSING OF FACE IMAGE DATABASES

DUO XU, MING DONG[1], FARSHAD FOTOUHI, and LOREN SCHWIEBERT

Department of Computer Science, Wayne State University, Detroit, MI 48202

ABSTRACT

The advent of large image databases has created a need for tools which can organize images automatically. This paper focuses on the study of different clustering techniques that can be used to derive the hierarchy in image databases. Specifically, we propose a novel framework to organize the image database based on users' relevance feedback. By treating each feedback as a weak classifier and combining them together, we are able to capture the concepts in the users' minds and build a hierarchy in the database to support effective and efficient search and browsing. We present here an image retrieval system, which has an integrated user interface to collect relevance feedback and supports hierarchical browsing and search. Based on a face image database with 1328 images, we compare our approach with typical top-down and bottom-up clustering methods. Superior retrieval accuracies have been achieved by our framework when compared with the clustering methods.

1. INTRODUCTION

The recent development of digital media technology has generated a huge amount of images in applications such as medical image database, criminal suspect tracking, travel image gallery, and personal or family picture collections. Unlike text information, image data have their own characteristics. Traditional keyword based indexing and retrieval does not work well for them. Content-based image retrieval has emerged as an important area in computer vision and multimedia computing.

Most of the current content-based image retrieval systems concentrate on search-by-query [1]. Those systems typically require that users provide an image example. The most similar images to the query are then retrieved and presented to users. However it is usually difficult to produce query images with good quality. In addition, repetitive queries sometimes tend to become trapped among a small group of undesirable images [2]. On the other hand only a few system support image browsing. Unlike textual documents, the content of an image can be taken in at a quick glance, and a large number of them can be presented to the users at once, using thumbnail versions. Browsing allows users to simply go over a large image set in order to find what they want. In general, a browsing environment seeks to logically and predictably organize the image database for such a purpose.

Recently, several researchers have found that computer vision techniques can be used to make a set of image/video collections easier to browse, giving a representation of the set a noticeable structure by clustering together mutually similar items. By organizing images into similar groups and forming a tree hierarchy in the image database, both efficient search (by a search algorithm) and browsing (by users) are supported. Yeung et al. studied the application of bottom-up clustering methods to the organization of video key frames [3]. Milanes et al. applied hierarchical clustering to organize an image database into visually similar groups [4]. Zhang and Zhong proposed a hierarchical Self-Organizing Map that used the SOM algorithm to organize an image database into a two-dimensional grid [5]. In [2], both top-down clustering based on K-means algorithm and hierarchical bottom-up clustering have been used to support fast search-by-query and effective browsing on large image databases.

Despite the great efforts that have been spent on building hierarchy in image databases, challenges remain.

- **Robust clustering:** Current technology for content-based image interpretation is limited by the fact that low-level features cannot represent high-level concepts accurately. Clustering based on color, texture, and shape features may not provide meaningful organization in the database. The design of a robust clustering system under such a situation is still a challenging problem.
- **Imprecise and Emergent Semantics:** An image usually has an imprecise meaning. To place the images that are alike together, we need to define their mutual similarity. There are, of course, many possible definitions of "similarity" depending users' interpretations [6]. Even for the same user, the similarity measure will change depending on his goal of search or browsing. In order to support effective retrieval and browsing, the user's interest has to be reflected in the hierarchical structure of the database. Building a complete offline hierarchy without including the users in the loop could not meet this requirement.

[1] Corresponding Author

In this paper, we present a novel framework to organize image databases based on multiple-user relevance feedback. A relevance feedback mechanism is one way to incorporate the user's perception during image retrieval. By treating each user as an independent weak classifier, we show that combining multi-users' feedback is equivalent to the combinations of weak, independent classifiers. The latter system has been shown to have good generalization performance both theoretically and empirically in the pattern recognition literature [7, 8]. Furthermore, the hierarchy is also able to reflect the interest of most users in the system by including them in the loop. We test our algorithm based on a face image database with 1328 images. Our results show that our system supports effective and efficient search and browsing.

We have laid out the rest of the paper as follows. In Section 2, we briefly review different clustering algorithms that have been used to organize image databases and establish the need to include the users in the loop. In Section 3 we present our multiple-user relevance feedback based framework to organize image databases. We present our experimental results on a face image database in Section 4. Section 5 concludes the paper.

2. TOP-DOWN AND BOTTOM-UP CLUSTERING ALGORITHMS

A natural way to organize an image database is through an indexing scheme that performs clustering based on color, texture and shape features. Two conventional approaches are top-down clustering and bottom-up clustering [2].

2.1 Top-down Clustering

Top-down clustering splits the node successively from the root to leaves to build up a tree. One way to split is using a k-means algorithm to recursively partition each node of the tree into its k children. The following sequence of steps illustrates a typical k-means algorithm.

Randomly select k seed patterns.
For other unselected patterns, cluster them to their nearest seed.
Compute each cluster's mean as new seed patterns.
Go to the second step until the seeds remain the same or stop if a predefined number of loops has been achieved.

2.2 Bottom-up Clustering

Bottom-up clustering is essentially a hierarchical clustering approach. It clusters the nearest two image patterns in the feature space. Then a function is used to produce a new image pattern on the two image patterns in the presence of the clustered set of images for further clustering. This process continues until the last two image patterns are clustered to a root node in the hierarchy.

The success of the above clustering-based schemes is often limited, largely due to two challenging issues mentioned in Section 1. In addition, clustering algorithms will usually generate a tree-like structure to represent the semantics between images, which is far from enough to meet the requirements of the real world. It is obvious that one class may have more than one parent class and this kind of semantic relationship can not be reflected in a tree-like structure. Furthermore, sometimes the clustering approaches put more importance on some prominent features in the feature space. In the case those features are not of user's interest, the corresponding nodes in the hierarchy may be senseless.

It has been shown that user interaction is essential to successfully organize an image database [6]. In the following, we present our relevance feedback based organizing framework for image databases.

3. RELEVANCE FEEDBACK BASED ORGANIZING FRAMEWORK

3.1 Relevance Feedback

A relevance feedback mechanism is an inexpensive way to incorporate users' perceptions [9]. Typically, the system returns a set of possible matches for a query, and users give feedback by marking items as relevant or not relevant. Sometimes the click sequences of users could be used as indirect relevance feedback. We propose a relevance feedback based framework to organize image databases by noticing the following two facts:

- Each user can be treated as a classifier. The database may be partitioned by the user into relevant and non-relevant items. However the partition usually has low accuracy due to the fact that low-level features can not represent the image content accurately and users' feedback usually has inevitable noise. Hence each user is a weak classifier. By saying weak here we mean that the classification accuracy is typically low.
- Users are usually independent of each other.

Therefore, combining multi-user feedback is equivalent to the combination of weak, independent classifiers, which is a classification system that has been shown to have good generalization performance both theoretically and empirically [7, 8]. In fact, if we make the assumption that the errors of weak classifiers have zero mean and are uncorrelated, we can relate the system error to the average error of the classifiers acting separately as follows [10],

$$E_{sys} = E_{ave} / L \tag{1}$$

Where L is the number of classifiers in our system.

3.2 Relevance Feedback Based Organizing Framework

As is evident from http://www.google.com/press/zeitgeist.html, people do often search for the same things. In other words, there are some image searches which are clearly popular. The corresponding concepts behind these popular searches provide us with a good opportunity to organize the image database based on users' interest. The proposed multi-user relevance feedback based framework comprises of 4 modules (see Fig. 1): Feedback Expansion Module (FEM), Feedback Clustering Module (FCM), Committee Vote Module (MVM), and Hierarchy Construction Module (HCM).

Based on the few positive and negative examples provided by the users' relevance feedback, FEM classifies the entire database into a relevant group and an irrelevant group through a Naive Bayes classifier [11]. FCM find out the similar feedbacks (feedbacks corresponded to the same concept) by improved k-means clustering algorithm. In the improved k-means algorithm, only an estimation of cluster number is required before clustering. A voting committee is then formed in CVM based on those feedbacks in one cluster. The committee vote is introduced to build the corresponding concept. In our experiment, voting rate is set at 0.7 to achieve high precision. In other words, only those images who received over 70% votes will be elected into the current concept in the hierarchy. Finally the relations between concepts (nodes in the hierarchy) are established in HCM. Assuming we have two concepts, C_1 with n_1 images and C_2 with n_2 images, and their intersection $C_{inter} = C_1 \cap C_2$ is of size n_{inter}. Let T_1 and T_2 indicate the overlap between the intersection and the two concepts, $T_1 = n_{inter} / n_1$, $T_2 = n_{inter} / n_2$. C_1 is considered as C_2's ancestor if $T_1 < T_2$ and $|T_1 - T_2| < T$, where T is a threshold and is set at 0.4 in our experiment. Any pair of concepts built in CVM is evaluated to find out their relationship and a directed graph is built. Finally, the directed graph is simplified by preserving only the relation between parents and children.

In summary, our system will collect users' relevance feedback and build a hierarchy based on that to improve the search and browsing in the image databases. In the following, we present our experimental results based on a face image database with 1328 face images.

4. EXPERIMENT RESULTS

4.1 Feature Extraction

Our database has 1328 136-pixel by 94-pixel grey scale face images. We selected the faces from two face databases, AR face database [12] and FERRET database [13]. We divide the entire database into 10 concepts based on visual inspection and the ground truth is shown in Tab. 1. All images are warped manually by fixing each face's left eye and right eye at position (31, 70) and (61, 70). For those faces with sunglasses, the location of the eyes is estimated. To make some critical parts of the face prominent, we cut the rectangular area of the mouth (34 by 51), the lip (15 by 35) and the eyes (29 by 59). In this way, each face image is represented by a vector of size 16754 (136*94+34*51+15*35+29*59). Eigenface [14, 15] is calculated through singular value decomposition and each image is projected to reduce the dimension from 16754 to 35. The most significant components are chosen by setting the threshold of corresponding eigenvalue to be 1/50 of the maximum eigenvalue. Each face image is projected to a new 35-dimension feature space and each feature x is normalized into the range of [0 1] as follows,

$$x = \frac{x - \min(x)}{\max(x) - \min(x)} \tag{2}$$

4.2 User Feedback Collection

A user feedback collection interface is developed with Java Servlet (Fig. 2). Face images are randomly chosen from the database and presented to users. Users are asked to select the faces belonging to the same concept in their own opinion by checking the box below each image. The number of face images that have already been chosen is shown in the shopping cart. When a user decides that he has checked enough images, he can submit by clicking the "submit" button. Both the positive examples (selected images) and negative examples (unselected images) are recorded. Assuming the users know the ground truth of the database, we collected 84 feedbacks in total from several students in our experiment.

4.3 Experimental Results

4.3.1 Expanded relevance feedback vs. Concept after voting

Our results show that combining multi-user feedback by the committee voting mechanism essentially increases the accuracy of each concept. For example, Tab. 2 shows the comparison of the classification accuracy of 7 expanded feedbacks for concept "Smile & Longhair" and the final concept after committee voting (Recall that each expanded feedback is a weak classifier that classifies the database into relevant and irrelevant groups.). The accuracy is obtained by comparing the classification results with the ground truth. It is obvious that the classification accuracy of this concept is greatly improved.

4.3.2 Class Hierarchy and Hierarchical Browsing

The class hierarchy built by our framework is shown in Fig. 3. For illustration purposes, each node is manually labeled with the corresponding concept. Based on the class hierarchy, our system provides hierarchical browsing of the image database. Users could move through the database in a natural manner following the hierarchy. By clicking any node in the hierarchy, users could see all the images in that category.

For comparison purposes, we also show the class hierarchy generated by top-down clustering (k-means algorithm with k =3 and depth=2) in Fig. 4. Visual inspection shows that most of the nodes have specific meaning except nodes 2, 6, and 7. Light direction seems to be a very important component in feature space because the root node is split mainly by it.

Fig. 5 shows the class hierarchy built by bottom-up clustering. After hierarchically clustering all face images in our database, only nodes within the top 4 levels and with at least 70 images are kept. This hierarchy is similar to the top-down result but only one node (node 2) is meaningless. The splitting of the root node also depends greatly on the direction of light.

4.3.3 Query Precision

We compare the class hierarchy built by the proposed framework with the other two in terms of query precision. For a face image belonging to a concept in the ground truth, we retrieve its k-nearest neighbors in all three class hierarchies. The retrieval is done in every node of the hierarchies and the precision is calculated by comparing the retrieved images with the images in the ground truth. Only the maximum precision in each hierarchy is recorded. For each concept in the ground truth, we make 20 random queries. The mean and variance of query precision for concept "Longhair" and "Shorthair" are shown in Fig. 6 as a function of k (the number of retrieved images). We get similar results for all other concepts. It is obvious that the proposed framework greatly improves the retrieval accuracy over other clustering methods.

4.3.4 Major Advantages of Our System

Our system has several major advantages over typical clustering approaches:

- Our class hierarchy support search and browsing with high accuracy as shown in Section 4.3.2 and 4.3.3.
- Our class hierarchy can truly reflect user's interest. Those concepts with great user interest are made prominent in our hierarchy. For example, a very popular concept for face images, "Smile", is included in our hierarchy, but not in the other two hierarchies. At the same time, our hierarchy will not generate meaningless concepts (nodes). In the other two clustering approaches, the images are grouped together naively in the feature space, which may produce some nodes with no obvious physical meaning for users.
- The class hierarchy in our system is a directed graph instead of a tree structure built by both top-down and bottom-up clustering. Because some concepts may have more than one parent concept in reality, directed graph is a better way to show the structure in the database.

CONCLUSION

In this paper we study the different techniques that can be used to organize image databases. We propose a novel framework to organize the image database based on multiple users' relevance feedback. By treating each feedback as a weak classifier and combining them together, we are able to capture the concepts in the users' mind and build a hierarchy in the database to support effective and efficient search and browsing. We present an image retrieval system, which has integrated user interface to collect relevance feedback and supports hierarchical browsing and search. Based on a face image database with 1328 images, we show that our system has several advantages over typical top-down and bottom-up clustering methods.

We are in the final step to put our system online and publicly available. In the future, we plan to build an image search engine and use the proposed framework to index the huge amount of images on the Internet.

REFERENCE

[1] M. Flickner, H. Sawhney, W. Niblack, and et al., "Query by image and video content: The qbic system," IEEE Computer, vol. 28, no. 9, pp. 23–32, 1995.

[2] J. Chen, C.A. Bouman, and J.C. Dalton, "Hierarchical browsing and search of large image databases," IEEE Trans. on Image Processing, vol. 9, pp. 442–455, 2000.

[3] M. Yeung and B. Yeo, "Video visualization for compact presentation and fast, browsing of pictorial content," IEEE Trans. on Circuits and Systems for Video Technology, Vol.7 pp. 771-785, 1997

[4] Milind Naphade and Thomas Huang, "Extracting semantics from audiovisual content: The final frontier in multimedia retrieval," IEEE Trans. on Neural Networks, vol. 13, no. 4, pp. 793–809, 2002.

[5] D. Zhong, H. J. Zhang, and S. F. Chang, "Clustering methods for video browsing and annotation," in Proc. SPIE Conf. Storage Retrieval Image Video Databases IV, San Jose, CA, February 1996, pp. 239–246.

[6] S. Santini, A. Gupta, and F. Jain, "Emergent semantics through interactuion in image databases," IEEE Trans. on Knowledge and Data Engineering, vol. 13, no.3, pp. 337–351, May/June 2001.

[7] Chuanyi Ji and Sheng Ma, "Combinations of weak classifiers," IEEE Transactions on Neural Networks, vol. 8, pp. 32–42, 1997.

[8] L. Kuncheva, "A theoretical study on six classi.er fusion strategies," IEEE Trans. on Pattern Analysis and Machine Intelligence, vol. 24, no. 2, pp. 281–286, February 2002.

[9] Y. Rui, T. Huang, M. Ortega, and S. Mehrotra, "Relevance feedback: A power tool in interactive content-based image retrieval," IEEE Trans. on Circuits and Systems for Video Technology, vol. 13, no. 4, pp. 811–820, 2002.

[10] C. M. Bishop, Neural Networks for Pattern Recognition, Oxford University Press, 1995.

[11] P. Langley,W. Iba, and K. Thompson. An analysis of Bayesian classifiers. In Proceedings of the Tenth National Conference on Artificial Intelligence, pages 399–406, San Jose, CA, 1992. AAAI Press.

[12] A.M. Martinez and R. Benavente, *The AR Face Database*, CVC Technical Report #24, June 1998.

[13] P. J. Phillips, H. Wechsler, J. Huang, and P. Rauss, "The FERET database and evaluation procedure for face recognition algorithms," Image and Vision Computing J, Vol. 16, No. 5, pp 295-306, 1998.

[14] Matthew A. Turk and Alex P. Pentland, "Face Recognition Using Eigenfaces", Proc. IEEE Conf. on Computer Vision and Pattern Recognition, pp. 586-591, 1991.

[15] Alex Pentland, Baback Moghaddam, Thad Starner, "View-Based and Modular Eigenspaces for Face Recognition", Proc. of IEEE Conf. on Computer Vision and Pattern Recognition, 1994

TABLES AND FIGURES

Concept	1	2	3	4	5	6	7	8	9	10	Total
# of images	446	672	192	83	237	124	105	28	43	42	1328

Table 1. Ground truth of our face image database. 10 concepts in total. 1: Longhair 2: Shorthair 3: Smile 4: Longhair & Smile 5: Sunglass 6: Longhair & Sunglass 7: Shorthair &Sunglass 8: Shorthair & Sunglass & Leftlight[2] 9: Shorthair & Sunglass & Rightlight[3] 10: Longhair & Sunglass &Leftlight

	1	2	3	4	5	6	7	Average	Committee vote
Accuracy	.6016	.5417	.6000	.5714	.4713	.7065	.4658	.5655	.7263

Table 2. The classification accuracy of seven expanded relevance feedbacks of concept "smile & long hair" and the final concept after committee voting.

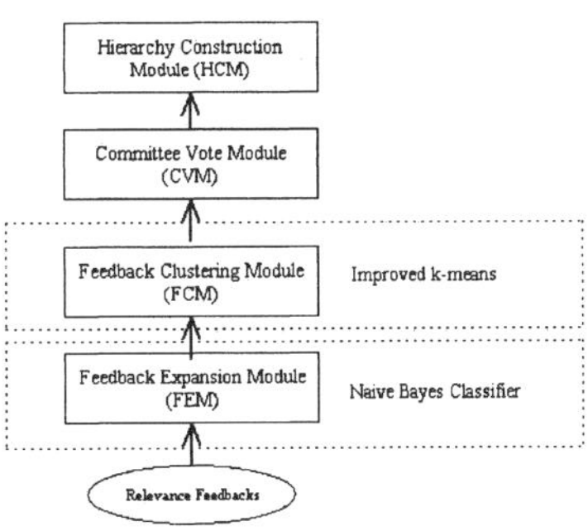

Figure 1. Proposed multi-user relevance feedback framework

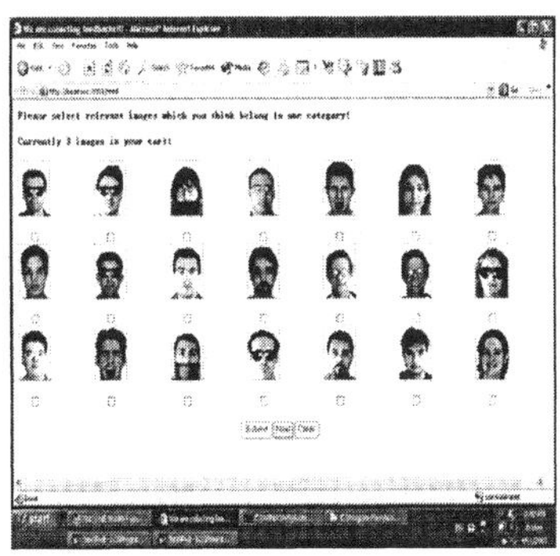

Figure 2. Web based interface for user feedback collection

[2] Leftlight means there is light on the left side of the face.

[3] Rightlight means there is light on the right side of the face.

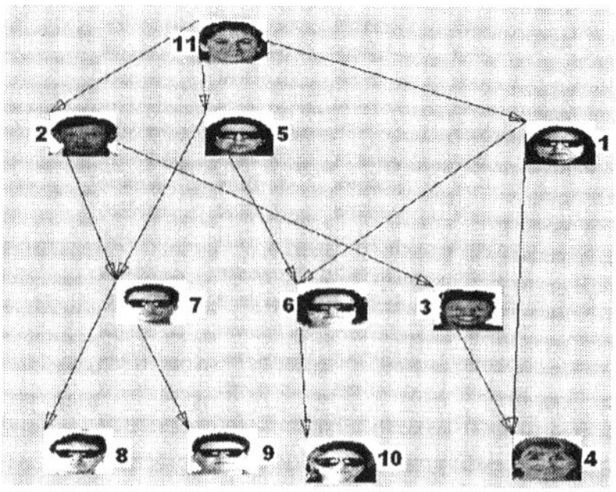

Figure 3. Class Hierarchy built up by proposed system. 1: Longhair 2: Shorthair 3: Smile 4: Longhair & Smile 5: Sunglass 6: Longhair & Sunglass 7: Shorthair &Sunglass 8: Shorthair & Sunglass & Leftlight 9: Shorthair & Sunglass & Rightlight 10: Longhair & Sunglass &Leftlight 11: All face images

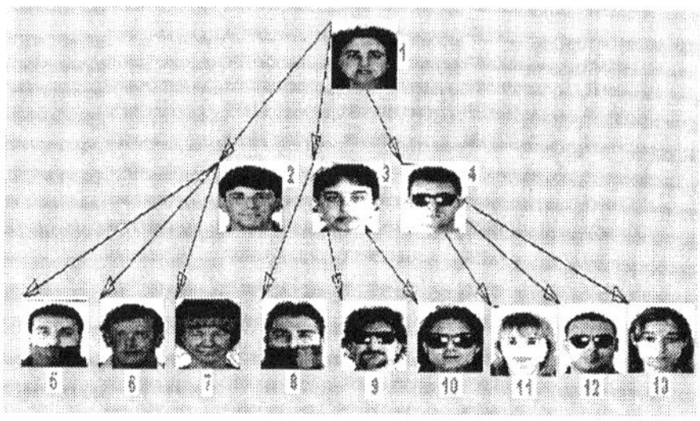

Figure 4. Class hierarchy built up by top-down clustering 1:All face images 2:Meaningless 3: Rightlight 4:Leftlight and Bothlight[4] 5:Scarf 6:Meaningless 7:Meaningless 8:Rightlight & Nosunglass 9:Rightlight & Sunglass 10:Shorthair 11: Rightlight & Sunglass&Longhair 12: Bothlight 13:Sunglass&Lleftlight 14: No-sunglass & Leftligh

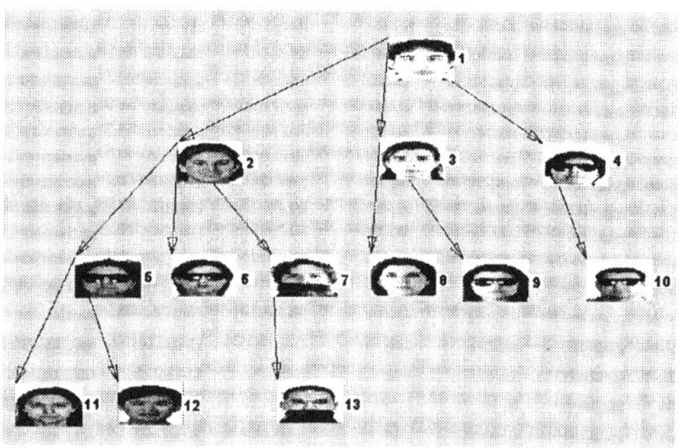

Figure 5. Class hierarchy built up by bottom-up clustering 1:All face images 2:Meaningless 3:Leftlight & bothlight 4:Rightlight 5:Nolight[5] & No-sunglass 6:Nolight & sunglass 7:Scarf 8:No-sunglass & Leftlight & Both light 9:Sunglass & Leftlight 10:Sunglass & Rightlight 11: Nolight & Nosunglass & Longhair 12: Nolight& Nosunglass & shorthair 13:Scarf& Nolight

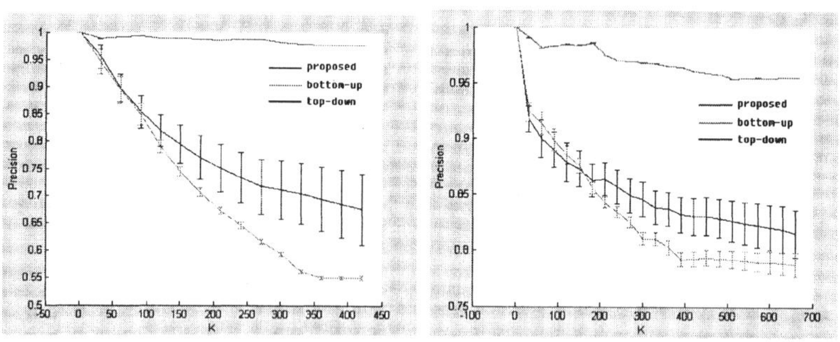

Figure 6. Comparison of mean and variance of query precision for Concept "Longhair" (left panel) and Concept "Shorthair" (right panel) in three different class hierarchies.

[4] Bothlight means there is light on both left and right side of the face.

[5] Nolight means there is no light on the face.

A NOVEL MULTI-POINT LUMINESCENT COATED ULTRA VIOLET FIBRE SENSOR UTILISING PATTERN RECOGNITION

COLIN FITZPATRICK
University of Limerick,
Dept. Electronics and Computer Engineering,
Limerick,
IRELAND.

WILLIAM B LYONS
University of Limerick,
Dept. Electronics and Computer Engineering,
Limerick,
IRELAND.

COLIN FLANAGAN
University of Limerick,
Dept. Electronics and Computer Engineering,
Limerick,
IRELAND.

ELFED LEWIS
University of Limerick,
Dept. Electronics and Computer Engineering,
Limerick,
IRELAND.

ABSTRACT

This paper reports on a novel multi-point (3 sensors) luminescent-coated fibre which utilises Artificial Neural Network (ANN) pattern recognition techniques to interpret its output. Sections of the fibres cladding are removed and replaced with a phosphor doped epoxy that emits visible light when stimulated with ultra-violet (UV) radiation. The various sensing locations utilise phosphors with different emission wavelengths resulting in individual spectral peaks for each of the sensing points. The system described is therefore capable of measuring levels of UV exposure at multiple points on a single fibre loop. The optical sensor system is interrogated using a fibre optic spectrometer. The signal analysis is performed using ANN, which allows interpretation and classification of the resulting complex spectral patterns containing spectral cross coupling due to the phosphor emission overlap from the separate sensing points. Initial measurements were trained using a feed forward ANN, and the system was 100% successful in classification of all subsequent test samples analysed on the trained network, some of which are presented.

INTRODUCTION

In optical fibre sensing, the potential for multiple sensing points along a single fibre is one of the core advantages of the technology. It is possible to monitor several points simultaneously on a loop of fibre with just a single optical output for interrogation. The information from these numerous points may be obtained using a multitude of different techniques based on the networking capabilities of optical communication systems (Grattan and Sun, 2000). The sensor used in this paper is based on the principle of wavelength division multiplexing, whereby, utilising different phosphors, the measurement points on the fibre loop generate different wavelength emissions when stimulated with the measurand – in this case exposure to UV. The resulting generated light travels through the fibre to the interrogating optical spectrometer (Oceanoptics S2000) where it is broken into its constituent parts and a spectral output is produced representing the wavelengths of the different sensing points. The spectrum is sent to the host P.C. for subsequent analysis uting ANNs.

The spectral responses in such systems are often complex and difficult to interpret due to phosphor emission overlap. This spectral overlap can be a limiting factor in the number of sensors possible on a single fibre. It has been proposed that for many applications of optical fibre multipoint sensors, ANN pattern recognition techniques may be used to resolve the problems arising from cross-coupling and cross-sensitivity to other parameters (Lyons and Lewis, 2000, Lyons *et al*, 1999, Lyons *et al*, 2000).

In order to overcome this, interrogation of the sensor system data is achieved using ANN Pattern Recognition techniques applied to the output spectral response, which allows the interpretation and classification of the resulting complex spectral patterns containing spectral cross coupling due to the phosphor emission overlap from the separate sensing points which have been exposed to UV. The software used to implement the neural network is Stuttgart Neural Networks Simulator (SNNS).

THE SENSOR

The multi-point ultra violet sensor described in this paper is an expansion of a single probe version which has been previously developed and tested at the University of Limerick (McSherry *et al*, 2003). These sensors operate by generating photoluminescence from incident ultra violet radiation in the cladding of the fibre. Part of this luminescent emission is then coupled to the fibres core through imperfections on the core-cladding interface which causes non-specular transmission and also by evanescent wave coupling. A diagram showing the sensor and its principle of operation is shown in Fig. 1.

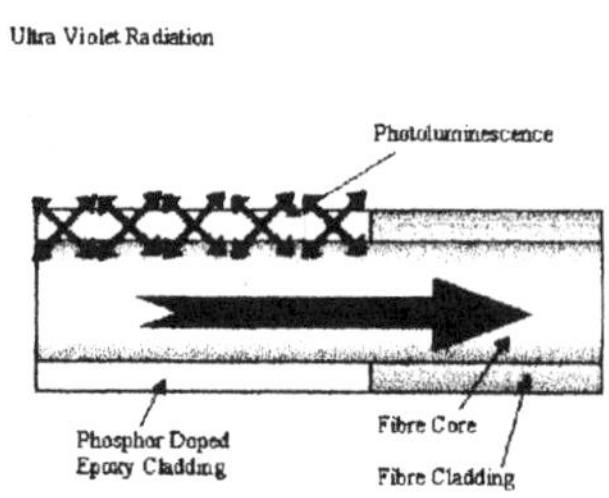

Figure 1. Luminescent Clad Optical Fibre UV Sensor

The photoluminescent coating is achieved by removing the regular cladding from the fibre and replacing it with a phosphor doped epoxy. The phosphor is the photoluminescent agent, and an important design consideration is the overlap of its absorption spectrum with the ultra violet radiation being monitored in order to achieve maximum sensitivity. The epoxy is required to have good transmission at the wavelengths under investigation in order to allow maximum penetration into the coating and also have a suitable refractive index in order to construct a positively guiding waveguide from the fibre. It has been shown that the intensity at the fibres end is directly proportional to the intensity of the incident ultra violet radiation.

The multi-point configuration of this luminescent clad sensor is based not solely on the intensity received at the output of the fibre but also on the spectral distribution. Coatings with the same absorption properties but different emission wavelengths are applied to different regions along the fibre. By monitoring the different peaks on the spectrum it is possible to gain information about localised areas of higher or lower ultra violet intensity.

SENSOR FABRICATION

The sensors were fabricated on a 1mm core diameter polycarbonate fibre. This fibre was chosen for its ease of handling, ease of coating and high refractive index. As explained in the previous section the high refractive index is necessary to create a positively guiding fibre as the refractive index of suitable epoxies is quite high. The cladding was removed from the fibre using solvents. The photoluminescent coating was applied using a rig specifically designed at the University of Limerick [3(6)]. The process involves the coating being pneumatically dispensed along the fibre as the fibre is spinning axially. This process creates smooth reproducible coatings with uniformity of thickness. The process is applied to separate sections of the fibre for the different coatings.

EXPERIMENTAL SETUP

The experimental set-up is shown in Fig. 2. A fluorescent ultra violet tube with peak emission wavelength of 370nm is used as the source for stimulating the photoluminescent process in the cladding of the fibre sensor.

The experiment was run for three different sensor stimulation conditions; all of the sensor stimulated equally; stimulation removed from the green section; and stimulation removed from the red section. Some non-ultra violet transmitting tubing was placed over the specific areas of the fibre that were to have its stimulation removed.

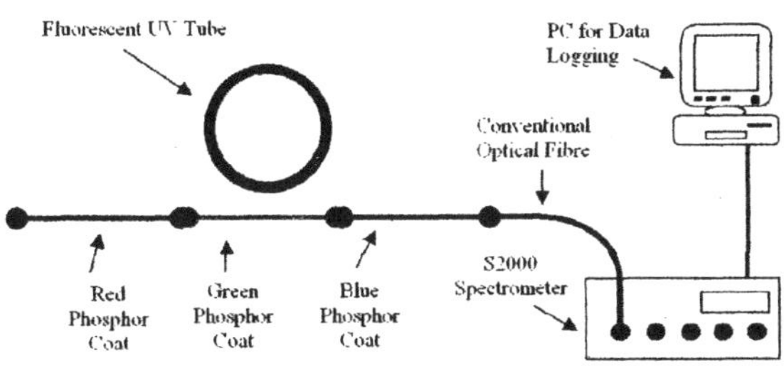

Figure 2. Experimental Set-up

A notebook PC was configured as a virtual instrument for control of the S2000 spectrometer, the resulting data capture, and pre-processing. This was implemented using a LabVIEW-based virtual instrument (VI) developed by the authors specifically for this purpose. The data output from this VI was then made available for analysis by the artificial neural network which was implemented using Stuttgart Neural Network Simulator (SNNS) version 4.1 (SNNS-url, 2003).

The spectra received from the Spectrometer range from approximately 190 nm to 880nm. This spectral range is represented by a total of 2048 analogue voltage levels received from the CCD (Charged Coupled Device) detector within the spectrometer. Fig. 3 illustrates a typical distributed spectral output from the S2000 spectrometer due to UV exposure over the whole fibre (covering red, green, and blue photoluminescent coated regions of the multipoint fibre sensor).

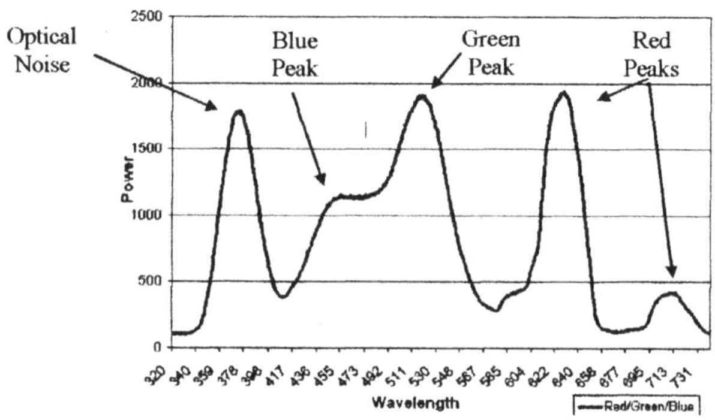

Figure 3. Spectral output showing UV exposure on the complete fibre sensor.

From Fig. 3, one can observe five peaks on the spectrum. The first peak (on the far left) at 370nm is the wavelength peak associated with the ultra violet stimulation used. Part of this stimulation is scattered into the fibres core and makes its way through the fibre and is detected. It is considered to be "optical noise".

The second (456nm) and third (530) peaks are resultant from the emissions associated with the blue and green phosphors respectively. The red phosphor used in the sensor manufacture has a characteristic two peak emission spectra, which are illustrated by the forth and fifth peaks at 630nm and 710nm respectively.

RESULTS AND ANALYSIS

Results

Numerous spectrometer readings were taken for each of the sensing stimulation conditions under test (7 in total; Red, Green, Blue, Red/Green, Red/Blue, Green/Blue, and Red/Green/Blue).

In order to maximise the efficiency of the computer algorithm, prior to application to the ANN software, the data from the S2000 spectrometer was pre-processed in order to extract the features of interest and hence reduce the amount of input data required for the network. Feature extraction involved a two-step process: first, a windowing of the data trace was carried out in order to remove both the UV simulated peak and the second red phosphor peak (the far left and far right peaks illustrated in Figure 3), and then the resulting trace was 'sampled' at every tenth point in order to further reduce the data set whilst still maintaining overall pattern integrity.

The feature extraction processes was implemented in LabVIEW using in-house developed virtual instruments (VIs) designed specifically for the task. An example selection of the resulting patterns (for all the sensing simulation conditions) used for ANN training and testing after feature extraction is illustrated in Figure 4. Each pattern is represented by a 92-point array.

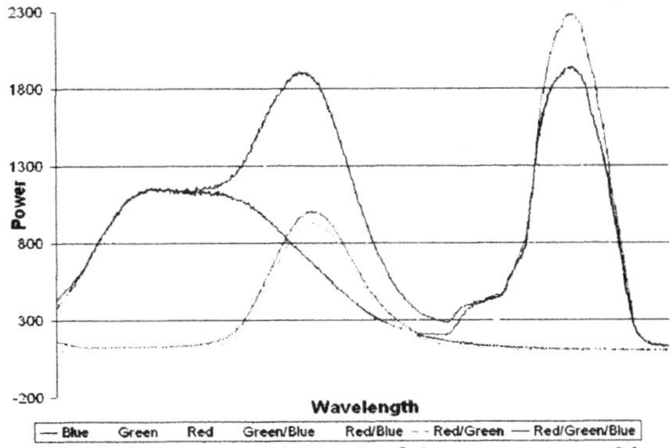

Figure 4. Training patterns for each of the exposure combinations.

Analysis

The saved spectrometer sensor data were then normalised between ± 1 using the standard LabVIEW Scale 2D array VI (LabView, 1998) prior to application to the ANN. The ANN consisted of a feed-forward three-layer network, with the number of input nodes representing the number of points required to represent the extracted feature from the sensors spectral response (in this case 92).

Although there is no set method for choosing the number of nodes to be used in a hidden layer, there are limitations associated with selecting too few nodes, which would result in the inability of the network to train successfully, and too many nodes would result in a poor generalisation by the network.

A number of different feed-forward networks were examined with various numbers of nodes in the hidden layer, and a total of 10 nodes were chosen, as it was found that it performed best. The output layer consisted of three nodes – one for each of the excitations possible (Green , Blue, and Red).

A total of 1960 result patterns were used in training the network, 280 each for the various excitation conditions under test. For training the feed-forward network, 150 epochs were required with a back-propagation algorithm utilising a learning rate of 0.8. The network was initialised with randomised weights. The testing of the trained network was carried out using 700 test patterns generated from an independent set of data from which had been applied in the training of the network. All 700 test patterns were correctly classified. Table 1 shows a selection of test results indicating the Sensor Excitation, Ideal Output, and Observed Output.

CONCLUSION

A novel multi-point (3 sensors) luminescent-coated fibre has been investigated and proven to be capable of detection the presence of UV at each of the three sensing points using a fibre optic spectrometer. The sensors were developed using a 1mm core diameter polycarbonate fibre where the cladding was removed in various sensing locations and replaced by different photoluminescent coatings. The length of the fibre was 15 cm, although longer and shorter lengths may be used as required.

Artificial neural network signal processing techniques have allowed the resulting spectral signatures to be characterised using pattern recognition. Extending this technique, it is envisaged that a more sophisticated multipoint luminescent based fibre optic sensor can be developed utilising more than three phosphors. Such a 'smart sensor' will capable of interpreting and separating relevant spectral response data from the data received containing spectral overlap, interfering parameter cross coupling, as well as faults or defects detected in the fibre.

ACKNOWLEDGMENTS

We wish to thank the "Irish Research Council for Science, Engineering and Technology" for the Government of Ireland Post-Doctoral Fellowship funding & EC Framework V GRD2-2000-30203 (MICROPRO) 'Novel Techniques Using High Power, High Frequency Microwaves for Material Production and Material Processing'

REFERENCES

Grattan, K.T.V., and Sun, T., 2000, "Fiber optic sensor technology : an overview," *Sensors and Actuators*, Vol. 82, pp. 40-61.

LabVIEW Function and VI Reference Manual Jan 1998 P/N 321526B-01

Lyons, W.B., and Lewis, E., 2000a, "Neural networks and pattern recognition techniques applied to optical fibre sensors," *Transactions of the Institute of Measurement and Control*, Vol. 22 (5), pp. 385-404

Lyons, W.B., Ewald, H', Lewis, E., 1999, "A Multi-point Optical fibre distributed Sensor Based On Pattern Recognition," *Proc. 2nd Int Automatisierungssymposium*, Wismar, Germany. Pp. 54-57.

Lyons, W.B., Ewald, H., Flanagan, C., and Lewis, E., 2000, "*An Optical Fibre Multipoint U Bend Sensor Based On Artificial Neural Network Pattern Recognition*" Proc. Artificial Neural Networks in Engineering Conference 2000 (ANNIE 2000), Vol. 10, pp. 663-670.

McSherry M., Fitzpatrick C., Lewis E., Pandithas I., Al-Shamma'a A.I., Lucas J., 2003, "A narrow band photoluminescent optical fibre sensor for the detection of high intensity germicidal UV radiation (254nm) from a microwave plasma UV lamp", *Journal of Optics A, Pure & Applied Optics*, July 2003.

O'Donoghue C., Fitzpatrick C., Lewis E., Grout I. 2002, "A Fabrication Process for Luminescent Clad Optical Fibre Sensors" 3rd International Conference on Optics-photonics Design & Fabrication, Oct 30th –Nov 1st 2002, Tokyo, Japan.

http://www.informatik.uni-stuttgart.de/ipvr/bv/projekte/snns/snns.html

Table 1: Sample of 7 results from the ANN test Set (from 700)

Sensor Excitation	Ideal Output	Observed Output		
Red	0 0 1	0.098	0.045	**0.931**
Green	0 1 0	0.045	**0.968**	0.025
Blue	1 0 0	**0.905**	0.018	0.085
Red/Green	0 1 1	0.006	**0.902**	**0.903**
Red/Blue	1 0 1	**0.898**	0.122	**0.994**
Blue/Green	1 1 0	**0.999**	**0.917**	0.078
Red/Green/Blue	1 1 1	**0.999**	**0.993**	**0.908**

GOAL DIRECTED VISUAL SEARCH BASED ON COLOR CUES: CO-OPERATIVE EFFECTS OF TOP-DOWN & BOTTOM-UP VISUAL ATTENTION

ROGER S. GABORSKI,
VISHAL S. VAINGANKAR, ROXANNE CANOSA
Laboratory for Applied Computing, Rochester Institute of Technology,
102 Lomb Memorial Drive, Rochester, NY 14623.

ABSTRACT

Focus of attention plays an important part in our perception of the world around us. Visual search is a combined effort of the top-down (cognitive cue) and bottom-up (low-level feature conspicuity) processes. Often during visual search our attention involuntarily gets directed to some irrelevant conspicuous objects, such as a bright object, regardless of the cued object. Objects that share similar characteristics with the cued object also influence our attention.

In this paper we analyze the mechanism of visual search based on color cues in natural images as well as computer generated images. We demonstrate the characteristics of the visual scene search by tracing the focus of attention path using data from RIT's human eye tracking system. The results are compared with the computer simulation results of our model in which we capture the interaction between the two systems and explain the behavioral dependencies. The top-down system is implemented using a neural network mimicking the working memory area of the brain and the bottom-up system is implemented using saliency maps.

1. INTRODUCTION

Focus of attention is a crucial mechanism we adopt while observing our surroundings. We as humans do not perceive every aspect of the surroundings, but instead focus on the interesting aspects and ignore the non-interesting ones. Thus attention acts as a gating mechanism to the higher level processing in the visual cortex, processing only the attended locations of the scene. Saccadic movements of the eye get the region of interest under the direct focus of the fovea, an area of maximal acuity in the retina. The filtering theory of attention (Broadbent, 1958) stated the need for attention for processing required information and ignoring the rest. We focus on certain aspects of the image with a higher priority than others. This prioritizing of objects leads us to the concept of saliency of low-level features (image based features) of objects. These low-level properties of objects are extracted by the bottom-up focus of attention mechanism that was initially described by Koch and Ullman (1985), where a saliency map reflects the relative saliency of objects from their surrounding. A top-down cognitive process modulates this early visual attention. For instance, given a visual search task, the knowledge about the target object's features biases the perception of the scene. High-level knowledge about the target object can be its shape, color or motion information. The guided search model (Wolfe, 1994) captures visual search in a computational paradigm. Their model created an activation map, which was a weighted summation of activity in the pre-attentive feature maps. Feature maps with the target object's features are assigned higher weights. The activation map represented the task based relevance of objects in the scene. Psychology based research in color visual search by D'Zmura (1991) explains the serial/parallel nature of color search. Above-mentioned research in color search has been confined to synthetic images where the research goal has been to find relationship between target and distracter colors. We report results for color search in natural images.

2. MODEL

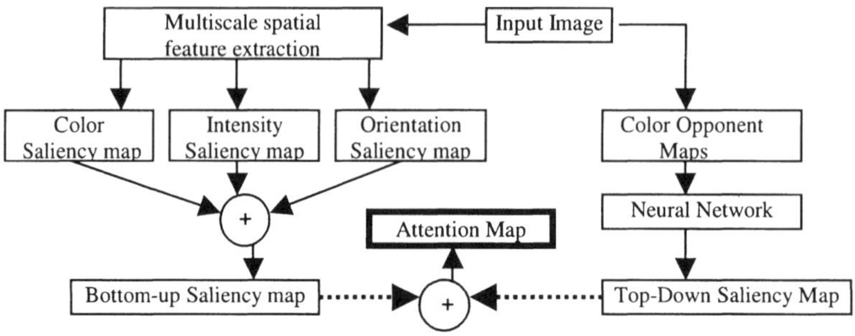

Figure 1: Block diagram of the attention system

2.1 Bottom-Up attention

Conspicuous sensory features of a scene are instrumental in directing our attention. A bright source of light in a dark room instantly catches our attention due to the light's intensity contrast with the dark surrounding. Computational modeling of early visual mechanisms is implemented in intensity, color and orientation feature channels. Multiresolution feature extracting spatial filters are inspired from the biological vision principles simulating the function of the retina, lateral geniculate nucleus and the primary visual cortex. Intensity contrast in the input image is calculated using Difference of Gaussian filters that simulate the center-surround receptive field structure of the retinal ganglion cells. Color opponent cells in the lateral geniculate nucleus are modeled using the center-surround receptive field for color, where an on-center red, off-surround green filter extracts the red regions of the image and an on-center green and off-surround red extracts the green regions of the image. Blue-yellow color opponent filter extracts blue and yellow regions of the image. Color opponent cells that are excited by red and inhibited by green are denoted as having R+G- response. Cells with opposite response characteristics are G+R-. Similar behavior is associated with the blue-yellow color opponent cells. Processing in the color channel produces four different color opponent feature maps, which are as follows, red excitatory/green inhibitory (R+G-), red inhibitory/green excitatory (R-G+), (B+Y-), (B-Y+).

Orientation information is extracted using directional Gabor wavelets of 0°, 45°, 90° and 135° orientations. Gabor filters are convolved with the input image to extract edges of above-mentioned orientations. Convolution of spatial filters with the input image produces a set of topographical feature maps of the input image. Since the center-surround mechanism is used for extraction of the features, the topographical maps obtained are called feature salience maps since they contribute to the spatial competition within a map during feature extraction. Previous approaches (Itti and Koch, 2001) constructed feature pyramids for each feature channel and applied spatial competition using Difference of Gaussian filters on the feature maps to generate saliency maps.

Multi resolution feature saliency maps are summed together within the respective feature channels to generate the final three feature saliency maps. The three feature saliency maps are linearly summed to form a final saliency map called the bottom-up saliency map as shown in Fig.1. This inter-feature channel summation contributes to the inter-feature competition in the bottom-up saliency map. The bottom-up saliency map is representative of the combined effects of the three feature channels with varying strength from each channel.

2.2 *Top-Down attention (Neural Network Architecture)*

During a visual search for objects of interest, high-level information about the searched-for object usually guides our focus of attention. We have experimented with color as a high-level feature for visual search. The neural network in our system mimics the working memory area of the brain. The inputs to neural network are the color opponent cell output maps (R+G-, R-G+, B+Y-, B-Y+). Four different neural networks are trained on the four prototype colors (Red, Yellow, Green, Blue). Assuming that a neural network is trained on pure saturation of red, this network gives an output of 1 for regions containing a high value of red. The output response reduces with the reduction in amount of the trained color. Thus orange which is a combination of Red and Yellow, produces a lower response than a pure red does. Orange generates response in both the red and yellow trained neural networks.

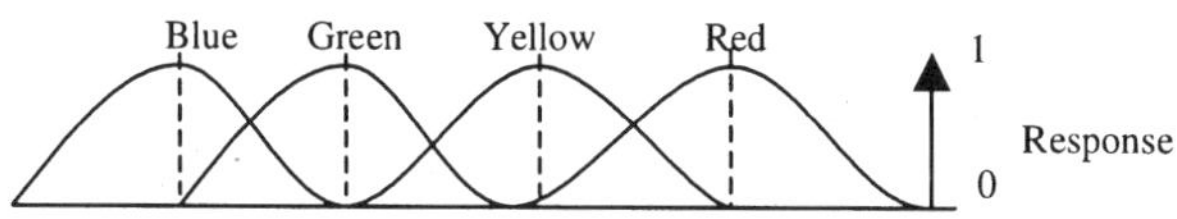

Figure 2: Color opponent cell responses.

As observed from the response curve Fig. 2, there is no overlap between the red and green response curves, indicating the mutually exclusive nature of the two colors. Similar behavior is true for blue and yellow color opponent cells. This behavior can be simulated using a neural network. The interaction between the responses of different color opponent cells gives the highest response for the prototype colors, marked with vertical dotted lines on the response curve. A reduced response for the deviation colors (orange, violet) is indicated by the fall in the response (to the right and left of the vertical dotted lines).

While training the neural network, the training image is an image containing red, yellow, green, and blue regions. The target matrix is a binary image of the training image with the region of the color to be learned as 1 and the remaining three color regions as 0. Thus while training the neural network on red as the target color, the red region in the training image is 1 and distracter colors, yellow, green and blue regions as 0. The output of the neural network processing is an output map (top-down saliency map) with relative saliency values with respect to the trained prototype color. Saliency values of the output map close to 1 signify high target color and values closer to 0 lack the target color properties.

A two layer feed forward neural network is used, with the first layer using a hyperbolic tangent sigmoid transfer function and the second layer using a linear transfer function. The back propagation learning technique is employed in which a training vector and target vector are provided. A maximum of 200 epochs were allowed for the network while training. The number of epochs selected for the network gives expected outputs without over training the network. This is proved by the reduced response generated for deviation colors after being trained on the prototype colors.

Another neural network architecture proposed by De Valois et al. (1966) explained the interaction between the three cone types (short, medium and long wavelength) to explain the opponent responses of the color opponent cells that produce a myriad of color perception. Using the neural network as a model for capturing the interaction between the color opponent cells we are able to efficiently predict the regions of the input image with task-relevant saliency.

2.3 Interaction between Top-down & Bottom-up attention

Research conducted by Parkhurst et al. (2002) explains the effects of bottom-up and top-down attention on perception of a scene. The bottom-up and the top-down saliency maps in our system are summed to generate a final attention map. Summation of the maps enhances the saliency values of the relevant search task regions, and simultaneously maintains the saliency of task irrelevant conspicuous features. Figure 1 shows the interaction between two attention systems. The salient regions in the attention map are the regions that have a high probability of being fixated when tested with human subjects. The results section of this paper addresses this topic.

3. EXPERIMENTAL METHODS

Ten Rochester Institute of Technology students were selected as subjects for conducting the eye tracking experiments. Since color was the basis for the visual search, only subjects with normal color vision were chosen. Subjects were naive about the nature of experiments. Subjects were seated at a distance of 38" from the 50" Pioneer plasma display on which the images were displayed. Experiment images consisted of indoor scenes, bookshelf scenes and computer-generated scenes.

3.1 Procedure

The first part of the experiment displayed images for free viewing. Subjects were instructed to freely view images displayed on the screen, and proceed to the next image by pressing the space bar key. This part of the analysis gives us an understanding of the non-goal directed perception of a scene. The next part displayed images from a different image set for the search task. Search task consisted of searching for objects of a particular color in the image and subjects were instructed to proceed onto the next image after locating all the target colored objects.

3.2 Eye tracking

Rochester Institute of Technology's Visual Perception Laboratory uses an Applied Science Laboratory model 501 head-mounted eyetracker to monitor subject's eye movements while viewing a scene. Recent research (Pelz and Canosa, 2001) conducted on the eyetracker shows the scope of the eyetracker for predicting human behavior in different visual perception tasks. The eyetracker monitors the subject's pupil and the first corneal reflection of an infrared illuminator. The fixation coordinates and fixation duration of the scanpath are collected for data analysis. Fixation duration is the amount of time spent at the fixation points.

4. RESULTS

In order to estimate how well the fixations correspond to various saliency maps in the system, a correlation measure was calculated. The saliency map mean is indicative of the strength of individual maps. The scanpath is overlaid on the saliency map and the saliency values at the location of fixations are extracted. The mean of all the extracted saliency values is termed the fixation mean (fix_mean). A ratio of fix_mean to the map mean is calculated. If this ratio is close to 1, it indicates that the fixations were randomly distributed around the map since the fix_mean and map mean are almost similar. A ratio significantly higher than 1 indicates the majority of fixations occurred on the high saliency regions. If the ratio is less than 1, it indicates the fixations tend to be focused on the low saliency regions. Fixation duration also can be used for calculating the correlations values by multiplying the extracted fixation salience with the corresponding fixation duration. The ratio of fix_mean_duration and map mean generates the correlation value based on duration.

4.1 Relationship between the target and distracter prototype colors in visual search.

The correlation for various prototype colors is calculated by overlaying the search scanpath onto the R+G-, R-G+, B+Y- and B-Y+ maps. Pattern bars in the graphs of Fig.3 are the correlation values without considering the fixation durations. Black bars are correlation values with fixation durations as a weighting factor. Y-axis represents correlation values normalized to the highest of the four correlations.

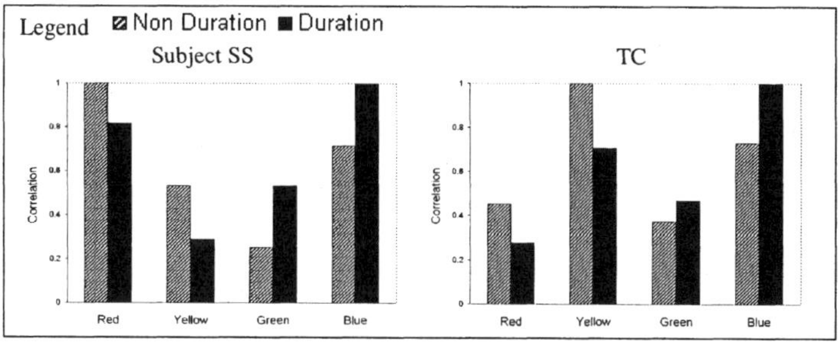

Figure 3. The target color is Blue. Graphs for two subjects are shown

In Fig. 3 the target color (blue) has a consistently high correlation value in the non-duration analysis, but other distracter colors tend to have comparatively high correlation values too. This behavior is because the subjects fixated the distracter colors while searching for the target colors. For subject SS, red distracter attains highest correlation in non-duration analysis and for subject TC, yellow was a prominent distracter. When applying fixation durations as weighting factors, the target color emerges with the highest correlation value, and the distracter colors attain more decorrelation with the target color. The graphs suggest that the fixation durations reflect important information about the nature of search.

4.2 Graphs explaining the relative effects of top-down and bottom-up attention on eye movements during visual search.

Analyses of our results show that bottom-up factors do affect the visual search task. The graphs in Fig. 4 are obtained by overlaying the search scanpath on the top-down and bottom-up saliency map. The y-axis shows the normalized correlation values, and the x-axis the sequential fixation numbers of the scanpath for the search conducted by subjects. The correlation values are normalized with respect to the maximum correlation possible for the respective saliency maps. Maximum correlation is calculated by taking a ratio of 1 (global maxima of the map) to the map mean. Correlation values in top-down curve are normalized with the maximum correlation of the top-down map, obtained by dividing 1 by the mean of top-down map. The normalized correlation values lie between 0 and 1, with 1 indicating the maximum correlation. This normalization allows us to compare the fixation strengths of the two curves. The top-down and bottom-up correlation curves are depicted in the graphs.

When not focusing on a target object, the low-level conspicuous features catch the subject's attention, which are predicted by the bottom-up saliency map. This is evident in the graphs of Fig.4 where the fixations with low top-down correlation values are compensated with high bottom-up correlation value, suggesting a potential bottom-up influence on the fixation seen in subject JS, NB. The results indicate that search task is influenced by intermediate task irrelevant bottom-up fixations in process of fixating on search relevant cognitive cues (top-down features). Another interesting observation is, even after the target object is located and if the search is in progress, the final fixations are primarily on high salient regions of the bottom-up map evident for subject NB, TC.

Legend: —o— Bottom-up correlation curve, -----*----- Top-down correlation curve

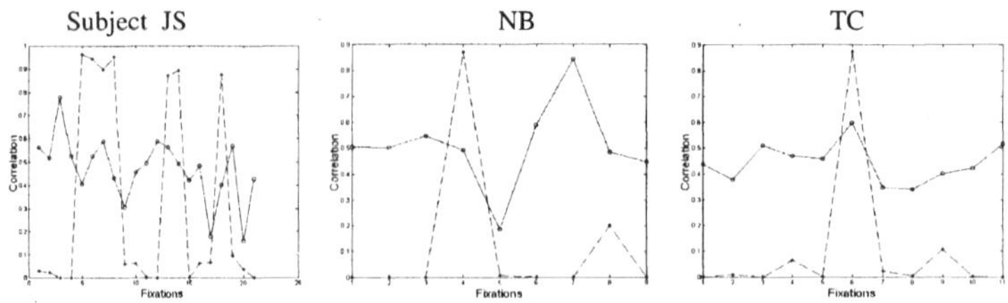

Figure 4: Search scanpath overlaid on top-down and bottom-up saliency maps.

5. CONCLUSION

As an engineering model the system is able to predict the regions of the scene that have a high probability of being fixated when confirmed with human data. This is evident from the high correlation values obtained from correlation curves in the graphs that are obtained from the eye movements of subjects. Neural network is efficient as a predictor for search task relevant regions in a scene. Color spectrum response to changing wavelength is captured using neural network. The system is not intended to be complete representation of the human visual search mechanism. Psychophysical factor like reduction in spatial resolution with increasing visual angle from the point of gaze has not been attempted.

ACKNOWLEDGEMENTS

We would like to thank Dr. Jeff B. Pelz, Center for Imaging Science, RIT, for granting us access to the eyetracker equipment in the Visual Perception Lab. We appreciate the support of all the students who participated as subjects in the experiments.

REFERENCES

Broadbent D. E. (1958) *Perception And Communication*. London: Pergamon

De Valois, Abramov, & Jacob, (1966). Analysis of response patterns in LGN cells. *Journal of the Optical Society of America,* 56, 966-977

D'Zmura M. (1991) Color in Visual Search. *Vision Research.* 31(6) pp 951-966.

Itti, L. and Koch, C. Computational Modeling of Visual Attention *Nature Neuroscience Review* (2001) 2, 194-204

Koch C, Ullman S. (1985). Shifts in selective visual attention: towards the underlying neural circuitry. *Human Neurobiology.* 4, 219-227

Parkhurst, Law, and Neibur (2002). Modeling the role of salience in the allocation of overt visual attention. *Vision Research,* 42(1), 107-123.

Pelz, J.B. & Canosa, R. (2001) Oculomotor Behavior and perceptual strategies in complex tasks. *Vision Research.* 41: 3587-3596.

Wolfe, J.M (1994) Guided search 2.0: A revised model of visual search. *Psychonomic Bulletin & Review* 1(2), 202-238.

An Investigation of the Effects of Correlation in Classifier Fusion

SUSAN A. STORM
Department of Operational Sciences
Air Force Institute of Technology
Wright-Patterson AFB Ohio

MARK E. OXLEY
Department of Mathematics and Statistics
Air Force Institute of Technology
Wright-Patterson AFB Ohio

KENNETH W. BAUER, JR.
Department of Operational Sciences
Air Force Institute of Technology
Wright-Patterson AFB Ohio

ABSTRACT

The effects of data correlation on three classifier fusion techniques were examined. The three fusion methods tested were the ISOC fusion method (Haspert, 2000), the ROC "within" Fusion method (Oxley and Bauer, 2002) and the simple use of a Probabilistic Neural Network (PNN) as a fusion tool. Test situations were developed to allow the examination of various levels of correlation between feature sets. We consider a two-class problem in which we simplify the target determination to friendly or hostile.

INTRODUCTION

Tumer and Ghosh (1996) provided expressions for quantifying the effects of using correlated classifiers. The classifiers they considered were correlated due to the sharing of a common feature set. Hashem (1993) discussed similar effects for linear combinations of neural networks. Duin and Tax (2000) examined combining classifiers under the conditions of shared and different (perhaps correlated) feature sets. In this paper we are concerned with controlling the levels of correlation between different feature sets as applied in a classifier fusion context. This paper documents the first step towards the creation of a synthetic classifier fusion-testing environment. The purpose of such an environment is to allow for the testing of various fusion algorithms in a controlled fashion. The effects of data correlation on three classifier fusion techniques were examined. The three fusion methods tested were the ISOC fusion method (Haspert, 2000), the ROC "within" Fusion method (Oxley and Bauer, 2002) and a Probabilistic Neural Network (PNN). Test situations were developed to allow the examination of various levels of correlation between feature sets. We considered a two-class problem in which we simplify the target determination to friendly or hostile. This paper is organized as follows. First, the ISOC, ROC "within" and PNN fusion techniques are briefly summarized. Next, Intra- and Inter- correlation are defined and methodology is established for generating experimental feature sets. Finally, an experimental design is described, executed, and interpreted.

FUSION METHODS INVESTIGATED

Consider a situation where there are two candidate classifiers for fusion in a two-class situation. The ISOC fusion method (Haspert, 2000) takes the class indications of the two classifiers and finds the optimal Boolean fusion rule; e.g., declare class A if the 1st

classifier indicates it is from class A and the 2nd classifier indicates either that it is from class A or class B. This method uses fixed thresholds for each classifier. Storm (2003) investigated a subsequent optimization over a threshold space and found that significant improvements were possible. Here ISOC is applied in it original formulation.

The ROC "within" Fusion method (Oxley and Bauer, 2002) uses a (fixed) logical "or" rule and develops an optimal fused ROC curve from the individual classifier ROC curves. Optimal thresholds are calculated for each individual classifier. This is in contrast to the ISOC method that finds an optimal rule but with a fixed threshold.

PNN (Probabilistic Neural Network) fusion simply entails using the posterior probability estimates from the two classifiers as features in a PNN.

CORRELATION

Both the ISOC and the ROC "within" method assume independence between the feature sets as presented to the two classifiers. PNN fusion makes no assumption of independence. We are interested in assessing the effects that correlation has on all three fusion schemes.

We will consider successive observations of a feature or a set of features. In this setting we will define two types of correlation: intra-correlation and inter-correlation. Intra-correlation refers to the autocorrelation of a specific data stream. Inter-correlation is the correlation between features in a given feature set. Correlation of the intra-correlation type is not considered in this paper but is the subject of current efforts.

METHODOLOGY

DATA GENERATION

From this point forward "within" correlation refers to intra-correlation and "across" correlation refers to inter-correlation. Let $F = F_1 \times F_2 \subset \Re^4$ where F_1 is feature set 1 and F_2 is feature set 2. Assume the correlation of the data is given by

$$\Sigma^i = \begin{bmatrix} \Sigma^i_{F_1,F_1} & \Sigma^i_{F_1,F_2} \\ \Sigma^i_{F_2,F_1} & \Sigma^i_{F_2,F_2} \end{bmatrix}$$

where

$$\Sigma^i_{F_1,F_1} = \Sigma^i_{F_2,F_2} = \begin{bmatrix} 1 & 0 \\ 0 & 1 \end{bmatrix} \text{ and } \Sigma^i_{F_1,F_2} = \begin{bmatrix} 0 & \rho \\ \rho & 0 \end{bmatrix}$$

where $\rho \in \{0, 1/5, ..., 4/5\}$ and $\Sigma^i_{F_j,F_k}$ is the correlation matrix between the features contained in the feature set F_j and F_k in class i (i,j,k = 1,2). If F^i_j designates feature set *j* for class *i*, let $F_1 = F_1^1 \cup F_1^2$ where

$$F_1^1 \sim N_2(\mu_1^1, \Sigma_{F_1,F_1}^1) \text{ and } F_1^2 \sim N_2(\mu_1^2, \Sigma_{F_1,F_1}^2)$$

and where $\mu_1^1 = (0,0)^T$ and $\mu_1^2 = (0.95, 0.95)^T$.

Let $F_2 = F_2^1 \cup F_2^2$ where

$$F_2^1 \sim N_2(\mu_2^1, \Sigma_{F_2,F_2}^1) \text{ and } F_2^2 \sim N_2(\mu_2^2, \Sigma_{F_2,F_2}^2)$$

and where $\mu_2^1 = (0,0)^T$ and $\mu_2^2 = (1.15, 1.15)^T$.

In our experiments, we simulate the fusion of two classifiers under different levels of correlation. This situation is portrayed in Figure 1 below. Multivariate realizations of F are generated and split into F_1 and F_2. These features sets are used to train and test two separate classifiers, C_1 and C_2. Hence, the classifier C_1 was trained and tested with realizations of F_1 (TR_1 and T_1) while classifier C_2 was trained and tested with realizations of F_2 (TR_2 and T_2) as shown in Figure 1. The two classifiers were: C_1- a linear discriminant and C_2 – a quadratic discriminant. Once these classifiers were trained, the validation set was applied to the results, and the posterior probabilities from this validation set were fused using the ISOC fusion method, the ROC "Within" Fusion, and the PNN fusion method.

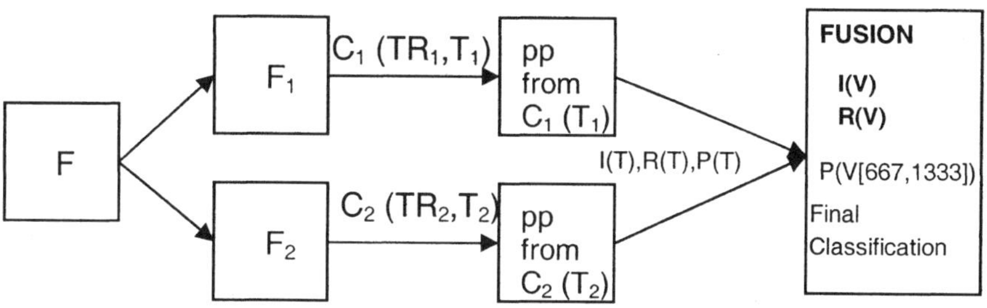

Figure 1: Fusion Flowchart.

In Figure 1 $F_1 \equiv$ Feature Set 1 with two features, $F_2 \equiv$ Feature Set 2 with two features, TR $\equiv$ 2000 training exemplars with two features (f_1, f_2 in F_1 and f_3, f_4 in F_2), T $\equiv$ 2000 test exemplars with two features, V $\equiv$ 2000 validation exemplars with two features, I $\equiv$ ISOC Fusion, R $\equiv$ ROC "within" fusion, and P $\equiv$ PNN Fusion. The symbol C_1(TR,T) signifies that classifier 1, the linear discriminant, was trained on data set TR and tested on data set T. The symbol I(V) shows that the posterior probabilities from the validation set V were fused using the optimal ISOC rule. The symbol P(V(667,1333)) denotes that in the PNN fusion, 667 posterior probability exemplars from the validation set were used for training the neural net, while 1333 exemplars were used for application of the PNN.

EXPERIMENTAL DESIGN

For each fusion method, posterior probabilities were calculated for every exemplar in the training sets as presented to the two classifiers. These posterior probabilities enabled the application of the three fusion methods. In the case of the ISOC method, an optimal Boolean rule is generated from which an idealized (from the training data) ROC curve can be generated. The ROC "within" method fuses the individual ROC curves (forming an idealized ROC curve) but in this case optimal thresholds can be calculated for the fixed "or" rule. This is in contrast to the ISOC method that finds an optimal rule with a fixed threshold. PNN fusion simply entails using the posterior probabilities from the two classifiers as features in the PNN. In the case of ISOC and the ROC "within" fusion methods, a final ROC curve is generated from independent validation data. A final ROC curve is generated for PNN fusion by training the PNN on 1/3 of the validation data and applying the PNN to the remaining 2/3. Curves were generated at 6 levels of correlation (i.e., 0, .2, .4, .6, .8, .9). Below only final or "applied" ROC curves are presented.

RESULTS

The final fusion process step is to apply the ISOC, ROC "within", and PNN fusion rules to the validation data set. These applications are shown in Figure 2. Here we have plotted common FP(false positive), TP (true postive) points for comparative purposes.

We can see that the ISOC model is more robust relative to correlation than the other fusion tools. The PNN maintains a higher performance at low correlation; however this performance is degraded at a high level of correlation. The within ROC curve has a similar performance to the ISOC curve at low correlation, but is degraded slightly at a higher correlation. As discussed earlier, the ISOC method finds the best rule for fixed thresholds, while the ROC "within" fusion method finds optimal thresholds for a given rule, in this case logical "or".

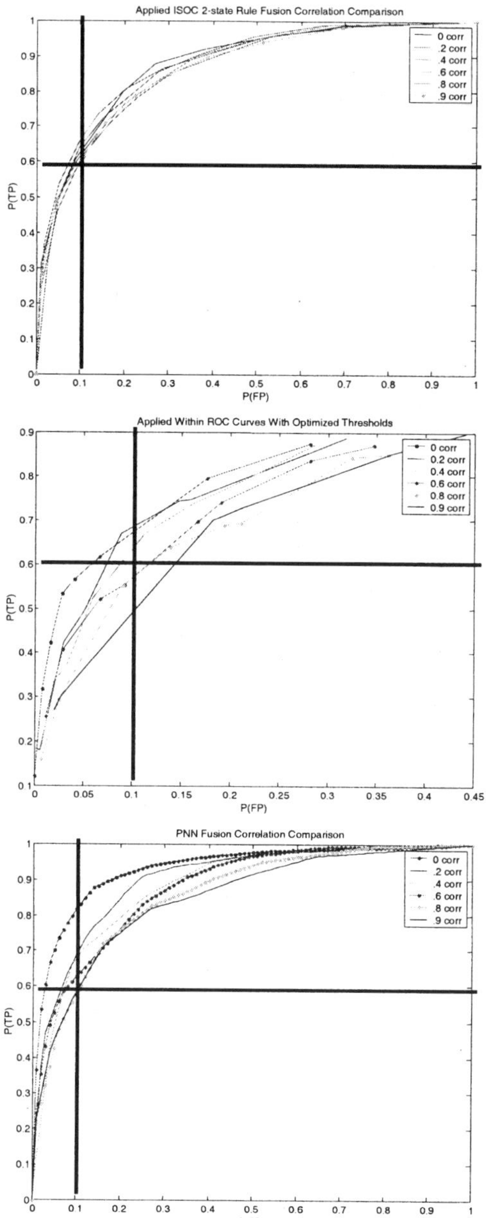

Figure 2: Comparison of Three Fusion Models

SUMMARY

The effects of correlation on three fusion schemes were explored. Test situations were developed to allow the examination of various levels of correlation between feature sets. The ISOC method appears to be the most robust to correlation. However, the PNN is an extremely simple, easy to apply method that outperforms the other fusion methods at low correlation levels. This research is the first step towards the creation of a synthetic classifier fusion-testing environment.

References

Duin, R.P.W. and Tax, D.M.J., "Experiments with combining Classifier Rules", Multiple Classifier Systems: *Proceedings of the First International Workshop*, MCS 2000, Cagliari, Italy, June 2000.

Hashem, S., Optimal Linear Combinations of Neural Networks. PhD thesis, School of Industrial Engineering, Purdue University, December 1993.

Haspert, J.K., "Optimum ID Sensor Fusion for Multiple Target Types." *IDA Document D2451*, 2000.

Oxley, M.E., and K. Bauer, "Classifier Fusion for Improved System Performance." *AFIT/ENS Working Document 02-02*, 2002.

Ralston, J.M., "Bayesian Sensor Fusion for Minimum-Cost I.D. Declaration." *1998 Joint Service Combat Identification Systems Conference on Requirements, Technologies and Developments*, (CISC-98), Volume 1 – Technical Proceedings.

Storm A. Susan. *An Investigation of the Effects of Correlation in Sensor Fusion.* Masters Thesis, WPAFB, OH, March 2003, Air Force Institute of Technology.

Tumer, K. and Ghosh J., "Error Correlation and Error Reduction in Ensemble Classifiers," *Connection Science, Special issue on combining artificial neural networks: ensemble approaches*, volume 8, numbers 3 & 4, pp 385-404, December 1996.

IMAGE RECONSTRUCTION FROM PART OF SUBBAND IMAGES

XIAO ZHOU
Master student of The Research Inst. Of Radio and Automation, SCUT.
Guang Zhou 510640 China

YINGLIN YU
Professor of The Research Inst. Of Radio and Automation, SCUT.
Guang Zhou 510640 China

ABSTRACT
To increase the compression ratio in image communication channels is one of the most challenging research issues .In this paper, we discuss how to restore the original image from its narrow-band version both in black and white image and color image. In the experiments with black and white image we use the given low frequency components of an image to estimate the high frequency components, then fuse add the estimated high frequency components with the given low one. Because the parameter of the filter is less than one, so through finite iterations, we can restore the original image. If we get the high frequency first, we can also restore the original image with the similar method. We analyze the theoretical background and informatics background of our method in detail.In the experiments with color image, we decompose the image into S、H、V, components, we use the above method to restore H component with its low frequency component. Since the S、V components are mainly consisted with low frequency components, so we can directly use the low frequency components of S、V components and the estimated H component to restore the whole original color image. Experimental results show that it can efficiently compress image in both lossy and lossless cases.So using our method can communicate a large amount of image data with less frequency band. It's an efficient way to save frequency band resource and it's a attractive method with bright research prospect.

INTRODUCTION

For further increasing the compression ratio in image communication channels, we propose a novel method of pre-narrowing the bandwidth of an original image so that we can further save the bandwidth resources of a channel, and reduce the coding rate of the encoder. The key problem is how to restore the original image from its narrow–band version. The other motivations of this research subject include how to overcome the effect of the frequency distortion problem on the image quality. In some practical usages the low frequency components or the high frequency

components will suffer heavy loss during transmission, some of them will be lost completely.

In order to restore an image from its incomplete frequency components we propose a scheme and try to solve this problem in a general way.

BODY OF PAPER

1.THEORETICAL BACKGROUND

The basic principle of the image analysis and synthesis shown in Fig 1(a) and Fig 1(b) ,where f_{sl} is the low frequency subband image, f_{sh} is the high frequency subband image, LPF and HPF are the low pass filter and high pass filter respectively. R is a synthesis operator. The figs above are not only fit for the black and white images but also fit for the H,S,V components of color images.By Fig 1,we can easily derive a reconstruction scheme shown in Fig 2(a) and Fig 2(b).

In Fig 2(a) we show a reconstruction scheme of $\hat{f}_0$ (the estimate of f_0)from f_{sl} component only, correspondingly we give the reconstruction scheme from f_{sh} in Fig 2(b).The key problem of the reconstruction is that how we can estimate the lost frequency components from the narrow-band version. The theoretical background may be based on the Forier Transform, we show the detailed theoretical analysis as follows:

In Fig 1(a),through low pass filter we can get the low frequency subband image, in theory, we suppose the filter has abrupt edge (as shown in Fig 3),but in practical experiments the edge of the low pass filter may not be abrupt and the real model may as shown in Fig 4. Then let the low frequency subband image pass a high pass filter ,like the low pass filter the edge of the high pass filter is *also not abrupt(as shown in Fig 5).*

We take Fig 1(a) and Fig 2(a) for example. We design the parameters of the two filters to let the filters have a certain overlap region(Fig 6),in this way we can get the remainder high frequency information of the low frequency subband image, then through several times of iteration in addition to blowup the estimated high frequency component we can get the result of the estimated original images. The whole process of our method is shown in Fig 7 which is based on the Forier Transform characteristic.

So in theory it's possible for us to estimate the lost frequency components from the narrow-band version.

2.SYSTEM ANALYSIS

I. Now ,we use Fig 1(a) and Fig 2(a) for system analysis ,then we can get the same result with Fig 1(b) and Fig 2(b) using the same method .First, let the matrices $[H_l]$, $[H_h]$, [R] are the operator of LPF ,HPF, R respectively, and also denote the image matrix $[f_0]$,$[f_{sl}]$,$[f_{sh}]$,$[\hat{f}_0]$ by the single column vector $\vec{f}_0$, $\vec{f}_{sl}$,$\vec{f}_{sh}$,$\vec{\hat{f}}_0$ respectively. From Fig1(a),

we have: $\vec{f}_{sl}=[H_l]\vec{f}_0$ .... Eq(1)

FromFig2(a),wehave: $\vec{\hat{f}}_0=R[\vec{f}_{sl},[H_h\vec{\hat{f}}_0]]$ ....Eq(2)

In our experiment the synthesis operator [R] is simply addition, in practical use we can use different operation of [R] according to image characteristic and the different requirements of the reconstructive quality , for example , we can *use fuzzy addition to instead of the simply addition.*

The reconstruction scheme shown in Fig2(a) forms a closed feedback loop, and can also be viewed as an adaptive filter, we can adaptively adjust the parameters in operators HPF,LPF,R to get an optimal or suboptimal estimate of $\hat{f}_0$. Because we don't exactly or approximately know what the original image is, so we can use the difference between the adjacent iteration results to get the best parameters and to decide when to stop the iteration so that to have an optimal estimate of $\hat{f}_0$.Another question is whether the result $\hat{f}_0$ is convergent by using our reconstructive method, the article bellow will show our discussion of the convergence of $\hat{f}_0$.

If we use Eq(2) to estimate the $\hat{f}_0$ (here, the synthesis operator [R] is simply addition), we have:

$$\overrightarrow{\hat{f}_0}^{(n)} = \vec{f}_{sl} + [H_h]\overrightarrow{\hat{f}_0}^{(n-1)}$$ ($\overrightarrow{\hat{f}_0}^{(n)}$ is the estimate result after n^{th} iteration)

$$\because \overrightarrow{\hat{f}_0}^{(0)} = (1+[H_h])\vec{f}_{sl}$$

$$\therefore \overrightarrow{\hat{f}_0}^{(n)} = [(1-[H_h])^{-1}+[H_h]^{n-1}+[H_h]^n]\vec{f}_{sl}$$

Generally, we design the filter to let the parameters satisfy that: $0<[H_h]<1$, so from the above analysis we can draw a conclusion that we manage to have a convergent estimate result of $\hat{f}_0$ by our method.

From Fig 2(a) , it's clear that the closed feedback loop will enforce the frequency component in the overlap region in Fig 6.The nonlinear effect of the window function in Fig 7 will help us to get some frequency component which has completely lost.

Ⅱ.Then we take the color images for analysis. We decompose the image into S 、H、V, components, where V stands for the intensity component ,the characteristic of V component is similar with the black and white images so now we focus on the S and H components which contain the most of the color information.

Here, we let a vector $f_c(x,y)$ to stand for the S and H components , where

$$F_c(u,v)=\iint s(x,y)e^{-jH(x,y)}e^{-j2\pi(ux+vy)}dxdy \qquad \text{……………Eq(3)}$$

For our analysis we divide the color images into many subsets, if the subset is small , we can describe the color information of it in two ways. In one case, the color of the subset is uniform, most of the plot color is the same. In the other case, there are color change edge in the subset .If we take(x_{0i}, y_{0j})as the center of the subset (i , j) then the Forier Transform of it is :

$$F_c^{(x_{0i},y_{0j})}(u,v)=\iint_{(x,y)\in D(x_{0i},y_{0j})} s(x-x_{0i},y-y_{0j})e^{-jH(x-x_{0i},y-y_{0j})}e^{-j2\pi(ux+vy)}dxdy \qquad \text{……………Eq(4)}$$

In the area $D(x_{0i},y_{0j})$ the Taylor expansion of $H(x-x_{0i},y-y_{0j})$ is:

$$H(x-x_{0i},y-y_{0j})=H(x_{0i},y_{0j})+\left.\frac{\partial H}{\partial x}\right|_{\substack{x=x_{0i}\\y=y_{0j}}}(x-x_{0i})+\left.\frac{\partial H}{\partial y}\right|_{\substack{x=x_{0i}\\y=y_{0j}}}(y-y_{0j}) \qquad \text{Eq (5)}$$

In the first case which the color of the subset is uniform,then $\left.\frac{\partial H}{\partial x}\right|_{\substack{x=x_{0i}\\y=y_{0j}}}(x-x_{0i})+\left.\frac{\partial H}{\partial y}\right|_{\substack{x=x_{0i}\\y=y_{0j}}}(y-y_{0j})$ is very small so the Forier Transform is :

$$F_c^{(x_{0i},y_{0j})}(u,v)=\iint_{(x,y)\in D(x_{0i},y_{0j})} s(x-x_{0i},y-y_{0j})e^{-jH(x-x_{0i},y-y_{0j})}e^{-j2\pi(ux+vy)}dxdy$$

$$=e^{-j2\pi(x_{0i}u+y_{0j}v)}e^{-jH(x_{0i},y_{0j})}s(u,v)^{(x_{0i},y_{0j})}$$ ……………Eq(6)

In the second case which contains color edge $\left.\frac{\partial H}{\partial x}\right|_{\substack{x=x_{0i}\\y=y_{0j}}}(x-x_{0i})+\left.\frac{\partial H}{\partial y}\right|_{\substack{x=x_{0i}\\y=y_{0j}}}(y-y_{0j})$ may can't be ignored so the Forier Transform is :

$$F_c^{(x_{0i},y_{0j})}(u,v)=e^{-j2\pi(x_{0i}u+y_{0j}v)}e^{-jH(x_{0i},y_{0j})}s(u-\frac{m_x}{2\pi},v-\frac{m_y}{2\pi})$$ ……..Eq(7)

From the above analysis we can see that in the first case H component make a certain change to the phase of S component in frequency space and in the other case H component not only make a certain change to the phase of S component but also make a spectrum shift of S(u,v) .

In conclusion, in the SHV system of color images the information of H component is very important , so the loss in H component may cause the distortion of color images. Using the reconstruction method introduced above we can estimate the lost information of H component and get a good restoration of the color images.

3.THE EXPERIMENT RESULT

Basing on the scheme of this paper, computer simulations are performed.We design low pass filter and high pass filter shown in Fig 8.

Experiment 1:

First we use black and white images to do the experiment, we perform the image reconstruction from $\vec{f}_{sl}$ only. The $\vec{f}_{sl}$ is obtained by feeding a standard test image through the LPF. It is noted that we hereafter don't use the standard test image at all. We insert an adjustable amplification factor k into the closed loop of Fig 2(a) .In the case of linear feedback loop, the experiment results are shown in Fig9.We also discuss the effect of value k on the reconstruction results. Furthermore, we insert a finite support gate into the closed loop of Fig 2(a) as a nonlinear element ,then form a nonlinear loop. We hope that it is possible to regenerate some lost high frequency component by the nonlinear effect. This gate is produced by a contour extraction operator. The experiment result is shown in Fig 10,by this scheme, we restore the small leaves more clearly. This effect can not achieved by the above linear loop method.

Experiment 2:

Now we use color images to do the experiment. As we have pointed out previously , in color images H component contains the most color information and S and V components don't have so much high frequency information most of their information concentrate in the low frequency domain. So we can use the low frequency component of S and V to instead of S and V ,in this way the bandwidth resources can be greatly saved. In Fig 11 , (a)is the original images,(b) is the images which use the low frequency component to instead both S and V.We can see that only use the low frequency of S and V don't make too much distortion to the original images while the bandwidth are greatly reduced.

If we want to save the bandwidth further it's necessary to save the bandwidth of H, but as we know H plays an important part in the color image so if we reduce the bandwidth of the H and get the low frequency component we should estimate the high frequency of H, we can use the method introduced above to estimate the high frequency component of H and restore the original color images. The result of the experiment is shown in Fig 12.

In Fig12(a),because of the loss of high frequency component of V, the resolution of the restored image is not very good ,so we can use the same way to estimate the high frequency component of V to improve the quality of the restored image.(the result is shown in Fig13).In Fig 12(b),because we can't completely restore the H component, so there is color distortion in some place. To improve the quality of the restored image further we use an adaptive algorithm to adjust the H component in the restored image,then we get a better experiment result(shown in Fig 14).In table 1,it shows the number of bit for the compressed images,table 2 shows the PSNR for experiment result images.

4.Figures and tables

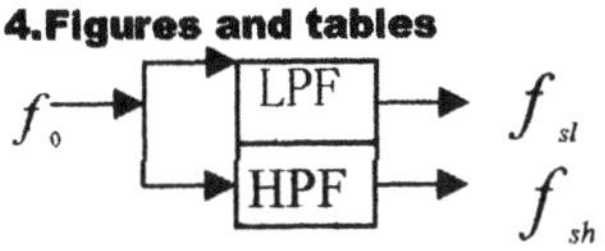

Fig1(a)

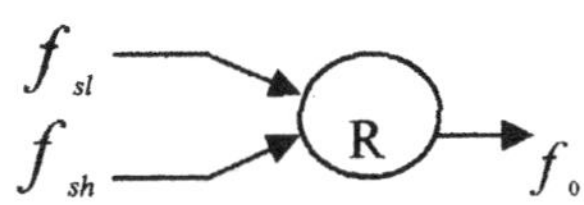

Fig1(b)

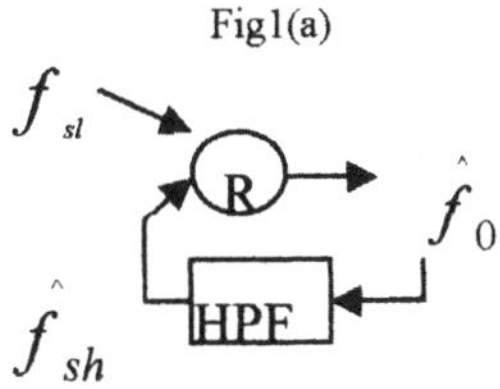

Fig 2(a)

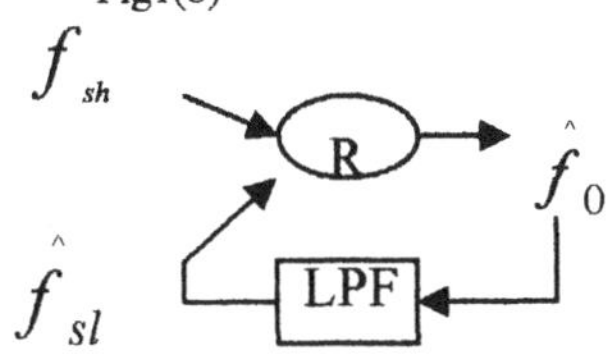

Fig2(b)

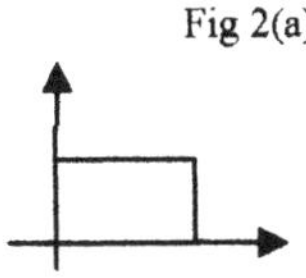

Fig 3

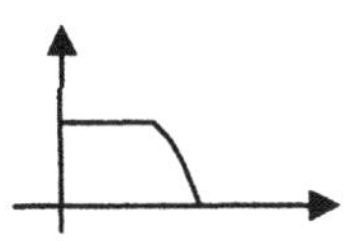

Fig 4

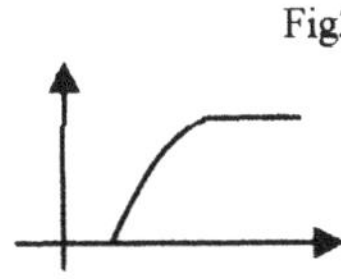

Fig 5

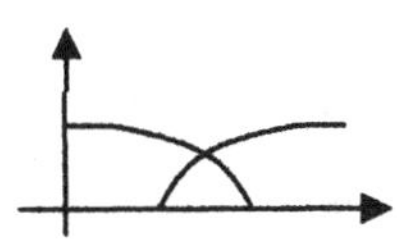

Fig 6

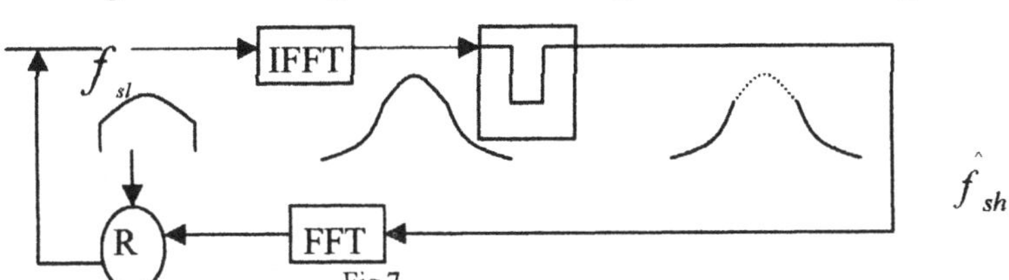

Fig 7

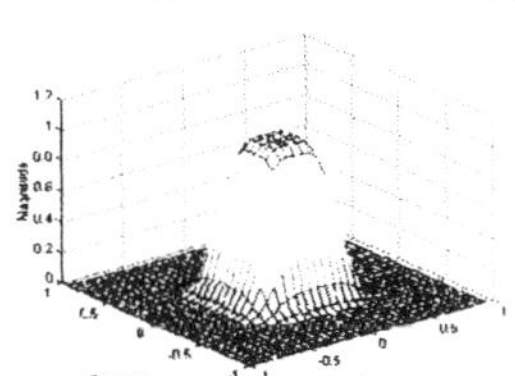

Fig 8

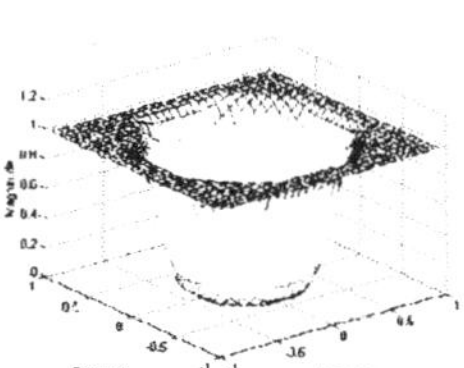

Fig 8

Fig 9(original image)

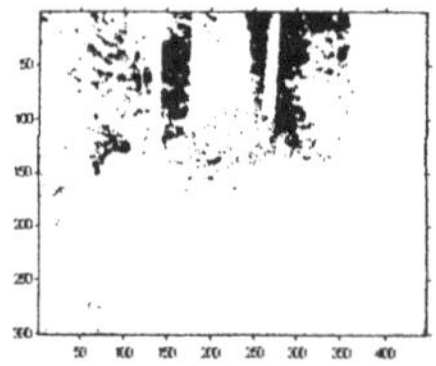

Fig 9 (restored image)

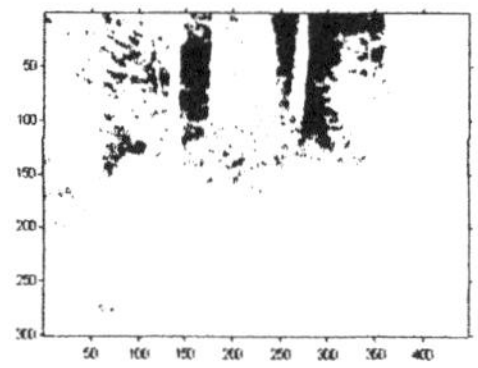

Fig 10

Fig 11(a)

Fig 11(b)

Fig11(a)

Fig11(b)

Fig 12(a)

Fig 12(b)

Fig 13

Fig 14

Image number	Fig 9	Fig13	Fig 14
Bit/pel	6. 6425	6..5466	6. 6444

Table 1

Image num.	Fig 9	Fig 12(a)	Fig 12(b)	Fig 13	Fig 14
PSNR	38.5997	33. 5709	35. 3534	34. 0344	36.7295

Table 2

APPENDICES

1. This research is supported by the GuangDong province natural science fund.(#011628)
2. The reference is elided due to the limit for the length of the paper.

A STATISTICAL NEURAL NETWORK MODEL FOR OBJECT DETECTION

MING-JUNG SEOW and VIJAYAN K. ASARI
Department of Electrical and Computer Engineering
Old Dominion University, Norfolk, VA 23529

ABSTRACT
A statistical neural network model for detection of objects is proposed in this paper. A new learning algorithm based on the statistical properties of the object for a recurrent neural network is developed. The learning is performed based on a mathematical model describing the probabilistic behavior of the object under consideration. The method of least squares estimation is utilized to approximate the relationship between the relative magnitudes of the output of a neuron with respect to the output of another neuron. The dynamics of the neural network functions as a line attractor. The convergence characteristics of the neural network are defined by the statistical properties of the input patterns. The performance of the network on object detection is evaluated by conducting several experiments on face images and the results are encouraging.

INTRODUCTION

The property that is of primarily significant for a neural network is the ability of the network to learn from its environment, and to improve its performance through learning. For more than a decade, neural network research has been considered important not only because it makes efforts towards explaining our intelligence, but also because it provides effective working models for solving a wide range of practical problems. There are too many activities associated with the notion of "learning" to define it in a precise manner. Moreover, the process of learning is a matter of viewpoint, which makes it all the more difficult to agree on a precise definition of the term. For example, numerous researchers such as Rumelhart, Grossberg, Kohonen, and Hopfield, established the theoretical background of "learning" in this field. The model they have developed: the multilayer perceptron trained with backpropagation learning algorithm (Rumelhart, 1986), the ART & ART-MAP (Grossberg, 1982), the self-organizing Kohonen Maps (Kohonen, 1984), and the Hopfield networks (Hopfield, 1982) were theoretically (Hornick, Stinchcombe, and White, 1997) and experimentally (Lawrance, Giles, Fong, 1995) proven to be capable of learning and solving many complex tasks (Meddsker, Jain, 2000).

One of the major goals of both biological neural network modeling and artificial neural network research is to discover better learning rules to yield networks that can learn more difficult tasks, such as tasks that the brain can handle (Seow, Asari, 2002). The principle of association formation has dominated our understanding of learning and memory since William James published "Principles of Psychology" at the end of the nineteenth century (James, 1890). A fundamental problem in associative memory concerns how memory is encoded by the neural network learning algorithm (Haykin, 2001).

In memory-based learning, all (or most) of the past experiences are explicitly stored in a large memory of an input – output system, as shown in Fig. 1. The input to the system can be an N dimensional vector $a \in R^N$ called the memory stimuli, and the output can be an L dimensional vector $b \in R^L$ called the memory response. The relation between the stimuli and the response is given by $b = f(a)$, where $f: R^N \rightarrow R^L$ is the associative mapping of the memory. Each input – output pair or memory association (a, b) is said to be stored and recalled in the memory.

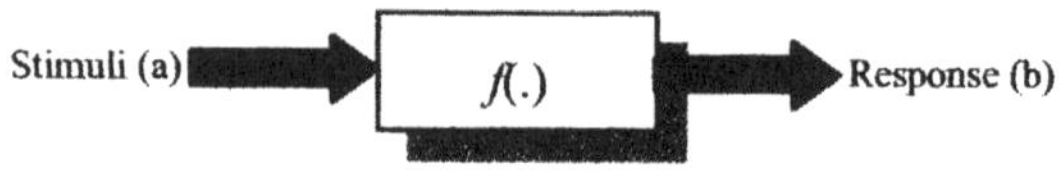

Figure 1: Block diagram of associative memory

Memories are usually thought of as an attracting fixed point at discrete locations in the state space in most of the associative memory models (Hopfield, 1982). These assumptions need not be true for patterns, which exhibit similar characteristics such as the face images that manifest the same pattern (or faces) with varying expression (or emotion). As a result, it may be more appropriate to represent the face expression images using a line attractor network.

In this paper, we propose a novel learning algorithm for a single layer recurrent neural network suitable for multiple valued (or gray scale) pattern association. The proposed learning algorithm involves minimizing the maximum distance of the probabilistic models of the state of two neurons to represent the synaptic weights between them. The dynamics of the neural network can function as a line attractor, which represents the statistical characteristics of the stored patterns.

LEARNING MODEL

The relationship of each neuron with respect to every other neuron can be expressed as:

$$b_i^s = \frac{1}{N}\sum_{j=1}^{N}\left[W_{ij}^s a_j^s + B_{ij}^s\right] \tag{1}$$

for stimulus-response (or input-output) pair (a^s, b^s) corresponding to patterns s, the bias value B(s), and the resultant memory W(s). W(s) and B(S) can be expressed as:

$$W(s) = \begin{bmatrix} w_{11}(s)\, w_{12}(s) \cdots w_{1N}(s) \\ w_{21}(s)\, w_{22}(s) \cdots w_{2N}(s) \\ \vdots \quad \vdots \quad \vdots \quad \vdots \\ w_{N1}(s)\, w_{N2}(s) \cdots w_{NN}(s) \end{bmatrix} \text{ and } B(s) = \begin{bmatrix} B_{11}(s)\, B_{12}(s) \cdots B_{1N}(s) \\ B_{21}(s)\, B_{22}(s) \cdots B_{2N}(s) \\ \vdots \quad \vdots \quad \vdots \quad \vdots \\ B_{N1}(s)\, B_{N2}(s) \cdots B_{NN}(s) \end{bmatrix} \tag{2}$$

To combine W(s) and B(s) of an experience (or pattern) to form a memory matrix W and bias matrix B; we need to utilize probabilistic methods.

The least squares estimation approach to this problem can determine the best approximation line when the error involved is the sum of the squares of the differences between the expected outputs and the approximated outputs. Hence, the weight matrix and the bias matrix must be found that minimize the total least squares error:

$$Err(W_{ij}, B_{ij}) = \sum_{j=1}^{N}\left[b_i - \left(W_{ij}a_j + B_{ij}\right)\right]^2 \tag{3}$$

The least squares estimation method is the most convenient procedure for determining best linear approaximation, and there are also important theoretical considerations that favor this method. For example, the least squares approach puts substantially more weight on a point that is out of line with the rest of the data but will not allow that point to dominate the approximation. The general problem of fitting the best fit line to a collection of data $\{(x_i, y_i)\}_{i=1}^{m}$ involves minimizing the total error described in Eq. (3) with respect to the parameter W_{ij} and B_{ij}. For a minimum to occur, we need:

$$\frac{\partial}{\partial W}\sum_{j=1}^{N}\left[b_i - \left(W_{ij}a_j + B_{ij}\right)\right]^2 = 2\sum_{j=1}^{N}\left(b_i - W_{ij}a_j - B_{ij}\right)\left(-a_j\right) = 0 \tag{4}$$

$$\frac{\partial}{\partial B}\sum_{j=1}^{N}\left[b_i - \left(W_{ij}a_j + B_{ij}\right)\right]^2 = 2\sum_{i=1}^{N}\left(b_i - W_{ij}a_j - B_{ij}\right)\left(-1\right) = 0 \tag{5}$$

Eq. (4) and Eq. (5) can be simplified to the normal equation

$$B_{ij}\sum_{i=1}^{N}a_i + W_{ij}\sum_{i=1}^{N}a_i^2 = \sum_{i=1}^{N}a_ib_i \tag{6}$$

$$B_{ij}N + W_{ij}\sum_{i=1}^{N}a_i = \sum_{i=1}^{N}b_i \tag{7}$$

The solution to this system is as follows:

$$W_{ij} = \frac{N\left(\sum_{i=1}^{N}a_ib_i\right) - \left(\sum_{i=1}^{N}a_i\right)\left(\sum_{i=1}^{N}b_i\right)}{N\left(\sum_{i=1}^{N}a_i^2\right) - \left(\sum_{i=1}^{N}x_i\right)^2} \quad \text{and} \quad B_{ij} = \frac{\left(\sum_{i=1}^{N}a_i^2\right)\left(\sum_{i=1}^{N}b_i\right) - \left(\sum_{i=1}^{N}a_ib_i\right)\left(\sum_{i=1}^{N}a_i\right)}{N\left(\sum_{i=1}^{N}a_i^2\right) - \left(\sum_{i=1}^{N}a_i\right)^2} \tag{8}$$

Based on the above theory, the memory matrix W and the bias matrix B for the P neuron network can be obtained by substituting N with the number of pattern P, a_i with the ith input of the sth pattern x_i^s and b_j with the jth input of the sth pattern x_j^s. Where $x_m \in \{1,256\}$ can be the magnitude of the input at m^{th} neuron.

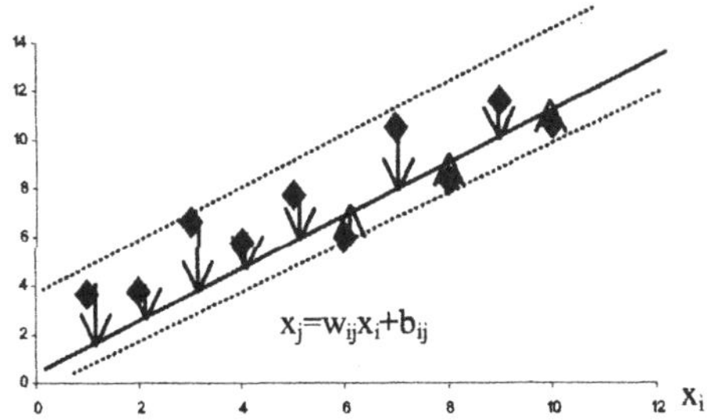

Figure 2: Weight graph

Fig. 2 shows the weight graph illustrating the concept of training based on Eq. (8). A weight graph is a graphical representation of the relationship between the ith neuron and the jth neuron for P patterns. Utilization of the weight graph may help visualize the behavior of one neuron pair.

The network architecture is a single layer fully connected recurrent neural network. The net output of the network can be computed as (Fig. 3):

$$\mathrm{Net}_i = \frac{1}{N}\sum_{j=1}^{N} w_{ij}x_j + B_{ij} \tag{9}$$

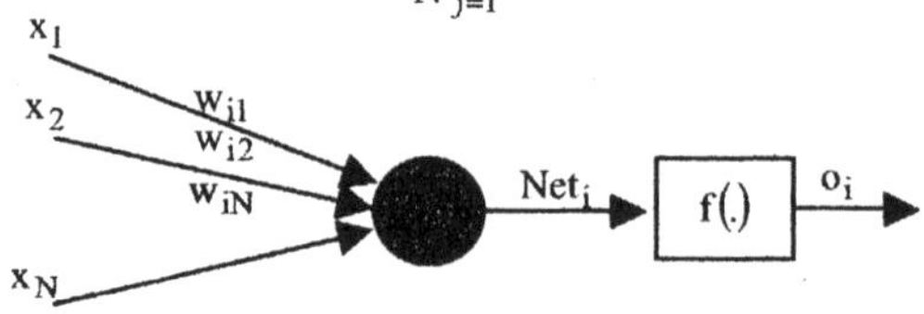

Figure 3: Simplified model of a neuron with threshold function

where N is the number of neurons. The Net_i can be thresholded considering the region of distance of the weight components in the weight graph shown in Fig. 2. That is, in order to preserve each pattern, we need to find the region where the threshold can preserve each pattern. The dashed line in Fig. 2 is such a distance. Mathematical representation of the threshold function can be expressed as:

$$o_i^{new} = f(\mathrm{Net}_i^{new}) = \begin{cases} o_i^{old} & \text{if } \mathrm{winL}_i \le \mathrm{Net}_i^{new} - o_i^{old} \le \mathrm{winR}_i \\ \mathrm{Max}\left(\mathrm{Net}_i^{new}, 256\right) & \text{otherwise} \end{cases} \tag{10}$$

where the window functions for thresholding, winL_i and winR_i, are given by,

$$\mathrm{winL}_i = \alpha \frac{1}{N}\sum_{j=1}^{N} \mathrm{Tl}_j \tag{11}$$

$$\mathrm{winR}_i = \alpha \frac{1}{N}\sum_{j=1}^{N} \mathrm{Tr}_j \tag{12}$$

where α is a constant and $0 < \alpha \le 1$ and Tl_j and Tr_j can be expressed as:

$$\mathrm{Tl}_j = \mathrm{Min}[(w_{ij}x_j + b_{ij}) - x_i] \quad \text{for } 1 \le i \le N \tag{13}$$

$$\mathrm{Tr}_j = \mathrm{Max}[(w_{ij}x_j + b_{ij}) - x_i] \quad \text{for } 1 \le i \le N \tag{14}$$

The window width between winL_i and winR_i decides if the o_i^{old} should be updated or not. That is, winL_i and winR_i calculate the mean of the maximum distance from the approximate functions, which is also the maximum error from the approximate function of the x_i and uses it as a function of threshold. Fig. 4 shows the activation function for each neuron.

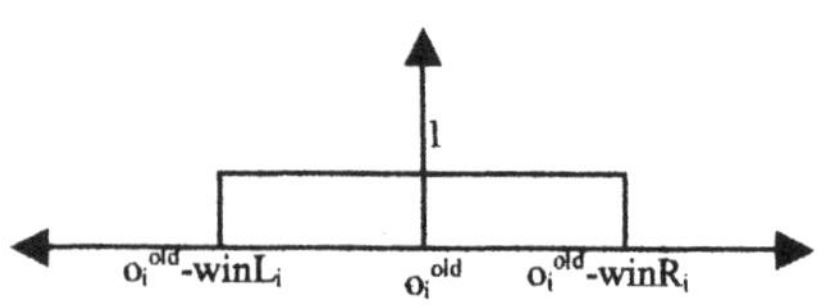

Figure 4: Activation function

RESULTS AND DISCUSSION

Simulations were performed on a 623 × 805 dimensional gray-level (256 level) face images, as shown on Fig. 5. Fig. 5(a) shows the image under consideration and Fig. 5(b) shows the images extracted from Fig. 5(a) for the first experiment.

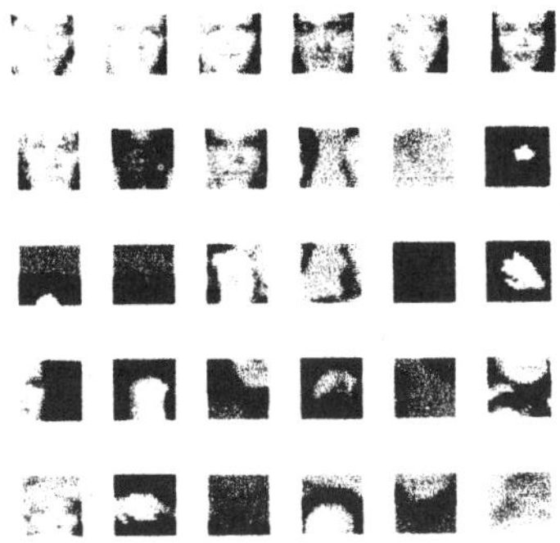

(a) 623 × 805 dimensional image (b) 64 × 64 extracted sub-images

Figure 5: Image under investigation

The first experiment tests the associability of the network by extracting 30 sub-images of size 64 × 64 from the 623 × 805 dimensional image. On this experiment, the network is initially trained using 9 face images as seen on Fig. 5(b). The performance of the network is tested by evaluating all the 30 sub-images in Fig. 5(b). It can be observed from Fig. 6 that the network is able to associate the image correctly, that is, it converges when it is a trained image and it diverges when it is an untrained image, which includes the partial face images.

Figure 6: Associability of the network

The second experiment tests the network detection capability by training the 9 face image as shown on Figure 5(b), and testing the 623 × 805 dimensional image. The detection of the network is designed in such a way that the whole image (623 × 805) is scanned with a moving network of size 64 × 64 at every possible position from left to right, top to bottom of the image and marked the face region with a bounding box. It can be observed in Figure 7 that the network is able to detect all the face images, which are trained to the network.

Figure 7: Face detection using the novel approach

CONCLUSION

A novel learning algorithm for image association and face detection has been proposed. The technique is based on the mathematical model of probabilistic nature of the object under consideration. The method of least squares estimation has been utilized to approximate the relationship between the relative magnitudes of the output of a neuron with respect to the output of another neuron. The dynamics of the neural network functions as a line attractor as opposed to a point attractor. Experiments show that the network has good associability and it is capable of face detection.

REFERENCES

D. E. Rumelhart, G. E. Hinton, and R. J. Williams, "Parallel distributed processes," Vol 1&2, MIT Press, Cambridge, MA, 1986

S. Grossberg, "Studies of mind and brain," Kluwer Academic Publishers, Boston, 1982

T. Kohonen, "Self-organization and associative memory," Springer-Verlag, New York, 1984

J. J. Hopfield, "Neural networks and physical systems with emergent collective computational abilities," Proc. Natl. Acad. Sci. USA, vol. 79, pp. 2554-2558. 1982.

K. Hornick, M Stinchcombe, and H. White, "Multilayer feedforward networks are universal approximators," Neural Network, 2, 1997

S. Lawrance, C. L. Giles, and S. Fong, "On the applicability of neural networks and machine learning methodologies to nature language processing," Technical Report UMIACS-TR-95-64 and CS-TR-3479, University of Maryland, 1995

L. R. Meddsker and L. C. Jain, "Recurrent neural network: Design and applications," CRC Press, New York, 2000.

M. J. Seow and V. K. Asari, "A modular architecture for a recurrent neural network for character recognition," *ANNIE 2002 Conference Proceedings vol. 12,* pp. 21-26, Nov. 10-13, 2002.

W. James, "The principles of psychology," New York: H. Holt and Co. BF121 .J2 1890 V. 1-2. 1890.

S. Haykin, "Neural network: A comprehensive foundation," New York: Macmillan, 2rd ed 2001.

SHAPE MORPHING USING SPHERICAL SOFMS

ARCHANA SANGOLE, GEORGE K. KNOPF AND PHILIP C. IGWE
Department of Mechanical & Materials Engineering,
Faculty of Engineering, The University of Western Ontario,
London, Ontario, Canada, N6A 5B9

ABSTRACT
Shape morphing begins by specifying the geometry and topology of the first object, and then transforming it into the second in order to visualize the intermediate shapes. Establishing a correct correspondence between the bounding shapes is critical in order to obtain a smooth transformation. The proposed metamorphosis technique utilizes a spherical self-organizing feature map (SOFM), with a fixed number of nodes, to fit a mesh around measured coordinate data points. Geometric information about the first bounding object is incorporated in the surface model as the spherical SOFM lattice deforms in the coordinate space. The resulting tessellated surface model is then used as the initial conditions for interpolating the intermediate non-rigid shape transformation. Topological consistency is therefore maintained as the tessellated form of the first shape evolves into the second bounding shape and correspondence between the two shapes is automatically established. The algorithm's performance is illustrated using two examples that involve a continuous transformation between multiple shapes.

INTRODUCTION

Shape morphing is the process of transforming one 3D object into another. Metamorphosis of 3D graphical objects includes the interpolation of their shapes as well as a reasonably smooth transition of other non-geometric object attributes such as colour, texture, and normal fields. These additional attributes are often considered to be part of the geometric model represented in a higher dimensional space (N-dimensional). The shape morphing procedure will consequently involve the transformation of one N-dimensional object to another. A variety of shape transformation algorithms have been proposed for industrial design and computer animation applications that enable the user to "fill-in" the sequence between key-framed objects or the blending of existing shapes to create new objects.

Developing a semi-automated or automated morphing procedure is a nontrivial exercise. The morphing sequence must be smooth and keep as much as possible to the two shapes during the transformation. This is purely an aesthetic requirement and as such highly subjective and task dependent. In general, a shape morphing technique involves establishing a correspondence between the bounding shapes, and then interpolating their geometry. For the purpose of simplification the two are often addressed separately. Correspondence is established by matching geometric attributes of the two shapes and then aligning them in a manner that ensures a smooth transformation from one shape to the other. It is of great importance while morphing shapes with different topologies. A general hypothesis in determining correspondence during shape metamorphosis is to construct a single mesh with two geometric forms. However, a common mesh implies that the bounding objects have the same global topology – a situation that is not always valid. For example, the assumption of identical topology will prevent the morphing technique from transforming a sphere into a torus.

The techniques used to establish a correspondence and then interpolating the shapes are primarily dependent on the representation of the freeform shape(s) prior to the morphing operation. Types of structural representations include surface boundary representations, volume-based approaches such as implicit function representations and secondary representations such as simplex meshes.

Surface boundary representations of 3D objects have data structures that contain both geometric and topological information. Geometry provides the precise coordinates of the vertices or control points that define the faces or parametric surface patches. Topology, on the other hand, gives the adjacency relationship between polygonal faces or surface patches. It is relatively easy to attach additional object properties such as colour, and texture to surface representation. Unfortunately, their model parameters are very sensitive and, even slight changes in data values due to random noise can significantly modify the location of the vertices or control points thereby limiting their implementation in interactive design.

Chen and Parent (1989) represent the 3D objects by a set of planar contours and apply a shape averaging method to corresponding pairs of contours. Lazarus and Verroust (1997) utilize boundary representations for morphing cylinder-like objects. An axis is defined for each object and a discretized representation of the objects is generated by constructing cylindrical meshes that approximate the object in the form of a star-shape around the axis. Shape transformation is achieved by interpolating between the two cylindrical axes of the objects, for each point along the local radius of the mesh. However, the method is largely topology driven and the complexity of the algorithm is proportional to the number of vertices and faces in the two polyhedra.

In order to morph objects with different topologies DeCarlo and Gallier (1996) represent the objects by a sparse control mesh of equal size and topology and, define a mapping between the respective meshes of the bounding shapes by identifying corresponding faces. The meshes then degenerate to the original models – maintaining a sense of correspondence. Alternatively, Kent *et al.* (1992) project the models onto a unit sphere and establish a mapping between their respective projections. Correspondence between the vertices is established by a process of subdivision and shape metamorphosis is achieved by interpolating between pairs of corresponding vertices. Gregory *et al.* (1999) extended it by providing an additional capability of interactively selecting important features between the two object meshes and then establishing a match between corresponding feature pairs thereby having more control on the transformation. These methods are more suited for polyhedral objects and do not require an explicit mapping for every point on the bounding objects.

Another approach is to define implicit functions for each shape and then interpolate between these two functions (Turk and O'Brien, 1999). A commonly used implicit function is the characteristic function, which is essentially a binary value that signifies the position (inside - outside) of the region with respect to the shape. Others include weighted Minkowski sums, wavelet decomposition of a characteristic function (Taosong *et al.*, 1994) and a signed distance function.

An alternative method to create a secondary representation of the objects is to approximate the shapes by simplex meshes (Delingette *et al.*, 1993; Hebert *et al.*, 1995). Mesh connectivity is established during transformation using a set of Eulerian operators. The shape morphing process between the bounding shapes is obtained by adding constraints that transform the intrinsic parameters. The operator must guide the approximation process by positioning and scaling the initial mesh for each object. Some problems may occur in rendering meshes of smooth surfaces because the simplex mesh is comprised of planar facets with an average degree of 6.

The proposed shape metamorphosis method utilizes a spherical self-organizing feature map (SOFM) to generate a closed representation of the freeform shape. Geometric information is incorporated in the model as the topologically ordered SOFM lattice deforms in the visual display space during the learning process. The deformed map generated for the first object is used as the initial condition for the second consequently maintaining topological consistency during the transformation process. The algorithm's performance is illustrated by continuously transforming the facetted surface model of one object into another.

SHAPE MORPHING USING A SPHERICAL SOFM

Prior to shape interpolation a tessellated representation of the reference freeform shape (**S**^{r}) is generated using the spherical self-organizing feature map (SOFM). The resulting tessellated form constitutes the neural architecture of the SOFM for the shape metamorphosis operation and serves as the initial conditions for morphing the reference object (**S**^{r}) into the target object (**S**^{t}). The weight vectors of the nodes in the tessellated representation of **S**^{r} constitute the initial weights for the shape morphing operation, while scattered coordinate data for the second object form the inputs to the network, Figure 1. An explanation of the spherical SOFM weight adaptation algorithm can be found in (Knopf and Sangole, 2002a).

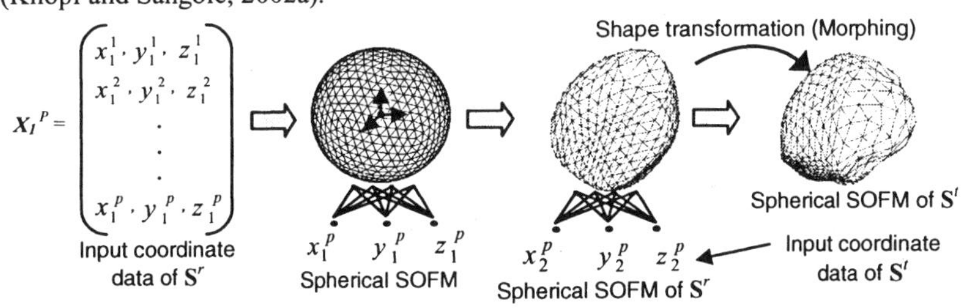

Figure 1 Non-rigid transformation from a reference shape (S^r) to a target shape (S^t).

While generating the tessellated form for **S**^{r} a relatively high learning rate 'μ' is selected so that the network quickly learns the topology of the reference coordinate data. During the morphing operation, 'μ' is reduced from 0.2 to 0.05 and the incremental change in the weights used in (A2b) is redefined as

$$\mathbf{S}^t: \ \alpha = 0.001 + \mu\,[1 - (NE_{i,j,k^*} / NE_{\text{initial}})\,] \ . \qquad (1)$$

in order to get a gradual transition from **S**^{r} to **S**^{t} thereby getting a sequence of intermediate shapes that exhibit a smooth metamorphosis between the bounding shapes. Shape transformation is observed by displaying the SOFM lattice at various time intervals during the weight adaptation process.

Due to a global mesh topology, a correspondence between the two shapes is automatically established within the SOFM while the nodal points are being dragged through the visual space during weight adaptation as the reference shape (**S**^{r}) morphs to the target shape (**S**^{t}). Eventually, **S**^{t} is represented by the same number of connected nodes as in **S**^{r} thereby retaining topological consistency throughout the transformation process. This simple operation is independent of the number of coordinate points in the bounding shapes. It not only eliminates the need to explicitly compute a correspondence between the two freeform shapes but also provides a fairly smooth transition from one shape to the other. Some shape morphing applications may however require preliminary alignment or registration data prior to morphing in order to ensure that a smooth and natural shape transition. Spherical SOFM representations of the shapes can be utilized for preliminary registration (Knopf and Sangole, 2002a) to align the two sets of coordinate data. Examples illustrating both cases have been presented in this paper.

EXPERIMENTAL STUDY

Examples include a set synthetic coordinate data of a cube and cylinder, and scattered coordinate data of two natural heads with distinct variations in facial characteristics. The tests were performed to demonstrate the topological continuity and geometric interpolation characteristics of the proposed shape morphing methodology.

Case 1: Shape transformation between a cube and a cylinder

The bounding shapes consist of scattered synthetic coordinate data of a cube, with 1500 points, and a cylinder, with 5137 points. A spherical SOFM network of 642 cluster units is used. The two shapes are aligned so that their centers-of-mass coincide. Figure 2 illustrates initialization of the map for the morphing operation, which is then used as the initial condition for morphing into the cylinder. The sequence of transformations between the two shapes is shown in Figure 3.

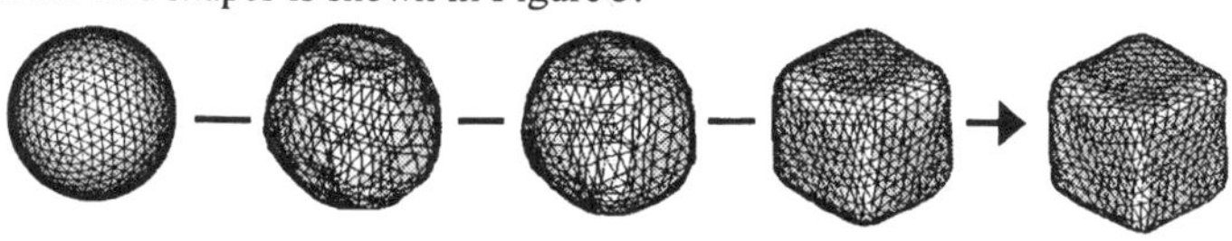

Figure 2 Initializing the spherical SOFM for the morphing operation by generating a tessellated representation of the reference shape (S^r), a cube.

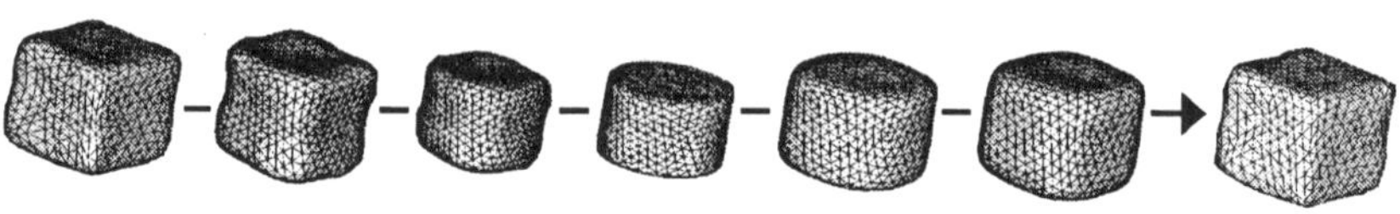

Figure 3 Sequence of shape transformation between a cube (S^r) and a cylinder (S^t).

Case 2: Shape transformation between two natural heads

The reference head consists of 18,882 scattered coordinate points while the target head consists of 21,474 points, Figure 4. It also illustrates the respective images and the initial maps for the two sets of coordinate data. A spherical SOFM network with 2562 cluster units is used. The mesh of the reference head $\mathbf{S}^r$ is transformed to take the shape of the target head $\mathbf{S}^t$ using the same weight adaptation algorithm of the spherical SOFM.

In this example, the shapes are pre-registered to align the facial characteristics in order to ensure a natural transition from one head to the other. In addition, it was observed that by using an initial neighbourhood size, NE_{initial}, of 1 unit radius for 80 % of the training process and later reducing it to just the winning unit towards the final stage ensures a natural transition between the two shapes and avoids a smoothing effect that would otherwise result due to a wide neighbourhood. The sequence of transformations as the reference head morphs into the target head and back is illustrated in Figure 5.

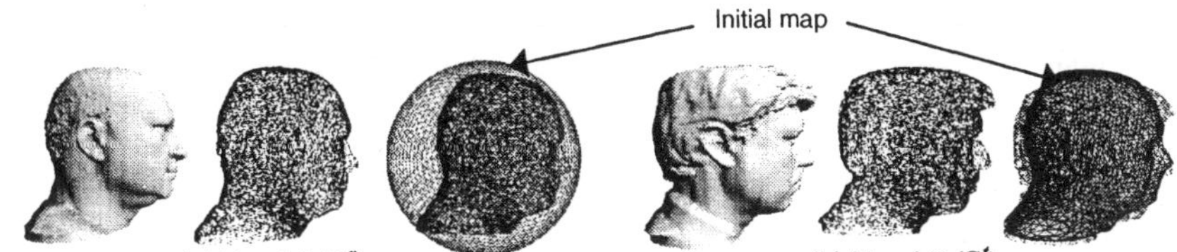

(a) Head-1 (S^r) (b) Head-2 (S^t)

Figure 4 The reference shape head-1 (S^r), the target shape head-2 (S^t) and their respective initial maps during the morphing operation.

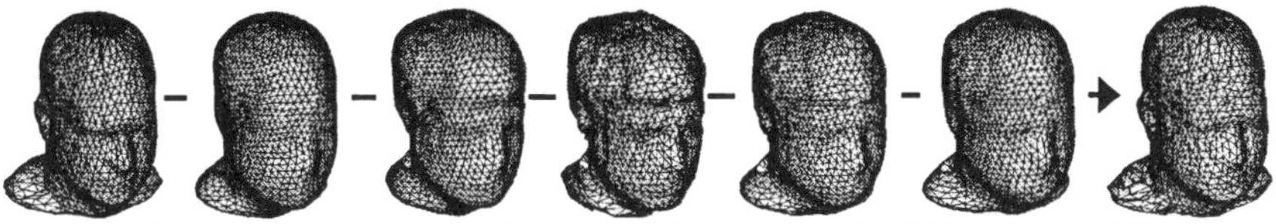

Figure 5 Sequence of shape transformations between head-1 (S^r) and head-2 (S^t).

DISCUSSION

In a general sense, shape morphing begins by specifying the geometry and topology of two objects, and then transforming the first reference model into the target object, with the primary goal of obtaining intermediate shape alternatives that exhibit a combination of features that are adopted from the bounding shapes. As demonstrated, the spherical SOFM provides a viable shape metamorphosis tool. Example one required minimal pre-registration while in the second, pre-registration of the heads was necessary to align their facial features in order to ensure a smooth transition from one head to the other and to get a range of intermediate heads that exhibit a combination of facial characteristics adopted from the bounding shapes, Figure 6.

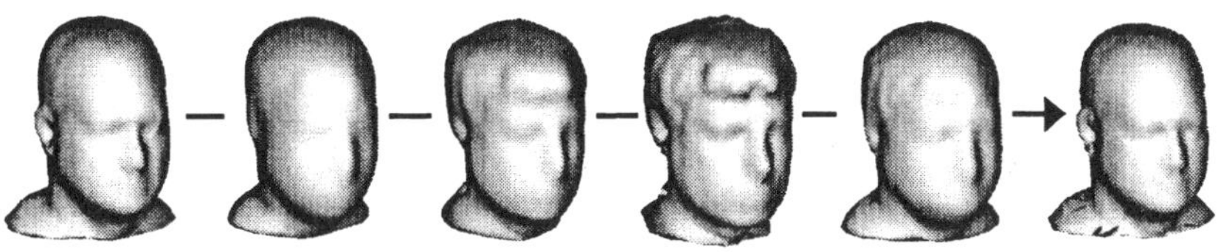

Figure 6 Sequence of intermediate rendered surface models while transforming head-1 (S^r) to head-2 (S^t) and back to head-1 (S^r).

The examples presented in this paper had geometry equivalent to that of a sphere and preliminary registration of the bounding shapes prior to morphing was not emphasized. While morphing objects with different topologies correspondence will be a critical issue and careful attention will have to be given on how to adapt inter-node connections during shape metamorphosis.

CONCLUSIONS

The spherical self-organizing feature map (SOFM) was introduced as a freeform shape metamorphosis tool. A novel feature of the algorithm is that correspondence between the bounding shapes is automatically established during shape transformation. Furthermore, only the tessellated form of the first bounding shape needs to be defined which serves as the initial conditions for shape metamorphosis. While morphing requires a clear definition of the bounding shapes, it is still one method of transforming between multiple shapes. The same concept may be extended to develop interactive software tools that enable the design engineer to intuitively create and visually assess alternative product concepts (Knopf and Sangole, 2002b) by manipulating their respective tessellated forms within a dynamic physics-based framework for virtual sculpturing. This application will however require an immersive virtual reality environment (IVR) that facilitates the display, reorientation, analysis and modification of the tessellated form of the object. By automatically generating topologically consistent geometries that exhibit smooth blending transitions from one shape extremity to another, the designer is provided with a versatile modeling tool that can help inspire new ideas and innovative concepts.

ACKNOWLEDGEMENTS

This work has been supported, in part, by the Natural Sciences and Engineering Research Council of Canada. The authors wish to express their sincere gratitude to Mr. Dennis Nancoo for his assistance in acquiring the range data necessary to test the algorithm.

REFERENCES

Chen, S.E., and Parent, R.E. (1989). "Shape averaging and its application to industrial design", *IEEE Computer Graphics*, vol. 9, pp. 47 – 54.

DeCarlo, D. and Gallier, J. (1996). "Topological evolution of surfaces", *Proceedings of Graphics Interface '96*, pp. 194-203.

Delingette, H., Watanabe, Y, and Seuenaga, Y. (1993). "Simplex based animation", *Computer Animation '93, Computer Graphics International*, Geneva, Switzerland, pp. 13- 28.

Gregory, A., State, A., Lin, M., Manocha, D. and Livingston, M.A. (1999). "Interactive surface decomposition for polyhedral morphing", The Visual Computer, vol. 15, pp. 453-470.

Hebert, M., Ikeuchi, K., and Delingette H. (1995). "A spherical representation for recognition of freeform surfaces", *IEEE Transactions on Pattern Analysis and Machine Intelligence,* vol. 17, no. 7, pp. 681 - 690.

Kent, J.R., Parent, R.E. and Carlson, W.E. (1992). "Shape transformation for polyhedral objects", *Proceedings of ACM SIGGRAPH Computer Graphics,* vol. 26, Issue 2, pp. 47-54.

Knopf, G.K., and Sangole, A., (2002a), "Registration of closed free-form surfaces using deformable maps", *Intelligent Engineering Systems Through Artificial Neural Networks ANNIE'2002 conference*, St. Louis Missouri, C.H. Dagli *et al.* (Eds.), ASME Press.

Knopf, G.K. and Sangole, A. (2002b). "Intelligent Systems for Interactive Design and Visualization", *28th Annual Conference of the IEEE Industrial Electronics Society* (IECON'2002), Sevilla Spain, Nov. 5-8, 2002, pp. 2995 – 3001.

Kohonen, T. (1997). *Self-Organizing Maps*. Springer, New York.

Krishnamurthy, A.K., Ahalt, S.C., Melton, D.E., and Chen, P. (1990). "Neural networks for vector quantization of speech and images", *IEEE Journal on Selected Areas in Communication*, vol. 8, no. 8, pp. 1449 – 1457.

Lazarus, F. and Verroust, A. (1997). "Metamorphosis of cylinder-like objects", *The Journal of Visualization and Computer Animation*, Vol. 8, pp. 131-146.

Taosong, H., Wang, S. and Kaufman, A. (1994). "Wavelet-based volume morphing", *Proceedings of Visualization '94*, Washington, D.C., D. Bergeron and A. Kaufman (Eds.), Oct. 17-21, pp. 85-92.

Turk, G. and O'Brien, J.F. (1999). "Shape transformation using variational implicit functions", *Proceedings of the SIGGRAPH '99 conference*, pp. 335-342.

SIGNATURE RECOGNITION AND VERIFICATION WITH ANN USING MOMENT INVARIANT METHOD

Cemil OZ
Sakarya University Computer Eng. Dep., Sakarya, Turkey,ozc@umr.edu,

Fikret Ercal
UMR Computer Science Department, Rolla, MO 65401,ercal@umr.edu

ABSTRACT
In this paper, we present an off-line signature recognition and verification system which is based on moment invariant method and ANN. Two separate neural networks are designed; one for signature recognition, and another for verification (i.e. for detecting forgery). Both networks use a four-step process. First step is to separate the signature from its background. Second step performs normalization and digitization of the original signature. Moment invariant vectors are obtained in the third step. And the last step implements signature recognition and verification.

Keywords: Signature verification, signature recognition, ANN, moment invariants, computer vision

1. INTRODUCTION

Signature is a special case of handwriting which includes special characters and flourishes. Many signatures can be unreadable. They are a kind of artistic handwriting objects. However, a signature can be handled as an image, and hence, it can be recognized using computer vision and artificial neural network techniques.

Signature recognition and verification involves two separate but strongly related tasks: one of them is identification of the signature owner, and the other is the decision about whether the signature is genuine or forged. Also, depending on the need, signature recognition and verification problem is put into two major classes: (i) online signature recognition and verification systems (SRVS) and (ii) offline SRVS. Online SRVS requires some special peripheral units for measuring hand speed and pressure on the human hand when it creates the signature. On the other hand, almost all off-line SRVS systems relies on image processing and feature extraction techniques.

In the last two decades, in parallel with the advancement in the sensor technology, some successful online SRVS are developed [1, 2, 3]. There are also many studies in the area of offline SRVS category [4, 5, 6, 7, 8, 9, and 10]. These studies are generally based on ANN [4,10], analysis of the geometry and topology of the signature [11], and its statistical properties [9].

In this study, we present an off-line signature recognition and verification system which is based on a moment invariant method. Two neural networks are designed; one for signature recognition, and another for verification (i.e. for detecting forgery). Both networks use a four-step process:

First step: this stage consists of basic image processing methods; converting color image to gray image, noise reduction, and background elimination.

Second step: different size signatures are normalized to have the same size.

Third step: moment invariant values at different angles are calculated for each signature.

Fourth step: signature recognition and verification is done at this stage. Two different ANN models are designed; one for recognizing the signature and the other for the verification purposes.

2. IMAGE PROCESSING

The camera-captured or scanned real world images containing human signatures are processed using several image processing algorithms before the calculation of the moment invariants. These processes are given below.

2.1 CONVERTING COLOR IMAGE TO GRAY SCALE IMAGE

In present technology, almost all image capturing and scanning devices use color. Therefore, we also used a color scanning device to scan signature images. A color image consists of a coordinate matrix and three color matrices. Coordinate matrix contains x,y coordinate values of the image. The color matrices are labeled as red (R), green (G), and blue (B). Techniques presented in this study are based on grey scale images, and therefore, scanned or captured color images are initially coverted to grey scale using the following equation [12]:

$$\text{Gray color} = 0.299*\text{Red} + 0.5876*\text{Green} + 0.114*\text{Blue} \quad (1)$$

2.2 NOISE REDUCTION

Noise reduction (also called "smoothing" or "noise filtering") is one of the most important processes in image processing. Images are often corrupted due to positive and negative impulses stemming from decoding errors or noisy channels. An image may also be degraded because of the undesirable effects due to illumination and other objects in the environment. Median filter is widely used for smoothing and restoring images corrupted by noise. It is a non-linear process useful especially in reducing impulsive or salt-and-pepper type noise. In a median filter, a window slides over the image, and for each positioning of the window, the median intensity of the pixels inside it determines the intensity of the pixel located in the middle of the window. Different from linear filters such as the mean filter, median filter has attractive properties for suppressing impulse noise while preserving edges. Median Filter is used in this study due to its edge preserving feature [13, 14, 15,16].

2.3 BACKGROUND ELIMINATION

Many image processing applications require the differentiation of objects from the image background. *Thresholding* is the most trivial and easily applicable method for this purpose. It is widely used in image segmentation [17, 18]. Thresholding is choosing a threshold value T and assigning 0 to the pixels with values smaller than or equal to T and 1 to those with values greater than T. We used thresholding technique for differentiating the signature pixels from the background pixels. Clearly, in this application, we are interested in dark objects on a light background, and therefore, a threshold value T, called the brightness threshold, is appropriately chosen and applied to image pixels f(x, y) as in the following:

$$\begin{array}{lll} \text{If } f(x,y) \geq T & \text{then} & f(x,y) = \text{Background} \\ & \text{else} & f(x,y) = \text{Object} \end{array} \quad (2)$$

Figure 1.a) Captured signature b) signature image with background removed

2.4 SIGNATURE NORMALIZATION

Signature dimensions may vary due to the irregularities in the image scanning and capturing process. Furthermore, height and width of signatures vary from person to person and, sometimes, even the same person may use different size signatures. First, we need to eliminate the size differences and obtain a standard signature size for all signatures. After this normalization process, all signatures will have the same dimensions. In this study, we used a normalized size of 40x40 pixels for all signatures that will be processed further. During the normalization process, the aspect ratio between width and height of a signature is kept intact.The normalization process is demonstrated in the following figure.

a) Original signature image b) Normalized signature

Figure 2. Signature normalization

3. MOMENT INVARIANT METHOD

Moment invariants are properties of connected regions in binary images that are invariant to translation, rotation and scaling. They can be easily calculated from region properties and they are very useful in performing shape classification and part recognition. One of the techniques for generating invariants in terms of algebraic moment was originally proposed by Hu [19]. The moment invariants used in our research are computed using the equations given in Table-1(b) for all signatures at various angles[20, 21, 22, 23].

Table 1 : (a) Formulas used for specific central moments b) List of the derived invariant moments

Central Moments
$\mu_{00}=m_{00}$
$\mu_{10}=0$
$\mu_{01}=0$
$\mu_{20}= m_{20}-x\, m_{00}$
$\mu_{02}= m_{02}- y\, m_{01}$
$\mu_{11}= m_{11}- y\, m_{10}$
$\mu_{30}= m_{30}-3x' m_{20} + 2x' m_{10}$

Derived Invariant Moments
$I_1=\eta_{20} + \eta_{02}$
$I_2=(\eta_{20} -\eta_{02})^2+4\eta_{11}{}^2$
$I_3=(\eta_{30} - 3\eta_{12})^2+ (3\eta_{21}-\eta_{03})^2$
$I_4=(\eta_{30} + \eta_{12})^2+ (\eta_{21}-\eta_{03})^2$
$I_5=(\eta_{30} - 3\eta_{12}) (\eta_{30} + \eta_{12})^2((\eta_{30} + \eta_{12})^2 -3(\eta_{21} + \eta_{03})^2) + (3\eta_{21} - \eta_{03}) (\eta_{21} + \eta_{03})(3 (\eta_{30} + \eta_{12})^2 -(\eta_{21} + \eta_{03})^2)$
$I_6=(\eta_{20} - \eta_{02}) ((\eta_{30} + \eta_{12})^2- (\eta_{21} + \eta_{03})^2)+ 4\eta_{11} (\eta_{30} + \eta_{21})(\eta_{21}+ \eta_{03})$
$I_7=(3\eta_{12} - \eta_{30}) (\eta_{30} + \eta_{12})((3\eta_{30} \eta_{12}) - 3(\eta_{21} + \eta_{03})^2) + (3\eta_{21}- \eta_{03}) (\eta_{21} + \eta_{03})((3\eta_{30} \eta_{12})^2 -(\eta_{21} + \eta_{03})^2)$

3.1 DIGITIZATION OF SIGNATURES WITH MOMENT INVARIANT METHOD

Feature vectors used for signature identification are generated using moment invariants. For this purpose, we produced six different sets of feature vectors for every signature where each set consisted of seven moment invariant values listed in Table 1(b). First, a feature vector for the normalized signature is produced using moment invariants, and these seven feature vector values are saved in a database. Then, the normalized signature is rotated using the two dimensional rotation equation given below:

$$\begin{bmatrix} x' \\ y' \end{bmatrix} = \begin{bmatrix} \cos\theta & -\sin\theta \\ \sin\theta & \cos\theta \end{bmatrix} \begin{bmatrix} x \\ y \end{bmatrix} \qquad (3)$$

Six different sets of feature vectors are calculated to correspond to six different reference angles. The normalized signature image is rotated from -15° to 90° with 15°-25° increments.

4. ANN DESIGN FOR SIGNATURE RECOGNITION AND VERIFICATION

We designed a multilayer feed forward artificial neural network for recognition of off-line digitized signatures. The proposed ANN consists of 7 input variables, 18 hidden neurons, and 30 output variables and it is designed to recognize one signature at a time. Backpropagation algorithm is used for training.

4.1. TRAINING FOR SIGNATURE RECOGNITION

First, an input/output database is created manually for training and testing the ANN. Each input vector consists of seven moment invariants obtained for a signature at a particular rotation angle. As explained earlier in section 3, six different moment invariant vectors each corresponding to an angle between -15° and 90° is produced for each signature. These six vectors are divided into two sets each containing three vectors. One of these sets (3 input vectors) is used in the training of ANN and the other set (remaining 3 input vectors) is used for testing. Each set is formed by picking the input vectors at alternate angles so that verification can be done in any given rotation of the signature image. The database contained a total of 30 different signature images which are used for both training and testing. Since 3 input vectors for each image is used for training purposes, there are a total of 90 (30*3) input vectors (data sets) in the training set. The remaining 90 data sets are used for testing. ANN contained 30 binary output values each corresponding to one signature being tested as shown in Table 2. Under normal (correct) operation of ANN, only one output is expected to take a value of "1" indicating the recognition of a signature represented by that particular output. The other output values must remain zero. In general, the number of outputs must be equal to the number signatures being considered. Table 2 shows a number of real input/output vectors used in the training set.which are obtained from a set of signatures.

Table 2. A sample input set for signature recognition NN.

		Inputs							Outputs					
	Signature images	I_1	I_2	I_3	I_4	I_5	I_6	I_7	O_1	O_2	O_3	O_4		O_{30}
1	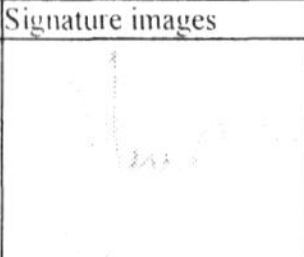	0.24498	0.01320	0.00311	0.01508	0.00007	0.00172	-.00003	1	0	0	0		0
		0.21825	0.06348	0.82200	0.24186	0.55277	0.79252	0.80334	1	0	0	0		0
		0.76491	0.12616	0.26356	0.99508	0.06991	0.75293	0.07709	1	0	0	0		0
2		0.51865	0.17235	0.13245	0.21363	0.01455	0.08733	0.00561	0	1	0	0		0
		0.49462	0.02351	0.09704	0.08364	0.04426	0.04839	0.00443	0	1	0	0		0
		0.68945	0.09235	0.03231	0.20112	0.01757	0.03869	0.01461	0	1	0	0		0
3	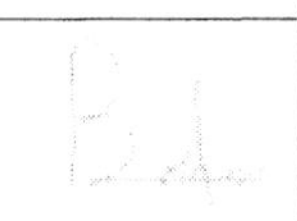	0.56872	0.16345	0.07310	0.22116	0.02386	0.08941	-.02038	0	0	1	0		0
		0.56872	0.16345	0.04132	0.12125	0.02121	0.08449	-.00234	0	0	1	0		0
		0.39925	0.10635	0.04071	0.02115	0.00098	-.06001	0.00000	0	0	1	0		0
30		0.22156	0.00396	0.00046	0.00842	-.00001	0.00052	0.00001	0	0	0	0		1
		0.37143	0.00296	0.00096	0.00242	-.00001	0.00082	0.00001	0	0	0	0		1
		0.21433	0.01591	0.00159	0.00359	0.00000	0.00013	0.00000	0	0	0	0		1

4.2. SIGNATURE VERIFICATION

In this part of the study, our purpose is to authenticate a signature i.e. to verify that the signature is not counterfeit and it really belongs to the person who is claimed to be the owner of the signature. The ANN used for this purpose is also a multilayer feed forward network which consists of 7 input variables, 8 hidden neurons, and 2 output variables indicating whether the signature is fake or true. Backpropagation algorithm is used for training. The training data set is obtained from three original (authenticated) signatures provided by the real owner and three fake signatures. As it was done for the preparation of the training data for the ANN used in recognition, three invariant vectors per signature is used in the training set. Therefore, a total of 18 moment invariant vectors are used in the training set. A sample set of three signatures belonging to the same person is shown in Figure 3.

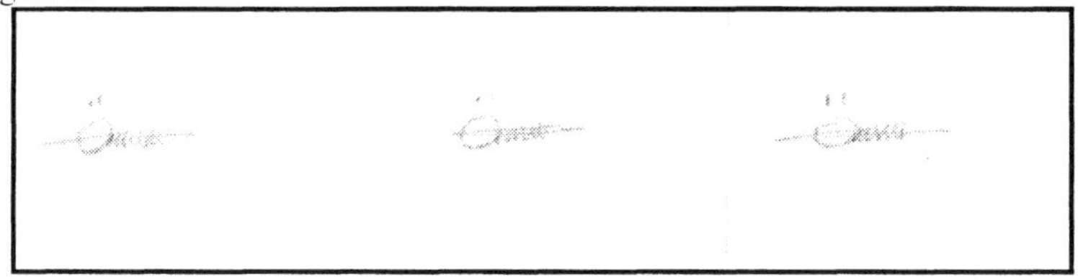

Figure 3. Tree signatures which belong to the true owner of the signature

5. IMPLEMENTATION AND TEST RESULTS

5.1. SIGNATURE RECOGNITION

We used a windows interface program. This software allowed the signature images to be loaded one at a time and used in training and testing. First, the signature image is captured using a CCD camera or a scanner, then, through several image processing operations, it is converted to binary and normalized to a 40*40 image as explained earlier. Moment invariant values are obtained from the normalized image which is then used as the input vector to the ANN. After the training of the ANN for signature recognition, the system is ready to recognize a given signature.

The signature recognition system is tested using 30 signatures chosen at random. The images were obtained using the following properties:

- Signatures were signed inside a special framed area. ANN T.
- Images were taken with a simple CCD camera and they were shot from a fixed distance.

As explained in Section 4.1., 30 images in our database belonging to 30 different signatures are used for both training and testing. Since 3 (out of 6) input vectors for each image were used for training purposes, there are only 90 (30*3) input vectors (data sets) left to be used for the test set. Under normal (correct) operation of the ANN, only one output is expected to take a value of "1" indicating the recognition of a signature represented by that particular output. The other output values must remain zero. The output layer used a logic decoder which mapped neuron outputs between 0.5-1 to a binary value of 1. If the real value of an output is less than 0.5, it is represented by a "0" value. The ANN program recognized all of the 30 signatures correctly. This result translates into a 100% recognition rate. We also tested the system with 10 random signatures which are not contained in the original database. Only two of these signatures which are very similar to at least one of the 30 stored images resulted in "false positives" (output > 0.5) while the remaining 8 are recognized correctly as not belonging to the original set (the output value was <= 0.5). Since recognition step is always followed by the verification step, these kinds of false positives can be easily caught by our verification system. In other words, the verification step serves as a safeguard against "false positives" as well as "false negatives".

5.2. TESTING THE VERIFICATION SYSTEM

Training for verification is explained in Section 4.2. Signatures used for testing the verification system are obtained the same way as in the recognition system. We tested the verification software using 10 signatures; 5 imitations (counterfeit signatures) and 5 true signatures. The program detected (classified) 4 true signatures and 5 counterfeits correctly. In other words, all counterfeit signatures are detected correctly. Only one signature is classified as a counterfeit while it was not (i.e. a "false negative"). Obviously, a "false negative" should be more acceptable in comparison to a "false positive", because the person can always be given a second chance to prove that the signature is his/hers. On the other hand, a false positive in verification carries a lot of risk.

6. CONCLUSION

In this study, we presented an off-line signature recognition and verification system which is based on image processing, moment invariants, and ANNs. Both systems used a four-step process. In the first step, the signature is separated from its image background. Second step performs normalization and digitization of the original signature. Moment invariants which are used as input features for the NN are obtained in the third step. Two separate ANNs are used; one for signature recognition and another for verification. Our recognition system exhibited a 100% success rate by identifying correctly all of the 30 signatures that it was trained for. However, it exhibited poor performance when it was presented with signatures that it was not trained for earlier. We did not consider this as a "high risk" case, because recognition step is always followed by the verification step and these kinds of false positives can be easily caught by the verification system. Indeed, the verification system did not miss any of the counterfeit signatures. However, its verification for true signatures lacked some accuracy. We think that this is also acceptable because a person can always be given a second chance to prove the ownership of a signature.

Generally, the failure to recognize/verify a signature was due to poor image quality and high similarity between two signatures. Recognition and verification ability of the system can be increased by using additional features in the input data set. In the future, we plan to expand the input feature vector by the addition of some grid information and features related to texture.

REFERENCES

1- Parizeu, M., Plamondon, R., " A comparative analysis of regional correlating dynamic time warping, and skeletal tree matching for signature verification", IEEE transactions on Pattern Analysis and Machine Intelligence 12, pp. 710-717, 1990.

2- Brault, J. Plamondon, R., "segmenting handwritten signatures at their perceptually important points." IEEE transactions on Pattern Analysis and Machine Intelligence 15, pp. 953-957, 1993.

3- Lee, L., Berger, T., Aviczer, e., "Reliable on-line human signature verification systems." IEEE transactions on Pattern Analysis and Machine Intelligence 18, pp. 643-647, 1996.

4- Xuhang, X., Graham, L., "Signature verification using a modified basian network", Pergamon Pattern Recognition 35, pp.983-995, 2002.

5- Yuan Y., T., Ernest C.M.L., "New method for feature extraction based on fractal behavior." Pergamon Pattern Recognition 35, pp.1071-1081, 2002.

6- Ismail M.A., Samia G., "Off-line Arabic signature recognition and verification", Pergoman Pattern Recognition 33, pp.1727-1740, 2000.

7- Qi, Y., Hunt, B.R., "signature verification using global and grid features." Pattern recognition 27, pp.1621-1629, 1994.

8- Yedekcioglu O.A., Akban M.B., Lim, Y.H., "off-line signature verification with thickened templates." COMCON5 proceedings of 5 th international conference on Advanced in communication and control, crete, Greece, pp. 131-142, 1995.
9- Han, K., Sethi I.K., "Handwritten signature retrieval and identification." Pattern Recognition Letters 17, pp. 83-90., 1996.
10- Baltzakis H., Papamorkos N., "A new signature verification technique based on a two-stage neurol network classifier.", PergomanEngineering Aplication of Intelligence 14, pp.95-103, 2001.
11- Droughord, J., Plamondon, R., Godbout, m., "A neural network approach to off-line signature verification using directional PDF.", Pattern Recognition 29, 415-424, 1996.
12- Luong, C. M., "Introduction to Computer Vision and Image Processing", web site: http://www.netnam.vn/unescocourse/computervision/comp_frm.htm
13- Lim. J.S., "Two-Dimensional and Image Processing", Prentice-Hall, 1990.
14- Yang, X., and Toh, P.S., " Adaptive Fuzzy Multilevel Median Filter", IEEE Transaction on Image Processing, Vol. 4, No. 5, pp.680-682, may 1995.
15- Hwang, H., and Haddad, R.A. "Adaptive Median Filters: new Algorithm and Results", Transactions on Image processing, Vol. 4, No. 4 pp.449-505, April 1995.
16- Rosenfeld, A., "Digital Picture Processing", Academic Press Inc., 1982.
17- Erdem, U.M., " 2D Object Recognition In Manufacturing Environment Using Implicit Polynomials and Algebraic Invariants", Master Thesis, Bogazici University, 1997.
18- Fu, K.S., Mui, J.K., "A survey On Image Segmentation", Pattern Recognition, Vol. 13, pp.3-16, Pergoman Press, 1981.
19- Hu, M., "Visual Pattern Recognition by Moment Invariants", IRE Trans. Inform. Theory 8, pp.179-187, 1962.
20- Koker, R., Oz, C., Ferikoglu, A., " Object Recognition Based on Moment Invariants Using Artificial Neural Networks", Proceedings of 3rd International Symposium an Intelligent Manufacturing Systems, August 30-31, Sakarya, 2001.
21- Awcock, G.J., and Thomas, R., "Applied image processing", McGraw Hill, Inc., 1996.
22- Reiss, T.H., "The Revised Fundamental Theorem of Moment Invariants", IEEE Transaction on Pattern Analysis and Machine Intelligence, 13(8):830-834, August 1991.
23- Ustun, A., "Cisim tanima Problemine Yapay Sinir Aglarinin uygulamasi", MSc Thesis, ITU, 1999

RECOGNITION OF THE OUTLINES OF AN IMAGE USING ANN

WAFI EL-CHAAR, CHAÏBAN NASR
Faculty of Engineering 1
Lebanese University - Lebanon

DENIS HAMAD
ULCO
France

ABSTRACT
The objective of the following work is to demonstrate the capacity of artificial neural networks in image processing, particularly in outlines detection problems. For that we used ANN combined with image treatment in order to optimize the structure of the architecture of the neural network. Moreover we have developed and experimented different convergence algorithms based on SVM in order to optimize the different structures obtained. We have developed a software tool based on MATLAB (Caweley, 2000) that converts an image into different forms (gray, RGB) that subdivides this image into blocs in the objective of recognizing the different outlines of a presented image.

INTRODUCTION

With the global trend towards building "Intelligent Machines" image processing is becoming a key point in machine-machine and man-machine communications. Therefore the need for a machine to "See" and "Understand" what it sees becomes one of the most important research domains nowadays (Barlett et al., 2002). It corresponds to the image processing domain and its applications that cover a wide range from object detection and recognition, to movement and path detection, in addition to image compression and querying.

On the other hand, human intelligence is based on the brain and its ability to "learn" and to "analyze" many complex situations in very short time. Now, for machines to be intelligent, they must have the ability to simulate parts or functions of the masterpiece of human brain, therefore the need for ANN. Too many ANNs were introduced and studied, some were designed for classification tasks, some were designed for data topology discovery (Fritzke, 1997), and some were designed for prediction, control, data mining and many other applications (Haykin, 1999).

Our goal is to study the possible contribution of artificial neural networks in the wide domain of image processing (Ma and Khorasani, 2002), particularly in image segmentation (Carson et al., 1999), in outlines recognition and in pattern recognition problems. On the intelligent path from "See" to "Understand" what you see, image segmentation and pattern recognition are the first steps that follow the "See" process. Afterwards, scene analysis, object detection, and movement detection (Burel, 1991) are based on the output of those first steps. Our approach was to extract the color definition of an image, and then analyze it in order to separate different color ranges, and then to group those colors ranges in objects or patterns.

The color definition of images is based on CIE's 1964 RGB color space, which defines a color to be the additive mix of 3 primary stimuli or channels (Red, Green, and Blue). Analysis of the image (Chi and Ersoy, 2002) involves decomposing the image into small blocks and studying the color ranges existing in each block by the mean of its color histogram. When two or more color ranges are found, we search for the separating surface between those ranges on the histogram by fitting a special polynomial function. And then, the calculated threshold is used for the classification of the image pixels. Once classified, those pixels will be fed to a SVM algorithm that will draw the separating surface between the two classes on the image, though realizing our goal of image segmentation and outlines recognition.

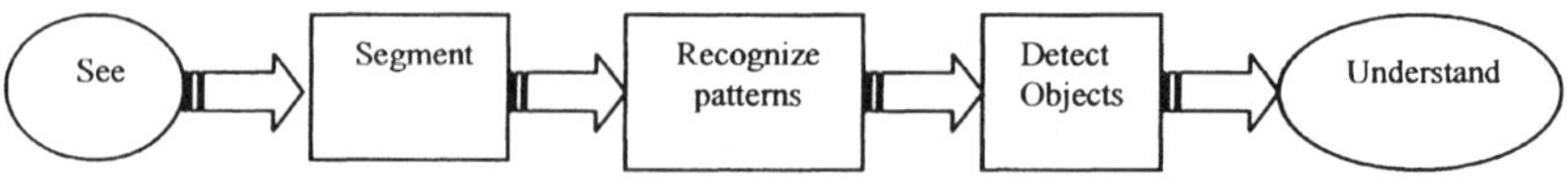

Future work will involve improvements to the color definition analysis and to grouping small blocks' patterns into macro patterns to initiate the process of Object Detection.

RGB COLOR DEFINITION OF AN IMAGE

Based in CIE's 1964 RGB Definition, an image pixel has three components: Red, Green, and Blue. Each is represented on a [0; 255] scale, where 0 represents the darkest side and 255 the brightest. In order to study an image, we first extract and store this RGB definition in three corresponding matrices and then we decomposed those matrices into 32x32 blocks corresponding to 32x32 pixels image blocks. Each one of those blocks will be treated separately; its threshold is calculated and then used for the classification of its pixels.

PATTERN COMPOSITION OF A BLOCK

We define a pattern to be a group or a set of groups of adjacent pixels having similar color definition. This definition of group can be represented on the histogram of the block. A histogram measures how many times a certain color has appeared in the block, and since we have 256x256x256=16,777,216 possible colors, the histogram size will be huge and will cause storage and manipulation problems. In order to avoid these problems, many data structures for color histograms were proposed (Brun, 2002). Among those propositions we chose the marginal histograms, where a separate histogram is calculated for each channel of the image. We propose to segment the image based on the best data from any channel: In certain zones of the image the red channel may present a clearer segmentation, on other zones the green channel is better and on others the blue channel is the best. Review Figure 1, Figure 2, Figure 3 and Figure 4. Hence, working on each channel separately presents better results than combining them and certainly much smaller data structures are needed.

MLP FOR PATTERN COMPOSITION DETECTION

In order to classify the blocks containing one pattern and those containing two patterns, we constructed and trained a MLP neural network. Our MLP has 256 inputs corresponding to the 256 shades of a color, its first and only hidden layer has 14 neurons (chosen after an optimization study on the size of the hidden layer), and the output layer has only 1 neuron. All 3 colors histograms are presented to the

MLP, but only the most certain histogram that is found to have 2 patterns will be taken into consideration in the next levels.

For the optimization study, we gathered 951 different histograms from blocks of different images in a training database and we used it to train different MLPs having each a different hidden layer size. Our training policy was to train the MLP on the first 256 training samples for 2000 epochs and requesting a training error (Mean Square Error MSE) goal of 0.001, then we use the second 256 inputs with the same parameters, and so on. Finally, in Figure 8 we show the results of the training of the MLP on the entire training database. This policy was adopted in order to make the trainings easier for MATLAB program since the size of the network and the training database is huge (an MLP with 256-14-1 has 3,613 weights to be calculated for each sample of the training database on each epoch). This was repeated for different hidden layer sizes. Finally using the graph on Figure 9 representing training errors encountered for each case, we chose 14 neurons on the hidden layer of the MLP as it gave the most satisfying results.

GAP DETECTION INSIDE COLOR HISTOGRAMS

When a block is found to have two patterns, it is essential to locate the best threshold between those two color patterns. To accomplish this task, we propose to search for the best 4th degree polynomial function that approximates, in a least squares way, the shape of the histogram (Figure 6). Based on the shape of the histogram with two zones, the best fitting 4th degree polynomial has to have two maximums (corresponding to the two zones) and one minimum corresponding to the gap. To initialize the search, we chose five points as follow:

- ❑ The barycenter of the histogram that we defined to be the mean point on the x-axis where each x coordinate is biased by the value of the histogram at that point.
- ❑ The maximum point on the histogram to the left of the barycenter.
- ❑ The maximum point on the histogram to the right of the barycenter.
- ❑ Two points that constitute the borders of the "useful" part of the histogram (limited between the first and the last point where the local histogram value is higher than the average value of the histogram).

Using those five points, we calculated a first 4th degree polynomial that initialized the search for the best polynomial. We used the least squares with constraints method, and stated that the polynomial should always have a maximum-minimum-maximum style. This condition is met when the polynomial tends to $-\infty$ when x tends to $\pm\infty$. Therefore the constraint had to be: $a_4 < 0$, *where* $p(x) = a_4 x^4 + a_3 x^3 + a_2 x^2 + a_1 x + a_0$

After calculating for each layer of a block the best fitting polynomial and its gap, we decide which layer is the best to be used in the next steps. This is done based on the mean square error (MSE) between the data points of each histogram and its best fitting polynomial: "The chosen layer will be the one having the least MSE among those already assigned by the MLP as having 2 patterns". All pixels on the chosen layer (R, G, or B) having a color value less than that of the calculated gap will be reported as class 1, the rest will be reported as class 2.

DRAWING THE SEPARATING SURFACE

For each block in the image, the separation surface is the surface that separates the pixels belonging to class one from the pixels belonging to the class two. One of the newest algorithms that are used to construct non-linear separating surfaces is the Support Vector Machines (SVM) algorithm. The SVM is a learning machine that is based on the theory of statistical learning. Its main idea is that the most important data on the data set are those belonging to class 1 and are the closest to class 2 and vice-versa. Those points are going to "support" the separating surface between the two classes (Burges, 1998).

If we consider the simplest case of linearly separable data, see Figure 7, and we construct a hyperplane H_1 on the boundary of class 1 and another hyperplane H_2 on the boundary of class 2, the best separating surface is the hyperplane H that lies in the middle of H_1 and H_2 and is parallel to each one of them. The SVM algorithm searches for the best separating hyperplane (Fritzke, 1997) that maximizes the margin between H_1 and H_2 (Or minimizes its inverse). "The problem is now to find the pair of hyperplanes H_1 and H_2 (and eventually H) that minimizes the margin between H_1 and H_2 subject to the constraints that the points belong the class 1 will be limited by H_1 and those belonging to class 2 will be limited by H_2" (Figure 7). Switching to the Lagrange formulation of the problem and multiplying each one of the positive constraints by a positive Lagrange multiplier α_i has the advantage that the data points will appear in the training algorithm only in the form of dot products. In the solution, those points for which $\alpha_i > 0$, are called "Support Vectors" and lie on one of the hyperplanes H_1, or H_2. All other training points have $\alpha_i = 0$ and hence do not participate in the construction of the separating surface. For these machines, the support vectors are the critical elements of the training set. They lie closest to the decision boundary; if all other training points were removed and training was repeated, the same separating hyperplane would be found (Burges, 1998).

In most practical cases (Christianini, 2001), the data points are not linearly separable and linear SVMs cannot perform well due the complex disposition of the data in the working space. This draws us to the case of non-linear SVMs. In non-linear SVM the original space $\Re^d$ is mapped into a higher order space $\aleph$ using a mapping functionΦ. Since that the data points figure in the training problem only in the form of dot product $x_i.x_j$, this product will be mapped in $\aleph$ as: $\Phi(x_i).\Phi(x_j)$. And more, a Kernel function is defined in $\aleph$ such that its represents this dot product: $K(x_i, x_j) = \Phi(x_i).\Phi(x_j)$ thus we need only to use this kernel function in the training algorithm wherever a dot product between data points is encountered (Burges, 1998).

- Polynomial kernel: $K(x, y) = (x.y + 1)^P$
- Gaussian RBF classifier: $K(x, y) = e^{\frac{-\|x-y\|^2}{2\sigma^2}}$
- Special two layers neural network: $K(x, y) = \tanh(kxy - \delta)$

For the RBF case, the number of centers, the centers themselves, the weights and the thresholds are all determined automatically by the SVM training and give excellent results compared to classical RBF. For that reason, we adopted a wide RBF kernel in the SVM training policy with $\sigma^2 = 10$ pixels and a low extra cost C = 0.5 only for points that do not abide by the constraints of the algorithm. This had the effect of relaxing the separating surface hence neglecting any small "noisy" regions. We used Cawley's support vector machine toolbox for MATLAB to detect and draw the separating surface between the 2 classes of each image block, given that we have already classified the pixels of the block using the threshold from the 4th degree polynomial.

RESULTS

To test the proposed system, we prepared a MATLAB tool that reads an image and extracts its RGB colors definition, and then uses a MLP, which we have already trained, to decide on the composition of each block of the image. Next the 4th degree polynomial is searched and the separating surface is drawn onto the image using the SVM algorithm. We defined for each image a Success Rate as the ratio of the blocks successfully classified by the MLP on the total number of blocks in an image, since all errors that our system may present are caused by the MLP misjudgments. If a block is successfully classified as having two patterns, the 4th degree polynomial implementation policy ensures a very good choice of the gap, as we initialize the search from a very close to optimum situation.

We tested our algorithm on a small database of six images and we achieved an Average Success Rate of 91.76%. The worst example is shown on Figure 10 where the test image "Frosty" was shown along with its separating surface and on Figure 11 where the separating surface was shown alone. We note that we did not include any block from this image in the training database of the MLP; its success rate was 86.22%.

We note finally that the MLP is still making errors with its decisions especially in uniform blocks i.e. sky where very few color shades are encountered. Thus more examples and more training are required to achieve a better performance of the system.

REFERENCE

Bartlett M., Movellan J., Sejnowski T., November 2002, "Face Recognition by Independent Component Analysis", *IEEE Transactions on Neural Networks*, Vol.13, pp.1450-1464

Brun L., January 2002, "Cours de Traitement d'Images Couleurs", Université de Reims – LERI, http://univ-reims.fr/lobos/LERI/member/luc/

Burel G., 1991, "Réseaux de Neurones en Traitement d'Images: Des Modèles Théoriques aux Application Industrielles", LEST-Université de Bretagne Occidentale, THOMSON-CSF, France

Burges C., 1998, "A Tutorial on Support Vector Machines for Pattern Recognition", Microsoft Research, Data Mining and Knowledge Discovery 2

Carson C., Belongie S., Greenspan H., Malik J., 1999, "Blobworld: Image Segmentation using Expectation-Maximization and its applications to image querying", University of California at Berkeley

Cawely G.,2000, "MATLAB Support Vector Machine Toolbox", University of East Anglia, School of Information Systems, Norwich, Norfolk, U.K

Chi H., Ersoy O., March 2002, "Support Vector Machine Decision Trees with Rare Event Detection", Smart Engineering System Design, Taylor & Francis

Cristianini N., 2001, "Support Vectors and Kernel Machines", ICML http://support-vector.net/tutorial.html

Fritzke B.,1997, "Growing Neural Gas", NeuroInfomatik Lab, Ruhr University Bochum, Germany

Haykin S., 1999, "Neural Networks, a Comprehensive Foundation", Prentice Hall, New Jersey

Ma L., Khorasani K., September 2002, "Application of Adaptive Constructive Neural Networks to Image Compression", *IEEE Transactions on Neural Networks*, Vol.13, pp. 1112-1126

FIGURES & RESULTS

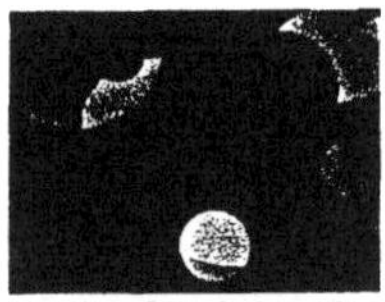

Figure 1 RGB Colors Image

Figure 2 Red Channel

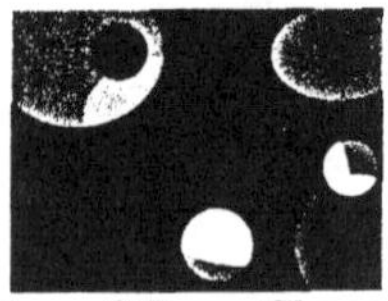

Figure 3 Green Channel

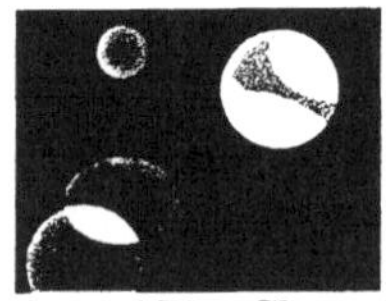

Figure 4 Blue Channel

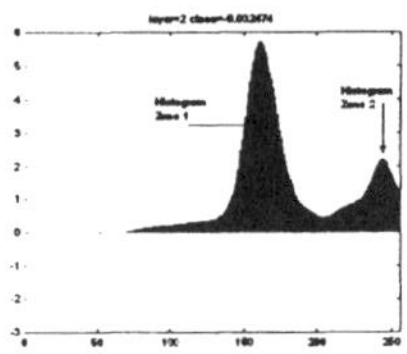

Figure 5 Histogram of an image block

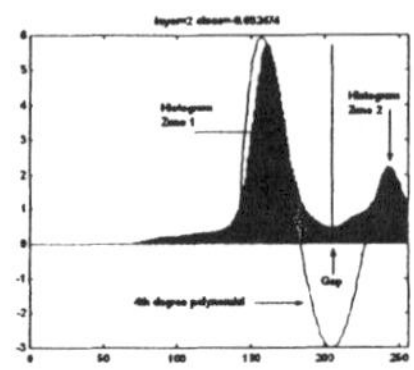

Figure 6 Histogram & its best 4th degree polynomial approximation.

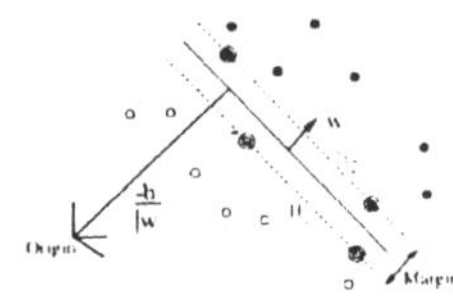

Figure 7 Linear SVM

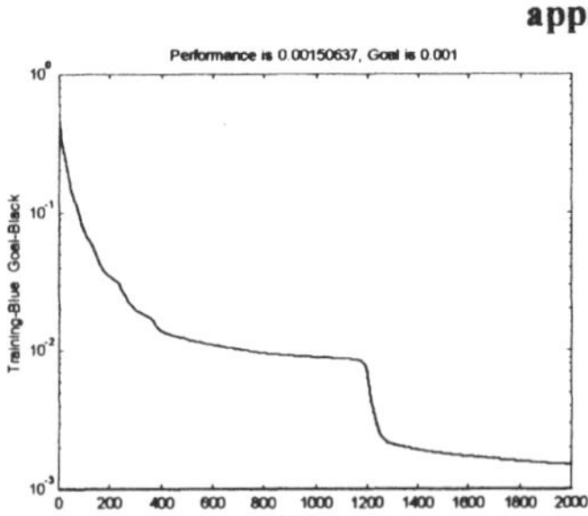

Figure 8 MLP Training: entire training database

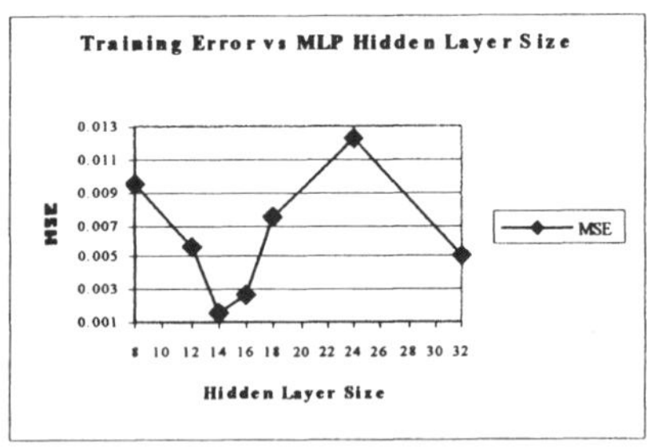

Figure 9 Optimization study for MLP hidden layer size

Figure 10 Test image "Frosty.

Figure 11 Separating surface for "Frosty"

LOCAL FEEDBACK IN THE EVOLUTION OF COMPEMENTARY FEATURE SUBSETS

DALE E. COURTE
University of Dayton
Computer Science, Dayton, Ohio

MATEEN M. RIZKI
Wright State University
Computer Science and Engineering,
Dayton, Ohio

LOUIS A. TAMBURINO
Wright State University
Computer Science and Engineering,
Dayton, Ohio

ABSTRACT

Recognition problems of high dimensionality, such as classification of 2-D Synthetic Aperture Radar (SAR) images, require determination of appropriate feature subsets. When such problems are addressed using a collection of voting classifiers, the choice of a feature subset for each constituent classifier is a considerable problem itself. Overlap of feature subsets should be minimized to avoid redundancy, and yet each subset of features must provide enough information to guarantee that the associated classifier contribute to the overall solution. An evolutionary system is presented that evolves complementary feature subsets for a collection of voting classifiers in parallel. A population of genotypes is maintained, each member of which uniquely defines a set of configurations for each of *n* classifiers. A novel performance feedback mechanism is integrated into the evolutionary system to prevent overtraining and encourage cooperation. The result is an optimized collection of cooperative classifiers evolved in parallel. An application of this system to the classification of SAR images is presented.

INTRODUCTION

Efficient classification of objects represented by large numbers of input features necessitates the discovery of effective subsets of those features. Ideally, classifications should be made using the smallest subset of available features that still result in a near-optimal classification rate. Classification of 2-D imagery is such a problem domain.

There has been considerable interest in recent years in the combination of classification systems into combined or fusion classifiers (Kuncheva and Jain, 2000) (Kittler et al, 1998). The Cooperative Hybrid Architecture for Multi-class Problems, or CHAMP (Courte, 2002) provides a platform for the optimization of combinations of cooperative classification systems. CHAMP utilizes a hybrid evolutionary computation (EC) (Bäck, et al, 2000) approach to optimize the cooperating classification systems, each of which is called an *elector* in this discussion. In a CHAMP system, individual classification modules (genes) can be altered through mutation and complete sets of voting modules (chromosome) can be recombined via crossover operations. As in all evolutionary systems, a population of such cooperating classifiers is maintained and allowed to evolve over a number of generations. The experiments discussed in this paper demonstrate the ability of the CHAMP architecture to encourage cooperation among electors, and establish the need for specialized feedback in such a cooperative evolutionary system.

OVERVIEW OF THE CHAMP SYSTEM

Figure 1 shows a CHAMP voting system assembled from m electors. An elector consists of a feature extraction unit γ and a classification unit ϕ. Each elector pair $<\gamma_i, \phi_i>$ is configured by a pair of parameter vectors $<a_i, b_i>$. Such parameter vector pairs are the *genes* of the evolutionary system, and a collection of such genes (one for each $<\gamma_i, \phi_i>$ pair) is a *chromosome* that represents a unique cooperative classifier. Each elector takes in an input vector $[x_1, x_2, ..., x_n]$ and outputs both a label ω_i and a confidence level s_i in the range $[0..1]$, with higher values indicating more confidence in the correctness of label ω_i. Such confidence levels can be estimated through supervised learning, assuming known training data is available. This label-confidence level pair is called a *proposal.*

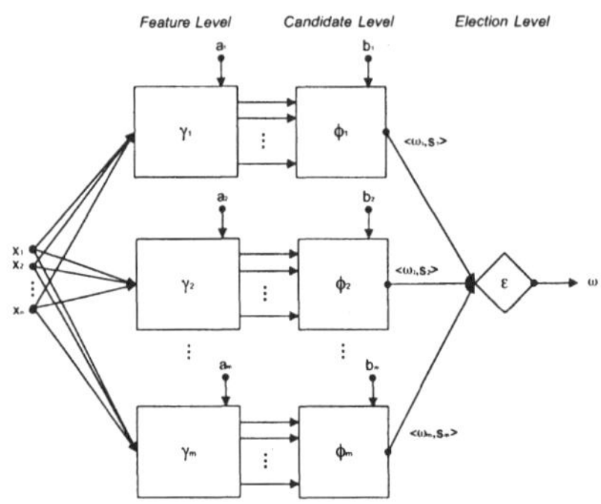

Figure 1: A CHAMP System with m Electors

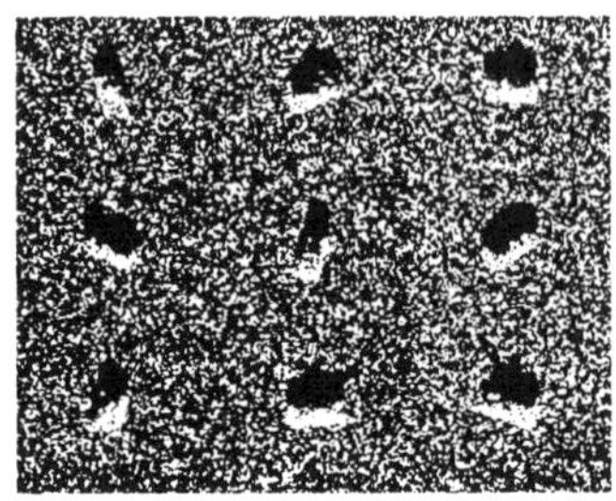

Figure 2: MSTAR Targets

Here, ε is a voting scheme that determines a final label assignment from the m proposals provided. Only one voting scheme is considered at this time:

$$\varepsilon : \left[\langle\omega_1, s_1\rangle, \langle\omega_2, s_2\rangle, ..., \langle\omega_m, s_m\rangle\right] = i, \text{ such that } v_i = \max_{j=1}^{k}(v_j) \qquad \text{(eq. 1)}$$

$$\text{where } \left(v_j = \sum_{\omega_a = j} s_a\right), j = 1..k \text{ and } k \text{ is the number of classes}$$

This is a variation on majority voting, which has been shown to be a very effective method of combining classifiers (Lam and Suen, 1997). Here, instead of each classifier casting an "all or nothing" vote for the label being proposed, the value of a vote is the confidence the classifier has in the associated proposal. Since the architecture encourages classifiers to specialize, this minimizes the influence of a classifier's vote for a label it has little confidence in assigning.

A *population* consisting of a number of chromosomes can be subjected to an evolutionary process through which a near-optimal configuration can be determined. Each generation of the evolutionary process, offspring can be created from existing chromosomes via either a mutation or crossover operation, and survivors are chosen based on a predetermined fitness criteria.

MSTAR EXPERIMENTS

The data chosen for these experiments is part of a collection of Synthetic Aperture Radar (SAR) images maintained at Wright Patterson Air Force Base. The collection is jointly sponsored by DARPA and Wright Laboratories and is accessible from the Sensor Data Management System (SDMS) website[1]. The images represent radar returns from aircraft circling a number of military targets. The images can be considered similar to low resolution photographic images, as they are 2-D collections of values that can be displayed as a grayscale image.

For these experiments, images represented three classes of vehicles were used. Each image consists of 128x128 pixels, and 128 images representing each vehicle class were used. The images represented depression angles of both 17 and 15 degrees and covered approximately 120 degrees of aspect. Half the samples were used for training, and half for testing. The training samples were randomly divided equally into training and validation sets for each experiment. Figure 2 shows nine grayscale representations of MSTAR targets. Each row of targets displayed depicts samples of a single class from different aspects. The data was smoothed using pixel averaging prior to processing.

Feature Reduction

The CHAMP system used here implemented a k-nearest neighbor (k-nn) classifier (Cover and Hart, 1967) for each elector, with $k=1$. The 128x128 images each provide 16,384 features per sample, making feature reduction not only desirable, but also necessary. Even in the unlikely event that a 1-nn classifier would perform well using this many features, the inefficiency of having to compare such large prototypes becomes prohibitive. It therefore was desirable to evolve smaller feature sets based on the training and validation samples.

Early attempts to select subsets of unrelated features indicated poor performance resulted when the number of selected features per elector was kept small, and resulted in overtraining when the number of selected features was allowed to grow. It was determined that allowing the selection of groups of neighboring features reduced the overtraining problem. These groups, or *patches* of features are of equal length and width, encompassing 4, 9, 16, 25, or 36 features each. A gene used to represent such a collection of patches in a CHAMP system can be defined as a list of 3-tuples:

$$g = \left[\langle x_1, y_1, n_1 \rangle, \langle x_2, y_2, n_2 \rangle, \ldots, \langle x_p, y_p, n_p \rangle\right] \qquad \text{(eq. 2)}$$

Where each x and y value represent the starting corner of one of p patches, while each n value represents the size of the corresponding patch. Note that since the classification scheme chosen here is fixed (not "tunable" by parameter settings), a gene here consists only of information needed by the feature extraction module of an elector.

Growing Genes through Random Mutation

The CHAMP system developed for these experiments allowed up to 20 patches per gene, resulting in a maximum of $20 \times 36 = 720$ features per elector. This is more than a 95% feature reduction per elector. Since a primary goal here was effective feature reduction, the decision was made to start with small genes and allow the number of selected features to grow via the evolutionary process, encouraging genes to grow no larger than necessary. To this end, each gene starts with just two patches, the size and location of which are randomly selected.

---

[1] http:// //www.mbvlab.wpafb.af.mil/public/sdms

A strict evolutionary programming (EP) (Fogel et al, 1965) approach was used in these experiments in that mutation is the only reproductive operation used. Mutation can take one of three forms, chosen with equal probability: 1) patches can be added, 2) patches can be removed, or 3) patches can be replaced. When patches are added or replaced, new patches are randomly generated to replace them. There is no attempt to "fine tune" genes by growing or shrinking patches, or by shifting the location of a patch a small amount. This was decided to promote diversity in the population, and encourage exploration throughout the feature space. The number of patches to be added, removed or replaced was determined by drawing a number between 1 and 1/2 the current number of patches, subject to size limits.

Local Feedback in the Determination of Fitness

In all evolutionary systems, the choice of the appropriate fitness function is among the toughest decisions to be made. When evolving a collection of cooperative classifiers, two types of feedback can be used in determining fitness. The overall performance of the combined classifier is an example of *global* feedback, while information about the performance and form of individual classification modules is considered *local* feedback. Such local feedback has been used in non-evolutionary cooperative systems, such as *boosting* (Schapire, 1999). The work described here indicates there is also value in considering local feedback in the determination of fitness in an evolutionary system. Since fitness is evaluated with each generation using a single population, the inclusion of both local and global information in a CHAMP system is continuous. This provides more immediate local feedback than alternative methods that evolve cooperative subcomponents, such as the *cooperative coevolution* systems of (Potter and DeJong, 2000).

Results from three series of experiments illustrate this point. Each series consists of 20 experiments involving systems of four electors, each experiment evolving a population of size 50 over 500 generations. Every generation, a number of offspring equal to the population size are created. These offspring then compete with their parents for survival, based solely on their fitness. Each experiment begins with a draw of training and test data samples. Training and test sets are of equal size, and the sampling method guarantees an even distribution of aspect (poses) in each set. The three series differ only in the computation of fitness.

Results and Observations

As a baseline, the first series of tests is based fitness solely on the combined performance of the evolved classifiers. The system exhibited considerable overtraining, as is evident in the leftmost column of results shown in Table 1. For each experiment, the table lists the terminal accuracy of the most-fit member of the population during training and the resulting performance against hold-out test data, as well as the difference between these two values. While terminal training performance is very high, even reaching 100% accuracy in 7 of 20 tests, the test performance is lagging by over 14%. Examination of individual experiments revealed the evolution of a dominant elector within each combined classifier, with the remaining electors making only marginal, or even non-existent, contributions. It appeared the evolutionary process resulted in "minimizing the damage" done by many electors, rather than rewarding their contribution. Clearly, global performance information is not sufficient feedback for this evolutionary process.

In order to insure that each elector makes a contribution to the overall accuracy of a combined classifier, the mean and variance of the performance measures of each individual classifier can be considered. In the second series of experiments, this local feedback was incorporated in a fitness measure $\mu-\sigma$, where μ is the average of the training accuracies of the electors when each is considered as a standalone classifier, and σ is the standard deviation of those accuracies. This rather restrictive measure insures that not only the average elector performs well but also that the variance between electors be minimized. The results shown in the center of Table 1show a considerable positive effect, both in terms of overall accuracy (improved by 6%) and in terms of preventing overtraining. However, terminal training accuracy was reduced considerably.

While the local feedback used in the experiments described above improved both test set accuracy and reduced overtraining, there are some intuitive drawbacks to this approach that imply further improvements could be achieved. By minimizing the variance in performance between electors while maximizing the mean accuracy may cause the system to converge prematurely resulting in very similar electors. The resulting lack of diversity between electors could inhibit the emergence of specialization – limiting the ability of the CHAMP system to evolve complementary electors, each making unique contributions to the overall accuracy. The third set of experiments adds a diversity measure to the fitness calculation. For each elector in a combined classifier, a *center* is defined as the average position of the patches for that elector. These centers represent points in the coordinate space of the images where the electors are considered to be *focused*, and the diversity measurement encourages separation of the centers of the elector set. To relax constraints on performance late in the evolutionary process, the effect of both this diversity measure and of the variance measure linearly decreased with time. The fitness measure used in the remaining series of experiments was:

$$\mu - \left(\sigma - \frac{d}{c}\right) \times \left(1 - \frac{g}{g_{\max}}\right) \qquad \text{(eq. 3)}$$

where μ and σ are as before, d is the mean distance between the centers defined above, c is a scaling factor (determined experimentally), and g and g_{max} are the current generation number and the terminal number of generations respectively.

As shown in the rightmost columns in Table 1, this fitness measure allowed training accuracy to approach that achieved using global feedback, but without the overtraining exhibited in the first experiments. In fact, both the training and test accuracy improved over the second series of tests discussed.

test	global accuracy			local accuracy			local accuracy + diversity		
	training	test	difference	training	test	difference	training	test	difference
1	100.00%	85.42%	14.58%	94.79%	90.63%	4.17%	96.88%	91.67%	5.21%
2	100.00%	84.90%	15.10%	94.79%	92.19%	2.60%	98.96%	93.23%	5.73%
3	98.96%	91.15%	7.81%	95.83%	91.67%	4.17%	98.96%	92.71%	6.25%
4	98.96%	86.46%	12.50%	94.79%	91.15%	3.65%	98.96%	95.31%	3.65%
5	98.96%	87.50%	11.46%	94.79%	92.71%	2.08%	95.83%	93.23%	2.60%
6	98.96%	81.25%	17.71%	94.79%	93.75%	1.04%	98.96%	89.58%	9.38%
7	98.96%	88.54%	10.42%	92.71%	90.10%	2.60%	98.96%	88.02%	10.94%
8	97.92%	86.46%	11.46%	95.83%	87.50%	8.33%	98.96%	89.06%	9.90%
9	98.96%	86.46%	12.50%	96.88%	91.67%	5.21%	97.92%	94.79%	3.13%
10	95.83%	76.56%	19.27%	92.71%	86.46%	6.25%	97.92%	95.31%	2.60%
11	98.96%	88.54%	10.42%	96.88%	91.15%	5.73%	98.96%	93.75%	5.21%
12	100.00%	85.42%	14.58%	96.88%	95.31%	1.56%	98.96%	89.58%	9.38%
13	100.00%	81.77%	18.23%	98.96%	86.98%	11.98%	98.96%	96.88%	2.08%
14	97.92%	81.25%	16.67%	97.92%	92.71%	5.21%	96.88%	92.19%	4.69%
15	98.96%	88.02%	10.94%	96.88%	91.15%	5.73%	97.92%	91.15%	6.77%
16	100.00%	76.56%	23.44%	94.79%	91.15%	3.65%	97.92%	91.15%	6.77%
17	100.00%	90.10%	9.90%	92.71%	91.15%	1.56%	98.96%	86.46%	12.50%
18	100.00%	85.42%	14.58%	95.83%	90.63%	5.21%	98.96%	95.83%	3.13%
19	98.96%	80.21%	18.75%	94.79%	90.63%	4.17%	98.96%	91.67%	7.29%
20	97.92%	84.90%	13.02%	96.88%	90.10%	6.77%	96.88%	92.71%	4.17%
mean	99.01%	84.84%	14.17%	95.52%	90.94%	4.58%	**98.28%**	**92.21%**	**6.07%**

Table 1: Performance of three series of experiments

Additional support for the value of local feedback can be found in (Courte et al, 2002) and further abilities of the CHAMP architecture are demonstrated in (Courte, 2002) and (Courte et al, 2003).

CONCLUSIONS

Inclusion of local feedback is an essential consideration when evolving systems of cooperative classifiers. In these experiments, local information relating to both individual performance and diversity were incorporated into fitness measures that exhibited considerable performance improvement over a measure based purely on global feedback. Further work may focus on identifying commonalities of such feedback over diverse problem sets, resulting in guidelines for determining appropriate feedback for future systems. The CHAMP architecture provides a suitable platform for such work.

ACKNOWLEDGEMENTS

Funding for this research was provided in part by the U.S. Air Force (AFRL/SNAT grant F33615-99-C-1441, "Evolving Pattern Recognition Systems").

REFERENCES

Bäck, T., Fogel, D.B., and Michalewicz, T., eds., *Evolutionary Computation I*, Institute of Physics, 2000

Courte, D.E., *Evolutionary Optimization of Voting Classifiers*, Ph.D. Dissertation, Wirght State University, Dayton, OH, 2002.

Courte, D.E., Rizki, M.A., and Tamburino, L.A., "Evolving Cooperative Recognition Systems", *Intelligent Engineering Systems through Artificial Neural Networks,* vol. 12 (ANNIE 2002), ASME Press, 2002.

Courte, D.E., Rizki, M.A., and Tamburino, L.A., "Evolutionary Optimization of Heterogeneous Classifiers", Proceedings of the International Conference on Artificial Intelligence, IC-AI'03, Vol. 2., CSREA Press, 2003.

Cover, T.M., and Hart, P.E., "Nearest Neighbor Pattern Classification", *IEEE Transactions on Information Theory*, IT-13, 21-27, January 1967

Fogel, L.J., Owens, A.J., and Walsh, M.J., "Artificial Intelligence through a Simulation of Evolution", *Biophysics and Cybernetic Systems: Proceedings of the 2nd Cybernetic Sciences Symposium*, Spartan Books, pp. 131-155, 1965

Kittler, J., Hatef, M., Duin, R.P.W, and Matas, J., "On Combining Classifiers", *IEEE Transactions on Pattern Analysis and Machine Intelligence*, Vol. 20:3, pp. 226-239, 1998

Kuncheva, L.I., and Jain, L.C., "Designing Classifier Fusion Systems by Genetic Algorithms", *IEEE Transactions on Evolutionary Computation*, Vol. 4:4, 2000

Lam, L., and Suen, C.Y., "Application of Majority Voting to Pattern Recognition: Am Analysis of its Behavior and Performance", *IEEE Transactions on Systems, Man, and Cybernetics-Part A: Systems and Humans*, Vol. 27:5, pp. 553-567, 1997

Potter, M.A., and DeJong, K.A., "Cooperative Coevolution: An Architecture for Evolving Coadapted Subcomponents", *Evolutionary Computation* 8(1):1-29, MIT Press, 2000

Schapire, R.,E., "A Brief Introduction to Boosting", Proceedings of the Sixteenth International Joint Conference on Artificial Intelligence, 1999

PATTERN MATCHING IN COLLECTIONS OF JAVA BYTECODE: MAXMISING REUSE POTENTIAL

ANGUS MACLEAN

JOHN A.W. MCCALL

DERYCK F. BROWN

School of Computing
The Robert Gordon University
Aberdeen, UK

ABSTRACT

Realising the benefits of systematized software reuse is non-trivial, and particularly problematic in the context of small to medium sized development environments. This paper outlines an approach to supporting object-oriented reuse by identifying and maximising reuse potential. It describes a process of intelligent pattern matching that identifies common structure present in collections of object-oriented code. It builds on the definition of an attributed, relational model, which captures the inherent structure of class-based code. Established graph-theoretic techniques are adapted and applied via this model to the problem of pattern matching within class collections. The approach is based on a vector-space model of global similarity, and an attributed, graph-match model of local similarity that incorporates a hybrid genetic algorithm. The overall approach is evaluated in the context of object-oriented development based on direct analysis of executable Java bytecode.

INTRODUCTION

The benefits of systematized reuse are undeniable, particularly in the context of large-scale development environments (Sametinger, 1998). Large organizations are increasingly committed to retaining and reusing development experience (Nada et al, 1998). However, realising the benefits of systematized software reuse is non-trivial. It is particularly problematic in the context of small to medium sized development environments where a reuse process is often informal or absent. The majority of software development organisations are small, produce non-trivial products, but arguably differ from their larger-scale counterparts (Fayad et al, 2000). This includes their limited capacity to accommodate development with and for reuse. It is also clear that complex, resource-intensive, formal approaches do not translate effectively to reuse "in the small", where *awareness* and *sharing* of reuse experience are more important enabling factors (Morisio et al, 2002).

Informal reuse refers to such development practices as "cutting-and-pasting" design or code within and between projects. It is based on an individual developer's prior knowledge of existing artifacts and recognition of their reuse potential. Re-implementing retained mental models, and recurring idiomatic coding conventions, are further, less tangible sources of undocumented yet potentially reusable code. Anecdotal evidence suggests that in the absence of a structured reuse process informal reuse is common. Baxter et al (1998) echo this belief when identifying code "cloning", Kreuger (1992) describing it as a process of design and code "scavenging". In reviewing both formal and anecdotal studies, Maiden and Sutcliffe (1993) reinforce the significance to programmers of a copying / modifying reuse philosophy, based on examples and past use / reuse experience. A significant problem with informal reuse is the lack of *shared* and *retained* awareness of the reused code and development experience. We aim to maximise reuse potential by searching for undocumented instances of reused, or re-implemented, versions of the same or similar classes. Existing approaches that rely on establishing degrees of code similarity include the automated identification of code clones (Baxter et al, 1998), source code plagiarism detection (Prechelt et al, 2000; Hislop, 1998) and code re-factoring (Balazinska et al, 2000). These have been shown to be efficient and effective. However, they are brittle when faced with classes that vary in fine detail, which nonetheless have strong similarity of purpose and / or physical functionality.

Whitmire emphasises the declarative, relational nature of class-based development: "I have not seen a case in which two classes participate in a similar set of relationships where they did not also serve the same purpose and were not also internally structurally similar." (Whitmire, 1997) The relationships inherent in the structure of object-oriented code are more important than method content, due to the methods being smaller and less complex than their procedural, or function-oriented counterparts. In effect, object-oriented development conveniently carries an *explicit structural formalism*, in the form of the *class as an organising principle*. We argue that this structure provides a valid means of supporting the assessment of class-based similarity.

We aim to test the hypothesis that an attributed relational model of object-oriented code structure can help identify useful instances of undocumented class-based reuse. Our approach is based on code-level analysis, relying *solely* on the structural characteristics of the code and the peculiarly object-oriented features of class-based development: classes, those entities comprising a class, and the intra and inter-class relationships existing between them. This is independent of any information contained in associated documentation, and "lightweight" in that it requires no additional external knowledge sources. By representing classes as attributed relational graphs, established graph-theoretic techniques drawn from the domains of chemoinformatics and computer vision are adapted and applied via this model to the problem of determining similarity between classes. The combinatorial complexity inherent in our approach is addressed by incorporating a genetic algorithm, which is used to solve the associated maximum clique problem.

AN ATTRIBUTED, RELATIONAL MODEL OF OBJECT-ORIENTED CODE

In computer vision, objects are modeled as graphs, where attributed primitives and relationships respectively equate to vertices and edges (Tsai and Fu, 1979). Such attributed, relational graphs (ARGs) are also fundamental in determining molecular similarity (Willett, 1999). Both domains provide a template for the comparison of structures as graphs. As our approach emphasizes the declarative, relational aspects of object-oriented code, a class can be analogously abstracted by an ARG. Vertices represent the class, its constituent and associated parts (e.g. internal fields and methods; externally called methods and accessed fields), and edges model their interrelationships (e.g., ownership; inheritance; invocation; usage; parameterisation). Additionally, vertices and edges can be assigned attributes based on their various metric properties.

Figure 1 shows an example of a class as represented by its ARG, simplified for the sake of illustration. (The full model currently defines 11 vertex and 33 edge types, with up to 20 attributes each.) The current approach generates ARGs for a collection of Java classes based on parsing and disassembling their executable bytecode files. Our particular domain generates directed ARGs, organized on three levels: they are rooted at level 0 (L0), the designated class vertex; have internal methods and fields, superclass and interfaces at level 1 (L1); external methods and fields, parameters and returns at level 2 (L2). The model is independent of identifier semantics and associated code documentation, being solely reliant on structure and structural typing.

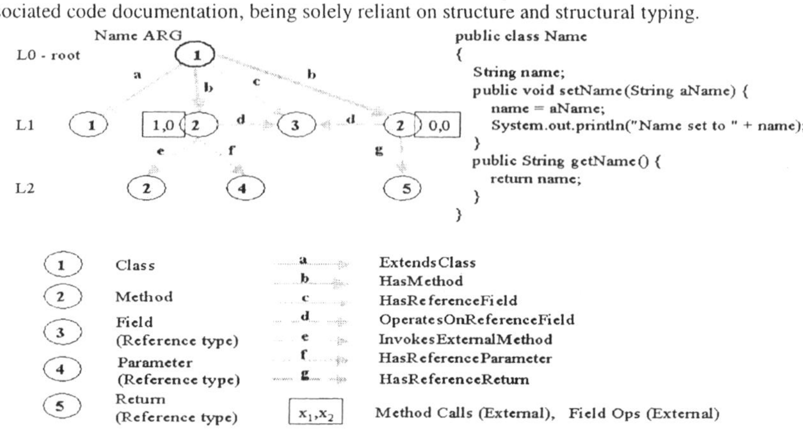

Figure 1: ARG abstraction of a simple class

CLASS-BASED PATTERN MATCHING

Two graphs are isomorphic if there is a one-to-one correspondence between their vertices, and between the edges incident to corresponding vertices. The largest pair of isomorphic vertex subsets, or subgraphs, one drawn from each of two graphs (G_1 and G_2), identifies a maximum common subgraph (MCS_{12}) of these two factor graphs. Similarity between the two factor graphs can be expressed in terms of the number of nodes in the MCS and each factor graph as: $MCS_{Sim} = | MCS_{12} | / \max (| G_1 | , | G_2 |)$.

Based on the ARG representation of a class, pair-wise similarity can be formulated in terms of identifying the MCS. This in turn can be reduced to identifying the maximum clique in the correspondence graph (CG) formed from the modular product of the underlying factor graphs. This concept is well known in computational chemistry (Gardiner et al, 1997). The CG vertices are formed by creating the set of all pairs of compatible vertices, one from each of the factor graphs. Two CG vertices are connected if the edges between the corresponding vertices in the underlying factor graphs are also compatible. In the case of our ARGs, compatibility is based on threshold matching of vertex and edges, in terms of both type and attributes. The largest clique, or fully connected subgraph, identifies an MCS of the two graphs.

Unfortunately, clique extraction, and consequently MCS identification, is a notoriously difficult NP-Complete combinatorial optimization problem. Preliminary experiments were carried out based on applying Bron and Kerbosch's (1973) exact clique detection algorithm to the comparison of a range of classes drawn from the Java Development Kit. These indicated that correspondence graph size and density would be prohibitive in terms of the time and space required by exact, deterministic MCS extraction. A reduction in CG size can be achieved by noting that our vertices and edges are typed and attributed (inc. edge direction), and can be further categorized by their level in the ARG. Our domain constraints also require that the common subgraphs identified in the underlying factor graphs are connected, and rooted at L0. In addition, by identifying equivalence classes based on indistinguishable pendant (leaf) vertices, we exploit symmetry in the factor graphs in order to reduce the search space. However, irrespective of these measures, time and space constraints render an exact solution entirely impractical except for small-order factor graphs.

The key to unlocking the class match problem is the adoption of a sub-optimal approach based on a hybrid genetic algorithm (HGA) (Marchiori, 1998). Marchiori's combination of a simple GA and a naïve greedy heuristic procedure has shown to be an effective approach to the maximum clique problem. The GA uses a standard bit-string representation where each bit corresponds to a vertex in the CG, the 1's defining the currently selected subgraph. A post-hybridising heuristic is applied to the individuals in the current population, which do not necessarily represent cliques due to the disruptive action of the genetic operators. The heuristic is a stochastic process that ultimately "repairs" subgraphs that are not fully connected, in addition to ensuring maximality, i.e., establishing a clique. Fitness equates to the number of vertices in the represented clique, and so to the size of the MCS, i.e., the number of 1's in the bit-string. The GA is generational, elitist (best two), uses uniform crossover (0.8) and swap mutation (0.1).

Any heuristic approach to MCS extraction is likely to remain computationally expensive, and the majority of class-pairs in a collection are likely to be dissimilar. In order to limit unnecessary comparisons, a low complexity screening process is introduced. Each ARG is characterized by a vector of features F[1,...,n] representing counts of its constituent vertex types; all single edge-unit[1] paths (vertex-edge-vertex); double edge-unit paths (vertex-edge-vertex-edge-vertex); and triple edge-unit paths (vertex-edge-vertex-edge-vertex-edge-vertex), irrespective of edge-unit path overlap. For example, referring to Figure 1, there are three occurrences of vertex type "2", one of type "4". There is one occurrence of double edge-unit path "2,-e,2,d,3", and two of triple edge-unit path "1,b,2,d,3,-d,2", the –ve signs indicating traversal against the directed edge. (The same features could be represented as "3,-d,2,e,2" and "2,d,3,-d,2,-b,1" respectively but we adopt the convention that the lexicographically lesser representation is used.) In practice, the edge-unit features are simplified to include only the edge types, as edge types are chosen to implicitly determine the adjacent vertex types. This Structure Path (SP) representation aims at approximating MCS_{Sim} using the SP_{Sim} similarity coefficient defined over two feature vectors F_1 and F_2 as:

$$SP_{Sim} = \frac{\sum_{i=1..n} \left(\frac{\min(F_1[i], F_2[i])}{\max(F_1[i], F_2[i])} \cdot \omega_i \right)}{\sum_{i=1..n} \left(\delta(F_1[i], F_2[i]) \cdot \omega_i \right)}$$

(Where $\delta(F_1[i], F_2[i])$ returns 0 if both $F_1[i]$ and $F_2[i]$ are 0 and 1 otherwise. ω_i is the weight associated with feature i, currently $2^{(1-v)}$, where v is the number of vertices in feature i.)

[1] An *edge-unit* comprises an edge and its two incident vertices. It is labeled according to its constituent edge and vertex types. When part of a multiple edge-unit path, each edge-unit has an additional direction label.

EXPERIMENTAL EVALUATION

Class-based similarity: a quantitative analysis

The base hypothesis assumes a continuous, inter-dependent, "similar task" development environment. In the absence of readily accessible commercial data, our initial data sets were random samples of classes drawn from two undergraduate and one graduate-level object-oriented design assignment, as characterized in Table 1.

Data Set	Sample Size	Project Specification	Bytecode Size
A	30	Constrained	Small (<2K)
B	30	Loose	Medium (<6K)
C	30	Loose	Larger (<12K)

Table 1: Data sets

Seta A and B were undergraduate-level Java projects, and set C a graduate-level Java assignment. The data sets are not independent in that the undergraduate students followed the same course of instruction, presented by the same course team. Additionally, in all cases, the common learning environment provided scope for both sharing of experience – legitimate and otherwise – and "cut and paste" reuse based on the course materials. A "constrained" specification limited the potential variability in the delivered application, whilst a "loose" specification provided the programmers with a greater degree of freedom during the interpretation of requirements and eventual implementation. These data sets provide an opportunity to determine the effects of implementation variability on class similarity. However, it should be clear that they can provide only a limited insight into the effects of transference of previous individual and shared experience; idiomatic development techniques; and retained mental models, within and across projects.

In the absence of an established reference method, our MCS_{Sim} measure of class similarity was compared against the "JPlag" approach to plagiarism detection (Prechelt et al, 2000). "JPlag" operates by tokenizing Java source code and applying a sophisticated string-matching algorithm over the tokenized representation of two classes, thereby establishing degrees of code similarity (JP_{Sim}). As an effective plagiarism engine, in principle, this is not unrelated to our current objective. "Plagiarism, after all, is simply a socially unacceptable form of reuse." (Hislop, 1998).

Marchiori's HGA was implemented using a randomly generated initial population of 50 individuals. Our MCS is constrained by requiring subgraphs identified in the underlying factor graphs be connected. Enforcing this penalty increases the execution time. Consequently, the number of iterations was set at 50. The pair-wise similarity of a typical combined data set is plotted in Figure 2. The rank correlation between MCS_{Sim} and JP_{Sim} is significant (Spearman: rc 0.56; $p<0.0001$). On closer inspection, variations in similarity reflect the differing emphases of the two models:

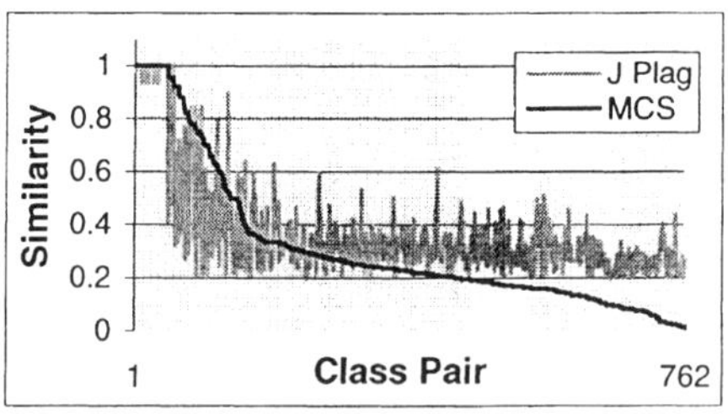

Figure 2: MCS_{Sim} and JP_{Sim} pair-wise similarity. Combined data set (A + B + C). (JP_{Sim} values < 0.2 were un+available.)

- $JP_{Sim} << MCS_{Sim}$: JP_{Sim} fails to match non-contiguous but similar blocks of code. In some cases, JP_{Sim} values could be increased by relocation of order-independent blocks of the class' source code.
- $JP_{Sim} >> MCS_{Sim}$: JP_{Sim} fails to localise the matching of class features, e.g., JP_{Sim} allows string matching *across* methods.
- $MCS_{Sim} << JP_{Sim}$: MCS_{Sim} enforces intra and inter-class relationships, e.g., JP_{Sim} can breach relational constraints and inconsistently match methods called; fields accessed; parameters and types.
- $MCS_{Sim} >> JP_{Sim}$: differences in method detail identified by JP_{Sim} but not reflected in the MCS_{Sim} measure.

A further experiment was carried out by grouping the classes based on abstract, high-level functionality, i.e., each class-pair was labeled as representative, or not, of classes broadly similar in purpose and/or physical functionality. Although MCS_{Sim} and JP_{Sim} both demonstrate a significant difference in median values between similar and dis-similar groupings (Mann-Whitney: $p<0.0001$), the difference is more pronounced for MCS_{Sim}, i.e., MCS_{Sim} appears to better discriminate between similar and dis-similar classes. This is illustrated in Table 2. A comparison of similarity values for class sub-groups within each data set is given in Table 3. In general, within-group similarity values are higher for MCS_{Sim}. A greater degree of implementation variability is reflected in the lower within-group similarity of groups B-SN2 and C-JB: a consequence of the looser specification and larger class sizes, exaggerated in this case by the individuality of the developers, not reflecting reuse in the typical target context of *the same individual, or group, working within and across projects*. Group A-TI was identified as a tutor-supplied utility that was reused largely unchanged.

	Class Group	Similarity	
		Median	IQRange
MCS_{Sim}	Similar	**0.625**	0.568
	Dis-similar	**0.158**	0.194
JP_{Sim}	Similar	0.425	0.523
	Dis-similar	0.255	0.089

Table 2: Comparison of MCS_{Sim} and JP_{Sim} similarity values within similar and dis-similar class groups in the combined data set (A + B + C). (5 sample avg.)

Data	Class	Similarity			
		MCS_{Sim}		JP_{Sim}	
		Median	IQRange	Median	IQRange
A	H	**0.880**	0.063	0.432	0.088
A	WP	**0.621**	0.197	0.326	0.206
A	TI	**1.000**	0.000	**1.000**	0.069
A	HP	**0.632**	0.224	0.383	0.465
B	SN1	**0.667**	0.298	0.387	0.201
B	SN2	0.478	0.065	**0.498**	0.054
C	JB	**0.329**	0.043	0.318	0.079

Table 3: Comparison of MCS_{Sim} and JP_{Sim} similarity values for similar class sub-groups within each data set. (5 sample avg.)

Although the results are encouraging, clique extraction times remain significant in the context of supporting interactive analysis. Figure 3 shows the results of applying SP_{Sim} screening as a means of limiting the number of expensive MCS_{Sim} analyses. SP_{Sim} values correlate well with MCS_{Sim} providing a good approximation to an upper bound. Table 4 shows the results of a ROC analysis using an SP_{Sim} cutoff value of 0.5 as a predictor of MCS_{Sim} being ≥ 0.4. Clearly, we can achieve very high values of sensitivity and specificity, i.e., SP_{Sim} predicts almost all significantly similar class-pairs while rejecting a high proportion that are potentially of little interest.

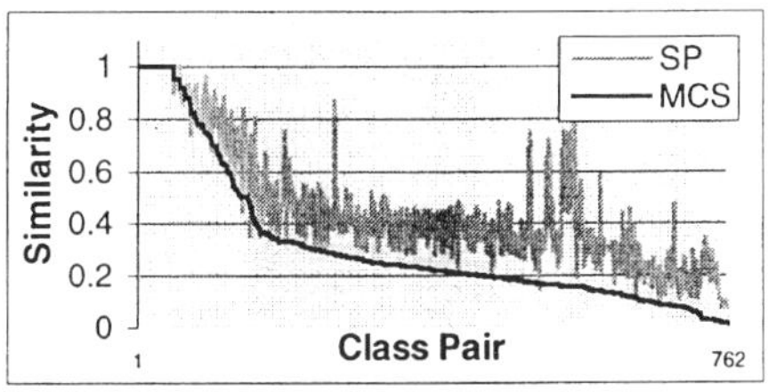

Figure 3: MCS_{Sim} screening using SP_{Sim}.

SP_{Sim} Cutoff	True +ve Fraction (Sensitivity)	True -ve Fraction (Specificity)	True +ve	True -ve	False +ve	False -ve
0.5	99.3%	90.2%	141	559	61	1

Table 4: ROC analysis using an SP_{Sim} cutoff value of 0.5 as a predictor of MCS_{Sim} being ≥ 0.4

Utility: a qualitative assessment

The developed similarity model currently provides the basis for a classification framework that identifies clusters of similar classes. This focuses attention on those potentially generalisable and reusable components associated with the current domain. In addition, the framework provides a searchable case-base for use with newly introduced classes. The "templates" provided by partially complete, iteratively constructed and refined classes associated with ongoing development are being used as "queries" against the classified code-base. This is being further explored and exploited in the context of *late life-cycle activated reuse*, similar in principle to the active approach of Ye and Fischer (2001). The underlying strategy is dependent on the code-compile-test cycle in which a class is iteratively developed until it meets a given, potentially fluid, specification. This is particularly relevant when one considers the emergence of "rapid", highly iterative and iterated development practices (Beck, 1999). By monitoring the production of classes and comparing them in real time with the existing code-base, the development process is being actively informed. Preliminary experiments based on an iterative build process applied to selected classes from the previous data sets are encouraging. Partially implemented classes comprising only fields and method signatures have been sufficient to retrieve good quality[2] clusters of existing code containing examples, where they exist, satisfying the required specification fully or in part. Again, this contributes to identifying and actively promoting foci for potential reusable assets. By improving awareness of other contexts in which similar code is being or has been deployed, this further

[2] In terms of recall and precision.

benefits the development process via reinforcement and knowledge transfer. In addition, where a code revision system is employed, insights into potential problems with code currently being developed can be coupled with the opportunity for immediate improvement, through heightened awareness of previous code re-engineering or refactoring.

CONCLUSIONS

The experimental evidence suggests that an attributed, relational, graph-theoretic model can be reasonably applied to the identification of recurring, useful structure present in established collections of object-oriented code. The key, enabling factor is the adoption of an evolutionary approach to the solution of an otherwise intractable problem. Preliminary work indicates that in principle the presented approach is capable of supporting both a reuse-oriented classification framework, and an automated, active approach to promoting code reuse late in the development life-cycle. However, exploring the utility of such an approach in a typical operational setting has still to be completed. This includes establishing a balance between the benefits and possible drawbacks, e.g., the level of acceptable intrusion, alongside an assessment of the role of the developer as the final arbiter of useful, class-based similarity. Continuing work seeks to validate the notion of late life-cycle activated reuse and further explores the extension of the pattern matching approach to incorporate identification of recurring class combinations. Again this is only possible given the use of the HGA, as CG sizes are further increased.

This lightweight, code-oriented, graph-theoretic pattern matcher can be used to promote reuse in situations where a reuse process is either absent or informal: the tools and techniques described here provide support for enabling and improving shared awareness of reuse opportunity, based on mining similarity in past and ongoing development. They can in turn be seen as part of a process of domain analysis, capable of stimulating the evolution of a more systematic reuse ethic and as a basis for enabling late life-cycle activated reuse as introduced here.

REFERENCES

Balazinska, M., Merlo, E., Dagenais, M., Lague, B. and Kontogiannis, K., 2000, "Advanced Clone Analysis to Support Object-Oriented System Refactoring.", Proceedings of 7th Working Conference on Reverse Engineering (WCRE'2000), IEEE Computer Society Press, pp98-107.

Baxter, I. D., Yahin, A., Moura, L., Sant'Anna, M., and Bier, L., 1998, "Clone detection using abstract syntax trees.", Proceedings of International Conference on Software Maintenance, ACM Press, Bethesda, Maryland, November 16-19, pp368-377.

Beck, K., 1999, "Extreme Programming Explained: Embrace Change.", Addison-Wesley.

Bron, C. and Kerbosch, J., 1973, "Algorithm 457, Finding all cliques of an undirected graph.". Communications of the ACM, 16, pp575-577.

Fayad, M. E., Laitinen, M., Ward, R.P., 2000, "Software Engineering in the Small.", Communications of the ACM, 43(3), pp115-118.

Gardiner, E.J., Artymiuk, P.J. & Willett, P., 1997, "Clique-detection algorithms for matching three-dimensional molecular structures.", Journal of Molecular Graphics and Modelling, 15, pp245-253.

Hislop, G., 1998, "Analysing Existing Software for Software Reuse.", The Journal of Systems and Software, 41, pp33-40.

Kreuger, C.W., 1992, "Software Reuse.", ACM Computing Surveys, 24, pp131-184.

Maiden, N.A. and Sutcliffe, A.G., 1993, "People-Oriented Software Reuse: the Very Thought.", Proceedings of the Second International Workshop on Software Reuse, Frakes, W.B. (Ed.). IEEE Computer Press, Lucca, Italy, March 24-26, pp176-185.

Marchiori, E. A., 1998, "Simple Heuristic Based Genetic Algorithm for the Maximum Clique Problem.", ACM Symposium on Applied Computing (SAC98), pp366-373.

Morisio, M., Ezran, M., Tully, C., 2002, "Success and Failure Factors in Software Reuse.", Transactions on Software Engineering, 28(4), pp340-357.

Nada, N., Rine, D. and Tuwaim, S., 1998, "Practices in Organizational Structure for Software Reuse.", Proceedings of International Conference on Information Systems Analysis and Synthesis.

Prechelt, L., Malpohl, G. and Philippsen, M., 2000, "JPlag: Finding plagiarisms among a set of programs.", Technical Report 2000-1, University of Karlsruhe, Germany.

Sametinger, J., 1998, "Software Engineering With Reusable Components.", Springer-Verlag.

Tsai, W.-H. and Fu, K.-S., 1979, "Error-correcting isomorphism of attributed relational graphs for pattern analysis.", IEEE Transactions on Systems, Man, and Cybernetics, 9, pp757-768.

Whitmire, S.A., 1997, "Object-Oriented Design Measurement. ", John Wiley & Sons.

Willett, P., 1999, "Matching of chemical and biological structures using subgraph and maximal common subgraph isomorphism algorithms.", in Truhlar, D.G., Howe, W.J., Hopfinger, A.J., Blaney, J.D. and Dammkoehler, R. (Eds.) "Rational Drug Design.", Springer Verlag, pp11-38.

Ye, Y. and Fischer, G., 2001, "An Active and Adaptive Reuse Repository System.", 34th Hawaii International Conference on System Sciences (HICSS-34), IEEE Press, Maui, Hawaii, Jan 3-6.

RADIOGRAPHIC IMAGE ESTIMATION FOR TWO MATERIAL ASSEMBLIES BY USING NEURAL NETWORKS

N. S. REEN, I. N. TANSEL, P. CHEN
Mechanical Engineering Department
Florida International University
10555 West Flagler Street, (EAS - 3473) Miami, FL 33174

C.V. KROPAS-HUGHES
Air Force Research Laboratory
AFRL/MLLP, Bldg. 655, R166
2230 Tenth Street
Wright-Patterson AFB
Ohio, 45433-7746

ABSTRACT

Radiographic image estimation of assemblies with parts made from different materials requires extensive information of the material and imaging system characteristics. Artificial neural networks are proposed to represent the relationship between the thickness of parts made of two different materials and the pixel value of the radiographic image. The part thicknesses were estimated at a grid from the STL files of parts and the image was generated by using the neural networks.

INTRODUCTION

Radiographic images are one of the most effective Nondestructive Testing (NDT) tools for inspection of hidden parts. The popularity of industrial radiographic techniques is expected to increase in the future for inspection of composite structures, complex assemblies, multilayer boards, and MEMs. The digital imaging of X-rays, solid modeling based CAD programs, gigantic data storage units, fast data transfer rates, and tremendous speed of microprocessors will influence the common inspection practices in near future. Instead of using the radiographic image of a perfect part as a reference, the same image can be estimated from the CAD file. In addition to detecting cracks, and large broken segments of the parts small material losses can be recognized by comparing the gray levels of actual and simulated images. In this study, the use of neural networks is proposed to prepare the radiographic images when only two materials exist in the assembly.

Researchers have been working on simulation of X-Ray images from the CAD files (Inanc and Gray, 1990), STL files generated by the solid modeling CAD programs (Duvauchelle et al., 2000; Reen, 2002), three-dimensional CT data (Riepert et al., 1995; Riepert, 2000), and MR images (Mizowaki, 2001). To improve the accuracy of the simulations, scattering of photons were considered by using the integral transport equation (Inanc and Gary, 1997) and Monte Carlo radiation transport simulation (Mathy et al., 2003). Some of the available radiographic image simulation programs are XRSIM (Inanc and Gary, 1997), FoXSIS (Riepert et al., 1995; Riepert, 2000), and Sinbad . In addition to industrial and medical applications, agricultural and food industry has also investigated X-ray simulation techniques for their applications (Zwiggelarr et al., 1996). It is possible to estimate the X-ray images of complex assemblies with several parts if extensive information is available about the material and imaging system. In the simulation of multi-material systems, work is easiest when only two materials exist (Wojcik et al., 1996).

When a commonly used material is considered, it is possible to find information on their characteristics and prepare the simulation. However, the conventional process to obtain information about a new material is time consuming. In this study, the thickness of materials will be determined by using the ray tracing method from the STL files and the pixel value of the final image will be estimated by using backpropagation (BP) type neural networks (Rumelhart et al., 1986). This approach simplifies necessary tests to obtain information about the characteristics of the materials.

THEORETICAL BACKGROUND

Transmissions of the X-rays up to about 200 keV can be represented with the following equation [10]:

$$T(E) = \mu(E)\, L \tag{1}$$

$$T(E) = \ln (Io(E)/I(E)) \tag{2}$$

Where, Io(E) and I(E) are the original and the transmitted X-ray beam intensities. μ(E) is the linear attenuation coefficient of the material at the considered (E) energy level. L is the thickness of the material. When two materials are on each other T(E) will be calculated by considering the length and attenuation of both materials with the following equation:

$$T(E) = \mu 1(E)\, L1 + \mu 2(E)\, L2 \tag{3}$$

When a common material is considered, linear attenuation can be calculated by using two materials and two energy dependent functions.

In this study BP type neural networks were used [11]. The number of the input and output nodes are determined according to the model. The user determines the number of hidden layers and nodes to obtain the most compact model, which satisfies the targeted accuracy.

PROPOSED PROCEDURE FOR ESTIMATION OF RADIOGRAPHIC IMAGE

The proposed approach is presented in Fig.1. The STL files of the parts were prepared by the CAD/CAM package (ProE). The thickness of the each part was calculated with constant intervals on a grid by using the ray tracing method. The BP type neural network estimated the pixel values of radiographic images by evaluating the thickness of both parts as an input.

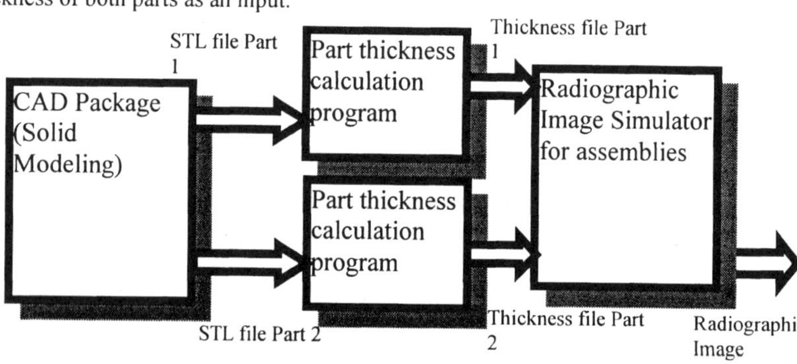

Figure 1. The proposed BP based radiographic image simulation program.

The BP type neural networks were trained by using the semi-experimental data. Brass and aluminum wedges, and flat pieces were used to prepare the radiographic images. The pixel values were taken from the radiographic image and interpolated by using the following equation to represent multiple combinations:

$$G'(L_1, L_2) = G(L_{1t}) - L_2 ((G(L_{1tb}, L_{2tb}) - G(L_{1tb})) / L_{2tb}) (R(L_{1t}, L_{1tb})) \quad (4)$$

Where G' (L1, L2) are the estimated gray level of the image when the material 1 has L1, and material 2 has L2 thickness. G (L1t) is the experimentally obtained gray level of the image when the thickness of the step wedge (material 1) is L1t. G(L1tb, L2tb) is the gray level of the image when two blocks are put on each other with the thickness of L1tb (material 1) and L2tb (material 2). G (L1tb) corresponds to the gray level when the block of material 1 with L1tb thickness is exposed to X-ray. This equation is written by assuming that material 1 is a very dense material (brass) compared to material 2 (aluminum). In this case, the relationship between the gray level and part thickness can be kept relatively linear if the maximum thickness of material 2 is selected reasonably. A nonlinear relationship is observed between the thickness of the material 1 and the gray level. R (L1t, L1tb) represents this characteristic. A constant was used for each L1t value to represent this function. The constant was selected bigger than 1 when L1t < L1tb and smaller than 1 when L1t > L1tb.

EXPERIMENTAL PROCEDURE

The experimental setup is presented in Fig.2. The aluminum and brass step wedges were located at two opposite sides. The brass wedge is on the right. Radiographic images were obtained by using an ARACOR Tomoscope in the preview mode. This system had a Fein Focus 225 KeV X-ray source, and fiber optic scintillators with four CCD image sensors. The digital radiographic images were obtained at 45 KeV and presented in Fig.3.

RESULTS AND DISCUSSION

The training file for the BP type neural network was prepared by using the Eq.4. The gray levels of the pixels at different part thicknesses were obtained from the digital images presented in Fig.3. The semi-experimental gray levels for different material combinations are presented with the 3-D graph in Fig.4 and used for the training of BP type neural network. The network had one hidden layer with 5 nodes.

Once the training of the neural network is complete, it is possible to estimate the X-ray simulation of any assembly with parts made of aluminum and brass, as long as their thicknesses are within the range of the test pieces. The 3-D view of parts for an assembly, and the estimated radiographic image of the system are presented in the Fig.5 and Fig.6 respectively.

The part on the top is an aluminum piece with 0.3-inch maximum thickness. The height of first step on the left is 0.14 and that of second and third step is 0.08 inch each. The thickness of the middle section is 0.03 inches and there is a square hole in the middle of it. The piece at the bottom is brass. It has 0.05-inch thickness at the thick section while the rest is 0.03 inches.

Figure 2. The experimental setup. Aluminum (on the left) and brass (on the right) step wedges and test blocks are located on the rotary table.

Figure 3. Radiographic images of the blocks at 45 KeV.

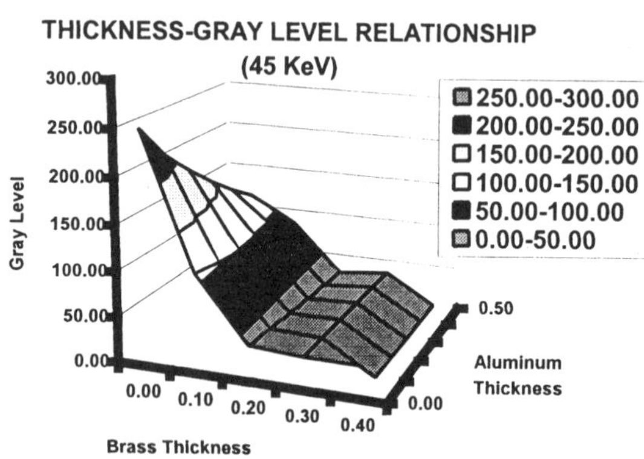

Figure 4. Material thickness versus gray level when aluminum and brass materials overlap.

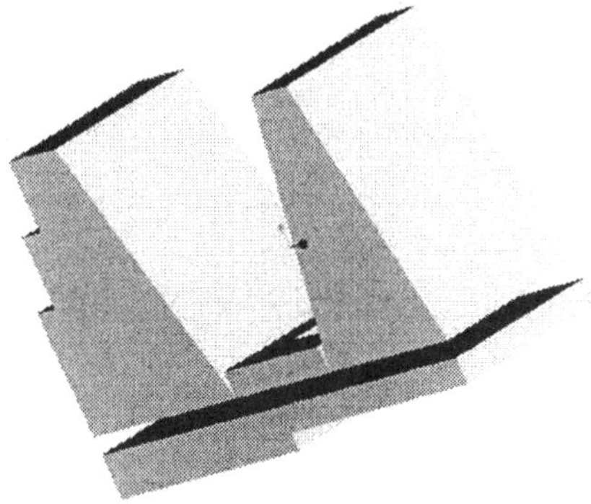

Figure 5. ProE generated 3-D views of parts and the proposed assembly. Aluminum part is at the top. Brass plate is at the bottom.

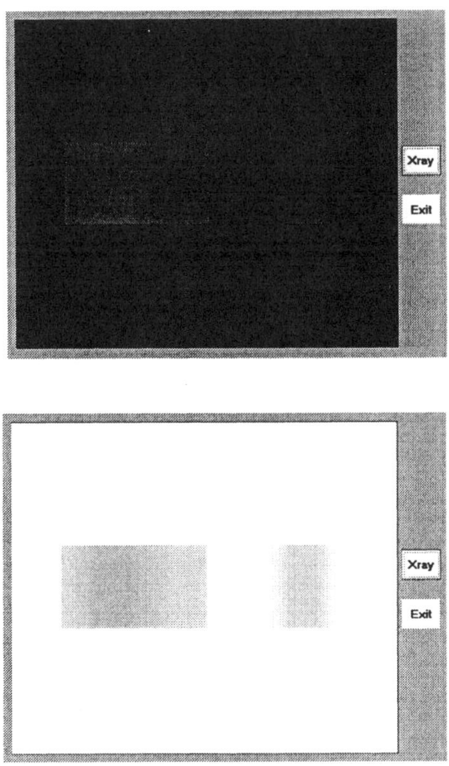

Figure 6. The radiographic image simulation for the assembly in Fig.5. The top image is the radiographic image simulation. The bottom one is the negative of the top image for easy visual inspection of the gray level variation.

CONCLUSION

The use of backpropagation type neural networks is proposed to simulate the radiographic image of an assembly with parts made of two materials. Simulation of the radiographic image of assemblies with different materials can be easily accomplished by using the analytical model as long as the characteristics of the materials and the imaging system are known. However, determination of these characteristics is time consuming. The proposed approach conveniently establishes a relationship between the part thicknesses and the pixel value of the radiographic image when two materials are used in the assembly. The necessary data could be collected from a single image of two wedges made of different materials and located with 90o angle. The radiographic image of one assembly was simulated by using the proposed method. Accuracy of the simulations depends on the training of the neural network.

ACKNOWLEDGEMENT

Mr. Ed Porter and Mr. Dan Daniels of Wright Patterson AFB prepared the digital radiographic images. Authors appreciate contributions of both engineers to the study.

REFERENCE

Duvauchelle, P., Freud, N., Kaftandjian, V., and Babot, D., 2000, "A Computer Code to Simulate X-ray Imaging Techniques", Nuclear Instruments and Methods in Physics Research: Section B, 170, Vol.1-2, pp.245-258.

Inanc, F., and Gray, J.N., 1990, "A CAD interfaced simulation tool for X-ray NDE studies", In Review of Progress in QNDE (Thomson D.O. and Chimenti D.E. Eds), Vol.9, Plenum, New York.

Inanc F. and Gray J. N., 1997, "Scattering simulations in radiography, Applied Radiation and Isotopes", Vol. 48, No. 10-12, Pages 1299-1305.

Mathy F., Guillemaud R. and Picone M., 2003, " Experimental validation of a coupled photon Monte Carlo and CAD software", Nuclear Instruments and Methods in Physics Research Section A: Accelerators, Spectrometers, Detectors and Associated Equipment, Volume 504, Issues 1-3, 21 May 2003, Pages 317-320

Mizowaki T., Araki N., Nagata Y., Negoro Y., Aoki T. and Hiraoka M., 2001, "The use of a permanent magnetic resonance imaging system for radiotherapy treatment planning of bone metastases", International Journal of Radiation Oncology*Biology*Physics, Volume 49, Issue 2, Pages 605-611.

Reen N. S., 2002, "Improvement of existing complexity estimation method", M.S. Thesis, Florida International University, Miami, FL.

Riepert T., Ulmcke D., Jendrysiak U. and Rittner C., 1995, "Computer-assisted simulation of conventional roentgenograms from three-dimensional computed tomography (CT) data -- an aid in the identification of unknown corpses (FoXSIS)", Forensic Science International, Volume 71, Issue 3, Pages 199-204.

Riepert T., Ulmcke D., Schweden F. and Nafe B., 2001, "Identification of unknown dead bodies by X-ray image comparison of the skull using the X-ray simulation program FoXSIS", Forensic Science International, Volume 117, Issues 1-2, Pages 89-98.

Rumelhart, D.E., Hilton, G., Williams, R.J., 1986, "Learning Internal Representations by Error Propagation, Parallel Distributed Processing:Explorations in the Microstructure of Cognition", Vol.1, Ed. E. Rumelhart, J.L. McClelland, MIT Press, pp. 319-362.

Wojcik, R., Majewski, S., Parker, F.R., Winfree, W.P., 1996, "Single shot dual energy reverse geometry X-radiography", Nuclear Science Symposium. IEEE Meeting: November 2, 1996, pp. 811 – 815.

Zwiggelaar R., Bull R. C. and Mooney J. M., "X-ray Simulations for Imaging Applications in the Agricultural and Food Industries", Journal of Agricultural Engineering Research, Volume 63, Issue 2, 1996, Pages 161-170.

PART VI: PREDICTION

TORNADO DETECTION WITH KERNEL-BASED METHODS

THEODORE B. TRAFALIS
Budi Santosa
School of Industrial Engineering
University of Oklahoma
202 West Boyd, Suite 124,
Norman OK, 73019
405/325-4347
ttrafalis@ou.edu,bsant@ou.edu

MICHAEL B. RICHMAN
School of Meteorology
University of Oklahoma
1100 East Boyd, Suite 1310
Norman, OK 73019
405/325-1853
mrichman@ou.edu

ABSTRACT

There are several statistical based methods used to detect tornadoes, such as learning networks (neural networks and radial basis function networks) and linear discriminant analysis (LDA). Different types of learning networks, such as Support Vector Machines (SVM) and minimax probability machine (MPM), are applied. The last two approaches utilize kernel methods to address nonlinearity of the data in the input space. These kernel-based methods are applied, using radar derived velocity data, to detect tornadoes. These approaches are based on maximizing the margin between two different classes: tornado and non-tornado. Results show that these approaches are successful for the tornado detection problem.

1. INTRODUCTION

Detection of tornadoes with ample warning times is a goal of severe weather forecasters. With modern weather radar, high speed computing and advanced signal processing algorithms, progress is being made increasing their average lead-time. Contemporaneously, there is a need to lower the false alarm rate (FAR), as a high FAR can lead to the public ignoring warnings. The Mesocyclone Detection Algorithm (MDA) was created by the National Severe Storms Laboratory and uses the Weather Surveillance Radar 1998 Doppler (WSR-88D) outputs to detect storm–scale circulations associated with regions of rotation in thunderstorms. Meteorologists use the MDA as an input in their decision to issue tornado warnings. Marzban and Stumpf (1996) show that the MDA performance can be improved by ANN post-processing.

SVM and MPM are applied herein to detect tornado circulations sensed by the WSR-88D radar. These kernel based methods do not make any assumptions about the data distributions. Their application is useful to address the problem of nonlinearity of the data in the input space since the data are mapped into a higher dimensional space where the problem becomes linear separable. By exploiting these methods, it is possible to find nonlinear classifiers that separate the data into tornado and non-tornado cases. MPM is a new technique that seeks the best classifier minimizing the maximum probability of misclassification error in a binary problem. By using the theorem introduced by Marshall and Olkin, generalized by Popescu and Bertsimas (2001), the problem of minimax can be formulated as an unconstrained second order conic programming problem (Lanckriet et al., 2002).

The paper is organized as follows. In section 2, the definition of the problem is provided. In section 3, description of the data is given. In section 4, the basics of the learning machines used and our methodology are discussed. In section 5, the experimental setting is described. Section 6 provides analysis of the results and, finally, section 7 concludes the paper.

2. PROBLEM STATEMENT

There are two challenges for improving tornado warnings. The first is tornado detection as, of those tornadoes that do occur, the number detected is smaller. The second one is false alarms, since algorithms detect tornado circulations more often than such circulations are confirmed. Accordingly, it is desirable to develop a statistical algorithm that will maximize detection and minimize false alarms. Despite the challenges, lead times for tornadoes have increased from a few minutes (a decade ago) to 11 minutes (with current radar), largely due to improvements in algorithms that use the radar data as inputs.

The second research problem is to develop an intelligent system that can generalize well. ANN are considered robust classifiers in terms of input noise. An alternative to ANN is SVM where the learning optimization problem is convex. Vapnik (1995) shows that SVM has better generalization properties than ANN. More recent work by Lanckriet et al. (2002) has shown that MPM is competitive with SVM.

3. DATA AND ANALYSIS

Any circulation detected on a particular volume scan of the radar data can be associated with a report of a tornado. If there is a tornado reported between the beginning and ending of the volume scan, and the report is within reasonable distance of a circulation detection (input manually), then the ground truth value is flagged. If a circulation detection falls within the prediction "time window" of -20 to +6 minutes of the ground truth report duration, then the ground truth value is also flagged. The key behind these timings is to determine whether a circulation will produce a tornado within the next 20 minutes, a suitable lead time for advanced severe weather warnings by the National Weather Service. Any data with the aforementioned flagged values are categorized as tornado cases (1). Other circulations, not associated with a tornado, are categorized as 0, corresponding to a non-tornado case. The predictor pool employed in this study consists of 23 MDA attributes based on Doppler velocity data. These same attributes have been used successfully by Marzban and Stumpf (1996).

4. METHODOLOGY

4.1 Support Vector Machines (SVM)

Consider a problem with two classes, where a classifier is sought to separate two classes of points. The SVM formulation follows Haykin (1999). SVM map a given set of binary labeled training data into a high-dimensional feature space and separate the two classes of data linearly with a maximum margin hyperplane in the feature space. In the case of nonlinear separability each data point x in the input space is mapped into a higher dimensional feature space using a feature map φ. In the new space, the dot product $\langle x, x' \rangle$ becomes $\langle \varphi(x), \varphi(x)' \rangle$. A nonlinear kernel function, *k(x,x')*, can be used to substitute the dot product $\langle \varphi(x), \varphi(x)' \rangle$. The use of a kernel function allows the SVM to operate efficiently in a nonlinear high-dimensional feature spaces without being adversely affected by the dimensionality of that space. In general, this hyperplane corresponds to a nonlinear decision boundary in the input space.

4.2 Minimax Probability Machines (MPM)

The MPM approach was introduced by Lanckriet et al. (2002). The objective of this approach is that the problem of binary classification is solved by minimizing the maximum probability of misclassification of the future data points. The problem can be defined as follows. Let x and y denote random vectors in a binary class, with means and covariance matrices given by $(\bar{x}, \Sigma_x)$ and $(\bar{y}, \Sigma_y)$ respectively, where x, $\bar{x}$, y, $\bar{y} \in \mathbb{R}^n$ and $\Sigma_x, \Sigma_y \in \mathbb{R}^{nxn}$ and both are symmetric and positive semidefinite. Then, a hyperplane $H(a,b) = \{z \mid a^T z = b\}$, where $a \in \mathbb{R}^n \setminus \{0\}$ and $b \in \mathbb{R}$ which separates the two classes of points with maximal probability with respect to all distributions having these mean and covariance matrices is determined. Mathematically the problem can be formulated as:

$$\begin{aligned} &\max_{\alpha, a \neq 0, b} \alpha \\ &s.t \inf_{x \sim (\bar{x}, \Sigma_x)} \Pr\{a^T x \geq b\} \geq \alpha \\ &\quad \inf_{y \sim (\bar{y}, \Sigma_y)} \Pr\{a^T y \leq b\} \geq \alpha \end{aligned} \tag{1}$$

After some manipulations by exploiting the powerful theorem by Popescu and Bertsimas (2001) that states that:

$$\sup_{y \sim (\bar{y}, \Sigma_y)} \Pr\{ y \in S\} = \frac{1}{1 + d^2}, \text{ with } d^2 = \inf_{y \in S} (y - \bar{y}) \Sigma_y^{-1} (y - \bar{y}),$$

(Lanckriet, 2002) the above problem can be simplified as:

$$\begin{aligned} &\kappa_*^{-1} = \min_{\alpha} \| \Sigma_x^{\frac{1}{2}} a \|_2 + \| \Sigma_y^{\frac{1}{2}} a \|_2 \\ &st.\ a^T (\bar{x} - \bar{y}) = 1 \end{aligned} \tag{2}$$

In the case of nonlinear separability the kernel method is used. In the high dimensional feature space we want to find a hyperplane $H(a,b) = \{\varphi(z) \mid a^T \varphi(z) = b\}$ that corresponds to the linear decision boundary $D(a,b) = \{z \in R^n \mid a^T \varphi(z) = b\}$ in the input space $\mathbb{R}^n$ ($a \in \mathbb{R}^n \setminus \{0\}$ and $b \in \mathbb{R}$).

4.3 Linear Discriminant Analysis (LDA)

Linear discriminant is applied herein. This method projects high-dimensional data onto a line and a binary classification task is performed in this one-dimensional space (in two dimensional space). The projection maximizes the distance between the means of the two classes while minimizing the variance within each class.

4.4 Neural Network (NN)

A feedforward neural network with two layers is utilized, in which the first is a hidden layer and the second is an output layer. The number of nodes in the hidden layers is varied from 2, 5 and 10. The experiments show that 5 hidden nodes provide the best results. The network is trained by using a gradient descent with momentum algorithm.

4.5 Forecast Evaluation Indices for Tornado Detection

In the detection paradigm, the forecasts are evaluated by using a suite of forecast evaluation indices based on a contingency table. The forecast evaluation indices used are: accuracy, false alarm rate (far), bias, probability of detection (pod), probability of false detection (pofd), critical success index (csi) and Heidke skill and are defined in Wilks (1995). They describe various aspects of tornado forecasts being observed and the number of those forecasts that verify compared to some reference (e.g., a random forecast).

5. EXPERIMENTS

For the training set, the ratio between tornado and non-tornado observations is about the same. In the *testing* sets, the ratio is varied from 5% to 50% in 5% increments. The cases used for training are different with those used in the testing set. The same *training* and *testing* sets are applied to all methods. The SVM, MPM approach, LDA and ANN experiments were performed in the MATLAB environment. The OSU SVM Classifier Matlab Toolbox by Jushui et al. was used to run experiments of SVM for classification. These codes are MATLAB versions of Chih-Chung Chang and Chih Jen Lin's LIBSVM algorithm (Chang and Lin, 2002). For the MPM approach, the MATLAB codes from Lanckriet are used. Only the linear MPM approach is implemented due to the limitation of memory. The ANN model is run 10 times. For each run the testing set output is examined. The predicted labels are obtained by discretizing the average of the 10 sets of outputs into –1 (non-tornado) or +1 (tornado). The data are preprocessed before being applied to each method. For each column, each data point (observation) is divided by the norm of the column. For ANN, a preprocessing is applied such that the data are located in the range between –1 and 1.

6. RESULTS

By preprocessing the data, the performance of LDA improves significantly. For the testing sample with a small ratio of tornadoes to non-tornadoes (5%), improvement in misclassified error (misses and false alarm) is from 21% to 4% (Table 1). Similar results were found for SVM (Table 2) and MPM (Table 3). For all three techniques, the misclassification rate was lowered by preprocessing to 4% for the lowest ratio and 7 or 8% for the second lowest ratio. In contrast, this preprocessing step does not have a significant impact on the ANN misclassification performance (Table 4).

Table 1. Misclassification rates for LDA before and after preprocessing.

% tornadoes	5	10	15	20	25	30	35	40	45	50
Before	22	21	24	25	26	27	25	24	23	21
After	4	8	11	13	17	19	20	20	20	20

Table 2. Misclassification rates for SVM before and after preprocessing.

% tornadoes	5	10	15	20	25	30	35	40	45	50
Before	95	90	87	83	80	77	74	71	69	67
After	4	7	10	12	15	18	18	19	19	19

Table 3. Misclassification rates for MPM before and after preprocessing.

% tornadoes	5	10	15	20	25	30	35	40	45	50
Before	21	21	23	24	25	26	24	24	22	21
After	4	8	11	13	16	18	19	20	19	19

Table 4. Misclassification rates for NN before and after preprocessing.

% tornadoes	5	10	15	20	25	30	35	40	45	50
Before	21	23	24	29	25	29	25	23	21	20
After	22	24	26	28	28	29	24	24	22	19

Figure 1(a-g) depicts different forecast evaluation indices for each classification method. Each graph describes the relation between a specific statistic and percentage of tornado observations in the *testing* set. All the methods are trained with a *fixed* training set. Tuning parameters are determined in the training phase and these are used to make predictions for different *testing* sets based on the ratio of tornado to non-tornado observations. By investigating this ratio, the performance for each forecast validation statistic can be observed. A climatological average of tornadoes close to 10% is reasonable.

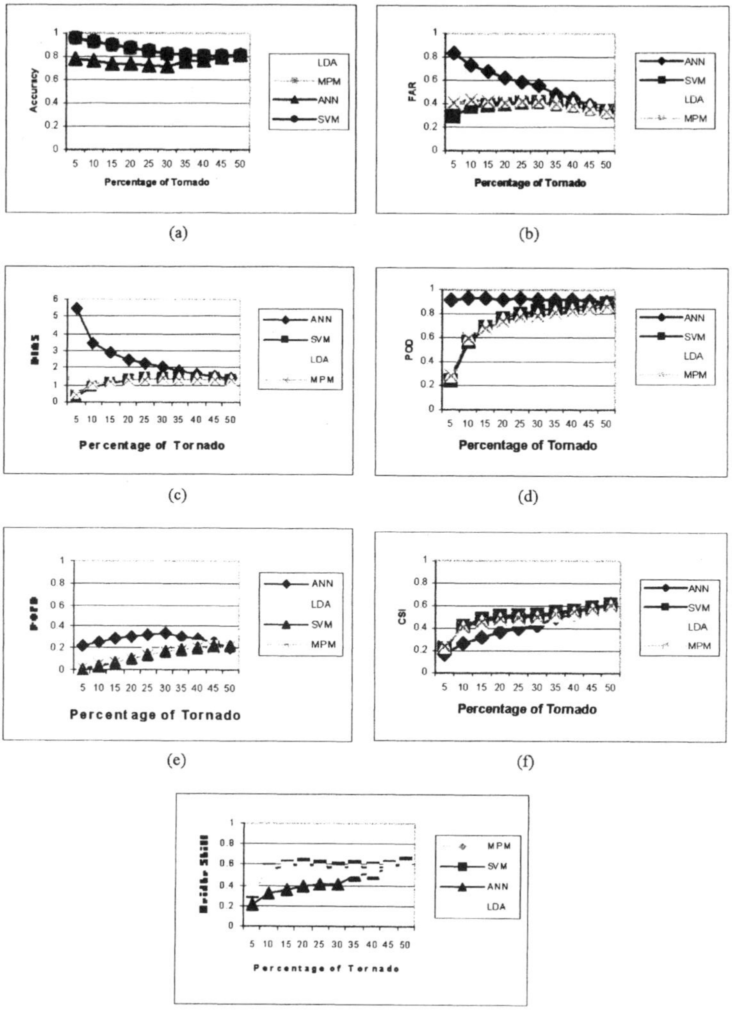

(g)

Figure 1 (a-g). Forecast indices versus percentage of tornadoes used in the test set. Panels (a) Accuracy, (b) FAR, (c) Bias, (d) POD, (e) POFD, (f) CSI, and (g) Heidke Skill

The accuracy of LDA, MPM and SVM are closely related (Fig. 1a). The ANN results differ with a lower accuracy than the other three methods at all ratio percentages under 50. Much of the explanation of the behavior can be explained by looking at the FAR (Fig. 1b). The SVM values are close to those for the MPM and LDA and the ANN values have a much larger FAR. This predisposition of ANN to forecast tornadoes when none are observed (overforecasting) can be seen in the BIAS values (Fig. 1c). These are large until the percentage of tornadoes to non-tornadoes exceeds 35. As in the previous evaluation indices, the MDA, SVM and MPM BIAS values are all clustered together and have an optimum bias value of 1.0 near the climatological probability of tornadoes. The POD (Fig. 1d) indicates ANN had an edge over the other techniques as it could detect over 90% of the tornadoes in the database. At the ten percent level, the remaining techniques had close to a 0.6 POD. Another way of examining the False Alarms is through POFD (Fig. 1e). The POFD should be as close to zero as possible. For the techniques evaluated, the ANN results have a high POFD. At the 10% tornado level, the POFD was close to 3% for SVM, LDA and MPM compared to 25% for ANN. CSI results (Fig. 1f) show the familiar pattern of the ANN exhibiting a different behavior than the other three techniques. ANN has a distinctly lower (poorer) CSI until the percentage of tornadoes reaches 35. At the level close to the climatological probability, the CSI is about 0.27 for ANN and about 0.41 for the other techniques. The Heidke skill results (Fig. 1g) show lower skill scores for ANN, compared to LDA, SVM and MPM. At the ten percent ratio, the ANN skill is about 32% and the other techniques have skills of about 54%. Taken collectively, these results suggest that ANN behaves differently then the other methods. This can be explained as ANN working better in detecting a tornado when there is tornado, overpredicts tornadoes in all circumstances (high false alarm rate) and is inferior in detecting no tornado when there is no tornado, compared to the other methods.

7. CONCLUSIONS

Based on the performance improvements of almost all the methods, except for ANN, after preprocessing of the data, it is concluded that the preprocessing step is very important here. These improvements are due to the preprocessing that makes the range of magnitude of each attribute considerably smaller.

Linear MPM shows comparable results to SVM and LDA. Due to the large memory requirement to store the kernel matrix as the data sets get larger, only linear MPM, which does not need a kernel mapping, is applied. For small data set (e.g., 50% ratio for tornadoes to non-tornadoes) kernel MPM show promising performance.
Global optimization techniques for selecting the best ANN architecture and optimal weights will be investigated. Reducing the kernel matrix size or applying a decomposition method such that MPM can be applied for large-scale problems is an alternative open research. Overall, the SVM technique shows a slight edge over the LDA and MPM methods for most forecast evaluations. However, the key finding is that LDA, MPM and SVM are more accurate, have lower false alarm rate, less bias and more skill than the NN technique. This is particularly noteworthy since the current tornado detection algorithms are processed through ANN.

From the meteorological viewpoint, the SVM and MPM techniques need to be tested in an operational setting in order to assess if lead times are better than the currently employed algorithms. Such experimentation provides valuable cross-validation that will reveal if these techniques remain stable as prediction tools in near real time. In the future, the use of multiple radars as sources of predictors is a good direction to investigate to obtain more accurate prediction.

Acknowledgements. The present work has been partially supported by the NSF grant EIA-0205628.

References

Chang, C. C and Lin, C.J, *LIBSVM: A Library for Support Vector Machines*, 2001, http://www.csie.ntu.edu.tw/~cjlin/libsvm.

Haykin, S. *Neural Networks: A comprehensive Foundation*, 2nd edition, Prentice-Hall, Upper saddle River, NJ, 1999.

Junshui, M., Y. Zhao, and S. Ahalt, "OSU SVM Classifier Matlab Toolbox", Available at http://eewww.eng.ohio-state.edu/~maj/osu_svm/

Lanckriet, G.R.G., El Ghaoui, L., Bhattacharyya, C., Jordan, M.I., A Robust Minimax Approach to Classification, *Journal of Machine Learning Research*, 3, 555-582, 2002.

Marzban, C. and Stumpf, G. J., A Neural Network for Tornado Prediction Based on Doppler Radar Derived Attributes, *Journal of Applied Meteorology*, 35,617-626, 1996.

Popescue, I. and Bertsimas, D., Optimal Inequalities in Probability Theory: A Convex Optimization Approach, *Technical Report TM62*, INSEAD, 2001.

Vapnik, V.N., *The Nature of Statistical Learning Theory*. John Wiley & Sons, New York, 1995

Wilks, D. S. *Statistical Methods in the Atmospheric Sciences*. Academic Press, London, 1995.

SPATIAL CHARACTERIZATION OF RAINFALL PATTERNS FOR USE IN A CLASSIFICATION SYSTEM

MICHAEL E. BALDWIN
Cooperative Institute for Mesoscale Meteorological Studies, University of Oklahoma, Norman, OK
also affiliated with: NOAA/NWS/Storm Prediction Center and NOAA/National Severe Storms Laboratory

and

S. LAKSHMIVARAHAN
School of Computer Science, University of Oklahoma, Norman, OK

ABSTRACT

The quality of the spatial structure and continuity of rainfall forecasts is of great interest to certain users of weather information. To assist in the determination of this type of verification information, a rainfall system classification procedure has been developed. This paper will focus on the feature extraction process and the experiments for discovering useful attributes that describe the spatial variability of rainfall systems. Results from expert classification of a training data set are compared to hierarchical cluster analysis using trial attributes. When the objective classification results agree with the results of the expert classification, the attributes are considered to be useful. The attributes are taken from aspects of the two-dimensional autocorrelation function, along with other statistics previously determined to be useful measures of rainfall intensity. An automated classification procedure is developed using the objective classification results from the training data set. To validate this procedure, a representative sample from a large data set is taken. Results from the automated classification procedure agree with an expert classification for 85% of the validation sample.

INTRODUCTION

The overall goal of this work is to develop an automated rainfall system classification procedure. This is part of an ongoing project of weather-related decision support system development at the National Severe Storms Laboratory (NSSL) and Storm Prediction Center (SPC) in Norman, OK. This paper follows previous work that used attributes related to the intensity of rainfall (Baldwin and Lakshmivarahan 2002, hereinafter referred to as BL02). To develop an automated classification system, the discovery of attributes that allow for automated characterization of rainfall patterns is required. The determination of useful attributes is based upon a relatively small training data set, comparing the results of automated classification experiments with a human expert classification. This data set consists of hourly accumulated rainfall analyses from an operational analysis system. A set of 48 separate precipitation events which occurred at various times and locations across the United States were selected for inclusion in this data set. Cases were included based upon the existence of typical rainfall patterns that are often found, in particular, *lin-*

ear, *cellular*, and *stratiform* precipitation events. In the *linear* class, there is larger-scale organization of smaller-scale elements of heavy precipitation. The smaller-scale elements (usually called "cells") appear to be arranged approximately along a line. For the *cellular* class, there is very little large-scale organization of smaller-scale heavy precipitation elements. The rainfall field typically consists of "cell" features somewhat randomly positioned in a disorderly fashion. In the *stratiform* class, there is large-scale organization of light rainfall, the precipitation field shows little variation in any direction over a large area. Linear and cellular events are considered to be related, due to the existence of smaller-scale heavy rainfall elements. These events fall under the *convective* precipitation class, which consists of rainfall produced by small-scale (wavelengths on the order 100km and smaller), convectively-driven atmospheric circulations. Stratiform events fall under the general *non-convective* precipitation heading, where rainfall is produced by upward vertical motion resulting from large-scale (wavelengths on the order 1000km and larger) forcing mechanisms. Note that the linear and cellular classes are sub-classes of the parent convective class, while the stratiform class is a sub-class of the non-convective class in the overall classification hierarchy.

In order to obtain summary information on the spatial organization of rainfall, statistical measures that are a function of the *location* as well as the amount of rainfall are required. Such measures can be found in the field of geostatistics. There is a long history of research using geostatistical tools to examine the characteristics of spatial radar/rainfall data. For example, Kessler and Russo (1963) noted how the ellipticity of the auto-correlation was an objective measure of the "systematic bandedness in the pattern." By using features of the auto-correlation as attributes in this work, the classification system is able to separate the linear and cellular events. Therefore, we conclude that a useful set of attributes have been obtained, and these results are used to develop an automated classification system.

An automated classification system is developed using the results from previous classification experiments with the training data set. A large data set covering a full year is classified by the automated system. In order to validate the automated classification system, a random, representative sample of rainfall events from 2002 is taken. Results from the automated classification procedure agreed with an expert classification for 85% of the validation sample, providing independent confirmation of the accuracy of this classification system.

SPATIAL CHARACTERISTICS

In BL02, results from an automated classification using attributes related to the overall distribution of rainfall across an object were presented. These results showed that useful information on the intensity of rainfall within each object was obtained through the use of these attributes. However, it was determined that these attributes do not provide information on the spatial continuity and variability of the rainfall within an object. In order to provide information on aspects of the spatial continuity and variability within rainfall objects, additional attributes related to the shape and structure of the spatial patterns are required. There are several measures of spatial variability and continuity to choose from (Isaaks and Srivastava 1989; Deutsch and Journel 1988), for this work two-dimensional plots of the *correlogram* were used. The correlogram $\rho(h)$ is also known as the auto-correlation, which is the correlation between all possible pairs of values separated by vector h.

A series of synthetic rainfall patterns and associated correlograms were generated and analyzed in order to determine the sort of information that could be obtained from the correlogram. Rainfall patterns of various sizes, shapes, and numbers of objects were inspected (not shown). Overall, these results demonstrated the potential of using several summary measures from the correlogram to characterize the rainfall field. The ellipticity and areal coverage of various correlation contours from the central maxima (connected to the origin) seem to represent the degree of organization in the original field. These aspects of the correlogram will be estimated by fitting an ellipse to various correlation contours, by measuring the lengths of approximate semi-major and semi-minor axes. The *ellipticity* of the correlation contour will be approximated by computing the ratio between the semi-major and semi-minor axes. The product between them will be used as an approximation to the *area* of an ellipse.

ANALYSIS OF TRAINING DATA SET

Each case of a training data set containing 48 rainfall objects (see BL02 for a complete description of the training data set) was classified by an expert into three main event classes; *linear, cellular,* and *stratiform*. The expert classification of the 48 cases was based entirely upon the rainfall pattern, no other information (such as meteorological conditions, location, time of year, time of day) associated with the events was provided to the expert analyst. Table 1 shows the distribution of the expert classification across the three main classes, note that the events were not uniformly distributed as the majority of the events were convective in nature. For each event in the training data set, a correlogram was computed using GSLIB, a freely available library of software packages for geostatistics developed at Stanford University (Deutsch and Journel 1988). The area and ellipticity of various contour values (0.2, 0.4, 0.6, and 0.8) in each correlogram from the training data set were estimated. The ratio of the semi-major and semi-minor axes will be used as a summary measure of the ellipticity or eccentricity of the ellipse. For a circle, this ratio will be equal to 1.0, the ratio will increase as the ellipticity of the contour increases. The product of the two axis lengths will also be used as a summary measure approximating the area of the ellipse.

Table 1: Hierarchy of rainfall classes (case numbers).

Convective	Convective	Non-convective
Linear (1-18)	Cellular (19-37)	Stratiform (38-48)

To determine the usefulness of these features in an automated classification system, the attributes corresponding to the 0.6 correlation contour were selected for thorough analysis. As in the analysis described in BL02, the classification system uses a hierarchical cluster analysis algorithm, specifically Ward's method (Ward 1963). Validation of the automated classification will be performed in the same way as in BL02. The hierarchical cluster analysis results must be analyzed by hand to determine a number of clusters. In this work, the dendrograms were cut into four or five clusters containing the majority of objects in the training data set. Outliers were not classified, and no more than six outlier cases were allowed. Every attempt was made to cut each dendrogram at a point where there was substantial separation between the intra-cluster and inter-cluster variance. Each major cluster is considered a separate class of objects. The definition of each class will be determined by the highest percentage of expertly classified cases detected for that particular cluster. The percentage of objects that are correctly classified by their membership in the dominant expert class for each cluster will be used as the metric for determining the skill of the classification experiments.

Now that the set of attributes and the classification algorithm have been selected, the question now becomes whether the attributes require standardization and what combination of these attributes will produce the best classification results. The question of whether to transform the attributes prior to the cluster analysis was investigated by testing the raw attributes, normalizing each attribute vector to produce zero mean and unit variance, and standardizing each attribute by dividing by its maximum value. For each of these transformations, different combinations of subsets of the four attributes (gamma distribution shape and scale parameters, correlogram contour area and eccentricity) were used. The results of these 33 experiments involving different combinations of attributes and standardization (not shown) demonstrated that correlogram contour eccentricity and gamma scale parameter were the attributes with the most discriminating power. The best results were obtained when these were used by themselves. When additional attributes were added, results degraded slightly. When only one of these were used in combination with other attributes, results were also degraded. Standardization of these attributes had little impact upon the classification results, since the ranges of these attributes are quite similar (typically 0-10). Therefore, the gamma scale parameter and eccentricity from all four correlation contours (0.2, 0.4, 0.6., and 0.8) were tested for their effectiveness.

Figure 1 shows the cluster analysis results using these five attributes. A cut has been made on the dendrogram separating the objects into five main clusters with six outliers. The first cluster is unanimously populated with all eleven stratiform events. The second cluster is dominated by nine cellular cases, with one linear case included. The third cluster is also cellular dominant, with six cases. The fourth cluster contains five linear cases, and the fifth cluster is split between seven linear cases and three cellular events. Validating these clusters, in the two-class case there are no incorrectly classified events, resulting in 100% correct classification. In the three-class case the clusters correctly placed 38 of 42 cases into the dominant class, for a 90.5% percent correct. The automated classification using these five attributes have successfully separated the cellular, linear, and stratiform events with over a 90% accuracy rate. This level of success demonstrates that useful attributes for an automated rainfall pattern classification system have been discovered.

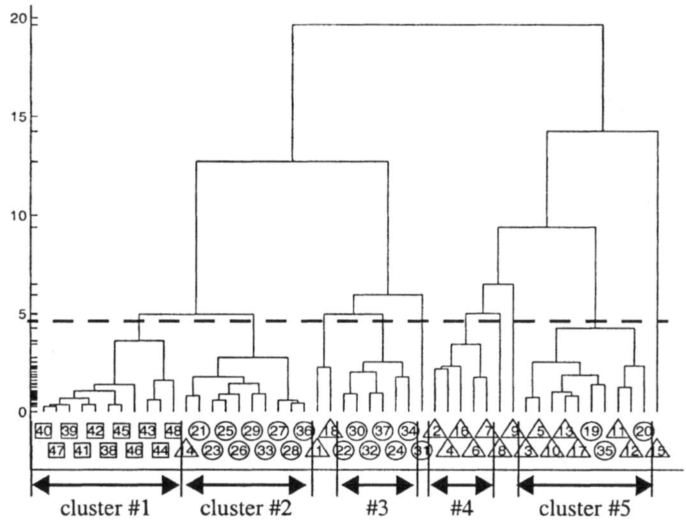

Figure 1: Dendrogram produced by Ward's method using gamma scale parameter and correlogram contour eccentricity from 0.2, 0.4, 0.6, and 0.8 contours as attributes. Cases of the training data set are symbolized by their expert classification, squares indicate stratiform, circles indicate cellular, and triangles indicate linear events. The cut level is represented by the thick dashed line, the resulting five clusters are idenitified at the bottom.

AUTOMATED RAINFALL OBJECT CLASSIFICATION

Rainfall objects in the training data set were located and identified by hand. An automated system for rainfall object identification is required. Basically, contiguous areas of rainfall are considered *systems* or separate objects for identification. A simple threshold (=0.05mm) was used to convert the rainfall image into a binary image. A connected component labeling algorithm (Klette and Zamperoni 1996) is applied to this binary image to locate individual objects within the full image and identify them with a separate label. This algorithm labels pixels that are connected to other pixels with the same label. Since the atmosphere is diffusive, it is physically sensible to expect a significant fraction of locations receiving non-zero precipitation over a region will receive less than a measurable amount of rain (also known as a *trace* amount). The size of the area receiving trace amounts of precipitation was specified to be 15% of the total area receiving detectable precipitation. The areal extent of each object was increased by a integer number of pixels in each direction such that the object's area was increased by as close to 15% as possible. In addition, it is not unusual to find small gaps between nearby regions of measurable rain. Therefore, the definition of *connected* pixels was expanded so that pixels that were within 5 pixels (~20km) of one another were considered connected. The rainfall objects are then analyzed so that statistical features described previously can be extracted.

Now that rainfall systems can be identified and features can be extracted in an automated fashion, these systems can be classified. The automated classification system will place objects into one of five classes, defined by the classification results of the training data set previously described. Each of these classes will be defined by the mean of the attribute vectors for the members of the five main clusters shown in figure 1. Similar to partitional/K-means clustering, the automated classification system will proceed as follows. The raw values of the five attributes for a given object will be compared to the five cluster means. The Euclidean distances between the object and each of the five cluster means will be computed. The object will be placed into the class represented by the nearest cluster mean in terms of the smallest Euclidean distance, in other words, the nearest-neighbor cluster.

VALIDATION

This automated classification system has been built upon the cases found in the training data set. A large data set covering a full year (2002) was analyzed using the automated rainfall object identification and analysis system. In order to validate the automated classification system, an independent random sample of 100 rainfall events from 2002 was taken. The distribution of attributes from objects belonging to the random sample was compared with that from the 2002 data to confirm that the validation data set was representative of the population. The sample was classified both by an expert and objectively via the automated classification system. Comparison of these results provides an independent confirmation of the accuracy of this classification system. In comparing the five classes of the automated system (figure 1) to the three classes of the expert classification, class 1 in the automated system is equivalent to the stratiform class, classes 2 and 3 are equivalent to cellular, and classes 4 and 5 are equivalent to linear. Overall, 89% of the objects were correctly classified into the convective/non-convective classes (2-class case), and 85% of the objects were correctly classified in the 3-class case (stratiform, linear, cellular).

The addition of spatial characteristics to the intensity-related attributes has successfully refined the rainfall system classification procedure, allowing for discrimination between linear and cellular events. Future work involves further refinement of the classification procedure, and application of this system to rainfall patterns forecasted by numerical weather prediction systems in order to provide more meaningful forecast verification information.

REFERENCES

Baldwin, M. E., and S. Lakshmivarahan, 2002: Rainfall classification using histogram analysis: An example of data mining in meteorology. *Intelligent Engineering Systems Through Artificial Neural Networks*, Volume 12, C.H. Dagli, A.L. Buczak, J. Ghosh, M.J. Embrechts, O. Ersoy, S.W. Kercel, Eds., ASME Press, 429-434.

Deutsch, C. V. and A. G. Journel, 1998: *GSLIB: Geostatistical software library and user's guide. Second edition.* Oxford University Press, 369pp.

Isaaks, E. H., and R. M. Srivastava, 1989: *An Introduction to Applied Geostatistics.* Oxford University Press, 561pp.

Kessler, E., III, and J. A. Russo Jr., 1963: Statistical properties of weather radar echoes. *Preprints, Proc. of 10th Weather Radar Conf.* Washington, DC, Amer. Meteor. Soc., 25-33.

Klette R. and P. Zamperoni, 1996: *Handbook of image processing operators.* John Wiley and Sons, 397pp.

Ward, J. H., 1963: Hierarchical grouping to optimize an objective function. *J. Amer. Stat. Assoc.*, **58**, 236-244.

NEURAL NETWORK AND STATISTICS-BASED SYSTEMS FOR PREDICTIONS IN COASTAL STUDIES

Alexey L. Sadovski
Philippe Tissot
Carl W. Steidley
Gary Jeffress

Department of Computing and Mathematical Sciences
Department of Physical and Life Sciences
Texas A&M University - Corpus Christi
6300 Ocean Dr. Corpus Christi, Texas 78412,
USA

ABSTRACT

This paper deals with developing and comparing tools, such as neural networks and a combination of regression and harmonic analysis, for the prediction of water levels in estuaries and shallow waters of the Gulf of Mexico. Harmonic analysis, as expressed in the form of published tide charts cannot provide reliable predictions in these waters. In the future these new methods of prediction will be applied to different regions with estuaries and shallow waters.

INTRODUCTION

The goal of our on-going research is to develop effective and reliable tools for predicting water levels in the shallow waters of the Gulf of Mexico. Here we will discuss two methodologies for the prediction of water levels: statistical models of multivariate regression integrated with harmonic analysis and neural networks [1]. Both of these approaches are based on the real-time data collected by the Texas Coastal Ocean Observation Network (TCOON). These methods are under development at the Division of Nearshore Research in cooperation with the Department of Computing and Mathematical Sciences both of Texas A&M University-Corpus Christi.

TCOON, developed and managed by the Division of Nearshore Research of the Center for Coastal Studies at Texas A&M University - Corpus Christi, consists of approximately 50 data gathering stations located along the Texas Gulf coast from the Louisiana to Mexico borders. Data sampled at these stations include: precise water levels, wind speed and direction, atmospheric and water temperatures, barometric pressure, and water currents. The measurements collected at these stations are often used in legal proceedings such as littoral boundary determinations; therefore data are collected according to National Ocean Service standards. Some stations of TCOON collect parameters such as turbidity, salinity, and other water quality parameters. All data are transmitted back to A&M-CC at multiples of six minutes via line-of-sight packet radio, cellular phone, or GOES satellite, where they are then processed and stored in a real-time, web-enabled database. TCOON has been in operation since 1988.

Coastal users will be interested in the model which best predicts water levels for their particular field(s) of interest but different users may have different priorities. Coastal populations and emergency management entities, worried about flooding, will favor models with good performance when the water levels are considerably higher than normal. Port authorities and other interests regulating navigation will be mostly interested in models performing well when the water levels are considerably lower than normal to avoid ship grounding. In this paper we are presenting a brief summary of our models and a comparison of their predictions.

STATISTICAL MODEL

Tide charts, based on harmonic analysis, are generally the method of choice for the forecast of water levels. However there are limitations to their use. Tide charts are mostly based on astronomical forcing or the influence on water levels of the respective motions of the earth, the moon, and the sun. There are locations around the world, including the Gulf of Mexico, where other factors such meteorological forcing often dominate tides [2] and limit significantly the application of tide charts. In such cases other models must be developed to forecast water levels.

We have considered three different models for next hour predictions, and two of these produced quite reliable predictions. The first of these models is a multi-regression model in which the hourly prediction is based on the levels of water, speed and direction of wind for the previous 48 hours with a step of 2 hours. This model did not produce expected results. The coefficient of correlation for these predictions was less than 0.5.

The second approach is also a multi-regression model in which two-hour predictions of water level are based on the levels of water during the previous 48 hours, using 2-hour steps. Here we now believe that information about weather (pressure, wind, temperature, etc.), used in the model previously described, is hidden in the levels of water. This model worked remarkably well: R squared for all stations was greater than 0.95. To make further predictions we used the previously determined levels of water. Such a step by step approach produced quite good predictions. Table 1 below presents statistical data for differences between predicted and real levels of water for 6, 12, 18, 24, 30, 36, 42, and 48 hours:

Table 1. Statistical characteristics of prediction errors (in meters)

	Mean	Median	Std. Dev.	Min. range	Max. range
Error 6hr	0.0124	0.0121	0.310	-0.858	0.796
Error 12hr	0.0129	0.0117	0.105	-0.421	0.442
Error 18hr	0.0155	0.0108	0.313	-0.951	0.866
Error 24hr	0.00924	0.0023	0.177	-0.580	0.622
Error 30hr	0.0176	0.0062	0.297	-0.748	0.803
Error 36hr	0.0140	0.0198	0.184	-0.653	0.641
Error 42hr	0.0156	- 0.0034	0.293	-0.746	0.828
Error 48hr	0.0265	0.0289	0.193	-0.568	0.593

The third approach is based on linear multi-regression of the levels of water, first differences, and second differences for such levels for the previous 48 hours with the step equal to two hours. This approach produces the same quality of water level prediction as the second approach, i.e. $R^2 > 0.95$. These results are quite understandable, since in both cases we have to deal with linear combinations of previous water levels. The difference in these two models is that the following: third approach has between four (4) and eight (8) significant variables in a linear regression while in the second model of linear regression we use all twenty four (24) variables which are water levels for the previous 48 hours.

A statistical model is also useful as a means to fill gaps, due to equipment failure, in the observed water level data. To fill gaps in tidal data, we used the following procedure. First, we found backward and forward linear regressions for the predicted water levels, and then evaluated lost data as a linear combination of forward and backward predictions with weights proportional to the distances from the edges of the gap.

INTEGRATION OF REGRESSION AND HARMONIC ANALYSIS

To determine why our regression models that do not include wind and atmospheric pressure data provide us with a much better prediction than the models that include such data, we performed a Factor Analysis. The analysis of the major components has shown that 5 factors explain 95% of variance for water levels. In deep water the first three components are periodical. In the shallow water the major component is not periodical, while the other components are periodical. Our conclusion is that the prime factor is weather. It is well known that the weather affects tides much more in shallow waters than in deep waters [1], [2]. Linear regression models for different locations have different coefficients for the same variables. This difference may be explained by the geography of the location where the data are collected.

There is the method to improve the model by integrating the regression approach with harmonic analysis. Namely, let us denote

$$x_n = w_n - h_n$$

x_n is the difference between water level w_n and harmonic water level h_n at the moment n

Then we can apply a technique, which is similar to what we have just done earlier in the paper. That is, we can predict next hour difference between water level and harmonic level

$$x_1 = a_0 x_0 + a_{-1} x_{-1} + \dots + a_{-n} x_{-n}$$

and step by step

$$x_k = a_0 x_{k-1} + a_{-1} x_{k-2} + \dots + a_{-n} x_{k-n}$$

Now we can find prediction for the water levels as follows

$$pw_t = h_t + x_t$$

ANN MODELING AND PREDICTIONS

The Artificial Neural Network (ANN) modeling approach is also based in forecasting future water level differences based on past water level differences. Other inputs have also been tested for the ANN models. For example, past wind squared is included in the model discussed below; as it has been recognized that wind forcing is well correlated with water anomalies. Other inputs, such as barometric pressure, have been tested but models which included: past water level differences, past wind measurements and wind forecasts have been shown to be optimal [3]. It has also been shown that simple neural networks with one hidden layer and one output layer have the best performance [2], [3]. With one input neuron with a tansig function and one output neuron with a purelin function and a number of inputs ranging from 10 to 30 an ANN forecasting a water level n hours past the time of forecast can be expressed as follows:

$$x(t_o + n) = a + \left(\frac{2b}{\left(1 + e^{-\left(c + \Sigma d_i y_i\right)}\right)} \right) - 1$$

In the expression above, the additive parameters (a, c) are identified as the model biases and the multiplicative parameters (b,d_i) are referred to as the model weights, and y_i is the input to the model. The exponential terms in the in the ANN model provide a non-linear modeling capability. The training of ANN models is different in nature as compared to the previous methods. There is typically no demonstrated method to identify a global optimum. The goal of the training process is therefore to find a suitable local optimum. To identify a good local optimum ANNs are trained over past data sets starting with a random guess of the model parameters and using the repeated comparison between the output of an ANN and an associated set of target vectors to optimize the weights of the neurons and biases of the model. All the ANNs discussed in this work were trained using the Levenberg-Marquardt backpropagation algorithm and implemented within version 4.0 of the Matlab Neural Network Toolbox and the MATLAB 6.0 Release 12 computational environment [5] running on a Pentium PC.

The performance of the ANN for the prediction of water levels was tested at the Bob Hall Pier, Texas, TCOON station. The model was trained and tested using three data sets composed of 3600 hourly measurements of water levels, wind speeds and wind directions. The data sets covered the Spring seasons of 1998, 2000, and 2001 from Julian day 21 to Julian day 182. The model was successively trained on each data set and applied to the other two data sets. This procedure provided a set of six time series of predicted water levels to be used for validation. For each time series the average absolute error between predicted and measured water levels was computed. Averages and standard deviations were then computed for the results of the six validation time series

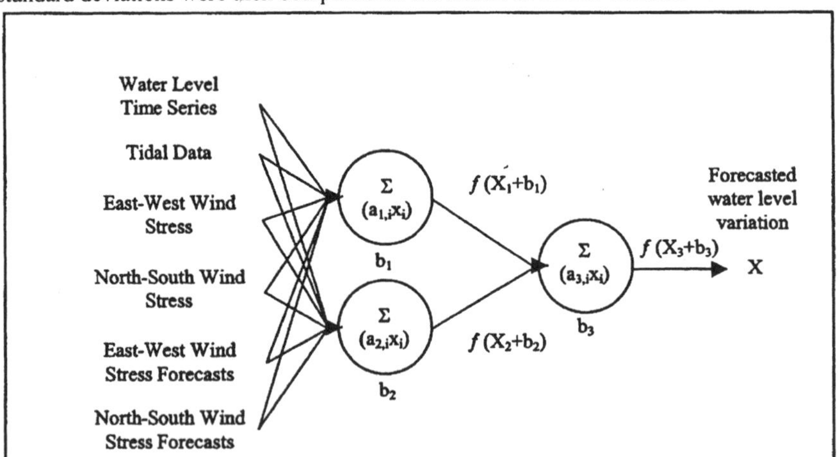

Figure1. Schematic of the type of neural network applied to the problem of water level

for these two parameters. The standard deviation gives an overall measure of the variability due to the differences between training sets as well as the differences resulting from the training process. The inputs to the model were selected as the previous12 hourly water level and wind measurements based on experience gathered during the modeling for other locations [4]. One model was trained without wind predictions while in the second case wind measurements were used to simulate wind forecasts. These wind forecasts consisted of future wind measurements at 3 hour intervals up to 36 hours. A database of wind forecasts is presently being constructed and models based on wind forecasts are expected to be more representative of future model performance. Figure 2 displays a comparison between a 36-hour water level hindcast, the tide tables, and TCOON measurements. As the figure indicates, the ANN model captures a large fraction of the water anomaly and improves significantly on the tide tables. The performance of the models with and without wind forecasts is compared with the performance of the tide tables for forecasting times ranging from 6 to 36 hours. Both ANN models improve significantly on the tide tables for forecasting times up to 24 hours. Improvements for 30-hour and 36-hour predictions are still measurable. The addition of wind forecasts improves the model performance although not significantly as compared to the improvement over the tide tables.

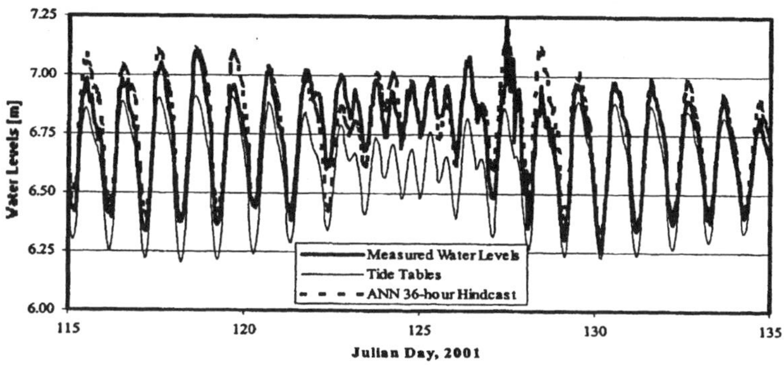

Fig. 2. Comparison of a 36-hour ANN forecast, tide table readings and TCOON measurements of water levels for the Bob Hall Pier Station located near Corpus Christi, Texas during the Spring of 2001.

CONCLUSIONS

The methods of tidal water level predictions described in this paper consistently outperform harmonic analysis which is expressed in the form of tide charts as could be seen in the Table 2 and Figure 3.

The next step of this on-going research [6] is the selection of expert scientists and representatives of port authorities with the goal of obtaining their real evaluation of the relative importance of different measurements. This will allow us to use multi-criteria decision-making to develop an aggregate goodness criteria for evaluation of the quality of water level predictions. Such goodness criteria may be used for determining the quality of the methods we are using to fill gaps of lost data in the data collected by Texas Coastal Ocean Observation Network (TCOON) [2].

Table 2. Comparison of 24 and 48 hours predictions by different methods

Rockport (015)

Model	RMSE	CF	POF	NOF	MDPO	MDNO
Harmonic	0.11+/- 0.02	85.02+/- 4.12	0.21+/- 0.19	1.90+/- 2.51	16+/-16	73+/-81
Pers 24hr	0.069+/- 0.006	95.75+/- 1.19	0.249+/- 0.231	0.023+/- 0.029	14+/- 19.17	0.6+/- 1.342
LR 24hr	0.106	97.18	0.261	0.027	9	1
NN-1 24hr	0.0588+/- 0.0085	97.848+/- 1.515	0.132+/- 0.114	0.104+/- 0.232	8.5+/- 8.7	7.6+/- 18.5
NN-2 24hr	0.053+/- 0.0079	98.563+/- 1.284	0.124+/- 0.115	0.08+/- 0.204	8.4+/- 8.7	6.2+/- 17.1
Pers 48hr	0.101+/- 0.009	87.18+/- 2.22	0.785+/- 0.528	0.424+/- 0.255	25.4+/- 17.813	13.8+/- 10.232
LR 48hr	0.122	91.05	0.466	0.409	16	19
NN-1 48hr	0.0889+/- 0.0123	91.396+/- 2.768	0.199+/- 0.158	0.57+/- 0.937	9.6+/- 7.6	26.4+/- 37.7
NN-2 48hr	0.0779+/- 0.0108	94.500+/- 2.616	0.123+/- 0.162	0.299+/- 0.575	6.8+/- 9.6	16.3+/- 30.7

ACKNOWLEDGEMENTS

The work presented in this paper is funded in part by the following federal and state agencies:

- National Aeronautic and Space Agency (NASA Grant #NCC5-517)
- Texas General Land Office
- National Oceanic and Atmospheric Administration (NOAA)
- Coastal Management Program (CMP).

The views expressed herein are those of the authors and do not necessarily reflect the views of NASA, TGLO, NOAA, CMP or any of their sub-agencies.

References.

[1]. Sadovski, A. L., P.Tissot, P. Michaud, C. Steidley, "Statistical and Neural Network Modeling and Predictions of Tides in the Shallow Waters of the Gulf of Mexico", WSEAS Transactions on Systems", Issue 2, vol. 2, WSEAS Press, pp.301-307.

[2] Tissot P.E., Cox D.T., Michaud P. 2002. "Neural Network Forecasting of Storm Surges along the Gulf of Mexico". Proceedings of the Fourth International Symposium on Ocean Wave Measurement and Analysis (Waves `01), ASCE, 1535-1544.

[3] P.E. Tissot, D.T. Cox, and P.R. Michaud, "Optimization and Performance of a Neural Network Model Forecasting Water Levels for the Corpus Christi, Texas, Estuary", 3rd Conference on the Applications of Artificial Intelligence to Environmental Science, Long Beach, California, February 2003.

[4] Campolo, M., Andreussi, P., and Soldati, A. 1997. "River Flood Forecasting with a Neural Network Model". *Water Resources Research*, 35 (4), 1191-1197.

[5] The MathWorks, Inc. 1998. Neural Network Toolbox for use with Matlab 5.3/version 3, The MathWorks, Natick, MA.

[6] Sadovski, A. L., C. Steidley, P. Tissot, P. Michaud, "Developing a Goodness Criteria for Tide Predictions Based on Fuzzy Preference Ranking", Developments in Applied Artificial Intelligence, pp. 391-401, Lecture Notes in Artificial Intelligence 2718, Springer, June 2003.

Lidar Detection of Underwater Objects using Neural Networks with Linear Prediction and Fourier Transform for Feature Extraction

Vikramjit Mitra	Chia-Jiu Wang	Satarupa Banerjee
University of Denver	University of Colorado	Visveswaraiah Tech. University
Dept. of Engineering	ECE Department	CS Department
Denver, CO	Colorado Springs, CO	Bangalore, India
:vmitra@du.edu	cwang@eas.uccs.edu	satarupa_b@rediffmail.com

Abstract: Light Detection And Ranging (Lidar) data is currently in extensive use for remote sensing and target detection. In this paper we present a neural network for detection of fish from Lidar data. The Lidar data is gathered by storing a sequence of laser backscatter amplitudes, obtained from laser shots at various heights above the earth surface. The data is preprocessed to eliminate noise and a window of the data is selected for further processing. We have used Linear Prediction Coding (LPC) and fast Fourier transform (FFT) for feature extraction and a multi-layer perceptron (MLP) to distinguish among different classes by feature detection, where classes include fish, water-bottom and water-layer. In this paper we have proposed a neural network for detection of fish from Lidar data and the results show that the detection technique is effective and can perform the required classification with high degree of accuracy.

1. Introduction

Lidar is a remote sensing system that transmits and receives laser light at a high frequency and operates in the ultraviolet, visible and infrared region [1,2]. The Lidar instrument transmits light to a target and some of this light is reflected and scattered back to the instrument where it is analyzed. The change in the properties of the light helps in determining some properties of the target. Lidar technique is well known for its high efficiency in classification of remote objects [1,2], high area coverage coupled with high spatial data density and rapid response [3]. Airborne Lidar has high measurement accuracy compared to conventional acoustic methods [4]. Neural network is generally accepted due to its high efficiency in classification of data, parallel computation potentials, high adaptability and great error tolerance [5,6].

In this paper, we present the results of using multi-layer perceptron (MLP) for classification of under water fish, water layer and waterbed. The differences in the Lidar backscatter pattern for different cases help in distinguishing them by analyzing the reflected pattern. The features extracted from the backscatter are used to train the network and these features are used to distinguish among different cases. Our previous work [7] shows that the neural network is an effective tool for fish detection from Lidar data, provided that there is an efficient feature extraction process. In this paper we have enhanced the feature extraction technique, in order to increase the desired accuracy.

1.1 Lidar

Lidar systems are used to collect topographic data. This data is collected by a Lidar system mounted on an aircraft. The Lidar system is capable of recording elevation measurements at a rate of 2,000 to 5,000 pulses per second with a vertical precision of 15 centimeters (6 inches). During the flight, the Lidar sensor pulses a narrow, high frequency laser beam toward the earth through a port opening in the bottom of the aircraft's fuselage. The Lidar sensor records the time difference between the emission of the laser beam and the return of the reflected laser signal to the aircraft. The Lidar operates on the principle that water depth can be calculated from the time difference between laser returns reflected from the sea surface and seabed. Generally, an infrared channel (wavelength = 1064 nm) is used for surface detection, while bottom detection done by a blue-green channel (wavelength = 532 nm) [3]. Figure 1.1(a) shows the characteristc plot of the intensity of backscatter versus time for one laser shot. After being shot from an aircraft, the laser beam has to travel through the air, where due to lack of reflecting particles the amplitude of reflection is minimun. But some reflections might occur due to the presence of fog or cloud in air. As the leaser beam moves towards the earth surface it first hits the air-water interface. Due to change in density of the medium, a huge amount of backscatter occurs, resulting in a peak in the backscatter plot as shown in Figure 1.1(a). As the laser beam penetrates deeper, three cases i.e., presence of water layer only, presence of under water object and presence of water-bottom may occur.

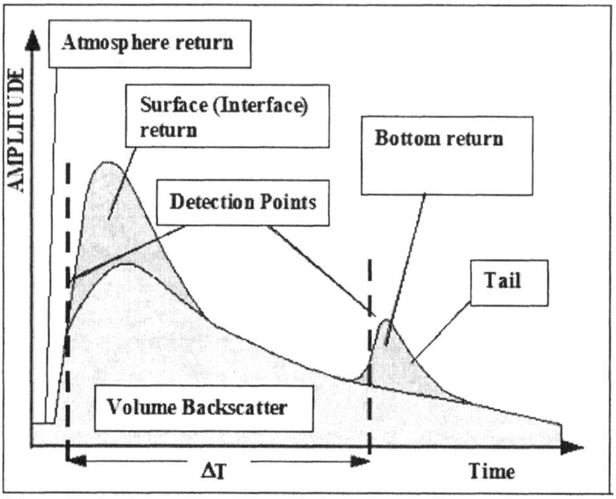

Figure 1.1(a). Back scatter of Lidar [4]

The Lidar backscatter characteristics in case of water layer are shown in Fig. 1.1(b). The peak represents contact of the laser beam with the air-water interface that results in a huge backscatter due to change in medium. The reflection dies steadily as the laser beam penetrates deeper into water. Some stray reflections may occur as the laser beam penetrates through the water, as shown in Fig. 1.1(c), implying that there are some objects underneath.

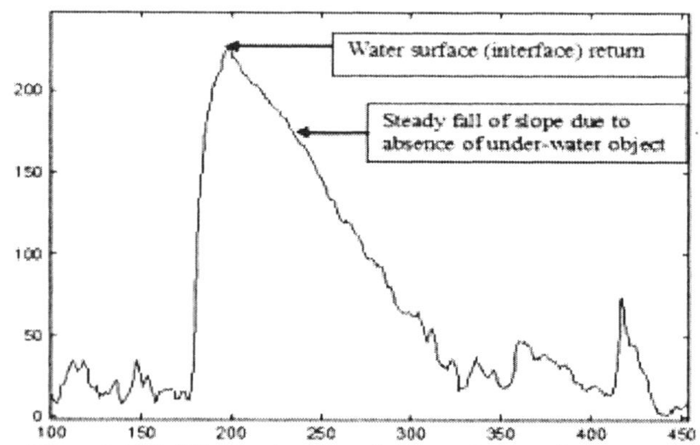

Figure 1.1(b) Lidar back scatter for water layer only
(Y axis: back scatter amplitude; X-axis: time in millisecond)

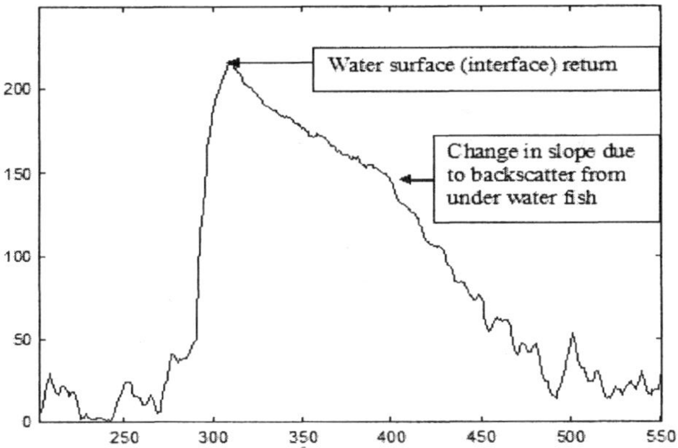

Figure 1.1(c) Lidar back scatter for water with under water fish (Y axis: back scatter amplitude; X-axis: time in millisecond)

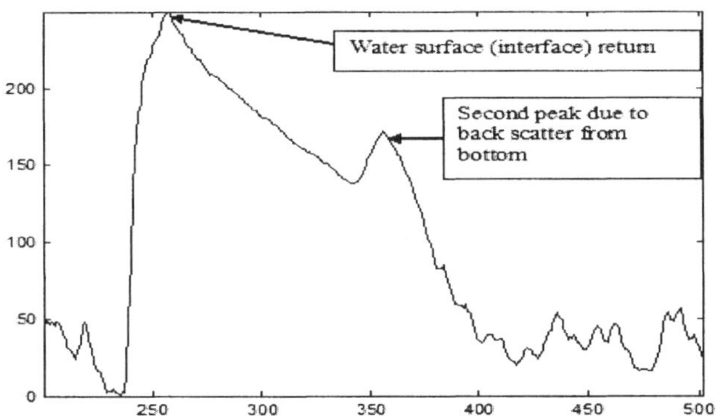

Figure 1.1(d) Lidar back scatter for bottom close to water surface (Y axis: back scatter amplitude; X-axis: time in millisecond)

2. Data Analysis

In this section we will analyze some samples and explain in detail the feature extraction process. Before the feature extraction process, the data is pre-processed to remove noise and then a window of the data is selected to eliminate all unwanted information. The data window is then used to extract features which will help in the classification process. We have used linear prediction coding [8] and fast Fourier transform techniques [9,10] to analyze the Lidar data sets. Linear Predictive Coding (LPC) [8] is a signal encoding technique, primarily used for speech signal encoding. LPC breaks a signal into specific characteristics and encodes only these characteristics instead of the whole sample. It is based on the fact that every signal has unique attributes. By breaking each sample into these components to generate prediction coefficients, LPC uses fewer bits to transfer the sample. Thus a significant amount of signal compression is achieved by building a predictive model of the waveform [8]. The predictive coefficients obtained from LPC coding are a linear combination of all the previous samples *S(n)*, as represented in the following equation:

$$S_{LPC}(n) = \sum_{i=1}^{p} a_i S(n-i) \qquad (1)$$

where a_i are the coefficients of the predictor. The coefficients can be found in many ways. The commonly used technique is the Least Square Error (LSE) principles where the error is minimized.

3. Feature Extraction

A study of the backscatter characteristics for the three cases: water-layer, fish and water-bottom, shows that the slope of the backscatter plot varies for the three mentioned categories. In case of water layer we have only one peak depicting the air-water interface backscatter and then the characteristic curve falls sharply with almost a constant slope. In case of bottom, the backscatter-curve shows two distinct peaks. The major peak depicts the air-water interface and the minor peak represents the water-bottom interface. Lastly for fish or presence of under-water object, the backscatter curve is almost similar to that of water layer. However, the curve does not fall sharply with a constant slope after the first peak instead the slope gets reduced at some points, implying the presence of some stray backscatters from objects below the water surface. These changes in slope of the characteristic curves resemble the change in frequency of the laser backscatter. Thus significant information about the slope changes is present within the spectral content of these laser backscatters.

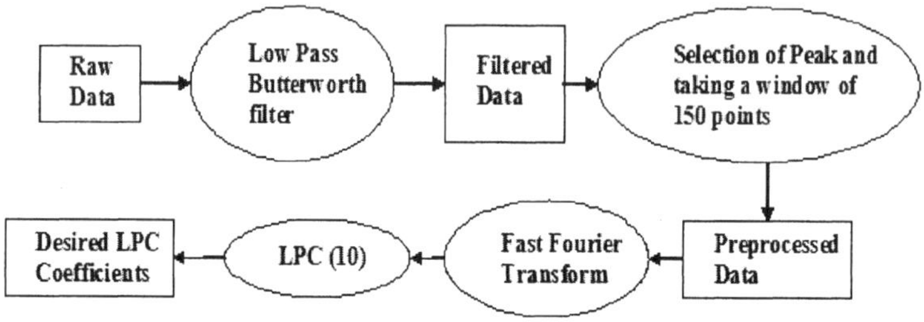

Figure 2. Block diagram of feature extraction process

Figure 2 presents the data analysis flow to extract features from a raw data set. After taking the FFT, the spectral data is broken into LPC coefficients. Thus the specific characteristics of the FFT data are now encoded as LPC coefficients. The number of coefficients determines the number of input nodes to be selected for the neural network. The Lidar data is preprocessed before obtaining the LPC coefficients to feed the input nodes of the neural network. The data is preprocessed by a low-pass butter-worth filter to filter off the high frequency signal components. These high frequency components add noise to the signal and hence hinder the process of feature extraction. After filtering the data, we select the peak value of backscattered signal, presenting the air-water interface point and take a window of 150 points after it.

4. Artificial Neural Networks

In our work, a multi-layer perceptron (MLP) is used with a rule-based decision logic module as shown in Figure 3. Figure 3 depicts a MLP with input nodes containing the LPC coefficients and output node leading to a decision logic module. Of the 10 LPC coefficients, one coefficient with a value of zero is ignored in training. The MLP has 9-input nodes, corresponding to the 9 nonzero LPC coefficients and two cases are considered for the output nodes: 1-output node and 2-output node, with sigmoid as the activation function. The desired output values corresponding to each of the three classes are shown in Table1.

Desired output (1 output node)	0	1	0.5
Class	Fish	Water Layer	Water Bottom

Desired output (2 output nodes)	0, 0	1, 1	0, 1
Class	Fish	Water Layer	Water Bottom

Table 1 Desired outputs for 1 output node and 2 output nodes

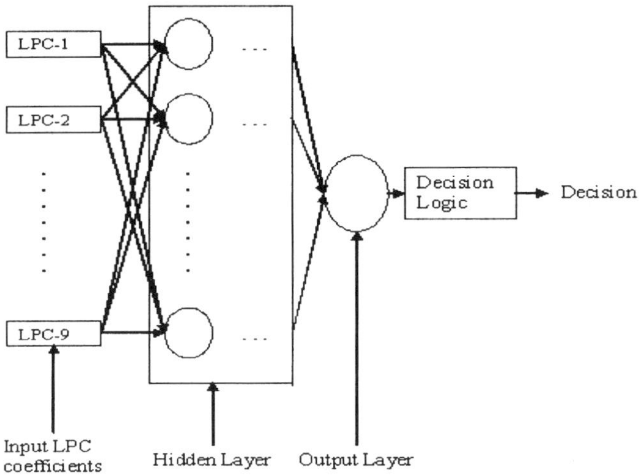

Figure3. An Artificial Neural Network with a decision module.

The actual outputs are close to these values but not exactly equal to them. To make a clear decision, we have used a rule-based decision logic module, which has a strong decision threshold that helps to make the final decision, as depicted in Table 2 and Table 3.

	Decision Boundary
Water Layer	0.75 < output value < 1
Water Bottom	0.25 < output value < 0.75
Fish	0.0 < output value < 0.25

Table 2 Decision boundaries for 1-output node

	Decision Boundary	
	output node 1	output node 2
Water Layer	0.5 < output value < 1.0	0.5 < output value < 1.0
Water Bottom	0.0 < output value < 0.5	0.5 < output value < 1.0
Fish	0.0 < output value < 0.5	0.0 < output value < 0.5

Table 3 Decision boundaries for 2-output node

5. Simulation results and Discussions

In this study, 9-input 1-output and 9-input 2-output neural networks with one to four hidden layers were considered. The results for all eight networks are presented below. We have used 210 samples to train the network (70 samples from each category) and 30 samples to test the network. Test results are presented in Table 4 and Table 5.

Output		
Desired	Actual	After decision logic
1.0	0.9956	1.0
1.0	0.8384	1.0
1.0	0.9855	1.0
0.5	0.4064	0.5
0.5	0.4988	0.5
0.5	0.4678	0.5
0.0	0.0001	0.0
0.0	0.0002	0.0
0.0	0.0001	0.0

Table 4. MLP with one hidden layer (one output node)

Desired		Actual		After decision logic	
1	1	0.9942	0.9929	1	1
1	1	0.9387	0.9996	1	1
1	1	0.9566	0.9949	1	1
0	1	0.0005	1.0000	0	1
0	1	0.0007	1.0000	0	1
0	1	0.0006	0.9989	0	1
0	0	0.0122	0.0108	0	0
0	0	0.0134	0.0121	0	0
0	0	0.0112	0.0109	0	0

Table 5. MLP with one hidden layer (two output nodes)

From the results above it is evident that the neural network works efficiently in detecting the required classes. From the data analysis, the LPC coefficients for all the three classes were found to be very close to each other without any strong decision boundary separating them. However the neural network worked efficiently with 100% accuracy. The final decision was inferred by using rule-based decision logic. A better approach may be to incorporate a fuzzy logic module to perform inference. As the output of the neural network is in form of numbers, closely resembling fuzzy membership functions, hence a better solution would be to implement a neural-fuzzy network. As the Lidar samples for fish and water-layer were very close to each other, hence extracting features to distinguish among them was a major challenge. The method presented in this paper meets this challenge.

Acknowledgement: The authors would like to acknowledge NOAA for providing the real experimental data.

References

[1] Dr. R. Sica, "Exploring the Atmosphere With Lidars". University of Western Ontario, Department of Physics and Astronomy, http://pcl.physics.uwo.ca/pclhtml/introLidar/introLidarf.html, © 1999 by Dr. R. J. Sica

[2] Dr. James E. Arnold, Dr. Michael J. Kavaya, NASA "LIDAR Tutorial", http://www.ghcc.msfc.nasa.gov/sparcle/sparcle_tutorial.html, 1999.

[3] Guenther, G.C., T..J. Eisler, J.L. Riley, and S.W. Perez, "Obstruction detection and data decimation for airborne laser hydrography," Proc. 1996 Canadian Hydro. Conf., June 3-5, Halifax, N.S.

[4] Geraint R. West, W. Jeff Lillycrop, "Feature Detection and Classification with Airborne Lidar- Practical Experience" http://shoals.sam.usace.army.mil/pdf/West_Lillycrop_99.pdf, Joint Airborne Lidar Bathymetry Technical Center of Expertise, AL.

[5] Principe, Euliano, Lefebvre "Neural and Adaptive Systems: Fundamentals through Simulations" © John Wiley & Sons, Inc. (2000).

[6] Rongrui Xiao, Robert Wilson and Richard Carande. "Neural Network Classification with IFSAR and Multispectral Data Fusion" 1998 IEEE transaction. http://www.vexcel.com/papers/nn_class.pdf, Vexcel Corp. Boulder, CO 80301.

[7] Vikramjit Mitra, Chia-Jiu Wang, George Edwards, "Neural Network for LIDAR detection of fish". Proceedings of International Joint Conference of Neural Networks (IJCNN), pp. 1001-1006, 2003.

[8] N. S. Jayant, P. Noll, "Digital Coding Of Waveforms" Prentice-Hall Signal Processing Series. © Prentice-Hall, Inc., (1984).

[9] Elliot, D.F. and Rao, "Fast Fourier Transforms: Algorithms" © K.R. Academic Press, New York, (1982).

[10] Oppenheim, A. and Shaffer, R. , "Discrete-Time Signal Processing", © Prentice-Hall, Inc., Englewood Cliffs, NJ, (1989).

INFLOW FORECASTING OF THE DAM BY THE NEURAL NETWORK

EIJI SENO
Ehime University, Matsuyama, JAPAN

MASANORI IZUMIDA
Ehime University, Matsuyama, JAPAN

KENJI MURAKAMI
Ehime University, Matsuyama, JAPAN

SUSUMU MATSUMOTO
Ehime University, Matsuyama, JAPAN

ABSTRACT

At the multipurpose dam, it is necessary to forecast inflow well. In this paper, we propose a method of inflow forecasting by neural network. The neural network system can forecast the inflow by learning inflow data and rain data inside of the dam basin. However, there is a problem in this system. As forecasting time becomes future, the error is increasing rapidly. We suppose one of the reasons that it has lacked important information of rain data. If weather moves to constant direction, this lacked information exists in outside of the dam basin. Therefore, we try to simulate using rain data outside of the dam basin in addition to data inside of the dam basin. Then, results of inflow forecasting show that our proposed method is effective.

INTRODUCTION

At the multipurpose dam, we must manage the water level to satisfy purposes such as flood control, hydraulic power plant, and so on. First, for the flood control, when a cloudburst come, we must lower the water level and store up the flood peak temporarily and discharge. The inflow forecasting becomes the basis of discharge gate control. Second, for the purpose of the hydraulic power plant, we must keep the water level high for efficient power supply, but we must avoid wasteful discharge when it rains. To satisfy these purposes, we need the precise inflow forecasting system. So, we try to forecast by the neural network with rain data of wide area.

INFLOW FORECASTING BY THE NEURAL NETWORK

There are papers of inflow forecasting using mathematical models. But there are two problems. One of them is generality. The mathematical model can not be applied to the other dam easily, because each dam differs in natural characteristics. Another problem is year change. If a good mathematical model is adapted to the dam, the year change occurs in the inflow by erosion of the coast and sand accumulation of the lake.

Recently, researchers try to forecast by neural network. The neural network can solve non-linear problems by learning input and output data. Therefore, about the generality problem, the neural network system can generalize easily by learning another dam data. About the year change problem, system can solve by learning new data. However, there is a new problem in the neural network system. As the forecasting time becomes future, the error is increasing rapidly.

We use data of Kanogawa Dam which was constructed at the midstream of Hijikawa River in Japan. There is Nomura Dam in upstream. Usually, Hijikawa River is gentle, but in case of cloud burst, it invite flood to downstream towns. The inflow changes in wide range, so inflow forecasting is difficult.

INFLOW FORECASTING USING WIDE AREA RAIN DATA

To find the reason why the error is increasing rapidly, we study about the elements which are related to the present inflow. Fig.1 shows correlations with the present inflow and related elements.

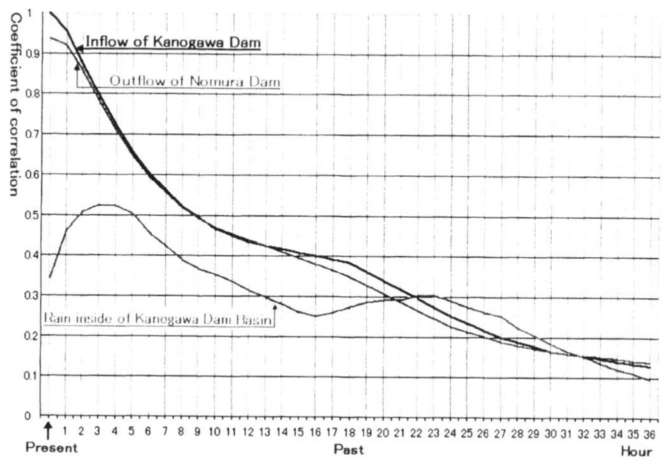

Fig.1. Correlations with the present inflow and related elements

Fig.1 shows as follows.

(1) The correlation of the inflow and outflow are very high near the present time, and decrease gently.
(2) The tendency of the inflow line and the outflow line is very close.
(3) The correlation of the rain, 2 hours later from 6 hours later is highest.
(4) The correlation of the rain has two peaks.

Fig.2 shows 1 hour later forecasting. We do not have the information about in 1 hour, shown by dotted line. The system can not use the highest correlative part of the inflow and outflow. But, a lot of information is still left in the data, system can forecast well.

Fig.3 shows 6 hours later forecasting. When the forecasting time becomes future, the mainly information decreases mainly. Especially, the rain data has lacked information in the part of highest correlation. We don't have data of high correlation, so, it is difficult to forecast the inflow of future. If we can add the lacked information in some way, forecasting ability will improve.

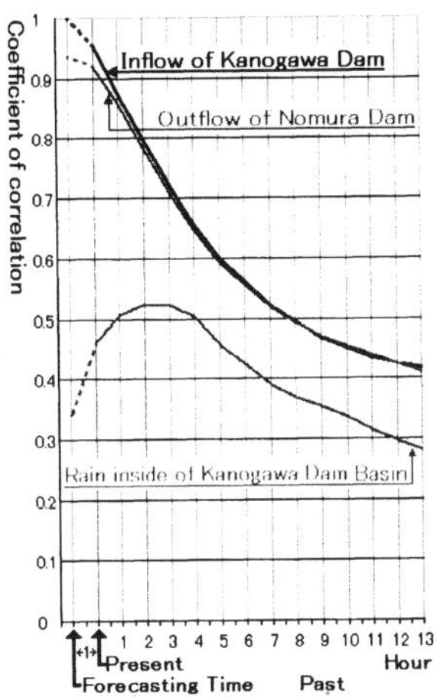

Fig.2. 1 hour later inflow forecasting

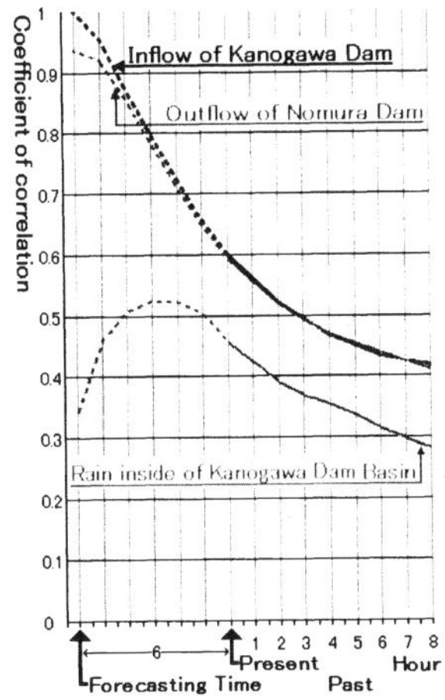

Fig.3. 6 hours later inflow forecasting

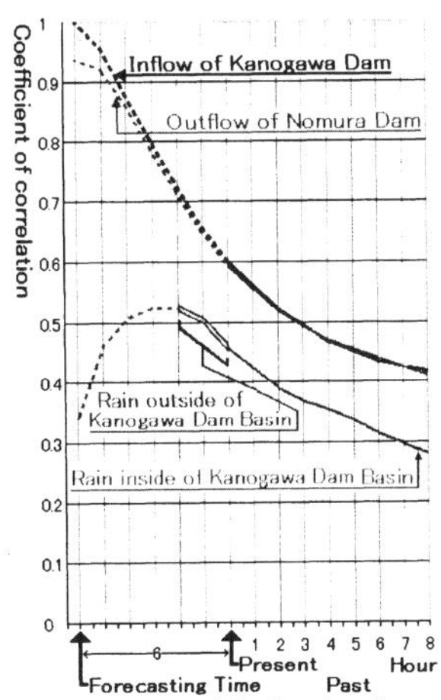

Fig.4. 6 hours later forecasting added rain data outside of the basin

Fig.4 shows 6 hours later forecasting added rain data outside of the basin. We want to get 2 hours rain data inside of dam basin. But it is impossible. Then we aim in characteristic of weather in Japan.

Basically, the weather moves from west to east. We decide weather movement assumption. If rain clouds move to constant direction at constant speed, there is rain information in outside of the dam basin in future. So, we use rain data outside of the dam basin, and we can guess this lacked information.

EXPERIMENT

In proposed method, we need rain data of wide area. We use "Radar-AMeDAS Data" which Japan Meteorological Business Support Center issues. This is rain data base which covers Japanese whole country with 6 km meshes.

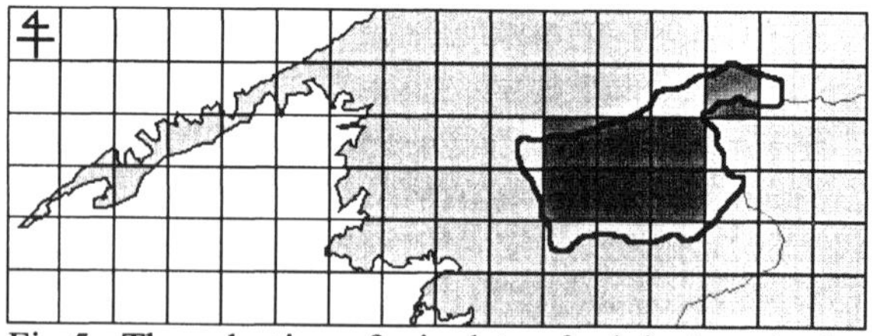

Fig.5. The selection of rain data of original method

First, we experiment original method of inflow forecasting to compare. In original method, the selection of area is rain data inside of the dam basin. As shown in fig.5, we use 7 meshes which represent the dam basin.

Original inflow forecasting system is composed as follows.

(1) Network type: 3-layered perceptron / Sigmoid function.

(2) Learning method: Back propagation / Learning times: 5,000 times

(3) Number of neurons: Output layer = 1 Middle layer = 5 Input layer = 216

(4) Input data: Outflow data of the upstream dam = 24 hours
Inflow data of the own dam = 24 hours
Rain data inside of the dam basin = 7 meshes × 24 hours

Table 1. Forecasting error using original method

Forecasting time	MSE $(t/s)^2$
1 hour later	149
2 hours later	424
3 hours later	726
4 hours later	1329
5 hours later	2684
6 hours later	4435

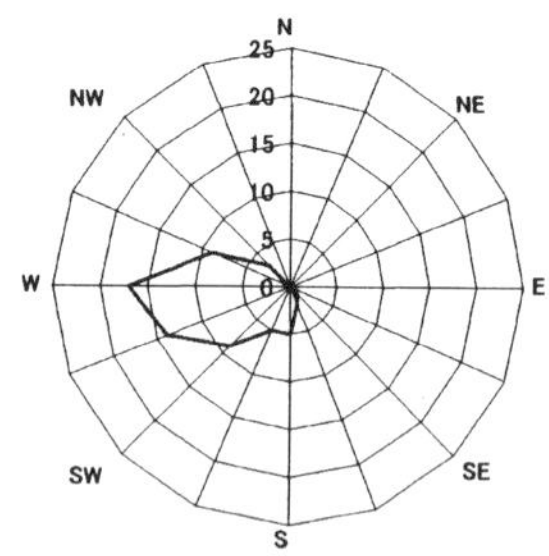

Fig.6. Counts which direction the rain clouds come from

On the experiment, we are choosing best parameters.

Table 1 shows forecasting error using original method. Table 1 shows the following.

(1) The system can forecasts 1 hour or 2 hours later inflow well.

(2) As the forecasting time becomes the future from 3 to 6 hours later, the error is increasing rapidly.

About proposed method, we must select good quality data from wide area.

Fig.6 shows counts which direction the rain clouds come from. It finds that many clouds come from the west. Next, about the speed, we get the value of under 35 km per hour.

Based on this results, we decide to use rain data in the area of the western side of the dam. The distance of the area is 70 km. It is the distance which cloud moves within 2 hours. Therefore, we use three kinds of area data as the input.

Proposed inflow forecasting system is composed as follows.

(1) Network type: 3 layered perceptron / Sigmoid function

(2) Learning method: Back propagation / Learning times: 5,000 times

(3) Number of neurons: Output layer = 1 Middle layer = 5 Input layer = 216

<Pattern A>

Input data: Outflow data of the upstream dam = 24 hours
Inflow data of the own dam = 24 hours
Rain data inside of the dam basin =3 meshes × 24 hours
Rain data outside of the dam basin =4 meshes × 24 hours

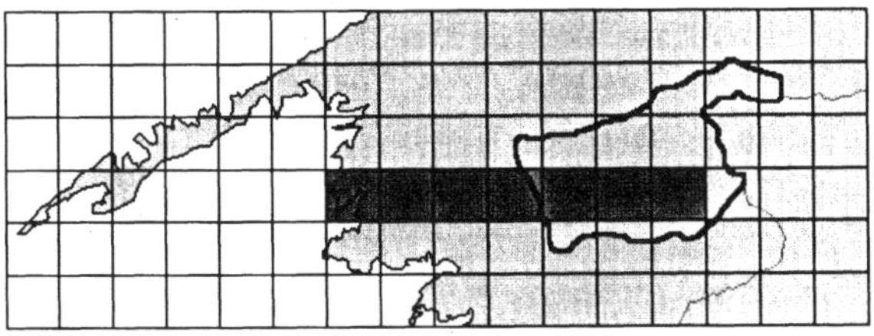

Fig.7. The rain data area of pattern A (near the dam basin)

<Pattern B>

Input data: Outflow data of the upstream dam = 24 hours
Inflow data of the own dam = 24 hours
Rain data inside of the dam basin =0
Rain data outside of the dam basin =7 meshes × 24 hours

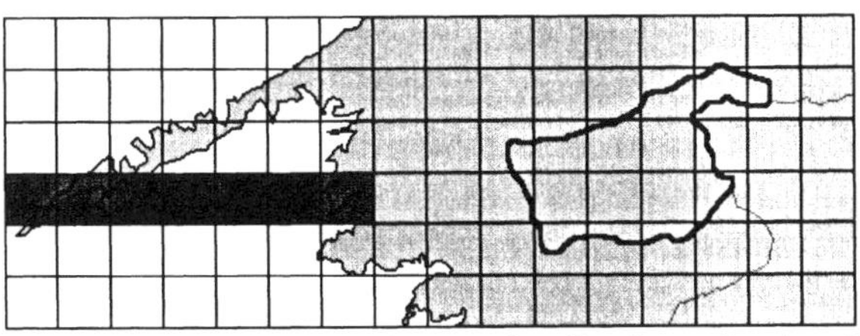

Fig.8. The rain data area of pattern B (far from the dam basin)

<Pattern C>

Input data: Outflow data of the upstream dam = 24 hours
Inflow data of the own dam = 24 hours
Rain data inside of the dam basin =2 meshes × 24 hours
Rain data outside of the dam basin =5 meshes × 24 hours

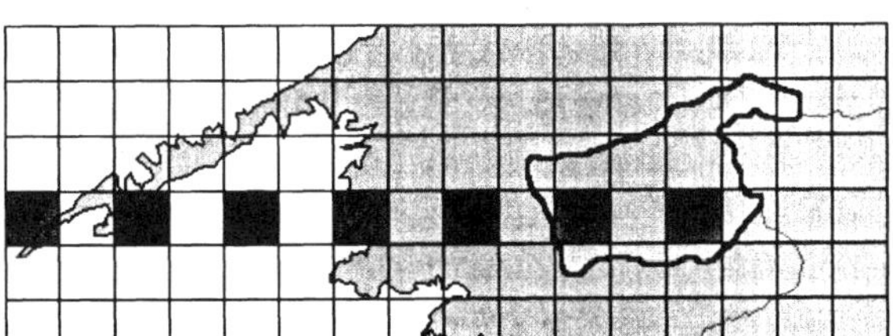

Fig.9. The rain data area of pattern C (wide area)

Table 2. Forecasting error using proposed method

Forecasting time	Pattern A		Pattern B		Pattern C	
	MSE (t/s)^2	Improvement Rate of MSE (%)	MSE (t/s)^2	Improvement Rate of MSE (%)	MSE (t/s)^2	Improvement Rate of MSE (%)
1 hour later	226	-51.9	223	-50.2	200	-34.7
2 hours later	506	-19.4	506	-19.4	432	- 1.9
3 hours later	670	7.7	779	- 7.4	632	12.9
4 hours later	1094	17.7	1363	- 2.5	951	28.5
5 hours later	2452	8.6	1870	30.3	1330	50.4
6 hours later	4051	8.7	1942	56.2	2689	39.4

These forecasting results are shown in Table 2.

Table 2 shows as follows.

(1) In case of the 1 to 2 hours later, the forecasting error is worse than the original method.

(2) In case of the 3 to 6 hours later, the forecasting error is better than the original method.

(3) Total result is better when we use rain data of the wide area. (Pattern C)

We experiment many patterns except these patterns, and Pattern C is the best.

From the results in Table 2, we conclude that the forecasting error is improved by our proposed method. In the proposed method, we use a very commonplace weather movement characteristic. So, if weather data of the dam neighborhood is given, we can apply to the inflow forecasting.

CONCLUSIONS

We proposed the method for inflow forecasting of the dam using the rain data outside of the dam basin in addition to the rain data inside of the dam basin. We show our method is effective by simulations. Forecasting errors are improved about 30 % for 3 to 6 hours later forecasting.

From now, we must try to choose best combination of data area.

REFERENCES

F.Imoto, M.Nakamura: "Online prediction method of several-hours-ahead inflow rate based on rainfall information", T.IEE Japan, Vol.113-B, No.3, pp.213-220, 1995.

F.Imoto, M.Nakamura, T.Taniguchi: "Model construction of inflow rate into a dam considering accuracy of rainfall prediction", T.IEE Japan, Vol.115-8-B, pp.869-874, 1995.

K.Mizuno, N.Hayashi, Y.Goto, K.Yukida, K.Ichiyanagi, Y.Yokomizu, T.Matsumura: "Investigation of input data for estimation of ground rainfall distribution by using Artificial Neural Network", T.IEE Japan, Vol.120-B, No.5, pp.665-671, 2000.

T.Iisaka, T.Matsui, Y.Ueki: "The development of the water flow forecasting system into the dam by neural networks and fuzzy theory", T.IEE Japan, Vol.119-B, No.10 pp.1020-1024, 1999.

A TEMPORAL PATTERN APPROACH FOR PREDICTING WEEKLY FINANCIAL TIME SERIES

DAVID H. DIGGS
Department of Electrical and Computer Engineering
Marquette University
P.O. Box 1881,
Milwaukee, WI 53201-1881, USA
Email: david.diggs@marquette.edu
Ph: 414.288.6046

RICHARD J. POVINELLI
Department of Electrical and Computer Engineering
Marquette University
P.O. Box 1881,
Milwaukee, WI 53201-1881, USA
Email: richard.povinelli@marquette.edu
Ph: 414.288.7088

ABSTRACT

Discovering patterns and relationships in the stock market has been widely researched for many years. The goal of this work is to find hidden patterns within stock market price time series that may be exploited to yield greater than expected returns. A data mining approach provides the framework for this research. The data set is composed of weekly financial data for the stocks in two major stock indexes. Experiments are conducted using a technique designed to discover patterns within the data. Results show that these methods can outperform the market in longer time ranges with bull market conditions. Results include consideration of transaction costs.

INTRODUCTION

Data mining is the process of discovering hidden patterns in data. Due to the large size of databases, importance of information stored, and valuable information obtained, finding hidden patterns in data has become increasingly significant. The stock market provides an area in which large volumes of data are created and stored on a daily basis, and hence an ideal dataset for applying data mining techniques.

Statistical analysis has been widely used for many years to make predictions on the future values of a security price and study its behavior over time. Times series such as the stock market are often seen as non-stationary which present challenges in predicting future values. The focus of this research is in analyzing and predicting weekly financial time series. This work will show the advantage of using a weekly trading strategy, which is an extension of the daily trading strategy, to overcome the transaction cost associated with trading.

The proposed method is a data mining approach that uses time-delay embedding and temporal patterns to characterize events. The method is designed to analyze non-stationary time series and provides the basis for this work. The paper is broken into five sections, which describe the goal of this work, overview of the Time Series Data Mining method, financial applications, results, and research conclusions.

PROBLEM STATMENT

Previously, we have applied our Time Series Data Mining (TSDM) approach to making one-step daily price predictions (Povinelli, 2000). We have shown that a simple trading strategy based on these predictions can yield greater returns than expected by the efficient market hypothesis. However, these returns are adversely affected by transaction costs. Hence, the goal of the current research is to develop an approach that increases returns and overcomes transaction costs. This is achieved by studying weekly stock price time series and making one-step weekly predictions.

Li and Tsang (2002) previously researched stock prediction using genetic programming in order to do financial forecasting. Results showed they were able to trade with better accuracy on long positions in indexes and individual stocks. Sauer (1994) also proposed a method similar to TSDM, in which time-delay embedding with interpolation and weighted regression were used to make time series predictions.

TIME SERIES DATA MINING

The proposed approach is based on TSDM (Povinelli, 2000). This method discovers hidden temporal structures predictive of sharp movements in price, using a time-delay embedding process that reconstructs the time series into a phase space that is topologically equivalent to the original system under certain assumptions. These assumptions are discussed in (Sauer et al., 1991). The TSDM technique was developed to make one-step predictions (Povinelli, 2000; Povinelli and Feng, 1998). The approach is used to make a prediction, which determines the buying and selling of stocks in a given time period. Figure 1 below shows two possible examples of the types patterns for which the TSDM method searches. The illustrated patterns are predictive of sharp increases in a time series.

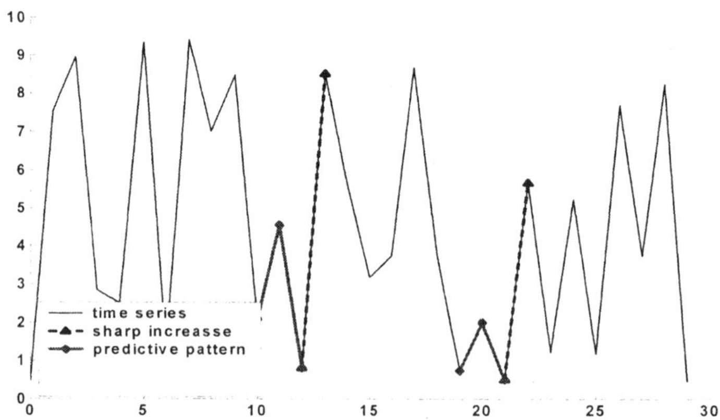

Figure 1 – Time series example

The patterns are determined by the three previous points to make a prediction for the next time step. The method involves clustering patterns, which are used to detect these sharp increases in the stock price. To find these temporal patterns the time series is embedded into a reconstructed phase space with a time delay of one and a dimension of three (Sauer et al., 1991). Once the data is embedded, temporal structures are located using a genetic algorithm search. Clusters are made of points within a fixed distance of the temporal structures. A percent change function

$$g(t)=(x_{t+1}-x_t)/x_t \tag{1}$$

determines the value given to the prediction made from the clustering using the temporal structures. This value is the percent change in the security price for the next week. The temporal structures are next ordered by how well each predicts the stock price movements. A ranking function is defined as the average value within a temporal structure, and it is used to order the structures for optimization. The optimization is a search to find the best temporal structures and is done with a simple genetic algorithm (sGA) that finds fitness value parameters that maximize the ranking function $f(P)$. The sGA uses a combination of Monte Carlo search for population initialization with roulette selection and locus crossover to find P^*, and a criteria of fitness values halt the genetic algorithm. See reference (Povinelli, 2000) for more details on the algorithm.

APPLICATION TO FINACIAL TIME SERIES

The stock market is a platform for millions of investors to interact through the buying and selling of securities on various equity markets, such as the NASDAQ, AMEX, and NYSE. The goal is to use the TSDM method to achieve small and unexpected returns that are greater than transaction costs associated with trading. The data is in the form of weekly stock price data comprised of the Dow Jones 30 and NASDAQ 100 component stocks.

The method is applied to the stocks of a given market index. Weekly buy or do nothing signals are generated by the TSDM method for each stock in the index. The TSDM returns are compared the index as a benchmark. Transaction costs are also considered to show the adverse effects of actually making these weekly trades. Transaction costs are calculated based on the number of stocks in the weekly portfolio selected by the TSDM model. A transaction cost of $20 is used for each buy and sell transaction.

METHOD RESULTS

Four experiments with various time ranges were run on each data set. Training the data was done using the TSDM method to find predictive structures. Testing for each experiment used a weekly trading strategy for buying and selling a set of stocks that are output from our model. Each week a new set is traded based on the model predictions. The time ranges for each data set were current year (1/01/2003-5/01/2003), previous year (1/01/2002-1/01/2003), previous 5 years (1/01/1998-1/01/2003), and previous 10 years (1/01/1993-1/01/2003). The training period for the four experiments was 25 weeks because previous test experiments with this training period gave the best results. The genetic algorithm parameters were the same for each experiment, consisting of population size 30, and fitness convergence value of 0.9, which is used as a stopping criterion.

The total portfolio values are set to $10,000 and $100,000 to calculate transaction costs and returns for the model. The weekly geometric average rate of return is calculated for the model and both indexes for each experiment, shown below in Table 1. The average rate of return for the model is based on the returns calculated from each stock in the portfolio, which are bought and sold equally on a weekly basis. The table also shows the adjusted return with transaction costs and the average transaction cost per week. Figures 2 through 5 plot the previous year and previous five years of TSDM model weekly returns and adjusted model returns based on a $10,000 portfolio value. The figures plot model and adjusted returns for both the Dow Jones 30 and NASDAQ 100 data sets along with a comparison against the index benchmarks.

Table – 1 Weekly Returns Comparison

Rate of Return	*TSDM Model*	*TSDM Model With Transaction Costs ($10,000)*	*TSDM Model With Transaction Costs ($100,000)*	*Index*
Dow Jones 30 ytd	-0.119 %	-0.126 %	-0.120 %	-0.000768 %
Dow Jones 30 1 Year	-0.0098 %	-0.017 %	-0.011 %	-0.000127 %
Dow Jones 30 5 Years	0.730 %	0.720 %	0.729 %	0.0000616 %
Dow Jones 30 10 Years	1.440 %	1.429 %	1.439 %	0.00000324 %
NASDAQ 100 ytd	-0.172 %	-0.193 %	-0.174 %	0.00586 %
NASDAQ 100 1 Year	-0.017 %	-0.042 %	-0.019 %	-0.00887 %
NASDAQ 100 5 Years	1.018 %	0.989 %	1.015 %	-0.00011 %
NASDAQ 100 10 Years	1.873 %	1.845 %	1.871 %	0.00201 %

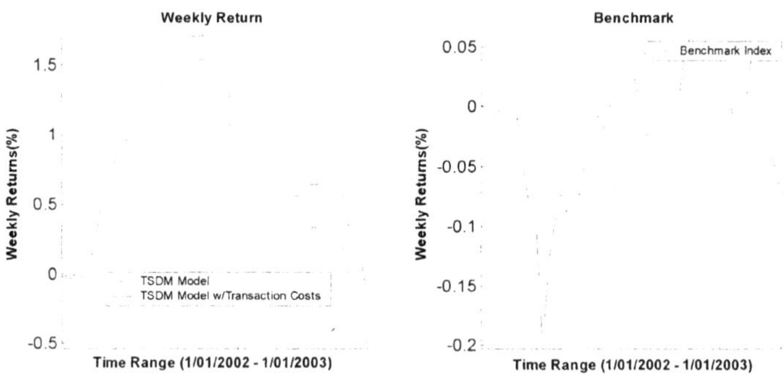

Figure 2 - Dow Jones 30 Experiment 2

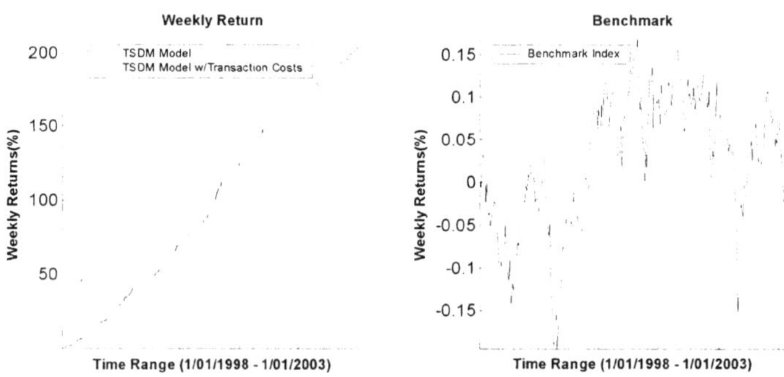

Figure - 3 Dow Jones 30 Experiment 3

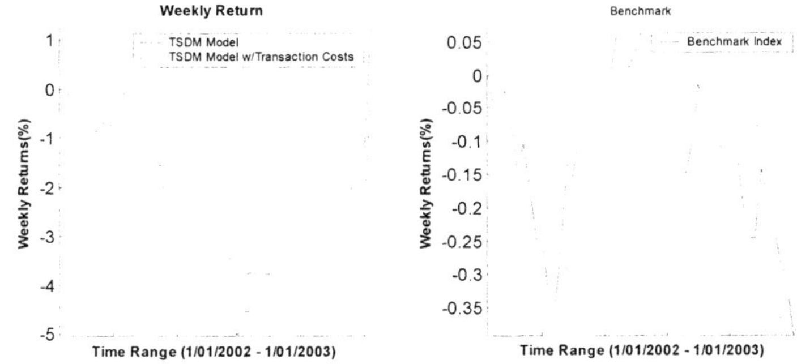

Figure - 4 NASDAQ 100 Experiment 2

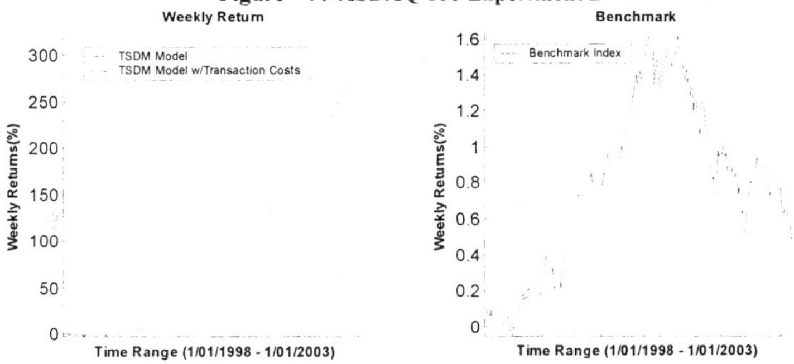

Figure - 5 NASDAQ 100 Experiment 3

CONCLUSIONS

Our data mining approach combined with a weekly trading strategy is used to overcome trading cost and compared against market index used as benchmarks. The benchmarks used in comparing results were two major stock market indexes, the Dow Jones 30 and NASDAQ 100. The model produced positive returns in the five and ten year experiments when transaction costs were both taken into account and ignored. When transaction costs were calculated, the results from the NASDAQ experiments were more affected due to larger portfolio sizes than the Dow Jones experiments. The different time ranges enabled us to see how are model reacted to the changes in the market setting. The first two experiments were set directly in the recent bear market, which has taken both major indexes down significantly over the past two years. Our model was unable to do better than the market due to a lack of diversity in the weekly portfolios. In the five and ten year time ranges our model was able to overcome transaction costs and outperform the market, showing a strong predictive stock selection process. In comparison to the Dow Jones experiments, the NASDAQ produced better results due to the larger selection of stocks to choose from to create better weekly portfolios With this trading strategy, a larger portfolio value with the ability to purchase more shares reduces the adverse impact of trading costs on the returns. As shown in the Table 1 the difference between the model return and the adjusted return decreases with a higher portfolio value.

This work leads itself into an optimization problcm for future work. The optimization is to be done in the form of portfolio management that will optimize the weekly portfolio predictions made using the TSDM method. The future directions of this work will include the use of Markowitz Portfolio Theory and the Capitol Asset Pricing Model (CAPM) (Reilly and Brown, 1997) to find optimal portfolios while meeting specific goals such as maximizing return and minimizing risk. This addition focuses on increasing the returns of the TSDM method over time by altering the weights of the weekly portfolios. In addition to portfolio management techniques the transaction cost model can be modified to include the bid-ask spread.

ACKNOWLEDGEMENTS

This material is based on work supported by the Department of Education GAANN Fellowship.

REFERENCES

Li, J., and Tsang, E. P. K., 2002, "Eddie for Financial Forecasting," Genetic Algorithms and Genetic Programming in Computational Finance,, S.-H. Chen, ed., Kluwer Academic Publishers, Norwell, Massachusetts, 161 -174.

Povinelli, R. J., 2000, "Identifying Temporal Patterns for Characterization and Prediction of Financial Time Series Events," *Temporal, Spatial and Spatio-Temporal Data Mining: First International Workshop; TSDM2000*, Lyon, France, 46-61.

Povinelli, R. J., and Feng, X., 1998, "Temporal Pattern Identification of Time Series Data Using Pattern Wavelets and Genetic Algorithms," *Artificial Neural Networks in Engineering*, St. Louis, Missouri, 691-696.

Reilly, F. K., and Brown, K. C., 1997, *Investment Analysis and Portfolio Management*, Dryden Press, Fort Worth, Texas.

Sauer, T., 1994, "Time Series Prediction Using Delay Coordinate Embedding," Time Series Prediction: Forecasting the Future and Understanding the Past, A. S. Weigend and N. A. Gershenfeld, eds., Addison-Wesley, 175-194.

Sauer, T., Yorke, J. A., and Casdagli, M., 1991, "Embedology," *Journal of Statistical Physics*, Vol. 65, 579-616.

New Methods to Optimize Stock Price Prediction with Neural Networks

Hua Jiang
Department of Mechanical Engineering
Ryerson University Toronto ON M5B2K3 Canada Email: hjiang@ryerson.ca

Abstract:
This paper provides an empirical result on developing neural networks trading system base on neural network nonlinear model and online trading technology. Through 250 days modeling and testing, we concluded that the forecasting model is highly efficient where CAPM is rejected by empirical result according to past research. It applies to the stock market where not all individuals hold all the stocks in the index, where market index is not universal and where market point is not efficient on CAPM model.

Key words:
Dynamic Beta, Pattern Recognition, Prediction, Stock market, Neural Network

Introduction:
In the past decade, a neural stock trading system has in fact already taken shape at online trading platform. The earliest study was done in 1990. Researchers concluded that neural network model outperformed the benchmark models in return. We anticipate that many online trading service providers will adopt and incorporate neural network in their service in the near future because of the following factors:

- Inter listing across countries and global capitalization.
- Financial market and institutional deregulation.
- Phenomenal growth in IT (cheap information, real time).
- New methods to minimize risk and maximize return.
- New trading methods (floor, telephone, internet, wireless, intelligence).

In this paper, we created a multi-method that begins with three methods to forecast the stock price and optimize the three results to select the one among the three. There are two core components. The first one is to determine economy macro factors and company micro factors in input variables adding noise to the training set. The second one is to forecast stock prices using optimized multi-methods. The multi-method includes *predict day close (PDC)* -using adjusted close price of the passing 60 days; *predict one se t(POS)* method-using first hour's trading data to forecast the next hour's trading data and run regression on the result to find out the trading day close price; *cyclic price prediction (CPP)* - beginning with first hour's trading data and predict next hour's data and then using the next hour's trading data to predict one more hour further till it reaches the trading day closing point. In multi-method, two type of prediction price are achieved. When short term factors such as overnight news occurs, PDC deviates more from POS and CPP. We use the average of POS and CPP as the final prediction when the deviation is beyond 2% from PDC. We use the average of

PDC, POS and CPP as the final prediction when the deviation is within 2%. The best forecasted price is recorded and corresponding investment decisions are made base on this price. The following figures and tables present in details the optimized results using multi methods on highly capitalized US and Canadian stock market, portfolio composition from different sectors, how neural network pattern day trader makes profit, and possible prediction of market failure. Factors considered are as follows:

Abbreviation	Variable
IR	Interest Rate
CPI	Consumer Price Index for Five Years
MS	Money 2 Supply
MIA	Market Index Average
UR	*Unemployment Rate*
MD	Monthly Trading Data for Five Years
FHTD	First Hour Trading Data
HHTD	Half Hour Trading Data
DD	60 Day Trading Data
TO	Target Output

Table1. List of input factors

CPP method sample:

We begin with first hour's trading data and predict next hour's data and then using the next hour's trading data to predict one more hour further till it reaches the trading day closing point.

IR	CPI	MS	MIA	UR	MD	FHTD	HHTD
0.0396	0	-0.1468	0.2161	0.1075	-0.05	0.987433	0.982362
-0.0403	0.0931	0.194	0.6785	-0.1064	0.0971	-0.96527	0.053537
...	...	...	...	...	...	...	...
0.1047	0.6843	-0.1331	0.2802	-0.2532	0.0103	0.016722	0.048202

Table 2. CPP training pattern

CPP weights:

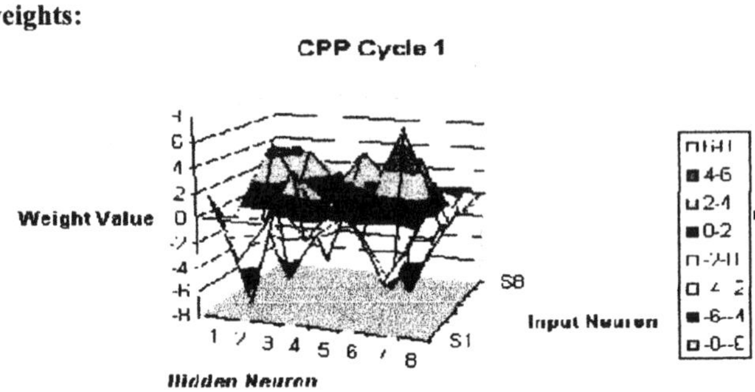

Figure 1. CPP cycle 1 analysis

We find that there are two connections that share the weights between 6 and 8 and they are the most important factors in determine the final forecast result.

PDC method sample:

The price change is defined as $\frac{P_t - P_{t-1}}{P_{t-1}}$ where P_t is the adjusted close price of day t . In this paper, we chose to transform data to neural network ready decimals. In this way, we can apply our neural network to stock of any type, i.e. *from penny stocks under the $5 value to stocks of $500 in value. The following* is sample input data sets for PDC daily cycle in the training process.

IR	CPI	MS	MIA	UR	MD	DD	TO
0.039646	0	-0.14684	0.216077	0.107527	-0.04996	-0.07563	0.029545
-0.04034	0.09311	0.193996	0.678459	-0.10638	0.097079	0.029545	-0.03091
...	...	...	...	...	...	...	...
0.104668	0.684346	-0.13312	0.280157	-0.25316	0.010329	0.056161	-0.02778

Table 3. PDC training pattern

PDC weights:

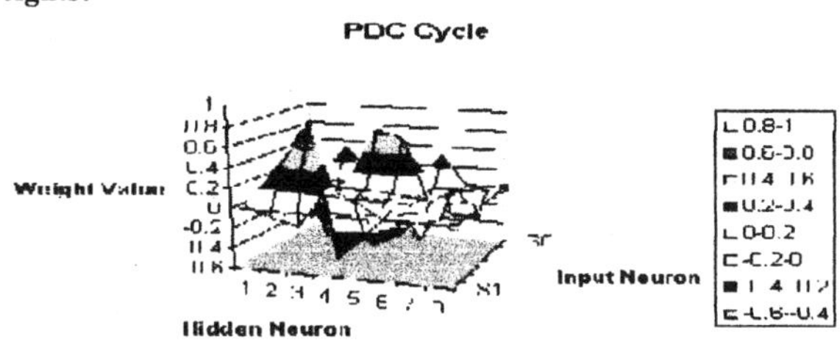

Figure 2. PDC cycle analysis

POS method sample:

We use first hour's trading data to forecast the next hour's trading data and run regression on the result to find out the trading day close price.

IR	CPI	MS	MIA	UR	MD	FHTD	HHTD
0.0396	0	-0.1468	0.2161	0.1075	-0.05	0.987433	0.982362
-0.0403	0.0931	0.194	0.6785	-0.1064	0.0971	-0.96527	0.053537
...	...	...	...	...	...	...	...
0.1047	0.6843	-0.1331	0.2802	-0.2532	0.0103	0.016722	0.048202

Table 4 POS training pattern

Sample regression result on forecasted data that is correlated with MIA.

OUTPUT OF SINGLE STOCK REGRESSION

Regression Statistics	
Multiple R	0.053052
R Square	0.002814
Adjusted R Square	-0.01438
Standard Error	0.106034
Observations	60

	df	*SS*	*MS*	*F*	*Significance F*
Regression	1	0.001841	0.001841	0.1637	0.687261
Residual	58	0.652104	0.011243		
Total	59	0.653944			

	Coefficients	*Standard Error*	*t Stat*	*P-value*	*Lower 95%*
Intercept	0.008393	0.013709	0.612263	0.542757	-0.01905

X Variable 1	0.238281	0.588931	0.404599	0.687261	-0.94059

Table 5. POS Regression

POS weights:

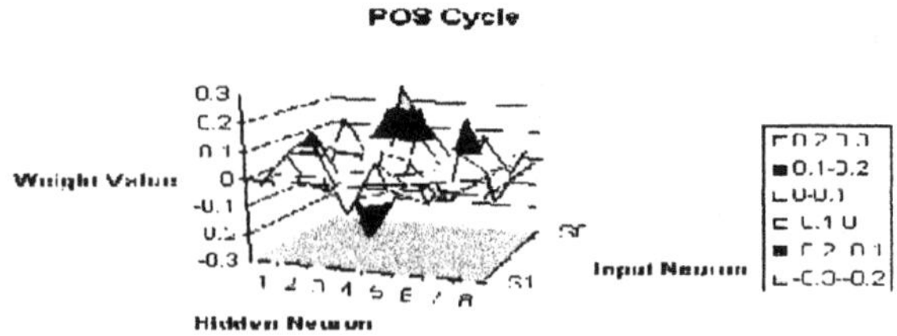

Figure 4. POS Analysis

Input neuron analysis:
For this stock, we find the following correlation result of input factors.

	IR	CPI	MS	MIA	UR	MD	FHTD
IR	1	0.04283	0.15333	0.31679	-0.07359	0.36008	0.10617
CPI	0.04283	1	0.27006	0.00507	0.23449	0.07319	0.11526
MS	0.15333	0.27006	1	0.01183	0.2418	0.22965	0.34612
MIA	0.31679	0.00507	0.01183	1	0.18609	0.44684	0.3257
UR	-0.07359	0.23449	0.2418	0.18609	1	0.2179	0.07268
MD	0.36008	0.07319	0.22965	0.44684	0.2179	1	0.49101
FHTD	*0.10617*	*0.11526*	*0.34612*	*0.3257*	*0.07268*	*0.49101*	1

Table 6. Input correlation

We conclude that overall importance level of input neurons from high to low sequence is MD, MS, MIA, CPI, IR, and UR.

Optimization of three methods:
In this section, we use the portfolio as an example to describe our optimization method using dynamic beta. Three methods produce different intra day close price. Which one to choose? We find the optimized result based on the following selection process. From the regression result of each stock on a testing day, we can find the dynamic beta as listed in table below.

SYMBOL	TLC	BR	AC	COR	HUM	BRA	RCMB	SCC	PCA	MFI
BETA	0.301	0.673	1.993	0.868	0.824	1.817	1.402	1.226	0.545	0.588

Table 7. Summary of dynamic beta

A higher beta means the stock is has more systematic risk, a lower beta means the stock has less systematic risk. For example AC with beta 1.993, we can assume because of political and macro economic situation, AC change 1.993 percent when MIA change every one percent. Therefore we choose PDC result as the optimized final result for all the stocks with POS beta smaller than 1. We choose PDC result as the optimized final result for all stocks with POS beta above 1. We choose POS result as the optimized final result for all stocks with zero betas.

The following figure illustrates our overall performance on randomly selected portfolio, neural network portfolio, and optimized portfolio using the dynamic beta. It also contains the market index history of the testing period. With the optimized multi-method neural network methods and the 7 input neurons, we did 83% better than the market. During the period of this project, no market failure was predicted. The neural network portfolio did 31% better than the market.

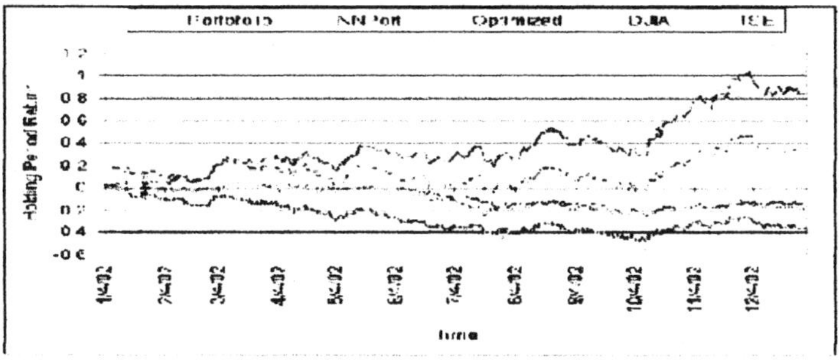

Figure 5. Return comparison

Conclusion:

During one-year period of research, Dow Jones Industrial Average lost 18% TSE300 index lost 13%. With the precise prediction, investors can short sell or buy and sell stocks using the forecasted result. In January 2, DJIA closed at 10073.4 TSE300Index closed at 7646.8. At the end of year 2002, DJIA closed at 8341.63 TSE300Index closed at 6614.5. Ten stocks and three portfolios holding period return were compared. The single neural network portfolio did significantly better than initial portfolio and the market indices. The optimized neural network portfolio with multi-methods achieved highest return among the three. Through analysis, we find that investors can double the forecast effect using data of different period and we recommend using first hours trading data as the training set.

The forecasting model is highly efficient in capitalized free trade market i.e. US and Canada. It applies to any stocks in all sectors. It increases the profit for institutional traders. On the retail side, it benefits the day traders and pattern day traders. On the regulatory side, it can be used to test the stock market efficiency. Basing on above findings in this paper, we proposed the elastic efficient market hypothesis base on CAPM model. The next paper we are working on proves the existence of dynamic beta β_i that is in the cubic relationship with the portfolio's risk σ_p. We found the equation $\psi = C(\sigma_p{}^{-3})$ where ψ is the relationship between $d\beta i$ and $d\sigma_p$, $C = -2Cov(r_i, r_p)$ is the constant coefficient because given the same mean time period, $Cov(r_i, r_p)$ is constant. The empirical result of this paper successfully recognized this pattern.

References:

1. Ahmadi, H. (1990). Testability of the arbitrage pricing theory by neural networks. Proceedings of the International Conference on Neural Networks (pp. 385 –393). San Diego, CA
2. Choi, J.H. Lee, M.K., & Rhee, M.W. (1995). Trading S&P500 stock index futures using a neural network. Proceedings of the Third Annual International Conference on Artificial Intelligence Applications on Wall Street (pp. 63-72). New York
3. Kamijo, K., & Tanigawa, T. (1990). Stock price pattern recognition: A recurrent neural network approach. Proceedings of the International Joint Conference on Neural Networks (pp. 215-221). San Diego, CA.
4. Kimoto, T., Asakawa, K., Yoda, M., & Takeoka, M. (1990). Stock market prediction system with modular neural network. Proceedings of the International Joint Conference on Neural Networks (pp. 1-6). San Diego, CA.
5. Trippi, R. R., & DeSieno, D. (1992). Trading equity index futures with a neural network. The Journal of Portfolio Management, 19, 27-33.
6. White, H. (1994). Can neural networks forecast in the bit leagues? Comparing network forecasts to the pros? Proceedings of International Symposium of Forecasting (p.24). Stockholm, Sweden.
7. Kohara, K., Ishikawa, T., Fukuhara, Y., & Nakamura, Y. (1997). Stock price prediction using prior knowledge and neural networks. International Journal of Intelligent Systems in Accounting, Finance and Management, 6, 11-22.
8. S. Haykin. Neural Networks A Comprehensive Foundation. Prentice-Hall, Upper Saddle River, NJ, second edition, 1999.
9. C. M. Bishop. Neural Networks for Pattern Recognition. Oxford University Press, Oxford, 1995.
10. G. J. Gibson and C. F. N. Cowan. On the decision regions of multi-layer perceptrons. Proceedings of the IEEE, 78(10):1590–1594, 1990.
11. S. Tamura and M. Tateishi. Capabilities of a four-layered feedforward neural network: Four layers versus three. IEEE Transactions on Neural Networks, 8(2):251–255, 1997.
12. K. J. Lang and M. J. Witbrock. Learning to tell two spirals apart. In T. Sejnowski D. Touretzky, G. Hinton, editor, Proceedings of the 1988 Connectionist Models Summer School, pages 52–59, San Mateo, CA, 1988.

A PRACTICAL ARTIFICIAL INTELLIGENCE-BASED PREDICTOR FOR REAL TIME STOCK INDEX FORECASTING PROBLEM

HAILIN LI
Department of Engineering Management
Smart Engineering Systems Laboratory
University of Missouri-Rolla
Rolla MO 65409-0370

CIHAN H DAGLI
Department of Engineering Management
Smart Engineering Systems Laboratory
University of Missouri-Rolla
Rolla MO 65409-0370

ABSTRACT

Ample research mainly based on supervised learning-oriented approaches already has been done in the real time financial series forecasting area. Although some of such predictors can perform better than conventional models consistently, they suffer from inherent drawbacks due to the characteristic of supervised learning. Reinforcement Learning (RL) methods embody a general Monte Carlo approach to dynamic programming for catching system dynamics under uncertain environments. The integrated architecture combined RL and artificial neural networks (ANNs) is the powerful tool for both control and prediction purpose. Relatively fewer studies were made in terms of predicting the stock market data by means of such hybrid framework. This paper presents the novel Neuro – Q Learning prediction model for forecasting real time stock index. The encouraging experimental results for NASDAQ market illustrate the effectiveness of the presented approach.

INTRODUCTION

The real time financial series forecasting applications of soft computing methods have attracted a lot of interest for many years. Although many academics and practitioners have criticized technical analysis for years, it had become clear that soft computing tools, especially those from the field of neurocomputing were frequently finding relevance in the financial markets.

In the stock markets, stock indices forecasting with supervised learning-based approaches has already been accepted as one of viable analytical options mainly due to the nonlinear relationship between the financial variables and stock returns [1] [2] [3] [4]. Although some of such prediction models can perform better than conventional stock market forecasting method (i.e. random walk model) consistently, several issues prevent them to be applied for real world application. First of all, they usually require very time-consuming work not only for selecting neural network input variables and data preprocessing but also for network structure optimization. More importantly, there is no solid support for such pure supervised learning-based approaches in terms of why historical patterns of stock indices should be expected to repeat. These challenges stimulate the research for many hierarchical and hybrid artificial intelligence-based financial prediction models. Among them, some interesting developments include the synthesis of genetic algorithm and ANNs [5], Neuro-fuzzy architecture [6] and combination of qualitative and quantitative data [7].

Real time financial performance depends upon sequences of interdependent decisions and is thus path-dependent. Such characteristic calls for the mechanism that can provide rational decision-making under uncertain environment. RL is attracting attention as the promising machine learning approach to solve the difficult problems in which external environment cannot be significantly modeled in the system, or environment is dramatically changing. Though much progress for various real-world applications has been made in recent years by using RL algorithms, relatively fewer studies were made for the purpose of forecasting the stock indices by means of RL.

Inspired by the fact that future financial target value can be viewed as the result influenced by the combination of historical market patterns and the actions taken by investors in the immediate previous market state, a new hybrid forecasting architecture for stock indices is proposed in this paper. Our goal is to produce the practical and convenient way to overcome the problem that traditional supervised learning–oriented models face, that is, complex neural network configuration procedure and weak ability to handle "unexpected" event. In such hybrid-learning framework, typical three-layer feedforward perceptron trained by backpropagation algorithm is used as approximation model to represent historical market patterns roughly and the Q-learning algorithm is used in order to generalize the possible actions of investors under different market "state". The final predicted value is the sum of neural network output and the action of reinforcement learning agent. The proposed hybrid forecasting method is applied to the NASDAQ market and the experimental results are presented to show its effectiveness.

REINFORCEMENT LEARNING

Similar to dynamic programming, Reinforcement Learning is concerned with rational decision-making process under uncertain environment. The goal of RL is to analyze how decisions ought to be made in the light of clear objectives so that agent can generate a series of actions to influence the evolution of a stochastic dynamic system. Compared with the classical dynamic programming algorithm, RL can handle the situation that no priori model information such as MDP state transition probability is available. This salient characteristic makes RL more flexible than dynamic programming. Much research work already has been done in this area. Among them the major development include the temporal-difference learning algorithm and Q-learning [8].

In this paper, our effort is restricted to the discrete-time dynamic system that is suitable for most financial series prediction environment. The system evolution at time t, action takes on a state x_t can be shown as:

$$x_{t+1} = f\left(x_t, a_t, w_t\right)$$

where w_t is a disturbance and a_t is a control decision generated by policy μ.

The value function J^{μ} is the basic criteria for decision-making of agent. Each $J^{\mu}\left(x\right)$ can be viewed as the assessment of long term rewards given that we start in state x and control the system using a policy μ. If the optimal value function J^{*} is available, the generation of optimal action requires computing one expectation per element of the decision space. Q-Learning algorithm is the popular variant of temporal-difference learning. Q value function $Q(x,a)$ is introduced in order to avoid the complex computation. Therefore, the optimal actions can be obtained without the requirement for evaluating the system function.

In the absence of the model of the MDP ($\Re$ and P^{u} are unknown), Q-Learning algorithm uses sample data $\{x_t, a_t, r_t, x_{t+1}\}$ generated by the interaction between agent and environment. r_t is the immediate reward.

NEURO-Q LEARNING VIEW OF FINANCIAL SERIES

Future financial target value can be viewed as the result influenced by the combination of inherent market inertia (normal market patterns generalized from historical data) and the actions taken by investors at the immediate previous market state which result in the unexpected market behavior.

It is well known that supervised-learning methods are the extremely useful tool to catch the system patterns that keep repeating themselves. The hardest part of financial prediction work is about how to handle "unexpected" event. For example, the future stock indexes not always follow the rule generated from historical data. As a matter of fact, there are wide day-to-day changes in the prices quoted on most stock exchanges. Although such high market volatility often creates the difficulties for existing theories to make clear explanation, essentially it is the direct consequence results from the behavior of people. The most recent investment actions of investors change the predicted market trajectory that is generated from historical data. Inspired by such observation, the proposed schema includes RL strategy into the problem of stock indices prediction regarding the process of market price changes as a Markov process.

Basically, a standard three-layer feedforward ANNs model is constructed to obtain the normal market pattern or normal value trajectory. Q-Learning algorithm is used for training Reinforcement Learning agent so that the agent can imitate the extra behavior (policies) of investors that result in unexpected value change other than just follow normal patterns. It is noted that all model configuration work is totally base upon the time-sequence values of target series itself. No other market indictors will be used.

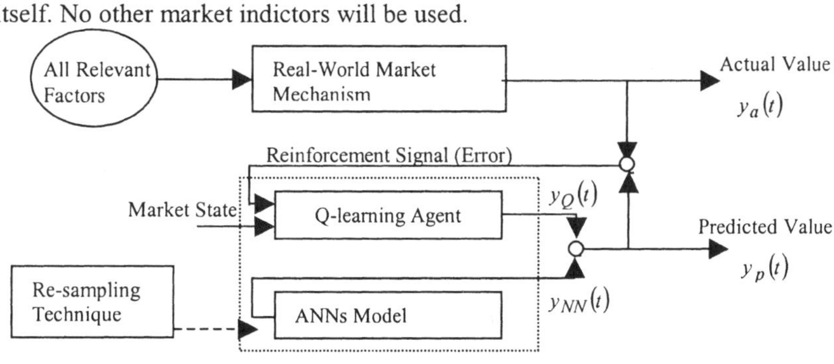

Figure 1. Hybrid Prediction Architecture for Stock Price

Figure 1 shows the proposed Neural-Q Learning forecasting architecture for financial series. Since the task was to forecast the future values of stock index simply based upon its historical values, the training of a hybrid time series model is involved. Let $y_a(t)$ be the current actual target value and $y_p(t)$ be the current predicted value. The final forecasting result of $y_p(t)$ is composed of two different parts, that is, the output of ANNs model $y_{NN}(t)$ generated from normal historical market pattern and the output of Q-Learning agent that depicts the "abnormal behavior" of investors in accordance with different market state.

Let f be the underlying function through which the "rational" part of $y_a(t)$ can be mapped from lagged values $(y_a(t-1),\ y_a(t-2),\ \cdots,\ y_a(t-k))$. f can be fairly accurately approximated by a three-layer feedforward ANNs model trained by various back-propagation algorithms. To overcome the overtraining phenomenon that is common in supervised-learning domain, a cross-validation approach is used to choose the optimal combination of hidden neuron number n and the number of lagged values k.

As to the RL agent training, the system parameters include market state, reinforcement learning signal and actions that agent can take. A market state at a given time is the basis for policy-generation. There should not exist any strict rule about how to pick up valid indicators. Various kinds of technical indicators as well as other derived data can be considered as potential candidates. The set of continuous or discrete actions A can be determined from the range of "abnormal" values within the training set. Typically a reinforcement signal can be designated as fixed integers base upon whether the agent output achieves the desired output. In this paper, the feedback signal R at time step t is determined by the squared error between $y_a(t)$ and $y_{NN}(t)$ plus a term proportional to the square of the action change from one time step to the next. The action change term is introduced to reduce the unnecessary fluctuation in the action.

The rolling-training process rely on the re-sampling procedure is only applied to the ANNs model. The optimal network structure is likely changed from period to period to account for possible changes of general normal market rules. Basically, the new value of financial series is added to the training set and the validation set when the next day actual value is available. The oldest one is then eliminated in order to maintain the size of training data. The new ANNs structure is thus generated from the new data set so that the model can follow the market dynamics timely.

For Q-learning agent, although the policies of investors do change even the market state is the same, the agent only trained within the fixed time period in the experiment of this paper. The assumption here is that the final convergence polices is stable enough if the market state cover the most range of series values.

EXPERIMENT

The NASDAQ Stock Market is adopted as the simulation environment in this paper. Its inherent high volatility provide excellent platform to test the performance of any stock price prediction model. Daily history data of NAS/NMS COMPSITE started from 1-Mar-1993 as well as its closing price on the previous trading day is used to assess the performance of Neural-Q learning prediction architecture. The objective of the problem is to make accurate forecasting for next day IXIC closing price simply base upon the historical daily closing price itself so that the time-consuming work for selecting suitable neural network inputs and the complex data preprocessing procedure can be neglected.

For a typical three-layer feedforward ANNs trained by back propagation algorithm, the perfect data fitting for training set can be obtained if sufficient lagged values are included and if the number of neurons in the hidden layer is also large enough. However, the optimal balance for the combination is hard to get and the ANNs is prone to overfitting. In order to construct relatively accurate ANNs model without introducing too much extra work, we experimented with different values of k ranging from 3 to 10 with an increment of 1, and n ranging from 3 to 7 with an increment of 2. Such combination resulted in 24 candidate models. As mentioned earlier, the way for choosing k and n is the standard cross-validation approach. In order to let ANNs model reflects market patterns changing as timely as possible, from the very beginning, 10 input-target pairs are sequentially picked up each time as the training data for the construction of networks model. Such 10 pairs were divided into training set (7) and validation set (3). Once the ANNs model was set up and the prediction for next day was completed, the new actual value of stock closing price is added to the training data once it became available and the oldest input-target pair is eliminated. Then the training process restarted. Throughout the experiment period, most time the selected model structure has 3 hidden layer units and 7 lagged values.

The Q-Learning agent plays the key role in the presented hybrid architecture. The inputs to agent are market state and reinforcement signal at each time step. Base upon the hybrid view of stock price, the actual stock market value reflects not only market inertia (modeled by ANNs model) but also the "abnormal" investment behavior according only to a given state. The market state can be viewed as Markov chain. In the proposed architecture, the financial series forecasting work is totally rely on the historical data of the series itself. Accordingly, the value of series is the only data in defining the market state. Two issues need to be considered for the generation of state. Firstly, the mapping relationship from the amount of unexpected action of investors (implicitly showed by difference between actual value and ANNs predicted value) to market state should be carefully selected. Secondly, one state should cover certain range so that the state visiting frequency can be promised.

The output of the Reinforcement Learning agent is directly added to the output of ANNs model. In principle, the allowed output actions of the agent should be the set of continuous actions in order to completely offset the error of simple ANNs model. However, in this paper, we adopt the discrete actions set as the simplified case to show the effectiveness of the approach. The Q function is implemented in a quantitized, or table look-up, method.

In the experiment, the Q-Learning agent is trained for 1699 time-step interaction between the NASDAQ market and the combination of the Q-Learning agent and Neural Networks predictor. At the time step t, the training parameters (state x_t and actions set A) are defined as following:

$$x_t = Round\{Aver[y_a(t-1) - y_{NN}(t-1)), (y_a(t-2) - y_{NN}(t-2)] / 5\}$$

$$A = \{-200,-100,-70,-50,-20,-10,0,10,20,50,70,100,200\}$$

The prediction performance of the presented hybrid architecture is beginning from 2-Dec-1999 and ended in 30-May-2003 (875 points totally).

All relevant results are illustrated by two figures. Figure 2 plots the real time actual NASDAQ price, the forecasting results of simple ANNs model and the prediction error. Although the model structure is changed dynamically, the error of single ANNs model is still pretty high especially from the year 1999. Apparently the high market volatility creates more possibility for stock price to departure from the "normal" patterns. Figure 3 clearly show the function of Reinforcement Learning agent. After learning the behavior of investors for more than six years, the outputs of agent started from 2-Dec-1999 catch the amount of "abnormal" market behavior amazingly. The most forecasting error of the hybrid architecture is due to the fact that a discrete action set is used.

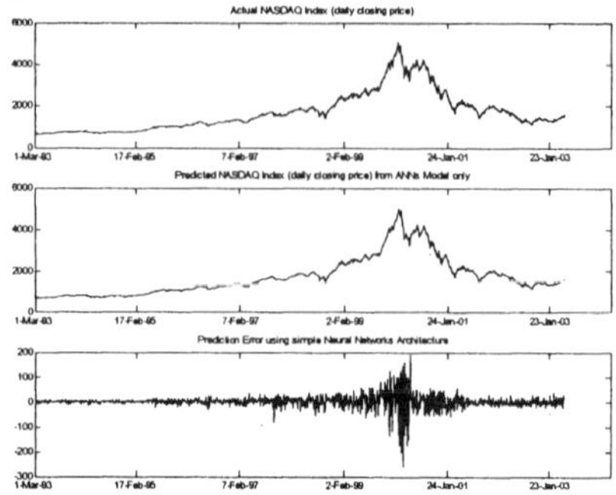

Figure 2. Single ANNs Forecasting Model

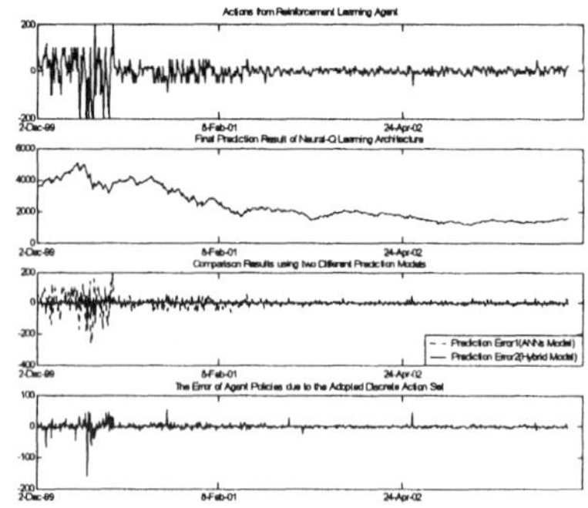

Figure 3. Hybrid Architecture

CONCLUSIONS

This paper presents the hybrid Neuro-Q Learning view for financial series. The proposed stock price forecasting architecture integrates the Reinforcement Learning agent with traditional ANNs model to generalize not only normal market patterns but also "abnormal" investment behavior under different market state. Such method for representing target financial series values might be utilized as a more practical method for financial trading. The experimental results demonstrate the effectiveness of the novel forecasting approach. More research base upon this schema is now in progress.

REFERENCES

[1] Donaldson, R.G., Kamstra, M., "Forecast Combining With Neural Networks", Journal of Forecasting, vol. 15, No. 1, pp. 49-61, (January 1996).

[2] Grudnitzky, G., Osburn, L., "Forecasting S&P and Gold Futures Prises: An Application of Neural Networks", Journal of Futures Markets, vol. 13, No. 6, pp. 631-643, (September 1993).

[3] Kryzanowski, L., Galler, M., Wright, D.W., "Using Artificial Networks to Pick Stocks", Financial Analyst s Journal, pp. 21-27, (August 1993).

[4] Refenes, A.N., Zapranis, A., Francis, G., "Stock Performance Modeling Using Neural Networks: A Comparative Study with Regression Models", Neural Networks, vol. 7, No. 2, pp. 375-388, (1994).

[5] P.G McCluskey, "Feedforward and recurrent neural networks and genetic programs for stock market and time series forecasting", Technical Report CS-93-96, Brown University (1993).

[6] R.J. Kuo, "A decision support system for the stock market through integration of fuzzy neural networks and fuzzy Delphi", Applied Artificial Intelligence, 6: 501-520 (1998).

[7] Marco Costantino, Russell J. Collingham and Richard G. Morgan. "Qualitative Information in Finance: Natural language Processing and Information Extraction", University of Durham, UK (1996).

[8] R.S.Sutton, A.Barto, "Reinforcement Learning: An Introduction", MIT Press, Cambridge, MA (1998).

A HYBRID OPTION PRICING MODEL USING A NEURAL NETWORK FOR FORECASTING VOLATILITY

SUNISA AMORNWATTANA, DAVID ENKE, AND CIHAN DAGLI
Smart Engineering System Lab
Intelligent System Center
University of Missouri - Rolla

ABSTRACT
A great deal of research in the financial domain involves the development of models for pricing financial derivatives, such as options. The Black-Scholes model is the standard approach used for pricing financial options. However, although being theoretically strong, option prices valued by the model often differ from the prices observed in the financial markets. This paper applies a hybrid neural network which preprocesses financial input data for improving the estimation of option market prices. This model is comprised of two parts. The first part is a neural network developed to forecast volatility. The second part is an additional neural network developed to value the difference between the Black-Scholes model result and the actual market option prices. The resulting option price is then a summation between the Black-Scholes model and the network response. The hybrid system with a neural network for estimating volatility provides better performance than either the Black-Scholes model with historical volatility, or the Black-Scholes model with volatility valued by the neural network.

INTRODUCTION

Option pricing has been of great interests for many years. Predicting the correct value of an option can be beneficial for both hedgers and speculators. The Black-Scholes model [1] has become a standard method to value option prices. Nonetheless, even though it has been widely accepted, actual market prices often differ from the value obtained from the Black-Scholes model. Part of this difficulty may result since the Black-Scholes model was developed for European options on non-dividend-paying stocks.

To improve the Black-Scholes model, many option pricing techniques, such as those using GARCH models [2], have been developed. Corrado and Su [3] modified the Black-Scholes model by using skewness and kurtosis adjustment terms. In the recent decade, nonparametric approachs or a neural network are often used for option pricing and volatility estimation [4],[5]. Chen and Lee [6] applied genetic algorithms as an alternative approach for option pricing with European call options. Schittenkopf and Dorffner [7] developed a method of risk-neutral density extraction from option prices based on mixture density networks. Lajbcygier et al. [8] introduced a hybrid system based on a combination of the standard Black-Scholes model and a neural network for estimating historical volatiliy. Karaali et al. [9] and Malliaris and Salchenberger [10] applied neural networks for volatility prediction benchmarked with conventional techniques in time series, digital signal processing, historical volatility, and implied volatility.

THE BLACK-SCHOLES MODEL

In 1973, Black and Scholes [1] proposed a major breakthrough in the pricing of stock options. The model has great influence on the growth and success of financial engineering. This model is derived for valuing European options on non-dividen-paying stocks. It was developed and based on the assumptions of geometric Brownian motion for the underlying asset price, constant variance of the stock return, and no transactions costs. This model uses five input parameters: volatility of the underlying asset, exercise price of the option, price of the underlying asset, number of days until the expriation date, and the risk-free interest rate. The Black-Scholes formulas for valuing call and put options are shown in equations (1) and (2), respectively:

$$C = S_0 N(d_1) - Xe^{-rt} N(d_2) \tag{1}$$

$$P = Xe^{-rt} N(-d_2) - S_0 N(-d_1) \tag{2}$$

where

$$d_1 = \frac{\ln(S_0 / X) + (r + \sigma^2 / 2)T}{\sigma\sqrt{T}} \tag{3}$$

$$d_2 = \frac{\ln(S_0 / X) + (r - \sigma^2 / 2)T}{\sigma\sqrt{T}} \tag{4}$$

and C and P are the European call and European put option price, S_0 is the stock price at time zero, X is the strike price, r is the continuously compounded risk-free rate, σ is the stock price volatility, and T is the time to maturity. The function $N(x)$ is the cumulative probability distribution function for a standardized normal distribution. This model has been extended to consider early exercise and has been improved for calculating values of American call options on dividend paying stocks [11].

NEURAL NETWORK ARCHITECTURE

For this research the general regression neural network (GRNN) [12] is applied. The GRNN architecture is chosen because it provides comparable accuracy to the backpropagation (BP) architecture with substantially less training time. Actually, it is significantly different from the BP model in both algorithm and architecture. It obtains the estimate in a single pass, therefore no iterative training is needed. The network also performs well under noisy environments when given sufficient data.

The GRNN is a supervised feed forward network, subsuming the radial basis-function method. It is based on nonlinear regression theory and computes its output using a variation of the "nearest-neighbor" approach. The forecast value for an input vector is a weighted average of the outputs in the training sample. The closer an input vector in the training sample, the larger the weight of its corresponding output. The GRNN is usually used for function approximation. It has been shown that the GRNN can approximate a continuous function when given a sufficient amount of training data.

HYBRID NEURAL NETWORK

A hybrid neural network model for options pricing was introduced by Lajbcygier et al. [7]. This approach is a combination of the standard Black-Scholes model and a neural network. The concept is to use the conventional Black-Scholes model as the core calculation and then apply a neural network to value the difference between the actual market price and the Black-Scholes model value. The inputs of this model are the risk-free interest rate, time to maturity, volatility, and the ratio of the stock price and strike price (S_0/X). The volatility is obtained from the historical volatility that is calculated from past market data. The target is the difference between the Black-Scholes model and the ratio of call option price and strike price (C/X), as shown in Figure 1.

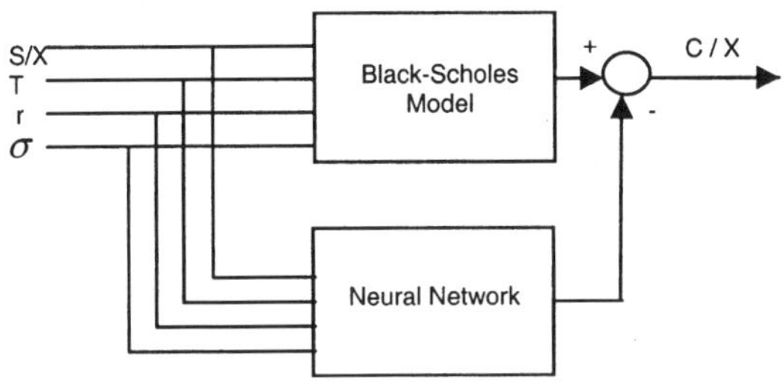

Figure 1. Hybrid Neural Network

NEW HYBRID NEURAL NETWORK MODEL

For this research, the hybrid neural network model was used, although the stock price and strike price are applied individually, instead of using the ratio of the stock price and strike price. Futhermore, the actual market option price replaces the ratio of the call option price and the strike price to obtain the actual difference between the Black-Scholes model and the market option price.

The new model also uses a neural network to estimate volatility. Historical volatility estimates are typically poor since they tend to equally weight each data point. In fact, the more recent the data points, the more influence on the predicted volatility. This assumption provides more accurate estimates of volatility. Several techniques are often applied to solve this problem, such as an exponentially weighted moving average (EWMA) and generalized autoregressive conditional heteroscedasticity (GARCH). Some studies have shown that neural networks for estimating volatility outperform other commonly used techniques [13]. The relationship between asset price and volatility is nonlinear. Neural networks have the characteristic of being able to solve nonlinear problems.

The new hybrid model is divided into two parts, as shown in Figure 2. The first part of the model applys a GRNN to predict volatility. This part has four inputs: stock price, strike price, time to maturity, and the risk-free interest rate. The implied volatility is the target output of this model. The system then combines the predicted volatility obtained from the first network with the first four parameters to provide five inputs to the second part of the model. The second part applies these inputs to obtain a value from the Black-Scholes model. The same five inputs are also applied to a second GRNN, which is also used to predict the price and to value the discrepancy of the Black-Scholes model. For the model, the difference between the Black-Scholes model and actual price are represented as targets. The estimated call or put option prices (C/P) are obtained from the combination of the Black-Scholes model and the second GRNN. The standard Black-Scholes model with historical volatility and the Black-Scholes model with volatility calculated from GRNN are compared as benchmarks.

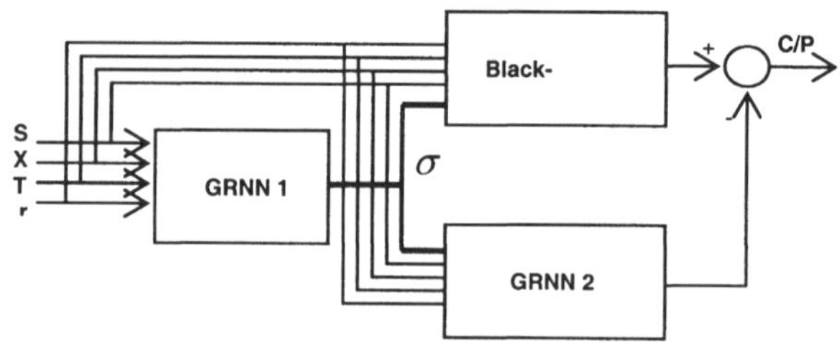

Figure 2. Option Pricing Model

DATA SET

This research applies the call option data from six different underlying primitive stock assets to train and test the model. This model is trained on the period of July 1, 2002 to September 30, 2002. The model is tested on the period October 1, 2002 to October 15, 2002. This study is focused on the closing prices for each day. Details of the data set are shown in Table 1. Each underlying asset is tested under three situations: in-the-money, out-of-the-money, and all (in and out). This experiment focuses on the 90 day historical volatility since the networks are trained under a three month period, allowing each model to be compared during the same time frame.

Underlying Stock	Number of data records					
	All (in and out)		In-the-money		Out-of-the-money	
	Train	Test	Train	Test	Train	Test
KO	3526	649	1676	345	1850	304
MCD	2678	579	915	164	1763	415
BA	4985	887	1755	194	3230	693
C	5678	944	2087	386	3591	558
IBM	6452	1118	2399	292	4053	826
CSCO	5075	889	1522	188	3553	701

Table 1. Amount of training and test data for each stock

IMPLEMENTATION

Data preparation is one of the most important steps for obtaining an effective result when modeling financial data. This is especially true for neural networks, were network training can be made more efficient if preprocessing steps are performed on the network inputs and targets. This study uses a mean and standard deviation approach to normalize the data set. After preprocessing, inputs and targets have zero mean and unity standard deviation. The mean squared error (MSE) and mean absolute deviation (MAD) are used to compare the performance of each model.

RESULTS

Once again, the research tested three models. The first model uses the 90 day historical volatility with the Black-Scholes model (HV-BS). The second model attempts to improve the volatility estimate by using a neural network to predict volatility, before providing this value to the Black-Scholes model (NN-BS). The third model is the proposed new hybrid model (NN-HB). Table 2 gives the results of the mean squared error of the test set. Table 3 gives the results of the mean absolute deviation calculation of the test set.

Stock	All (in and out)			In-the-money			Out-of-the-money		
	HV-BS	NN-BS	NN-HB	HV-BS	NN-BS	NN-HB	HV-BS	NN-BS	NN-HB
KO	0.3553	0.0388	0.0306	0.1279	0.0993	0.0535	0.6134	0.0257	0.0229
MCD	0.0845	0.0049	0.0028	0.0222	0.0089	0.0064	0.1092	0.0042	0.0020
BA	0.1695	0.0197	0.0057	0.1756	0.0571	0.0262	0.2215	0.0156	0.0161
C	3.5470	0.0580	0.0150	1.9558	0.1076	0.0128	4.6478	0.0319	0.0205
IBM	1.2099	0.0493	0.0184	0.7930	0.0847	0.0213	1.3572	0.0329	0.0165
CSCO	0.0358	0.0066	0.0020	0.0457	0.0098	0.0028	0.0331	0.0071	0.0027

Table 2. Mean squared error results

Stock	All (in and out)			In-the-money			Out-of-the-money		
	HV-BS	NN-BS	NN-HB	HV-BS	NN-BS	NN-HB	HV-BS	NN-BS	NN-HB
KO	0.4329	0.1464	0.1025	0.2328	0.2233	0.1427	0.4654	0.1252	0.1050
MCD	0.1846	0.0554	0.0329	0.1132	0.0757	0.0479	0.2095	0.0482	0.0290
BA	0.2638	0.0812	0.0575	0.3317	0.1805	0.0992	0.2996	0.0872	0.0834
C	1.1378	0.1644	0.0810	0.9424	0.2332	0.0749	1.2343	0.1257	0.0959
IBM	0.7146	0.1451	0.0844	0.5478	0.2120	0.0896	0.7494	0.1171	0.0782
CSCO	0.1199	0.0580	0.0324	0.0921	0.0584	0.0409	0.1193	0.0574	0.0355

Table 3. Mean absolute deviation results

The results from table 2 and table 3 show that the Black-Scholes model with volatility calculated from the neural network estimates the option price significantly better than when historical volatility is used as an input to the Black-Scholes model. Futhermore, the neural network trained on the difference between the Black-Scholes model and the neural network result presents an improvement in option pricing estimation for the tested data set. This demonstrates the usefulness of combining these two methods.

CONCLUSION AND FUTURE WORK

This research demonstrates the benefits of using neural networks for option pricing, since neural networks can be used to learn from the past data to predict volatility and account for non-ideal conditions that cause problems for the Black-Scholes model. The hybrid neural network is shown to improve the performance of the Black-Scholes model to capture deviations from the strict assumptions. When combining a neural network for estimating volatility with the hybrid neural network, superior performance is achieved.

Further research should examine different option pricing models to investigate the impact on the results. Dividends should also be considered as an additional parameter. Other volatility estimating approaches should also be explored.

REFERENCES

[1] Black, F. and Scholes, M., (1973) "The Pricing of Options and Corporate Liabilities", Journal of Political Economy, vol.81, pp. 637-654.

[2] Duan, J.C., Gauthier, G., Sasseville, C. and Simonato, J.G., (2003) "Approximating American Option Prices in the GARCH Framework", (to appear in Journal of Future Markets).

[3] Corrado, C.J. and Su, T., (1996) "Skewness and Kurtosis in S&P 500 Index Returns Implied by Option Prices", Journal of Financial Research, vol. 19, pp. 175-192.

[4] Hutchinson, J.M., Lo, A. W. and Poggio, T., (1994) "A Nonparametric Approach to Pricing and Hedging Derivatives Securities via Learning Networks", Journal of Finance, vol. 49, pp. 851-889.

[5] Hanke, M., (1999) "Neural Networks Versus Black-Scholes: An Empirical Comparison of the Pricing Accuracy of two Fundamentally Different Option Pricing Methods", Journal of Computational Intelligent Finance, vol.5, pp. 26-34.

[6] Chen, Shu-Heng and Lee, Who-Chiang, (1997) "Option Pricing with Genetic Algorithms: a Second Report" , Neural Network , International Conference, vol. 1, pp.21-25.

[7] Schittenkopf, C. and Dorffner, G., (2001) "Risk-Neutral Density Extraction from Option Prices: Improved Pricing with Mixture Density Networks", Neural Network, IEEE Transactions on, vol. 12, July 2001, pp. 716-725.

[8] Lajbcygier, P., Flitman, A., Swan, A. and Hyndman, R., (1997) "The Pricing and Trading of Options Using a Hybrid Neural Network Model with Historical Volatility", Neurovest Journal, vol. 5, no. 1, pp. 27-41.

[9] Karaali, O., Edelberg, W. and Higgin, J., (1997) "Modeling Volatility Derivatives Using Neural Networks", Computational Intelligence for Financial Engineering, pp. 280-286.

[10] Malliaris, M. and Salchenberger, L., (1994) "Do-Ahead Replaces Run-Time: a Neural Network Forecasts Options Volatility", Artificial Intelligence for Applications, Proceeding of the Tenth Conference on, 1994 pp. 480-481.

[11] Whaley, R. ,(1981) "On the Valuation of American Call Options on Stock with Known Dividends", Journal of Financial Economics, vol. 9, pp. 207-211.

[12] Specht, D. ,(1991) "A General Regression Neural Network", IEEE Transactions on Neural Networks, vol. 2, no. 4, pp. 568-576.

[13] Refenes, A.P. and Holt, W., (2001) "Forecasting Volatility with Neural Regression: A Contribution to Model Adequacy", IEEE Transactions on Neural Networks, vol.12, no.4.

New Methods to Test Elastic Market Efficiency Theory with Neural Networks

Hua Jiang
Department of Mechanical Engineering
Ryerson University Toronto ON M5B2K3 Canada Email: hjiang@ryerson.ca

ABSTRACT

In this paper, we proposed elastic efficient market theory followed by neural network experiments. Three neural network methods was used the optimization process is done by applying the new parameter dynamic beta to forecasting results. Mathematical induction of new parameter dynamic beta β_i is presented and the relationship ψ between dynamic βi and $d\sigma_p$ is defined with $\psi = C(\sigma_p^{-1})$ where C is constant. The empirical result successfully recognized the pattern of a ten stock portfolio.

KEYWORDS

Neural Network, Dynamic Beta, Pattern Recognition, Prediction, Stock Market

INTRODUCTION

We propose the elastic efficient market theory and derived our neural testing basics by comparing it with traditional efficient market theory in three levels, each of which is explained as below and the sample testing data are including in the section after. Lastly, we completed this paper by describes the mathematical induction of dynamic beta we found through the neural network experiments and formulated the final equation. The three neural network testing methods includes *predict day close (PDC)* -using adjusted close price of the passing 60 days; *predict one set (POS)* method-using first hour's trading data to forecast the next hour's trading data and run regression on the result to find out the trading day close price; *cyclic price prediction (CPP)* -beginning with first hour's trading data and predict next hour's data and then using the next hour's trading data to predict one more hour further till it reaches the trading day closing point. Taking one stock T (AT&T Corp.) as an example, we describe the optimization process as follows.

NEW METHODS

The following tables present in details the results obtained by using dynamic beta to optimize neural network testing result of the CPP method, PDC method and POS method. Factors considered are as follows:

Abbreviation	Variable
IR	Interest Rate
CPI	Consumer Price Index for Five Years
MS	Money 2 Supply
MIA	Market Index Average
UR	Unemployment Rate
MD	Monthly Trading Data for Five Years
FHTD	First Hour Trading Data
HHTD	Half Hour Trading Data
DD	60 Day Trading Data
TO	Target Output

Table1. List of input factors

INPUT NEURONS ANALYSIS

For this stock, we find the following correlation result of input factors.

	IR	CPI	MS	MIA	UR	MD	FHTD
IR	1	0.04283	0.15333	0.31679	-0.07359	0.36008	0.10617
CPI	0.04283	1	0.27006	0.00507	0.23449	0.07319	0.11526
MS	0.15333	0.27006	1	0.01183	0.2418	0.22965	0.34612
MIA	0.31679	0.00507	0.01183	1	0.18609	0.44684	0.3257
UR	-0.07359	0.23449	0.2418	0.18609	1	0.2179	0.07268
MD	0.36008	0.07319	0.22965	0.44684	0.2179	1	0.49101
FHTD	0.10617	0.11526	0.34612	0.3257	0.07268	0.49101	1

Table 2. Input correlation

We conclude that overall importance level of input neurons from high to low sequence is MD, MS, MIA, CPI, IR, and UR.

OPTIMIZATION THROUGH DYNAMIC BETA

In this section, we use the portfolio as an example to describe our optimization method using dynamic beta. Three methods produce different intra day close price. Which one to choose? We find the optimized result based on the following selection process. From the regression result of each stock on a testing day, we can find the dynamic beta as listed in table below.

SYMBOL	TLC	BR	AC	COR	HUM	BRA	RCMB	SCC	PCA	MFI
BETA	0.301	0.673	1.993	0.868	0.824	1.817	1.402	1.226	0.545	0.588

Table3. Summary of dynamic beta

A higher beta means the stock is has more systematic risk, a lower beta means the stock has less systematic risk. For example AC with beta 1.993, we can assume because of political and macro economic situation, AC change 1.993 percent when MIA change every one percent. Therefore we choose PDC result as the optimized final result for all the stocks with POS beta smaller than 1. We choose PDC result as the optimized final result for all stocks with POS beta above 1. We choose POS result as the optimized final result for all stocks with zero betas.

ELASTIC EFFICIENT MARKET THEORY

LEVEL 1. STRONG FORM

The strong form of market efficiency theorizes that the current price reflects all information available. It does not matter if this information is available to the public or privy to top management; if it exists at all, then it is already reflected in the current price. To this point, we agree with the efficient market theory. Because all possible information is already reflected in the price, investors and traders will not be able to find or exploit inefficiencies based on fundamental information. We can't, however, agree here. Our neural network-forecasting model actually predicted the inefficiencies when short-term market behavior went beyond control and when demands and supply rules are broken as is seen in September 11 event and timesaving clock shift period. Therefore we summarize one theory, elastic efficient market theory, which covers a short-term inefficient market that can be quickly recovered and changed to efficient market.

LEVEL 2. SEMI STRONG FORM

The semi-strong form of market efficiency theorizes that the current price reflects all readily available information. This information will likely include annual reports, SEC filings, earnings reports, announcements and other relevant information that can be readily gathered. However, there is other information not readily available to the public that is not fully reflected in the price. This could be information held by insiders, competitors, contractors, suppliers or regulators, among others. Anomalies exist when information is withheld from the public and the only way to profit is by using information not yet known to the public. This is sometimes called insider trading. Once this information becomes public knowledge, prices adjust instantaneously and it is virtually impossible to profit from such news. Academics and Random Walkers believe that the markets are semi-strong efficient. Prices reflect public information and it is virtually impossible to profit from this information. We cannot agree with it. Neural Networks has the ability to detect hidden patterns including insider trading. Base on our prediction, elastic efficient market theory successfully explained that investors could profit from insider news even if it is unknown to them. The reason is because insiders are generally limited to trade in terms of volume and time. For example, the director of a public corporation wishes to sell stock of that company in compliance with Rule 144, unregistered securities must be held for 1 year, fully paid, before they can be sold, if the sale is for more than 500 shares, worth more than $10,000, SEC filing is required, and any short swing profits (within 6 months) from trading the stock must be returned to the corporation. As a result, insiders trading could rarely occur. In some cases, even if it does happen, neural networks can detect the reflected market price changes and suggest public investors respond instantaneously. Another reason is that to reach price where insider news targeted, it takes a long time. If not, the news would have been circulated to the public already. The time value can be covered by securities and other financial instruments such as options, thus the general investors are free to get hold of the options and profit. Correspondingly, elastic efficient market theory proves that semi strong form doesn't hold. Prices reflect both public information and privy information. While prices works on other financial instruments i.e. options, other financial instruments prices reverse the same effect and secure the fair trading environment, which we call it an elastic efficient market.

LEVEL 3. WEAK FORM

The weak form of market efficiency theorizes that the current price does not reflect fair value and is only a reflection of past prices. Furthermore, the future price cannot be determined using past or current prices (sorry technical analysts). Fundamental analysts are champions of weak-form market efficiency and believe that the true value of a security can be ascertained through financial models using information readily available. The current price will not always reflect fair value and these models will help identify anomalies. To understand the word fair here, we'd better use "Price As If" to avoid the weak form anomalies. Weak form supporters blame that "As If" price is far from the current price. They think "As If" price is the true value of the stock. However, I have to argue that to cover these two prices difference, value of information flow should be counted. I.e. the amount of price difference equals the value we need to pay to move the information from where it is to the public and to be generally accepted. The flow of information has become faster with the Internet and surprises are factored in instantly. However quick arrival of information doesn't necessarily mean quick acceptance. Just like mass manufacturing sometimes can only cause excess inventory and reverse the original purposes. Since the boom of IT sector in the 90s, we all experience the difficulties of handling garbage information and filter the valuable information. Many emails, commercial letters and news are ignored and deleted without being reviewed. Thus the deviated market price is a "fair" value and we call it "As it is" value. Thus our elastic efficient market theory explained all the three levels forms and trims them into one. This has been proved by our experiment. The neural networks model has recognized the abstract rules of each.

FORMULATE DYNAMIC BETA EQUATION

TERMS

$E(r_c)$	*Expected return of combined portfolio C*
$E(r_i)$	*Expected return of individual asset I*
$E(r_p)$	*Expected return of tangency portfolio P*
$E(r_m)$	*Expected return of market portfolio*
	The fraction of individual asset i in combined portfolio C
$(1-\ )$	*The fraction of tangency portfolio P in combined portfolio C*

$_c$	*The variance of combined portfolio C*
$_i$	*The variance of individual asset I*
$_p$	*The variance of tangency portfolio P*
$Cov(r_i, r_p)$	*Covariance of return on individual asset i and return on portfolio P*
r_f	*Return on risk free asset i.e. T-Bill rate*

DYNAMIC BETA INDUCTION

$E(r_c) = \quad E(r_i) + (1- \quad) E(r_p)$ (1)

${}_c^2 = {}^2 \ {}_i^2 + (1- \ )^2 \ {}_p^2 + 2 \ (1- \ ) Cov(r_i, r_p)$ (2)

$E(r_c) = y E(r_p) + (1-y) r_f$

$_c = y \ _p$

$y = \ _c / \ _p$

$E(r_c) - r_f = y [E(r_p) - r_f] = (\ _c / \ _p) [E(r_p) - r_f]$ (3)

We are interested in how would expect return of the combined portfolio change for each unit increase in individual asset in the combined portfolio.

Use (1) $\frac{dE(r_c)}{d\alpha} = E(r_i) - E(r_p)$ (4)

Use (3) $\frac{dE(r_c)}{d\alpha} = \frac{d\sigma_c}{d\alpha}[\frac{E(r_p - r_f)}{\sigma_p}]$ (5)

Use (2) $2\sigma_c \frac{d\sigma_c}{d\alpha} = 2\alpha\sigma_i^2 - 2(1-\alpha)\sigma_p^2 + 2(1-\alpha)Cov(r_i, r_p) - 2\alpha Cov(r_i, r_p)$ (6)

$2\sigma_c \frac{d\sigma_c}{d\alpha} = -2\sigma_p^2 + 2Cov(r_i, r_p)$

$\frac{d\sigma_c}{d\alpha} = \frac{-\sigma_p^2 + Cov(r_i, r_p)}{\sigma_c}$ (7)

Substitute (4) & (7) to (5)

$E(r_i) - E(r_p) = \frac{-\sigma_p^2 + Cov(r_i, r_p)}{\sigma_c}[\frac{E(r_p) - r_f}{\sigma_p}]$

From (2) ${}_c^2 = {}_p^2, \quad {}_c = {}_p$

$\sigma_p^2[E(r_i) - E(r_p)] = -\sigma_p^2[E(r_p) - r_f] + Cov(r_i, r_p)[E(r_p) - r_f]$

$\sigma_p^2[E(r_i) - r_f] = Cov(r_i, r_p)[E(r_p) - r_f]$

$E(r_i) - r_f = \frac{Cov(r_i, r_p)}{\sigma_p^2}[E(r_p) - r_f]$ (8)

Since $\beta i = \frac{Cov(r_i, r_p)}{\sigma_p^2}$ (9)

Hence $E(r_i) - r_f = \beta_i[E(r_p) - r_f]$ (10)

$E(r_i) = \beta_i[E(r_p) - r_f] + r_f$

When p = m, this becomes

$E(r_i) = r_f + \beta_i[E(r_m) - r_f]$ (11)

Where $\beta i = \frac{Cov(r_i, r_m)}{\sigma_m^2}$ (12)

We solve the problem of inefficient market as follows

(A) When the market is inefficient, CAPM model doesn't hold. Traditional explanation on why CAPM model is rejected by empirical data is that the portfolio must be efficient and $\beta i, E(r_m)$ has to be efficient. In this thesis, we find that Neural Network model complies with elastic efficient market theory all the time. I.e. in strong form, semi-strong form and weak form. We find that the portfolio react to market in every minute and $\beta i, E(r_m)$ that we use are never constant, when we use minute by minute data, $E(r_m)$ is the actual forecasted market return and it is highly efficient. We are then interested in how the clastic efficient market theory applies to each variable of different mean time period.

Taking derivatives of time t on each side of (9), we have

$$\frac{d\beta i}{dt} = d\left(\frac{Cov(r_i, r_p)}{\sigma_p^2}\right)/dt$$

Because Covariance measures of the variance of two random variables r_i, r_p that are observed in the same mean time period, $Cov(r_i, r_p)$ is constant for each dt.

$$\frac{d\beta i}{dt} = Cov(r_i, r_p)d(\sigma_p^{-2})/dt$$

$$\frac{d\beta i}{dt} = -2Cov(r_i, r_p)(\sigma_p^{-3})d(\sigma_p)/dt$$

Let us call this factor ψ, we have $\psi = -2Cov(r_i, r_p)(\sigma_p^{-3})$ (13)

Hence $\frac{d\beta i}{dt} = \psi d(\sigma_p)/dt$ (14)

Also $\psi = \frac{d\beta i}{d\sigma_p}$, this is the term we use to describe how individual asset's β shift with portfolio P's risk factor. Thus we find that public traded firm's beta changes in proportion to the third power of portfolio P's risk reciprocal given the same mean time period. Special case occurs when the market portfolio M replaces P. In elastic efficient portfolio theory, public traded firm's beta changes in proportion to the third power of market portfolio's risk reciprocal given the same mean time period. These find is so significant that it proves that given the same mean time period, sets of data can be used to forecast dynamic beta movement and thus the dynamic price prediction.

This induction supports our three neural network methods which are based on minute by minute data in the cyclic price prediction method and predict one set method, day by day data in the predict day close method.

(B) Another explanation on why CAPM is rejected by empirical data is that P=m should for all individuals in order to validate CAPM. However, the reality is that not all individuals hold all the stocks in the index.

From (13) we know that the market is relatively efficient on the same mean time period, it is absolutely efficient when all individuals hold all the stocks in the index that could never possibly occur. Through empirical Neural Network testing result, we find that NYSE, NASDAQ, TSE are absolute efficient generally, relatively efficient when September 11 and timesaving occurred. However the relative status can be recovered after one mean time period say one month. This is how elastic efficient market theory overcome CAPM model in proving market efficiency.

(C) The third explanation on why CAPM is rejected by empirical data is that market index is not universal and market point is not efficient. I.e. DJIA is based on 30 stocks movement only.

Let's prove that our theory is valid by looking at the following index. DJIA is the most widely used indicator of the overall condition of the stock market, a price-weighted average of 30 actively traded blue chip stocks, primarily industrials. The 30 stocks are chosen by the editors of the Wall Street Journal (which is published by Dow Jones & Company), a practice that dates back to the beginning of the century. Charles Dow officially started the Dow in 1896, at which time it consisted of only 11 stocks. The Dow is computed using a price-weighted indexing system, rather than the more common market cap-weighted indexing system. From (13) we know that portfolio P can be any composition of publicly traded stocks and it of course include the 30 stocks in the DJIA index. This problem can be solved in the same way as in problem (B).

CONCLUSION

To summarize the above problems encountered by tradition CAPM model, we can say that the key to solve them is to find a model that can quickly derive rules out of the smallest divisible time frame. Neural Network forecasting data is no doubt the first candidate to accomplish this task. As described above, we use minute-by-minute trading data to look at the dynamic beta effect on portfolio risk and targeted the market efficiency with a new approach by proposing the elastic efficient market theory. We are looking forward to more industry support to our further research on bringing elastic efficient market theory and dynamic beta applied to neural network to more complex portfolio, index portfolio and testing whole market as one.

REFERENCE

1. *M. A. Arbib. Part I: Background. In M. A. Arbib, editor, The Handbook of Brain Theory and Neural Networks, pages 1–25. MIT Press, Cambridge, MA, 1995.*
2. *N. J. Nilsson. Learning Machines. McGraw-Hill, New York, 1965.*
3. *B. Widrow and S. D. Stearns. Adaptive Signal Processing. Prentice-Hall, Englewood Cliffs, N.J, 1985.*
4. *M. A. Arbib. Part I: Background. In M. A. Arbib, editor, The Handbook of Brain Theory and Neural Networks, pages 1–25. MIT Press, Cambridge, MA, 1995.*
5. *B. M"uller, J. Reinhardt, and M. T. Strickland. Neural Networks An Introduction. Physics of Neural Networks. Springer-Verlag, Berlin, second edition, 1995.*
6. *C. M. Bishop. Neural Networks for Pattern Recognition. Oxford University Press, Oxford, 1995.*
7. *S. Geman, E. Bienenstock, and R. Doursat. Neural networks and the bias/variance dilema. Neural Computation, 4(1):1–58, 1992.*
8. *S. Haykin. Neural Networks A Comprehensive Foundation. Prentice-Hall, Upper Saddle River, NJ, second edition, 1999.*
9. *C. M. Bishop. Neural Networks for Pattern Recognition. Oxford University Press, Oxford, 1995.*
10. *G. J. Gibson and C. F. N. Cowan. On the decision regions of multi-layer perceptrons. Proceedings of the IEEE, 78(10):1590–1594, 1990.*
11. *S. Tamura and M. Tateishi. Capabilities of a four-layered feedforward neural network: Four layers versus three. IEEE Transactions on Neural Networks, 8(2):251–255, 1997.*
12. *K. J. Lang and M. J. Witbrock. Learning to tell two spirals apart. In T. Sejnowski D. Touretzky, G. Hinton, editor, Proceedings of the 1988 Connectionist Models Summer School, pages 52–59, San Mateo, CA, 1988. Morgan Kaufmann.*
13. *D. L. Chester. Why two hidden layers are better than one. In International Joint Conference on Neural Networks, volume I, pages 265–268. Lawrence Erlbaum, 1990.*
14. *A. G. Barto. Learning as hill-climbing in weight space. In Michael A. Arbib, editor, The Handbook of Brain Theory and Neural Networks, pages 531–533. The MIT Press, Cambridge, MA, 1995.*
15. *D. E. Rumelhart, G. E. Hinton, and R. J. Williams. Learning internal representations by error propagation. In Parallel Distributed Processing, volume 1, chapter 8, pages 318–362. MIT Press, Cambridge, MA, 1986.*
16. *J. Hertz, A. Krogh, and R. G. Palmer. Introduction to the Theory of Neural Computation, volume I of Santa Fe Institute studies in the sciences of complexity. Addison-Wesley, Redwood City, CA, 1991.*

SHORT TERM FORECASTING WITH SUPPORT VECTOR MACHINES AND APPLICATION TO STOCK PRICE PREDICTION

HUSEYIN INCE
Faculty of Business Administration
Gebze Institute of Technology
Gebze,Kocaeli, TURKEY
ince@yahoo.com

THEODORE B. TRAFALIS
School of Industrial Engineering
University of Oklahoma, Norman
OKLAHOMA, USA
ttrafalis@ou.edu

ABSTRACT

Financial time series are complex, non stationary and deterministically chaotic. Therefore, it is impossible to forecast with parametric models such as regression. Instead of parametric models, we propose two techniques and compare those with each other. They are data-driven non parametric models. Two different models are assumed with different inputs. Our assumption is that the future value of a stock price depends on the financial indicators although there is no parametric model to explain this relationship. This relationship comes from the technical analysis. Comparison shows that SVR over performs the multi layer perceptron (MLP) networks for a short term prediction.

INTRODUCTION

One of the goals of financial methods is asset evaluation which means determining the market price of an asset, predicting what this price will be in the future, and how it will move with other indicators. The behavior of an asset can be analyzed by using technical tools, parametric pricing methods or combination of these methods. Prediction of financial time series is one of the most challenging applications. Since the financial market is a complex, non stationary and deterministically chaotic system, it is very difficult to forecast by using deterministic (parametric) techniques because of the assumptions behind the parametric techniques. In this case, nonparametric techniques such as support vector regression (SVR), and neural networks (NN) are good candidates for financial time series forecasting. Over the past decades, neural networks and radial basis function networks have been used for financial applications ranging from option pricing, stock index trading to currency exchange (Chen et al 2003, Galindo,1998, Leigh, 2002,Trafalis et al, 2003).

The SVR algorithm that was developed by Vapnik (1995) is based on statistical learning theory. In the case of regression, the goal is to construct a hyperplane that lies "close" to as many of the data points as possible (Cortes and Vapnik, 1995, Pontil and Verri, 1997). This determines the trend line of training data, hence the name support vector regression (SVR).

Technical analysis, also known as charting, is used widely in real life. It is based on some technical indicators. These indicators help the investors to buy or sell the underlying stock. Our goal is to try to understand the relationship between the stock price and these indicators. These indicators are computed by using stock price and volume overtime. Therefore, we cannot use parametric techniques such as regression. We will investigate the prediction of individual stock prices by applying neural networks and support vector regression methods. Several technical indicators are used as inputs to these models.

Section 2 explains the theory of support vector regression SVR. In section 3 experiments and comparison of the methods will be discussed. Section 4 concludes the paper.

SUPPORT VECTOR REGRESSION (SVR)

The ε -insensitive support vector regression will be explained. In the ε -insensitive support vector regression, our goal is to find a function *f(x)* that has ε deviation from the actually obtained target y_i for all training data and at the same time is as flat as possible. Suppose *f(x)* takes the following form,

$$f(x) = w \cdot x + b, \quad w \in X, b \in \Re \tag{1}$$

If we have a ***w*** with a small norm, then we can say that *f* is flat. One way to do this is to minimize $||w||^2$ using the euclidean norm (Smola and Schölkopf, 1998) subject to the constraints of the so-called ε-insensitive band. One has to solve this problem in order to obtain an ε-insensitive SVR solution. Usually, we need to allow for some errors. We introduce slack variables ξ_i, ξ^*_i to cope with this situation. This case is called soft margin formulation. Specifically, we solve the following problem:

$$\begin{aligned} & \min \ \frac{1}{2}\|\mathbf{w}\|^2 + C\sum(\xi_i + \xi_i^*) \\ & \text{Subject to} \\ & \mathbf{y}_i - \mathbf{wx}_i - b \le \varepsilon + \xi_i \\ & \mathbf{wx}_i + b - \mathbf{y}_i \le \varepsilon + \xi_i^* \\ & \xi_i, \xi_i^* \ge 0, C \ge 0 \end{aligned} \tag{2}$$

where *C* determines the trade-off between the flatness of the *f(x)* and the amount up to which deviations larger than ε are tolerated.

The lagrangian function will help us to formulate the dual problem, which will give us a quadratic programming problem formulation. Next we construct the dual problem. The reason is that solving the primal problem is difficult due to too many variables. If we use the dual problem formulation, we can decrease the variables and the size of the problem becomes smaller. Specifically the dual problem is as follows:

$$\begin{aligned} & \max \ -\frac{1}{2}\sum_{i=1}^{l}\sum_{j=1}^{l}(\lambda_i - \lambda_i^*)(\lambda_j - \lambda_j^*)\mathbf{x}_i\mathbf{x}_j - \varepsilon\sum_{i=1}^{l}(\lambda_i + \lambda_i^*) + \sum_{i=1}^{l} y_i(\lambda_i - \lambda_i^*) \\ & \text{Subject to} \ \sum(\lambda_i - \lambda_i^*) = 0 \\ & \qquad \lambda_i, \lambda_i^* \in (0, C) \end{aligned} \tag{3}$$

Next we consider the non-linear case. First of all, we need to map the input space into the feature space and try to find a regression hyperplane in the feature space. We can accomplish that by using the kernel function k(x,y). In other words we replace k as follows:

$$k(x,y)=\phi(x)\cdot\phi(y)$$

$$\max\ -\frac{1}{2}\sum_{i=1}^{l}\sum_{j=1}^{l}(\lambda_i-\lambda_i^*)(\lambda_j-\lambda_j^*)K(\mathbf{x}_i,\mathbf{x}_j)-\varepsilon\sum_{i=1}^{l}(\lambda_i+\lambda_i^*)+\sum_{i=1}^{l}y_i(\lambda_i-\lambda_i^*)$$

$$\text{Subject to}\quad \sum(\lambda_i-\lambda_i^*)=0 \qquad (4)$$

$$\lambda_i,\lambda_i^*\in(0,C)$$

At the optimal solution, we obtain

$$w^*=\sum_{i=1}^{l}(\lambda_i-\lambda_i^*)K(x_i),\quad \text{and}$$

$$f(x)=\sum_{i=1}^{l}(\lambda_i-\lambda_i^*)K(x_i,x)+b \qquad (5)$$

The main difference between the linear and the non-linear case is that in the non-linear case ***w*** is not given explicitly anymore. On the other hand, by using dot products, it is defined uniquely (Smola and Schölkopf, 1998). Also, we work in the feature space using linear techniques. In order to solve the optimization problem given in equation (4), one needs an efficient algorithm. Recently, several decomposition techniques have been developed to solve this large scale problem (Collobert and Bengio, 2001, Lee and Mangasarian,2001, Osuna et al,1997, Trafalis and Ince, 2002,Cortis and Vapnik, 1995).

TECHNICAL INDICATORS OF FINANCIAL TIME SERIES

Some of the financial indicators which have been used in this research will be explained briefly.

Exponential Moving Average (EMA): Exponential moving averages reduce the lag by applying more weight to recent prices relative to older prices. The weighting applied to the most recent price depends on the length of the moving average. The shorter the exponential moving average is, the more weight will be applied to the most recent price. The important thing is that the exponential moving average puts more weight on recent prices.

Relative Strength Index (RSI): The Relative Strength Index (RSI), developed by J. W. Wilder, is an extremely useful and popular momentum oscillator. The RSI compares the magnitude of a stock's recent gains to the magnitude of its recent losses and turns that information into a number that ranges from 0 to 100. It takes a single parameter, the number of time periods to use in the calculation.

Bollinger Band (BB): Developed by John Bollinger, Bollinger Band is an indicator that allows users to compare volatility and relative prices levels over a period of time. The indicator consists of three bands designed to encompass the majority of a security's price action, a simple moving average in the middle, an upper band (SMA plus 2 standard deviations), and a lower band (SMA minus 2 standard deviations).

Moving Average Convergence Divergence (MACD): It is one of the simplest and most reliable indicators available. MACD uses moving averages, which are lagging indicators, to include some trend-following characteristics. These lagging indicators are turned into a momentum oscillator by subtracting the longer moving average from the shorter moving average. The resulting plot forms a line that oscillates above and below zero, without any upper or lower limits. MACD is a centered oscillator and the guidelines for using centered oscillators apply. The most popular formula for the "standard" MACD is the difference between a security's 26-day and 12-day exponential moving averages.

Chaikin Money Flow (CMF): The CMF oscillator is calculated from the daily readings of the Accumulation/Distribution Line. The basic premise behind the Accumulation Distribution Line is that the degree of buying or selling pressure can be determined by the location of the close relative to the high and low for the corresponding period (Closing Location Value). There is buying pressure when a stock closes in the upper half of a period's range and there is selling pressure when a stock closes in the lower half of the period's trading range.

EXPERIMENTS

There have been several papers discussing the use of non parametric techniques for financial forecasting. For example neural networks, time series (ARIMA) models have been applied to stock index and exchange rate forecasting (Tsaih, 1999, Leigh 2002). SVR has been applied to stock price forecasting and option price prediction (Trafalis et al, 2003).

We have used the daily stock prices of Microsoft and Yahoo for 3 years span. The training set contains 670 observations and the testing/validation set consists of 20 examples. We propose two simple models. In the first model, we assume that the current stock price depends on the previous EMA, Volume, RSI, BB, MACD and CMF.

$$S_t = f(EMA_{t-1}, V_{t-1}, RSI_{t-1}, BB_{t-1}, MACD_{t-1}, CMF_{t-1}),$$

The second model is

$$S_t = f(EMA_{t-1}, EMA_{t-2}, V_{t-1}, V_{t-2}, RSI_{t-1}, RSI_{t-2}, RSI_{t-3}, BB_{t-1}, BB_{t-2}, MACD_{t-1}, MACD_{t-2}, CMF_{t-1}, CMF_{t-2})$$

The relationship between stock price and these indicators is highly nonlinear. They are also correlated with each other. Therefore, it should be wise not to use any parametric techniques such as multiple regression models. Instead of parametric models, non parametric data driven model, SVR and MLP will be applied with different settings. We have chosen three different techniques to apply these two models. Comparison of MLP and SVR will be given in terms of the following statistics; normalized mean square error (NMSE) and root mean square error (RMSE) respectively.

We have looked at the effect of the radial basis kernel function with different σ's for MSFT and YAHOO models. The experiments show that σ has different effect on training and testing sets in terms of RMSE and NMSE.

Table 1 shows the RMSE on the testing set. As it can be seen from table 1, when σ increases, the RMSE is getting smaller.

Table 1: Effects of the different σ in the radial basis kernel function for validation set in terms of RMSE.

Models	σ							
	0.5	1	2.5	5	10	50	75	100
Yahoo(Model 1)	19.34	21.63	31.62	21.16	4.211	1.056	0.869	0.725
Yahoo(Model 2)	18.60	18.64	20.34	26.69	15.858	1.485	1.608	0.892
MSFT(Model 1)	5.577	5.112	2.028	2.040	1.232	0.765	0.813	0.824
MSFT(Model 2)	5.588	5.576	4.811	2.968	1.639	0.728	0.672	0.753

Different network structures have been applied to these models. There are two problems with the backpropagation (BP) algorithm. The BP algorithm converges to local minima. Secondly, the overfitting problem may be seen if the numbers of neurons are increased more than the network needs. This can be seen from table (2). When the number of neurons in the hidden layer is increased, the RMSE is decreased.

Table 2: When the number of neurons in hidden layer is increased, the RMSE decreases for the training set. On the other hand, after a point, the RMSE for testing set starts increasing. This can be explained with over fitting.

	(15-7-1)	(15-12-1)	(15-14-1)	(15-21-1)	(15-30-1)
Testing(RMSE)					
Yahoo (Mod. 1)	1.1069	1.0278	1.6312	0.9666	1.5854
Yahoo(Mod. 2)	1.3676	1.6722	1.4945	1.8418	3.6541
MSFT(Mod. 1)	0.7616	0.7988	0.8569	1.6214	1.7575
MSFT(Mod. 2)	0.9336	1.1179	0.9758	1.3890	1.7656
Training(RMSE)					
Yahoo (Mod. 1)	2.6056	2.0594	1.8035	1.0927	0.8622
Yahoo(Mod. 2)	1.8767	1.2232	0.8491	0.6370	0.3395
MSFT(Mod. 1)	0.8704	0.7452	0.7362	0.6183	0.5204
MSFT(Mod. 2)	0.7554	0.6771	0.5514	0.4119	0.1905

Table 3 shows the comparison of the best models for SVR and MLP in terms of the RMSE and NMSE statistics. As it can be seen from table 4, SVR outperforms the MLP.

Table 3: Comparison of RMSE and NMSE for SVR and MLP networks. For each technique, the best configuration is determined in terms of testing data(RMSE) from table 1 and table 2.

	RMSE		NMSE	
	SVR	MLP	SVR	MLP
Yahoo(Model 1)	0.7249	0.9667	0.0009	0.0016
Yahoo(Model 2)	0.8918	1.3676	0.0014	0.0033
MSFT(Model 1)	0.7651	0.7617	0.5787	0.5745
MSFT(Model 2)	0.6718	0.9337	0.4479	0.8562

CONCLUSIONS

We have presented a different approach for stock price forecasting. Two different models are used to explain the relationship between stock price and technical indicators. Usually, these indicators help the investors/traders to identify a pattern in a stock price chart. In this paper, we look at them in different perspective. They are the input variables for MLP and SVR methods.

Our results show that SVR outperforms the MLP networks. The next step is to develop a classification model for the stock price. By doing this, one can identify when to buy and sell the underlying stock. This kind of model would be more appropriate than stock price forecasting.

REFERENCES

Chen, A., Leung, M. T., and Daouk, H., 2003, Application of neural networks to an emerging market: forecasting and trading the Taiwan Stock Index, *Computers & Operation Research (30)*: 901-923.

Collobert, R., and Bengio, S., 2001, Svmtorch: Support vector machines for large-scale regression problems. *Journal of Machine Learning Research*, 1:143–160.

Cortes, C., and Vapnik, V.,1995, Support vector networks. *Machine Learning*, 20:273–297.

Galindo, J., 1998, A framework for comperative analysis of statistical and machine learning methods: An application to the black scholes option pricing equations. Technical report, Banco de Mexico, Mexico, DF, 04930.

Lee, Y.-J., and Mangasarian, O.L.,2001, Rsvm: Reduced support vector machines. CD Proceedings of the First SIAM International Conference on Data Mining, 89.

Leigh, W., Paz, M., and Purvis, R., 2002, An analysis of a hybrid neural network and pattern recognition technique for predicting short-term increases in the NYSE, *The International Journal of Management Science (30)*: 69-76.

Osuna, E., Freund, R., and Girosi, F.,1997, Training support vector machines: An application to face detection. *Proc. Computer Vision and Pattern Recognition'97*, pages 130–136.

Pontil, M., and Verri, A.,1997, Properties of support vector machines. Technical report, Massachusetts Institute of Technology, Artificial Intelligence Laboratory, 1997.

Smola, A., and Schölkopf, B., 1998, A tutorial on support vector regression. *Statistics and Computing*, Invited paper, in press.

Trafalis, T.B., Ince, H., and Mishina, T., 2003, Support vector regression in option pricing, *Procedings of Conference on Computational Intelligence and Financial Engineering (CIFer 2003)*, March 20-23, Hong Kong.

Trafalis, T.B., and Ince, H.,2002, Benders decomposition technique for support vector regression. In *Neural Networks, 2002. IJCNN '02. Proceedings of the 2002 International Joint Conference on*, volume 3, pages 2767– 2772. IEEE.

Tsaih, R., 1999, Sensitivity analysis, neural networks and, the finance. pages 3830–3835.

Vapnik, V., 1995, *The Nature of Statistical Learning Theory*. Springer Verlag.

SHORT-TERM STOCK FORECASTING BASED ON A NEURAL NETWORK MODEL INCORPORATING FUZZY LOGIC AND FIBONACCI RATIOS

GARY R. WECKMAN
Industrial and Manufacturing
Systems Engineering
Ohio University

SRIRAM LAKSHMINARAYANAN
Industrial and Manufacturing Systems
Engineering
Ohio University

ABSTRACT
Financial Forecasting is a vital tool when considering investments in the stock market. The decision of a typical investor at any point in time is to either buy or sell a particular stock based on the short-term trend. Trading systems based on neural networks have become useful in this decision making process. This paper focuses on developing a neural network model incorporating Fuzzy Logic techniques for discrete symbolic inputs to determine an output decision to buy/sell. The input signals to the neural network are various technical, Fibonacci based Elliott Wave and short-term trend indicators. This methodology will integrate the use of wave patterns associated with stock market activity into the learning mechanism of the neural network model. Various neural network architectures are tested on the same dataset to derive the best model. The best model is further tested for stability and convergence of the output.

INTRODUCTION

The economic environment seems to be affected by almost every major world event, therefore perceived as very volatile. These events also make it difficult to predict the movements of the stock market. To be successful in any business there is a need to forecast the financial scenario of the near future [Gately 1996, Shadbolt and Taylor, 2002, Strand, 1999]. The process of forecasting future price movements in the stock market involves technical analysis, in which there is an abundance of historical data available. Technical analysis can be classified into two main types. The first is based on mechanical indicators such as price, volume, and technical indicators to aid in the forecasting of price movements. The other type is based on the recognition and interpretation of chart patterns using techniques such as trend lines, Japanese candlestick charts [Lee and Jo, 1999], Elliot Wave Theory [Hadik, 1999, Prechter, 1994], Dow theory and others.

Several researchers have developed applications using artificial Neural Networks (NN) to help make more informed financial decisions [Kohara and Fukuhara, 1997, Saad, et al, 1998, Schadbolt and Taylor, 2002, Yao and Tan, 2000]. These NN models have often been successfully used, receiving much attention in trading and financial forecasting of problems, such as price forecasting, stock performance, selection and buy-sell decisions [Smith and Gupta, 2000]. However, most studies have shown that NN have some limitations in learning the patterns since stock market data has immense noise, complex dimensionality, and is affected by various factors, most of which may not have been considered. Many studies on stock market prediction using NN intelligence techniques have been attempted in the past decade. These studies have used various types of NN to predict the stock index and its direction of change [Kyoung-jae and Ingoo, 2000, Thawornwong, et al, 2001]. Recently, in efforts to improve the NN model's capability to include the more subjective interpretation of technical indicators, the use of fuzzy logic to decide when to buy and sell stocks was incorporated [Bogullu, et al, 2002].

INDICATORS

A number of indicators were considered within the initial model such as the Relative strength Index (RSI), Money Flow Index (MFI), Moving and Stochastic Oscillator (SO), but proven to be insignificant. For further details on these indicators see [Agarwala et.al, 2002]. Since the only financial indicators considered in the final model were Moving Average (MA), and Moving Average Convergence/divergence (MACD). The following provides a brief description of technical trading criteria used to perform the technical indicator calculations and stock trading exercises [Thawornwong, et al, 2001]: The MA function returns the moving average of a field over a given period of time. The MA is calculated by averaging together the previous values over the given period, including the current value.

$$MA = \frac{\sum_{1}^{n} Closing\ Price}{n} \tag{1}$$

The MACD function calculates the MACD, which is the difference between a short and a long term moving average for a field. The MACD is a specific instance of a Value Oscillator and is typically used on the closing price of a security to detect price trends. EMA is the exponential moving average.

$$\begin{aligned} &EMA = [\alpha \times Today's\,Close] + [(1-\alpha) \times Yesterday's\,EMA] \\ &MACD = [0.075\,EMA\,of\,Closing\,Prices] - [0.15\,EMA\,of\,Closing\,Prices] \\ &Signal\,Line = 0.20\,EMA\,of\,MACD \end{aligned} \tag{2}$$

WAVE THEORY

Since the discovery of the Elliott Wave Theory in the 1930's, the world of financial forecasting was exposed to a new approach in stock market trading. Ralph Nelson Elliott, the creator of the theory discovered patterns within the price charts of the liquid markets. Elliott concluded that observing and analyzing the repetitive patterns in which the stock market waves unfold would forecast the movement of the stock market. The basic principles of his research identified numerous patterns or waves that occur in the stock markets, divided into trend and corrective patterns. The trend patterns may be constituted of 5 waves and the corrective pattern of 3 waves. Thus any wave moving up will have to be composed of 5 waves and any wave moving down of 3 waves regardless of the cycle that they are part of. In addition, smaller patterns of the trend and corrective waves fit within larger patterns (see Figure 1). The study of wave patterns and wave counts in Elliott Wave Theory is essential because it tells the pattern in which the market is going to move rather than the direction of movement alone. The Fibonacci series is the mathematical backbone of the Elliott's Wave theory. The wave counts of the impulsive and corrective patterns (5 + 3 = 8 total) are Fibonacci numbers, and breaking down wave patterns into their respective sub waves produces Fibonacci numbers indefinitely. Since Fibonacci ratios are found to occur in the proportions of one wave to another, waves are often related to each other by the ratios [Hadik, 1999, Prechter, 1994]. This theory is currently used by a large number of stock traders for financial forecasting, but academicians have found little research value in this concept.

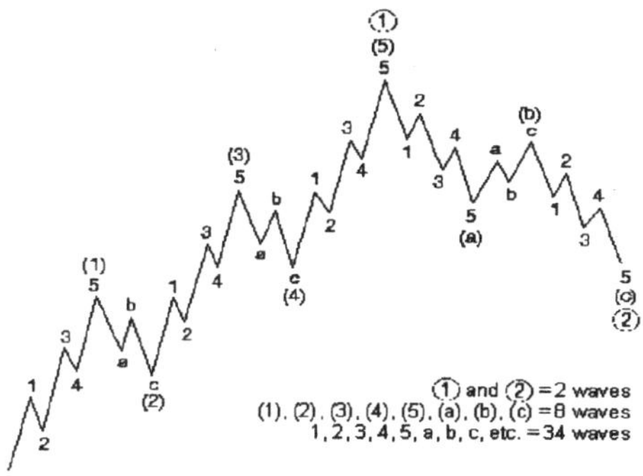

Figure 1: Trend and corrective waves

FUZZY LOGIC

Fuzzy Logic (FL) is a technique that was originally developed to mimic humanlike decision making involving uncertainty. FL simulates how a normal human decsion is made through reasoning by allowing the computer to determine outputs in a less precise manner. In a real life continuous data set, a definitive threshold cannot be established, but each data point does have a certain degree of membership to a belief function [Bogulla, et al, 2002, Turban 2001]. This concept is used to define the membership of Elliott Wave points in the current data set. A Fibonacci Ratio based percentage is calculated for a given data set and the membership functions for the Elliott Wave indicators are defined. Three fuzzy classes of high, medium and low are set indicating the probability of the point being an Elliott Turning point (see Figure 2). The fuzzy classes used are rectangular binding functions. The area depicts the probability of that point being a Elliott Turning point.

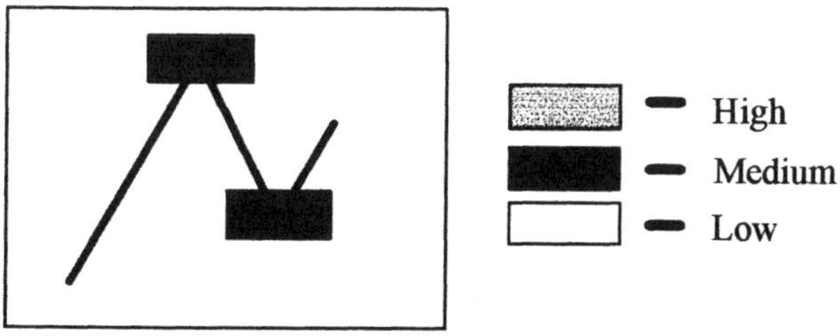

Figure 2: Probability of data point being a turning point

METHODOLOGY

This paper focuses on two types of neural networks which are the Multilayer Perceptrons (MLPs) and the Generalized Feedforward Networks (GFN). The MLPs are layered feedforward networks typically trained with backpropagation (learning algorithm). The GFNs are a generalization of the MLP, which contains connections that can jump over one or more layers. The data set consisted of 1527 points of the closing prices of Hewlett Packard over a 5 year period. 1377 of these data points were used for training the network and 100 data points were used for cross validation. The data set was randomized before training and cross validation. 50 data points were used for final testing of the model. The original model used 5 technical indicators (MACD, RSI, MFI, MA, SO) along with a fuzzy logic based Elliott Wave indicator (EWFL) and Elliot Wave turning point indicator (EWTP). The MACD, trading volume technical indicators along with Elliot Wave indicators (EWFL & EWTP) were used as the first four inputs. In addition, 40 inputs were added for the RSI, MFI, MA, SO indicators that were computed for periods of 5, 10, 15, 20, 25, 30, 35, 40, 45 and 50 days resulting in a total number of 45 inputs to the NN Model. The least significant inputs were dropped based on sensitivity analysis after 6 iterations. For more specifics on sensitivity see Weckman and Agarwala (2002). The analysis, data and model were collected using two different software packages; TradingSolutions and NeuroSolutions V4.2 from NeuroDimension Inc (see http://www.nd.com/). Different models were created reducing the number of inputs by removing the least sensitive variables from the model. The selection of inputs (PEs) was based on the results of the Sensitivity Analysis. The final GFN model is summarized below (see Figure 3):

Input Layer:	Input Processing Elements = 8
	MACD, MA(5, 10, 20, 25, 30 ,40), EWFL, and EWTP
Hidden Layers:	Processing Elements (PEs) = 4 each layer
	PE Transfer Function = TanhAxon
Output Layer:	Output Processing Elements = 1 (Closing Sales Price)
	PE Transfer Function = TahnAxon
	Maximum Number of Epochs = 2000

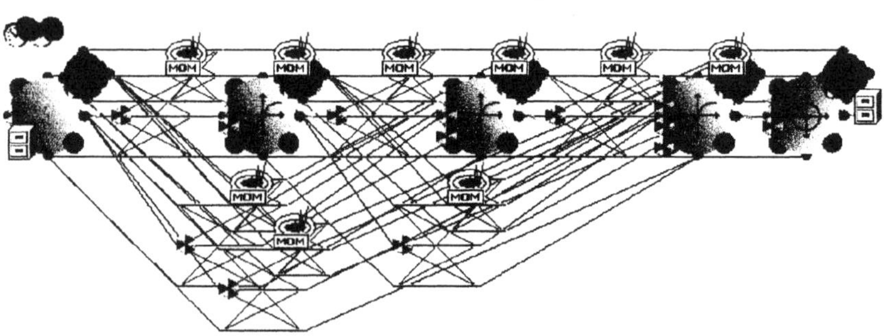

Figure 3: GFN Model Representation in NeuroSolutions

The Neurofuzzy logic and wave patterns associated with stock market activity were integrated into the learning mechanism of the NN Model, along with the technical indicators. The output of the model and the actual closing price were compared to generate buy-sell decisions. The buy-sell decisions were short-term indicators and these decisions were compared to the results from the actual and the moving average analysis. To generate the decisions for the neurofuzzy model, the forecast of the next day was compared to the actual of the current day. For decisions based on moving averages, a buy signal is generated when the stock price rises above the moving average and a sell signal is generated when the stock price dips below the moving average. The economic impact of these decisions in financial protfolio analysis could be incorporated in further research.

RESULTS AND CONCLUSIONS

The GFN network was found to give superior results when compared to the MLP network (see Table 1 and Figure 4). The results of the GFN network were used for computing the buy-sell indicators. These results were then compared to the indicators determined by the analysis of the moving averages. The buy-sell signals generated by Moving Averages analysis and the signals generated by the Neurofuzzy model were compared to the original decisions based on known closing price value. The neurofuzzy model was found to generate decisions that were correct 78.4% of times when compared to 18.8% of times by the moving Average analysis (see Table 2).

Table 1: Comparison of MLP and GFN Neural Network Models

	MLP	GFN
Performance	Close	Close
MSE	8.350437	4.339483
MAE	2.343918	1.517511
Min Abs Error	0.002981	0.024118
Max Abs Error	8.138203	9.032437
r	0.981729	0.989098

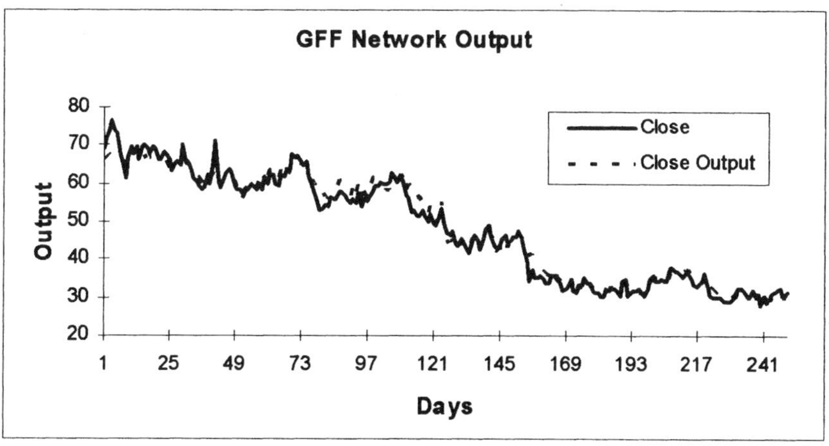

Figure 4: GFN Closing Price Output (Forecast) versus Actual

Table 2: Comparison of Moving Average and Neurofuzzy Buy/Sell Signals

	MA Vs. Actual	Neurofuzzy Vs Actual
Decision Mismatch	204	53
Score	46	197
Percentage Accuracy	18.4	78.80

A lower score on the decision mismatch indicates fewer erroneous decisions that can be obtained by using that method. The higher percentage of accuracy indicates that the Neurofuzzy model outperforms the decisions based on simple moving average.

REFERENCES

Agarwala, R, Weckman, G. R., and McLauchlan, R., 2002, "Optimization of Data Collection and Training Time for Forecasting Stock Market Trends Using Artificial Neural Network", Proceedings of the 2002 Artificial Neural Networks in Engineering Conference (ANNIE '02), ASME, St. Louis, Missouri, pp. 707-712.

Bogullu, V.K., Enke, D., and Dagli, C., 2002, "Using Neural Networks and Technical Indicators for Generating Stock Trading Signals", Proceedings of the 2002 Artificial Neural Networks in Engineering Conference (ANNIE '02), ASME, St. Louis, Missouri, pp. 721-726.

Gately, Edward, 1996, Neural Networks for Financial Forecasting, New York, Wiley.

Hadik, E. S., 1999, "Fibonacci squared", Futures, 28 (7), pp. 38-42.

Kohara, Ishikawa, and Fukuhara, Nakamura, 1997, "Stock price prediction using prior knowledge and neural networks", International Journal of Intelligent Systems in Accounting, Finance and Management, 6 (1), pp. 11-22.

Kyoung-jae, Kim, and Ingoo, Han, 2000, "Genetic algorithms approach to feature discretization in artificial neural networks for the prediction of stock price index", Expert System with Applications, 19 (2), pp. 125-132.

Lee, K. H., and Jo, G. S., 1999, "Expert System for predicting stock market timing using candlestick chart", Expert Systems with Applications, 16, pp. 357-364.

Prechter Jr., .R. R., 1994, R.N. Elliott's Masterworks: The Definitive Collection, New Classics Library, Gainesville, Georgia.

Saad, E. W., Prokhorov, D. V., and Wunsch, D. C., 1998, "Comparative study of stock trend prediction using time delay, recurrent and probabilistic neural networks", IEEE Transactions on Neural Networks, Vol. 9, No. 6, pp. 1456-1470.

Shadbolt, J., and Taylor, J.G., 2002, Neural Networks and the Financial Markets: Predicting, Combining and Portfolio Optimization, Springer, New York.

Smith, K. A., and Gupta, J. N. D., 2000, "Neural networks in business: techniques and applications for the operations researcher", Computer& Operations Research, 27, pp. 1023-1044.

Strand, Sverre, 1999, "Forecasting the future: pitfalls in controlling for uncertainty", Futures, 31, 333-350.

Thawornwong, S., Enke, D., and Dagli, C., 2001, "Using Neural Networks and Technical Analysis Indicators for Predicting Stock Trends", Proceedings of the 2001 Artificial Neural Networks in Engineering Conference (ANNIE '01), ASME, St. Louis, Missouri.

Turban, E., and J. E. Aronson, 2001, Decision Support Systems and Intelligent Systems, Sixth Edition, Prentice Hall, New Jersey.

Weckman, G. R., and Agarwala, R., 2003, 'Identifying Relative Contribution of Selected Technical Indicators in Stock Market Prediction" Proceedings of the IIE Annual Conference 2002, May 18-21, Portland, Oregon.

Yao, J., Li, Y., and Tan, C. L., 2000, "Option price forecasting using neural networks", Omega, 28, pp. 455-466.

FORECASTING FATIGUE ALLOWANCES USING NEURAL NETWORKS

KAMBIZ FARAHMAND
Mechanical & Industrial Engineering
Texas A&M University-Kingsville
Kingsville, Texas

ABSTRACT
The world of Methods Engineering and or Work Measurements have debated the setting of fatigue allowances for time studies since the time of Fredrick Taylor, the father of modern time studies. The current methods of setting allowances are still being used and engineers are being handed down this information as time passes by. This paper will show a new method for setting fatigue allowances using ANN. The ANN model will be taught to recognize patterns from an assigned time standard with an allowance and compare to actual hours worked for a given task using multilayer back-propagation method. The ANN model will then train itself and forecast a new standard for an allowance on a monthly basis. Finally, a comparison will be made to analyze the overall performance of the ANN model comparing it to the traditional methods of setting allowances.

INTRODUCTION

The world of Methods Engineering and or Work Measurements have debated the setting of fatigue allowances for time studies since the time of Fredrick Taylor, the father of modern time studies. Often allowances have been set by crude estimations, referenced from tables and graphs or have been negotiated between labor unions and management. Engineers are understanding the nature of fatigue and its effects, however, it has also shown how much farther they are from the final solution for this problem. Hence, the current methods of setting allowances are still being used and engineers are being handed down this information as time passes by. These methods have range from logical analysis to pure guesses.

The field of methods engineering (or work measurements) aids the analysis of a manufacturing process to accurately capture a task. "It involves the technique of establishing an allowable time standard to perform a given task based on measurement of the work content of the prescribed method with due allowance for fatigue and personal delay" (Niebel et al., 1993). Knowing or calculating the exact time to perform a task can give any company the competitive edge to accurately determine the contracts man-hour requirements and to predict production output to meet demand.

The current method of setting and assigning fatigue allowances is not sufficiently accurate for the current operation. The allowance system evaluates and accounts for breaks, special clothing, job and environmental conditions. Personal delay allowances and pace ratings are already factored in the standard. The information and guidelines used to set the allowances are not very accurate and may vary from person to person.

Therefore, this system can be bias by individual preference and is subject to human error. The current system does not account for complex undefined relations when setting fatigue allowances and an ANN would be the best application for such application.

The current method of setting and assigning fatigue allowances is not sufficiently accurate for the current operation. The allowance system evaluates and accounts for breaks, special clothing, job and environmental conditions. Personal delay allowances and pace ratings are already factored in the standard. The information and guidelines used to set the allowances are not very accurate and may vary from person to person. Therefore, this system can be bias by individual preference and is subject to human error. The current system does not account for complex undefined relations when setting fatigue allowances and an ANN would be the best application for such application.

DEFINING ALLOWANCES

Time and labor standards do not accurately reflect the true nature of an operation without accounting for an operators work environment. According to the Webster's dictionary, fatigue is defined as, "The exhaustion of physical and mental strength." Areas where a work environment may change day by day might cause work delays of even more costly work stoppages.

When completing a time standard, an allowance is commonly added. The categories or parts of an allowance can be classified as personal allowance, unavoidable or delay allowance, individual pace ratings and fatigue allowances. These categories are often called personal, fatigue and delay (PF&D) allowances. According to Bostion, "The inclusion of this allowance is common practice in the development of a labor standard" (Bostion et al., 1980). Setting a good PF&D is key when developing time standards.

PACE RATING

While performing a time study, an analyst typically chooses an "average" operator. This operator should have proper training, good skill level and aptitude. When performing a time study, that may not always be the case. A normal pace rating of an operator or mechanic is rated at 100%. When an operator works too slow, they are usually rated less than 100%. However, if the mechanic or an operator works faster than normal, the person is rated more than 100%. But, how does an analyst know when an operator is working at an average pace?

PERSONAL AND UNAVOIDABLE DELAY ALLOWANCES

In the manufacturing field, delays are a part of the process. Delays are usually divided into two categories, personal delays and unavoidable delays. Allowing for such delays also contribute to computing the proper standard time for an operation. Throughout the working day, personal time allowances must be set. Scheduled breaks are usually accounted for as a percent of a standard workday.

FATIGUE ALLOWANCE

For a final standard time, one more allowance that must be accounted for is fatigue. The fatigue allowance has been the most debated part of a time standard and has challenged industrial engineers for decades. Immense interest and studies have been developed to accurately account for fatigue allowances. According to Payne "It can easily cause the success of failure of the time of this positions" (Payne et al., 1949). To understand the nature of the fatigue allowance one must understand the nature of fatigue. When studying fatigue, three major factors affect human fatigue. The three factors are mental/psychological fatigue, fatigue from the environment and fatigue from motion.

MENTAL AND PSYCHOLOGICAL FATIGUE

The mental and or psychological fatigue plays a significant factor in an industrial job setting. There are many factors that attribute to mental and psychological fatigue. Jobs that take great mental effort, such as working with fine controls like working on circuit boards or with chemicals in a laboratory may be more mentally draining than simple mechanical of repetitive assemblies. However, repetitive operations such as assembly line work can cause boredom and fatigue. Most mental and psychological allowances are estimated and usually overlooked while setting the allowances.

PHYSICAL FATIGUE

Physical fatigue is the second factor that plays into setting an allowance. There are two basic types of physical fatigue, static fatigue and dynamic fatigue. Static fatigue is defined as fatigue caused by working in a set position for a long period. For example, when a worker has his/her hands above their head for long period, fatigue will eventually set in. The workers shoulder, back and arms will start to fatigue will eventually have to take multiple breaks to complete the task.

Dynamic fatigue occurs while the body is in working motion. This fatigue is usually caused with repetitive motions of jobs associated with excessive movement. Some examples of this are:

- Continuous assembly line work
- Torque multiple bolts
- Commercial construction
- Road Construction

ENVIRONMENTAL FATIGUE

The third and final factor that accounts for fatigue is the surrounding environment. Personal fatigue is also attributed to the working conditions around the worker. Temperature, wind speed (if work is outdoor), precipitation, humidity, etc. All play a factor in work performance. Knowing and understanding how climate affects the body is challenging and difficult to account for when establishing a fatigue allowance. Many studies and methods have been developed to measure the effects of fatigue on the body.

EFFECTS OF CLIMATE

Many workers have been exposed to heat and/or cold stress at some point during their work. When working in hot or cold areas, the core body temperature will significantly change. To dissipate this heat, the body utilizes its natural cooling system, sweating. When temperature is cold, the body will respond by vasoconstriction and shivering. The effects of heat stress have also been equally complex and contradictory. Therefore, setting a fatigue allowance would nearly be impossible to set using a set method.

NEURAL NET ARCHITECTURE FOR FORECASTING ALLOWANCES

A three-layer backpropagation ANN model was designed using NeuroSolutions version 3.0. One of the three layers was hidden and the software program allowed different processing elements and learning rules to create the best forecasting system. Setting up the ANN model required three basic functions, stating assumptions to set up the model, deciding the inputs for the model's parameters and selecting the desired output.

INPUT VARIABLE

The input variable selected highly was dependent on the working environment for the operator. Since the task was performed outdoor, the environment plays a drastic role in worker performance. The following variables were chosen which directly correlate to the work performed:

- Temperature
- Precipitation
- Wind Speed
- Humidity (Maximum and Minimum)
- Illumination (Sunshine)

Once the data was set, different ANN models were constructed to determine the best forecasting model. The best models were chosen during the training phase by the lowest mean square error. A three-month and twelve-month moving average was performed and set on an MS Excel spreadsheet. Both sets of data were trained using the different training options from NeuroSolutions; the twelve-month moving average produced the lowest MSE.

Using different combinations to build an ANN forecasting model, the top four models were chosen. Table 1 shows the final Four Combinations of Foecasting ANN models. The selected models were then run with the test data to examine how the ANN forecasted data would compare to the original. Once the data was collected, Model A produced the lowest mean square error as shown in Table 2. Figure 1 shows the training graphs for the models

CONCLUSIONS

This technique predicts or forecasts fatigue allowances for outdoor operations. A three-layer ANN with backpropagation learning has been used to predict fatigue allowances. Fatigue allowances allow an operator to complete an operation compensating for environmental conditions and difficulty of the task. Many companies that have to submit bids for specific jobs such as commercial building or road construction may not include allowances for these conditions. Using ANN to forecast an allowance typically can help businesses bid for work more accurately.

Forecasting fatigue allowances can help companies schedule a job more accurately using proper environmental delay allowances. This can aid companies to minimize the risks in bidding for work and aid them to compete in a global economy by reducing labor cost. Using ANN to forecast the monthly fatigue allowances was significantly better than the traditional method. The model will also help secure bids that are close to the actual work performed. This can be used to help provide a better and more accurate scheduling system that will also help create efficient methods to manage tools and materials. As a result, the ANN forecasting model will help reduce labor costs and increase the profit margin for an organization with narrow profit margin. To confirm the results, the models were also compared using a percent error analysis and depicted in Figure 2.

It is obvious that model B performs the best; however, model A has the least mean square error. Either model A or B would be ideal to use to forecast the fatigue allowance. The system would automatically update the weather information via the World Wide Web or be input manually in on a monthly basis. The program would perform a twelve-month moving average. The software would also be connected to the accounting system to acquire the time charged to perform an operation. The software would then use the ANN model to forecast the monthly standard with the fatigue allowance built in.

Since both models A and B performed the best, model A with the lowest MSE and Model B with the lowest percent error, can both be run by the expert system. Each test result can be compared to actual data and the best result can be chosen based on the lowest percent error. The expert system can be an ideal tool for any job shop operation to help forecast work to secure bids with proper allowance.

REFERENCES

Niebel, Benjamin W. Motion and Time Study. Irwin, Burr Ridge, Illinois. Ch 16, 21. 1993.

Bostion, W. H. "Development of personal Fatigue and Delay (PF&D) Allowances". Proceedings of the Fall Industrial Engineering Conference. Institute of Industrial Engineers, Atlanta, 1980, pp. 457-464.

Payne, Matthew A. The Fatigue Allowance in Industrial Time Study. 1949.

Table 1: Top Four Models (ANN Training)

Models	MSE
A	0.0105
B	0.0106
C	0.0107
D	0.0107

Table 2: Top Four Models (ANN Testing)

Models	MSE
A	77.92
B	79.3
C	82.64
D	87.38

Model A Training Graph

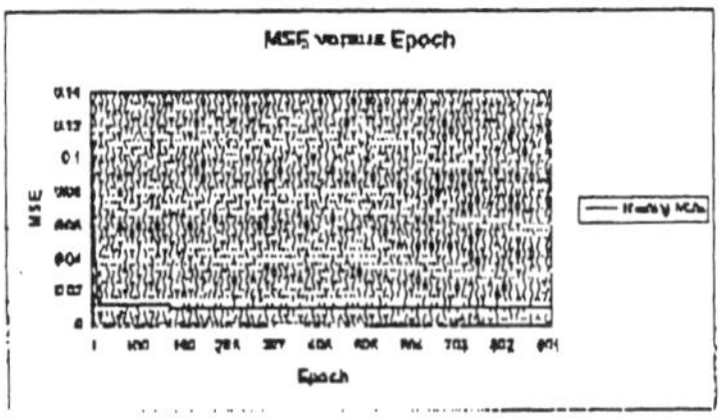

Model A Testing Graph

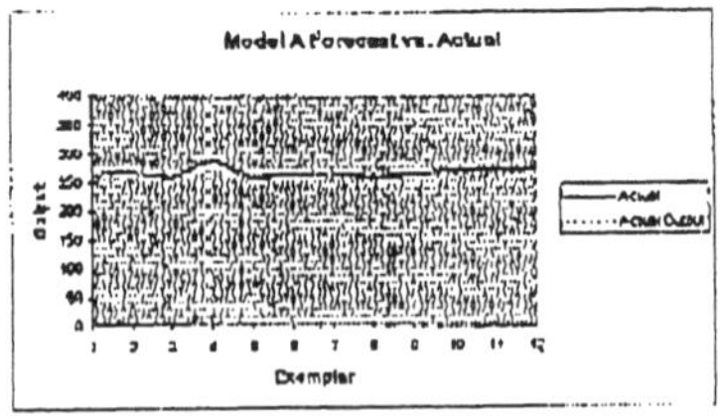

Model B Training Graph

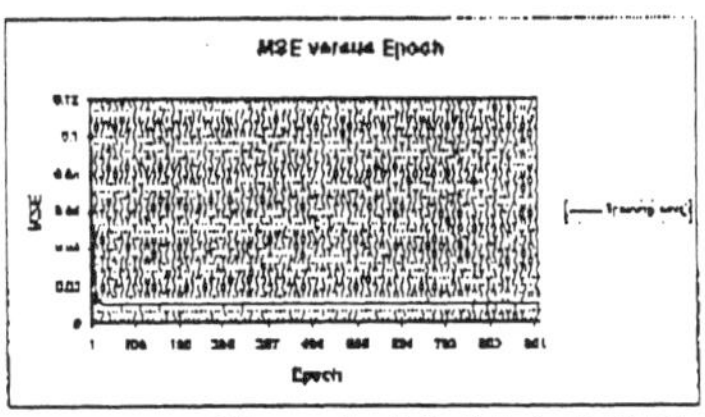

Model B Testing Graph

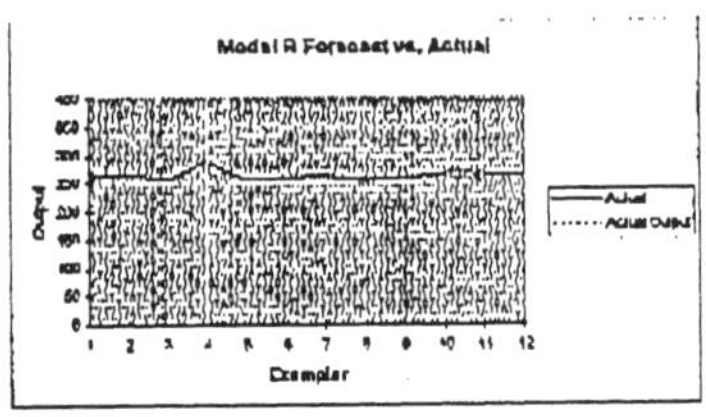

Figure 1. Model A & B Training and Testing Graphs

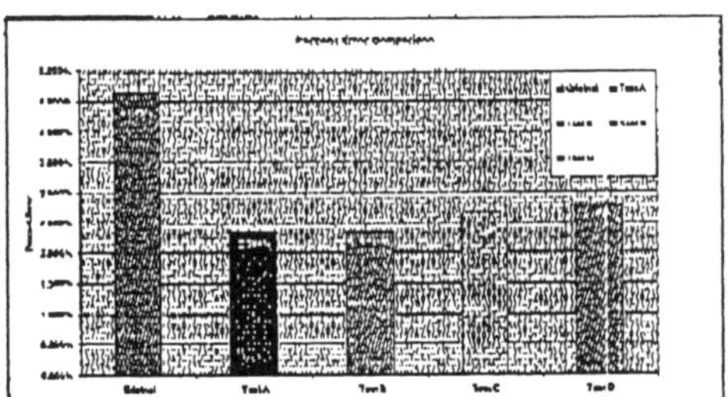

Figure 2: Percent Error Comparison of Actual vs. ANN Models

PREDICTING STRENGTH OF HIGH PERFORMANCE CONCRETE USING ARTIFICIAL NEURAL NETWORK

RANJEET AGARWALA
Texas A&M University-Kingsville
Mechanical Engineering
Kingsville, Texas.

ROBERT A. MCLAUCHLAN
Texas A&M University-Kingsville
Mechanical Engineering
Kingsville, Texas.

MOHAMMED FARUQI
Texas A&M University-Kingsville
Civil Engineering
Kingsville, Texas.

BHAVESH PATEL
Pune University
Mechanical Engineering
Pune, India

ABSTRACT

There are no set formulas for mixing concrete constituents to obtain desired properties of the final product. Concrete mixing is predominantly a qualitative knowledge based approach prone to quality variations. Reliance on such an approach compromises the precision and accuracy of concrete properties and hence necessitates the development of a reliable quantitative mixing formulation. Statistical modeling techniques like Multiple Linear Regression Analysis have been used in the past. All these methods have failed to accurately predict the properties due to the highly nonlinear relationship between the concrete proportions and its properties. In this paper, we will develop neural network (NN) models for predicting, with a significant degree of accuracy, the compressive strength of concrete for different mix-design parameters. We will also compare and conduct sensitivity analysis for different neural network models to obtain satisfactory results.

INTRODUCTION

A recent increase in the concrete ingredients has made concrete more and more complicated and therefore its performance varies vastly. For a long time concrete engineers have been designing mixes by trials and their personal experiences. Some of these mix designs have been successful while others were discarded. A question then arises that how can researchers obtain useful information from the successful mixed design and where can researchers find a quick way of designing a mix to give desired properties?

Zong, Gung, and Yun (Zong et al., 1999) utilized an automatic knowledge-acquisition system based on neural networks to design concrete mixes. In a later work Gung, and Zong (Guang et al., 2000) proposed a method to predict 28-day compressive strength by using multi layer feed forward neural networks. Lai, and Serra (Lai et al., 1997) developed a model based on neuro computing for predicting with sufficient approximation, the compressive strength of cement conglomerates. Yeh (Yeh , 1998) developed a strength based artificial neural network (ANN) model and found that the model was more accurate than the one based on regression analysis. He also found that his ANN model gave detailed effects of the proportions of each variable from the concrete mix. Dias and Pooliyadda (Dias et al., 2001) used back propagation neural network models to predict the strength and slump of ready mixed ordinary concrete and high strength concrete, in which the chemical admixtures where used.

Although trial mixes (Kosmata et al.; 1994) are important and will remain so for a long time, they are not always the answer. In this work we have developed compressive strength based neural network models for the concrete. Previous works have successfully modeled concrete behavior, but did not adequately address the relative contribution or sensitiveness of every proportion of concrete towards the compressive strength.

In our work, we have developed different neural network models with varied hidden layers and subjected them to varied training cycle. Finally, we chose a neural network model based on the strength of prediction and conducted sensitivity analysis to observe the relative contribution of each input.

CONCRETES MIXES

Concrete is a mixture of fine aggregate, coarse aggregates, or other aggregates held together in a rocklike mass with a paste of cement and water. Sometimes one or more admixtures are added to change certain characteristics of the concrete, such as its workability, durability and time of hardening.

The Five types of cements are:

Type I: The common all-purpose cement used for general construction work.

Type II: A modified cement, which has a lower heat of hydration than, does Type I cement and one which can withstand some exposure to sulfate attack.

Type III: A highly early strength cement, which will produce in the first 24 hours, a concrete with strength about twice that of Type I cement.

Type IV: Low heat cement, which produces a concrete that dissipates heat very slowly.

Type V: Cement used for concretes, which are to be exposed to high concentration of sulfate.

Fine aggregates: Fine aggregates generally consist of natural sand or crush stones with most particles smaller than 5.08 mm.

Coarse Aggregates: Coarse aggregates consists of one or a combination of gravels or crush aggregates with particles predominantly larger than 5.08 inches and generally between 9.525 and 12.7 mm.

Compressive strength may be defined as the measured maximum resistance of a concrete or mortar specimen to axial loading. It is generally expressed in N/mm^2 at an age of 28 days. The compressive strength tests are made on specimens of motor or concrete. In United States, unless otherwise specified; compression tests of concrete are made on cylinders of 152.4 mm diameter and 304.8 mm in height.

NEURAL NETWORK MODELING, TRAINING AND PREDICTION

Concrete is a highly complex material and therefore, an attempt to model its behavior is a great challenge. The basic strategy for developing a neural network model for the concrete behavior is to train a neural network, on the results of series of experiments. Such a trained network will be able to reproduce the experimental results and approximate results from other experiments. It is known that by controlling some parameters of fresh concrete, the properties of concrete can be significantly improved.

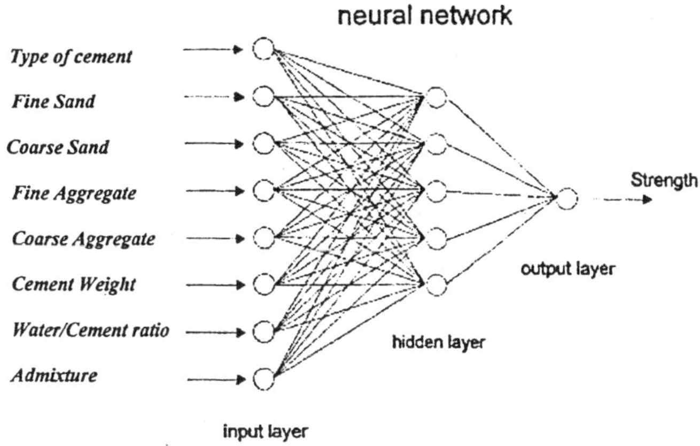

Figure 1: Proposed Neural network Model

There exists a complex non-linear relationship between these factors and the concrete strength as shown in Fig 1. These factors or variables, which serve as input parameters to the neural network, are:

- ***The grade of cement***: The class of cement helping in producing a 28 day strength concrete.
- ***The fine module of sand***: Proportion per m^3 of fine sand (0-2 mm).
- ***The coarse module of sand***: Proportion per m^3 of coarse sand (2-6 mm).
- ***Fine Aggregate***: Proportion per m^3 of fine aggregate (3-18 mm).
- ***Coarse Aggregate***: Proportion per m^3 of coarse aggregate (15-30 mm).
- ***Cement Weight***: Proportion of cement per m^3 of coarse aggregate
- ***Water Cement Ratio***: w/c ratio
- ***Fluidificant***: The effect of admixtures or Plasticizer.

SIMULATION RESULTS

Various NN models based on the eight inputs were used in predicating the twenty-eight day concrete strength. The neural models were run under varied training cycles and varied hidden layers. Ultimately, the Generalized Feedforward Networks (GFN) with two hidden layers and a training cycle of 1000 epochs, was chosen based on the strength of output.

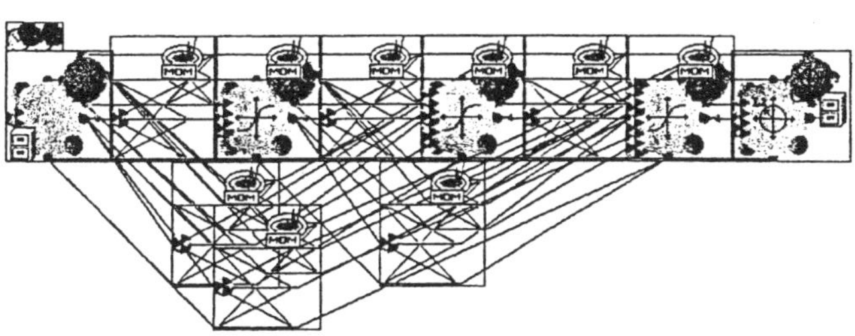

Figure 2: Simulated GFN Model with two hidden layers

The data below summarizes the attributes of the network model used.

Final GFN Model:	Input Processing Elements = 8 Output Processing Elements = 1 Maximum Number of Epochs = 1000 No of hidden layers = 2
Hidden Layer 1:	Processing Elements (PEs) = 4 PE Transfer Function = TanhAxon Learning rule = Momentum Step Size = 0.1 Momentum=0.7
Hidden Layer 2:	Processing Elements (PEs) = 4 PE Transfer Function = TanhAxon Learning rule = Momentum Step Size = 0.001 Momentum=0.7
Output Layer:	PE Transfer Function = TahnAxon Learning rule = Momentum Step Size = 0.1 Momentum=0.7
Training :	Minimum Mean Square Error(MSE) = 0.0048 Final Mean Square Error (MSE) = 0.0048
Testing:	Nominal Mean Square error = 0.0940 Minimum Absolute error = 2.9211 Maximum Absolute error = 39.0035 Root Mean Square Error = 0.9565

Fig. 3 is representative of neural network training and Fig. 4 represents the ability of the neural network model to predict compressive strength.

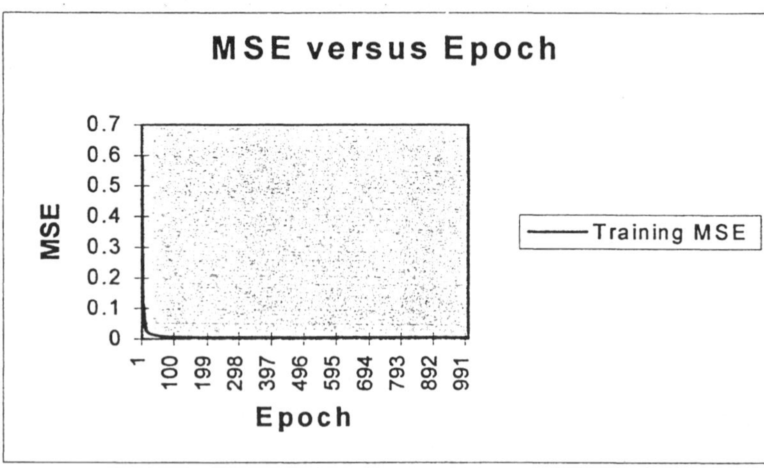

Figure 3: GFN Training Results of Mean Square Error vs. Number of Epochs

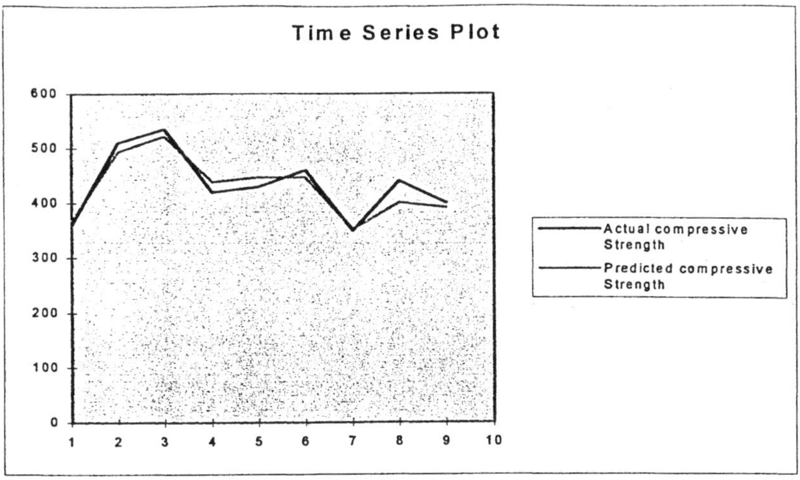

Figure 4: Actual vs. Predicted 28-day Compressive Strength

After having determined the suitable neural network model and the no of training cycles, the relative contribution of each input for a particular neural network model was evaluated through sensitivity analysis. The data below represents the relative sensitivity of each input in predicting 28-day compressive strength.

Fig. 5 is a graphical representation of the relative sensitivity of each input.

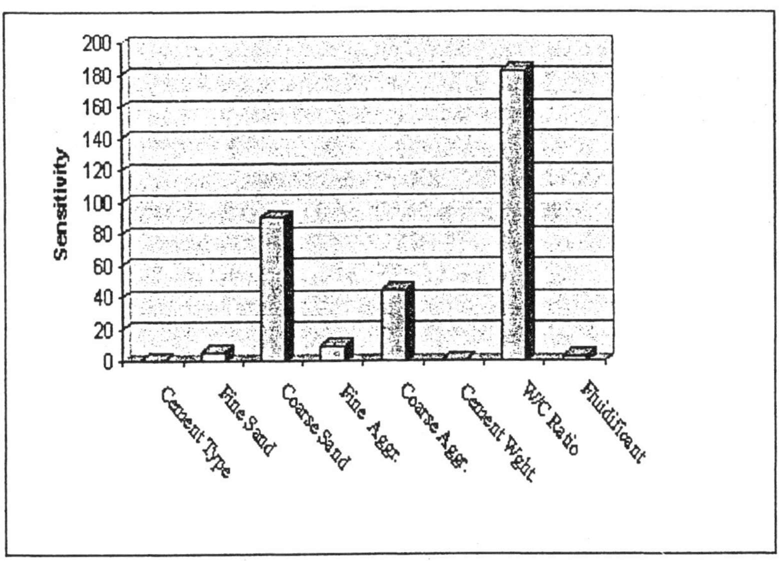

Figure 5: Sensitivity of each input to the neural net.

INPUT	SENSITIVITY
CEMENT TYPE	0.436218113
FINE SAND	5.519771576
COARSE SAND	90.3889389
FINE AGGREGATE	9.266704559
COARSE AGGREGATE	44.15232468
CEMENT WEIGHT	0.260421127
W/C RATIO	182.5594788
FLUIDIFICANT	2.822932005

CONCLUSIONS AND FUTURE WORK

It was observed that the neural network model performed well in the prediction of 28-day compressive strength. It was much easier to determine the relative contribution of each input by performing Sensitivity Analysis. The model was reduced in size taking into consideration the best five contributing inputs. This reduced the size of the model and the amount of data, but the accuracy of prediction was lost significantly. It would be of further interest to modify the model by incorporating additional factors such as the effects of silica fume, sand-aggregate ratio, aggregate-cement ratio and a small allowance for imprecision to reflect human behavior. One of the future endeavors would be to establish a neural network model to predict thermal properties such as conductivity, diffusivity, coefficient of thermal expansion and the heat of hydration of the concrete.

REFERENCES

Kosmatka and Panarese, 1994, " Design and Control of Concrete Mixtures", *Portland Cement Association.* Thirteenth edition.

McCormac, 1998, "Design of Reinforced concrete", *Addison Wesley*; Fourth Edition

Zong, Guang, Yun, 1999, "The application of automatic acquisition of knowledge to mix design of concrete", *Cement and Concrete research;* Volume 29 Pages 1875-1880.

Guang and Zong, 2000, "Prediction of the compressive strength of concrete by Neural Networks", *Cement and Concrete research;* Volume 30 Pages 1245-1250.

Lai and Serra, 1997, "Concrete strength prediction by means of neural networks", *Construction and building materials;* Vol 11, 2, pp 93-98.

Yeh, 1998, "Modeling of Strength of high performance concrete using Artificial Neural Networks", *Cement and Concrete research;* Volume 28, No-12, pp 1797-1808.

Dias and Pooliyadda, 2001, " Neural networks for predicting properties of concrete with Admixtures", *Construction and building materials;* Vol 15, pp 371-379.

Principe, J.C., Euliano, N.R., and Lefebvre, W. C., 2000, "*Neural and Adaptive Systems: Fundamentals Through Simulations*", John Wiley & Sons, Inc., New York.

PART VII: DATA MINING AND DECISION SUPPORT SYSTEMS

DYNAMIC DETERMINANTS FOR DECISION-MAKING IN SMART ENGINEERING SYSTEMS

THOMAS KOKUMO YESUFU
Obafemi Awolowo University,
Department of Electronic and Electrical Engineering,
Ile-Ife, Nigeria.
E-mail: tyesufu@oauife.edu.ng
Telephone: +234 80 3472 0266

OLUFEMI ADEDOTUN YESUFU
Obafemi Awolowo University,
Department of Agricultural Economics,
Ile-Ife, Nigeria.
E-mail: olufemi_yesufu@yahoo.com
Telephone: +234 80 3472 0266

ABSTRACT
The complexity of the Information Age attracts a lot of concern as to whether there is a way for maximizing its potentials without generating a terminal crisis of unimaginable proportions in the fast shrinking global village. This paper presents a tool that can be used to obtain the key determinants for improving the art of decision-making in smart engineering systems. The method employed is based on the fact that no true social revolution can be excised from the past, present and future of a society. Furthermore, a case is made by illuminating the way to maximizing the potentials of the information superhighway through the market paradigm, a pragmatic tool that embraces the developmental principles entrenched in an exemplary window and an analytical shadow of reasoning.

INTRODUCTION

The societal status in the Agricultural Age was determined by how much land was owned whilst in the Industrial Age, one could say status was largely defined in economic terms – how much wealth an individual possesses (Beach (1997) and Yanoff (1997)). The question is what tool can be used to obtain the key determinant in the Information Age? The new and emerging information society therefore leaves no one in doubt about the sustainability inherent in the gains of a system that is driven along the information superhighway. However, the complexity of the Information Age attracts a lot of concern as to the understanding of the dynamic determinants of an ideal decision-making process and the developmental principles of systems necessary to averting a terminal crisis in decision-making.

The art of decision-making, which normally proceeds from an intuition/vision, could take place in the form of action (adding reason to logic), thought (adding learning to logic) and by proxy machinations (adding belief to logic). On a comprehensive note, intuition leads to logic, and logic leads to learning, and learning leads to belief, and belief leads to reason and reason leads to action (Fig. 1). Thought provides an unbounded framework for decision-making in the form of the global manifestations of a little star of the rhyme: *Twinkle, twinkle, little star!* Action, on the other hand, provides a bounded framework through an exemplary window for decision-making in the form of the local manifestations of a little bird in a window, depicted by the rhyme: *Once I saw a little bird.* Also, puppet regimes, simulations or proxy machinations/animations are derived from the randomized combination of action and thought in the form of the irrational manifestations of an egg-laying black hen of the rhyme: *Higgledy-piggledy, my black Hen!* Since reason, learning and belief have a common ground to logic, they could contribute to reconstruct the logic and meet the dare need for silence (a successful performance of risk assessment and control following a given charge) in smart engineering system (SES). Moreover, the autonomy of a SES is based on its ability to employ trust, discretion and resources successfully in the respective form of belief, reason and learning. There are three ways that we may consider for the understanding of this system:

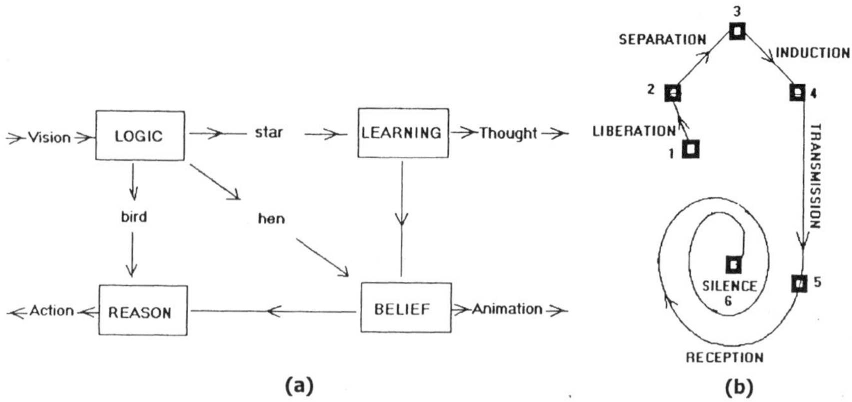

Fig. 1: Decision-Making Processes

- Newton's laws of motion
- Charge (command) communication laws of Maxwell and the
- Control (combination of the charge and motion laws) law of Schrödinger.

This is exemplified by the respective components of the Internet in the resolution of the US Federal Networking Council (FNC) unanimously passed on October 24, 1995. Subsequently, we state that reason, learning and belief are mutually expressed in the laws of Maxwell and Newton. Hence,

- reason takes the form of Newton's first law of motion and the Maxwell's equation given by

$$\nabla \bullet \vec{D} = \rho \tag{1}$$

- learning (bounded induction), in a similar vein, takes the form of the Newton's second law of motion and the Maxwell's equation:

$$\nabla \times \vec{E} = -\frac{\partial \vec{B}}{\partial t} \tag{2}$$

- belief (in the form of action at a distance), in its own right, is also expressed as the Newton's third law of motion, and as a consequence of the role it played in the following Maxwell's equation, which stated the existence of a displacement current before Hertz experimentally observed it:

$$\nabla \times \vec{H} = \vec{J} + \frac{\partial \vec{D}}{\partial t} \tag{3}$$

Another form of Eq. (2) is the Maxwell's equation:

$$\nabla \bullet \vec{B} = 0 \tag{4}$$

where H and E are the respective magnetic and electric field intensities, J is the current density, and B and D are the respective magnetic and electric flux densities).

On the other hand, the control law is found by the rearrangement of Maxwell's equations to link charge (q) to current (i) or flows (logic at maturity) as stated in Yesufu and Yesufu (1999). Hence, this form of the Schrödinger's equation appears as the only existing proof as follows:

$$\frac{dq}{q\,dt}\frac{di}{i\,dt} = j\omega^2\left(\sigma/_{\omega\varepsilon}\right) \tag{5}$$

The intuition-silence evolution of a SES makes it mandatory to state that the dynamic determinants of decision-making can be expressed through the prediction of the behavior of a falling charge in order to reconstruct the SES on each occasion; the dynamic determinants σ, ω and ε are respectively the conductivity, the frequency of energy transfer and the permittivity of the medium. This requires the use of an exhaustive set of risk assessment parameters and control instructions derived from the dispersion expressed either in terms of the closed form q-i spatio-temporal data or the dynamic determinants of decision-making in SES.

DISPERSION, COMMUNICATION AND TRANSPORT

Dispersion, θ, is a closed form property indicating the adaptation of a system right from emotion to societal action (integrated engineering values) through randomly processed data, gaussian distribution of counsel, uniformity of knowledge and binomially distributed wisdom using

a) the past or inertia (I), characteristic of the flip-flop based (modulo counter in a) logic machine or motion,

b) the present or appreciation (A), characteristic of change or a control system's transfer function and

c) the motivation or future (M), characteristic of market development or modulation in a communication.

We consider that communication is fundamental and hence shows the expression for motivation in a communication generating random numbers $x_1, x_2, \ldots x_t$ up to t seconds before computing the appreciation and inertia using the market paradigm, which is a quantitative and qualitative expression of the behavioral systems approach stated in Kohls and Uhl (1980). Accordingly, it models the marketing system as being made up of the power, communication, adaptation and input/output sub-systems. The random walk model defines an integrated series that is gaussian; other series must be differenced in order to become stationary (impart information or experience different from random walk) or possess statistical or structural properties invariant with time. For instance, the series, x_t, is a general form for this stationary series and it is given by

$$x_t = r_t\, x_{t-1} + e_t \tag{6}$$

where e_t, the error term has zero mean and covariance, and a constant variance and r_t, the root of the series, has a value of unity for random walk.

The value of r_t affords us the opportunity to generate standard test statistics through the participatory and simultaneous processes of regression and co-integration. The value of r_t can, thus, indicate whether another series y_t is stationary or not relative to the ideal random walk model in Eq. (6) by co-integrating y_t with x_t. The value of r_t for y_t as we regress towards the mean is, therefore, such that convergence occurs when

$$e_t = 0 \tag{7}$$

and

$$y_t = x_t \tag{8}$$

According to Hallam et al. (1992), variables are co-integrated if there is a linear combination of their data series that is stationary (as indexed globally), even though the individual series are non-stationary. Hence,

$$z_t = y_t - d_t x_t \tag{9}$$

In this case, z_t indicates the extent of any divergence from that relationship and it is given by

$$z_t = x_t \tag{10}$$

And d_t measures the degree of integration and represents the long run or equilibrium relationship between the two variables on the circular flow involving x or y series based on Eq. (5) and it is given by

$$d_t = 0.5 x_t x_{t-1} \tag{11}$$

If the relationship between the variables is positive, z_t being stationary implies that they cannot drift apart. This shows Granger causality in at least one direction between the two variables concerned.

On rearranging Eqs. (6) to (11), we have

$$r_t = 2 \frac{x_t - x_{t-1}}{x_t x_{t-1}^2} \tag{12}$$

The value of r_t indicates the motivation or amount of information present by adding a new term, x_t, to the series after the term x_{t-1}. In other words this is the characteristic root of the series at the time t. Moreover, the direction of causation (appreciation, A) is in favor of the series that has a greater motivation whilst the inertia, I, is the cumulative sum of the appreciation up to the time in focus. It is worth noting that appreciation can take three range of values: $A = 0$, $A < 0$ and $A > 0$. On the other hand, I takes either positive or negative values. The values of A and I from this market paradigm give the basis for finding the θ, which specifies the normative and positive optima in marketing systems (Yesufu and Yesufu, 2002).

The Market Paradigm And Motion

Decision-making is bound to determine the evolution, gender, marketing and conflict scenarios in a system. This implies a general modification of the simple planar illustration of the Cobweb effect by introducing the important dimension of time variation. Furthermore, the cumulative effect of the appreciation of the system explains the composition of its inertia and, hence, the visionary component for decision making in the market. The normative marketing scenarios vis-à-vis the states of motion are as follows:

(i) When $A = 0$, this implies that price (supply) will maintain a perpetual oscillation resulting in specialization ($I > 0$) or extinction ($I < 0$) given a respective positive or negative climate of incentives. The equivalent motion scenario for specialization and extinction are respectively steady motion and stop.

(ii) When $A > 0$, the marketing system will exhibit explosive oscillation resulting in diversification ($I > 0$) or mutation ($I < 0$) given a respective positive or negative climate of incentives. The equivalent motion scenario for diversification and mutation are respectively slow down and turn-about.

(iii) When $A < 0$, the system will exhibit damped oscillation resulting in hawking ($I > 0$) or touting ($I > 0$) given a respective positive or negative climate of incentives and thus, the equivalent motion scenario for hawking and touting are respectively bending and U-turn.

Table 1: Dispersion, Communication, Marketing and Transport Scenarios

Dispersion	Communication	Marketing	Transport	Viewgraph
$\theta = 0°$	Ducting	Specialization	Steady Motion	
$\theta \approx 0°$	Attenuation	Diversification	Slow Down	
$0° < \theta < 90°$	Refraction	Hawking	Bending	
$90° < \theta < 180°$	Reflection	Touting	U-Turn	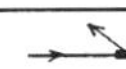
$\theta \approx 180°$	Extinction	Extinction	Stop	
$\theta = 180°$	Diffraction	Mutation	Turn-About	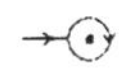

DISCUSSION

The engineering values in society are self-evident and continuously evolving in real operating systems/networks over seen by people that plan and/or perform acts of liberation, separation, induction, transmission and reception (Fig 1). It is therefore quite remarkable to state that the US took an anticipatory posture in maximizing the potentials that have so far greeted the development of the Internet. Subsequently, the scenarios marking the multi-faceted development of the Internet constitute an exemplary window for decision-making in smart engineering systems. This was possible by making the network to absorb new streams of innovations along its revolutionary path via the six milestones marking the very beginning and the ends of liberation, separation, induction, transmission, reception and silence. It is worth noting that the military played the initial role of liberation, the universities and research centers played the role of separation, a national agency played the role of induction, the service providers or private sector played the role of transmission, the commercial users played the role of reception and the group of individual men and women to be networked played the role of silence. Hence, the motion coordinates (scenarios) of a SES must imbibe the developmental principles of the Internet. In addition, a conformity is sought with the evolution, gender, conflict and marketing scenarios of an electron in order to respectively define the when, what, how and where decisions are to be made. The consistency of the coordinates of a smart engineering system (SES) is often compromised when there is a significant degree of uncertainty in its motion and/or cost factors referred to as dynamic determinants. Subsequently, the behavior of electrons in an atom gives examples of the various potentials related to the decision-making processes in the developmental hierarchies of the information superhighway. Subsequently, the electron has been considered as an ideal SES that has so far not suffered inconsistencies due to its fundamental, rational and universal role in suggesting closed form solutions for material composition/dispersion and transport, communications and other distributional consequences of electrons, as Table 1 shows. In this regard, communications is seen to be fundamental and useful for carrying out risk assessment and control in order to track the process of decision-making and, hence, improve on the total cost parameters due to the interactions of a SES. There has been remarkable progress in the study of Maxwell's equations and the Schrödinger's equation in the understanding of the basic modes of interactions causing the electron to produce (and relate with) the effects of forces: translation, vibration, rotation and heat. Maxwell's equations (including the Newton's and the Schrödinger's equations) therefore offer SES an analytic shadow of reasoning for decision-making based on the market paradigm.

CONCLUSION

Comparative analysis with the normative typologies offered by the market paradigm is required to understand the reactions expressed in operating systems. This is with a view to making it possible for improvements to be made through an unbiased or neutral analysis. Subsequently, the market paradigm combines the exemplary window in the developmental principles of the Internet with the analytic shadow of reasoning in the behavior of the electron with a view to having an exhaustive risk analysis and control of engineering systems in motion. The market paradigm, therefore, makes possible the most meaningful appraisal of an existing system both in terms of the delineation of problem areas and of the indications of the magnitude and importance of the distortions in decision-making applications. Other applications envisaged range from the understanding of natural phenomena, business mergers and design of early warning systems to the design of smart engineering systems and the safe navigation of special crafts in unexplored force fields. This is possible through the market paradigm's unification of the separate methodologies for achieving soft and hard computing respectively defined as an exemplary window and an analytic shadow of reasoning.

NOMENCLATURE

A	:	appreciation
B	:	magnetic flux densities
D	:	electric flux densities
E	:	electric field intensities
H	:	magnetic field intensities
I	:	inertia
J	:	current density
M	:	motivation
d_t	:	degree of integration of any two series at any time t
e_t	:	error term
i	:	current
j	:	complex operator
t	:	time
r_t		root of a series
q	:	charge
x_t	:	value of series x in a time t
y_t	:	regression towards the mean of a series
z_t	:	extent of any divergence from a relationship
ε	:	permittivity
σ	:	conductivity
ω	:	angular frequency
θ	:	dispersion

REFERENCES

Beach, G.J., 1997, "Editorial: Information Age", *ALCATEL Telecommunications Review*, 4th Quarter, pp. 243 – 245.

Kohls, R.L. and J.N. Uhl, 1980, "Marketing of Agricultural Products", Macmillan Publishing Co. Inc., 5th Edition, 612p.

Hallam, D., Machado, F. and Rapsomanikis, G., 1992, "Co-integration Analysis and the Determinants of Land Prices", *Journal of Agricultural Economics*, Vol. 41, No. 3, pp. 28 – 37.

Yanoff, L., 1997, "Building the Next Generation Internet", *ALCATEL Telecommunications Review*, 4th Quarter, pp. 246 – 255.

Yesufu, T.K. and O.A. Yesufu, 1999, "Pattern Recognition in Time series Data", *Proceedings of the International Symposium on Nonlinear Theory and Applications (NOLTA '99)*, Hawaii (USA), pp. 543 - 544.

Yesufu, T.K. and O.A. Yesufu, 2002, "Development of a Methodology for Studying Nonlinear Dynamics", Klonowski, W., ed., "Attractors, Signals and Synergetics", *Proceedings of the 1st European Interdisciplinary School on Nonlinear Dynamics for System and Signal Analysis (EUROATTRACTOR 2000)*, Warsaw, June 06 to June 15, 2000, pp. 231 – 242.

GENERIC BI-LAYERED NET MODEL CONSERVATION AND INFORMATION PROCESSES

BÉLA CSUKÁS
University of Kaposvar
Dept. of Information Technology
Kaposvar, Hungary

GYÖNGYI BÁNKUTI
University of Kaposvar
Dept. of Applied Mathematics
Kaposvar, Hungary

ABSTRACT
This paper starts from the idea of Direct Computer Mapping and introduces the formal description of a new conceptual modeling methodology called Generic, Bi-layered Net GBN). It is a novel framework for process simulation, as well as an adequate software technology and a hypothetical hardware architecture that make possible the common implementation of the hybrid (discrete & continuous) models. Next a very simple example for a Switched Server System illustrates how we can describe a "hybrid automaton" with this tool. Finally we summarize the existing applications and outline the GBN representation, suggested for the problem class of the supply / demand networks.

1. INTRODUCTION

The computer modeling of the continuous and discrete processes have been evolving in two different ways. Continuous processes are usually described by a set of algebraic, differential and/or integral equations (IPDAE). The 'a priori' (white box) models are derived from the simple elementary balances of the characteristic measures, and then they are transformed into sophisticated mathematical constructs, that cannot be solved. Consequently they are discretisized, and finally, the computer executes simple arithmetical operations, again. The 'a posteriori' (black box) models differ only in the origin, however, the numerical solution of the identified mathematical equations are similar. The discrete processes can be modeled by abstract automata, as well as by various nets. Many net models originated from the Petri-net and the State-Transition Nets. The solution of the hybrid, discrete / continuous models is a big issue, because the usual integrators do not tolerate the discrete events, while the usual representation of the continuous processes cannot be embedded into the discrete models.

In the Generic, Bi-layered Net based Direct Computer Mapping (Csukás and Bánkuti, 2003), the simple building blocks of the conservational and informational processes are mapped onto the generic "passive" and "active" elements of the program. The "active" elementary transitions may be either balance-based primitive "stoichiometric" processes or any kind of rules. The "passive" elements may be either the finite balance elements, describing the state of the physical system, or any kind of signs. In this knowledge representation, the model is organized by the elementary processes, which gives a common structure for the continuous transitions and discrete events.

2. GENERIC, BI-LAYERED SOFTWARE AND HARDWARE ARCHITECTURE

The generic elements of the proposed hybrid simulator are brief programs, organized in a bi-layered architecture, illustrated in Fig. 1 (left). The programs in the layers and the communication between them have different functionalities, which makes possible to execute the continuous changes and the sequential events with the same kernel.

The "active" elements (A), with the knowledge of the output of the "passive" ones (P), calculate the changes or rules, and then modify the input of the "passive" layer.

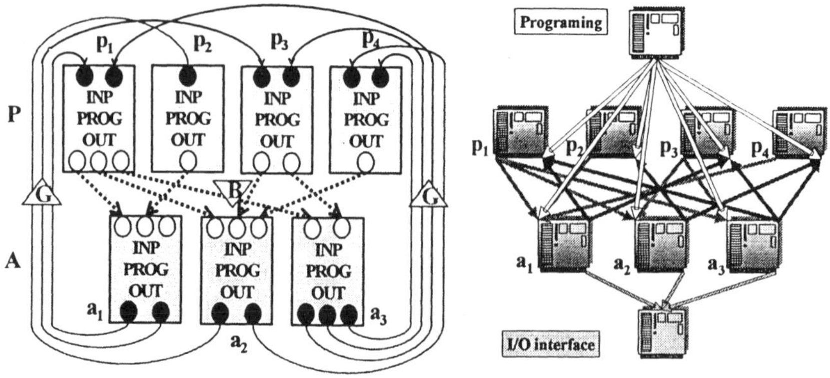

Figure 1. Generic, bi-layered software and hardware architectures

The hardware implementation of the Generic Bi-layered Net can be solved by a hypothetical multiprocessor machine (see Fig. 1 right), consisting of two classes of processors. Programming means the configuration of processor network, the declaration of the initial conditions and the distribution of the brief elementary programs amongst the units. Execution is based on the wired or wireless communication between the two layers of processors, having their own unique broadcast and multiple receiving frequencies. The four steps of the run are: the P→A broadcasting, the execution of the 'active' units, the A→P broadcasting and the execution of the 'passive' units, respectively.

3. FORMAL DESCRIPTION OF THE GENERIC BI-LAYERED NET MODEL

The generic bi-layered net can be defined by the ten-tuple of

$$\langle \mathbf{P},\mathbf{A},\mathbf{B},\mathbf{G},\mathbf{X},\mathbf{Y},\mathbf{\Phi},\mathbf{\Psi},\underline{\mathbf{r}},\mathbf{t}\rangle \tag{1}$$

The communication channels (B and G) of the di-digraph determine the passive→active

$$B(\tau)\subset P(\tau)\times A(\tau) \qquad {}_j b_i(\tau)=\left(p_j(\tau),a_i(\tau)\right)\in B(\tau) \tag{2}$$

$$\exists_j\underline{b}(\tau)\mid \underset{i}{\forall}\left(p_j(\tau),a_i(\tau)\right)\in {}_j\underline{b}(\tau) \qquad \exists\underline{b}_i(\tau)\mid \underset{j}{\forall}\left(p_j(\tau),a_i(\tau)\right)\in\underline{b}_i(\tau) \tag{3}$$

and active→passive data flows

$$G(\tau)\subset A(\tau)\times P(\tau) \qquad {}_i g_j(\tau)=\left(a_i(\tau),p_j(\tau)\right)\in G(\tau) \tag{4}$$

$$\exists_i\underline{g}(\tau)\mid \underset{j}{\forall}\left(a_i(\tau),p_j(\tau)\right)\in {}_i\underline{g}(\tau) \qquad \exists\underline{g}_j(\tau)\mid \underset{i}{\forall}\left(a_i(\tau),p_j(\tau)\right)\in\underline{g}_j(\tau) \tag{5}$$

respectively. Index j designates the ordered sets of the existing output (${}_j\underline{\mathbf{b}}$ (Eq. 3)) and input ($\underline{\mathbf{g}}_j$ (Eq. 5)) connections for the j-th passive element. Similarly, index i defines the ordered sets of the existing output (${}_i\underline{\mathbf{g}}$ (Eq. 5)) and input ($\underline{\mathbf{b}}_i$ (Eq. 3)) connections for the i-th active element. Variable r designates the geometrical and property coordinates of the model. the Variable τ denotes the optional points or intervals of the continuous or discrete time t, declaring the existence of the respective elements and relations.

The passive elements **P** are associated with state variables $\mathbf{X_j}$ and with operators ψ_j, describing the change of the state:

$$\underset{j}{\forall} p_j \rightarrow X_j \in X; \quad \psi_j \in \Psi \qquad \psi_j = \begin{bmatrix} \underline{y}_j[g_j] \\ {}_j x[{}_j b] \end{bmatrix} \tag{6}$$

where $\mathbf{X_j}$ contains any structured data set, and operator ψ_j describe how the $\underline{y}_j$ change via channels g_j modifies the state. The active elements A are characterized by an operator φ_i, providing a mapping. This determines how the output changes ${}_i y_j \in \underline{y}_j$, carried by ${}_i g$ are calculated from the coordinated input readings ${}_j x_i \in \underline{x}_i$ that comes from the passive elements through the channels b_i:

$$\underset{i}{\forall} a_i \rightarrow \varphi_i \in \Phi; \quad \varphi_i = \begin{bmatrix} \underline{x}_i[b_i] \\ {}_i\underline{y}[{}_i g] \end{bmatrix} \tag{7}$$

The net structure determines also the network (ring) structures of the influence routes and flux routes. The alternating, connected, ordered set

$$\left\{{}_{j_1}b_{i_1}, {}_{i_1}g_{j_2}, {}_{j_2}b_{i_2}, {}_{i_2}g_{j_3}, \ldots {}_{j_n}b_{i_n}, {}_{i_n}g_{j_{n+1}}\right\} \tag{8}$$

of the communication channels is called influence route, which determine a special network structure. The influence routes carry the influence, e.g. a perturbation of the content $\mathbf{X}_{j_1}$ of the element $\mathbf{p}_{j_1}$ affects the content $\mathbf{X}_{j_{n+1}}$ of the element $\mathbf{p}_{j_{n+1}}$, according to the influence:

$$\left\{{}_{j_1}\Delta x_{i_1}(t_1), {}_{i_1}\Delta y_{j_2}, {}_{j_2}\Delta x(t_2)_{i_2}, \ldots, {}_{i_n}\Delta y_{j_{n+1}}\right\} \tag{9}$$

where ${}_j\Delta x_i$ and ${}_i\Delta y_j$ refer to the perturbation of the state and the change. Obviously, sensitivity and its special forms, such as observability and controllability can be studied by means of the influence route network. The minimal (generating) influence routes are the basic edges. The maximal influence routes are the transferring routes and the complete loops. The simplified structure of the influence routes is a special ring, where the two algebraic operations are the concatenation and the common part.

The formal analysis designates typical process classes. An important special case of net (1) is the class of balance processes, where the basic part of the state $\mathbf{X_j}$ is a measure and the operator ψ_j summarizes the simultaneous rates. Depending on the discrete or continuous time, operator ψ_j generates also the appropriate

$$\psi_j(\underline{y}_j) = \frac{\Delta X_j[p_j]}{\Delta t} = \sum_i {}_i y_j \tag{10}$$

difference or differential equations, called balance equations.

4. GENERIC, BI-LAYERED NET REPRESENTATION OF A SIMPLE HYBRID AUTOMATON

We shall illustrate the GBN method with the simplest discrete / continuous, switched, hybrid dynamic system (Perkins and Kumar, 1989; Agarwal et al., 2002; see Fig. 2). The suppliers send materials in the buffer tanks A, B and C at constant flow rates of S_A, S_B and S_c, respectively. Machine can process any one material at a time at rate P_A, P_B and P_c. The level of materials changes in the buffers, however there are determined minimal and maximal levels. A material specific setup time is incurred each time machine switches to different material. The goal is to design a switching strategy, which satisfies various single or multiple objectives, e.g. maximal production, minimal setup and waiting time, minimal buffer levels, etc. There is an obvious interaction between the continuous and discrete (logic) components, i.e. it is a simple prototype of the hybrid dynamic systems.

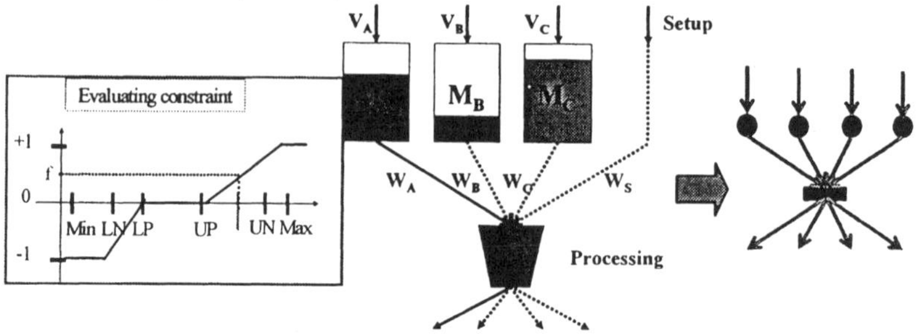

Figure 2. The Switched Server System and the applied "fuzzy" constraint

In the usual formal modeling approach, the main feature is that the system switches between various sets of differential equations. One cannot start emptying the next buffer tank immediately, because there is a material specific setup time, and during that time, none of the tanks are being emptied. The GBN model of example is shown in Fig. 3.

Passive elements sre: p_1 = buffer tank A, p_2 = buffer tank B, p_3 = buffer tank C, p_4 = setup supply (conservational); and p_5 = state of buffer tank A, p_6 = state of buffer tank B, p_7= state of buffer tank C, p_8 = state of setup supply, p_9 = state of setup, p_{10} = state of machine, p_{11} = proposed change (informational).

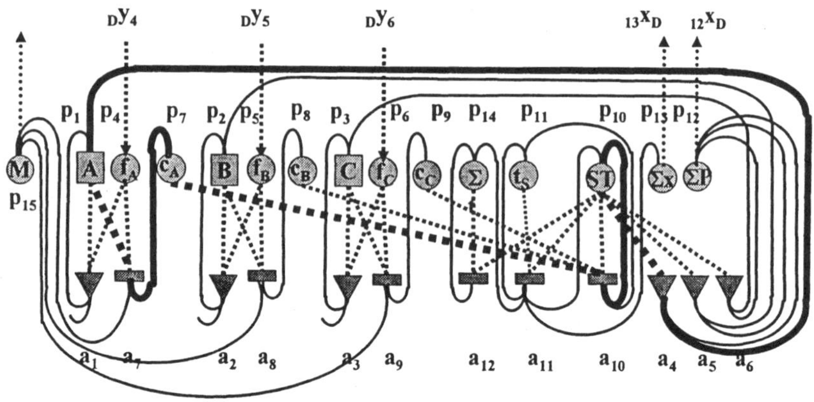

Figure 3. The Generic, Bi-layered Net model of the example

Active elements are: a_1 = supply A, a_2 = supply B, a_3 = supply C, a_4 = process A, a_5 = process B, a_6 = process C, a_7 = setup (conservational); and a_8 = characterize buffer tank A, a_9 = characterize buffer tank B, a_{10} = characterize buffer tank C, a_{11} = characterize setup, a_{12} = evaluate system, a_{13} = decide change (informational). As an example for the functioning of an active element please consider the mapping

$$a_{10} \rightarrow \varphi_{10} = \begin{bmatrix} \{f_A, T_{S,A}\} \models_7 x_{10},\ \{f_B, T_{S,B}\} \models_6 x_{13},\ \{f_C, T_{S,C}\} \models_7 x_{13},\ ST =_{10} x_{10} \\ {}_{10}y_{10} = ST,\ {}_{10}y_{11} = T_S \end{bmatrix} \tag{11}$$

while the actual rule is

if ST = OK and $f_X = \max(f_A, f_B, f_C)$
then ST =: Process_X and $T_S =: T_{S,X}$

where $f_X = f(M_X$, Min,LN,LP,UP,UN,Max) is the evaluation of the level in the buffer tank X, based on the level M_X and on the six characteristic points of the trapezoidal fuzzy variable; and

$T_{S,X}$ is the setup time of tank X.

The mapping and functioning of a passive element are the followings:

$$p_1 \rightarrow \psi_1 = \begin{bmatrix} \Delta M_{A1} = {}_1y_1,\ \Delta M_{A4} = {}_4y_1 \\ {}_1x_1 = M_A,\ {}_1x_7 = M_A \end{bmatrix} \tag{12}$$

$$M_A =: M_A + \sum_i {}_iy_1 \tag{13}$$

As an example for an influence route follow the thick line in Fig. 3, designating the

$$\{{}_1b_7,\ {}_7g_7,\ {}_7b_{10},\ {}_{10}g_{10}\ ,{}_{10}b_4,\ {}_4g_1\} \tag{14}$$

feedback loop that carries the influence of

$$\{{}_1\Delta x_7(t_1),\ {}_7\Delta y_7(t_2),\ {}_7\Delta x_{10}(t_3),\ {}_{10}\Delta y_{10}(t_4),\ {}_{10}\Delta x_4(t_5),\ {}_4\Delta y_1(t_6)\} \tag{15}$$

This describes how the level of p_1 can effect on the change of its next state. The manifold feedback loops of the Generic Bi-Layered Net model determine the skeleton of a Recurrent Neural Network structure.

For the time being, we use a multiobjective genetic algorithm (Csukás and Balogh, 1998) for optimal process design. In the applied method, the dynamic models to be generated and executed are proposed by a genetic algorithm, while the "fitness" evaluation of the variants comes from the result of the detailed dynamic simulation.

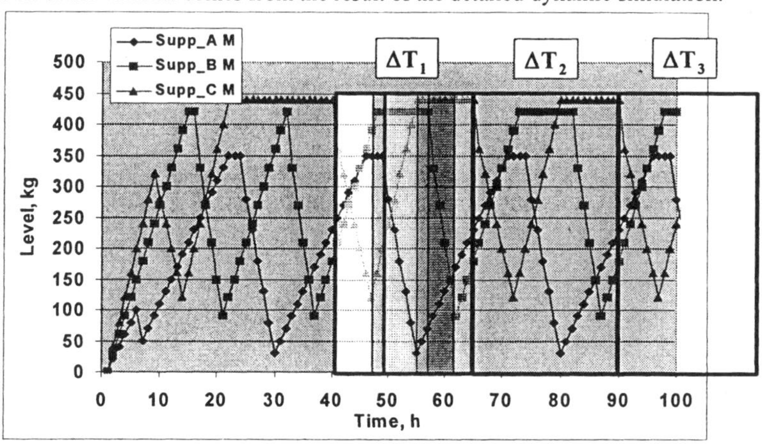

Figure 4. Fast development of a robust, periodic solution

Actually, the genetic algorithm tunes the parameters of the above mentioned fuzzy constraints f(Min,LN,LP,UP,UN,Max), while the objective functions of the evaluation may be the maximization of the produced materials, the minimization of the setup, the minimization of the buffer levels, etc. A compromise good solution can be seen in Fig. 4.

5. PRACTICAL APPLICATIONS OF THE METHODOLOGY

The methodology has been successfully applied to several complex practical examples, as follows: Simulated Moving Bed preparative chromatographic processes (industrial applications); batch, controlled co-polymerization of products with directed structures (industrial application); metabolic network simulation (Csukás, et al., 2003, ongoing experiment); dynamic simulation based planning and scheduling of an agricultural farm (including cost analysis and year long cash-flow calculation); quantitative health risk analysis of multi-product, batch pharmaceutical plants (industrial application). Present work has been continued toward the simulation based design and control of realistic supply / demand networks. Figure 5 shows, that the typical local decision blocks can be generated from the simplified flowsheet, automatically. The local decisions (d) are made with the knowledge of the fuzzy constraints (f_X), associated with the adjacent resources (X). The constraints (f_X) are controlled by any (e.g. genetic, neural or fuzzy) supervisory algorithm, according to the prescribed objectives.

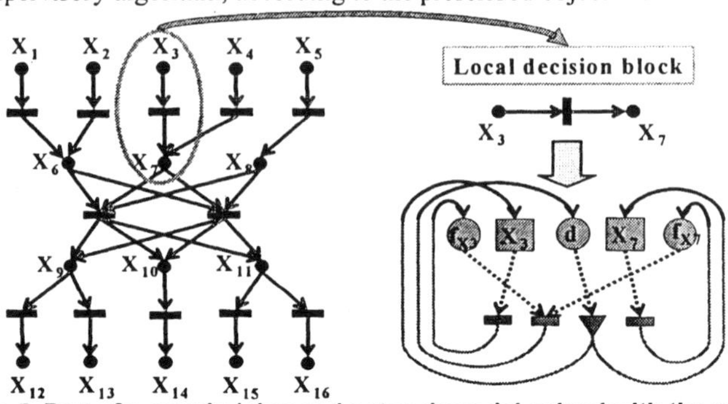

Figure 5. Part of a supply / demand network model solved with the method

CONCLUSIONS

The recently developed Generic Bi-layered Net model seems to be a robust and flexible common background for the simulation based design and control of the various processes. Considering the hybrid, discrete / continuous systems the novel method can be used as an adequate representation of the so-called hybrid automata. Consequently, the list of successful, previous applications can be supplied with the local decision based operation of the supply / demand networks.

REFERENCES

Agarwal, A., Ydstie, B. E., Grossmann, I., 2002, "Stability, Performance and Control of Switched Systems", *Annual Meeting of Center of Computer Aided Process Design*, Carnegie Mellon University, Pittsburgh.

Csukas B., Bankuti Gy., 2003, "Direct Computer Mapping of Process Models", *Proceedings, Foundations of Computer Assisted Process Operations*, I. E. Grossmann and C. McDonald, eds., AIChE INFORMS, pp. 577-581.

Csukas B., Balogh S., 1998, „Combining Genetic Programming with Generic Simulation Models in Evolutionary Synthesis", *Computers in Industry*, Vol. 36, pp. 181-197.

Perkins, R. and Kumar, P. R., 1989, "Stable, distributed, real-time scheduling of flexible manufacturing assembly disassembly systems", *IEEE Transactions on Automatic Control*, Vol. 34, pp. 139-148.

Csukas B., Debelak, K. A., Prokop, A., Balcarcel, R. R., Tanner, R. D., Bankuti Gy., Balogh S., 2003, „Generic Bi-layered Net Model-Based Discrimination of Chemical and Biological Warfare Agents", *AIChE Annual Meeting*, San Francisco, 2003 (ongoing).

ACKNOWLEDGEMENT

The work has been supported by Hungarian Scientific Research Fund T 037 297.

KNOWLEDGE-BASED SMART PROTOCOL FOR BIOMETRIC AUTHENTICATION

V. K. Murthy
School of Information Technology and Electrical Engineering,
UNSW@ADFA, University of New South Wales,
Canberra, ACT 2600, AUSTRALIA
email: murthy@cs.adfa.edu.au

ABSTRACT
A smart biometric-authentication scheme is described for real-time security control in distributed on-line transaction processing systems (OLTP).This scheme uses a probabilistic quorum protocol to validate users using their biometrical characteristics,e.g., speech, finger and hand scan, handwriting. This multi-modal-input authentication scheme is well suited for smart card based template and agent-based distributed implementation.

1. INTRODUCTION

Biometric technology has emerged as a viable solution for a range of application where a person's identity should be established. It is no longer a science fiction myth ,but is being deployed as a major tool in many organizations to authenticate employees, customers and citizens, Nanavati et al [13], Jain [6]. The design of user authentication schemes has been a very active area of research , Abraham et al [1]. Landwehr and Jajodia [9] . Real-time database authentication schemes are required in a variety of applications, especially in sensitive applications such as defence control systems. Most of the authentication schemes described in the literature use exact arithmetic based on finite fields for designing cryptographic algorithms and protocols Brickell [2] , Guillou [4] , Mitchell [11], Preneel[14] , Simmons[16] . These schemes are user independent and are well suited for keyboard modal input from the user.

However, other modes of input such as speech, mouse and pen are gaining importance and users are showing increasing preferences for these modes Martin[10] , Rudnicky [15],Nanavati et al [13]. Rudnicky [15] shows that a high percentage of users prefer the speech mode of input with short duration utterances. Also, multi-modal biometric testing devices would become more popular than the cryptographic schemes, since the users' profiles can be monitored unobtrusively, if required.

However,since the biometrical feature verification schemes are statistical in nature they are prone to errors in judgement. To obviate this difficulty, we propose an authentication scheme which uses a probabilistic quorum protocol that minimizes the error in verification arbitrarily close to zero. This scheme can be implemented using smart card and agent-based techniques.

The quorum protocol can also be used for unobtrusive monitoring and at the same time improve the performace rating. Thus, an On Line Transaction Processing system (OLTP) can detect an intrusion by a masquerader who gained access on the basis of the masquerader's self-declared false identity. In such a case, the OLTP would abort the user transaction, alert the security manager, and initiate recovery , Krishnamurthy and Murthy [8], [12].

The user monitoring can be performed by an agent -based cooperative transaction called the `**Scrutinizing transaction**', [12] which can run concurrently ,if required, with the user transactions. An efficient and trusted service can be provided to users by integrating the actions of the scrutinizing transactions (that ensure dependability) with the various concurrency control and recovery schemes of the user transactions (that guarantee consistency). Blackboard architectures (Engelmore [3]) widely used in distributed artificial intelligence would be required to coordinate the scrutinizing transactions at different sites in distributed databases.

2. BIOMETRIC AUTHENTICATION

A user initially enrols in a biometric system by providing biometric data (finger scan , facial scan, iris scan, voice scan, hand scan ,retina scan,finger print, signature or keystroke feature) which is converted into a template called "Enrolment template" which is a small file derived from the distinctive features of the data to enable matching and verification The user carries this template in the form of a smart card. The template is also stored at the main database of the concerned organization and can be stored and deployed through a trusted agent anywhere in the networked world , adding appropriate matching and other computational procedures for scrutiny .

When a user is to be verified, the biometric systems store and compare the enrolment templates (not data) with the newly created "Presentation template" of the user under verification .To generate the presentation template, the agent can prompt the user to carry out a particular task in the domain of enrolment template, e.g. uttering a particular word. This is called a challenge . A presentation template is then created and compared with the enrolment template . If the comparison is rendered as a score, this score is compared against a preset threshold for security clearance [13]. In practice,the final score has to be arrived at as a cumulative sum over of scores , by using a number of challenges that minimize false acceptance and true rejection. Finally, if the cumulative sum of scores exceeds a specified threshold then the clearance is given, else an alarm is raised.

3 . AUTHENTICATION PROTOCOLS

A protocol consists of an ordered sequence of message exchanges that take place between the user and the machine (system). In an authentication protocol, the machine provides an authentication certificate to the user for access to the OLTP as well as the commitment of the user transaction. The machine's prompting message to the user for performing a specified action is called a challenge and the user's message is called a response. In a typical authentication protocol, we will have 2k (where k is a positive integer) challenges by the machine and 2k accepted responses from the user in order to gain accreditation and receive an authentication certificate.

3.1 Quorum Authentication Protocols

In a quorum authentication protocol, the machine provides an authentication certificate to a user if the user's response is accepted at least k+1 times out of 2k challenges.

3.2 Probabilistic Protocols

Probabilistic protocols are protocols that find the right answers to decision problems within a certain error probability. When combined with a quorum protocol, a probabilistic protocol can reduce the error in verification of a user arbitrarily close to zero, provided the initial probability of error is less than 1/2 and the ratio of probability of errors in successive steps of the protocol is less than one.

4. QUORUM PROTOCOL

The conventional zero-knowledge cryptographic protocols Brickell[2] , Imai [5] Preneel [14] , Simmons [6] use exact modular arithmetic for challenge and response protocols. Here, the system (verifier) asks the user (prover) to prove that the latter is a genuine user without revealing secrets. Our aim in this paper is to devise an authentication scheme based on biometrical characteristics of users. It is assumed that the smart card as well as agents ,Guillou [4], can store reference patterns of the biometric characteristics that can be delivered to the local/global processor when challenges take place and responses are provided. Presently, self-programmable one-chip microcomputers (SPOM) are manufactured by several companies including Motorola and Hitachi, Guillou [4] .They can permit the storage of reference patterns of biometric characteristics, as well as executing conventional cryptographic protocols.

As earlier mentioned, the biometrical characteristics are statistical in nature and so it is not possible to ensure exact matches with the template stored . That is, matching can only be carried out within certain confidence limits. This matching process can be performed by a local processor or server under the direction of an agent.

4.1 Probabilistic Quorum Authentication

The quorum authentication protocol based on biometrical characteristics (such as speech, handwriting, fingerprints and keyboard characteristics) is as follows:

1.When the user initiates a transaction by presenting the smart card, the agent selects a random task stored in it that is representative of its user's characteristics and sends it to the local processor or server (the verifying entity).

2. The agent then challenges the user (the prover) to perform certain actions, in order to extract the user's biometrical characteristics.

3.The user then responds with the appropriate actions so that the verifier can proceed to certify. (For instance, if speech characteristics are used, the user pronounces the chosen word; if handwriting characteristics are used, the user writes the word using a pen-based interface; if keyboard characteristics are used, the user types the word.)

4. The local processor then computes the user's features for each of the user's responses and compares it with the features of the word stored in theagent, as well as the smart card. (For instance, the local processor computes the features of the pronounced word and compares it with the speecific features of the word stored in the smart card). For each response of the user, the matching only succeeds if all or a quorum of the required features stored in the smart card and those computed agree within certain well-defined threshold of taxonomic distances, resulting in the acceptance of the response;otherwise, the response is rejected.

5. There are 2k challenges and responses in this protocol. If the user's response is accepted at least k +1 times out of 2k responses (i.e., there is a majority or quorum of accepted responses), the user is verified and gains accreditation by obtaining an authentication certificate. It is advisable that a user is not informed as to which of his responses are not accepted in order to minimize security breaches. In this authentication protocol, it is extremely difficult to reconstruct the chosen word from its features; thus, this protocol can be thought of as a statistical zero-knowledge protocol. Note that after k+1 accepted responses, the user is verified and no further challenges/ responses are required. In particular, if all the initial k+1 responses are accepted, then the user is verified after k +1 challenges; thus, the minimum number of challenges required for verification is k+1 . Also, if all the initial k+1 responses are not accepted, the user is rejected and the user transaction is aborted.However, in agent-based sytem continued unobtrusive monitoring can be carried out even for a genuine user.

Each response can come from either a true (genuine) user or a false user (masquerader). Each such response can be accepted or rejected. Thus, the possible events are:

TA: True user is accepted (true acceptance).

TR: True user is rejected (true rejection).

FA: False user is accepted (false/erroneous acceptance).

FR: False user is rejected (false rejection).

Let P be the intrinsic probability of correct verification (or certification) for the authentication scheme, i.e,P is the probability for false rejection (FR) and true acceptance (TA) and 1-P represents the maximum initial knowledge error in the verification of the user. In probabilistic protocols, it is required that

$$P > 1/2 \qquad (1)$$

in order to achieve rapid improvement in verification probability.

Let 1 - u be the desired probability of correct verification, i.e,u is the probability of incorrect verification or the final knowledge error.

Our aim is to minimize the final knowledge error u in the authentication scheme (protocol), or maximize the probability of correct verification 1 - u , so that it is greater than P and is as close to unity as possible. This can be achieved by increasing the number of challenges and responses in the authentication scheme (protocol). The proof of this protocol is based on the use of the binomial distribution function. This distribution is applicable whenever a series of responses (trials) satisfy the following conditions , Trivedi [17] :

1. Each response has exactly two mutually exclusive outcomes - acceptance or rejection.
2. The probability of acceptance for each response is a constant P where $P = 1/2 + v$. The probability of rejection is $1-P$. This means that abnormal variations in users' characteristics do not happen.
3. The outcomes of successive responses are mutually independent, i.e., they do not influence each other.

Subject to the above conditions, the probability that we shall not perform correct verification is:

$$\sum_{i=0}^{k} \binom{2k}{i} (1/2+v)^{i} (1/2-v)^{2k-i} \qquad (2)$$

That is, we sum over all the insufficient numbers of accepted responses, i = 0 through i= k , the probability of that number of accepted responses occurring. This sum is no larger than ,Ullman [18]:

$(k + 1)\, 4^{k} (1/2 + v)^{k} (1/2 - v)^{k} = (k + 1)(1 - 4v^{2})^{k}$

which is an approximant larger than the expression specified in (2). Hence, an upper bound on the probability of incorrect verification or knowledge error u is given by :

$u = (k + 1)(1 - 4v^{2})^{k}$

Substituting $v = P - 1/2$, we have $u = 4^{k} (k + 1) P^{k} (1 - P)^{k}$ or

$1 - u = 1 - 4^{k} (k + 1) P^{k} (1 - P)^{k}$ (3)

We can therefore achieve the desired probability of correct verification 1-u (improving it from P) by a suitable choice of k in (3). Thus, we can increase the probability of correct verification 1 - u (closer to 1) by choosing larger values for k in the authentication scheme, i.e., by increasing the number of challenges and responses.

While (3) gives a quick estimate for the value of k , this estimate can be poor (giving a high value of k) for values of $P < 0.7$. In practical situations where a binomial distribution table is available and P is known, it is better to use the table to determine the actual value of k to obtain the desired value of 1-u. The probability of incorrect verification u (based on the error bound) for k+1 accepted responses out of a maximum of 2k challenges and responses is:

$u = 4^{k} (k + 1) P^{k} (1 - P)^{k}$

For each successive step, we have two more challenges and responses. Hence, the probability of incorrect verification u' (based on the bound) for the next step where we have k+2 accepted responses out of a maximum of 2(k+1) challenges and responses is:

$u' = 4^{k+1} (k + 2) P^{k+1} (1 - P)^{k+1}$

Since it is required in a probabilistic protocol that the probability of incorrect verification or knowledge error (based on the bound) decreases for successively increasing steps, it is necessary that the ratio $u' / u < 1$. In other words,

$u'/u = 4P (1 - P) (k + 2) / (k + 1) < 1$ (4)

Equation (4) holds only if :

$P > 0.5 + 0.5/\sqrt{k+2}$ (5)

Also, it is desirable that the ratio u'/u is less than 1 from the initial step (k=1) of the process onwards. Note that if (5) holds for k=1 then it holds for all other k >1 since P is a maximum for k=1 . Thus,

$P > .5 + 0.5/\sqrt{3} = 0.79$ (6)

Note that the requirement $P > 0.79$ is based on the approximation to and this leads to a loose upper error bound. If the sum is calculated directly using (2}) rather than its approximant, we will get the actual value of P (lower than 0.791) that satisfies the condition $u'/u < 1$. This is achievable with currently available speaker verification system technology, Nanavati et al [13] .

If we want to reduce the error quadratically, however, as in probabilistic algorithms we require that $u'/u < 0.5$. This condition holds only if:

$P > 0.5 + 0.5\ \sqrt{k+3}/\sqrt{2(k+2)}$. For k=1 , we have $P > 0.9$.

For highly secure systems, the probability of correct verification 1 - u should be very close to 1 and thus k would have a relatively large value, i.e., there would be a large number of challenge and response steps and response times would be increased.

In less secure systems, we could reduce response times by choosing a smaller k at the expense of increased probability of incorrect verification u , i.e., there is a tradeoff between the probability of correct user verification and response times. Thus, we should choose optimum values of k and u according to our security and response time requirements.

5. EXAMPLE

Let the intrinsic probability of correct verification $P = 0.8$. Let k=10 i.e., we have a maximum of $2k = 20$ challenges and responses out of which there are at least $k + 1 = 11$ accepted responses. Then the probability of correct verification $1-u = 0.9999$ after 11 acceptances (out of a maximum of 20 responses), using the binomial tables Trivedi [17].

6. IMPLEMENTATION ISSUES

A possible implementation strategy is to use a conventional number theoretic cryptographic protocol and a probabilistic quorum protocol for user authentication during the authentication. Then, the quorum protocol can be used to perform unobtrusive key-hole monitoring of a user during the whole OLTP session until commitment by providing accreditation using several challenges and evaluating the responses. Thus, the quorum protocol provide an additional layer of security control.

If a user is verified by the quorum protocol (user is genuine), then the system allows the transaction to complete and commits the transaction. If the quorum protocol detects intrusion by a masquerader, the system aborts the transaction and raises an alarm.

An audit log can record the actions of a transaction and the probabilities associated with correct verification, false acceptance and true rejection. The arbitrator can inspect the audit log to resolve disputes, Mitchell [11] e.g., the probabilities would indicate if a user is (falsely) denying that he/she performed some action even though he/she actually carried it out. Also, we can ensure that committed transactions were carried out by genuine users from an inspection of the log.

7. USE OF TRUSTED MOBILE AGENTS

A mobile agent is ideally suited for performing biometric authentication tasks in distributed systems since it has the following properties (Knabe[7]):

1. It is a code containing object along with data and execution context that is able to migrate autonomously and purposefully within a computer network .
2. It can react to external events
3 It can be persistent in the sense that it can suspend execution and keep local data in stable storage. After resuming activity an agent's execution is continued, but not necessarily at the same location.
4. A user or a program can delegate tasks to an agent and vest it with an authority (power of attorney) to act on their behalf by providing terms of reference and time dead-lines.
5. Agent can make decisions, based on rules, goals, policies and preferences set by its owner.

6.Agents can interact with the host environment and owners.
7. Agents can cooperate with each other to achieve common goals.
8. Agents can support peer-peer model for distributed computing.
9. Agents can vary their roles-may act as clients or servers, observers, seekers.
10.Ability to reason and learn from interactions with other agents, owners and environment.

A large number of trusted agents can be deployed across the network to provide additional support for generating challenges and verifying. Also a trusted agent can combine fetures of digital cryptography anlong with biometric authentication schemes.

Further,once the authentication is successful ,agents can learn to modify or update their rule base with the new inputs received from a genuine user and enlarge its knowledge base to accomodate for changes in user's biometric data and also environmental changes for modifying the challenges. This feature of dynamic expansion of the knowledge base of an agent will be useful for reducing false acceptance and true rejection rates. Also agents can collect information about the impostor for future criminal invstigation., and transmit across the network to appropriate security agencies.

8. CONCLUSION

An agent and smart- card based biometric authentication using a probabilistic quorum protocol is described. The performance of this protocol is alaso analysed.

REFERENCES

[1] D.G.Abraham et al,Transaction security system, IBM Systems Journal,206-229,30, 1991.
[2] E.F. Brickell (Ed.), Advances in Cryptology - CRYPTO '92}, Lecture Notes in Computer Science 740, Springer-Verlag, New York, 1993.
[3]R. Engelmore and T. Morgan, Blackboard Systems, Addison-Wesley, Reading, Mass., 1989.
[4] L.C. Guillou, M. Ugon and J. Quisquater, The Smart Card}, in Contemporary Cryptology, (Ed.) G.J. Simmons, pp 561-613, IEEE Press, New York, 1992.
[5]H. Imai and R.L. Rivest (Eds.), Advances in Cryptology , Lecture Notes in Computer Science 739, Springer-Verlag, New York, 1993.
[6]A.Jain et al, Biometric Authentication, Lecture Notes in Computer science, Vol,2359, Springer Verlag, New York, 2002.
[7]F. Knabe, An overview of Mobile Agent Programming. Lecture Notes in Computer Science, Vol. 1192, pp.100-115, Springer Verlag, New York,1996
[8] E.V. Krishnamurthy and V.K. Murthy, Transaction Processing Systems , Prentice-Hall, New York, 1991.
[9] C.E. Landwehr and S.Jajodia (Eds.), Database Security , IFIP Proceedings, Elsevier, New York, 1992.
[10] G. Martin, The utility of speech input in user-computer interfaces , International Journal of Man-Machine Studies, pp. 355-376, 29,1989.
[11]C.J. Mitchell, F. Piper and P. Wild, Digital signatures , in Contemporary Cryptology, (Ed.) G.J. Simmons, pp 325-378, IEEE Press, New York, 1992.
[12]V.K. Murthy and E.V. Krishnamurthy, Knowledge-Based Security Control for On-Line Database Transaction Processing Systems , ACM SIGSAC Review, Vol. 12, pp. 7-15, 1994.
[13]S. Nanavati , M.Thieme and R.Nanavati,Biometrics, John Wiley , New York, 2002.
[14] B. Preneel and R. Govaerts (Eds.), Computer Security and Industrial Cryptography, Lecture Notes in Computer Science 741, Springer-Verlag, New York, 1993.
[15] A.I. Rudnicky, Mode preference in a simple data-retrieval task}, in Proc. Arpa Workshop on Human Language Technology, Morgan Kaufmann, Calif., 1993.
[16] G.J. Simmons (Ed.), Contemporary Cryptology , IEEE Press, New York, 1992.
[17]K.S. Trivedi, Probability and Statistics with Reliability, Queuing, and Computer Science Applications}, Prentice-Hall, Englewood Cliffs, N.J., 1982.
[18]J.D. Ullman, Computational Aspects of VLSI ,Computer Science Press , Rockville, MD, 1984.

A DATA REDUCTION TECHNIQUE FOR WEB MININIG APPLICATIONS

LATRONIA D LOVELL
The University of Alabama
Department of Computer Science
Tuscaloosa, Alabama

HUI-CHUAN CHEN
The University of Alabama
Department of Computer Science
Tuscaloosa, Alabama

ABSTRACT
Although web mining is a relatively new subject, it is widely recognized that there are many benefits to such application of data mining algorithms. As with any data mining task, data reduction is a critical step in the mining process. However, the structure of data collected by many web services prevents the application of current mining algorithms. Therefore, in this paper, we not only discuss data reduction techniques, but we also propose a new method to handle the semi-structured or unstructured data and various data types that may be present in web mining applications. A data set collected by Blue Martini Software from Gazelle.com is used to illustrate and test the proposed method. The results show that our method generates comparable, if not better, results in addition to being computationally more efficient than traditional methods.

INTRODUCTION

Our ability to generate and collect data has lead to the incredible need and popularity of data mining. Although current data mining techniques have enjoyed success, there are still many areas where improvement is crucial to implement the mining techniques effectively. For example, typically data mining algorithms are employed on massive data sets, thus the computations required may grow exponentially. This rapidly growing computation time may soon far exceed our ability to manage. This is evident especially in web mining applications mainly because of the difficulty in processing the tremendous amounts of raw data that may be stored on the web (Mulvenna et al., 2000). While the potential benefits of such mining are great in areas such as e-business, the application of current data mining techniques to web data is still a very challenging problem. Since the data was not stored or collected with the purpose of data mining in mind, much preprocessing is necessary. Therefore, before applying data mining techniques to web-based data, data reduction methods should be performed. Unfortunately, most existing data reduction techniques are not designed to handle the unstructured, or semi-structured, data produced by web services. Consequently, existing reduction methods must be modified or new methods must be developed.

Although we do mention the rule extraction technique used to evaluate the performance of the proposed approach, the primary focus of this paper is to introduce an effective way to transform web-based data into an appropriate format and reduce the dimensionality of the data while retaining the useful knowledge imbedded in the database. In order to evaluate our proposed approach, a real world e-commerce data set is used.

WEB-LOG DATA

Web mining can be broadly categorized into three areas: web content mining, web structure mining, and web usage mining (Mulvenna et al., 2000). In this paper, the scope of discussion is limited to web usage mining where the regularity of the data collected by the web user's sessions will be addressed.

The data set used to implement our approach is from the 2000 KDD Cup competition (KDD, 2000). It's real world e-commerce data involving web clickstreams and purchase transactions collected by Blue Martini Software from Gazelle.com, a web legwear and legcare retailer that closed its online store in August of 2000. In an attempt to gain insight for determining business directions and marketing decisions, knowledge of the navigational behavior of customers is needed. To test our approach in the e-commerce domain, we focused on the task of characterizing customers who spent more than $12 on an average order at the site. Clearly, being able to understand such customer characteristics will not only help in providing better customer access, but also enable web designers to place advertisements in the most cost and profit beneficial places. However, the available data consists of mixed data types in various formats. In our situation, for each of the 1781 cases provided there are 517 features. The feature values may be numeric, categorical, and, in some cases, an undefined number of strings. The number of distinct values for each feature range from 2 to 1716.

FEATURE REDUCTION

One of the main themes of data reduction is dimension reduction by feature selection. For our specific task, there were 517 features. The feature labeled "Spend over $12 per order on average", which has a value of either "TRUE" or "FALSE", has been designated as the target feature. It is noted that many of the given features are insignificant to the target. Therefore, we used an approach which combines the chi squared test and an information gain analysis to assess feature relevance.

First of all, we used the chi squared test of independence to determine if a given feature was statistically significant to the target feature (Nukoolkit et al., 2001b). With a significance level of 1%, we reduce the number of features from 517 to 300. To further reduction, we used an information gain analysis based on the entropy of the remaining features to select the features that provide information gain greater than a pre-determined threshold value (Jaynes, 1957). In our study, the 45 features with the largest information gained were selected. This method of feature subset selection can be much faster than methods where various combinations of features are examined (Kautardzic, 2001).

DATA TRANSFORMATION

As mentioned previously, data collected on the web may be unstructured, semi-structured, or in different formats. Therefore, we take an intermediate step to transform all feature values, regardless of data type, into a numerical form that will not only be a more suitable format for various data mining tasks, but also can represent its relationship to the target variable.

We propose to represent a feature value (v_i) with its relative frequencies to target outcomes. For example, if the target outcomes for a feature k are $c_1, c_2, \dots c_s$, then the new transformed numerical form will be $[p_{1i}{}^k, p_{2i}{}^k, \dots, p_{si}{}^k]$ with $p_{ji}{}^k$ defined as

$$p_{ji}{}^k = N_{ji}/M_i \qquad \text{for } j = 1, 2, \dots, s \qquad (1)$$

where N_{ji} is the number of cases where feature $k = v_i$ and c_j occurred and M_i is the total number of occurrences of feature $k = v_i$.

The obtained $p_{ji}{}^k$ is a conditional probability of the target outcome c_j given feature k equal to v_i occurred. Thus, $0 \leq p_{ji}{}^k \leq 1$ and $\sum i\ p_{ji}{}^k = 1$ for all v_i of a given feature k. This representation shows the likelihood of each target outcome for a given feature equal to v_i. Hence, if we want to find the most likely cause for an interesting outcome (or a particular subset of target outcomes) the given representation can serve as the basis of this search. For example, in our case study, the interesting outcome is C_1 signaling that more than $12 was spent during this session. Thus we want to find the likelihood of C_1 when the feature k equal to v_i is observed. The transformed value implies that a customer with feature k equal to v_i has a probability of $p_{1i}{}^k$ to spend more than $12. Therefore the potential customers will be the group of people who possess feature values with high $p_{1i}{}^k$. These high $p_{1i}{}^k$ feature values are the factual knowledge we are seeking to uncover.

Another important probability in this data mining task is the total number of occurrences of feature k with value equal to v_i. In some features, the occurrences of a particular value may be so rare that it is statistically underrepresented while other feature values may be overrepresented (Parrish et al., 2002). Thus the probability of occurrences, denoted by $q_i{}^k$, is computed as follows:

$$q_i{}^k = M_i/ N_k \qquad (2)$$

where N_k is the total number of occurrences of feature k.

Now we have transformed a feature into the form $[p_{1i}{}^k, q_i{}^k]$ and the corresponding $p_{2i}{}^k$ ($p_{2i}{}^k = 1 - p_{1i}{}^k$) is eliminated. This idea can be generalized to any conditional probability vector which describes the causal relationship between v_i and c_j (Nukoolkit 2001a).

This transformed vector, which can be viewed as a point in an n dimensional Euclidean space, provides a unified way to solve the problem of inconsistent data types by using a simple counting scheme.

VALUE REDUCTION

The mining task may become more manageable if the number of distinct values can be reduced. For example, in our data set, the maximum number of distinct values for one feature was 1716. Attaching a value reduction process will obviously render the mining task less difficult.

In the previous section, we transformed all feature values into a point in an n-dimensional Euclidean space. Now existing clustering algorithms can be applied to group these data points into clusters in multidimensional space. The cluster centers, medians, or majority values are typical representatives of each cluster. With the data transformation that we have presented, we group points with similar conditional probabilities.

In order to evaluate the proposed algorithm with our case study a K-Means clustering technique is applied to either the original values or the transformed values. Before applying any value reduction, there were 1781 cases each having 45 features accounting for a total of 3041 possible values per case. After using the K-Means clustering algorithm on the original values, the total number of values is reduced to 431 per case, hereby creating an 86% reduction. However, first using the data transformation technique discussed above reduces the number of values to 465; subsequently applying K-Means further reduces the number of values to 345, accounting for an 89% reduction.

PERFORMANCE EVALUATION

Since we have seen that our data reduction technique produces great dimension reduction for our case study, a time reduction is obvious. The next question would be whether some of the preprocessed data could be discarded without sacrificing the quality of results. In order to evaluate the quality of results, two measures are used:

a) an error measure of the relative purities of K-Means clustering using the original data value compared to using the transformed (new) value

b) generating association rules from both the original and new data and comparing the respective rule coverage and accuracy

Instead of transforming original values into the conditional probability vector as discussed previously, existing clustering algorithms seek to distribute values in clusters to minimize the average distance of a value to its cluster mean. The K-Means clustering algorithm is one of the most popular of such methods. In order to obtain a good approximation to the original values, the purity of a cluster, called the tth cluster, is measured by entropy computed by

$$\text{Entropy}(t) = \Sigma_j -P(c_j)*\log(P(c_j)) \quad (3)$$

where $P(c_j)$ is the probability of c_j in the tth cluster.

In general, a perfect cluster is one containing only one class of target outcomes. We thus measure the overall degree of mixture (error) of a feature by

$$Err = \Sigma_t \text{Entropy}(t)*n(t)/N \quad (4)$$

where n(t) is the number of values in t^{th} cluster and N is the total number of cases. Mathematically, features with less error, as calculated by Eq. (4), are prime candidates of interesting features in data mining. The question is whether the proposed transformation degrades the overall purity of a given feature. In other words, we wish to determine if the transformation will result in a higher Err value for a given feature thus making the identification of prime candidates harder. To test this situation, the same K-Means algorithm is applied to both the original and new data. Table 1 shows the computed Err values for some features in its original form and its new form. The results show that the average error for the original data is 0.434 while the average error for the new data is 0.482. However 53% of the time, the new data generated a smaller Err value.

Table 1: Feature Error Measures

FEATURE	ORIGINAL Err	NEW Err
Num articles Template Views.txt	0.502968	0.492383
Num Brands Assortment Views.txt	0.507218	0.501345
Num Energizers & Relaxers Views.txt	0.498573	0.478656
Num Fashion Product Views.txt	0.494233	0.499705
Num Legwear Product Views.txt	0.455074	0.486809
Num main Template Views.txt	0.456311	0.489213
Num mainhome Template Views.txt	0.455074	0.496275
Num mainlifestyles Template Views.txt	0.504119	0.466724
Num Reinforced Toe Views.txt	0.5058	0.494812
Num Sandal Foot Views.txt	0.49969	0.478985
Num Ultra Sheer Look Product Views.txt	0.467403	0.456783
Num UniqueBoutiques Assortment Views.txt	0.469198	0.461957
Order Line Quantity Average.txt	0.0469426	0.470356
Order Line Quantity Maximum.txt	0.0286812	0.498904
Order Line Quantity Sum.txt	0.443054	0.440094
Pack Average.txt	0.477659	0.51545

To compare the quality of mining results, in this paper only rules with one feature variable are extracted and the rule quality will be compared in terms of rule accuracy and the number of C_1 cases covered. Since 23% of the records have a target value of "TRUE," for quality assurance purposes, we maintain rules with 40% or higher accuracy. Table 2 and table 3 show the best 7 rules generated from the original and new data respectively. The results show that the average accuracy of the original data is 74% while the average accuracy of the new data is 81%.

Table 2: Significant Association Rule Features Using Original Values

Original Feature	Accuracy
Order Amount Sum Percent Having Discount Range (5...10]	75%
Order Amount Sum Percent Having Discount Range 0	82%
Order Discount Amount Average	82%
Order item Quantity Sum Percent Having Discount Range (5...10]	75%
Order item Quantity Sum Percent Having Discount Range 0	82%
Percent Order Promotion FRIEND	83%
Percentage Legwear size One Size Purchased	41%
Average	74%

Table 3: Significant Association Rule Features Using New Value

New Feature	Accuracy
Order Amount Sum Percent Having Discount Range (5...10]	75%
Order Amount Sum Percent Having Discount Range 0	82%
Order Discount Amount Average	82%
Order item Quantity Sum Percent Having Discount Range (5...10]	75%
Order item Quantity Sum Percent Having Discount Range 0	82%
Percent Order Promotion FRIEND	83%
Request Processing Time Average	94%
Average	81%

In order to eliminate the possible redundancy in the discovered rules, a sequential search (Mitchell, 1997) is employed to produce a set of rules that cover at least 85% of the cases belonging to class C_1 in the given data set. Table 4 and table 5 show the results of the sequential search association rule generation. We found that, the original data generated a set of three rules with 85% coverage of class C_1 cases with 75% accuracy. On the other hand, the new data yields a set of two rules with 100% coverage and 82% accuracy.

Table 4: Association Rule Set for Original Values

	Feature	Total Records	TARGET=TRUE
Rule 1	Percent Order Promotion FRIEND	390	324
Rule 2	*Order Discount Amount Average*	*75*	*26*
Rule 3	Order item Quantity Sum Percent Having Discount Range (5...10]	3	2
	TOTAL	468	855
		Accuracy	75%
		Coverage	85%

Table 5: Association Rule Set for New Values

	Feature	Total # Records	# TARGET = TRUE
Rule 1	Request Processing Time Average	433	409
Rule 2	Percent Order Promotion FRIEND	70	4
	TOTAL	503	413
		Accuracy	82%
		Coverage	100%

CONCLUSIONS

Applying current data mining techniques to web mining applications is a difficult problem due to the structure of web data. In this paper, we introduce an effective way to transform web usage data that is successful not only in creating a unified representation of data regardless of data type but also successful in value reduction and feature smoothing. In addition, the simplicity of the proposed method does not introduce a high time/complexity factor that may be seen with other value reduction techniques. The results of our case study analyzing the navigational behavior of customers from Gazelle.com show that the proposed method accounts for the greatest data reduction and the best rule generation with minimal increase in error.

REFERENCES

Jaynes, E.T., 1957, "Information Theory and Statistical Mechanics, Physical Review, Vol. 106, pp. 763-768

Kautardzic, M., 2001, "Data Mining: Concepts, Models, Methods and Algorithms," IEEE Press

KDD Cup Competition, 2000, http://www.ecn.purdue.edu/KDDCUP, Sixth ACM SIGKDD International Conference on Knowledge Discovery & Data Mining

Mitchell, T.M., 1997, Machine Learning, Web/McGraw-Hill

Mulvenna, M.D., et al, eds, 2000, "Personalization on the Net Using Web Mining: A Collection of Articles," CACM Vol. 43, No. 8, pp. 122-125

Nukoolkit, C., 2001a, "Knowledge Discovery in Databases with Categorical Data", Ph.D Thesis, University of Alabama, Tuscaloosa, AL

Nukoolkit, C., Chen, H.C., Brown, D., 2001b "Improving Accuracy of Nearest Neighbor Algorithm in Highway Accidents Prediction", ANNIE '01 for the Proceedings of Smart Engineering System Design, Vol. 11, pp. 763-768

Parrish, A., Dixon, B., Cordes, D., Vrbsky, S., Brown, D., "CARE: A Tool to Analyze Automobile Crash Data," August 2002

Zheng, Z., Kohavi, R., Mason, L., 2001, "Real World Performance of Association Rule Algorithms," Proceedings of the Seventh ACM SIGKDD International Conference on Knowledge Discovery and Data Mining, pp. 401-406

DISCRETE-EVENT SIMULATION MODELS WITH JAVA/JDBC/RMI

DARUSH DAVANI AND GORAN TRAJKOVSKI

Computer and Information Sciences Department
Towson University, 8000 York Road
Towson, MD 21252, USA
E-mail: {ddavani, gtrajkovski}@towson.edu

ABSTRACT
The paper provides with a framework for a system to assist users in designing single-run simulation experiments with arbitrary systems for goal seeking and optimization analysis, what-if analysis, interpolation, extrapolation, integration and Monte-Carlo Simulation. Using the object-oriented, and the client/server technologies, we have developed an expert system for the design of discrete event simulation models using the JAVA/JDBC/RMI technologies. The choice of these tools is due to their popularity, flexibility, transparency, ease-of-use, and integration with standard database management systems. This approach differs from the traditional ones in that it is a step-forward toward the implementation of the thin-client paradigm, and optimizes critical resources, such as the network traffic. The paper discusses the theoretical framework, and the implementation details for the expert system.

INTRODUCTION

Although the fields of Simulation and Artificial Intelligence (AI) are both concerned with the development and the use of models, these fields evlolved almost independently. Now the potential contributions of expet systems to the development of simulation models are being considered more widely. AI can play a variety of roles in the simulation process, including: knowledge representation of the model, decision making within a simulation, rapid prototyping ofmodels, data analysis of the outputs, model modifications and maintenance, etc.

The growth of Simulation in several areas, such as Engineering, and its applications to manufacturing has created a demand that exceeds the supply of simulation experts. This shortage could be relieved by an expert system for simulation.

This paper provides a framework for such an expert system to assist users in designing 'single-run' simulation experiments for goal-seeking, optimization analysis, incorporating 'what-if' analysis, interpolation, (Davani, 1992a), area under the curve, extrapolation, and Monte-Carlo simulation, (Davani, 1995).

This paper is organized as follows. The next section presents the models that we used in the development of our expert system. The development environmnet is then presented. In the following section, the architectural overview is given further. The last section concludes the paper.

MODELS

A detailed description of the models can be found in (Davani, 1992a), (Davani, 1992b); (Davani, 1995) and (Davani, 1996).

Sensitivity Analysis

The approach for obtaining sensitivities information combines the Monte Carlo estimation with finite differencing. In recent years, two or more efficient methods, infinitesimal Perturbation Analysis (PA), (Abell and Judd, 1991) and Score Function (SF), known as Likelyhood Rato (LR), have been proposed, (Davani, 1996).

PA basically considered the nominal simple path, which can be obtained from a simulation or from an experiment on the actual system. From this nominal part, with a litlle extra calculations, one can approximate perturbed paths and estimate sensitivity.

In the SF method, the gradient is expressed as expectation with respect to the "same" distribution as the performance measure function itself, (Arsham and Davani, 1992). These estimates are generally unbiased and consistent. One big advantage of this method is that implementation details are similar for all models, whereas PA is problem-dependent.

Sensitivity analysis in its own right gives us the range for which a solution is valid in dealing with the stability analysis.

Optimization Algorithms

The single-run sensitivity estimation is inncorporated in the stochastic algorithms using PA. In the SF sensitivity estimate because all the performance and all the sensitivities are from the same sample, they are dependent; therefore one must be satisfied with slower procedures. One can incorporate the single-run extrapolation methods, such as LR extrapolation, in an adaptive random search optimization technique.

Goal Seeking Problem

The Goal Seeking Problem considers the inverse question in simulation, namely what must be the controllable input parameter level to achieve the desired output value, (Davani, 1995). Explicit use of the inverse problem is for quality control and robust design. This provides a framework of arriving at a targetvalue for a product, and process and service attributed through a set of experiments, which include Monte-Carlo simulation experiments.

An indirect approach to solve the Goal Seeking Problem is by inverse interpolation based on regression analysis. In this treatment we simulate the system for many different values of parameters, and then approximate the response surface function by using goodness fit criteria. Finally, the fitted function is used to int6erpolate for the unknown parameter.

Another method for approaching the Goal Seeking Problem is by estimation of the derivative of the output function, with respect to the input parameter for a nominal system in a single simulation run. The estimated derivative is used in the first order local approximation of the stochastic version of the Newton's root-finding algorithm to estimate the necessary controllable input parameter of a desired accuracy. Under fairly mild conditions, this algorithm normally almost always converges.

'What-If' Analysis

Simulational models are often subjects to serious errors propagated from statistical errors in estimating the parameters of the input distribution. Post-prescriptive analysis allows us to establish confidence in the prescriptive resuts with respect to small changes in parameters in the input distributions. The routine approaches are sensitivity analysis, and 'what-if' analysis for controllable and uncontrollable input parameters.

To solve the 'what-if' proble, for several input values requireds a separate simulation run for each input value. The estimator requires one sample path of the system. As the likelyhood ratio extrapolation estimator estimates for any number, estimations are computed simultaneously from a single simulation run, while simulating the nominal system.

Two other general methods for estimating the performance for several scenarios using a single simulation run based on sensitivities are the Exponential Tangential extrapolation, and the Linear Tangential extrapolation.

Area Under the Curve

This particular program gets the data (the lower and the upper limits from the user), then validates the data, and if the data is not valid, they are fetched from the database using prepared SQL statements. Once the data is fetched, the program gets the function from the GUI, calculates the area under that curve, and displays the data to the user.

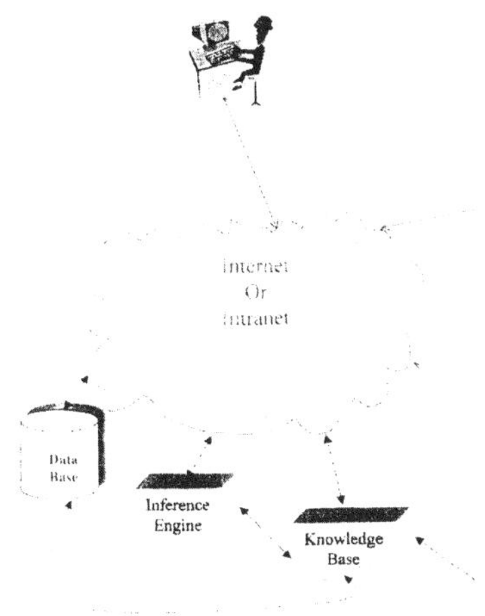

Figure 1. Graphical representation of the approach taken in the development of the tool.

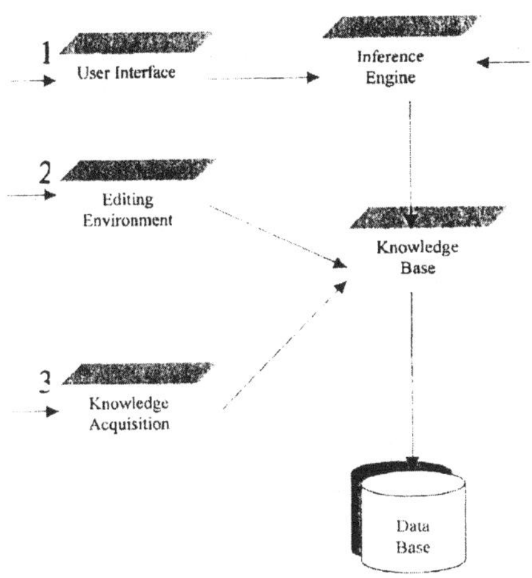

Figure 2. The structure of the interaction with the product (1- User, 2-Developer, 3- Expert)

THE DEVELOPMENT ENVIRONMENT FOR THE SOFTWARE TOOL

The approach for the system that we have been developing is a little bit different from the traditional ones. The reason the tradition is broken is the flexibility in solving the problem, the use of object-oriented technology, popularity of the JAVA programming language, the demand for thin clients, and the advantages of the client/server approach in distributed systems., (Davis et al., 1996), (Hughes et al., 1999), (SAMS.net, 1996); (Whalen and Schroeter, 2002).

To reduce the proble of network traffic, we have taken the thin client approach. The tablesin the database that this expert system uses are

- WHAT, for the 'what-if' analysis
- SENSITIVE, for the sensitivity analysis
- AREA, for the Area under the curve, and
- INTERPOLATION, for the interpolation tool of the expert system.

The Managing Programming Environmnet discusses about how to create packages, how to name and manage source directory structure, and how to create and maintain the Makefile source tree. There is a minimal number of things that need to be performed in order to assure proper operations of the code with the JAVA compiler and interpreter.

The programming platform provides for a system-indepenedent programming interface allowing the programs to use them without compromising portability. Often, a program requires access to system resources such as properties, standard input and output streams, or the current time. The program could use these resources directly from the current runtime environmnet.

To solve the problem of database access in Java, Sun has introduced the Java Database Connectivity package (JDBC). JDBC enables Java applications to connct to any database using various drivers and a new set of Java API objects and methods.

The JDBC provides various objects for connecting to a database, executing SQL statements, and executing stored procedures. JDBC also provides various objects and methods for getting information about objects within the database.

Why JDBC? One of the alternative that we sidered was the CGI scripting language. By calling a CGI script from JAVA, the application allows the call of separate programs to access the database and return results. Using this approach is slow and also lets more bugs creep into the applications. Developing two different development languages requires the knowledge of the two technologies. Another reason to perefer JDBC over CGI is the fact that JDBC is noticeably faster than the CGI approach. CGI requires multi-level of processing of requests, and thus the waiting time increases.

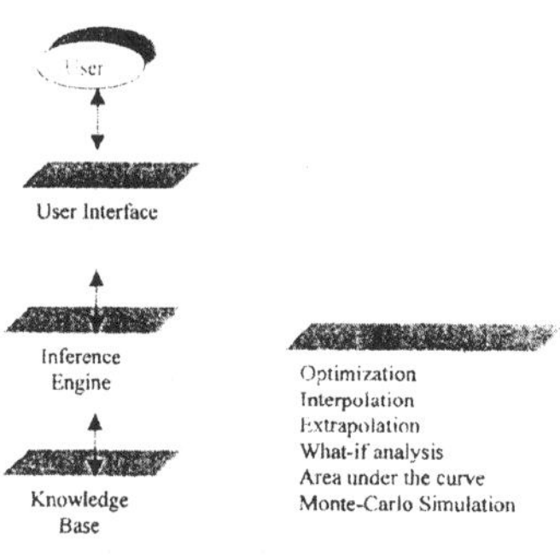

Figure 3. The user's perspective while using the expert system.

The major advantage of using JDBC drivers is that one can write their applications without regard for the database that has beein used. JDBC provides application developers with a single API that is unform and database independent. The driver translates the standard calls into specific calls required by the database it supports. The application can be written once and moved to various drivers. It allows developers to retain specific functionality that their database offers. The design can be either two or three-tier architecture.

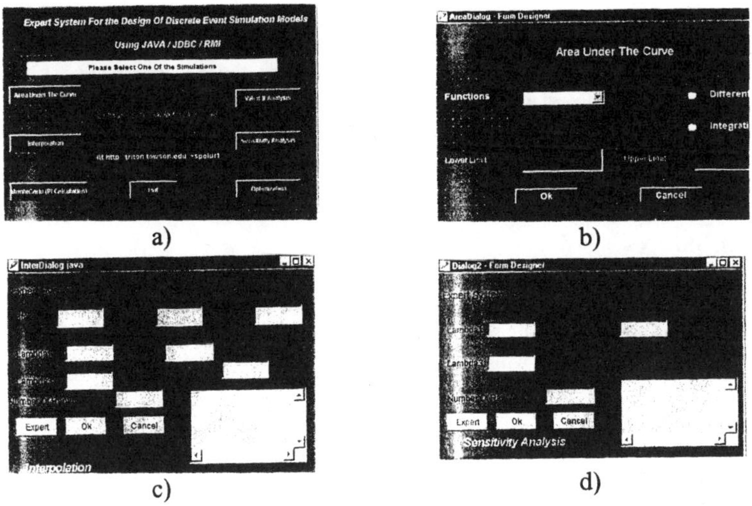

Figure 4. Screenshots of the GUI: a) The front screen with the options; b) The form for the Area under the Curve; c) Interpolation component; d) Sensitivity Analysis screen.

ARCHITECTURAL OVERVIEW

The RMI sysytem consists of three levels: the stub/skeleton layer (client-side stubs (proxies) and server-side skeletons), the remote reference layer (remote reference behavior-e.g. invocation to a single object or to a replicated object), and the transport layer (connection set up and management and remote object tracking).

A remote method invocation from a client to a remote server object travels down through the layers of the RMI system to the client-side transport, then up through to the server-side transport to the server. The return value of a call is sent back through the skeleton, remote reference layer and transport on the server side, and then up through the transport, remote reference layer, and the stub on the client side.

In the sever environmnet, where a JAVA process is being used to serve the RMI requests, the server may need to use a security manager to isolate itself from stub misbehavior. The server functions will usually be implemented by classes loaded from the local system and therefore not subject to the restriction of the security manager. If the remote object interfaces allow objects, either local or remote, then those objects classes must be accessible to the server. Usually those classes will be built-in classes or will be deined by the server. As long as the classes are available locally, there is no need for a specialized security manager or stub loader.

In remote procedural call systems, client-side stub code must be generated and linked into a client before a remote procedure can be done. This code may be either statistically linked into the client, or linked in at run-time via dynamic linking with libraries available locally or over a network file system. In either the case of of static or dynamic linking, the specific code to handle an RPC must be available to the client machine in a compiled form.

For an RMI sever to be able to pass on to clients the location of the stub classes, the server must configure its own stub code base with a URL that is used to retrieve the needed classes. As in applications, classes are always loaded from the local file system if they are available. When loaded by the core class loader, those classes are considered to be trusted and not subject to the security manager. Once a class is loaded by a class loader, the same loader is used to resolve and load all other classes it references, even if they are loaded from the local file system.

CONCLUSIONS

In this paper we presented an expert system that we developped for discrete-event simulations. The GUI is universal, and can be used for other simulation engines, to compare the performance. The structural overview of the product is given, as well as the options it provides the users with.

ACKNOWLEDGEMENT

The authors want to thank Sridhar Poluri for the technical assistance during the work on this project.

REFERENCES

Abell, A., Judd, R.P., 1991, "Perturbation Analysis for Systems with Structural Changes," Proc. Summer Computer Simulation Conf, pp. 104-109.

Arsham, H., Davani, D., 1992, "Monte-Carlo gradient estimators' Some experiments with the score function method.," Simulation in Engineering Education. Society for Computer Simulation.

Davani, D., 1992a, "Parametric What-if analysis in MTTF: A single run Monte-Carlo-based approach," Proc. ACM Annual Conference on Communications, Kansas City, Missouri.

Davani, D., 1992b, "Performance Interpolation for Computer Simulation Models: A Simulation-base Approach", Simulation in Engineering Education, Society for Computer Simulation.

Davani, D., 1995, "A Secant Based Approach to Goal Seeking Problem arising in Monte-Carlo Experiments, " Proc. IASTED Modeling and Simulation, Pittsburgh, PA.

Davani, D., 1996, "Optimization of Discrete Event Static Computer Simulation Models: Score Function (Likelihood Ratio Method)," Proc. 5th SIAM Conf. Optimization. Victoria, BC.

Davis, O., McGinn, T., Bhatiani, A., 1996, Instant Java Applets. Ziff Davis Press.

Hughes, M., Shoffner, M., Hamner, D., Hughes, C., 1999, Java Network Programming: A Complete Guide to Networking, Streams, and Distributed Computing. Manning.

SAMS.net Development Group, 1996, . JAVA Unleashed, Sams Plus Publ.

Whalen, E., Schroeter, MM., 2002, Oracle Performance Tuning, Addison-Wesley.

MULTI-ATTRIBUTE LEARNING MECHANISM FOR CUSTOMER SATISFACTION ASSESSMENT

AKIYA INOUE
NTT Service Integration
Laboratories, Tokyo, JAPAN

KEN NISHIMATSU
NTT Service Integration
Laboratories, Tokyo, JAPAN

SHOKO TAKAHASHI
NTT Service Integration
Laboratories, Tokyo, JAPAN

ABSTRACT

This paper presents a multi-attribute learning mechanism called *LMDCM* (*Learning Mechanism using Discrete Choice Models*) for assessing customer satisfaction with various types of Internet services and IT systems. This mechanism can indicate the customer satisfaction level of each service being provided under given situations. Discrete choice models expressing customer choice behavior are applied to the learning module. The feature of this model is that it considers satisfaction or dissatisfaction as a choice set of the discrete choice model. Customer satisfaction assessment for various Service Providers (*xSPs*) is shown as an application example.

1. INTRODUCTION

The number of Internet users exceeded 50 million last year in Japan. The Internet diffusion rate of business users having more than 300 employees is now over 98%. Stewart Personick mentioned in his paper (Personick, 2002) that "Even in the context of the current dramatic slowdown in telecommunications investment, the Internet and its supported applications are continuing to revolutionize almost all aspects of our business and personal lives". On the other hand, almost all business users are trying to reduce the total cost invested in their networks and IT systems, their management costs and labor costs. A major trend is the effective use of outsourcing such as Internet data centers (*IDCs*), Application service providers (*ASPs*) and Management service providers (*MSPs*). The rate of using outsourcing among business users is over 70% in Japan. However, due to the increase of providers of various services (*xSPs*) and the variety of their service menus, the *xSP* market is experiencing severe competition.

Needless to say, the main objective of service providers is to increase revenue. Providers must give careful consideration to the trade-off between getting an advantage over other providers and reducing network and system investment to achieve the objective. Not only getting new customers but also keeping existing customers continuously for a long period is very significant. We have proposed service choice models to predict the service choice probability under competitive conditions (Inoue and Yamamoto, 1990). Although this choice model is effective for analyzing *xSP* choice behavior, it is not enough to understand the satisfaction level of existing customers and predict churning, that is, changing to other providers. We propose a new analysis method to assess customer satisfaction and to predict churning.

This paper presents a new learning mechanism called *LMDCM* (*Learning Mechanism using Discrete Choice Models*) to assess customer satisfaction for various types of Internet services and IT systems. It also demonstrates that the proposed mechanism is available for customer satisfaction assessment for *xSPs*.

2. REQUIREMENTS FOR CUSTOMER SATISFACTION ASSESSMENT

In various types of Internet services and IT systems, a service level agreement (*SLA*) is a significant service menu. *SLA* is defined in the book (Lewis, 1999) written by Lundy Lewis as a contract between a supplier and a consumer that identifies (1) the service supported by the enterprise network, (2) service parameters for each service, (3) service levels, and (4) liabilities on the part of the supplier and the consumer when service levels are not met. Service level management is a necessary process to guarantee service levels based on the *SLA*. However, the customer satisfaction for a service is not necessarily the same as the guarantee provided by an *SLA*.

Customer satisfaction for conventional network services basically depends on simple values such as average blocking probability, delay time, down time, and throughput. The appearance of various types of service providers such as *IDCs* and *xSPs* has resulted in service selection becoming a complicated and difficult process. A customer can choose the most appropriate service from among many different kinds of services. The primary decision-making factor and the priority among factors vary from customer to customer. Moreover, the number of factors expressing service specifications differs according to the service. In general, a decision to change the service provider under contract depends on the customer satisfaction level. Service providers require powerful support capabilities to assess customer satisfaction level and to predict churning.

A decision to change the service provider under contract is based on the service evaluation factors. These consist of quantitative factors such as charge and performance and qualitative factors such as brand image, help-desk quality and maintenance-personnel attitudes. It is difficult to understand the relationships among service evaluation factors. Moreover, high performance is not necessarily the same as the high satisfaction level. The satisfaction level generally changes over time. For example, a high-performance service is perceived as ordinary after a few months.

A new analysis method should be able to:

(1) handle both quantitative factors and qualitative factors,

(2) analyze the weight of factors in relation to decision-making,

(3) analyze the relationship between each factor and satisfaction level,

(4) learn satisfaction level changes,

(5) predict the satisfaction level of new services,

(6) predict churning.

An expert system such as an equipment-failure diagnosis system, in which the "if-then rule" is used, is not enough to fulfill them. This is because some of factors are probabilistic or uncertain. AI methods such as fuzzy logic and neural networks have been proposed to assess customer satisfaction (Kuo, et al., 1996), (Temponi, et al., 1999). Although these methods are effective, the satisfaction level is expressed as the score obtained on the basis of factors. It is difficult to predict churning by using the score. The approach of the component-to-service mapping problem (Lewis, 1999) is effective to analyze the relationship between each factor and the satisfaction level.

Our multi-attribute learning mechanism *LMDCM* is an effective way to fulfill these requirements.

3. MULTI-ATTRIBUTE LEARNING MECHANISM USING DISCRETE CHOICE MODELS

3.1 Architecture of *LMDCM*

The basic architecture of *LMDCM* is shown in Fig. 1. The main objectives of *LMDCM* are to assess customer satisfaction level and to predict churning for various types of services. *LMDCM* indicates the customer-satisfaction probability or churning probability of each customer segment under given service specifications. A service provider can choose an action pattern on the basis of these outputs. For example, the provider may modify the service menu such as a threshold value in the *SLA*, or renew the server systems.

The *LMDCM* consists of three modules: i) the service evaluation module, ii) the learning module, and iii) the customer satisfaction evaluation module. The *LMDCM* is defined as $\{S, X, f, E, R, G, P\}$. Here:

$S_a = \{s_1, s_2, \dots, s_m\}$: The set of service specifications of service a, where s_i is a service specification vector expressed by several service attribute parameters, and m is the total number of service specifications.

$X_a = \{x_{a1}, x_{a2}, \dots, x_{ak}\}$: The set of service attribute parameters of service a; X_a is the service attribute vector for service a.

$E_a = \{e_{a1}, e_{a2}, \dots, e_{ak}\}$: The set of evaluation results for service attribute parameters, where e_{ai} is the evaluation result for service attribute parameter x_{ai}.

f : A conversion rule of a service attribute parameter on the basis of the evaluation result or a predetermined value x_0,

$X^*_a = \{x^*_{a1}, x^*_{a2}, \dots, x^*_{ak}\}$: The set of converted service-attribute parameters of service a.

$$x^*_{ai} = f(x_{ai}, e_{ai}, x_0). \tag{1}$$

$R_a = \{0, 1\}$: The set of responses for S_a; r is an evaluation result of S_a such that

$r = 1$: satisfaction, $r = 0$: dissatisfaction, (in the case of the customer-satisfaction-probability model), or

$r = 1$: change to other services, $r = 0$: continuation, (in the case of the churn-probability model).

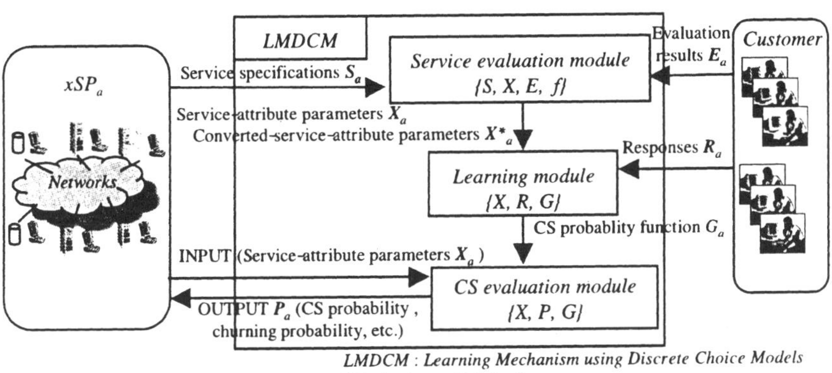

Figure 1 Architecture of *LMDCM*.

$P_a = \{p_{a1}, p_{a2}\}$: $p_{a1} = Prob\{\, r = 1\,\}$, $p_{a2} = Prob\{\, r = 0\,\}$, $p_{a1} + p_{a2} = 1$.
G_a : The customer-satisfaction probability function of service a , where

$$p_{a1} = G_a (X^*_a). \tag{2}$$

In the case of the customer-satisfaction-evaluation model, probability p_{a1} is the customer-satisfaction probability of service a. Probability p_{a2} is the customer-dissatisfaction probability of service a. In the case of the churning prediction model, p_{a1} indicates the churning probability of service a. G_a indicates the churning-probability function.

In the *LMDCM*, discrete choice models (Ben-Akiva and Lerman, 1987) are applied to determine the form of G_a and to estimate G_a, so as to take into account many different service-attribute-parameters. In this paper, *Binary Choice models* (*BC* models), in which the choice set consists of exactly two alternatives, are applied to the learning module. The learning module substitutes responses for decision j (alternative j). If the response is satisfaction, r =1; otherwise, r=0. In *LMDCM*, the attribute values are represented by service-attribute parameters X_a. A sample includes service-attribute parameters X_a and the response r. Consequently, the probability p_{a1} indicates the customer-satisfaction probability of each customer segment under given service specifications.

3.2 Modeling with *LMDCM*

The *LMDCM* consists of seven components. There are various types of modeling methods using these components.

1) Definition of Service Specifications

The service specifications of Internet services and IT systems are classified into several categories such as quality of service, reliability, security, optional service menu, charge, and maintenance service. Each category has several service-attributes. For example, the quality of service is expressed by network throughput, delay time, packet loss, etc. Each attribute such as delay time has several parameters: average value, maximum value, measured data, catalog data, etc. In the first step of the modeling process, the categories, the attributes of each category, and the parameters of each attribute are defined as the service specifications S_a.

2) Definition of Service Attribute Parameters

It is necessary to extract only significant parameters from among a large number of parameters on the basis of the modeling objective. In general, parameters with the correlation should be checked and omitted as well as the conventional multivariate regression analysis used so far. In the second step of the modeling process, the set of service-attribute parameters X_a is defined. The way to extract significant parameters is omitted in this paper. The rule for converting service-attribute parameters is described below. It is classified into three types:

(1) No conversion

A service-attribute parameter is used without converting the parameters in order to evaluate the relationship between the absolute value and the satisfaction level. When the parameter is a quantitative factor such as delay time,

$$x^*_{ai} = x_{ai}\,. \tag{3}$$

When the parameter is a qualitative factor such as help-desk service,
$x^*_{ai} = 1$: service function is provided, $x^*_{ai} = 0$: no function.

(2) Conversion based on reference value

In order to evaluate the degree of deviation of the service level, a service-attribute parameter is converted based on the predetermined threshold or reference value x_0 such as the service level of the *SLA*.

$$x^*_{ai} = x_{ai} - x_0, \quad \text{or} \quad x^*_{ai} = (x_{ai} - x_0) / x_0. \tag{4}$$

In order to evaluate the differences among competing providers, a service-attribute parameter is converted based on the service-attribute value of the competitive provider.

(3) Conversion based on evaluated value

When a service-attribute parameter is a qualitative factor such as brand image, help-desk quality or maintenance-personnel attitude, the evaluation result e_{ai} for the service attribute parameter x_{ai} is used.

$$x^*_{ai} = e_{ai} \tag{5}$$

Generally, the evaluation result e_{ai} is expressed by three to five levels. An example of five levels is shown: $e_{ai} = 1$: very dissatisfied, 2: dissatisfied, 3: neutral, 4: satisfied, 5: very satisfied.

When the sensitivity to the service-attribute parameter differs according to customer segmentation, this conversion is effective. There are two cases to determine the value of the evaluation result e_{ai}. One is to determine the value based on the difference between the reference value and measured value. In this case, the rule to convert the difference into the evaluation result e_{ai} such as a satisfaction level is given. An example of this conversion rule is shown below. Suppose that a service attribute parameter x_{al} is average delay time, and the reference value is x_0,

$e_{ai} = 1$: if $x^*_{al} \geq 0.1$ (very dissatisfied),
$e_{ai} = 2$: if $0.1 > x^*_{al} \geq 0.5$ (dissatisfied),
$e_{ai} = 3$: if $0.05 > x^*_{al} > -0.05$ (neutral),
$e_{ai} = 4$: if $-0.05 \geq x^*_{al} > -0.1$ (satisfied),
$e_{ai} = 5$: if $-0.1 \geq x^*_{al}$ (very satisfied),

where

$x^*_{al} = (x_{al} - x_0) / x_0$.

The other is to determine the value directly by the questionnaire survey.

3) Estimation of Customer-Satisfaction Probability Function G_a

The learning algorithm of *LMDCM* is a method for estimating the customer-satisfaction probability (or churning-probability) function G_a which corresponds to the "Learning Module" function in Fig. 1.

In *LMDCM*, a *BC* model (Ben-Akiva and Lerman, 1987) is used for estimating G_a. In the case of the customer-satisfaction-evaluation model, the utility U_{ij} of decision j with respect to service a_i is defined as the satisfaction level. In the case of the churning-prediction model, the utility U_{ij} of decision j with respect to service a_i is defined as the utility obtained by changing to other services from service a_i. In random utility theory, U_{ij} is a random variable and is composed of two parts as follows:

$$U_{ij} = V_{ij} + \varepsilon_{ij}, \qquad (i=1, 2, ..., j=0, 1). \tag{6}$$

V_{ij} is called the *systematic (or representative)* component of utility U_{ij} and is assumed to be deterministic; ε_{ij} is the random part and is called *disturbance*. Provided that the functional form for V_{ij} has linear variables, the difference between the systematic component V_{i1} and V_{i0} can be defined as follows:

$$V_{i1} - V_{i0} = \beta_0 + \beta_1 x_1 + \beta_2 x_2 + \ldots + \beta_K x_K, \tag{7}$$

where β_k (k=1, 2, $\cdots$, K) is the coefficient denoting the weight for the value x_k of service-attribute parameter k, and $\beta_0$is a constant related to the other attributes with the exception of $x_1, x_2, \cdots, x_K$. Assuming that the disturbances are independent and identically Gumbel distributed (i.e., logistically distributed), the customer-satisfaction probability, that is G_a, is given by (Ben-Akiva and Lerman, 1987)

$$p_{i1} = 1/(1+exp(V_{i0} - V_{i1})). \tag{8}$$

In conventional *BC* models, the decision (j=1 *or* 0) is then made using the threshold value p_0, as shown below.

$$j=1, \text{ if } p_{i1} \geq p_0, \quad j=0, \text{ if } p_{i1} < p_0. \tag{9}$$

Equation (8) is a function of the unknown coefficients $\beta_1, \beta_2, \cdots, \beta_K$. These coefficients are estimated from a set of N samples. In conventional *BC* models, each sample which is used to estimate the unknown coefficients consists of a decision and observed attribute values. In *LMDCM*, the attribute values are converted-service-attribute vector $\boldsymbol{X}^*_a(t)$ at time t when response r is obtained. Therefore, a sample consists of the set of converted-service-attribute parameters $\boldsymbol{X}^*_a$ and the response r. The learning module substitutes response r for decision j (alternative j). Consequently, the probability p_{i1} in the customer-satisfaction-evaluation model indicates the customer-satisfaction possibility of service a. And that in the churning-prediction model indicates the churning possibility of service a.

In the third step of the modeling process, the customer-satisfaction probability (or churning-probability) function G_a is estimated. Generally, G_a is constructed for each customer segment under given service specifications. After response r has been obtained N times, which is determined by the number of samples in the sample set, a vector is estimated for the unknown coefficients $\beta_0, \beta_1, \cdots, \beta_K$ by using maximum likelihood estimation (Ben-Akiva and Lerman, 1987). G_a, which is a vector of the unknown coefficients $\beta_0, \beta_1, \cdots, \beta_K$, is updated every learning cycle N_0 ($N_0 \leq N$).

4) Customer Satisfaction Evaluation

This section describes the customer-satisfaction-evaluation process corresponding to the "CS Evaluation Module" function in Fig. 1. The probability p_{i1} is obtained periodically by using estimated function G_a and service-attribute parameters $\boldsymbol{X}_a$ at the time. When a service provider understands that customer-satisfaction probability decreases every update-cycle, it is necessary to think about countermeasures. *LMDCM* can estimate the customer-satisfaction probability under virtual service specifications. A service provider can choose the most appropriate renewal plan based on the estimated results for several virtual service specifications.

4. CHURNING MODEL AND EVALUATION EXAMPLES

An example model to predict churning with respect to an ASP and the evaluation results are shown in this section. The service specifications are classified into six categories: charge, quality of service, reliability, security, customer-support services, and optional services. Each category of the service specifications includes several service-attributes shown in Table 1.

The following conditions are assumed in connection with the definition of service attribute parameters.

- Six significant parameters are extracted from among a number of parameters.
- Each evaluation result e_{ai} for service attribute parameter x_{ai} is obtained periodically by the questionnaire survey.
- A conversion rulc based on evaluated value e_{ai} for service attribute parameter x_{ai} is used.

$$x^*_{ai} = e_{ai} \tag{10}$$

The evaluation result e_{ai} is expressed by five levels: e_{ai} = 1: very dissatisfied, 2: dissatisfied 3: neutral, 4: satisfied, 5: very satisfied.

Therefore, the set of service specifications $\boldsymbol{S}_a$, the set of converted-service-attribute parameters $\boldsymbol{X}^*_a$ and the set of responses $\boldsymbol{R}_a$ are expressed as below:

$$\begin{aligned} \boldsymbol{S}_a &= \{s_1, s_2, \dots, s_6\}, \\ \boldsymbol{X}^*_a &= \{x^*_{a1}, x^*_{a2}, \dots, x^*_{a6}\}, \\ \boldsymbol{R}_a &= \{0, 1\}, \end{aligned} \tag{11}$$

(r =1: change to other providers, r =0: continuation).

Equation (8) is presented as:

$$V_{i1} - V_{i0} = \beta_0 + \beta_1 x_1 + \beta_2 x_2 + \dots + \beta_6 x_6. \tag{12}$$

The number of samples N in a sample set for estimating G_a is 100. In this model, six service-attribute parameters do not correspond to six service-specification categories. However, for reasons of confidentiality, we cannot describe the service-attribute parameters in this paper.

Table 1 Service specifications example

Category	*Service-attribute*
Charge	- Initial charge - Monthly charge - Discount menu ·····
Quality of service	- Throughput - Delay time - Packet loss rate ·····
Reliability	- Mean time to repair - Mean time to failure - Availability ·····
Security	- Firewall function - Virus check function - Authentication function ·····
Customer-support service	- Help-desk service - Training service - Installation service ·····
Optional Service	- SLA (Service Level Agreement) menu - Load balance control function ·····

Figure 2 shows the churn probabilities against the service level of the primary service-attribute. It shows that there is a strong possibility that the customer will change to another provider when the service level of the primary service-attribute is less than "dissatisfied" or worse. Thus, it can be said that it is effective to evaluate service specifications and business strategies under competitive conditions. We will show the experimental results in our presentation.

5. FUTURE STUDY

These days, a customer can choose the most appropriate service from among not only service providers but also many different kinds of services. It is very important to understand customer-satisfaction levels under highly competitive conditions. *LMDCM* has an effective function for learning changes of service specifications and customer's decision-making process. And it can be used in various fields including marketing. Evaluation regarding the convergence and accuracy is under way.

REFERENCES

Personick, S. D., 2002, "Evolving toward the Next-Generation Internet: Challenges in the Path Forward", *IEEE Communications Magazine*, Vol. 40, No. 7, pp. 72-76, July.

Inoue, A., and Yamamoto, H., 1990, "Evaluation of New Telecommunications Services using Stated Preference Techniques", *IFIP International Conference Modeling the Innovation: Communications, Automation and Information Systems*, Rome, pp. 71-78, March.

Lewis, L., 1999, *Service Level Management for Enterprise Networks*, Artech House, Inc. MA, U.S.A.,.

Kuo, Y. F., Temponi, C., and Corley, H. W., 1996, "Customer Satisfaction Assessment through a Fuzzy Neural Controller", *Fuzzy Information Processing Society, NAFIPS, Biennial Conference*, pp. 195-199, June.

Temponi, C., Kuo, Y. F., and Corley, H. W., 1999, " A fuzzy neural architecture for customer satisfaction assessment", *Journal of Intelligent and Fuzzy Systems*, 7, pp. 173-183,.

Ben-Akiva, M., and Lerman, S. R., 1987, *Discrete Choice Analysis*, MIT Press, MA, U.S.A.,.

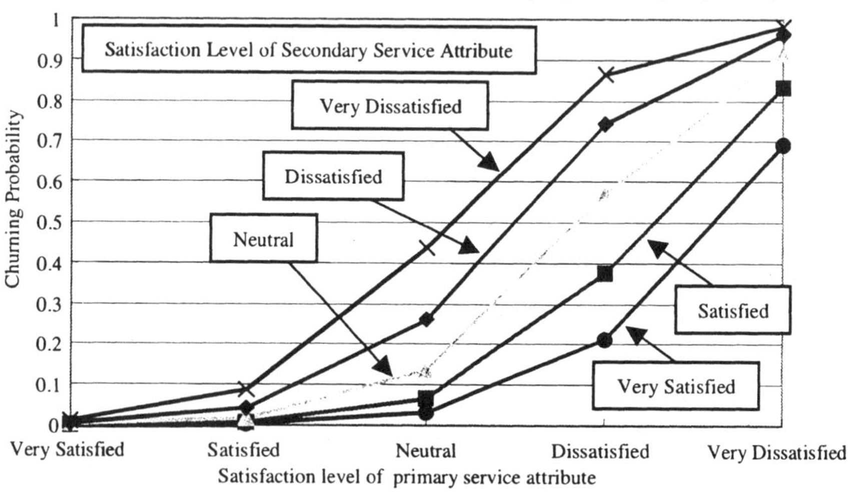

Figure 2 Churning probability against satisfaction level of primary service attribute.

NEURAL NETWORKS IN EMBEDDED LINUX

XING LIU
Gonzaga University
Department of Electrical & Computer Engineering
Spokane, Washington

ABSTRACT
This paper introduces the research in developing a neural network based software program that runs on an embedded Linux operating system named BlueCat. The program is used for dynamic system identification. A multilayer perceptron neural network is used as the identifier. Details on how the software is implemented and how it is exported to the embedded Linux system are provided. A design strategy that can accommodate system dynamics change is also proposed. Potential applications of this research are embedded controllers and instruments where the system in question is nonlinear and its model is required. Examples of such systems can be found in chemical and petroleum industries.

INTRODUCTION

Neural networks are computational models developed from the studies of the brain. Although research on neural networks can be dated back to early 1940s, recent research prospered due to the publication of the book by Rumelhart and McClelland (1986). Since then tremendous amount of research has been conducted on theoretical and applied neural networks.

Due to their ability in approximating arbitrary nonlinear input-output relationships (mappings) (Cybenko, 1989; Huang, 2003), neural networks have found extensive applications in system modeling and identification (Chen and Billings, 1990; Narendra and Parthasarathy, 1990; Pham and Liu, 1996). The use of neural networks leads to a uniform approach in obtaining models for dynamic nonlinear systems.

There are many different types of neural networks that are in use. A widely used neural network in system identification is the multilayered perceptron (MLP) network (Rumelhart and McClelland, 1986). An MLP network can be trained by a simple and effective algorithm called backpropagation (Rumelhart and McClelland, 1986).

On the other hand, as an operating system, Linux provides a powerful application development environment, with extensive support for standard I/O hardware. Due to its stability, scalability and zero (or low) cost, Linux is increasingly finding applications in the embedded systems world (Abbott, 2003; Webb et al; 2002). Progress in microprocessor technology has made it possible to implement computationally heavy-duty technologies such as neural networks on an OS such as Linux in terms of both speed and memory requirements.

This paper introduces the development a software program that is used for system identification and the experience in exporting it to BlueCat - a commercial embedded Linux system made by Lynuxworks, Inc. The advantages in having a software based on Linux are: (i). standard tools for software development; (ii). extensive support for various hardware platforms and I/O devices; (iii). easy information exchange between different systems; (iv). no or very low cost; (v) increasingly wide adoption by the embedded industry. (iii). easy multitask programming - multiple programs (tasks) can run at the same time, with each serving a different client. For example, when a trained neural network model is in use in a Linux program, another program can be initiated to update the model without affecting the normal operation of the principal program if a change in the environment is detected.

MULTILAYER PERCEPTRON BASED SYSTEM IDENTIFICATION

This paper employs a multilayer perceptron (MLP) network for system identification. It is known that, with one or more hidden layers in the network, an MLP can approximate arbitrary nonlinear static functions.

Research of non-neural network types indicates that discrete time dynamic systems can be described by the following equation:

$$y(k+1)=g(y(k),y(k-1),\dots,y(k-n+1),u(k),u(k-1),\dots,u(k-m+1)) \qquad (1)$$

where $y(.)$ is the system's output, $u(.)$ is the system's input. $g(.)$ is a function relating the current/past outputs and current/past inputs with the upcoming output. Function $g(.)$ is usually nonlinear. For a system with fixed dynamics, function $g(.)$ does not vary with time.

Because a feedforward MLP does not have internal dynamics, the current/past inputs and outputs of the system have to be fed into the MLP in order for it to be able to represent a dynamic system. The sole purpose of the MLP is to approximate the nonlinear function $g(.)$. Figure 1 shows how an MLP can be used in the identification of a dynamic system. In this paper the MLP neural network is used to identify the following dynamic system that has "strong nonlinearity" (Narendra and Parthasarathy, 1990):

$$y(k+1)=y(k)/(1+y^2(k)) + u^3(k) \qquad (2)$$

NEURAL NETWORKS IN BLUECAT LINUX

BlueCat Linux is a product by LynuxWorks, Inc. It is an open source distribution based on the Linux kernel 2.4.2. In addition to the basic Linux kernel, the BlueCat Linux package also includes tools and utilities for creating Linux that runs on different embedded targets. The package is open source and trial versions can be downloaded from LynuxWorks' web site. The following introduces how a neural network software is created and set up to run in BlueCat.

To prepare the neural network application software program to run in BlueCat Linux, a few items need to be produced. The first is a Linux kernel that is suitable for the embedded application. The second is the neural network executable program. The third is a root file system. All the three items will be exported to the embedded Linux system.

To develop the above three items, the BlueCat package needs to be installed on top of a standard Linux operating system, which is RedHat Linux 7.0 in this research. The running RedHat Linux 7.0 can be considered to be the "host". Executing the "install" program of BlueCat uncompresses BlueCat and installs it on the host. The uncompressed BlueCat file system tree looks very similar to that of the standard Linux. Under this tree there are directories that hold binary files to be downloaded to embedded targets, plus files supporting cross development (for building and debugging target systems). BlueCat should be installed in its own directory. Although BlueCat becomes part of the file system of the host after installation, paths should be set up so that the correct BlueCat files are used in building the needed items for the target system. This is achieved by running the shell script `SETUP.sh`.

The first item to build is the target's kernel. Before the kernel is built, it has to be configured so that it meets the target's requirements. Configuration options can be selected/deselected for the kernel to have different functionalities and sizes and to support different hardware devices. The procedure for configuration is the same as that of the standard Linux (for example, run command `make xconfig`). After the configuration, the kernel can be built using the BlueCat tool `mkkernel` or `make kernel`.

The next item to develop is the neural network program. The program is written using the C programming language. The dynamic system model used is described in Equation (2). The neural network used is an MLP that has two input units (one for $u(k)$, the other for $y(k)$), several hidden units (four used in this work), one output unit (for the predicted $y(k+1)$). The activation functions for the hidden units are hyperbolic tangent. Backpropagation is used in training. The `gcc` command was used to build the neural network program into an executable. Because this build operation takes place in the BlueCat environment, the neural network program is able to run on the desired target.

The third item is the BlueCat Linux root file system. The root file system helps organize data on the embedded target. It specifies the file system structure of the target: what directories and files are going to be generated on the target. When the target boots up, these directories and files will exist in the Linux environment on the target device. The file for the root file system is created using command `mkrootfs` or `make rootfs`.

In this work, the shell project files provided by Abbott (2003) are modified to have an embedded Linux system that boots from an $x86$ computer with a floppy disk drive. The booted embedded Linux system has a standard Linux shell interface. The root file system includes a few standard Linux commands such as `cd`, `ls` and `cat`, plus the neural network program. It should be very interesting to see a small Linux OS running on an embedded device and the user can communicate with this device via the standard Linux command interface such as `ls`.

The next thing to do is export the above three items to an embedded system which in this work is a Pentium computer (can be an old 386 machine) that boots from a floppy disk. The `mkboot` command is used to make the floppy disk bootable and copy the prepared file items to the disk.

Now a floppy disk that contains the embedded Linux kernel and the root file system (including the neural network program) is ready to be used to boot the embedded target where the neural network program can execute.

To test the embedded Linux system, the floppy disk is placed in the Pentium machine and the machine is booted by pressing the reset button. It was observed that the BlueCat Linux booted up, with the kernel image and root file system loaded and uncompressed, and finally a shell prompt appeared. It was able to use the Linux `ls` command to navigate through the file system. Then the command `./mlp` (`mlp` was the name of the neural network program) was used to start the neural network program. It was observed that the training of the neural network converged very quickly. Table 1 shows the parameters used in the test. Fig. 2 shows the expected output and neural network output obtained from the test.

IDENTIFYING TIME VARYING DYNAMIC SYSTEMS - PROPOSED DESIGN

The previous section discussed how a neural network program could be placed in an embedded Linux system. The neural network program deals only a dynamic system with fixed dynamics. If there is significant change in the dynamics, the neural network model will generate large errors. In this case the neural network has to be re-trained in order to represent the new dynamics.

If the existing Linux task has to use the neural network to calculate an estimated output of the system at regular intervals, a new task can be created to update the weights of the neural network because Linux is a multitasking system.

Assume the task that uses the trained neural network is called *main*. This task uses the neural network to predict the outputs of the dynamic system, and in the meantime, it samples the actual outputs of the dynamic system, all at specified time instants. A number of such sampled actual outputs can be saved in memory and can be used later in training. If this task finds that the error is too large between the predicted and actual outputs, it creates an *update* task that is responsible for updating the weights of the neural network to accommodate the system dynamics change. Linux allows the two tasks to execute in parallel. On the other hand, *main* and *update* can use "signals" to communicate with each other and can exchange numeric data using "shared memory". "Semaphores" can be used to prevent collision in accessing memory. The overall design discussed above can be represented by Figs. 3(a) and (b). This design is still to be implemented and tested.

CONCLUSIONS

Linux is increasingly finding applications in embedded systems. Many embedded applications require the modeling and control of nonlinear systems. This is where neural networks can fit in. This paper has discussed how to implement neural network programs in a typical embedded Linux environment. Experience shows that it is feasible to have an embedded system that runs both Linux and a neural network program. A software design strategy is also proposed for time-varying nonlinear systems in the paper. If implemented, it should be useful to applications in various industrial processes.

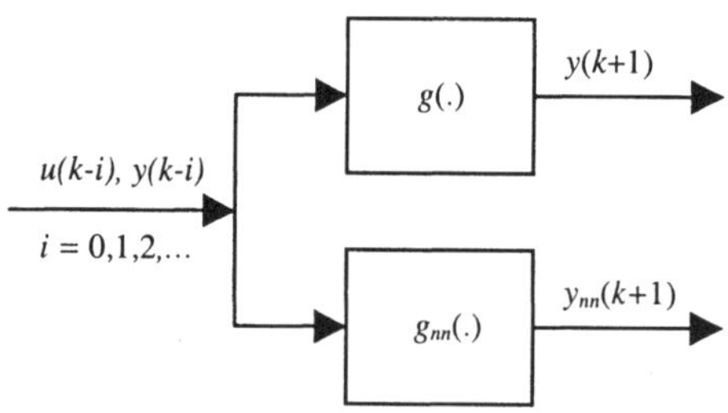

Fig. 1 Neural network based dynamic system identification

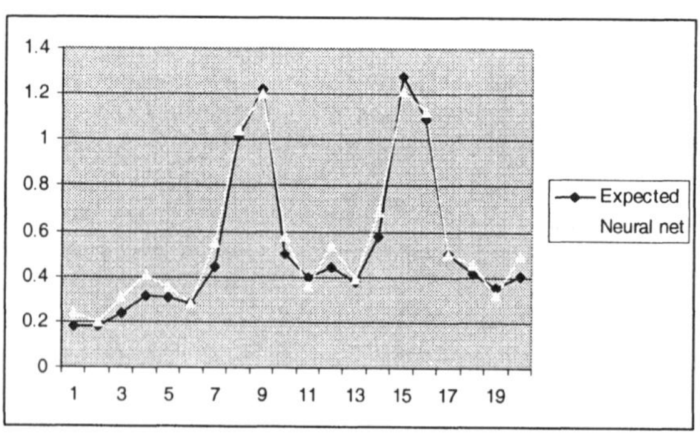

Fig. 2 Outputs of dynamic system and neural network

Table 1 Parameters Used in Test

Number of input neurons	2
Number of hidden neurons	4
Number of output neurons	1
Input $u(k)$	Random signal; $0 \leq \lvert u(k) \rvert \leq 1$; Seed for random number generator (training set) : 1 Seed for random number generator (testing set) : 3
Initial weights	Small random values: $0 \leq w \leq 0.1$; Seed for random number generator: 5
Learning coefficient	0.1
Momentum value	0.9
Epochs trained	40
Training data size	30
Test data size	20

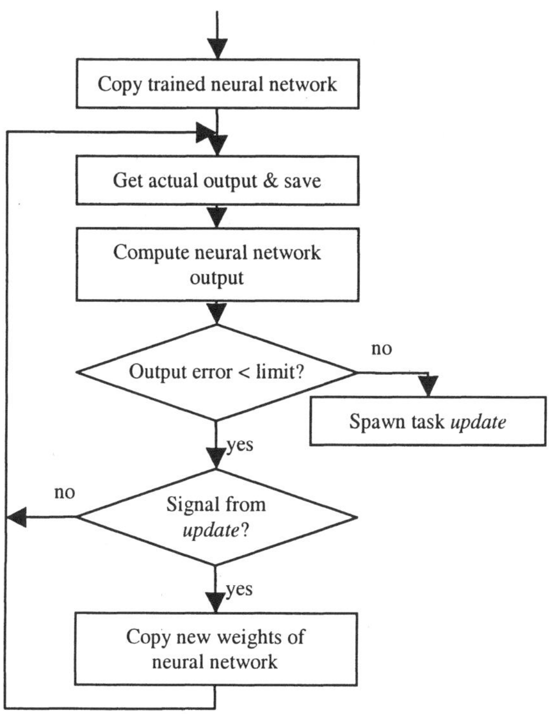

Fig. 3(a) The *main* task

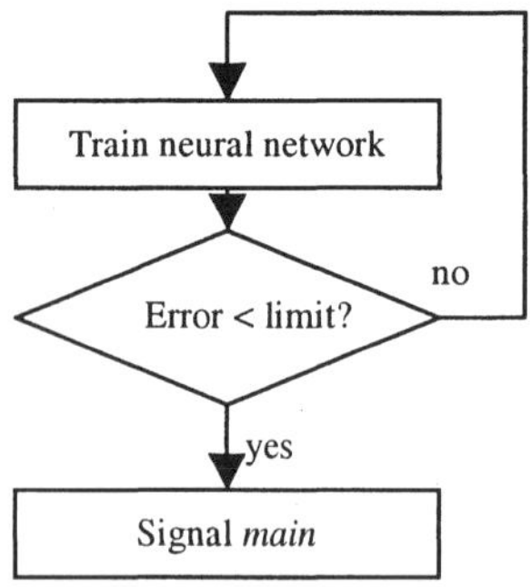

Fig. 3(b) The *update* task

REFERENCES

Abbott, Doug (2003), Linux for Embedded and Real-Time Applications, Elsevier Science (USA), New York.

Chen, S. and Billings, S.A. (1990), Neural networks for nonlinear dynamic system modelling and identification, ***International Journal of Control***, **56**(2), 319-346.

Cybenko, G. (1989), Approximation by superpositions of a sigmoidal function, ***Mathematics of Control, Signals, and Systems***, **2**, 303-314.

Huang, G.B. (2003), Learning capacity and storage capacity of two-hidden-layer feedforward networks, ***IEEE Trans. on Neural Networks***, **14**(2), 274-281.

Narendra, K.S. and Parthasarathy, K. (1990), Identification and control of dynamic systems using neural networks, ***IEEE Trans. on Neural Networks***, **1**(1), 4-27.

Pham, D.T. & Liu, X. (1996), Training of Elman Networks and Dynamic System Modeling, International Journal of Systems Science, **27**(2), 221-226.

Rumelhart, D. and McClelland, J. (1986), ***Parallel distributed processing: exploitations in the micro-structure of cognition***, volumes 1 and 2, Cambridge: MIT Press.

Webb, Warren, Pick and Place (2002), Linux grabs the embedded market, the EDN Magazine, October 31, 41-46

WEB STRUCTURE REORGANIZATION FOR ADAPTIVE NAVIGATION THROUGH CONCEPTUAL CLUSTERING

SANG C. SUH
SAM I. SAFFER
VAN N. KINCAID
BIN YU
Department of Computer Science
Texas A&M University, Commerce
Commerce, TX 75429, U.S.A

ABSTRACT
In this paper, we propose a method that combines both web structure and usage mining called conceptual clustering. First, a conceptual cluster repository is created by dynamically classifying and grouping all the pages of a web site into different categories subdivided by goal and purpose of the page contents. Second, we capture the user navigation information from click streams, using cookies or log files. User behavior patterns are compared to a specified threshold of the number of pages visited. We can then determine if a user belongs to an appropriate cluster in the conceptual repository to predict the user's intention more accurately. At that point, we will construct an adaptive web site that dynamically improves its organization and presentation by guiding the user's navigation in a predicted manner according to the cluster previously determined. If the user cannot be classified into any existing cluster predetermined, a new concept is created and added into the repository, if the number of users with similar pattern reaches the threshold with great certainty.

INTRODUCTION

Most of the web sites these days are often statically designed so that they exhibit the same structural behavior for all visitors. However visitors' navigation behaviors are often dynamic and variant. Making a web site automatically adaptive is not an easy task because each visitor has different browsing patterns and purposes at different times. As a result, visitors' navigation prediction is the key factor to shorten the navigation time and improve visitor satisfaction by dynamically customizing the web site's organization and structure.

My Yahoo [1] is an example of adaptive web site that supports fairly simple adaptations and customizations based on each user's personal profile. However, this technique does not generate very sophisticated adaptations. It is dependent on person-to-person basis and there is not any general prediction for all users. Likewise, Web Watcher [2], Letizia [3] and Fink [4] rely on client side personal profiles and content similarity of web document to these profiles to dynamically present the information to a visitor or a class of visitors.

Perkowitz [5] in 1997 introduced the concept of adaptive web sites and proposed various approaches to make this feasible. In his further research on this topic in 1998 [6], he formally defined the index page synthesis problem and presented PageGather algorithm. As the research went on, in [7] in 1999, Perkowitz introduced a new approach called conceptual cluster mining to find a small set of cohesive clusters that correspond to concepts expressible in a given conceptual language.

Mobasher, Cooley and Srivastava [9] described an approach to usage-based Web personalization taking into account both offline tasks related to the mining of usage data, and the online process of automatic Web page customization. The clustering technique called Association-Rule Hypergraph Partitioning is used to discover clusters of transactions. However, all these methods discussed so far rely heavily on users click stream behavior patterns and do not take into account the structure of the web site.

In this paper we describe an approach that combines both of the issues together. We use a conceptual clustering repository to implement adaptive web navigation system. This repository consists of a set of different clusters representing different concepts. Each cluster contains a set of pages that are related to a topic (that is, concept) where each page is given its own weight value. We first introduce the technique to create the base repository by using the site's metadata to construct conceptual clusters. Next, web log access entries are collected and transformed into meaningful semantic of units of navigation. Based on the click stream information gathered during the initial user interaction session with the system, we can then try to determine if a user is interested in a concept(s) that corresponds to one of the clusters in the repository. Once the concept is determined, navigational behavior of the web site will automatically become dynamic by avoiding visits to unnecessary pages and guiding the user through the sequence of web pages he needs to visit to find the exact information that is desired. Finally an experimental result is presented to reflect the effectiveness of our approach in web navigation.

PRELIMINARIES

On the web site, we assume that there are n static pages, $P=\{P_1, P_2, P_3, ..., P_n\}$ pages and m predefined concepts, $C=\{C_1, C_2, C_3, ..., C_m\}$. Each page consists of a number of concepts from which we can construct a concept chain to represent a page. Therefore, a web site can be viewed as a collection of pages represented with a concept chain. For each page P_k its corresponding concept chain is represented as $C_i (W_i) + C_j (W_j) + ... + C_l (W_l)$, where W_i represents a weight associated with a concept C_i. Conversely, a concept may be present in several pages. Those pages that cover the concept constitute a page chain. For each concept C_m, its corresponding page chain is represented as $P_a (W_a) + P_b (W_b) + P_c (W_c) + ... + P_z (W_z)$. Each page in the page chain is given a weight associated with it. The weight Wi shows the degree of significance of the concept in the page which is represented as the ratio $f(C_i)$ / $F(C_i, C_j, C_l)$, where f is a significance frequency of C_i and F is the sum of significance frequency of all concept in a page.

ID	*P1*	*P2*	*P3*	*P4*	*P7*	*P8*	...	*Pk*
C1	*0.0*	*0.2*	*0.5*	*0.0*	*0.0*	*0.2*	...	*0.0*
C2	*0.0*	*0.0*	*0.3*	*0.1*	*0.0*	*0.0*	...	*0.0*
C3	*0.5*	*0.1*	*0.0*	*0.0*	*0.2*	*0.1*	...	*0.1*
...	...	...	...	...	...	...	...	...
Cm	*0.3*	*0.0*	*0.0*	*0.5*	*0.1*	*0.0*	...	*0.0*

Table 1. Conceptual Clustering Repository Structure

Based upon the identified concepts and all the static pages in the web, both concept and page chains are constructed. These chains constructed are stored in a conceptual clusters repository. Its structure is shown in Table 1. Processes that initiate the dynamic navigation of web pages use this repository to obtain relevant information for concept classification.

SYSTEM OVERVIEW

Concept names will first be automatically generated following the keyword frequency method [10]. Using these concept names, the concept chains and page chains are constructed and placed into the conceptual clusters repository. Figure 1 gives brief sketch of the role of conceptual clusters repository in adaptive web navigation. The web server will save the users' click streams into a log file and at the same time provide this information to the User Behavior Track Module (UBTM). UBTM will provide the click streams to the Conceptual Clusters Repository (CCR) to determine a concept(s) that the user may be interested in. When the click stream closely matches with any concept in CCR, the concept is selected as a target concept and the web navigation is focused on the specific target concept. The Adaptive Navigation Engine (ANE) will help the web server to dynamically generate web pages that are related to the selected concept(s).

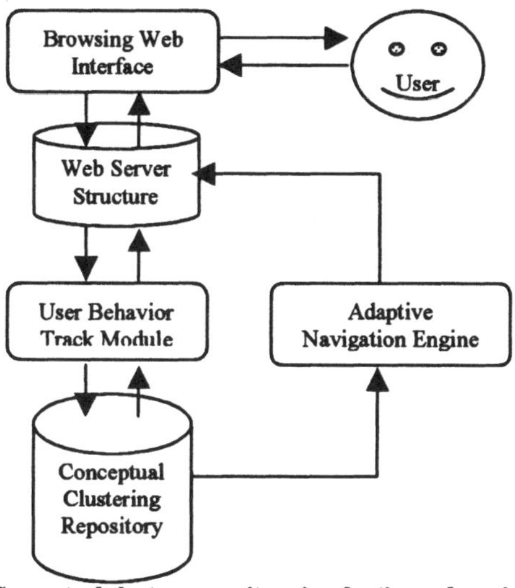

Figure 1. Conceptual clusters repository in adaptive web navigation

CONCEPTUAL CLUSTERING REPOSITORY

Jenkins and Inman [10] proposed an approach called "*adaptive automatic classification on the web*". This method uses a recursive discovery tools to generate vocabularies from a set of documents based on keyword occurrence frequency. Our method is a variation and an extension of their approach in that we consider relative significance of each and every keyword on each page. We are not concerned with characterizing a web page with a representative key word. Rather we consider all possible concepts represented by keywords on all pages.

Generally, every web page contents some tags like <TITLE>, <H1>, <H2>. If a word appears in the <TITLE> tag, which means it should be given more weight than the words appearing in <H1> tag. Likewise, <H1> tag has more weight that <H2> tag and <H2> tag has more weight than Content occurrences. As a result, we adopt the following formula to be used to calculate the weight of a concept in a page.

$$\mathbf{W_i = f(C_i) / F(C_1, C_2, \ldots C_l)}$$

where: W_i = the weight of a concept C_i in a page

$f(C_i) = t_i*2 + h1_i*1 + h2_i*0.5 + ct_i*0.25$

t_i = the number of the TITLE occurrences of the concept C_i

$h1_i$ = the number of the H1 occurrences of the concept C_i

$h2_i$ = the number of the H2 occurrences of the concept C_i
ct_i = the number of the Content occurrences of the concept C_i

$$F = \sum_{i=1}^{n} f_i$$, where we assume that there are n concepts in the page.

For example, let's assume that a page, P_1 consists of 2 concepts, C_1 and C_2:
C_1 has $t_1 = 1, h1_1 = 2, h2_1 = 1, ct_1 = 30$ in P_1
C_2 has $t_2 = 0, h1_2 = 1, h2_2 = 2, ct_2 = 40$ in P_1
Then, $F = f_1 + f_2$
$= (1*2+2*1+1*0.5+30*0.25) + (0*2 + 1*1+ 2*0.5 + 40*0.25)=13+12 = 25$
Therefore, we get the following weights:
Weight $(C_1, P_1) = W_1(P_1) = 13 / 25 = 0.52$
Weight $(C_2, P_1) = W_2(P_1) = 12 / 25 = 0.48$

Once the conceptual clusters repository is constructed, the number of concepts in it may be very large. In an effort to reduce the size of the repository, we may identify those concepts that have the same or almost the same page chain and merge them

AUTOMATIC ADAPTATION

Several adaptive navigation algorithms have been proposed in previous work. Mobasher, Colley, and Srivastava [9] developed an approach that adds recommended links to the last page of the session before the page is sent to the browser. Perkowitz and Etzioni [6,7,8] developed the IndexFinder algorithm to generate the index page automatically. However, these methods are not directly based on the user's dynamic/adaptive need for specific information in the web site but are rather based on the patterns of the majority of users' navigational activities obtained from web logs for the web site in consideration.

Our approach, on the other hand, makes use of the web site structure and organization, which is reflected in the conceptual clusters repository structure (Table 1) to meet the dynamic and adaptive need of navigators to search for information. After each page click, as discussed in the previous section to discover the user-interested concepts, Adaptive Navigation Engine (ANE) as depicted in Figure 1 takes charge for the task of adaptive deployment of dynamically created web pages. ANE uses the concept candidate list to identify all relevant pages to the user-interested concept. If the list is not empty, obtain all the page links inside the page chain of each candidate concept in the descending order of the weight values. If the links selected by the current page overlaps with those previously selected, highlight them to get user's attention. If these links are not yet in the page, add them to the end of the current page group by concepts on the recommendation list in the descending order of the weight values before sending the page for the browser to open. This process is summarized below as a 2-step process.

Step 1: [Find the recommended links]
Suppose the current page contains the following page links:

$P_1 \rightarrow \{P_2(0.5), P_4(0.2), P_6(0.1), P_{11}(0.2)\}$

Suppose that the major concepts and their corresponding page chains are follows:

$C_{11} \rightarrow \{P_3(0.2), P_4(0.1), P_{11}(0.3)\}$
$C_{16} \rightarrow \{P_2(0.4), P_8(0.5)\}$
$C_{14} \rightarrow \{P_4(0.5), P_{10}(0.25), P_{11}(0.15), P_{15}(0.34)\}$

By comparing the links in the current page P_1 with the page chains in the candidate concept list, we can acquire the following recommended links that are not in current page: P_3, P_8, P_{10}, P_{15}. These links are then sorted in the descending order of the weight values and prepared for the next step.

Step 2: [Add the recommended links to the current page]

The recommended links can be added to the end of the current page as related links for recommendation to the navigators. These links are grouped by each concept and listed in the descending order of overall weight values. Threshold values can be used to limit the number of recommended links per concept. This dynamically updated page is then sent to the browser as the next page to open. The dynamically created page structure is shown in Figure 2.

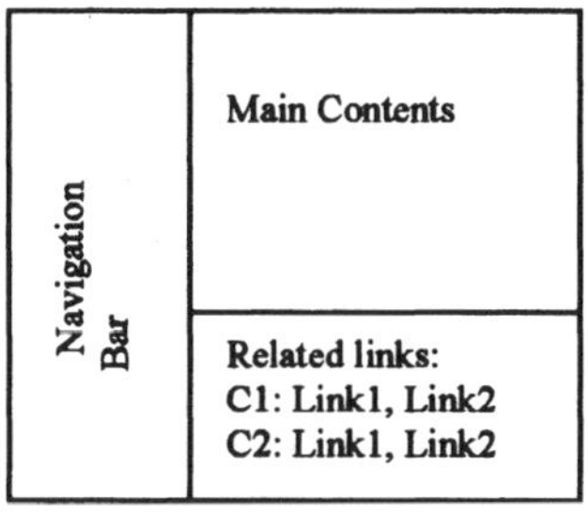

Figure 2. The dynamically updated page structure

EXPERIMENTAL RESULTS

Similar to the criteria presented in [11] and [12] for a navigation structure evaluation, we will evaluate our system based on the usability:

The summary of our experiments is given in Figure 3. The result shows that the adaptive system is more helpful to the users than non-adaptive system in that it takes a fewer number of pages to visit to find the desired piece of information. The figure compares the number of visited pages using the adaptive system with that of non-adaptive system to search for ten different pieces of information. Each column consists of two bars representing the average numbers of clicks the users have to make in each session before he or she can find one particular piece of information. The left bar represents the non-adaptive system, whereas the right bar represents the adaptive system. The adaptive system is shown to be more effective since for most of the cases it takes a fewer number of page clicks for the users to find the same piece of information in the adaptive system. In adaptive system, the number of visited pages that the users may not be interested in is significantly reduced, thus achieving significant time saving on web navigation. Therefore, our conclusion is that the adaptive system is more efficient than the usual non-adaptive system in that it is quicker to find the desired piece of information.

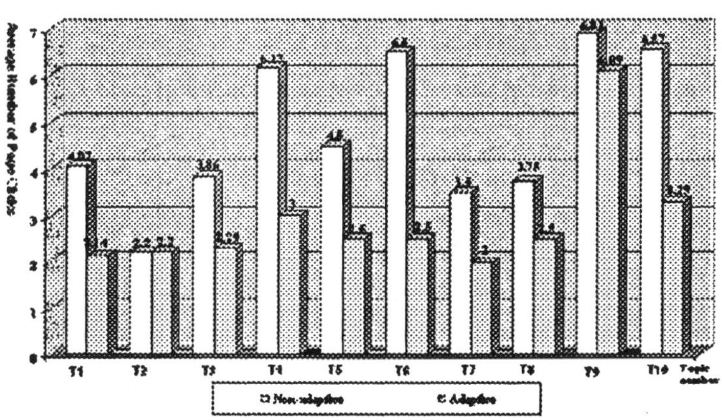

Figure3. Comparison between non-adaptive and adaptive systems

CONCLUSION AND FUTURE WORK

In this paper we have presented an approach to produce an adaptive web sites using concept repository. Adaptive navigation works by making popular pages easily accessible, highlighting interesting links, connecting related pages and clustering similar documents together. The web site's organization is reflected in our conceptual clusters repository. The adaptive navigation is based on this data structure combined with the users' behavioral patterns and interests. Our experimental results indicate that the techniques presented in this paper are very promising and bear further investigation and development.

Our future work in this area involves the automatic expansion of the conceptual repository by adding new concepts formed by user transactions that are not matched with any of the existing concepts. This requires identifying new criteria to determine the membership of a concept. A secondary repository might be used to store the candidates for new concepts to be identified and added into the primary repository. This will enable the dynamic modification capability of the repository. Furthermore, the time factors may be added when calculating weight values for both page chains and concept chains. The longer time a user spends on actively browsing a page, the bigger weight value for that page should be given.

ACKNOWLEDGEMENT

This research is funded by a grant from Texas A&M University – Commerce, grant # 142001-20300. We would like to express our special thanks to Ankit Africawala and Naren Surampalli for their help with the web site construction and experimentation. Our thanks also go to Mau-shing Lee who helped in tuning the ideas for the adaptive navigation.

REFERENCES

[1] My Yahoo: http://my.yahoo.com

[2] Joachims, T., Freitag, D., Mitchell, T., 1997, "WebWatcher: A Tour Guide for the World Wide Web," Proceedings of the International Joint Conference in AI (IJCAI97).

[3] Lieberman, H., 1995, "Letizia: An Agent that Assists Web Browsing," Proceedings of the 14th International Joint Conference in AI (IJCAI95), AAAI Press, Menlo Park, California.

[4] Fink, J., Kobsa, A., and Nill, A., 1996, "User-oriented Adaptivity and Adaptability in the AVANTI Project, " Designing for the Web: Empirical Studies.

[5] Perkowitz, M. and Etzioni, O., 1997, "Adaptive Web Sites: an AI challenge," Proceedings of the Fifteen International Joint Conference on Artificial Intelligence.

[6] Perkowitz, M. and Etzioni, O., 1998, "Adaptive Web Sites: Automatically Synthesizing Web Pages," Proceedings of the Fifteen International Joint Conference on Artificial Intelligence.

[7] Perkowitz, M. and Etzioni, O., 1999, "Adaptive Web Sites: Conceptual Cluster Mining.," Proceedings of the Fifteen International Joint Conference on Artificial Intelligence.

[8] Perkowitz, M. and Etzioni, O., 1999, "Towards Adaptive Web Sites: Conceptual framework and case study.," Artificial Intelligence 118(2000) 245-275

[9] Mobasher, B., Cooley, R., Srivastava, J., 1999, "Creating Adaptive Web Site Through Usage-Based Clustering of URLs," Workshop on Knowledge and Data Engineering Exchange

[10] Jenkins, C. and Inman, D., 2000, "Adaptive Automatic Classification on the Web," Proceedings of the 11th International Workshop on Database and Expert System Applications 2000 (DEXA'00)

[11] Masthoff J., 2002, "Automatic Generation of a Navigation Structure for Adaptive Web-Based Instruction," 2nd International Conference on Adaptive Hypermedia and Adaptive Web Based Systems

[12] Nielsen, J., Heuristic evaluation. In Nielsen, J., and Mack, R.L.(Eds.),1994, "Usability Inspection Methods," John Wiley & Sons, New York, NY.

PART VIII: SMART ENGINEERING SYSTEMS

AN APPLICATION OF CLUSTERING AND CLUSTER VALIDITY MEASURES TO EPIDEMIOLOGY DATA

DEBORAH STACEY, NEIL HARVEY
ZEYAD AZEM
Computing and Information Science
University of Guelph
Guelph, Canada

SHANNON MAJOWICZ
Population Medicine
Ontario Veterinary College
University of Guelph
Guelph, Ontario

ABSTRACT

The analysis of survey data that is essentially unlabelled poses an interesting problem for both pattern classification professionals and for epidemiologists. The epidemiological community approaches the problem by using expert opinion to derive case definitions that are then used to classify the collected data. This research proposes that various clustering techniques and methodologies for evaluating the clusters can be used to supplement the traditional case definition. Validating the number of clusters selected by a clustering technique allows the analyst to partition the survey data into classes that can then be analyzed with respect to the class definitions.

INTRODUCTION

Infectious gastrointestinal illness (GI) is an important public health issue around the world, and surveillance of such illness is central to effective control. Passive surveillance information has traditionally been the source of data used to estimate the magnitude and distribution of illness in the population. However, under-reporting is known to be a major limiting factor associated such information. Population-based surveys are a better way to obtain more accurate estimates of the magnitude, burden, and distribution of illness.

From February 2001 to February 2002, a retrospective, cross-sectional telephone survey of randomly selected residents of a city in southern Ontario, Canada was administered. To identify those individuals with self-reported GI, respondents were asked if they experienced any vomiting or diarrhea in the 28 days prior to the interview. Additional questions explored secondary symptoms, medical history, health care use, and self-reported causes of illness. The response rate for this survey was 37%, and of the 3,496 respondents, 12% (n=428) reported experiencing an illness that included vomiting or diarrhea.

The case definition of GI adopted in an epidemiology study can radically alter the magnitude and distribution estimates obtained. The absence of a universally accepted case definition of acute or infectious GI is noticeable in the literature; however, it may be that an internationally standardized case definition is not feasible, and that the best option is to understand how different case definitions relate in order to compare results between studies. Applying published case definitions from multiple studies to one data set would allow for such a comparison.

Alternately, it may be possible to examine the underlying data structure to determine if (a) natural case groupings occur in the data, and (b) if those groupings are clinically or epidemiologically meaningful. Since this survey collected information on self-reported acute GI using very liberal inclusion criteria, the data represented all GI which occurred in the population (including chronic, and non-infectious). Thus, application of clustering techniques to evaluate the underlying cluster structure of the data was possible.

CLUSTERING

A variety of algorithms are available for discovering natural groupings or clusters in data. For this study we used k-means and Kohonen's Self-Organizing Map (SOM).

Both k-means and the SOM rely on a measure of distance between two data points. We treated each survey response as a point in a high-dimensional space and used the Euclidean distance between points.

All survey questions allowed missing values. One way to handle missing values is to fill in the mean or median value for that variable. Ahmad and Tresp (1993), who have written extensively on training neural networks on data with missing attributes, criticise this *usual practice* for filling in missing features and offer alternatives based on estimating the distribution of the variables (Tresp et al. (1994, 1995)). Rather than choose a single rule for missing values, we drew on survey practices to fill in reasonable, conservative values. Because we lacked *a priori* knowledge of *natural* clusters in the data, we computed many clustering solutions and looked to cluster validity measures and cluster profiling to choose the best one(s).

There is no universally agreed-upon definition of what a cluster *is*, just an intuitive sense that two points from the same cluster should be somehow more similar to each other than to a point from a different cluster. Many researchers have translated this notion into numerical measures of the quality of a clustering solution. In 1985, Milligan and Cooper (1985) compared 30 measures from the literature and exposed many of them as very poor ones. Of the ten top-ranked measures from Milligan and Cooper's study, we used the Calinski and Harabasz measure, the C-Index created by Hubert and Levin (1976), the cubic clustering criterion (CCC) created by Sarle (1983), a point-biserial correlation, and the Davies and Bouldin index (DB-Index).

These measures must be combined with a certain amount of intuition since they do not always show a clear peak, they do not always agree among themselves, and they do not clearly *fail* if no meaningful clusters actually exist. We examined the clustering solutions by hand and chose ones that scored well on all five validity measures.

RESULTS

SOMs and k-means were used on the available data set to produce clusters that could provide a better understanding of the primary differences between acute and non-acute GI cases. Different clustering schemes were produced where

Table 1: Two demographics clusters

	Clusters		
Variable	1	2	Sample
urban_rural (% urban)	82.5	90.7	86.2
education_yrs (% HSD)	32	42	36.5
cultural_grp (% 1)	88	82.4	85.5
Income	2.86	1.53	2.26
people_hhold	3.6	1.46	2.63
workers_hhold	1.92	1.21	1.6
Age	29.47	50.3	38.88
sex (% Female)	65.8	67.9	66.7
nghbrhd_avg_income	55961	29774	44125
Frequency	234	193	427

each had a unique number of clusters. The selection of the best clustering scheme in each methodology is based on a set of cluster quality measures along with expert field knowledge.

The goal of the following experiments was to extract a set of acute GI diagnosis rules that would assist a field expert in classification of new cases; in effect, a new case definition. It was known that the dataset contained very few known, or *labelled* GI cases (1 acute and 12 non-acute). Both the SOM and the k-means algorithms were applied to Demographics variables, Symptoms variables, and Demographics plus Symptoms variables.

DEMOGRAPHICS CLUSTERS

There are 9 demographics variables in the data set. Several clustering schemes were generated using different number of clusters. Due to the size of the data set, the number of clusters was limited to range from 2 to 10 clusters. This range was selected to allow for the examination of different clustering schemes that would contain clusters of significant frequencies (i.e. number of records within each cluster). It is very unlikely that clusters based on demographics alone would be sufficient to detect acute GI cases, however, the objective here is to get an understanding of the demographic profiles of the cases and to identify the primary demographic variables that differentiate well between the various clusters. Table 1 presents some of the results obtained.

Table 1 contains the profiles of a 2-cluster scheme. The values represent the means of the variables within the clusters. A comparison of variable means among the clusters provides an understanding of the overall profile and highlights the variables that differentiate the clusters well. The evaluation of variable differentiability is based on comparison between the clusters as well as the data sample as a whole.

Further analysis of other clustering schemes (3, 4 and 6 clusters) reveals a set of primary explanatory variables consistent with those of the previous clustering scheme (2 clusters). The set of primary demographic variables identified above will be used in combination with symptom variables later on.

The SOM clustering and the cluster validity measures used also supported a similar results in that a low number of clusters were favoured (usually 2 to 3, but not more than 4). Figure 1 shows how a point-biserial correlation appears to recommend 3 to 4 clusters for the demographics variables.

SYMPTOMS CLUSTERS

Many of the symptom variables in the data sample had high rate of missing values. The problem of missing values was compounded more by the relatively small size of the data sample. To avoid using variables with many missing values and in consultation with a field expert, it was decided that only the following symptom variables (5) should be considered for cluster analysis : Number of consecutive days vomiting, Maximum number of vomiting in a 24-hour period, Consecutive days of diarrhea, Maximum number of loose stools in a 24-hour period, Blood in stool. Several cluster schemes were generated based on these symptom values alone with the number of clusters varying between 2 and 10. It was decided that such a range is appropriate considering the impact of sample size bias as the number of clusters increases. The 2-cluster scheme in the Table 2 summarizes the profiles of 2 clusters that differ significantly from one another as well as from the overall sample profile. If the sample profile is an accurate estimate of the actual distribution profile then it could be stated that cluster 2 represents people who are unlikely to be GI positive. Cluster 1 on the other hand, which has average symptom values higher than those of the sample, is likely to contain positive GI cases. Cluster 1 contains 48 cases (11%) that would require further field expert analysis.

We also used clustering schemes of 3, 4, 6, and 8 clusters. As the number of clusters is increased, cluster frequency decrease. For example, there are 3 clusters (1, 3, and 5) in the 8-cluster scheme containing a total of 5 cases. Although such cases may indeed be so unique justifying a separate cluster, it is difficult to validate the quality of such clusters given the limited amount of data available. Experimenting with different number of clusters highlights interesting relationships. For example, cluster 1 in the 3-cluster scheme is similar to cluster 1 in the 2-cluster scheme. It still has above sample average for all the symptoms but not as extreme. Also, cluster 3 in the 3 cluster case has above average values for some of the symptoms, differentiating it from cluster 1 in the same scheme.

The various clustering schemes confirm that all 5 selected symptom variables were significant in differentiating among the clusters. Next, these symptom variables were combined with the primary demographic variables identified in the previous experiment to produce a more comprehensive clustering scheme.

DEMOGRAPHICS AND SYMPTOMS CLUSTERS

Various clustering schemes were produced with number of clusters up to 10. The following results are focused on the overall best clustering scheme which has 2 clusters as shown in Table 3. Cluster 1 has a symptom profile which is less severe

Table 2: Two symptoms clusters

	Clusters		
Variable	1	2	Sample
cons_days_vom	6.94	0.46	1.19
max_#_vom_24hrs	5.67	0.5	1.08
cons_days_diar	10.65	2.55	3.46
max_#_lstl_24hrs	6.1	3.84	4.09
bld_stl (% Yes)	10.4	5	5.6
Frequency	48	379	427

Figure 1: A point-biserial correlation appearing to recommend 3-4 clusters for demographics data.

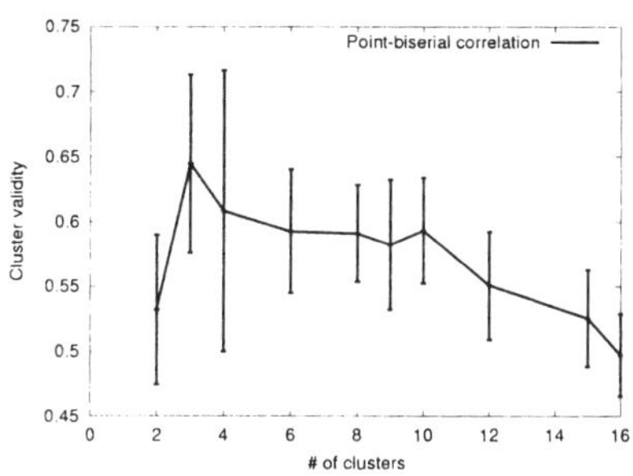

Figure 2: The Calinski and Harabasz measure appearing to recommend 2 clusters for combined symptoms and demographics data.

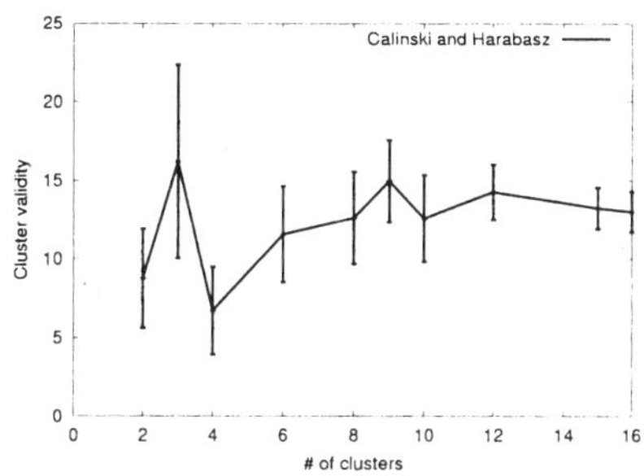

than the overall sample average. In addition, this cluster includes all 12 known cases of non-acute GI. Cluster 2 has a symptom profile that can be considered as severe, above average values for all symptoms. This cluster also includes the only known case of positive GI. Cluster 2 has a frequency of 26 (6% of sample). Increasing the number of clusters might lead to a more refined description of positive GI cases, however, the combination of small sample size and the lack of known positive GI cases (only 1 case) will increase decision boundary bias. On the other hand, 6% positive GI rate might be too high an estimate which may justify increasing the number of clusters.

Once again the SOM and its validity measures favoured a 2 cluster solution for the combined symptoms and demographics data as can be seen in Figure 2 that shows the Calinski and Harabasz validity measure applied to the clustering of this data set.

CONCLUSION

This study has produced results that lead us to believe that clustering is producing an interesting and valuable view of the data. The fact that the clus-

Table 3: Two symptoms/demographics clusters

	Clusters		
Variable	1	2	Sample
Cons_days_vom	0.7	8.65	4.26
Max_#_vom_24hrs	0.95	3.08	2.44
Cons_days_diar	2.5	18.19	6.11
Max_#_lstl_24hrs	3.89	7.23	3.72
Bld_stl (% Yes)	4.75	19.4	5.6
Education_yrs (% HSD)	37.4	23	36.5
People_hhold	2.58	3.5	1.5
Workers_hhold	1.57	2.12	0.71
Age	39.17	34.42	21.01
Nghbrhd_avg_income	44090	44656	25823
GI (Yes)	0	1	1
GI (No)	12	0	12
Frequency	401	26	427

tering is favouring lower numbers of clusters that are of a reasonable size (the positive infectious GI is much smaller than the non-acute) and that the clustering always separates the known positive case from the known non-acute cases is also promising. We believe that this work will open up a new way of looking at epidemiological survey data that is useful for the field expert who is searching for a new way to validate and use case definitions and classification.

REFERENCES

Ahmad, S. and Tresp, V. 1993, Some Solutions to the Missing Feature Problem in Vision, in *Advances in Neural Information Processing Systems 5*, ed. S. J. Hanson, J. D. Cowan, and C. L. Giles (San Mateo, CA: Morgan Kaufmann), pp. 393–400

Hubert, L. J. and Levin, J. R. 1976, A General Statistical Framework for Assessing Categorical Clustering in Free Recall, *Psychological Bulletin*, 83, pp. 1072–1080

Milligan, G. W. and Cooper, M. C. 1985, An Examination of Procedures for Determining the Number of Clusters in a Data Set, *Psychometrika*, 50, pp. 159–179

Sarle, W. S. 1983, Cubic Clustering Criterion, Tech. Rep. A-108, SAS Institute Inc., Cary, NC

Tresp, V., Ahmad, S., and Neuneier, R. 1994, Training Neural Networks with Deficient Data, in *Advances in Neural Information Processing 6*, ed. J. D. Cowan, G. Tesauro, and J. Alspector (San Francisco: Morgan Kaufmann), pp. 128–135

Tresp, V., Neuneier, R., and Ahmad, S. 1995, Efficient Methods for Dealing with Missing Data in Supervised Learning, in *Advances in Neural Information Processing 7*, ed. G. Tesauro, D. Touretzky, and T. Leen (Cambridge, MA: MIT Press), pp. 689–696

DETECTION OF MULTIPLE SCLEROSIS USING NEURAL NETWORK ENSEMBLES

GREG EASTMAN
Department of Math and Computer Science
Drury University
Springfield, Missouri

DANIEL C. ST. CLAIR
Department of Computer Science
University of Missouri-Rolla
Rolla, Missouri

ABSTRACT
Event related potentials, or ERPs, are averaged waveforms of EEG cerebral activity measured using non-invasive electrodes attached to the scalp. Certain diseases resulting in neuropsychological impairment seem to result in changes in the waveform and latency of the P300 ERP. One example of this type of disease is multiple sclerosis. Experiments using artificial neural networks to analyze ERP data sets and discriminate between multiple sclerosis patients and a control population have been performed in previous research. These experiments yielded classification accuracies of 81% for single channel (electrode) training and 90% based on a vote of 2 out of 3 single channel outputs. This research describes a series of experiments on the same data sets undertaken with the goal of seeking higher accuracies. Single channel training using inputs that were pruned to isolate the P300 resulted in classification accuracies approaching 86%. By constructing a multi-channel ensemble network that used raw outputs of the single channel networks as inputs, classification accuracies of 96% were achieved.

INTRODUCTION

Electroencephalography is the measurement of neural activity by electrodes placed on the scalp of a human subject. Neurological activity in the brain results in the presence of electrical potentials that can be recorded by these electrodes. The electrodes are labeled according to the position they occupy on the skull. These electrodes detect fluctuations in the electrical potential of the underlying brain tissues.

The event related potential, or ERP, is a change in potential at a particular time related to an event in the physical world or to some psychological process. The P300 is the name given to one of several long latency ERP components that are observed in the measured potentials induced by an experimental method where a sensory stimulus is presented in an oddball paradigm. In this method of presentation, an infrequent stimulus is interspersed with a more frequent stimulus (Slater et al., 1993) (Picton and Hillyard, 1988). Long latency components are visible as a sequence of positive and negative peaks. The P300 is so named because the maximum amplitude of the measured positive (P) potential occurs at some time approximately 300 msec after the presentation of the stimulus. The P300 itself consists of two peaks called the P3a and P3b respectively

It is theorized that diseases or drugs that affect cortical function might affect the P300 in a manner that could be detected by analysis of the structure of this waveform (Slater et al., 1993). Of interest in this research is whether one might detect the presence of multiple sclerosis in subjects based on an analysis of the ERPs collected for control and patient groups. Previous work using artificial neural networks to analyze the data sets used in this research obtained classification accuracies of 81% for single channel analysis (Wu et al, 1993), and accuracies as high as 90% for a simple jury of three networks (Slater et al., 1993).

This paper describes methods used to improve the classification accuracy attained for this data set. The experiments undertaken to this end include single channel training using inputs that have been pruned to isolate the P300 and the construction of a multi-channel ensemble network trained using the raw outputs of multiple single channel networks as inputs. The accuracies obtained with the multi-channel ensemble were on the order of 96%, a significant improvement over the simple jury investigated in previous work (Slater et al., 1993).

BASELINE EXPERIMENTS

To establish a benchmark for assessing improvements in classification accuracy, baseline experiments were performed in order to replicate the results reported in previous work. The data were obtained from a pool of subjects consisting of two groups. The first was a control group of individuals with no history of central nervous system illness. The second consisted of patients who were diagnosed as having multiple sclerosis. The clinical diagnosis of subjects was made by neurologists at the University of Miami Department of Neurology prior to selection for inclusion in this data set (Slater et al., 1993).

Raw EEG signals were measured at 20 scalp positions. These signals were digitized over a 512 msec epoch consisting of 256 data points. The data obtained for this research consisted of two runs per patient, each containing the averaged response of an individual to 100 infrequent tones. The data from midline electrodes are used in this research. A midline electrode is one labeled with a lower case 'z', such as Cz, Fz, Oz, or Pz. These electrodes are attached along the line that would divide the left and right hemisperes of the brain.

Previous work employed a simple pruning method whereby the original 256 data points of each instance were uniformly sampled at a fixed interval (Slater et al., 1993), (Wu et al, 1993). They selected an interval of 10 data points thereby reducing each 256 element input vector to a 25 element input vector. For the purposes of these experiments a pruning factor of 8 was selected, yielding input vectors of 32 elements. It was thought that a smaller pruning factor would reduce the likelihood that important information would be discarded during the pruning process.

The baseline networks used for the present research are all multilayer, feedforward architectures using a backpropagation algorithm for training. The input layer consists of 32 input nodes. This layer is fully connected to a hidden layer of 8 nodes. The output layer consists of 2 nodes. The logarithmic sigmoid, or *logsig* function, was selected as the activation function for both the hidden and output layers. This architecture is identical to that used by Slater et al except that there are more input nodes (32 vs. 25) due to the selection of a smaller pruning factor. The training algorithm used for training the networks is the Levenberg-Marquardt algorithm[1]. The performance of the network during training is monitored by the Average Square Error (ASE). The output targets are [0 1] for a classification of multiple sclerosis, or MS, and [1 0] for a classification of a control, or normal. The network classification is considered correct if the error of both

---

[1] It is not clear what specific training algorithm was employed by Slater et al.

output nodes with respect to the correct targets is less than 0.5. Rounding the output values to the nearest integer and then comparing the rounded values to the expected values determine the network classification. Rounded outputs of [0 0] or [1 1] are considered to be unknown. The basic experiments described here use *v-fold* cross-validation for error rate estimation where $v = 5$. The original work reported by Slater et al used a value of $v = 6$. These experiments use 5 in order to increase the relative size of the set of unseen test cases.

Before training begins, several preprocessing steps are performed on the raw data instances. The first of these is the pruning described earlier. Next, the instances in the training set are normalized so that the values fall in the interval [1 −1]. This is done in order to reduce computational rounding that might be caused if some of the inputs differ from others by several orders of magnitude. Finally, the data must be partitioned into 5 sets as required by the 5-fold cross validation. Due to the nature of the data sets there is a special consideration that affects this partitioning. The is the fact that each individual subject is represented by two instances per electrode. Each set in the partition will contain 13 subjects, with two instances per subject, for a total of 26 instances. The subjects are distributed across the folds to maintain about the same ratio of controls to MS patients in each fold. Both instances for a given subject are placed in the same partition.

Since the convergence of the network, all other variables held constant, is completely a function of the initial input and layer weights, it is best to try several different initial conditions to ensure that an optimal solution has been found (Hagan et al., 1996). In these experiments, this is accomplished by reinitializing the input and layer weights at the beginning of each training session. The input and layer weights are reinitialized using the default algorithm provided by the software used for these experiments, in this case the Nguyen-Widrow algorithm (Nguyen and Widrow, 1990). The 'best result' for a given fold comes from the session where the initial conditions led to the most accurate performance on unseen test conditions. By saving information about the performance on unseen test cases after each epoch of training, it is possible to identify the point where the network shows optimal behavior.

The results obtained from the Cz baseline experiments performed for the purposes of this research are given directly below in Table 1. The line labeled MS (%) represents the number of correctly classified MS instances divided by the total number of actual MS instances, likewise for the line labeled Control (%). The line labeled All (%) represents the number of correct classifications divided by the total number of test instances in the fold. Due to the contents of the data sets, the numbers of MS and Control instances in each fold are not equal. This explains why the numbers listed for All(%) are not simply (MS(%) + Control(%)) / 2.

Fold	1	2	3	4	5	Average	Std
MS (%)	85.71	75.0	66.67	83.33	83.33	78.81	7.91
Contol (%)	100	92.86	85.71	71.43	92.86	88.57	10.83
All (%)	92.31	84.62	76.92	76.92	88.46	83.85	6.88

Table 1 – Cz Baseline Experiment – Best Classification Accuracies

The results obtained for the baseline experiment are very similar in kind to those reported by Wu et al., (1993). Not only is the overall classification accuracy similar, but the average accuracies for MS and controls are also similar. Using these observations as the basis for an assumption that a suitable baseline experiment has been constructed, training was carried out for 3 other midline electrodes, namely Fz, Oz, and Pz.

As noted, the raw outputs from the networks consist of a tuple in which each of the two elements varies between 0 and 1. These outputs were rounded to the nearest integer in order to decide the network classification for a particular instance. This method obscures information about the certainty of classification. For example, an output of [0.49 0.51] would be rounded to [0 1] and result in a classification of MS. However, such an output might be regarded as less certain than an output of [0.1 0.9]. In order to get an idea of how brittle the classification decisions are, an uncertainty interval u was defined to represent the distance of a number from the value of 0.5. Outputs that fall within this distance from 0.5 are considered to be so weak as to be indeterminate. When $u = 0$ an output of [0.4 0.6] would be classified as MS. However, if we let $u = 0.15$, a network output of [0.4 0.6] now falls into the region of uncertainty and is not counted as a classification success. Table 2 shows the effect of increasing the uncertainty interval on the classification accuracies obtained for the Cz folds.

Cz Electrode

u	Fold 1	Fold 2	Fold 3	Fold 4	Fold 5	Avg
0.00	92.3	84.6	76.9	76.9	88.5	83.8
0.10	76.9	65.4	73.1	73.1	76.9	73.1
0.20	34.6	65.4	69.2	73.1	73.1	63.1

Table 2 – Uncertainty Interval Effects on Cz

EXTENSIONS TO BASELINE EXPERIMENTS

This section describes two distinct extensions to the baseline experiments. The first extension was based on the hypotheses cited in previous work that cognitive impairment may be evident in the structure of the P300. To test this hypothesis, experiments were performed that used data points from 200 msec and beyond, the portion of the data where the P300 resides. The second extension was designed to improve previous classification accuracies of 90% that were achieved by combining the decisions of 3 single channel networks and deciding by a 2 out of 3 majority vote (Slater et al., 1993). To obtain this improvement, an ensemble jury network was constructined using the raw outputs of each of the four single channel networks, Cz, Fz, Oz, and Pz, as inputs. Rather than using a simple vote, the ensemble is trained in the hope that it can learn how much weight should be attributed to the outputs of each of the single channel electrodes. Using this method, accuracy on unseen test cases was about 96%, higher than previously published results using these data.

To examine the possibility that the most important information resides in the P300, network training was performed using instances that were pruned to remove data points that precede the P300 feature. Using the methods and protocols described in the baseline experiments, networks were trained using instances with the first 100 points removed. If important information resides in the first 100 data points one would expect the training to be less successful. In fact, as shown in Table 3, the prediction accuracy of these networks appears to be slightly better than that of the baseline networks.

Fold	1	2	3	4	5	Average	Std
MS (%)	85.71	91.67	100.00	83.33	83.33	88.81	7.13
Contol (%)	91.67	85.71	71.43	85.71	85.71	84.05	7.51
All (%)	88.46	88.46	84.62	84.62	84.62	86.15	2.10

Table 3 – Cz P300 Experiment – Best Classification Accuracies

Figure 1 compares the cross fold average accuracy for increasing values of the uncertainty interval *u* for both the baseline and P300 Cz electrode training. The shallow slope of the P300 line and the fact that it remains above the baseline plot are indicators that P300 networks have stronger discrimination abilities. Similar results were noted for th Fz and Pz electrodes, however this was not the case for the Oz electrodes (Eastman, 2001).

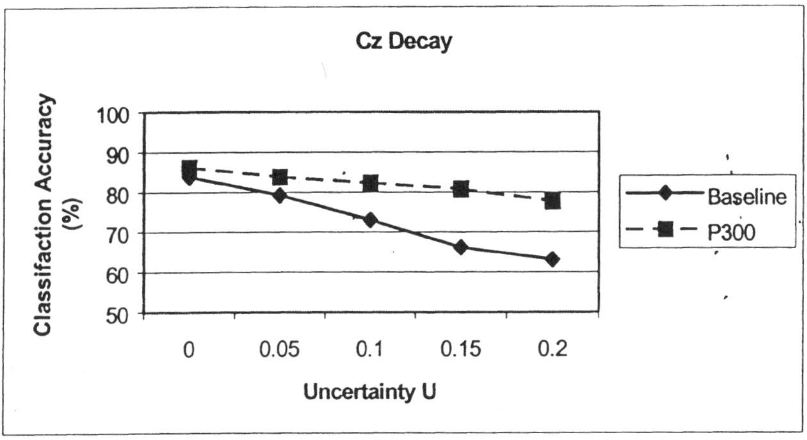

Figure 1 – Cz Classification Accuracy Decay

In conclusion, while the better outcomes may be due to the slight variance in results that can be seen from experiment to experiment, confidence is high that pruning the first 100 data points from each instance did not degrade the performances of these networks. In fact, the networks appear to provide better discrimination between control and MS subjects. The more robust decisions rendered indicate a real improvement as opposed to a statistical fluke.

The success encountered using a simple voting Jury in the previous work by Slater et al. (1993) led to the idea that an ensemble network might increase the classification accuracy still further. The ensemble network was trained to render decisions based on the values of the raw outputs from each of the four midline electrode networks trained using data from Cz, Fz, Oz, and Pz. This means that each input vector is now an 8-element column vector. The architecture used is the same as that for the previous experiments except that the number of nodes in the hidden layer has been reduced to 4 from 8. This reduction was made knowing that our input vectors are much smaller than they are for the single channel networks. The inputs were partitioned in the same manner as in both the baseline and P300 experiments. The outputs of the P300 networks for unseen test cases were used to construct inputs since they seemed to show higher accuracies and more decision confidence than the baseline networks.

Table 4 shows the results of the jury ensemble training. The average accuracy across folds is 96%! Overall, the standard deviation is lower than in the previous experiments indicating a high consistency between folds. Not only are the ensemble networks very accurate in their predictions, but they also render much stronger decisions than the single electrode networks. Even with the uncertaintly factor at 0.2, the overall accuracy for the ensemble remains above 90% (Eastman, 2001).

Fold	1	2	3	4	5	Average	Std
MS (%)	92.7	100.0	100.0	91.7	100.0	97.0	4.3
Contol (%)	100.0	85.7	92.9	100.0	100.0	95.7	6.4
All (%)	96.2	92.3	96.2	96.2	100.0	96.2	2.7

Table 4 – Jury Ensemble – Best Classification Accuracies

CONCLUSIONS

These experiments show that any diagnostic tool based on examination of ERPs will probably perform better with information provided by more than one electrode. By sampling potentials from numerous scalp positions, a more complete picture of the neurological activity is obtained. One thing that makes this type of analysis attractive is the relative simplicity and low cost of the data collection mechanisms. Other types of brain activity monitoring are available, however the cost of obtaining and operating such equipment could be prohibitive. If this type of ERP analysis could be turned into a production quality diagnostic tool it would be more accessible to smaller medical facilities.

REFERENCES

G. Eastman,, 2001, "Detection of Multiple Sclerosis Using Neural Network Ensembles", MS Thesis, University of Missouri – Rolla, Rolla, Missouri.

M. Hagan, H. Demuth and M. Beale, 1995, *Neural Network Design*, PWS Publishing Company, Boston, MA, pp. 11-21.

D. Nguyen and B. Widrow, 1990, "Improving the Learning Speed of 2-Layer Neural Networks by Choosing Initial Values of the Adaptive Weights," *International Joint Conference on Neural Networks*, Vol 3, pp. 21-26.

T. Picton and S. Hillyard, 1988, "Endogenous Event-Related Potentials," in T. W. Picton, ed. *Human Event-Related Potentails, Handbook of Electroencephalography and Clinical Neurophysiology*, Revised Series, Volume 3, Elsevier Science Publishers, pp. 361-426.

J. D. Slater, F. Y. Wu, L. S. Honig, R. E. Ramsay, and R. Morgan, 1993, "Neural Network Analysis of the P300 Event-Related Potential in Multiple Sclerosis," *Electroencephalography and Clinical Neurophysiology*.

F. Y. Wu, J. D. Slater, L.S. Honig and R. E. Ramsay, 1993, "A Neural Network Design for Event Related Potential Diagnosis," *Computers in Biology and Medicine*, Vol. 23, No. 3, pp 251-264.

APPROXIMATE REASONING IN TRADITIONAL CHINESE MEDICAL DIAGNOSIS AND TREATMENT FOR THE ABNORMAL RHYTHM OF HEART-BEATS

YUAN, LUNCI
Lcy_999@yahoo.com
48th Building, #103
3rd Yanji New Town
Shanghai, 200093,
P. R. China

FENG, CHU*
Chuf_99@yahoo.com
42-97 Crommelin St.
Flushing, NY 11355
USA

ABSTRACT

By means of the brand-new technique of decision-making and evaluation based on Approximate reasoning[5,6] a computer diagnosis and treatment system of the excellent Doctor Yuan Lunci of traditional Chinese medicine has been created in this paper. This technique can truly sum up the expert's experience in computer.

Key words: Approximate reasoning; Decision-making; Diagnostic expert system; Fuzzy logic; Implication;Medical diagnosis; Multiple criteria evaluation; Multicriteria decision; Possibility theory

INTRODUCTION

So far in the Western medicine there are no specifics for some kinds of the abnormal rhythm of heart-beats. On the basis of ***Yin***, ***Yang***, ***Qi***, ***Blood***, however, the excellent Doctor of traditional Chinese medicine can diagnose and find the treatments, which is unique and effective. In this paper, many factors, types and criteria model based on approximate reasoning has been applied successfully to the traditional Chinese medical diagnosis and treatment systems, which is more in accordance with the thinking and reasoning processes of the excellent Doctor of traditional Chinese medicine in diagnosing and finding the treatments for very difficult and complicated cases.

MODEL[1,2,3,4]

The set of diagnosis factor is

$$F = \{F_i\}, \qquad I = 1, 2, \ldots, I, \qquad I = 25 \tag{1}$$

where

F_1 - thready, rapid and forceful pulse

F_2 - floating, slippery and forceful pulse

F_3 - thready weak and feeble pulse

F_4 - weaker and feeble pulse

F_5 - thready faint and feeble pulse

F_6 - deep, slow and feeble pulse

F_7 - crimson tongue

F_8 - red tongue

*Feng, Chu is the corresponding author

F_9 - reddish tongue
F_{10} - pale tongue
F_{11} - flushed cheeks
F_{12} - unhealthy flush face
F_{13} - pale and sallow complexion
F_{14} - pale face
F_{15} - brightly pale complexion
F_{16} - constipation
F_{17} - frequency of bowel movement
F_{18} - loose stool
F_{19} - night sweats
F_{20} - abnormal sweating due to general debility
F_{21} - spontaneous perspiration
F_{22} - sweatiness and short of breath
F_{23} - polydipsia
F_{24} - thirst.
F_{25} - no thirst.

The set of the disease type is

$$T = \{T_p\}, \qquad p = 1, 2, \ldots, P, \qquad P = 4 \tag{2}$$

Where T_1 - deficiency of ***Yin,*** T_2 - insufficiency of ***Yang,***
T_3 - deficiency of ***Q,*** T_4 - deficiency of ***Blood.***

The diagnosis criterion set is

$$C_p = \{c_{pj_p}\}, \qquad j_p = 1, 2, \ldots, J_p. \tag{3}$$

Let p=1, J_1=5. The criteria diagnosed the illness as the type of deficiency of ***Yin,*** T_1, are stated as follows.

c_{11} : If the patient's pulse is thready, rapid and forceful, F_1, gravely, $H_2 F_1$, then the patient is considered possible.

c_{12} : Apart from thready, rapid and forceful pulse, F_1, with crimson tongue, F_7, - more possible.

c_{13} : Besides c_{12}, the patient has flushed cheeks, F_{11}, or unhealthy flush face, F_{12}, and night sweat, F_{19}, it can be considered much more possible.

c_{14} : Apart from c_{13}, with constipation, F_{16}, and polydipsia, F_{23}, - definite.

c_{15} : If the pulse is deep, slow and feeble, F_6, or tongue is pale, F_{10}, - impossible.

Let p=2, J_2=5. The following is the criteria diagnosed the illness as the type of insufficiency of ***Yang,*** T_2.

c_{21} : If the patient's pulse is thready, faint and feeble, F_5, then the patient is considered possible.

c_{22} : Apart from c_{21}, with gravely pale tongue, $H_2\ F_{10}$,
- more possible.

c_{23} : Besides c_{21}, and pale tongue, F_{10}, the patient has brightly pale complexion, F_{15}, and no thirst, F_{25},
- much more possible.

c_{24} : If the patient has deep, slow and feeble pulse, F_6, frequency of bowel movement, F_{17}, and abnormal sweating due to general debility, F_{20} ,
- definite.

c_{25} : If the pulse is thready, rapid and forceful, F_1, and the tongue is crimson, F_7, or constipation, F_{16},
- impossible.

Let p=3, J_3=5. The criteria diagnosed the illness as the type of deficiency of ***Qi*** , T_3, are stated as follows.

c_{31} : If the pulse is weaker and feeble, F_4,
then the patient is considered possible.

c_{32} : Apart from c_{31}, with pale tongue, F_{10},
- more possible.

c_{33} : Besides c_{32}, the patient has pale face, F_{14}, no thirst, F_{25}, and abnormal sweating due to general debility, F_{20}, or spontaneous perspiration, F_{21},
- much more possible.

c_{34} : Besides c_{32}, the patient has pale face, F_{14}, no thirst, F_{25}, loose stool, F_{18}, and sweatiness and short of breath gravely, H_2F_{22},
- definite.

c_{35} : If the pulse is thready, rapid and forceful, F_1, or floating, slippery and forceful, F_2,
- impossible.

Let p=4, J_4=4. The following is the criteria diagnosed the illness as the type of deficiency of ***Blood***, T_4.

c_{41} : If the pulse is thready, weak and feeble, F_3,
then the patient is considered possible.

c_{42} : Apart from c_{41}, with pale and sallow complexion gravely, H_2F_{13},
- more possible.

c_{43} : Besides c_{41}, the patient has pale and sallow complexion, F_{13}, pale tongue, F_{10}, and thirst, F_{24},
- definite.

c_{44} : If the patient has floating, slippery and forceful pulse, F_2, and red tongue, F_8, or reddish tongue, F_9 ,
- impossible.

We define the appraisal set

$A = \{A_k\}, \quad k = 1,2,\ldots,K, \quad K = 5$ (4)

and the appraisal function A(v) is defined as

A_1 ------- possible $A_1(v) = v$ (5)

A_2 ------- more possible $A_2(v) = v^{3/2}$ (6)

A_3 ------- much more possible $A_3(v) = v^2$ (7)

A_4 ------- definite $A_4(v) = \begin{cases} 1 & v=1 \\ 0 & v \neq 1 \end{cases}$ (8)

A_5 ------- impossible $A_5(v) = 1-v$ (9)

where $v \in V$,

$V = \{ v_l \} = \{0, 0.1, 0.2, 0.3, 0.4, 0.5, 0.6, 0.7, 0.8, 0.9, 1\}, l = 1,2,\ldots,L, L=11.$ (10)

The diagnosis criteria are described as:

$p = 1, \quad J_1 = 5,$

If $c_{11} = H_2F_1$ then A_1

If $c_{12} = F_1 \cap F_7$ then A_2

If $c_{13} = F_1 \cap F_7 \cap (F_{11} U F_{12}) \cap F_{19}$ then A_3

If $c_{14} = F_1 \cap F_7 \cap (F_{11} U F_{12}) \cap F_{19} \cap F_{16} \cap F_{23}$ then A_4

If $c_{15} = F_6 \text{ U } F_{10}$ then A_5

$p = 2, \quad J_2 = 5,$

If $c_{21} = F_5$ then A_1

If $c_{22} = F_5 \cap H_2F_{10}$ then A_2

If $c_{23} = F_5 \cap F_{10} \cap F_{15} \cap F_{25}$ then A_3

If $c_{24} = F_6 \cap F_{17} \cap F_{20}$ then A_4

If $c_{15} = (F_1 \cap F_7) \text{U } F_{16}$ then A_5

$p = 3, \quad J_3 = 5,$

If $c_{31} = F_4$ then A_1

If $c_{32} = F_4 \cap F_{10}$ then A_2

If $c_{33} = F_4 \cap F_{10} \cap F_{14} \cap F_{25} \cap (F_{20} U F_{21})$ then A_3

If $c_{34} = F_4 \cap F_{10} \cap F_{14} \cap F_{25} \cap H_2F_{22} \cap F_{18})$ then A_4

If $c_{35} = F_1 U F_2$ then A_5

$p = 4, \quad J_4 = 4,$

If $c_{41} = F_3$ then A_1

If $c_{42} = F_3 \cap H_2F_{13}$ then A_2

If $c_{43} = F_3 \cap F_{13} \cap F_{10} \cap F_{24}$ then A_4

If $c_{44} = F_2 \cap (F_8 U F_9)$ then A_5

where H_2 is the mood concentrator "gravely" defined by

$H_2F_1 = (F_1)^2$, etc.

Suppose the set U is built up by all patients

$$U = \{u_m\}, \qquad m = 1,2,\ldots,M . \tag{11}$$

The single-factor evaluation for patient u_m is a fuzzy mapping from F to U, $f : F \sim\!\rightarrow U$, and the fuzzy mapping implies a fuzzy relation which can be represented by a fuzzy matrix

$$R = (r_{im}) = (fR_1 \;\; fR_2 \;\; \ldots \;\; fR_i \;\; \ldots \;\; fR_I)^T \tag{12}$$

where fR_i is a fuzzy subset of U.

We use R as input. Processing it according to the diagnosis criteria , we can get fuzzy mapping from C_p to U, $f : C_p \sim\!\rightarrow U$, which can be described by a fuzzy matrix

$$C_pR = (fc_{p1} \;\; fc_{p2} \;\; \cdots \;\; fc_{pj_p} \;\; \cdots \;\; fc_{pJ_p})^T \tag{13}$$

where fc_{pj_p} is a fuzzy subset of U .

We can make the following approximate reasoning

$p = 1, \quad J_1 = 5,$

If $x_1 = fc_{11}$ then $y_1 = A_1$

If $x_1 = fc_{12}$ then $y_1 = A_2$

If $x_1 = fc_{13}$ then $y_1 = A_3$

If $x_1 = fc_{14}$ then $y_1 = A_4$

If $x_1 = fc_{15}$ then $y_1 = A_5$

$p = 2, \quad J_2 = 5,$

If $x_2 = fc_{21}$ then $y_2 = A_1$

If $x_2 = fc_{22}$ then $y_2 = A_2$

If $x_2 = fc_{23}$ then $y_2 = A_3$

If $x_2 = fc_{24}$ then $y_2 = A_4$

If $x_2 = fc_{25}$ then $y_2 = A_5$

$p = 3, \quad J_3 = 5,$

If $x_3 = fc_{31}$ then $y_3 = A_1$

If $x_3 = fc_{32}$ then $y_3 = A_2$

If $x_3 = fc_{33}$ then $y_3 = A_3$

If $x_3 = fc_{34}$ then $y_3 = A_4$

If $x_3 = fc_{35}$ then $y_3 = A_5$

$p = 4, \quad J_4 = 4,$

If $x_4 = fc_{41}$ then $y_4 = A_1$

If $x_4 = fc_{42}$ then $y_4 = A_2$

If $x_4 = fc_{43}$ then $y_4 = A_4$

If $x_4 = fc_{44}$ then $y_4 = A_5$.

Using implication as translation rule[5],

$$d_{pj_p}(m,l) = 1 \wedge (1 - fc_{pj_p}(u_m) + A_k(v_l)). \tag{14}$$

From this, we can get a fuzzy mapping from U to V, $f : U \sim\!\!\rightarrow V$, which can be represented by a fuzzy matrix

$$D_{pj_p} = (d_{pj_p}(m,l)). \tag{15}$$

And then, we get the fuzzy multicriteria evaluation matrix

$$D_p = \bigcap_{j_p=1}^{J_p} D_{pj_p} = (fE_{p1}\ \ fE_{p2}\ \ \cdots\ \ fE_{pm}\ \ \cdots\ \ fE_{pM})^T \tag{16}$$

where fE_{pm} is a fuzzy subset of V, which represents the possibility of suffering from the disease for patient u_m.

Assume $E_{pm\alpha}$ is the α level set of fE_{pm}, $\alpha \in [0,1]$. It should be noted that the set $E_{pm\alpha}$ is an ordinary subset of V.

The mean value of the elements in $E_{pm\alpha}$ is

$$H_l(E_{pm\alpha}) = \frac{1}{N_\alpha} \sum_{n=1}^{N_\alpha} Z_n(\alpha) . \tag{17}$$

where α is the level of the level set, $Z_n(\alpha)$ is the element in $E_{pm\alpha}$, $Z_n(\alpha) \in E_{pm\alpha}$, and N_α is the cardinality of the finite set $E_{pm\alpha}$.

The point value[3] of fE_{pm} is

$$S_p(m) = \frac{1}{\alpha_{max}} \sum_{l=1}^{L} H_l(E_{pm\alpha}) \,\Delta\, \alpha_l \tag{18}$$

where α_{max} is the maximum membership grad of fE_{pm},
$\Delta\alpha_l = \alpha_l - \alpha_{l-1}$, $\alpha_0 = 0$.
Compare $S_p(m)$ with the threshold defined by the authors and the patient u_m is diagnosed and the treatment is found.

EXAMPLE

This example taken from the experience of an existing patient is one of very difficult and complicated cases, which the author, Doctor Yuan, accurately diagnosed a patient's abnormal rhythm of heart-beats as the type of deficiency of ***Yin***, T_1, thus successfully treating this disease.

Input is

$R = [.8\ .2\ 0\ 0\ 0\ 0\ .6\ .4\ 0\ 0\ .7\ .3\ 0\ 0\ 0\ 1\ 0\ 0\ .7\ .3\ 0\ 0\ .6\ .4\ 0]^T$.

Processing it according to the above model, we get

$C_1R = [0.6400\ \ 0.4800\ \ 0.3360\ \ 0.2016\ \ 0.0000]^T$

$C_2R = [0.0000\ \ 0.0000\ \ 0.0000\ \ 0.0000\ \ 1.0000]^T$

$C_3R = [0.0000\ \ 0.0000\ \ 0.0000\ \ 0.0000\ \ 1.0000]^T$

$C_4R = [0.0000\ \ 0.0000\ \ 0.0000\ \ 0.0800]^T$

p=1

D_{11}=[0.3600 0.4600 0.5600 0.6600 0.7600 0.8600 0.9600 1.0000 1.0000 1.0000 1.0000]

D_{12}=[0.5200 0.5516 0.6094 0.6843 0.7730 0.8736 0.9848 1.0000 1.0000 1.0000 1.0000]

D_{13}=[0.6640 0.6740 0.7040 0.7540 0.8240 0.9140 1.0000 1.0000 1.0000 1.0000 1.0000]

D_{14}=[0.7984 0.7984 0.7984 0.7984 0.7984 0.7984 0.7984 0.7984 0.7984 0.7984 1.0000]

D_{15}=[1.0000 1.0000 1.0000 1.0000 1.0000 1.0000 1.0000 1.0000 1.0000 1.0000 1.0000]

P=2

D_{21}=[1.0000 1.0000 1.0000 1.0000 1.0000 1.0000 1.0000 1.0000 1.0000 1.0000 1.0000]

D_{22}=[1.0000 1.0000 1.0000 1.0000 1.0000 1.0000 1.0000 1.0000 1.0000 1.0000 1.0000]

D_{23}=[1.0000 1.0000 1.0000 1.0000 1.0000 1.0000 1.0000 1.0000 1.0000 1.0000 1.0000]

D_{24}=[1.0000 1.0000 1.0000 1.0000 1.0000 1.0000 1.0000 1.0000 1.0000 1.0000 1.0000]

D_{25}=[1.0000 0.9000 0.8000 0.7000 0.6000 0.5000 0.4000 0.3000 0.2000 0.1000 0.0000]

P=3

D_{31}=[1.0000 1.0000 1.0000 1.0000 1.0000 1.0000 1.0000 1.0000 1.0000 1.0000 1.0000]

D_{32}=[1.0000 1.0000 1.0000 1.0000 1.0000 1.0000 1.0000 1.0000 1.0000 1.0000 1.0000]

D_{33}=[1.0000 1.0000 1.0000 1.0000 1.0000 1.0000 1.0000 1.0000 1.0000 1.0000 1.0000]

D_{34}=[1.0000 1.0000 1.0000 1.0000 1.0000 1.0000 1.0000 1.0000 1.0000 1.0000 1.0000]

D_{35}=[1.0000 0.9000 0.8000 0.7000 0.6000 0.5000 0.4000 0.3000 0.2000 0.1000 0.0000]

P=4

D_{41}=[1.0000 1.0000 1.0000 1.0000 1.0000 1.0000 1.0000 1.0000 1.0000 1.0000 1.0000]

D_{42}=[1.0000 1.0000 1.0000 1.0000 1.0000 1.0000 1.0000 1.0000 1.0000 1.0000 1.0000]

D_{43}=[1.0000 1.0000 1.0000 1.0000 1.0000 1.0000 1.0000 1.0000 1.0000 1.0000 1.0000]

D_{44}=[1.0000 1.0000 1.0000 1.0000 1.0000 1.0000 1.0000 1.0000 1.0000 1.0000 0.9200]

P=1

D_1 = [0.0992 0.1365 0.1918 0.2719 0.3865 0.5482 0.7548 0.7984 0.7984 0.7984 1.0000]

P=2

D_2 = [1.0000 0.9000 0.8000 0.7000 0.6000 0.5000 0.4000 0.3000 0.2000 0.1000 0.0000]

P=3

D_3 = [1.0000 0.9000 0.8000 0.7000 0.6000 0.5000 0.4000 0.3000 0.2000 0.1000 0.0000]

P=4

D_4=[1.0000 1.0000 1.0000 1.0000 1.0000 1.0000 1.0000 1.0000 1.0000 1.0000 0.9200].

Finally, we have
S_1=0.7609
S_2=0.2250
S_3=0.2250
S_4=0.4960.
The calculated result corresponds to Doctor Yuan's diagnosis, because the biggest is S_1=0.7609.

CONCLUSIONS

The Traditional Chinese Medicine is an ancient science, which is unique and effective but extremely hard to learn, particularly diagnosis based on ***Yin, Yang, Qi, Blood***.

In this paper through applying the brand-new theory, Approximate Reasoning, to the ancient science, a computer diagnosis and treatment system of the excellent Doctor Yuan Lunci of traditional Chinese medicine has been created.

It is quite evident that we shall be able to create a series of diagnosis systems, which will enable better propagating of the traditional Chinese medicine to The West.

REFERENCES

[1] Chu Feng, Quantitative evaluation of university teaching quality ----- an application of fuzzy set and approximate reasoning, Fuzzy Sets and Systems 37(1990)1 – 11
[2] Chu Feng, Fuzzy multicriteria decision-making in distribution of factories: an application of approximate reasoning, Fuzzy Sets and Systems 71(1995)197 – 205
[3] R.R.Yager, Multicriteria decision with soft information: an application of fuzzy set and possibility theory (I) (II), Fuzzy Math. 2 (2) (3) (1982) 21 – 28, 7 – 16
[4] Lunci Yuan Fuzzy multicriteria evaluation in traditional Chinese medical diagnosis of cerebral apoplexy: an application of approximate reasoning, the Proceedings of 19th International Conference of the NAFIPS 2000, pp.211-214. http://www.gsu.edu/~dscthw/PeachFuzz/
[5] L.A.Zadeh, A theory of approximate reasoning , in : J.Hayes, D.Michie and L.I.Mikulich, Eds., Machine Intelligence, Vol.9 (Halstead Press, New York, 1979)149 - 194
[6] L.A.Zadeh, The concept of a linguistic variable and its application to approximate reasoning --- I, II, III, in R.R.Yager, S.Ovchinnikov, R.M.Tong, H.T.Nguyen, Eds., Fuzzy Sets and Applications: Selected Papers by L.A.Zadeh (John Wiley & Sons, New York, 1987 219 - 366

EXPERIMENTS USING AN ADAPTIVE NETWORK-BASED FUZZY INFERENCE SYSTEM (ANFIS) FOR THE DIAGNOSIS OF BREAST CANCER

WALKER H. LAND, JR.,
Research Professor,
Dept. of Computer Science,
Binghamton University
Binghamton, New York

HAROLD W. LEWIS, III, Professor
and Director of Center for Intelligent
Systems, Dept. of Systems Science,
Binghamton University
Binghamton, New York

ELIZABETH A. VERHEGGEN,
Graduate Student,
Dept. of Systems Science,
Binghamton University
Binghamton, New York

ABSTRACT
This paper describes a series of neural network experiments which contribute decision support information, aimed at reducing the misclassification rate of screening mammograms and improving recommendations regarding breast biopsy. Features were extracted from digitized film images, each with pathologically definite diagnosis, from a large public domain database. Decision trees were used to illicit separate approaches to classify different mammographic structures, which were then tuned for each structure as part of the fuzzy rule set in an Adaptive-Network-based Fuzzy Inference System (ANFIS). By comparing several neural network based computer-aided diagnosis (CAD) approaches for breast cancer classification as benign or malignant, our experiments indicate that separating discriminatory features into categories with predominant mammographic structures imparts substantial classification improvements of 50% to 151% at the clinically imperative levels of high sensitivity.

Introduction
Breast cancer is the second leading cause of death from cancer among women in the United States. More than 180,000 new cases of invasive breast cancer are diagnosed each year, with more than 40,000 deaths from the disease (Dronkers, 2002 and Nass, 2001). Prior to treatment by radical mastectomy at the end of the 19th century, breast cancer was considered incurable and treatments were actually palliative (Dronkers, 2002). Additional improvements in treatment reduced breast cancer mortality to approximately 40-45%. Since research has yet to discover methods which will provide a 100% effective treatment which is independent of the stage of tumor detection, prevention assumes primary importance for reducing mortality further. While prevention covers a range of modalities, diagnosing breast cancer earlier currently offers the best hope of reducing mortality (Nass, 2001 and Olson, 2002). Mammography is currently the mainstay for diagnosis of breast cancer. Despite its utility to detect breast cancer early, there are inherent limitations of the technology which result in false positive results. This undesirable consequence of mammography's low percentage of true positive classifications or low positive predictive value (PPV) means that 60 to 85 percent of detected lesions are actually benign (Karssemeijer, 1998 and Nass, 2001). The additional medical interventions necessary to evaluate false positives represents one third of the total cost of screening for all women. In addition to increasing healthcare costs, the associated physical and psychological costs of false positive results may serve as a

barrier for widespread screening of appropriate age groups. These issues serve as a driving force behind computer-aided diagnosis (CAD) efforts to reduce the number of benign outcome biopsies associated with mammography.

In this paper we introduce a neuro-fuzzy modeling approach as a CAD tool to resolve the issue of the low PPV of mammography. Our neuro-fuzzy approach is based on the belief that fuzzy logic's capability of capturing uncertainty due to non-randomness and ambiguity in the problem domain is well suited to mammography's diagnostic challenges. In the following section we provide a brief overview of the experimental frame of the classification problem. We then describe the extraction of fuzzy rules from data using subtractive clustering. These rules are implemented in an Adaptive Neuro-Fuzzy Inference System called ANFIS. Following that we propose a divide-and-conquer approach to classification based on partitioning the feature space using the tree-based method called CART®. We conclude our study of breast cancer classification using ANFIS with a discussion of the results obtained by separating the mass mammographic structures for improved diagnostic performance, and outline future directions.

EXPERIMENTAL FRAME OF THE CLASSIFICATION PROBLEM

The databases used to evaluate ANFIS as a classifier of benign and malignant lesions originated from two sources, Duke University and the University of South Florida. Together they provide a retrospective study of approximately 2000 film screen mammogram cases. We refer to the Duke data as RDU, as it has undergone variable simplification in previous studies, reducing its set of discriminators from 16 to 7. Similarly we refer to the other dataset by abbreviation as USF. Both databases contain cases of non-palpable, mammographically-suspicious breast lesions which were characterized by biopsy, the gold standard for definitive diagnosis. Three lesion types which describe the main categories of breast cancer were represented: mass lesions, calcification lesions, and a mixed category referred to as other lesions. These types of breast cancer were approximately evenly split between the two main categories of mass and calcification, with about 45% in each, with the mixed category of other lesions comprising the remaining 7 to 9%. Malignant and benign classes are represented similarly (60/40).

With the exception of an input variable for age as a measured feature in years on [24, 91], the types of breast cancer have qualitative feature descriptors which reflect standardization by the American College of Radiology called the BIRADS lexicon (BIRADS, 1993). Therefore, descriptors for mass and calcification take on specific definitions given by their BIRADS designation. Two BIRADS descriptors for each type of breast cancer were chosen as model input variables based upon earlier senstivity studies. Mass margin and mass shape, and calcification description and distribution form the BIRADS discriminator set. Each varaible is described by 5 to 14 states which are ordered in terms of suspicion of breast cancer. However, as these variables are qualitative and linguistic (categorical) variables, their order is arbitrary, and reflects no measureable unit of severity. Given these methodolgocial distinctions, the datasets were divided into training (80%) and validation (20%) sets, maintaining equal proportions of outcome, for 5-fold cross validation of the models, to prevent overfitting and ensure generalization to new patterns (Haykin, 1999 and Stone, 1974).

ANFIS EXPERIMENTS USING SUBTRACTIVE CLUSTERING FOR RULE INDUCTION

ANFIS was chosen to model the breast cancer databases because the neural portion of this hybrid capitalizes on the data driven capabilities of neural networks while the fuzzy portion allows the input to the neural network to be related to fuzzy terms more suitable for categorical, linguistic (qualitative) variables. This is accomplished using the Tagaki-Sugeno-Kang (TSK) class for a fuzzy inference system (FIS). The FIS portion plays the role of the 'gray box' approximating an unknown nonlinear mapping via a smooth combination of simple and generally linear and local models (Jang, 1997).

In a TSK fuzzy model the equations express as separate fuzzy IF/THEN rules all possible combinations of the inputs. These rules contain linguistic information in the antecedent and numerical information in the consequent, such as:

IF X_1input is $O_{1(obscured)}$ and X_2 input is $L_{2(lobulated)}$ THEN $y=c_{10}+c_{11}X_1+c_{12}X_2$.

ANFIS uses these fuzzy rules as local mappings, facilitating the minimum disturbance principle (Jang, 1997). This rule states that adaptation should not only reduce the output error for the current training pattern but should minimize disturbance to the response already learned. In an ANFIS implementation then, for the input BIRADS diagnostic and patient history (age) features, the fuzzy relations which relate linguistic terms to numerical values, as well as the consequent coefficients, are determined and trained as part of the neural network. A brief overview of the architecture of ANFIS follows (Lewis and Yang, 1998). Layer 0 distributes system inputs while layer 1 finds the membership grades of each term, for each of the inputs using any appropriate parameterized membership function. Here, Gaussian Bell was used. Layer 2 acts as the conjunction operator. Layers 3 and 4 normalize rule antecedents. Layer 5 applies the rule consequents, while layer 6 computes the overall output as the summation of all incoming signals. A diagram is shown in fig. 1.

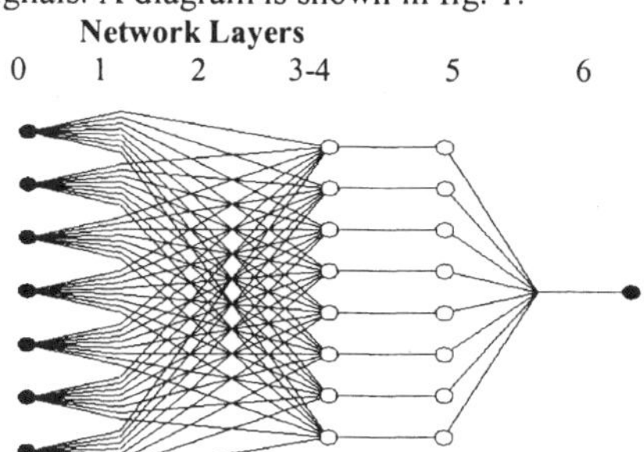

Fig1. ANFIS Structure

Fuzzy rules were extracted from the RDU and USF data sets by a variant of mountain clustering called subtractive clustering. In subtractive clustering divisions of the input space are done by defining cluster centers from the data, and assigning behaviors or outputs to those clusters that carry the greatest weight. Input states with few or no data are largely ignored. Subtractive clustering thus partitions the input space so that the centers of the data become the nucleus of the fuzzy rules. Based on this method of rule induction, several ANFIS models were run with an average of 8 fuzzy rules. The output threshold was varied to achieve the clinical imperative of high sensitivity and specificity, as given by the standard definitions, and these metrics were used as the performance measures. With sensitivity as the probability of predicting disease, given the true state is disease and specificity as the probability of predicting non-disease, given the true state is non-disease, Stolley (1995), the following results for both datasets were achieved as shown in table 1.

Objective	Performance	RDU	USF
100% Sensitivity	Specificity PPV	32.2 49.6	5.0 51.0
98% Sensitivity	Specificity PPV	37.4 51.4	20.2 52.0
95% Sensitivity	Specificity PPV	39.8 51.5	32.4 58.4

Table 1. ANFIS Results Using Subtractive Clustering

The results of these first experiments indicate that although the PPV values exceeded mammography's PPV, which is typically 25 to 40%, specificities are low at 100% sensitivity, where no cancers are missed, particularly for the USF dataset. Based upon these trials, rule induction was augmented by using CART® to ascertain whether rules might be improved by estimating a simple tree-based structure from the data.

ANFIS EXPERIMENTS USING CART FOR RULE INDUCTION

The tree structured classifier derived by the CART® algorithm reveals where the most accurate classification can be achieved, within the limits of the data, and provides some insight into the predictive structure of the data. Whereas subtractive clustering determines data groupings based on the actual centers of the data, the CART® algorithm reveals where the most accurate classification can be achieved by determining the decision boundaries which partition the data into disjoint subsets. Using the full training set, by exhaustive search the optimal split is identified which reduces an error measure. This error or impurity function (Gini index) computes the cost of the node, such that growing the tree is optimized and cases are split in the node resulting in the greatest cost reduction (Brieman 1998). Splitting continues at the new 'child' nodes, terminating when the node error measure falls below a tolerance level or when the error reduction from further splitting does not exceed a threshold. A distinct advantage of a recursive binary tree is its interpretability, which provides a simple view stratifying the population into partitions of benign and malignant outcome, on the basis of the patient characteristics. This can be seen in the CART diagram which follows in fig. 2. The feature mass margin splits the data set. Its states of "well circumscribed and obscured" go to the left terminating at an 85% benign node, while the first terminal node on the right is malignant (92%) for the state "spiculated". "Microlobulated", and "ill defined" masses are further split by age, and result in poorer classification. Based upon this analysis, three additional ANFIS experiments were run to determine whether performance improvements could be garnered with new induced fuzzy rules, and by modeling mass lesions separately.

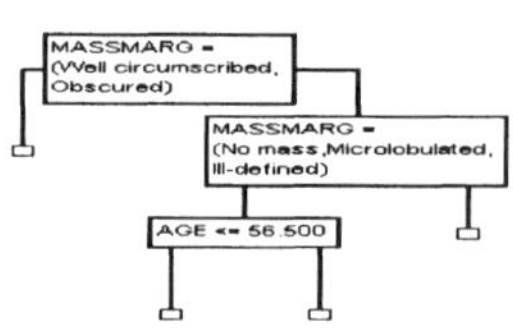

FIG. 2 CART DIAGRAM OF USF DATASET

DISCUSSION OF NEW ANFIS RESULTS

For this series of experiments the data set was cleansed, removing the 'other' lesion category of mixed features which did not contribute to classification accuracy based on the decision tree results. Additionally, three fuzzy rules based on mass margin, with age less than or equal to 63.5 were extracted from the tree of the full data set: IF well-circumscribed, THEN benign; IF obscured, THEN benign; and IF spiculated, THEN malignant. These fuzzy rules were used in place of the 8 subtractive clustering induced rules. Following these procedures, three ANFIS models were constructed. The first used the cleansed data set of both mass and calcification features, and the second isolated mass mammographic features (margin and shape), and together with age, comprised a mass model based on the insights from the tree partitioned feature space which indicated superior classification performance. For the larger USF data set, these results were compared to a benchmark of the less challenging masses, where the margin features yielded 85% to 94% classification accuracy in the decision tree. We refer to these as 'mass subset' for short. The results follow in tables 2 and 3.

Performance		Mass and Calcification	Mass	Mass Subset Benchmark
100% Sensitivity	Specificity	12.56%	25.39%	29.73%
	PPV	50.53%	55.20%	50%
98% Sensitivity	Specificity	25.99%	39.19%	41.89%
	PPV	53.94%	59.24%	54.26%
95% Sensitivity	Specificity	37.77%	55.37%	82.43%
	PPV	57.24%	65.27%	79.03%

Table 2. ANFIS Results on Cleansed USF Data Using 3 Fuzzy Rules

Performance		Mass and Calcification	Mass
100% Sensitivity	Specificity	48.31%	84.21%
	PPV	50.12%	72.61%
98% Sensitivity	Specificity	54.98%	84.21%
	PPV	52.65%	72.61%
95% Sensitivity	Specificity	57.57%	84.21%
	PPV	53.02%	72.24%

Table 3. ANFIS Results on Cleansed RDU Data Using 3 Fuzzy Rules

CONCLUSIONS

Our experiments using tree-based rule induction as an initial starting point for further development of fuzzy rules in ANFIS indicate that significant improvements in the desired performance measures can be achieved for both types of breast cancer modeled. While missing no mass or calcification cancers, the application of three fuzzy mass rules was sufficient to garner improvements in specificity of 50% for RDU and 151% for the USF data set. Furthermore, the data cleansing which removed the mixed lesion category (approximately 130 cases) was extended to focus modeling approaches to the region of the feature space where discriminatory features showed better classification – mass mammographic structures. The results of ANFIS mass models suggest a foundation for splitting the data sets by mammographic structure in a statistical manner, to achieve the performance needed for a CAD system used as a viable decision support tool in a clinical environment. Taken as an intermediary step, modeling mass mammographic structures imparts a specificity improvement of 102% over the mass and calcification model, at 100% sensitivity (25.39% versus 12.56%). PPV values also surpass those of mammography (50 -57% versus 25 – 40%). Additionally, these results can be benchmarked against other CAD implementations by Markey (2002) which utilized the USF and RDU data, along with an additional data set from the University of Pennsylvania. Performances measured in terms of area under the ROC curve or A_z were reported for a Back-Propagation Network (BPN), Case-Based Reasoning (CBR), CART, Linear Discriminant Analysis (LDA) and Support Vector Machine (SVM). For brevity, considering only the results for the partial area under the curve in the critical 90% region or A_z 90, ANFIS yields superior performance by at least 29% when compared to the best performing model, the BPN (44.63% versus 34.7%) on both lesion features. CBR, CART, LDA, and SVM models yielded poorer A_z 90 results (32.40, 28.60, 26.10, 23.70, respectively). Similarly, the ANFIS model surpassed these CAD benchmarks for masses modeled separately, where an A_z 90 of 62.42% was achieved compared to 48.3, 46.3, 41.5, and 46.8 for the BPN, CBR, CART and LDA. Other paradigms modeled in Verheggen (2003) also did not surpass ANFIS results in specificity or ROC metrics. These include mass CAD models with specificities at 100% sensitivity, missing no cancer, which ANFIS at 25.39% surpassed by at least 68%, and include: an Evolutionary Programmed Neural Network with Adaptive Boosting (15.08%), a Generalized Regression Neural Network (9.52%), a Radial Basis Function Network (1.38%), and a Levenberg-Marquardt MLP (4.30%). Ongoing work focuses on additional benchmarking and the derivation of fuzzy rules for the more challenging calcification cancers, which have the higher percentage of benign–outcome biopsies.

REFERENCES

BIRADS, Breast Imaging – Reporting and Data System, *American College of Radiology*, 1993

Breiman, L.,J. H. Friedman, R.A. Olshen, and C.J. Stone, *Classification and Regression Trees*. CRC Press, 1998.

Haykin, S., *Neural Networks: a comprehensive foundation*, 2d Edition. Prentice Hall, 1999.

Jang, J.-S.R, C.-T.Sung, and E. Mizutani, *Neuro-Fuzzy and Soft Computing: A Computational Approach to Learning and Machine Intelligence*. Prentice Hall, 1997.

Karssemeijer, N., M.Thijssen, J. Hendriks, and L. van Erning, *Digital Mammography*. Kluwer Academic Publishers, Nijmegen, 1998, pp. 457-460.

Lewis, H. W., and M. -C. Yang, "Neural Genetic Fuzzy Models," *The Shogaku Ronshu*, vol. 66, no. 3, pp. 1-20, 1998.

Markey, M. K., *Modular Machine Learning Methods for Computer-Aided Diagnosis of Breast Cancer* PhD Thesis, Duke University, 2002

Nass, S. J., I. C. Henderson, and J. C. Lashof, National Cancer Policy Board – Institute of Medicine and National Research Council, *Mammography and Beyond: Developing Technologies for the Early Detection of Breast Cancer*. National Academy Press, Washington, DC, 2001.

Olson, J. S., *Bathsheba's Breast – Women, Cancer, and History*. Johns Hopkins University Press, 2002.

Salford Systems, *CART® Classification and Regression Trees*, [On-line] http://www.salford-systems.com,(accessed 12/5/2002).

Stolley, P. D., Lasky, T., *Investigating Disease Patterns: The Science of Epidemiology*, Scientific American Library, 1995.

Stone, M., Cross-Validatory Choice and Assessment of Statistical Predictions, *Journal of the Royal Statistical Society*, Series B, Vol. 36, Issue 2, pp. 111-147. 1974.

Verheggen, E. A., *Neuro-Genetic and Neuro-Fuzzy Machine Learning Paradigms for the Diagnosis of Breast Cancer*. Master's Thesis. Binghamton, NY: Binghamton University, 2003

An Integrated Simulated Annealing/Genetic Algorithm Approach to Synthesis of Digital Circuits

GARY GREWAL
Dept. of Computing and
Information Science
University of Guelph
Guelph, Ontario, Canada
N1G 2W1

Michael O'Cleirigh
Dept. of Computing and
Information Science
University of Guelph
Guelph, Ontario, Canada
N1G 2W1

ABSRACT
This paper presents a novel approach to the high-level synthesis problems of scheduling, module allocation, and binding for behavioral descriptions. A very general version of the problem is considered where modules may perform different operations in different numbers of control steps. The three inherently interdependent problems are solved using the hierarchical application of two meta-heuristics: Simulated Annealing (SA) and Genetic Algorithm (GA). A GA is used to select functional units from a hardware library, while an algorithm based on SA uses the allocations to perform scheduling and binding. The SA algorithm performs the role of fitness function for the GA. In this way, we achieve a truly integrated solution as both algorithms run concurrently and cooperate with each other until a satisfactory solution is found.

1. Introduction

High-Level synthesis of digital systems begins with a behavioral description of the target design. From this, a register-transfer level (RTL) description of circuit behaviour is derived. The behavioral description is transformed into intermediate graph-based representations depicting both control flow and data flow. The scheduling process assigns operations in the data-flow graph (DFG) to control steps. Module allocation selects specific numbers and types of modules from a module library to execute the operations specified in the DFG. Instance binding assigns constituent operations of the DFG to specific module instances.

It is well known that the problems of scheduling, module allocation, and instance binding are inherently independent. Thus, for guaranteed optimal results, all three problems must be considered simultaneously. However, the problem of scheduling alone is NP-complete. As such, most current approaches incorporate heuristic methodologies in an attempt to produce results in reasonable amounts of time. However, the quality of results varies an may not be optimal, since heuristic approaches often fail to address the interdependence between all three problems.

In contrast to heuristic approaches, Integer Linear Programming (ILP) formulations have proven effective in handling the interdependence between the three problems. Some formulations even support complex module libraries where modules can perform different operations in different numbers of control steps. However, their run-times become unacceptable as the problem size increases. (An excellent survey of recent developments in high-level synthesis can be found in [1].)

To compensate, we present an integrated solution to the HLS problems of scheduling, allocation, and binding based on the hierarchical application of two meta-heuristics: Simulated Annealing (SA) and Genetic Algorithm (GA). The main advantage of this hybrid search over other approaches is a reduction in time to obtain optimal solutions. We use two algorithms in a hierarchical manner to solve he problems in an integrated manner. A GA is used to select functional units from a hardware library, while an algorithm based on SA uses the allocations to perform scheduling and binding. The SA algorithm performs the role of fitness function for the GA. In this way, we achieve a truly integrated solution as both genetic algorithms run concurrently and cooperate with each other until a satisfactory solution is found.

2. Basic Concepts

We now begin with a closer look at the three problems of scheduling, allocation, and binding. In Fig. 1 a simple data-flow graph and hardware library are shown. The DFG contains five operations: 2 multiplication operations, 2 addition operations, and 1 subtraction operation. The hardware library contains three types of functional units, each with various execution times and areas. More than one instance of each functional unit may be chosen in the final allocation. Notice that all of the functional units present in the library are complex, allowing different operations to be performed in different numbers of control steps. The faster functional units also tend to consume more space (chip area) and thus cost more to instantiate. Here each functional unit is capable of performing each operation in the DFG; in general, this may not be true.

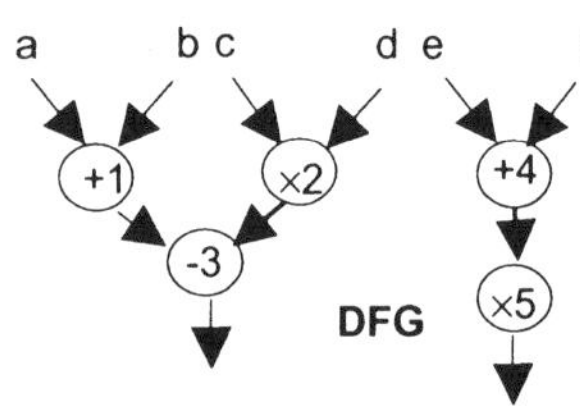

	Time Delay ×	(+,-)	Area
ALU 1	2	1	600
ALU 2	3	1	400
ALU 3	3	2	300

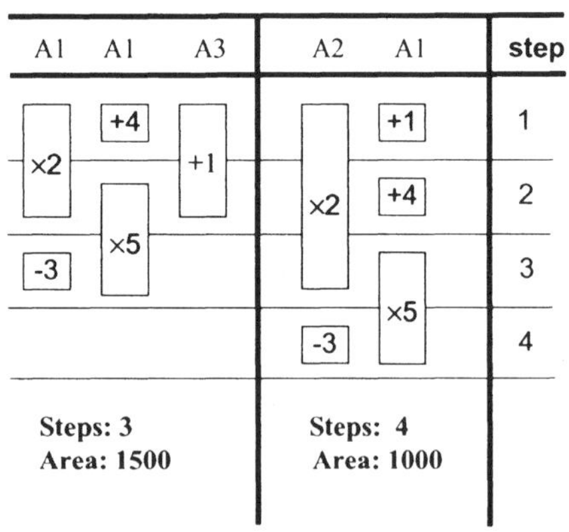

Figure 1: Data flow graph and module library.

Given a behavior specified as a DFG, as well as a hardware library, the problem is to find a suitable schedule such that: the operations are performed by the units having the required capability, a unit can execute at most one operation at a time, the operations are executed in the order specified by the DFG, the total area is limited, and the execution occurs with a number of steps that is limited.

Figure 1 also illustrates two different schedules and allocations for the behavior specified in Fig. 1. Each schedule incorporates different combinations of units. The short 3-step schedule requires 2 instances of the faster – but larger – multiplier (ALU1), resulting in a 1500 unit area design. By extending the schedule length to 4 steps, only 1 instance of the fast multiplier is required to satisfy the timing requirement using a smaller 1000 unit design. Notice how the issues of schedule and area restrictions, availability of units, and unit allocation and binding are all interdependent.

3. Our Approach

In practice, the total *time* and *area* available on a chip are often restricted to lie within some prespecified limits. Consequently, a designer is required to verify the existence of some feasible solution that meets both time and total area constraints; this requires no objective function at all. By progressing through a series of plausible time and area limits, the sequence of solutions clearly display the area/time tradeoffs to the designer. Our integrated approach to the scheduling, allocation, and binding problems is based on the hierarchical application of two meta-heuristics. They are hierarchical because the second meta-heuristic (SA) acts as the fitness function for the first (GA) as shown in Fig. 2.

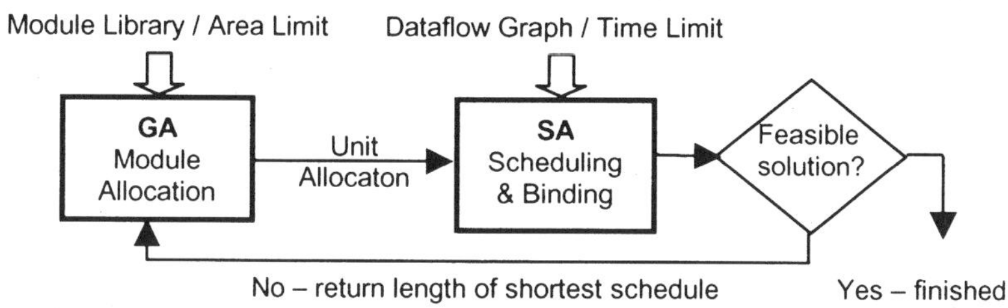

Figure 2: Hierarchical application of two meta-heuristics.

The genetic algorithm, GA, performs allocation; that is, it selects functional units from a general hardware library to implement the operations in the original specification. Each unit instance is assigned a unique unit number, and multiple instances of the same unit type can be selected. The only restriction is that the total area required for a given allocation cannot exceed the area limit specified by the designer.

The second simulated annealing algorithm, SA, performs scheduling and binding using the allocation provided by the GA. Thus, potential allocations provided by the GA are evaluated on the basis of their ability to satisfy the timing constraint imposed by the designer. If a schedule is found that satisfies the timing constraint, a feasible solution has been found and the process terminates. Otherwise, SA returns the length of the best (shortest) schedule found after a predetermined amount of time. The point is that both algorithms will continue to run concurrently and cooperate with each other until a solution is found that satisfies both timing and area constraints imposed by the designer (assuming, of course, the original constraints provided by the designer are feasible).

3.1 GA – Allocating Functional Units

We now take a closer look at the genetic algorithm used to allocate functional units. The total area available on a chip is restricted to lie within a prespecified limit. More formally, $\sum_m U_m A_m \leq AreaLimit$,where variable m indexes each unit instance being considered in the current allocation, A_m is the chip area consumed by including unit m

in the design, and U_m is a 0-1 variable. If $U_m = 1$, unit m is selected to appear in the final design.

We employ a bit-string representation where the values of all variables U_m are directly stored in a string of length n, where n is the maximum number of functional units the designer is willing to consider using. At first glance, this representation seems to be the most direct and easiest, but infeasible solutions containing constraint violations need to be considered. One way to do this is based on the application of a special repair algorithm to "correct" any infeasible solutions generated. For the allocation problem, this can be done by setting some U_m variable(s) to 0. The repair procedure is outlined in Fig. 3.

procedure repair (x)
begin
  area-exceeded = false
  $x' = x$
  if $\sum_{i=1}^{n} x'[i] \cdot A[i] > AreaLimit$
    then area-exceeded = true
  while (area-exceeded) **do**
  begin
    i=randomly **select** a unit and remove it from the current allocation: i.e., $x'[i] = 0$
    if $\sum_{i=1}^{n} x'[i] \cdot A[i] \leq AreaLimit$
      then area-exceeded = false
  end
end

Figure 3: The repair procedure.

The repair procedure transforms an infeasible allocation (x) into a feasible one (x') by removing (setting $x[i]'=0$) randomly selected units from the current allocation until the area constraint is satisfied. Thus, all candidate allocations generated by the GA are guaranteed to satisfy the area constraint imposed by the designer.

The remainder of the GA (not presented here) is based on the simple GA introduced by Holland [2], and runs with the following parameter settings: Population size = 20, crossover rate = 75%, and mutation rate = $1/m$.

3.2 SA – Scheduling and Binding Operations

The role of the simulated annealing algorithm is to schedule the operations in the data-flow graph under the precedence and resource restrictions implied by the GA. Scheduling operations using only the units provided by the GA involves making a choice as to the order in which operations can be executed and assigned (bound) to units. The problem is to find a schedule and binding that satisfies the timing constraint imposed on the final circuit by the designer.

The SA algorithm uses a modified version of List Scheduling [3] to construct feasible schedules. In List Scheduling, all "ready" operations of the DFG are placed into a list ordered by relative priorities and whenever a unit becomes available the operation with the highest priority in the list is chosen and assigned to that unit. An operation is considered "ready", if and only if all of its predecessors have completed executing. Figure 4 shows the general list-scheduling algorithm.

1. Assign a **priority** to each operation in the DFG.
2. Initialize a priority queue for ready operations by inserting every operation that has no immediate predecessors. Sort the operations in increasing order of priority.
3. While the priority queue is not empty do the following:
   a. Get the operation at the front of the queue.
   b. Select an idle unit to perform the task.
   c. After all immediate predecessors of an operation are executed, insert that operation's successor(s) into the priority queue.

Figure 4: List-Scheduling Algorithm.

It is clear from Fig. 4 that the particular assignment of priorities to operations (step 1) results in different schedules (with possibly different lengths) being generated. This is because operations are selected for execution in different orders. Various heuristics have been proposed for prioritizing the operations in the DFG. However, no one heuristic has emerged as the clear winner. Rather, recent results [4] suggest that the performance of a list scheduler can be improved by using a mixture of several heuristics.

In the approach presented here, we employ SA to determine the priority of the operations to be scheduled. This is a *permutation* problem by nature. We employ a priority-based encoding where the *position* of each operation in a vector is used to identify the operation to be scheduled and the *value* of each entry in the vector is used to denote the priority associated with the operation. This value is an integer exclusively within $[1,p]$, where p is the number of operations in the original DFG. The larger the integer (p), the higher the priority.

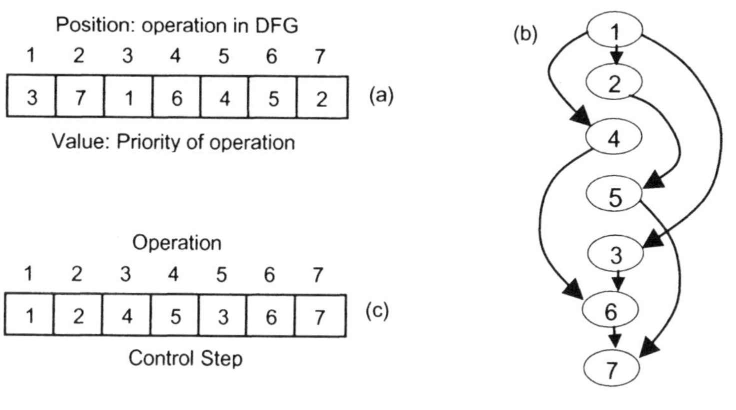

Figure 5: (a) Priority-based encoding; (b) dataflow graph; and (c) final schedule.

Figure 5(a) shows a possible (vector) encoding for the operations of the DFG in Fig. 5(b). If we assume the existence of a single functional unit capable of performing all of the operations in the DFG in a single control step, a feasible schedule can be constructed as follows. At the beginning, only operation 1 is ready to be scheduled and is therefore assigned to control step 1. Then three operations 2, 3, and 4 become ready and compete for the second control step. Their priorities are 7, 1, and 6, respectively. Operation 2 wins, because it has the highest priority. After scheduling operation 2 on control step 2, the candidates for control step 3 are operations 3, 4, and 5. Operation 4 wins and is scheduled on control step 3. The process simply repeats until all of the operations are scheduled. The final schedule is shown in Fig. 5(c).

In the previous example, a single unit capable of executing all of the operations in the DFG was used. Most designs, however, will include multiple units. It should be clear, therefore, that the decision regarding whether or not an operation can be scheduled not only depends on whether the operation's predecessors have executed, but also on the availability of a unit to perform the operation. Thus, a vector does not directly represent a schedule, but rather a way of resolving conflicts when more than one operation is ready for scheduling.

The SA algorithm we use here is based on the original SA algorithm by Fitzpatrick et. al. [5]. With SA, one proceeds from the current solution (state) to a neighbouring solution (state) by performing a *move* operation in the *neighborhood* of the current solution. For the scheduling problem discussed here, we define the neighborhood to consist of all solutions that can be reached by randomly selecting two vector values and exchanging them. This has the effect of changing (swapping) the priorities of two (randomly selected) operations. The remainder of the algorithm (not presented here) conforms to the original algorithm proposed by Fitzpatrick et. al., and runs with the following parameter settings: Initial temperature = 100, the number of times the temperature is lowered according to the cooling parameter = 25, the number of moves to make at a given temperature = 100, the rate at which the temperature will be cooled = 0.25, where we assume a geometric cooling schedule.

The SA algorithm terminates when a solution (schedule and binding) is found that satisfies the timing constraint imposed by the designer. If no such solution can be found, the SA algorithm terminates once the final temperature is reached and returns the length of the best (shortest) schedule found to the GA.

4.0 Results and Discussion

The previously described Hierarchical Algorithm (HA) is implemented in the *C* programming language in a UNIX environment. Using well known benchmarks, the HA will be compared to other related works, including the HAL system [6], SYMPHONY [7], TASS [8], and ADPS [9]. In selecting comparison systems, a wide variety of approaches have been chosen. HAL uses the popular force-directed scheduling heuristic, while SYMPHONY uses an integrated ILP formulation which is sensitive to the interdependence between the 3 problems. TASS uses a variant of simulated annealing. ADPS first uses force-directed scheduling to obtain a schedule; the resulting schedule is then used for allocation and binding, all within an iterative process that attempts to refine the design. All times given represent CPU time expressed in seconds.

4.1 The Elliptical Wave Filter

An important facet of any HLS systems is the ability to determine a lower bound performance for execution of a behavioral specification, for a fixed number of resources. This first example demonstrates the ability of the HA we are proposing to quickly minimize schedule length given a fixed bound on area.

The elliptical wave filter consists of 34 addition and multiplication operations. The hardware library contains two unit types: adders and multipliers. Each adder has a delay of 1 control step and an area of 100 units, while each multiplier has a delay of 2 control steps and an area of 250 units.

Initially, we limit area to 1050 units and seek a minimum time solution. Table 1 reveals that the HA found an optimal 17-step schedule, requiring 3 adders and 3 multipliers. When the area limit was reduced to 700 units the HA found an optimal 18-step schedule requiring 2 adders and 2 multipliers. When the area limit was further reduced to 450 units an optimal 21-step schedule was found requiring 1 multiplier and 2 adders. Finally, when the area limit was reduced to just 350 units an optimal 28-step schedule was found requiring 1 adder and 1 multiplier. In each case, the comparison with HAL, TASS and SYMPHONY show a reduction in time to obtain the same optimal results.

Table 1: Execution Times for Elliptical Wave Fitler.

Area Limit	Units Allocated	Control Steps	Time (s) HAL	TASS	SYMPHONY	HA
1050	*,*,*,+,+,+	17	120	10.0	2	0.015
700	*,*,+,+	18	240	10.1	233	0.020
450	*,+,+	21	360	10.2	Not Reported	0.005
350	*,+	28	480	Not Reported	Not Reported	0.001

4.2 Bandpass Filter

We now demonstrate the HA's ability to minimize area given constraints on execution time. For purposes of comparison, we choose the Bandpass Filter benchmark and complex unit library presented in [7], and compare the HA with SYMPHONY and ADPS. The DFG consists of 29 operations including multiplication, addition, and subtraction operations. The cumulative area of the units present in the library is 1620, with each unit having a delay of 1 control step.

Initially, time was limited to 8 control steps and a minimum area solution sought. Table 2 reveals that the HA and SYMPHONY found an optimal allocation (675), while ADPS produced a sub-optimal (685) solution. With the time limit relaxed to 9 control steps, the HA and SYMPHONY again found an optimal allocation (625), but ADPS could only find a sub-optimal solution (650). This trend continued for the 11 step schedule. In every case, HA and SYMPHONY produced better results than ADPS. However, the HA required less time than SYMPHONY to obtain equivalent (optimal) solutions.

Table 2 Execution Times for Bandpass Filter.

Control Steps	Model	Units Allocated	Area Constraint	CPU Time (s)
8	SYMPHONY	+*,+-,+-,*	675	4.8s
	ADPS	+-,*,*,+,-	685	9.0
	HA	+*,+-,+-,*	675	2.07
9	SYMPHONY	+-,*,*,+	625	17
	ADPS	+-,*,*,+-	650	64
	HA	+-,*,*,+	625	2.67
11	SYMPHONY	+*,+-,*	600	42
	ADPS	+-*,*,+,-	630	197
	HA	+*,+-,*	600	0.04

5.0 Conclusions

In this paper we have presented an integrated solution to the high-level synthesis problems of scheduling, allocation, and binding. Our approach is based on the hierarchical application of two meta-heuristics, simulated annealing and genetic algorithms, that handle all the interactions among the three subproblems. The solution permits the use of complex functional units and allows operations to be implemented by a variety of functional units, possibly requiring different execution times. Our results show a reduction in time to obtain optimal solutions to standard benchmarks compared with other systems.

In a straightforward way, our model can be extended to effectively use *pipelined* units or *chained* operations whenever there is an opportunity to do so. Furthermore, our model can also be extended to find optimal schedules and module allocations for multiple-block designs, not block-by-block or just along critical paths, but for all the blocks of a design simultaneously; we plan to publish the details of multi-block synthesis in a separate manuscript.

References

1. Y. Lin, "Recent developments in high-level synthesis," ACM Transactions on Design Automations of Electronic Systems, Volume 2, Issue 1, January 1997.
2. J. Holland, "Adaptation in Natural and Artificial Systems," University of Michigan Press, Ann Arbor, 1975.
3. A. Zomaya, Parallel and Distributed Computing Handbook, McGraw-Hill, 1996.
4. Auyeung, I. Gondra, and H. Dai, Multi-heuristic list scheduling genetic algorithm for task scheduling, *The Eighteenth Annual ACM Symposium on Applied Computing*, pp.721-724, 2003.
5. S. Kirkpatrick, C. Gelatt, and M. Vecchi, "Optimization by simulated annealing," Science, vol. 220, pp. 671-680, May 1983.
6. P. Paulin, High-level synthesis of digital circuits using global scheduling and binding algorithms, PhD Thesis, Carleton University, Ottawa, (January 1989).
7. G. Grewal, Scheduling, allocation, and binding in multiple-block synthesis, Master's Thesis, University of Guelph, Department of Computing and Information Science, Ontario, Canada, April 1993.
8. S. Amellal and B. Kaminska, Functional synthesis of digital systems with TASS, *IEEE Transactions on Computer-Aided Design Conference*, *13*, May 1994.
9. C. Papachristou and H. Knouk, A linear program driven scheduling and allocation method followed by an interconnect optimization algorithm, *Proceedings of the 27th European Design-Automation Conference*, pp. 193-199, 1992.

IMPROVED NETWORK DESIGN WHEN CONSIDERING s-t RELIABILITY USING AN EXACT RELIABILITY BLOCK DIAGRAM

MEHMET SAHINOGLU
Computer and Information Science
Troy State University Montgomery

ALICE SMITH
Industrial & Systems Engineering
Auburn University

BERNA DENGIZ
Industrial Engineering
Gazi University

ABSTRACT
Using an actual communications network design problem with eleven nodes at a major Turkish university in Ankara, three possible designs were previously developed by use of a memetic algorithm by Altiparmak et al. (2003) and an objective of maximizing all terminal reliability. This paper modifies one of those designs by considering s-node (source) to t-node (target) reliability. To choose which network paths should be reinforced for best s-t reliability, all node to node path reliabilities are calculated using a novel exact method which reduces any complex system to a simple series reliability block diagram as by Sahinoglu and Libby (2003). First, all possible paths are found and then trimmed of redundant or protruding component duplications. Then, the exact s-t availability is calculated. This calculation favorably compares with conventionally employed upper bound calculations with respect to path set and cut set formulations.

INTRODUCTION

Interest in considering reliability during design of computer communications networks with a large number of nodes and connecting links, such as those found in hospitals, universities, electricity distribution, gas pipelines, military or internet, has increased in recent years. Due to geographical and physical constraints in such critical systems, designers at the initial or improvement stages usually base their decisions on approximate or upper-bound estimates of reliability to compute a given ingress (source) to egress (target) reliability. This practice may be deceptive, erroneous and overly optimistic due to computational complexity when reliability remains of a crucial importance that requires human life and health. This paper shows a new exact s-t reliability calculation and its use in network redesign.

MOTIVATION

Conventionally, a system is defined as an entity with components, parts or blocks with lines of communications or links between them. An electric circuit, a piece of hardware or software or a managerial organization are examples of systems, which can best be described by a diagram illustrating the constituent parts and their interconnecting links, usually referred to as a reliability block diagram (RBD). With the two basic types of system structure, series and parallel, a series path works if all the constituent components work (intersection event) and a parallel path works when any one of the components operates (union event) using Boolean Algebra. This can be summarized as follows: Rel=Pr(system operates)=Pr(at least one successful path)=1-Pr(at least one cut failed). A minimal tie set or minpath contains paths where Rel= Pr(∪ T_i, i=1,..,m) consist of the smallest number of branches. Note that ∪ denotes "the union of". A minimal cut set (mincut) contains a minimal number of components which when removed from the system would cause a system failure, where Rel=1-Pr(∪ C_j, j=1,..,n).

However, the exponential size of enumeration of such tie sets or cut sets quickly becomes unmanageable as size (links and nodes) of the network grows (Colbourn, 1981; Aggarwal, 1981; Dengiz et al. 1997; Ball et al., 1977). All current exact computational algorithms for general networks are based on enumeration of states, minpaths or mincuts (Cavers, 1975). Hence, network reliability bounds, estimation or simulation is commonly used for non-trivial sized networks (Jan, 1993; Yeh et al., 1994; Fishman, 1986). The method used in this paper, Exact Reliability Block Diagram Calculation (ERBDC), is an exact calculation of s-t reliability but tractable for large networks. As an example of this method, a Java applet (Screen 1 of the Appendix) examines Example 6.3 of Figure 6.4 given on pp. 106-7 of *Reliability Modeling* by Linda Woltensholme, 1999. The node failures are all q=0.1 except for q_1=q_3=2q=0.2. (Note, links have zero failure probability for simplicity.) If q=failure probability for all components, then Pr(T_1 ∪ T_2 ∪ T_3) =Pr(164) + Pr(1234) + Pr(1254) - Pr(12364) - Pr(12654) - Pr(12354) + Pr(123654) = (1-2q) (1-2q+q^2) (1+q-2q^3) = 0.8 * 0.81* 1.098 = 0.711504, which can be observed in the array 1,4 of Screen 1. The ingress-egress relationship is also tabulated in Screen 1 by Polish, or postfix, notation where postfixes * and + denote two-at-a–time series and parallel components, respectively by Sahinoglu and Libby (2003). The upper-bound of system reliability is calculated by treating the three paths in parallel, Rel = 1-(1-0.0648)(1-0.5184)(1-0.5832) = 0.92934, as shown in Screen 1.

USING ERBDC TO CALCULATE NETWORK S-T RELIABILITY

Design 1 of the Gazi University network was identified by the memetic algorithm in Altiparmak et al. (2003) using all terminal network reliability, that is, the probability that all nodes can communicate with all other nodes. The topology of the Gazi University network is described in terms of its links and nodes later in the paper. The information of topology, link distances, the type of links, their costs, the type of servers and server costs are given in Table 1-4. The method is as follows:

In a parallel set composed of *i, j, k, l, m*... paths; at each item *i*, compress for each following item *j*. If *i* can combine with *j*; then do so, and remove *j*. If not, then keep it and then compress again with the next k^{th} path, until all of the parallel paths have been exhausted. Finally, one should have a single compressed path RBD from ingress to egress node. A line between two nodes is treated as a series component between the two, such as in Screen 1. Line -1 connecting the nodes 1 and 2, is expressed as in 1,-1,*,2,*. Two components in parallel are designated as 1,2, + and in series are denoted as 1,2,* by Sahinoglu et al. (2001, 2003). Considering path reliability, where the source (ingress) is node 5 and the target (egress) is node 9, the paths or tie-sets are as follows when all links are assumed to operate with full reliability of unity.

Path #	Reliability	Polish Notation
1	0.98210778	5, 1, *,9, *,
2	0.97621513	5, 1, *,10, *,9, *, +,
3	0.95874848	5, 2, *,3, *,4, *,11, *,1, *,9, *, +,
4	0.95299599	5, 2, *,3, *,4, *,11, *,1, *,10, *,9, *, +

The + signs at the end of each serial path denote that those paths will be combined in parallel to calculate the upper bound. Otherwise, each path has all its components connected in series denoted by a succession of * postfixes. One will take each path in ascending or descending sequence as convenient to compare and contrast as follows:

5, 1, *,9, *
5, 1, *,10, *,9, *

```
5 _____________1_____________9,  to merge with
               |-----10--------|
```

5, 2, *,3, *,4, *,11, *,1, *,9, *   to converge to:

```
5 _________________________1 _______________9, which will finally merge with
                           |----------10--------|
 |2----3-------11------------|
```

5, 2, *,3, *,4, *,11, *,1, *,10, *,9, *   to converge to:

```
5 _________________________1 _______________9
                           |------------10-----|
 |2--- 3-------11----------- |
        |--4----|
```

Since the straight line (with a reliability of unity as assumed) between nodes 5 to 1 and 1 to 9 dominates, the rest of the branches are ineffective. Therefore, the system reliability is the series connection of three nodes 5, 1 and 9; i.e., 5, 1, *, 9,* using Polish notation. That is, $0.994^3 = 0.982107784$. If links have non-unity reliability, then links have to be multiplied in series.

When the links have failures operating with non-unity reliability, then it is a different scenario. The following table is utilized to compose the finalized single-path RBD to calculate the exact reliability. Note, the negative digits denote links and sole digits show the nodes.

Path #	Reliability	Polish Notation
1	0.97622004	5, -1, *,1, *,-11, *,9, *,
2	0.95583166	5, -1, *,1, *,-13, *,10, *,-12, *,9, *
3	0.93687685	5, -2, *,2, *,-3, *,3, *,-4, *,4, *,-5, *,11, *,-6, *,1, *,-11, *,9, *
4	0.91731016	5, -2, *,2, *,-3, *,3, *,-4, *,4, *,-5, *,11, *,-6, *,1, *,-13, *,10, *,-12, *,9, *

The algorithm works as follows:

1) Take paths 4 and 3 from the reverse order usually choosing from longest to shortest. Enumerate those common elements, shown in series, in the center to branch out to those legs which are not common, shown in parallel to enable all paths successfully from the source (5) to the target (9) node. That is, merge

5 -2 2 -3 3 -4 4 -5 11 -6 1 -13 10 -12 9, with

5 -2 2 -3 3 -4 4 -5 11 -6 1 -11 9, to converge to:

```
                                   |-13  10  -12 |
5 -2  2 -3  3  -4  4  -5  11  -6  1               9
                                   |     -11     |
```

2) Then take the next path backwards 5 -1 1 -13 10 -12 9 and merge it with the RBD in **1)** by following the same rule of thumb:

```
                                   |-13  10  -12 |
5 -2  2 -3  3  -4  4  -5  11  -6  1               9,  with
                                     |   -11      |

5                 -1               1  -13  10 -12   9,   to converge to:

 |-2  2  -3  3  -4  4  -5  11  -6|  |-13   10  -12|
5|                                1               9
```

so as to enable the passage for all paths from 5 to 9.

```
 |              -1              |  |     -11     |
```

3) Finally take the last path in the reverse order to merge with the RBD in **2)**

```
 | -2  2  -3  3  -4  4  -5  11  -6|  |-13   10  -12|
5|                                 1               9,  with
 |              -1              |  |     -11     |

5               -1               1       -11      9,
```

to converge to the same tableau as in **3)** as the last path already existed in tableau of **2)**

```
 |-2  2  -3  3  -4  4  -5  11  -6|  |-13   10  -12|
5|                                1               9
 |              -1              |  |     -11     |
```

The final table can be expressed as in the following reliability block diagram

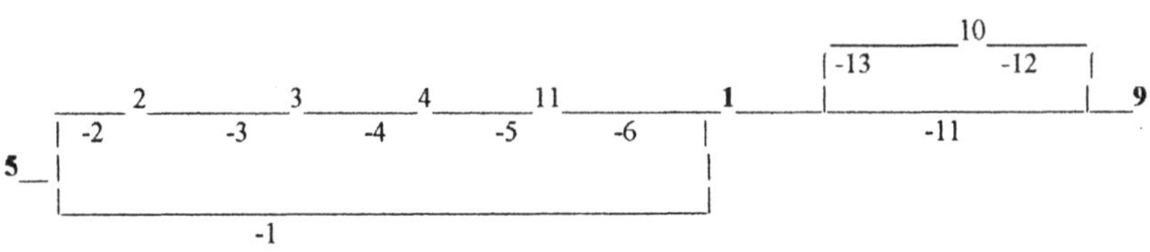

which results in an exact single-path reliability of 0.981940669581914, with its Polish Notation: 5, -1, -2, 2, *, -3, *, 3, *, -4, *, 4, *, -5, *, 11, *, -6, *, +, 1, *, -11, -13, 10, *, -12, *, +, *, *, 9, * that describes the above relationship in the final RBD.

APPLICATION TO AN ACTUAL NETWORK REDESIGN PROBLEM

The backbone network of Gazi University is located in downtown Ankara, Turkey, with nodes described in the text and links in Table 1. The discontinuous representation of some links represents that some facilities are located outside of the main campus. Considering the technological improvements in the telecommunications area and also the increasing load on its backbone network, Gazi University administration has decided to renew all links and equipment in order to increase performance, considering both cost and reliability. The geographical locations of the facilities require a "star" network topology, which is known to have poor reliability. In order to help overcome this weakness, some of the closer facilities will be connected by additional links. The network has eleven nodes and thirteen links with distances given in Table 1. While the dominant node choice is server type 1, links varied. Note the following nodes and attached buildings: Rector (1), Education (2), Science (3), Pharmacy (4), Technical Education (5), Law (6), Dentistry (7), Medicine (8), Economics & Administrative Sciences (9), Engineering (10), Fine Arts (11).

In this paper, the first recommended design (design 1 given in Table 3) of Gazi University backbone network was chosen to examine all s-t nodes reliabilities and all path reliability upper bounds by the newly proposed ERBDC method. This method will be very convenient for improvement purposes so as to reinforce the weakest connections. One can observe that the s-t pairs of (5-6), (5-8), (5-9), (6-11), (8-11), (9-11) have lower reliability than others. These (s-t) nodes and their weakest path are given in Table 4. After determining the nodes that have the weakest connections, the link types between them are upgraded as given in Table 4. As shown in Table 4, changing the type of the two links numbered with 12 and 13 from type 3 (with the lowest reliability) to type 1 (with the highest reliability) in the recommended design of the Gazi University backbone network, both s-t reliabilities and weakest path reliabilities are improved. All terminal reliability of the system is also improved at an additional cost of $54,400 = 1900m.($16) + 1500m.($16), where 1900 m is the distance of link 12, 1500m is the distance of link 13 *and $16 is the unit cost difference to replace type 3 link with type 1.* The new all terminal reliability of this reinforced network is 0.994254 whereas the design from Altiparmak et al. (2003) had a reliability of 0.9940.

Table 1. Links (negatives are redefined as in Screen 2) and their distances

Links Between Entities	Link Number	Distance (Meters)
1-5	1 (-1)	220
1-6	2 (-7)	2420
1-8	3 (-10)	700
1-9	4 (-11)	850
1-10	5 (-13)	2300
1-11	6 (-6)	120
2-3	7 (-3)	150
2-5	8 (-2)	350
3-4	9 (-4)	200
4-11	10 (-5)	450
6-7	11 (-8)	150
7-8	12 (-9)	1900
9-10	13 (-12)	1500

Table 2. Alternatives for fiber optic cables (links)

Type	Description	Reliability	Unit Cost*
1	SC to SC connector, multimode, PVC 62.5/125 duplex	0.9990	28
2	SC to ST connector, multimode, PVC 62.5/125 duplex	0.9950	17
3	ST to ST connector, multimode, PVC 62.5/125 duplex	0.9900	12

* Includes switch, router and connector costs.

Note the alternatives for servers as follows: Each server consists of Intel Pentium III processors with 512KB L2 Cache and 10000 RPM Ultra 160 SCSI Hard Disk Drives. Server 1 is configured with dual processor 1.4 GHz, 2GB (2@1-GB DIMMS), one 36GB disk drive, costing $5300 with reliability at 0.99995. Server 2 is configured with dual processor 1.4 GHz, 1GB (2@512-MB DIMMS), one 72GB disk drive, costing $4300 with reliability at 0.9909. Server 3 is configured with single processor 1.4 GHz, 512MB (2@256-MB DIMMS), one 36GB disk drive, costing $2800 with reliability at 0.9950.

Corresponding to the table below, the first original design investment amount is $218,635, with a reliability of 0.994 while the second improved design investment amount is 283,035 with a reliability of 0.994254. Note, D stands for Design type.

Table 3. The recommended first design and improved new design

D	1	2	3	4	5	6	7	8	9	10	11	12	13	1	2	3	4	5	6	7	8	9	10	11
1	1	3	2	2	3	1	1	2	1	3	1	3	3	1	1	1	1	1	1	1	1	1	1	1
2	1	3	2	2	3	1	1	2	1	3	1	**1**	**1**	1	1	1	1	1	1	1	1	1	1	1

Table 4. Sensitivity analyses for the weakest paths

s-t reliability for current design	Weak-est path #	Current recommended path design	Reliability of the weakest path	Link # to alter	Results s-t reliability	Results upgraded-path reliability
.981795	(5-6)4	8,7,9,10,6.3,12,11	0.91549497	12	.981881	.923817
.981907	(5-8)3	8,7,9,10,6,2,11,12	0.91089449	12	.981950	.919175
.981194	(5-9)4	8,7,9,10,6,5,13	0.91731016	13	.981984	.925649
.981795	(6-11)3	11,12,3,1,8,7,9,10	0.91549497	12	.981881	.923817
.981907	(8-11)1	12,11,2,1,8,7,9,10	0.91089449	12	.981195	.919175
.981940	(9-11)3	13,5,1,8,7,9,10	0.91731016	13	.981984	.925649

CONCLUSIONS

In this study the novel ERBDC method is introduced as a tool both to compute exact (s-t) reliability for a network, which is important in the design of even moderately sized networks, and to redesign an existing network. The work was first initiated by Sahinoglu (2000, 2001) and brought to its present graphically convenient and advanced stage.

The recommended first design of the Gazi University backbone communications network is analyzed and weakest links between certain nodes among all paths associated with specific (s-t) nodes are identified. This method allows designers to see and investigate all possible paths between all s-t nodes in a network as well as computing exact s-t reliabilities to make sensitivity analysis on any design obtained by a heuristic design algorithm.

Thus, designers can reinforce their designs considering both reliabilities and costs of all components and links in an existing network being redesigned. The proposed ERBDC method opens a new avenue to overcome the impasse of calculating the exact s-t reliabilities of very large complex systems, hence paving the way to conveniently and accurately redesigning and improving network reliability. This will consequently improve the quality of distributed network systems.

REFERENCES

Colbourn, C. J., 1981, "Combinatorial aspects of network reliability," *Annals of Operations Research*, Vol. R-30, pp. 32-35.

Aggarwal, K. K., and Rai, S., 1981, "Reliability evaluation in computer communication networks," *IEEE Transactions on Reliability*, R-30, No.1, pp. 32-35.

Cavers, J. K., 1975, "Cutset manipulations for communication network reliability estimation," *IEEE Transactions on Communications*, Com-23, No.6, pp. 569-575.

Jan, R. H., 1993, "Design of Reliable Networks," *Computers and Operations Research*, Vol.20, No.1, pp. 25-34.

Yeh, M. S., Lin, J. S. and Yeh, W. C., 1994, "New Monte Carlo Method for Estimating Network Reliability," *Proceeding of 16th International Conference on Computers and Industrial Engineering*, pp.723-726.

G.S. Fishman, 1986, "A Comparison of Four Monte Carlo Methods for Estimating the Probability of s-t Connectedness," *IEEE Trans. on Reliability*, Vol R-35,No 2, pp.145-155.

Sahinoglu, M, Libby D., 2003, "Sahinoglu-Libby (SL) Probability Density Function-Component Reliability Applications in Integrated Networks," *Proceedings Annual Reliability and Maintainability Symposium*, pp.280-287

Dengiz B., Altiparmak F., Smith A. E, 1997, "Efficient Optimization of all-Terminal Reliable Networks Using Evolutionary Approach," *IEEE Transactions on Reliability*, Vol.46 (3), No.1, pp. 18-26.

Dengiz B., Altiparmak F., Smith A. E, 1997,"Local Search Genetic Algorithm For Optimal Design of Reliable Networks," *IEEE Transactions on Evolutionary Computation*, Vol 1, No.3, pp. 179-188.

Altiparmak F., Dengiz B., Smith A., 2003, "Optimal Design of Reliable Computer Networks: A Comparison of Meta-Heuristics," Rev. for *Journal of Heuristics*.

Ball M., Van Slyke R. M., 1977, "Backtracking Algorithms for Network Reliability Analysis," *Annals of Discrete Math*, vol. 1, pp. 49-64

Woltenshome L.C., 1999, Reliability Modelling- A Statistical Approach, Chapman & Hall/CRC.

Sahinoglu M., 2000, "Reliability Index Evaluations of Integrated Software System (INTERNET) for Insufficient Software Failure and Recovery Data", *Lecture Notes in Springer Verlag, 1st International Conference*, ADVIS-2000, Izmir, Turkey, pp. 25-27.

Sahinoglu M., Munns W. 2001, "Availability Indices of a Software Network," *Proceedings of IX Brazilian Symposium on Fault Tolerant Computing*, Florianopolis, Brazil, pp. 123- 131.

ACKNOWLEDGEMENTS

The authors thank Justin Larson from TSUM, for his efforts toward the coding of the ERBDC Java software and type setting this paper.

Screen 1. Node to node reliability matrix, RBD and s-t reliability results for sample network

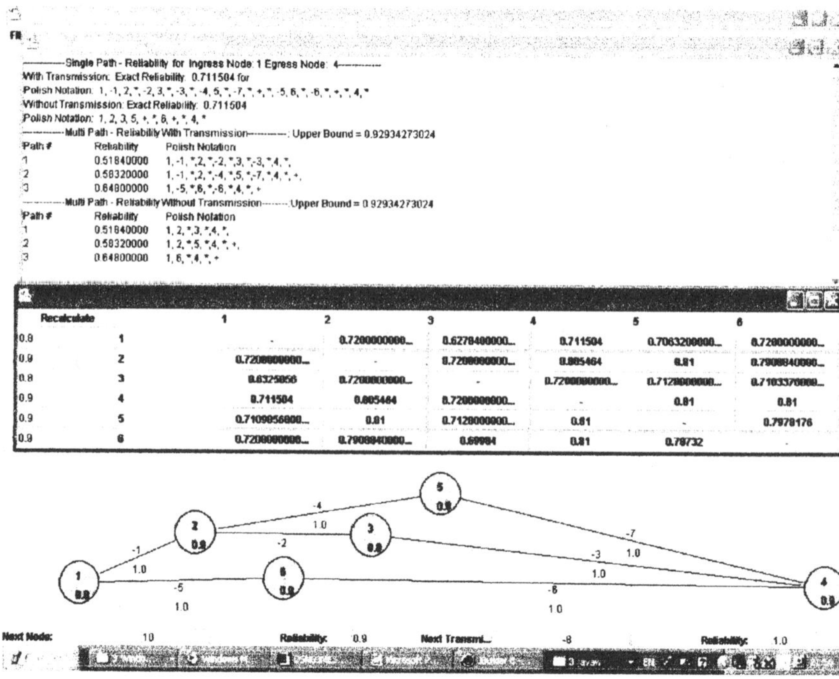

Screen 2. Node to node reliability matrix, RBD and s-t reliability results for the Gazi University network problem

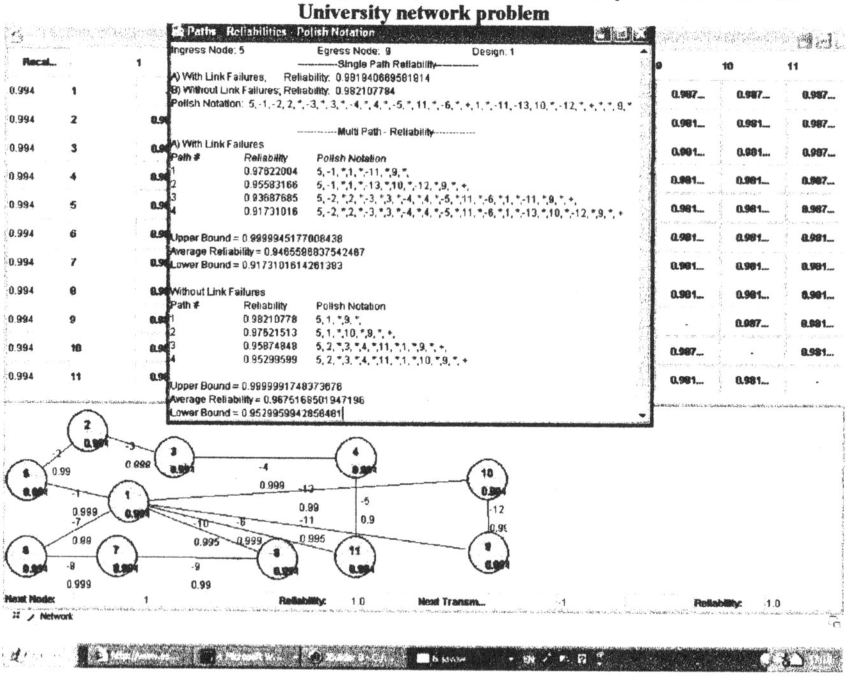

STRATEGIC PRODUCT PLANNING USING NEURAL NETWORKS

Dhananjay Trichinapally, Aravind Sankaran, Venkat Allada
Engineering Management Department
Sustainable Design Laboratory
University of Missouri-Rolla
Rolla, MO 65409

ABSTRACT

Strategic product planning is a challenging entrepreneurial task. It supplies input to product design and development and creates a business plan for profitability. One of the strategic product planning considerations is whether to develop a single product line or a product family consisting of several product variants. However at present, there are not many quantitative tools to determine whether to pursue the single product line option or a product family option. This paper proposes a four-force model that describes the factors influencing the product planning decision. Using the four-force framework, a neural network model has been developed to help strategic product planning.

1 INTRODUCTION

Global markets and shorter product cycles has made the strategic product planning a very challenging task (Gausemeier and Binger, 2001). The important strategic issues in establishing a new industry are as follows (Abell, 1980):

- Who is the customer?
- What products should be offered to the chosen customers?
- How can we offer products or services in a cost effective manner?

The issue of customer selection is based on various marketing factors such as market size, market type and growth potential (Montoya-Weiss and Calantone, 1994). The specific product portfolio offered by a company is based on factors such as financial capability, profitability, and demand. (Mcmath and Forbes, 1998, Freeman and Golden, 1997, Adler and Houghto, 1997). The issue of how to offer the product is based on the specific core competencies of the company (Barnett and Clark, 1996).

The success of establishing a new product is highly correlated with the above-mentioned strategic issues, which in turn, are dependent on the type of product introduction strategies namely, single product development and product family development. A single product design and development approach employs a distinct product for serving a specific market segment. A product family can be defined as a group of products that share similar features and functions and can easily be adapted to satisfy a wide variety of customer requirements or to target specific market niches (Allen, 2000). In the past decade, the aspect of discussion was on developing derivates for the product by using existing technology and minimizing the cost incurred in the development process. Result of this research has led to the development of various product family design strategies.

This paper investigates one of the main issues of strategic product planning i.e., selection of appropriate strategy. Specifically, we address the following issues

- Determination of various factors, which influence product planning decision-making.

- Using these factors in a neural net model to determine most appropriate design strategy to be followed by an industry.

2 TAXONOMY OF VARIOUS DESIGN STRATEGIES

Various product planning strategies have been described in the literature (Maier and Fadel, 2001, Yu, et al. 1998, Zamirowski and Otto 1999). These strategies could be defined using the following three attributes (Maier and Fadel, 2001):

1) Number of products: It can be defined as the number of different products, which belong to the same family and manufactured by a single manufacturer.

2) Degree of commonality: It can be defined as the number of common components or features that exists among products in the same product family.

3) Time of introduction: It is defined as the time required to introduce a new product to the market.

In the remainder of this section, eight types of product planning strategies proposed by Maier and Fadel (2001) are discussed. The product design strategies are represented in Figure 2 (a-d) (where each dot represents a single product). The product family design strategies (5-7) are represented in Figure 2(e-g).

2.1 Strategy 1: Single product design

A single product design strategy is defined as a strategy in which a distinct product is manufactured for serving a specific market segment. Refer to Figure 2 (a).

2.2 Strategy 2: Evolving single product design

A single evolving product can be defined as a product designed at one time and periodically updated. Refer to Figure 2 (b).

2.3 Strategy 3: Mutating single product design

A mutating single product design can be defined as a product that is designed at one time but is periodically updated into a new product. (i.e., to serve a different market niche than the preceding generation), Figure 2 (c) shows that for every new design increment, the product moves into a new market segment.

2.4 Strategy 4: Punctuated equilibrium in single product design

This strategy can be described as a combination of both mutating and evolving single product strategies. Refer to Figure 2 (d).

2.5 Strategy 5: Concurrent product family design

Concurrent product family designs are defined as a group of products that are designed at one time and are never intended to be modified in the near future. Refer to Figure 2 (e).

2.6 Strategy 6: Concurrent evolving product family design

Concurrent evolving product family designs can be defined as a group of products designed at one time and updated as a group periodically. Refer to Figure 2 (f).

2.7 Strategy 7: Mixed evolving/ mutating product family design

This kind of product family can be described by using the descriptions of mutating and evolving products. Here a family of products, which is not only evolving but also changing market, is a mixed evolving/mutating product family. Refer to Figure 2 (g).

2.8 Strategy 8: Customized products

Customized product design can be defined as a product design methodology in which products are designed to suit needs of an individual customer, where needs vary across the customers.

3 FOUR FORCES MODEL

The four-forces model proposed in the paper captures the various factors that influence the strategic product planning. Michael Porter (1980) provided a five forces model to depict the competitive forces that act on any business unit. We derive our basic inspiration from the Porter's model and attempt to create our own four forces model for the product planning strategy selection problem. We attempt to measure the four forces using linguistic variables. The four forces are as follows:

1) Core capability
2) Competition
3) Marketing
4) Product Characteristics

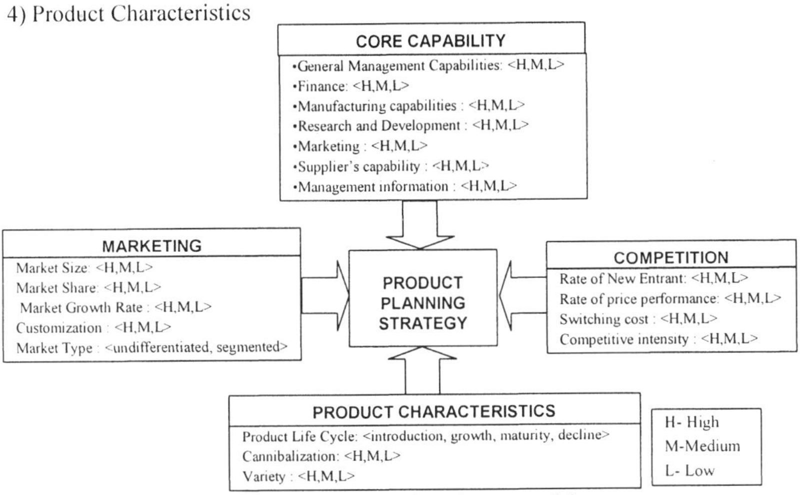

Figure 1: The four-forces model

3.1 Core capability

The core capability of a company can be described using the following factors:

- General management capabilities: Efficiency with which the management performs functions like production, planning, scheduling and implementation of manufacturing operations.
- Finance: Financial capabilities of an industry are described in terms of capital structure, liquidity, and cash flow of an industry.
- Manufacturing capabilities: Production capability of an industry and their flexibility in manufacturing different products.
- Research and development: Ability in the field of research and development, which helps in company's strategic planning.
- Management information: Ability of an industry to gather and process internal and external data to generate useful information to keep up with the changes in technology and market trends.
- Marketing: Ability of an industry to distribute, sell, advertise, and properly price the products offered.
- Supplier capability: Effectiveness with which the supplier network can provide raw material and services to the company.

3.2 Competition

The important factors related to competition are defined below:

- Rate of new entrant: Rate at which a new manufacturer would enter into the existing market.
- Rate of price performance: Ratio of cost and performance of the product varies among competitors due to variations in cost and quality of the products offered by different companies.
- Switching cost: Minimum price differential needed for a consumer to switch to competing products of some other company.
- Competition Intensity: Degree of competition in an industry as all the business competitors perceives it.

3.3 Market characteristics

The various market characteristics are briefly described below:

- Market size: Market size contains information about the size of a market, industry, product group, or commodity. Size may be measured in any of various ways such as the dollar value of sales, number of units produced/manufactured, value of items imported/exported, unit volume of shipment or transactions.
- Market share: Total sales of a given product for the entire market to the total of all sales of that specific product made by a company for a given period of time.
- Market growth rate: Rate at which the market size grows.
- Customization preference: Power given to the customer by virtue of which the customer can alter the degree of customization.
- Market type: Market type can be categorized into two broad categories, which are defined below:
  - i) Undifferentiated market: In this market type, all consumers have similar needs for a specific kind of product, or the customer demand is so diffused that it cannot be differentiated.
  - ii) Segmented market: Market segmentation is a process of dividing a total market into market groups consisting of consumers who have relatively similar product needs. Two types of market segments can exist:
    - a) Concentrated market: The industry looks into a single market segment. Example of a concentrated market is a car manufacturer focusing on sports cars.
    - b) Multi-segmented market: In this case, the industry looks at more than one market segment. Example of a multi-segmented market is a car manufacturer who manufactures sports cars as well as family cars.

3.4 Product characteristics

Some of the important product characteristics are as follows:

1. Product life cycle characteristics (PLC): The PLC characteristic determines the life cycle stage of a given product. There are four stages in PLC:
   - a) Introduction: It is a stage in which a new product is introduced to the market.
   - b) Growth: This refers to the period following the introduction stage during which the annual growth rate in sales exceeds some norm.
   - c) Maturity: The period following growth during which the annual sales rate does not fall more than some percentage (say 75 percent) below the maximum possible demand.
   - d) Decline: The period following maturity during which the annual sales fall more than some percentage (say 65 percent) below the maximum

2. Cannibalization: Cannibalization describes the degree of competition among products manufactured by the same manufacturer catering the same market.
3. Variety: Product variety describes the number of different models of the same product having a significant amount of market differentiation.

4 NEURAL NETWORK MODEL FOR PRODUCT PLANNING

Neural networks are distributed processing systems that can continuously improve their performance via dynamic learning (Hagan, 1996, Lin 1996, Zurada, 1992). Neural networks can provide a feasible approach to the design and implementation of pattern recognition based system for intelligent control and model identification. The product planning strategy selection problem can be modeled as a supervised learning input-output mapping problem, where the various factors influencing the product family selection strategy are mapped onto various design strategies.

4.2 Training and Data Set

Data was collected by analyzing products from 106 industries. For each of these industries, we looked at 19 different factors (described in the four forces model). All the factors except the market type and the product life cycle were described using variables such as high, medium or low corresponding to 1, 2 and 3 respectively. The three market types, viz, undifferentiated market, multi-segmented market and concentrated market were given values 1, 2, and 3 respectively. Similarly, the product life cycle stages, introduction, growth, maturity, decline were given values 1, 2, 3, and 4 respectively.

After compiling the dataset, a feature vector set was created for each industry by product type. Each feature vector set consists of 19 factors. Each feature vector was assigned a target vector representing the product planning strategy employed by the company.

4.3 Neural Network Specification

The feed forward back propagation architecture iteratively feeds the network with inputs and the weights are continuously updated depending on the error. The architecture used in this paper has one input layer, two hidden layers and one output layer. The hidden layers have 10 neurons each and the output layer has 8 neurons. The first or the input layer serves as a hold site for the input applied to the network. The last or the output layer is where the overall mapping of the network is available. The Levenberg Marqaurdt algorithm is used for the training purpose. The advantage of using the Levenberg Marquardt algorithm is to attain the second-order training speed without having to calculate the Hessian matrix. This helps in achieving a fast convergence. The network was trained for 2500 epochs. The number of epochs refers to the number of times the complete input patterns are presented to the network for training. Thus, the error 'e' represents the overall or accumulated correction to the weight set after each sweep of all vectors in the training set.

5 RESULTS

The neural network is trained using data set collected from the 106 industries representing various products. After training the network for the given data set, untrained data set consisting of 8 data points representing the four forces acting on a particular industry was presented to the neural network. In each of the 8 data sets, the neural network was able to classify the products into a product family. The network achieved 90% accuracy for the test data. The results indicate that each of the product families had unique characteristics, which was easily detected by the neural network. The input space was this mapped to the target space enabling the neural network to classify these product families into different categories based on the product planning strategy. This type of classification may be useful for companies contemplating to enter new markets or repositioning themselves in existing markets.

In order to illustrate and validate the results obtained, we have considered the Cannon and Olympus companies manufacturing cameras. Table 1 gives the values of 19 attributes for each of the manufacturer. The neural net gives the appropriate product design strategy based on the core competencies captured in the attribute vector.

Industry	General Mgmt.	Finance	Operations	R&D	Mgmt info.	Marketing	Supplier cap.	Cannibalization	Product variety	rate of new	Rate of price	Switching cost	Market size	Market share	Market growth	Competition	Life cycle	customization	Market type
Ca	H	H	H	H	H	H	H	H	H	L	H	L	H	H	M	H	Mt	L	D
Ol	H	H	H	H	H	H	H	H	H	L	H	H	H	L	M	H	Mt	L	D

Table 1: Test data vector for Ca - Canon and Ol - Olympus industries (H- high, L – Low, Mt – Maturity, D – Differentiation)

For the canon industry, a mixed evolving/ mutating product family design was suggested by the neural net whereas for the Olympus a concurrent evolving product family design was recommended. To validate these results obtained, we have looked into the history[1,2] of these industries and analyzed their product innovation strategies which confirm the output of the neural network.

6. CONCLUSIONS AND FUTURE WORK

We have proposed a four-forces model to address an important issue the product planning strategy. A back propagation network proposed in this paper is used to for the classification of product planning strategies. The data set data set from industries was synthetically constructed based on the limited information publicly available from various company websites. More real life data set will help in validating the result of this study. Also, data mining techniques could be used for determining the associated relationship between the strategic product planning decisions.

REFERENCES

Abell, D., 1980. Defining the Business: The starting point of Strategic planning. Englewood Cliffs, New Jersey: Prentice-Hall.

Adler, W., Houghton, J. 1997. America's Stupidest Business Decisions. William Morrow and Company Inc., New York.

Barnett, B.D., Clark, K.B., 1996. Technology newness: An empirical study in the process industry, J Tech Management, vol:13, pp.263-283.

---

[1] http://www.canon.com/camera-museum/history/

[2] http://www.star.ucl.ac.uk/~rwesson/esif/om-sif/camhistory/camhistory.htm

Beltratti, A, S., Margarita, Terna, P., 1996. Neural Networks for Economic and Financial Modeling, International Thomson Computer press, London, U.K.

Constantinos, M.,1998. Strategic Innovations in established companies Sloan Management review, vol: 39(3), pp.31-41.

Seipersad, C. C., and Hernandez, G,2000. A quantitative approach to determine product platform extent, DETC2000/DAC-14288, Baltimore, Maryland.

Demuth, Howard, Mark, B., Neural Network Toolbox, 2001 Math works

Freeman, Allyn, Bob Golden.1997. Why didn't I think of it? Johan Wiley & Sons, Inc., New York.

Gausemeier, J., Binger.V., 2001. Strategic product planning the development of products and markets of tomorrow as an entrepreneurial challenge. DETC 2001/CIE-21244, Pittsburgh, PA.

Krishnan, V and Gupta, S., 2001. Appropriateness and impact of platform-based product development, Management science, vol 47, No: 1.pp52-68.

Maier, J.R.A. and Fadel, G.M., 2001. Strategic Decisions in the early stages of product family design, DETC2001/DFM-21200, Pittsburgh, PA.

Martin, N., Mild. A., Feurstein, M., Dorffner, G., Taudes. A., 2001. The effects of incentive schemes and organizational arrangements on the new product development process, Management Science, vol.47, No. 8, pp.1029-1045.

Mcmath, M. R., Thom, F., 1998. What were they thinking? Times Business-Random House, New York.

Montoya, W., Mitzi, M., Roger C., 1994. Determinants of new product performance: A review and meta-analysis. Journal of Product Innovation Management, vol: 11, pp. 397-417.

Paul, A, G., 1999. Early Warning of New rivals. Sloan Management review, vol: 40, No: 3, pp.107-116.

Porter, M.E., 1980. Competitive Strategy: Techniques for Analyzing Industries and Competitors (New York: The Free Press).

Zamirowski, J. E and Otto, .K. N., 1999. Product portfolio architecture definition and selection, ICED1999, Munich.

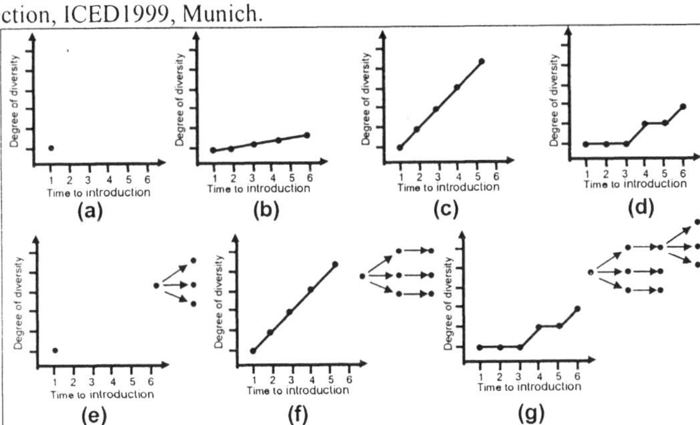

Figure 2: Various design strategies

RECEIVER OPERATING CHARACTERISTIC CURVES AND CLASSIFIER FUSION

JUSTIN M. HILL
United States Air Force Academy
Dept. of Mathematical Sciences
USAF Academy, Colorado

MARK E. OXLEY
Air Force Institute of Technology
Dept. of Mathematics and Statistics
Wright-Patterson AFB, Ohio

KENNETH W. BAUER
Air Force Institute of Technology
Dept. of Operational Sciences
Wright-Patterson AFB, Ohio

ABSTRACT
A classifier typically has parameters that can vary; thus a classifier is, in fact, a family of classifiers. Varying the parameters and graphing the probability of false positive verse probability of true positive yields the Receiver Operating Characteristic (ROC) curve. Given a collection of classifiers and their corresponding Receiver Operating Characteristic (ROC) curves, can the ROC curve for the fused classifier be determined? We address this question in this paper. This paper also presents a unique representation of the fusion of multiple classifiers where the fusion is based on Boolean rules.

INTRODUCTION

Receiver operating characteristic (ROC) curves are commonly used to summarize the performance of imperfect diagnostic systems, especially in automatic target recognition (ATR) and in biomedical research, when classification accuracy alone is not sufficient. A ROC curve is the graph of a relation which summarizes the performance of a signal detection system faced with the task of detecting a signal (target) in the presence of background noise (clutter). ROC curves usually relate the detection or "hit" rate (i.e., probability of true positive (P_D)) to the false alarm rate (i.e., probability of false positive (P_{FA})) as an internal decision threshold is varied. For a typical ROC curve, the decision threshold is varied from a very conservative value, i.e., a value that results in zero detection rate and zero false alarm rate, to a very aggressive value resulting in 100% detection rate and 100% false alarm rate.

These values are actually conditional probabilities such that P_D is the probability of correctly concluding that a target is present and P_{FA} is the probability of incorrectly concluding that a target is present. ROC curves have their foundation in statistical decision theory (Wald, 1947) and were originally developed as tools for electronic signal detection (Peterson, 1954). ROC analysis has been extensively applied to human perception and decision-making problems (Green, 1966) and is also commonly used in biomedical research (Metz, 1986; Metz, 1989). Alsing (2000) provides a comprehensive review of the use of ROC curves in ATR and medical research. For an in-depth technical discussion of ROC curves, consult Egan (1975) and Swets & Pickett (1982).

Consider a collection of classifiers, a multiple classifier system (MCS), in which the individual decisions are to be combined to make one overall decision. Given that one knows the ROC curves of the individual classifiers and the fusion rule, can the ROC curve of the fused classifier be determined without further experimentation? First we define some matrices that can be used to quantify the performance of a single classifier. Then we explain how to use these matrices to estimate the probabilities of all possible states for an MCS. Then, with a fusion rule and an overall system goal, we can use this information to estimate the ROC curve for the system. Reference (Hill, 2003a) and (Hill, 2003b) for a more in-depth discussion of this work.

SINGLE CLASSIFIER PERFORMANCE

Conditional Performance Matrix

One can summarize the performance of a classifier operating at a particular decision threshold θ in terms of the conditional probabilities $P_D(\theta)$ and $P_{FA}(\theta)$ by recording these values and their respective complements in a matrix equivalent to the "performance matrix" defined by Ralston (1999). For specificity, we call this matrix the Conditional Performance Matrix (CPM). One might also consider a CPM in which $P_D(\theta)$, $P_{FA}(\theta)$ and their complements are functions of θ rather than simply values for a given θ. CPMs would normally be computed using empirical evidence. To be consistent, the first row should correspond with the *clutter* label, and the last row should correspond with the *target* label. Similarly, the first and last columns should correspond with instances of clutter and targets, respectively. The set of CPMs for all θ can be used to construct the ROC curve for that classifier. We denoted the CPM for a particular classifier k with C_k.

Prior Probabilities Matrix and Joint Performance Matrix

One can compute the joint probability of two events (A and B) by multiplying the conditional probability of event A given event B by the *a priori* probability of event B. In the context of the detection problem, the joint probability of detecting a target (A) when a target is present (B) can be computed by dividing P_D by the *a priori* probability of seeing that type of target P(B). Thus, one can multiply the columns of the CPM by appropriate *a priori* probabilities to arrive at a matrix similar to the CPM, but with joint rather than conditional probabilities. We arrange these *a priori* probabilities in a diagonal matrix we call the Prior Probabilities Matrix (PPM). The result of post-multiplying the CPM by the PPM is the Joint Performance Matrix (JPM). We denote the PPM and JPM for a classifier k with ρ_k and J_k, respectively.

CLASSIFIER SYSTEM PERFORMANCE

We define similar (larger) matrices for an MCS. Each cell in these new matrices also indicates the probability of a certain event defined by the intersection of a particular row with a particular column. In these matrices, each row corresponds with a combination of output labels from the individual classifiers in the form of ordered pairs for a two-classifier system, ordered triplets for a three-classifier system, etc. Each column corresponds with a combination of possible true events (e.g., no targets are present, target A is present but target B is not, etc.). There is one row for each possible combination of output labels and one column for each possible combination of true events. Thus, the cells in these larger matrices give the probabilities of all possible states that the system might enter.

The Conditional State Probabilities Matrix (CSPM) is one of these larger matrices. Each cell in the CSPM gives the conditional probability of a combination of labels given a combination of true events (e.g., the probability that the system concludes that target types A and B are present but C is not given that all three types of target are actually present), and the CSPM is analogous to the CPM. When a CSPM is post-multiplied by the Combined Prior Probabilities Matrix (CPPM), a diagonal matrix containing the a priori probabilities of each possible combination of true events, the result is a Joint State Probabilities Matrix (JSPM). The JSPM gives the joint probability of a combination of labels coinciding with a combination of true events (e.g., the probability that the system concludes that target types A and B are present but C is not when all three types of target are actually present) and is denoted S_J. The CPPM and JSPM are analogous to the PPM and JPM, respectively.

Oxley and Bauer defined the difference between fusion *within* and *across* target types. A system of classifiers that are fused *within* is a system in which all classifiers are trained to detect a particular type of target. Thus, they share the same prior probability of detection. Moreover, there are only two possibilities for *truth* in such a system. Either the target is present, or it is not. A system that is fused *across* target types includes classifiers trained to detect a number of different types of targets. Each of these types of targets may have a different prior probability of detection, and since the system seeks different types of targets, it can *accidentally* arrive at the correct conclusion if a classifier seeking target type A incorrectly detects a target when a target type B is present. For reasons such as these, an *across* system may be more difficult to analyze than a *within* system.

Decision Fusion *Across* Target Types

If one assumes statistical independence between the individual classifiers, one can easily compute the CSPM of an MCS fusing *across* target types by taking the Kronecker product of each of the CPMs.

$$CSPM = C_1 \otimes C_2 \otimes \ldots \otimes C_n \qquad (1)$$

One can also compute the CPPM with the Kronecker product of the PPMs.

$$CPPM = \rho_1 \otimes \rho_2 \otimes \ldots \otimes \rho_n \qquad (2)$$

Some fundamental properties of the Koneke product are given in (Graham, 1981). Van Loan notes the widening use of the operation and lists some areas where Koneke product research is thriving: signal processing, image processing, semi definite programming, quantum programming, and fast Fourier transform (Van Loan, 2000). It is worth noting that the Koneke product lacks the commutative property; therefore, the order in which the Koneke product is applied to the Cams should be the same order applied to the Pomes. One can find the JSPM with the following formula:

$$S_J = (CSPM)(CPPM) = (C_1 \otimes C_2 \otimes \ldots \otimes C_n)(\rho_1 \otimes \rho_2 \otimes \ldots \otimes \rho_n). \qquad (3)$$

Decision Fusion *within* Target Types

Again assuming statistical independence, one can also use the Koneke product to compute the CSPM of a system fusing *within*. While there may be many possible scenarios in an *across* system, there are only two possibilities in a *within* system (i.e., target type A is present or not). Thus, the CSPM for a *within* system will have only two columns. A simple way of removing the extraneous (and meaningless) columns is to post-multiply the Koneke product of the Cams by an appropriately dimensioned matrix ($d \times 2$) containing ones in the (*1, 1*) and (*d, 2*) cells. The result is a $d \times 2$ matrix consisting of the first and last columns of the Kronecker product term.

In a *within* fusion system, computation of the CPPM is trivial. Note that the CSPM has only two columns. Therefore, one needs to post-multiply the CSPM by a 2×2 CPPM to arrive at a properly dimensioned JSPM. Since each classifier seeks the same type of target, the *a priori* probability of the system being in the *target* state is the same as the *a priori* probabilities for all the classifiers in the system. Thus, the 2×2 matrix required is simply the PPM shared by the individual classifiers in the system. The JSPM could then easily be computed by post-multiplying the CSPM by the CPPM.

$$S_J = (C_1 \otimes C_2 \otimes \ldots \otimes C_n)\begin{pmatrix} 1 & 0 & \cdots & 0 & 0 \\ 0 & 0 & \cdots & 0 & 1 \end{pmatrix}^T \rho \tag{4}$$

SYSTEM CONDITIONAL AND JOINT PERFORMANCE MATRICES

Truth Matrix (T)

Recall that the last column of each JPM and CPM corresponds with the presence of a target. Consequently, all but the first column of the JSPM correspond with the presence of at least one type of target. Note that the first column gives the probabilities of certain labels given that no targets are present. If the goal of the MCS is to determine if *any* targets are present, we can add together the remaining columns to arrive at the probability of a target being present under each possible label set. We accomplish this by post-multiplying the JSPM by a "truth matrix." The truth matrix will usually have two columns and an appropriate number of rows. The *(1, 1)* cell will contain a one, and the remaining cells in the first column will be zero. The second column will be the complement of the first column. After multiplying, the result is a $d\times2$ matrix in which the rows correspond with each possible combination of labels, the first column corresponds with the absence of targets, and the second column corresponds with the presence of at least one target.

$$T^T = \begin{pmatrix} 1 & 0 & 0 & 0 & 0 & 0 & 0 & 0 \\ 0 & 1 & 1 & 1 & 1 & 1 & 1 & 1 \end{pmatrix} \tag{5}$$

Fusion Rule Matrix (R)

Recall that a logical fusion rule selects combinations of labels for which the system concludes that a target is present. For example, a system using the *OR* rule will conclude that a target is present if any of the classifiers detect a target. Thus, the *OR* rule corresponds with all rows of the JSPM except the first. All other Boolean rules can be defined in this manner.

If we want to determine the probability of identifying a target under a particular fusion rule we can add the appropriate cells from the second column of the matrix resulting from post-multiplying the JSPM by the Truth Matrix. If we want to determine the probability of identifying a target when none exists we can add the corresponding cells from the first column. Note that the *OR* rule corresponds with all label combinations but the one represented by the first row of the JSPM. Now we can define a fusion vector as a column vector of zeros and ones, the ones corresponding to the rows appropriate for that rule. The fusion vector for the OR rule for our example contains a zero in the first cell and a one in all other cells (i.e., $r_{OR}^T = (0,1,1,1,1,1,1,1)$). This fusion vector is similar to the "rule of engagement" defined by Ralston (1999) and the rule vectors defined by Liggins (2000).

Pre-multiplying the previous result by the transpose of the fusion vector (to preserve dimensionality) gives a vector containing the probabilities of correctly identifying a hostile target and misclassifying a friendly object; and pre-multiplying by the complementary vector, gives the probabilities of misclassifying a hostile target and correctly identifying a friendly object. If one augments the fusion vector with its complimentary vector, the result is a fusion rule matrix.

$$R_{OR}^T = \begin{pmatrix} r_{OR} \\ \bar{r}_{OR} \end{pmatrix} = \begin{pmatrix} 0 & 1 & 1 & 1 & 1 & 1 & 1 & 1 \\ 1 & 0 & 0 & 0 & 0 & 0 & 0 & 0 \end{pmatrix} \tag{6}$$

System Joint Performance Matrix (S_J)

If the classifiers are statistically independent, the system JPM can be computed with the formula

$$J_S = R^T S_J T \text{ ,} \tag{7}$$

where S_J, R and T are computed with the formulas presented earlier.

System Prior Probabilities Matrix and Conditional Performance Matrix

Recall that the JPM for a classifier can be computed by multiplying the CPM and the PPM. Thus, one could post-multiply the system JPM by the inverse of the system prior probabilities matrix to compute the system CPM, but the system PPM has not yet been computed. Recall the CPPM matrix, and note that each row/column of the diagonal CPPM corresponds with a particular combination of true events defined by the columns of the CSPM and JSPM. All but the first column correspond with any event where a target is present, and the first column corresponds with instances where no target is present. Pre- and post-multiplying by the truth matrix give the sums of appropriate mutually exclusive probabilities that correspond to the appropriate probabilities summarized in the PPM. One can now compute the system CPM by inverting the PPM and post-multiplying the JPM by the result.

ESTIMATING ROC CURVES

Logical AND and OR

One can employ Oxley and Bauer's method (Oxley, 2002) to analytically estimate the optimal system ROC curve for an MCS of any size if the system outputs only two labels (*target* or *clutter*). The key to their formula for the *AND* rule was the observation that, under an *AND* rule, the system probability for assigning a *target* label is equal to the product of the probabilities that each of the individuals assign a *target* label.

$$\Pr\{t_S\} = \Pr\{t_1\} \cdot \Pr\{t_2\} \cdots \Pr\{t_n\} \tag{8}$$

Maximizing both sides of the separable equation with respect to the individual threshold values and manipulating the result allows one to derive a formula for the maximum P_D for a given P_{FA}.

Oxley and Bauer derived a similar formula for the *OR* rule. The key to this formula was the observation that, under an *OR* rule, the system probability for assigning a *clutter* label is equal to the product of the probabilities that each of the individuals assign a *clutter* label.

$$\Pr\{c_S\} = \Pr\{c_1\} \cdot \Pr\{c_2\} \cdots \Pr\{c_n\} \tag{9}$$

Minimizing both sides of the separable equation with respect to the individual threshold values and manipulating the result gives a formula for *min* ($-P_{D,S}$), which is equivalent to *max* $P_{D,S}$.

Other Boolean Rules

The ROC Fusion method can also be used to estimate ROC curves for Boolean rules other than the simple *AND* and *OR* under certain conditions. Specifically, it can be implemented in a piecewise manner for any rule that can be represented with a *simple* Boolean sentence (i.e., one in which the decision for each classifier is listed only once). For classifiers A, B and C, the rule "*A* or (*B* and *C*)" (A is a dominant sensor) is a simple sentence, but "(*A* and *B*) or (*A* and *C*) or (*B* and *C*)" (majority vote) is not.

CONCLUSIONS

The primary contribution of this paper is a matrix algebraic formula for computing the Conditional and Joint Performance Matrices for a Boolean Multiple Classifier System. This work is also the first known use of the Kronecker product in evaluating classifier system performance. Also presented were definitions and/or derivations for the following: Conditional Performance Matrix (CPM), Prior Probabilities Matrix (PPM), Joint Performance Matrix (JPM), Combined Prior Probabilities Matrix (CPPM), Conditional State Probabilities Matrix (CSPM), Joint State Probabilities Matrix (JSPM), Truth Matrix, and Fusion Rule Matrix. Furthermore, the ROC fusion method was presented for estimating an optimal ROC curve for classifier system using the system CPM. The individual CPMs were used previously to determine optimal fusion rules (Ralston, 1999); however, that work did not take into account the possibility of varying the decision thresholds for the individual classifiers, nor did that work provide a methodology for analyzing systems fusing *across* target types.

REFERENCES

Alsing, S. G. *The Evaluation of Competing Classifiers.* Ph.D. Dissertation, Air Force Institute of Technology, Wright-Patterson AFB, OH, March 2000.

Egan, J. P. *Signal Detection Theory and ROC Analysis*, Academic Press, New York, 1975.

Graham, A. *Kronecker Products and Matrix Calculus*, Ellis Horwood Limited, Chichester UK, 1981.

Green, D. M. and Swets, J. A. *Signal Detection and Psychophysics*, John Wiley & Sons, New York, 1966.

Hill, J. M. "Evaluating the Performance of Multiple Classifier Systems: A Matrix Algebra Representation of Boolean Fusion Rules," Master's Thesis, Air Force Institute of Technology, Wright-Patterson AFB, OH, March 2003.

Hill, J. M., Oxley, M. E. and Bauer, K. W. "Evaluating the Fusion of Multiple Classifiers via ROC Curves," in *Proceedings of SPIE AeroSense*, **5096**, 2003.

Liggins, M. E., Nebrich, M. A., Berlin, M. S. and Lazaroff, M. "Adaptive Multi-Image Decision Fusion Analysis," 2000.

Metz, C. E. "ROC Methodology in Radiologic Imaging," *Investigative Radiology* **21**, pp. 720-733, 1986.

Metz, C. E. "Some Practical Issues of Experimental Design and Data Analysis in Radiological ROC Studies," *Investigative Radiology* **24**, pp. 234-245, 1989.

Oxley, M. E. and Bauer, K. W. "Classifier Fusion for Improved System Performance," AFIT/ENS Working Paper 02-02, Air Force Institute of Technology, Wright-Patterson AFB, OH, January 2002.

Ralston, J. M. "Bayesian Sensor Fusion for Minimum Cost ID Declaration," Tech. Rep. IDA Paper P-3441, Institute for Defense Analyses, June 1999.

Peterson, W. W., Birdsall, T. G., and Fox, W. C. "The Theory of Signal Detectability," *Tranactions of the IRE Professional Group on Information Theory* **PGIT-4**, pp. 171-212, 1954.

Swets, J. A. *Signal Detection and Recognition by Human Observers*, John Wiley & Sons, New York, 1964.

Swets, J. A. and Pickett, R. M. *Evaluation of Diagnostic Systems: Methods from Signal Detection Theory*, Academic Press, New York, 1982.

Van Loan, C. F. "The Ubiquitous Kronecker Product," *Journal of Computaional and Applied Mathematics* **123**, pp. 85–100, 2000.

Wald, A. *Sequential Analysis*, John Wiley & Sons, New York, 1947.

ON-LINE PREDICTION OF STOCHASTIC SYSTEMS

M. A. FKIRIN
Professor of Automatic Control and Computer Systems
Department of Industrial Electronics and Control Engineering,
Faculty of Electronic Engineering , Menoufia University,
Menouf , Egypt.

ABSTRACT

This paper describes the application of the exponential-forgetting least-squares algorithm in prediction of stochastic systems. It is shown to be exponentially stable. The validity of the algorithm is confirmed by comparing the experimentally observed solar circuit response with the predicted values. The results are of interest for control design of real systems.

INTRODUCTION

In recent years a remarkable research effort has been directed towards the study of the prediction of stochastic systems (Ljung and Gunnarsson 1990, Hui and Zak 1993, Weigend and Gershenfeld 1994 , Benveniste et al. 1998) , and the dynamic performance of solar energy systems (Duffie and Beckman 1997). A prediction model should be judged solely on the accuracy of its predictions.

Solar thermal energy systems operate by allowing solar radiation to heat a solar collector panel, which is continuously cooled by a circulating fluid. This fluid gives up its energy to a thermal storage unit which is tapped as required by the user. The storage unit (boiler) is a tank of water to which the fluid transmits its energy by mixing. The solar energy system (shown in figure 1) is designed for collector testing and/or thermal transient evaluation. The energy is achieved from the solar heated storage tank by refilling the store with an equal amount of cold water.

The units DT1 and DT2 represent a differential thermostat controlling the circulation of liquid through the solar collector panel and an auxiliary electronic heater, respectively. The other control is due to the operation of a

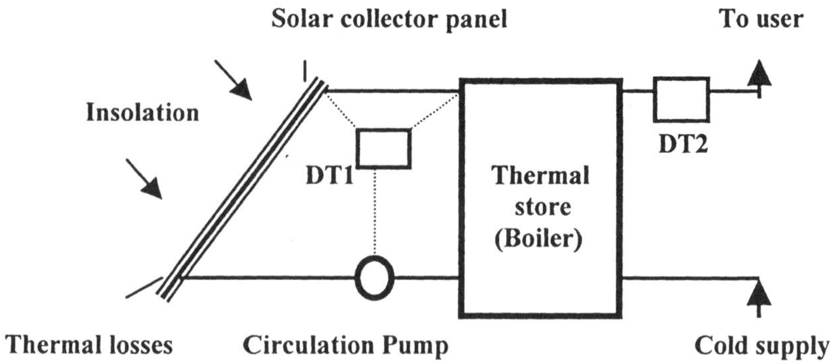

Figure 1 Experimental solar water heating system

differential thermostat between the panel and the store, for it is only advantageous to operate the panel when to do so would result in a net energy gain at the store. This is because the panel is in two-way thermal communication with the environment- gaining by radiation and losing by convection/ conduction.

SYSTEM MODEL

Consider the discrete-time stochastic auto-regressive moving average ARMA model:

$$A(z^{-1})\ y(k) = B(z^{-1})\ u(k) + C(z^{-1})\ e(k) \tag{1}$$

where u(k) and y(k) are the input and output at sample time k. e(k) is a sample from a zero-mean white sequence un-correlated with the input, with variance δ^2. $A(z^{-1})$, $B(z^{-1})$, $C(z^{-1})$ are polynomials in the unit backward shift operator z^{-1}.

Equation (1) can be written as an observation form (Fkirin 1985):

$$y(k) = h^T(k)\ x(k) + e(k) \tag{2}$$

with a state p-vector of system and noise parameters

$$x^T(k) = [\ a_1(k)\ \ a_2(k)\ \ldots\ a_{na}(k)\ |\ b_0(k)\ \ b_1(k)\ \ldots\ b_{nb-1}(k)\ |\ c_1(k)\ \ c_2(k)\ \ldots\ c_{nc}(k)\ ] \tag{3}$$

and observer vector

$$h^T(k) = [\ -y(k-1)\ \ -y(k-2)\ \ldots\ -y(k-na)\ |\ u(k)\ \ u(k-1)\ \ldots\ u(k-nb+1)\ |\ e(k-1)\ \ e(k-2)\ \ldots\ e(k-nc) \tag{4}$$

where p is na+nb+nc. Equation (2) represents a time-varying stochastic linear model.

The parameter variation at time k can be given by:

$$x(k) = x(k-1) + \gamma\, w(k) \tag{5}$$

where w(k) can be arbitrary or centered random variables with a weak dependence condition. The noise-excitation matrix γ is usually taken to be unit matrix in the absence of better information.

EXPONENTIAL-FORGETTING LEAST-SQUARES ALGORITHM

The recursive least-squares identification algorithm with exponential-forgetting ρ can be used to estimate the time-varying parameters x(k). It is defined by (Juditsky and Priouret 1994):

$$\hat{x}(k) = \hat{x}(k-1) + \mu\, P(k)\, h(k)\, \hat{e}(k) \tag{6}$$

$$P(k) = (\mu \rho^{-1} R(k) + S^{-1})^{-1} \quad , \quad S = S^T > 0 \tag{7}$$

$$R(k) = (1-\rho)\, R(k-1) + \rho\, h(k)\, h^T(k) \quad , \quad R(0) = 0 \tag{8}$$

The best model structural form can be establish using the one-step-ahead prediction error (Fkirin 1991 , Weigend and Gershenfeld 1994):

$$\hat{e}(k) = y(k) - \hat{y}(k) \; . \tag{9}$$

The term S is introduced to grantee the invertibility of P(k). $\bar{\mu}' << 1$ is a regularization parameter , and $0 < \rho < 1$.

The prediction value $\hat{y}(k)$, in equation (9) , can be obtained from:

$$\hat{y}(k) = h^T(k)\, \hat{x}(k-1) \quad . \tag{10}$$

Equations (6)-(10) are the exponential-forgetting least-squares identification algorithm.

STABILITY ANALYSIS OF THE EXPONENTIAL-FORGETTING LEAST-SQUARES ALGORITHM

The stability and convergence of the proposed algorithm can be proved. It can be seen that there exists a critical value $\mu_0 < 1$ such that the algorithm is unstable if $\mu >= \mu_0$ is chosen .

The main advantage of the regularized algorithm (6)-(10) is that it remains stable for any $\mu > 0$ and $0 < \rho < 1$. The forgetting-factor algorithm , under assumptions studied by Chen and Guo 1991 , also retain this property of uniform stability.

If the distribution of h(k) has a continuous density, the algorithm is exponentially stable for any $0 < \rho < 1$. Also, the values of μ and γ are very small. Therefore, the algorithm has good asymptotic properties, which implies that, under rather standard conditions, the mean-squared error (Clarke and Gawthrop 1975 , Chao and Hershauer 1993):

$$e_s(k) = E\{ | x(k) - x(k) |^2 \} \tag{11}$$

is of the order of $(\mu + \gamma^2)/\mu$ if μ and γ are sufficiently small. This indicate that:

$$\mathit{Lim}\ e_s(k) = 0 \tag{12}$$

as k gets larger, which is the sufficient condition for stability (Fkirin 1985) of the proposed algorithm.

PREDICTION MODEL

To illustrate the potential use of the exponential-forgetting least-squares identification algorithm to prediction, a solar circuit system in Egypt (shown in Figure 1) , is presented. Hourly solar energy records, of Menoufia University station are used. They are covered 13 hours, over a period from 6 a.m. to 6 p.m., in first of October 2002.

A second-order model is chosen because many high-order systems can be represented by second-order system models with good accuracy. Input temperature records (deg c) are shown in Figure 2.

The prediction output energy records (w/m^2) , $\hat{y}(k)$ of equation (10), are shown in Figure 3, with the actual value for comparison.

In Figure 3, $\hat{y}(k)$ has corresponding rise during the effective peaks of the input temperature of Figure 2. The one-step-ahead prediction error $\hat{e}(k)$, of equation (9), is shown in Figure 4. The largest errors of the prediction model, occur on the rising limb and peak of the records.

It is concluded that $\hat{y}(k)$ and $\hat{e}(k)$ are feasible indictors for identifying the model structure goodness.

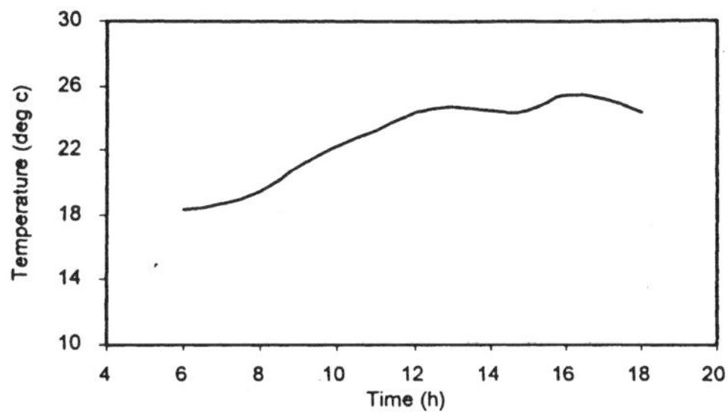

Figure 2 Input records of the solar energy system

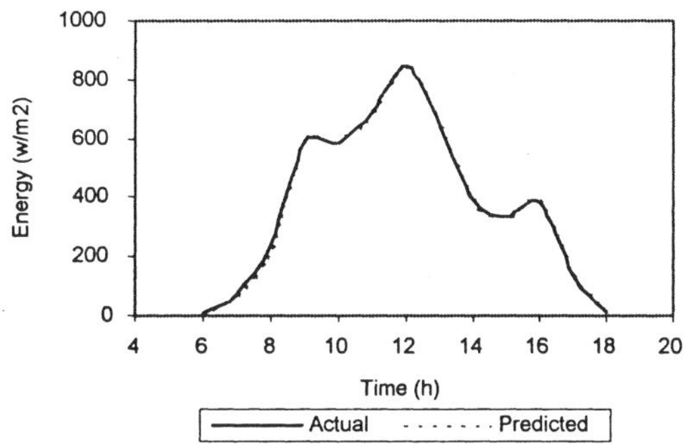

Figure 3 Actual and prediction output records of the solar energy system

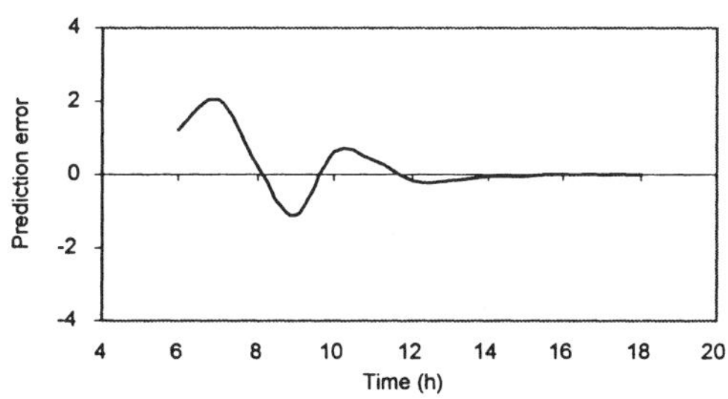

Figure 4 One-step-ahead prediction error of the solar energy system

CONCLUSIONS

In this paper, the exponential-forgetting least-squares identification algorithm has been demonstrated for prediction the stochastic systems, in the ARMA mode. stability of the proposed algorithm is proved. Prediction output model and one-step-ahead prediction error, obtained from the proposed algorithm, are tested using solar energy records. The flexibility of the identification algorithm to constructing the predict model is strong motivation for widely applied in many real systems.

ACKNOWLEDGMENTS

The provision of computing facilities and solar energy station records by the Menoufia University, Egypt, is gratefully acknowledged.

REFERENCES

Benveniste A., Metivier M. and Priouret, 1998, "Adaptive Algorithms and Stochastic Approximations", Berlin : Springer-Verlag.

Chao J. C. and Hershauer C., 1993 , "Modeling data communications controls", *Computers & Industrial Engng.*, vol. 24, pp. 91-107.

Chen H. F. and Guo L., 1991 , " Identification and Stochastic Adaptive Control" , Berlin : Birkhauser.

Clarke D. W. and Gawthrop R. J., 1975, "Self-tuning controller", *Proc. IEE*, vol. 122, pp. 929-934.

Duffie J A and Beckman W A, 1997, "Solar energy thermal", London:Wiley

Fkirin M. A., 1985 , "On the stability of time-varying models obtained by extended recursive least-squares identification algorithm", *Int. J. Systems Sci.*, vol. 16, pp. 1555-1558.

Fkirin M. A. , 1991 , "Identification of dynamic processes at various system time constants", *Cybernetics and Systems: An Int. J.*, vol. 22, pp. 349-356.

Hui S. and Zak S. H., 1993 , "Robust output feedback stabilization of uncertain dynamic systems with bounded controllers", *Int. J. Robust and Nonlinear Control*, vol. 3, pp. 115-132.

Juditsky A. and Priouret P. , 1994 , "A robust algorithm for random parameter tracking", *IEEE Trans. Automat. Contr.*, vol. 39, pp. 1211-1221.

Ljung L. and Gunnarsson S., 1990 , "Adaptation and tracking in system identification- A survey", *Automatica*, vol. 26, pp. 7-21.

Weigend A. S. and Gershenfeld N. A. , 1994 , " Time series prediction: Forecasting the future and understanding the past" , Reading , MA : Addison-Wesley Pub. Co. .

DEVELOPMENT OF AUTONOMOUS SNOWPLOW

TETSUSHI SATO
Complex System Engineering
Department, Hokkaido University
tetsushi@complex.eng.hokudai.ac.jp

KEITARO NARUSE
Complex System Engineering
Department, Hokkaido University
naruse@complex.eng.hokudai.ac.jp

YUKINORI KAKAZU
Complex System Engineering
Department, Hokkaido University
kakazu@complex.eng.hokudai.ac.jp

TSUTOMU MAKINO
Electronic and Information
Engineering, Graduate School of
Science and Engineering, Hokkaido
Tokai University
gu000105@gus.htokai.ac.jp

HIROSHI YOKOI
Complex System Engineering
Department, Hokkaido University
yokoi@complex.eng.hokudai.ac.jp

ABSTRACT
This paper presents the development of a robot that can perform snow-removal work in both an autonomous control mode and a remote control mode. In this paper, first, difficulties of practical autonomous snowplow are described. An autonomous snowplow is developed, which includes a sensor module, a vehicle control module, an emergency stop module and work planning module. The sensor module of the snowplow is evaluated and their characteristics are shown through experiments. Finally, the emergency stop module is tested and it is shown that the snowplow can stop even when a person comes up suddenly.

INTRODUCTION

In a chill snowfall zone, people perform daily snow-removal work at roads, railroads, ordinary houses, factorys, and so on. The snow-removal work is very hard work, and it gives too much burden to a worker. Moreover, an old person, a physically handicapped person, etc. are difficult to perform such work. Therefore, it is desired to mitigate the snow-removal worker's burden.

The snow-removal work treated in this study includes outdoor work robot's research domain. However, an environment where the snow removal is performed changes considerably in short time as compared with the other outdoor work environment. Moreover, sensors may not work properly in snowfall because snow can adhere to the sensors(Tatumi, 1992). We can say that the snow-removal work is more difficult than other outdoor work robots.

There are some research projects which robotize the snow-removal work. For example, in Kitami Institute of Technology, Japan, and the Hokkaido Industrial Research Institute, Japan, a remote control snowplow(Tada, 1992)(Fukuzawa, 2000)(Suzuki, 2000) has been developed. Moreover, at AHMCT in University of California, USA, an operation support system of a snowplow(AHMCT website) for highways has been developed. However, these are a support system for the snow removal work, and an autonomous snow removal robot have not been achiered, yet.

Then, in this paper, it is tried to automate the snow removal by the following techniques. First, sensors are installed in both a snowplow and a work place. Next, two information processing modules, an emergency stop module and a work planning module, are built.

This paper is organized as follows. First, difficult points about the automation of a snowplow are summarized. Next, in order to make a safety autonomous snowplow, the emergency stop module treated in this paper is shown. Then the manufacturing of an experimental model of a snowplow are performed, and the basic performance of the experimental model is evaluated. Finally, the emergency stop module is verified.

DIFFICULT POINTS WHEN DEVELOPING AUTONOMOUS SNOWPLOW

A snowplow is a large-sized machine used outdoors. Therefore, in order to robotize the snow-removal work, many difficult points exist, which are common to other outdoor work robots, such as irregular ground adaptation, environmental recognition, reliability of functions and safety. In addition to these, difficult points unique in snow-removal work are listed below.

(a) Sensor: Field of view can be bad; snow quality is uneven; snow quality changes considerably in short time; objects can be covered by snow.

(b) Actuation: Environment is uneven; restriction of operation by snow coverage state; where snow should be moved to; environment changes considerably in short time.

(c) Safety: A snowplow rotates a blade for taking in and efecting snow. Damage is serious when people and thing contact with the blade. Therefore, high safety is needed for when developing autonomous snowplow.

As shown in the above, there are many difficult points which should be solved in the development of the autonomous snowplow. The following section describes the autonomous snowplow system proposed in this study.

SNOW-REMOVAL WORK SYSTEM THAT IS NEEDED FOR AUTONOMOUS SNOWPLOW

In it is difficult to establish the reliability of sensors attached to a snowplow under a noise that the snowplow generates. Therefore, it should be assumed that information on a whole work place cannot be acquired by the sensors. On the other hand, the environment where the snowplow works in tends to change considerably. Therefore, it is necessary to acquire the information on the whole work place in real time.

In order to develop the autonomous snowplow, it is necessary to install auxiliary equipment in the work place. Then, we propose the following system, as shown in the Fig.1. First, sensors are installed in both the snowplow and the work place.

The emergency stop module is installed in the snowplow. It runs only using the sensors of the snowplow. For example, the module considers a process of avoiding collision with a man in the field.

On the other hand, the work planning module should model the whole work place. This module includes a database(Okuno, 2001) about the work place. The database is updated in real time by the sensors installed in the work place. The work planning module generates a work plan using the database and restriction conditions.

Also in the environment where there are many noises and they tend to change, the snow-removal work can be performed safely, if this system is used. In this paper, the emergency stop module in the proposed snow-removal work system is examined.

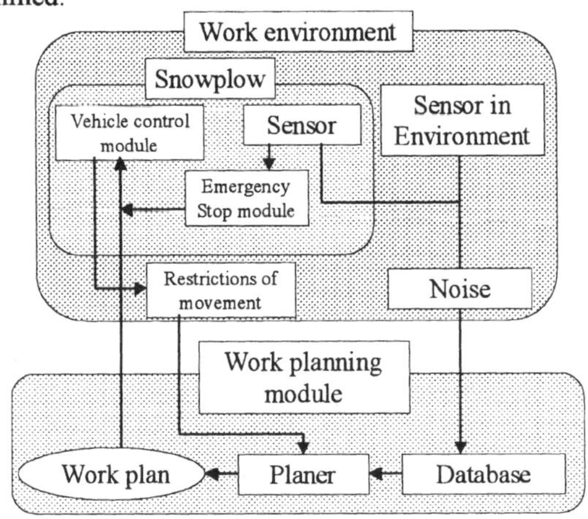

Fig.1 Proposed system

EXPERIMENTAL MODEL

We have developed the autonomous snowplow of a medium size rotary one, "Type 11-22HST (manufactured by Ishikarizouki, Japan)". The experimental model is shown in Fig.2.

The outline of the sensors attached in the snowplow is shown in a Table.1. The system configuration of the manufactured experimental model is shown in Fig.3.

Fig.2 The experimental snowplow

Table.1 Setup position of sensors

sensor type	setup position	number of sensors
Pyroelectric Infrared Sensor	front: 2, both ends of front: 2	4
Reflected Infrared Sensor	center of front: 1	1
Ultrasonic Sensor	front: 2	2
Iron Sensor	front: 1, both ends of front: 2	3
GPS	center of front: 1	1

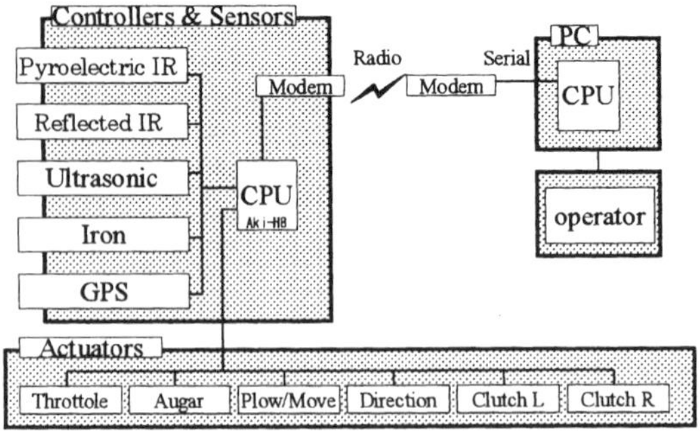

Fig.3 Sensors and controller for experimental snowplow

PERFORMANCE EVALUATION OF THE EXPERIMENTAL MODEL'S SENSORS

In this section, the results of basic examination of the obstacle sensor attached to the snowplow, i.e., a pyroelectric sensor, a reflected infrared sensor, an ultrasonic sensor, and an iron sensor, are shown. The weather conditions of all experiments are fine weather in summer, fine weather in winter, and snowfall. Moreover, the state of the snowplow is either in stop, idling, or in motion.

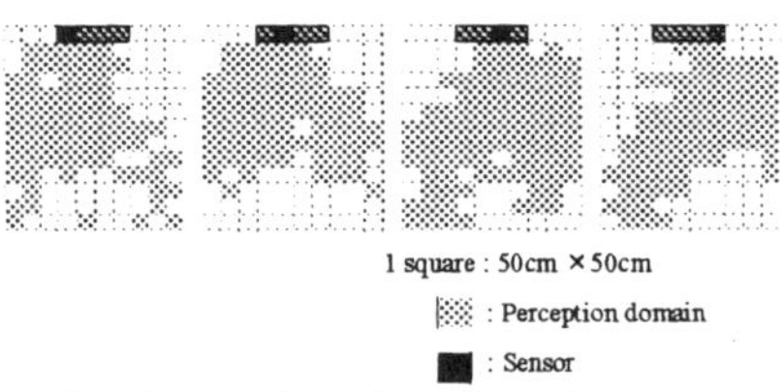

Fig.4 Pyroelectric infrared sensor (Idling in summer)

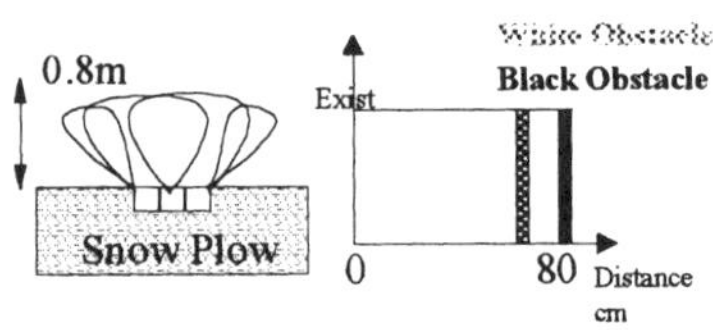

Fig.5 Reflected infrared sensor (Moving in winter)

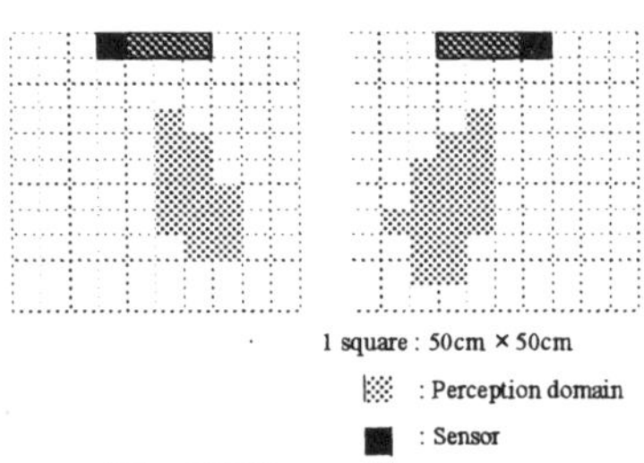

Fig.6 Ultrasonic sensor (Idling in summer)

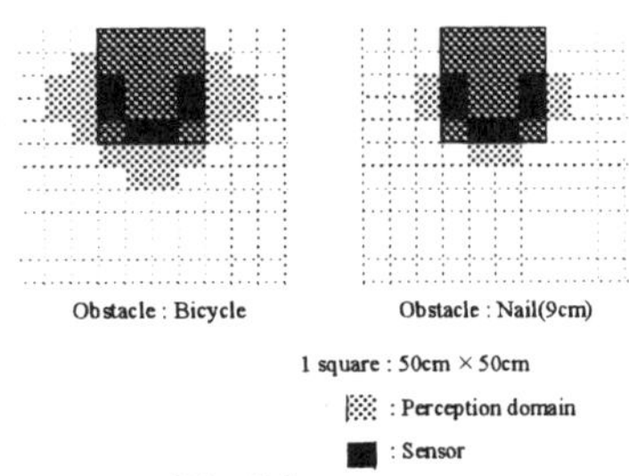

Fig.7 Iron sensor (Engine is stopped)

Fig.4, Fig.6 and Fig.7 expresse the sensor location and the front space of the snowplow. One rectangle is 0.5m * 0.5m, and the sensor is depicted as the black rectangle. The gray rectangle shows that the sensor can detect an obstacle at the position.

In Fig.5, the left figure expresses the sensor location and the front space of the snowplow, and the right figure expresses the maximum detection distance of the sensor. The reflected infrared sensor can detect the obstacle inside the curve in the left figure.

The features of each sensor are summarized in Table.2.

Table.2 Features of each sensor

Pyroelectric IR	A sensor value is not stabilized at the time the snowplow in motion.
Reflected IR	The recognition range is extremely narrower than spec.
Ultrasonic Sensor	A perception domain changes a lot with environmental atmosphere.
Iron Sensor	A sensor value is not stabilized while engine is operated

EVALUATION OF EMERGENCY STOP MODULE

We tested the emergency stop module using the manufactured experimental model.

First, the experiment for basic operation of the module is conducted. It sees whether the emergency stop module operates correctly, when a person approaches from the front of the snowplow in a stop state. The weather of the experiment was fine, and a road surface was in a hardened snow state. The experiment is conducted 10 times for each sensor. The average and standard deviation of the distance between the person and the snowplow when the emergency stop module operates are shown in Fig.8. In this experiment, when the pyroelectric infrared sensor is used, it reacts to the convection by the heat of a searchlight currently installed, and there is a case where the emergency stop module operated.

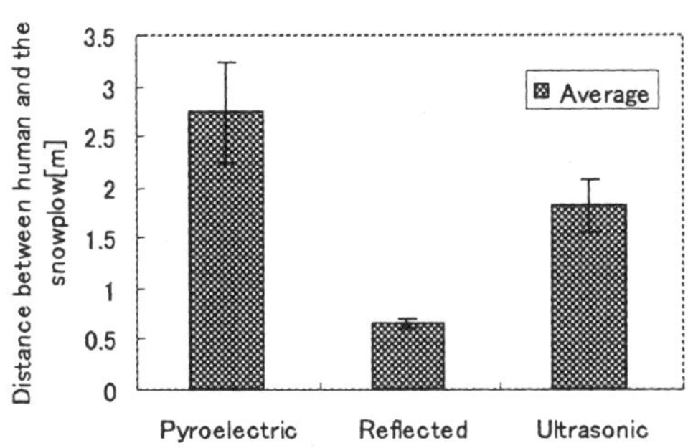

Fig.8 Experiment Result

Then, it is experimented a sensor integrated module using the pyroelectric infrared sensor as the assistance of the ultrasonic sensor. This module sees a value of the pyroelectric infrared sensor, when there is the reaction of the ultrasonic sensor.

By this sensor integrated module, it has verified that the emergency stop module could be operated correctly. It can distinguish a heating element like a human being, and the other obstacle in some extent by this module.

Furthermore, it is experimented the operation of the emergency stop module under actual snow-removal work. The weather of the experiment was fine, and amount of snow cover was about 10cm. The experiment was conducted by the ultrasonic sensor and the reflected infrared sensor.

As the result of this experiment, the emergency stop system operated correctly under the real environment under snow-removal work. In order to perform snow-removal work safely, the following points should be examined; in future, braking distance of a snowplow, an obstacle must be recognized in at least 1 meter before.

CONCLUSIONS

In this paper, first, we have summarized the difficult points when developing an autonomous snowplow actually. Based on these, we have proposed the autonomous snowplow system including the sensor module, the vehicle control module, the emergency stop module and the work planning module.

We have manufactured the experimental model includeing the sensor module and the vehicle control module.

We have performed the evaluation of the sensor attached in the experimental model. The results of the experiments have showed that sufficient performance was not obtained by each independent sensor. It is thought that techniques, such as sensor fusion, are needed.

At the end, using the manufactured experimental model, the emergency stop module has been built and the validity has been verified.

REFERENCES

Advanced Highway Maintenance and Construction Technology, http://www.ahmct.ucdavis.edu/

FUKUZAWA. S, TADA. T, TAKAHASHI. H and ISODA. K, 2000, "Development of an autonomous snow-removing machine – part 2 (On performance test in the snow environment)", *In Proc, 16th Symposium on Cold Region Development, (In Japanese)*

OKUNO. T, TAKAMATSU. K, SUZUKI. K and KAKAZU. Y, 2001, "FIELD INFORMATION SYSTEM FOR MAP-BASED PRECISION FARMING", *In Preprints of The Fourth IFA Symposium on Intelligent Autonomous Vehicles, September 5-7, pp.213-218*

SUZUKI. S, KUMAGAMI. H, TSUNEMOTO. H and HANIU. H, 2001, "Experiment of a practical automatic snowplow", *In Preprints of The Fourth IFA Symposium on Intelligent Autonomous Vehicles, September 5-7, pp.225-229*

TADA. T, KITAMURA. K, KOBAYASHI. M and MITO. M, 1992, "Intelligent Snow Remover Project –Environment Recognition of Snow Remover-", *In Report of Hokkaido Industrial Research Institute, No.291, (In Japanese)*

MODELING THE TIME-DEPENDENT ROUGHNESS PERFORMANCE OF KANSAS PCC PAVEMENTS

YACOUB NAJJAR[1] & VICTORIA FELKER[2]

[1]Professor & [2]Doctoral Candidate
Department of Civil Engineering
Kansas State University
Manhattan, Kansas 66506
USA

ABSTRACT
In this study, dynamic Artificial Neural Network (ANN) approach was used to develop a reliable and accurate time-dependent roughness prediction model for newly constructed Kansas Jointed Plain Concrete Pavements (JPCP). To achieve this objective, relevant data was obtained from the historical Kansas pavement condition database. The database included construction and materials data as well as other inventory items such as traffic and climatic related data. Utilizing a two-stage training approach and through numerous training, testing, and validation processes, a three-layer (19-10-1) time-dependent ANN-based roughness prediction model was developed. The developed model produced output values that are very close to the actual (measured) roughness IRI (International Roughness Index) values. Accordingly, it was able to project the time-dependant roughness behavior with a reasonably high $R^2 = 0.90$.

INTRODUCTION

A significant portion of funds is spent on pavement rehabilitation, resurfacing, and new construction. The new construction program for Jointed Plain Concrete Pavements (JPCP) in Kansas is in the range of $150 to $200 million (Hancock, 2000). Therefore, accuracy of the pavement performance prediction model is essential to maximize funds expended on these pavements. The pavement surface type selection process in the Kansas Department of Transportation (KDOT) for alternate strategies uses Life-Cycle-Cost (LCC) analysis. Performance prediction of JPCP over a longer analysis period (30 years as minimum) is required in order to compare the competing surface types. Performance prediction models for JPCP are now lacking since new construction practices started in the early 1990's. Identifying and selecting appropriate strategies that can potentially perform better than others would result in high benefit return. Accurate prediction of pavement performance over longer periods of time represents a critical issue in the Pavement Management System (PMS) of KDOT (Hass et al., 1994).

In this study, the dynamic Artificial Neural Network (ANN) methodology was utilized to establish an efficient, rational, and practical (easily implementable) time-dependent roughness prediction model for the newly constructed JPCP in Kansas. Inputs for the developed model take into consideration various parameters related to pavement design factors, concrete paving material used, prevailing traffic loadings, construction quality, subgrade soil properties, and prevailing climatic conditions.

DATABASE DEVELOPMENT

All data types for 25 constructed projects (total 125 of 1-mile sections) for newly built JPCP on I-70, I-435, K-15, K-96, US-50, US-56, US-75, and US-77 were collected from the Kansas Department of Transportation (KDOT) historical pavement database. Each project has a different construction date. The oldest project has a construction date of year 1995, while the newest project has a construction date of year 1997. Therefore, the age of all sections considered was varied between 3 and 5 years since the last available roughness IRI data is for year 2000. All the Portland Cement Concrete (PCC) pavement sections considered are Jointed Plain Concrete Pavements (JPCP) with 4.6 m (15 ft.) joint spacing and doweled joints. The JPCP slab thickness varied from 229 mm (9 in.) to 292 mm (11.5 in.). The database for each section included annual roughness values measured during each year after construction. Longitudinal profile measurements were done on right and left wheel path with a South Dakota-type Profilometer. From these profile data, the International Roughness Index (IRI) values were calculated with the RoadRuf software developed by the University of Michigan Transportation Research Institute. The IRI value used in this study represents only the right wheel path IRI.

The data was categorized and assembled into an Excel spreadsheet to prepare for the analysis using ANNs methodology. Initially, fifty-six different potential input variables were listed. However, after data cleansing and thorough examination, the database was limited to twenty six practical input variables. The detailed database development of the twenty six input variables included pavement profile (initial smoothness represented by initial roughness IRI value), pavement section, pavement layer, time-series traffic, subgrade type and treatment, climatic conditions, shoulder type, concrete materials test data, and concrete mixture data. Pavement layer data included PCC slab thickness and type of base drainage, such as Portland Cement Treated Base (PCTB) and Bound Drainable Base (BDB). Time-series traffic data refers to the cumulative 80-kN (18-kip) Equivalent Single-Axle Loads (ESAL's) and is used herein to represent traffic loadings. Climatic information was included in order to develop a model, which can directly account for section-specific climatic conditions. The output variable was the time-series (i.e., yearly) right wheel path roughness IRI value, which is used to quantify the long-term pavement roughness performance.

MODEL DEVELOPMENT

According to Najjar and co-workers (Basheer, 1998; Itani and Najjar, 2000; Ali, 2000; Najjar et al., 2000; Najjar et al., 1997), for proper ANN modeling, a two stage training approach may be needed. In the first stage, the full database is divided into training, testing, and validation sub-bases, and then, the least-error-structure is determined. If accuracy measures of the least-error structure are comparable for training, testing, and validation data sets, then the second stage of training is not warranted. On the other hands, if the accuracy measures are not comparable, then as a compromise, the least-error structure determined from stage one is then retrained on the entire database. In this study, during the first stage of training, the database containing a total of 415 data sets was divided into 225, 93, and 97 data sets for training, testing, and validation, respectively.

Dynamic ANN-based training technique adopted by Najjar (1999) and Najjar & Zhang (2000) is utilized herein to model the time-dependent pavement roughness performance. This technique is utilized within the framework of the conventional feed-forward error-back propagation neural network approach. According to the feedback approach, the futuristic [i.e., year (n+1)] roughness IRI value [i.e., $(IRI)^{n+1}$] is determined from a number of previously determined input parameters. This logic can mathematically be represented in the following compact form:

$$\{(\mathbf{IRI})^{n+1}\} = \mathbf{ANN}_{(m+1)\text{-}k\text{-}1}\{x_1, x_2, \ldots, x_m, (\mathbf{IRI})^n\} \quad (1)$$

where ANN denotes the neural network model that best relates a given number of inputs (m+1) [i.e., x_1, x_2, ... , x_m, $(IRI)^n$] to the desired output [i.e., $(IRI)^{n+1}$]. Note that $\{x_1, x_2, \ldots, x_m\}$ is a vector of *(m)* parameters used herein to represent all static input parameters that are believed to affect the desired output. The *(m+1)-k-1* notation denotes the architecture of the selected network. In this case, *(m+1)* represents the *(m)* static inputs, and the one additional feed-back parameter, *k* is the optimal number of hidden nodes, which needs to be determined through training and online testing process, and *(1)* is the desired number of output, namely; the futuristic roughness IRI value [i.e., $(IRI)^{n+1}$].

In order to filter out the most influential static input parameters, various ANN modeling trials were performed. In the final optimal network achieved (utilizing the previously mentioned two stage training approach), the input parameter vector contained the following 19 variables:
1) PCC slab thickness (SLTH), 2) Drainable base (BTYD), 3) Non-drainable base (BTYN), 4) Concrete unit weight (UW), 5) Cement factor (CFA), 6) Non-treated subgrade (SUBTRT_N), 7) Lime treated subgrade (SUBTRT_L), 8) Subgrade modification (SUBTRT_M), 9) % of natural subgrade soil material passing No. 4 sieve (FSI), 10) % of natural subgrade soil material passing No. 200 sieve (TSI), 11) Plasticity index of natural subgrade soil material (PI), 12) Cumulative yearly ESAL values (ESAL), 13) Average number of freeze-thaw cycles per year (FTC), 14) Cumulative total number of days below 32 °F/yr (DB), 15) Cumulative total number of days above 90 °F/yr (DA), 16) Cumulative number of wet days/yr (WET), 17) Initial right wheel path IRI values (IIRI), 18) Age of pavement (AP), and 19) IRI value at age (n) year $(IRI)^n$.

As noted earlier, the input layer includes a dynamic variable [i.e., $(IRI)^n$] that represents the IRI value at age (n) year. Accordingly, all nineteen input parameters (i.e., eighteen static and one dynamic) are used via the developed ANN-based model to predict the IRI value [i.e., $(\text{IRI})^{n+1}$] at age (n+1) year. Note that, the initial value for (n) is zero. Based on the stage one training, the optimal network architecture was found at ten (10) hidden nodes. Accordingly, the final form of Eq. (1) can be represented by the following expanded form:

$$\{(\mathbf{IRI})^{n+1}\} = \mathbf{ANN}_{19\text{-}10\text{-}1}\{\text{SLTH, BTYD, BTYN, UW, CFA, SUBTRT\_N, SUBTRT\_L, SUBTRT\_M, FSI, TSI, PI, ESAL, FTC, DB, DA, WET, IIRI, AP, }(\mathbf{IRI})^{n}\} \quad (2)$$

Based on the stage two training, the 19-10-1 ANN-based IRI prediction model yielded a coefficient of determination, R^2 of 0.90. To further validate the developed ANN-based model, it was used to predict IRI values for years 2001 and 2002 data. This new data was provided by KDOT after the 19-10-1 ANN-based IRI prediction model, described in Eq. (2), was fully developed. This was done in order to obtain a truly independent validation of the developed model. Accordingly, R^2 values of 0.8 and 0.74 were obtained for 2001 and 2002 year's data, respectively. As expected, model prediction accuracy reduces as extrapolation time is increased. For this reason, it is imperative that such model be annually updated on newly acquired data. This update, will allow the modified ANN-based model to better project/predict IRI value for future years.

SENSITIVITY ANALYSIS

In order to assess the impact of each independent input variable on the time-dependent IRI profile, a sensitivity analysis was performed. To accomplish this objective, a PCCP section with average input values was chosen for this purpose. Accordingly, when logically possible, each input variable was then varied within its applicable range (keeping all other input variables stationary). The time-dependent IRI profile for up to seven years (two years more than the maximum actual age of the available database) was projected via the developed 19-10-1 ANN-based model. Selected plots for variables showing the maximum impact on IRI time-dependent profile are presented in this paper. Factors that, so far, have shown the greatest impact on the roughness are BTYD, BTYN, IIRI, SLTH, SUBTRT_N, SUBTRT_L, and SUBTRT_M.

The choice of using drainable or non-drainable bases significantly influences the roughness as illustrated in Figure 1. This factor seems to produce the largest impact on IRI profiles, observed in this study, for Kansas JPCP pavements constructed after 1995. Figure 1 demonstrates that non-drainable bases will result in PCC pavements having higher roughness. Drainable bases tend to have overall lower roughness profiles. Note that, bases that are non-drainable, are also impermeable. In this case, trapped water can cause swelling of the bases (or subgrades) and eventually, leads to an uneven pavement surface. On the other hand, proper drainage will decrease the swell potential of the subgrades or bases. Numerically, for the case considered herein, the use of drainable base will decrease dramatically the roughness by 48 in./mile. In other words, after seven years of service, a drainable base tends to produce a roughness, which is about 48 in./mile lower than non-drainable base. This clearly indicates that a drainable base will help retain the smoothness longer compared to non-drainable bases.

Figure 2 identifies three different natural subgrade treatment types as a significant variable affecting the roughness. A subgrade modification may be used as a stabilized base for lower volume roads or detours. The aggregate, soil, lime, and water are mixed either in-place or off-site in a borrow area. The subgrade modification of an existing pavement usually involves the milling of a part of or all of the existing pavement prior to commencement of the subgrade modification operations. The aggregate for the subgrade modification may be either the remainder of the existing pavement left in-place or some of the milled material. As noted in Figure 2, pavements placed on lime treated subgrades are substantially smoother than those placed on non-treated subgrades. Use of 6 in. lime treated subgrade will tend to decrease the roughness by about 63 in./mile after 7 years. This indicates that subgrade treatment would be the most beneficial to sustain smoothness since a large percentage of subgrade material consists of silt and clay particles.

CONCLUDING REMARKS

The dynamic ANN modeling strategy was used in this study to develop an efficient time-dependent PCC pavement roughness prediction model. Inputs for the developed model take into consideration various parameters related to pavement design factors, concrete material parameters, subgrade treatment, foundation soil properties, prevailing traffic loadings, prevailing climatic conditions, and construction quality. Developed model produced output values that are very close to the actual IRI values. The developed model could project the time-dependant roughness behavior with a reasonably high $R^2 = 0.90$.

ACKNOWLEDGEMENTS

The authors gratefully acknowledge the continued financial support and guidance received from KDOT to conduct this research study.

REFERENCES

Hancock, J. D., 2000, *A Study of Factors Responsible for Roughness Progression on Portland Cement Concrete Pavements,* M.S. Thesis, Department of Civil Engineering, Kansas State University, Manhattan.

Hass, R., Hudson, W. R., Zanieweski, J., 1994, *Modern Pavement Management.* Krieger Publishing Company, Florida.

Basheer, I., 1998, *Neuromechanistic-Based Modeling and Simulation of Constitutive Behavior of Fine-Grained Soils,* Ph.D. Dissertation, Department of Civil Engineering, Kansas State University, Manhattan.

Itani, O., Yacoub, N., 2000, "3-D Modeling of Spatial Properties via Artificial Neural Networks," *Transportation Research Record,* No. 1709, pp. 50-59.

Ali, H., 2000, *Neuroses-Based Liquefaction Potential Assessment and Stress-Strain Behavior Simulation of Sandy Soils,* Ph.D. Dissertation, Department of Civil Engineering, Kansas State University, Manhattan.

Najjar, Y., Basheer, I. A., Ali, H., McReynolds, R., 2000, "Swelling Potential of Kansas Soils: Modeling and Validation Using ANN-Reliability Approach," *Transportation Research Record,* No. 1736, pp. 141-147.

Najjar, Y., 1999, *Quick Manual for TR-SEQ1,* Department of Civil Engineering, Kansas State University, Manhattan.

Najjar, Y., Basheer, I. A., McReynolds, R., 1997, *Modeling the Durability of Aggregate Used in Concrete Pavement Construction: A Neuro-Reliability-Based Approach,* Final Report No.: KS-97-3, Kansas Department of Transportation.

Najjar, Y. M., Zhang, X, 2000, *Characterizing the 3D Stress-Strain Behavior of Sandy Soils: A Neuro-Mechanistic Approach,* Geotechnical Special Publication No. 96, ASCE.

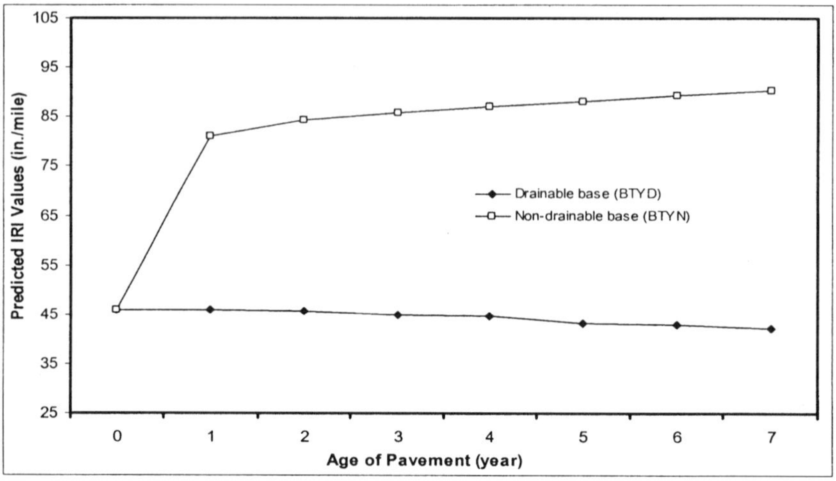

Fig. 1 Predicted IRI values for drainable and non drainable base.

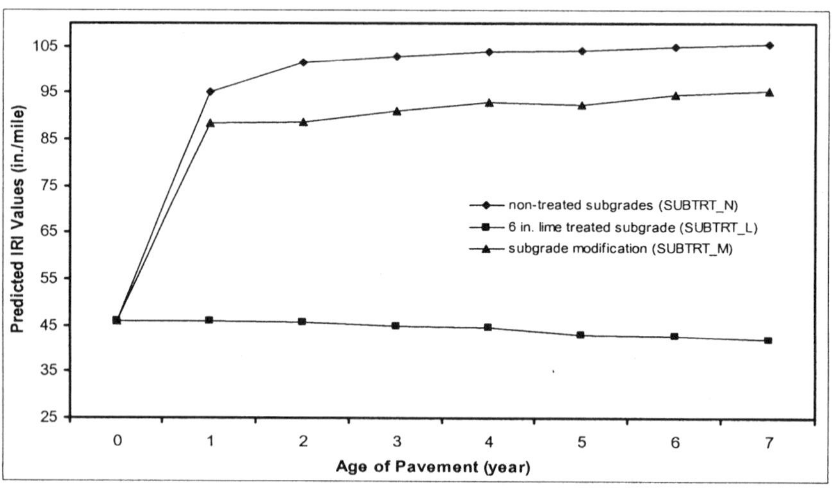

Fig. 2 Predicted IRI values for various subgrade treatment cases.

MULTI-OBJECTIVE GENETIC ALGORITHMS FOR SENSOR LAYOUT OPTIMIZATION IN STRUCTURAL DAMAGE DETECTION

ANNE M. RAICH
Department of Civil Engineering
Texas A&M University
College Station, Texas

TAMAS R. LISZKAI
Material and Structural Analysis
Framatome ANP
Lynchburg, Virginia

ABSTRACT

To maximize the performance of structural damage identification methods (SDIM), a methodology is defined for optimizing the excitation and sensor locations used to collect measurement information from a structure. The multi-objective optimization problem had two objectives: maximize the amount of diagnostic information contained in the vibration signatures and minimize the number of sensors used to obtain the measurements. The information measure was based on the sensitivities of frequency response functions with respect to the indicators that identify structural damage. The Pareto front was determined using a Nondominated Sorting Genetic Algorithm. Beam and frame case studies investigated the benefit of using the optimum sensor layouts on SDIM performance. Using the optimal layouts increased the amount of measurement information available to the SDIM, which improved the accuracy of the SDIM results, even when noise was present in the measurements.

INTRODUCTION

Many engineering tools depend on the quantity and quality of diagnostic information contained in measurements. The information captured using sensors varies with the domain. In most cases, however, the number of possible sensor locations is much larger than the actual number of sensors used to collect measurements. In damage detection, the number of sensors is limited by the available access and equipment. Therefore, it is vital to place the limited sensors in a near-optimal way to maximize the diagnostic information with respect to a SDIM. Sensor layout design therefore poses an optimization problem with an objective of maximizing the amount of information. This objective, however, contradicts the need to minimize the number of sensors used. A significant amount of research has been conducted in recent years in structural damage identification. When damage influences dynamic properties such as stiffness, vibration data can be used for damage identification. Frequently used vibration data are natural frequencies, mode shapes, and frequency response functions (FRF). The sensor layout method proposed by Shi et al. (2000) focused on improving the diagnostic information for a *modal*-based SDIM. Worden and Burrows (2001) compared several methods, including genetic algorithms (GA), for sensor optimization for a *modal*-based SDIM.

This paper discusses optimizing sensor layouts to improve the performance of a *FRF*-based SDIM (Raich and Liszkai, 2003). A damage indicator, x_j^s, was introduced for each finite element in the model. Using the damage indicators, the stiffness matrix $\mathbf{k}_j^{dam}$ for the jth damaged finite element in the model is obtained: $\mathbf{k}_j^{dam} = (1 - x_j^s)\mathbf{k}_j$, $0 = x_j^s = 1$, where $\mathbf{k}_j$ is the stiffness matrix of the jth finite element without damage. An error function, which was defined as the difference between the FRFs obtained from measurements on the damaged structure and the FRFs obtained from the structural model defined using the proposed damage indicators, was minimized using an IRR GA (Raich and Ghaboussi, 1997) to obtain the vector of damage indicators, $\mathbf{x}^s$, for the model.

Satisfying the first objective of the optimum sensor layout problem required an information measure for FRF data. The sensitivity of the FRF matrix with respect to the damage indicator of the jth finite element can be written as:

$$\frac{\partial \mathbf{H}(\omega)}{\partial x_j^s} = -\mathbf{H}(\omega)\frac{\partial \mathbf{Z}(\omega)}{\partial x_j^s}\mathbf{H}(\omega) \tag{1}$$

where $\mathbf{H}(\omega)$ is an FRF matrix and $\mathbf{Z}(\omega)$ is the inverse of the FRF matrix. Equation 1 is computationally prohibitive for large structures and numerical differentiation is used to calculate the FRF sensitivities. The information, ι_{km}, contained in the H_{km} FRF function for a pair of excitation, k^{th}, and measurement, m^{th}, DOFs is defined as:

$$\iota_{km} = \sum_{j=1}^{n_e}\left(\int_{\varpi_0}^{\varpi_1}\left|\frac{\partial H_{km}(\omega)}{\partial x_j^s}\right| d\omega\right)^2 \tag{2}$$

where ϖ_0 and ϖ_1 define the measured frequency range and n_e is the number of finite elements. The total information, I, contained in n_{meas} sensors is defined as:

$$\mathrm{I} = \sum_{m\in J}^{n_{meas}} (\iota_{km})^2 \tag{3}$$

SENSOR OPTIMIZATION USING MULTIOBJECTIVE GAS

The excitation and sensor layout optimization problem can be stated as:

$$\begin{aligned} \min f_1(\mathbf{x}) &= n_{meas} \\ \max f_2(\mathbf{x}) &= \mathrm{I} \end{aligned} \tag{4}$$

The objectives contradict each other, since reducing the number of sensors usually decreases the amount of information and vice versa. Therefore it would be unlikely to find a solution that optimizes both objectives. A multi-objective (MO) optimization problem, however, can be defined that generates multiple solutions called "trade-off" or Pareto optimal solutions. A Nondominated Sorting Genetic Algorithm (NSGA) proposed by Srinivas and Deb (1994) is used in this research. NSGA evolves a set of Pareto optimal solution in a single run. NSGA employs a different selection procedure then used in a simple GA. A dummy fitness value is used in the selection process that is based on a pareto ranking of the population that considers nondomination of individuals. Fitness sharing is used to maintain population diversity across the Pareto front. An overview of MO Gas, including the NSGA, is provided by Coello (2001). The sensor decision variables, x_j, are logical variables that act as an "on/off" switch related to each DOF in the finite element model where a sensor can be placed (n is the total number of DOFs):

$$n_{meas} = \sum_{i=1}^{n} x_i, \qquad x_i = \{0,1\} \tag{5}$$

There is an additional decision variable, x_0, that defines the excitation location DOF.

A simple GA encoding was used to represent the excitation and sensor layouts. The excitation and sensors can be placed at any DOF in the horizontal, x, or vertical, y, direction. The first m-bits of each GA individual encode the excitation DOF. The rest of the GA individual consists of n_{sens} number of bits ("on/off" switches) that define the DOFs where sensors are placed. The only problem constraint is a limit on the number of sensor used in the measurements. An individual was repaired if the number of sensors was zero or exceeded the maximum limit of 10 sensors.

RESULTS OF SENSOR LAYOUT OPTIMIZATION TRIALS

Two-span continuous beam (Fig. 1) and unbraced frame (Fig. 2) trials were performed. The two-span continuous beam model was defined by 20 finite elements (W12x65). In the frame model, the columns consisted of 3 finite elements (W14x132), while the beams were defined by 5 finite elements (W12x65), for a total of 76 nodes and 81 elements. The beam to column elements were 1-ft long. (E=207 GPa, υ=0.3, ρ=7780 kg/m$^{3)}$. For the two-span continuous beam trial, there were 38 possible excitation and sensor locations, while for the frame trial there were 144 possible locations. The GA parameters included: population size = 300; mutation rate = 0.5%; crossover rate = 80%; and tournament size = 3. Figure 3 shows the final population in objective space for the two-span beam problem. After 34 generation, 60 % of the population defined the Pareto front. The individuals are well distributed across the Pareto front (the number of individuals in each group is noted). Table 1 indicates that the relative increase in information decreases with increasing number of sensors; with 10 sensors showing a 400% increase in information over one sensor. The optimal layout for one sensor is to place both the sensor and the excitation at the vertical DOF 13 of the two-span beam.

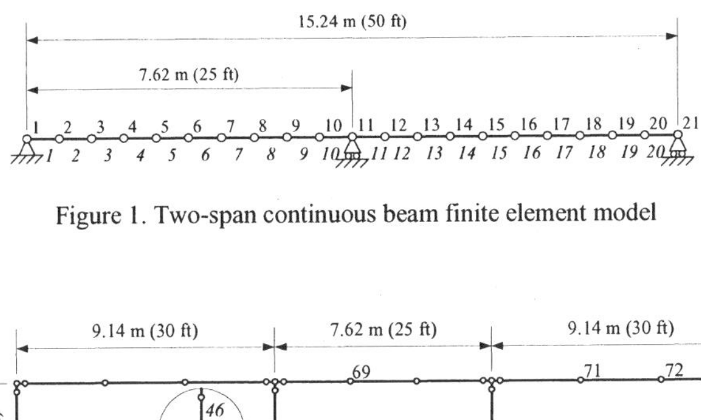

Figure 1. Two-span continuous beam finite element model

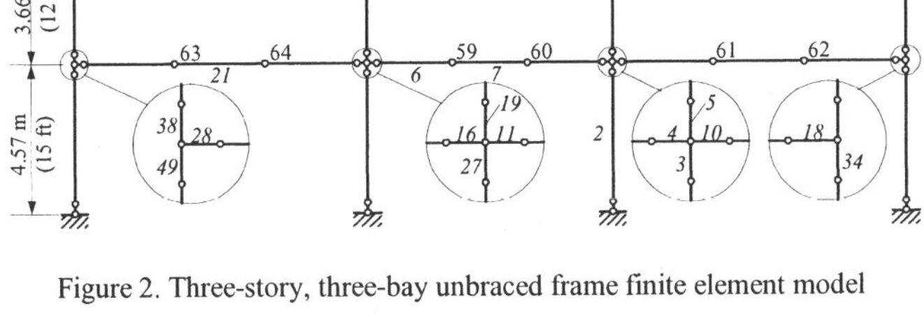

Figure 2. Three-story, three-bay unbraced frame finite element model

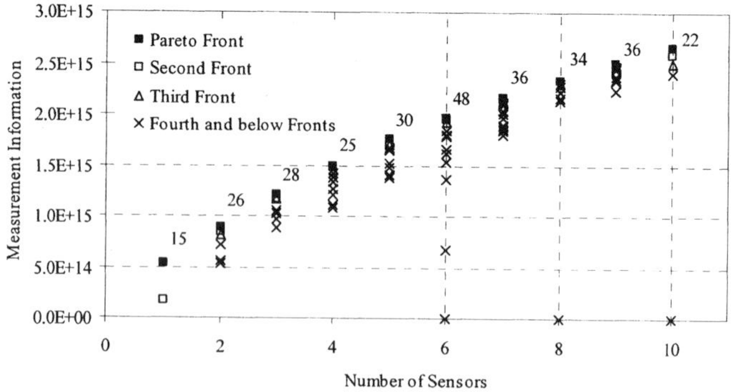

Figure 3. Individuals in the Pareto front and the final population for the two-span beam

Table 1 Pareto optimal excitation and sensor locations for two-span continuous beam

# of sensors	Node Excitation	Sensor(s) at Node	Measurement Information Value	Measurement Information Percent
1	13	13	$5.3574 \cdot 10^{14}$	100.0%
2	13	13, 19	$8.9212 \cdot 10^{14}$	166.5%
3	13	9, 13, 19	$1.2077 \cdot 10^{15}$	225.4%
4	13	3, 9, 13, 19	$1.4877 \cdot 10^{15}$	277.7%
5	13	3, 9, 13, 19, 20	$1.7648 \cdot 10^{15}$	329.4%
6	13	2, 3, 9, 13, 19, 20	$1.9772 \cdot 10^{15}$	369.1%
7	13	2, 3, 9, 13, 17, 19, 20	$2.1658 \cdot 10^{15}$	404.3%
8	13	2, 3, 9, 13, 15, 17, 19, 20	$2.3366 \cdot 10^{15}$	436.1%
9	13	2, 3, 9, 12, 13, 15, 17, 19, 20	$2.5027 \cdot 10^{15}$	467.1%
10	13	2, 3, 9, 12, 13, 14, 15, 17, 19, 20	$2.6658 \cdot 10^{15}$	497.6%

Table 2. Pareto optimal excitation and measurement locations for the unbraced frame

# of sensors	Node Excitation	Sensor(s) at Node	Measurement Information Value	Measurement Information Percent
1	61	61	$9.8223 \cdot 10^{11}$	100.0 %
2	61	61, 62	$1.8670 \cdot 10^{12}$	190.1 %
3	61	61, 62, 72	$1.8850 \cdot 10^{12}$	191.9 %
4	61	61, 62, 71, 72	$1.9029 \cdot 10^{12}$	193.7 %
5	61	61, 62, 66, 71, 72	$1.9066 \cdot 10^{12}$	194.1 %

Table 2 summarizes the optimum sensor layout results obtained for the unbraced frame problem. All sensors were placed at vertical DOFs. Placing a sensor at the excitation DOF was the most beneficial, which is similar to the results of the two-span beam trial. Table 2 indicates that using two sensors increases the information by about 90%, but adding more sensors does not increase the diagnostic information significantly.

EFFECT OF OPTIMAL LAYOUTS ON SDIM PERFORMANCE

Trials for the two-span beam and the unbraced frame problem domains were performed to evaluate the benefit of using the optimal layouts defined to collect diagnostic information for a FRF-based SDIM (Raich and Liszkai, 2003). Table 3 summarizes the two damage cases investigated. The GA parameters included: population size = 200; seeding the initial population with the zero damage individual (Liszkai,

2003); tournament sizes = 4; mutation rate = 0.005; adaptive and equal probability crossover with primary crossover rate of 0.9; gray coding; and elitism. After GA trial convergence, local hillclimbing was applied to the fittest GA individual. For the SDIM performance trials, a seeded IRR GA encoding was represented the damage indicators used to identify the location and severity of damage in the structure (Liszkai, 2003).

The two-span beam (TWSP) case study simulates a damage scenario with two damaged elements with different severities. SDIM results using the optimal and non-optimal sensor layouts are compared in Fig. 4. Both the optimal and non-optimal sensor layouts used only one sensor. A 10%, zero-mean Gaussian noise with standard deviation of 1 was added to the simulated measurements to account for typical measurement noise. Trials using the optimum sensor layout converged faster, which reduced the number of hillclimbing iterations. The average initial population fitness increased by an average factor of 3.9 using the optimal sensor layout, which indicates that more diagnostic information is contained in the measurements. The results presented in Fig. 4 show that the SDIM's damage predictions using the optimal sensor layout are better correlated with the imposed damage scenario. With the non-optimal sensor layout, the SDIM predicted the following damages in the true damaged elements: 5.9% in Element 6 and 9.5% in Element 10. In comparison, the SDIM using the optimal sensor layout predicted 4.9% damage in Element 6 and 10.0% damage in Element 10. The damage-balancing phenomenon (i.e. predicted damage indicators of unequal damages tend to approach the average value of the inflicted damage severities) observed with the non-optimal sensor layout trials is not present in the optimum sensor layout trials. Using the optimal sensor layout also reduced the number and severity of falsely identified elements. When the non-optimal sensor layout was used, two elements were falsely identified with damages less than 1%. No false predictions were obtained using the optimal sensor layout.

The unbraced frame (FRM I) case study simulates damages associated with soft story mechanism. All four damaged elements were located on the first floor columns. The damage severity for the perimeter columns was 10% (Elements 34 and 49), while the damage severity for the interior columns was 5% (Elements 34 and 49). Noise free measurements were used and seeded IRR GA trials are compared. The SDIM results obtained using the non-optimal and optimal sensor layouts are presented in Fig. 5. Using the non-optimal sensor layout, only the two interior column elements out of the four damaged elements were correctly identified and the damage indicator values for these elements were overestimated (6.7% damage in Element 3 and 11.6% damage in Element 27 instead of the imposed 5% damage). Furthermore, four finite elements were falsely identified as damaged with smaller damage indicator values. In comparison, the SDIM results obtained using the optimal sensor layout show that all four true damaged elements were identified with the correct damage indicator values, which is the global optimum.

Table 3. Definition of damages imposed for the case studies

Case	Imposed Damage		Excitation		Sensor(s)	
	Element	Percent	Non-optimal	Optimal	Non-optimal	Optimal
TWSP	6, 10	5, 10 %	18-*y*	13-*y*	14-*y*	13-*y*
FR I	49, 27, 3, 34	10, 5, 5, 10 %	71-*x*	61-*y*	59, 67, 69	61, 62, 72

CONCLUSIONS

A multi-objective optimization method was presented to obtain near-optimal excitation and sensor layout designs for damage identification problems. An information measure based on FRF sensitivities was used by the NSGA to evolve the Pareto optimal set. Two-span beam and unbraced frame trials using non-optimal and optimal sensor layouts were performed to identify the benefit of using the optimal sensor layout designs for a FRF-based SDIM. The increased fitness values for the optimal sensor layouts indicate an increase in the diagnostic information collected. The amount of diagnostic information, however, did not increase proportionally with increasing number of sensors. Therefore, it is appropriate to use a minimal number of sensors. Significant improvement in SDIM prediction accuracy was obtained using the optimal sensor layout designs.

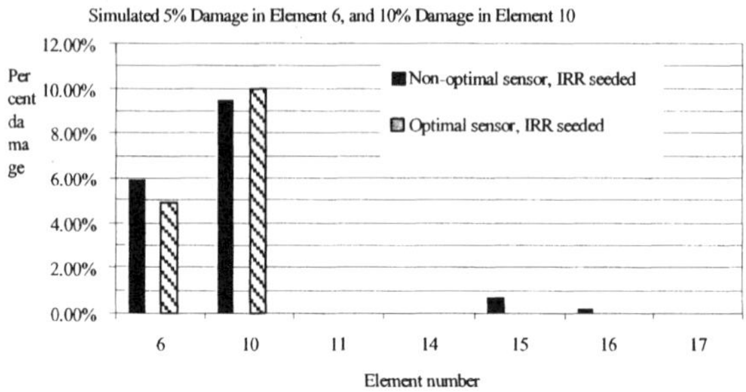

Figure 4. SDIM results for the two-span beam with *10%* measurement noise imposed

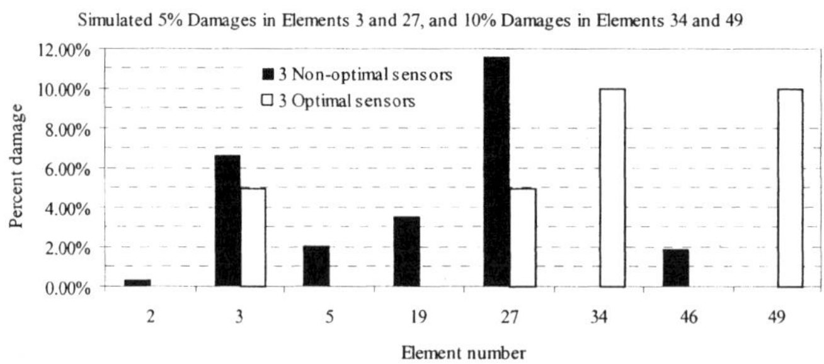

Figure 5. SDIM results for unbraced frame soft story mechanism damage case (*0% noise*)

REFERENCES

Coello, C.A.C., 2001, "A short tutorial on evolutionary multiobjective optimization," *First International Conference on Evolutionary Multi-Criterion Optimization, EMO 2001*, LNCS 1993, Springer-Verlag, Berlin Heidelberg, pp. 21-40.

Liszkai, T.R., 2003, "Modern heuristics in structural damage detection using frequency response functions." Doctoral dissertation, Texas A&M University, College Station, TX.

Raich, A.M., and Ghaboussi, J., 1997, "Implicit redundant representation in genetic algorithms," *Evolutionary Computation*, 5(3), pp. 277-302.

Raich, A.M., and Liszkai, T.R., 2003, "A model parameter updating damage detection technique using genetic algorithm," *Proceedings of the International Conference on Metal Structures*, April 3-5 2003, Miskolc, Hungary, pp. 293-300.

Shi, Z.Y., Law, S.S., and Zhang, L.M., 2000, "Optimum sensor placement for structural damage detection," *Journal of Engineering Mechanics*, 126(11), pp. 1173-1179.

Srinivas, N., and Deb, K., 1994, "Multiobjective optimization using nondominated sorting in genetic algorithms," *Evolutionary Computation*, 2(3), pp. 221-248.

Worden, K., and Burrows, A.P., 2001, "Optimal sensor placement for fault detection," *Engineering Structures*, 23, pp. 885-901.

SIMULATING THE 3D STRESS-STRAIN BEHAVIOR OF SANDY SOILS

YACOUB M. NAJJAR, Professor
Department of Civil Engineering
2118 Fiedler Hall
Kansas State University
Manhattan, Kansas 66506
USA

ABSTRACT
An adaptive dynamic BackPropagation artificial Neural Net (BPNN) approach is used in this study to characterize the 3-D monotonic stress-strain behavior of Canadian River Sand. To accomplish this objective, four consecutive modeling stages were conducted. In the first stage, mechanics-based knowledge relating to the cause and effect process (i.e., stress-strain) was used to design an appropriate stress-controlled general-purpose ANN model. In the second stage, the model with appropriate input categories was trained and tested on 3-D monotonic stress-strain data of the Canadian River sand using enforced training technique. In the third stage, statistical and graphical accuracy outcomes on training and testing stress-strain responses were used to arrive at the optimal ANN-based model. In the final stage, the selected ANN-based model was combined with appropriate mechanics-based concepts to create a hybrid neuro-mechanistic simulator (NMS). The NMS developed was found to be efficient in characterizing the 3-D monotonic stress-strain behavior of Canadian River sand subjected to shear loading stress paths.

INRODUCTION

In the development of tractable and practical constitutive models, mechanics-based models usually adopt idealization and assumptions that result in the inability to describe the complex nature of soil deformation behavior. On the other hand, some classical mechanics-based constitutive models are handicapped for practical use by the model sophistication in terms of increasing number of required material parameters. Moreover, due to the complex nature of the stress-strain relation, a specific classical model is usually applicable to limited material types and loading conditions.

During the last decades, many investigations have focused on developing ANN-based constitutive models for geomaterials. Using the sequential (feed-back) backpropagation training approach, Ellis et al. (1995) developed appropriate ANN models to simulate both monotonic and cyclic undrained triaxial stress-strain response of sand at various relative densities, uniformity coefficients, overconsolidation ratios and effective confining pressures. Realizing the benefits of the feed-back approach, Basheer and Najjar (1998) used it to simulate the uniaxial stress-strain constitutive behavior of fine-grained soils under cyclic loading. Subsequently Najjar and Ali (1999) and Ali and Najjar (2000) have used this feed-back approach to characterize the drained/undrained stress-strain response of Nevada sand subjected to both triaxial compression and extension stress paths.

DATA DESCRIPTION

A series of comprehensive truly triaxial cubical experiments (Oyesanya, 1992) were performed on the prepared Canadian River sand samples using a fluid cubical device. The truly triaxial or multi-axial cubical testing device has the advantage of applying three independent principal stresses to the six sides of a cubical specimen and is thus able to simulate more complex but realistic stress paths. A total of four stress paths were conducted. Three CTC (Conventional Triaxial Compression), TC (Triaxial Compression), CTE (Conventional Triaxial Extension) and TE (Triaxial Extension) tests were performed at different initial confining pressure values ranging from 35 kPa to 138 kPa. Additionally, one Hydrostatic Compression (HC) and one Simple Shear (SS15) tests were carried out. The HC test was performed up to a maximum confining pressure of 310 kPa while the SS15 test was performed at an initial confining pressure of 103 kPa.

MODEL DEVELPMENT STAGES

In this study, an adaptive feed-back backpropagation ANN approach is used along with proven mechanics-based knowledge/concepts (i.e., hybrid appproach) to characterize the complex 3D monotonic stress-strain behavior of a sandy soil. To successfully implement this hybrid approach, four consecutive modeling stages were conducted. In the first stage, mechanics-based knowledge relating to the cause and effect process (i.e., stress-strain) was used to design an appropriate general purpose stress-controlled ANN model. In the second stage, the designed model with appropriate input categories (stress invariants and Lode angle) dictated by mechanics-based knowledge was trained and tested on 3D monotonic stress-strain experimental data of a sandy soil under different stress paths (i.e., compression, extension and simple shear). During the third stage, statistical and graphical accuracy outcomes on training and testing stress-strain responses were used to arrive at the optimal neuro-mechanistic based model. In the last stage, the selected neuro-mechanistic based model was combined with appropriate mechanics-based concepts to create the hybrid neuro-mechanistic simulator (NMS).

Stage one: model architecture design

Based on our knowledge of geomechanics (Najjar 1990 and Najjar and Zaman 1993) and previous ANN research (Najjar and Basheer 1996, Basheer and Najjar 1998 and Najjar and Zhang (2000) related to constitutive behavior modeling, it is envisioned that a general purpose ANN-based model for monotonic and cyclic stress-strain behavior can be expressed in the following compact form.

$$\left\{\varepsilon_j^{\,i+1}\right\} = ANN_{20-NH-3}\left\{\sigma_j^{\,i}, \Gamma^i, \sigma_j^{\,i+1}, \Gamma^{i+1}, \lambda_k, \varepsilon_j^{\,i}\right\} \tag{1}$$

According to this notation, ANN represents the neural network models that best relate given inputs (i. e., $\left\{\sigma_j^{\,i}, \Gamma^i, \sigma_j^{\,i+1}, \Gamma^{i+1}, \lambda_k, \varepsilon_j^{\,i}\right\}$) to desired outputs (i.e., $\left\{\varepsilon_j^{\,i+1}\right\}$). The 20-NH-3 label denotes the architecture of the selected network. In our case, the network architecture is made up of 20 inputs, NH optimal number of hidden nodes which needs to be determined through training and online testing processes, and three desired outputs. Accordingly, all outputs at state $(i+1)$ are fed back to the network inputs to predict outputs at the $(i+2)$ state. This scheme is referred to as the one-point scheme. This implies that the stress- and strain-relating vectors at current state i plus the stress-relating vectors at one futuristic state $(i+1)$ are used as inputs to predict the strain-relating vectors at futuristic state $(i+1)$. In this case, strain-relating vector (outputs, ε_j^{i+1}, j=1,2,3) refers to the three strain components corresponding respectively to vertical Z-axis, horizontal X-axis and horizontal Y-axis at futuristic state $(i+1)$. On the other hand, our input nodes are made up of six major components namely: (i) stress-relating vectors (σ_j^i, j=1,2,3) representing vertical Z-axis, horizontal X-axis and horizontal Y-axis at state i, (ii) stress invariant-related vector $\left\{\Gamma^i\right\}$ at state i which contains various stress related parameters that are believed to affect the stress-strain response of the material under consideration, (iii) stress-relating vectors ($\sigma_j^{i+`1}$, j=1,2,3) representing vertical Z-axis, horizontal X-axis and horizontal Y-axis at futuristic state $(i+1)$, (iv) stress invariant-related vector $\left\{\Gamma^{i+1}\right\}$ at futuristic state $(i+1)$, (v) state of loading condition vector (λ_q, q = 1,2,3) that allows the network to differentiate between virgin, unloading and reloading regions, and (vi) strain-relating vectors (ε_j^i, j=1,2,3) vertical Z-axis, horizontal X-axis and horizontal Y-axis at state i. The stress invariant-related vector $\left\{\Gamma\right\}$ is made up of four major components: first stress invariant J_1, second deviatoric stress invariant J_{2D}, third deviatoric stress invariant J_{3D} and stress-path Lode angle θ ($\theta = 0^\circ$ for compression tests, $\theta = 30^\circ$ for simple shear tests, and $\theta = 60^\circ$ for extension tests).

Stage two: model training and testing

In this study, monotonic shear-loading ANN-based model was trained and tested. The architecture of the model suited for this loading condition can be extracted from Eq. (1). Accordingly it can be expressed in the following compact form:

$$\left\{\varepsilon_j^{\,i+1}\right\} = ANN_{17-NH-3}\left\{\sigma_j^{\,i}, \Gamma^{i}, \sigma_j^{\,i+1}, \Gamma^{i+1}, \varepsilon_j^{\,i}\right\} \quad (2)$$

Note that by eliminating the λ_q vector, the number of input parameters is reduced from 20 in the general purpose model to 17 for this model. To accomplish this objective, the monotonic shear stress-strain data reported by Oyesanya (1992) were used to train and test this model.

Stage three: model selection

Both statistical and graphical criteria were adopted to select the desired optimal network model. The selection process was composed of two consecutive steps. In the first step, statistical accuracy measures such as Sum-Squared-of-Error (SSE), Coefficient of Determination (R^2) and the Mean Absolute Relative Error (MARE) on both training and testing data sets were used to filter out the most promising optimal networks. Statistically, an optimal network is defined as the one with the best overall accuracy measures. In the second step, the predicted and actual stress-strain graphical responses for both training and testing experiments were produced for the most promising networks. Based on the overall graphical evaluation of each model's performance, the absolute optimal network was selected.

Stage four: neuro-mechanistic simulator implementation

In order to properly apply the ANN-based model to practical application, it is necessary to develop a simulation model that completely overcomes the need for experimental testing. As a result, a properly designed neuro-mechanistic simulator (NMS) should be able to efficiently predict all strain components for any specified stress path. To accomplish this task, the desired NMS was assembled by integrating the following five components: i) initial conditions and update module (ICUM), ii) monotonic shear loading model (MSLM), iii) monotonic HC loading model (MHLM), iv) linear elastic mechanics-based model (LEMM), and v) neuro-simulator activation/update manager (NSA/UM). Details of the step-by-step operation of the NMS are given in Zhang (2000) and Najjar and Zhang (2000).

COMARSION OF SIMULATED AND ACTUAL RESPONSES

To verify the capability of the developed neuro-mechanistic simulator, a series of compression, extension and simple shear monotonic stress-strain loading tests were simulated and compared with their corresponding actual experimental responses. Due to space limitations, only simulated response for CTE15 is presented and discussed in this section. Fig. 1a shows the agreement between simulated and experimental stress-strain responses in the ($\tau_{oct} - \varepsilon$) space. In this figure, the Y-axis represents the octahedral shear stress (τ_{oct}) while the X-axis denotes the axial strain (ε) for each direction. Similarly agreement between simulated and experimental volumetric response (negative sign means contraction, while positive sign indicates dilation) for the same test is presented in Fig. 1b. As can be noted from these direct comparisons of both stress-strain and volumetric responses, it is apparent that the developed shear-loading model predicts the stress-stain behavior of training and testing responses with high degree of accuracy. Moreover, the onset of dilation in the volumetric response space is nicely realized by the model. It is worthy to mention herein that the slight deviations between predicted and actual responses are not mainly due to modeling deficiencies.

CONCLUDING REMARKS

The developed neuro-mechanistic based simulator (NMS) was found efficient in characterizing the 3D monotonic stress-strain behavior of Canadian River sand subjected to both shear and HC-loading stress paths. Inclusion of the linear elastic model within the NMS assures that predictions at early loading shearing stress states are consistent with proven mechanics concepts. As a result, the developed NMS can easily be used to create in real time any desired monotonic stress-strain response for given stress path and known initial stress and strain conditions. The general purpose ANN-based model proposed in this paper allows modelers the flexibility to easily formulate specialized forms of this model to accommodate any limitations in the available experimental data in regard to stress path variations and loading conditions. The proposed neuro-mechanistic based methodology can also be applied to characterize the deformational behavior of other geomaterials such as rocks, concrete and soils.

ACKNOWLEDGEMENT

Financial support received from National Science foundation under grant CMS-98-00569 to conduct this research study is gratefully acknowledged.

REFERENCES

Ali, Hossam E. and Najjar, Yacoub M. 2000, *Simulating the CD Behavior of Nevada Sand Via Feed-Back Network Approach*. Proceedings (CD-ROM) of the *14th-ASCE Conference on Engineering Mechanics*, Austin, Texas, May 21-24.

Basheer, I. A. and Najjar, Y. 1998, *Modeling Cyclic Constitutive Behavior by Neural Networks: Theoretical and Real Data*. Proceedings of the *12th-ASCE Speciality Conference on Engineering Mechanics*, La Jolla, California, May 17-20, 1998. pp. 952-955.

Ellis, G.W., Yao, C., Zhao, R., and Penumadu, D. 1995, *Stress-Strain Modeling of Sand Using Artificial Neural Networks*. *J. Geotech. Engrg.*, ASCE, Vol. 121, No. 5, pp. 429-435.

Najjar, Y.M. 1990, *Constitutive Modeling and Finite Element Analysis of Ground Subsidence Due to Mining*. PhD dissertation, Univ. of Oklahoma, Norman, Oklahoma.

Najjar, Y.M., and Zaman, M.M. 1993, *Surface Subsidence Prediction by Nonlinear Finite Element Analysis*. *J. Geotech. Engr.*, ASCE, Vol. 119, No. 11, pp. 1790-1804.

Najjar, Y.M., and Basheer, I. A. 1996, *Stress-Strain Modeling of Sands Using Artificial Neural Networks*. Discussion, *J. Geotech. Engrg., ASCE*,Vol. 122, No. 11, pp. 949-951.

Najjar, Y. M. and Ali, H. E. 1999, *Simulating The Stress-Strain Behavior of Nevada Sand by ANN. Proc. of the 5th U.S. National Congress on Computational Mechanics (USACM)*, Boulder, Colorado, August 4-6, 1999, Book of Abstracts, pp.500.

Najjar, Yacoub and Zhang, Xiaobin. 2000, *Characterizing the 3D Stress-Strain Behavior of Sandy Soils: A Neuro-Mechanistic Approach*. In Numerical Methods in Geotechnical Engineering, *ASCE Geotechnical Special Publication No. 96*, Filz, G. and Griffiths, D. (Editors), pp. 43-57

Oyesanya, A. I. 1992, *Laboratory Testing and Constitutive Modeling of Canadian River Sand*. MS thesis, Univ. of Oklahoma, Norman, Okla.

Zhang, Xiaobin. 2000, *Characterizing the 3-D Stress-Strain Behavior of Canadian River Sand: A*

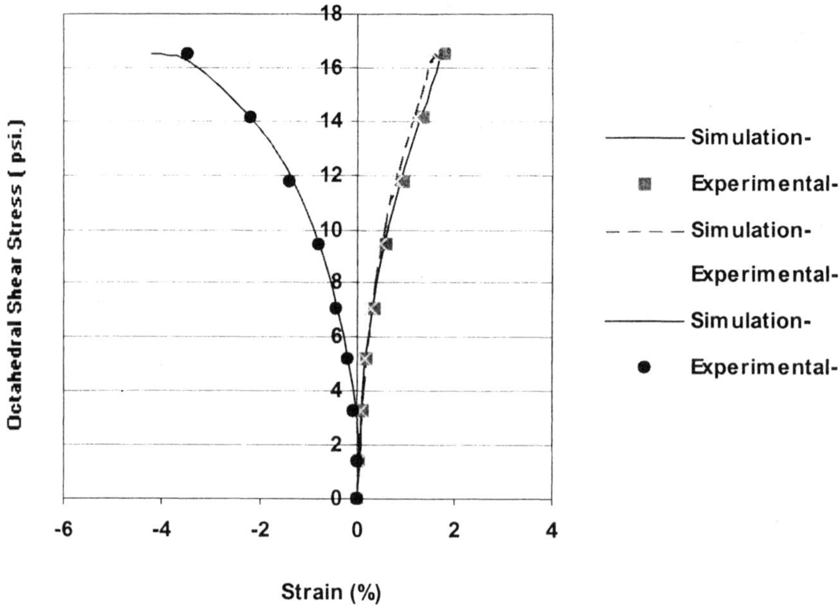

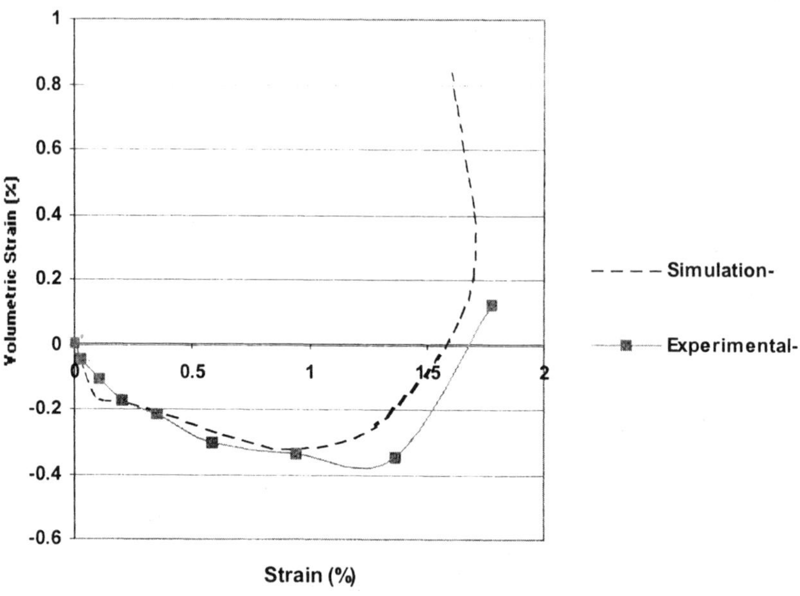

Fig. 1 Agreement between simulated and experimental responses of conventional triaxial extension test @ confining pressure of 103 kPa (CTE15): (a) stress-strain response; (b) volumetric response

SENSITIVITY ANALYSIS OF KNOWLEDGE DISCOVERY FOR FATAL CAR ACCIDENTS

WAN HUANG
Department of Computer Science
The University of Alabama
Tuscaloosa, Alabama, USA

HUI-CHUAN CHEN
Department of Computer Science
The University of Alabama
Tuscaloosa, Alabama, USA

DAVID CORDES
Department of Computer Science
The University of Alabama
Tuscaloosa, Alabama, USA

ABSTRACT

In this paper we investigate the sensitivity of the association rules generated for fatal car accidents by applying a distance-based rule extraction approach to the dataset from the Baldwin County of the state of Alabama. The applicability and portability of those rules are evaluated by analyzing datasets from other counties within the state of Alabama. Specifically, we study the relationship between the datasets and the effects of the dataset used for training in an effort to improve the reliability of the knowledge discovery process.

INTRODUCTION

Discovering underlying rules from existing traffic datasets often aids in better understanding the problem and identifying significant factors that contribute to major car accidents. However, discovering rules for fatal car accidents is not easy since most of the accidents reported are only car-damage (or other property-damage) or injury accidents, with very few (less than 1%) involving fatalities. Considering that current data mining methods (Agrawal and Srikant, 1994; Adamo, 2001; Kantardzic, 2003) often discount those improbable events, a distance-based rule extraction approach (Huang el al., 2003) has been designed that generates rules related to fatal car accidents using existing CARE (Critical Analysis Reporting Environment, developed at the University of Alabama) databases. The six best rules extracted from 1998 dataset of Baldwin County (the largest county in terms of area) in Alabama were compared with the rules extracted by the Apriori algorithm and decision tree algorithm. Rule accuracy analysis on the test set verified that the distance-based rule extraction approach outperformed the other methods when the target class is fatality.

In this paper, the sensitivity of the generated rules is investigated. The applicability and portability of those rules are evaluated by analyzing rule accuracy for datasets from other counties within Alabama. Public data, such as geographical location, size, and vehicle-related statistics, are considered in an effort to capture the characteristics of each county. This helps to determine the portability of the rules and to improve the reliability of the knowledge discovery process. Rule lifetime study is also considered.

Rule 1: If estimated speed (V064) is over 70 mph and roadway character (V077) is straight or curve hillcrest, curve up or down grade, then accident severity is fatal (confidence=28.5%, accuracy=50%)
Rule 2: If estimated speed (V064) is over 70 mph and roadway character (V077) is straight up grade or curve level, then accident severity is fatal (confidence=11.8%, accuracy=33.3%)
Rule 3: If estimated speed (V064) is over 70 mph and Traffic Control at Unit Causal (V079) is no passing zone, then accident severity is fatal (confidence=20%, accuracy=33.3%)
Rule 4: If other circumstances of driver causal (V038) is misjudged stopping distance, or over speed limit, or fail to yield roadway, and roadway character (V077) is straight or curve hillcrest, curve up or down grade, then accident severity is fatal (confidence=11.6%, accuracy=33.3%)
Rule 5: If condition of driver causal (V032) is unknown and roadway character (V077) is straight up grade or level curve, then accident severity is fatal (confidence=15.4%, accuracy=25%)
Rule 6: If causal driver's location (V040) is on roadway and estimated speed (V064) is over 70 mph, then accident severity is fatal (confidence=20.5%, accuracy=25%)
Rule 7: If causal driver claimed unseen object or was not in control (V038), and estimated speed (V064) is over 70 mph, then accident severity is fatal (confidence=11.1%, accuracy=25%)
Rule 8: If event location (V016) is off roadway and estimated speed (V064) is over 70 mph, then accident severity is fatal (confidence=20%, accuracy=22.2%)

Figure 1. The best eight rules extracted from Baldwin 98 dataset

Variable	Name	Variable	Name
V016	Event Location	V064	Estimated Speed of Causal Vehicle
V032	Condition of Causal Driver	V077	Roadway Character of Causal Unit
V038	Other Circumstances of Causal Driver	V079	Traffic Control of Causal Unit
V040	Event Location of Causal Driver		

Table 1. Variables appeared in the rules

PORTABILITY ANALYSIS OF EXTRACTED RULES

As mentioned previously, fatal car accidents are improbable (less than 1%) yet critical events among all types of car accidents, thus we may expect to see rules with support and confidence levels that are much lower than those of other dominant events. That is, if the rules we identify have confidence levels of greater than 10%, they actually increase our prediction ability by more than a factor of ten. Thus, the rules are considered valuable given that the event only occurs with a probability of 1%. Given this, we assume an acceptable rule confidence level of 10% for the analysis performed below. In this study, several aspects of the rules extracted from Baldwin County are selected for sensitivity analysis. We apply the eight rules with the highest accuracy (obtained from the test set) and acceptable confidence levels (>=10%, obtained from the training set) to each case. The new accuracy of each rule is obtained, allowing us to evaluate the portability of those rules. Figure 1 lists the eight rules to be studied.

Basically, these eight rules relate to the seven variables as listed in Table 1. Among those variables, V064 (Estimated Speed of Causal Vehicle), and V077 (Roadway Character of Causal Unit) are shown to be highly significant to the fatality level of the accidents. They appear in 75% and 50% of the extracted rules, respectively.

Sensitivity of Portability of Baldwin 98 Rules on County Statistics

There are 67 counties in the state of Alabama, with each county having its own unique characteristics. To study the portability of the above rules to counties other than Baldwin, a regression analysis was employed. In this regression analysis, a set of variables that are relevant when determining county characteristics are selected. They are area (AR), persons per vehicle (PV), and vehicles per mile (VM) (Remington, 2000). We first studied the relationship

Rule 1 Regression Analysis: accuracy versus VM, PV, AR				
The regression equation is: Accuracy = 0.216 – 0.000590 VM – 0.208 PV + 0.000116 AR				
Predictor	Coef	SE Coef	T	P
Constant	0.2159	0.1239	1.74	0.087
VM	-0.0005901	0.0003204	-1.84	0.071
PV	-0.2076	0.1167	-1.78	0.081
AR	0.00011593	0.00008306	1.4	0.168
S = 0.1394 R-Sq = 11.0%		R-Sq (adj) = 6.3%		

Figure 2. Results of regression analysis for Rule 1

Rule No.	Area (AR)	Persons per vehicle (PV)	Vehicles per road mile (VM)
1	0.168	0.081*	0.071*
2	0.856	0.505	0.346
3	0.867	0.323	0.653
4	0.000*	0.119	0.109
5	0.919	0.289	0.302
6	0.346	0.062*	0.267
7	0.880	0.073*	0.416
8	0.473	0.035*	0.239

Table 2. P value of each variable in each rule

between rule accuracy (dependent variable) and the selected relevant variables (independent variables) to determine how well these rules applied to the other counties. A linear multiple-variable regression model was assumed, and regression analysis was performed for all eight rules. Figure 2 illustrates the result of regression analysis for rule #1. The regression equation is given and each independent variable's prediction ability is represented by its corresponding P value. If P value is less than a pre-selected level of significance $\alpha = 0.1$, we consider that variable to be statistically significant in the prediction of rule accuracy. Thus Fig. 2 suggests that rule #1 is highly related to the variables PV (persons per vehicle) and VM (vehicle per mile) since their corresponding P values are 0.081 and 0.071, respectively.

In the regression equation of rule #1 given in Fig. 2, the coefficient of VM is negative, which suggests that the accuracy of rule #1 will decrease if the rule is applied to a county with a larger VM value. However, if the rule is applied to a county with a small VM value, the accuracy may increase. For example, we applied rule #1 to counties with small VM values, such as Marion (VM=27) and Randolph (VM=22), the accuracy was 0.5 for both counties. But when the same rule was used in counties with large VM values, such as Jefferson (VM=257) and Montgomery (VM=251), the accuracy was 0.0507 and 0, respectively.

Table 2 listed the computed P values of the three independent variables for all eight rules. Figures marked with an asterisk ("*") are considered statistically significant at $\alpha = 0.1$. Regression results indicate that area (AR) is not significant for most of the rules (except rule #4), thus we can remove this factor from consideration when applying these rules to other counties. Basically, the results of regression analysis indicate the portability of these rules and where these rules can be applied.

Sensitivity of Portability of Baldwin 98 Rules on Geographical Sections

Assuming that counties can be grouped into different sections based on their geographical information (such as relative geographical location, number of state highways crossing that county, or population), it is expected that counties

Rule No.	Number of counties with accuracy >=0.3	Distribution of counties among sections
1	6	3 in section 7, 2 in section 3, 1 in section 6
2	4	1 in section 7, 2 in section 5, 1 in section 6
3	10	5 in section 7, 2 in section 3, 2 in section 6, 1 in section 5
4	0	
5	2	1 in section 7, 1 in section 6
6	7	4 in section 7, 2 in section 3, 1 in section 6
7	4	2 in section 5, 2 in section 3
8	8	4 in section 7, 2 in section 3, 1 in section 4, 1 in section 6

Table 3. Number of counties with accuracy >=0.3 in each section for each rule

within the same geographical section tend to share similar rules. Thus a threshold clustering algorithm (Looney, 1997) was used to divide the 67 counties into seven sections based on the counties' geographical data (Remington, 2000). A portability study was then conducted on each section by applying the eight rules. For each rule, we counted the number of counties whose accuracy levels were greater than or equal to 0.3, as shown in Table 3.

It can be seen from Table 3 that, for each rule, section 7 takes around 50% of all the counties with accuracy greater than or equal to 0.3, while sections 1 and 2 have no rule with accuracy greater than or equal to 0.3. This means that sections 1 and 2 may have quite different accident causal patterns. In fact, the three counties in section 1 (Jefferson, Madison and Montgomery) and the only county in section 2 (Mobile), all have far greater VM values (above 250) than do the other counties (below 130). On the other hand, rules #1, #3, #6, and #8 have higher portability than the other four rules since they have more counties satisfying the accuracy threshold, while rule #4 (whose regression analysis P value of 0.0000 indicates that it is highly sensitive to the area of the county) is the least portable of the eight rules.

LIFETIME ANALYSIS OF EXTRACTED RULES

As information fed into the CARE system grows daily, the accident database should be mined periodically to capture new accident causal patterns. The discovered rules for car accidents have a limited lifetime. For example, when a new highway is built, or lighting conditions are improved, or even when a new mall is opened, the traffic patterns may change accordingly and thus the number of accidents may also change. The rules extracted before these changes occur may become inaccurate. A rule lifetime study is necessary in the sense that its results usually shed some light on what kind of rules (and thus which variables) are relatively sustainable over time. This is helpful for traffic professionals since the sustainability suggests that some factors are taking effect over time and thus discloses possible negligence of countermeasures.

To analyze the trend of accuracy change over time, a linear regression was employed. The eight rules, which were extracted from dataset of year 1998, were applied to datasets of years 1999, 2000, and 2001 of Baldwin County. Twelve datasets (each contained accident records in a 3-month period) were used, and the relationship between rule accuracy and time was sought . Figures 3 and 4 plot (in fitted regression equation) the rule accuracy relative to time. It is worth noting that the accuracy is consistently decreasing for all eight rules, although with different slopes. If the minimum accuracy is 0.1 for a rule to be valid, it is noted that rule #3 has the longest length of lifetime (approximately 4

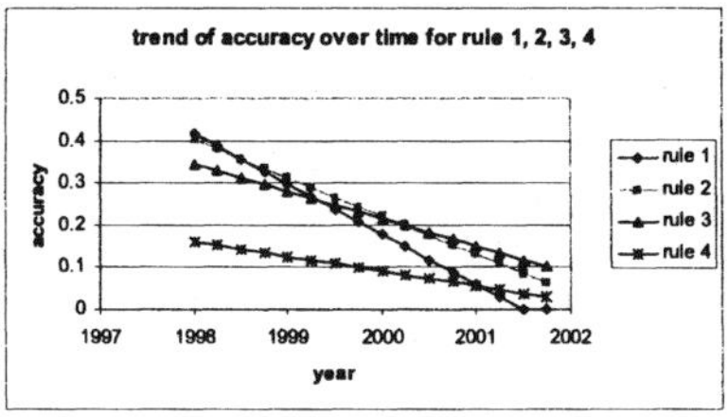

Figure 3. Trend of accuracy over time for rule 1, 2, 3, 4

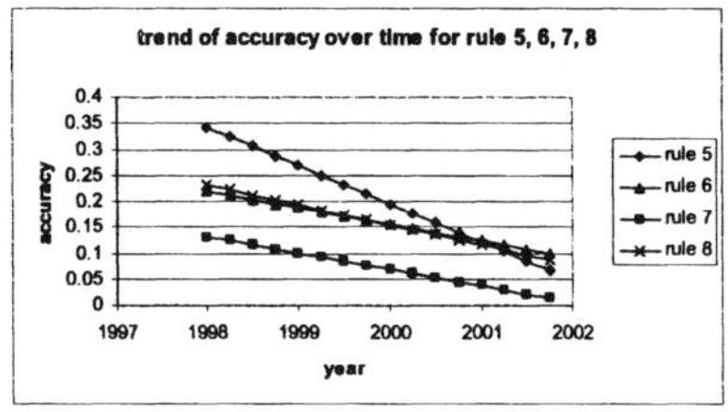

Figure 4. Trend of Accuracy over time for rule 5, 6, 7, 8

Figure 5. Lifetime comparison of rule 3 and rule 6

years). Rule #6 starts with a relatively lower accuracy than rule #3, however, it decreases much slower than rule #3, and turns out to have only a slightly shorter lifetime (approximately 3.5 years), as shown in Fig. 5. Rule #7 has the shortest lifetime (approximately 1 year) among the eight rules. On average, the lifetime is approximately three years for the eight rules. The results suggest the need for mining rules annually in Baldwin County, and that previously discovered rules may become invalid after 3 or 4 years. In addition, the decreasing accuracy trend implies that some other causal factors may take more effective role over time than those discovered in the previous year's dataset.

To verify this intuition, we applied the same distance-based rule extraction approach to datasets of years 1999, 2000, 2001 of Baldwin County and retrieved the rules as described in Figs. 6, 7 and 8. It is noted that some variables, which were not significant from the 1998 dataset, become more significant and appear in the later years' rules. This includes variables such as V032, condition of causal driver. Other variables become less significant and disappear from the previous year's rules. It is noted that variable V077 is still a very significant variable for 1999 dataset, which means that roadway characteristics are still an important factor in the fatalities of car crash for Baldwin County in 1999. However, for the datasets of 2000 and 2001, variables V033 (Officer's opinion of sobriety of causal driver), V098 (Officer's opinion of sobriety of driver 2), and V032 (Condition of causal driver) surface and become more significant variables. This change suggests that in recent years, although roadway condition has been improved, drivers generally do not pay enough attention to possible fatalities in car accidents. Moreover, they are ignorant of basic safe driving guidelines, such as not consuming alcohol or drugs and getting sufficient rest. Thus, some countermeasures targeting these problems may be considered.

CONCLUSION

In this paper, we conducted sensitivity studies on the portability of rules

Rule 1: If Condition of Causal Driver (V032) is apparently asleep, and roadway character (V077) is Straight Hillcrest, or Curve Down Grade, or Curve Up Grade, or Curve Hillcrest, then accident severity is fatal (confidence=11.1%, accuracy=100%).
Rule 2: If Condition of Causal Driver (V032) is apparently asleep, and there is no Traffic Control at Unit Causal (V079), then accident severity if fatal (confidence=13.3%, accuracy=50%)
Rule 3: If roadway character (V077) is Straight Hillcrest, or Curve Down Grade, or Curve Up Grade, or Curve Hillcrest, and light condition (V238) is darkness not lit, then accident severity is fatal (confidence=12%, accuracy=25%)

Figure 6. Rules from Baldwin 1999 dataset

Rule 1: If Officer's opinion of Sobriety of Driver 2 (V098) is Alcohol only, and light condition (V238) is darkness not lit, then accident severity is fatal (confidence=18.2%, accuracy=100%).
Rule 2: If Officer's opinion of Sobriety of Driver 2 (V098) is Alcohol only, and Local (V240) is Open county, then accident severity is fatal (confidence=15.4%, accuracy=100%)
Rule 3: If Highway class is County (V010), and Officer's opinion of Sobriety of Driver 2 (V098) is Alcohol only, then accident severity is fatal (confidence=33.3%, accuracy=50%)
Rule 4: If Highway class is County (V010), and Estimated Speed of Causal Driver (V064) is Over 70 MPH, then accident severity is fatal (confidence=35%, accuracy=50%)
Rule 5: If License status of Causal Driver (V029) is Suspended, and Condition of Causal Driver is Unknown (V032), then accident severity is fatal (confidence=20%, accuracy=50%)
Rule 6: If Condition of Causal Driver (V032) is Unknown, and Officer's opinion of Sobriety of Causal Driver (V033) is Alcohol only, then accident severity if fatal (confidence=18%, accuracy=50%)

Figure 7. Rules from Baldwin 2000 dataset

Rule 1: If Condition of Causal Driver (V032) is Apparently asleep, and Officer's opinion of sobriety of Causal Driver (V033) is Drugs or Both Alcohol and Drugs, then accident severity is fatal (confidence=14.3%, accuracy=100%)
Rule 2: If Condition of Causal Driver (V032) is Fatigued or ill, and Other Circumstances of Causal Driver (V038) is Driver not in control, then accident severity is fatal (confidence=12.5%, accuracy=100%)
Rule 3: If Condition of Causal Driver (V032) is Fatigued or ill, and Event Location of Causal Driver (V040) is Off-Roadway, then accident severity is fatal (confidence=11.1%, accuracy=50%)
Rule 4: If Condition of Causal Driver (V032) is Fatigued or ill, and Event Location (V016) is Off-Roadway, then accident severity of fatal (confidence=11.1%, accuracy=50%)

Figure 8. Rules from Baldwin 2001 dataset

associated with fatal car accidents. The results suggest that the variables contributive to fatalities vary with geographical characteristics and vehicle-specific statistics (persons per vehicle, vehicles per mile). Meanwhile, lifetime analysis of the rules extracted suggests that factors related to fatalities change over time, and frequent rule mining is necessary to capture properties of fatal car accidents and take effective countermeasures. The approach presented in this paper can also be employed for other applications where a set of association rules is mined.

REFERENCES

Adamo, J., *Data Mining for Association Rules and Sequential Patterns*, Springer, Berlin, 2001.

Agrawal R., and Srikant R., Fast Algorithms for Mining Association Rules, *Proceedings of International Conference on Very Large Data Base (VLDB)*, 487-499, Santiago, Chile, 1994.

Huang W., Chen H. and Cordes D., Discovering Causes for Fatal Car Accidents Using Data-Mining Techniques, submitted for publication, 2003.

Looney C. G., *Pattern Recognition Using Neural Networks*. Oxford University Press, NY, 1997.

M. Kantardzic. Data Mining-Concepts, Models, Methods, and Algorithms. IEEE Press, A John Wiley & Sons, Inc., NY, 2003

Remington W.C. (editor). *The Atlas of Alabama Counties*. University of Alabama, Tuscaloosa, 2000.

A METHOD FOR OBTAINING ROBUST MODULATION TEMPLATES FOR LASER PRINTERS

AKARAPHUNT VONGUNGHAE
Electrical & Computer Engineering
Department, University of Idaho.

RICHARD B. WELLS
MRC Institute, University of Idaho

ABSTRACT
High print density laser printers require pattern-dependent modulation of the laser beam to obtain best print quality. In a previous work we reported a neural network scheme for optimizing modulation templates in laser printers. Because the training method for this problem involves adaptation over a spatially-distributed output, it is observed that often the best neural network mapping from input bit-map to modulation pattern produces an equivocation, i.e. a one-to-many position-dependent and pattern-dependent optimum response. In this paper we introduce a novel method for mapping one-to-many solutions to an optimized one-to-one solution using statistical techniques. The technique is a two-pass training algorithm. The first pass finds a set of optimized templates for each input bit-map allowing position-dependent one-to-many mappings to occur. In the second pass, a statistical procedure is used for obtaining a robust one-to-one template mapping. The neural network is then re-trained with this mapping, using a previously reported variation of the well-known Madaline II rule, to generalize to a final mapping. The re-trained neural network is then used to obtain the final modulation templates. The method is applied to an experiment from the previous work for doubling the resolution of a laser printer. The experimental print quality resulting from this method compares favorably to that obtained using human-generated templates.

I. INTRODUCTION

Printing a high quality image requires laser beam modulation. The laser beam inside printers is modulated by scanning the input image with a certain size window i.e. 4 pixels by 3 pixels, and mapping those scanned bitmap window patterns into pulse patterns to modulate the laser beam. Finding a set of modulation templates mapping scanned bitmap window patterns into modulation patterns usually is a time consuming task. In this paper we discuss neural network generation of templates for Resolution Enhancement Technology (RET) [1] – [3]. In a previous work [4] an automatic system for finding a set of the modulation templates by using a feed-forward neural network as a compensator was proposed. The system in [4] and the method in the next section could generally be applied to many cases of RET. In this paper finding templates for doubling the resolution of a laser printer is demonstrated via an experiment in order to present and explain the proposed method. The system block diagram from [4] is shown in Figure 1. For resolution doubling a set of bitmap training images are scanned using a 4x3 pixel window. The 4x3 pixel scanned bitmap is the input to the neural network compensator, and the NNC produces the modulation pattern as the output. The modulation pattern feeds into a signal processing model (SPM1X) to model laser exposure [5] on the photoconductor drum inside the printer [6], [7]. Another signal processing model (SPM2X) models the desired response i.e. the exposure of a higher resolution printer with twice the resolution. The adaptation algorithm that is established in [4] uses special patterns in the algorithm as an advisor. The NNC seeks advice from the

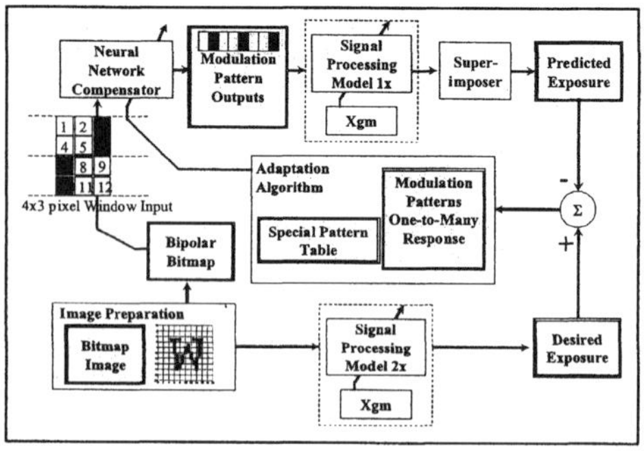

Figure 1: Model – Reference Neural Network Scheme Block Diagram

special pattern table when it does not produce an improved exposure response for the 4x3 pixel scanned bitmap input. The algorithm in [4] is simply explained as three cases in the following,

- If the NNC produces a pattern that has a higher squared error than the previous modulation pattern in the one-to-many response table at the corresponding position, the NNC is adapted toward the previous modulation pattern from the 'one-to-many' response table.
- If the NNC produces a modulation pattern with the same error as the previous modulation pattern, the algorithm seeks the best modulation pattern directly from the special patterns and the NNC is adapted once using that pattern as the desired result. If the modulation pattern of the adapted NNC with the direct search pattern has a lesser squared error, that pattern of the adapted NNC will be inserted into the one-to-many response table.
- If the NNC has already produced the modulation pattern that has a lesser squared error than the previous modulation pattern, the NNC – produced pattern is inserted into the one-to-many response table.

The algorithm in [4] allows one-to-many mapping optimization. The same 4x3 pixel scanned bitmap input at a different position on the page might have a different modulation output because the best optimized pattern at a local position depends on the combination of modulations and pixels around itself. After the optimization is converged and settled, the one-to-many response table's updating will stop. Because of this one-to-many adaptation, the NNC exhibits an equivocation. To eliminate the equivocation and generalize the template solution, a statistical procedure to map the one-to-many response into a one-to-one mapping response is proposed in the section II.

II. METHOD FOR OBTAINING ROBUST MODULATION TEMPLATES

To illustrate the general idea of this method, Figure 2 shows a graphical result calculated by this method. After a first pass using the adaptation algorithm and the system in [4] has been done, the probability of turning-on each pulse bit from all modulation patterns of a particular group of scanned bitmap inputs from all location in the training images is calculated using the stored one-to-many table. In Figure 2(a), the modulation pattern is divided into 16 pulse bits; the probability curve reveals which pulses are likely to be turned on for this particular group of scanned bitmap inputs. The scanned bitmap inputs

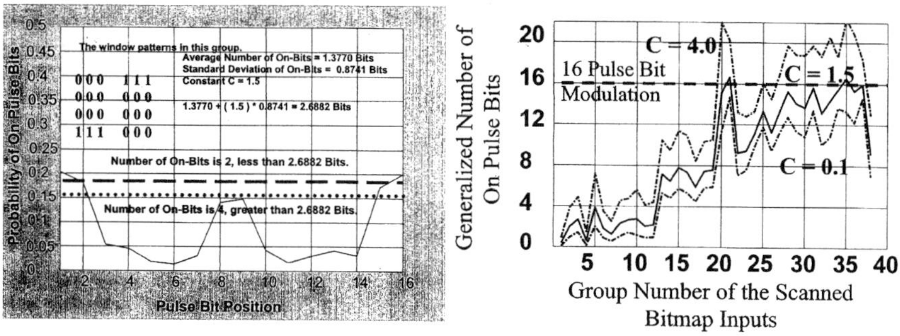

(a) (b)

Figure 2: (a) Probability of turning on of each modulation pulse
(b) Finding bit saturation parameter C

are grouped using a set of flipping operations. Three flipping operations are defined as follows,

$$wUD = flipUpDown(w) \quad (1)$$

$$wLR = flipLeftRight(w) \quad (2)$$

$$wUDLR = flipLeftRight(wUD) \quad (3)$$

where,

w	is scanned bitmap input,
wUD	is the result of flipping the input w up-down,
wLR	is the result of flipping the input w left-right
$wUDLR$	is the result of flipping the result of wUD left-right.

The functions *flipUpDown( )* and *flipLeftRight( )* are defined as fowllows,

$$\left(\begin{bmatrix} a11 & a12 & a13 \\ a21 & a22 & a23 \\ a31 & a32 & a33 \\ a41 & a42 & a43 \end{bmatrix}\right) \xrightarrow{flipUpDown(\,)} \left(\begin{bmatrix} a41 & a42 & a43 \\ a31 & a32 & a33 \\ a21 & a22 & a23 \\ a11 & a12 & a13 \end{bmatrix}\right)$$

$$\left(\begin{bmatrix} a11 & a12 & a13 \\ a21 & a22 & a23 \\ a31 & a32 & a33 \\ a41 & a42 & a43 \end{bmatrix}\right) \xrightarrow{flipLeftRight(\,)} \left(\begin{bmatrix} a13 & a12 & a11 \\ a23 & a22 & a21 \\ a33 & a32 & a31 \\ a43 & a42 & a41 \end{bmatrix}\right)$$

From the three operations above, the other scanned bitmap inputs are compared to the results of these 3 operations. If any scanned bitmap pattern is the same as one of the results, *wUD*, *wLR*, *wUDLR*, that scanned bitmap pattern is assigned to the group of the scanned bitmap pattern *w*. As an example, in Table 1 the bitmap input *w* at the row 2 and column 4 is the input to the set of flipping operations above. The results *wUD*, *wUDLR* are the scanned bitmap at the row 2, column 5 and column 6 respectively. The result *wLR* is row 9, column 5. These scanned bitmap inputs are grouped as the group B in Table 1.

To make a decision how many and which pulses should be turned on for a particular group of scanned bitmaps, the average and standard deviation of the number of on pulses in modulation patterns for that particular scanned bitmap group are calculated. With the

Table 1: Example of Calculations

1	Group Index j		A	B	B	B	C
2	4x3 Pixel Scanned Bitmap Inputs		0 0 0	0 0 0	0 1 1	1 1 0	0 0 0
			0 0 0	0 0 0	0 0 0	0 0 0	1 1 1
			0 0 0	0 0 0	0 0 0	0 0 0	0 0 0
			0 0 0	1 1 0	0 0 0	0 0 0	0 0 0
3	(Mod. Pattern), number of the patterns (b3 b2 b1 b0) i , n i	i = 1	(0000), 60	(0100), 5	(0010), 4	(1000), 10	(0110), 8
		i = 2	(0010), 2	(1000), 6	(0100), 2	--	(0100), 5
		i = 3	(1000), 1	(0001), 1	--	--	--
		i = 4	(1010), 1	(0101), 1	--	--	--
		i = 5	(1100), 1	(1101), 1	--	--	--
		i = 6	(0001), 2	--	--	--	--
		i = 7	--	--	--	--	--
4	Sum of On Pulse Bits	(s3,s2,s1,s0)	(3,1,3,2)	(7,7,0,3)	(0,2,4,0)	(10,0,0,0)	(0,13,8,0)
5	Total Number of Mod Patterns	N	67	14	6	10	13
6	On-Bit Average and Standard Dev of Mod Patterns	AvOnBit	0.134		1.100		1.615
		StdOnBit	0.423		0.403		0.506
7	Generalized Number of On Pulse Bits	=AvOnBit + C*StdOnBit where C = 1.5	0.768		1.704		2.374
8	Probability of On-Modulation Bit	bit 3	0.045		0.567		0.000
		bit 2	0.015		0.300		1.000
		bit 1	0.045		0.133		0.615
		bit 0	0.023		0.100		0.000
9	Additional Bitmap Flip Inputs		None		0 0 0		0 0 0
					0 0 0		0 0 0
					0 0 0		1 1 1
					0 1 1		0 0 0
10	Modulation Pattern Mapping Result	(b3 b2 b1 b0)	(0 0 0 0)		(1 0 0 0)		(0 1 1 0)

average and standard deviation the generalized number (GOn_j) of on-pulse bits for each group of scanned bitmap inputs is calculated by using an empirical relationship,

$$C_j = \left(\frac{(Number\ of\ Modulation\ Bits\ + 1 - AvOnBit_j)}{StdOnBit_j} \right) \quad (4)$$

$$C = Min(C_j) \quad (5)$$

$$GOn_j = AvOnBit_j + C \times StdOnBit_j \quad (5)$$

where,

- for each group j of scanned bitmap inputs,

  j is the group index of scanned bitmap inputs,

  GOn_j is the generalized number of on-pulses in group j,

  $AvOnBit_j$ is the average number of on-pulses in group j,

  $StdOnBit_j$ is the standard deviation of on-pulses in group j,

- for all groups of scanned bitmap inputs,

  C is a bit saturation parameter.

This GOn_j value is used as an upper bound for finding the modulation pattern. After the probability curve and GOn_j are calculated, the position that has a highest probability is assigned to be on-pulse first. For the curve in Figure 2(a), the possible modulation patterns for this group of scanned bitmap inputs are ‘1000000000000000’, ‘1000000000000001’, ’1100000000000001, ’1100000000000011’, ‘1100000010000011. Each modulation bit has a corresponding on probability. Because the total number of on pulses for modulation pattern should be equal or less than GOn_j (2.6882), the pattern for this scanned bitmap input group is '1000000000000001, i.e. the on-pulse bits are the 2 bits having the highest on probability. The range of values GOn_j is controlled by the bit saturation parameter *C*. Equations 4 and 5 provide the value of *C* for the given number of bits in the modulation pattern. Figure 2(b) shows an example of the range of GOn_j with the modulation pattern divided into 16 bits. The solid line is the result GOn_j when *C* is equal to 1.5 (calculated by Equation 4 and 5). The dashed lines are other GOn_j curves with *C*'s equal to 4.0 and 0.1. With $C = 4.0$ the templates would produce images that are too dark because GOn_j exceeds the available 16 bits of modulation. With $C = 0.1$ the template would produce white – saturated images.

The detailed stepwise procedure is as follows. Table 1 shows a calculation example for the resolution doubling templates; the number of pulse bits for the laser modulation is 4 pulse bits. The calculation steps are:

1. Run the optimization with the adaptation algorithm in [4] until the convergence is satisfied or the rate of one-to-many updating is lower than a preset value.
2. Collect mapping data (all pairs of 4x3-pixel scanned bitmap input and modulation pattern) from one-to-many response table in the adaptation algorithm. Rows 2 and 3 of the table show the results of this step.
3. Group the scanned bitmap inputs as defined and explained above. The results of this step are shown in row 1. The scanned bitmap inputs are classified into 3 groups, A, B and C. This step also adds additional bitmap input patterns to the training set for increasing the symmetry in the response of the NNC. Row 9 shows the results of additional bitmap inputs. Note that these "symmetry bitmaps" are obtained from operation (1) – (3).
4. Calculate the statistical values of modulation patterns of each group of scanned bitmap inputs. The averages, standard deviations, and probabilities are shown in rows 4, 5, 6 and 8.
5. Calculate the generalized number of on-pulse bits for each group of scanned bitmap inputs (using Equation 6) and the bit saturation parameter C (using Equations 4 and 5). In Table 1, this calculation step corresponds to the row 7 when C is equal to 1.5.
6. Calculate the one-to-one modulation template mapping using the generalized number as the upper bound value. The on pulses are assigned to the bit that has the higher probability first, and are followed by the lower ones until the number of on pulses satisfies the generalized number. This step is the same as the previous explanation for Figure 2, corresponding to the rows 7, 8, and 10.
7. Re-train the NNC with the one-to-one solution and generalize the NNC using the Madamant Rule [8] (a Madaline II-like algorithm [9] with improved generalization).

III. EXPERIMENTS AND RESULTS

For the experiments in this paper, a set of vertical and horizontal lines comprise the training images. The space between the lines in the training images is 7 pixels. The line thicknesses of the training images are varied from 1 to 6 pixels and the size of training images is 30×30 pixels. Two test images, "man" and "stingray", are used to test the modulation patterns. The NNC is never shown the test images during training. Figure 3 shows the test images generated by a native 1200 dpi printer. SPM2X modeled this native 1200 dpi printer. This is the model reference printer that the 600 dpi printer with modulation templates is to mimic. SPM1X modeled a 600 dpi printer that was used for

Figure 3: Printed with the Native 1200 dpi Printer

Figure 4: Printed with the 600 dpi Printer and the Templates Obtained by the Method Presented in this Paper

Figure 5: Printed with the 600 dpi Printer and Human Generated Templates

Figure 6: Printed with the 600 dpi Printer and the Templates Obtained by Omitting Step 3 of Procedure in Section II

printing the hard copies of Figure 4, 5 and 6. Figure 4 shows the result of the 600 dpi printer with the templates of the re-trained NNC optimized by the method presented here.

Figure 5 shows the images printed with the 600 dpi printer using a set of human generated templates. Comparing the result Figure 4 against the real 1200 dpi Figure 3 and the human generated template image Figure 5, differences in the "man" image are not easily discernable. If we look at the "stingray" images, one is able to see differences in the results because human eyes have a low pass filter characteristic. Most areas of the "man" image have higher spatial frequencies than the "stingray" image.

Figure 6 shows the result if we omit step 3 of the procedure in the section II where scanned bitmap inputs are grouped by the flipping operations. The result is poorer than that in Figure 4, especially on the "stingray" image. This shows that the eye is sensitive to symmetrical patterns and grouping scanned bitmap inputs by the flipping operations is necessary. The NNC has no sense of symmetry. The training algorithm has to provide symmetry information to the NNC as done in the step 3. Flipping scanned bitmap inputs, comparing them, and grouping them together provide the symmetry information to the re-trained NNC.

VI. CONCLUSIONS

This paper reports a new method for solving the equivocation problem encountered due to temporal-spatial optimum weight variability in large printed images. Good quality images, comparable to those obtained from a native-mode 1200 dpi printer, have been demonstrated on a 600 dpi printer using NNC generated modulation templates. The result in Figure 4 compares well with Figure 3. Symmetry information has to be provided to the NNC because human eyes are sensitive to symmetry, but NNC lacks that sense.

ACKNOWLEDGMENT

This work was supported by a grant from the Hewlett Packard Co, the NSF – Idaho EPSCoR Program, and by the National Science Foundation under award number EPS – 0132626.

REFERENCES

[1] DP-Tek, Inc., Resolution transforming raster-based image system, United States Patent # 5,134,495, Jul. 28,1992.

[2] DP-Tek, Inc., Interleaving vertical pixels in raster-based laser printers, United States Patent # 5,193,008, Mar. 9,1993.

[3] DP-Tek, Inc., System and method for enhancing graphic features produced by making engines, United States Patent # 5,515,480, May. 7,1996.

[4] Richard B. Wells, and Akaraphunt Vongkunghae, "A Model – Reference Neural Network Scheme for Obtaining Laser Printer Modulation Templates", Proc. 28th Ann. Conf. Indus Electron. (IECON'02), Sevilla, Spain, Nov 5 -- 8 2002, pp. 1872 – 1875.

[5] Richard B. Wells, Akaraphunt Vongkunghae and Jang Yi, "A Signal Processing Model for Laser Print Engines", Proc. 28th Ann. Conf. Indus Electron. (IECON'02), Sevilla, Spain, Nov 5 – 8 2002, pp. 1514 – 1519.

[6] J. Yi, "A xerographic simulation model," M.S. thesis, University if Idaho, Moscow ID, May 1999.

[7] E. M. Williams, *The physics and technology of xerographic processes*. Malaba, Florida: Krieger Publishing Company, 1993

[8] Hua Lin, "An image extraction and recognition system using neural networks," M.S thesis, University of Idaho, Moscow, ID, 1999.

[9] B. Widrow and M. A. Lehr, "30 years of adaptive neural networks: perceptron, madaline, and backpropagation," *Proc. IEEE*, vol. 78, no. 9, pp. 1415-1442, Sept. 1990.

A MAX-PLUS, DIOID BASED NEURAL NETWORK FOR DISCRETE EVENT SYSTEM MODELING

JEFF DALTON
Department of Mechanical and Aerospace Engineering and Engineering Mechanics University of Missouri-Rolla

MATT INSALL
Department of Mathematics
University of Missouri-Rolla

DONALD WUNSCH
Department of Electrical and Computer Engineering University of Missouri-Rolla

ABSTRACT
Discrete event system simulation models provide an alternative methodology for analyzing behavior of certain types of systems where we wish to emphasize the reaction of components within the system to events that occur either within the system or in the system environment. The implementation of discrete event system simulations has been primarily algorithmic rather than computational. Specific non-traditional algebraic structures have been used to provide a mathematical foundation for modeling discrete event systems in a more formal and rigorous setting. In this paper we present a tutorial introduction to the use of so called max-plus algebras formulated in a universal algebraic setting for discrete event system simulation. We make the observation that slightly modified versions of standard feed-forward neural networks can be used to implement discrete event system simulations in the max-plus algebraic setting.

INTRODUCTION

The logical and mathematical modeling of the behavior of systems has historically been a corner stone for the engineering process. We model systems to better understand their behavior, to analyze and improve their performance, and to modify their behavior to achieve desired results. Traditional approaches to mathematical modeling of systems have been based on the development of differential or difference equations that involve the evolution of a system state that is dependent on time and external influences in the system's environment. For these approaches the system state is a set of one or more real valued functions. In a discrete time model composed of difference equations, states change at discrete points in time. Systems modeled by linear ordinary differential equations have solutions that are continuous in time. There is a large class of problems that are not adequately modeled using either difference or differential equations. Examples include planning and scheduling problems, discrete part manufacturing systems, supply chain management, queuing problems, multitasking operating systems and others. These types of problems are characterized by a finite, discrete set of states with transitions between states that are dependent upon asynchronous events that occur in the system (Casandra 1993). State trajectories are piecewise constant functions.

Several approaches to modeling discrete event systems have been studied extensively over the past 20-30 years. For example, Cassandra (1993) discusses both timed and untimed petri nets as methodologies for modeling discrete event systems.

Stochastic methods and dioid algebras are also discussed by Cassandra (1993). Implementations of these methods are typically algorithmic in nature and use software simulation. They are difficult to describe mathematically. Ziegler et. al. (2000) have developed a formal, set theoretical specification of discrete event systems. Again, implementations are algorithmic. Although definitions in the formalism are based on algebraic concepts, mathematical descriptions and formal mathematical proofs are difficult. In both Cassandra (1993) and Zeigler, et. al (2000) the dynamics of discrete event systems are considered in conjunction with traditional dynamic system theory in the form of hybrid systems and they make a clear case that the most general models must include each approach for general system modeling.

Baccelli, et. al. (1992), develop discrete event system modeling and a system theory that is based on algebraic structures specifically chosen for their application to discrete event systems. Their development depends on an algebraic theory surrounding the definition of a dioid which is an algebraic structure composed of operations with specific properties defined on product spaces of an underlying set. We will use specific instances of the dioid structure where we replace the usual addition operation with the *max* function and the usual multiplication operation with addition. We (and others) refer to this particular dioid structure as a max-plus algebra. Using the tools developed by Baccelli, et. al. (1992), we have a formal theoretical basis for the development and study of discrete event system models that is founded in abstract algebra. The theory developed by Baccelli et. al (1992) parallels that of conventional system theory with required adaptations made in the theory so that a formal mathematical structure exists specifically for the purpose of modeling discrete event systems. We propose to exploit that foundation and use it in the context of neural network systems.

By slightly modifying the presentation of the theory by Baccelli et. al. (1992) we can take advantage of several key features in the theory of universal algebra as presented by Burris and Sankappanavar (1981). Universal algebra has been developed and widely disseminated in the mathematical community over the last 20 years. The theoretical development of universal algebra is based in an equational logic that is particularly suited for computation. Definitions, axioms and proofs in universal algebra can be studied using theorem proving software that has been developed. Goals for the study of universal algebra include 1) the comparison of structures that are very different in presentation but may have common algebraic features, 2) the generalization and unification of properties and characteristics in large classes of problems, and 3) a deeper understanding of system concepts and interrelationships through a higher level of abstraction (Burris and Sankappanavar, 1981).

The use of neural networks for system modeling and control offers several advantages over conventional approaches. Neural networks provide the ability to decompose large centralized model representations into representations that are distributed and amenable to parallel computations. Static network structures exist that can represent complex functional relationships between input and output spaces and recurrent structures can be used for modeling dynamic behavior of complex systems. In this paper we show a direct mapping of algebraic structures suitable for discrete event system simulation into a neural network for parallel and distributed processing. This work can be used in conjunction with prior work to develop hybrid dynamical systems that are described using neural networks. The universal algebraic context can be used as a basis for comparison between max-plus neural networks described in this paper and neural networks in the traditional algebra.

In the following sections we give a formal definition for the dioid algebraic structure that is used to develop a theory for discrete event system modeling. Examples illustrate the basic operations on the max-plus dioid which we then extend to the matrix case. Next we show how the matrix dioid can be mapped to a max-plus neural network. The use of the max-plus neural network structure in a discrete event system is demonstrated and we conclude with directions for further research.

DIOID THEORY AND MAX-PLUS UNIVERSAL ALGEBRAS

In the universal algebraic setting we will use operations defined on product spaces created by taking direct products of an underlying set. We will use D^0 to denote the singleton zero, $D^0 = \{\phi\}$. An operation defined on D^0 (a nullary operation) simply chooses a constant from the underlying set D. Unary operations are defined on $D^1 = D$, binary operations are defined on $D^2 = D \times D$ and n-ary operations are defined on D^n. With these ideas in mind we can now formalize the notion of a dioid structure that will be used below. The following definition is based on that given in [3].

A dioid is an algebraic structure, $\mathcal{D} = (D; \varepsilon; e; \oplus; \otimes)$, with underlying set, D, two nullary operations, zero element ε: $D^0 \rightarrow D$ and identity element e: $D^0 \rightarrow D$, two binary operations, $\oplus: D^2 \rightarrow D$ that is associative, commutative and idempotent and $\otimes: D^2 \rightarrow D$ that is distributive over $\oplus$, is associative and is absorbing with respect to ε.

As an example of a dioid consider the underlying set consisting of the reals extended on the left, $D = R \cup \{-\infty\}$. Take constant bottom element, $\varepsilon = -\infty$, and constant identity element, $e = 0$. Choose the *max* function as the first operation and the usual addition operation as the second operation. We further stipulate that $\varepsilon + a = \varepsilon$ for any a in D. We can concisely write these requirements in a universal algebraic structure as

$$\mathcal{M} = (D; \varepsilon; e; \max(\,); +) \tag{1}$$

The following simple calculation demonstrates the use of max-plus algebraic concepts

$$(\varepsilon \oplus 5) \otimes (e \oplus 3) \oplus 7 = \max(\max(\varepsilon, 5) + \max(e, 3), 7) = 8 \tag{2}$$

MATRIX DIOIDS IN THE MAX-PLUS SETTING

Next we consider the extension of the max-plus universal algebraic structure above to a matrix setting. In the dioid matrix multiplication, addition replaces scalar multiplication and then the maximum operator is applied over all elements of the "product" of rows and columns. Examples of operations in the matrix dioid structure follow:

$$\begin{bmatrix} 2 & 5 \\ 1 & \varepsilon \end{bmatrix} \otimes \begin{bmatrix} 5 \\ 3 \end{bmatrix} = \begin{bmatrix} \max(2+5, 5+3) \\ \max(1+5, \varepsilon+3) \end{bmatrix} = \begin{bmatrix} 8 \\ 6 \end{bmatrix} \tag{3}$$

$$\begin{bmatrix} 2 \\ \varepsilon \end{bmatrix} \otimes 1 = \begin{bmatrix} 2+1 \\ \varepsilon+1 \end{bmatrix} = \begin{bmatrix} 3 \\ \varepsilon \end{bmatrix} \tag{4}$$

$$\begin{bmatrix} 8 \\ 4 \end{bmatrix} \oplus \begin{bmatrix} 3 \\ \varepsilon \end{bmatrix} = \begin{bmatrix} \max(8,3) \\ \max(4, \varepsilon) \end{bmatrix} = \begin{bmatrix} 8 \\ 4 \end{bmatrix} \tag{5}$$

$$\begin{bmatrix} 1 & 2 \end{bmatrix} \otimes \begin{bmatrix} 5 \\ 3 \end{bmatrix} = \max(1+5, 2+3) = 6 \tag{6}$$

MAX-PLUS NEURAL NETWORKS

The reader may be familiar with matrix formulations for both linear systems of difference equations in discrete time simulation models and linear systems of first order differential equations in continuous time systems. The following matrix formulation for a linear system in the max-plus matrix dioid structure described above provides a method for describing recursion expressions of particular interest in modeling discrete event systems.

$$\begin{aligned} x(k+1) &= A \otimes x(k) \oplus B \otimes u(k) \\ y(k) &= C \otimes x(k) \end{aligned} \tag{7}$$

With the following selection of matrices, initial state, and input function values we see that Eq. (3) – (6) are calculations of a linear system of the form given in Eq. (7).

$$A = \begin{bmatrix} 2 & 5 \\ 1 & \varepsilon \end{bmatrix},\ B = \begin{bmatrix} 2 \\ \varepsilon \end{bmatrix},\ C = \begin{bmatrix} 1 & 2 \end{bmatrix},\ x(1) = \begin{bmatrix} 5 \\ 3 \end{bmatrix},\ u(1) = 1 \tag{8}$$

The calculations shown in Eq. (7)–(8) can be mapped into a neural network structure as shown in Fig. 1. The input layer represents elements of the state vector, x, and the input vector u, both at index k. Weights in the interconnection matrix between the input and hidden layers have elements that appear in the A and B matrices in Eq. (7) and are added to the input layer nodes creating a vector that appears at the hidden layer nodes. We use the *max* function as the activation function in the hidden layer. The output of the hidden

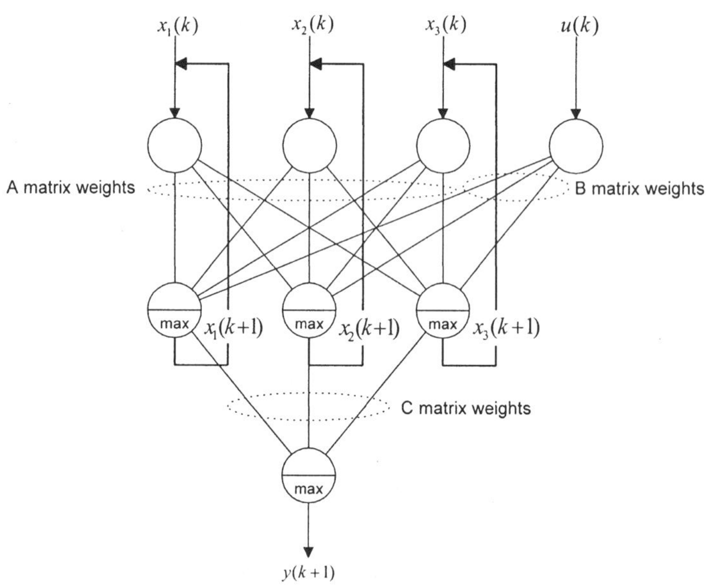

Figure 1: Max-Plus Recurrent Neural Network

layer is then the vector, $x(k+1)$. From Fig. 1,

$$x_i(k+1) = \max\left\{\max_j[x_j(k) + a_{ij}], \max_l[u_l(k) + b_{il}]\right\} \tag{9}$$

$$y_i(k) = \max_j[x_j(k) + a_{ij}] \tag{10}$$

Interconnection weights between hidden layer and the output node shown in Fig. 1 represent values of the C matrix in Eq. (7). Hidden layer node values are added to their respective connection weights and appear as a vector at the output node. The *max* operator is again applied to all values at the output node to produce the final output value, $y(k+1)$. We now demonstrate how this neural network structure based on the max-plus matrix diode can be applied in a simplified manufacturing system model.

AN ILLUSTRATIVE EXAMPLE

The following example illustrates a way that a max-plus neural network may be developed for the purpose of modeling a problem in discrete event systems. Consider the graph shown in Fig. 2. Nodes of this graph may represent places in a manufacturing process where operations are performed. Edges in the graph describe the movement of discrete parts from one place in the manufacturing system to another place. Edge weights of the graph define the length of time that it takes for a part to be processed at the originating node and be moved to the destination node for processing. The $u(k)$ and $y(k)$ nodes in the graph represent an input node (with no predecessor) and an output node (with no successor), respectively. In a more general setting there may be more than one input or output node for the system. We consider the single input, single output case here for clarity. The input sequence, $u(k)$, represents the time intervals at which parts appear at the input (receiving area) of the manufacturing system relative to a fixed initial time. The output sequence, $y(k)$ represents the time intervals when finished parts emerge from the manufacturing system (shipping area) relative to the fixed initial time. Based on the graph topology, we can formulate the matrices of a max-plus linear system of the form given in Eq. (7) as follows:

$$A = \begin{bmatrix} a_{11} & \varepsilon & \varepsilon \\ \varepsilon & a_{22} & \varepsilon \\ a_{31} & a_{32} & a_{33} \end{bmatrix}, B = \begin{bmatrix} b_{11} \\ b_{21} \\ \varepsilon \end{bmatrix}, C = \begin{bmatrix} \varepsilon & \varepsilon & c_{13} \end{bmatrix}, x(0) = \begin{bmatrix} \varepsilon \\ \varepsilon \\ \varepsilon \end{bmatrix}, u(1) = 1 \tag{11}$$

Elements of A represent processing and transition times between internal places in the manufacturing system, elements of B represent transitions between receiving areas and the initial processing places and elements of C represent transitions between final processing places and the shipping area.

Mapping these matrices to the neural network structure as described above and shown in Fig. 1 provides a way to use the neural network to carry out the recursive steps of Eq. (7). The output trajectory from a simulation predicts delivery times for our final product and with this we can, for example, ask the question: given an expected receiving schedule can we meet our desired shipping schedule subject to the processing times encoded in the neural network? Components of the state vector trajectory, $x_i(k)$, give times when discrete parts arrive for processing at place i in the manufacturing system. By analyzing the state trajectory we can assess manufacturing system balance, bottleneck nodes, and flow of material through the system.

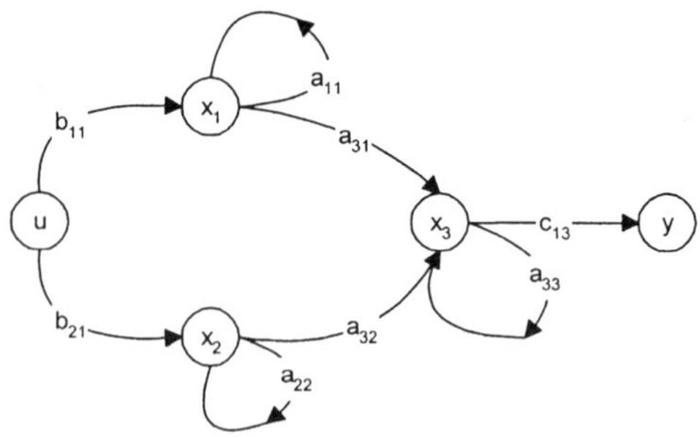

Figure 2: Discrete Event System Graph

CONCLUSIONS

We have presented a tutorial introduction of the use of dioid theory in the context of a max-plus algebra for developing discrete event system simulations and make the observation that the max-plus algebraic system theory can be mapped to a recurrent neural network structure. Using this, identification and control problems in discrete event systems are mapped to neural networks. Modified neural network training can now be developed to train neural network models of complex systems in the application areas noted. There are several extensions to this work that our group is considering. Can we incorporate neural network training algorithms with constraints on interconnection weights to suggest changes in a manufacturing system that will enable us to meet delivery schedules given receiving schedules? If we know the structure of the manufacturing system and we know shipping and receiving schedules, can we predict the times and utilization of places within the manufacturing system? How do we manage the receiving schedules in order to meet specific delivery schedules? If we can collect material flow data at the various places in a manufacturing system can we use this data to develop a max-plus neural network model of the manufacturing process?

REFERENCES

Baccelli, F., Cohen, G., Olsder, ,G.J., and Quadrat, J. P. (1992). *Synchronization and Linearity: An Algebra for Discrete Event Systems.* NewYork, Wiley.

Burris, S. and Sankappanavar, H.P., (1981) *A Course in Universal Algebra*, Graduate Texts in Mathematics, Springer-Verlag.

Cassandra, C.G., (1993) *Discrete Event Systems: Modeling and Performance Analysis*, Aksen Associates Inc., Publishers.

Zeigler, B., Praehofer, H. and Kim, T.G., (2000) Theory of Modeling and Simulation: Integrating Discrete Event and Continuous Complex Dynamic Systems, 2nd ed., San Diego, Academic.

A DECISION TREE APPROACH FOR SUBCIRCUIT EXTRACTION PROBLEM

[1]NIAN ZHANG [2]DONALD C. WUNSCH II
[1,2]Applied Computational Intelligence Lab
Dept. of Electrical and Computer Engineering
University of Missouri-Rolla
1870 Miner Circle
Rolla, MO 65409 USA
[1]nzhang@umr.edu [2]dwunsch@ece.umr.edu

ABSTRACT
The layout versus schematic (LVS) analysis is an essential part of topology design verification, and subcircuit extraction is one of the operations in the LVS testing. We provided an efficient decision tree approach to the subcircuit extraction problem based on the solution to the graph isomorphism problem. We construct the decision trees for the pattern circuit and the main circuit. And then we perform a classification process. To evaluate its performance, we compare it with the neural networks based heuristic dynamic programming algorithm (SubHDP) which is by far one of the fastest algorithms for subcircuit extraction problem.

Keywords: Decision Tree, Neural Networks, Adaptive Critic Designs, Layout versus Schematic (LVS), VLSI Layout Verification

1. INTRODUCTION

The subcircuit extraction is a very important application in many fields of computer aided design, especially for the present advanced VLSI design. For example, as VLSI design gets larger and larger every year, semiconductor industries have to ensure that there are no bugs in the design and that the manufactured chips are defect-free. Hence, it is desirable to identify the complete or partial set of invalid states before test generation [1]. In addition, subcircuit extraction is one of the operations in the Layout vs. Schematic (LVS) testing, which is a useful means of minimizing mask errors. Moreover, it is also an important application in circuit partition, and so on.

Several methods have been proposed to solve the subcircuit extraction problem [2] – [9]. The SubHDP algorithm proposed by Wunsch and Zhang has been implemented on a circuit up to 100,000 transistors in 16.002 seconds. It is by far one of the fastest algorithms for subcircuit extraction problem [10].

2. DEFINITIONS AND NOTATIONS

In this section, we give the basic definitions and notations that will be used throughout the paper.

Definition 2.1. A labeled directed graph G is a 6-tuple, $G = (V, E, \mu, v, L_V, L_e)$, where

- V is the set of vertices,
- $E \subseteq V \times V$ is the set of edges,
- $\mu : V \rightarrow L_V$ is a function assigning labels to the vertices,

- $\nu : E \to L_e$ is a function assigning labels to the edges,
- L_V and L_e are finite sets of symbolic labels.

Definition 2.2. Given a graph $G = (V, E, \mu, \nu, L_V, L_e)$, a *subgraph* of G is a graph $S = (V_S, E_S, \mu_S, \nu_S, L_V, L_e)$ such that

1. $V_S \subseteq V$,
2. $E_S = E \cap (V_S \times V_S)$,
3. $\mu_S(v) = \begin{cases} \mu(v) & \text{if } v \in V_S, \\ \text{undefined} & \text{otherwise} \end{cases}$
4. $\nu_S(e) = \begin{cases} \nu(e) & \text{if } e \in E_S, \\ \text{undefined} & \text{otherwise.} \end{cases}$

Let $G = (V, E, \mu, \nu, L_V, L_e)$ be a graph with $V = \{v_1, v_2, \cdots, v_n\}$. Then G can also be represented by its *adjacency matrix* $M = (m_{ij}), i, j = 1, \cdots, n$, where $m_{ii} = \mu(v_i)$ and $m_{ij} = \nu((v_i, v_j))$ for $i \neq j$. Clearly, the matrix M is not unique for a graph G. If M represents G, then any permutation of M is also a valid representation of G.

Definition 2.3. A $n \times n$ -matrix $P = (p_{ij})$ is called a *permutation matrix* if

1. $p_{ij} \in \{0,1\}$ *for* $i, j = 1, \cdots, n,$
2. $\sum_{i=1}^{n} p_{ij} = 1$ *for* $j = 1, \cdots, n,$
3. $\sum_{j=1}^{n} p_{ij} = 1$ *for* $i = 1, \cdots, n$

Definition 2.4. Let G_1 and G_2 be two graphs and M_1 and M_2 their corresponding adjacency matrices. G_1 and G_2 are isomorphic if there exists a permutation matrix P such that

$$M_2 = PM_1P^T \tag{1}$$

Notice that the matrix P can be understood as a bijective function f that maps the vertices of G_1 to G_2, and vice versa. That is, $f(v_j) = v_i$ *iff* $m_{ij} = 1$. We will call both P and f a graph isomorphism between G_1 and G_2. Thus, the problem of finding a graph isomorphism between two graphs G_1 and G_2 is equivalent to finding a permutation matrix P for which (1) holds true.

3. SUBCIRCUIT EXTRACTION

Basically, the problem of subcircuit extraction can be deemed as a subgraph isomorphism problem. A graph G consists of a finite nonempty set V = V (G) of p points together with a prescribed set X of q unordered pairs of distinct points of V [11]. Each pair x = {u, v} of points in X is a line of G, and x is said to join u and v. A subgraph of G is a graph having all of its points and lines in G. The subgraph isomorphism problem is defined as: *Given a graph S and another larger graph T, to find all the subgraphs of T which are identified with S*. Similarly, the subcircuit extraction problem is to determine whether one given pattern circuit has any isomorphic subcircuits in another larger model circuit. A 2-input NAND gate serving as the subcircuit (i.e. pattern circuit) is shown in Fig. 1 (a). Its corresponding circuit graph is shown in Fig. 1 (b). In Fig. 1 (b), device is represented as a square, while terminal is represented as a circle. In Fig. 2, a netlist is shown as the main circuit, in which we will find the instances of the pattern circuit. Our goal is to verify that the netlist composed of M8, M9, M10 and M11 in the main circuit is isomorphic to the pattern circuit.

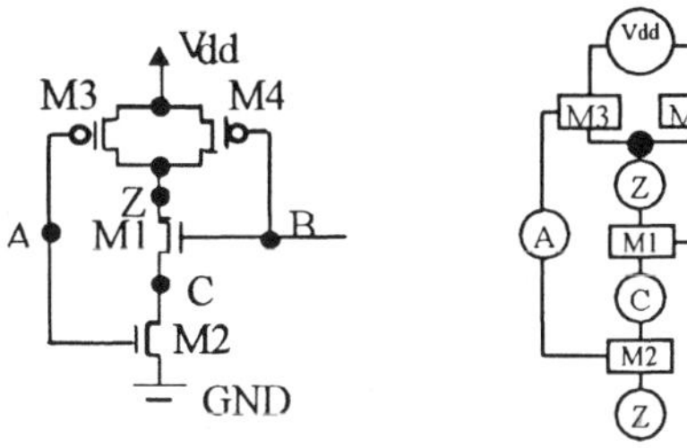

(a) (b)

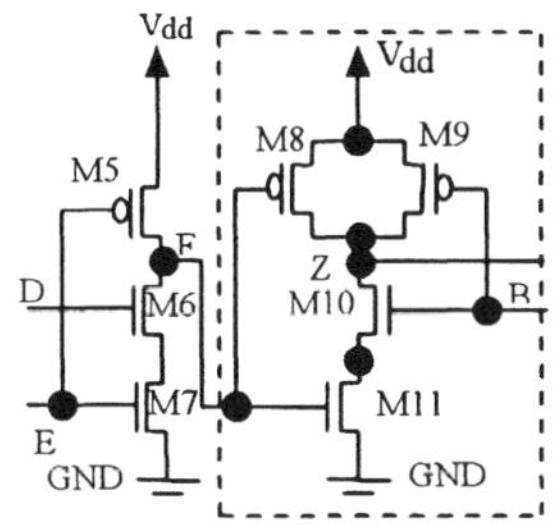

Fig. 1 (a) Pattern circuit. (b) The corresponding graph representation. Fig. 2 Main Circuit

The method of how to select candidate subcircuits can be found in [8]. By comparing the pattern circuit graph and the input circuit graph, the subcircuit extraction problem is then transformed from subgraph isomorphism problem to graph isomorphism problem.

4. DECISION TREE APPROACH TO GRAPH ISOMORPHISM

The decision tree approach to graph and subgraph isomorphism detection was introduced by Messmer and Bunke [12]. It presents a deterministic method for the problems of graph and subgraph isomorphism detection from an input graph to a set of a priori known pattern graphs.

For the pattern graph we compute all possible permutations of its adjacency matrix, and transform these adjacency matrices into a decision tree. The matrix of the input graph is then used to find those adjacency matrices in the decision tree that are identical to it at run time. The permutation matrices that correspond to these adjacency matrices represent the graph or subgraph isomorphisms that we are looking for.

Let $G = (V, E, \mu, v, L_v, L_e)$ be a pattern graph and M its corresponding $n \times n$-adjacency matrix. Furthermore, let A(G) denote the set of all permuted adjacency matrices of G,

$$A(G) = \{M_P \mid M_P = PMP^T \text{ where P is a } n \times n \text{ permutation matrix}\}.$$

Now the graph and subgraph isomorphism problem can be restated in terms of A(G). For a pattern graph G with corresponding $n \times n$-adjacency matrix M and an input graph G_1 with an $m \times m$-adjacency matrix M_1 and $m \le n$, we want to determine whether there exists a matrix $M_P \in A(G)$ such that $M_1 = S_{m,m}(M_P)$. If such a matrix M_P exists, the permutation matrix P corresponding to M_P describes a subgraph isomorphism from G_1 to G, i.e. $M_1 = S_{m,m}(M_P) = S_{m,m}(PMP^T)$. If G and G_1 are of equal size, the permutation matrix P represents a graph isomorphism between G_1 and G, i.e. $M_1 = PMP^T$.

Apparently, the set A(G) can be organized in a decision tree. The features that will be used for the classification process are the individual elements in the adjacency matrices, arranged in a particular way. We say that matrix M consists of an array of so-called *row-column* elements a_i, where each a_i is a vector of the form

$$a_i = (m_{1i}, m_{2i}, \cdots, m_{ii}, m_{i(i-1)}, \cdots, m_{i1}).$$

The matrix can then be written as

$$M = (a_1, a_2, \cdots, a_n), i = 1, \cdots, n.$$

5. SUBCIRCUIT EXTRACTION USING DECISION TREE APPROACH

NAND gate will be served as our pattern circuit. We generate a graph for each of the pattern circuit and input circuits (i.e. candidate subcircuits).

As we know, the circuit graph is a bipartite graph, which means device vertices only connect to terminal vertices, and terminal vertices only connect to device vertices. Thus, for a graph of a node and its neighbors, it has only two possible forms, as shown in Fig. 3 (a) and (b). The vertex *a* is called the *top node*. If it is a device, vertex *a* and its neighbors can only construct a graph like Fig. 3 (b); if it is a terminal, vertex *a* and its neighbors can construct a graph either like Fig. 3 (a) or Fig. 3 (b).

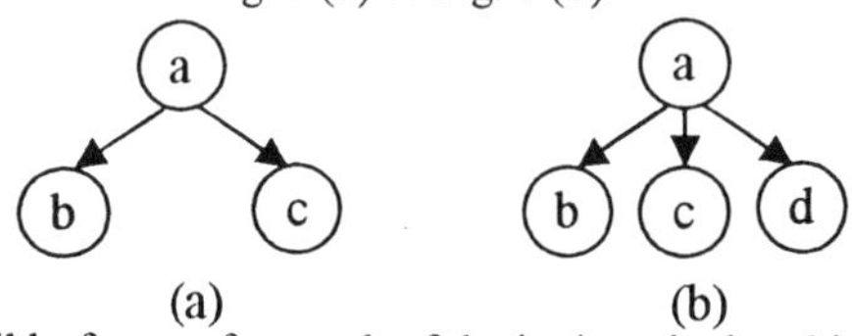

Fig. 3 Two possible forms of a graph of device/terminal and its neighbors. The vertex *a* is called the *top node*. If it is a device, vertex *a* and its neighbors can construct a graph like Fig. 3 (b); if it is a terminal, vertex *a* and its neighbors can construct a graph either like Fig. 3 (a) or Fig. 3 (b).

The subcircuit identification algorithm is as follows:

Step 1 We start from Vdd, and make it as the top node;

Step 2 We construct a graph, g consisting of the top node and its neighbors;

Step 3 List the set $A(g)$ of permuted adjacency matrices of g by its row-column elements. For example, the set of permuted adjacency matrices of Fig. 3 (a) is shown in Fig. 4 (a). The corresponding row-column representation of the matrices is shown in Fig. 4 (b).

Step 4 Build a decision tree for g with adjacency matrices according to the row-column elements of each adjacency matrix $M_P \in A(G)$. For example, the corresponding decision tree for Fig. 3 (a) is shown in Fig. 5. In Fig. 5, the nodes of the decision tree are represented by circles. Each directed branch from one node to another has a row-column element associated with it;

Step 5 Perform a classification process. At the top of the decision tree there is a root node. Its direct successors constitute the first level of the decision tree. On the first level, the classification is done by comparing the first row-column element of the input graph (i.e. candidate subcircuit graph) to the first row-column element a_1 of each matrix $M_P \in A(G)$ of the pattern graph. The element $a_1 = (m_{11})$ represents the label of the first vertex in each matrix in A(G), with $m_{11} \in L_V$. Thus, on the first level of the decision tree, we simply check if the label of the first vertex of the input graph matches any of the vertex labels of the pattern graph. Next, on the second level of the decision tree, the second row-column element a_2 of each matrix is used for the classification. This means that we check if the label of the second vertex of the input graph and the edges between the second and the first vertex have identical occurrences in the pattern graph.

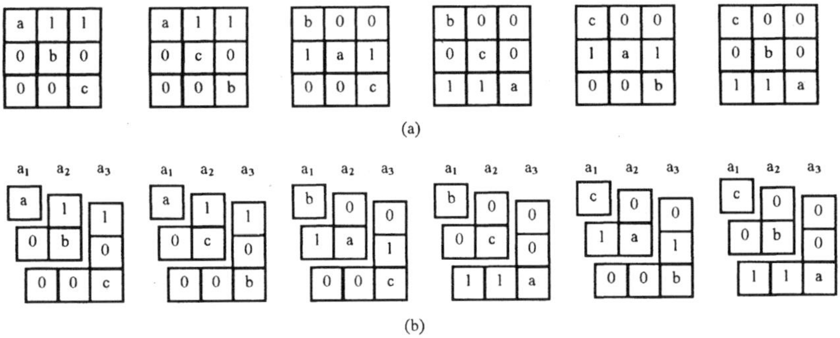

Fig. 4 (a) The adjacency matrices of Fig. 3 (a). (b) The corresponding row-column representation of the matrices.

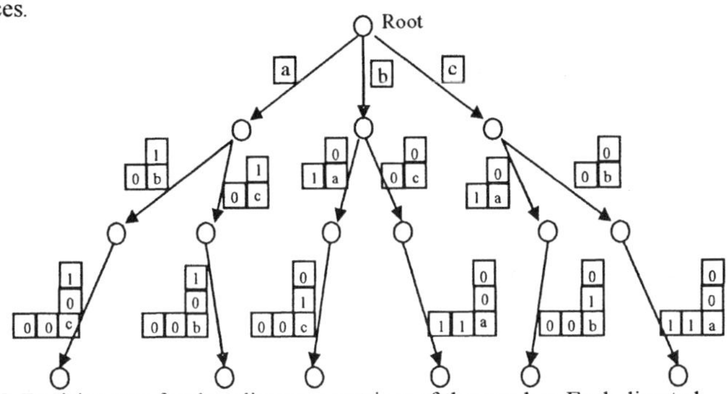

Fig. 5 Decision tree for the adjacency matrices of the graph g. Each directed branch from one node to another has a row-column element associated with it

This process is continued on the subsequent levels in the decision tree. With each matrix Mp that is represented by some node N on level k, the corresponding permutation matrix P is also stored. If at some point no classification is possible, then the input graph G_1 is not isomorphic to any subgraph of the pattern graph. If each row-column element of G_1 has been used in the classification process and some node N in the decision tree has been reached, then each permutation matrix that is associated with N represents a subgraph isomorphism from the input graph to the pattern graph. If node N is a leaf node, the input graph and the pattern graph are of the same size, each permutation matrix associated with N describes a graph isomorphism between those two graphs.

Step 6 If there is a graph isomorphism, then choose one of the top node's neighbors as the top node, and repeat step 2 to 5; otherwise, make a conclusion that there is no graph isomorphism between the two graphs, and then quit.

6. EXPERIMENTAL RESULTS

In this experiment, we want to investigate the run time efficiency of the new algorithm. We use the Dell Precision Workstation 420, Intel Pentium III processors up to 800MHz, and 256MB memory to do the experiments. We try to identify the 2-input NAND gate from different size of RAM circuits. Identification time is the time to find out all the subcircuits. Total run time includes circuit setup time and subcircuit identification time. Run time is measured in seconds.

We compare the total run time of decision tree approach and SubHDP approach, as shown in Fig. 6. The x-axis denotes the number of transistors in the main circuit, and the y-axis denotes the total run time. The solid line represents the SubHDP algorithm, and the dashed line represents the decision tree approach. From Fig. 6, we can see that the decision tree approach identifies the NAND gates from the main circuit not as faster as the SubHDP algorithm. In addition, the total run time of the decision tree approach is a sharp curve with the main circuit size exceeding 50,000 transistors. However, the total run time of the SubHDP algorithm is approximately a smooth linear line versus the number of transistors. The experiments also show that both the algorithms can find out all the NAND gates from all the main circuits correctly.

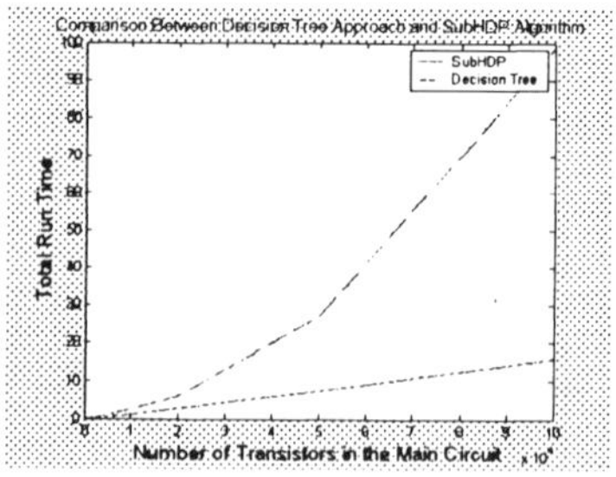

Fig. 6 Run time comparison between the decision tree algorithm and the SubHDP.

7. CONCLUSIONS

From the experiments, we prove that the decision tree approach is an efficient approach to implement the subcircuit extraction problem. It can successfully find out all the instances of the pattern circuit (i.e. NAND gate) from the main circuit. In this paper, we use the NAND gate as our pattern circuit, however, it can also be NOR gate, OAI gate, or RAM cell etc.

ACKNOWLEDGEMENT

The authors would like to acknowledge the support of NSF and the MK Finley endowment.

REFERENCES:

[1] Michael John Sebastian Smith, "Application-Specific Integrated Circuits," Addison-Wesley, ISBN 0-201-50022-1, 1997.

[2] G. Yokomizo, et al, "A New Circuit Recognition Method for Pattern Based Circuit Simulation," *Proceeding of CICC*, May 1990.

[3] Miles ohlrich, et al, "SubGemini: Identifying Subcircuit using a Fast Subgraph Isomorphism Algorithm," *30th ACM/IEEE Design Automation Conf.*, Dallas, Texas, June 14 - 18, 1993.

[4] N. Vijaykrishnan, et al, "SUBGEN: A Genetic Approach for Subcircuit Extraction," *9th International Conference on VLSI Design*, January 1996.

[5] Zong Ling and David Y.Y.Yun, "An Efficient Subcircuit Algorithm by Resource Management," *Proceeding of the Second International Conference of ASIC*, Shanghai, P.R.China, 1996.

[6] Wei-Hsin Chang, et al, "A Novel Extraction Algorithm By Recursive Identification Scheme," *2001 IEEE International Symposium on Circuits and Systems*, Australia, 2001.

[7] Nian Zhang and Donald C. Wunsch II, "A Novel Subcircuit Extraction Algorithm Using Heuristic Dynamic Programming (HDP), " *2002 International Conference on VLSI*, Las Vegas, Nevada, June 24 – 27, 2002.

[8] Nian Zhang and Donald C. Wunsch II,"A Fuzzy Attributed Graph Approach to Subcircuit Extraction Problem," *The IEEE International Conference on Fuzzy Systems*, St. Louis, MO, May. 25 - 28, 2003. (Best Student Paper Award)

[9] Nian Zhang, Frank Harary and Donald C. Wunsch II, "CMOS IC Topology Design Verification By Heuristic Dynamic Programming, " *Proceedings of ANNIE '02*, St. Louis, MO, Nov.10 - 13, 2002.

[10] Nian Zhang, Donald C. Wunsch II, and Frank Harary "The Subcircuit Extraction Problem," *IEEE Potentials*, August/ September Issue, 2003.

[11] Frank Harary, "Graph Theory," Narosa Publishing House, 1998.

[12] B.T. Messmer and H. Bunke, "A Decision Tree Approach to Graph and Subgraph Isomorphism Detection, " *Pattern Recognition*, vol. 32, pp. 1979 – 1998, 1999.

INTELLIGENT CONDITION-BASED MAINTENANCE OF REACTIVE ION ETCHING PROCESS

SHAIKH, N.I.
The Pennsylvania State University
University Park, Pennsylvania,
16802

PRABHU, V.V.
The Pennsylvania State University
University Park, Pennsylvania,
16802

ABSTRACT
The semiconductor industry is facing increasing competition; motivating the need for condition-based maintenance in conjunction with process control. Condition-based maintenance can reduce the downtime of expensive equipment, production costs, and improve yield. Moreover, it can potentially reduce operation cost of semiconductor fabs by lowering the number of expensive spare parts that need to be in stock. This paper proposes an intelligent condition-based maintenance approach wherein the operating parameters for the process are selected while being constrained both by the process and maintenance requirements. Reactive ion etcher is selected as the target equipment: it is widely used and is a critical equipment in semiconductor industry. Based on real-time process and equipment condition data, artificial neural networks are used for assessing the current condition of the equipment and predicting the remaining time before the etcher needs to be shut down for maintenance. The proposed data driven approach is specially suited for the semiconductor industry, which relies heavily on statistical techniques for process control and optimization.

1. INTRODUCTION

Semiconductor manufacturing is a complex and highly integrated industry involving intricate processes like lithography and reactive ion etching (RIE). Increasing competition and frequent technology upgrade make high efficiency and accuracy of processes equipment mandatory for the industry. Several factors combine to make effective production control of wafer fabrication facilities a complex task: variability due to unscheduled equipment downtime being a critical factor (Mosley, et al, 1998). The focus of this paper is on combining maintenance scheduling with process level controls to synchronize the equipment downtime with maintenance crew availability and low-load periods of production of that machine.

The lack of understanding of the processes and interactions in the semiconductor equipment and heavy reliance on data driven approaches to control and fault diagnosis make Neural Networks (NN) a good tool for mapping (process emulator) and estimation (remaining lifetime of a part/equipment). NN based process control and fault prognosis have been combined in an attempt to develop intelligent condition-based maintenance programs for semiconductor manufacturing processes such as RIE. Maintenance crew availability information has been added as an additional parameter in the NN based controller to select operating parameters based on the machine's current condition, estimates of the remaining useful lifetime of the machine, and the maintenance crew visiting time.

The rest of the paper is organized as follows: A brief introduction to operating parameter selection and its effects on process and machine health is presented in Section 2. Section 3 looks into the unified approach for process parameter selection and condition based maintenance scheduling for RIE. The simulation results are presented in Section 4, and conclusions in Section 5.

2. BACKGROUND

Studies have shown that the actual machine deterioration due to a specific job is a function of both the operating parameters used for the job and the machine's condition at the time of the job (Chakravatry and Balakrishnan 1995). By taking process control decisions in conjunction with condition monitoring and machine maintenance decisions, potential benefits can be reaped. RIE has been selected as the target process to demonstrate the efficacy of the approach. It is representative of most of the equipments that are available in the fabs: complex and difficult to model, with repairs that are usually complex and time consuming and repair crew that is scarce (Mosley, et al, 1998).

RIE: RIE is a critical technology for modern VLSI circuit fabrication at many steps of the manufacturing process (details are available in Stokes and May, 2000). The physical and chemical mechanisms in RIE are known to be complex and are currently not well understood, making the process difficult to control in manufacturing applications. Himmel et al, (1995) proposed a model-based control that was primarily founded on the non-linear mapping capabilities of neural networks. The three components of the control scheme are the Neural Controller (NC), Plant Emulator, and the plant. A modification of the proposed control architecture, that will be used in this work is presented in Figure 1.

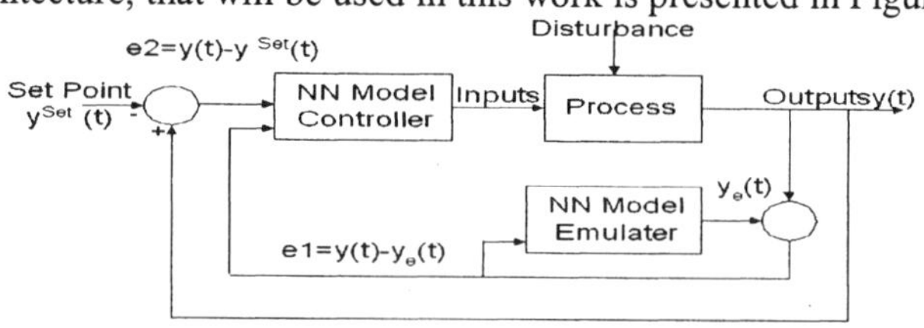

Figure 1 Model based Real Time Control Architecture for RIE

Control and Manipulated Variables: Quality of plasma etch is characterized by its selectivity, anisotropy, uniformity and etch rate- etch rate being a key indicator of machine health. The inputs/outputs of the system can be represented as shown in Figure 2. Voltage bias and gas concentration are the observable variable for monitoring these parameters (Rashap et al, 1995-Figure 3).

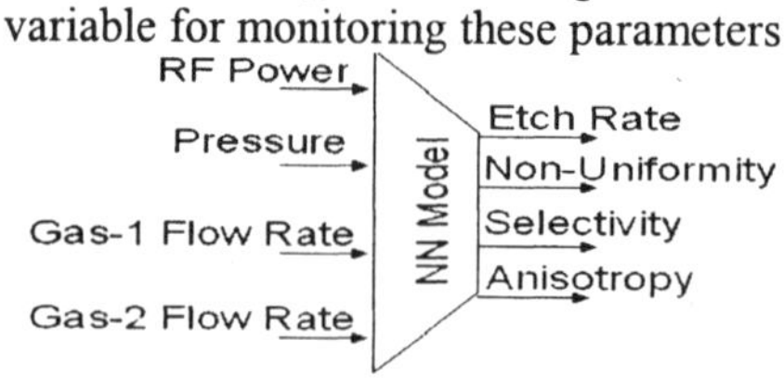

Figure 2 Control and Manipulated Variables for RIE

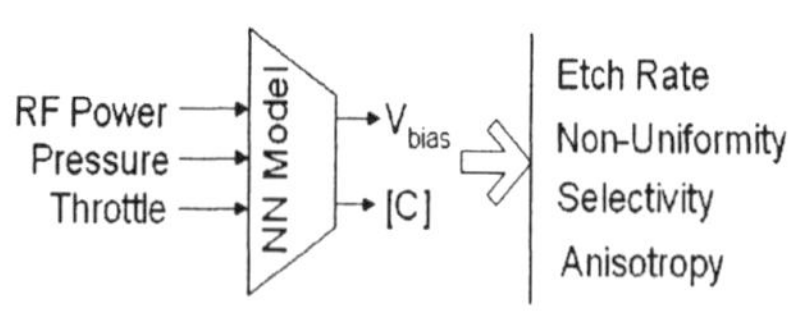

Figure 3 Indirect Control and Manipulated Variables for RIE

NN based Plant Emulator for RIE: The forward NN based process model defines a functional relationship between RIE process conditions (inputs) such as RF power or gas compositions and responses (outputs) such as etch rate or uniformity. The NN based emulator used by Kim and May (1997) is presented in Figure 4.

NN based Controller for RIE: For the implementation of a control action, an inverse of the process model is required. Figure 5 displays the structure of the inverse model proposed by Stokes and May, (2000), where e_1 and e_2 have been defined in Figure 1.

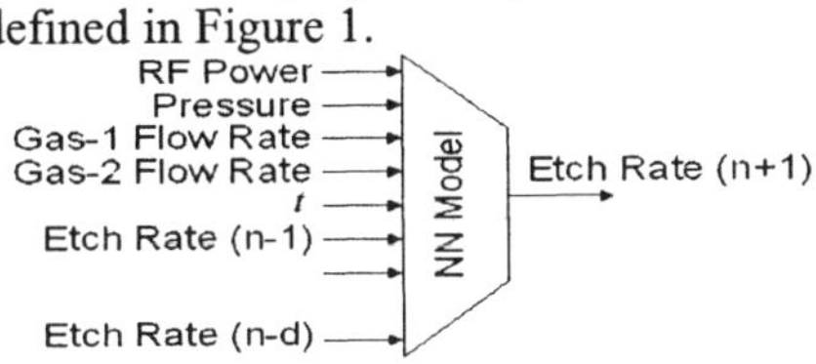

Figure 4 General structure of RIE Process Emulator

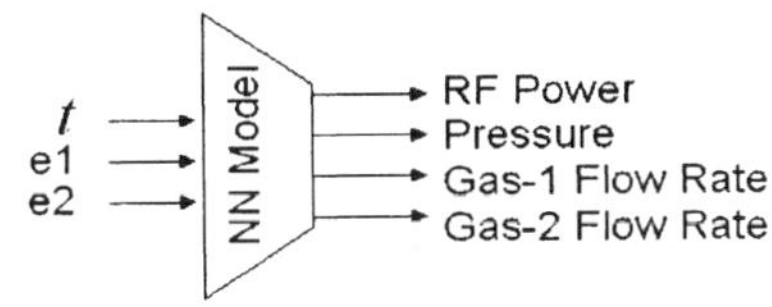

Figure 5 General structure of RIE Inverse Process Model

Predictive Maintenance: Condition based maintenance scheduling has three major parts: condition monitoring, remaining lifetime estimation, and maintenance scheduling.

Condition Monitoring: Condition monitoring comprises of identification of the possible faults, identification and measurement of the indicator variables of the faults, and mapping the measured indicator variables to the machine condition. Though the framework for condition monitoring can be generalized, the faults, the indicator variables, and the mapping itself are specific to the machines. The major faults for RIE and their plausibility, as identified by Kim and May (1997) are presented in Table 1.

Lifetime Estimation: Three relevant time periods for evidence collection in semiconductor manufacturing have been proposed: during equipment maintenance periods (before processing), during on-line equipment operation (during processing), and during in-line, post process physical and/or electrical inspection. In this work, we will be focusing on on-line diagnosis. Kim and May have proposed a NN based parameter estimation technique as well (Figure 6), a modification of which will be used for lifetime estimation in the current work.

Serial No	Components	Support	Plausibility
1	Capacitance-manometer	0.353	0.508
2	Pressure Switch	0.353	0.507
3	Electrode Assembly	0.113	0.267
4	Exhaust Valve Controller	0.005	0.160
5	Throttle Valve	0.003	0.159

Table 1 Identified Faults for RIE (from Kim and May, 1997)

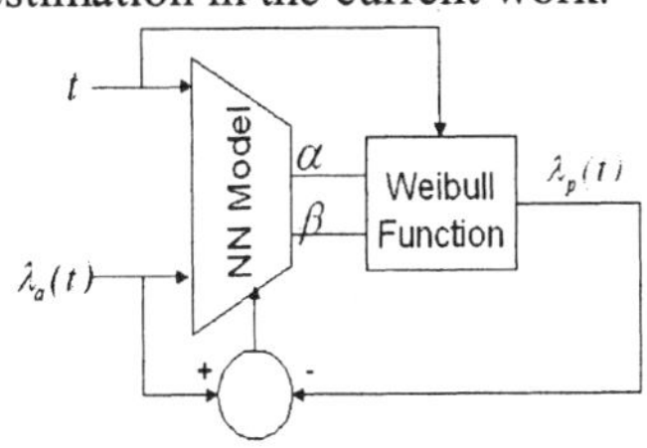

Figure 6 Weibull Distribution Parameter Estimator

Maintenance Scheduling: Typically, 20-30 etchers are used in a fab, and manufacturers have to allocate maintenance resources to service these effectively. The maintenance resource-scheduling problem, in general, can be considered to be similar to the production-scheduling problem.

3. CONDITION BASED MAINTENANCE ARCHITECTURE

Based on the information from an assigned machine and/or the process and arrival time of maintenance crew, this condition-based maintenance controller outputs a maintenance service request time and recommends operating parameters for the machine or other machine execution commands. The logic for the condition based maintenance controller is presented in Figure 7. The main components of the architecture are the remaining lifetime estimator, NN based process emulator, the NN controller, and the maintenance scheduler.

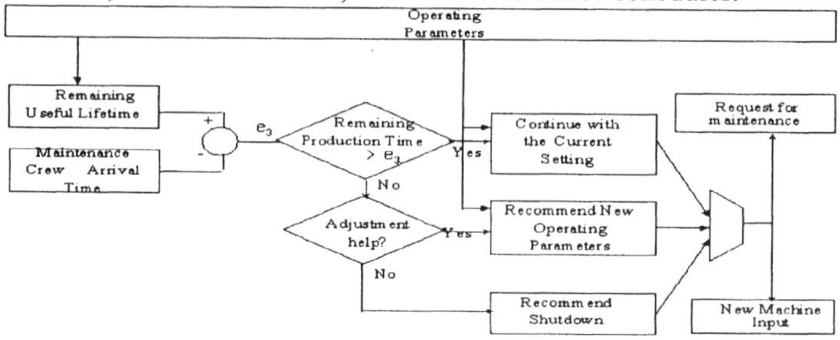

Figure 7 The Condition Based Maintenance Logic

Lifetime Estimation: NN are used for assessing the current condition of the equipment and predicting the remaining time before the etcher needs to be shut down for maintenance. The basic logic is similar to that presented in Figure 6, however, the system is MISO, as against MIMO in earlier works.

Process Emulator: An empirical nonlinear model of the RIE process (Equation (1)) proposed by Rashap et al (1995) has been used to simulate and generate realistic process data to test the condition-based maintenance approach. It may be noted that this is a 2 input 1 output system.

$$[V_{\text{bias}}] = \left[\frac{2.72e^{-0.5s}}{s+0.17} \quad \frac{0.444}{s+1.25} \right] \begin{bmatrix} \text{Throttle} \\ \text{RF Power} \end{bmatrix} \tag{1}$$

Inverse Process Model: The proposed inverse process model is a modification of the model presented in Figure 5. An additional input parameter has been introduced that incorporates machine health information into operating parameter selection decisions. Error e_3 is threshold-based input, i.e., it is triggered only in certain scenarios. When e_3 is not triggered, the process control decisions are made independent of the maintenance requirements.

Maintenance Scheduling: In a service request process, the maintenance-scheduling model first sends the remaining useful lifetime to a maintenance crew model as a maintenance service request time, and waits for the crew model to respond with a crew visiting time. Then it calculates a cost that corresponds to the response and sends the result to the crew model for a record. The parameters used for simulating the maintenance requirements in a fab are based on those determined by Mosley, et al (1996), and are presented in Table 2. Detailed description and simulations of the system are available in Koomsap et al, 2003.

Table 2 Range for The RIE process variables

No of Etchers	Failure distribution Exponential (Mean)	Repair distribution Gamma (beta, alpha)	Required Availability
20	2160	48.6,11	80%

4. SIMULATION

The empirical model presented in Equation (1) was developed for an Applied Materials 8300 Hexode Reactive Ion Etcher that uses a simple CF_4 chemistry to etch unmasked PolySi/SiO_2/Si wafers. The recipe parameters include etch time (seconds), gas flow rates (SSCM), chamber pressure (mtorr), incident RF Power (W), and temperature. The operating region that was used is presented in Table 3. The throttle angle corresponding to these settings was 12.5% open. Three networks, corresponding to the remaining lifetime estimator, the plant emulator, and neural controller were developed. The details of the structure and training of these networks (done using NN toolbox in MATLAB) is presented in Table 4. The performance of the three networks is graphically displayed in the following sub-sections.

Serial No	Parameter	Nominal Value	Range
1	CF_4	30	10-40 sccm
2	Pressure	20	8-30 mtorr
3	RF Power	1000	800-1300 watts

Table 3 Range for The RIE process variables

Serial No	Neural Network	Structure (In-hidden-out)	Learning Rate	RMS Error
1	Plant Emulator	2-4-1	0.01	7%
2	Neural Controller	4-6-2	0.01	9%
3	Lifetime Estimator	2-3-1	0.001	9%

Table 4 NN Structure and training parameters

NN Based Lifetime Estimator: The inputs and the outputs to the NN based remaining lifetime estimator are presented in Figure 8. This is a 2 input, single output system, with initial lifetime of the machine (throttle valve of the RIE) taken as 2160 minutes.

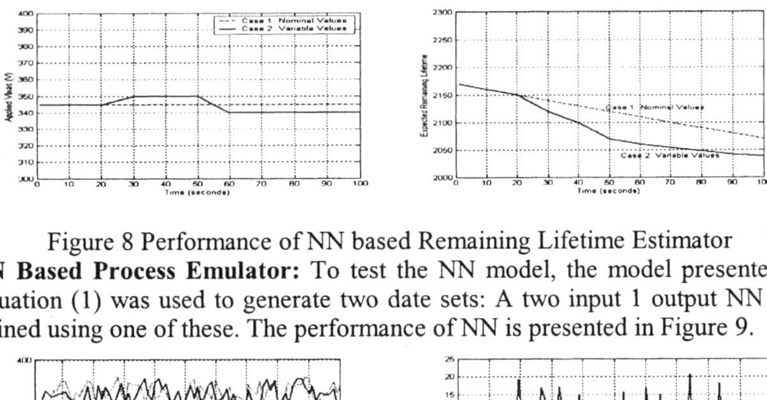

Figure 8 Performance of NN based Remaining Lifetime Estimator

NN Based Process Emulator: To test the NN model, the model presented in Equation (1) was used to generate two date sets: A two input 1 output NN was trained using one of these. The performance of NN is presented in Figure 9.

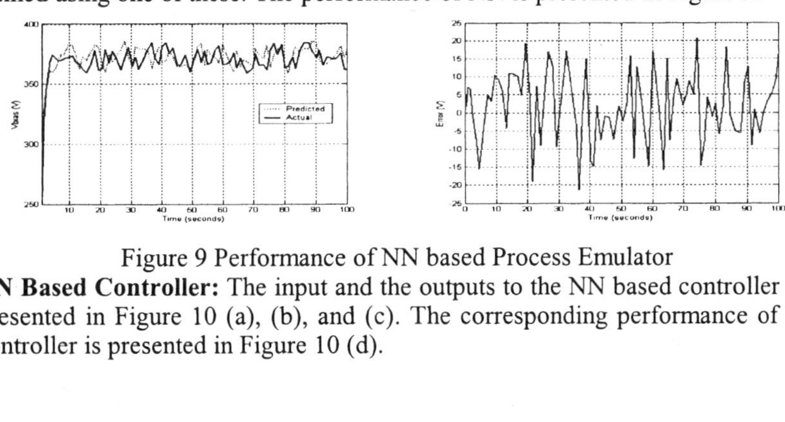

Figure 9 Performance of NN based Process Emulator

NN Based Controller: The input and the outputs to the NN based controller are presented in Figure 10 (a), (b), and (c). The corresponding performance of the controller is presented in Figure 10 (d).

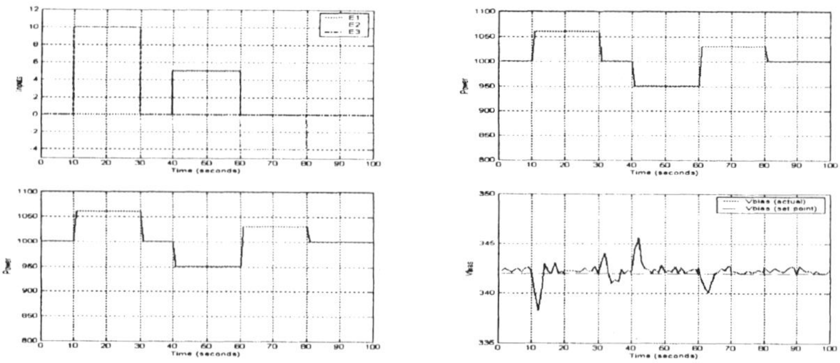

Figure 10 Performance of NN based Controller

5. CONCLUSIONS

The focus of this work was on developing architecture for unifying machine level process control and condition-based maintenance scheduling. A condition-based maintenance controller has been developed to monitor the machine's condition and to recommend the operating parameters to the process controller, The contributions of this development is prolonging the machine's lifetime and sustaining the productivity by adjusting operating parameters and providing a data driven ready-to-run condition-based maintenance controller. This is specifically important for semiconductor manufacturing process because of the poor understanding of the processes and high costs of operation. Initial results show a promising prospect of real-time integrated maintenance scheduling and process control.

Acknowledgments: Work was partially supported by NSF Grants DMI-9908267 and DMI-0075572, and Ben Franklin Technology Partnership through Penn State's CMEI.

REFERENCES

Chakravarty, A.K. and Balakrishnan, N. (1995), "Impact of Job-Sequence on the Down-time of a Deteriorating Flexible Machine," European Journal of Operational Research, v.87, p.299-315.

Himmel, C.D., Kim, T.G., Krauss, A., Kamen, E.W., and May, G.S., (1995), "Real time Predictive Control of Semiconductor Manufacturing Processes Using Neural Networks", Proceedings of the American Control Conference, Seattle, Washington.

Kim, B., and May, G.S., (1997), "Real Time Diagnosis of Semiconductor Manufacturing Equipment Using a Hybrid Neural Network Expert System", IEEE Transactions on Components, Packaging, and Manufacturing Technology-Part C, v-20, no-1.

Koomsap, P., Prabhu, V.V., Shaikh, N.I., Schriempf, J.T., and McDermott, J.H., (2003), "Condition monitoring and lifetime estimation of a CO_2 laser", Journal of Laser Applications.

Mosley, S.A., Teyner, T., and Uzsoy, R.M., (1998), "Maintenance Scheduling and Staffing Policies in a Wafer Fabrication Facility", IEEE Transactions on Semiconductor Manufacturing, v-11, no-2.

Rashap, B., Elta, M., Etemad, H., Fournier, J., Freudenberg, J., Giles, M., Grizzle, J, Kabamba, P., Khargaonekar, P., Lafortune, S., Moyne, J., Teneketzis, and Terry Jr., F., (1995), "Control of Semiconductor Manufacturing Equipment: Real time Feedback Control of a Reactive Ion Etcher", IEEE Transactions on Semiconductor Manufacturing, v-8, pp. 286-297.

Stokes, D., and May, G.S., (2000), "Real Time Control of Reactive Ion Etching Using Neural Networks", IEEE Transactions on Semiconductor Manufacturing, v-13, no-4.

MONITORING SINGLE FORCE VERSUS RESULTANT FORCE WHEN A GENETIC TOOL MONITOR IS USED

I.N. TANSEL, W.Y. BAO, D. RINCON
Mechanical Engineering Department
Florida International University
Center for Engineering & Applied Science
10555 West Flagler Street, (EAS - 3473) Miami, FL 33174

C.V. KROPAS-HUGHES
Air Force Research Laboratory
AFPL/MLLP, Bldg. 655, R166
2230 Tenth Street
Wright-Patterson AFB
Ohio, 45433-7746

KEYWORD LIST:
Genetic Algorithms, Manufacturing Applications, Process Monitoring, Artificial Neural Networks, Backpropagation, Machine Monitoring.

ABSTRACT
A Genetic Tool Monitor (GTM2) was developed to evaluate tool condition by estimating the key parameters from cutting forces. Genetic algorithm of the GTM2 is capable of estimating a large number of parameters with desired resolution. Theoretically, the method should work with the resultant force or any force component data with the same accuracy. This concept was tested on the simulated cutting force data. Simulations showed that any force component could be used to estimate the critical parameters with the same accuracy of the resultant cutting force. Use of the resultant force is recommended in the experimental studies to avoid significant estimation errors and false alerts caused by the noise and modeling errors.

INTRODUCTION

The consistency of the machining quality is very important in automated manufacturing. Many monitoring techniques have been developed to detect tool breakage and to estimate tool wear by evaluating the most important characteristics of the signals coming from the sensors such as dynamometer, accelerometers, acoustic emission sensors, thermocouples, and microphones. The first approach was to evaluate these characteristics at different tool conditions and to set the limits or to relate them to tool wear. Later, intelligent computational algorithms such as neural networks were used to automate this task. The operating conditions such as speed, feed rate, depth of cut, and tool-material couple should be identical or very close to the test conditions to use these methods effectively. The Genetic Tool Monitor (GTM) was developed to interpret the cutting force signals by using analytical expressions. Theoretically, once the key parameters are estimated in a single test from the data of any cutting force component, cutting operations can be monitored by measuring any cutting force at any operating condition. However, GTM or improved version GTM2 should be used very conservatively to avoid problems created by the noise, modeling error, and necessary computational time. In this study, the performance of GTM2 is evaluated on simulated cutting force data when the resultant or only one force component is used.

The genetic algorithm (Goldberg, 1989; Holland, 1992; Carroll, 1996) estimates multiple parameters of an analytical expression by using a search method, which emulates the evolution in the nature. It works with linear and non-linear analytical expressions and very complex conditional statements can be included in the objective function. The GTM concept was developed to estimate the key parameters such as cutting force component and runout of the analytical model (Tansel et. al.,1998; Tansel et. al.,2002; Bao and Tansel, 2000a; Bao and Tansel, 2000b) of micro-end-milling operations. Other researchers have used GA to determine the optimal cutting conditions (Cus and Balic, 2003; Davim and Antonio, 2001; Shunmugam et. al., 2000; Khan et. al, 1997) in machining, and monitoring of turning (Szecsi, 1999; Achiche, 2002)operations.

Generally, continuous monitoring of the machining operations while the cutting conditions change is the main goal of the academic studies. However, industry prefers a robust system with minimal false alarms even it works at the strictly limited conditions. In this study only two parameters will be monitored. Cutting force coefficient and runout will indicate the dullness and reduction of the contact length of one of the cutting edges. Although, the genetic algorithm of the GTM2 (Li, 2002) may identify most of the cutting parameters theoretically, the computational time will be unacceptable in the practical applications and will not be considered in this study.

THEORETICAL BACKGROUND

The analytical cutting force model developed by Bao and Tansel (Tansel et. al.,1998; Tansel et. al.,2002) was used in this study. The model calculates the cutting forces by considering the trajectory of the tips of the end mill and the tool run-out.

Cutting condition was monitored by evaluating the estimated parameters of the analytical model. Genetic algorithm was used to fit the analytical cutting model to experimental data. Genetic algorithms (Goldberg, 1989; Holland, 1992; Carroll, 1996) use a repetitive process to find the optimal solution by using the natural selection rule of the genetic evaluation. The user assigns a certain number of binary elements for each parameter, and selects the population size, mating pool size and the numbers of the children from each couple according to the problem, application and resources. During the repetitive process, the mating couples (parents) are selected, and the hereditary chromosomes of the next generation are determined. The process continues with gene crossover, gene mutation, creation of next generation and evolution. Hereditary chromosomes can be selected from the stronger one of the mating couple, or one of them in turn, or selected randomly in the second step. Fixed-point or uniform crossover is used in the third step according to the problem. Jumping or creeping mutation can also be used. The process is repeated until the fitness reaches the desired level.

GENETIC TOOL MONITOR 2

The diagram of the Genetic Tool Monitor (GTM2) is presented in the Figure 1. The user gives the program all the known parameters including the operating conditions, and tool geometry. Depending on the operating conditions, one or more parameters can be estimated. The objective is mostly to estimate the key cutting parameters such as cutting force coefficient, and tool runout. Depth of cut, entry and exit angles, and other parameters may also be estimated if on-line monitoring of the machining operation is considered. However, computational time will increase while the accuracy of the estimations is compromised.

The genetic algorithm estimates the selected parameters. Cutting forces are estimated by using the analytical model and estimated parameters. The accuracy of the estimated parameters is evaluated by comparing the sampled and estimated cutting force(s). The genetic algorithm updates the estimated parameters to minimize the error. The process is repeated until the allowed iteration number is reached. The value of the cutting force coefficient increases when the cutting edges wear out and get dull. The chipped away segments of the cutting edges reduce the effective cutting length of the tips. Tip and runout estimations start to vary with time.

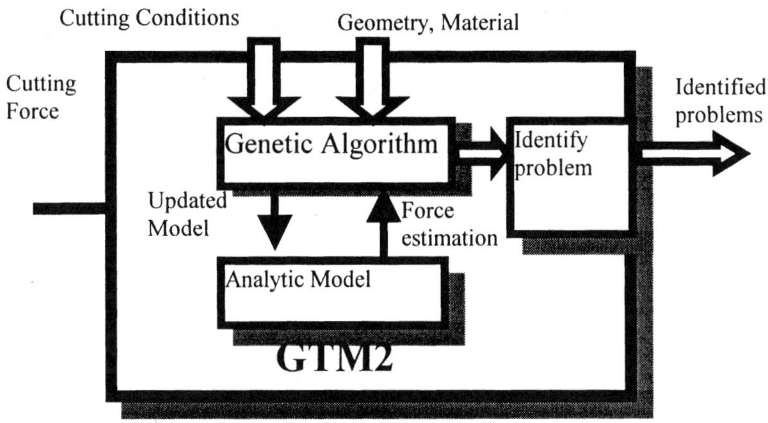

Figure 1. Block diagram of the GTM2.

GENERATION OF THE SIMULATED DATA

The simulated data was generated by using the analytical cutting force model (Tansel et. al.,1998; Tansel et. al.,2002)[4,5]. Micro-end milling operations were considered. The considered trajectory of the cutting tips and the estimated cutting force is presented in Figures 2 and 3 respectively.

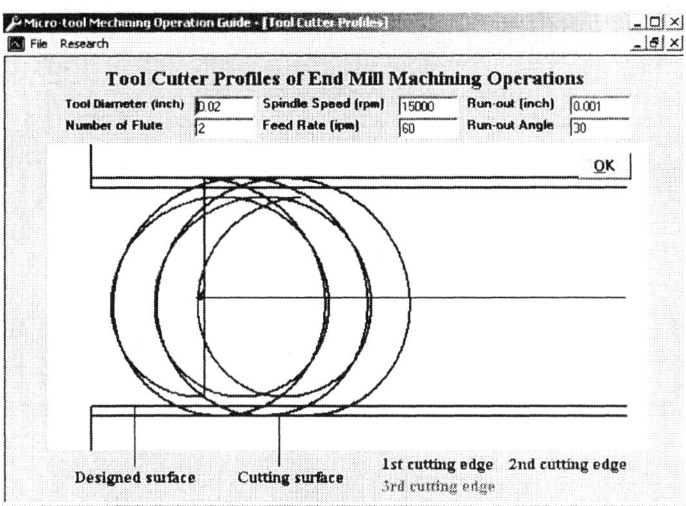

Figure 2. Trajectory of the tool tip of the analytical model

Feed, thrust and resultant forces of worn tools were simulated in this study by considering three different conditions. In the simulations, the cutting force coefficient and contact length of one of the cutting edges was kept constant. The other edge was assumed to get dull and/or partially chiped away when the tool wears out. The contact length of a cutting edge is the section of a tooth which actively removes material from the workpiece. The contact length of the perfect cutting edge was taken as 0.001". Different levels of dullness and contact length loss were simulated to evaluate if the GTM2 would be able to recognize the difference. The cutting force coefficient of one of the edges was increased first 10% in three simulations with partial cutting edge loss at three different levels. In the other three simulations, cutting force coefficient was increased by 20%. For partially chipped away cutting edges, the contact length was reduced to 0.0009", 0.0008", and 0.0007". The center of rotation (run-out) and diameter of a tool were changed to simulate the cutting force of a tooth with partially chipped away cutting edge.

RESULTS AND DISCUSSION

The cutting force coefficient and contact length of one of the cutting edges was estimated from the resultant (Figure 4-5), feed (Figure 6-7), and thrust direction(Figure 8-9), cutting forces. To demonstrate the performance of the proposed approach, simulated and estimated values of the Dullness Index (DI) and contact length estimations were compared in the figures. The genetic algorithm was stopped after 500 iterations in all the test cases. The dullness index (DI) is the ratio of the estimated cutting force coefficient over the theoretical value of the same coefficent for the perfect tool. The contact length is the section of the cutting edge which removes the material.

The cutting force coefficient was estimated with less than 1% error in all the studied cases either the resultant force or single force component was used. The contact length estimation error was less than 10% when the GTM2 made the estimations from the resultant force. The estimation error was less than 2% when a single force component was used to estimate the contact length. These results indicated that GTM2 was capable to distinguish the variation of the cutting force coefficient and the change of contact length.

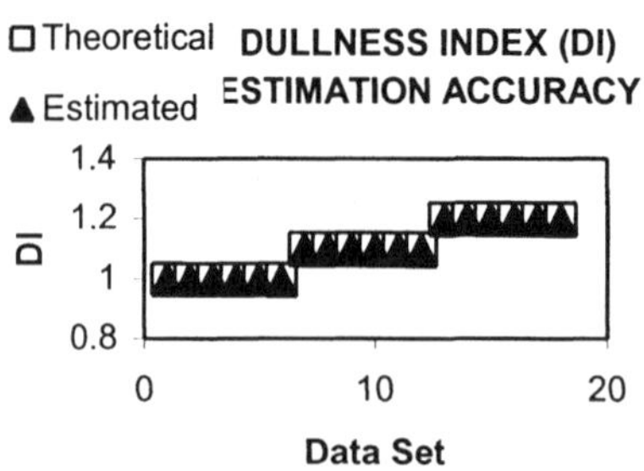

Figure 4. Cutting force coefficient estimation accuracy of the GTM2 when the resultant cutting force is used for estimations. (Values were taken from (Li,2002))

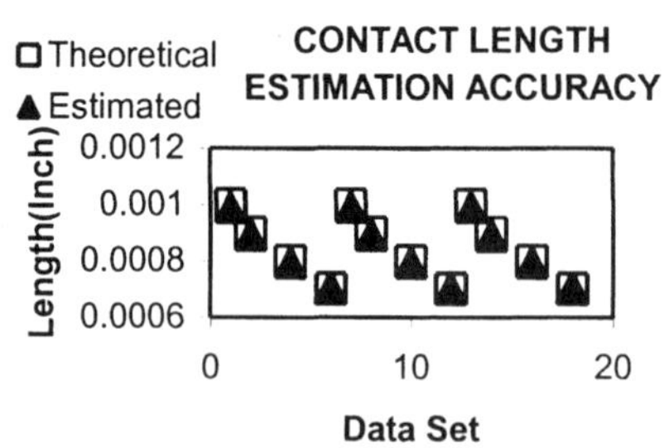

Figure 5. Contact length estimation accuracy of the GTM2 when the resultant cutting force is used for estimations. (Values were taken from (Li,2002))

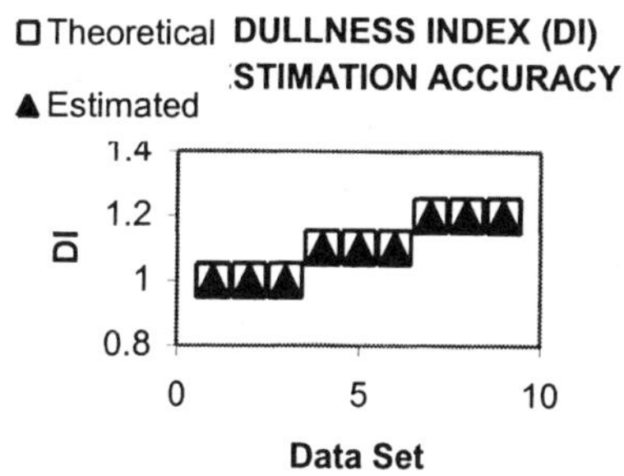

Figure 6. Cutting force coefficient estimation accuracy of the GTM2 when the feed direction cutting force is used for estimations. (Values were taken from (Li,2002))

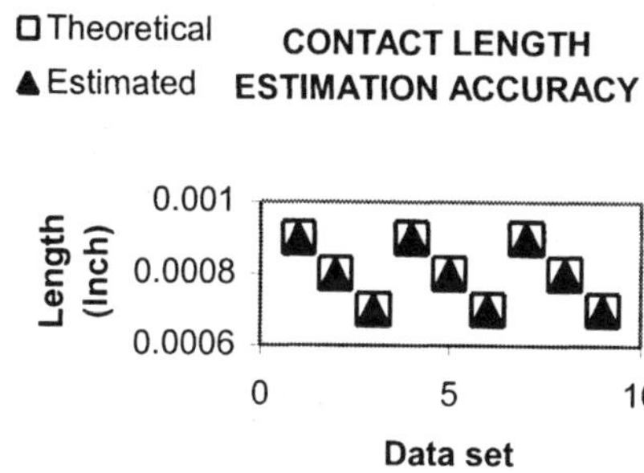

Figure 7. Contact length estimation accuracy of the GTM2 when the feed direction cutting force is used for estimations. (Values were taken from (Li,2002))

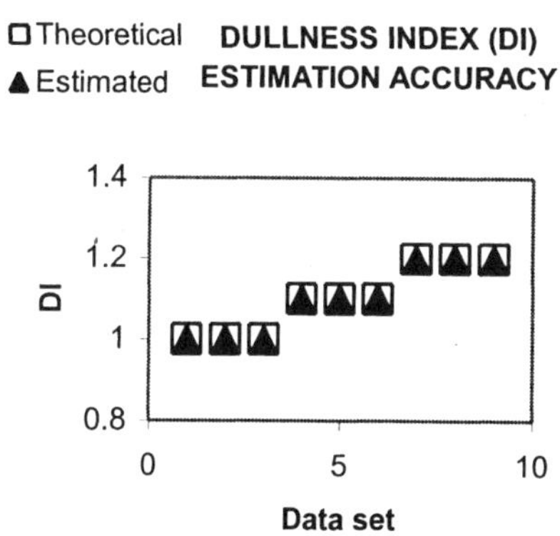

Figure 8. Cutting force coefficient estimation accuracy of the GTM2 when the thrust direction cutting force is used for estimations. (Values were taken from (Li,2002))

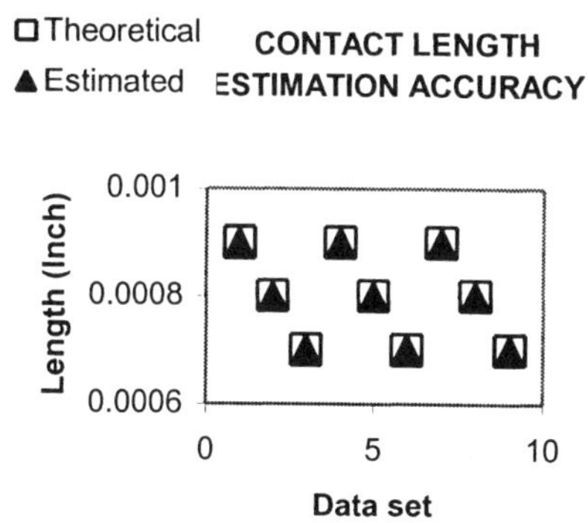

Figure 9. Contact length estimation accuracy of the GTM2 when the thrust direction cutting force is used for estimations. (Values were taken from (Li,2002))

CONCLUSION

The cutting force coefficient and contact length between the cutting edges and the workpiece were estimated by using the Genetic Tool Monitor (GTM2). The GTM2 estimated both the parameters with less than 10% error when the resultant force was used. Estimation error was better than 2% when the estimations were made by inspecting a single force component.

Although, the results indicate that GTM2 will work effectively when the resultant force or any other force component is monitored, we suggest the monitoring of the resultant cutting force. Variation of the cutting force coefficient, and contact length influenced the characteristics of the single force components more than the resultant force. Most probably, noise and modeling mistakes will create significant errors when single force components are used.

The distinction of the cutting force coefficient variation from the contact length variation and accurate estimations in less than 500 generations indicated that genetic algorithm was a good choice for the task. It is possible to increase the number of the estimated coefficients by including the depth of cut and contact angles to monitor the entire cutting operation on-line. However, each additional parameter will increase the computational time and reduce the accuracy of the estimations.

ACKNOWLEDGEMENT

Different versions of the simulation and GTM programs were prepared by W.Y. Bao and Mr. L. Li. The values for tables and figures were taken from Mr. L. Li's M.S. thesis. This project was partially supported by the Motorola Corporation and Wright Patterson AFB. Mr. Robert Shisler made valuable contributions to the project while he managed the project for Motorola Corporation. Authors appreciate contribution of all the parties.

REFERENCES

Achiche, S., Balazinski, M., Baron ,L.,and Jemielniak, K., 2002, "Tool wear monitoring using genetically-generated fuzzy knowledge bases," Engineering Applications of Artificial Intelligence, Volume 15, Issues 3-4, Pages 303-314 (June-August 2002)

Bao, W. Y., and Tansel, I. N., 2000a, "Modeling micro-end-milling operations. Part I: analytical cutting force model," International Journal of Machine Tools and Manufacture, Volume 40, Issue 15, December 2000, Pages 2155-2173

Bao ,W. Y., and Tansel, I. N., 2000b,"Modeling micro-end-milling operations. Part II: tool run-out," International Journal of Machine Tools and Manufacture, Volume 40, Issue 15, December 2000, Pages 2175-2192

Carroll, D.L., 1996, "Chemical Laser Modeling with Genetic Algorithms", AIAA Jour., Vol.34, No.2, pp.338-346.

Cus, F., and Balic, J., 2003,"Optimization of cutting process by GA approach, Robotics and Computer-Integrated Manufacturing," Volume 19, Issues 1-2, Pages 113-121 (February-April 2003).

Davim, J. P., and António, C. A. C., 2001, "Optimisation of cutting conditions in machining of aluminium matrix composites using a numerical and experimental model," Journal of Materials Processing Technology, Volume 112, Issue 1, Pages 78-82 (3 May 2001).

Goldberg, D.E., 1989, "Genetic Algorithms in Search, Optimization and Machine Learning", Addison-Wesley, Reading, MA.

Holland, J.H., 1992, "Adaptation in Natural and Artificial Systems", A Bradford Book, The MIT Press, Cambridge, MA.

Khan, Z., Prasad, B., and Singh, T., 1997, "Machining condition optimization by genetic algorithms and simulated annealing," Computers & Operations Research, Volume 24, Issue 7, Pages 647-657 (July 1997).

Li, L., 2002,"Genetic Algorithm Based Open Architecture Cutting Tool Monitoring System", M.S. Thesis, Florida International University.

Shunmugam, M. S., Bhaskara Reddy, S. V, and Narendran, T. T., 2002, "Selection of optimal conditions in multi-pass face-milling using a genetic algorithm," International Journal of Machine Tools and Manufacture, Volume 40, Issue 3, Pages 401-414 (February 2000).

Szecsi, T, 1999, "A DC motor based cutting tool condition monitoring system", Journal of Materials Processing Technology, Volumes 92-93, Pages 350-354 (30 August 1999).

Tansel, I., Bao,W.Y., Tansel, B., Shisler, R., Smith, D., and Murray, J., 1998, "Identification of Cutting Conditions by Using an Analytical Model and Genetics Algorithms for Micro-End-Milling Operations," Smart Engineering Systems: Neural Networks, Fuzzy Logic, Evolutionary Programming, Data Mining, and Rough Sets, edited by Drs. Dagli, Akay, Buczak, Ersoy, and Fernandez, ASME Press, New York.

Tansel, I.N., Bao, W.Y., Arkan, T.T., and Kropas-Hughes, C.V., 2002, "Genetic algorithm versus neural network in health monitoring applications," Intelligent Engineering Systems Through Artificial Neural Networks v 12 p 255-260.

Thermal Agents: Learning Thermal Profiles for Rapid Design

Daniel A. Ashlock
Mathematics Department
Iowa State University
Ames, Iowa 50011
danwell@iastate.edu

Kenneth M. Bryden
Mechanical Engineering
Iowa State University
Ames, Iowa 50011
kmbryden@iastate.edu

Abstract

The temperature profile across an object is easy to compute by iterative methods. The time spent waiting for iterative solutions to converge for multiple objects in a complex configuration is an impediment to exploratory analysis of engineering systems. A rapidly computed initial guess can speed convergence for an iterative algorithm. We describe and test a system for creating thermal agents that supply such initial guesses. Thermal agents are specific to an object but general across different thermal boundary conditions. During an off-line training phase, genetic programming is used to locate a thermal agent by training on several sets of boundary conditions. In use, thermal agents transform boundary conditions into rapidly converged initial values on a cellular decomposition of an object.

1 Introduction

In a virtual engineering environment exploratory analysis demands rapid recalculation when changes are made by a user of the system. In this study we describe and demonstrate a type of *thermal agent* that can be used to accelerate convergence of an iterative solver for temperature distribution. Thermal agents supply initial estimates of the temperature distribution of an object that permit iterative solvers to converge rapidly. The thermal agents are specific to particular objects but general across different thermal boundary conditions. The two objects used in this study are shown in Figure 1.

In order to simulate a system involving many objects with temperature profiles that vary over time a thermal agent is trained for each object. The creation of thermal agents is an off-line process in which the space of possible thermal agents is searched. Once the thermal agents are located for each object, these agents permits rapid recalculation of thermal profiles as the boundary values for objects are updated by a higher level simulation system.

Genetic programming [2] permits an algorithm modeled on biological evolution to search a space of formulas. This study uses genetic pro-

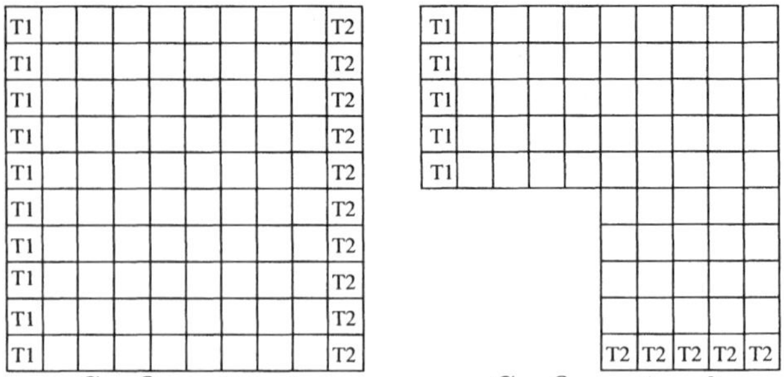

Figure 1: **The two configurations of objects used in this study. The values T_1 and T_2 represent fixed boundary temperatures in any given problem case.**

gramming to locate effective thermal agents. Given information about the boundary value temperatures, a thermal agent maps the (x, y)-coordinates of a cell onto an initial temperature estimate for a cell. The quality of a thermal agent is judged by the amount of computational work required to converge the temperature profile of its associated object starting from its initial temperature approximations.

2 Training Thermal Agents

Feature	Case 1	Case 2
Grid Size	20x20	20x20
Number of boundary conditions	2	2
Boundary block 1	X=0..0,Y=0..19	X=0..0,Y=0..9
Boundary block 2	X=19..19,Y=0..19	X=10..19,Y=19..19
Number of fitness cases	2	1
Fitness case 1	$T_1 = 100, T_2 = 300$	$T_1 = 100, T_2 = 500$
Fitness case 2	$T_1 = 100, T_2 = 500$	n/a
Adiabatic blocks	n/a	X=0..9,Y=10..19

Table 1: **Descriptions of problem cases used in this study.**

An object used to train thermal agents is specified by a partition into discrete cells in a grid. A given cell may be *active* or *adiabatic* with the active cells further divided into *boundary cells* and *free cells.* Boundary cells have fixed temperatures, adiabatic cells do not participate in heat flow, and free cells are those whose temperature is to be initially approximated by the thermal agent. In this study the objects and grids are two dimensional though the technique can easily by generalized to three dimensions. Figure 1 and Table 1 describe the experimental cases used in this study.

A simple iterative solver is used to evaluate the thermal agents. This solver updates its temperatures in the free cells by computing the average temperature of the cell and is active grid neighbors. If $\Gamma(C)$ is the set of grid neighbors of a cell C, including C, then the temperature updating rule for C is given by

$$T_{new}(C) = \frac{1}{|\Gamma(C)|} \sum_{D \in \Gamma(C)} T_{old}(D). \tag{1}$$

This updating rule is applied synchronously to all free cells.

A thermal agent τ produces initial approximations to the temperatures in free cells given the x, y coordinates of the cell and the temperatures $T_1, T_2, \ldots, T_k$ for the boundary blocks. The initial temperatures in cell (x,y) are given by

$$T_{initial}(x, y) = \tau(x, y, T_1, T_2, \ldots, T_k) \tag{2}$$

Different fitness cases for a thermal agent are specified by choices of the fixed boundary temperatures. For a given fitness case the score of a thermal agent is determined by first loading the agent's initial temperature approximations into the grid and then running the iterative solver until the largest temperature change during an iteration drops below a threshold t_{stop}. The score is the number of iterations required to converge the initial approximation. The fitness of a thermal agent is the average of its scores over all fitness cases.

Details of the genetic programming system used to locate thermal agents are given in Section 3. The operations used by the system are standard arithmetic and transcendental functions from a pocket calculator together with a numerical $If-Then-Else$ of the sort described in [2]. Terminals are cell coordinates, boundary temperatures, and ephemeral constants in the range $0 \leq t \leq 10000$ degrees centigrade. Using convergence time as a fitness function in a genetic programming system creates a problem that suggests a modification of the way thermal agents are evaluated. Sub-tree crossover, one of the search operators used in genetic programming, is known to be highly disruptive [1, 2] and is sometimes referred to as a "macromutation". In the case of thermal agents, this disruption results in bad agents. Finding out that an agent is bad is computationally costly. If an agent produces initial temperature estimates in the millions of degrees the iterative solver requires many iterations to converge. To quickly eliminate such agents, we define a convergence time threshold L_{stop}. If the best fitness found thus far is B, then any agent whose initial estimates have not converged in $L_{stop} \cdot B$ iterations is assigned a score of $L_{stop} \cdot B + 1$ for the current fitness case. Some variation of this technique may be applicable in any system with disruptive crossover and expensive fitness evaluation.

3 Experimental Design

Two sets of experiments were performed to test the thermal agent location system. The first is a simple problem involving a square with a hot side and a cool side. Solutions to this problem are one dimensional. Fitness was evaluated for two sets of boundary temperatures for this experimental case. The second experimental case used an L-shaped object with a hot and cool boundary on orthogonal axes that are spatially separated. A collection of 100 runs were done for each experimental case. For each population the steady state evolutionary algorithm was run until either an agent with fitness of at most 1 was located or until 50,000 tournaments had taken place. The best agent from each run was saved, yielding a set of 100 agents for each problem case. The stopping temperature variation was set to $t_{stop} = 1$ and the convergence time boundary was set to $L_{stop} = 3$.

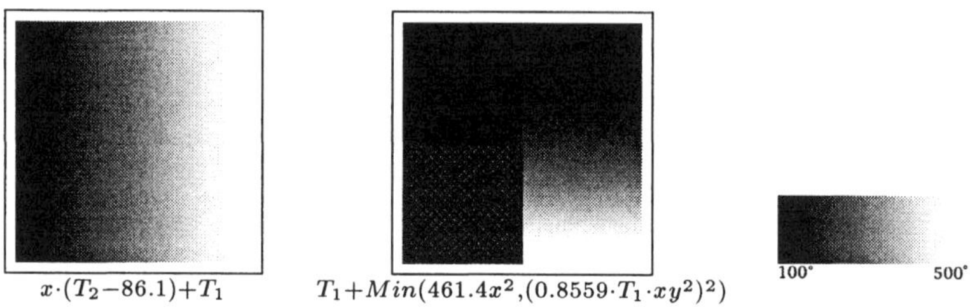

Figure 2: **Initial thermal guesses supplied by the best thermal agents located for the square and L-shaped objects. The formulas for the thermal agents are given beneath their thermal profiles.**

The genetic programming system used to search for thermal agents operates on a population of 400 parse trees. An initial set of trees with six operations and terminals each are generated at random. The values of ephemeral real constants in the initial population was in the range 0-1000 degrees, but constants were permitted to move outside this range by constant mutation, described below. The evolutionary algorithm is a steady state [3] algorithm using single tournament selection with tournament size seven. Reproduction consists of copying the two parent trees, performing subtree crossover [2] on the copies, and then performing a subtree mutation on one of the new trees and a constant mutation on the other. We augment subtree crossover with an operation called *chopping*. If a tree is produced with more than twenty total operations and terminals, we iteratively promote a randomly selected argument of the root node of the tree to become the new tree until the trees size drops to twenty or less. Subtree mutation picks a node of a parse tree uniformly at random and replaces it with a random tree the same size. Constant mutation picks an ephemeral constant from the tree uniformly at random and adds to it a Gaussian random number with mean zero and standard deviation 10.

4 Results

The best thermal agent found for the first experimental case is

$$\tau_1(x, y, T_1, T_2) = x \cdot (T_2 - 86.1) + T_1, \tag{3}$$

which posses the correct form for an analytic solution to the problem. This thermal agent misses being an analytic solution only by the adjustment of its ephemeral constants. It requires one iteration of the solver to converge this agent's initial estimates for both fitness cases.

The second experimental case was more challenging, and the best thermal agent located has a fitness of 3, requiring that the solver iterate 3 times to converge it. The form of this agent is

$$\tau_2(x, y, T_1, T_2) = T_1 + Min(461.4x^2, (0.8559 \cdot T_1 \cdot xy^2)^2). \tag{4}$$

This agent uses a minimum operation to break the object into two pieces, an upper area with dependence of temperature only on x and a lower region with dependence on both x and y to provide the curved temperature profile.

For both experimental cases the fitness of members of the initial population varied in the range 25-300 with means well above 100. Best final fitnesses of 1 and 3 respectively show that the evolutionary algorithm is successfully locating agents that perform well. In order to check that the convergence times for these thermal agents was objectively low, the fitness was computed for a thermal agent that filled in the value $\frac{(T_1+T_2)}{2}$ in all free cells. A summary of evolutionary and baseline results is given in Table 2.

Experiment Number	Best Fitness	Mean Fitness	Standard Deviation	Worst Fitness	Baseline Fitness
1	1.0	1.89	1.16	3	33
2	3.0	13.2	3.7	18	47

Table 2: **Performance summary for thermal agents. These statistics are drawn from 100 runs of the evolutionary algorithm for each experimental case. The baseline fitness is the fitness of an agent that fills in the average of the boundary temperatures on the free cells.**

The first experimental case, an essentially one-dimensional problem, yielded solutions that converged in one iteration in 43 of 100 runs and never produced solutions requiring more than three iterations. This experimental case is an easy problem and was included not because we wish to solve it. Rather, if the performance of the thermal agent locating system was not significantly better then the system itself is suspect. The best solution located had the form of a correct analytic solution and, as such, showed they system will function well if it is given an easy problem.

The second experimental case was a non-trivial problem in which a perfect solution must depend on both spatial variables as well as the boundary temperatures. The best solution located in this case was *not* an analytic solution, depending as it did on a non-differentiable minimum operation. This minimum operator acted to decompose the space and then provided adequate approximating functions in each part of the decomposition.

5 Future Work

Working in three dimensions is an obvious generalization and not a difficult one, save for the increase in the number of cells required with a consequent impact on run-time. Shifting the work to multiprocessor machines should compensate for such increases in run-time and three-dimensional versions of this work are planned.

A improvement of this system that might help substantially is to perform a preliminary evaluation for dependence on relevant variables. A thermal agent that does not depend on x, y, and at least one of the boundary temperatures is unlikely to have good fitness.

The eventual goal of this work is to use multiple thermal agents to model multiple objects undergoing thermal interaction. Several steps that are required for this. The thermal agents should be generalized to work with more complex thermal solvers than our simple averaging solver [4]. Such solvers are both faster than simple averaging and can handle non-homogeneous materials.

References

[1] Peter J. Angeline. Subtree crossover causes bloat. In John R. Koza *et. al.*, editor, *Genetic Programming 1998: Proceedings of the Third Annual Conference*, pages 745–752, University of Wisconsin, Madison, Wisconsin, USA, 22-25 July 1998. Morgan Kaufmann.

[2] Wolfgang Banzhaf, Peter Nordin, Robert E. Keller, and Frank D. Francone. *Genetic Programming : An Introduction.* Morgan Kaufmann, San Francisco, 1998.

[3] Gilbert Syswerda. A study of reproduction in generational and steady state genetic algorithms. In *Foundations of Genetic Algorithms*, pages 94–101. Morgan Kaufmann, 1991.

[4] J. C. Tannehill, D. A. Anderson, and R. H. Pletcher. *Computational Fluid Mechanics and Heat Transfer.* Taylor and Francis, Washington D.C., 1997.

A METHOD FOR ASSESSING QUALITY OF FOOD PRODUCTS BASED ON OPTICAL SPECTRA USING INTELLIGENT PATTERN RECOGNITION IN A FULL SCALE PRODUCTION ENVIRONMENT

M. O'FARRELL
ECE Department,
University of Limerick
Castletroy, Limerick,
Ireland

E. LEWIS
ECE Department,
University of Limerick
Castletroy, Limerick,
Ireland

C. FLANAGAN
ECE Department,
University of Limerick
Castletroy, Limerick,
Ireland

W. B. LYONS
ECE Department,
University of Limerick
Castletroy, Limerick,
Ireland

N. JACKMAN
Food Design
Applications Ltd.
Newtown, Castletroy,
Limerick, Ireland.

ABSTRACT

An Optical Fibre Sensor System has been developed for the online measurement of food quality both internally and externally as it cooks in a large-scale industrial oven. The quality of the food is determined by its colour as it cooks and by automatically classifying the food at different stages it is possible to determine whether the food is cooked to an optimum colour for profits. Developments have been made on previous work by the authors [3] by further examining the internal colour of the food and also looking at food that does not have an even colour externally. Spectroscopic techniques are employed to determine the colour and this signal is interrogated using an Artificial Neural Network. The resultant output spectral patterns can therefore be quantitatively classified and thus categorized.

INTRODUCTION

With an ever-increasing volume in the throughput of the food industry, profit margins dictate that it is no longer acceptable for large errors to occur, as the result would be significant. This is evident in the cooking of large batches of various food types in industrial ovens of lengths up to 20 metres Fig.1. The conveyor belts of these ovens may be fully loaded with food for maximum efficiency and the cost of an error in one batch would be intolerable for the producer. Automation in the cooking of the food is critical in order to control oven temperature, conveyor belt speed, height of steamers etc. so that the food is cooked to the highest quality. There is a need for monitoring the food as it passes through the oven and determining that all the processing parameters are correct, a critical one of these being the food's colour of the food (both internal and external).

Grading the quality of a product e.g. in terms of colour [1] or shape and size [2], ensures consistency in the product. Factors such as colour have a huge influence over a customer decision in purchasing a product, as they often have preconceived notions of how a product should look; roast chickens must be golden, flame-grilled hamburgers must have black lines and cooked sausage must be a rich brown.

Colour is determined in the visible spectrum. Using a broadband tungsten halogen lamp, the food sample is illuminated through an optical fibre. The reflected light is returned to the spectrometer (Oceanoptics S2000) and the spectrum is sent to the host P.C. for subsequent analysis using Artificial Neural Networks. The software used to implement the neural network is Stuttgart Neural Networks Simulator (SNNS).

This paper focuses on taking both internal and external readings of the same product. It is also investigated what the effects of something like herbs being scattered on top of the food. The cooking-stages of pastry are examined as a product that colours in an uneven manner.

EXPERIMENTAL SET-UP

Food producers often employ a technique to colour the food by using surface air-blowing towards the end of the belt to colour the food. It is clear from the diagram (Fig. 1) that there are five main stages involved in the cooking, represented by the five stages in the oven. The colour sensing is performed between the fourth and fifth stages, after "colouring by roasting" has taken place.

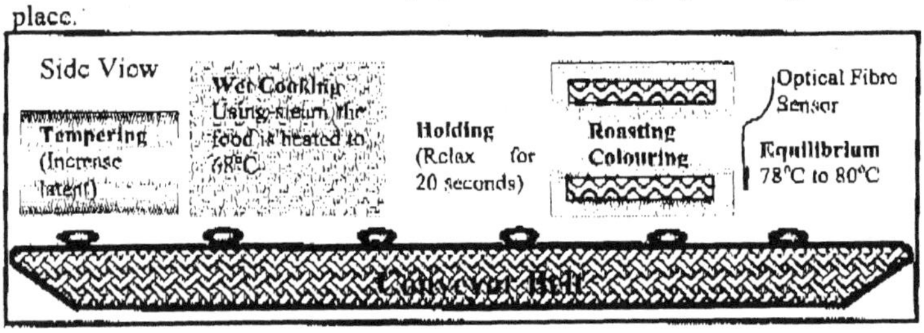

Figure 1: Industrial oven used in on-site testing

The colour is measured using a reflective probe and recording the light reflected in the visible spectrum, which is representative of the colour. Illumination of the food is performed with 6 fibres attached to a tungsten halogen white light source. One fibre returns the reflected light to an Ocean Optics s2000 spectrometer, controlled by a host P.C., Fig. 2(a). The data is then saved for subsequent analysis using the Stuttgart Neural Networks Simulator (SNNS) the Neural Network Software [4]. Examples of the use of artificial intelligence for decision-making in the food industry are plentiful [5], [6].

The probe for measuring the external readings is required to have a diameter that is as large as possible in order to cover a greater area. The fibres used are 400μm in diameter. For testing a single probe was used, as for initial measurements, it was important for the investigators to have full and free control over where it is placed in order to train the Neural Network with correct data; it is envisaged to have 5 or 6 separate probes placed around the surface of the food sample to read the colour over a greater area. The internal readings need to be unobtrusive so it is important that the diameter is kept to a minimum. The fibres for this probe are 200/220 μm. The tip of this probe is polished at 45 Deg. This gives the tip a pointed edge for insertion and increases the area of the fibre exposed to the food, Fig. 2(b).

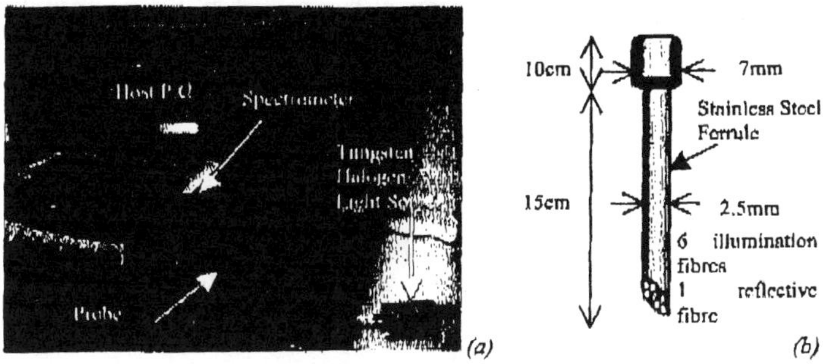

Figure 2: (a) Photo of the System Set-up (b) Probe for internal colour readings.

Both probes are encased in stainless steel. This material was chosen to allow the probe to be regularly sterilised, something that is essential in the food industry. Its robustness allows the probe to be injected at high speed into the product and puncture without causing damage to the product or itself.

RESULTS

Results were obtained in conjunction with Food Design Applications Ltd.

Puff Pastry

Firstly tests were carried out on puff pastry products, i.e. sausage rolls and cornish pasties. These products colour quite quickly and can become burnt in a matter of seconds. It is important to monitor the edge of the food, as burnt patches will appear there to begin with. The graph of Fig.3 represents the variations in 985 spectra, i.e. 197 per class- raw, too light, correct colour, too dark and burnt.

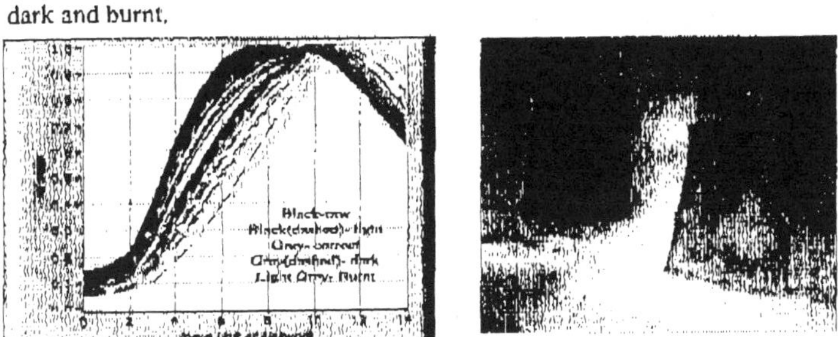

(a) (b)

Figure 3: (a) Spectral Variations of Puff Pastry (b) Cornish Pasty; Overcooked (left) and Raw (right)

Marinated Chicken

Testing was also done on roasted marinated chicken thighs and wings. The internal colour was examined using an unobtrusive optical fibre probe of diameter 2.5mm. Fig 4(a) shows a photo of the food when it is cut open it is possible to observe pink appearance of the uncooked meat in the centre. Fig. 4(b) shows the spectral variations between 375 of each undercooked and correctly cooked samples.

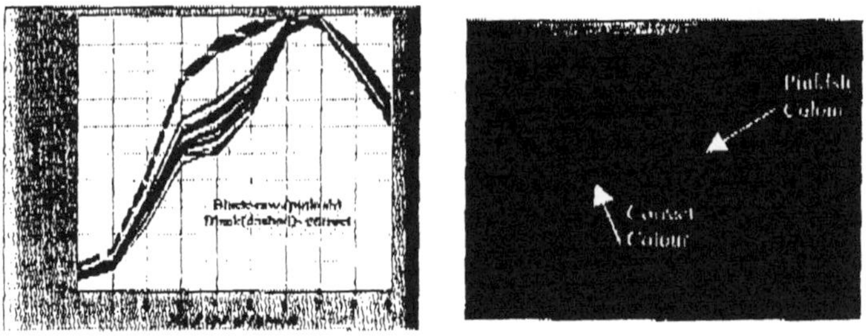

(a) (b)

Figure 4: (a) Spectral Representation of Internal Readings of Chicken. (b) Photo showing example of pinkish undercooked meat and correctly cooked meat

The surface colour of these products is quite varied, Fig. 5(b), with dark spots and herbs from the marinade. The graph of Fig. 5(a) represents variations in 1875 independently acquired spectra, i.e. 375 per colour class- raw, light, correct colour, dark, and herbs.

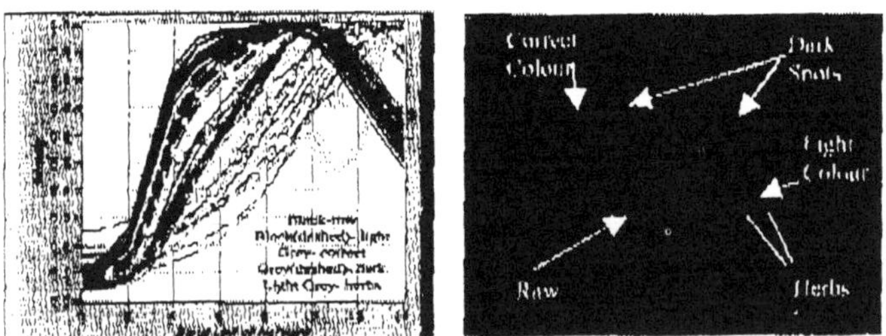

(a) (b)

Figure 5: (a) Spectral Representation of External Readings of Chicken. (b) Photo showing example of raw to dark meat with herbs scattered on top

Data Conditioning

The x-axes of each of the graphs represent the amount of data points that were used to represent the spectra. Each spectral reading from the s2000 spectrometer has 2048 data samples, resulting in high resolution, in the range between 200nm and 900nm. It was possible to reduce the amount of data points whilst still preserving the original spectral shape. For the pastry 15 points were taken between the wavelengths of 400nm and 850nm. For the marinated chicken 25 points were needed for the external readings and 10 data points for the internal readings.

THE NEURAL NETWORKS

Three Feed Forward 3 layer neural networks were configured with the number of input nodes representing the number of data points required to represent the spectra of each cooking stage, i.e. 15 for the puff pastry and 25 for the external chicken readings and 10 for the internal chicken readings. At the output layer of each network one node was chosen for each condition to be sensed. This represents 5 nodes for the puff pastry and external chicken neural networks and 2 for the internal chicken neural network.

Neural Network for Marinated Chicken

1. External Readings

Following a comparison of different configurations, a hidden layer of 5 nodes was chosen for the neural network for external chicken readings. Using the standard training algorithm of Backpropagation with Momentum the network was trained using the Neural Network software, SNNS. The following parameters as required by η(learning rate) = 0.8, μ (momentum term) = 0.2, c (flat spot eliminator) = 0.1 and δ (maximum error) = 0.1. A total of 985 result patterns were used to train the network, 197 of each condition; raw, too light, optimum colour, dark spots, and herbs. The trained network was then tested using an independent set of 370 data patterns (74 of each condition) to that which had been applied previously in the training of the network. The network correctly classified all test patterns. A sample of the results obtained from the test set is shown in Table 1.

2. Internal Readings

A hidden layer of 2 nodes was chosen for the neural network for internal chicken readings. Using the standard training algorithm of Backpropagation (η = 0.8, δ = 0.1), a total of 596 result patterns were used to train the network, 298 of each condition; pinkish and optimum colour. The trained network was then tested using an independent set of 370 data patterns (74 of each condition). The network correctly classified all test patterns. A sample of the results obtained from the test set is shown in Table 1.

Neural Network for Puff Pastry

Various network configurations were tested and it was decided that a hidden layer of 12 units was most suitable. Using Backpropagation with Momentum (η = 0.8, μ = 0.2, c = 0.1 and δ = 0.1) the Network trained successfully, distinguishing between 5 conditions: raw, undercooked, correctly cooked, over cooked and burnt. 985 test patterns, 197 for each condition, were used to train the network. When the network was tested using a test set of 370 test patterns, 74 of each condition, it correctly classified each cooking stage. A sample of the results obtained from the test set is shown in Table 1(b).

Colour	Ideal O/p	Observed Activation Output **Pastry**	Observed Activation Output **External Chicken**	Ideal O/p	Observed Act. Output **Internal Chicken**
Raw	10000	0.92 0.08 0.04 0.02 0.00	0.93 0.07 0.00 0.02 0.04	10	0.89 0.10
Raw	10000	0.91 0.08 0.04 0.02 0.00	0.92 0.08 0.00 0.02 0.04	10	0.89 0.06
Light	01000	0.03 0.85 0.06 0.01 0.00	0.02 0.98 0.09 0.02 0.00		
Light	01000	0.03 0.85 0.05 0.001 0.00	0.03 0.94 0.08 0.02 0.00		
Correct	00100	0.00 0.00 0.80 0.19 0.01	0.00 0.00 0.80 0.08 0.04	01	0.10 0.89
Correct	00100	0.00 0.00 0.70 0.11 0.01	0.00 0.06 0.89 0.12 0.03	01	0.06 0.93
Dark	00010	0.00 0.00 0.02 0.92 0.08	0.01 0.00 0.00 0.90 0.08		
Dark	00010	0.00 0.00 0.02 0.90 0.07	0.01 0.00 0.00 0.92 0.08		
Burnt	00001	0.00 0.00 0.00 0.00 1.00			
Burnt	00001	0.00 0.00 0.00 0.00 0.89			
Herb	00001		0.00 0.00 0.04 0.08 0.93		
Herb	00001		0.00 0.00 0.04 0.008 0.94		

Table 1: Sample Results using Test Sets on the trained ANN for each food type.

CONCLUSION

A fibre optic based spectroscopic technique, coupled with Artificial Neural Network signal processing techniques, has been developed for the purpose of the recognition and classification of the various cooking stages of pastry and marinated chicken. This paper further expands on previous research done by the authors [3]. It explores other food types both internally and externally, thus strengthening the viability of the system.

REFERENCES

[1] McConnell RK Jr., Blau HH Jr. "Colour classification of non-uniform baked and roasted foods." *Proc. Food Processing Automation IV, pp.40 –46, 1995*

[2] Aleixo N., Blasco B., Molto E., Navarron F., "Assessment of citrus fruit quality using a real-time machine vision system", *Pattern Recognition, 2000. Proc. 15th International Conference pp. 482 –485 vol. 1, 2000*

[3] O'Farrell M., Lewis E., "An Intelligent Optical Fibre based System for Monitoring Food Quality" *IEEE Sensors 2002. Proc. 1st International Conference pp. 308-311 vol. 1, 2002*

[4] SNNS, Stuttgart Neural Network Simulator, User Manual version 4.1 or http://www-ra.informatik.uni-tuebingen.de/SNNS/

[5] Shiranita K., Hayashi K., Otsubo A., Miyajima T., Takiyama R., "Determination of Meat Quality by image processing and Neural Network Techniques" *The ninth International Conference on Fuzzy System, pp 989 –992, vol.2, 2000*

[6] Abdesselam A., Abdullah R.C., " Pepper Berries Grading using Artificial Neural Networks" *TENCON 2000. Proceedings, pp 153 –159, vol.2, 2000*

Waveform Recognition of Back-Electro Motive Force for Real-time Embedded System Applications Using Support Vector Machines

Jeff Stewart and Chia-Jiu Wang
University of Colorado at Colorado Springs
Department of Electrical and Computer Engineering
P.O. Box 7150
Colorado Springs, CO 80933

Abstract

This paper investigates the possibility of using support vector machines to recognize distinct patterns produced by a back electro motive force (back-emf) in a sampled analog waveform. The intended use of the design would be an embedded controller using the support vector machine output as feedback to control a device. The back-emf waveforms contain many hidden distinct features that "describe" the motion of the device that generated the signal. This research, therefore, is aimed to use support vector machines to analyze a batch of samples of back-emf data and thereby measure several aspects about the corresponding motion of the device. The results show great promise that support vector machines are able to analyze the back-emf data properly. The outputs of the support vector machines can be used to control the system accurately.

1. Introduction

A microcontroller system can be used to control a device that moves when a current is applied. The motion of the device is discrete; that is, after a current is applied, the device moves a fixed distance in one direction, and when another current is applied, the device will move back to its starting position. In most systems, the timing of the motion is critical down to the microsecond. Since the devices are sensitive to temperature, and also to other forces acting on them, albeit to a lesser degree, a significant challenge is posed to the microcontroller to maintain the timing tolerances required. If this challenge is to be met, the microcontroller must receive feedback about the motion of the device so that it may compensate for the outside disturbances.

As well as correcting for timing variances, the controller must also have the ability to detect and, to the best of its ability, correct failures. The most common failure types result in complete loss of motion which is relatively easy to detect. On the other hand, some failures may occur where the device may not stay in one position because it bounces when it lands in that position. This mode of failure requires a more detailed examination of the back-emf signal.

2. Back-EMF Characteristics

By looking at a back-emf trace as shown in Figure 1, one can determine many things about the operation of the device. For instance, different shapes of the trace may determine the speed, position, and stability of the device. Figure 1 shows a picture of four different back-emf traces. The key features in this trace have been exaggerated for clarity. The first noteworthy characteristic to discuss is the general motion. Note that three of the traces are moving to the top (positive direction) while the green trace moves to the bottom (negative direction). Any motion to the negative direction is an indication that the device did not move. Sometimes motion in the positive direction with significantly less amplitude can also indicate loss of motion. This is not illustrated here

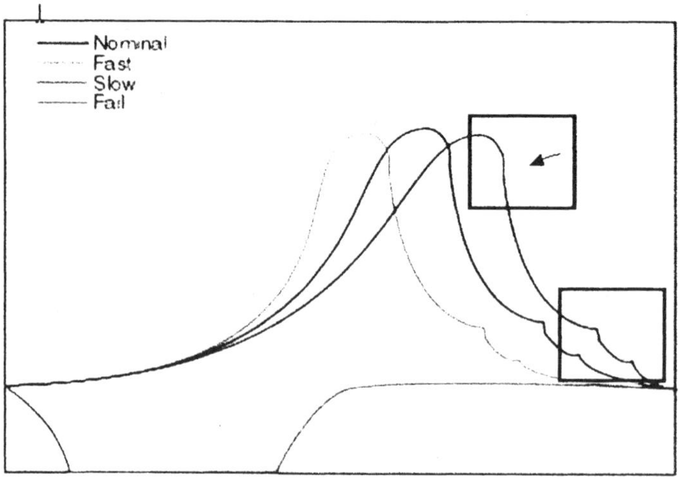

Figure 1. Example back-emf traces with exaggerated features

The next feature of importance to this study is the "cusp" at the top of each trace (highlighted by the blue square on the red trace). This cusp or change in slope of the trace is an indication that the device has reached the opposite position. In a sense, this inflection marks the end of the travel time for the device. This can give a controller an indication of whether the device is operating nominally as is shown in the black trace, or whether the device is fast or slow as shown in the cyan and red traces respectively. The last feature of real interest for a controller is the inflection points at the end of the trace (highlighted in the magenta square). These inflections represent bouncing; that is, when the device moves into a position, rather than just coming in and landing, it bounces off the wall and lands again. The number of inflections corresponds to the number of bounces.

3. Proposed Solution

Real-time digital signal processing and pattern recognition techniques have been used to remove noise and recognize the system's operating conditions. In this paper we propose a new solution using support vector machines to identify and predict the nonlinear operation conditions of a back-emf device. From the back-emf characteristics, it is evident that there is some valuable feedback to be gained from analyzing the back-emf data and adjusting the control appropriately. There are several basic features that a controller could be expected to implement given the trace data. A system is proposed to handle the following functions efficiently. These functions are:

(1) Failsafe Mechanism – A controller could determine whether the device moved or not and take any action possible to correct the situation or shut down.

(2) Speed Detection – The time that the device takes to travel can vary by several microseconds. If this is critical to the overall system, the controller could take this time variance into account, and correct other factors accordingly.

(3) Stability Analysis – If the controller sees evidence of the device bouncing, there might be corrections that are possible if the device was known not to be landing consistently.

3.1 Proposed System Overview

Figure 2 shows the block diagram of the proposed system. Support vector machines [1, 2, 3] are used to perform classification and regression functions. Because the system must identify or measure three conditions, the overall design will contain three SVM's: one for the failure detection, one for the stability analysis, and one for the measurement of the device speed. The failure detection unit will simply have a pass or fail type of output and it will only classify whether or not the device moved. Similarly, the stability machine will only classify whether or not the device bounced as it landed with a pass or fail criteria. The speed measurement unit, on the other hand, is somewhat more complicated, more than two outputs are required; instead of a simple pass/fail, the system needs to formulate a different output depending on the range of speed that the device moved in. For instance, in a simple case, there could be three bands each two hundred microseconds wide, 100-299,300-499,and 500-699. These bands could be labeled as fast, nominal, and slow respectively. In this case, the SVM would determine which band the speed of a particular back-emf trace resided in and classify the particular actuation with one of the labels. Finer resolution could also be achieved by having bands that were ten or even one microsecond apart. Now, the SVM would be classifying the actuation into actual time down to whatever the resolution of the bands.

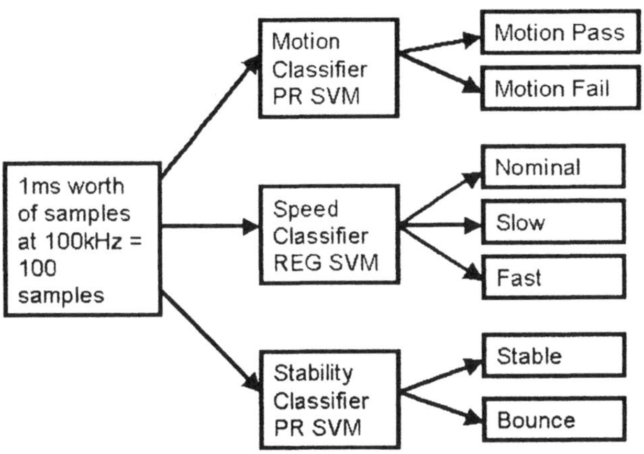

Figure 2 Block Diagram of the proposed SVM based back-emf analysis system.

4. Results Analysis

4.1 Speed Classifier SVM

Figure 3 shows 100 time-shifted back-emf traces. In training process we use about 20 traces. The rest of 80 traces are used for tesing. Along with the testing data, a set of theoretical "answer" values had to be calculated so that the training process would know the desired result of each trace. For this study, a simple level threshold was used. Since all of the traces are about the same amplitude, the cusp also occurs at about the same height. Thus, the test script simply calculated the sample at which the trace crossed the threshold. If the threshold occurred between two samples, an interpolated value was given. An example of four plots with the calculated cusp point highlighted is shown in Figure 4.

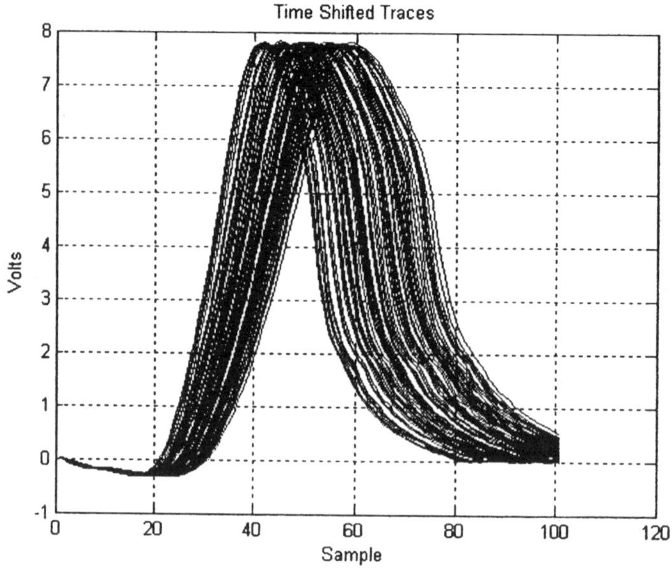

Figure 3: Plot of 100 time-shifted testing traces. The training data is identical to this except that there are fewer traces.

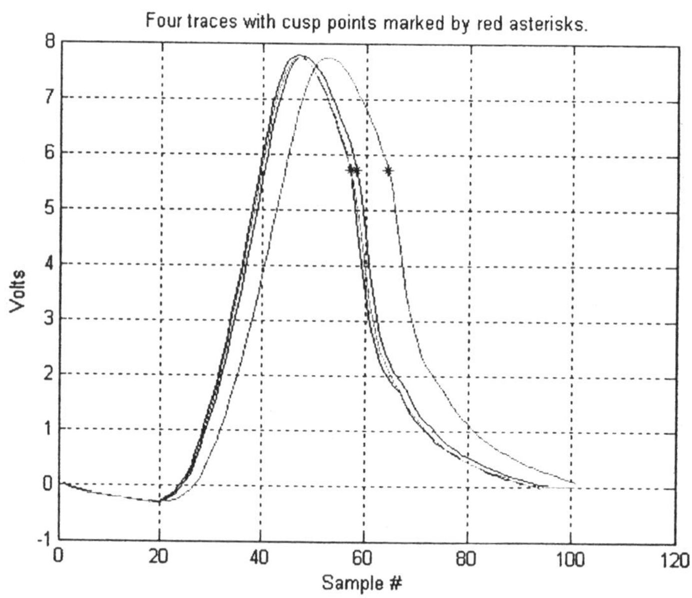

Figure 4: Plot of four back-emf traces with the cusp points highlighted by red asterisks. In this trace, the cusp threshold was set to 5.7.

The plot in Figure 5 shows the results of the prediction. This plot is shown with the SVM complexity constant C =101. We use mySVM developed by Stefan Ruping [4]. The predicted speed is about 95% same as the actual speed.

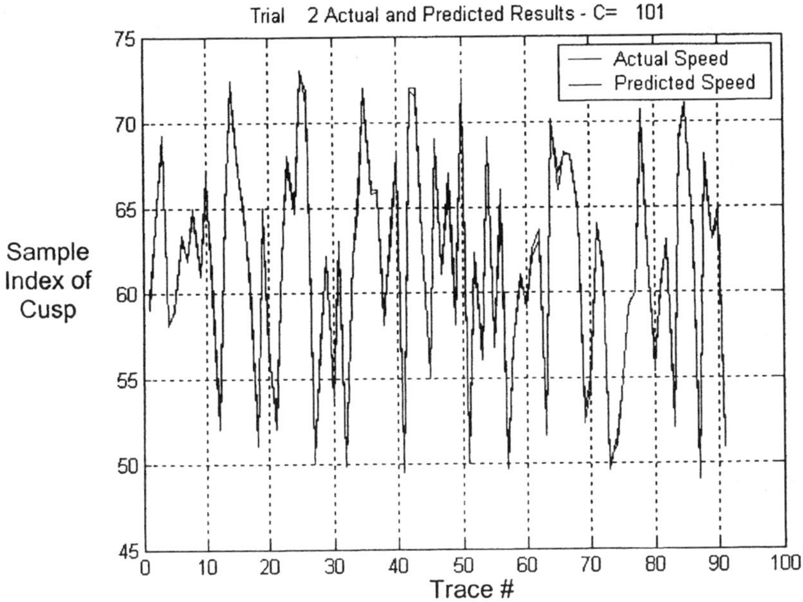

Figure 5: Plot of the prediction results of the trained speed measurement SVM.

4.2 Integrated System Results

After testing all three subsystems, an integrated system test is conducted. The test data was generated exactly the same way that the other three SVM test sets were developed, but this time all of the possible variances, speed, failures, and bouncing were placed in the system. Figure 6 shows the test results.

5. Future work

In this study, the SVM showed outstanding promise for back-emf feature extraction. In order to get this system into a microcontroller, there are several steps that need to be taken in the training data development, and SVM refining realms.

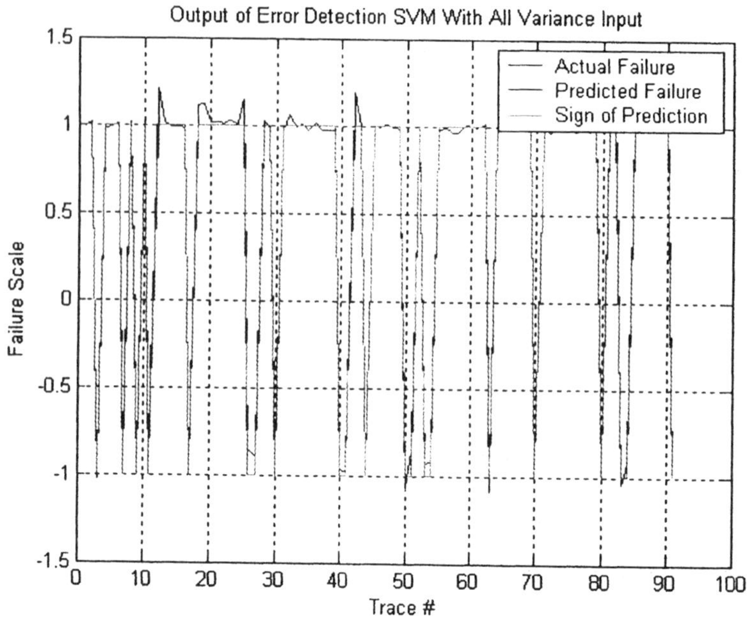

Figure 6: Plot of the actual, predicted, and sign(predicted) results for motion classifier SVM. Percent Classified Correctly=100.00%

The training data used in this study was taken from a real device, but it was a device that worked very well. Therefore, no true failures or bounces were recorded. Also, since the device had been well developed, its motion speed was very stable. All of the variances used for training had to be artificially put into place. This is acceptable to determine the capabilities of a SVM as they relate to this problem, but there are many errors that may be introduced because of artificial features placed into the back-emf signature that may not be present on an actual system. It is critical that an improved method of creating broad and encompassing training sets is developed to ensure that the accuracies that were seen in this study can be reached or even exceeded on a real system. All three machines only had between twenty and thirty support vectors. In order to run in real time, it would be imperative to strip out the unused inputs keeping only the support vectors. This would reduce the processor power consumption to a minimum and it is possible to have a SVM system on a single silicon chip in the future.

References

[1] Bernhard Schölkopf, *Support Vector Learning,* Microsoft Research Ltd, 1999

[2] Nello Cristianini, *Support Vector and Kernel Machines,* Online Tutorial, 2001 http://www.support-vector.net/tutorial.html

[3] Pike & Mustard, *A neural network for the evaluation of hemodynamic variables.,* Handbook of Neural Computation, Oxford 1997

[4] S. Ruping, "mySVM – Manual", October 30, 2000.

VIEW-BASED OBJECT RECOGNITION GUIDED BY INTELLIGENT VISUAL EXPLORATION

QIANG LU
Computer Science and Engineering
University at Buffalo
Buffalo, NY 14260-2000

PETER D. SCOTT
Computer Science and Engineering
University at Buffalo
Buffalo, NY 14260-2000

ABSTRACT
We present an active 3-D object recognition scheme in which, when an ambiguous object is viewed, a new imaging viewpoint is computed to efficiently reduce the predicted object class ambiguity. The camera moves toward that viewpoint and a new image of the object is acquired. The data from the set of images of the object acquired to-date is combined, and the consistent hypothesis set reduced as the number of these views increases, until recognition is achieved. Detailed approaches are presented and discussed along with preliminary experimental results.

1. INTRODUCTION

3-D object recognition on the basis of a single image can be challenging since (outside of toy problems) there normally is ambiguity, i.e. uncertainty upon matching the recovered image features against the nominal characteristic of object models in a database. The traditional approach to this problem is to provide all consistent recognition hypotheses in a list ordered by confidences or cost functions. Active recognition addresses this ambiguity by deferring a recognition decision until sufficient images at selected viewpoints are captured. By intelligently changing the intrinsic and extrinsic parameters of the camera, views from which additional information can be gathered so that classification ambiguity can be efficiently reduced is determined by the view planning module of the system.

Effective active recognition requires the interplay of several major modules, the feature extraction module, the view planning module, the motion control module, the view integration module, and the decision module. Each will be introduced here and described in subsequent sections.

Regions characterized by the shape of their bounding contours are selected as the primary recognition features for the system to be described. They are gathered from a new hybrid color image segmentation scheme developed to cope with lighting situations, such as strong shadings and highlights. A shift in viewpoint is requested by the motion control module to resolve the skew condition where the unreliable classification may be derived. With a reliable initial view captured, if the object cannot be recognized, selection of the next view by the view planning module is required. We define the aspect transition graph of an object as the aspect graph with edges attributed by directional information between the nodes (aspects) each edge connects. The algorithm searches within the aspect transition graph of each candidate and uses a min-max rule to determine an efficient next viewpoint, one predicted to lead to the most discriminate aspect i.e. the aspect with the least number of remaining consistent hypotheses. View integration is accomplished by mapping the newly acquired aspect into a developing recognition tree, and the recognition decision made on the basis of the remaining consistent hypotheses.

Following a brief review of related work and a short description of database construction, we will describe each of these modules in more detail.Then, we will show preliminary experimental results with a testbed and conclude with a brief discussion.

2. RELATED WORK

Active object recognition has been studied in various contexts, with different recognition schemes and view planning approaches [1]-[5] . Unique to active recognition strategies are the tasks of identifying sets of view points from which objects can be unambiguously identified, and determining how to move the camera through these viewpoint sets in the most efficient manner.

The active vision system described by Dickinson et al. [5] is the most closely related to that reported here. They describe a system for recognizing 3-D objects from 2-D images in a bottom-up fashion by searching through a hierarchical aspect graph. Viewpoint planning is based on the aspect prediction graph which connects all aspects of an object associated with camera transitions. The camera is moved to a position such that the most informative aspect is in view. A concern with this strategy is that it might not lead to a viewpoint which has high discrimination power, as we will discuss in section 6.

The active vision system reported here differs from [1]-[5] and other active recognition systems we are aware of by including appearance-based active elements into the motion control module as well as the view planning module. This provides more robust object face recognition which leads to more reliable view planning and makes the vision exploration process a more systematic, reliable goal-directed process.

3. TRAINING SET DATABASE CONSTRUCTION

Training of the system consists in capturing an aspect hierarchy for each object to be recognized. The initial test dataset in the system contains nine objects, of which eight are constructed with superquadrics and the ninth object is selected such that only small portion of this object differs from one of the other objects (object 1 in Figure 1).

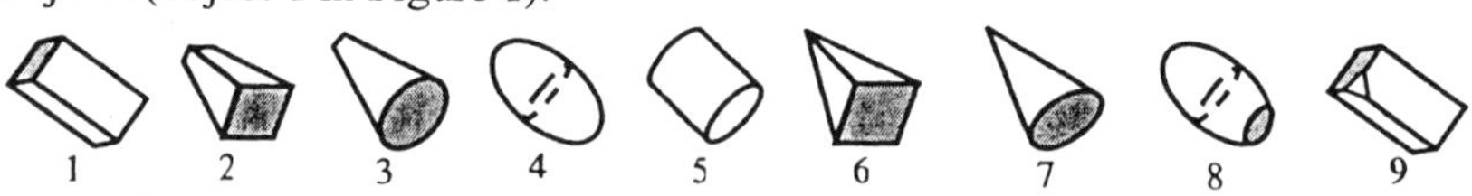

Figure 1 3-D objects in the database

These objects are each characterized by a data structure similar to Dickinson's aspect hierarchy [5]. It consists of four layers: 3-D objects, face aspects, faces, and boundary groups. Based on this image dataset, upward and downward conditional probabilities are computed between adjacent layers of the training set to link the recognition and view planning procedures.

4. FEATURE RECOVERY

The hue cue in the HSI color model has been chosen as a starting point for the new hybrid color image segmentation algorithm since it can discount the influence of shading, and highlight assuming reasonable material properties and lighting conditions [6]. Ohta's color features [7] are applied in a second step for further segmentation. To deal with strong shading and/or specular highlight regions, a principle component analysis approach of the color distributions of the image, similar to the method of Klinker et al. [8], is used.

After candidate regions have been segmented, we identify their bounding contours. Then these contours are divided into boundary segments based on curvature maxima computed from the chain code [9]. To compensate for boundary noise and the digitalization error, the discrete contour evolution scheme [10] with segment merge is employed.

A region graph is constructed whose nodes represent contour segments and edges represent spatial adjacency between contour segments after the boundary segmentation. Image region labeling then consists of graph comparison between the region map and the corresponding face maps which characterize the face layer of the aspect hierarchy. If a graph isomorphism is not found, a subgraph isomorphism is next considered. The result is a hypothesis (face within an object) for each image region.

5. MOTION CONTROL FOR ROBUST FACE CHARACTERIZATION

When a region face is visible but highly skewed (oriented nearly perpendicular to the optical axis), robust characterization of its shape is not possible. Movement of the camera within the characteristic view region to eliminate skewed faces is required.

To identify these regions, we first compute a region's confidence based on its size, eccentricity and recognition probability. A small camera movement is then executed for each visible region with insufficient region confidence. The direction is based on the position of the region relative to the adjacent region with the highest region confidence.

After these refined movements to expose all visible faces adequately, a recognition attempt is made by first matching each image region graph with training set face graphs and then verifying its parent aspect graph in the hierarchy. The verification involves two steps, aspect instantiation which generates all the face aspect hypotheses consistent with the face-to-aspect mappings, and the aspect covering step based on consistency with the face adjacency graph.

6. VIEW PLANNING MODULE

During the aspect recovery process, we may not be able to recover a unique object, since, from certain viewpoints, multiple objects in the training set may display the same aspect. The key to disambiguate such objects is a planned movement of the camera to a new view position from which the objects appear distinct. This view position does not need to be the position where the most likely aspect of each ambiguous candidate can be recovered, as Dickinson et al.

[5] chose. What needed is that the new view disambiguates the object, that is, the new view together with the previous ones is consistent with just one object.

Therefore the goal of the view planning is to guide the camera to a useful characteristic view region with high discrimination. Our approach is to employ the aspect transition graph, which attributing the usual aspect graph edges with directional information for transit between aspects, to determine the direction of translational motion. Beginning with the current aspect, we list all consistent candidates based on the training set aspect-to-object mappings then use a breath-first search on the adjacent nodes (aspects) in the aspect transition graph of all candidates. Among them, a target aspect will be chosen using a min-max rule. Finally, the planned camera movement would be the one from the current aspect to the selected target aspect in the aspect transition graph.

When moving along this path, the large scale motions implied by view planning must be implemented with motion control, i.e. the camera in the system should have the ability to fixate the test object (assume static) from each selected viewpoint. After each step taken by the motion control module, the object is re-fixated along the optical axis by rotational movements of the camera. Meanwhile, we need to track aspects from view to view so that we can monitor the visual events specified in the aspect transition graph. The two visual events need to be captured are the appearance and the disappearance of a face. A modified region adjacency graph with its match algorithm is used for this purpose.

7. FEATURE INTEGRATION

Our strategy of feature integration is to first construct a recognition tree whose branches represent all candidate objects derived from the original recovered set of aspect. For each candidate object, we check if the camera action and the newly recovered aspect pair are consistent with its aspect transition graph, i.e. whether the new aspect can be recovered in the candidate object if the current camera action is taken. If the pair is consistent, we retain that candidate object, otherwise the candidate and its branch is deleted from the recognition tree. If a unique candidate is left, the system terminates with the corresponding recognition decision, else the view plan module is queried for the next viewpoint.

Thus feature integration is implemented by aspect integration. As more views add more aspects to the recognition tree, fewer consistent candidate branches remain. If the remaining tree has a path which leads to a single consistent candidate, a recognition can be declared.

8. EXPERIMENTAL RESULTS

We tested this system in a simulated 3-D working space containing the test objects depicted in Figure 1. The simulation environment was implemented in C++ with the OpenGL™ graphics library for 3-D object rendering. 256x256 simulated color images served as inputs to our active OR system.

In Figure 2, we present a demonstration of how the motion controller tries to manipulate the virtual camera to gather the robust face information. Figure 2(a) shows an initial view which contains three visible faces among which the top and right face have low region confidence. They both are highly skewed and would not yield accurate features. Therefore, the camera is guided to positions where these regions will show better. Figure 2(c) is an improved viewpoint for the top face, and 2(d) the acceptable appearance for the right face. The bounding contours of three faces from different viewpoints (the front face from (a), the top face from (c), and the right face from (d)) would yield accurate aspect in the hierarchy and thus lead to a good recognition result.

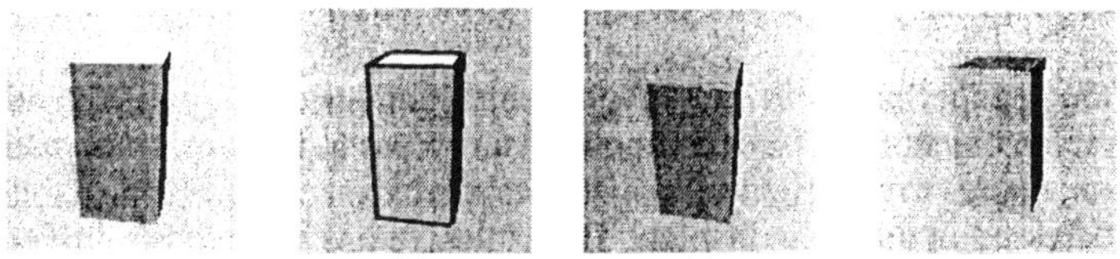

(a) (b) (c) (d)

Figure 2 Viewpoint tuning by the motion controller

Figure 3 illustrates how the view planner guides the camera to move to a characteristic view capable of disambiguating an aspect. The object shown is actually object 1 and the initial view shows three faces. There are three candidates, object 1, object 2 under perspective projection, and object 9 (a cut corner may be on the left side of the back-bottom face). Our view planner produced two steps. The first is to move the camera to the left so that the optical axis is orthogonal to the main visible face such that object 1 and object 2 can be distinguished (object 2 would show four faces). After feature integration, objects 1 and 9 are still both consistent so another view is planned. The camera is moved further to the left so that the left face of the object is visible. Lacking the cut corner, recognition as object 1 follows.

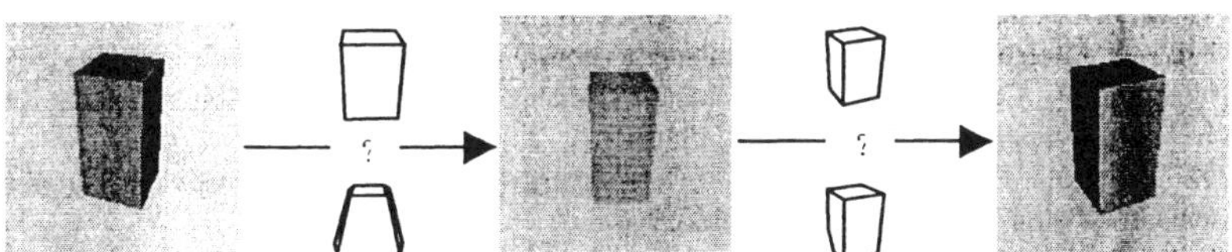

(a) (b) (c)

Figure 3 A example of viewpoint control

To evaluate the efficiency of the proposed gaze control algorithm, a comparison system was also tested. This system has all the same functional modules as the original system except the view planning module which simply selects next views randomly. This random comparison has been adopted by other active object recognition researchers [3] [11] for performance evaluation. Comparison experiments have been conducted with the same initial view positions and recovered aspects to both systems. Results have shown significant advantage of the proposed gaze control algorithm in terms of number of views to recognition.

We have not acquired sufficient results to date to determine if the comparison system has a larger error rate, so far both systems have continued visual exploration until sufficient information to reach the correct decision was acquired.

CONCLUSIONS

We have described an active object recognition system for 3-D object scenes and a test simulation environment for such system. The primary recognition features are the visible faces, derived through an aspect hierarchy. At the bottom of the hierarchy, a new hybrid color image segmentation algorithm has been developed for robust extraction of image regions from images with difficult lighting and reflective properties. Active vision is introduced in an early stage, motion control as well as a later stage, view planning.

Although experiments to date involve only geometrically simple objects, more complex objects can be decomposed using a parts-decomposition module. In addition, scenes involving multiple objects, perhaps occluding one another, can be addressed in the same fashion. The exclusive use of local primitives (bounding contours) and features (faces) is consistent with the capability to recognize under partial occlusion. These extensions are being addressed.

REFERENCES

[1] H. Borotschnig et al., Appearance-based active object recognition, Image and Vision Computing, Vol. 18, pp. 715-727, 2000.

[2] T. Arbel and F. P. Ferrie, Entropy-based gaze planning, Image and Vision Computing, Vol. 19, No. 11, pp.779-786, 2001.

[3] R. D. Rimey and C. M. Brown, Control of selective perception using Bayes nets and decision theory, International Journal of Computer vision, Vol. 12, No. 2/3, pp. 173-207, 1994.

[4] P. Gvozdjak and Z-N. Li, From Nomad to Explorer: Active Object Recognition on Mobile Robots, Pattern Recognition, Vol. 31, No. 6, pp. 773-790, 1998.

[5] S. Dickinson et al., Active Object Recognition Integrating Attention and Viewpoint Control, Computer Vision and Image Understanding, Vol. 67, No. 3, pp. 239-260, 1997.

[6] F. Perez and C. Koch, Toward Color Image Segmentation in Analog VLSI: Algorithm and Hardware, International Journal of Computer Vision, Vol. 12, No. 1, pp. 17-42, 1994.

[7] Y.I. Ohta et al., Color information for region segmentation, Computer Vision, Graphics, and Image Processing, Vol.13, pp.222-241, 1980.

[8] G. Klinker et al., A Physical Approach to Color Image Understanding, International Journal of Computer Vision, Vol. 4, No. 1, pp. 7-38, 1990.

[9] F. Leymarie and M. Levine, Curvature Morphology, Technical Report TR-CIM-88-26, Center for Intelligent Machines, McGill University, December, 1998

[10] L. Latecki and R. Lakamper, Convexity Rule for Shape Decomposition Based on Discrete Contour Evolution, Computer Vision and Image Understanding, Vol. 73, pp. 441-454, 1999.

[11] F. G. Callari and F. P. Ferrie, Active Object Recognition: Looking for Differences, International Journal of Computer Vision, Vol. 43, No. 3, pp. 189-204, 2001.

Intelligent Music Classification with Support Vector Machines

Chris Felton
Department of Electrical and Computer Engineering
University of Colorado at Colorado Springs
P.O Box 7150, Colorado Springs, CO 80933

Chia-Jiu Wang
Department of Electrical and Computer Engineering
University of Colorado at Colorado Springs
P.O Box 7150, Colorado Springs, CO 80933

ABSTRACT

Larger storage systems and efficient audio compression together create a platform for a large collection of digital music. These massive collections of digital music can be classified into different genres for information retrieval, automatic playback or categorization. Intelligent signal processing (ISP) can be applied to automatically classify the digital songs. This paper will explore the efficiency of Support Vector Machines (SVM) to classify the different genres of music. This information can help to determine if there are enough statistical data points between similar genres of music to be differentiated. Sets of musical data will be used to train the SVMs to classify the following types: Alternative, Blues, Classical, Country, Electronica, Hip-Hop/Rap, Jazz, R&B/Soul, Rock/Pop, and World/Folk.

INTRODUCTION

The collections of digital music need to be easily accessible. Automated music classification will aid in digital music retrieval and organization. This paper will explore music classification using Support Vector Machines (SVMs). SVMs have shown great success in other classification problems such as, face detection, verification and recognition, speech recognition, text detection [1, 2, 3, 4, 5, 6, 7]. The goal will be to use multi-class SVMs to classify music into 10 different types: Alternative, Blues, Classical, Country, Electronica, Hip-Hop/Rap, Jazz, R&B/Soul, Rock/Pop, and World/Folk.

Music can automatically be classified into genres based on the audio content. An efficient means for classifying the data is needed. Support vector machines provide a means to extend the high dimensional data space into separable classes. This paper presents methods used in feature extraction and music classification results by a system with multi-class classification SVMs.

FEATURE EXTRACTION

First to classify music, information will be extracted from each sample. These "features" will be used to distinguish one type of music from another. In this paper all the features are spectral based. Figure 1 shows the system block diagram. Figure 2 shown below, is the time domain plot of four music samples.

The first two samples represent the Alternative genre and the last two samples represent the Classical genre. They are thirty-second samples recorded from online streaming music stations. The streaming audio is of comparable quality to 128 kbit MP3 samples, which is considered CD quality. A large number of the music samples used in this work for training and testing are 30-second samples recorded from online stations.

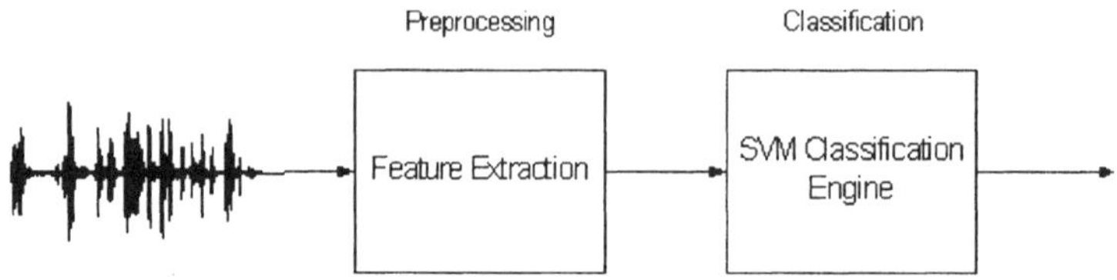

Figure 1 System Block Diagram.

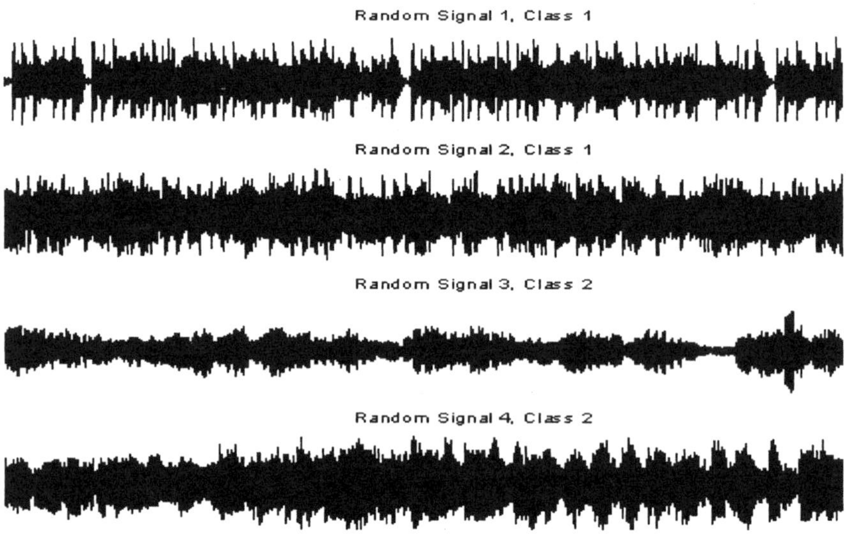

Figure 2. Music Samples

The fast Fourier transform (FFT), Mel-Frequency Cepstral Coefficients (MFCC) and Spectrogram are used to process music data samples. The MFCC is a short-term spectral based feature, successful in speech recognition. The spectrogram is the spectrum during short periods of time. The spectrogram gives a good representation of the sample. It can be shown that an alternative sample has a lot of energy in frequencies below 8k and it is flat in the higher frequencies.

The classical sample energy is spead out across a majority of the frequencies. The spectrogram experimentally showed improvement in classifying songs. The FFT, MFCC, and Spectrogram will be combined to generate the input vector to classify the songs.

Support Vector Machines

Support vector machines are a relatively new type of learning algorithm, originally introduced by Vapnik and co-workers [6] and successively extended by a number of other researchers. Their remarkably robust performance with respect to sparse and noisy data is making them the system of choice in a number of applications. We apply SVM to classify songs. This is a multi-class classification problem. Currently there are two types of approaches for multi-class SVM. One is by constructing and combining several binary classifiers while the other is by directly considering all data in one optimization formulation. In our work we use the former method.

One vs. Rest Multi Classification

One vs. rest architecture has a SVM each class and a arbitrator. The one vs. rest SVMs are a binary classifier trained to identify one class. The class that is to be identified is one class, denoted as 1, and all the other classes are combined to create the other class denoted as –1, for a specific SVM. Figure 3 presents the multi-class classification architecture. In this work the total number of music classes is 10.

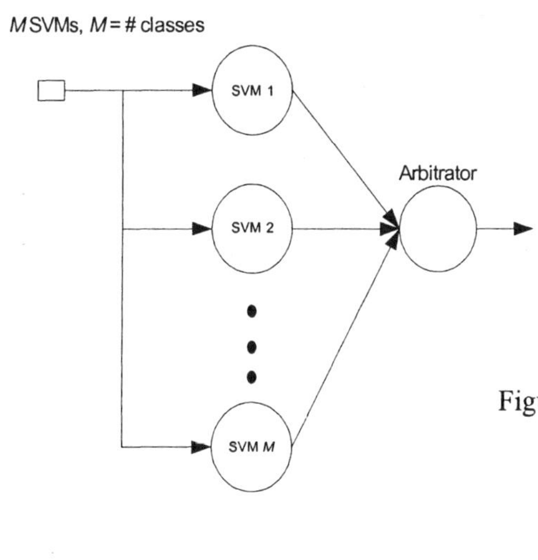

Figure 3. One vs. Rest SVM

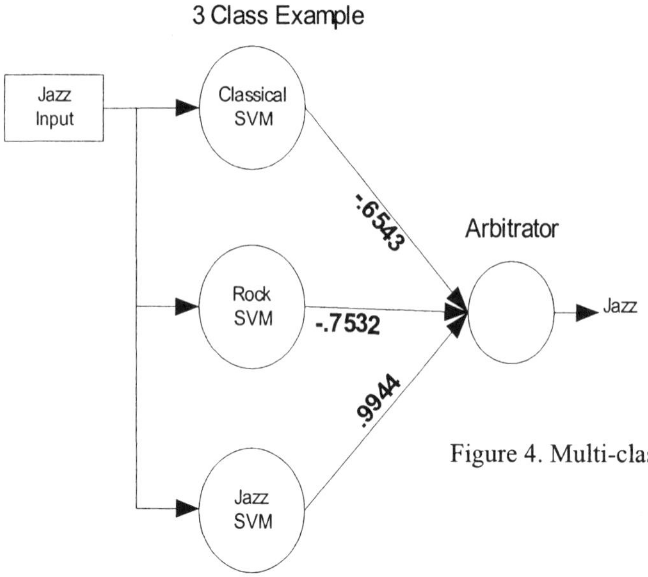

Figure 4. Multi-class Arbitrator

Input Vector Construction

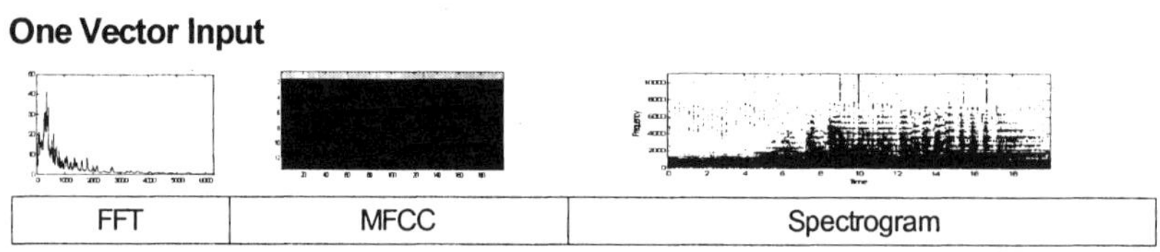

All three feature extraction techniques are combined to create the above input. From the graph these data points are easily separable. But it is not known yet if more samples will end up in similar areas as the previous samples. This is our current on-going research. The input vector's organization is illustrated in Figure 4. Figure 4 shows the layout for each input vector and also shows an example how the different samples are used to train a Jazz one vs. rest binary SVM.

Training data sets

Two training sets were used to train the SVMs. The first set with 10 samples from each music type was used to train the binary one vs. rest SVMs. All 10 samples from the identifying SVM class and 2 samples from the rest of the genres were used for training. An additional 10 samples from each class was used to train the arbitrator. Figure 4. is an example of the training sets as input vectors to the SVMs.

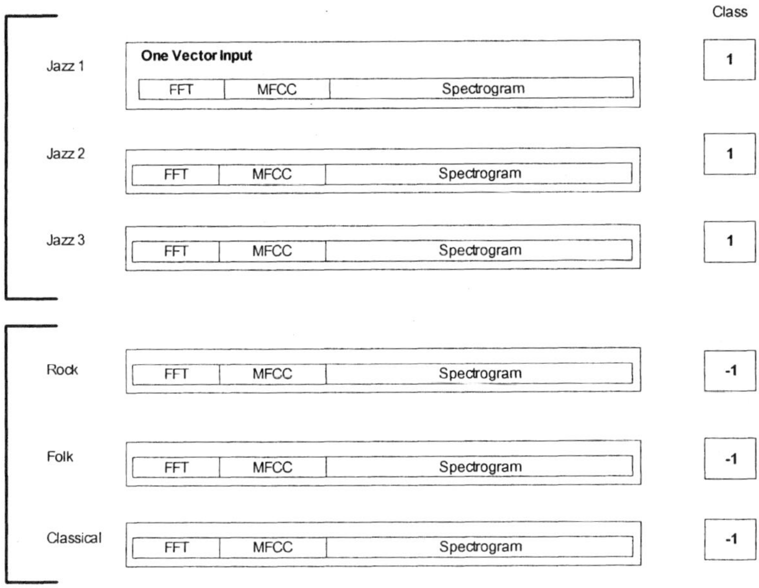

Figure 4 training examples

Results and Future Work

The first experiment was to classify a set of music using only the binary one vs. rest SVMs. Table 1 shows the results from classification with out an arbitrator. Only one class (or none) was recognized per sample. The following table is the result from a larger collection of 898 songs. The results were manually determined to be correctly or incorrectly classified.

Table 1 Confusion matrix using no arbitrator testing individual SVM classifiers

Out of 898	Alternative	Blues	Class.	Country	Electronica	Jazz	R&B-Soul	Rap/Hip-Hop	Rock-Pop	World/ Folk
Alternative	283	1	0	0	0	0	0	3	2	0
Blues	0	3	0	0	0	0	0	0	0	0
Class.	3	1	69	0	0	0	0	6	5	9
Country	1	0	0	0	0	0	0	0	0	2
Electronica	0	0	0	0	0	0	0	0	0	0
Jazz	0	0	0	0	0	0	0	0	0	0
R&B/Soul	1	0	0	0	0	0	6	0	0	1
Rap/Hip-Hop	1	0	0	0	0	0	0	1	0	0
Rock/Pop	2	0	0	0	0	0	0	0	1	0
World/Fold	7	0	0	0	0	1	0	1	1	17

The available larger collection of music used above did not have equal representation of all music types. It was heavily biased to alternative type music. As seen in the above table, the majority of the classified music landed in the Alternative genre.

The next experiment was to test with an arbitrator. Currently this is limited to 4 music types. Table 2 is the confusion matrix for a 4-class One vs. Rest multi-class SVM. The arbitrator is a winner-take-all (W-T-A) arbitrator. Table 3 presents the same case but using a SVM as an arbitrator which has better results.

Table 2 Confusion matrix using a W-T-A arbitrator

Classified / Actual	**Alternative**	**Classical**	**Rap-HipHop**	**World-Folk**	**Total Classified**
Alternative	**11**	0	3	2	16
Classical	9	**20**	5	6	40
Rap-HipHop	1	0	**5**	0	6
World-Folk	7	0	5	**2**	14
TotalTest Samples	28	20	18	10	76
Percentage Correct	39.29%	100.00%	27.78%	20.00%	

Table 3 Confusion matrix using SVM as an arbitrator

Classified / Actual	**Alternative**	**Classical**	**Rap-HipHop**	**World-Folk**	**Total Classified**
Alternative	**15**	1	5	2	23
Classical	2	**19**	0	2	23
Rap-HipHop	5	0	**11**	2	18
World-Folk	6	0	2	**4**	12
Total Test Samples	28	20	18	10	
Percentage Correct	53.57%	95.00%	61.11%	40.00%	

Conclusion

The results are promising because there were small numbers of misclassifications. But there were a lot of no classifications present. To remove the no classifications a better arbitrator based on the decision value should be added. This would force one classification for each genre. We are now still working on this project to achieve over 90% music classification by using balanced training data set and bootstrapping techniques.

References

[1] Scholkoppf and Smola, *Learning with Kernels.* Massachusetts Inst. of Tech., 2002

[2] Principe, Euliano, and Lefebvre, *Neural and Adaptive Systems.* New York: John Wiley & Sons, 2000.

[3] Haykin, *Neural Networks a Comprehensive Foundation.* New Jersey: Prentice Hall.

[4] Cristianini and Shawe-Taylor, *An Introduction to Support Vector Machines and other kernel-based learning methods.* United Kingdom: Cambridge University Press, 2000.

[5] Lee, *Pattern Recognition with Support Vector Machines.* Germany: Springer, 2002.

[6] Gunn, "Support Vector Machines for Classification and Regression" University of Southampton, May 1998.

[7] Lee and Hwang, "Selecting Good Speech Features for Recognition" ETRI Journal, volume 18, number 1, April 1996.

INTELLIGENT TWO-SIDED VOICE COMMUNICATION SYSTEM BETWEEN THE MACHINING SYSTEM AND THE OPERATOR

WOJCIECH KACALAK
Department of Mechanical Engineering
Technical University of Koszalin
Koszalin, Poland

MACIEJ MAJEWSKI
Department of Mechanical Engineering
Technical University of Koszalin
Koszalin, Poland

ABSTRACT
In the future, speech unquestionably will become the primary means of communication between human beings and machines. By primary it is meant the one most often used. In this paper we present a two-sided voice communication system between the machining system and the operator using artificial intelligence. The introduced intelligent communication system is composed of the following subsystems: voice communication, command syntax and effect analysis, safety assessment, and command execution. The voice communication between the machining system and the operator contains an application of intelligent and indirect layer, which is equipped in neural networks and fuzzy logic methods. The voice communication system is a part of the complex system of technological process supervision that is also composed of the following subsystems: subsystems of process diagnostic, prognosing of process state, analyses of causes and results of inaccuracy in process, analysis of syntax, as well as a subsystem of estimation of results and safety of commands given by the operator.

INTRODUCTION

There is no need of an operator being present at the work stand by the machine in the modern supervising systems of technological process. According to the new conception, the operator may stay away from the machining system in another room, or even in any distance from it (Kacalak, 2003). It is made possible by using a voice communication system between the operator and the machine (O'Shaughnessy, 2000). The conditions of the effectiveness of that system are as the followings:

- automatic recognizing the typical condition of the technological process with the intelligent data analysis system,
- permanent self-learning while working,
- informing the operator by voice about the results of automatic analysis of the process states,
- recognizing the operator's voice commands in continuous speech (Jurafsky, 2000) filtering with the command syntax and meaning analysis module,
- checking if the operator's commands are understood and not contradictory with the safe process state system - assessment of a grade of affiliation of the command to the correct command category,
- correcting the operator's incorrect commands with an optional condition of confirmation (Weinschenk, 2000).

ARCHITECTURE OF THE INTELLIGENT TWO-SIDED VOICE COMMUNICATION SYSTEM BETWEEN THE MACHINING SYSTEM AND THE OPERATOR

Creation of the modern supervising systems of technological process using the assumption mentioned above will make it possible to eliminate scarcities of the typical co-operation between the operator and the machining system (Kacalak, 2001).

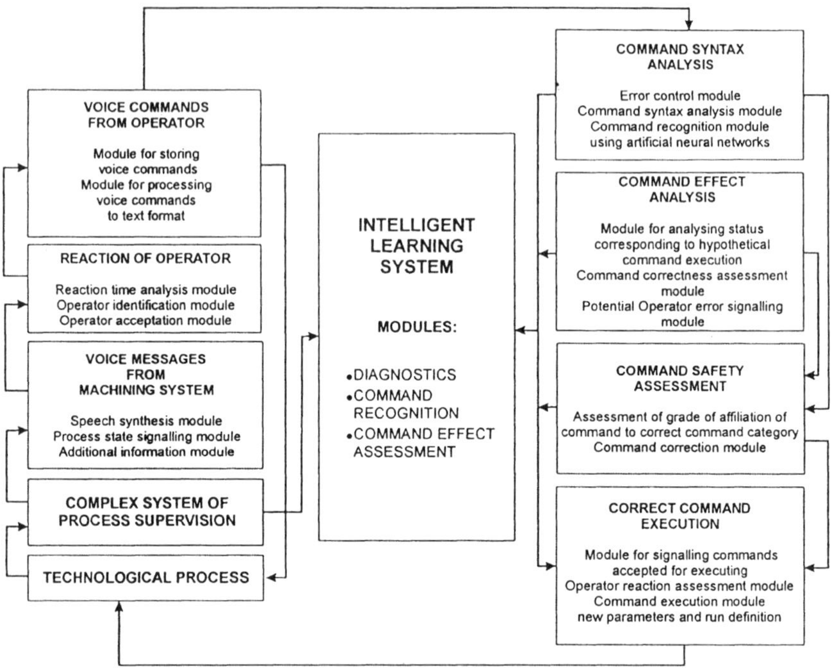

Fig. 1. General scheme of technological system with the voice communication system with the operator

The intelligent two-sided voice communication system between the machining system and the operator consists of the following subsystems (fig. 1):

A. Subsystem of passing voice messages from the machining system to the operator consists of:

- process status signalling module,
- additional information module.

B. Subsystem of analysis of the operator's reaction uses:

- reaction time analysis module,
- operator identification module,
- operator acceptation module.

C. Subsystem of passing voice commands from the operator works with:

- module for storing voice commands,
- module for processing voice commands to text format.

D. Subsystem of analysis of command syntax contains:

- error control module,
- command syntax analysis module,
- command recognition module using artificial neural networks.

E. Subsystem of analysis of command effect consists of:
- module for analyzing the status corresponding to the hypothetical command execution,
- command correctness assessment module.

F. Subsystem of assessment of command safety contains:
- assessment of a grade of affiliation of the command to the correct command category,
- command correction module.

G. Subsystem of execution of the correct command uses:
- module for signalling commands accepted for executing,
- operator reaction assessment module,
- command execution module - new parameters definition.

H. Subsystem of intelligent learning may consist of:
- diagnostic module,
- command recognition module,
- command effect assessment module.

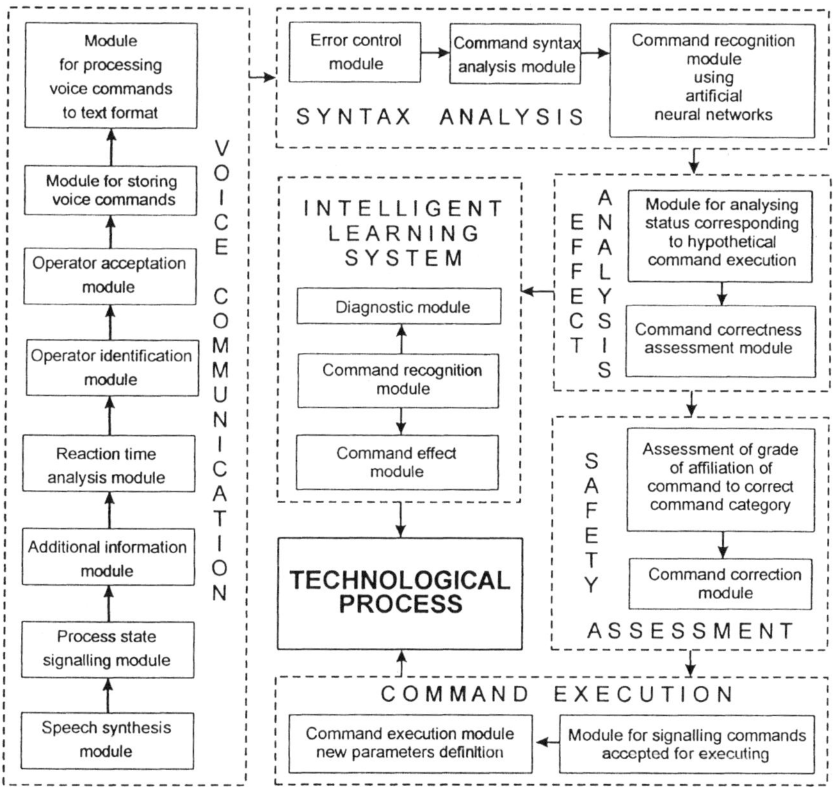

Fig. 2. Scheme of the voice communication system between the machining system and the operator

The task of the voice communication system between the operator and the machining system (fig. 2) is generating voice messages to the operator about the process state, storing the operator's voice commands, analyzing reaction time of the operator, analyzing syntax and assessing safety of the operator's commands,

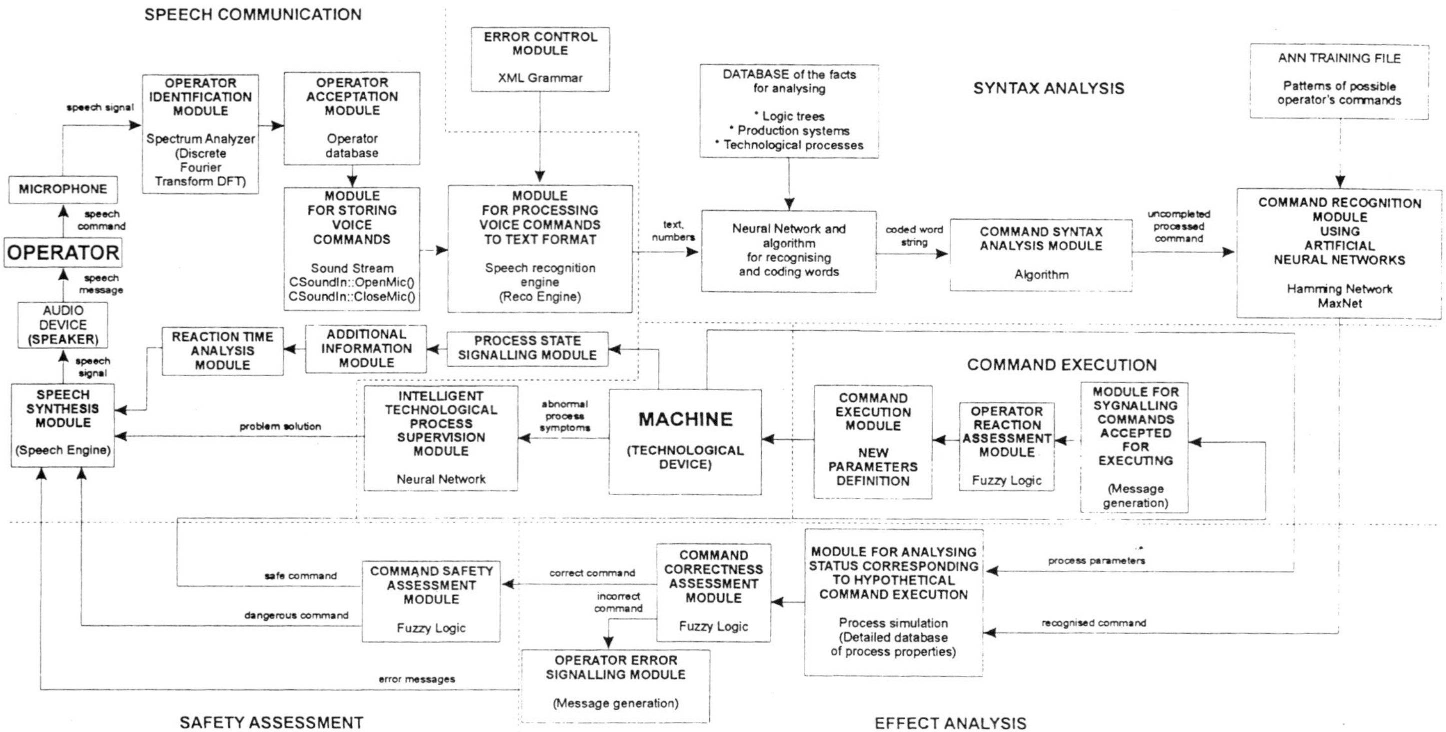

Fig. 3. Detailed architecture of the intelligent two-sided voice communication system

and also estimating the operator's reaction. That system in the co-operation with the comprehensive diagnostic system will also make it possible to analyze the state of the process corresponding to the hypothetical execution of the operator's command.

The scheme of the intelligent two-sided voice communication system (fig. 3) presents a practical example of the system applied in supervising of the technological process. The operator produces a voice command, which is recognized by the speech recognition module and processed to the text format. Next the syntax of the command is analyzed by an algorithm that returns a command in the numeric vector format. The processed command is sent to the command recognition module using artificial neural network in order to recognize the command. The next step is to provide the command effect analysis module with the information on the process parameters from the machine and the vector format of the command. The module does its work with a neural network. Then a fuzzy logic algorithm is used to analyze the command safety and produce the assessment of the command safety that can be either safe command to execute or dangerous command. If the command is safe it is executed immediately. If the operator's command is classified as a dangerous one, the system produces a voice message with a speech synthesis module and the command is never executed.

The functioning of the automatic command recognition of the intelligent two-sided voice communication system between the machining system and the operator is based on neural networks and algorithms for text analysis (fig. 4).

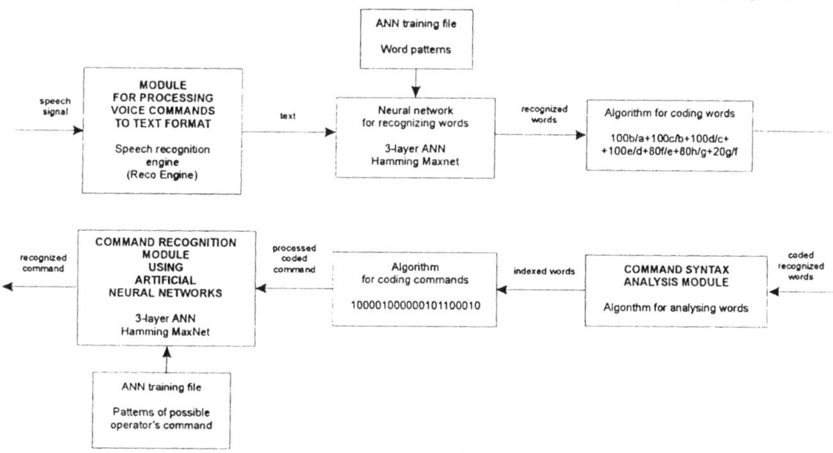

Fig. 4. The automatic command recognition system

The speech signal is processed to text and numeric values with the module for processing voice commands to text format using the speech recognition engine. The separated words of the text are the input signals of the neural network for recognizing words. The applied network is the 3-layer Hamming neural network with the additional layer named Maxnet. The network is capable of optimal classification of binary images. It represents a heteroassociative memory, and it has a high rate of processing signals. It also works in parallel. The recurrent network Maxnet performs functions of vector classifier, it contains neurons connected in feedback. The network has a training file containing word patterns. As the work result, the network recognizes words as the operator's command components, which are represented by its neurons. The recognized words are sent to the algorithm for coding words. Next the coded words are transferred to the command syntax analysis module. It is equipped with the algorithm for analyzing and indexing words. The module indexes words properly and then they are sent to the algorithm for coding commands. The commands are coded as vectors and they are input signals of the command recognition module using artificial neural network. The module uses the 3-layer Hamming neural network either to recognize the operator's command or to produce the information that the command is not recognized. The neural network is equipped with a training file containing patterns of possible operator's commands.

SPEECH AND COMMAND RECOGNITION RESULTS

The speech recognition module recognizes 85-90% of the operator's words right. As more training of the neural networks is done, accuracy rises to around 95%. The command recognition module using 3-layer neural networks is capable of recognizing different commands of the same meaning in the natural speech.

CONCLUSIONS

Creating the complex system of technological process supervision becomes a more important task. It is because of growing demands involving the reliability of the process and stability of the results. The aim of the research carried out is to reach a higher level of the intelligent process supervision systems provided with the intelligent two-sided voice communication system between the machining system and the operator. The research aiming at developing an intelligent and indirect communication layer is very difficult, but the prognosis of the technology development and its first use shows a great significance in efficiency of supervision and production humanization.

ACKNOWLEDGEMENTS

The support of the Polish State Committee for Scientific Research (KBN grant 5 T07D 036 24) is highly appreciated.

REFERENCES

Jurafsky D., Martin J. H., 2000, "Speech and Language Processing: An Introduction to Natural Language Processing, Computational Linguistics, and Speech Recognition", Prentice Hall, New Jersey 2000.

Kacalak W., Majewski M., 2003, "Supervising of Technological Process Using Two-Sided Voice Communication Between the Machining System and the Operator", Modern Trends In Manufacturing CAMT 2003, 20-21 February 2003, OW PW, Wroclaw 2003, pp.175-182.

Kacalak W., Krzyżyński T., Lipiński D., Kamiennik T., 2001, "Intelligent Systems of Inaccuracies Minimization and Disturbance Compensation in Process of Fine Grinding", 15th European Simulation Multiconference, June 6-9 2001, Prague.

O'Shaughnessy D., 2000, "Speech Communications: Human and Machine", IEEE Press, New York 2000.

Weinschenk S., Barker D. T., 2000, "Designing Effective Speech Interfaces", John Wiley & Sons Inc., New York 2000.

Interpreting Complex Data From A Dual Element Multipoint Optical Fibre Sensor System For Use In Process Water Systems

D. King, W.B. Lyons, C. Flanagan & E. Lewis.

Department of Electronic and Computer Engineering, University of Limerick, Castletroy, Limerick, Ireland.

Tel: +353-61-213558 / +353-61-213409. Fax: +353-61-338176. E-mail: Damien.King@ul.ie

Keywords: Pattern Recognition, ANN, Backpropagation Algorithm, Discrete Signals Fourier Transform, Fast Fourier Transform, Feedforward Neural Networks, Multi-Layer Perceptron, Optimisation, Sensor Recognition, Signal Processing, Supervised Training.

ABSTRACT

A dual element multipoint optical fibre sensor system capable of detecting ethanol in water supplies is reported. A U-bend configuration is used for each sensor element in order to maximize sensitivity and the sensor system is interrogated using Optical Time Domain Reflectometry (OTDR). It has been proposed to apply Artificial Neural Network Pattern Recognition Techniques to the optical fibre sensor system to accurately classify each sensor element test condition. Novel signal processing techniques utilising Fourier Transform methods are applied to the resulting OTDR data, with the aim of reducing the computational resources of the ANN required to accurately classify the sensor system data.

INTRODUCTION

Optical fibre sensors possess a number of advantages over conventional electronic sensing techniques, which make them attractive for use in a wide range of application areas. These advantages include safety in chemically hostile environments, immunity to electromagnetic interference, electrically passive operation, sensitivity, weight and versatility. As a result of this, the use of optical fibre sensors for the purpose of environmental monitoring has expanded rapidly over the past decade [1-2]. In this investigation, two sensor elements were incorporated into a 1km length of 62.5μm core polymer clad silica (PCS) optical fibre cable. In order to maximise the sensitivity of the sensor system, a U-bend configuration was used for each of the sensor elements and the core of the fibre exposed directly to the measurand under test.

The interrogation of such a sensor uses a technique known as Optical Time Domain Reflectometry, OTDR. OTDR is capable of detecting attenuation as a function of distance along the fibre and therefore is able to locate position and changes in the sensor signals along the fibre [3-4]. As a result of this, OTDR has found many applications in both single and multipoint sensors where the OTDR instrument is used to monitor the fluctuation in the optical fibre attenuation caused by an external parameter induced by a measurand.

Optical fibre sensor signals can often be complex and cross coupling of signals from external parameters e.g. temperature (the true measurand) and strain or microbending (interfering parameters in this case), adds to the difficulty interpreting data from such systems. It has been proposed that for many applications of optical fibre sensors, artificial neural network pattern recognition techniques may be used to resolve the problems arising from cross-sensitivity to other parameters [5].

In this investigation, it proposed to apply novel signal processing techniques to the obtained OTDR data, with the aim of reducing the required computational resources of the implemented artificial neural network. Previous work by *King et al* has shown it is possible to train a feed-forward ANN based on the frequency domain response of a single element optical fibre sensor system [6]. In this investigation the sensor system has been expanded to contain two U-bend sensor elements and the sensors are exposed to varying combinations of air, distilled water and 50% ethanol.

SENSOR FABRICATION

In order to produce the sensor elements along the fibre cable, the buffer and cladding were chemically removed from a 2cm length located at 665m and 756m respectively from the launch end of the fibre. In order to achieve the desired U-bend configuration for each sensor element, the exposed fibre was cleaned using acetone and slowly bent into a U-shape using heat from a flame. The bending procedure was controlled using an in house developed fixture to improve the repeatability of the sensor manufacturing and hence improve the reproducibility characteristics between successive sensors. The final bend radius was measured to be 1mm using a conventional optical laboratory microscope.

In order to minimise optical power loss, the sensor was fabricated from the same fibre as that used to transmit light through the system (i.e. 62.5μm PCS), removing the need to splice the sensor into the transmitting fibre and the resulting power losses.

The operation of the sensor is based on the modulation of the light intensity propagating in the fibre by the measurand as a result of the interaction with the evanescent field penetrating into the absorbing measurand. Much experimental work has already been reported [7-8] for a single U-bend sensor detailing resulting sensitivity gains from evanescent wave increases from the curving of the sensing fibre [9]. It has been shown by *Gupta et al* that the sensitivity of the sensor increases with decreasing bend radius of the probe and also with the increase in refractive index of the fluid under test [7-8].

MEASUREMENT SYSTEM CONFIGURATION

The system configuration for this investigation is shown in figure 1. It comprises of the optical fibre sensor elements, an EXFO IQ7000 (0.85μm) OTDR and a Pentium MMX 200 MHz PC running LabVIEW [10] Virtual Instrument (VI) programs for data capture and pre-processing. The LabVIEW VI programs were developed by the authors specifically for this investigation, and the resulting data output were made available for analysis by the signal processing technique and the artificial neural network implemented using SNNS V3.2 (Stuttgart Neural Network Simulator) [11].

RESULTS

In order to successfully train and test an ANN pattern recognition system; it is necessary to obtain a large number of patterns. For this reason, numerous OTDR readings were taken for each of the sensing conditions under test, see table 1.In order to allow the sensor system to fully stabilise, the first OTDR reading for each test case is ignored and the remaining patterns are used in the training and testing of the ANN.

Figure 2 shows an example of a resulting OTDR output trace. The sensing area of interest on the overall trace is quite small (256 data points out of a total of 12,000) and therefore to maximise the efficiency of the computer algorithm, it was necessary to design an in-house LabVIEW VI that would locate the sensor peak, select the required window width and save this data for analysis by the signal processing and the SNNS software. Prior to signal processing analysis, the extracted sensor peaks are normalised between –1 and +1 using the standard LabVIEW scale 2D array VI, see figure 3.

SIGNAL PROCESSING ANALYSIS

Previous work by *W.B. Lyons et al* [12] has shown it is possible to train a multi layer perceptron using the data obtained from a U bend sensor interrogated by an OTDR. In this investigation it has been proposed to apply a digital signal processing technique to the obtained sensor data, prior to the training of the artificial neural network, with the aim of reducing the computational resources of the implemented artificial neural network, i.e. fewer nodes required in the input and hidden layers.

It can be observed from the extracted OTDR peaks shown in figure 3, that it is relatively low frequency information that is of interest in this classification. Due to the low frequency nature of the information, a discrete Fourier transform (DFT) using an FFT algorithm can be directly applied to the OTDR peaks without having to apply any windowing transform. The low frequency nature of the information also reduces concerns over filtering and aliasing in the frequency domain [13].

The signal processing analysis in this investigation is performed using an in-house developed MATLAB program. Once the extracted OTDR sensor peak is inputted into MATLAB, a 256-point DFT of the peak is calculated and from the resulting Fourier transform the power spectral density (PSD) of the OTDR output is calculated, see figure 4.

As anticipated, the main PSD area of interest is located in the low frequency region. As a result of the application of the discrete Fourier transform, the OTDR peak information is now more explicit and easier for the user to access in comparison to time-domain based results which require all of the extracted OTDR peak data points. An empirical decision was made to select the first twelve points of the PSD plot as the main area of interest on the trace and these six points form the input layer of the ANN implemented in SNNS. These twelve points adequately represent the area of interest of the PSD trace whilst minimising the required number of nodes in the input layer of the implemented ANN.

ARTIFICIAL NEURAL NETWORK PATTERN RECOGNITION

Using SNNS, a three layer feed forward neural network was implemented, see figure 5, consisting of a twelve node input layer, a five node hidden layer and a six node output layer. The latter includes one node to represent each sensor test condition. To determine the optimal size of the hidden layer used, multiple trials using hidden layers consisting of one node up to eight nodes were performed. In this investigation, a hidden layer of five nodes was found to perform best. The use of a feed forward network is a robust way of building a non-parametric classifier.

The application of the discrete Fourier transform to the extracted output peaks from the OTDR has achieved it aim as it has significantly reduced the computational resources of the artificial neural network in comparison to time domain based results. The network resources required with the DFT applied were an input layer of twelve nodes, a hidden layer of five nodes and an output layer of six nodes in comparison to previous work by *Lyons et al* which required ANN resources of an input layer of 253 nodes, a hidden layer of 30 nodes and an output layer of 9 nodes, based on *time domain* OTDR results obtained for a multipoint optical fibre sensor system [14].

A total of 360 patterns are used to train the ANN, 40 patterns for each test condition listed in table1. To train the implemented feed-forward ANN, 500 epochs were required using a *backpropagation* algorithm with a momentum term. The momentum term of the backpropagation algorithm introduces the old weight change as a parameter for the computation of the new weight change. The new weight change is calculated using the formula

$$\Delta w_{ij}(t+1) = \eta * \delta_j * o_i + \alpha \Delta w_{ij}(t)$$

α is a constant term specifying the influence of the momentum term. This helps to avoid common oscillation problems when the error surface has a very narrow minimum area. The main advantage of a back-propagation algorithm is that flat spots of the error surface are traversed relatively rapidly with a few big steps. In this investigation the learning function used a learning parameter η of 0.9, a momentum term μ of 0.1 and a flat spot elimination value c of 0.1.The network was initialised with randomised weights and trained with a *topological order* update function [11].

In order to test the trained ANN, an independent set of data to that which was used to train the ANN is used. This data consists of the remaining 10 patterns for each of the test conditions listed in table 1. The resulting 90 patterns were applied to the trained ANN and all patterns were classified correctly, see table 2.

From the results obtained in table 2, it can be seen that the application of the discrete Fourier transform has achieved its aim. The computational resources of the ANN have been significantly reduced by the application of the DFT, in comparison with previous work by *Lyons et al* [14], without affecting the accuracy of the ANN's classifications. The ANN, using the frequency domain response of the sensor as its input layer, accurately classifies all of the sensor's test conditions. Based on the results shown in table 2, the example of the ANN shown in figure 5 shows sensor 1 detecting air and sensor 2 detecting 50% ethanol.

CONCLUSION

A reliable measurement system based on optical fibre sensors for the purpose of fluid monitoring has been presented. A dual element optical fibre sensor system based on a 1km length of 62.5μm core PCS fibre has been investigated and proven to be capable of detecting varying combinations of air, distilled water and 50% ethanol at each of the sensor elements, using OTDR based techniques. Although the length of fibre used in this investigation was 1km, longer or shorter lengths may be used as required.

Artificial neural network techniques have allowed the resulting OTDR signals to be accurately determined using pattern recognition. The ANN implemented in SNNS successfully classified each sensor test condition correctly based on the frequency domain response of the sensor. Due to the application of the Fourier transform to the OTDR peaks, refinements have been made to the resources used by the feed forward ANN, with a result of a reduction in the training time of the ANN.

REFERENCES

[1] Grattan K.T.V., Sun T., "Fiber Optic Sensor Technology: An Overview", *Sensors and Actuators A* **Vol. 82**, pp40-61, 2000.
[2] Kersey A.D., "A Review of Recent Developments in Fiber Optic Sensor Technology", *Optical Fiber Technology* **Vol2**, pp291-317 1996.
[3] Barnoski, M.K., Jensen S.N., "Fiber Waveguides: a Novel Technique for Investigating Attenuation Characteristics", *Applied Optics* **15**, pp2112-2115, 1976.
[4] Tateda M., Horiguchi T., "Advances in Optical Time Domain Reflectometry", *IEEE Journal of Lightwave Technology* **7(8)**, pp1217-1224, 1989.
[5] Lyons W.B., Lewis E., "Neural networks and pattern recognition techniques applied to optical fibre sensors" *Transactions of the Institute of Measurement and Control* **22 (5)**, pp385-404 2000a.
[6] King D., Lyons W.B., Flanagan C., Lewis E, "An Optical Fibre Sensor For Use In Process Water Systems Utilising Signal Processing Techniques and Artificial Neural Network Pattern Recognition", *Proc Artificial Neural Networks in Engineering (ANNIE) 2002*, Intelligent Engineering Systems Through Artificial Neural Networks, **12**, St.Louis, Missouri, USA, pp593-599. 2002.
[7] Gupta, B.D., et al, "Fibre optic evanescent field absorption sensor based on a U-shaped probe" *Optical and Quantum Electronics,* **28**, pp1629-1639 1996.
[8] Khijwania, S.K., Gupta B.D., "Maximum achievable sensitivity of the fibre optic evanescent field absorption sensor based on a U-shape probe" *Optics Communications,* **175**, pp135-137 2000.
[9] Soichi Otsuki et al, "A novel fibre optic gas-sensing configuration using extremely curbed optical fibres and an attempt for optical humidity detection" *Sensors and Actuators* B **53**, pp91-96 1998.
[10] LabVIEW Function and VI Reference Manual, P/N 321526B-01, Jan 1998.
[11] Stuggart Neural Network Simulator (SNNS), User Manual, Version 4.1, Report No. 6/95, 1995. http://www-ra.informatik.uni-tuebingen.de/SNNS/
[12] Lyons W.B., Lewis E., Ewald H., "An Optical Fibre Multipoint U Bend Sensor Based on Pattern Recognition", *Proc Irish Signals & Systems Conference (ISSC) 2000, UCD, Dublin, Ireland*, pp412-417 2000.
[13] Press W.H., Vetterling W.T., Teukolsky S.A., Flannery B.P., "Numerical Recipes in C: The Art of Scientific Computing", Cambridge University Press, 1990.
[14] Lyons W.B., King D, Flanagan C, Lewis E, "A Three Sensor Multipoint Optical Fibre Water Sensor Utilising Artificial Neural Network Pattern Recognition", *Proc OFS2002 – 15th Optical Fiber Sensor Conference Technical Digest*, Portland, Oregon, USA, pp463-466 2002.

TABLE 1: SENSOR TEST CONDIDITONS INVESTIGATED

Sensor 1	Sensor2	No. of Readings Taken
Air	Air	51
Air	50% Ethanol	51
Air	Water	51
50% Ethanol	Air	51
50% Ethanol	50% Ethanol	51
50% Ethanol	Water	51
Water	Air	51
Water	50% Ethanol	51
Water	Water	51

TABLE 2: ANN VALIDATION PATTERN RESULTS

Sensor Test Conditions		Expected Output	Observed Output
Air	Air	0 0 1 0 0 1	0.007 0.081 **0.919** 0.007 0.081 **0.919**
Air	50% Ethanol	0 0 1 0 1 0	0.007 0.081 **0.919** 0.023 **0.968** 0.009
Air	Water	0 0 1 1 0 0	0.007 0.081 **0.919** **0.955** 0.031 0.014
50% Eth	Air	0 1 0 0 0 1	0.004 **0.973** 0.023 0.054 0.032 **0.914**
50% Eth	50% Ethanol	0 1 0 0 1 0	0.004 **0.973** 0.023 0.012 **0.981** 0.007
50% Eth	Water	0 1 0 1 0 0	0.004 **0.973** 0.023 **0.921** 0.012 0.067
Water	Air	1 0 0 0 0 1	**0.943** 0.045 0.012 0.004 0.087 **0.909**
Water	50% Ethanol	1 0 0 0 1 0	**0.943** 0.045 0.012 0.012 **0.981** 0.007
Water	Water	1 0 0 1 0 0	**0.943** 0.045 0.012 **0.957** 0.041 0.002

FIGURE 1: MEASUREMENT SYSTEM CONFIGURATION

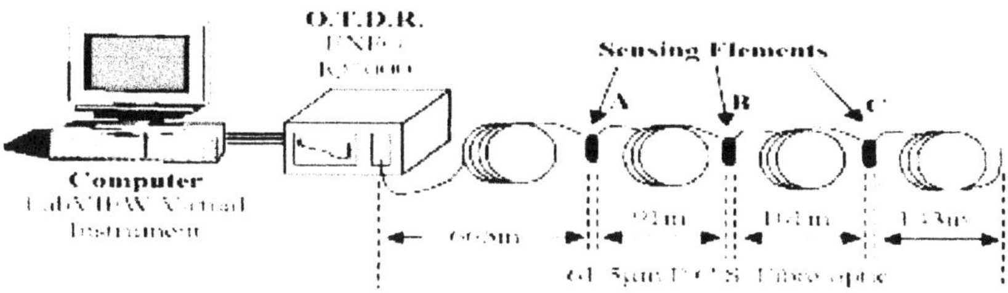

FIGURE 2: OTDR OUTPUT TRACE

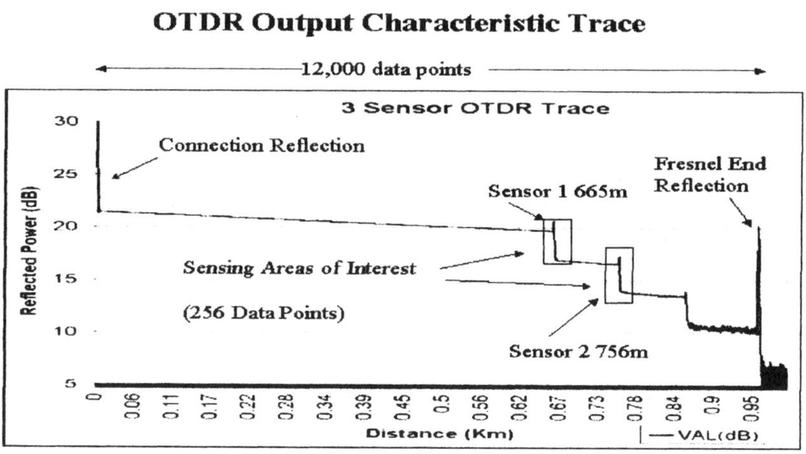

FIGURE 3: EXTRACTED OTDR SENSOR PEAKS

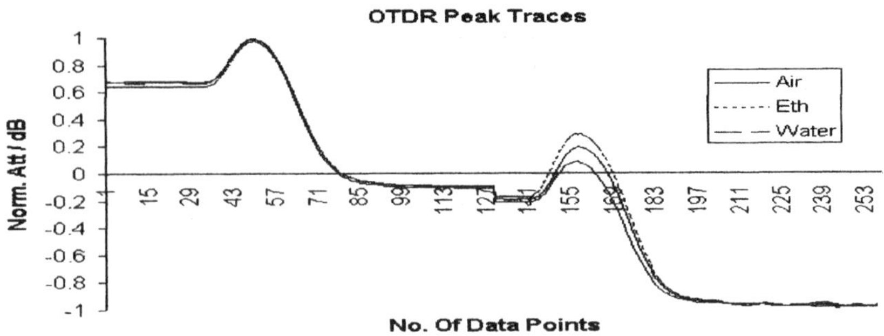

FIGURE 4: CORRESPONDING OTDR PEAKS PSD TRACE

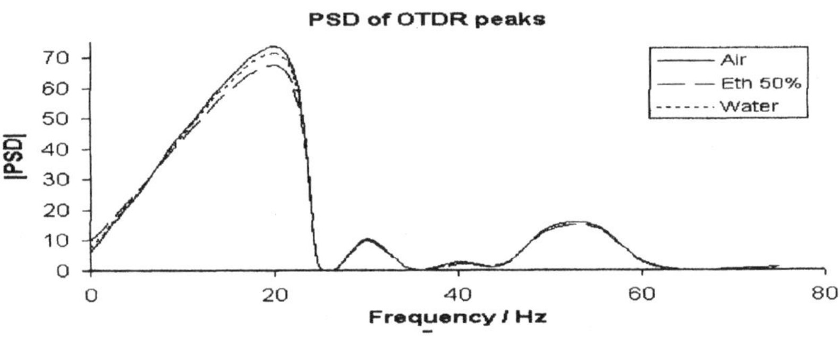

FIGURE 5: ANN IMPLEMENTED IN SNNS

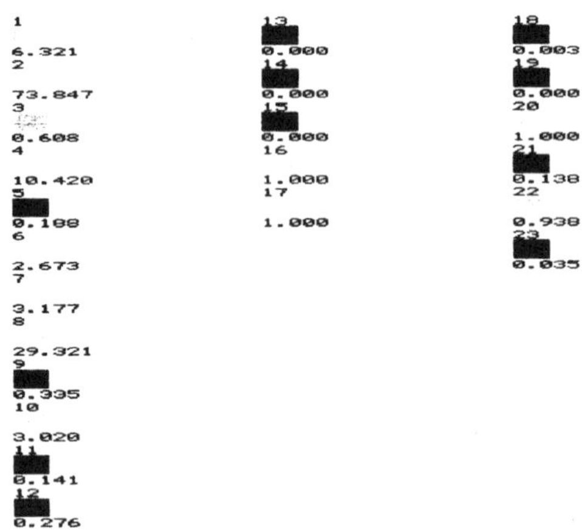

SMART VEHICLE CATEGORIZATION TECHNIQUES

CARLOS SUN
University of Missouri-Columbia

ABSTRACT
Vehicle type categorization is the separation of vehicles into predefined classes and can be useful for improving transportation efficiency, cost, environmental sustainability, enforcement, safety, and education. Two different approaches to vehicle type categorization will be presented. Both approaches use inductive loop signature data which is one of the most common forms of traffic detector. One approach is ICAN (Inductive Classifying Artificial Network) which is a self-organizing feature map (SOFM). ICAN mainly focuses on the challenging task of differentiating between two-axle vehicles such as passenger car, sports utility vehicle (SUV), van, truck, and bus. This is in contrast to systems that classify according to the number of axles. The overall categorization results of 87% (dataset 1) and 82% (dataset 2) for 7 categories coupled with consistent performance across all vehicle categories was significant and encouraging. Another system uses a heuristic discriminant algorithm for categorization and multi-objective optimization for training the heuristic algorithm. The heuristic algorithm yielded encouraging results of 81%-91% overall categorization rates. The results demonstrate the potential of collecting network-wide vehicle categorization data from inductive loops.

INTRODUCTION

The potential long term benefit areas of Intelligent Transportation Systems can be summarized as efficiency, economy, environment, enforcement, enhanced safety, and education. A major component of ITS is the improvement in the collection and transmission of transportation data. One type of data that is valuable in the effort to improve data quality is vehicle type categorization. Vehicle type categorization data is information on the types of vehicles that are on the roadways, such as passenger car, van, bus, truck, sports utility vehicle (SUV), etc… This type of data can be useful in for many areas of transportation planning and operations. Light "trucks" or LTVs (including SUVs) and heavier trucks are of special interest to agencies in terms of transportation safety and efficiency.

Heavy trucks and other oversized vehicles have different operating characteristics than passenger cars. Generally, heavy trucks travel at slower speeds, have longer deceleration times and distances, require more maneuvering times, and occupy more space on the roadway. The assumption that all vehicles have similar performance characteristics leads to errors in the modeling of traffic flow on networks.

There are many aspects of LTVs that differentiate them from other vehicles. One aspect that has been studied by the National Highway Traffic Safety Administration (NHTSA) is rollover crashes related to all classes of LTV's but especially light trucks such as pickups, sport utility vehicles, and vans. Even though LTVs are in 68 percent as many crashes per registered vehicle as are passenger cars, LTVs are in 127 percent as many rollover crashes per registered vehicles (Garrott, 1999).

There are many other classes of vehicles that merit identification. One example of an application that results from the unique identification of buses and emergency vehicles is pre-emption at traffic signals. By pre-empting or giving these vehicles the green phase, buses can be kept closer to published schedules and collisions with emergency vehicles can be reduced. Another application of vehicle categorization is the automatic identification of the vehicle class for toll collection purposes since many agencies charge different toll amounts for different classes of vehicles. Vehicle class information can also help traffic enforcement agencies in maximizing their resources by identifying concentrations of commercial vehicles. Since the types and amounts of noise and airborne pollutants emitted by vehicles differ significantly among vehicle classes, it is important to perform vehicle categorization in order to obtain an accurate assessment of area-wide traffic.

This paper presents two related methodologies for performing vehicle categorization using inductive loop detectors. In contrast to the some of the newer types of traffic detectors, the inductive loop detector has seen wide implementation and is still popular despite its long history. Figure 1 shows examples of the signatures obtained from inductive loop detectors and the corresponding ground truth video. One methodology for vehicle categorization is the use of Self-Organizing Feature Maps for automatically categorizing images obtaining form loop detectors. Another methodology is the use of a heuristic discriminant algorithm for categorization and multi-objective optimization for training the heuristic algorithm.

LITERATURE

Early vehicle type categorization systems consisted of axle counters used in combination with inductive loop detectors. One example was ALICE (Automatic Vehicle Categorization System) developed by the Transport and Road Research Laboratory in the U.K. (Davies, 1986). This system consisted of a preferred configuration of one inductive loop with two axle detectors per lane and the associated processing hardware and software. Speed, vehicle length, and wheelbase length(s) can all be derived using this system.

More recently, there have been investigations with detectors that produce images which can be evaluated using pattern recognition techniques. These systems are increasingly more sophisticated in terms of the categorization algorithms and automated software tools. These approaches include video image processing (Wei, 1996 and Yuan, 1994), infrared detection (Lu, 1989), acoustic waveform analysis (Nooralahiyan, 1997), and inductive signature recognition (Pursula, 1994).

There are also commercial products available that use infrared detector or multi-beam vertical light curtain for vehicle categorization. Companies such as Schwartz Electro-Optics (2000) and Transport Data Systems (2000) use 4 and 5 category schemes with subcategories. Details on the algorithms and performance for such commercial systems are often unavailable.

ARTIFICAL NEURAL NETWORK APPROACH

ICAN stands for Inductive Classifying Artificial Network and is used to conveniently describe a self-organizing feature map (SOFM) for vehicle type categorization using inductive signatures as input. A Self-Organizing Feature Map is an Artificial Neural Network that forms clusters of neurons which reflect similarities in the input vector. SOFM is also known as the Kohonen network.

The choice of the particular SOFM configuration is not a precise science and involves engineering judgement. The two major design parameters are the dimension of the network and the number of neurons. As noted by Schalkof (1992), powerful results have been obtained by just using 1- and 2-dimensional topologies. In selecting the dimensionality of the network for vehicle categorization implementation, the computational intensity and the categorization rate are both important. The input vector was composed of ninety-three components corresponding to the ninety-three equally spaced interpolations of the vehicle signature from the inductive loop detector. This input format was chosen to match directly the output of the detector cards with very little processing. One characteristic of ICAN is the simplicity of the 13 neuron 1-dimensional neural network, and the employment of a small training set of 13 signatures. The thirteen outputs were mapped to two vehicle categorization schemes, one with 9 vehicle classes and the other with 7 vehicle classes. The choice of these two vehicle categorization schemes is somewhat arbitrary since it was not dictated by any particular transportation application. However, the authors felt that the schemes chosen were general enough to accommodate several useful transportation applications such as the examples mentioned in the beginning of this paper. The two schemes are listed in Table 1.

The classification rate using 9 vehicles classes was 71% for the first data set. Most of the error is due to ICAN's inability to distinguish between 3 vehicle classes. These vehicle classes are classes 2, 3, and 4. In other words, cargo vans, SUV's, pickups, and full-size pickups exhibit very similar signatures. This is not surprising, since smaller SUV's share platforms with pickups, and bigger SUV's share platforms with full-size pickups. By expanding the SUV/pickup cluster to include cargo vans and full-size pickups, a 7 class scheme results. The overall classification rate for that scheme is 87%. The overall test result from using the second data set was 82%.

HEURISTIC APPROACH

A second vehicle categorization system used a heuristic discriminant algorithm for categorization and multi-objective optimization for training the heuristic algorithm. A heuristic discriminant algorithm is a hierarchical combination of discriminant functions, similar to a decision tree, that separates the vehicle pattern space in a piece-wise linear fashion. Feature vectors obtained by processing inductive signatures were used as inputs into the algorithms. The use of feature vectors reduced data storage and communications requirements, and enabled a hierarchical approach for categorization. Some examples of feature vectors include the electronic length, inductive magnitude, energy content, signature variance, skewness, kurtosis, Discrete Fourier Transform (DFT) coefficients, and first and second order gradients of the signature. Since some feature vectors contained similar information only a subset of feature vectors was used in algorithms. The heuristic algorithm employed seven categorization classes and yielded encouraging results of 81%-91% overall categorization rates using different datasets. The results by class is presented in Table 2. Note that the performance is consistent among all category types.

CONCLUSIONS

The results from both vehicle categorization systems demonstrate the potential of collecting network-wide vehicle categorization data from inductive loops. From the limited amount of testing, no generalized statement can be made regarding the superiority of one method over the other. Since both algorithms require fairly low computation, they can both be implemented on common computer hardware. The primary difference between the two methods is that no class can be specified aprioi using the SOFM, while the heuristic needs to have each class defined apriori. The availability of vehicle categorization data helps to improve traffic surveillance and paints a more detailed picture of dynamic traffic networks.

REFERENCES

Davies, P. (1986) Vehicle Detection and Classification, Information Technology Applications in Transport, VNU Science Press, 11-40.

Garrott, W.R., Howe, G.J., and Forkenbrock, G. An Experimental Examination of Selected Maneuvers That May Induce On-Road Untripped, Light Vehicle Rollover. National Highway Traffic Safety Administration. Washington, D.C. July. 1999.

Lu, Y. Vehicle Classification Using Infrared Image Analysis. ASCE Journal of Transportation Engineering. Vol. 118. No. 2. March/April. 1989. Pp. 223-240..

Nooralahiyan, A.Y. et al. A Field Trial of Acoustic Signature Analysis for Vehicle Classification. Transportation Research-C. Vol. 5. No.3/4. 1997. Pp. 165-177.

Pursula, M. and Pikkarainen, P. A Neural Network Approach to Vehicle Classification with Double Induction Loops. Proceedings of the 17th ARRB Conference. Part 4. 1994. Pp. 29-44.

Schwartz Electro-Optics (SEO). (2000) AUTOSENSE for automated highway systems, Product literature, http://www.seord.com/

Schalkoff, R. (1992) Pattern Recognition: Statistical, Structural, and Neural Approaches, John Wiley and Sons, New York.

Transport Data Systems. (2000) Automatic Vehicle Classification System, Product literature, http://www.transportdatasystems.com.

Wei, C. et al. Vehicle Classification Using Advanced Technologies. Transportation Research Record 1551. November. 1996. Pp. 45-50.

Yuan, X. et al. Computer Vision System for Automatic Vehicle Classification. ASCE Journal of Transportation Engineering. Vol. 120. No. 6. November/December. 1994. Pp. 861-876.

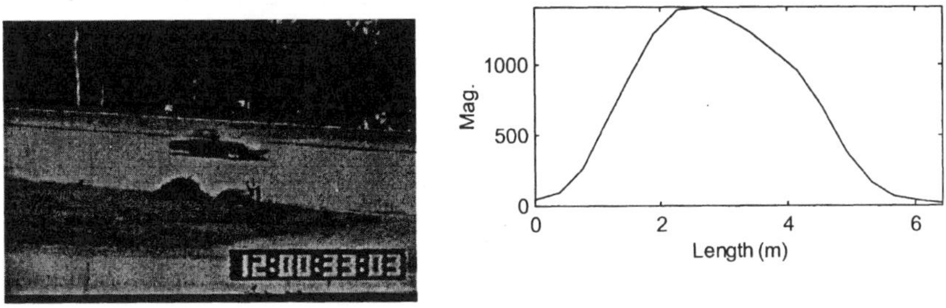

Figure 1a. Sports Car

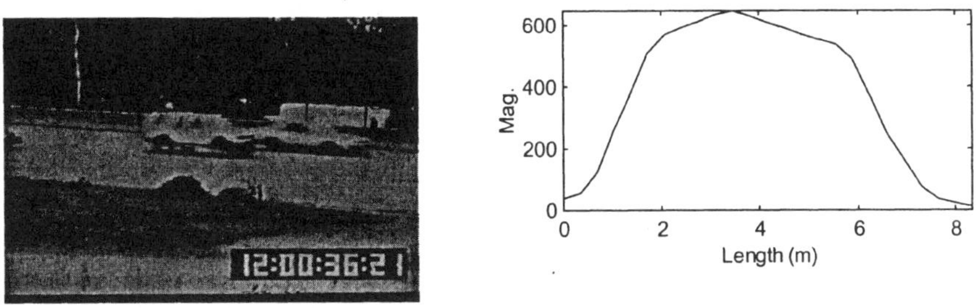

Figure 1b. Sports Utility Vehicle

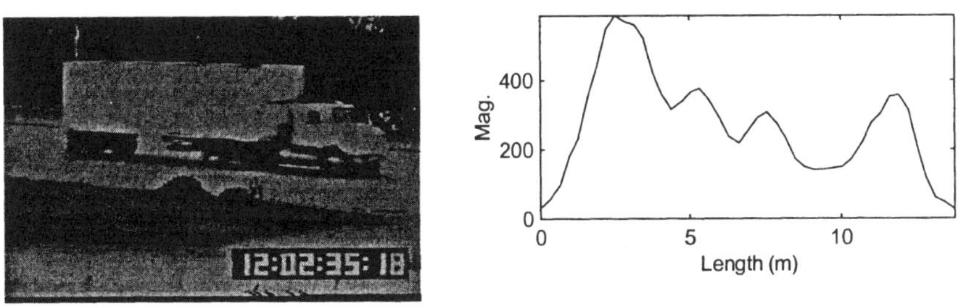

Figure 1c. Semi-trailer Truck

Figure 1 Signatures and Images from Representative Vehicles

Table 1 ICAN Categorization Schemes

Nine-Vehicle Categorization Schemes	
Class	Vehicle Types Represented
1	passenger car, minivan, sports car, station wagon
2	van
3	sports utility vehicle, pickup
4	full-size pickup
5	truck
6	> 2-axle truck
7	bus
8	vehicle with trailer
9	limousine
Seven-Vehicle Categorization Schemes	
Class	Vehicle Types Represented
1	passenger car, minivan, sports car, station wagon
2	van, sports utility vehicle, pickup, full-size pickup
3	truck
4	> 2-axle truck
5	bus
6	vehicle with trailer
7	limousine

Table 2 Heuristic Classification Results

		Heuristic Classification for Training Set						
		1	2	3	4	5	6	7
Correct Class	1	64	4	1	0	0	0	0
	2	2	22	1	0	0	0	0
	3	4	2	20	0	0	0	0
	4	0	0	0	2	1	0	0
	5	0	0	1	0	7	0	0
	6	0	0	0	0	0	2	0
	7	0	0	0	0	1	0	3
		Heuristic Classification for Test Set						
		1	2	3	4	5	6	7
Correct Class	1	66	3	0	0	0	0	0
	2	1	18	6	0	0	0	0
	3	3	2	20	0	1	0	0
	4	0	0	0	2	1	0	0
	5	0	0	1	0	6	1	0
	6	0	0	0	0	0	2	0
	7	0	0	0	0	1	0	3

QUANTITATIVE ANALYSIS OF RELATION BETWEEN SUBJECTIVE IMPRESSIONS AND CHARACTERISTICS OF ARM MOTION IN FREE-STYLE DANCE

SHIHOKO KAMISATO,
SATORU ODO,
KOJI YAMADA
Faculty of Engineering, University of the Ryukyus

KIYOSHI HOSHINO
Institute of Engineering Mechanics and Systems,
University of Tsukuba

ABSTRACT
The purpose of this study is to examine the relationship between physical motion characteristics and the subjective impressions produced in dancing. Motions of Okinawan folk dance"Kachaasii" were used for the experiment, because the dancers can produce various kinds of movement with no regular rules of operation, which suggests that the structures of subjective impressions for dance motions are not limited, and that quantitative analyses are easily adoptable to in the frequency or time domain. Questionnaire with adjectives and adverbs for subjective estimation were the target for principal component analysis. Response variables were each principal component scores which are equivalent to subjective impressions, and explanatory variables were kinematics features of arm's movement in multiple regression analysis.

1. INTRODUCTION

The human motions produce various kinds of impressions. For example "graceful movements " and "skillful movements" expressed an impression of motions by words. Over the past few years, several studies have been made on impressions structure by sign language and dance as a means of communication [1]-[3]. Furthermore, the concern with the study of humanoid robotics and sensitivity of the humans who used computer graphics has been growing for the last several years [4]-[6]. However, there has been no study that tried to make relationship clear quantitatively about motion characteristics of humans and subjective impressions. So, in this study, the purpose is to identify the relationship between motion kinematics or dynamics and the impressions. Here, we tried approach using dance motions. Because it examines relationship with subjective impressions by various motions so that it is represented in a creation dance. The structure of subjective impressions for dance motions with no regular rule of operation was investigated using Okinawan folk dance "Kachaasii". First, questionnaires with adjectives and adverbs for subjective estimation were adopted for principal component analysis using semantic differential method. Second, response variables were each principal component scores which are equivalent to subjective impressions, and explanatory variables were kinematics features of arm's movement in multiple regression analysis [7]-[8]. This study is important in order to make clear the mechanism of humans and motion control of robot accompanied with sensitivity.

2. SUBJECTIVE IMPRESSION EXPERIMENT

2.1 METHODS

The purpose of the impression evaluation experiment here in this chapter is to know the impressions to various patterns of dance motion and to examine the impression structure of free-style dance Kachaasii with principal component analysis. The video image was prepared for the evaluation in order to carry out the impression evaluation of Kachaasii. At first a lot of dance patterns were collected without relationship in presence of dance experience, age and sex. Dance motions of 26 patterns were selected by motion characteristics pattern. The various dance patterns performed by the same dancer were recorded at the front with video camera. The size and width of dancer's body or arm were uniformed in each video sequence. The face was hidden with dark-color veil so as to avoid the impression deviated by dancer's sexuality, face or facial expression. The questionnaire was carried out with thirty pairs of adjectives and the antonyms. The adjectives are shown in Table1 and fig.1. Twenty-six patterns of dance were prepared for subjective evaluation as video image. The example is shown in Fig.2. Sufficient trial time was allowed for subjects to watch and answer in each evaluation of dance.

Table 1 Adjectives for principal component analysis.

(1)	speedy	(16)	accented
(2)	regular	(17)	straight
(3)	stabilized	(18)	rising
(4)	strong	(19)	heavy
(5)	dynamic	(20)	cheerful
(6)	beautiful	(21)	skillful
(7)	graceful	(22)	harmonious
(8)	like	(23)	wide
(9)	magnificent	(24)	good
(10)	profound	(25)	rhytmical
(11)	smooth	(26)	complex
(12)	large	(27)	balanced
(13)	toward	(28)	delight
(14)	hard	(29)	thick
(15)	forceful	(30)	peculiar

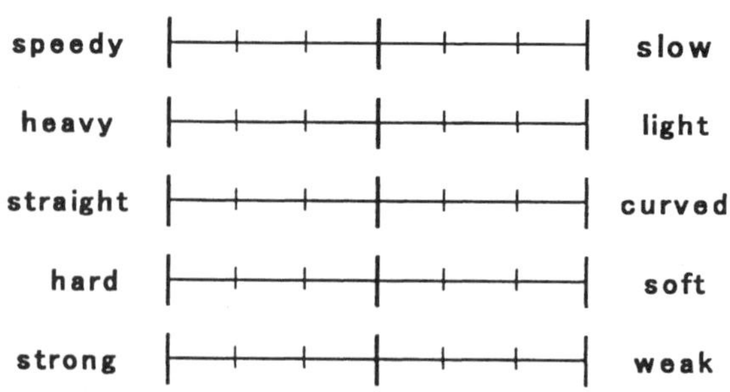

Fig.1 An example questionnaire

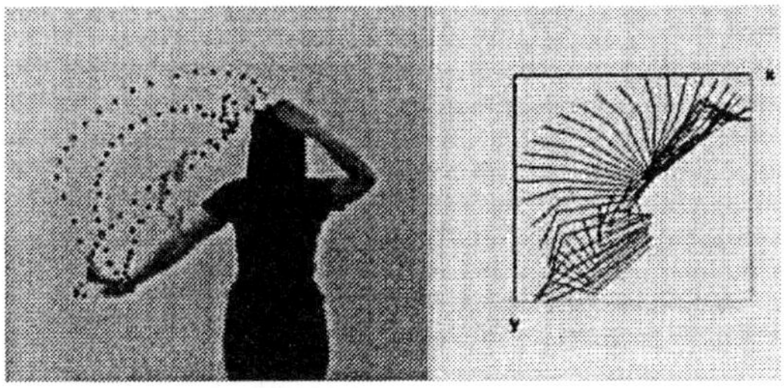

Fig.2 An example of video image

2.2 RESULTS

The analytical results by principal component analysis were as follows: eigenvalue, proportion and cumulative proportion of each component are shown in Table 2. The first through five components seemed to be empirically meaningful, mainly from the view point of eigenvalue. Then, the values of each factor loading were examined in first through five components. Impression characteristics of each component are shown in Fig.3. Each component have impressions characteristics as follows: the first component exhibited "preference", the second "hard - soft ", the third "dynamic - static ", the fourth "monotonous - peculiar ", and the fifth "balanced - large ".

Table2. Eigenvalues and propotions of principal component analysis

principal component	eigenvalue	propotion [%]	comulative_ propotion [%]
1st	9.00	30.01	30.01
2nd	3.53	11.77	41.77
3rd	2.33	7.75	49.53
4th	1.91	6.38	55.91
5th	1.39	4.63	60.53
6th	1.05	3.49	64.03

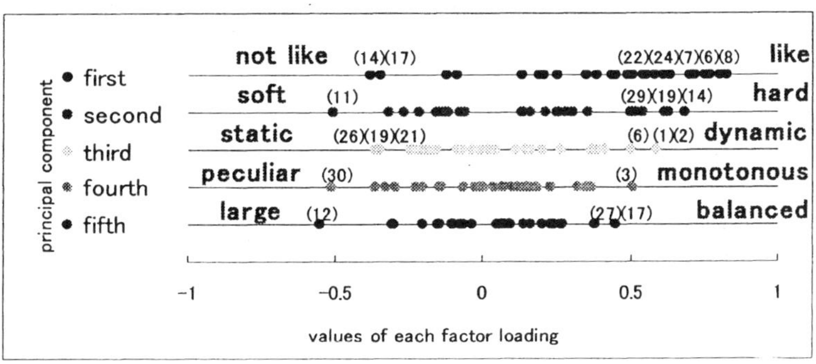

Fig.3. Adjectives selected in each component.

3. EXTRACTION OF KINETIC CHARACTERISTICS

A statistical measure as a principal component score is also obtained by the principal component analysis, which is an averaged score in each dance pattern weighted by each value of factor loadings obtained at word pairs in each principal component, through which the degree of a certain kind of impression of a dance pattern can be clarified. The larger a principal component score is, the greater impact a dance pattern has in a certain kind of impression represented by each principal component. In this chapter, therefore, kinematics features were parameterized in each dance pattern, and the relationship between the motion features as explanatory variables and subjective impressions as response variables were quantified by the multiple regression analysis. Kinetic characteristics which seemed to affect the impression of each component were extracted from the trajectories of hand. The examples of different patterns of extracted trajectories are shown in Fig.2. Kinetic characteristics of 15 patterns were obtained from the trajectories of hand. Kinetic features of the arm adopted for explanatory variables were as following: area of movement as the horizontal and vertical center of gravity coordinates x and y, direction of rotation as the clockwise or counterclockwise circumferential trajectory, size of movement as the projected area of trajectory, complexity of movement as the square of circumference length divided by gross area, directivity of movement as the angle of inclination of principal axis, expansibility of trajectory as the ratio of length between principal axes and its orthogonal axes, smoothness of movement as the square of integrated acceleration, horizontal or vertical symmetry of speed distribution as kurtosis of trajectory distribution in x or y axes, size of angular variation as the maximum curvature, flexuosity of a wrist joint at flexion or extension as the difference of normalized areas of trajectory between fingertip and wrist, existence of wrist winding, existence of standstill in movement and expandability of trajectory by ratio of horizontal and vertical. Furthermore, a time series transformation of degree was measured for movement of 6 joints. As for the motion characteristic of each joint, it was each extracted average of degree, variance, kurtosis, square integration of angular acceleration transformation, angular velocity plus side maximum, angular velocity minus side maximum, angular acceleration plus side maximum, angular acceleration minus side maximum and phase difference between two joints next to each other. The relationship of impression and motion characteristics was analyzed by multiple regression analysis of stepwise method.

4. RESULTS AND DISCUSSION

The result of multiple regression analysis is shown by table 3. The results of multiple regression analysis between motion features as explanatory variables and principal component scores as response variables were as follows: in the first component, winding and rotation of the wrist joint was important component the impression of "preference"; in the second component, the "hard" impressions were produced when existence of standstill in movement and motion of shoulder extension degree was big, and the "soft" impression tended to be produced when motion of wrist extension degree was big; motion features necessary in the third principal component as "dynamic" impression was quick movement to raise above of shoulder flexion degree, and as "static " impression was phase difference from elbow extension to wrist rotation and phase difference from elbow extension to wrist abduction; in the fourth component, the"monotonous" impression was produced when keep the state that inclined of wrist abduction, and the "peculiar" impression tended to be produced when motion to raise the arms in the inside with high position; in the fifth component, the "balanced" impressions were produced when existence of standstill in movement and Arm moves with the position which is near to a body, and "large" impression tended to be produced when extend by an elbow and a wrist to make trajectory big. The results of this study maybe supply a method for quantitatively describing the relationship between impression and motion.

Table3. Result of the multiple regression analysis

first principal component: like-not like

motion chatacteristics		Comulative propotion	Regression coefficient	F Value	p
no.13	existence of wrist winding	50.05%	0.63	24.04	**
no.39	the wrist rotation angular velocity - said maximum	67.81%	-0.88	8.44	**
no.57	the shoulder adduction angular acceleration - side maximum	76.74%	-0.83	12.69	**

* p < 0.05 ** p < 0.01

second principal component: hard-soft

motion chatacteristics		Comulative propotion	Regression coefficient	F Value	p
no.14	existence of standstill in movement	49.59%	0.70	23.61	**
no.29	the variance of shoulder extension degree	71.93%	0.391	10.66	**
no.17	the variance of wrist extension degree	55.97%	-0.59	6.88	*

* p < 0.05 ** p < 0.01

third principal component: dynamic-static

motion chatacteristics		Comulative propotion	Regression coefficient	F Value	p
no.69	phase difference from elbow extension to wrist rotation	42.48%	-0.38	17.72	**
no.71	Phase difference from elbow extension to wrist abduction	48.18%	-0.54	10.70	**
no.42	the shoulder flexion angular velocity + side maximum	48.28%	0.77	10.17	**

* p < 0.05 ** p < 0.01

fourth principal component: monotonous-peculiar

motion chatacteristics		Comulative propotion	Regression coefficient	F Value	p
no.21	the kurtosis of wrist abduction degree	32.37%	0.96	11.49	**
no.31	the average of shoulder extension degree	56.37%	-0.30	12.65	**
no.45	the shoulder adduction angular velocity - side maximum	68.64%	-0.59	8.61	**
no.42	the shoulder flexion angular velocity + side maximum	78.59%	-0.34	9.76	**

* p < 0.05 ** p < 0.01

fifth principal component: balanced-large

motion chatacteristics		Comulative propotion	Regression coefficient	F Value	p
no.7	expansibility of trajectory	39.14%	-0.71	15.44	**
no.1	center of gravity of the X-axis	57.83%	0.47	10.19	**
no.14	existence of standstill in movement	65.21%	0.08	4.67	*
no.23	the variance of wrist rotation degree	75.59%	-0.22	8.93	**
no.25	the average of elbow extension degree	82.22%	-0.35	7.46	*
no.19	the average of wrist extension degree	87.72%	-0.33	8.5	**

* p < 0.05 ** p < 0.01

5. CONCLUSIONS

The purpose of this study is to identify the relationship between motion kinematics or dynamics and the impressions produced in dancing. The structure of subjective impressions for dance motions was investigated with Okinawan dance "Kachaasii" which has no regular rule of operation. First, the questionnaire method and the principal component analysis were adopted for investigation of the mental structure of the dance. Second, the motion dynamics and kinematics were examined from the view point of the areas of the trajectory, directions of rotation, the spatial location, maximum acceleration, center of gravity, the complexity, the gradient of main axis, the direction, the ratio of quadrature axis and vertical line, the sum of acceleration, range of trajectories of hand. Furthermore, a time series transformation of degree was measured for movement of 6 joints. As for the motion characteristic of each joint, it was each extracted average of degree, variance, kurtosis, square integration of angular acceleration transformation, angular velocity plus side maximum, angular velocity minus side maximum, angular acceleration plus side maximum, angular acceleration minus side maximum and phase difference between two joints next to each other. The results showed that such factors were involved in the dance as follows: "preference", "hard - soft ", the third " dynamic - static ", " monotonous - peculiar ", and " balanced - large ". Then, the relationship between the impressions and kinetic characteristics of motions were quantitatively analyzed from the view point of spatial and temporal features of trajectories of finger, hand and time series transformation of joints degree. The results showed that each kind of impressions tended to be produced by specific motion features.

REFERENCES

[1] Akiko Zukawa and Yoshiyuki Matsuura, The image of dance relating to dance movement pattern in semantic space: Health & Sports sciences, Univ. of Tsukuba, Vol.9, pp.103-114, 1986.

[2] Mariko Shiba, Extraction of kansei information in dance movements: ATR workshop on virtual communication environments, pp.66-88, 1998.

[3] Satoru Odo, Shihoko Kamisato, Kiyoshi Hoshino, Effects of movement durations in different words with similar motions in the sign language discrimination: The journal of the institute of image information and television engineers, Vol.56, No.2, pp.302-306, 2002.

[4] Takeshi Arakaki and Kiyoshi Hoshino, Dance CG animation tool producing subjective impressions: Intl.Sumposium on KANSEI Technology 2001, pp.169-172, 2001.

[5] Christopher G. Atkeson, Josha G. Hale, Frank Pollick, Marcia Riley, Using a humanoid robot to explore human behavior and communication: JRSJ, Vol.19, No.5, pp.42-47, 2001.

[6] Shihoko Kamisato, Tomohiro Shiroma, Kiyoshi Hoshino, Estimation of subjectivity in the dancing movement of arm motion and generation of dance movements by humanoid robotic arm: Technical report of IEICE, HIP2001, Vol.101, No.594, pp.61-66, 2002.

[7] Shihoko Kamisato and Kiyoshi Hoshino, Motion characteristics of finger and wrist trajectories and subjective impression in a free-style dance: IEICE HIP, vol.100, no.138, pp.47--51, 2000.

[8] Shihoko Kamisato and Kiyoshi Hoshino, Evaluation of subjective impressions and motion characteristics in a free-style dance: Proc. of ORSJ, vol.2, pp.309--314, 2000.

APPROXIMATING ECONOMIC DATA SETS FROM THEIR UNDER-SAMPLED WEYL-HEISENBERG EXPANSIONS

ANTHONY JOSEPH, PH.D.
Computer Science Dept.
Pace University
One Pace Plaza
New York, New York 10038
Telephone No.: 212 346 1492
Fax No.: 212 346 1863
Email : ajoseph2@pace.edu

AHMAD ALNIMRAT, PH. D.
Electrical Engineering Dept.
The City College of New York
Convent Avenue at 138th Street,
New York, New York 10031

ABSTRACT

This paper focuses on an experiment designed to show that orthogonally projected sales and inventory data sets onto integer under-sampled Weyl-Heisenberg expansions in Zak space can yield comparable or better results in comparison to conventional spectral analysis in terms of the number of spectral coefficients needed to adequately represent the data sets. The algorithm used to perform the projection of these economic data sets from a critically sampled Weyl-Heisenberg subspace onto an under-sampled Weyl-Heisenberg subspace is based on periodization-decimation operations using Poisson summation formula in Zak space. Decimating in frequency by a factor of two, the projection of the sales and inventory data contained 79.55% and 82.61% respectively of the original data energies. Additionally, ten and eight of the under-sampled Weyl-Heisenberg coefficients from the sales and inventory data sets respectively contained 74.14% and 75.03% of the original energies.

INTRODUCTION

An important feature of many time series, including sales and inventory, is that they are periodic. Therefore, these series can be represented in the spectral domain where each component's independent contribution to the total energy (or power) of the time series can be readily analyzed. The questions that now arise are the following:

- How many spectral components significantly contribute to the total energy contained in the time series?
- Does one need to study all the spectral components in a time series to determine those that highly contribute to the total energy?

Larrain (2002) examined the macroeconomic interrelationship between nominal interest rates, real sales, and inventories. He used 372 data points for each of these business cycle variables and discovered that over 70% of the total energy in each was concentrated in 15 Fourier coefficients - a mere 4% of the spectral components. In this study, we seek to obtain more highly concentrated sets of spectral components from under-sampled subspaces of the respective original subspaces of sales and inventory data sets consisting of 400 data points. We expect to find comparable energies in relatively fewer numbers of integer under-sampled Weyl-Heisenberg coefficients obtained from the orthogonal projection of the critical sampled Weyl-Heisenberg expansions of sales and inventory data sets in Zak space. There will be no focus on the exact frequency locations of the time-frequency cells that contain these dominant coefficients.

Inventory plays a prominent role in a nation's economy and is closely related to sales. In general, inventory investment is positively correlated to sales (Blinder and Maccini, 1991) and it is dominant feature in business cycles.

Weyl-Heisenberg expansions allow for high time-frequency localization of data sets. They have the ability to evaluate time-varying spectra, to discriminate data in noise, and to compact information. They represent generalized forms of spectral analysis that have bases functions are that generally nonorthogonal.

The algorithm used in this experiment to perform the orthogonal projection of the economic data sets from their critically sampled subspace into integer under-sampled Weyl-Heisenberg subspaces is based on periodization-decimation operations via Poisson summation formula in the convenient and computationally efficient Zak space. Decimation in frequency by a factor of two projects the sales and inventory data sets onto smaller respective subspaces containing 79.55% and 82.61% of the original total energies with only ten and eight of the under-sampled Weyl-Heisenberg coefficients from the sales and inventory data sets, respectively, containing 74.14% and 75.03% of the original total energies. Decimation in time by a factor of two yielded under-sampled Weyl-Heisenberg coefficients that were less highly concentrated in energy distribution.

MATERIALS and METHODS

The raw data (monthly total business sales and total business inventory) used in this experiment were equivalent to those Larrain (2002) used in his work. However, the data used here covered the period from January 1965 to April 1999. Before subjecting the data to time-frequency digital signal processing techniques, the data were preprocessed to remove some of their nonlinearity (Pena et al, 2001) and some of the low frequency components as well as some of the nonstationarity (Koopmans, 1995 and Pena et al, 2001). Twelve-month backward first difference was used for the high pass filtering and nonstationarity removal. Each of the preprocessed sales and inventory data sets, with means removed, consisted of 400 samples.

Spectral analysis using Fourier analysis is a customary technique for studying time series data sets including economic ones. It involves the decomposition of a data set into its Fourier coefficients. A coefficient is associated with a single frequency. If the data set is real as in this case of sales and inventory, only half of the Fourier coefficients are unique. Often one is interested in the energy that each coefficient independently contributes to the total energy of a data set. The cumulative sum of the energies in the Fourier coefficients equals the total energy of the data set. Therefore, it is desirable to have the total energy in a data set concentrated in as few spectral coefficients as possible. This can facilitate analysis and reduce processing time if the analysis tool has the ability to compact data set energy in relatively few coefficients.

Generalized and alternate forms of spectral analyses are performed using Weyl-Heisenberg expansions and Zak transform (Auslander et al, 1990; Joseph et al, 2001; and Tolimieri and An, 1998). These transforms are linear with nonorthogonal bases functions. In particular, a finite energy data set does not have a unique Weyl-Heisenberg expansion over its domain of definition.

Mathematical Framework

The foundation upon which the orthogonal projection used in this experiment is based is detailed in Joseph et al (2001) and Tolimieri and An (1998). Some of the most salient equations include the following:

- For $\mathbf{m} \in Z/N \times Z/N$ and $g \in L(Z/N)$, define the time-frequency translates of g as

$$g_{\mathbf{m}} = g(a - m_1)x_{m_2}(a), \quad x_{m_2}(a) = \exp(2\pi i m_2 a / N), \quad a \in Z/N, \quad g_m \in L(Z/N) \quad (1)$$

- A Weyl-Heisenberg expansion of $f \in Z/N$ over $[g, \Delta]$ has the form

$$f = \sum_{\mathbf{m} \in \Delta} c(\mathbf{m}) g_{\mathbf{m}} \quad (2)$$

where $c \in L(\Delta)$ is the Weyl-Heisenberg coefficient set of f over $[g, \Delta]$. $[g, \Delta] = \{g_{\mathbf{m}} : \mathbf{m} \in \Delta\}$ is Weyl-Heisenberg system over subgroup $\Delta \in Z/N \times Z/N$ with window function g for a collection of data sets in $L(Z/N)$. A Weyl-Heisenberg system $[g, \Delta_0]$ is critically sampled if Δ_0 is the critically sampling subgroup of $Z/N \times Z/N$. The critically sampling subgroup may have the form $\Delta_0 = B \times B_*$ for a subgroup $B \in Z/N$ (B_* is the dual of B). If $\Delta_0, \Delta_1, \Delta_2 \in Z/N \times Z/N$ are subgroups satisfying the condition $\Delta_1 \subset \Delta_0 \subset \Delta_2$, then $(\Delta_2)_* \subset \Delta_0 \subset (\Delta_1)_*$, $(\Delta_0)_* = \Delta_0$.

- Zak transform, a simple and computationally efficient tool for the analysis and construction of time-frequency algorithms (Tolimieri and An, 1998), of a data set $f \in L(Z/N)$ over a subgroup $B \in Z/N$ is defined as

$$F(\mathbf{a}) = \sum_{m \in B} f(a_1 + m)\overline{x}_m(a_2), \quad F \in L(Z/N \times Z/N) \quad (3)$$

where $F = Z(B)f$ and $\mathbf{a} \in Z/N \times Z/N$. It satisfies the functional relationship for the space of all $F \in L(Z/N \times Z/N)$, $\forall k \in \Delta_0$:

$$F(\mathbf{a} + \mathbf{k}) = x_{k_1}(a_2)F(\mathbf{a}), \quad \mathbf{a} \in Z/N \times Z/N \quad (4)$$

- The relationship between Zak transform and time-frequency translates is the following: $G_{\mathbf{m}}(\mathbf{a}) = \psi_{\mathbf{m}}(\mathbf{a})G(\mathbf{a})$ with $\mathbf{m} \in \Delta_0$ and $\mathbf{a} \in Z/N \times Z/N$.
- The main computational step in the orthogonal projection algorithm in Zak space is the computation of the $\Delta -$ periodizations:

$$X(\mathbf{a}) = \sum_{\mathbf{m}\in\Delta} |G(\mathbf{a}+\mathbf{m})|^2, \quad Y(\mathbf{a}) = \sum_{\mathbf{m}\in\Delta} |G(\mathbf{a}+\mathbf{m})|^2 P(\mathbf{a}+\mathbf{m}), \quad \mathbf{a} \in Z/N \times Z/N \qquad (5)$$

EXPERIMENT

The experiment, which was conducted in Zak space, consisted of orthogonally projecting sales and inventory data sets from their critically sampled subspaces of Weyl-Heisenberg coefficients into integer under-sampled subspaces of Weyl-Heisenberg coefficients. The two types of periodization-decimation operations consisted of decimation in frequency by a factor of two and decimation in time by a factor of two. In both cases the Gaussian window was used. The lower limit of the energy threshold of the under-sampled Weyl-Heisenberg coefficients that was considered to be sufficiently highly contributory to the total energies in the projections was 0.02.

RESULTS

The experiment showed that for decimation in frequency by a factor of two, the inventory projection retained 82.6% of the total energy in the inventory data set while the sales projection retained 79.6% of the total energy in the sales data set. Of these, 75.0% of the inventory and 74.1% of the sales total energies were concentrated in eight and ten Weyl-Heisenberg coefficients respectively (Table 1) on the integer under-sampled subspace. The projection data sets of sales and inventory, recovered from the 200 nonzero Weyl-Heisenberg coefficients, retained many of the salient features of the original data sets that had a critically sampled Weyl-Heisenberg representation with 400 nonzero coefficients. Further evidence of the projections close approximation to the original sales and inventory data sets was seen in the overlap of their dominant power spectral densities at their frequency indices. The results obtained from decimating the sales and inventory Weyl-Heisenberg coefficients in time by a factor of two were less favorable (Table 1). Moreover, the percentage of the inventory and sales projection energies contained in the aforementioned eight and ten under-sampled Weyl-Heisenberg coefficients with each having energy content of 0.02 or greater was over 90% for decimation in frequency: 90.9% for inventory and 93.1% for sales. A single coefficient for inventory contained 31.5% of the total energy and 38.1% of the projection energy while for sales the corresponding values were 16.1% and 20.3%. For decimation in time, six and five under-sampled Weyl-Heisenberg coefficients with energies equal to or exceeding the 0.02 threshold contained 66.8% and 62.2% respectively of the sales and inventory projection energies.

DISCUSSION

Decimation in frequency yielded more favorable results than decimation in time. This appeared to be consistent with the fact that economic time series such as sales and inventories have most of their energies concentrated in relatively few low frequency spectral components. On the contrary, the energy distributions in the temporal components of these time series variables seemed to be spread out over a relatively larger number of components thereby accounting for smaller contributions to a larger number of Weyl-Heisenberg coefficients when the data sets were decimated in time. The results obtained for decimation in frequency compared favorable with results obtained by

Larrain (2002). Larrain recovered over 70% of the sales and inventory energy distributions from 4% of the spectral components while in this experiment, there was over a 70% recovery rate of the sales and inventory total energies from 2.5% and 2% respectively of the spectral components. Additionally, over 90% of the total energy distributions in the sales and inventory projections were concentrated in 5% and 4%, respectively, of the spectral components in Zak space. These higher concentrations of energies in fewer spectral components suggests that Weyl-Heisenberg expansions are a viable tool for studying the visible and hidden periodicities in sales and inventory data sets and, perhaps more generally, any time series.

The orthogonal projection algorithm uses a process wherein it removes decimated time-frequency cells of the Weyl-Heisenberg coefficients, and then encodes their information into adjacent non-decimated cells. Thus, with a better choice of window and or the under-sampling rate, it may be possible to achieve relatively higher percentages of the sales and inventory total energies in the projections. These improvements in energies could be such that the mean square error between the original data sets and their orthogonal projections are negligible. Approximated data sets with relatively few Weyl-Heisenberg coefficients that are almost replicas of the originals have implications in reduced data storage space, transmission time, computation speed, and resultantly lower cost.

If the orthogonal projection algorithm for Weyl-Heisenberg expansions of a data set in Zak space onto a smaller subspace were to be used for data mining, it should provide better results than traditional data mining techniques like statistical analysis and artificial intelligence for time varying data sets. However, these two approaches could be combined to automatically detect and extract the significant features of both stationary and nonstationary data sets. More specifically, orthogonal projection algorithm for Weyl-Heisenberg expansions could be included in a preprocessor stage to a principal component analysis based neural network to improve the neural network's general efficiency and applicability to different types of data sets of arbitrarily large sizes.

CONCLUSION

With 200 of the original 400 Weyl-Heisenberg coefficients on an under-sampled Weyl-Heisenberg subspace, the resulting projections from the sales and inventory data sets compared favorably, in terms of energy distributions, with results obtained by Larrain (2002) using 372 Fourier coefficients. Although, there was no determination of the exact frequency locations of the Weyl-Heisenberg coefficients, it was observed that the dominant under-sampled Weyl-Heisenberg coefficients were located at no more than two different frequencies with most located at a single frequency. Future investigations will include determining the under-sampling rate that yields the best approximation of the original sales and inventory data sets in the least mean square error sense while concentrating energies in fewer Weyl-Heisenberg coefficients, finding the exact time and frequency locations of these coefficients, and comparing the exact relationship between the approximated aggregate sales and inventory with results obtained from classical spectral analysis.

TABLE

The table shown below describes the energy contribution of the under-sampled

Weyl-Heisenberg coefficients to the sales and inventory data. The sales energy contribution is enclosed within parenthesis

Table 1. Energy under-sampled Weyl-Heisenberg coefficients contributed to inventory (sales) data

Coefficient	Decimation in Frequency by 2		Decimation in Time by 2	
	Energy Contribution	**Percentage of total Energy**	**Energy Contribution**	**Percentage of total Energy**
1	0.217 (0.024)	31.47% (2.70%)	0.020 (0.027)	2.84% (3.06%)
2	0.020 (0.085)	2.92% (9.62%)	0.133 (0.123)	16.36% (13.92%)
3	0.066 (0.083)	9.57% (9.43%)	0.056 (0.120)	8.07% (13.68%)
4	0.028 (0.052)	3.99% (5.96%)	0.026 (0.086)	3.71% (9.76%)
5	0.078 (0.065)	11.36% (7.33%)	0.086 (0.050)	12.53% (5.70%)
6	0.024 (0.020)	3.44% (2.23%)	(0.028)	(3.21%)
7	0.035 (0.124)	5.07% (16.15%)		
8	0.050 (0.054)	7.21% (6.17%)		
9	(0.068)	(7.68%)		
10	(0.061)	(6.88%)		
Total	**0.518 (0.652)**	**75.03% (74.14%)**	**0.300 (0.434)**	**43.51% (49.32%)**

REFERENCES

Auslander, L. et al, 1990, "A Comparison of the Gabor and Short-Time Fourier Transform for Signal Detection and Feature Extraction in a Noisy Environment," in *SPIE Proceedings: Advanced Signal Processing Algorithms, Architectures, and Implementations*, F. Luk, ed., Vol. 1348, San Diego, pp.230-240.

Bils, M. and Kahn, J., 2000, "What Inventory behavior Tells us about Business Cycles," *American Economic Review*, Vol. 90, pp. 458-481.

Blinder, A. and Maccini, L., 1991, "Taking Stock: a Critical Assessment of Recent Research on Inventories," *Journal of Economic Perspective*, Vol. 5, Winter, pp. 73-96.

Christiano, L., 1988, "Why does Inventory Investment Fluctuate so Much?" *Journal Of Monetary Economics*, Vol. 21, pp. 247-280.

Joseph, A. et al, 2001, Under-Sampled Weyl-Heisenberg Expansions via Orthogonal Projections in Zak Space," *Signal Processing*, Vol. 81, March, pp. 2383-2402.

Koopmans, L., 1995, *The Spectral Analysis of Time Series*, Academic Press, New York, pp. 165-208.

Larrain, M., 2002, "Do Interest Rates Lead Real Sales and Inventories? A Spectral Analysis Approach" *Business Economics*, pp. 33-43.

Manolakis, D. et al , 2000, *Statistical and Adaptive Signal Processing: Spectral Estimation, Signal Modeling, Adaptive Filtering, and Array Processing,* Mc Graw Hill, New York, pp. 395-498.

Pena, D. et al, 2001, *A Course in Time Series Analysis*, John Wiley & Sons, New York, pp. 25-85.

Ramsey, V., 1989, "Inventories as Factors of Production and Economic Fluctuations," *The American Economic Review*, Vol. 79, June, pp. 338-354.

Tolimieri, R. and An, M. 1998, *Time-Frequency Representations*, Birkhauser, Boston, pp. 47-92.

MONITORING THE OZONE HOLE USING NEURAL NETWORKS AND THE MIXTURE-OF-EXPERTS MODEL

DIEGO G. LOYOLA R.
German Aerospace Center (DLR)
Remote Sensing Technology Institute (IMF)
Oberpfaffenhofen, D-82234 Wessling, Germany

ABSTRACT
This paper presents the retrieval of ozone from satellite data using an algorithm based on neural networks combined with the mixture-of-experts model. Single neural networks are trained to approximate sub-regions of a complex multidimensional function; the neural networks are then combined with a gating network that identifies the expert or blend of experts that most likely approximate the target function. The resulting multi-network is being used for the operational processing of satellite data, including the reprocessing of historical measurements. The evolution of the Antarctic ozone hole from 1995 to 2002 and the ozone mini-holes over Europe in 1999 and 2002 are presented.

INTRODUCTION

Ozone is an important constituent of the Earth atmosphere. It absorbs the potentially harmful ultra-violet radiation from the sun which can cause skin cancer and damage vegetation. Stratospheric ozone has been depleted over Antarctica the last decades at certain times of the year forming the well know ozone hole. The destruction of the ozone layer is mainly due to the release of manmade chemicals, and therefore it is very important to monitor its evolution.

This paper presents the determination of ozone total column from the GOME sensor using an algorithm based on neural networks and the mixture-of-experts model. Single neural networks are trained to approximate sub-regions of a complex multidimensional function; the neural networks are then combined with a gating network that identifies the expert (neural network) or blend of experts that most likely approximate the target function.

A geophysical validation campaign demonstrated that the ozone total columns determined with this novel neural network and mixture-of-experts algorithm is more accurate than ozone obtained from GOME with classical algorithms. The resulting system is being used for the operational processing of GOME data, including the reprocessing of historical measurements.

The next sections present the motivation of the work, the modular combination of neural networks used for function parameterization and finally the evolution of the ozone hole.

MOTIVATION

Radiative transfer models (RTM) are needed for the retrieval of atmospheric constituens from satellite measurements. The operational use of such complex and computational intensive models is limited by the computer power available.

Analysis of the time needed by the main algorithms of the GOME Data Processor system (Loyola et al., 1997), the ground segment of the GOME sensor flying on-board the ERS-2 satellite, show that radiative transfer model computations are by far the most time consuming task, taking almost 2/3 of the total processing time. RTM computations must be accelerated for the operational (re)processing of satellite data on a global scale and, what is more difficult, for near-real-time applications.

The most commonly used techniques for speeding up radiative transfer models is the parameterization based on look-up tables. The main problems with look-up tables is the usually huge size of the required database and the relatively slow recovery of RTM values.

On the last years neural networks have been combined into multi-net systems with either redundant or modular elements, see for example the work of Naftaly et al. (1997) and Lu B. and Ito (1998). The combination of neural networks can result in dramatically performance (accuracy) improvement (Ji and Ma, 1997) and it is a way of more effectively solving complex tasks (Trentin and Giuliani, 1997). This work uses a novel RTM parameterization technique based on the combination of neural networks.

MODULAR COMBINATION OF NEURAL NETWORKS

A complex problem could be better solved when it is broken down into simpler modules or subtasks. Each subtask could be then solved with a neural network. The 'divide and conquer' principle is familiar to engineering design, making the overall system easier to understand and to change.

Modular components can be created in a top-down or bottom-up way. The bottom-up technique is used when the inputs to the system coming from different sources are combined to form a overall solution. This approach is used in this work, following the statistical model Mixture-of-Experts presented in Jacobs et al. (1991).

Mixture models provide a principled way of combining two or more Gaussian distributions into a single distribution that summarizes the entire data set. In other words, mixture models are piecewise approximators where different components are used to summarize sub-regions of the input data. The regions may have 'soft' boundaries; hence data points may lie simultaneously in multiple regions. The mixture-of-experts model consists of a number of experts combined by means of a gating network which identifies the expert or blend of experts that most likely approximate the desired function. In this work the experts are neural networks, and in opposition to Jacobs et al. (1991), the experts are not mutual exclusive (each input point is assigned to one or more experts).

Let $T = \{x^{(t)}, y^{(t)}\}$ for $t=1$ to N be a set of data items or patterns, then for each input $x^{(t)}$:

i) at least one label i is chosen from a multinomial distribution with probability $P(i|x^{(t)}, V)$ where $V=[v_1, ..., v_I]$ is a matrix of parameters underlying this distribution.

ii) a response $y^{(t)}$ is generated with probability $p(y^{(t)}|x^{(t)}, u_i, \phi_i)$ where u_i is a parameter vector associated with expert i, and ϕ_i is a dispersion parameter.

in other words the output $y^{(t)}$ corresponding to an input $x^{(t)}$ is generated by first selecting one or more experts based on the input $x^{(t)}$, and then the selected experts map the input $x^{(t)}$ to the response $y^{(t)}$.

The total probability of generating $y^{(t)}$ from $x^{(t)}$ is given by the mixture density

$$p(y^{(t)} \mid x^{(t)}, \Theta) = \sum_i p(y^{(t)} \mid x^{(t)}, u_i, \phi_i) P(i \mid x^{(t)}, V) \tag{1}$$

where Θ is the matrix of all parameters. The probability of the data set T is the product of m such densities, with the likelihood given by

$$L(\Theta \mid x^{(t)}) = \prod_t \sum_i p(y^{(t)} \mid x^{(t)}, u_i, \phi_i) P(i \mid x^{(t)}, V) \tag{2}$$

Mixture-of-experts are a type of distributions in which the mixing weights $P(i|x^{(t)}, V)$ and mixed distributions $p(y^{(t)}|x^{(t)}, u_i, \phi_i)$ are dependent on the input $x^{(t)}$.

A graphical representation of the mixture-of-experts model is given in Fig. 1. The modules or experts are neural networks where different networks are appropriated in different regions of the input space. Expert network i maps its input x to an output y_i. A special module, referred as a gating network, identifies the expert or blends of experts whose output is most likely to approximate the corresponding response y. The output of the gating network for an input x is a vector g of scalars g_i that weight the contributions of the various experts according to the input x. The final output of the model is a combination of K expert outputs calculated as

$$y = \sum_{i=1}^{K} g_i \cdot y_i \tag{3}$$

where for each possible input x, the gating weights g_i are greater than or equal to zero and their sum is equal to one.

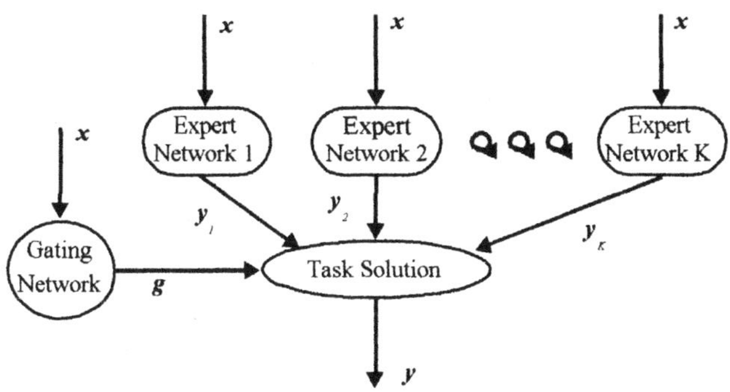

Fig. 1. Mixture-of-experts model. The gating network identifies for any input x the expert (neural network) or blend of experts whose output is most likely to approximate the corresponding output y. The vector g weights the contribution of the experts as function of the input.

FORWARD MODEL PARAMETERIZATION WITH NEURAL NETWORKS

The RTM parameterization process has the following steps: node points are generated and validated, generated data is transformed and separated in independent sets, then the multi-network of mixture-of-experts is designed trained and validated, and finally a detail error analysis is performed. If the performance of the multi-network is not satisfactory, then the previous stages should be iterated. The single stages are described in the following subsections.

RTM Node Points Generation and Validation

The first step is the definition of the minimum and maximum values for each RTM input dimension, covering the expected geophysical and viewing scenarios of the target sensor. The sample points must be properly balanced, additional sample points are needed in regions where the output function present non-monotonic changes or regions with high curvature.

The RTM is used to compute the output corresponding to the combination of all input parameters. The input/output node points of the RTM function are the training patterns.

Data Preprocessing and Separation

The RMT node points are selected, coded and scale and the data set is shuffled so that successive training examples rarely belong to the same class. The resulting data is separated into three independent sets: the training set T with 75% of the data, the testing set S with 15% of the data, and the validation set V with the remaining 10% of the data.

Multi-network Design, Training and Validation

A hierarchically multi-network architecture is designed for the parameterization of RTMs following the approaches presented in the previous chapter. A mixture-of-experts network forms the first level and an ensemble of networks implementing each expert in the second level of the hierarchically multi-network.

The resulting multi-network is tested on the independent validation set V to check its accuracy and generalization ability.

Error Analysis

The commonly used rms error is not sufficient to characterize the network performance, error historgrams showing the statistical distribution of the errors produced by multi-net compared to the original RTM nodal points are additionally used.

A network error or poor accuracy may be caused by wrong network models or deficient training data. The first case can be solved reviewing the network topology and/or the activation functions. In the second case, the training data does not represent appropriately the target function and additional RTM node points must be generated.

Parameterization Results

The final mixture-of-experts have a total of 12 experts (neural networks). The goal of reaching an average relative error of less than 1% was achieved by networks with two hidden layers. The errors are normal distributed with a mean in the order of 10^{-3} and a standard deviation in the order of 10^{-2}.

MONITORING THE OZONE HOLE

Very accurate ozone can be now retrieved out of the GOME sensor using an iterative algorithm based o the mixture-of-experts model (Lambert, 2002).

The development of ozone during the period 1995 to 2003 is presented in Fig. 2. The average global total column amount for this period is 290,42 Dobson Units (DU). Therefore, the global total ozone remains approximately 3% below the pre-1980 averages.

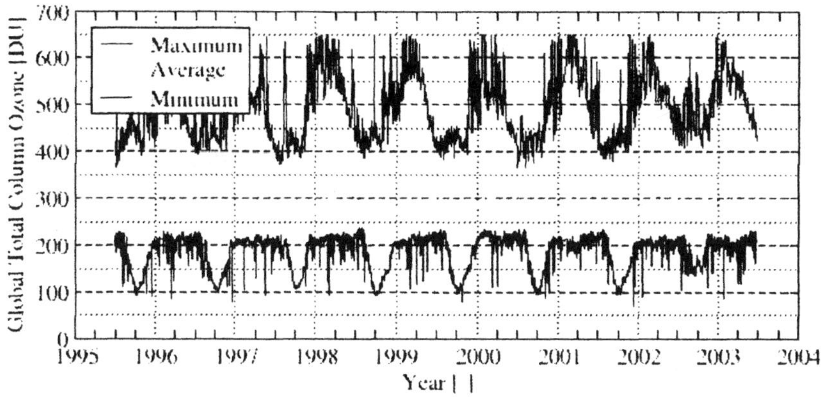

Fig. 2 Total column ozone measured by GOME from July 1995 to July 2003.

Figure 3 shows the development of the Antarctic ozone hole (caused by chemical reactions and CFCs) in the month of September from 1995 to 2002. The ozone in September 1995 had a depletion of around 50% (175 DU) compared with August. The ozone values in 1995, 1998 and 2000 reach a minimum below 100 DU. The ozone hole in 2000 was the largest ever observed, *the ozone hole expanded to approximately 26 million km^2 (6422 million acres).* The ozone hole in 2002 behaved very differently, it became elongated and finally broke up into two parts.

GOME also detected the occurrence of ozone mini-holes over Europe caused by unusual atmospheric circulation flow. Ozone losses of around 25% were observed in November 1999 and January 2002 with ozone total column amounts below 200 DU.

CONCLUSIONS

This paper presented the combination of neural networks with the mixture-of-experts model for the determination of ozone out of satellite data. The results obtained with this algorithm are very precise while the computation time and memory demands are reduced several orders of magnitude.

The resulting system is being used for the operational processing of GOME data. The retrieved ozone is public available at http://auc.dfd.dlr.de/ngong .

The gain in processing speed maintaining an excellent accuracy opens new perspectives for the operational utilization of neural networks for the retrieval of ozone. This is specially important for the next generation of European atmospheric remote sensing missions that will provide long-term key trace gas observations to the scientific community.

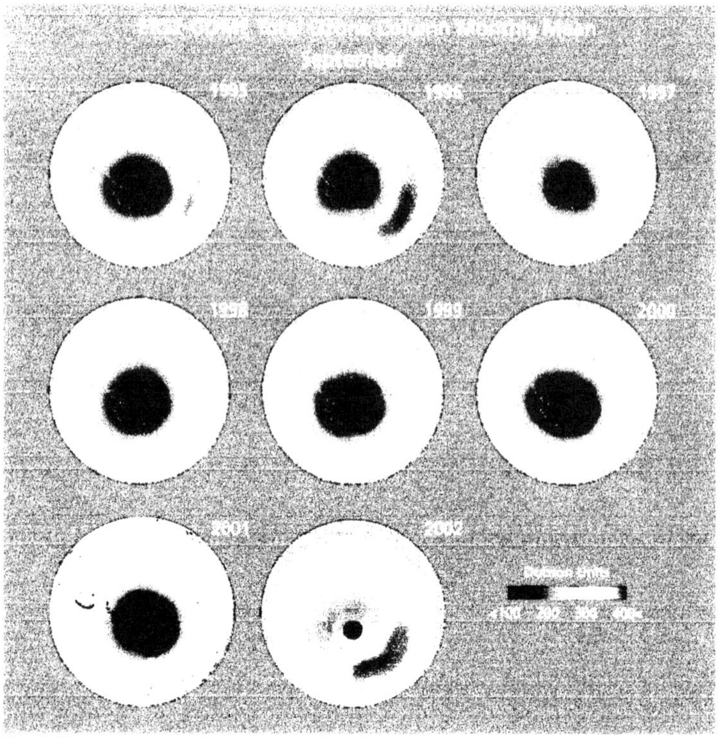

Fig. 3. Development of the Antarctic ozone hole in September 1995 to 2002 ().

ACKNOWLEDGEMENTS

The author would like to thank to the DLR colleagues Werner Thomas, Thomas Ruppert and Thilo Erbertseder for their support.

REFERENCES

Jacobs R.A., Jordan M.I., Nowlan S.J., and Hinton G.E., 1991, "Adaptive mixtures of local experts," Neural Computation, vol 3, pp. 79-97.

Ji C., and Ma S., 1997, "Combination of week classifiers," Special Issue of Neural Networks and Pattern Recognition, IEEE Trans. Neural Networks, vol 8, pp. 32-42.

Lambert J.-C. (ed), 2002, "ERS-2 GOME GDP 3.0 Implementation and Delta Validation, " ESA/ESRIN, Italy, Tech. Rep. ERSE-DTEX-EOAD-TN-02-0006.

Loyola D., Balzer W., Aberle B., Bittner M., Kretschel K., Mikusch E., Muehle H., Ruppert T., Schmid C., Slijkhuis S., Spurr R., Thomas W., Wieland T., and Wolfmueller M., 1997, "Ground segment for ERS-2/GOME sensor at the German D-PAF," 3rd ERS Scientific Symposium, Florence, Italy, pp. 591-596.

Lu B. and Ito M., 1998, "Task decomposition and module combination based on class relations: A modular neural network for pattern classification," Bio-Metric Control Research Center, Japan, Tech. Rep. BMC TR-98-1.

Naftaly U., Intrator N., and Horn D., 1997, "Optimal ensemble averaging of neural networks," Networks, vol. 8. no. 3, pp. 283-296.

Trentin E., and Giuliani D., 1997, "Speaker normalization with a mixture of recurrent networks," Proc. Of ESANN97, Belgium.

KERNEL PLS FEATURE SELECTION WITH SENSITIVITY ANALYSIS FOR IN-SILICO DRUG DESIGN

MARK J. EMBRECHTS
Department of Decision Sciences & Engineering Systems
MUHSIN OZDEMIR
Department of Decision Sciences & Engineering Systems
KRISTIN P. BENNETT
Department of Mathematical Sciences

FABIO A. ARCINIEGAS
Department of Decision Sciences & Engineering Systems
CURT M. BRENEMAN
Department of Chemistry

Rensselaer Polytechnic Institute,
Troy, NY 12180-3590, USA

ABSTRACT

Kernel Partial Least Squares (K-PLS) regression is a nonlinear extension of the PLS regression method, which has certain similarities to principal component regression (PCR). This paper introduces Kernel PLS regression as an alternative to artificial neural networks. Kernel PLS regression is a nonlinear statistical method that, contrary to neural networks, requires few heuristics. This paper introduces and applies K-PLS and K-PLS sensitivity analysis for in-silico drug design and compares the results with neural network models. Two standard benchmark datasets for drug design are used in this paper.

INTRODUCTION

Training multi-layered perceptrons with backpropagation presents a difficult challenge for small datasets with a relative large number of features (Kewley et al. 2000) and generally relies on heuristics and experience. This paper presents Kernel Partial Least Squares (K-PLS, first introduced by Rosipal and Trejo 2001) as an alternative to neural networks for nonlinear regression that requires less heuristics.

When training an artificial neural network (ANN), one is usually interested in obtaining a network with optimal generalization performance. However, it is well known that neural networks are prone to over-fitting unless proper precautions are taken (Geman et al. 1992): while the network gets better and better for the training set, the overall generalization performance for the test set does not follow the same trend. There generally is a point for which the generalization error starts to increase while the training error keeps decreasing. Ideally, training should be halted at this point (Prechelt 1998).

The two most common approaches for preventing overfitting during the training ANNs are early stopping (Morgan and Bourlard 1990) and weight decay (Plaut et al. 1986). Early stopping has the advantage of being a quick trick, since it reduces the training time. On the other hand, the disadvantages of early stopping are: i) the ad-hoc heuristic for selecting the right stopping point for training, ii) the partial use of the available training data, and iii) the relatively linear model. A further disadvantage with early stopping is that the generalization error is typically estimated from a validation set which forces the splitting of the original dataset into three sets (training set, validation set, and external test set). When the number of observations is limited, as for the datasets used in this paper, creating an additional validation set becomes unpractical. Weight decay, on the other hand, has the advantage of being based on regularization theory, but has the disadvantage of being time consuming and difficult to tune for determining an appropriate weight decay constant.

A third approach for avoiding overfitting is to halt training by a policy (i.e., using multiple training/test set models, all trained to the same policy error and averaging the performance of the test sets for evaluating the test error). Stopping by policy is a fast method in comparison with weight decay and early stopping. It also does not require a third dataset for the estimation of the generalization error (Kewley et al. 2000). This method is conservative in the sense that it does not halt at the minimal generalization error for each split, but is heuristic in determining the stopping point for the policy. On the other hand, because we generally undertrain rather than overtrain, the neural network results are not very sensitive to the number of neurons in the hidden layers. These ANN models are also more linear than models based on small optimally tuned ANNs.

K-PLS will be introduced in this paper as an alternative nonlinear modeling methodology to neural networks. K-PLS is based on Partial Least Squares regression (PLS). PLS modeling is a powerful multivariate regression tool developed by Herman Wold in the 1970s (Wold 1970) that has found utility in solving a variety of chemometrics problems with a large number of descriptive variables (Wold 1970, and Geladi and Kowalski 1986). PLS regression has many similarities with Principal Component Regression (PCR) and has been applied recently to datasets that exhibit multi-collinearity and a low observation/variable ratio.

This paper is organized as follows. First, the concept of partial least squares (PLS) regression is presented, followed by a nonlinear extension based on the use kernels (K-PLS). While this introduction of (Direct) Kernel PLS or (D)K-PLS shares many similarities with the K-PLS algorithm original K-PLS introduced by Rosipal and Trejo, it is in fact a very different and more transparent algorithm that is closer to the classical PLS method. Feature selection methods for PLS are reviewed and a new algorithm for feature selection based on sensitivity analysis algorithm is introduced. Finally, results for two in-silico drug design related datasets and the conclusions are discussed.

PARTIAL LEAST SQUARES REGRESSION

PLS regression is a multivariate regression method, which seeks to uncover a small number of almost orthogonal "latent" variables (scores) from a much larger set of correlated descriptors. PLS is built on the properties of a modified NIPALS algorithm (first used in PLS by Wold et al. in 1984). PLS usually has superior prediction properties compared to PCR, because unlike PCR it is a supervised learning method. PLS is similar to principal component regression in the sense that it uses just a few latent variables (or scores), which can be considered as synthetic features that are made up by a linear combination of the original descriptive variables. The latent variables or scores in PLS are similar to the projections on a set of eigenvectors or basis vectors in PCR that span the data space. PLS differs from PCR in the sense that the basis vectors in PLS are not orthogonal but span a Krylov space. These basis vectors are "peeled of" one-by-one while reducing the rank of the data matrix by applying the NIPALS algorithm and (contrary to PCR) are optimally aligned with the response of interest.

A simplified PLS model for just one response variable **y** consists of a regression between the scores for the data matrix **X** and the response **y**, and has as one of the main relations an outer relation linking the latent variables or scores **T** to **X** according to $\mathbf{X}_{nm} = \mathbf{T}_{nh}\mathbf{P}^T_{hm} + \mathbf{E}$ (Equation (1)), where $\mathbf{X}_{nm}$ is the data matrix consisting of n observed data records with m descriptive features or variables. $\mathbf{T}_{nh}$ is a matrix that consists of h latent variables or scores for each of the n data records and **E** is an error measure resulting from a truncation into just a few latent variables (typically 3-6 for QSAR studies). The matrix $\mathbf{P}^T_{hm}$ forms a set of intermediate or internal bases also called the loading matrix and is calculated in the NIPALS algorithm presented below. The **T** scores can be calculated in principle by projecting the data matrix **X** on a set of h basis vectors **W*** according to $\mathbf{T}_{nh} = \mathbf{X}_{nm}\mathbf{W}^*_{mh}$. Note that $\mathbf{W}^*_{mh}$ represents a set of h basis vectors, similar to the principal components in PCR. The basis vectors $\mathbf{W}^*_{mh}$ are usually not calculated explicitly in PLS and contrary to PCR do not form a set of orthogonal bases: i.e. $\mathbf{W}^{*T}_{hm}\mathbf{W}^*_{mh} \neq \mathbf{I}$. On the other hand the bases for PLS do span a Krylov space, which can be considered as a softer form of orthogonality and form conjugate gradient directions with respect to the correlation matrix according to $\mathbf{W}^*_{mh}(\mathbf{X}^T_{mn}\mathbf{X}_{nm})\mathbf{W}^*_{mh} = \mathbf{I}$.

Just as is standard practice in neural networks, PLS modeling proceeds in two stages: modeling and prediction. In the modeling stage we estimate the response variable of the training set with PLS according to

$$\hat{y}_{n1} = X_{nm}W_{mh}\left(P^T_{hm}W_{mh}\right)^{-1}T^T_{hn}\vec{y}_{n1} = X_{nm}\vec{b}_{m1} \qquad (2)$$

where n is the number of samples, m is the number of features, and h is the number of latent variables. For the prediction stage on an external test set or a validation set we use:

$$\hat{y}_{k1} = X_{km}W_{mh}\left(P^T_{hm}W_{mh}\right)^{-1}T^T_{hn}\vec{y}_{n1} = X_{km}\vec{b}_{m1} \qquad (3)$$

where it is now assumed that k is the number of data for the test set. Note that the b vector was obtained from the training set, and plays a similar role to the weights in a neural network. Equation 3 is obviously a linear model that can also be viewed as a neural network with just one neuron and no transfer function. While this simplistic neural network doesn't seem to be impressive, equation 3 has to be interpreted in reality as a multilayered network with linear transfer functions. For more details about the PLS algorithm see Swante Wold et al. (2001). The calculation of **T**, **W**, and $\mathbf{P}^T$ proceeds by using a modified NIPALS algorithm (See Geladi and Kowalski 1986 for details).

KERNEL PARTIAL LEAST SQUARES REGRESSION

Kernel Partial least squares (KPLS) regression is a nonlinear extension of the PLS regression method. This extension was first proposed by Rosipal and Trejo (2001), based on the PLS regression $\mathbf{Y} = \mathbf{XB} + \mathbf{F}$, where **B** has the form

$$\mathbf{B} = \mathbf{X}^T\mathbf{U}(\mathbf{T}^T\mathbf{X}\mathbf{X}^T\mathbf{U})^{-1}\mathbf{T}^T\mathbf{Y} \qquad (4)$$

In this formula, $\mathbf{XX}^T$ is a linear similarity measure (e.g., association matrix or the variance-covariance matrix X^TX), which in principle, can be replaced by any similarity or dissimilarity measurement. Kernel PLS uses a kernel representation for $\mathbf{XX}^T$ instead of $\mathbf{XX}^T$, which is a nonlinear similarity measure. This kernel representation, **K**, is a nonlinear transformation, of the input variables into a feature space F; or mapping $\Phi : x_i \in R^N \rightarrow \Phi(x_i) \in F$.

In this paper, a different approach for KPLS than originally proposed by Rosipal is used. Instead of applying the kernel trick on equation (4), it is applied on the original data matrix in equation (3). Here $\mathbf{X}_{nm}$ is replaced by $\mathbf{X}_{nn}$, which is a larger matrix that consists of artificial features created by projecting the features on themselves. This projection is a nonlinear kernel based correlation matrix, which contains nonlinear similarities of data points rather than descriptor similarities. This modification in the PLS algorithm makes possible for PLS to construct nonlinear regression models; it also made the NIPALS PLS algorithm to resemble PCA more. In PLS regression, the weight matrix **W** and the score matrix **T** are the eigenvectors of the matrices $\mathbf{X}^T\mathbf{YY}^T\mathbf{X}$ and $\mathbf{XX}^T\mathbf{YY}^T$, respectively. In KPLS, the score matrix **T** contains the eigenvectors of the matrix $\mathbf{K}(\mathbf{XX}^T)\mathbf{YY}^T$, where $\mathbf{K}(\mathbf{XX}^T)$ is the kernel matrix.

Due to its elegant calibration algorithm, the Kernel PLS is a very fast statistical alternative for nonlinear regression problems. A drawback about the kernel PLS method is that the kernel matrix is of the order N^2 and can be quite large, and that all the training data are needed for making a prediction. However, the weight vector in the Kernel PLS algorithm might be used to either detect bad samples (outliers) or to select a small set of "relevant" data points for the model calibration. This reduces the amount of training data to carry on, and improves the scalability of the algorithm. Note also that by making a kernel from a sparser dataset, the kernel matrix will not be square anymore. Finally, KPLS requires both the training and test data to be centered (see Rosipal and Trejo 2001 for details). For square matrices these centering routines need to be modified.

FEATURE SELECTION AND DATASETS

A challenge for kernel partial least squares regression (KPLS) relates to the fact that traditional feature selection techniques might not apply. In this paper, sensitivity analysis is used for feature selection for predictive data mining problems. Sensitivity analysis determines the saliency of each of the features in a machine-learning model and reduces the number of descriptive features to predict an activity of interest (Kewley et al. 2000, and Zurada et al. 1994).

Sensitivity analysis uses a trained machine learning model to determine the sensitivities for the features of the model. The sensitivity of a feature is defined as the output range for that feature from the machine learning model when all the descriptive features are held constant (i.e., at their average or mean value) and the feature of interest is changed within its allowable range. For 1-D feature selection sensitivity analysis, features can be dropped by comparing its sensitivity with the sensitivity of an additional random variable that is added as an additional phantom input feature to the model.

Two datasets were used in this paper: The Lombardo blood-brain barrier partitioning and the HIV (Human Immunodeficiency Virus) reverse transcriptase inhibitors data sets, which are well-known and challenging datasets for QSAR benchmark studies. The Lombardo dataset contains a diverse set of molecules (62 molecules and 698 descriptors) for which blood-brain barrier penetration data is available. The HIV data set contains 64 molecules and 620 features, and it is significant since it represents one of several ways that the spread of HIV and AIDS may be slowed or halted (See Embrechts et al. 2001 for details about the datasets)

Next section will summarize the results of feature selection for Kernel PLS with sensitivity analysis for the HIV and Lombardo datasets and the results will be compared with the feature selection for SVM with sensitivity analysis. It will be shown that the selected feature subset for the Kernel PLS is as good as the SVM feature subset and therefore, that Kernel PLS is a feasible statistical alternative to machine learning algorithms for nonlinear regression problems.

RESULTS

In the QSAR literature, the prediction quality is often reported via $\mathbf{q^2}$ or $\mathbf{Q^2}$, which is called as ***predictive*** $\mathbf{R^2}$ or ***cross-validated*** $\mathbf{R^2}$. It is calculated in the same way as in $\mathbf{R^2}$ but the predicted activities ($\hat{y}_i$) are always obtained from the leave-one-out-cross-validation. In this paper $\mathbf{Q^2}$ is always calculated on the test or validation sets and are different than the ***predictive*** $\mathbf{R^2}$ (***cross-validated*** $\mathbf{R^2}$) used in the QSAR literature. $\mathbf{Q^2}$ is defined for a test set by $Q^2 = 1 - R^2$, which is given by

$$Q^2 = \frac{\sum (y_i - \hat{y}_i)^2}{\sum (y_i - \bar{y})^2}$$

For both datasets, four algorithms were used for the 1-D sensitivity analysis: ANN, SVM, PLS, and Kernel PLS. For the PLS models, the number of latent variables used was set at 4. The neural network models are multilayered perceptron with two hidden layers, with 13 and 11 neurons respectively (for details about training, see Embrechts et al. 2001). All results are based on 100-bootstrap validation. The 1-D feature selection sensitivity analysis results for each algorithm are given in Table 1 and Table 2.

Table 1. Feature selection results for HIV dataset

# Descriptors	Method	Q^2	CV R^2
230	SVM	0.39	0.61
39	SVM	0.19	0.81
24	ANN	0.26	0.74
45	PLS	0.23	0.77
47	KPLS	0.23	0.77

Table 2. Feature selection results for Lombardo dataset

# Descriptors	Method	Q^2	CV R^2
308	SVM	0.38	0.62
31	SVM	0.10	0.90
26	ANN	0.24	0.76
13	PLS	0.11	0.89
36	KPLS	0.12	0.88

The best K-PLS model obtained for the HIV dataset yields a $Q^2 = 0.23$, with 47 features; the best model for Lombardo yields a $Q^2 = 0.12$, with 36 features. The best model obtained with ANN yield a $Q^2 = 0.26$ (24 features) and $Q^2 = 0.24$ (26 features) for the HIV and Lombardo datasets, respectively. Even though the predictive powers of the feature subsets obtained using ANNs are lower, the resulting models still display a good performance. This behavior is due to the use of oversized neural networks and the early stopping by policy.

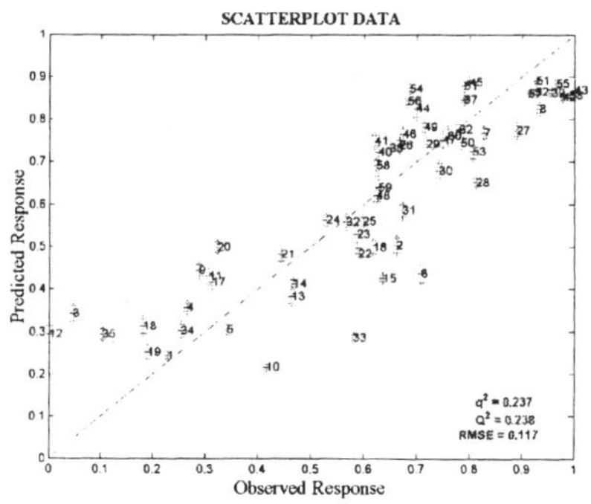

Figure 1. ANN prediction for the Lombardo dataset

Now, if one compares the Kernel PLS results with the subset obtained by using a SVM, it was found that for both datasets, almost 90% of the features were common (it is expected that a small number of features will be unique to each method). Moreover, the predictive power for the HIV selected feature subset with SVM ($Q^2 = 0.19$), and the Lombardo selected feature subset with SVM ($Q^2 = 0.10$), was not very different from the Kernel PLS subsets (see Table 3 and Table 4).

Table 3. Summary predictive power for the HIV dataset

Table 4. Summary predictive power for the Lombardo dataset

Subset	Validation Algorihtm (Q^2)			
	ANN	*SVM*	*PLS*	*KPLS*
ANN	0.24	0.21	0.25	0.15
SVM	0.16	0.10	0.13	0.14
PLS	0.16	0.11	0.11	0.12
KPLS	0.16	0.13	0.13	0.12

Subset	Validation Algorihtm (Q^2)			
	ANN	*SVM*	*PLS*	*KPLS*
ANN	0.26	0.29	0.28	0.30
SVM	0.21	0.19	0.20	0.21
PLS	0.27	0.30	0.23	0.23
KPLS	0.22	0.20	0.21	0.23

CONCLUSIONS

As it was explained in this paper, fully training a neural network not only requires lot of resources, but also involves the use of heuristics (architecture selection). With respect to the problem of training individual ANNs, this research proposed a training schema by early stopping by policy. This approach has the following setbacks: first, it requires the use of more heuristics (e.g., the stopping error is obtained from expertise knowledge); and second, this approach uses oversized neural networks, which in combination with the early stopping by policy schema, makes the trained neural networks more linear. The PLS, and KPLS regression models used in this paper are implemented in the StripMiner™ code, developed by Mark J. Embrechts.

REFERENCES

R. Kewley, M. J. Embrechts, and C. M. Breneman, "Data Strip Mining for the Virtual Design of Pharmaceuticals with Neural Networks," *IEEE Transactions on Neural Networks*, vol. 11, 3, pp. 668-679, 2000.

R. Rosipal and L. J. Trejo, "Kernel Partial Least Squares Regression in Reproducing Kernel Hilbert Space," *Journal of Machine Learning Research*, vol. 2, pp. 97-123, 2001.

S. Geman, E. Bienenstock, and R. Doursat, "Neural Networks and the Bias/Variance Dilemma," *Neural Computation*, vol. 4, pp. 1-58, 1992.

L. Prechelt, "Early Stopping - But When?," in *Neural Networks: Tricks of the Trade*, vol. 1, G. B. Orr and K.-R. Muller, Eds., First ed. New York: Springer, pp. 54-69, 1998.

N. Morgan and H. Bourlard, "Generalization and Parameter Estimation in Feedforward Nets: Some Experiments," presented at Advances in Neural Information Processing Systems 2, D. Touretzky Eds., Morgan Kaufman Publishers, Inc., pp. 630-637, San Mateo, CA, 1990.

D. Plaut, S. Nowlan, and G. Hinton, "Experiments on Learning in Backpropagation," Carnegie Mellon University, Pittsburgh, PA, Technical CMU-CS-86-126, 1986.

M. J. Embrechts, F. Arciniegas, M. Ozdemir, M. Momma, R. Kewley, C. M. Breneman, L. Lockwood, and K. P. Bennett, "Strip Mining for Molecules," presented at International Joint Conference in Artificial Neural Networks - IJCNN, Hawaii, 2001

H. Wold, "Estimation of Principal Components and related Models by Iterative Least Squares," in *Multivariate Analysis*, P. R. Krishnaiah, Ed. New York: Academic Press, pp. 391-420, 1966.

P. Geladi and B. R. Kowalski, "Partial Least-Squares Regression: A Tutorial," *Analytica Chimica Acta*, vol. 185, pp. 1-17, 1986.

S. Wold, A. Ruhe, H. Wold, and W. J. Dunn III, "The Collinearity Problem in Linear Regression. The Partial Least Squares (PLS) Approach to Generalized Inverses," *SIAM Journal of Scientific Statistical Computing*, vol. 5, 3, pp. 735-743, 1984.

S. Wold, M. Sjöström, and L. Erikson, "PLS-Regression: A Basic Tool of Chemometrics," *Chemometrics and Intelligent Laboratory Systems*, vol. 58, pp. 109-130, 2001.

J. M. Zurada, A. Malinowski, and I. Cloete, "Sensitivity Analysis for Minimization of Input Dimension for Feedforward Neural Networks," presented at International Symposium on Circuits and Systems, vol. 6, pp. 447-450, London, 1994.

A NEW INFORMATICS METHOD FOR MEASURING SYNONYMOUS CODON USAGE BIAS

XIUFENG WAN
wanx@ornl.gov
Environmental Sciences Division
Oak Ridge National Laboratory
Oak Ridge, TN 37831

DONG XU
xud@ornl.gov
Life Sciences Division
Oak Ridge National Laboratory
Oak Ridge, TN 37831

JIZHONG ZHOU
zhouj@ornl.gov
Environmental Sciences Division
Oak Ridge National Laboratory
Oak Ridge, TN 37831

ABSTRACT
Most of the current codon usage bias computational approaches are only suitable for the comparison of codon usage bias within a single genome. Here we introduce a new informatics method, referred to as synonymous codon usage order (SCUO), to measure synonymous codon usage bias. In this method, we used Shannon informational theory to describe the SCUO of each gene using a value ranging from 0 to 1, with larger values associated with greater codon bias. We compared our method with the codon adaptation index (CAI) method for measuring codon bias in *Escherichia coli* and *Sacharomyces cerevisiae*. We also studied the correlation between SCUO and CAI, and the relation of SCUO with gene length and gene function. Finally, we explored the correlation between SCUO and mRNA abundance in *Sacharomyces cerevisiae* using SAGE expression data.

INTRODUCTION

All amino acids except Met and Trp are coded by more than one codon. DNA sequence data from diverse organisms clearly show that synonymous codons for any amino acid are not used with equal frequency, even though choices among codons should be equivalent in terms of protein sequences (Grantham et al., 1980; Aota and Ikemura, 1986; Murray et al., 1989; Sharp et al., 1988; Shields et al., 1988; D'Onofrio et al., 1991). The relative frequency of synonymous codons varies with both the genes and the organisms. In *Escherichia coli*, genes of highly expressed proteins use codons corresponding to the most abundant tRNAs (Ikemura, 1985). Proteins expressed at low level use synonymous codons in rough proportion to the abundance of the corresponding tRNA, and have weaker codon preference. In contrast, non-coding regions of *E. coli* DNA showed no pronounced preference for any codon. Recently, the constraint of tRNA contents on synonymous codon choice were confirmed in 18 different unicellular organisms (Kanaya et al., 1999).

Three approaches were used to quantify the degree of synonymous codon usage bias. One approach is to devise a measure for assessing the degree of deviation from a postulated impartial pattern of synonymous codon usage, such as the codon usage preference bias measure (CPS), scaled χ^2 (McLachlan et al., 1984) and correlated χ^2 (Shields and Sharp 1987). Another approach is to use a reference set of gene sequences to assess the relative merits of different codons. The codon preference statistic and codon adaptation index (CAI) proposed by Sharp and Li (1987) used a group of highly expressed genes as a reference dataset. Karlin et al. (1998a) computed codon bias (CB) based on the codons of a relative group of other genes such as the average genes or another group of functional genes. CPS is useful for locating genes in sequenced DNA and for detecting certain sequencing errors. CAI is similar to CPS, but it employs a different normalization. A disadvantage of CPS and CAI is that they depend on the choice of the reference set of sequences for highly expressed genes. Although CB does not have this kind of disadvantage, it is not straightforward to compare and analyze codon usage bias within and across genomes. Recently, Zeeberg (2002) successfully applied Shannon informational theory to compute the synonymous coding bias in the human and mouse genomes. This method allows the comparison of codon usage bias between different organisms.

In this paper, we propose a simple informational index, also based on Shannon's information theory, to characterize the patterns of synonymous codon usage. Different from Zeeberg (2002), our informational index applies the maximum entropy techniques (Cosmi et al., 1990) to normalize the synonymous codon usage orderliness (SCUO), which allows us to compare codon usage bias across genomes as well as within a single genome. To evaluate our method, we studied the correlation between SCUO and CAI, and the relation of SCUO with gene length and function. We also explored the correlation between SCUO and mRNA abundance in yeast using SAGE expression data.

THE INFORMATICS METHOD

To implement the informatics method, we created a codon table for the amino acids that have more than one codon, indexed in an arbitrary way, so that we may unambiguously refer to the j-th (degenerate) codon of amino acid i, $1 \leq i \leq 18$. In mycoplasmas, Trp was also included into the codon table since a standard stop codon TGA encodes Trp in this specific species so that $1 \leq i \leq 19$. To simplify the explanation, the following description of the method is only based on the standard genetic codon table although the actual SCUO computation considered special cases for different organisms. Let n_i represent the number of degenerate codons for amino acid i, so $1 \leq j \leq n_i$; for example, $1 \leq j \leq 6$ for leucine, $1 \leq j \leq 2$ for tyrosine, etc. For each sequence, let x_{ij} represent the occurrence of synonymous codon j for amino acid i, $1 \leq i \leq 18$, $1 \leq j \leq n_i$. Normalizing the x_{ij} by their sum over j gives the frequency of the j-th degenerate codon for amino acid i in each sequence.

$$p_{ij} = \frac{x_{ij}}{\sum_{j=1}^{n_i} x_{ij}} \quad 1$$

According to information theory, we define the entropy H_{ij} of the i-th amino acid of the j-th codon in each sequence by

$$H_{ij} = -p_{ij} \log p_{ij} \quad 2$$

Summing over the codons representing amino acid i gives the entropy of the i-th amino acid in the each sequence

$$H_i = -\sum_{j=1}^{n_i} p_{ij} \log p_{ij} \quad 3$$

If the synonymous codons for the i-th amino acid were used at random, one would expect a uniform distribution of them as representatives for the i-th amino acid. Thus, the maximum entropy for the i-th amino acid in each sequence is

$$H_i^{\max} = -\log \frac{1}{n_i} \quad 4$$

If only one of the synonymous codons is used for the i-th amino acid, i.e., the usage of the synonymous codons is biased to the extreme, then the i-th amino acid in each sequence has the minimum entropy:

$$H_i^{\min} = 0 \quad 5$$

Unlike Shannon's definition of information, Gatlin (1972) and Layzer (1977) define the information as the difference between the maximum entropy and the actual entropy as an index of organization. The greater the information is, the more organized the sequence will be (Brooks and Wiley, 1988). In our case, this information measures the nonrandomness in synonymous codon usage and therefore describes the degree of organization for synonymous codon usage for the i-th amino acid in each sequence.

$$I_i = H_i^{\max} - H_i \quad 6$$

Let O_i be the normalized difference between the maximum entropy and the observed entropy for the i-th amino acid in each sequence, i.e.

$$O_i = \frac{H_i^{\max} - H_i}{H_i^{\max}} \quad 7$$

Obviously, $0 \le O_i \le 1$. When synonymous codon usage for the i-th amino acid is random, $O_i = 0$. When this usage is biased to the extreme, $O_i = 1$. Thus, O_i can be thought as a measure of the bias in synonymous codon usage for the i-th amino acid in each sequence. We designate the statistics O_i as the synonymous codon usage order (SCUO) for the i-th amino acid in each sequence.

Let F_i be the composition ratio of the i-th amino acid in each sequence:

$$F_i = \frac{\sum_{j=1}^{n_i} x_{ij}}{\sum_{i=1}^{18} \sum_{j=1}^{n_i} x_{ij}} \quad 8$$

Then the average SCUO for each sequence can be represented as

$$O = \sum_{i=1}^{n_i} F_i O_i \quad 9$$

VALIDATION AND EVALUATION OF THE INFORMATICS METHOD

The bacteria and archaea genomes and annotations were downloaded from ftp://ftp.ncbi.nlm.nih.gov/genome/Bacteria/ in August, 2002. The codonW program was downloaded from ftp://molbiol.ox.ac.uk/cu/ codonW.tar.Z (Peden J. F., 1999). The values of CAI of *E. coli* and *Saccharomyces cerevisiae* (yeast) were computed through codonW. The SAGE data (Velculescu et al. 1997) was downloaded from ftp://genome-ftp.stanford.edu/pub/ (Nov, 2002).

To validate this informatics approach, we measured the correlation between SCUO and CAI in *E. coli and S. cerevisiae*. The short sequences (less than 100 codons) were ignored during the comparisons. The mitochondrial genes in yeast and the plasmid genes in *E. coli* were also ignored. As a result, 6109 genes in yeast and 3887 genes in *E. coli* were included in the comparison.

We also measured the correlation between gene size and SCUO. *E. coli* genes were separated into 6 groups: 101-200 codons, 201-300 codons, 301-400 codons, 401-500 codons, 501-600 codons, and above 600 codons. We calculated the mean SCUO for each group of genes.

To evaluate the impact of codon usage bias on gene function, *E. coli genes* were divided into 18 groups according to the COG functional annotations (Tatusov et al. 1997 and 2000): C (energy production and conversion), D (cell division and chromosome partitioning), E (amino acid transport and metabolism), F (nucleotide transport and metabolism), G (carbohydrate transport and metabolism), H (coenzyme metabolism), I (lipid metabolism), J (translation, ribosomal structure and biogenesis), K (transcription), L (DNA replication, recombination and repair), M (cell envelope biogenesis, outer membrane), N (cell motility and secretion), O (posttranslational modification, protein turnover, chaperones), P (inorganic ion transport and metabolism), Q (secondary metabolites biosynthesis, transport and catabolism), R (general function prediction only), S (function unknown), and T (signal transduction mechanisms). We also included those genes undefined in the COG categories in an additional U (undefined) group. We compared the average SCUO values for the genes in different functional groups.

We compared the mRNA expression levels with their associated codon usage bias measured by our informatics method. We extracted the effective genes for measurement based on the following criteria: 1) Genes with at least one tag; 2) genes having syn names; 3) genes expressing mRNA transcripts >0 during at least one of the examined growth stages. If one gene has two or more assoicated tags, we summed the mRNA copies. From the downloaded SAGE datasets, we obtained 861 effective genes. Then we averaged the mRNA copies at L, S, and GM growth stages to the associated mRNA copies for this gene. Based on SCUO values, we divided them into 7 groups: 6 groups with a uniform interval of 0.1 for codon usage bias between 0 to 0.6 and one group with a codon usage bias larger than 0.6. We assigned the genes with codon usage bias larger than 0.6 into a single group because the associated gene number is very small.

RESULTS

The correlation of SCUO with CAI

The comparison between SCUO and CAI demonstrated that the SCUO values are positively correlated with CAI (Fig. 1). We also compared SCUO and CBI, Fop and N_C (data not shown). Our results show that SCUO gives similar features to measure codon usage bias as other indices.

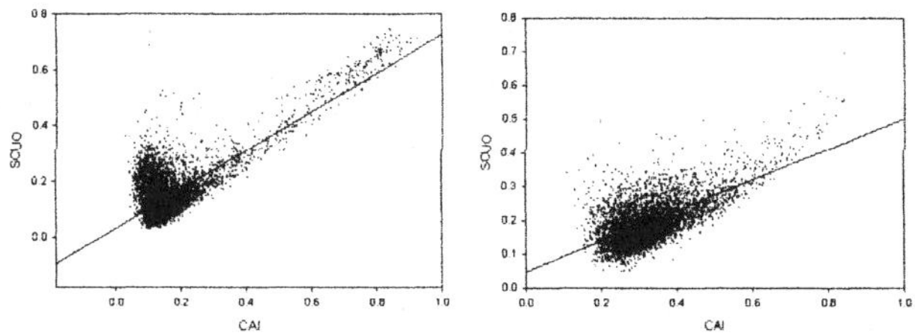

(a) (b)

Fig. 1. Comparison between SCUO and CAI. (a) relationship between CAI and SCUO in *S. cerevisiae*; (b) relationship between CAI and SCUO in *E. coli*.

Gene size and SCUO

We compared SCUO between different gene groups with different sizes in *E. coli*. Fig. 2 shows the mean values and standard deviation of each gene group. The genes with 101-200 codons exhibit about 0.05 SCUO unit higher than the genes with different sizes. The difference between the other five groups of genes is less than 0.02 SCUO unit. To ensure that the following comparison is independent of the effects of gene size, we only compute those genes with sizes over 200 codons in other parts of the paper.

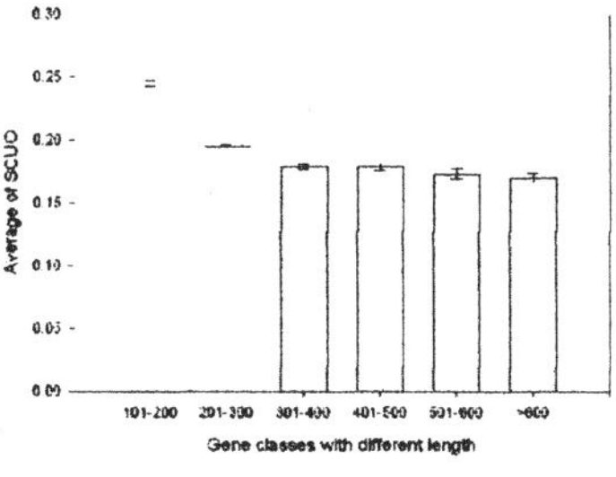

Fig. 2. Average SCUO of genes with different lengths. The *E. coli* genes were separated as 6 groups: 101-200 codons (n = 961), 201-300 codons (n = 984), 301-400 codons (n = 808), 401-500 codons (n = 723), 501-600 codons (n = 242), and above 600 codons (n = 349).

SCUO varies with gene functions

Fig. 3 shows SCUO varies with COG gene functions. The translation genes, class J (translation, ribosomal structure and biogenesis; 0.249±0.0104) and O (posttranslational modification, protein turnover, chaperones; 0.210±0.0079) have the highest SCUO values. The class U (undefined genes, 0.154±0.0030) has the lowest SCUO. But the majority of the genes in class U are hypothetical ORFs (57.7%) and putative genes (22.6%). Our results were consistent with previous reports (Karlin et al., 1998a; VanBogelen et al., 1996). Most ribosomal proteins have a higher codon bias and are considered as containing the "optimal" codons in bacteria (Karlin et al., 1998b). Our results showed that ribosome proteins have a relatively higher SCUO than tRNA synthesases in *E. coli* (data not shown), and these results are similar to previous reports (Karlin et al., 1998a).

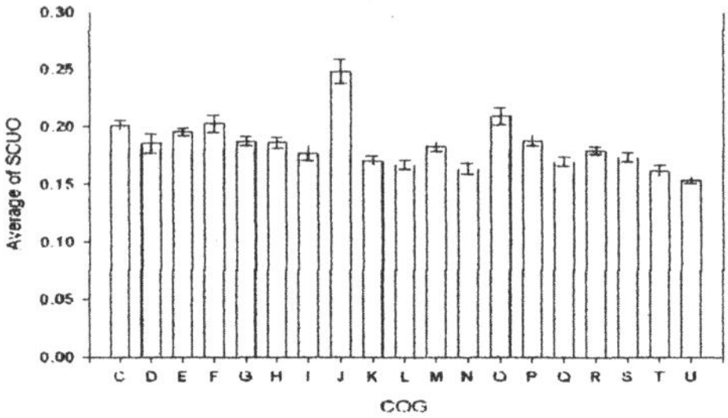

Fig. 3. Average SCUO of genes in different functional groups. The *E. coli* genes were grouped into 18 COG functional groups and an additional undefined subgroup.

The correlation between mRNA abundance and SCUO in yeast

Figure 4 illustrates the positive correlation between mRNA abundance and codon usage bias. These results demonstrated that larger SCUO is associated with more mRNA copies. These results were consistent with previous reports (Coghlan and Wolfe 2000). The positive correlation of SCUO with mRNA abundance is more persistent than that of CAI with mRNA abundance. For example, the mRNA copies of genes with 0.4-0.5 CAI unit is less than those with 0.3-0.4 unit.

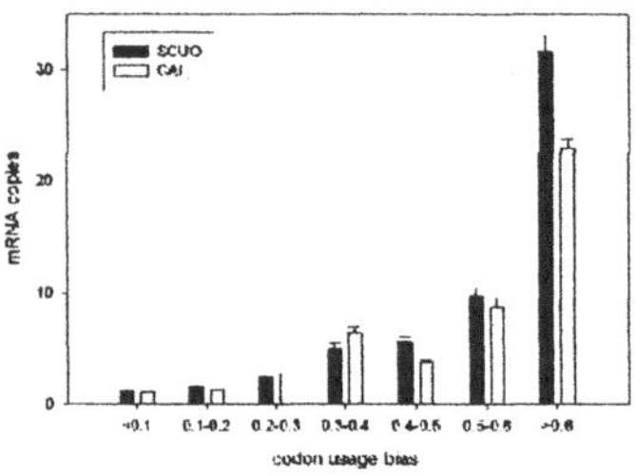

Fig. 4. The correlation between codon usage bias and mRNA copies in yeast. In group 1 (<0.1), $n_{SCUO}=227$, $n_{CAI}=50$; in group 2 (0.1-0.2), $n_{SCUO}=443$, $n_{CAI}=598$; in group 3 (0.2-0.3), $n_{SCUO}=106$, $n_{CAI}=218$; In group 4 (0.3-0.4), $n_{SCUO}=29$, $n_{CAI}=29$; In group 5 (0.4-0.5), $n_{SCUO}=12$, $n_{CAI}=15$; In group 6 (0.5-0.6), $n_{SCUO}=21$, $n_{CAI}=13$; In group 7 (>0.6), $n_{SCUO}=23$, $n_{CAI}=38$.

CONCLUSION

In summary, we present in this paper a simple informational method based on Shannon information theory and entropy theory to compare the synonymous codon usage across different organisms. To validate the proposed informatics method, we compared this method with a well-known codon usage measurement, CAI, which is only suitable for the comparison of codon usage bias within a single organism. We also explored the relationship between gene size and SCUO, gene function and SCUO. Finally, we confirmed the positive correlation between codon usage bias and mRNA abundance in yeast using our information method. All of the above results demonstrate that SCUO is an effective method for measuring codon usage bias. We plan to apply this informatics method to explore the comparison of codon bias across genomes, in particular, to examine the underlying mechanism between GC composition and codon usage bias across genomes.

ACKNOWLEDGMENTS

The authors wish to acknowledge Dorothea Thompson for her critical review. The research of XW and JZ was supported by the US DOE Office of Science as part of its Biological and Environmental Research Programs in Genome To Life and Microbial Genome Programs. DX's work was funded by the US Department of Energy's Genomes to Life program (www.doegenomestolife.org) under project, "Carbon Sequestration in Synechococcus Sp.: From Molecular Machines to Hierarchical Modeling" (www.genomes-to-life.org). Oak Ridge National Laboratory is managed by the University of Tennessee-Battelle LLC for the Department of Energy under contract DE-AC05-00OR22725.

REFERENCES

Aota, S. and Ikemura, T. (1986) Diversity in G + C content at the third position of codons in vertebrate genes and its cause. *Nucleic Acids Res.*, **14**, 6345-6355.

Brooks, D. R. and Wiley, E. O. (1988) *Evolution as entropy: toward a unified theory of biology.* 2nd edition. The University of Chicago Press, Chicago.

Coghlan, A., and K. H. Wolfe. (2000) Relationship of codon bias to mRNA concentration and protein length in *Saccharomyces cerevisiae*. *Yeast*, **16**, 1131-1145.

Cosmi, C. C., Ragosta, V., and Macchiato, M. F. (1990) Characterization of nucleotide sequences using maximum entropy techniques. *J. Theor. Biol.*, **147**, 423-432.

D'Onofrio, G., Mouchiroud, D., Aissani, B., Gautier, C. and Bernardi, G. (1991) Correlations between the compositional properties of human genes, codon usage, and amino acid composition of proteins. *J. Mol. Evol.*, **32**, 504-510.

Gatlin, L. L. (1972) *Information Theory and the Living System.* Columbia University Press.

Grantham, R., Gautier, C. and Gouy, M. (1980) Codon frequencies in 119 individual genes confirm consistent choices of degenerate bases according to genome type. *Nucleic Acids Res.*, **8**, 1893-1912.

Gouy, M. and Gautier, C. (1982) Codon usage in bacteria: correlation with gene expressivity. *Nucleic Acids Res.*, **10**, 7055-7074.

Ikemura, T. (1985) Codon usage and tRNA content in unicellular and multicellular organisms. *Mol. Biol. Evol.* **2**, 13-34.

Kanaya, S., Yamada, Y., Kudo, Y. and Ikemura, T. (1999) Studies of codon usage and tRNA genes of 18 unicellular organisms and quantification of *Bacillus subtilis* tRNAs: gene expression level and species-specific diversity of codon usage based on multivariate analysis. *Gene*, **238**, 143-155.

Karlin, S., Mrazek, J. and Campbell, A. M. (1988a) Codon usages in different gene classes of the *Escherichia coli* genome. *Mol. Microbiol.*, **29**, 1341-1355.

Karlin, S., Campbell, A. M. and Mrazek, J. (1998b) Comparative DNA analysis across diverse genomes. *Annu. Rev. Genet.*, **32**, 185-225.

Layzer, D. (1977) Information in cosmology, physics and biology. *Int. J. Quantum Chem.*, **12** (suppl. 1), 185-195.

McLachlan, A. D., Staden, R. and Boswell, D. R. (1984) A method for measuring the non-random bias of a codon usage table. *Nucleic Acids Res.*, **12**, 9567-9575.

Murray, E. E., Lotzer, J. and Eberle, M. (1989) Codon usage in plant genes. *Nucleic Acids Res.*, **17**, 477-498.

Peden, J. F. (1999) *Analysis of codon usage*, Ph.D. Thesis, University of Nottingham.

Sharp, P. M., Cowe, E., Higgins, D. G., Shields, D. C., Wolfe, K. H. and Wright, F. (1988) Codon usage patterns in *Escherichia coli, Bacillus subtilis, Saccharomyces cerevisiae, Schizosaccharomyces pombe, Drosophila melanogaster* and *Homo sapiens*; a review of the considerable within-species diversity. *Nucleic Acids Res.*, **16**, 8207-8211.

Sharp, P. M. and Li, W. H. (1986) An evolutionary perspective on synonymous codon usage in unicellular organisms. *J. Mol. Evol.*, **24**, 28-38.

Sharp, P. M., Tuohy, T. M., and Mosurski, K.R. (1986) Codon usage in yeast: cluster analysis clearly differentiates highly and lowly expressed genes. *Nucleic Acids Res.* **14**, 5125-5143.

Sharp, P. M. and Li, W. H. (1987) The codon Adaptation Index--a measure of directional synonymous codon usage bias, and its potential applications. *Nucleic Acids Res.*, **15**, 1281-1295.

Shields, D. C., Sharp, P. M., Higgins, D. G. and Wright, F. (1988) "Silent" sites in Drosophila genes are not neutral: evidence of selection among synonymous codons. *Mol. Biol. Evol.*, **5**, 704-716.

Shileds, D. C. and Sharp, P. M. (1987) Synonymous codon usage in *Bacillus subtilis* reflects both translational selection and mutational biases. *Nucleic Acid Res.*, **15**, 8023-8040.

Smith, N. G. C. and Eyre-Walker, A. (2001) Why are translationally sub-optimal synonymous codons used in *Esherichia coli*? *J. Mol. Evol.*, **53**, 225-236.

Tatusov, R. L., Koonin, E. V. and Lipman, D. J. (1997) A genomic perspective on protein families. *Science*, **278**, 631–637.

Tatusov, R. L., Galperin, M. Y., Natale, D. A. and Koonin, E. V. (2002) The COG database: a tool for genome-scale analysis of protein functions and evolution. *Nucleic Acids Res.*, **28**, 33-36.

VanBogelen, R. A., Olson, E. R., Wanner, B. L., and Neidhardt, F. C. (1996) Global analysis of proteins synthesized during phosphorus restriction in *Escherichia coli*. *J. Bacteriol.*, **178**, 4344-4366.

Velculescu, V. E., L. Zhang, W. Zhou, J. Volgelstein, M. A. Basrai, D. E. Jr. Bassett, P. Hieter, B. Vogelstein, K. W. Kinzler. (1997) Characterization of the yeast transcriptome. *Cell*, **88**, 243-251.

Zeeberg, B. (2002) Shannon information theoretic computation of synonymous codon usage biases in coding regions of human and mouse genomes. *Genome Res.*, **12**, 944-955.

THE DIRECT-KERNEL SELF-ORGANIZING MAP (DK-SOM)

Mark J. Embrechts (embrem@rpi.edu), **R. Bress, T. Naenna***
Department of Decision Sciences and Engineering Systems
*Department of Engineering Science
Rensselaer Polytechnic Institute
Troy, New York 12180-3590

ABSTRACT
This paper introduces the Direct-Kernel Self-Organizing Map (DK-SOM) as a novel learning/visualization technique. DK-SOM is a self-organizing map just as the original SOM, but operates on a kernel transformation of the data, where the kernel transform is applied in a pre-processing phase. DK-SOM has the advantage over SOM that it operates in a latent feature space just as support vector machines, but otherwise keeps the SOM algorithm intact. For problems with more features than data, DK-SOM can offer a distinct speed-up over SVMs. DK-SOMs are capable of incorporating distance measures that are more natural to the problem domain via the definition of an appropriate kernel. The prime application areas for DK-SOM are data visualization and multi-class data mining.

INTRODUCTION

Support Vector Machines or SVMs have proven to be formidable machine learning tools because of their elegance, model flexibility, predictive power, and theoretical transparency [1-3]. The purpose of this paper is to incorporate some of the inherent advantages of SVMs into Kohonen's Self-Organizing Map or SOM [4] by applying the "kernel trick" as a data transformation in a the data pre-processing stage. While the nonlinear properties of SVMs can be exclusively attributed to the kernel trick, SOMs are inherently nonlinear because they incorporate various elegant neighborhood-based manipulations. This way of accounting for nonlinearity effects is in a sense similar to the way how K-nearest neighbor algorithms incorporate nonlinearity. Unlike SVMs, the prime use for SOMs is often as a visualization tool [5, 7] for revealing the underlying similarity/cluster structure of high-dimensional data on a two-dimensional map. Because SOMs are trained with an unsupervised learning algorithm, they show a poorer performance for regression or classification predictions than is the case with models based on supervised learning methods. SOMs have the inherent advantage that they incorporate class ordinality in a natural way: i.e., via their self-organization properties that preserve the topology of a high-dimensional space in the two-dimensional SOM. SOMs are therefore quite powerful for multi-class classification, especially when the classes are not ordinal: a problem that is far from trivial.

The SOM can be turned into a kernel method by explicitly transforming the data into a kernel matrix and then to let the SOM directly operate on these kernel-transformed data. The DK-SOM now has the many of the same advantages of SVMs: e.g., the possibility for choosing a kernel distance metric that is more natural for a specific problem. In addition to these advantages the DK-SOM has the advantage over the SOM that the execution is extremely fast for small datasets with a large number of features, as is commonly the case with pharmaceutical or QSAR data, where it is the aim to predict biological activities for molecules based on molecular descriptors.

This paper is organized as follows: the next section will briefly review the kernel trick, we will then discuss the extension of the SOM algorithm on kernel transformed data, explain semi-supervised learning for regression and classification predictions on test data, introduce semi-supervised learning and the cellular automata model to color nodes in the SOM for predictive purposes. To benchmark DK-SOM we will use the olive oil data for predicting the region of origin. This non-ordinal, nine-class classification problem, was introduced by Forina [6] and studied in detail by Zupan and Gasteiger [7].

WHAT IS IN A KERNEL?

The kernel transformation goes back to the early 1900's, when Hilbert addressed the use of kernel transformations in the mathematical literature. A kernel can be described as a matrix which contains similarity measures of data records: either directly between the data themselves or with other data (e.g., support vectors). A classical use of a kernel is the correlation matrix in principal component analysis, which contains linear similarity measures between features. In support vector machines the kernel entries are similarity measures between data rather than features and these similarity measures are usually nonlinear.

There is a variety of possible nonlinear similarity measures, but in order to be mathematically tractable the kernel has to satisfy certain conditions, the so-called Mercer conditions [1-3, 9].

$$\vec{K}_{nn} = \begin{bmatrix} k_{11} & k_{12} & \cdots & k_{1n} \\ k_{21} & k_{22} & \cdots & k_{2n} \\ k_{i1} & k_{i2} & k_{ij} & k_{in} \\ k_{n1} & k_{n2} & \cdots & k_{nn} \end{bmatrix}$$

The formula above introduces the general stricture of a kernel matrix, K_{nn}, for n data. The kernel matrix is a symmetrical matrx where each entry contains a (linear or nonlinear) similiraty between two data vectors. There are several different possibilities for defining similarity metrics such as the dot product, which is a linear similarity measure, and the Radial Basis Function kernel or RBF kernel, which is a nonlinear similarity measure. The RBF kernel is the most popular nonlinear kernel and the kernel entries are defined by

$$k_{ij} \equiv e^{-\frac{\left\| \vec{x}_j - \vec{x}_l \right\|_2}{2\sigma^2}}$$

Note that in the kernel definition above, the kernel entry contains the Euclidean distance between data points, which is a dissimilarity, in a negative exponential. The negative exponential also contains a free parameter, σ, which is the Parzen window width. The proper choice for selecting the Parzen window is determined by hyper-tuning on an external test set. There is often a relatively broad window for σ for which the model quality is fairly stable. For our learning models we will consider a data matrix X_{nm} which has n data entries and each data vector contains m descriptive attributes or features. The purpose of a learning model in the context of this paper is to apply a "wisdom vector" or weight vector, w, to the data in order to make reasonable predictions for y according to:

$$\hat{\vec{y}}_n = \vec{X}_{nm}\vec{w}_n$$

One way different learning methods distinguish themselves is the way how to derive the weights or the wistom vector. Linear models, such as the one in the equation above, cannot capture inherent nonlinearities present in the data. The kernel trick can overcome this handicap. Nonlinearities can now be accounted for, because the linear model above now does not directly operate on the data, but on the nonlinear kernel transform of the data. This is very similar to what is done in principal component analysis, where the data are substitued by their (linear) principal components before building a model.

The kernel transformation is introduced as a data transformation in a pre-processing step. We will refer to this procedure as a direct kernel method. The beauty of such a direct kernel method is that nonlinearities are captured and hidden in the kernel. Nonlinear models can be build applying a linear model on kernel transformed data. A similar procedure will be applued to thet test data: the test data are now be projected via the kernel transfomation on the same training data, leading to a rectangular kernel with the proper dimensionality. The weights that resulted from a training process, can now be applied on the kernel transform of the test data. This idea is illustrated in fig. 1.

Training Mode

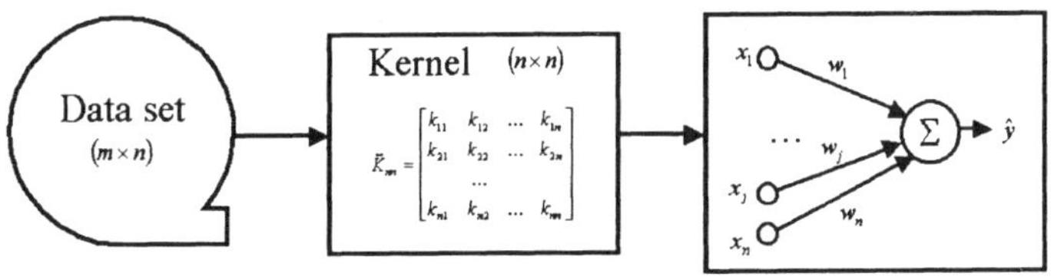

$$K_{nn} \otimes \vec{w}_n = \hat{\vec{y}}_n$$

Prediction Mode

$$K_{tn} \otimes \vec{w}_n = \hat{\vec{y}}_{predicted}$$

Figure 1. Operation schematic for direct kernel methods as a data pre-processor

DIRECT-KERNEL SOM

The direct kernel SOM, or DK-SOM, is the traditional SOM applied to kernel-transformed data. An efficient SOM module was implemented in the Analyze/StripMiner code [9] with as goal to keep user decisions minimal by incorporating robust default parameter settings. In order to do so the SOM is trained in the usual two-stage procedure: an ordering phase and a fine-tuning phase. The weights are trained by competitive learning where the winning neuron and its neighboring neurons in the map are updated according to

$$\vec{w}_m^{new} = (1-\alpha)\vec{w}_m^{old} + \alpha\, \vec{x}_m$$

$\vec{x}$ is a pattern vector with m features, α is the learning parameter and $\vec{w}$ represents the weight vector. Data patterns are presented at random and the learning parameter α is gradually (e.g., linearly) reduced from 0.9 to 0.1 during the ordering phase of the training. During this ordering phase, the neighborhood size is reduced from six to one on a hexagonal grid in a linear fashion. The number of iterations can be user specified, Kohonen suggests al least 100*the number of samples [4]. During a second fine-tuning stage, α is reduced from 0.1 to 0.01 in a linear fashion. The code defaults to 20,000 iterations for the ordering phase and 50,000 iterations in the fine-tuning phase. The map size defaults to a 9*x*18 hexagonal grid. Following Kohonen's suggestions the initial weights are random data samples and a supervised LVQ algorithm can be implemented in the fine-tuning stage [4]. The cells in the Kohonen map are colored by applying semi-supervised learning: i.e., the weight vector is augmented by a color code entry, where the color code is note used to measure distance metrics, but is otherwise updated in the same way as the weights are updated. A second optional cell-coloring scheme implemented in the code is based on a cellular automata rule. The data are pre-processed prior to applying the SOM by Mahalanobis scaling (i.e., the data are centered by subtracting the average for each feature and dividing each entry by the variance for the corresponding feature).

Figure 2 shows the results by applying a Kohonen map to order the growth region for 572 Italian olive oil samples. These same data were introduced Forina [6] and extensively analyzed by Zupan and Gasteiger [7]. The nine growth regions, or classes, are shown in Fig 3. The olive oil data, containing the contents of eight fatty acids, were split in a training set of 250 and a test set of 322 samples. This data set is challenging because there are 9 different classes or growth regions, but the classes are not entirely ordinal: a neural network or SVM with one single output neuron for all 9 classes can do no better than 83% correct classification. A 9*x*18 Kohonen map with the 322 test data mapped onto it is shown in Figure 2. 93% of the test samples are correctly classified and the class results are detailed in Table II. An overall comparison between different kernel methods with single output class regression encoding is shown in Table I (least-squares SVMs, LibSVM, Kernel Partial Least-Squares or K-PLS, Direct Kernel Partial Least Squares or DK-PLS [8], SOM and DK-SOM). Note that in this case the performance of the DK-SOM is not as good as the SOM in classification accuracy and speed of the algorithm. seconds training for the SOM. This is not entirely surprising because the DK-now SOM operates on data with 250 kernel features while the SOM operates on the eight original features.

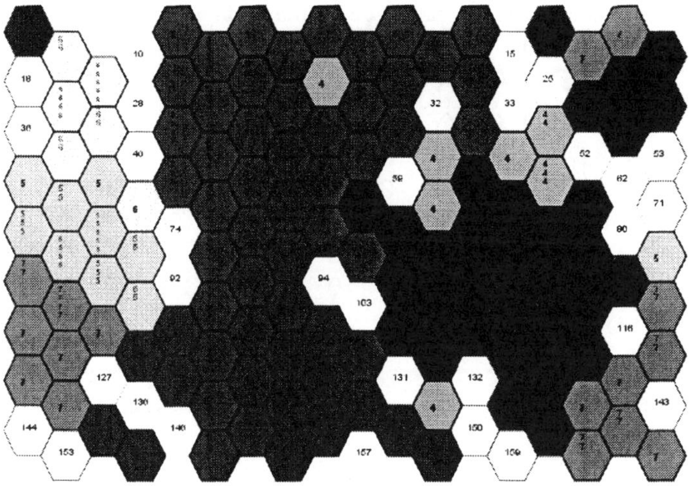

Figure 2. Direct-Kernel SOM for 250 training data for olive oils (σ = 2)

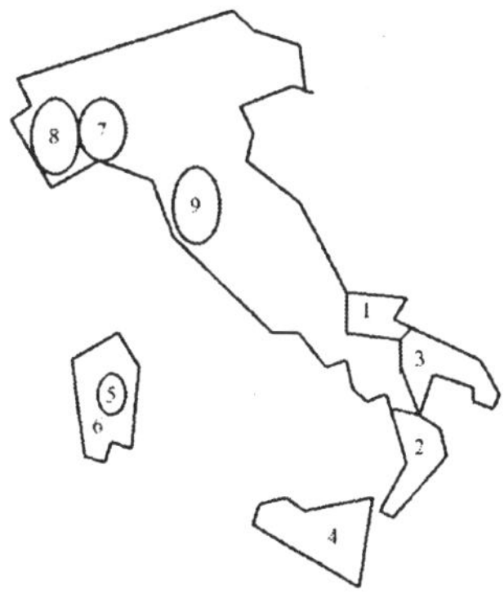

Class	Region	#samples
1	North Apulia Calabria	25
2	Calabria	56
3	South Apulia	206
4	Sicily	36
5	Inner Sardinia	65
6	Coastal Sardinia	33
7	East Liguria	50
8	West Liguria	50
9	Umbria	51
		572

Figure 3. Regions of origin for 572 olive oils

Table I. Comparison of classification results for olive oils (single encoding)

method	r^2	R^2	% correct	time
LS-SVM	0.03	0.03	83.23	0.58
LibSVM	0.033	0.033	83.54	0.9
K-PLS	0.03	0.03	86.02	0.79
DK-PLS	0.037	0.037	81.06	0.91
SOM	0.04	0.041	92.55	3
SOM-cell	0.068	0.071	89.75	3
DK-SOM	0.053	0.055	87.89	136
DK-SOM-cell	0.1	0.11	86.65	136

Table II. Typical confusion matrix for 332 test data for nine non-ordinal classes

Region										tot test
North Apulia	10	1	1	1	0	0	0	0	0	13
Calabria	0	25	6	0	0	0	0	0	0	31
South Apulia	0	1	108	0	0	0	0	0	0	109
Sicily	1	5	3	13	0	0	0	0	0	22
Inner Sardinia	0	0	0	0	40	1	0	0	0	41
Coast. Sardinia	0	0	0	0	0	16	0	0	0	16
East Liguria	0	0	0	0	1	0	23	1	0	25
West Liguria	0	0	0	0	0	0	0	33	0	33
Umbria	0	0	0	0	0	0	0	0	32	32
										322

CONCLUSIONS

This paper introduces direct kernel self-organizing maps or DK-SOMs. This method was applied to a challenging nine-class non-ordinal classification problem. While the predictive performance of DK-SOM is generally similar to their SOM counterpart, DK-SOM can have a distinct advantage in computing time for datasets with more features than data and has many of the advantages of support vector machines.

ACKNOWLEDGEMENTS

The authors hereby thank Dr. Kristin Bennett for interesting discussions on direct kernel methods. The authors gratefully acknowledge the National Science Foundation for support of this work (IIS-9979860).

REFERENCES

[1] Nello Cristianini and John Shawe-Taylor [2000] *Support Vector Machines and Other Kernel-Based Learning Methods*, Cambridge University Press.

[2] Vladimir Vapnik [1998] *Statistical Learning Theory*, Wiley.

[3] Bernhard Schölkopf and Alexander J. Smola [2002] *Learning with Kernels*, MIT Press.

[4] Teuvo Kohonen [1997] *Self-Organizing Maps, 2nd Edition*, Springer.

[5] Guido Deboeck and Teuvo Kohonen (Eds.) [1998] *Visual Explorations in Finance with Self-Organizing* Maps, Springer.

[6] M. Forina and C. Armanino, "Eigenvector Projection and Simplified Non-Linear Mapping of Fatty Acid Content of Italian Olive Oils," *Ann. Chim. (Rome)* Vol. 72, pp. 127-155.

[7] Jure Zupan and Johann Gasteiger [1999] *Neural Networks in Chemistry and Drug Design*, Wiley-CH.

[8] Bennett, K. P., and Embrechts, M. J. [2003] *"An Optimization Perspective on Kernel Partial Least Squares Regression,"* in *Advances in Learning Theory: Methods,* Models and Applications, Suykens J.A.K. et al., Eds., NATO-ASI Series in Computer and System Sciences, IOS Press, Amsterdam, The Netherlands.

[9] The Analyze/StripMiner code is available on request or can be downloaded from www.drugmining.com.

Design and Analysis of Sequential Function Charts Using Sensor-Based Stage Petri Nets

ShihSen Peng
Department of Mechanical Engineering, Chinese Military Academy, Taiwan, ROC
Email: pengss@cc.cma.edu.tw

ABSTRACT
Sequential Function Chart (SFC) is currently one of the design diagrams of automated control systems to capture the sequence of operations executed by the system's control software. This method grows so complex that locating the cause when a problem is detected becomes extremely difficult. Researchers are constantly pursuing to develop tools that eliminate the limitations and increase the design performance analysis using SFCs. Petri net resides among the most widely used tools to model and evaluate the behavior of automated control systems. Petri net control of real-world systems is achieved by mapping sensors signals and actuator commands to Petri net transitions. Being a well-defined and well-developed tool based on mathematical formulation, PN models are ideal as a method to analyze and implement real-time control for discrete industrial systems. This paper discusses a methodology to convert the SFC of an automated control system for design and analysis using the sensor-based stage Petri nets (SBSPN). Appropriate rules are suggested to convert all the I/O sensors of SFC into stages Petri net picture. A pneumatic control system is illustrated as an example to demonstrate the practical use of this technique.

Keywords: Sequential function chart, Petri nets, Stage Control Net, Real-time Petri nets, Sensor-based stage Petri nets.

INTRODUCTION

Programmable logic controllers (PLCs) are widely used for real-time control of automated industrial processes. They are well suited for discrete event control because they are able to directly implement the control sequence specified by means of standard languages such as Ladder Diagram (LD), IL (Instruction List), FBD (Function Block Diagram), Sequential Function Chart (SFC), and Grafcet defined in the IEC 61131-3. However, all the languages defined in the IEC 61131-3 standard are unable to analyze the designed ladder logic for evaluation purpose. Therefore, the developments of tools arming at modeling and analysis prior to implementation for the industrial sequential control have to be powerful and easy to use.

Sequential function chart (SFC) is the most popular design method of automated control systems used to capture the sequence of operations executed by the system's control software. The sequence control specifications are given by the ON/OFF sequence of switches, sensors and actuators, as well as the related combinational logic. They specify the I/O procedures of the controllers that drive and perform the repetitive operations of the system. The SFC diagram grows so complex that locating the cause when a problem is detected becomes extremely difficult. Moreover, its usage is limited only to control the systems and not to analyze and evaluate their qualitative and performance characteristics. In addition, SFC usually needs significant changes after a modification is made to the systems specifications, resulting in difficult maintenance problems. Therefore, researchers are constantly pursuing to develop tools that eliminate the limitations of the sequential function charts.

Petri net (PN) models reside among the most widely used tools to model and evaluate the behavior of a discrete event control system. PN models help us to understand the interactions and relations of stochastic and asynchronous events, to visualize conflicts and problems with buffers, to detect system deadlocks, etc. They also provide quantitative and qualitative analysis methods for resources utilization, consequences of system failure, systems throughput rates and more. Being a well-defined and well-developed tool based on mathematical formulation, PN models are ideal as a method to implement real-time control for the discrete event control system (DECS).

The work of this paper will discuss the methodology to convert the sequential function chart for design and analysis using the sensor-based stage Petri nets (SBSPN). Appropriate rules are suggested to convert all the I/O sensors of the SFC into a staged Petri net model. A pneumatic control industrial system is illustrated as an example to demonstrate the practical use of this technique. The paper is divided into five sections. In Section II, an illustrated example of a pneumatic control automated system is briefly described. In Section III, the technique of sensor-based stage Petri net is introduced and the integrating method of SFC stages for the model into a complete SBSPN representation is discussed. In Section IV, the design analysis of the model for liveness, boundedness, consistent, and reversibility are discussed. Finally, in Section V, the results and future work of this research are proposed.

SEQUENTIAL FUNCTION CHART FOR THE MODEL

A pneumatic control industrial system is illustrated as an example to demonstrate the practical use of this technique. As shown in Figure 1 and Figure 2, the pneumatic control system consists of two pneumatic pistons (A, B) that execute the push/pull actions to achieve a sequential stage processes and cyclic operation. Each piston is controlled by a 5/2 control valve with double-sided solenoid (A+, A–, B+, B–) to activate either push or pull action. The designed operation performs six stages in a cycle. Two flow rate valves adjust time duration of push/pull action at each stage. As shown in Figure 3, the I/O configuration of control program requires four limit switches (a_0, a_1, b_0, b_1), a start pushbutton (Pb1), and a toggle switch (Sw1) as input sensors and four solenoids (A+, A–, B+, B–) as output actuators. The SFC diagram for the MITSUBISHI FX2N-32MR PLC is designed as shown in Figure 4.

To implement the SFC diagram, the M8002 pulse starts the initial step state S0 for the sequential operations. The system starts by pushing Pb1 to ON to perform the sequential operations stage-by-stage and repeats both A– and A+ actions four times at stage 3 and 4 during a cycle. Whenever required, a jumping switch (Sw1=ON) is toggled to skip stages 3, and 4 during the cyclic operations.

Limit switches a_0, and b_0 are two normally open holding closed (NOHC) contacts, which are initially holding closed and will be released by piston A and B to execute A+ and B+ during the operations. The control logic is designed for cyclic operation but an emergency pushbutton can be installed to stop the power-initialized pulse M8002 at any time during the operation. All the limit switches signal the piston push/pull actions. Therefore, piston A is activated to operate four different strokes such as [A+] at stage 1, [A–] at stage 3, [A+] at stage 4, and [A–] at stage 5 individually, and piston B is activated to operate [B+] at stage 2 and [B–] at stage 6. After A– completed at stage 5, Piston B is activated to draw back to complete the processing in a cycle. According to the designed logic of SFC shown in Figure 4, three alternatives operation sequences for this model that starts from the push action of piston A and ends at the pull back of piston B are depicted as follows:

(1) Operation cycle with no jump/repeat action.

$[A+ \Rightarrow B+ \Rightarrow A- \Rightarrow A+ \Rightarrow A- \Rightarrow B-]$

(2) Operation cycle with four repeated actions only.

$\{A+ \Rightarrow B+ \Rightarrow [A- \Rightarrow A+]_4 \Rightarrow A- \Rightarrow B-\}$

(3) Operation cycle with jump action at stage 3 and 4.

$[A+ \Rightarrow B+ \Rightarrow A- \Rightarrow B-]$

The SFC with different input sensors designed in alternative logic combinations normally activates each output action stage-by-stage without considering the interlocking controls relays. In this model, the immediate binary conditions of six input sensors (Pb1, Sw1, a_0, a_1, b_0, b1), the deterministic timed output actuators (A+, A–, B+, B–, C+, C–), and a preset internal counter are the only related control elements required for the logic design stage-by-stage to achieve the sequential pistons push/pull actions shown in Figure 4. Therefore, all these elements are selected to construct different logic combinations stage-by-stage in the form of Boolean expressions as shown follows:

Stage 1: $A+ = b_0 * Pb1$

Stage 2: $B+ = a_1$

Stage 3: $A- = (b_1 * Sw1) + [a_1 * \overline{C0(K4)}]$

Stage 4: $A+ = C0 = a_0$

Stage 5: $A- = [a_1 * C0(K4)] + (b_1 * \overline{Sw1})$

Stage 6: $B- = a_0$

The Boolean expressions represent the sensor-based control logic at each stage for the required sequential operation of the illustrated model. We can then simply use these logic expressions to design the SFC diagram without considering the ladder logic-interlocking problem. Nevertheless, once the logic control of the SFC is increased for more complex, it is more difficult to analyze or evaluate the program for modification and cause difficult maintenance problem. The sensor-based stage Petri net technique is thus developed to help analyzing, modification, and evaluation the complex control logic effectively.

DESIGN OF SENSOR-BASED STAGES PETRI NETS FOR THE SFC OF THE ILLUSTRATED MODEL

(a) The Basic Rules of SBSPN

A sensor-based stage Petri net (SBSPN) [1] is the extension of real-timed Petri net (RTPN) [7-8] by adding a self-loop, priority stage place in the stage control net (SCN) to the corresponding timed transitions at that stage and follow the behavior of a CPU scanning process in PC/PLC. This scanning process can be represented by a token flow cyclically through the simultaneous stage places and transitions of SCN. The flow of tokens is used to drive the associated elements of RTPN at each stage. A SBSPN is a union of SCN and RTPN that can be formally defined as follows:

SBSPN $= (SCN) \cup (RTPN)$

$= (P_{Si}, T_{Si}) \cup (P, T, I, O, m_0, D, X, Y)$

Where:

- $\{P, T, I, O, m_0, D, X, Y\}$ is RTPN as defined before [7-8]; and
- $SCN = \{P_{Si}, T_{Si}\}$

Where:

P_{Si} is the set of stage places and T_{Si} is the set of stage transitions in the SCN

$\forall$ SCN $\in$ RTPN;

Let $O(P_{Si}, T_{Si}) = I(P_{Si}, T_{Si}) = 1$

Since SCN is established as a stage-enabled consistent Petri net simply to monitor the sequential status of RTPN, there is only one token allowed to flow sequentially and cyclically through SCN. The I/O status of RTPN at each stage is checked by the priority places of SCN first once its token moves up to that stage. The feedback token in the priority place then moves back to its transition for further stage scanning processes. A double directed arc in the SCN represents the priority choice, self-loop characteristic of a stage token to enable the stage I/O combination logic. The execution of SBSPN is thus controlled by the token flow of SCN through the priority place at each stage. If no SCN token presents at the stage, no action in that stage is enabled to take place. On the other hand, even if the scanning tol en presents at the priority place, no action of RTPN is executed if any input sensor is not simultaneously activated to achieve the combinational logic at the stage. This is the reason we can call SCN as a stage-control net to monitor and also integrate the design of RTPN for SFC or PLC stage programming. The proposed SCN in this study can simplify the SFC program design process. Once the control variables for a discrete event process are identified, a function between inputs/outputs at each stage can then be easily defined. The individual logic relationship at each stage is thus expressed as the Boolean equation of the stage without considering the interlocking problem from other variables at different stages.

(b) SBSPN Representation of SFC stage for the Model

(1) Segmented Timed Transition

In the designed SFC, different logic combinations of input switches and sensors activate each output actuator or solenoid. In this study, an extended segmented timed transition is used to represent the output actuator in the SFC. Each segment of the timed transition only accepts one set of ANDed input sensor as places in logic combination. Thus, the number of ORed combinations in SFC logic Boolean expression is the number of segments drawn for the segmented timed transition in the SBSPN representation. For example, the output actuator Y0 shown in Figure 5 is defined as a segmented timed transition and the Boolean equation can be expressed as:

$Y0 = S1 \cup [(X2 * X4) + (X3 * \overline{X4})]$

This Boolean expression depicts that once token moves to the stage place S1, then only the first enabled segment of the timed transition is allowed to fire the transition Y0 at this stage. If no any condition matches the logic combinations, no output action is performed at this stage. The token in S1 then moves automatically to the next stage S2 for further set of I/O logic combinations monitoring. A priority arc is double directed line connected between the stage place S1 and the deterministic timed transition Y0. Each timed transition is segmented to accept different set of input logic combinations and also connected by a priority arc for the CPU scanning process of controller at this stage. For example, the output logic expression Y0 at S1 stage consists of two input logic combinations, $[(X2 * X4) + (X3 * \overline{X4})]$, which are ORed together. Therefore, two segments are required to represent the output actuator as deterministic timed transition Y0. Each segment accepts two different input places ANDed together individually. Once the timed transition Y0 is fired one of the logic combinations, after the deterministic timed duration, an output token will be sent to the input sensor X0 to conclude the operation at this stage. Using this rule and method, all the stages of the SFC are easily represented by either a stand-alone or several segmented timed transitions. By using this method, all the combination logics of SFC diagram are converted into the stage logic combinations.

After integrating both the SCN and RTPN together, a complete sensor-based stage Petri net is established for the model design, analysis, and implementation.

(2) Integration of SFC Stages for the Model using SBSPN

The complete sensor-based stage Petri nets requires a cyclic scanning process (SCN) and the real-timed Petri net (RTPN) stage-by-stage to view all the I/O status during system operation. For the illustrated example, the SBSPN is a union of SCN and RTPN defined as:

SBSPN = (SCN) ∪ (RTPN); where

SCN= (Stage 20, Stage 21, Stage 22, Stage 23, Stage 24, Stage 25)
= (S20, S21, S22, S23, S24, S25); and

RTPN is real timed Petri net of all I/O logic combinations.

According to Figure 4, once the system power is initialized by the special internal contact M8002, the design model normally performs the process from the stages S20 to S25 in a sequential and cyclic operations. To view the system operation behavior, the complete sensor-based stage Petri net for the model is represented stage-by-stage as the following different sets of logic expressions:

Stage 20:

(RTPN) ⇒ $Y0 = X2 * X4$

(SCN)∪(RTPN)⇒ $S20 \cup Y0 = S20 \cup (X2 * X4)$

Stage 21:

(RTPN)⇒ $Y2 = X1$

(SCN)∪(RTPN)⇒ $S21 \cup Y2 = S21 \cup X1$

Stage 22:

(RTPN)⇒ $Y1 = (X3 * X5) + (X1 * \overline{C0})$

(SCN)∪(RTPN)⇒ $S22 \cup Y1 = S22 \cup [(X3 * X5) + (X1 * \overline{C0})]$

Stage 23:

(RTPN)⇒ $Y0 = C0 = X0$

(SCN)∪(RTPN)⇒ $S23 \cup Y0 = S23 \cup C0 = S23 \cup X0$

Stage 24:

(RTPN)⇒ $Y1 = (X1 * C0) + (X3 * \overline{X5})$

(SCN)∪(RTPN)⇒ $S24 \cup Y1 = S24 \cup [X1 * C0) + (X3 * \overline{X5})]$

Stage 25:

(RTPN)⇒ $Y3 = X0$

(SCN)∪(RTPN)⇒ $S25 \cup Y3 = S25 \cup X0$

All the stage logic combinations shown above consist of the related I/O sensors to establish the real-timed Petri net (RTPN) at each stage. The RTPN is then integrated with stage control net (SCN) in a union format. These logic combinations are converted into an extended timed Petri net called a sensor-based stage Petri net (SBSPN) for the design performance analysis. Since the SBSPN consists of immediate places, immediate transitions, and segmented deterministic timed transitions. A complete SBSPN diagram for the illustrated model in this study is then viewed as a more detailed picture to depict the stage-by-stage SFC programming behavior as shown in Figure 6.

DESIGN PERFORMANCE ANALYSIS

Petri net control of real-world systems is achieved by mapping sensors signals and actuator commands to Petri net transitions. To view the operation behavior of the illustrated model, all the Boolean equations are converted into a complete SBSPN graph as shown in Figure 6. The stage scanning control process is a live and safe circuit since only a token is allowed to pass through the six stage places (S20, S21, S22, S23, S24, S25) cyclically and reversibly. The operation normally starts from stage S20 by pushing the pushbuttons PB1 in Figure 4. This means a token resides in S20 of SFC starts to perform the scanning control process that is represented by the stage control net (SCN). The SCN is thus defined as an immediate timed Petri net that embedded in the controller for monitoring all the I/O logic combinations that is defined as real-timed Petri net in the complete SBSPN implementation and find the possible conflict during the RTPN operation stage-by-stage. Once a conflict message is found at any stage, the SCN sends a signal to the controller to solve the complex maintenance problem and also prevent system deadlock.

In RTPN, all the control relay outputs and actuators are defined as real-timed transitions, which are associated with deterministic processing time. When a transition is enable, the transition fires after the deterministic amount of time. Thus, the RTPN is a deterministic timed Petri net that is a clear mapping picture to show the SFC diagram logic for deadlock identification, analysis, and performance evaluation if the design order of program logic does not conflict with each other. The liveness, safeness, and reversibility properties of the RTPN to represent the SFC logics are guaranteed by the stage-by-stage algorithm, the stage logic of SFC for the PLC programming using SBSPN is live, safe and reversible if and only if the programming logic designed is live, safe, and the operation causality condition is satisfied. The evaluation and analysis of SBSPN technique can use the entire I/O sensors binary status to establish mapping incidence matrix and determine the behavior properties at each stage using Petri net technique.

Currently, PLCs are the most popular devices for implementing logic controllers. SFC is one of the IEC 61131-3 programming languages that are inspired from Petri nets structure. SBSPN technique is a live, safe, and reversible Petri net which has mutually exclusive transitions can be implemented directly in SFC of PLC programming. Once a system causes deadlock due to conflict in process implementation, the SCN in SBSPN representation can verify stage matrix and stop the process at that stage for system maintenance. The SCN capability added to the RTPN to establish the SBSPN technique does help to solve the system troubleshooting and modification difficulty.

CONCLUSION AND FUTURE WORK

This paper discusses the SBSPN technique applied to SFC for analysis and evaluation. The approach is simple, clear and strongly approved. It concludes:

(1) The related logic function of inputs and outputs sensors of a SFC for the complex automated control system can be designed using the stage combined logic expressions directly without considering the interlocking problem.

(2) The automated control logic can be easily converted into the sensor-based stage Petri net stage-by-stage for analysis and evaluation prior to the design of SFC.

(3) In addition to the analysis capability of Petri net technique, the SBSPN representation is the union of both SCN and RTPN that shows a clearer picture of CPU scanning process of controller implementation compared to the branching graph of the sequential function chart programming.

(4) SBSPN technique is a live, safe, and reversible Petri net, which has mutually exclusive transitions can be implemented directly in SFC of PLC programming.

(5) Future work is to develop the union incidence matrix of SCN and RTPN for performance evaluation of the SBSPN technique.

REFERENCES:

[1] Peng, S. S., and Zhou, M. C., "Sensor-Based Stage Petri Net Modeling of PLC Logic Programs for Discrete-Event Control Design, " *International Journal of Production Research* 41(3), 629-644, Feb.2003.

[2] Peng, S. S., and Zhou, M. C., "Sensor-based Petri net modeling for PLC stage programming of discrete-event control design," in *Proc. of the SCI2002, The 6th World Multiconference on Systematics, Cybernetics and Informatics*, July 14 -18, 2002, Orlando, Florida, USA. pp. 140-145.

[3] Peng, S. S., and Zhou, M. C., "Sensor-based stage Petri net modeling of PLC logic programs for discrete-event control design," in *Proc. of the IEEE2002 Int. Conf. on Robotic & Automation*, May 10-14, 2002, Washington DC, USA. pp. 1907-1912.

[4] Peng, S. S., and Zhou, M. C., "Petri net based PLC stage programming for discrete-event control design," in *Proc. of the IEEE2001 Int. Conf. on Systems, Man, and Cybernetics*, October 7-10, 2001, Tucson, Arizona, USA. pp. 2706-2710.

[5] Peng, S. S., and Zhou, M. C., "Conversion between ladder logic diagrams and Petri nets in discrete-event control design---a survey," in *Proc. of the IEEE2001 International Conference On Systems, Man, and Cybernetics*, October 7-10, 2001, Tucson, Arizona, USA. pp. 2682-2687.

[6] Sato, T. and Nose, K., "Automated generation of sequence control programs via Petri nets and logic table for industrial application," in *Petri Nets in Flexible and Agile Automation*, Edited by MengChu Zhou, Kluwer Academic Publishers, 1995. pp.93-107.

[7] Zhou, M. C., and Venkatesh, K., *Modeling, Simulation, and Control of Flexible Manufacturing Systems – A Petri Net Approach*. World Scientific Publishers, Singapore, 1998.

[8] Zhou, M. C., and Twiss, E., "Design of industrial automated systems via relay logic programming and Petri nets," *IEEE Tran. On Systems, Man, and Cybernetics*. vol.28, no.1, pp.137-150, Feb. 1998.

FIGURES:

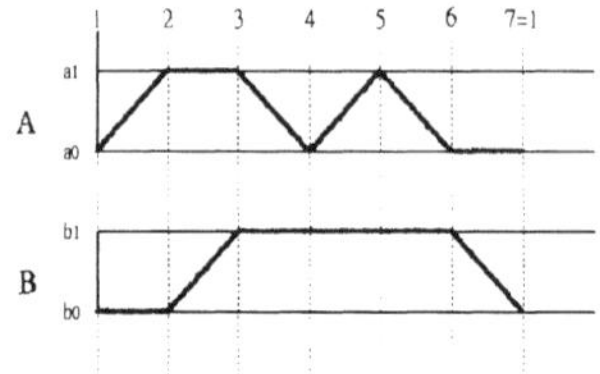

Stage 1 Stage 2 Stage 3 Stage 4 Stage 5 Stage 6
[A+] → [B+] → [A–] → [A+] → [A–] → [B–]

Figure 1. Operations stages for the model

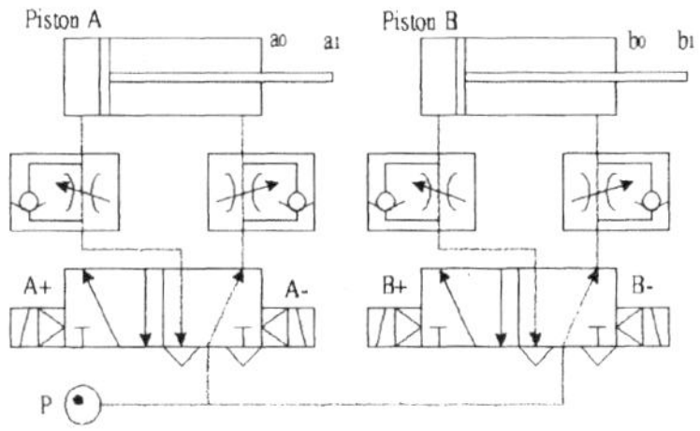

Figure 2. Power Schematic for the Model

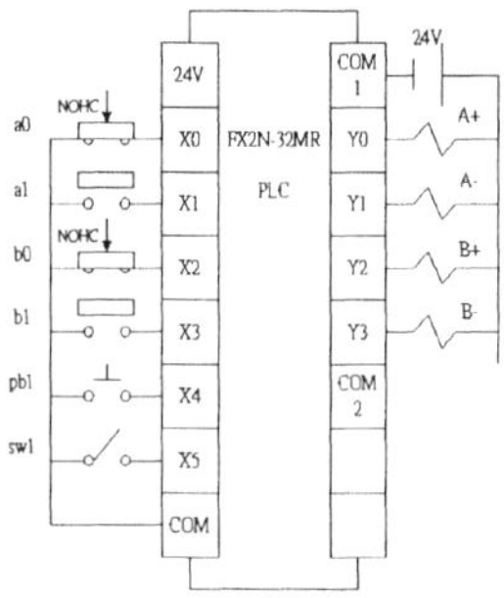

Figure 3. Inputs/Outputs Configuration of PLC

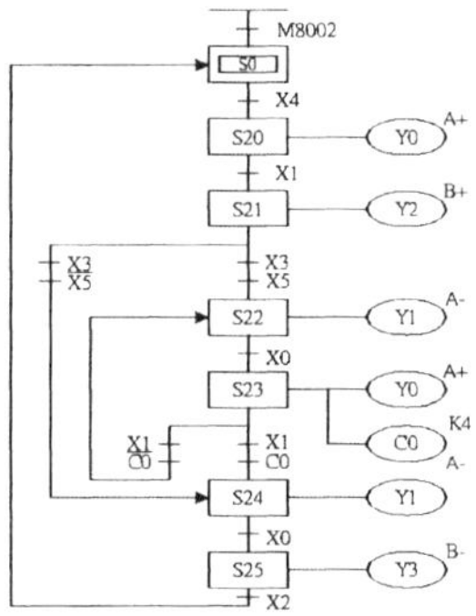

Figure 4. Sequential Function Chart for the Model

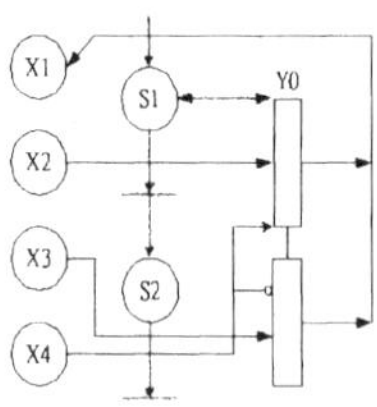

Figure 5. A Segmented Timed Transition

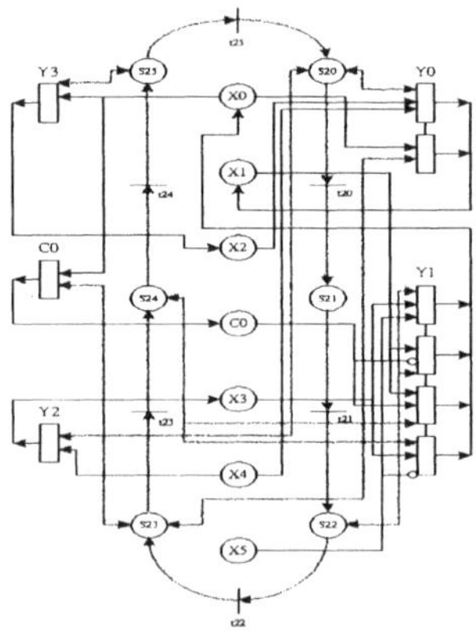

Figure 6. A Complete Sensor Based Stage Petri Nets for the illustrated Model

SUBJECT INDEX

A

B

C

D

E

F

Q

R

S

AUTHOR INDEX